Springer Series on
ATOMIC, OPTICAL, AND PLASMA PHYSICS 40

Springer Series on

ATOMIC, OPTICAL, AND PLASMA PHYSICS

The Springer Series on Atomic, Optical, and Plasma Physics covers in a comprehensive manner theory and experiment in the entire field of atoms and molecules and their interaction with electromagnetic radiation. Books in the series provide a rich source of new ideas and techniques with wide applications in fields such as chemistry, materials science, astrophysics, surface science, plasma technology, advanced optics, aeronomy, and engineering. Laser physics is a particular connecting theme that has provided much of the continuing impetus for new developments in the field. The purpose of the series is to cover the gap between standard undergraduate textbooks and the research literature with emphasis on the fundamental ideas, methods, techniques, and results in the field.

27 **Quantum Squeezing**
By P.D. Drumond and Z. Ficek

28 **Atom, Molecule, and Cluster Beams I**
Basic Theory, Production and Detection of Thermal Energy Beams
By H. Pauly

29 **Polarization, Alignment and Orientation in Atomic Collisions**
By N. Andersen and K. Bartschat

30 **Physics of Solid-State Laser Physics**
By R.C. Powell (Published in the former Series on Atomic, Molecular, and Optical Physics)

31 **Plasma Kinetics in Atmospheric Gases**
By M. Capitelli, C.M. Ferreira, B.F. Gordiets, A.I. Osipov

32 **Atom, Molecule, and Cluster Beams II**
Cluster Beams, Fast and Slow Beams, Accessory Equipment and Applications
By H. Pauly

33 **Atom Optics**
By P. Meystre

34 **Laser Physics at Relativistic Intensities**
By A.V. Borovsky, A.L. Galkin, O.B. Shiryaev, T. Auguste

35 **Many-Particle Quantum Dynamics in Atomic and Molecular Fragmentation**
Editors: J. Ullrich and V.P. Shevelko

36 **Atom Tunneling Phenomena in Physics, Chemistry and Biology**
Editor: T. Miyazaki

37 **Charged Particle Traps**
Physics and Techniques of Charged Particle Field Confinement
By V.N. Gheorghe, F.G. Major, G. Werth

38 **Plasma Physics and Controlled Nuclear Fusion**
By K. Miyamoto

39 **Plasma-Material Interaction in Controlled Fusion**
By D. Naujoks

40 **Relativistic Quantum Theory of Atoms and Molecules**
By I.P. Grant

Vols. 1–26 of the former Springer Series on Atoms and Plasmas are listed at the end of the book

I.P. Grant

Relativistic Quantum Theory of Atoms and Molecules

Theory and Computation

Springer

I.P. Grant
Mathematical Institute
Oxford University
Oxford, UK
ipg@maths.ox.ac.uk

Library of Congress Control Number: 2006926733

ISBN-10: 0-387-34671-6 Printed on acid-free paper.
ISBN-13: 978-0387-34671-7

Printed in the United States of America. (MVY)

9 8 7 6 5 4 3 2 1

springer.com

For my wife, Beryl, my sons, Paul and David, and my grandchildren Jacob, Georgia, Joshua, and Imogen

Preface

Relativistic atomic structure theory has been widely used for many years for predicting atomic energy levels, transition rates, collision cross sections and many other properties. The results have been important for fundamental physics experiments as well as providing data for space science, controlled fusion, and industrial applications. The development of a relativistic theory of the electronic structure and properties of molecules and condensed matter is more recent because it is technically more complex and needs more elaborate mathematical machinery than nonrelativistic theory. The first attempts in the early 1980s at relativistic molecular electronic structure calculations based on Dirac's Hamiltonian were unsuccessful and were mostly abandoned in favour of semi-relativistic approximations for the bulk of applications to quantum chemistry. Now that the dust has settled, there is a need for a book on rigorous foundations that sets out workable and economical methods of fully relativistic calculations for atoms, molecules and clusters that can be used by both physicists and chemists.

Part I of the book, which is nonspecialist, aims to equip the reader to recognize the qualitative signature of relativistic effects in the electronic structure of atoms and molecules. Part II deals with the theoretical foundations of the field. The form of relativistic wave equations is determined by the geometry of Minkowski space-time and the structure of the Lorentz and Poincaré groups. Quantum electrodynamics describes the physics of the interaction of electrona and electromagnetic fields; the equations are too complicated to solve exactly, but we can write down systematic approximation schemes that have proved very effective for modelling electronic structure.

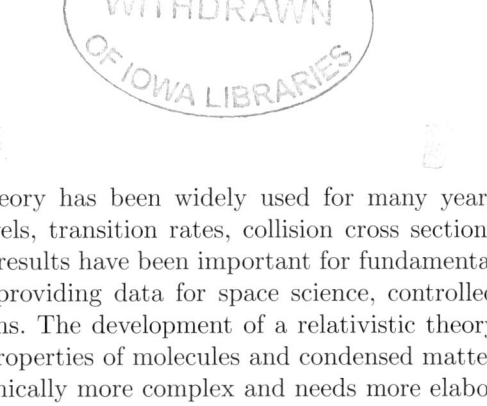

This provides the foundation for practical applications to atomic and molecular physics in Part III of the book. The electrostatic potential near each atomic nucleus is almost spherical, so that the *nonrelativistic* electron wavefunction has a characteristic *central field* character in that region, permitting factorization into radial and angular parts. Relativistic effects on electron dynamics are most marked in the strong electric field near the nucleus; the consequential coupling of the four components of Dirac central field spinors in that region therefore characterizes relativistic atomic wavefunctions. The long-range electron-electron interaction propagates relativistic effects right across the atom or molecule, so that it is essential to treat the whole many-electron system relativistically in order to get reliable results.

The coupled radial components of Dirac atomic spinors can be approximated numerically with finite differences, finite elements, or by expansion in analytic functions. Only the last approach is viable for polyatomic systems. Just as nonrelativistic molecular wavefunctions are often approximated as linear combinations of Gaussian functions (GTF), so relativistic molecular wavefunctions can be constructed from G-spinors – four-component generalizations of GTF having relativistic central field character. These incorporate all the internal relations between the components to satisfy the boundary conditions at the nuclei. Nonrelativistic GTF and G-spinors both represent the electron distribution between atoms and the resultant molecular bonding in much the same way. The failures of the 1980s were due to the incorrect assumption that the four components of a Dirac spinor can be considered independent and that variational calculations made with this assumption would reproduce the internal structure of orbital spinors with sufficient accuracy. Successful four component calculations need a spinor basis set.

The body of the book presents the technology needed for practical calculations; the appendices contain supplementary material. Chapters 6 to 8 set out the mathematical machinery of relativistic electronic structure calculations on atoms and ions, illustrated with output from a version of the GRASP computer package. Chapter 9, centred on the DARC relativistic R-matrix package, discusses mainly electron-atom/ion scattering, photo-excitation, and photo-ionization. The construction and application of the relativistic molecular structure code BERTHA are described in Chapters 10 and 11. Appendix A lists frequently used mathematical formulae, whilst Appendix B presents mathematical background material on linear operators in Hilbert space, Lie groups and Lie algebras, angular momentum theory (including diagram techniques), and various aspects of numerical approximation including the theory of variational methods for Dirac operators and iterative solution of (MC)DHF equations.

The book is primarily intended as a resource for research physicists and chemists, experimental or theoretical, who recognize that using the available relativistic electronic structure packages as "black boxes" is not always wise. These readers need to understand the physical and theoretical background in order to appreciate what can be done with existing codes and. what is just as important, what cannot be done. I have tried to include enough detail so that this book can be used by graduate students starting work in the field as well as by experts. Some material, especially Chapter 1, should be accessible to undergraduates. More difficult sections marked with an asterisk (∗), can be skipped on a first reading.

When the late Charles Coulson first suggested that I write a book on relativistic atomic structure theory more than 30 years ago, I had no inkling how much the subject would develop nor how long the writing would take. I regret I have not been able to find more space for some important topics such as relativistic many-body theory and the calculation of radiative corrections. The relativistic atomic structure package GRASP owes its present form to many collaborators: in particular Nicholas Pyper, Steven Rose, Neil Beatham, Bruce McKenzie, Jiro Hata, Ken Dyall, Patrick Norrington, and Farid Parpia. Charlotte Froese Fischer who, with Farid Parpia and others, has developed GRASP92 for modern multi-processor computers in order to study complex electronic structures in heavy atoms, continues to introduce further innovations, and her influence can be seen on both Chapters 7 and 8. I am also grateful to Stephan Fritzsche, whose RATIP procedures take relativistic atomic calculations in a new direction, and to illuminating correspondence with Steven Manson on photoionization theory and RRPA methods. The DARC code for relativistic R-matrix calculations, Chapter 9, much of which is based on GRASP, was mainly developed by Patrick Norrington and Wasantha Wijesundera. The BERTHA relativistic molecular structure code, Chapters 10 and 11, involved a close collaboration with Harry Quiney and Haakon Skaane. Harry Quiney's Oxford D. Phil. thesis laid the foundations of the relativistic basis set method in 1987 which were implemented in the DHF atomic code that preceded BERTHA. This book reflects the innumerable discussions on all aspects of relativistic atomic and molecular theory that we have had over the past 20 years. I am most grateful for his careful reading of much of the draft, which has resulted in many improvements. All errors and omissions are, of course, my responsibility alone.

Oxford, *Ian Grant*
August 2006

Contents

Part I Relativity in atomic and molecular physics

Part II Foundations

Part III Computational atomic and molecular structure

Part I

Relativity in atomic and molecular physics

1

Relativity in atomic and molecular physics

1.1 Elementary ideas

The standard quantum mechanical theory of atomic and molecular structure [1, 2, 3] assumes the constituent particles move nonrelativistically. It models atoms and molecules with a many-particle Schrödinger equation whose solutions enable us to predict physical quantities. The nuclei are represented as point masses; the a-th nucleus carries a positive electric charge $Z_a e$, say, whilst each electron, some 2000 times lighter than the individual nucleons making up the nucleus, carries a charge $-e$. The moving particles interact according to Coulomb's law: nucleus a attracts an electron at distance r with a force $-Z_a e^2/4\pi\epsilon_0 r^2$, electrons repel each other with a force $+e^2/4\pi\epsilon_0 r^2$, whilst nuclei a and b repel each other with a force $+Z_a Z_b e^2/4\pi\epsilon_0 r^2$. The electron has an intrinsic angular momentum, its spin $\mathbf{s}$, associated with an intrinsic magnetic moment. The spin state is characterized by the eigenvalues of the quantum mechanical operator $\boldsymbol{s}^2$ and (conventionally) its z-component s_z where $\boldsymbol{s} = \frac{1}{2}\hbar\boldsymbol{\sigma}$ and $\boldsymbol{\sigma} = (\sigma_x, \sigma_y, \sigma_z)$ denotes a vector whose Cartesian components are the Pauli spin matrices. Thus s_z can take values $\frac{1}{2}\hbar\sigma$, where $\sigma = \pm 1$. Each electron is described in configuration space by three position coordinates $\boldsymbol{r}$ and its spin label, σ, on the axis of quantization. Nonrelativistic theory does not couple the spatial degrees of freedom and the spin.

Let q_i denote the coordinates $(\boldsymbol{r}_i, \sigma_i)$ of the i-th electron. The quantum mechanical wavefunction for a system of N indistinguishable electrons, $\Psi(q_1, q_2, \ldots, q_N, t)$, is such that $|\Psi(q_1, q_2, \ldots, q_N, t)|^2$ can be interpreted as a probability density for finding the system in the configuration $q_1, q_2, \ldots, q_N$ at time t. Because the electrons are indistinguishable, the result of interchanging the coordinates q_i and q_j $(i \neq j)$ must give the same probability distribution, so that this operation can at most multiply Ψ by a phase factor. For fermions, particles such as the electron with intrinsic spin $\frac{1}{2}\hbar$, the N-electron wavefunction is totally antisymmetric with respect to such permutations. This has the effect that $\Psi(q_1, q_2, \ldots, q_N, t) = 0$ whenever $q_i = q_j$: no two electrons can be at the same point in configuration space at the same time. This is one mathematical expression of Pauli's exclusion principle.

The Schrödinger equation cannot be solved in closed form for anything more complicated than a one-electron system acted on by relatively simple external forces. Most interactions with external electromagnetic fields, light sources, charged and uncharged projectiles, or other atoms or molecules can only be treated by making approximations. Modern computational techniques for the calculation of atomic structures are described in, for example, such books as Fischer *et al.* [4], and in older books on atomic and molecular structure such as those by Pauling and Wilson [5], Murrell, Kettle and Tedder [6], and McWeeny [7]. These books generally present relativistic effects as a minor correction to models of atomic and molecular structures and processes. However, relativistic quantum electrodynamics (QED) provides a more fundamental starting point from which these nonrelativistic models, depending on electromagnetic forces, the spin of the electron, and the doctrine of the Pauli exclusion principle, emerge naturally.

Today, few people would question the evidence in favour of a relativistic model of the physical world as a more exact description than the Newtonian model that preceded it. Although Bohr's 1913 analysis of the hydrogen atom [8] successfully predicted the gross structure of its line spectrum, it was not long before Paschen's observations [9] of the fine structure of the hydrogenic spectrum of He^+ appeared in 1916. They were published in *Annalen der Physik* just before Sommerfeld's relativistic extension of Bohr's model [10] in which he predicted the fine structure independently. Schrödinger took the wave equation

$$\left(\partial^2/c^2\partial t^2 - \partial^2/\partial x^2 - \partial^2/\partial y^2 - \partial^2/\partial z^2 - m^2c^2/\hbar^2\right)\psi = 0, \qquad (1.1.1)$$

as a relativistic starting point in his first paper on wave mechanics in 1926 [11][1]. He then introduced electromagnetic interactions and attempted to calculate the hydrogen spectrum; he obtained the expected Balmer formula, but not the observed fine structure interval. This posed several problems: in particular, although one can write a continuity equation

$$\frac{\partial \rho}{\partial t} + \operatorname{div} \boldsymbol{j} = 0 \qquad (1.1.2)$$

in which the probability current

$$\boldsymbol{j} = -\frac{i\hbar}{2m}\left\{\psi^*(\nabla\psi) - (\nabla\psi^*)\psi\right\},$$

has the familiar nonrelativistic form, the associated expression for the probability density,

$$\rho = \frac{i\hbar}{2mc^2}\left(\psi^*\frac{\partial\psi}{\partial t} - \frac{\partial\psi^*}{\partial t}\psi\right),$$

[1] Pais [12, pp. 286 *et seq.*] has an illuminating historical discussion.

is not positive definite. Evidence that this serious defect was due to the presence of a second order time derivative in (1.1.1) provided motivation for Dirac's search for a first order wave equation with an unambiguously positive probability density [13, 14]. Dirac's free particle equation, in the notation we shall use in this book, reads

$$\{\gamma^{\mu}p_{\mu} - mc/\hbar\}\,\psi = 0, \tag{1.1.3}$$

where the *spinor* wavefunction ψ has four coupled components, and Einstein's summation convention over the space-time component indices $\mu = 0,\ 1,\ 2,\ 3$ applies. For a particle with charge $+e$ moving in an electromagnetic field with four-potential A^{μ}, this becomes

$$\{\gamma^{\mu}(p_{\mu} - eA_{\mu}) - mc/\hbar\}\,\psi = 0. \tag{1.1.4}$$

As usual, $p_{\mu} = i\hbar\partial/\partial x^{\mu}$ are the components of four-momentum and the γ^{μ} are 4×4 matrices. The charge density $j^0 = c\rho$ and current density $\boldsymbol{j}$ are now components of a four-vector $j^{\mu} = ec\psi^{\dagger}\gamma^0\gamma^{\mu}\psi$, so that the probability density $\rho = \psi^{\dagger}\psi$ is clearly positive definite. The current density three-vector is $\boldsymbol{j} = ec\psi^{\dagger}\boldsymbol{\alpha}\psi$. where $\boldsymbol{\alpha}$ has matrix valued components: $\boldsymbol{\alpha} = (\alpha^1, \alpha^2, \alpha^3)$ with $\alpha^i = \gamma^0\gamma^i$ for $i = 1, 2, 3$. In relativistic notation, the continuity equation (1.1.2) becomes

$$\partial_{\mu}j^{\mu} = 0. \tag{1.1.5}$$

Dirac's theory gives a very satisfactory account of the spectrum of atomic hydrogen, including the fine structure [15, 16]. The existence of electron spin emerges in a natural way, along with "spin-orbit coupling": in the absence of a preferred spatial direction, the total angular momentum of the electron, $\boldsymbol{j} = \boldsymbol{l} + \boldsymbol{s}$, is a constant of the motion, although neither $\boldsymbol{l} = \boldsymbol{r} \times \boldsymbol{p}$, the orbital angular momentum, nor $\boldsymbol{s}$ are constants as they would be in nonrelativistic theory.

The Dirac energy spectrum for a free electron has not only the expected continuum $E > mc^2$, but also another continuum of "negative energy" states with $E < -mc^2$. It took some time before Dirac was able to interpret the negative continuum in terms of the states of a new *anti-particle*, the positron, having a charge $+e$ and the same mass as the electron [17]. Cosmic ray positrons, which were observed by Anderson a year later [18], provided crucial support for this theoretical prediction. The hydrogenic atom has a point spectrum of bound states in the interval $-mc^2 < E < mc^2$, having an accumulation point at the series limit $E = mc^2$ as well as the two continua of scattering states. Although this is now well established, echoes of the confusion associated with the appearance of negative energy states can still be found in the literature.

In 1929, Breit [19] attempted to extend Dirac theory to the helium atom. This most memorably added a relativistic correction

$$B(\boldsymbol{r}_1, \boldsymbol{r}_2) = -\frac{e^2}{8\pi\epsilon_0}\left\{\boldsymbol{\alpha}_1.\boldsymbol{\alpha}_2 + \frac{\boldsymbol{\alpha}_1 \cdot \boldsymbol{R}\,\boldsymbol{\alpha}_2 \cdot \boldsymbol{R}}{R^2}\right\}, \quad \boldsymbol{R} = \boldsymbol{r}_1 - \boldsymbol{r}_2. \tag{1.1.6}$$

to the Coulomb interaction between electrons, now universally known as the Breit interaction. Various objections have been lodged against Breit's equation for helium, in particular that it is not Lorentz invariant and it is only an approximation to the full relativistic interaction between electrons [20, p. 170]. An expansion of Breit's equation in terms of the fine structure constant α adds a family of perturbation operators to the two- (or many-) electron Schrödinger Hamiltonian [20, p.181]. These perturbations, associated with the names of Breit and Pauli, are still what most people think of when they speak of "relativistic corrections" in atomic and molecular physics.

Relativistic self-consistent field equations for closed shell atoms – now termed Dirac-Hartree-Fock (DHF) equations – were first formulated by Swirles [21] in 1935.[2] Techniques of Racah's quantum theory of angular momentum [22], summarized in Appendix B.3, were not then available and Swirles's equations therefore appear somewhat complicated. Modern relativistic atomic and molecular structure theory has a more compact and transparent appearance [23] that relies on exploiting the underlying symmetries of Dirac central field spinors using Racah's methods (Appendix A.4). Although calculations for a small number of atoms were performed in the intervening decades, exploitation of DHF theory had to await the arrival of sufficiently powerful computers in the 1960s. Tables of nonrelativistic Hartree-Fock (HF) solutions for LS-coupled atomic ground states covering more or less the whole Periodic Table of the elements were first published in 1972 and 1973 [24, 25] at more or less the same time as similarly comprehensive tables of LS average of configuration DHF solutions [26, 27]. We shall see in §1.3.5 how these compilations have contributed to our understanding of relativistic effects in many-electron atoms. The steady growth in computer power since that epoch has been accompanied by related elaboration of the theoretical and computational machinery for relativistic modelling of many-electron atoms, which we describe in Part III of this book. Today, multi-configurational self-consistent field models (MCDHF), many-body perturbation theory (MBPT) and configuration interaction (CI) models can yield highly accurate predictions of atomic properties. The approximate energy levels and wavefunctions generated by these methods are the starting point for further investigations of radiative and collisional processes of interest for a range of applications in cosmology, astronomy, solar physics, controlled fusion, laser physics and other areas [28].

It has taken longer for relativistic effects to find a role in quantum chemistry. This is easy to understand: relativistic effects are small in the first two rows of the periodic table, and nonrelativistic models are a good first approximation for most descriptions of chemical processes involving first and second

[2] It has been common to use the notation DF and MCDF for the relativistic self-consistent fields, omitting the H for Hartree. This book uses DHF and MCDHF in recognition of Hartree's continuing influence. Swirles's original paper [21] was written at Hartree's suggestion and the input of his former students David F. Mayers and Charlotte Froese Fischer to today's software for relativistic atomic structure has been crucial to its success.

row atoms. Relativistic effects thus may appear unimportant for example in much of theoretical organic chemistry. Many accepted Dirac's 1929 dictum [29] that relativity "...give(s) rise to difficulties only when high-speed particles are involved and are therefore of no importance in the consideration of atomic and molecular structure and ordinary chemical reactions ...". We now recognize that this is a gross oversimplification. Reliable relativistic quantum mechanical models are essential for molecules containing heavier elements such as transition metals, lanthanides, or actinides [30]. Even in molecules containing only light elements, recent work has revealed that relativistic effects should not be forgotten [31].

Atomic calculations show that relativity most directly affects the dynamics of core electrons suggesting that it may be possible to model the combined effect of screening and relativity on outer electrons by an effective potential using nonrelativistic dynamics. This line of thought motivates *ab initio* the use of relativistic effective core potentials (RECP) in conjunction with various approximations to the Breit-Pauli Hamiltonian [32]. Spin-dependence is often averaged when constructing RECPs, so that some parts of the Breit-Pauli Hamiltonian have to be reintroduced later as a perturbation to account for fine structure effects. Other approaches use methods that approximate the relativistic wavefunctions in 2-component form, either by eliminating the "small" component of Dirac spinors from the energy expression or by using the generalized Douglas-Kroll transformation approach) [33, 34, 35, 36].

The main problem of principle affecting all these schemes is that it is very hard to estimate the errors associated with intuitive approximations and *ad hoc* parametrizations. Although nonrelativistic calculations are relatively cheap, relativistic perturbation operators are often difficult to handle, inviting approximating shortcuts. Benchmarking these approximations is a major chore. This book aims to formulate a practical approach to relativistic atomic and molecular structure in which the accuracy can be improved systematically to whatever level is desired.

1.2 The one-electron atom

1.2.1 Classical Kepler orbits

Hydrogenic atoms are the starting point for both relativistic and nonrelativistic theories of atomic and molecular structure. A single electron, mass m, electric charge $-e$, interacts with a stationary point charge $+Ze$ at the origin of a Cartesian reference frame. In nonrelativistic classical mechanics, the mathematics is the same as for Newtonian gravitation. [37, §3-7] The energy E and the orbital angular momentum vector $\boldsymbol{l}$ are *constants of the motion* that together serve to characterize the orbits. The general equation for the orbit in plane polar coordinates (r, θ) is

$$\frac{1}{r} = \frac{mk}{|\boldsymbol{l}|^2}\left\{1 + \epsilon\cos(\theta + \alpha)\right\}, \tag{1.2.1}$$

where $k = Ze^2/4\pi\varepsilon_0$ and ε_0 is the *electric constant* [38, Table XXIII]. This represents a conic section in a plane perpendicular to $\boldsymbol{l}$, with one focus at the origin and eccentricity

$$\epsilon = \sqrt{1 + \frac{2E|\boldsymbol{l}|^2}{mk^2}},$$

where E is the particle energy. The arbitrary constant α serves to define the direction in the orbital plane for which the polar angle θ is taken to be zero. The nature of the orbit depends on the energy:

- $-mk^2/2|\boldsymbol{l}|^2 \le E < 0$: In this case, $0 \le \epsilon < 1$ so that the right-hand side of (1.2.1) can never vanish, r lies between finite limits, and the orbit is an *ellipse*. The *closest approach* is at $r = |\boldsymbol{l}|^2/[mk(1+\epsilon)]$ and the maximum distance is $r = |\boldsymbol{l}|^2/[mk(1-\epsilon)]$.

  When $\epsilon = 0$, the orbit is a *circle* with radius $r = l^2/mk$ and energy $E = -mk^2/2|\boldsymbol{l}|^2$.
- $E = 0$: the orbit is a *parabola* and $\epsilon = 1$.
- $E > 0$: Then $\epsilon > 1$, and the orbit is a *hyperbola*; the two values of θ at which $r \to \infty$ determine the angle between its *asymptotes*.

As the radius is a minimum at $r_{min} = |\boldsymbol{l}|^2/[mk(1+\epsilon)]$, for all values of ϵ, the the velocity is perpendicular to the radius there, and the maximum speed is $v_{max} = |\boldsymbol{l}|/mr_{min}$. The electron is said to be *bound* to the centre of force when $E < 0$ and *unbound* when $E \ge 0$.

1.2.2 The Bohr atom

Bohr's first attempts to understand the spectrum of atomic hydrogen from 1913 onwards [8] naturally assumed that the electron described a closed classical orbit. The energy E of the particle, the orbital average of the kinetic energy $< T >$ and the potential energy $< V >$ satisfy the virial relation

$$E = \frac{1}{2} < V > = - < T > . \tag{1.2.2}$$

Analysis of the optical spectra of atomic hydrogen, the simplest of all atoms, revealed that the frequencies of the spectral lines could be fitted to Rydberg's formula

$$\nu = R\left(\frac{1}{n^2} - \frac{1}{m^2}\right) \tag{1.2.3}$$

where m, n are integers, inviting the hypothesis that radiation was emitted when the atom made a transition between two states whose energies fitted a formula

$$E_n = -\frac{R}{n^2}.$$

Bohr's analysis reproduced this formula with $R = 2\pi^2 me^4/ch^3$ by making the remarkable assumption that the angular momentum $|\boldsymbol{l}|$ of the classical orbit can take only the values

$$|\boldsymbol{l}| = n\hbar, \quad \hbar = h/2\pi,$$

where h is Planck's constant and n is an integer; the angular momentum was said to be *quantized.*

1.2.3 X-ray spectra and Moseley's Law

In 1913, Moseley [39] studied the K-shell X-ray lines of the metals from Ca (Z=20) to Zn (Z=30). His conclusion, that the square root of the frequency of each corresponding X-ray line was approximately proportional to Z, was crucial for settling the order with which elements are placed in the Periodic Table [40, §16-3]. Moseley's regularities were subsequently shown to hold for the rest of the Periodic Table, although screening and relativistic effects cause departures from Moseley's law at high Z. The elementary theory of X-ray lines is that of a one-electron spectrum [2, §9[13]], in which the frequency of each line is given by a formula similar to (1.2.3) [40, §16-3]. Relativistic effects modify the Z-dependence as Z increases.

1.2.4 Transition to quantum mechanics

Quantum mechanics describes the motion of a particle in terms of states whose *wavefunctions*, $\psi(\boldsymbol{r}, t)$, are labelled by constants of the motion. The probability of finding the particle in a volume element dv centred on $\boldsymbol{r}$ at time t is given by $|\psi(\boldsymbol{r}, t)|^2\, dv$. Schrödinger's wave equation

$$i\hbar\frac{\partial\psi}{\partial t}\psi(\boldsymbol{r}, t) = H\,\psi(\boldsymbol{r}, t), \tag{1.2.4}$$

where H is the quantum mechanical Hamiltonian operator, determines the evolution of $\psi(\boldsymbol{r}, t)$. This differential operator is derived from the classical Hamiltonian by making the subsitutions $\boldsymbol{p} \to -i\hbar\boldsymbol{\nabla}$, $\boldsymbol{r} \to \boldsymbol{r}$. For a particle of mass m moving in a conservative force field with potential energy $V(\boldsymbol{r})$ the classical Hamiltonian is

$$H(\boldsymbol{p}, \boldsymbol{r}) = \frac{1}{2m}\boldsymbol{p}^2 + V(\boldsymbol{r}).$$

In our model $V(\boldsymbol{r})$ is just the potential energy of an electron at a distance $r = |\boldsymbol{r}|$ from a fixed nucleus of charge $+Ze$,

$$V(\boldsymbol{r}) = -\frac{Ze^2}{4\pi\varepsilon_0 r},$$

so that the Schrödinger differential operator is

$$H = -\frac{\hbar^2}{2m}\nabla^2 - \frac{Ze^2}{4\pi\varepsilon_0 r}. \tag{1.2.5}$$

The solution of (1.2.4) with the Hamiltonian operator (1.2.5) is a standard student exercise [5], from which one deduces that the electron has *bound* states, replacing the classical elliptic orbits, whose probability density distribution vanishes exponentially as $r \to \infty$ and whose energies are given by the formula

$$\epsilon_{nl} = -\frac{mZ^2e^4}{32\pi^2\varepsilon_0^2\hbar^2 n^2} \tag{1.2.6}$$

where n is a positive integer. The electron is not confined to the neighbourhood of the nucleus when $E > 0$, so that these solutions correspond to the hyperbolic orbits. The orbital angular momentum vector $\boldsymbol{l}$, is a constant of the motion, as in classical mechanics, but its magnitude $\boldsymbol{l}^2$ takes values $l(l+1)\hbar^2$, where $l = 0, 1, \ldots, n-1$ and its projection on a fixed axis, conventionally taken as the z-axis, takes the $2l+1$ values $m\hbar$, where $m = -l, \ldots, l$. The bound state energies, which are independent of l, agree with Rydberg's formula (1.2.3) if $R = mZ^2e^4/32\pi^2\varepsilon_0^2\hbar^2$ where $\hbar = h/2\pi$. From the relation (1.2.2), we deduce that

$$< T_n >= -E_n,$$

and writing $T_n = mv_n^2/2$, giving the *root mean square velocity*

$$\frac{v_n}{c} = \frac{\alpha Z}{n} \tag{1.2.7}$$

where $\alpha = e^2/4\pi\varepsilon_0\hbar c$ is the dimensionless *fine structure constant* and c is the speed of light (see [38, Table XXIII] for the currently recommended numerical values of physical constants).

In spherical polar coordinates, the spatial part of the wavefunction of a bound state of a particle in a conservative central field factorizes into radial and angular parts,

$$\psi_{nlm}(\boldsymbol{r}, t) = \text{const.}\ \frac{P_{nl}(r)}{r}\, Y_l^m(\theta, \varphi), \tag{1.2.8}$$

where the angular amplitude reflects the spherically symmetry of the problem, and only the radial part depends on the details of the potential.

1.2.5 Sommerfeld's relativistic orbits and Dirac's wave equation

Elementary relativity theory tells us to expect deviations from Newtonian mechanics when speeds reach an appreciable fraction of the speed of light. In the Kepler problem, the particle speed attains its maximum at closest approach to the centre of force, namely

$$\frac{v_{max}}{c} = \frac{k}{c|\boldsymbol{l}|}(1+\epsilon), \quad \epsilon > 0. \tag{1.2.9}$$

Clearly, $v_{\rm max}$ is inversely proportional to $|\boldsymbol{l}|$, so that we can anticipate the largest effects in states with the lowest angular momentum.

Elementary relativistic kinematics [12, page 87] tells us that a particle moving in some reference frame with velocity $\boldsymbol{u}$ has four-momentum p^μ, where

$$p^0 = E/c = mc\gamma(u), \quad p^i = mu^i\gamma(u), \quad i = 1, 2, 3. \tag{1.2.10}$$

with $\gamma(u) = 1/\sqrt{1 - u^2/c^2}$ where u^i are the Cartesian components of $\boldsymbol{u}$ in the given reference frame, $u = |\boldsymbol{u}|$, E is the energy, and c is the speed of light. These equations are often said to express the relativistic variation of mass with velocity. For low speeds we can expand $\gamma(u)$ in powers of u/c giving

$$E = mc^2 + \frac{mu^2}{2} + \frac{3mu^4}{8c^2} + \ldots, \qquad p^i = mu^i\left(1 + \frac{u^2}{2c^2} + \ldots\right)$$

When $u/c \ll 1$, these reduce to the Newtonian energy (plus the *rest energy*, mc^2) and Cartesian components of momentum of a particle. We can expect to see relativistic effects easily when, say, $u^2/c^2 > 0.1$ where, for hydrogenic atoms, u is either v_n (1.2.7) or v_{max} (1.2.9). We conclude

- Electrons in hydrogenic atoms with low orbital angular momentum l are most likely to exhibit the largest relativistic effects.
- The maximum orbital speed ratio v_{max}/c given by (1.2.9) is inversely proportional to l. Like v_n/c it is roughly of order $\alpha Z \approx Z/137$. In the region of the actinides $(89 \leq Z \leq 103)$, $\alpha Z \sim 2/3$. Relativistic effects are thus likely to be quite important for understanding the properties of high-Z atoms and materials containing them.

Sommerfeld's investigations in 1914–15 [10] showed that the classical orbits were approximately ellipses whose axes precess about the normal to the plane of the motion. He concluded that the orbital energy was

$$E_{nk} = mc^2 - \frac{me^4}{32\pi^2\varepsilon_0^2\hbar^2}\left[\frac{Z^2}{n^2} + \frac{\alpha^2 Z^4}{n^3}\left(\frac{1}{k} - \frac{3}{4n}\right)\right] + O(\alpha^4). \tag{1.2.11}$$

where $n = 1, 2, 3, \ldots$ and $k = l + 1 = 1, 2, \ldots, n$. The first term on the right is the rest energy of the electron, the second is just Balmer's formula (1.2.6), and the next term gives the lowest order relativistic corrections. Dirac's relativistic wave equation for hydrogenic atoms, (1.1.4), is nowadays written (see §3.2)

$$i\hbar\frac{\partial\psi}{\partial t} = H\psi, \quad H = c\boldsymbol{\alpha}\cdot\boldsymbol{p} + \beta mc^2 - \frac{Ze^2}{4\pi\epsilon_0 r}. \tag{1.2.12}$$

Its solutions in spherical polar coordinates for energy E have the 4-component structure (3.2.4):

$$\psi_{E\kappa m}(\boldsymbol{r}) = \text{const.}\ \frac{1}{r}\begin{pmatrix} P_{E\kappa}(r)\chi_{\kappa m}(\theta,\varphi) \\ iQ_{E\kappa}(r)\chi_{-\kappa m}(\theta,\varphi)\end{pmatrix}. \tag{1.2.13}$$

$\kappa = \pm 1, \pm 2, \dots$ is the *angular number* and, (3.2.9),

$$\chi_{\pm\kappa m}(\theta, \varphi) = \sum_{\sigma} (l, m - \sigma, 1/2, \sigma \,|\, l, 1/2, j, m)\; Y_l^{m-\sigma}(\theta, \varphi)\phi_\sigma, \qquad (1.2.14)$$

where

$$\phi_{1/2} = \begin{pmatrix} 1 \\ 0 \end{pmatrix}, \quad \phi_{-1/2} = \begin{pmatrix} 0 \\ 1 \end{pmatrix},$$

are 2-component spin-orbitals. The angular amplitudes $Y_l^{m-\sigma}(\theta, \varphi)$ are vector coupled to the spin functions ϕ_σ to give a 2-component function that is a simultaneous eigenfunction of $\boldsymbol{j}^2$, $\boldsymbol{l}^2$, $\boldsymbol{s}^2$ and j_z. Clearly, the orbital angular momentum, $\boldsymbol{l}$, and spin angular momentum, $\boldsymbol{s}$, of the electron are not constants of the motion as in nonrelativistic quantum mechanics. Instead, Dirac central field spinors, (1.2.13), are eigenfunctions of $\boldsymbol{j}^2$ and j_z, where $\boldsymbol{j} = \boldsymbol{l} + \boldsymbol{s}$. The two possible coupling modes are defined by writing

$$\kappa = \eta(j + 1/2), \quad l = j + \eta/2, \quad \eta = \pm 1 \qquad (1.2.15)$$

which identifies the connection between the labels in (1.2.14) uniquely. We refer to κ as the *angular quantum number*. The Dirac spinors (1.2.13) can therefore be characterized by the value of κ associated with the upper components, or by the total angular momentum labels j and m, where $\boldsymbol{j}^2$ and j_z have eigenvalues $j(j+1)\hbar^2$ and $m\hbar$ respectively. The properties of the 2-component spin-orbitals, $\chi_{\kappa m}$, presented in Chapter 3 and elsewhere, are critical for the relativistic quantum mechanics of atoms and molecules. They ensure that the linear space spanned by the 4-component spinors $\psi_{E\kappa m}(\boldsymbol{r})$, and also by the 2-component spinors $\chi_{\pm\kappa m}$, for $m = -j, -j+1, \dots, +j$ is a representation space for a $(2j+1)$-dimensional representation of the rotation group. Notice that although the 2-spinor components of (1.2.13) are different, each one factorizes like the Schrödinger wavefunction (1.2.8) in a central field.

We define the *principal quantum number* by

$$n = n_r + |\kappa|. \qquad (1.2.16)$$

so that we label bound states with the triple $n\kappa m$ (or alternatively $nljm$). For low Z, ϵ_{nl} (1.2.6) is close to the shifted Dirac eigenvalue

$$\epsilon_{n\kappa} = E_{n\kappa} - mc^2 \qquad (1.2.17)$$

for the κ values of (1.2.15). It is therefore useful to retain the spectroscopic letters $s, p, d, f, g, \dots$ corresponding to $l = 0, 1, 2, 3, 4, \dots$, for labelling the Dirac states; when convenient we shall replace the label $n\kappa$ by nl_j. Thus the lowest Dirac states of hydrogenic atoms can be labelled $1s_{1/2}$, $2s_{1/2}$, $2p_{1/2}$, $2p_{3/2}$, ...The energy difference between the levels nl_j with $j = l \pm 1/2$ is referred to as the *fine structure splitting* of the nonrelativistic energy level. For *bound states*, the energies are given by the eigenvalue equation (3.3.7):

$$E_{n\kappa} = mc^2\sqrt{1 - \alpha^2 Z^2/N^2}. \qquad (1.2.18)$$

N is usually called the *apparent principal quantum number*

$$N = \sqrt{(n_r + \gamma)^2 + \alpha^2 Z^2},$$

in which the non-negative integer n_r is the *inner* or *radial quantum number* and the radial amplitudes $P_{E\kappa}(r)$, $Q_{E\kappa}(r) \sim r^\gamma$ as $r \to 0$, where $\gamma = +\sqrt{\kappa^2 - \alpha^2 Z^2}$. For electrons, $P_{E\kappa}(r)$ approximates the Schrödinger radial amplitude in the formal nonrelativistic limit $\alpha \to 0$ ($c \to \infty$), whilst $Q_{E\kappa}(r) = O(\alpha)$ and hence vanishes in that limit; the former is therefore often designated the *large* and the latter the *small* radial component.[3] The bound state radial quantum number n_r counts the number of nodes or zeros in the *large* component. When (1.2.18) is expanded in powers of αZ, we recover (1.2.11) with k replaced by $|\kappa| = j + 1/2$, so that the allowed energies of a hydrogenic atom now depend upon the angular momentum quantum number j (or on $|\kappa|$) as well as n.

Although Dirac theory was able to explain the gross structure as well as the fine structure of the spectra of hydrogenic atoms to quite high precision [41, Chapter VIII], the experiments suggested residual effects, which were firmly established by the experiments of Lamb and Retherford in 1947 [42]. The discovery of "Lamb shifts" led to a major re-examination of the theory of interactions of charged particles with electromagnetic fields and to dramatic advances in the relativistic theory of quantum electrodynamics (QED) which are, more than half a century later, still the subject of active experimental and theoretical research. The Lamb shift, which grows roughly like Z^4, is essential for the good agreement of theory with experiment in X-ray and inner-shell transitions in atoms with charges greater than $Z \approx 15$, or in the high-precision physics of atoms such as hydrogen and helium with small numbers of electrons [43].

1.2.6 Dirac and Schrödinger charge distributions

A comparison of hydrogenic radial density distributions obtained from analytic solutions of the Dirac and Schrödinger equations for hydrogenic atoms allows us to look at the primary dynamical effects of relativity [44] without the complications introduced by electron-electron interactions in many-electron systems. The electron density distribution in a hydrogenic atom given by the wavefunction (1.2.13) can be written as the product of a (scalar) angular density $A_{|\kappa|,m}(\theta) = \chi^\dagger_{\pm\kappa m}(\theta,\varphi)\chi_{\pm\kappa m}(\theta,\varphi)$,[4] see §3.2.5, and a radial density

[3] This labelling is inappropriate for negative energy states where $Q_{E\kappa}(r)$ is the large component and $P_{E\kappa}(r)$ the small component.

[4] The Hermitian conjugate of a 2-spinor $\chi = \begin{pmatrix} a \\ b \end{pmatrix}$ is $\chi^\dagger = (a^*, b^*)$ where the asterisk denotes complex conjugation.

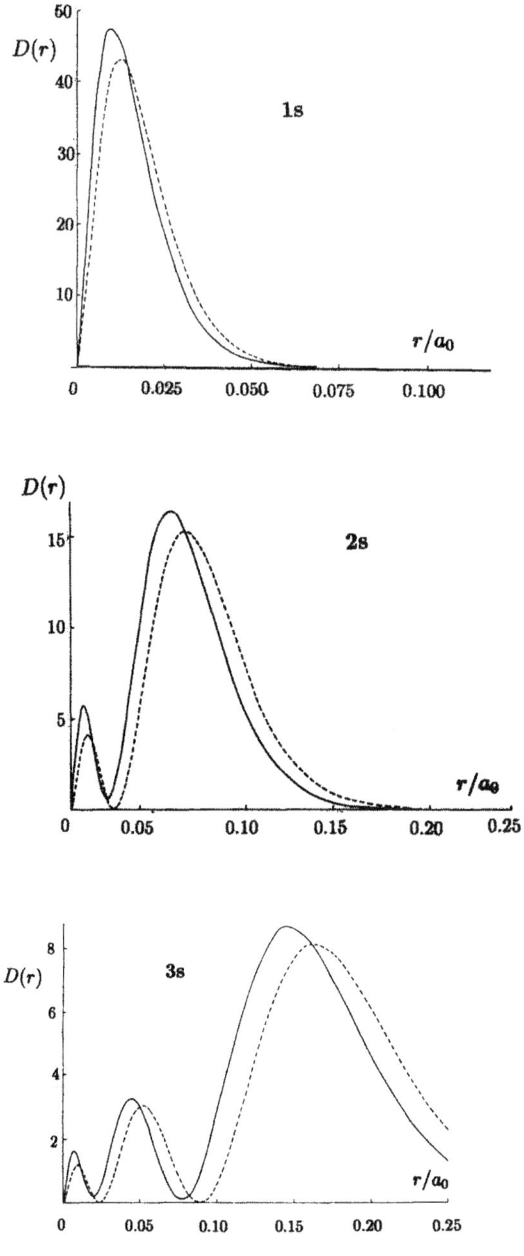

Fig. 1.1. Radial charge densities for *s*-orbitals in hydrogenic Hg^{79+}: Dirac, solid lines; Schrödinger, broken lines.

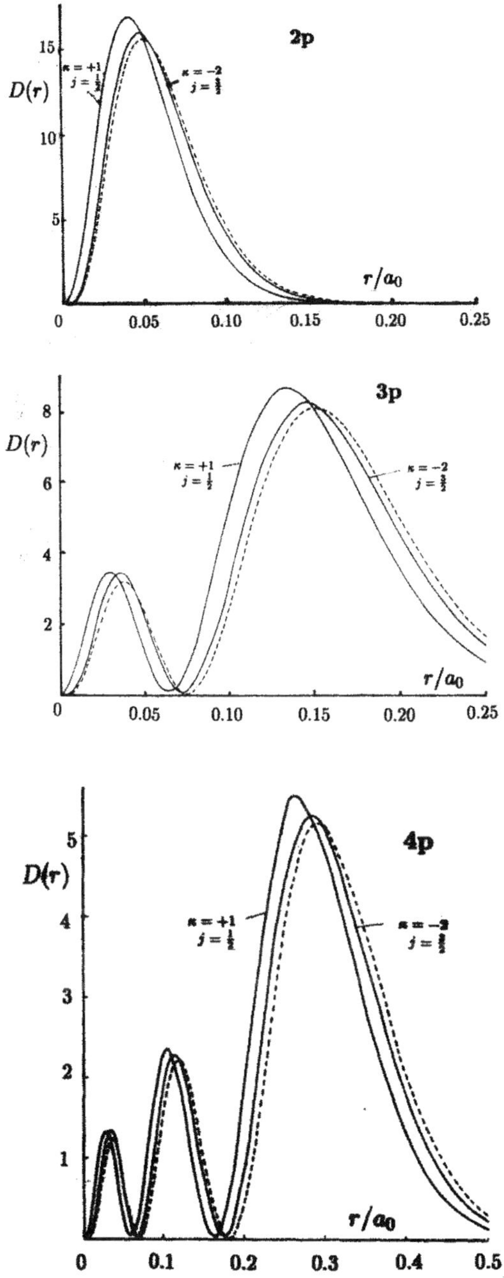

Fig. 1.2. Radial charge densities for p-orbitals in hydrogenic Hg^{79+}: Dirac, solid lines; Schrödinger, broken lines.

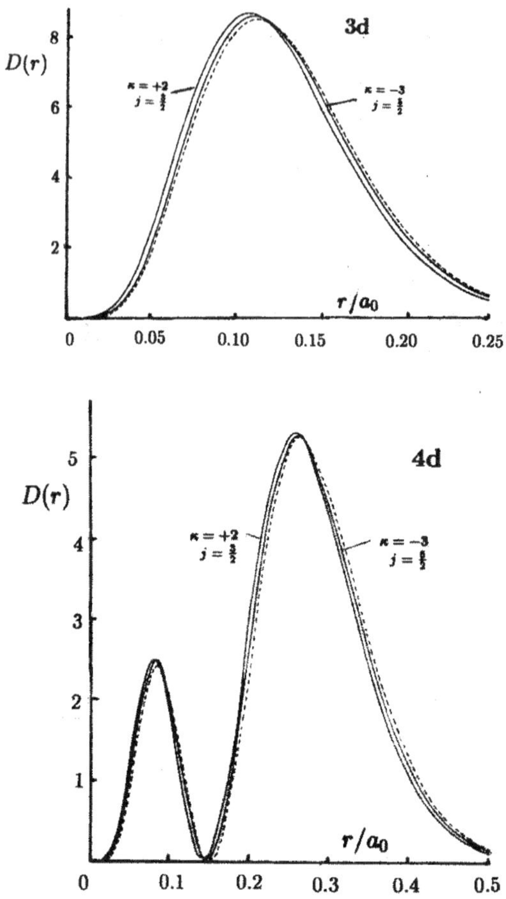

Fig. 1.3. Radial charge densities for *d*-orbitals in hydrogenic Hg^{79+}: Dirac, solid lines; Schrödinger, broken lines.

distribution $D_{n,\kappa}(r)$. The angular density is independent of atomic number and is the same for both upper and lower 2-spinor components. The radial density is therefore a sum of two terms

$$D_{n,\kappa}(r) = |P_{n,\kappa}(r)|^2 + |Q_{n,\kappa}(r)|^2. \tag{1.2.19}$$

The density plots for hydrogenic Hg^{79+} (where the nuclear charge $Z = 80$ is large enough to show appreciable relativistic dynamical effects) are presented in Figs. 1.1, 1.2 and 1.3. The main conclusions [44] are

- Dirac *radial density distributions* are more compact than their nonrelativistic counterparts. Using the inequalities such as (3.3.24) it is easy to show that the (scaled) mean radii, $\rho = \langle 2Zr \rangle$, satisfy

$$\rho_{n\kappa} < \rho_{nl} \tag{1.2.20}$$

where $\rho_{n\kappa}$ is the Dirac value, and ρ_{nl} the corresponding Schrödinger value. The relative difference is

$$\delta = (\rho_{n\kappa} - \rho_{nl})/\rho_{nl} = C(\alpha Z)^2 + O((\alpha Z)^4),$$

where $C < 0$ [44, Equations (10), (11) and Table 2].

- Dirac electrons are more tightly bound than their Schrödinger counterparts:

$$\epsilon_{n\kappa} < \epsilon_{nl} \tag{1.2.21}$$

where $\epsilon_{n\kappa}$, given by (1.2.18) and (3.3.25), is independent of the sign of κ, and ϵ_{nl} is given by by (1.2.6).

- *Spin-orbit splitting*. More precisely,

$$\epsilon_{n,\kappa=l} < \epsilon_{n,\kappa=-l-1} < \epsilon_{nl},$$

so that orbitals $n\kappa$ with $\kappa = +l$ are more tightly bound than those with $\kappa = -l-1$. The splitting is of order $O(\alpha^4 Z^4)$.

- The radial component $P_{n\kappa}(r)$ approaches the Schrödinger radial wavefunction in the nonrelativistic limit $\alpha \to 0, c \to \infty$, and in this case the Pauli formula (8.2.14)

$$Q_{n,\kappa} \approx \frac{1}{2c}\left(\frac{dP_{n,\kappa}}{dr} + \frac{\kappa P_{n,\kappa}}{r}\right)\{1 + O(1/c^2)\}$$

is a good first approximation. It follows that the zeros of the two components interlace; as a result there is no radius at which the density $D_{n\kappa}(r)$ (1.2.19) vanishes. The nonrelativistic zeros in the radial density are therefore replaced by positive minima,

Noticeable relativistic effects require the relative mean orbital velocity v_n^2/c^2 to be at least about 0.1, so that the criterion (1.2.7) from mean kinetic energy gives $Z/n \geq 40$. The mean speed is less than the maximum attainable so that a better indicator might be the fraction of the orbital charge density within the sphere inside which the (classical) electron moves at relativistic speeds, given by $Z/r > 0.1mc^2/E_h$. This gives $r/a_0 \leq 10^{-3}Z$, which for Hg^{79+} gives $r/a_0 \leq 0.08$. Figures 1.1, 1.2 and 1.3 shows that a reasonable proportion of the total charge density of electrons with $|\kappa| = 1$ ($s_{1/2}$ and $p_{1/2}$) is indeed located in that region. The proportion decreases as the angular momentum $j = |\kappa| - 1/2$ increases.

1.2.7 The Dirac hydrogenic spectrum at high Z

The properties of the Dirac hydrogenic states depart more and more from those of the corresponding nonrelativistic model as Z increases. Scattering solutions are asymptotically proportional to $\exp(\pm ipr)$ at large radii, where

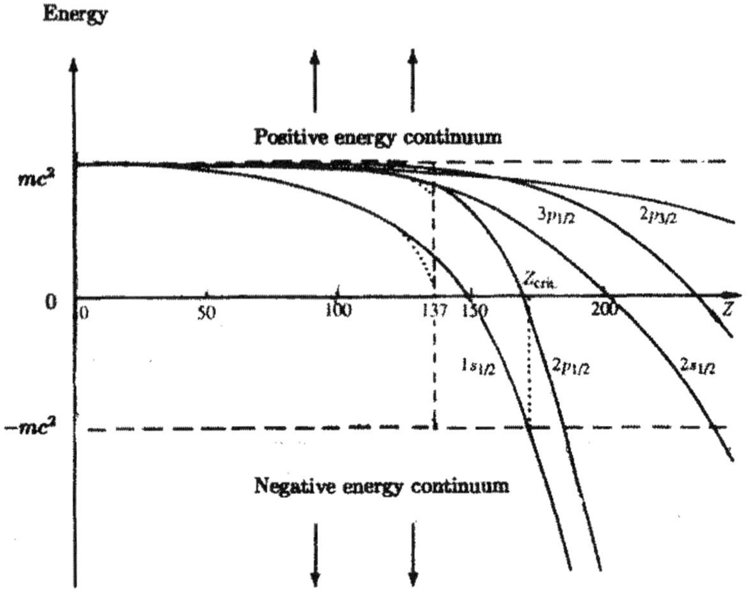

Fig. 1.4. Dependence of Dirac eigenvalues on atomic number.

$$E^2/c^2 - p^2 = m^2c^2, \tag{1.2.22}$$

as in classical relativistic mechanics (1.2.10). This Lorentz invariant equation has two solutions $\pm|E|$, one in each continuum when $|E| > mc^2$, so that p is real. When $|E| < mc^2$, $p = \pm i\sqrt{m^2c^2 - E^2/c^2}$ is pure imaginary so that there is only one square integrable bound solution asymptotically proportional to $\exp(-|p|r)$. For each value of κ, there is an increasing sequence of point eigenvalues in the bound state gap converging to a limit point at mc^2.

The dependence of the bound state energies from (1.2.18) on Z is depicted in Figure 1.4. The finite size of the nucleus becomes more important at high Z, because the proportion of the electron density penetrating the nuclear region increases signficantly. For a point nucleus with charge Ze, normalizable solutions cease to exist for $Z > |\kappa|/\alpha \approx 137|\kappa|$, as shown by the dotted curves for $|\kappa| = 1$ ($s_{1/2}$ and $p_{1/2}$). The corresponding solid lines were calculated using plausible model nuclear charge distributions with nuclear radii roughly proportional to $A^{1/3}$, where A is the nuclear mass number. Numerical eigenvalues can be found for more or less arbitrarily large Z, although the $1s_{1/2}$ curve descends into the lower continuum in the *supercritical* region $Z > Z_{crit} \approx 172$. There have been speculations that "superheavy" nuclei with lifetimes long enough to be detectable may exist with nucleon numbers in so-called "islands of stability" in the region shown which have fuelled research into the associated atomic properties [45]. This brief description raises a variety of questions of physical interpretation. The lower continuum appears to be an inescapable consequence of relativity theory. The most immediate problem is what effect this has on the stability of the hydrogenic ground state. In nonrelativistic quantum mechanics this is the lowest energy state available for a bound electron. The unavoidable appearance of negative energy states suggests that the $1s_{1/2}$ bound state cannot be stable. Dirac [46] proposed in 1929 a "vacuum" state in which each negative energy state has been assigned to an electron keeping all states with $E \geq -mc^2$ vacant. The $1s_{1/2}$ state is then the lowest available to the added atomic electron; this arangement is stable because the electron cannot decay to occupied lower states without violating the Pauli exclusion principle. The physics of a hypothetical system with $Z > Z_{crit}$ remains open.

If we wish to observe an electron that has been extracted from a negative energy state, it must have been given sufficient energy to be assigned to a state with energy $E \geq E_{1s}$. The negative energy "hole" left behind behaves like a particle with positive energy and charge $+e$, now recognized as a positron. The process is usually described as creating an electron-positron pair; such pairs, either real or virtual (when they appear as intermediate states of a perturbation expansion), can be created without affecting the total charge of the system. The appearance of negative energy states in the theory forces us to develop a formalism that can cope with an indefinite number of particles: quantum electrodynamics (QED).

Attempts have been made to realize high effective charges in superheavy quasi-molecules created by collision of heavy ions: for example, at high enough kinetic energy the nuclei of the Pb+Pb system can be made almost to touch. The combined nuclear charge of 164, just below the theoretical limit Z_{crit}, is high enough to test ideas on the physics of supercritical electric fields in QED [45, Chapter 11]. However, physics and chemistry is largely concerned with elements having much lower atomic numbers so that these fascinating issues are rather remote from the more everyday concerns with which this book deals.

1.3 Many-electron atoms

The relativistic changes in the electronic structure of hydrogenic atoms are modified by the mutual interactions of electron charges and currents in many-electron systems that propagate secondary relativistic effects across the atom or molecule. The relativistic "contraction" of hydrogenic orbitals naturally plays a prominent part: $1s$, $2s$, and $2p_{1/2}$ electrons, which are anyway the most compact, screen the nuclear charge more effectively in relativistic models so that electrons with higher angular momentum ($np_{3/2}$, $nd_{3/2}$, $nd_{5/2}$, $\ldots$) tend to expand and to have correspondingly lower binding energies. All bound electrons show evidence of the interaction between the *direct* dynamical relativistic contraction and the *indirect* relativistic expansion due to changes in electron screening of the nuclear charge and the interactions between subshells [47]. We shall look at this from two points of view. A comparison of eigenvalues from (nonrelativistic) Hartree-Fock (HF) and (relativistic) Dirac-Hartree-Fock (DHF) calculations is sufficiently realistic to model binding energy trends across the Periodic Table. Secondly a survey of measured successive atomic ionization potentials gives insight into the buildup of electronic shell structure in individual atoms and shows direct evidence of relativistic effects similar to those revealed by the HF and DHF models. This section utilizes a heuristic approach to theoretical atomic structure, which reflects the way in which the subject has developed historically in the last three-quarters of a century.

It will be convenient from now on to work in Hartree atomic units: the unit of mass is that of the electron $m_e \approx 9\ 10^{-31}$ kg.; the unit of length is the Bohr radius $a_0 = 4\pi\epsilon_0\hbar^2/m_e e^2 \approx 5.29\ 10^{-11}$ m ; the unit of energy is $E_h = e^2/4\pi\epsilon_0 a_0 = \alpha^2.m_e c^2 \approx 4.36\ 10^{-18}$ J where $\alpha = e^2/4\pi\epsilon_0\hbar c \approx 1/137.036$ is the fine structure constant. In atomic units, the unit of time is $\tau = \hbar/E_h \approx 2.42\ 10^{-17}$ s and c, the speed of light, is $\alpha^{-1} \approx 137.036$. See [38] for current recommended values of these physical constants.

1.3.1 Central field models of the atom

The central field model [2, Chapter 6] has given invaluable insight into the properties of many-electron atoms. The charge distribution in an isolated atom in its ground state is almost spherically symmetric, suggesting an atomic Hamiltonian of the form

$$H_0 = \sum_{i=1}^{N} \{t_i + U(r_i)\} \tag{1.3.1}$$

where t_i is the free particle Hamiltonian operator for the i-th electron ($m = 1$)

$$t_i = \begin{cases} \boldsymbol{p}_i^2/2 & \text{(Schrödinger case)} \\ c\boldsymbol{\alpha}_i \cdot \boldsymbol{p}_i + \beta_i mc^2 & \text{(Dirac case)} \end{cases}$$

and $U(r_i)$ is the effective potential energy of an electron at position $\boldsymbol{r}_i$, where

$$U(r) \sim \begin{cases} -\dfrac{Z}{r} + \text{const.} & r \to 0, \\ -\dfrac{(Z-N+1)}{r} & r \to \infty \end{cases} \tag{1.3.2}$$

and N is the number of electrons in the atom.

The electrons move independently in this model, so that the atomic wavefunction must be constructed from products of orbital wavefunctions, one for each physical electron. These orbitals, of central-field type, have the nonrelativistic form (1.2.8) or the relativistic form (1.2.13). The simplest conceivable wavefunction for the atom has the form of a product $\Phi := \psi_{a_1}(\boldsymbol{r}_1)\,\psi_{a_2}(\boldsymbol{r}_2)\ldots\psi_{a_N}(\boldsymbol{r}_N)$ where, for nonrelativistic electrons, $a_i = n_i l_i m_i \sigma_i$, $\sigma_i = \pm 1/2$ being the spin projection. Electrons are indistinguishable so that, when $\mathcal{P}$ is any permutation acting on the set $\{a_1, a_2, \ldots, a_N\}$, the function $\mathcal{P}\Phi$ is also a wavefunction for the system Pauli's exclusion principle allows at most one electron to have the orbital label a_i, and this is best accomodated by choosing a determinantal wavefunction

$$\Psi = \sum_{\mathcal{P}} (-1)^p \mathcal{P}\,\Phi \equiv \text{const.} \begin{vmatrix} \psi_{a_1}(\boldsymbol{r}_1) & \psi_{a_1}(\boldsymbol{r}_2) & \ldots & \psi_{a_1}(\boldsymbol{r}_N) \\ \psi_{a_2}(\boldsymbol{r}_1) & \psi_{a_2}(\boldsymbol{r}_2) & \ldots & \psi_{a_2}(\boldsymbol{r}_N) \\ \ldots & \ldots & \ldots & \ldots \\ \psi_{a_N}(\boldsymbol{r}_1) & \psi_{a_N}(\boldsymbol{r}_2) & \ldots & \psi_{a_N}(\boldsymbol{r}_N) \end{vmatrix} \tag{1.3.3}$$

where p is the parity of the permutation $\mathcal{P}$. The determinant vanishes whenever any two rows have the same label, $a_i = a_j$, or two columns the same argument: $\boldsymbol{r}_i = \boldsymbol{r}_j$ with $i \neq j$. The same construction is valid for Dirac electrons with the labelling $a_i = n_i \kappa_i m_i$. We shall usually assume when dealing with bound states that the orbitals are orthonormal.

1.3.2 Closed and open shells

We obtain a single electron density by integrating $|\Psi|^2$ over all particle coordinates save one; the resulting volume density of electrons is $\rho_{tot} = \sum_i \rho_{a_i}$ where $\rho_{a_i} = |\psi_{a_i}|^2$. The number of nlm particles contained in the spherical polar volume element $d\boldsymbol{r} = r^2\,dr\,d\Omega$ is

$$\rho_{nlm}(\boldsymbol{r})\,d\boldsymbol{r} = |P_{nl}(r)|^2 |Y_l^m(\theta,\varphi)|^2\,dr\,d\Omega \tag{1.3.4}$$

for a normalized nonrelativistic wavefunction of the form (1.2.8). From the properties of spherical harmonics,

$$\sum_{m=-l}^{l} \rho_{nlm}(\boldsymbol{r})\,d\boldsymbol{r} = |P_{nl}(r)|^2\,dr.(2l+1)\frac{d\Omega}{4\pi} \tag{1.3.5}$$

is spherically symmetric. The $2l+1$ particles of the set $nlm, m = -l, \ldots, +l$ are therefore said to form a *closed subshell*, with radial density

$$D_{nl}(r) = |P_{nl}(r)|^2. \tag{1.3.6}$$

The spatial density is the same for both spin projections and this *doubles* the total number in the subshell to $4l+2$ when spin is taken into account. The arrangement of electrons in subshells is called a *configuration*. If the atom has a configuration consisting only of closed subshells its electron distribution is spherically symmetric and the electric field is purely radial. Most of the electrons in a neutral atom in its ground state will belong to closed subshells, making the central field model a good starting point for atomic physics.

In the relativistic central field model, where the orbitals are 4-spinors of the form (1.2.13), the subshells are defined by the $2j+1$ states $n\kappa m$ (or $nljm$) with $m = -j, -j+1, \dots, +j$. The Dirac analogue of (1.3.4) is

$$\rho_{n\kappa m}(\boldsymbol{r})\, d\boldsymbol{r} = D_{n\kappa}(r) |A_{|\kappa|,m}(\theta, \varphi)|^2\, dr\, d\Omega \tag{1.3.7}$$

where $D_{n\kappa}(r)$ is defined by (1.2.19), and the angular density is given in §3.2.5. The subshell density is now given by

$$\sum_{m=-j}^{j} \rho_{n\kappa m}(\boldsymbol{r})\, d\boldsymbol{r} = D_{n\kappa}(r)\, dr.(2j+1)\frac{d\Omega}{4\pi}. \tag{1.3.8}$$

A closed nonrelativistic nl subshell with $4l+2$ electrons has a density equivalent to the sum of the densities of closed subshells nlj with $j = l - 1/2$ and $j = l + 1/2$ containing respectively $2l$ and $2l+2$ electrons.

A more realistic atomic model is based on the Hamiltonian

$$H = \sum_{i=1}^{N} t_i + \sum_{i<j} 1/r_{ij} \tag{1.3.9}$$

so that we can write

$$H = H_0 + V, \quad V = \sum_{i<j} 1/r_{ij} - \sum_i U(r_i). \tag{1.3.10}$$

The eigenvalues of the zero-order Hamiltonian, H_0, depend only on the way in which the electrons have been assigned to the central field orbitals. When there are open shells, any of the $\binom{2l+1}{q_{nl}}$ determinants that can be formed by selecting q_{nl} different values of m from the set $m = -l, \dots, +l$ will give energy $E_0 = (\Psi|\, H_0\, |\Psi)$.. The full set of determinants Ψ_i so obtained span a linear space; by diagonalizing H in this space we obtain a set of energy levels $E_\alpha = E_0 + V_\alpha$, where V_α is an eigenvalue of the secular matrix $\boldsymbol{V}$ whose elements are

$$V_{ij} = (\Psi_i|\, V\, |\Psi_j), \tag{1.3.11}$$

resolving the degeneracy, at least in part. There is in principle no reason to restrict ourselves to a space consisting of determinants spanning a single open subshell. The *superposition of configuration* (SOC) method employs wavefunctions that are linear combinations of determinants belonging to more than one configuration; the method is often said to be one of *configuration interaction* (CI), reflecting the fact that configurational wavefunctions of the same symmetry may have off-diagonal matrix elements in the Hamiltonian matrix [3, 4, 48]

Things are actually slightly more complicated: we need to take other symmetries into account. Thus there can be no preferred spatial direction for an isolated atom with no external interactions. The result is that the infinitesimal generators of spatial rotations, the operators $\boldsymbol{L} = \sum_i \boldsymbol{l}_i$ and $\boldsymbol{S} = \sum_i \boldsymbol{s}_i$ (or $\boldsymbol{J} = \sum_i \boldsymbol{j}_i$ in the relativistic case) commute with H. Thus eigenstates of the nonrelativistic H can be classified as simultaneous eigenstates of $H, \boldsymbol{L}^2, \boldsymbol{S}^2$. In some cases, the only remaining degeneracy will be with respect to the projections L_z and S_z, so that just one member of the set serves to represent them all. The theory of angular momentum in atomic structure and spectra is described in detail in many well-known texts including [2, 3, 48, 49, 50, 51]. With the exception of [51] these texts all deal with the field from a nonrelativistic viewpoint. Aspects of the theory of angular momentum required in relativistic atomic structure theory are treated in Appendix B.3 and Chapter 6.

1.3.3 Mean field potentials

A simple mean field *parametric potential* has been used extensively, amongst other things, for large-scale collisional-radiative calculations for highly ionized atoms involving hundreds of levels and millions of cross sections for various processes, [52, 53]. The radial charge distribution in nonrelativistic atom models has several peaks, one for each nl subshell; each peak can be modelled by writing [53, Equation (5)]

$$\rho(r) = -4\pi r^2 q \mathcal{N} r^{2l+2} \mathrm{e}^{-\alpha r} \tag{1.3.12}$$

where q is the number of electrons associated with this peak, $\alpha = (2l+3)/\langle\, r\,\rangle_{nl}$ where $\langle\, r\,\rangle_{nl}$ is the mean radius of the peak and $\mathcal{N}$ is a normalizing constant. It is simple to solve Poisson's equation for this charge density, leading to a mean field potential with parameters $\boldsymbol{\alpha} = \{\alpha_{nl}\}$

$$\begin{aligned} U(\boldsymbol{\alpha}, r) = -\frac{1}{r}\Big\{ & Z - N + 1 + \sum_s q_s\, g(L_s, \alpha_s, r) \\ & + \sum_t q_t\, f(l_t, \alpha_t, r) \Big\} \end{aligned} \tag{1.3.13}$$

where s runs over the closed shells, t over the open shells, and $N = \sum_s q_s + \sum_t q_t$ is the total number of electrons. The radial functions are

$$\begin{aligned} f(l, \alpha, r) &= \mathrm{e}^{-\alpha r} \sum_{j=1}^{2l+1} \left(1 - \frac{j}{2l+2}\right) \frac{(\alpha r)^j}{j!}, \\ g(L, \alpha, r) &= \sum_{l=0}^{L=n-1} \frac{4l+2}{2n^2} f(l, \alpha^{(l)}, r) \end{aligned}$$

where $\alpha^{(l)}$ takes account of the fact that the nl subshells have slightly different mean radii for different l values. Thus $f(l, \alpha, r)$ is the radial potential due to an electron in the nl subshell at radius r, and $g(L, \alpha, r)$ is the radial potential due to each of the $2n^2$ electrons in the complete n shell summed over the contributing nl subshells. Observable quantities are functions of the parameters α_s; they can be determined by, for example, minimizing the average configurational energy of the atom with resepct to these parameters or by fitting a chosen set of atomic energy levels. [53]

Another *ab initio* mean field potential is provided by Thomas-Fermi (TF) theory [2, §2[14]] [48, §7.8]; this is based on a statistical model of the electron distribution parametrized only by the values of the nuclear charge Z and the number of atomic electrons N. The nonrelativistic SUPERSTRUCTURE code [54, 55] constructs a mean field potential with a separately parametrized TF potential for each subshell.

Model potentials are less expensive than fully *ab initio* self-consistent field (SCF) models of the Hartree or Hartree-Fock type. The nonrelativistic wavefunction Ψ for a single closed shell configuration depends on the radial amplitudes $P_{nl}(r)$ used in its construction; the expectation value of the Hamiltonian (1.3.9) can be regarded as a functional $E[P_{a_1}, P_{a_2}, \ldots]$ so that standard methods of the calculus of variations can be used to derive nonlinear integro-differential (SCF) equations for the radial amplitudes P_a. This system of equations must be solved iteratively: we start with a trial set of P_a, from which we calculate subshell potentials enabling us to solve for a new set of trial orbitals. The cycle is repeated until the system is *self-consistent.* The method can be elaborated to include CI-type wavefunctions, and the resulting multi-configuration Hartree-Fock (MCHF) method is now widely used [4, 56, 57]. The relativistic DHF equivalent was first proposed by Swirles [21].

1.3.4 Comparison of Hartree-Fock and Dirac-Hartree-Fock models for ground states

The Periodic Table of the Elements, summarizing a huge body of information on trends in the physical and chemical properties of the elements, a simplified version of which appears in Figure 1.5, arranges the elements vertically in *groups* that can be correlated with the electronic configurations of atomic outer or *valence* shells ($ns^q, np^q, nd^q, \ldots$ according to the subshell l values) and in *periods* that reflect the principal quantum number n of the open shell that is being filled as q increases from left to right across the table. The lanthanides and actinides, in which the filling of the $4f$ and $5f$ subshells spoil this simple description, are shown at the bottom of the table.[5]

The mechanism of shell filling, Bohr's *Aufbau* (or *building up*) principle, involves constructing electron configurations for atomic ground states by assigning each electron to the orbital vacancy of lowest energy currently available

[5] Seaborg [61] describes the evolution of the modern Periodic Table including observed and predicted chemical properties of transactinide elements and possible extensions of the table to include undiscovered elements up to atomic number 168.

according to some central field model. The ordering, at least in the upper rows of the table, follows the ordering in hydrogenic atoms:

$$1s_{1/2}, 2s_{1/2}, 2p_{1/2}, 2p_{3/2}, 3s_{1/2}, 3p_{1/2}, 3p_{3/2}, 3d_{3/2}, 3d_{5/2}, \ldots$$

The Hartree-Fock (HF) and Dirac-Hartree-Fock (DHF) eigenvalues give a very simple picture of the way in which this happens. Koopmans' Theorem, §7.7, interprets the energy eigenvalue of an HF or DHF orbital as the energy required to remove an electron from a subshell without allowing the remaining electrons to relax. Whilst the neglect of relaxation and the details of interaction with the other electrons are needed for a fuller understanding of the *Aufbau* process, the HF and DHF eigenvalues provide a remarkably simple explanation of the main features. This can be seen from the extensive tables of HF eigenvalues and orbital expectation values for LS-configuration average atomic ground states published by Froese Fischer [58] and Mann [59] together with corresponding data for DHF average of configuration atomic ground states published by Desclaux [60] at about the same time.

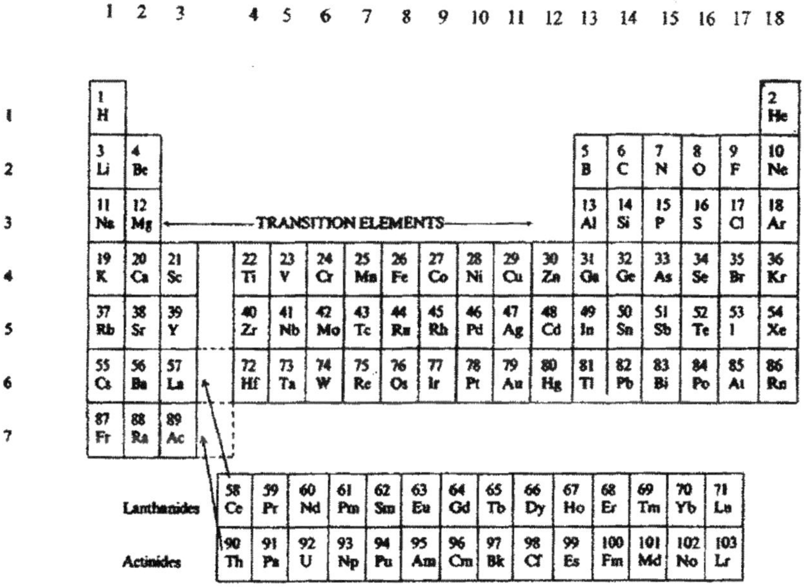

Fig. 1.5. Simplified version of the Periodic Table of the Elements.

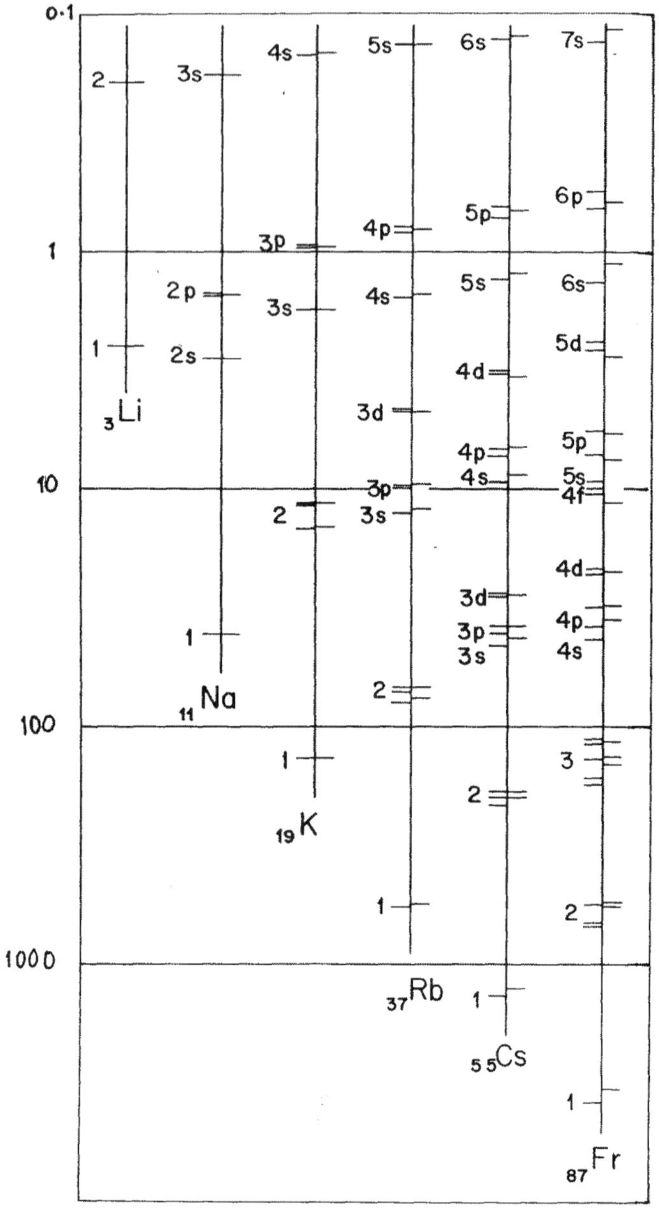

Fig. 1.6. Comparison of atomic relativistic subshell binding energies $-\varepsilon_{n\kappa}/E_h$ predicted by Koopmans' theorem from DHF jj-average of configuration calculations [60] (left-hand ticks on each vertical line) with corresponding nonrelativistic eigenvalues $-\varepsilon_{nl}/E_h$ (right-hand ticks) from HF LS-average of configuration calculations [58, 59] for alkali atoms of Group 1. The electron configuration is [Core] ns, where the core consists entirely of filled subshells. Binding energies increase down the page on a logarithmic scale.

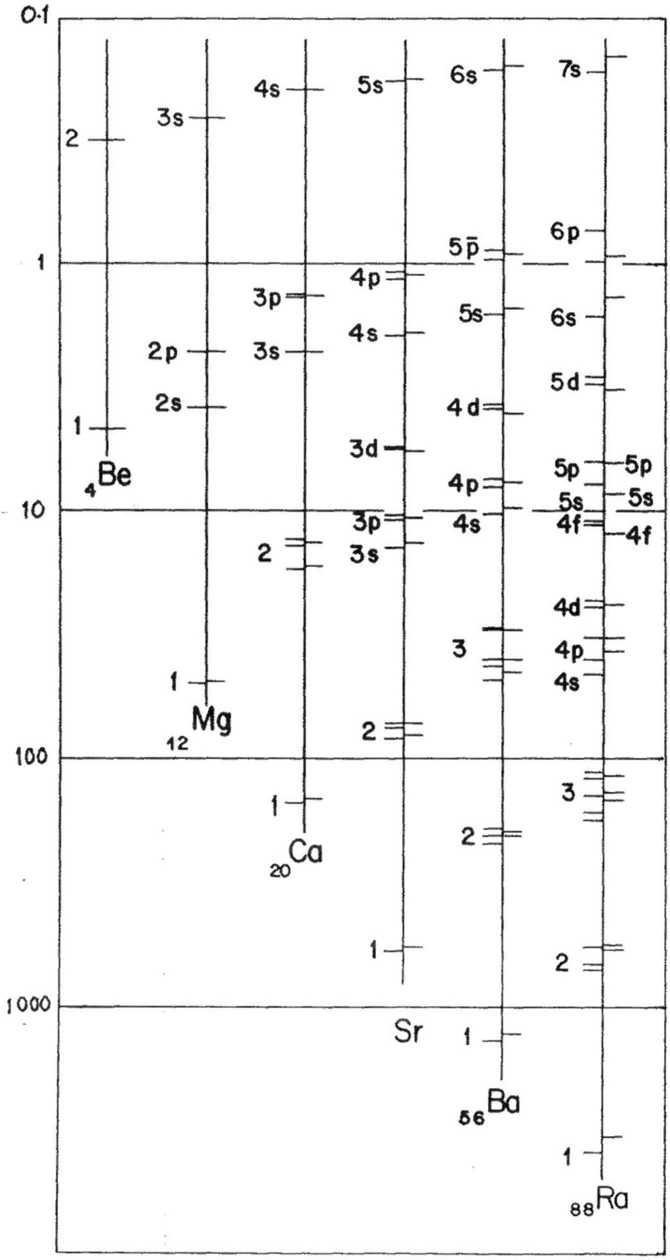

Fig. 1.7. Koopmans' theorem binding energies for alkaline earths, Group 2. The electron configurations are [Core] ns^2. The data are organized as in Figure 1.6.

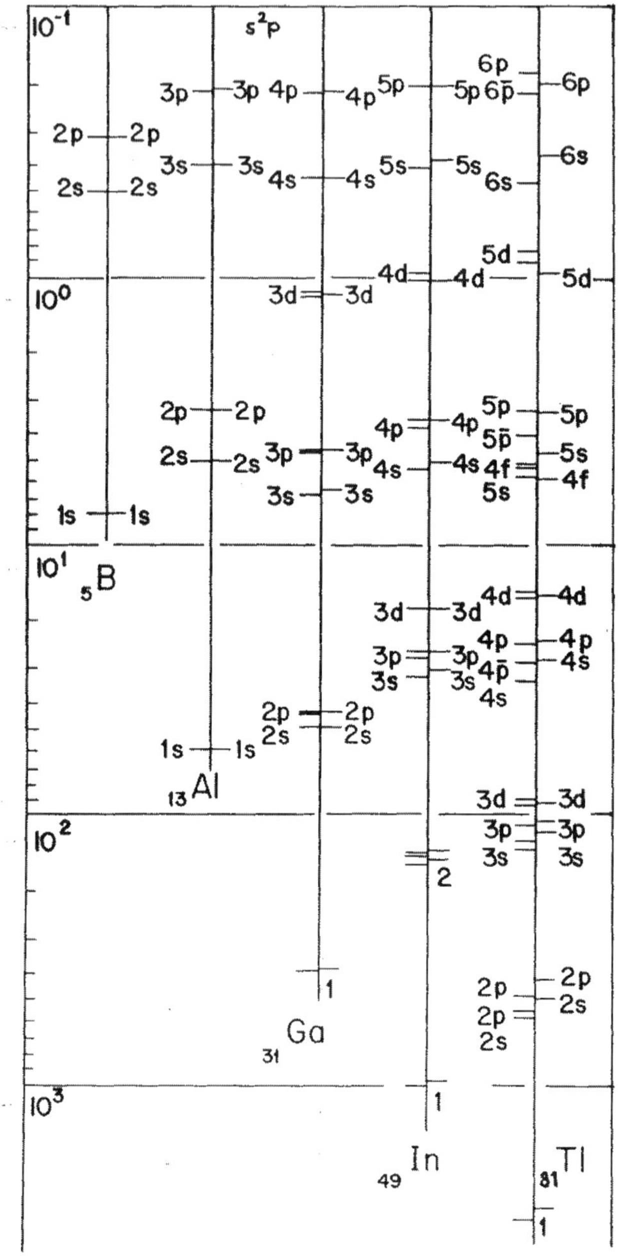

Fig. 1.8. Koopmans' theorem binding energies for Group 13. The electron configurations are [Core] ns^2np. The data are organized as in Fig. 1.6.

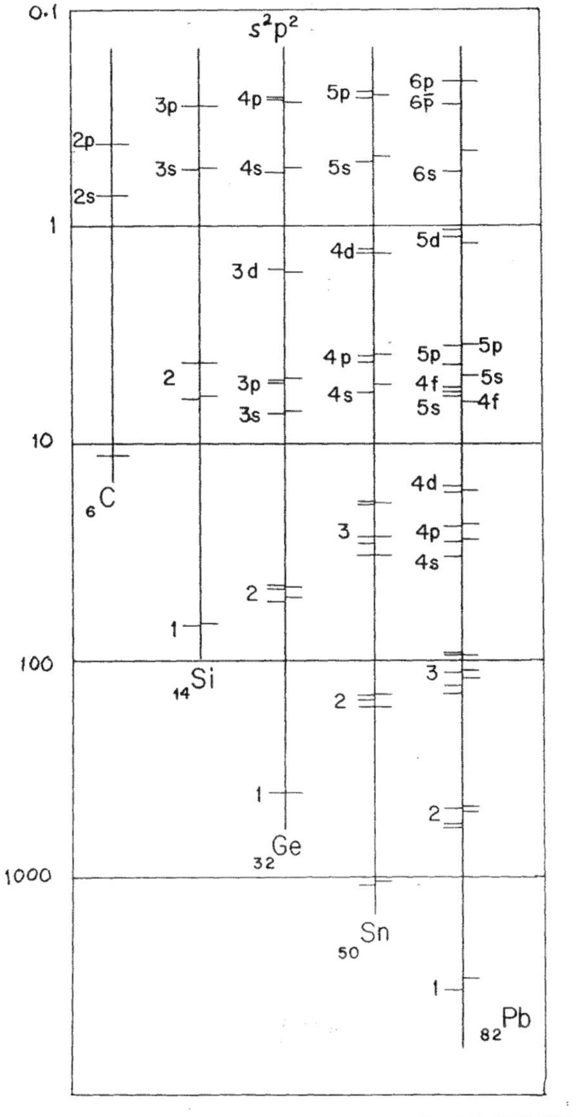

Fig. 1.9. Koopmans' theorem binding energies for carbon Group 14. The electronic configurations are [Core] ns^2np^2. The data are organized as in Fig. 1.6.

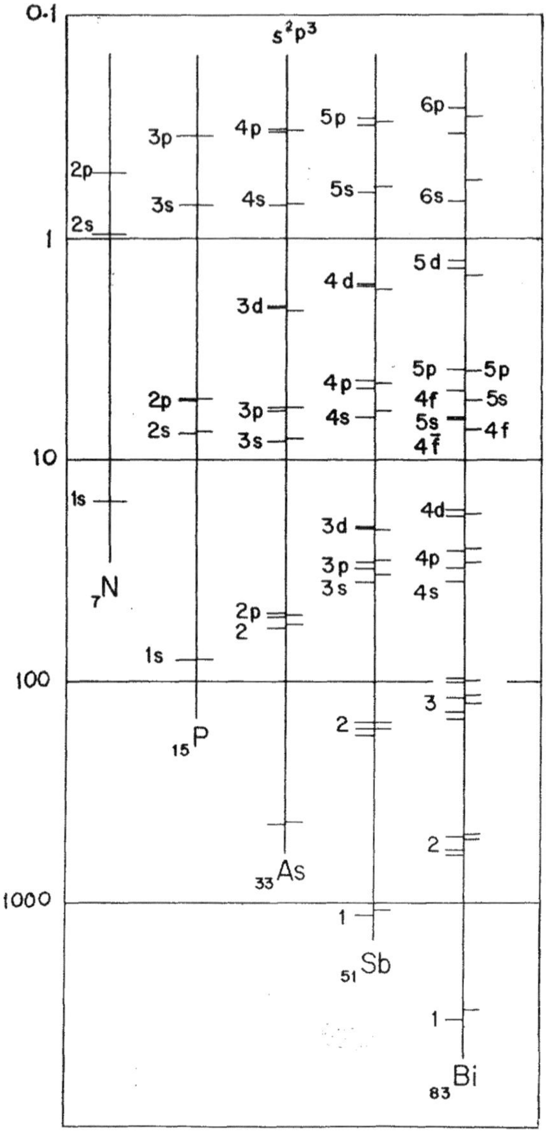

Fig. 1.10. Koopmans' theorem binding energies for nitrogen Group 15. The electronic configurations are [Core] $ns^2np^2np^3$. The data are organized as in Fig. 1.6.

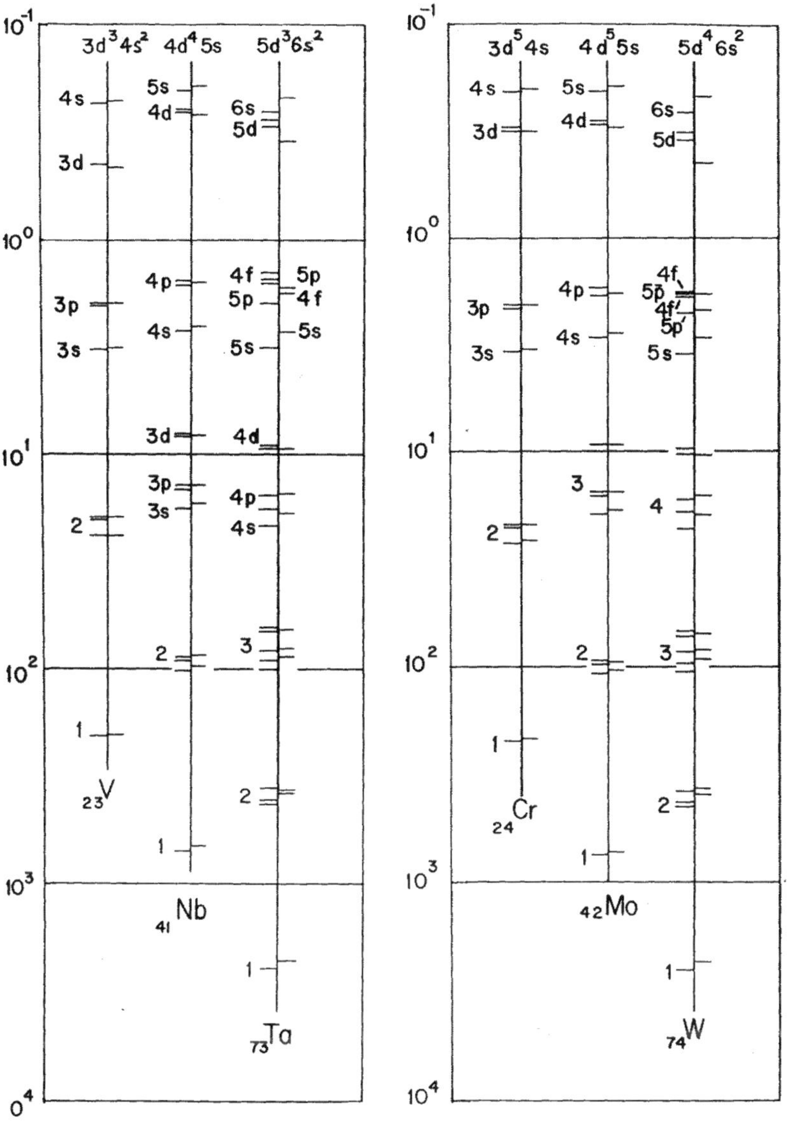

Fig. 1.11. Koopmans' theorem binding energies for elements of Groups 5 and 6. The electronic configurations are [Core] $(n-1)d^3ns^1$ or $(n-1)d^4ns$ for Group 5 elements and $(n-1)d^5ns$ or $(n-1)d^4ns^2$ for Group 6 elements. The data are organized as in Fig. 1.6.

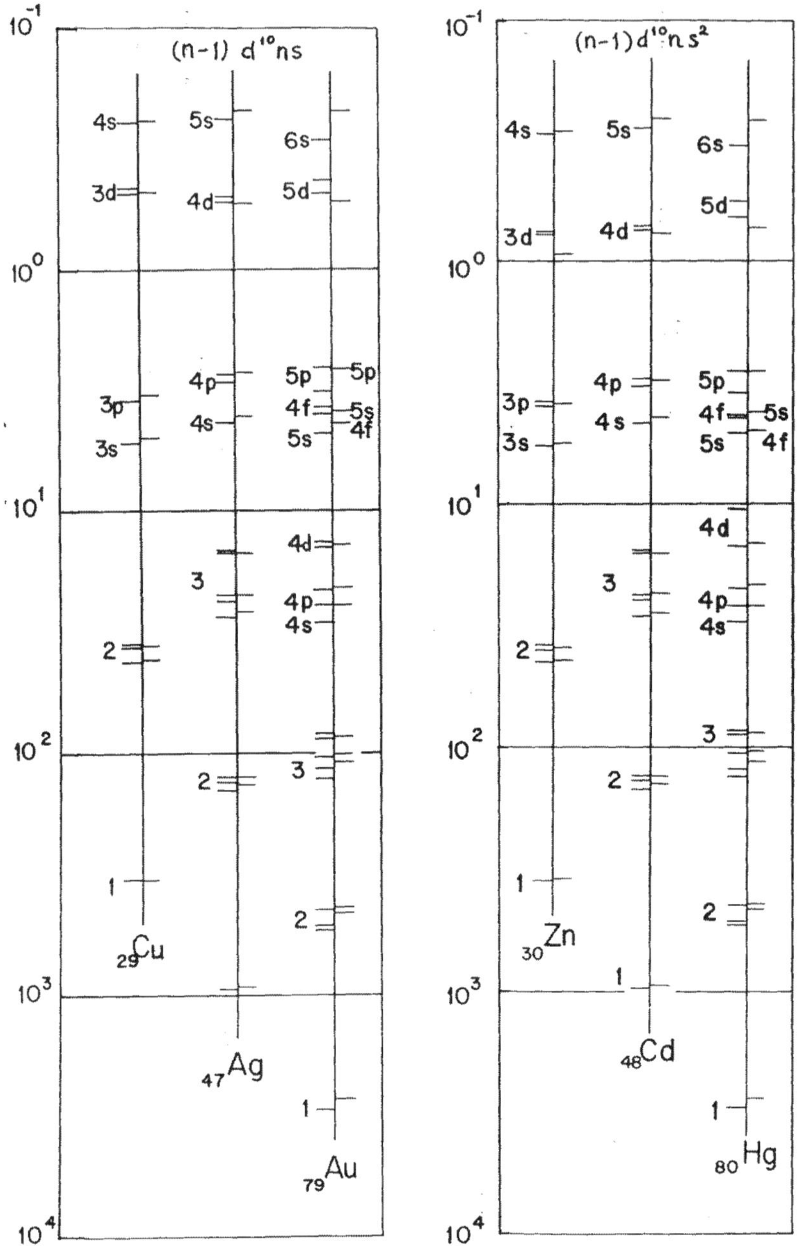

Fig. 1.12. Koopmans' theorem binding energies for coinage metals, Group 11, with electronic configuration [Core] $(n-1)d^{10}ns$ and Group 12, with electronic configuration [Core] $(n-1)d^{10}ns^2$. The data are organized as in Fig. 1.6.

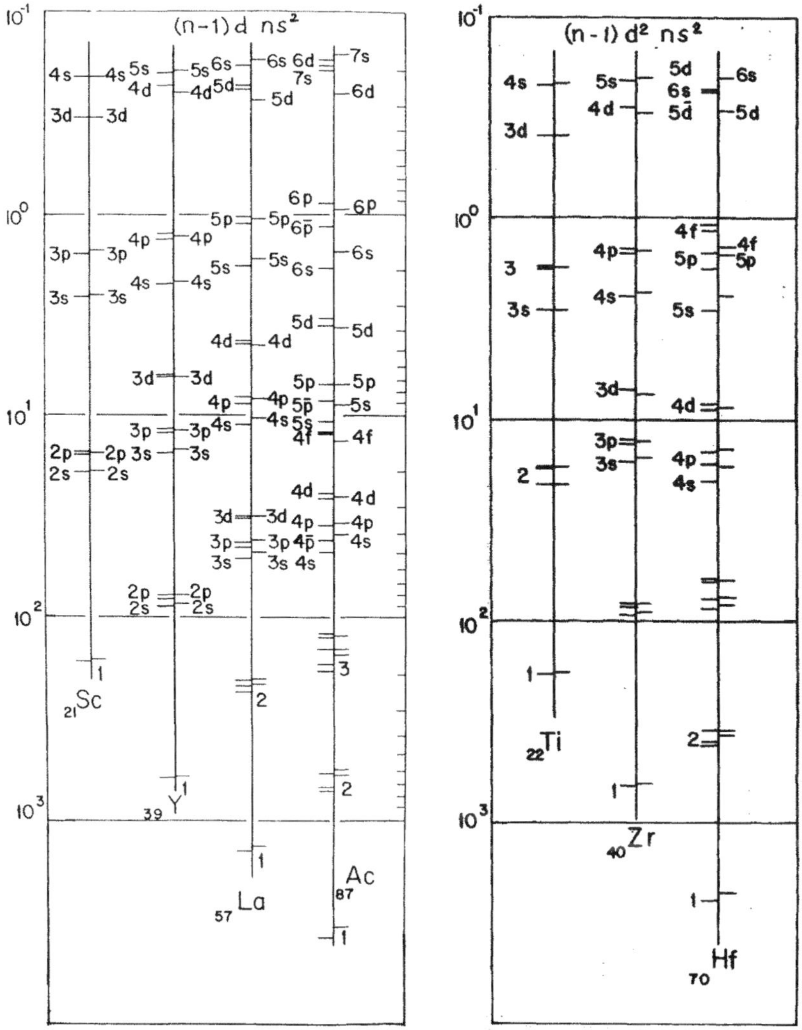

Fig. 1.13. Koopmans' theorem binding energies for Group 3, electronic configuration [Core] $(n-1)dns^2$ and Group 4, electronic configuration [Core] $(n-1)d^2ns^2$. The data are organized as in Fig. 1.6.

Results for main groups of the Periodic Table are displayed in Figs. 1.6–1.13. The ticks on the left-hand side of each vertical line in the diagram indicate the DHF Koopmans' ionization energy, whilst those on the right are from the equivalent nonrelativistic HF calculation. The ground electronic configurations of the alkali metals, Fig. 1.6, are [Core] ns: the valence subshell has a single ns electron outside a closed shell core. In elements such as Cs, the DHF eigenvalues of inner shells are below the HF eigenvalues, very much as expected from the hydrogenic calculations: for example $-\varepsilon_{2s}^{DHF}$ (a positive number) is larger than $-\varepsilon_{2s}^{HF} \approx -\varepsilon_{2p_{1/2}}^{DHF}$, whilst $-\varepsilon_{2p_{3/2}}^{DHF} \approx -\varepsilon_{2p}^{HF}$. Spin-orbit splittings are substantial in the high-Z members of the sequence. However, the valence ns electrons are progressively less tightly bound as n increases although $-\varepsilon_{7s}^{DHF}$ in Fr is greater than $-\varepsilon_{6s}^{DHF}$ in Cs, in line with observed trends in ionization energies (IE). Vertical distances in the logarithmic energy scale are proportional to the ratio of two energies, and it is worth noting that spin-orbit splittings and sometimes relativistic/nonrelativistic energy ratios may be larger for valence and sub-valence electrons than for the innermost shells, providing unequivocal evidence of the propagation of relativistic effects across the atom. The trends in the alkaline earth elements, Figure 1.7, are similar; the most notable new feature is the relativistic destabilization of the $4f_{5/2,7/2}$ and $5d_{3/2,5/2}$ subshells of Ra as a result of the more compact electron distribution in the inner core subshells. In the boron Group 13, with electronic configurations [Core] ns^2np, the adjustments are still more complicated, with the $4f_{5/2,7/2}$ subshells of Tl lying between the $5s$ and $5p_{1/2}$ subshells. We see the same effect in the carbon Group 14 where the electronic configuration is [Core] ns^2np^2.

The ordering of eigenvalues in DHF calculations gives some insight into the electronic structure of neutral atom ground states, but gives no information on the order in which shells fill in individual atoms. Some insight into the interactions determining into which shell an additional electron is likely to be placed can be obtained from the regularities of observed ionization energies [62] in conjunction with a heuristic model [63] employing a simplified expression for the energy of the atom.

The flavour of this analysis can be appreciated from Fig. 1.14. As the degree of ionization i steps to the right, each straight line segment sloping upwards to the right links points whose ordinates are the energies, I_i, of ionization of successive electrons from the corresponding atom. The transverse lines link points at which neighbouring atoms of the group have similar valence electronic configurations. Thus the ground configurations of the neutral atoms N, P, As, Sb, Bi, ... have been assigned to [Core] $2p^3$, $3p^3$, $4p^3$, ... We see that successive ionization energies of np electrons down to the closed subshell core increase in a roughly linear fashion: $I_i \approx iI_1$ for $i = 1, 2, 3$.

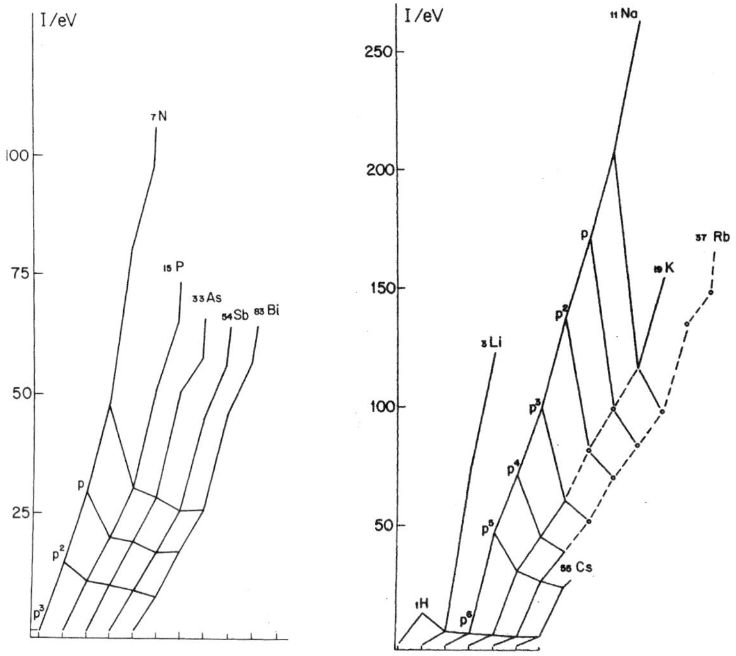

Fig. 1.14. Observed successive ionization energies, I_i/eV, against i for elements of Group 15 and Group 1. Source: NIST on-line database [62].

1.3.5 The mechanism of shell filling

A simple model due to Pyper and Grant [63] readily explains this behaviour. The first three ionization energies of nitrogen are

$$I_1 = 14.5eV, \qquad I_2 = 29.6eV \qquad I_3 = 47.5eV,$$

at which stage we are left with the C^{3+} Be-like ion core in the $1s^2 2s^2$ configuration. The central field approximation assumes that the $2p$ orbital wavefunctions is insensitive to the valence electron configuration, and we also assume that every $2p$ electron has the same effective interaction energy, $-C$, with the Be-like core. The first step in building up the valence $2p$ shell binds the electron with an energy

$$-C = -I_3 \approx -47.5eV. \tag{1.3.14}$$

Our assumptions require the next electron to be bound to the core with energy $-C$, but now its energy will be raised by an amount, $+F$, because of repulsion by the first electron. Thus

$$-I_2 \approx -C + F \approx 29.6eV. \tag{1.3.15}$$

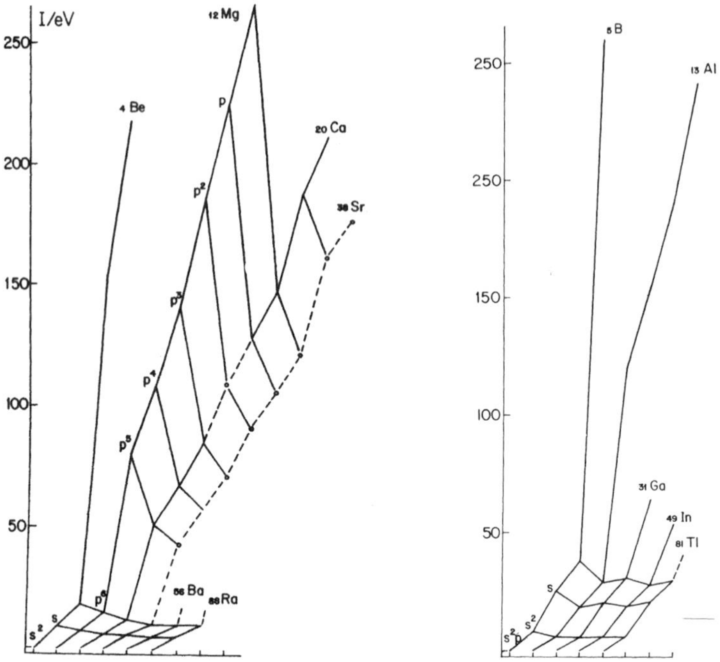

Fig. 1.15. Observed successive ionization energies, I_i/eV, against i for elements of Group 2 (left) and Group 13 (right). Source: NIST on-line database [62].

The third and final electron is repelled by the first two electrons and so

$$-I_1 \approx -C + 2F \approx 14.5eV. \tag{1.3.16}$$

Thus $F \approx 15eV$ and $C \approx 45eV$ consistent with the observation that the binding energy (or electron affinity) of an additional electron is very small. (On this model it is given by $I_0 = -C + 3F \approx 0$). The Pyper/Grant model thus approximates the total energy of an l^N configuration by a formula of the form

$$E_N = -NC + \frac{1}{2}N(N-1)\,F, \tag{1.3.17}$$

where C is the mean binding energy of an l electron to the core and F is the mean repulsion energy of a pair of l electrons. Because

$$I_i = E_{N-i} - E_{N-i+1}$$

the general formula [63, Equation (2.08)] is

$$\begin{aligned} I_i &= \frac{i}{N}C + \left(1 - \frac{i}{N}\right) I_0 \\ &= i\,I_1 - (i-1)\,I_0 \qquad (1.3.18)\\ &\approx i\,I_1. \qquad (1.3.19) \end{aligned}$$

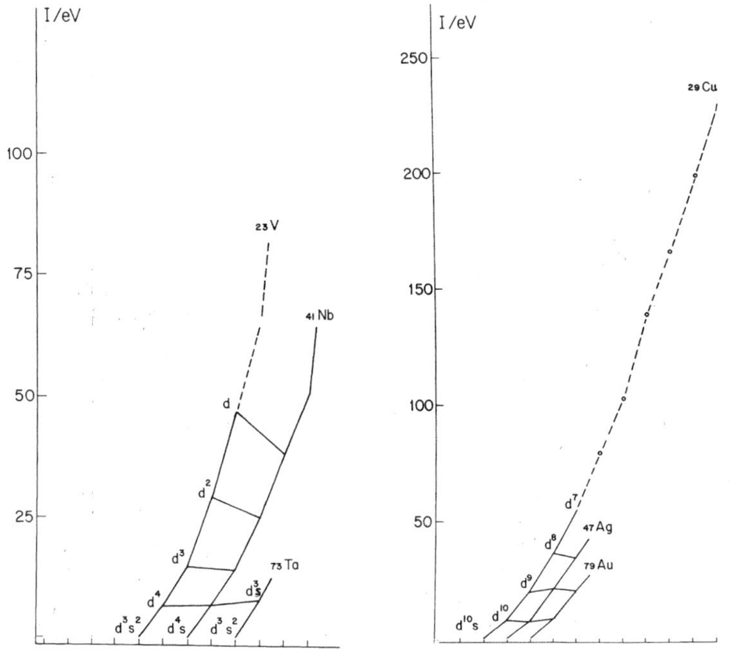

Fig. 1.16. Observed successive ionization energies, I_i/eV, against i for elements of Group 5 (left) and Group 11 (right). Source: NIST on-line database [62].

As the remaining figures, 1.15–1.18 confirm, this linear dependence on the degree of ionization is roughly fulfilled for ionization from s^n, p^n, d^n, and even f^n configurations in positive ions as well as neutral atoms.[6] Pyper and Grant [63] studied successive ionization from s^2 and p^N configurations from several atoms using theoretical single configuration DHF calculations. Table 1.1 shows results for the elements of Group 14, headed by carbon. The Pyper/Grant models (1.3.18) and (1.3.19) do better than one might have expected. The DHF energy differences in the last column follow experimental trends but the values are uniformly low; this is not surprising as the model takes no account of the phenomenon of electron correlation.

The plots of successive ionization reveal cases in which filling of configurations is somewhat irregular. Thus Group 5 neutral ground states are $3d^34s^2$ in vanadium and $5d^36s^2$ in tantalum but $4d^45s$ in niobium, whilst the ions V^+, Nb^+ and Ta^+ have the respective configurations $3d^4$, $4d^4$ and $5d^36s$. The lanthanides, Figure 1.18 are even more complicated, with most ground states having the form [Xe]$4f^n6s^2$, whilst La has [Xe] $5d6s^2$, Ce has [Xe] $4f5d6s^2$,

[6] Pyper and Grant [63] say that it fails for successive ionization from d^n and f^n configurations, but that assertion is not borne out by data of Fig. 1.18 that were not available at the time.

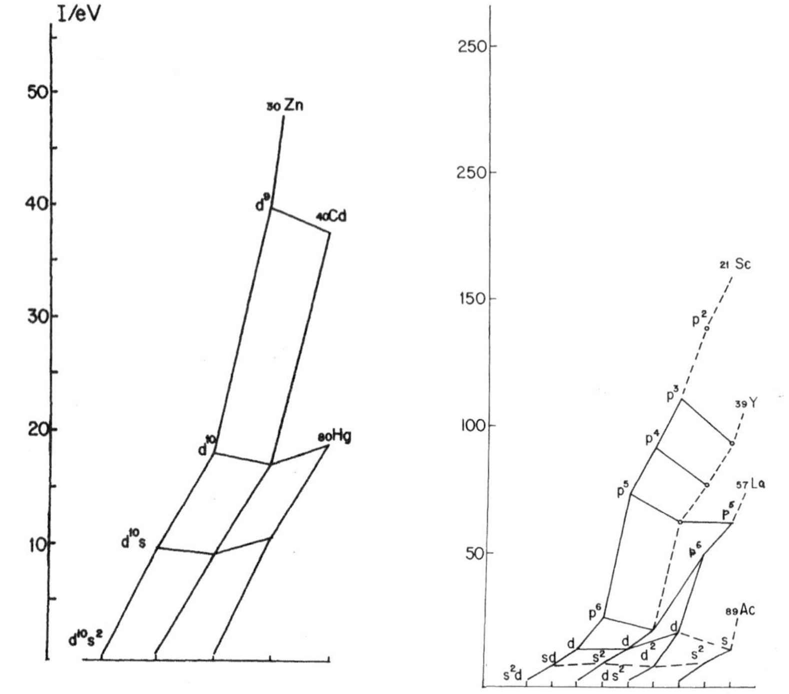

Fig. 1.17. Observed successive ionization energies, I_i/eV, against i for elements of Group 12 (left) and Group 3 (right). Source: NIST on-line database [62].

Gd has [Xe] $4f^7 5d6s^2$. The lanthanide $5d$ shell fills smoothly as Z increases once the $4f$ shell is complete at Yb. An extension of the simple Grant/Pyper model to include more than one valence subshell, allowing for different orbital properties of $nd_{3/2}$ and $nd_{5/2}$ subshells, for example, or $(n-1)d^x ns^{N-x}$ gives some idea of the energy balance involved in alternative modes of filling. The NIST database [62] shows that the more complicated configurations interact quite strongly so that whilst such extensions of the simple Pyper/Grant model may give some insight into the relative energies of different configurations, the differences may be too small to order configurations with certainty.

1.3.6 Other approaches

The original heuristic paper on relativistic self-consistent fields [21] started from the Dirac-Coulomb Hamiltonian (1.3.9). Z-expansion methods were suggested in the 1960s to study astrophysically important spectra from ions with only a small number of electrons. They rely on a formal double series expansion, in the parameters $\lambda = Z^{-1}$ and $\mu = \alpha^2 Z^2$, of the matrix elements of the Dirac-Coulomb-Breit Hamiltonian (in which the Breit interaction (1.1.6) has

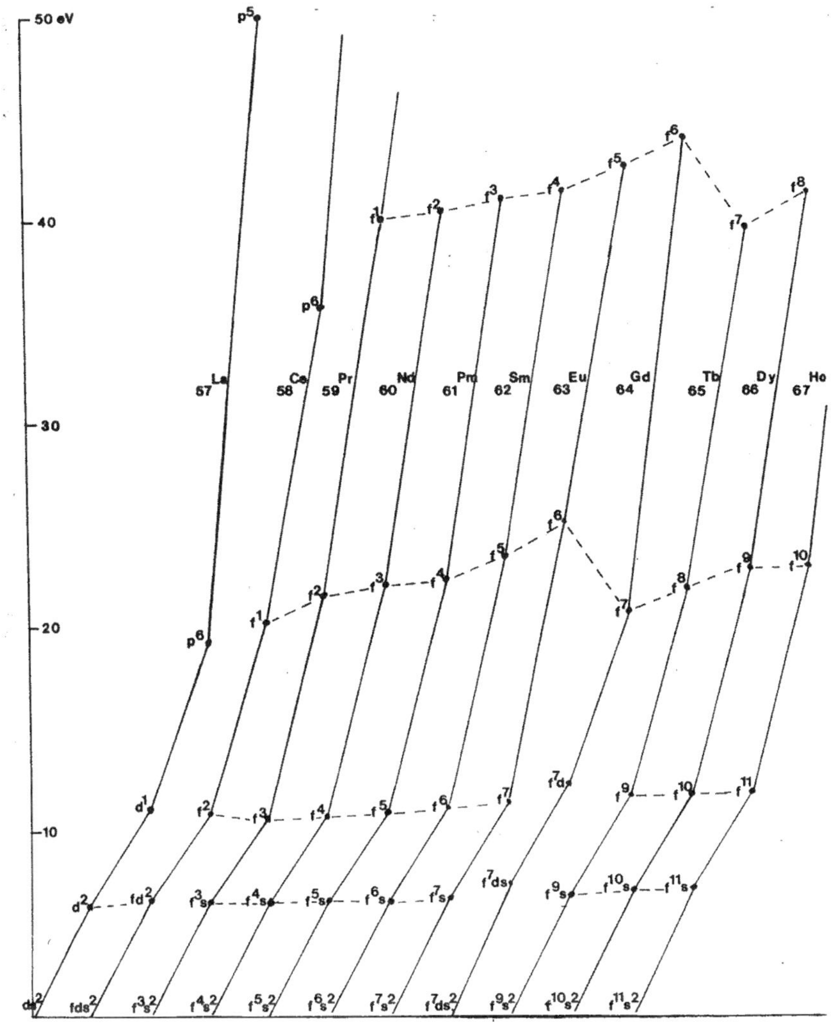

Fig. 1.18. Observed successive ionization energies, I_i/eV, against i for Lanthanides. Source: NIST on-line database [62].

been added to the DC Hamiltonian (1.3.9)) retaining only the lowest powers of μ [64]–[69]. At this level of approximation, correlation is neglected, so that it can only be applied to fine structure within a complex (that is to say configurations built from different assignments within the same shell). Better results are obtained by including nonrelativistic correlation terms of second order [69]. A semi-empirical screening model has also been used to estimate correlation corrections [66]–[68].

Table 1.1. Successive ionization energies (eV) in Group 14 (p^2 configurations)

	Expt.	(1.3.19)	(1.3.18)	ΔE_{DHF}
		Carbon		
I_1	11.26	12.49	11.90	10.77
I_2	24.38	24.97	24.97	24.05
		Silicon		
I_1	8.15	8.16	8.12	7.54
I_2	16.34	16.33	16.33	15.72
		Germanium		
I_1	7.88	7.79	8.04	7.35
I_2	15.93	15.58	15.58	15.29
		Tin		
I_1	7.34	7.16	7.37	6.82
I_2	14.63	14.31	14.31	14.00
		Lead		
I_1	7.41	7.31	7.07	6.79
I_2	15.05	14.68	14.68	14.40
		114		
I_1	–	8.48	8.24	8.03
I_2	–	16.96	16.96	16.59

Direct application of QED perturbation theory [70] has been attempted for a few systems.

1.4 Applications to atomic physics

The methods described later in this book have been used widely in atomic physics to interpret experimental atomic spectra, to evaluate cross sections for a range of continuum processes, and as an enabling technology for practical applications. Atomic structure calculations estimate energy levels directly, but the attainment of spectroscopic accuracy requires accurate modelling that can never be straightforward. The calculation of spectral line widths is often sensitive to the choice of electromagnetic gauge potentials; although gauge sensitivity can be mitigated for strong electric dipole transitions, it remains a challenge for forbidden lines and for intercombination lines, which also are sensitive to the treatment of electron correlation. There are further challenges in the treatment of continuum processes such as photo-ionization, radiative recombination, Auger transitions, and three-body recombination processes of interest, for example, in plasma physics.

Traditional spectroscopy records the emission and absorption of radiation by atoms in different spectral ranges, from which line positions and profiles can be measured. Laser spectroscopy and the invention of devices such as the electron beam ion trap (EBIT) [71], which can selectively produce and excite particular stages of ionization, make it possible to produce very high quality spectra demanding theoretical models with matching accuracy. Highly ionized atoms of heavy metals such as tungsten are present in high-temperature laboratory plasmas [72], and the modelling of their emission spectra requires a range of data on radiative transitions, Auger and Coster-Kronig transitions, and electron collisions. Laser interactions with matter and observations of astrophysical plasmas involving ions of all but the lightest atoms also benefit from calculations based on relativistic quantum theory.

Line positions, energy levels, and line widths are quantities that can be estimated fairly straightforwardly for atomic ions using a variety of *ab initio* schemes such as MCDHF. Neutral atoms and negative ions are generally more demanding than positive ions. The relative dominance of the electron-nucleus interaction in highly ionized atoms often makes a central field model a surprisingly good starting point. In this case, SOC wavefunctions built from a relatively small number of configurational states (CSF) can give excellent results. The power and memory of modern computers now permit calculations with wavefunctions containing several hundred thousand CSFs. Such time-consuming calculations are still uncommon, and they require computational technology that is still being developed.

1.4.1 X-ray spectra

A combination of relativistic atomic structure theory with critically evaluated experimental data provides the ingredients of a comprehensive tabulation of K-shell and L-shell X-ray transition and absorption edge energies for all the elements from neon to fermium and illustrates the use of many of the methods described in this book. The motivations for this ambitious 20-year programme coordinated by the US National Institute of Standards and Technology (NIST) [73] included the need to improve on prior compilations of X-ray data, to combine data from X-ray and optical interferometry to provide an accurate linkage of crystal spacings to optical wavelengths and to the SI definition of length, and also to provide reference lines for specific applications in X-ray crystallography.

The first step in the calculation of X-ray energies [73] is a Dirac-Hartree-Fock calculation for inner shell hole states. It is usually sufficient for this to use an average of configuration procedure that ignores open shell effects.

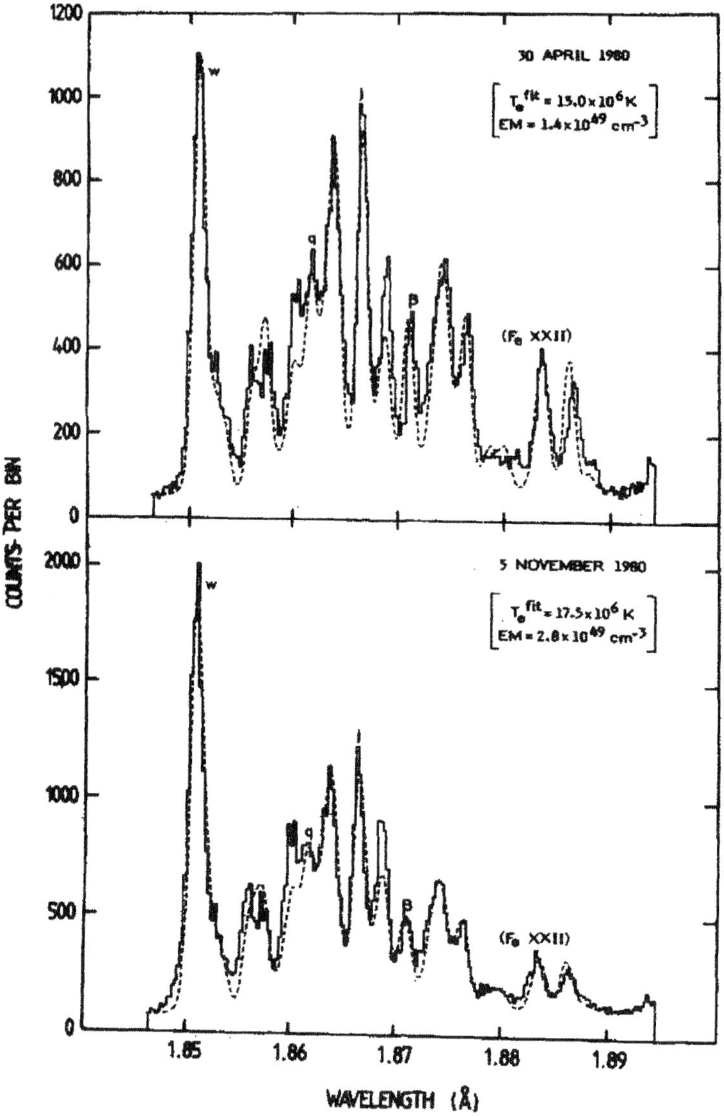

Fig. 1.19. The histogram records portions of the X-ray emission from solar flares taken on 30 April 1980 (4 minutes duration) and 5 November 1980 (about 1.5 minutes duration). The calculated spectrum (dashed line), comprising lines from Fe XXII – XXV spectra, is a fit for the single indicated temperatures. Reprinted, with permission, from [77].

Treatment of the magnetic part of the electron-electron interaction on the same footing as the Coulomb interaction in the DHF calculation accounts for higher order effects on the wavefunction. Good nuclear charge models including, where necessary, nonspherical effects in heavy nuclei, model finite nuclear size effects. Relaxation of spectator electrons can be handled by treating initial and final states independently. Electron correlation, which plays a significant role, can be modelled using relativistic many-body theory (RMBPT). Admixtures of configurations with two holes and a particle with the dominant hole states, producing so-called Auger corrections, can also be important. One-electron radiative corrections due to the electron self-energy and vacuum polarization scale roughly as Z^4/n^3, where n is the principal quantum number; radiative corrections to the electron-electron interaction scale roughly as Z^3/n^3. The X-ray database is available on-line [74].

1.4.2 Applications to astrophysics and plasma physics

The extreme ultraviolet (EUV) part of the solar spectrum exhibits spectra from multiply charged ions of the iron group [75]. Many previously unidentified lines have been classified from beam-foil measurements. However, theoretical predictions of oscillator strengths (proportional to radiative transition probabilities) and lifetimes for long-lived levels can be useful because most laboratory light sources yielding precise wavelength measurements do not allow measurement of lifetimes of excited levels. A series of MCDHF calculations [76] of level energies, transition probabilities and lfetimes of several phosphorus-like ions of the iron group ($22 \leq Z \leq 32$) in the $3s3p^4$ and $3s^2 3p^2 4d$ configurations was made after long-lived lines of phosphorus-, sulphur-, and chlorine-like ions were observed in beam-foil experiments [80]. The small probabilities of these long-lived intercombination lines are often difficult to predict accurately.

The methods of Chapter 8 underpin packages such as the GRASP92 relativistic atomic structure program [81]. The technology enables calculations of electron correlation with multiconfigurational SOC wavefunctions using CI, self-consistent fields or, in principle, MBPT methods. The configuration state functions (CSF), labelled by $\gamma \Pi J M$, are built from anti-symmetrized products of Dirac central field orbitals of the form (1.2.13). Here J, M are the quantum numbers of the coupled angular momenta of the electrons, Π is the parity, and γ denotes other data such as the configurational composition and the angular momentum coupling scheme, which are needed to identify the state. Atomic state functions (ASF) are linear combinations of n_c CSF of the same symmetry,

$$|\alpha \Pi J M\rangle = \sum_{r=1}^{n_c} c_r(\alpha) |\gamma_r \Pi J M\rangle \tag{1.4.1}$$

The atomic energy levels are approximated by the eigenvalues $E_{\alpha \Pi J}$ of the atomic Hamiltonian in the CSF basis and the ASFs are defined by the corresponding vector $\boldsymbol{c}(\alpha)$ of the coefficients of (1.4.1).

The phosphorous-like ions studied in [76] have five electrons in open shells outside a Ne-like closed shell core. An MCDHF calculation based on an average energy expression for the ground $3s^2 3p^3$ and excited $3s3p^4$ and $3s^2 3p^2 4d$ configurations generated the set of *spectroscopic orbitals*. Further MCDHF calculations in which the spectroscopic orbitals are frozen gave additional *correlation orbitals*. Together these constituted a common orthonormal orbital set from which all levels and radiative transition rates could be calculated. These are substantial calculations: 2,365 CSFs were needed to represent $3s^2 3p^3$ states at the $4l$ level and 7,917 CSFs at the $5l$ level whilst the $J = 5/2$ excited states needed a maximum of 3,698 at the $4l$ level and 16,633 at the $5l$ level. [76, Table A]. These calculations differ from those performed for the X-ray project in several ways. The MCDHF method was used to determine all orbitals and the fully retarded magnetic interaction was treated as a perturbation. QED corrections were ignored, The correlated wavefunctions were derived by using CI rather than RMBPT methods.

A disadvantage of this procedure is that the weaker transition rates are often far from gauge invariant. The results usually improve with systematic enlargement of the CSF set, but convergence is often slow. Evidence suggests that SOC wavefunctions converge better when the orbitals are optimized separately for each state. When this is done, the radiative transition calculation must allow for the use of different orbital sets for the initial and final states of each transition: see §8.4 and §8.5. This generally reduces the gauge dependence of the transition rates, but the cost may make state specific optimization impractical when information is needed for many atomic states.

1.4.3 Modelling atomic processes in plasmas

The modelling of atomic processes in high-temperature plasmas which are not in thermodynamic equilibrium (non-LTE) requires extensive data on ionic level populations, spectral line intensities, and cross sections for electron-ion collisions. Highly charged ions of heavy atoms are often found in fusion plasmas and X-ray laser experiments, as well as in ion traps, so that relativistic methods are essential and systems such as the HULLAC package [53] have been developed specifically for this purpose. Its treatment of electronic structure and the calculation of collisional and radiative rates is founded upon methods such as we discuss in this book; however much of the package is concerned with shortcut methods to handle the huge amount of atomic data that is typically needed in applications.

The X-ray emission spectra from solar flares recorded by NASA's Solar Maximum Mission (SMM) spacecraft in 1980 that contained strong lines of the Fe XXII–XXV spectra [77] provide a contrasting example of successful application of MCDHF-EAL calculations with a minimal CSF set. Fig. 1.19 compares two synthetic spectra (dashed curves) with spectra recorded by the SMM's bent crystal spectrometer (histograms). The line positions and radiative transition rates for the He-like (Fe XXV) and Li-like (Fe XXIV) ions were taken from MCDHF calculations [78, 79]. The w resonance line $(1s^2\,{}^1S_0 - 1s2p\,{}^1P^o_1)$ in the He-like Fe XXV spectrum at 1.8509 Å is from a

7 CSF calculation [79]. The lines labelled j, q are from 57 CSF calculations of the Li-like Fe XXIV satellite lines, respectively $1s^2 2p\ ^2P^o_{3/2} - 1s2p^2\ ^2D_{5/2}$ at 1.8666 Å and $1s^2 2s\ ^2S_{1/2} - 1s2p\,2s\ (^1P)^2P_{3/2}$ at 1.8617 Å [79]. The intensity ratio j/w was used for a first estimate of electron temperature, T_e, in the plasma which was later improved using a χ^2 test of the fit of the calculated spectrum to the observed spectrum in the neighbourhood of the two lines [77]. The β line is from the Fe XXIII spectrum. The observed spectrum contains overlapping lines from different ionization stages. The model used data acquired from a number of different theoretical atomic calculations; in particular radiative and autoionization rates for dielectronic excitation of $n = 2$ and $n = 3$ satellites in Fe XXIV and Fe XXIII were calculated using the RCN and RCG codes [48, Chapters 8, 16]. The RCN code generated orbital wavefunctions using a Hartree-Fock-Slater (HFS) scheme with a statistical model exchange potential whilst RCG calculated energy levels and transition rates in intermediate coupling using standard Slater-Condon methods in a single-configuration approximation ignoring spin-spin interactions, Zeeman and Stark effects. Each line is distributed over a wavelength interval according to a Voigt profile, (a convolution of Lorentzian component and Gaussian components). The Gaussian component represented Doppler broadening by ionic motion assuming some ion temperature T_{ion} roughly independent of T_e [77, §3].

1.5 Relativistic molecular structure

The use of relativistic methods to study the electronic structure of molecules and solids has expanded rapidly in recent years as shown by Pyykkö's series of bibliographies [82, 83, 84, 85], which now list some 12,700 papers on relativistic effects on atoms, molecules, and condensed matter and their chemical implications. The chemistry of compounds containing heavy elements has motivated calculations on: molecular geometry – bond lengths, bond angles and potential energy surfaces; electromagnetic properties – g-tensors, electric dipole moments, NMR shifts, and hyperfine interactions; and effects on chemical reactions. Much of this information has been obtained using methods that treat relativity approximately rather than from the more fundamental theoretical standpoint of this book. The calculations listed for the years 1993–99 in [85, Table 7.10] for compounds of elements ranging from hydrogen to the superheavy element 118 are dominated by calculations with relativistic effective core potentials (RECP); other methods are comparatively unusual. Although there is no reliable way to estimate the errors of these approximate methods, which make assumptions for which there can be no rigorous justification, they have still provided extensive and valuable insights into relativistic effects in quantum chemistry.

1.5.1 Relativistic interpretations of chemical anomalies

Pyykkö and Desclaux [86] asserted in 1979 that relativistic effects "...seem to explain some of the most conspicuous anomalies in the latter half of the periodic system." This review, along with other articles by Pitzer [87] and Pyykkö [88], has strongly influenced later research on relativistic effects in chemistry, especially in relation to heavy and superheavy elements. As always, relativistic effects on hydrogenic (§1.2) and many-electron atoms (§1.3) provide the starting point for investigation of the molecules of which they form part. The evidence cited in [86] was based on Dirac-Fock one-centre expansion (DF-OCE) calculations [89], supplemented by calculations using the local density discrete variational method (DVM) [90], the multiple scattering Dirac-Slater (MS-Xα) method [91], methods using relativistic effective core potentials (RECP) and relativistic extended Hückel (REX) models [92].[7] Pyykkö and Desclaux highlighted specific properties such as the yellow colour of gold and why it is so different from silver; why mercury is a liquid and has a strong tendency to two-coordination; and why, in Group 13, the valency changes from III for indium to I for thallium or in Group 14 from IV for tin to II for lead. Pitzer's article [87], published in the same issue of *Accounts of Chemical Research* also highlighted some of these questions as well as the mechanism of lanthanide contraction; the unique properties of gold compared with copper and silver and of U^{VI}, Np^{VI}, and Pu^{VI} as compared with the corresponding lanthanides and noted magnetic effects attributed to the large spin-orbit interaction in compounds of Os^{IV} and other substances.

The DHF calculations of [47] emphasized that the radial contraction of s and $p_{1/2}$ orbitals in many-electron atoms is mainly a relativistic dynamical effect whilst indirect effects generally cause orbitals of higher angular momentum, $p_{3/2}, d_{3/2}, d_{5/2}, \ldots$ to expand. The height of the angular momentum barrier increases with j, reducing the orbital electron density near the nucleus; the electron therefore spends less time in the high potential region in which it moves with relativistic speeds, reducing the dynamical relativistic contraction. The direct effect is thus more than counterbalanced by indirect effects for $d, f, \ldots$ orbitals: these are partly mutual adjustments in the electronic repulsion between shells and partly the effect of the increased screening of the nucleus by the more compact s and p orbitals. Clearly, the observed shell structure of individual atoms reflects a delicate and subtle balance between the direct and indirect effects of relativity. Relativistic adjustments propagate outwards through the electron shells to modify the wavefunctions of valence electrons, so affecting their bonding with electrons on neighbouring atoms. Pyykkö [88] built up a database that combined results of relativistic and nonrelativistic calculations with experimental data for a range of small molecules to illustrate the effect of relativity on various properties. For example, relativistic bond lengths are usually less than nonrelativistic bond lengths [88, Table IV]; attempts to explain these findings in simpler terms vary according

[7] Reference [88] has an extensive bibliography.

to the theoretical model from which the results were derived [88, §II.D]. The bond contractions, C, for H_2^+ and Group 11, 13, and 14 hydrides had originally been fitted to a formula $C = c_Z Z^2$, but the more comprehensive data of [88, Table IV] revealed that the "constant" c_Z varies substantially from group to group, with the smallest values in Groups 1 and 18 and a "gold maximum" for the coinage metals in Group 11.

As discussed in §1.3.4 and §1.3.5, the relativistic contraction of s and $p_{1/2}$ orbitals is accompanied by an increase in their IP whilst orbitals of higher angular momenta exhibit the opposite behaviour. Many physical and chemical properties have a saw-tooth behaviour, *secondary periodicity*, superimposed on the regular trend down the column of each group in the Periodic Table. Thus, in Group 15, N, As, and Bi are preferentially trivalent, whilst P and Sb are pentavalent. The outer shell configurations are respectively $2s^2 2p^3$, $3d^{10} 4s^2 4p^3$, $4f^{14} 5d^{10} 6s^2 6p^3$, $3s^2 3p^3$, and $4d^{10} 5s^2 5p^3$. The explanation suggested [88, p. 566] is that the anomalous trivalency of As in the fourth row arises from "an increase in the effective nuclear charge" seen by the $4s$ shell due to the filling of the $3d$ shell. HF and DHF calculations on a pseudo-atom with nuclear charge reduced by 10 and omitting the 10 electrons of the $3d$ shell were used to support this conjecture. In the case of Bi, the "lanthanoid contraction" associated with the filling of the $4f$ shell and the direct relativistic $6s$ stabilization contributed almost equally. A similar explanation was advanced for the "inert pair effect", namely the tendency of the $6s^2$ pair to remain formally unoxidized in compounds of Tl(I), Pb(II), and Bi(III). The HF and DHF atomic eigenvalues, although not directly applicable to molecules or bulk matter, suggest that atomic relativistic effects account for many of the experimental facts.

The different colours of silver and gold provide perhaps the most notorious phenomenon in which relativistic effects have been implicated [88, p. 583]. The observed reflectivity of gold is high for photon energies less than about 2.4 eV, in the middle of the visible light spectrum but drops abruptly at higher energies. This is attributed [93] to the onset of absorption by electrons in the $5d$ band due to excitation to the Fermi level (essentially $6s$ in character). This is consistent with a calculated relativistic interband gap of 2.38 eV, whereas the nonrelativistic estimate would have been much higher, pushing the absorption edge to higher photon energy in the ultraviolet. Silver, in the previous row of the Periodic Table, has a smaller relativistic rise in the $4d$ energy and a smaller reduction of the $5s$ energy, shown qualitatively by the DHF and HF eigenvalues of Fig. 1.12 and resulting in a relativistic absorption edge at around 3.7 eV in the ultraviolet. Thus "nonrelativistic gold" should appear white like silver, a conclusion supported by band structure calculations [88, References 359 b-k]. It seems reasonable to assign much of the difference in properties of the two coinage metals to relativistic mechanisms. Pyykkö [88] discusses similar examples in other groups of the Periodic Table.

1.5.2 Relativistic effective core potentials and other approximations

Effective core potentials (ECP) were first suggested about 70 years ago by Hellmann and Gombás [94, 95]. The valence electrons of an atom determine, at least qualitatively, its chemical behaviour so that a model treating only valence electrons moving in the field of a suitable ECP may be good enough to reproduce the same chemical behaviour as more eleborate models. This chemically intuitive approach reduces the size of the computation, enabling more effort to be put into the parts of the calculation that are of most interest to chemists. Because relativistic effects originate in the strong field region near the origin, the hope is that RECPs account for much of the relativistic effects due to the core, and that it is only necessary to treat the valence electrons nonrelativistically. Thus standard nonrelativistic codes can often be used with only minor modifications. Modern RECPs contain several approximations that can often be justified only empirically and that must be calibrated against more exact models. Dolg has comprehensively reviewed the theory of RECPs [32]; see also Balasubramanian [96].

The basic nonrelativistic valence-only ECP model has the form [32, eq. (23)]

$$\mathcal{H}_v = \sum_i^{n_v} h_v(i) + \sum_{i<j}^{n_v} g_v(i,j) + V_{cc} + V_{cpp}. \tag{1.5.1}$$

Subscripts c, v stand for core and valence, respectively; h_v, g_v for effective one- and two-electron operators; V_{cc} for the repulsion between all cores and nuclei of the system; and V_{cpp} is a core polarization potential (CPP). The number of valence electrons treated explicitly is given by $n_v = n - \sum_\lambda^N (Z_\lambda - Q_\lambda)$, where Z_λ is the nuclear charge on centre λ and Q_λ is the net charge of the associated core.

There are many possible choices for each of these operators [32, 96]. A popular scheme for constructing model potentials requires fitting a parametrized expression to the Fock operator $\mathcal{F}_v$ of a valence orbital φ_a^V:

$$\sum_\lambda^N \Delta V_{cv}^\lambda(\mathbf{r}_{\lambda i}) \approx -\frac{Z_\lambda - Q_\lambda}{r_{\lambda i}} + \sum_c (2J_c(i) - K_c(i)) \tag{1.5.2}$$

In the AIMP (*ab initio* model potential) scheme described in [32, §4.3], the right-hand side comes from an all-electron quasi-relativistic atomic HF calculation. The direct part is the more straightforward:

$$-\frac{Z_\lambda - Q_\lambda}{r_{\lambda i}} + \sum_c 2J_c(i) \approx \sum_\lambda^N \Delta V_C^\lambda(i) = \frac{1}{r_{\lambda i}} \sum_k C_k^\lambda \exp(-\alpha_k^\lambda r_{\lambda i}^2) \tag{1.5.3}$$

The parameters α_k^λ and C_k^λ are determined from a least-squares fit to the left-hand side subject to the constraint $\sum_k C_k^\lambda = Z_\lambda - Q_\lambda$. The nonlocal exchange terms are approximated in terms of a suitable basis set $\{\chi_p^\lambda\}$ on core λ as

$$-\sum_c K_c(i) \approx \Delta V_X^\lambda(i) = \sum_{p,q} |\chi_p^\lambda(i)\rangle A_{pq}^\lambda \langle \chi_q^\lambda(i)|. \qquad (1.5.4)$$

With these definitions, the core-like solutions of the valence Fock equation would still fall below the desired valence-like solutions. This has to be prevented by adding a shift operator

$$P^\lambda(i) = \sum_{c\in\lambda} D_c^\lambda \, |\varphi_c^\lambda(i)\rangle\langle\varphi_c^\lambda(i)| \qquad (1.5.5)$$

where the set φ_c^λ is localized on core λ must be represented by a sufficiently large basis set. It is common to set $D_c^\lambda = -2\epsilon_c^\lambda$, largely on grounds of convenience. Thus the construction of the model potential

$$\Delta V_{cv,av}^\lambda(i) = \Delta V_C^\lambda(i) + \Delta V_X^\lambda(i) + P^\lambda(i)$$

has not used any properties of the valence orbitals or their energies.

We now turn to the choice of one-electron operator h_v. The Dirac operator itself is usually rejected, partly because it requires new machinery to deal with Dirac 4-spinors, partly because of worries about the issues of variational collapse, continuum dissolution, and finite basis set disease discussed elsewhere in this book. The Douglas-Kroll transformation [33] as implemented by Hess [34] gives a quasi-relativistic one-electron Hamiltonian [32, §3.2]

$$\begin{aligned} h_{DKH}(i) = E_i - A_i\,[V(i) + R_i V(i) R_i]\,A_i \\ - W_1(i) E_i W_1(i) - \frac{1}{2}\{W_1(i)^2, E_i\} \end{aligned} \qquad (1.5.6)$$

where

$$\begin{aligned} E_i &\equiv E(\boldsymbol{p}_i) = c\sqrt{\boldsymbol{p}_i^2 + c^2}, \\ A_i &\equiv A(\boldsymbol{p}_i) = \sqrt{(E_i + c^2)/2E_i}, \\ R_i &\equiv R(\boldsymbol{p}_i) = c\,\boldsymbol{\sigma}_i \cdot \boldsymbol{p}_i/(E_i + c^2), \end{aligned}$$

and $W_1(i)$ is an integral operator with the momentum space kernel

$$W_i(\boldsymbol{p},\boldsymbol{p}') = A(\boldsymbol{p})\,[R(\boldsymbol{p}) - R(\boldsymbol{p}')]\,A(\boldsymbol{p}')\,V(\boldsymbol{p},\boldsymbol{p}')/(E(\boldsymbol{p}) + E(\boldsymbol{p}')),$$

in which $V(\boldsymbol{p},\boldsymbol{p}')$ is the kernel of the external potential $V(i)$ and $\{\ldots\}$ represents an anticommutator. The Coulomb interaction $g(i,j) = 1/r_{ij}$ is usually used unmodified; the changes made by replacing it by the transformed two-electron operator are usually not enough to justify the additional cost of calculating it. The model potential approach yields valence orbitals with the same nodal structure as all-electron orbitals, and the DKH Hamiltonian can therefore be used to approximate relativistic effects in the valence shell explicitly. The family of *regular approximation* Hamiltonians [36] in which

ZORA (the zero order regular approximation) closely resembles the DKH operator (1.5.7) has also been used in molecular calculations. A popular alternative to the DKH operator has been the Wood-Boring (WB) Hamiltonian [32, §3.3] [97], originally proposed for atomic self-consistent DFT calculations following Cowan [48, §7-14] and Cowan and Griffin [98](CG). In the WB form (there are minor technical differences with the CG Hamiltonian) the starting point is the elimination, in a manner similar to §3.7.1, of the lower pair of components from the Dirac one-electron equation giving the two-component energy dependent Hamiltonian operator

$$h_{WB}(i) = \frac{1}{2}\boldsymbol{\sigma}_i \cdot \boldsymbol{p}_i \left[1 + (E_i - V(i))/2c^2\right] \boldsymbol{\sigma}_i \cdot \boldsymbol{p}_i + \sum_{\lambda} V_{\lambda}(\boldsymbol{r}_{i\lambda}). \qquad (1.5.7)$$

In the central field approximation, this gives a radial wave equation for the large component

$$(\mathcal{H}_S + \mathcal{H}_{MV} + \mathcal{H}_D + \mathcal{H}_{SO})\, P_{n\kappa}(r) = \varepsilon_{n\kappa} P_{n\kappa}(r) \qquad (1.5.8)$$

where

$$\begin{aligned}
\mathcal{H}_S &= -\frac{1}{2}\frac{d^2}{dr^2} + \frac{l(l+1)}{2r^2} + V(r)\\
\mathcal{H}_{MV} &= -\frac{\alpha^2}{2}[\varepsilon_{n\kappa} - V(r)]^2\\
\mathcal{H}_D &= -\frac{\alpha^2}{4}\frac{dV}{dr}B_{n\kappa}\left(\frac{d}{dr} - \frac{1}{r}\right)\\
\mathcal{H}_{SO} &= -\frac{\alpha^2}{4}\frac{dV}{dr}B_{n\kappa}\frac{\kappa+1}{r}
\end{aligned}$$

with $B_{n\kappa} = \left[1 + \alpha^2(\varepsilon_{n\kappa} - V(r))/2\right]^{-1}$. The nonrelativistic Schrödinger Hamiltonian is denoted by $\mathcal{H}_S$, $\mathcal{H}_{MV}$ is the mass-velocity operator, $\mathcal{H}_D$ the Darwin operator, and $\mathcal{H}_{SO}$ the spin-orbit interaction. Usually the nonlocal HF potential is used in $\mathcal{H}_S$, and a local approximation to it in the remaining relativistic correction operators. Averaging over the two cases $\kappa = -l-1, +l$ gives a so-called *scalar-relativistic* equation.

The model potential scheme should, in principle, give the same valence wavefunctions as an all-electron calculation. The problem is that this requires compact basis functions to reproduce the right nodal structure in the core region of the atom. These are generally not needed for chemical bonding, and pseudo-potential approaches generate pseudo-valence orbitals with simplified nodal structure rendering the compact basis functions unnecessary. *Shape-consistent* pseudopotentials are derived by replacing valence orbitals $\varphi_{v,lj}(r)$ by pseudo-orbitals $\varphi_{p,lj}(r)$ such that

$$\varphi_{p,lj}(r) = \begin{cases} \varphi_{v,lj}(r) & \text{for } r \geq r_c,\\ f_{lj}(r) & \text{for } r < r_c, \end{cases}$$

where the nodeless polynomial $f_{lj}(r)$ is defined on $[0, r_c)$ to satisfy suitable continuity conditions at the core-boundary radius r_c as well as at the origin. The pseudo-potential $V^{PP}(r)$ is then evaluated pointwise so that the pseudo-orbital satisfies a pseudo-Fock equation with the correct eigenvalue $\varepsilon_{v,lj}$. It is usually fitted to an expansion of the form

$$V^{PP} = -\frac{Q}{r} + \sum_{lj} \left[\sum_{k} A_{lj,k} r^{n_{lj,k}-2} \exp(-\alpha_{lj,k} r^2) \right] \mathcal{P}_{lj}$$

where $\mathcal{P}_{lj}$ projects onto orbitals of lj symmetry. *Energy-consistent* pseudopotentials have also been proposed for which the free parameters are chosen to reproduce the relevant parts of the experimental atomic spectrum [32, §4.4].

1.5.3 Dirac four-component methods for molecules

The calculation of molecular structures is much more demanding than that of atomic structures, and relativistic effects exacerbate the difficulties. Whilst four-component calculations of atomic structure are now common, there have been very few such calculations for molecules. This has been partly due to worries, now dispelled, about the validity of four-component methods, and partly because of the relatively high computational cost. Calculations of electron repulsion integrals (ERI) for four-component wavefunctions are the biggest bottleneck. The earlier codes treated each component of the 4-spinor as if it were a nonrelativistic wavefunction in its own right suggesting that, if all symmetry is ignored, each relativistic ERI is a linear combination of up to 4^4 nonrelativistic ERI! This simplistic argument grossly overestimates both the memory and effort required whilst making it clear that it is necessary also to find ways to reduce relativistic computational overheads without compromising the accuracy of the model.

The most developed four-component machinery available for molecules includes the DIRAC [99, 100] and MOLFDIR systems [101], both of which are based on older nonrelativistic molecular structure packages. The codes developed by Hirao *et al.* [102, 103, 104] work with spherical Gaussian functions (SGTF) and exploit the ERI code SPHERICA [105], which uses generalized contraction ideas of Raffenetti [106] and coordinate expansion schemes due to Ishida [107], to make relativistic ERI generation more efficient. Grant and Quiney's BERTHA code [108], described in detail in Chapters 10 and 11, is based on a new ERI algorithm that is a relativistic generalization of the popular nonrelativistic McMurchie-Davidson algorithm [109]. BERTHA is based on G-spinor basis functions, §5.10, and achieves its efficiency by exploiting the internal symmetry relations between the four SGTF components of each G-spinor. In the nonrelativistic case, overlaps of GTF basis functions on different nuclear centres can be expressed as a linear combination of products of Hermite Gaussian functions (HGTF) with numerical coefficients which depend on the nuclear positions. The relativistic charge density and current density vector of Dirac theory can be written in much the same way in terms of G-spinor overlaps; the coefficients of their HGTF expansions then incorporate the spinor structure along with the nuclear geometry [110]. The result is that calculation of the Fock matrix for the Dirac-Coulomb operator, which requires only the charge density overlaps, has relatively small relativistic overheads. The current density vector, which is only required for the magnetic (or Breit) interaction part, is also relatively economical to construct.

Much use has been made of density functional theory (DFT) in quantum chemistry owing to its computational simplicity. Some methodological papers on relativistic DFT, surveyed in §4.15, have been applied mainly to atoms rather than to molecules. BERTHA has DFT modules [111] which are being developed [112, 113] to exploit both computer parallelism and other techniques for higher speed and accuracy. As with older codes, BERTHA has only been applied to relatively small molecules and atomic clusters, but this can be expected to change in the course of the next few years.

Most four-component electronic structure calculations so far have been for diatomic or polyatomic molecules with at most one heavy atom [85, Table 7.10]. The recent introduction [111] of DFT modules into BERTHA was tested with DHF/DFT calculations on small molecules like H_2O, NH_3, P_4, C_2H_4, CH_4, SiH_4, and $TiCl_4$. Near optimal parallelization strategies [112] implemented on a Beowulf workstation cluster with up to seven machines reduced the time of one SCF iteration for the HgF_2 molecule by a factor of four. Significant increases in speed can result from employing techniques such as fitting the electron density to an expansion in terms of a modest set of scalar basis functions. Calculations on closed shell gold clusters [113] show this scales like $O(N^3)$, reducing the computing time for the Coulomb matrix to under 3% of the normal value for Au_2 and under 1% for Au_5^+. Developments of this sort will greatly extend the range of problems that can be studied with four component methods.

The DHF and DHFB models give corrections to quantities of chemical interest such as bond lengths and bond energies, and nonrelativistic methods can be used straightforwardly to incorporate correlation corrections using perturbation theory [114, 115]. The DHF model can be made the starting point of a consistent relativistic many-body theory in which either relativistic many-body theory (RMBPT) or relativistic configuration interaction (RCI) methods are applied to molecular properties using spinor basis sets. Applications include magnetic properties [116, 117], molecular Auger effects [118], and bonding of compounds containing superheavy elements [119, 120].

1.5.4 Parity violation and hyperfine interactions

Relativistic atomic and molecular structure theory is also important for fundamental investigations into the forces of nature. Suggestions for studying parity-violating interactions by Purcell and Ramsey [121] were followed by the discovery, by Wu *et al.* [122] and Lee and Yang [123], of P-odd processes in nuclear β-decay. The $(V - A)$ theory of β-decay of Feynman and Gell-Mann [124] was followed by development of the theory of electroweak interactions [125, 126, 127], which gives a complete and renormalizable model of the P-odd interactions. This theory predicts an interaction of the electron with the weak neutral currents within the nucleus whose magnitude, scaling like Z^3, is determined by the Fermi constant, $G_F = 2.2\ 10^{-14}$ a.u. Physical consequences include optical rotation of polarized light transmitted by atomic vapours, energy differences between enantiomeric forms of chiral molecules, and nonvanishing probabilities for transitions that would otherwise be strictly forbidden [128]. Elaborate RMBPT calculations [129] of the electronic structure of atomic Cs taking P-odd effects into account combined with precision measurement of the tiny induced transition rates provided convincing evidence for the internal consistency of the electroweak theory and verified that the signal is proportional to the so-called weak charge of the nucleus. The detection of a nuclear anapole moment, due to a nuclear spin-dependent P-odd interaction, has provided further supporting evidence [130].

Only one example of a T-odd interaction, the decay of the neutral K^0-meson, is known. The standard electroweak model does not include such interactions and several particle physics theories have been proposed to account for it [131]. Several of these theories also propose PT-odd interactions, which would have an experimental signature attributing a non-zero electric dipole moment (edm) to the electron or to nucleons [132]. Sandars [133] was the first to recognize that such weak spin-dependent interactions would be enhanced by the strong internal electric field within polar molecules by mixing spin-rotational states that are nearby in energy but that have opposite parity. A high atomic number is needed for the necessary strong electric field, so that a relativistic treament of the molecule is essential [134].

PT-odd effects in the ground state, ${}^2\Sigma_{1/2}$, of the paramagnetic radical YbF have been investigated by various methods [135, 136] in order to set limits on any permanent EDM of the electron. The electronic matrix elements closely resemble those of the M1 hyperfine interaction of ^{171}Yb in YbF, so that predictions of the hyperfine interaction constant have been used to assess the reliability of the electronic PT-odd calculations. Similar calculations have been made for nuclear PT-odd effects in the TlF molecule [137], with the aim of setting limits on the EDM of the proton. The four component DHF calculations are demanding; large basis sets are needed: for example, $31s31p15d8f3g$ in the case of Tl. Highly contracted basis elements are essential to get reliable wavefunctions near the (suitably modelled) nucleus whilst long range functions are needed to model molecular bonding accurately.

There has been relatively little work on parity-violating energy differences between enantiomers of chiral molecules. It has been suggested that differences in the vibrational spectra of two enantiomers of handed molecues might be detectable. A study by Lærdahl *et al.* [138] found the biggest such differences, about 0.2 Hz, in chiral methane derivatives including an iodine substituent.

In the case of CHBrClF, the difference for the C-Cl stretching mode is only 7 mHz, whilst for the C-F stretching mode it is only 2 mHz, some 3 to 4 orders of magnitude smaller than could be measured in recent experiments.

1.5.5 High-precision spectroscopy of small molecules containing light elements

Although it is natural to think of relativistic methods as being essential for studying matter containing the heavier elements, some of the more striking recent results have come from applications to molecules containing only light elements. Calculations usually start from the Born-Oppenheimer approximation [139] in which the electrons move relative to a static nuclear skeleton. A sequence of such calculations in which the skeleton is deformed gives a potential energy hypersurface (PES) that can be used to determine the slow nuclear motions adiabatically. The precision of nonrelativistic calculations of the vibration-rotation spectra of light molecules is now so good that small corrections to the Born-Oppenheimer (BO) PES must be considered: these include adiabatic and nonadiabatic BO corrections and relativistic effects. A recent review [140] compared perturbation estimates of the relativistic mass-velocity and the Darwin one- and two-body corrections together with in small molecules with the corrections obtained from four-component calculations using BERTHA and MOLFDIR together with estimates of one-body and two-body Lamb shifts. BERTHA calculation [141] suggested that the two-electron relativistic contributions have a substantial effect on the rotation-vibration levels in H_2O, and similar results were obtained for H_2S [142]. The emission and absorption of light by water vapour is responsible for about 70% of the absorption of sunlight in the Earth's atmosphere and the majority of the greenhouse effect [31]. Water is a major product of combustion and a dominant constituent of the atmospheres of cool stars. Hence much effort has been devoted to constructing as complete and accurate a model as possible to predict the high-resolution water spectrum. Polyansky *et al.*'s study [31] stressed that inclusion of relativistic and other corrections had improved the accuracy of the *ab initio* PES of water by an order of magnitude compared with previous work, predicting individual line positions with a typical accuracy of 0.2 cm^{-1}. This somewhat unexpected conclusion suggests that there is continued scope for relativistic calculations on molecules with light element constituents, although the need to study molecules containing heavy atoms remains a major factor motivating development of relativistic methods.

References

1. Drake G W F ed. 2005 *Springer Handbook of Atomic, Molecular and Optical Physics* Second edn (New York: Springer-Verlag)

2. Condon E U and Shortley G H 1953 *The Theory of Atomic Spectra* (Cambridge: University Press).
3. Slater J C 1960 *Quantum theory of atomic structure*, 2 vols. (New York: McGraw-Hill).
4. Fischer C F, Brage T and Jönsson P 1997 *Computational Atomic Structure. An MCHF approach* (Bristol and Philadelphia: Institute of Physics)
5. Pauling L and Wilson Jr E B 1985 *Introduction to Quantum Mechanics with applications in chemistry* (Reprint of the 1935 edition) (New York: Dover)
6. Murrell J N, Kettle S F A and Tedder J M 1985 *The Chemical Bond* Second edn (New York: John Wiley)
7. McWeeny R 1989 *Methods of Molecular Quantum Mechanics* Second edn (London: Academic Press)
8. Bohr N 1913 *Nature* **73** 1109.
9. Paschen F 1916 *Ann Phys Lpz* **50** 901.
10. Sommerfeld A 1916 *Ann Phys Lpz* **51** 1, 44, 125.
11. Schrödinger E 1926 *Ann der Phys* **79** 361; see also [12, p. 288].
12. Pais A 1986 *Inward Bound* (Oxford: Clarendon Press)
13. Dirac P A M 1928 *Proc Roy Soc (Lond) A* **117** 610.
14. Dirac P A M 1928 *Proc Roy Soc (Lond) A* **118** 351.
15. Gordon W 1928 *Z f Phys* **48** 11.
16. Darwin C G 1928 *Proc Roy Soc (Lond) A* **118** 654.
17. Dirac P A M 1931 *Proc Roy Soc (Lond) A* **133** 60.
18. Anderson C D 1932 *Science* **76** 238.
19. Breit G 1929 *Phys Rev* **34** 353; — 1930 *Phys Rev* **36** 383; — 1932 *Phys Rev* **39** 616.
20. Bethe H A and Salpeter E E 1957 *Quantum Mechanics of One- and Two-Electron Atoms* (Berlin-Göttingen-Heidelberg: Springer-Verlag)
21. Swirles B 1935 *Proc Roy Soc (Lond) A* **152** 625; — 1936 *Proc Roy Soc (Lond) A* **157** 680
22. Racah G 1942 *Phys Rev* **61** 186; — 1942 *Phys Rev* **62** 438; — 1943 *Phys Rev* **63** 367; — 1949 *Phys Rev* **76** 1352
23. Grant I P 1961 *Proc Roy Soc (Lond) A* **262** 555; — 1965 *Proc Phys Soc* **86** 523
24. Fischer C F 1972 *Atomic Data* **4** 302; — 1973 *At Data Nucl Data Tab* **12** 87
25. Mann J B 1973 *At Data Nucl Data Tab* **12** 1
26. Desclaux J P 1973 *At Data Nucl Data Tab* **12** 311
27. Mann J B and Waber J E 1973 *Atomic Data* **5** 201
28. Mohr P J and Wiese W L (editors) 1998 *Atomic and Molecular Data and their Applications. ICAMDATA – First International Conference* (AIP Conference Proceedings 434) (Woodbury, NY: American Institute of Physics).
29. Dirac P A M 1929 *Proc Roy Soc A* **123**, 714. (The quotation is used with the permission of the Royal Society.)
30. Schwerdtfeger P (editor) *Relativistic Electronic Structure Theory. Part 1. Fundamentals*(Amsterdam: Elsevier).
31. Polyansky O L, Czászár A G, Shirin S V, Zobov N F, Barletta P, Tennyson J, Schwenke D W and Knowles P J 2003 *Science* **299**, 539; see also Polyansky O L, Zobov N F, Viti S, Tennyson J, Bernath P F and Wallace L 1997 *Science* **277**, 346; Tennyson J and Polyansky O L 1998 *Contemp. Phys.* **39**, 283.
32. Dolg M 2002 in *Relativistic Electronic Structure Theory. Part 1. Fundamentals* (ed. P Schwerdtfeger) Chapter 14. (Amsterdam: Elsevier).

33. Douglas M and Kroll N M 1974 *Ann Phys (NY)* **82** 89.
34. Hess B A 1986 *Phys Rev A* **32** 756; — *Phys Rev A* **33** 3742.
35. Wolff A, Reiher M and Hess B A 2002 in [30, Chapter 11].
36. Sundholm D 2002 in [30, Chapter 13].
37. Goldstein H 1980 *Classical Mechanics* (2nd edition) (Reading, Mass.: Addison Wesley)
38. Mohr P J and Taylor B N 2000 *Rev Mod Phys* **72** 351 (CODATA recommended values of the fundamental physical constants: 1998). These values are regularly updated: `http://physics.nist.gov/constants`.
39. Moseley H G J 1913 *Phil. Mag.* **26**, 1024; 1914 *Phil. Mag.* **27**, 703.
40. White H E 1934 *Introduction to Atomic Spectra* (New York: McGraw Hill).
41. Series G W 1988 *The Spectrum of Atomic Hydrogen - Advances* (Singapore: World Scientific)
42. Lamb W E and Retherford R C 1947 *Phys Rev* **72** 241.
43. Karshenboim S G, Pavone F S, Bassani F, Inguscio M and Hänsch T W 2001 *The Hydrogen Atom. Precision Physics of Simple Atomic systems* (Berlin: Springer-Verlag)
44. Burke V M and Grant I P 1967 *Proc Phys Soc* **90** 297.
45. Greiner W, Müller B and Rafelski J 1985 *Quantum Electrodynamics of Strong Fields* (Berlin: Springer-Verlag)
46. Dirac P A M 1929 *Proc Roy Soc A* **126**, 360; see also Dirac P A M 1930 *Nature* **126**,605.
47. Rose S J, Pyper N C and Grant I P 1978 *J. Phys. B: Atom. Molec. Phys.* **11**, 1171.
48. Cowan R D 1981 *The theory of atomic structure and spectra* (Berkeley: University of California Press)
49. Judd B R 1963 *Operator techniques in atomic spectroscopy* (New York: McGraw-Hill)
50. Lindgren I and Morrison J 1982 *Atomic many-body theory* (Berlin: Springer-Verlag)
51. Rudzikas Z B 1997 *Theoretical atomic spectroscopy* (Cambridge: Cambridge University Press)
52. Klapisch M, Schwob J L, Fraenkel B S and Oreg J 1977 *J. Opt. Soc. Am.* **67**, 148.
53. Bar-Shalom A, Klapisch M and Oreg J 2001 *J. Quant. Spectrosc. Rad. Transf.* **71**, 169.
54. Eissner W, Jones M and Nussbaumer H 1974 *Comput. Phys. Commun.* **8**, 70.
55. Nussbaumer H and Storey P J 1978 *Astron. Astrophys.* **70**, 37.
56. Hartree D R 1957 *The calculation of atomic structures* (New York: John Wiley)
57. Fischer C F 1977 *The Hartree-Fock method for atoms* (New York: John Wiley)
58. Fischer C F 1973 *At. Data Nucl. Data Tables* **12**, 87.
59. Mann J B 1973 *At. Data Nucl. Data Tables* **12**, 1.
60. Desclaux J P 1973 *At. Data Nucl. Data Tables* **12**, 311.
61. Seaborg G T 1996 *J. Chem. Soc., Dalton Trans.* 3899.
62. URL: http://physics.nist.gov/PhysRefData/contents.html
63. Pyper N C and Grant I P 1978 *Proc. Roy. Soc. A* **359**, 525.
64. Layzer D and Bahcall J N 1962 *Ann. Phys. (N.Y.)* **17**, 177.
65. Doyle H T 1969 *Adv. At. Mol. Phys.* **5**, 337.
66. Snyder R 1971 *J. Phys. B*, **4**, 1150.

67. Snyder R 1972 *J. Phys. B*, **5**, 934
68. Snyder R 1974 *J. Phys. B*, **7**, 335.
69. Goldsmith S 1974 *J. Phys. B*, **7**, 2315.
70. Labzowsky L, Klimchitskaya G L and Dmitriev Yu Yu 1993 *Relativistic effects in the spectra of atomic systems* (Bristol and Philadelphia: Institute of Physics Publishing)
71. Levine M A, Marrs R E, Henderson J R, Knapp D A and Schneider M B 1988 *Phys. Scr.* **T22**, 157.
72. Radtke R, Biedermann C, Schwob J L, Mandelbaum P and Doron R *Phys. Rev. A* **64**, 012720.
73. Deslattes R D, Kessler J G Jr., Indelicato P, de Billy L, Lindroth E and Anton J 2003 *Rev Mod Phys* **75**, 35.
74. Deslattes R D, Kessler J G Jr., Indelicato P, de Billy L, Lindroth E, Anton J, Coursey J S, Schwab D J, Chang C, Sukumar R, Olsen K and Dragoset R A 2005 *X-ray Transition Energies* (version 1.2). [Online] Available: `http://physics.nist.gov/XrayTrans` [2005, November 29]. National Institute of Standards and Technology, Gaithersburg, MD.
75. Dere K P 1978 *Astrophys. J.* **221**, 1062.
76. Fritzsche S, Froese Fischer C and Fricke B 1998 *Atomic Data and Nuclear Data Tables* **68**, 149.
77. Lemen J R, Phillips K J H, Cowan R D, Hata J and Grant I P, 1984 *Astron. Astrophys.* **135**, 313.
78. Hata J and Grant I P, 1981 *J. Phys B: At. Mol. Phys.* **14**, 2111.
79. Hata J and Grant I P, 1982 *Mon. Not. R. Astr. Soc.* **198**, 1081.
80. Träbert E, Brand M, Doerfert J, Granzow J, Heckmann P H, Meurisch J, Matinson I, Hutton R and Myrnäs 1993 *Phys. Scr.* **48**, 580.
81. Parpia F A, Froese Fischer C and Grant I P 1996 *Comput. Phys. Commun.* **94**, 249.
82. Pyykkö P 2005. *Database RTAM*: `http://www.csc.fi/rtam/`. This also contains the bibliographic parts of [83, 84, 85] and is updated regularly. Some explanations of the organization of the bibliography will be found in the printed books. There were about 12,700 references in August 2005.
83. Pyykkö P 1986 *Relativistic Theory of Atoms and Molecules I* (Berlin: Springer).
84. Pyykkö P 1993 *Relativistic Theory of Atoms and Molecules II* (Berlin: Springer).
85. Pyykkö P 2000 *Relativistic Theory of Atoms and Molecules III* (Berlin: Springer).
86. Pyykkö P and Desclaux J P 1979 *Acc. Chem. Res.* **12**, 276.
87. Pitzer K 1979 *Acc. Chem. Res.* **12**, 271.
88. Pyykkö P 1988 *Chem. Rev.* 563.
89. 1983 in *Relativistic Effects in Atoms, Molecules and Solids* (ed. G L Malli) pp. 213 – 225. (NATO ASI Series B 87) (New York: Plenum Press)
90. Rosén A and Ellis D E 1974 *Chem. Phys. Lett.* **27**, 595; see also Rosén A and Ellis D E 1975 *J. Chem. Phys.* **62**, 3039.
91. Yang C Y 1983 in *Relativistic Effects in Atoms, Molecules and Solids* (ed. G L Malli) pp.335. (NATO ASI Series B 87) (New York: Plenum Press)
92. Pyykkö P 1988 in *Methods in Computational Chemistry* Vol. 2 (ed. S Wilson) Chapter 4. ((New York: Plenum Press)
93. Christiansen N E and Seraphin B O 1971 *Phys. Rev. B* **4**, 3321.
94. Hellman H 1935 *J. Chem. Phys* **3**, 61.

95. Gombás P 1935 *Z. Phys.* **94**, 473.
96. Balasubramanian K 1998 *Encyclopedia of Computational Chemistry* 2471. (Chichester: Wiley).
97. Wood J H and Boring A M 1978 *Phys Rev B* **18**, 2701.
98. Cowan R D and Griffin D C 1976 *J. Opt. Soc. Am.* **55**, 1010.
99. Saue T, Fægri K, Helgaker T and Gropen O, 1997 *Mol. Phys.* **91**, 937.
100. Jensen H J Aa, *et al.* 2004 Online: `http://dirac.chem.sdu.dk`
101. Visscher L, Visser O, Aerts P J C, Merenga H and Nieuwpoort W C, 1994 *Comp. Phys. Commun.* **81**, 120.
102. Yanai T, Nakajima T, Ishikawa Y and Hirao K 2001 *J. Chem. Phys.* **114**, 6526.
103. Yanai T, Iikura H, Nakajima T, Ishikawa Y and Hirao K 2001 *J. Chem. Phys.* **115**, 8267.
104. Yanai T, Nakajima T, Ishikawa Y and Hirao K 2002 *J. Chem. Phys.* **116**, 10122.
105. Yanai T, Ishida K, Nakano H and Hirao K 2000 *Int. J. Quant. Chem.* **76**, 396.
106. Raffenetti R C 1973 *J. Chem. Phys.* **58**, 4452.
107. Ishida K 1996 *Int. J. Quant. Chem.* **59**, 209; — 1998a *J. Chem. Phys.* **109**, 981; — 1998b *J. Comput. Chem.* **19**, 923; — 1999 *J. Chem. Phys.* **111**, 4913.
108. Grant I P and Quiney H M 2002 in *Relativistic Quantum Chemistry – Theory*, Chapter 3. ed. P Schwerdtfeger (Amsterdam: Elsevier).
109. McMurchie L E and Davidson E R 1978 *J. Comput. Phys.* **26**, 218.
110. Quiney H M, Skaane H and Grant I P 1997 *J. Phys B: At. Mol. Opt. Phys.* **30**, L829.
111. Quiney H M and Belanzoni P 2002 *J. Chem. Phys.* **117**, 5550.
112. Belpassi L, Storchi L, Tarantelli F, Sgamellotti A and Quiney H M 2003 *FGCS* **20**, 739.
113. Belpassi L, Tarantelli F, Sgamellotti A and Quiney H M 2005, private communication.
114. Sjøvoll M, Fagerli H, Gropen O, Almlöf J, Saue T, Olsen J and Helgaker T 1997 *J. Chem. Phys.* **107**, 5496.
115. Lærdahl J K, Fægri Jr. K, Visscher L and Saue T 1998 *J. Chem. Phys.* **109**, 10806.
116. Quiney H M, Skaane H and Grant I P 1998 *Adv. Quant. Chem.* **32**, 1.
117. Aucar G A, Saue T, Visscher L and Jensen H J A 1999 *J. Chem. Phys.* **110**, 5208.
118. Ellingsen K, Matila T, Saue T, Aksela H and Gropen O 2000 *Phys. Rev. A* **62**, 032502.
119. Saue T, Fægri Jr. K and Gropen O 1996 *Chem. Phys. Lett.* **263**, 360.
120. Fægri Jr. K and Saue T 2001 *J. Chem. Phys.* **115**, 2456.
121. Purcell E M and Ramsey N F 1950 *Phys. Rev.* **78**, 807.
122. Wu C S, Ambler E, Hayward R W, Hoppes D and Hudson R P 1957 **105**, 1413.
123. Lee T D and Yang C N 1957 *Phys. Rev.* **105**, 1671.
124. Feynman R P and Gell-Mann M *Phys. Rev.* **109**, 193.
125. Glashow S L 1961 *Nucl. Phys.* **22**, 579.
126. Weinberg S 1967 *Phys. Rev. Lett.* **19**, 1264.
127. Salam A 1968 in *Nobel Symposium, No. 8* ed N Svartholm (Stockholm: Almquist and Wiksell).
128. Sandars P G H 1987 *Phys. Scripta* **36**, 904; *ibid.* 1993 *Phys. Scripta* **T46**, 16.

129. Johnson W R, Sapirstein J and Blundell S A 1993 *Phys. Scripta* **T46**, 184.
130. Wood C S, Bennett S C, Cho D, Masterson B P, Roberts J L, Tanner C E and Wieman C E 1997 *Science* **275**, 1759.
131. Commins E D and Bucksbaum P H 1983 *Weak interactions of leptons and quarks* (Cambridge: Cambridge University Press).
132. Schiff L I 1963 *Phys. Rev.* **132**, 2194.
133. Sandars P G H 1967 *Phys. Rev. Lett.* **19**, 1396.
134. Mårtensson-Pendrill A-M 1993 in *Methods in Computational Chemistry, Vol 5* ed. S Wilson pp. 99 –156. (New York: Plenum Press)
135. Quiney H M, Skaane H and Grant I P 1998 *J. Phys. B* **31**, L85.
136. Mosyagin N S, Kozlov M G and Titov A V 1998 *J. Phys. B* **31**, L763.
137. Quiney H M, Lærdahl J K, Fægri Jr. K, and Saue T 1998 *Phys. Rev. A* **57**, 920.
138. Lærdahl J K, Quiney H M and Schwerdtfeger P 2000 *Phys. Rev. Lett.* **84**, 3811.
139. Born M and Oppenheimer J R 1927 *Ann der Phys* **84** 457.
140. Tarczay G, Czászár A G, Klopper W and Quiney H M 2001 *Mol. Phys.* **99**, 1769.
141. Quiney H M, Barletta P, Tarczay G, Czászár A G, Polyansky O L and Tennyson J 2001 *Chem. Phys. Lett.* **344**, 413.
142. Barletta P, Czászár A G, Quiney H M and Tennyson J 2002 *Chem. Phys. Lett.* **361**, 121.

Part II

Foundations

2

Relativistic wave equations for free particles

The topics presented in this chapter are indispensible foundations for the relativistic theory of atomic and molecular structure that are often taken for granted by those whose main interest is in application of the theory. Section 2.1 gives a very brief account of the principles of the special theory of relativity that are applied throughout the book. The physical content of §2.2, the Lorentz group, and §2.3, the Poincaré group, unifies the three sections on the Klein-Gordon equation (for spin zero particles), the Dirac equation (for spin 1/2), and the Maxwell equations (for photons) that follow. Although §2.4–§2.6 can be read and largely understood without first reading the material on the Lorentz and Poincaré groups, the reader is likely to find that he will need it for a full understanding of the properties of these relativistic wave equations. Similarly, the reader may wish initially to accept unread much of the content of the two sections, §2.7 and §2.8, on local and global conservation laws as consequences of the precious sections. The final section §2.9 sets up machinery that will later become very familiar as we develop quantum electrodynamics (QED) and the formalism of relativistic atomic and molecular structure.

2.1 The special theory of relativity

The special theory of relativity [1, 2] is fundamental to our treatment of atomic and molecular structure. In flat space-time, *contravariant* vectors are written

$$x^\mu = (x^0, x^1, x^2, x^3) = (x^0, \mathbf{x})$$

with $x^0 = ct$ being the *time-like* component of the 4-vector and c the speed of light, so that x^0 has the dimension of length, putting it on the same footing as the other components . We use a boldface letter, $\boldsymbol{x}$ to denote a 3-vector whose components are the three *space-like* components of a contravariant 4-vector:

$$\boldsymbol{x} = (x^1, x^2, x^3) = (x, y, z).$$

We suppose that each freely moving particle, or *free particle*, is equipped notionally with some form of standard clock which measures time τ along the particle's space-time trajectory, its *worldline*, given by some equations of the form

$$x^\mu = x^\mu(\tau), \quad \mu = 0, 1, 2, 3. \tag{2.1.1}$$

Each event in the particle's history is therefore characterized by some value of τ. The concept of an *inertial observer*, who can move in the same way as a free particle, is convenient for setting down the principles of special relativity, namely

- Free particles and photons appear to inertial observers to travel in straight lines at constant speeds.
- Photons appear to inertial observers all to have the *same* constant speed, denoted by c.
- Each inertial observer's standard clock, from which he obtains the value of τ, appears to any other inertial observer to run at a constant rate. However the clocks of different observers do not necessarily run at the same rate.
- Free particles cannot travel faster than photons.

Thus *photons* are pictured in a similar fashion to free particles apart from their constant speed in the *inertial frame* associated with each inertial observer. To these assumptions we add the *Principle of Relativity*, which states that only relative motion of inertial observers is detectable.

Thus the worldline of a free particle is given by

$$x^\mu = v^\mu \tau + a^\mu, \quad \mu = 0, 1, 2, 3$$

where v^μ and a^μ are constants[1]. The requirement that free particles cannot travel faster than photons can be expressed as

$$(v^0)^2 > (v^1)^2 + (v^2)^2 + (v^3)^2.$$

It follows that two events, coordinates x^μ, y^μ, on the worldline of a photon are related by

$$(x^0 - y^0)^2 - (x^1 - y^1)^2 - (x^2 - y^2)^2 - (x^3 - y^3)^2 = 0. \tag{2.1.2}$$

The *metric coefficients* $g_{\mu\nu} = g^{\mu\nu}$ $(\mu, \nu = 0, 1, 2, 3)$ are the elements of the array

$$g = \begin{pmatrix} 1 & 0 & 0 & 0 \\ 0 & -1 & 0 & 0 \\ 0 & 0 & -1 & 0 \\ 0 & 0 & 0 & -1 \end{pmatrix} \tag{2.1.3}$$

so that (2.1.2) can be expressed succinctly in terms of Einstein's summation convention,

[1] We use Greek superscripts and subscripts to label space-time components.

$$g_{\mu\nu}\,(x^\mu - y^\mu)(x^\nu - y^\nu) = 0, \tag{2.1.4}$$

in which paired sub- and superscripts are summed over all possible values $0, 1, 2, 3$. So if O and O' are two inertial observers who have set up coordinate systems x and x', it can be shown that the most general transformation compatible with the assumptions is of the linear form

$$x^\mu = \Lambda^\mu{}_\nu\, x'^{\,\nu} + a^\mu. \tag{2.1.5}$$

For the moment we shall consider only *homogeneous Lorentz transformations* for which $a^\mu = 0$.

Define the *scalar product* of two 4-vectors U^μ, V^ν by

$$U \cdot V = g_{\mu\nu} U^\mu V^\nu = U^0 V^0 - \mathbf{U} \cdot \mathbf{V} = U^0 V^0 - U^i V^i, \tag{2.1.6}$$

where we use a summation convention over the space components, $i = 1, 2, 3$. This definition ensures that $U \cdot V$ is unchanged under Lorentz transformations. For example, because photons appear to travel with the same constant speed in every inertial frame, (2.1.4) must hold for the coordinates in every such frame, so that

$$x' \cdot x' = g_{\mu\nu} x'^\mu x'^\nu = g_{\mu\nu} \Lambda^\mu{}_\rho x^\rho \Lambda^\nu{}_\sigma x^\sigma = g_{\rho\sigma} x^\rho x^\sigma = x \cdot x.$$

This requires

$$g_{\mu\nu} = \Lambda^\rho{}_\mu g_{\rho\sigma} \Lambda^\sigma{}_\nu \tag{2.1.7}$$

or, in matrix notation,

$$g = \Lambda^T g \Lambda. \tag{2.1.8}$$

where superscript T denotes the matrix transpose. It follows that the determinant of the transformation matrix satisfies

$$(\det \Lambda)^2 = 1, \quad \det \Lambda = \pm 1$$

so that the transformation Λ is nonsingular, and its inverse is given by

$$\Lambda^{-1} = g^{-1} \Lambda^T g. \tag{2.1.9}$$

Covectors are defined by

$$x_\mu = g_{\mu\nu} x^\nu = (x_0, -\mathbf{x}); \tag{2.1.10}$$

from (2.1.9), covectors transform according to Λ^{-1} rather than Λ, and scalar products (2.1.6) take the simple form

$$U \cdot V = g_{\mu\nu}\, U^\mu\, V^\nu = U_\mu\, V^\mu = U^\mu\, V_\mu.$$

We shall also encounter 4-tensors, in particular second rank tensors $T^{\mu\nu}$, whose transformation law is

$$T^{\mu\nu} = \Lambda^{\mu}{}_{\rho}\,\Lambda^{\nu}{}_{\sigma}\,T'^{\,\rho\sigma}.$$

As with 4-vectors, we can define tensors with a mix of contravariant and covariant indices, for example the rank 2 tensor $T^{\mu}{}_{\nu} = g_{\rho\nu}\,T^{\mu\rho}$, in which the covariant indices transform using Λ^{-1} instead of Λ. Tensors of higher ranks with, say, p contra- and q covariant indices will also appear in this book.

The simple form of the Minkowski metric g, (2.1.3), means that the distinction between co- and contravariant indices is not very significant here, except for algebraic book-keeping. It is quite a different matter in curvilinear coordinate systems [3], which we need not consider in this book.

2.2 The Lorentz group

If Λ_1 and Λ_2 are two Lorentz transformations, then the matrix product, $\Lambda = \Lambda_1\Lambda_2$ is also a Lorentz transformation: for by (2.1.8),

$$\Lambda^T g\Lambda = (\Lambda_1\Lambda_2)^T g\Lambda_1\Lambda_2 = \Lambda_2^T\Lambda_1^T g\Lambda_1\Lambda_2 = \Lambda_2^T g\Lambda_2 = g,$$

and $\det\Lambda = \det(\Lambda_1\Lambda_2) = \det\Lambda_1\,\det\Lambda_2 = \pm 1$. Hence the set of Λ matrices forms a regular matrix group (the Lorentz group, $\mathcal{L}$) with respect to ordinary matrix multiplication, designated SO(3,1) [4, Chapter 3] [5, 6]. Equation (2.1.7) furnishes 10 constraints on the 16 components of the 4×4 matrix Λ; the result is that each Λ in $\mathcal{L}$ can be indexed in terms of 6 free parameters.

A *proper* Lorentz transformation is characterized as having $\det\Lambda = 1$; such transformations can be generated by a succession of infinitesimal changes in the parameters starting from the identity $\Lambda = 1$. They include (a) the *rotations*, forming a proper subgroup, SO(3), of $\mathcal{L}$, of the form

$$\Lambda(\mathbf{R}) = \begin{pmatrix} 1 & \mathbf{0} \\ \mathbf{0}^T & \mathbf{R} \end{pmatrix} \tag{2.2.1}$$

where $\mathbf{0} = (0,0,0)$ and $\mathbf{R}$ is a 3×3 orthogonal matrix; and (b) the *boosts* consisting of all transformations which relate one frame of reference to another moving with uniform relative speed v. In particular, a boost that leaves the x^2 and x^3 axes invariant has the form

$$\Lambda(b_1) = \begin{pmatrix} \cosh\beta & -\sinh\beta & 0 & 0 \\ -\sinh\beta & \cosh\beta & 0 & 0 \\ 0 & 0 & 1 & 0 \\ 0 & 0 & 0 & 1 \end{pmatrix} \tag{2.2.2}$$

where

$$\tanh\beta = \frac{v}{c}.$$

A general proper Lorentz transformation can be constructed by combining two or more of these elements.

The Lorentz transformations can be divided into four classes, characterized by the sign of $\det \Lambda$ and by whether $\Lambda^0{}_0 \geq 1$ or $\Lambda^0{}_0 \leq -1$, distinguished by the fact that it is impossible to pass from one class to another by smoothly varying the parameters:

1. $\mathcal{L}_+^\uparrow$: The *proper, orthochronous* transformations, which transform positive time-like vectors into positive time-like vectors, have $\det \Lambda = 1$, $\Lambda^0{}_0 \geq 1$.
2. $\mathcal{L}_-^\uparrow$: $\det \Lambda = -1$, $\Lambda^0{}_0 \geq 1$. These are obtained from the proper Lorentz transformations by a space inversion: $x^0 \to x^0, \boldsymbol{x} \to -\boldsymbol{x}$.
3. $\mathcal{L}_-^\downarrow$: $\det \Lambda = -1$, $\Lambda^0{}_0 \leq -1$. These are obtained from the proper Lorentz transformations by a time inversion: $x^0 \to -x^0, \boldsymbol{x} \to \boldsymbol{x}$.
4. $\mathcal{L}_+^\downarrow$: $\det \Lambda = 1$, $\Lambda^0{}_0 \leq -1$. These are obtained from the proper Lorentz transformations by both space and time inversions: $x^\mu \to -x^\mu, \mu = 0, 1, 2, 3$.

2.2.1 ∗ Spinor representation of Lorentz transformations

The group SL(2,C) of complex linear transformations of unit determinant in two-dimensional spinor space acts as the universal covering group for $\mathcal{L}_+^\uparrow$. We denote the Pauli spin matrices by

$$\sigma_1 := \begin{pmatrix} 0 & 1 \\ 1 & 0 \end{pmatrix}, \quad \sigma_2 := \begin{pmatrix} 0 & -i \\ i & 0 \end{pmatrix}, \quad \sigma_3 := \begin{pmatrix} 1 & 0 \\ 0 & -1 \end{pmatrix} \tag{2.2.3}$$

together with the identity matrix

$$\sigma_0 := \begin{pmatrix} 1 & 0 \\ 0 & 1 \end{pmatrix}$$

and define the Hermitian matrix

$$X := x^\mu \sigma_\mu = \begin{pmatrix} x^0 + x^3 & x^1 - ix^2 \\ x^1 + ix^2 & x^0 - x^3 \end{pmatrix} \tag{2.2.4}$$

The condition that this matrix be Hermitian is necessary and sufficient to ensure that the x^μ are real. It follows immediately that

$$\det X = (x^0)^2 - (x^1)^2 - (x^2)^2 - (x^3)^2 = x^\mu x_\mu \tag{2.2.5}$$

is Lorentz invariant, and because

$$x^\mu y_\mu = \frac{1}{2}\left[\det(X+Y) - \det X - \det Y\right],$$

so are all scalar products of four vectors. We conclude that given any complex 2×2 matrix U with unit determinant, $\det U = 1$, we can find a proper orthochronous Lorentz transformation $\Lambda(U)$ such that

$$X' = UXU^\dagger = [\Lambda(U)x]^\mu \, \sigma_\mu . \tag{2.2.6}$$

The neatest way to exhibit this correspondence uses the fact that every complex matrix U can be written

$$U = u^\mu \sigma_\mu, \quad u^\mu \in \mathbb{C};$$

so that

$$u^\mu u_\mu = \det U = 1,$$

by (2.2.5). Thus only 6 of the 8 real numbers comprising the elements of a complex 2×2 matrix are independent, corresponding to the number of parameters need to specify a Lorentz transformation. Because

$$\sigma_i \sigma_j = \delta_{ij} + i\epsilon_{ijk}\sigma_k, \text{ and } [\sigma_0, \sigma_i] = 0 \tag{2.2.7}$$

the Pauli matrices anticommute,

$$\{\sigma_i, \sigma_j\} = 2\delta_{ij}, \tag{2.2.8}$$

so that

$$\mathrm{tr}\ \sigma_\mu = 2\delta_{\mu 0}, \quad \mathrm{tr}\ \sigma_\mu \sigma_\nu = 2\delta_{\mu\nu}. \tag{2.2.9}$$

Applying (2.2.9) to (2.2.4) we see that

$$x^\mu = \frac{1}{2}\,\mathrm{tr}\ (\sigma_\mu X) \tag{2.2.10}$$

so that the result of a Lorentz transformation gives components

$$x'^\mu = \frac{1}{2}\,\mathrm{tr}\ (\sigma_\mu X') = \frac{1}{2}\,\mathrm{tr}\ (\sigma_\mu U X U^\dagger) = \frac{1}{2}\,\mathrm{tr}\ (\sigma_\mu U \sigma_\nu U^\dagger) x^\nu$$

from which we get

$$\Lambda^\mu{}_\nu(U) = \frac{1}{2}\,\mathrm{tr}\ (\sigma_\mu U \sigma_\nu U^\dagger). \tag{2.2.11}$$

It remains to verify that $\Lambda^\mu{}_\nu(U)$ satisfies the group multiplication law of $\mathcal{L}_+^\uparrow$. This follows directly from (2.2.11) with the aid of the result

$$\mathrm{tr}\ (U\sigma_\mu)\,\mathrm{tr}\ (\sigma_\mu U') = 2\,\mathrm{tr}\ (UU').$$

A convenient parametrization is

$$U = \sigma_0 \cos\frac{1}{2}\varphi + i\boldsymbol{n}\cdot\boldsymbol{\sigma}\sin\frac{1}{2}\varphi \tag{2.2.12}$$

where $\boldsymbol{n}$ is a unit vector in $\mathbb{R}^3$ and φ may be complex. Then

$$\begin{aligned} U^{-1} &= \sigma_0 \cos\frac{1}{2}\varphi - i\boldsymbol{n}\cdot\boldsymbol{\sigma}\sin\frac{1}{2}\varphi, \\ U^\dagger &= \sigma_0 \left(\cos\frac{1}{2}\varphi\right)^* - i\boldsymbol{n}^*\cdot\boldsymbol{\sigma}\left(\sin\frac{1}{2}\varphi\right)^*. \end{aligned} \tag{2.2.13}$$

The subgroup SU(2)⊂ SL(2,C) consists of all elements of SL(2,C) such that $U^\dagger = U^{-1}$, namely

$$U_{\boldsymbol{n}}(\varphi) = \sigma_0 \cos\frac{1}{2}\varphi + i\boldsymbol{n}\cdot\boldsymbol{\sigma}\sin\frac{1}{2}\varphi, \quad \varphi\in\mathbb{R},\ \boldsymbol{n}\in\mathbb{R}^3. \tag{2.2.14}$$

This corresponds to rotations through an angle φ about the real axis $\boldsymbol{n}$ in $\mathbb{R}^3$ as studied in Appendix B.3 and is double valued. Alternatively, we can require U to be Hermitian, $U = U^\dagger$, which leads to

$$U_{\boldsymbol{n}}(\beta) = \sigma_0 \cosh\frac{1}{2}\beta + i\boldsymbol{n}\cdot\boldsymbol{\sigma}\sinh\frac{1}{2}\beta \tag{2.2.15}$$

corresponding to a Lorentz boost (2.2.2). Examination of the dependence of $\Lambda(U)$ on the components u^μ shows that $\Lambda^0{}_0 > 0$, so that this construction covers only $\mathcal{L}_+^\uparrow$. However, the identity in $\mathcal{L}_+^\uparrow$ can be generated with $(u^0, \boldsymbol{u}) = (\pm1, \boldsymbol{0})$, showing once again the two-valued character of the homomorphism.

2.2.2 ∗ Infinitesimal Lorentz transformations and their generators

Infinitesimal proper Lorentz transformations are close to the identity in $\mathcal{L}_+^\uparrow$, and we can write

$$\Lambda^\mu{}_\nu = \delta^\mu{}_\nu + \varepsilon\lambda^\mu{}_\nu + \cdots, \tag{2.2.16}$$

where $\delta^\mu{}_\nu$ takes the value 1 if $\mu = \nu$ and zero otherwise, or

$$\Lambda_{\mu\nu} = g_{\mu\nu} + \varepsilon\lambda_{\mu\nu} + \cdots$$

where ε is a real parameter and (2.1.8) is satisfied to $O(\varepsilon^2)$ if $\lambda_{\mu\nu} = -\lambda_{\nu\mu}$. To determine these infinitesimal generators, consider first the 3-dimensional rotations (2.2.1). A rotation about the x^3-axis through an angle θ in the plane of x^1 and x^2 is defined by

$$\Lambda(\theta)^{12} = \begin{pmatrix} 1 & 0 & 0 & 0 \\ 0 & \cos\theta & \sin\theta & 0 \\ 0 & -\sin\theta & \cos\theta & 0 \\ 0 & 0 & 0 & 1 \end{pmatrix}$$

which can be expanded in powers of θ to give an expression of the form (2.2.16) with the *infinitesimal generator* matrix

$$(\lambda^\mu{}_\nu)^{12} = \begin{pmatrix} 0 & 0 & 0 & 0 \\ 0 & 0 & 1 & 0 \\ 0 & -1 & 0 & 0 \\ 0 & 0 & 0 & 0 \end{pmatrix} \tag{2.2.17}$$

where μ labels rows, ν labels columns, the superscripts 12 label the plane of rotation, and ε is replaced by θ. Similar matrices can be constructed for rotations about each of the other space axes. The standard boost (2.2.2) along the x^1 axis has the infinitesimal generator

$$(\lambda^\mu{}_\nu)^{01} = \begin{pmatrix} 0 & -1 & 0 & 0 \\ -1 & 0 & 0 & 0 \\ 0 & 0 & 0 & 0 \\ 0 & 0 & 0 & 0 \end{pmatrix} \tag{2.2.18}$$

where the superscripts 01 show that the boost operates in the plane of x^0 and x^1, and ε is replaced by β. If the collection of infinitesimal generators are designated $\mathcal{M}^\mu{}_\nu$, then we can define

$$\mathcal{M}^{\mu\nu} = -\mathcal{M}^{\nu\mu}$$

and an arbitrary infinitesimal Lorentz transformation can be expressed as

$$\Lambda(\varepsilon) = I + \frac{1}{2}\varepsilon^{\mu\nu}\mathcal{M}_{\mu\nu} \tag{2.2.19}$$

where the infinitesimal parameters satisfy $\varepsilon^{\mu\nu} = -\varepsilon^{\nu\mu}$ of which only 6 are independent.

The infinitesimal generators satisfy the commutation relations

$$[\mathcal{M}_{\mu\nu}, \mathcal{M}_{\rho\sigma}] = g_{\mu\rho}\mathcal{M}_{\nu\sigma} + g_{\nu\sigma}\mathcal{M}_{\mu\rho} - g_{\mu\sigma}\mathcal{M}_{\nu\rho} - g_{\nu\rho}\mathcal{M}_{\mu\sigma} \tag{2.2.20}$$

Whilst this looks complicated, the fact that the metric coefficient $g_{\mu\nu}$ vanishes unless $\mu = \nu$ means that the right-hand side of (2.2.20) is non-zero only when, say, $\mu = \sigma$; in this case the right hand side reduces to $\pm\mathcal{M}_{\nu\rho}$. All other commutators vanish.

2.2.3 * Representations of the Lorentz group

Let $\mathcal{G}$ be a group with elements $g_1, g_2, \ldots$ and identity e, and let $\mathcal{S}$ be a linear vector space. A set of linear operators $T(g_i) : \mathcal{S} \to \mathcal{S}$ is said to generate a *representation* of $\mathcal{G}$ if $T(e) = E$, where E is the identity in $\mathcal{S}$, and if for any pair of elements $g_1, g_2 \in \mathcal{G}$,

$$T(g_1)T(g_2) = T(g_1 g_2).$$

If the *representation space* $\mathcal{S}$ has finite dimension s, then the representation is said to be s-dimensional. In general, several group elements can map into the same operator, the extreme case being *the identity representation* when $T(g) = E$, $\forall g \in G$. The representation is said to be *faithful* when the representation is one-to-one.

Let $D(\Lambda)$ be a representation of the Lorentz group, and denote the infinitesimal generators of the representation by $M_{\mu\nu}$. When Λ has the form (2.2.19), it is represented by

$$D = I + \frac{1}{2} i \, \varepsilon^{\mu\nu} M_{\mu\nu} \tag{2.2.21}$$

where, following from (2.2.20), the operators $M_{\mu\nu}$ satisfy

$$[M_{\mu\nu}, M_{\rho\sigma}] = -i \left(g_{\mu\rho} M_{\nu\sigma} + g_{\nu\sigma} M_{\mu\rho} - g_{\mu\sigma} M_{\nu\rho} - g_{\nu\rho} M_{\mu\sigma} \right). \tag{2.2.22}$$

Thus to find the representations of the Lorentz group we must identify all possible realizations of these commutation relations.

The Lorentz group has irreducible representations that have both finite and infinite dimension. To construct finite dimensional irreducible representations, we note that the operators

$$\boldsymbol{J} = (M_{23}, M_{31}, M_{12}), \quad \boldsymbol{K} = (M_{01}, M_{02}, M_{03}), \tag{2.2.23}$$

have the commutation relations

$$[J_i, J_j] = i\epsilon_{ijk} J_k, \quad [K_i, K_j] = -i\epsilon_{ijk} J_k, \quad [J_i, K_j] = i\epsilon_{ijk} K_k, \tag{2.2.24}$$

where ϵ_{ijk} is the Levi-Civita symbol, taking the value +1 or -1 according as ijk is an even or odd permutation of 123, and zero otherwise. There are two operators, the Casimir operators, which commute with all operators of the representation, namely

$$C_1 = \frac{1}{2} M_{\mu\nu} M^{\mu\nu} = \boldsymbol{J}^2 - \boldsymbol{K}^2, \quad C_2 = \frac{1}{4} \epsilon^{\mu\nu\rho\sigma} M_{\mu\nu} M_{\rho\sigma} = 2\boldsymbol{J} \cdot \boldsymbol{K}. \tag{2.2.25}$$

These are therefore group invariants, and, by Schur's first lemma [6, §4.8], are multiples of the identity in any irreducible representation. We can therefore use the values of C_1 and C_2 of to label each irreducible representation.

The combinations

$$\boldsymbol{J}' = \frac{1}{2}(\boldsymbol{J} + i\boldsymbol{K}), \quad \boldsymbol{K}' = \frac{1}{2}(\boldsymbol{J} - i\boldsymbol{K}) \tag{2.2.26}$$

have commutation relations

$$[J_i', J_j'] = i\epsilon_{ijk} J_k', \quad [K_i', K_j'] = i\epsilon_{ijk} K_k', \quad [J_i', K_j'] = 0. \tag{2.2.27}$$

The operator sets $J_i', i = 1, 2, 3$ and $K_i', i = 1, 2, 3$ thus satisfy the standard commutation relations for angular momentum operators [6, §7.4]. The $J_i', i = 1, 2, 3$ have an irreducible representation of integer dimension $2j + 1$ spanned by the vectors $|j, m\rangle$, $m = -j, -j-1, \ldots, j$. If $J_\pm' = J_1' \pm iJ_2'$, $\boldsymbol{J}'^2 = J_1'^2 + J_2'^2 + J_3'^2$, then

$$\begin{aligned} \boldsymbol{J}'^2 \, |j, m\rangle &= j(j+1) \, |j, m\rangle \\ J_3' \, |j, m\rangle &= m \, |j, m\rangle, \\ J_\pm' \, |j, m\rangle &= \sqrt{j(j+1) - m(m \pm 1)} \, |j, m \pm 1\rangle, \end{aligned} \tag{2.2.28}$$

where j can take values $0, 1/2, 1, 3/2, \ldots$. Because J_i' and K_j' commute for all pairs of subscripts i, j, the infinitesimal generators possess irreducible representations of finite rank $(2j+1)(2j'+1)$ which we can label $D^{(j,j')}$. The Casimir operators for this representation have the values

$$C_1 = 2[j(j+1) + j'(j'+1)], \quad C_2 = -2i[j(j+1) - j'(j'+1)].$$

We consider the simplest cases:

- $D^{(0,0)}$: This has rank 1; the infinitesimal generators are all null. Objects belonging to this trivial representation are therefore *relativistic scalars.*
- $D^{(\frac{1}{2},0)}$ and $D^{(0,\frac{1}{2})}$: These are two-dimensional conjugate *spinor* representations. They are inequivalent, since the infinitesimal operators have the realizations

$$\begin{aligned} D^{(\frac{1}{2},0)}: \qquad & J_k = \tfrac{1}{2}\sigma_k \quad K_k = -\frac{1}{2}i\sigma_k \\ D^{(0,\frac{1}{2})}: \qquad & J_k = \tfrac{1}{2}\sigma_k \quad K_i = +\frac{1}{2}i\sigma_k \end{aligned}$$

  where the σ_k are Pauli matrices (2.2.3).
- $D^{(1,1)}$: This has rank 4, and the representation consists of four component vectors. The matrices represented by (2.2.17) and (2.2.18) generate the infinitesimal operators of this representation.

The improper operations of space inversion, I_s, and time inversion, I_t satisfy commutation $([\,\cdot\,,\,\cdot\,])$ and anticommutation $(\{\,\cdot\,,\,\cdot\,\})$ relations

$$\begin{aligned} [I_{st}, J_i] &= [I_{ts}, K_i] = 0, \\ \{I_t, K_i\} &= [I_t, J_i] \quad = 0, \\ \{I_s, K_i\} &= [I_s, J_i] \quad = 0. \end{aligned} \tag{2.2.29}$$

Consequently operators $\boldsymbol{J}', \boldsymbol{K}'$ defined in (2.2.26) satisfy

$$I_s\boldsymbol{K}' = \boldsymbol{J}'I_s, \quad I_t\boldsymbol{K}' = \boldsymbol{J}'I_t. \tag{2.2.30}$$

If we now adjoin I_s to the proper orthochronous Lorentz group to generate the set $\mathcal{L}_-^\uparrow$, we find that a basis vector $|jm, j'm'\rangle$ in $D^{(j,j')}$, with $j \neq j'$ is mapped by I_s into a multiple of $|j'm', jm\rangle$; for

$$J_3' I_s |jm, j'm'\rangle = I_s K_3' |jm, j'm'\rangle = m' I_s |jm, j'm'\rangle.$$

It follows that $|j'm', jm\rangle$ and $I_s|jm, j'm'\rangle$ transform under different irreducible representations. If $V^{jj'}$ denotes the vector space associated with $D^{(j,j')}$ and $j \neq j'$, then the space $V^{jj'} \oplus V^{j'j}$ of dimension $2(2j+1)(2j'+1)$ will be an irreducible vector space for the improper orthochronous Lorentz group.

This does not exhaust the catalogue of irreducible representations of the Lorentz group. Hermitian representations are necessarily infinite dimensional; they can be parametrized by setting

$$2j + 1 = k + ic, \quad 2j' + 1 = -k + ic$$

where k is an integer and c is real, so that the Casimir operators of the representation take the values

$$C_1 = c^2 + 1 - k^2, \quad C_2 = ck.$$

Unitary representations also exist with

$$2j + 1 = 2j' + 1 = c, \quad |c| \leq \frac{1}{2}.$$

These are of no interest in this book.

2.3 The Poincaré group

The Poincaré (or inhomogeneous Lorentz) group consists of all coordinate transformations $L = \{a, \Lambda\}$ of the form

$$x'^{\mu} = (Lx)^{\mu} = \Lambda^{\mu}{}_{\nu} x^{\nu} + a^{\mu} \tag{2.3.1}$$

where Λ is a homogeneous Lorentz transformation as described in §2.2. If $\bar{L} = \{\bar{a}, \bar{\Lambda}\}$ is a second such transformation, then we have the law of composition

$$\bar{L}\,L = \{\bar{\Lambda}\,a + \bar{a}, \bar{\Lambda}\Lambda\}. \tag{2.3.2}$$

Evidently $\{0, \mathbf{1}\}$, where $\mathbf{1}$ is the 4×4 identity matrix, plays the part of the identity, and it is easy to verify that these coordinate transformations generate a group.

Close to the identity, (2.3.1) may be written

$$x'^{\mu} = x^{\mu} + \varepsilon^{\mu}{}_{\nu}\, x^{\nu} + \varepsilon^{\mu}, \quad \varepsilon_{\mu\nu} = -\varepsilon_{\nu\mu}$$

so that the infinitesimal operators in a group representation have the general form

$$D = 1 + \frac{1}{2} i\, \varepsilon^{\mu\nu}\, M_{\mu\nu} + i\, \varepsilon^{\mu} P_{\mu} \tag{2.3.3}$$

where, if ∂_{μ} denotes the partial derivative with respect to x^{μ},

$$P_{\mu} = i\partial_{\mu}, \quad M_{\mu\nu} = x_{\mu} P_{\nu} - x_{\nu} P_{\mu} = -M_{\nu\mu}.$$

Thus the 6 independent operators $M_{\mu\nu}$ are supplemented by the 4 operators P_{μ}, so the the Poincaré group has 10 parameters. The Lorentz group commutators (2.2.22),

$$[M_{\mu\nu}, M_{\rho\sigma}] = -i \left(g_{\mu\rho} M_{\nu\sigma} + g_{\nu\sigma} M_{\mu\rho} - g_{\mu\sigma} M_{\nu\rho} - g_{\nu\rho} M_{\mu\sigma} \right), \tag{2.3.4}$$

must be augmented [7, 8, 9] by

$$[P_\mu, P_\nu] = 0, \quad [M_{\mu\nu}, P_\sigma] = i\,(g_{\nu\sigma}P_\mu - g_{\mu\sigma}P_\nu)\,. \tag{2.3.5}$$

Wigner [7] and Shirokov [9] introduce the vector

$$g_\mu = M_{\mu\nu}\, P^\nu \tag{2.3.6}$$

and the pseudovector

$$w_\mu = \frac{1}{2}\epsilon_{\mu\nu\rho\sigma}\, M^{\nu\rho} P^\sigma, \tag{2.3.7}$$

where $\epsilon_{\mu\nu\rho\sigma}$, takes the value +1 if $\mu\nu\rho\sigma$ is an even permutation of 0123, -1 if it is an odd permutation and zero otherwise. Because

$$g_\mu\, P^\mu = 0, \quad w_\mu\, P^\mu = 0, \tag{2.3.8}$$

only 6 of the 8 components of g_μ and w_μ are independent, so that we can replace the 6 independent operators $M_{\mu\nu}$ by these vectors, using

$$M_{\mu\nu} = (g_\mu P_\nu - g_\nu P_\mu + \epsilon_{\mu\nu\rho\sigma} P^\rho\, w^\sigma)\,/C_1, \tag{2.3.9}$$

where C_1 is one of the two Casimir invariants,

$$C_1 = P_\mu\, P^\mu, \quad C_2 = w_\mu\, w^\mu. \tag{2.3.10}$$

In terms of the original set of operators,

$$C_2 = M_{\mu\sigma}\, M^{\nu\sigma}\, P^\mu P_\nu - \frac{1}{2} M_{\mu\nu}\, M^{\mu\nu}\, P^\sigma P_\sigma.$$

The commutation relations are

$$[M_{\mu\nu}, w_\rho] = i\,(g_{\nu\rho}w_\mu - g_{\mu\rho}w_\nu)\,, \tag{2.3.11}$$

$$[w_\mu, P_\nu] = 0, \quad [g_\mu, w_\nu] = i w_\mu P_\nu, \quad [w_\mu, w_\nu] = i\epsilon_{\mu\nu\rho\sigma} w^\rho P^\sigma, \tag{2.3.12}$$

$$[g_\mu, P_\nu] = i\,(P_\mu P_\nu - g_{\mu\nu}C_1)\,, \quad [g_\mu, g_\nu] = i M_{\mu\nu} C_1. \tag{2.3.13}$$

These relations are easier to understand if we rewrite these equations in terms of the 3-vectors $\boldsymbol{J}$ and $\boldsymbol{K}$ defined in (2.2.23) along with the 3-vector $\boldsymbol{P} = (P_1, P_2, P_3)$ and $H/c = P_0$. This gives the set of commutators

$$[J_i, J_j] = +i\epsilon_{ijk} J_k, \quad [J_i, K_j] = +i\epsilon_{ijk} K_k \quad [K_i, K_j] = -i\epsilon_{ijk} J_k, \tag{2.3.14}$$

$$[J_i, P_j] = +i\epsilon_{ijk} P_k, \quad [K_i, P_j] = +i\delta_{ij} H, \tag{2.3.15}$$

$$[J_i, H] = 0, \quad [K_i, H] = i P_i. \tag{2.3.16}$$

For a free particle we can interpret the operator $\boldsymbol{P}$ as its *linear momentum*, $H = cP_0$ as its *kinetic energy*, $\boldsymbol{J}$ (the infinitesimal operator connected with spatial rotation) as its angular momentum, and $\boldsymbol{K}$ with boosts. The remaining generators are

$$w^\mu = (\boldsymbol{P}\cdot\boldsymbol{J},\ P_0\boldsymbol{J} + \boldsymbol{P}\times\boldsymbol{K}), \tag{2.3.17}$$

$$g^\mu = (-\boldsymbol{K}\cdot\boldsymbol{P},\ P_0\boldsymbol{K} - \boldsymbol{J}\times\boldsymbol{P}).$$

2.3.1 * Representations of the Poincaré group

Construction of the irreducible representations of the Poincaré group requires a complete set of six commuting operators. The two Casimir operators provide two of the set and, in view of (2.3.12), it is convenient to choose the others to be the three components of $\boldsymbol{P}$ and one of the components of w^μ, say w^3, whose eigenvalues then label the basis vectors of the representation. We write these vectors $|C_1, C_2; \boldsymbol{p}, \sigma\rangle$, where $\boldsymbol{p}$ denotes the eigenvalues of $\boldsymbol{P}$ and σ is proportional to the eigenvalue of w^3. It follows that when C_1 and C_2 have been specified, and

$$\boldsymbol{P}\,|C_1, C_2; \boldsymbol{p}, \sigma\rangle = \boldsymbol{p}\,|C_1, C_2; \boldsymbol{p}, \sigma\rangle$$

then, since $C_1 = P_\mu P^\mu = (P^0)^2 - \boldsymbol{P}^2$ and $H = cP^0$,

$$H\,|C_1, C_2; \boldsymbol{p}, \sigma\rangle = \pm c\sqrt{C_1 + \boldsymbol{p}^2}\,|C_1, C_2; \boldsymbol{p}, \sigma\rangle$$

Table 2.1 shows the three separate types of irreducible representation corresponding to the sign and magnitude of $C_1 = P_\mu P^\mu$ in keeping with the physical interpretation of P^μ as the 4-momentum of a particle. The first two

Table 2.1. Irreducible representations of the Poincaré group

Class		
P_m	P^μ time-like:	$C_1 > 0$
P_0	P^μ null:	$C_1 = 0,\ P^0 \neq 0$
P_π	P^μ space-like:	$C_1 < 0$

classes, $C_1 = m^2c^2 \geq 0$, are the most interesting, as we can interpret m as a particle *rest mass*. In the class P_m, we have

$$C_1 = H^2/c^2 - \boldsymbol{p}^2 := m^2c^2$$

so that

$$H = \pm c\sqrt{m^2c^2 + \boldsymbol{p}^2} \tag{2.3.18}$$

The sign of H is an invariant, and therefore we can define a third invariant operator

$$C_3 := \operatorname{sgn} H = \pm 1 \tag{2.3.19}$$

As C_2 is an invariant, we can calculate its value in any convenient frame, in particular the rest frame of the particle in which $\boldsymbol{p} = 0$ and so $H = mc^2$. From (2.3.17) we see that in this case

$$w^\mu = (0, mc\boldsymbol{J}), \quad C_2 = -w_\mu w^\mu = m^2c^2\boldsymbol{J}^2$$

so that the space-like part of w^μ is proportional to $\boldsymbol{J}$ in the particle rest frame. According to (2.3.14), the components of w^μ satisfy the usual commutation relations for angular momentum operators, so that the states can be classified in terms of the eigenvalues of $\boldsymbol{J}^2$ and of J_3. Thus the quantum numbers we need are

$$C_2 = m^2c^2 s(s+1), \quad s = 0, \frac{1}{2}, 1, \frac{3}{2}, \ldots \tag{2.3.20}$$

and

$$\sigma = -s, -s+1, \ldots, s-1, s.$$

Because the pair s, σ refer to a particle at rest, we interpret this as the *intrinsic angular momentum* or *spin* of the particle. In sum, a free particle with rest mass m and spin s has an irreducible representation space with basis vectors $|\boldsymbol{p}, \sigma\rangle$, $\sigma = -s, -s+1, \ldots, s-1, s$.

The class P_0 corresponds to rest mass $m = 0$ ($C_1 = 0$). If C_2 also vanishes, we have that both P^μ and w^μ are null vectors such that, (2.3.17), $P^\mu w_\mu = 0$. So we can set $w^\mu = \sigma P^\mu$; again using (2.3.17) we find that the eigenvalues of σ are

$$\sigma = \boldsymbol{p}\cdot\boldsymbol{J}/p_0. \tag{2.3.21}$$

Because p^μ is a null vector, $(p^0)^2 = \boldsymbol{p}^2$, so that (2.3.21) identifies σ as the *helicity* of the particle, defined as its component of the intrinsic angular momentum along the particle's momentum vector. When $\sigma \neq 0$, there are two independent states corresponding to two different polarizations, one parallel, the other anti-parallel to the motion. When $\sigma = 0$, there is only one such state.

The helicity of a massless particle is a Lorentz invariant quantity, having the same value in every inertial frame. We shall see below that particles of opposite helicity are related by space inversion. Because electromagnetic forces have space inversion symmetry, the massless particles with helicity ± 1 associated with electromagnetic phenomena are called *photons*. Similarly, the massless particles with helicity ± 2 believed to be associated with gravitation, which also has space inversion symmetry, are called *gravitons*. However, the supposedly massless particles emitted in nuclear beta decays have helicity $\pm\frac{1}{2}$. Apart from gravitation, they have no interactions that respect space inversion symmetry, and they are therefore distinguished by calling those with helicity $+\frac{1}{2}$ *neutrinos*, and those with helicity $-\frac{1}{2}$ *antineutrinos*.

There exist more representations when $C_2 > 0$. These are infinite-dimensional in the spin variable, so that the corresponding polarization is a continuous variable. Such representations appear not to have any counterpart in the real world, and we need not consider them further. Similarly, we can rule out the class P_π, having $C_1 < 0$, on the grounds that then we can always find an inertial frame in which the energy p_0 becomes arbitrarily large.

2.3.2 ∗ Space and time reflections

When the improper operations of space and time reflection are included, the full Poincaré group, like the Lorentz group, has four disjoint components only one of which is connected to the identity. The others are obtained from the continuous component by adjoining the discrete operators $\mathcal{P}$ and $\mathcal{T}$ of space and time reflection

$$\mathcal{P}^{\mu}{}_{\nu} = \begin{pmatrix} 1 & 0 & 0 & 0 \\ 0 & -1 & 0 & 0 \\ 0 & 0 & -1 & 0 \\ 0 & 0 & 0 & -1 \end{pmatrix}, \quad \mathcal{T}^{\mu}{}_{\nu} = \begin{pmatrix} -1 & 0 & 0 & 0 \\ 0 & 1 & 0 & 0 \\ 0 & 0 & 1 & 0 \\ 0 & 0 & 0 & 1 \end{pmatrix} \tag{2.3.22}$$

to the operations of the proper orthochronous Poincaré group $\mathcal{P}_{+}^{\uparrow}$. Let us denote a representation of the Poincaré group by $D(L)$, where $L = \{a, \Lambda\}$ satisfies the composition law (2.3.2). Then the operators of the representation must satisfy

$$D(\bar{L})D(L) = D(\bar{L}L).$$

Suppose that $\mathcal{P}$ and $\mathcal{T}$ have the representations

$$\mathsf{P} = D(\{0, \mathcal{P}\}), \quad \mathsf{T} = D(\{0, \mathcal{T}\}).$$

Then if $L = \{a, \Lambda\}$ is an arbitrary proper orthochronous transformation, we should expect that

$$\mathsf{P}D(L)\mathsf{P}^{-1} = D(\{\mathcal{P}a, \mathcal{P}\Lambda\mathcal{P}^{-1}\}) \tag{2.3.23}$$

and

$$\mathsf{T}D(L)\mathsf{T}^{-1} = D(\{\mathcal{T}a, \mathcal{T}\Lambda\mathcal{T}^{-1}\}) \tag{2.3.24}$$

if our description is to be invariant with respect to space and time reflections. When $D(L)$ is the infinitesimal transformation of (2.3.3) these relations give

$$\mathsf{P}\, iM^{\mu\nu}\, \mathsf{P}^{-1} = i\, \mathcal{P}_{\rho}{}^{\mu}\mathcal{P}_{\sigma}{}^{\nu}\, M^{\rho\sigma}, \quad \mathsf{P}\, iP^{\mu}\, \mathsf{P}^{-1} = i\, \mathcal{P}_{\rho}{}^{\mu}\, P^{\rho} \tag{2.3.25}$$

$$\mathsf{T}\, iM^{\mu\nu}\, \mathsf{T}^{-1} = i\, \mathcal{T}_{\rho}{}^{\mu}\mathcal{T}_{\sigma}{}^{\nu}\, M^{\rho\sigma}, \quad \mathsf{T}\, iP^{\mu}\, \mathsf{T}^{-1} = i\, \mathcal{T}_{\rho}{}^{\mu}\, P^{\rho}. \tag{2.3.26}$$

Before we can extract the commutation rules from these results, we have to decide whether P and T are linear and unitary or antilinear and antiunitary; definitions in Appendix B.1.7. Consider P first. Then, using (2.3.22) we see that

$$\mathsf{P}\, iP^{0}\, \mathsf{P}^{-1} = iP^{0}.$$

If P is unitary, then we can cancel i on both sides and we get $\mathsf{P}\, P^{0}\, \mathsf{P}^{-1} = P^{0}$, so that P and $P^{0} = H$ commute. If, on the other hand, P is antiunitary, then complex conjugation is required which would make P *anticommute* with H. Suppose that ψ is an eigenstate of H with energy $E > 0$; then if P and H anticommute we should have

$$\mathsf{P}\, E\psi = \mathsf{P}\, H\psi = -H\, \mathsf{P}\psi$$

so that $\mathsf{P}\psi$ would be an eigenstate of H with energy $-E < 0$. If ψ belongs to the particle representation characterized by mass m and spin s, then $\mathsf{P}\psi$ belongs to a different representation with the opposite sign of H. We prevent this by choosing P to be *unitary*. Applying the same argument to T forces us to choose it to be *antiunitary*. Taking account of (2.2.23), we find

$$[\mathsf{P}, H] = 0, \quad \{\mathsf{P}, \boldsymbol{P}\} = 0, \quad [\mathsf{P}, \boldsymbol{J}] = 0, \quad \{\mathsf{P}, \boldsymbol{K}\} = 0, \tag{2.3.27}$$

and

$$[\mathsf{T}, H] = 0, \quad \{\mathsf{T}, \boldsymbol{P}\} = 0, \quad \{\mathsf{T}, \boldsymbol{J}\} = 0, \quad [\mathsf{T}, \boldsymbol{K}] = 0. \tag{2.3.28}$$

This seems physically reasonable: P reverses the sense of $\boldsymbol{P}$ and $\boldsymbol{K}$ whilst $\boldsymbol{J}$, which must transform like the vector product $\boldsymbol{r} \times \boldsymbol{p}$, is left unchanged. Similarly, T reverses the sense of $\boldsymbol{P}$ and of $\boldsymbol{J}$, consistent with the observation that an observer would see bodies spinning in the opposite sense after a time reversal. It is easy to check the consistency of this choice with other commutation relations.

Suppose now that $\Psi(x)$ is a smooth function of the space-time coordinate x; then

$$\Psi(x + X) = \exp{(iX \cdot p)}\, \Psi(x)$$

which is formally equivalent to Taylor's theorem if we write, as usual in quantum mechanics,

$$p_\mu = i\frac{\partial}{\partial x^\mu}.$$

Because $p_0 = H/c$ and $x^0 = ct$, we see that

$$H\Psi(x) = cp_0\Psi(x) = i\frac{\partial \Psi}{\partial t} \tag{2.3.29}$$

which has the form of a Schrödinger equation. Our construction ensures that if $\Psi(x)$ satisfies (2.3.29), then because $D(L)\Psi(x)$ is in the same representation space when L is in the proper orthochronous Poincaré group $\mathcal{P}_+^\uparrow$, $D(L)\Psi(x)$ also satisfies (2.3.29). However, things are more complicated with the operators P and T.

Again we consider P first. Because it is a unitary operator, we must have in general

$$\mathsf{P}\Psi(ct, \boldsymbol{x}) = \mathrm{e}^{i\alpha}\Psi(ct, -\boldsymbol{x}) \tag{2.3.30}$$

where α is some real number. Clearly we are dealing with a *ray* representation in which Ψ and $\mathrm{e}^{i\alpha}\Psi$ represent the same state. It follows that $\Psi(ct, -\boldsymbol{x})$ and $\Psi(ct, \boldsymbol{x})$ both belong to the same representation (m, s) and satisfy the same Schrödinger equation (2.3.29). It follows from (2.3.30) that

$$\mathsf{P}^2 \sim 1 \tag{2.3.31}$$

since two inversions multiply the original $\Psi(ct, \boldsymbol{x})$ by a complex factor of unit modulus, denoted by the symbol $\sim$.

The treatment of time inversion is more complicated. The corresponding relation to (2.3.30) is

$$\mathsf{T}\Psi(ct, \boldsymbol{x}) = \tau\Psi^*(-ct, \boldsymbol{x}) \tag{2.3.32}$$

where τ is a matrix of dimension $(2s+1) \times (2s+1)$, such that

$$\tau \boldsymbol{s}^* \tau^{-1} = -\boldsymbol{s}. \tag{2.3.33}$$

Such a matrix exists because the matrices $-\boldsymbol{s}^*$ and $\boldsymbol{s}$ have the same commutation relations. There is only one irreducible representation of spin operators of dimension $2s+1$, so that the two representations must be equivalent. Thus ${\tau^*}^{-1}\tau$ commutes with all the components of $\boldsymbol{s}$, and by Schur's first lemma ([6, §4.8]) is a multiple of the identity. Hence

$$\mathsf{T}^2 \sim 1 \tag{2.3.34}$$

It can also be shown that τ is defined up to an arbitrary phase factor and must be a symmetric matrix when s is an integer, and anti-symmetric otherwise.

The scheme just outlined does not encompass the usual relativistic wave equations for which a representation space that admits a unitary representation of the full Poincaré group is needed. In addition to the Wigner time reflection operator T satisfying (2.3.28), we introduce the Pauli operator Z satisfying

$$\{\mathsf{Z}, H\} = [\mathsf{Z}, \boldsymbol{P}] = \{\mathsf{Z}, \boldsymbol{K}\} = [\mathsf{Z}, \boldsymbol{J}] = 0, \tag{2.3.35}$$

along with

$$\mathsf{ZP} \sim \mathsf{PZ}, \quad \mathsf{Z}^2 \sim 1.$$

We also need the antiunitary *charge conjugation* operator C, defined by

$$\mathsf{C} := \mathsf{ZT}, \tag{2.3.36}$$

which commutes with all the generators of $\mathcal{P}_+^\uparrow$:

$$[\mathsf{C}, H] = [\mathsf{C}, \boldsymbol{P}] = [\mathsf{C}, \boldsymbol{K}] = [\mathsf{Z}, \boldsymbol{J}] = 0. \tag{2.3.37}$$

If we also assume

$$\mathsf{TZ} \sim \mathsf{ZT}$$

then

$$\mathsf{CP} \sim \mathsf{PC}, \quad \mathsf{CT} \sim \mathsf{TC}, \quad \mathsf{CZ} \sim \mathsf{ZC}, \quad \mathsf{C}^2 \sim 1. \tag{2.3.38}$$

Because Z and H commute, the eigenvalues of H must now occur in pairs of equal and opposite sign. We can then accomodate all the improper operations in a unitary representation for (m, s) by doubling the dimension to $2(2s+1)$. Following Foldy [10] we choose a coordinate representation space on functions $\Psi(ct, \boldsymbol{x})$ with inner product

$$\int \Psi'^{\dagger}(ct, \boldsymbol{x})\Psi(ct, \boldsymbol{x})\, d^3x \tag{2.3.39}$$

on which the infinitesimal operators are realized by

$$H := cp_0 = \beta E \tag{2.3.40}$$
$$\boldsymbol{P} := \boldsymbol{p} = -i\boldsymbol{\nabla} \tag{2.3.41}$$
$$\boldsymbol{J} := \boldsymbol{L} + \boldsymbol{S} \tag{2.3.42}$$
$$\boldsymbol{K} := \beta\frac{1}{2c}(\boldsymbol{x}E + E\boldsymbol{x}) - \beta\frac{c\boldsymbol{S}\times\boldsymbol{p}}{mc^2 + E} - ct\boldsymbol{p} \tag{2.3.43}$$

where $\boldsymbol{L} = \boldsymbol{x}\times\boldsymbol{p}$ as usual and from (2.3.18) and (2.3.19)

$$E^2 = c^2\boldsymbol{p}^2 + m^2c^4.$$

All matrices have dimension $2(2s+1)\times 2(2s+1)$, and may be partitioned into $(2s+1)\times(2s+1)$ blocks, for example

$$\beta := \begin{pmatrix} I & 0 \\ 0 & -I \end{pmatrix}, \quad \boldsymbol{S} := \begin{pmatrix} \boldsymbol{s} & 0 \\ 0 & \boldsymbol{s} \end{pmatrix}.$$

The realization of the improper operators is

$$\mathsf{P}\Psi(ct, \boldsymbol{x}) = \sigma\Psi(ct, -\boldsymbol{x}) \tag{2.3.44}$$
$$\mathsf{Z}\Psi(ct, \boldsymbol{x}) = \zeta\Psi(-ct, \boldsymbol{x}) \tag{2.3.45}$$
$$\mathsf{T}\Psi(ct, \boldsymbol{x}) = \tau\Psi^*(-ct, \boldsymbol{x}) \tag{2.3.46}$$
$$\mathsf{C}\Psi(ct, \boldsymbol{x}) = \mathsf{ZT}\Psi(ct, \boldsymbol{x}) = \kappa\Psi^*(-ct, \boldsymbol{x}), \tag{2.3.47}$$

where the matrices have the block structure

$$\sigma = e^{i\theta_\pi}\begin{pmatrix} I_s & 0 \\ 0 & I_s \end{pmatrix}, \quad \zeta = e^{i\theta_z}\begin{pmatrix} 0 & I_s \\ I_s & 0 \end{pmatrix},$$
$$\tau = \begin{pmatrix} \tau_s & 0 \\ 0 & \pm\tau_s \end{pmatrix}, \quad \kappa = \zeta\tau = e^{i\theta_z}\begin{pmatrix} 0 & \pm\tau_s \\ \tau_s & 0 \end{pmatrix}$$

in which I_s is the $(2s+1)\times(2s+1)$ identity and τ_s is a matrix satisfying (2.3.33),(2.3.34). The phases and the signs may be assigned independently. It follows that

$$\{\zeta, \beta\} = [\zeta, \boldsymbol{S}] = 0, \quad \zeta\sigma \sim \sigma\zeta, \quad \zeta^2 \sim I$$

where I is the $2(2s+1)\times 2(2s+1)$ identity, whilst

$$\tau\beta^*\tau^{-1} = \beta, \quad \tau\boldsymbol{S}^*\tau^{-1} = -\boldsymbol{S}, \quad \tau\sigma^*\tau^{-1} \sim \sigma, \quad \tau\zeta^*\tau^{-1} \sim \zeta, \quad \tau\tau^* \sim 1.$$

The selection of a canonical form for strongly invariant wave equations requires the phases to be defined:

$$\beta = \beta^\dagger, \quad \beta^2 = \sigma\sigma^\dagger = \zeta\zeta^\dagger = I, \quad \sigma^2 \sim I, \quad \zeta^2 \sim I, \quad \zeta\sigma \sim \sigma\zeta \tag{2.3.48}$$

$$\boldsymbol{S} = \boldsymbol{S}^\dagger, \quad [S_i, S_j] = i\epsilon_{ijk} S_k \tag{2.3.49}$$

$$[\beta, \boldsymbol{S}] = [\sigma, \boldsymbol{S}] = [\zeta, \boldsymbol{S}] = 0, \quad [\beta, \sigma] = \{\beta, \zeta\} = 0 \tag{2.3.50}$$

It follows that β has eigenvalues ± 1 and that if ψ is an eigenvector belonging to eigenvalue $+1$, then $\zeta\psi$ is an eigenvector belonging to eigenvalue -1, so that the representation space always has even dimension. For T, Z we have

$$\tau\tau^\dagger = I, \quad \tau\beta^*\tau^{-1} = \beta, \quad \tau\boldsymbol{S}^*\tau^{-1} = -\boldsymbol{S},$$
$$\tau\tau^* \sim I, \quad \tau\sigma^*\tau^{-1} \sim \sigma, \quad \tau\zeta^*\tau^{-1} \sim \zeta$$

$$\kappa\kappa^\dagger = I, \quad \kappa\beta^*\kappa^{-1} = -\beta, \quad \kappa\boldsymbol{S}^*\kappa^{-1} = -\boldsymbol{S},$$
$$\kappa\kappa^* \sim I, \quad \kappa\sigma^*\kappa^{-1} \sim \sigma, \quad \kappa\zeta^*\kappa^{-1} \sim \zeta, \quad \kappa\tau^* \sim \tau\kappa^*.$$

The canonical equations are invariant under similarity transformations of the form

$$\xi \to U\xi U^{-1}, \qquad \xi = \beta, \sigma, \zeta, \boldsymbol{S}$$

and

$$\xi = \xi \to U\xi U^{*-1}, \qquad \xi = \kappa, \tau$$

where U commutes with both $\boldsymbol{x}$ and $\boldsymbol{p}$. This defines an equivalence class of representations in which any member can be transformed into the canonical form. An example is Dirac's equation for spin $\frac{1}{2}$ particles for which the relevant transformation was obtained by Foldy and Wouthuysen [11].

2.4 The Klein-Gordon equation

For a particle with mass $m > 0$ and spin $s = 0$, the relation

$$E^2 = c^2p^2 + m^2c^4 \tag{2.4.1}$$

combined with the correspondence principle identification $p^\mu = i\partial^\mu$ leads immediately to the equation

$$\left(\Box + m^2c^2\right)\phi(x) = 0, \tag{2.4.2}$$
$$\Box := p_\mu p^\mu = \frac{1}{c^2}\frac{\partial^2}{\partial t^2} - \boldsymbol{\nabla}^2.$$

Although this equation is associated with the names of Klein and Gordon, it was proposed independently by several other people. Indeed Schrödinger wrote it down along with his more familiar non-relativistic wave equation. The differential operator $\left(\Box + m^2c^2\right)$ is Lorentz invariant, and the amplitude $\phi(x)$ transforms under an inhomogeneous Lorentz transformation, $x' = \Lambda x + a$, to

$$\phi'(x') = \phi(x)$$

or

$$\phi'(x) = \phi\left(\Lambda^{-1}(x-a)\right) \tag{2.4.3}$$

The equation (2.4.1) makes it clear that solutions exist with both positive and negative energies: $E_{\boldsymbol{p}} = cp^0 = \pm\sqrt{p^2 + m^2c^2}$.

Every acceptable wave equation must admit the existence of a four-current density, j^μ, satisfying a *continuity equation*

$$\partial_\mu j^\mu = 0, \quad j^0 = c\rho, \quad \boldsymbol{j} = (j^1, j^2, j^3) \tag{2.4.4}$$

where $\rho(x)$ is the density at the space-time point x and $\boldsymbol{j}$ is the associated current. This equation expresses the conservation of matter and is clearly Lorentz invariant provided j^μ is a genuine four vector. In more familiar nonrelativistic notation (2.4.4) reads

$$\frac{\partial\rho}{\partial t} + \operatorname{div}\boldsymbol{j} = 0. \tag{2.4.5}$$

To find j^μ we proceed as in nonrelativistic wave mechanics to write down the equation

$$\phi^*\left(\Box + m^2c^2\right)\phi(x) - \phi\left(\Box + m^2c^2\right)\phi^*(x) = 0,$$

which can be simplified to read

$$\partial_\mu\left(\phi^*\partial^\mu\phi - \partial^\mu\phi^*.\phi\right) = 0,$$

suggesting that we define the four-current density as

$$j^\mu = \frac{i}{2m}\left(\phi^*\partial^\mu\phi - \partial^\mu\phi^*.\phi\right), \tag{2.4.6}$$

in which the pre-factor has been chosen so that the space-like components, $\boldsymbol{j}$, have the same form as in nonrelativistic theory. It follows that the density, ρ, is represented by

$$\rho = \frac{j^0}{c} = \frac{i}{2mc}\left(\phi^*\partial^0\phi - \partial^0\phi^*.\phi\right) = \frac{i}{2mc^2}\left(\phi^*\frac{\partial\phi}{\partial t} - \frac{\partial\phi^*}{\partial t}\phi\right) \tag{2.4.7}$$

The Klein-Gordon equation has plane wave solutions of the form

$$\phi(x) = A\,\mathrm{e}^{-ik_\mu x^\mu}, \quad k_\mu k^\mu = m^2c^2, \tag{2.4.8}$$

and then

$$ic\,\partial_0\phi(x) = E_{\boldsymbol{k}} = \pm\sqrt{c^2k^2 + m^2c^4}.$$

Substituting into (2.4.7), we see that

$$\rho = \frac{E_{\boldsymbol{k}}}{mc^2}\phi^*\phi \tag{2.4.9}$$

so that the sign of ρ depends on the sign of the energy.

The fact that expression (2.4.7) is not positive definite, and so can hardly represent a probability density, was a major obstacle to acceptance of the Klein-Gordon equation when it was first introduced. The indefinite sign of the energy E appears connected with the appearance of second order time derivatives in the Klein-Gordon equation (2.4.2). The initial value solution of the equation thus requires both ϕ and $\partial_0\phi$ to be given initially, whereas ordinary quantum theory, in which only a first order time derivative appears, just needs the value of ϕ. Dirac's equation [12, 13] for $s = 1/2$ grew out of a search for an acceptable relativistic wave equation that was first order in time. Although it was easy to define a positive definite probability density for Dirac's equation, it also possessed negative energy states that were difficult to understand until the discovery of the positron in 1932 [14] made it feasible to interpret them in terms of states of anti-particles.

We can put the Klein-Gordon equation into the Foldy canonical form (2.3.40) for $s = 0$ by considering the two-component expression

$$\chi(x) = \begin{pmatrix} \chi_1(x) \\ \chi_2(x) \end{pmatrix}, \tag{2.4.10}$$

where ψ satisfies (2.4.2), $E = +\sqrt{m^2c^2 + \boldsymbol{p}^2}$, $p^0 = i\partial_0$, and

$$\begin{aligned} \chi_1(x) &= \frac{-i}{\sqrt{2}}\left(E^{-1/2}ic\partial_0\psi + E^{1/2}\psi\right) \\ \chi_2(x) &= \frac{-i}{\sqrt{2}}\left(E^{-1/2}ic\partial_0\psi - E^{1/2}\psi\right) \end{aligned}$$

It follows that $\chi(x)$ is a basis vector for the 2-dimensional ($s = 0$) representation of the full Poincaré group since

$$\begin{aligned} H\chi &= ic\partial_0\chi = \beta E\chi \\ P_i\chi &= i\partial_i\chi, \quad i = 1, 2, 3, \\ J_i\chi &= L_i\chi \quad i = 1, 2, 3, \\ K_i\chi &= \left[\frac{1}{2c}\beta(x_iE + Ex_i) + ict\partial_i\right]\chi \quad i = 1, 2, 3, \end{aligned}$$

in which $L_i = (\boldsymbol{x} \times \boldsymbol{p})_i$, and

$$\beta = \begin{pmatrix} 1 & 0 \\ 0 & -1 \end{pmatrix}, \quad \sigma = \pm I, \quad \zeta = \kappa = \begin{pmatrix} 0 & 1 \\ 1 & 0 \end{pmatrix}, \quad \tau = I,$$

where I is the 2×2 identity matrix. Particles whose properties are represented by taking $\sigma = +1$ are said to be *scalar* whilst those whose amplitudes change sign under P, $\sigma = -1$ are said to be *pseudoscalar*. If we substitute (2.4.8) in (2.4.10), we find that plane wave solutions $\chi_+(x)$ for positive energy and $\chi_-(x)$ for negative energy are independent:

$$\chi_+(x) = A \begin{pmatrix} \mathrm{e}^{-ik_\mu x^\mu} \\ 0 \end{pmatrix}, \qquad \chi_-(x) = A \begin{pmatrix} 0 \\ \mathrm{e}^{-ik_\mu x^\mu} \end{pmatrix}. \tag{2.4.11}$$

Let

$$\chi_+(ct, \boldsymbol{x})_c := C\chi_+(ct, \boldsymbol{x}) = \kappa\chi_+^*(ct, \boldsymbol{x})$$

be the result of applying the charge conjugation transformation to $\chi_+(x)$ so that

$$\chi_+(ct, \boldsymbol{x})_c = A^* \begin{pmatrix} 0 \\ \mathrm{e}^{+ik_\mu x^\mu} \end{pmatrix}$$

which is the same as $\chi_-(x)$ with the signs of E and $\boldsymbol{k}$ reversed.

So far we have considered only neutral particles with zero spin. To describe charged particles, we make the usual *minimal coupling* substitution

$$P_\mu \to \varPi_\mu := P_\mu - qA_\mu \tag{2.4.12}$$

where A_μ is the four-potential of some external electromagnetic field and q is the particle's charge. This leads to the wave equation

$$\left(\varPi_\mu \varPi^\mu + m^2c^2\right)\phi(x) = 0. \tag{2.4.13}$$

It now makes more sense to replace the particle four-current density (2.4.6) with an electric *charge-current* density

$$j^\mu = \frac{iq}{2m}\left(\phi^*\partial^\mu\phi - \partial^\mu\phi^*.\phi\right) - \frac{q^2}{m}A^\mu\phi^*\phi, \tag{2.4.14}$$

which presents even more formidable problems of intepretation than (2.4.6). Thus the *charge density* becomes

$$\rho = \frac{j^0}{c} = \frac{iq}{2mc}\left(\phi^*\partial^0\phi - \partial^0\phi^*.\phi\right) - \frac{q^2}{mc}A^0\phi^*\phi, \tag{2.4.15}$$

which includes the A^0 component of the four-potential. Suppose that the space-like part, $\boldsymbol{A} = 0$, and that A^0 is independent of time in the chosen frame of reference. Then we expect to obtain stationary states, $\phi(x) = \psi(\boldsymbol{x})\exp(-iEt)$, for which we find

$$\rho(\boldsymbol{x}) = \frac{q}{mc^2}\left[E - qA^0\right]\phi^*\phi, \tag{2.4.16}$$

The implications of this sort of result become evident on choosing a simple model of a pionic atom, with a negative pion, charge $q = -e$, orbiting a massive nucleus, charge $+Ze$. We assume the nucleus is fixed in space, and ignore strong nuclear forces which should really be taken into account. Then

$$qA^0 = -\frac{Ze^2}{4\pi\epsilon_0 c}V(r),$$

where $V(r) \sim 1/r$ outside the nucleus, $r \gg R_{nuc}$, and $V(r) \to -V_0$ as $r \to 0$. Then (2.4.16) gives

$$\rho(\boldsymbol{x}) = \frac{e}{mc^2}\left[E + \frac{\alpha Z}{r}\right]\phi^*\phi, \quad r \gg R_{nuc}$$

where (recall $\hbar = 1$ in the units we are using) $\alpha = e^2/4\pi\epsilon_0\hbar c$ is the *fine structure constant* [15], and it is clear that the charge density changes sign whenever $E < -\alpha Z/r$. This compounds the difficulty of interpreting the Klein-Gordon equation as a single particle wave equation. The contradictions are resolved by quantum field theories such as quantum electrodynamics (QED).

Table 2.2. Fine structure coefficients F_{nl} and F_{nj}

	Klein-Gordon				Dirac		
$l:$	0	1	2	$j:$	1/2	3/2	5/2
n				n			
1	0.625			1	0.125		
2	0.1016	0.0182		2	0.0391	0.00781	
3	0.0324	0.0077	0.0028	3	0.0139	0.0046	0.0015

The solution of this model problem for a pure Coulomb attraction, $V(r) = 1/r$ has been given, for example, by Schiff [16, pp. 468–471], and is of interest for comparison with the corresponding solution of Dirac's equation (see later reference). The energy of the particle, when expanded in powers of the coupling parameter αZ, can be written

$$E_{nl} = mc^2\left[1 - \frac{(\alpha Z)^2}{2n^2} - \frac{(\alpha Z)^4}{2n^4}\left(\frac{n}{l+1/2} - \frac{3}{4}\right) + \dots\right],$$

where $n = n' + l + 1$, with n' and l both non-negative integers, so that n is a positive integer. The leading term is the rest energy, the second is the nonrelativistic energy, and the third predicts that the single nonrelativistic energy level is split into several *fine structure* levels depending on the value of l:

$$\Delta E_{nl} = -F_{nl}\, mc^2(\alpha Z)^4, \quad F_{nl} = \frac{1}{2n^4}\left(\frac{n}{l+1/2} - \frac{3}{4}\right).$$

Although this result predicts some fine structure, it did not fit Paschen's 1916 results for the fine structure of the He^+ $n = 3 \to n = 4$ line [17, p. 214], and a satisfactory explanation was not forthcoming until 1925 when Uhlenbeck and Goudsmit tried replacing the denominator $l+1/2$ by $j+1/2$, where $j = l \pm 1/2$ is the total angular momentum quantum number resulting from coupling the orbital angular momentum l to the electron spin 1/2 [17, p. 214]. We shall see in Chapter 3 that an expansion of the eigenvalues of Dirac's equation,

(3.3.25), confirmed Uhlenbeck and Goudsmit's guesswork. Table 2.2 shows that the splitting predicted by the Klein-Gordon equation is on average more than double that of the modified formula. There seems no way to escape the conclusion that the Klein-Gordon equation for a spin zero particle is not the right starting point for atomic and molecular modelling.

2.5 The Dirac equation

Dirac's theory of canonical transformations in quantum mechanics relied on a density ρ that was positive definite, and he believed in 1927 that this meant that the wave equation must be linear in the time derivative [17, p. 289]. He therefore searched for a free-particle equation of the form

$$\partial_0 \psi(x) + \alpha^k \partial_k \psi(x) + imc\beta\psi(x) = 0 \tag{2.5.1}$$

in which $\psi(x)$ is vector-valued, and α^μ and β are $n \times n$ matrices, for some as yet unknown value of n. Then it seems reasonable to take ρ to be the positive definite scalar

$$\rho = \psi^\dagger(x)\psi(x) \tag{2.5.2}$$

where $\psi^\dagger$ is the Hermitian adjoint (conjugate transpose) vector. This has to satisfy the continuity equation (2.4.5)

$$\frac{\partial \rho}{\partial t} + \mathrm{div}\boldsymbol{j} = 0.$$

Taking the adjoint of (2.5.1),

$$\partial_0 \psi^\dagger(x) + \partial_k \psi^\dagger(x)(\alpha^k)^\dagger - im\psi^\dagger(x)\beta^\dagger = 0$$

Multiply this equation on the right by $\psi(x)$, (2.5.1) on the left by $\psi^\dagger(x)$ and adding gives

$$\begin{aligned}\partial_0(c\rho(x)) + c\left\{\partial_k \psi^\dagger(x)(\alpha^k)^\dagger \psi(x) + \psi^\dagger(x)\alpha^k \partial_k \psi(x)\right\}& \\ +imc\psi^\dagger(x)\left(\beta - \beta^\dagger\right)\psi(x) = 0,&\end{aligned}$$

which reduces to the continuity equation if all the matrices are Hermitian,

$$\beta^\dagger = \beta, \quad (\alpha^k)^\dagger = \alpha^k$$

and we define

$$j^k(x) = c\psi^\dagger(x)\alpha^k\psi(x). \tag{2.5.3}$$

In the same way, if we operate on (2.5.1) from the left with

$$\partial_0 - \alpha^k\partial_k - imc\beta$$

we recover the Klein-Gordon equation

$$\left(\Box + m^2c^2\right)\psi(x) = 0$$

provided

$$\left\{\alpha^j, \alpha^k\right\} = 2\delta_{jk}, \quad \left\{\alpha^j, \beta\right\} = 0, \quad \left(\alpha^j\right)^2 = (\beta)^2 = I_n \tag{2.5.4}$$

where I_n is the $n \times n$ identity matrix. Dirac found that he could satisfy (2.5.4) by choosing $n = 4$. Whenever we need to be explicit, we shall use the *standard realization*

$$\alpha^j = \begin{pmatrix} 0 & \sigma^j \\ \sigma^j & 0 \end{pmatrix}, \quad \beta = \begin{pmatrix} I & 0 \\ 0 & -I \end{pmatrix}, \tag{2.5.5}$$

where the σ^j, $j = 1, 2, 3$ are the Pauli matrices (2.2.3) and I is the corresponding 2×2 identity matrix.

The success of this formulation is a matter of history. One of its triumphs is the prediction that a particle satisfying Dirac's equation has intrinisic spin $s = 1/2$, consistent with the four-dimensional spinor character of the Dirac wavefunction. (The term spinor, introduced by Ehrenfest [17, p.292], has stuck.) However, as formulated in this section, Dirac's equation is not in Foldy's canonical form [10], and we shall examine the relation between the representations in Section 2.5.3 below.

2.5.1 γ-Matrices and covariant form of Dirac's equation

Dirac's equation takes a more symmetric form that is very convenient for exhibiting its covariance properties if we introduce new 4×4 matrices

$$\gamma^0 := \beta, \quad \gamma^i = \beta\alpha^i, \quad i = 1, 2, 3. \tag{2.5.6}$$

Multiplying (2.5.1) on the left by $i\beta$, we find

$$\left\{\gamma^\mu p_\mu - mc\right\}\psi = 0. \tag{2.5.7}$$

where we have made the replacement $p_\mu := i\partial_\mu$. The anticommutation relations (2.5.4) are replaced by

$$\left\{\gamma^\mu, \gamma^\nu\right\} = 2g^{\mu\nu} I, \tag{2.5.8}$$

so that the matrices with space-like indices are *antihermitian*,

$$\gamma^i\gamma^i = -I_4, \quad (i \text{ not summed}) \tag{2.5.9}$$

whilst

$$\gamma^0\gamma^0 = I_4. \tag{2.5.10}$$

Using the standard realization of the α-matrices (2.5.5), we find

$$\gamma^0 = \begin{pmatrix} I & 0 \\ 0 & -I \end{pmatrix}, \quad \gamma^i = \begin{pmatrix} 0 & \sigma^i \\ -\sigma^i & 0 \end{pmatrix}. \tag{2.5.11}$$

See Appendix A.2 for other properties of Dirac matrices.

2.5.2 * Lagrangian formulation of Dirac's equation

Dirac's equation can be derived variationally along the lines of Appendix B.9.1 by requiring that the Lagrangian action, S, be stationary

$$\delta S = 0, \qquad S := \int_D \mathcal{L}\, d^4 x, \tag{2.5.12}$$

with respect to variations in the spinor ψ, its Dirac adjoint $\widetilde{\psi} = \psi^\dagger \gamma^0$, and their space-time derivatives. The Lagrangian density, $\mathcal{L}$, is defined by

$$\mathcal{L} := \frac{1}{2}\widetilde{\psi}\left(i\gamma^\mu \partial_\mu - mc\right)\psi + \frac{1}{2}\left(-i\partial_\mu \widetilde{\psi}\gamma^\mu - mc\widetilde{\psi}\right)\psi$$

so that for weak variations $\psi \to \psi + \delta\psi$, $\widetilde{\psi} \to \widetilde{\psi} + \delta\widetilde{\psi}$ we have

$$\begin{aligned}\delta\mathcal{L} = \delta\widetilde{\psi}\left[\frac{\partial\mathcal{L}}{\partial\widetilde{\psi}} - \partial_\mu\left(\frac{\partial\mathcal{L}}{\partial(\partial_\mu\widetilde{\psi})}\right)\right] + \left[\frac{\partial\mathcal{L}}{\partial\psi} - \partial_\mu\left(\frac{\partial\mathcal{L}}{\partial(\partial_\mu\psi)}\right)\right]\delta\psi \\ +\partial_\mu\left(\delta\widetilde{\psi}\frac{\partial\mathcal{L}}{\partial(\partial_\mu\widetilde{\psi})} + \frac{\partial\mathcal{L}}{\partial(\partial_\mu\psi)}\delta\psi\right)\end{aligned} \tag{2.5.13}$$

retaining only terms linear in $\delta\psi$ and $\delta\widetilde{\psi}$. Then since

$$\frac{\partial\mathcal{L}}{\partial\widetilde{\psi}} = \frac{1}{2}\left(i\gamma^\mu\partial_\mu - mc\right)\psi - \frac{1}{2}mc\psi, \quad \partial_\mu\left(\frac{\partial\mathcal{L}}{\partial(\partial_\mu\widetilde{\psi})}\right) = -\frac{1}{2}i\gamma^\mu\partial_\mu\psi,$$

for variations $\delta\psi$ and $\delta\widetilde{\psi}$ that vanish on the boundary, (2.5.12) gives the field equations

$$\left(\gamma^\mu p_\mu - mc\right)\psi = 0 \tag{2.5.14}$$

in agreement with (2.5.7), together with its adjoint. Thus for fields that satisfy (2.5.14), we have

$$\delta\mathcal{L} = \partial_\mu\left(\delta\widetilde{\psi}\frac{\partial\mathcal{L}}{\partial(\partial_\mu\widetilde{\psi})} + \frac{\partial\mathcal{L}}{\partial(\partial_\mu\psi)}\delta\psi\right), \tag{2.5.15}$$

from which we can derive conservation equations. A gauge transformation of the form

$$\psi \to \mathrm{e}^{i\alpha}\psi, \quad \widetilde{\psi} \to \mathrm{e}^{-i\alpha}\widetilde{\psi}$$

where α is a real infinitesimal constant leaves $\mathcal{L}$ invariant, so that in (2.5.15) $\delta\mathcal{L} = 0$. To lowest order in α,

$$\delta\psi = i\alpha\psi, \quad \delta\widetilde{\psi} = -i\alpha\widetilde{\psi};$$

substituting into (2.5.15), we see that

$$-\alpha\partial_\mu\left(\widetilde{\psi}\gamma^\mu\psi\right)=0, \tag{2.5.16}$$

consistent with the interpretation of

$$j^\mu=c\widetilde{\psi}\gamma^\mu\psi \tag{2.5.17}$$

as a conserved particle four-current vector, with time-like component given by (2.5.2) and space-like components given by (2.5.3). Integrating the continuity equation (2.4.5) over all space gives

$$\partial_0\int\widetilde{\psi}\gamma^0\psi\,d^3x=0, \tag{2.5.18}$$

because the surface integral of the current density must vanish. It follows that the total charge is a constant of the motion.

The canonically conjugate momenta are not independent of the variables ψ and $\widetilde{\psi}$ so that it is not possible to proceed to derive a Hamiltonian in the manner of Appendix B.9. However we can obtain the Hamiltonian indirectly from the (unsymmetrized) energy-momentum tensor

$$T'^{\mu\nu}=\frac{1}{2}i\left(\widetilde{\psi}\gamma^\nu\partial_\mu\psi-\partial_\mu\widetilde{\psi}\gamma^\nu\psi\right). \tag{2.5.19}$$

To derive this, we consider the effect of a translation

$$x\to x'=x+\epsilon a,$$

with infinitesimal ϵ, under which

$$\psi(x)\to\psi'(x')=\psi(x+\epsilon a)=\psi(x)+\epsilon a^\mu\partial_\mu\psi(x)+o(\epsilon).$$

so that we can put

$$\delta\psi=\epsilon a^\mu\partial_\mu\psi(x)$$

in (2.5.15). If $\mathcal{L}$ does not depend explicitly on the coordinates, we have

$$\delta\mathcal{L}=\epsilon a^\mu\partial_\mu\mathcal{L},$$

yielding

$$\epsilon a^\mu\left\{\partial_\mu\mathcal{L}-\partial_\nu\left(\partial_\mu\widetilde{\psi}\frac{\partial\mathcal{L}}{\partial(\partial_\nu\widetilde{\psi})}+\frac{\partial\mathcal{L}}{\partial(\partial_\nu\psi)}\partial_\mu\psi\right)\right\}=0.$$

Because a^μ is arbitrary, this equation gives

$$\partial_\nu T'^{\mu\nu}=0, \tag{2.5.20}$$

showing that $T'^{\mu\nu}$ gives conserved quantities. The energy-momentum tensor can be symmetrized, but there is no need to do this, as we can obtain everything we need from the momentum four-vector,

$$P^\mu = \int_\sigma d\sigma_\nu T'^{\mu\nu}, \tag{2.5.21}$$

Taking the space-like surface $d\sigma_\nu = (1,0,0,0)d^3x$ as usual, this definition gives

$$\begin{aligned} P^\mu = \int T'^{\mu 0} d^3x &= \frac{1}{2} i \int \left(\widetilde{\psi}\gamma^0 \partial_\mu \psi - \partial_\mu \widetilde{\psi}\gamma^0 \psi \right) d^3x \\ &= i \int \widetilde{\psi}\gamma^0 \partial_\mu \psi \, d^3x. \end{aligned} \tag{2.5.22}$$

We can show that the momentum four-vector is constant in time by integrating (2.5.20) over all space:

$$\partial_0 \int T'^{\mu 0} d^3x + \int \partial_i T'^{\mu i} d^3x = 0.$$

The second integral can be converted into a surface integral at spatial infinity, and if we assume the components of the energy-momentum tensor are such that this surface integral vanishes, the expression reduces to $\partial_0 P^\mu = 0$, proving the result.

We can interpret the time-like component as the particle Hamiltonian as in the Foldy representation (2.3.40),

$$H = cP^0 = c \int \widetilde{\psi}\gamma^0 p_0 \psi d^3x = \int \psi^\dagger \left(c\boldsymbol{\alpha} \cdot \boldsymbol{p} + \beta mc^2 \right) d^3x \tag{2.5.23}$$

where we have used equation (2.5.14), the equivalence $\gamma^\mu p_\mu = \gamma^0 p_0 - \boldsymbol{\gamma} \cdot \boldsymbol{p}$, and $\gamma^0 = \beta$, $\gamma^0 \boldsymbol{\gamma} = \boldsymbol{\alpha}$.

2.5.3 Foldy canonical form and the Foldy-Wouthuysen transformation

We can fix the representation of the infinitesimal operators of the Foldy algebra by writing

$$\begin{aligned} H &:= mc^2\beta + c\boldsymbol{\alpha} \cdot \boldsymbol{p}, \\ P_i &:= p_i, & i = 1,2,3, \\ J_i &:= L_i + \frac{1}{2} S_i, & i = 1,2,3, \\ K_i &:= \frac{1}{2c}(x_i H + H x_i) + ctp_i, \; i = 1,2,3, \end{aligned}$$

where $\boldsymbol{L} = \boldsymbol{x} \times \boldsymbol{p}$ and where

$$\boldsymbol{S} = \frac{1}{2}\boldsymbol{\Sigma} := \frac{1}{2} \begin{pmatrix} \boldsymbol{\sigma} & 0 \\ 0 & \boldsymbol{\sigma} \end{pmatrix}$$

The space and time reflection operators are as given by (2.3.44) *et seq.* with the 4×4 matrices appropriate to the case $s = 1/2$, with

$$\sigma := \beta, \quad \zeta = \beta\alpha_1\alpha_2\alpha_3 = \begin{pmatrix} 0 & I \\ I & 0 \end{pmatrix} \tag{2.5.24}$$

whilst

$$\begin{gathered} \tau = -\tau^t, \quad \tau\tau^\dagger = -\tau\tau^* = -I, \\ \tau\beta^*\tau^{-1} = \beta, \quad \tau\alpha_i^*\tau^{-1} = -\alpha_i, \\ \kappa = \kappa^t, \quad \kappa\kappa^\dagger = \kappa\kappa^* = I, \\ \kappa\beta^*\kappa^{-1} = -\beta, \quad \kappa\alpha_i^*\kappa^{-1} = \alpha_i. \end{gathered}$$

Clearly, these expressions are not quite in the Foldy canonical form. However, they can be transformed into it with the unitary transformation

$$\psi(x) \to \chi(x) := U\psi(x) \tag{2.5.25}$$

where

$$U = \exp\left\{\beta\frac{\boldsymbol{\alpha}\cdot\boldsymbol{p}}{2|\boldsymbol{p}|}\arctan\frac{|\boldsymbol{p}|}{mc}\right\} = \frac{E(\boldsymbol{p}) + mc^2 + c\beta\boldsymbol{\alpha}\cdot\boldsymbol{p}}{[2E(\boldsymbol{p})(E(\boldsymbol{p}) + mc^2)]^{1/2}} \tag{2.5.26}$$

with $E(\boldsymbol{p}) = (\boldsymbol{p}^2 + m^2c^2)^{1/2}$. It is straightforward to verify that U is indeed unitary, with

$$U^{-1} = \frac{E(\boldsymbol{p}) + mc^2 - c\beta\boldsymbol{\alpha}\cdot\boldsymbol{p}}{2E(\boldsymbol{p})(E(\boldsymbol{p}) + mc^2)},$$

so that, for example,

$$\begin{aligned} H &\to UHU^{-1} \\ &= \frac{(E(\boldsymbol{p}) + mc^2 + c\beta\boldsymbol{\alpha}\cdot\boldsymbol{p})\,(mc^2\beta + c\boldsymbol{\alpha}\cdot\boldsymbol{p})\,(E(\boldsymbol{p}) + mc^2 - c\beta\boldsymbol{\alpha}\cdot\boldsymbol{p})}{2E(\boldsymbol{p})(E(\boldsymbol{p}) + mc^2)} \\ &= \frac{(E(\boldsymbol{p}) + mc^2 + c\beta\boldsymbol{\alpha}\cdot\boldsymbol{p})\,(\beta E(\boldsymbol{p}))(E(\boldsymbol{p}) + mc^2 + c\beta\boldsymbol{\alpha}\cdot\boldsymbol{p})}{2E(\boldsymbol{p})(E(\boldsymbol{p}) + mc^2)} \\ &= \beta\frac{(E(\boldsymbol{p}) + mc^2 - c\beta\boldsymbol{\alpha}\cdot\boldsymbol{p})\,(E(\boldsymbol{p}) + mc^2 + c\beta\boldsymbol{\alpha}\cdot\boldsymbol{p})}{2(E(\boldsymbol{p}) + mc^2)} \\ &= \beta E(\boldsymbol{p}). \end{aligned}$$

The first step used the relations $-(\beta\boldsymbol{\alpha}\cdot\boldsymbol{p})^2 = +(\boldsymbol{\alpha}\cdot\boldsymbol{p})^2 = |\boldsymbol{p}|^2$, to simplify the product of the last two factors in the numerator, and the remaining steps depend on the anticommmutation relations (2.5.4). Foldy [10] showed that the other generators also take the canonical form, the only difference being the realization of the $\boldsymbol{\alpha}$ matrices, for which

$$\boldsymbol{\alpha} = -i\begin{pmatrix} 0 & \boldsymbol{\sigma} \\ \boldsymbol{\sigma} & 0 \end{pmatrix}$$

instead of (2.5.5). As with the Klein-Gordon equation, we see that the representation space decomposes into two disjoint subspaces corresponding to the sign of E:

$$H\chi_+ = +|E|\chi_+, \quad H\chi_- = -|E|\chi_-$$

with respective basis vectors

$$\chi_+ = \begin{pmatrix} u \\ 0 \end{pmatrix} \quad \chi_- = \begin{pmatrix} 0 \\ v \end{pmatrix}$$

where u, v each have two nonvanishing components. In the Dirac representation there is no such clean-cut separation of positive and negative energy states, and the two manifolds are both spanned by spinors which in general have four nonvanishing components.

2.5.4 $*$ Position operators in Dirac theory

Another problem of interpretation of Dirac's equation, which also afflicts wave equations for other spins such as the Klein-Gordon equation, is the definition of an operator representing the particle's position. Suppose, in the Dirac representation, we calculate a "velocity" operator, $\boldsymbol{v}$, using the usual expression $\boldsymbol{v} = i[H, \boldsymbol{x}]$. A short calculation gives $\boldsymbol{v} = c\boldsymbol{\alpha}$; the eigenvalues of all components of this operator are $\pm c$, which is physically unacceptable if we want to use $\boldsymbol{x}$ as a position variable and $\boldsymbol{v}$ as the corresponding velocity. However if we follow this by calculating the corresponding "acceleration" vector, we find

$$\frac{d\boldsymbol{v}}{dt} = 2iE(\boldsymbol{p})\left(\boldsymbol{v} - \frac{c^2\boldsymbol{p}}{E(\boldsymbol{p})}\right)$$

where $E(\boldsymbol{p}) = +\sqrt{c^2p^2 + m^2c^4}$, so that

$$\boldsymbol{v}(t) = \frac{c^2\boldsymbol{p}}{E(\boldsymbol{p})} + \boldsymbol{F}(t)$$

where $\boldsymbol{F}(t)$ oscillates undamped about the "classical mean velocity" of the particle, $c^2\boldsymbol{p}/E(\boldsymbol{p})$; this is known as *Zitterbewegung*. A fuller discussion may be found in the books of Thaller [18, §1.6], or Greiner [19, §2.2], where the Zitterbewegung is attributed to interference between the positive and negative energy components of the Dirac wavefunction.

Another problem with identifying $c\boldsymbol{\alpha}$ as the particle's "velocity" vector is that its components do not commute, $[\alpha_i, \alpha_j] \neq 0$, so that the "velocity" components are not simultaneously measurable. This suggests that there is something wrong with the identification, and we should look for another way of interpreting the formalism. Foldy and Wouthuysen [11] observed that the Zitterbewegung arises because the α_i matrices connect the upper and lower components of the Dirac spinor, and therefore sought a unitary transformation of the states in which these were decoupled. We have already seen that this is achieved by the Foldy Wouthuysen operator U of equation (2.5.26). If we interpret $\boldsymbol{x}$ as the position operator in the Foldy canonical representation, then the corresponding operator in the Dirac representation is

$$\begin{aligned}\boldsymbol{X} &:= U^{-1}\boldsymbol{x}U \\ &= \boldsymbol{x} + ic\frac{\beta\boldsymbol{\alpha}}{2E(\boldsymbol{p})} - ic^2\frac{\beta(\boldsymbol{\alpha}\cdot\boldsymbol{p})\boldsymbol{p} - i\boldsymbol{\Sigma}\times\boldsymbol{p}\,|\boldsymbol{p}|}{2E(\boldsymbol{p})(E(\boldsymbol{p})+mc^2)|\boldsymbol{p}|}.\end{aligned} \tag{2.5.27}$$

This operator has the expected commutation relations

$$[X^i, X^j] = 0, \quad [X_i, p_j] = i\delta_{ij}$$

and the corresponding velocity operator, $\boldsymbol{V}$, satisfies

$$\boldsymbol{V} := i[H, \boldsymbol{X}] = \frac{\boldsymbol{p}}{E(\boldsymbol{p})}\frac{\beta mc^2 + c\boldsymbol{\alpha}\cdot\boldsymbol{p}}{E(\boldsymbol{p})}. \tag{2.5.28}$$

The factor $\boldsymbol{p}/E(\boldsymbol{p},)$ is just the classical relativistic particle velocity whilst $(\beta mc^2 + c\boldsymbol{\alpha}\cdot\boldsymbol{p})/E(\boldsymbol{p})$ projects onto the states of positive energy, and vanishes otherwise. Thus this identification works for positive energy solutions. The position operator $\boldsymbol{X}$ was also derived by Newton and Wigner [20], who showed that it transforms like a vector under rotations and has eigenfunctions termed "localized wave functions" satisfying a list of desirable requirements. It also transforms positive energy wavefunctions into positive energy wavefunctions.

The Foldy-Wouthuysen transformation has interesting consequences for angular momentum variables. Neither the orbital angular momentum, $\boldsymbol{l} := \boldsymbol{x}\times\boldsymbol{p}$ nor the spin angular momentum $\boldsymbol{s} := \frac{1}{2}\Sigma$ are separately constants of the motion for the free particle in the Dirac representation, although their sum $\boldsymbol{j} := \boldsymbol{l} + \boldsymbol{s}$ is. However $\boldsymbol{L} := \boldsymbol{X}\times\boldsymbol{p}$ is a constant of the motion in the Foldy-Wouthuysen representation, and so is the *mean spin operator*,

$$\begin{aligned}\boldsymbol{\Sigma}_M &:= U^{-1}\boldsymbol{\Sigma}U \\ &= \boldsymbol{\Sigma} - ic\frac{\beta(\boldsymbol{\alpha}\times\boldsymbol{p})}{E(\boldsymbol{p})} - c^2\frac{\boldsymbol{p}\times(\boldsymbol{\Sigma}\times\boldsymbol{p})}{E(\boldsymbol{p})(E(\boldsymbol{p})+mc^2)}\end{aligned} \tag{2.5.29}$$

The Foldy-Wouthuysen transformation was a major step forward in understanding the nature of the solutions of Dirac's equation for the free electron. We shall discuss the application of this method to the motion of an electron in a hydrogenic atom in Section 3.7.2. Here the transformation to the canonical form requires the construction of a sequence of unitary transformations to decouple the positive and negative energy states, equivalent to a perturbation expansion on powers of $\boldsymbol{p}^2/c^2$. This introduces operators of order $(\boldsymbol{p}^2/c^2)^2$ and higher which have infinite expectation values on the nonrelativistic states and which are therefore no use for applications. Nevertheless the finite lowest order terms are often used to give "relativistic corrections" which give quite accurate results up to the second row of the Periodic Table. Thus the Foldy-Wouthuysen transformation is less useful than might have been hoped, especially for high-Z elements and many-electron systems. The Douglas-Kroll transformation [21], which also uses a sequence of unitary transformations to reduce the equations to a two-component form without introducing unusable perturbation operators, has been developed as a powerful tool for quantum chemistry by Hess and collaborators [22, 23, 24].

2.5.5 Dirac particles in electromagnetic fields

The minimal coupling *ansatz* has already been used in discussing the Klein-Gordon equation (2.4.12), and it is equally applicable to Dirac's equation. We assume that the electromagnetic field is defined in terms of a covariant *four-potential*

$$a^\mu = (A^0, c\boldsymbol{A}), \quad A_\mu = g_{\mu\nu}A^\nu = (A_0, -c\boldsymbol{A}), \tag{2.5.30}$$

satisfying Maxwell's equations, discussed in the next section, where ϕ is the scalar potential and $\boldsymbol{A}$ the vector potential.

Following the same procedure as in nonrelativistic quantum mechanics, we incorporate this in Dirac's equation for a particle with charge q by making the substitution

$$p^\mu \to \Pi^\mu := p^\mu - \frac{q}{c}a^\mu \tag{2.5.31}$$

The new equation has the form

$$\{\gamma^\mu \Pi_\mu - mc\}\,\psi = 0 \tag{2.5.32}$$

which we can rearrange in the form

$$i\frac{\partial\psi}{\partial t} = \{c\boldsymbol{\alpha}\cdot(\boldsymbol{p} - q\boldsymbol{A}) + \beta mc^2 + q\Phi\}\,\psi \tag{2.5.33}$$

where $\Phi = cA^0$.

The wave equation in the form (2.5.32) can be derived straightforwardly from a Lagrangian

$$\mathcal{L} = \mathcal{L}_0 - q\widetilde{\psi}\gamma^\mu a_\mu \psi \tag{2.5.34}$$

where $\mathcal{L}_0$ is now the free particle Lagrangian (2.5.12) and the second term couples the Dirac current to the electromagnetic field. The formalism of Section 2.5.2 is virtually unchanged, yielding the same expressions for the energy-momentum tensor and for the total momentum provided the four-potential is also translation invariant. In the hydrogenic case, where we assume a fixed centre of force, the system is no lon ger invariant with respect to spatial translations although it is still invariant with respect to translations in time. This leads to the Hamiltonian

$$H = cP^0 = \int \psi^\dagger \left\{ c\boldsymbol{\alpha} \cdot (\boldsymbol{p} - q\boldsymbol{A}) + \beta mc^2 + q\Phi \right\} \psi \, d^3x \tag{2.5.35}$$

appearing in (2.5.33). For an electron, we write $q = -e$.

The solution of Dirac's equation for the hydrogenic case, $\boldsymbol{A} = 0$, $\Phi = +Ze/4\pi\epsilon_0 r$ is a standard textbook problem (see, for example [4, Chapter 9], [18], [25]) which is treated in detail in Sections 3.2 and 3.3. The hydrogenic spectrum is the union of a continuous spectrum describing scattering states with two disjoint segments, $-\infty < E \leq -mc^2$ and $mc^2 \leq E < \infty$, along with a point spectrum describing bound states having energies in the range $0 < E < mc^2$. The point eigenvalues form a countable infinite set whose distribution is qualitatively much the same as in the nonrelativistic theory of the hydrogen atom. The eigenvalue sequence converges from below to a limit, $E = mc^2$, given by the energy of an electron at rest. The main difference is that the nonrelativistic terms, $E_n, n = 1, 2, 3, \ldots$, are split, by "spin-orbit coupling", into fine structure levels $E_{nj}, n = 1, 2, 3, \ldots, j = 1/2, 3/2, \ldots, n - 1/2$. Since $E_{nj} < E_n$, electrons are more tightly bound in the relativistic theory of the hydrogen atom than in the Schrödinger theory.

2.5.6 * Negative energy states

One of the basic assumptions of quantum mechanics, due originally to Bohr [26], is that an atom has a stable bound state of lowest energy, the *ground state*. It is therefore impossible to release energy from the atom through a spontaneous transition to a state of lower energy. Evidently relativistic wave equations predict a continuum of negative energy states with energies less than $-mc^2$. The bound state Dirac and Schrödinger spectra for hydrogen-like atoms are qualitatively similar and give a good account of the observations. How can we reconcile this with the presence of negative energy states?

Dirac [27] wrote that "an electron with negative energy moves in an external field as though it carries a positive charge", as emphasized in the discussion of charge conjugation above. He then postulated " ... that all the states of negative energy are occupied except perhaps a few of small velocity ... Only the small departure from exact uniformity, brought about by some of the negative-energy states being unoccupied, can we hope to observe ... " At first, Dirac identified the "hole" states with protons, although the proton mass is much bigger than that of the electron. At the time, it was thought that the only "elementary particles" in nature were electrons and protons, and it was hypothesized that the mass difference between them might be accounted for by electromagnetic interactions. It was not until Anderson's discovery of the positron [28], which has the same mass as the electron but charge $+e$, that some of the confusion was resolved. The "holes" were now identified as positrons, and the symmetry of the Dirac description of electrons and positrons could begin to be understood.

At this point, Dirac envisaged the vacuum as a region of space in its lowest possible energy state in which all negative energy states were occupied. This stabilized the ground state of the hydrogen atom, as there can be no "hole" in the "negative energy sea" into which an electron in the ground state can fall. Creating a "hole" state requires a minimum energy of order $2mc^2$ to put the ejected electron into a positive energy state. The "hole" can then be observed as a real positron. Dirac realised that this meant that even "The simple problem of the scattering of a photon on an electron is no longer a two-body problem. It is an infinitely-many particle problem." [17, p. 350]

The language of Dirac's "hole theory" remains useful for computational purposes in atomic and molecular physics as long as any excitations have energy low in comparison with mc^2. It implies that the presence of negative energy states can, for many purposes, be ignored. It is however an uncomfortable construction as the "negative energy sea" has both infinite mass and charge, both clearly unobservable. It is clearly preferable to work with a formalism in which electrons and positrons are put on the same footing without the need to invoke these infinite quantities. Feynman [29] showed that it was possible to write down a complete solution of the problem of electron and positron motion in an external electromagnetic field in terms of boundary conditions on the wave function. An electron in a positive energy bound or continuum state has a motion in which the particle's proper time increases along its path in space-time. As suggested also by Stückelberg [30], Feynman envisaged a positron as an electron for which coordinate time *decreases* along the particle's path; this is reflected in the negative frequency appearing in the time dependence of negative energy states. The formalism of quantum field theory, Chapter 4, is presented in terms of field variables with an indefinite number of particles. The numbers of electrons, N_e and positrons, N_p, are not conserved individually, although the total charge $Q = e(N_p - N_e)$ is a constant of the motion. Electrons and positrons can thus be created in pairs without affecting Q.

Each particle in quantum field theory is represented by a field amplitude, which will in general be a linear superposition of positive and negative frequency components. The fields propagate according to Huygens' principle, in which the solution at a given time can be regarded as the source of secondary waves whose propagation is described by a Green's function satisfying appropriate boundary conditions. The above argument requires that positive frequency components propagate forwards, negative frequency components backwards, in time. Section 2.9 shows how this can be accomplished.

2.6 Maxwell's equations

2.6.1 Covariant form of Maxwell's equations

Although it is possible to present the electromagnetic field equations due to Maxwell as an example of a massless field with spin $s = 1$ within the framework established in this chapter, we shall here follow a more traditional route.

Maxwell's equations relate the electric and magnetic fields $\boldsymbol{E}$ and $\boldsymbol{B}$ to the space-time distribution of electric charge ρ and current density $\boldsymbol{j}$ so that (in SI units)

$$\mathrm{div}\boldsymbol{B} = 0, \qquad \mathrm{curl}\ \boldsymbol{E} = -\frac{\partial \boldsymbol{B}}{\partial t}, \tag{2.6.1}$$

$$\mathrm{div}\boldsymbol{E} = \frac{\rho}{\epsilon_0}, \qquad \mathrm{curl}\ \boldsymbol{B} = \mu_0\, \boldsymbol{j} + \frac{1}{c^2}\frac{\partial \boldsymbol{E}}{\partial t}, \tag{2.6.2}$$

where ϵ_0 is the *electric constant*, related to the *magnetic constant* μ_0 and the speed of light *in vacuo* by, [15], $\epsilon_0\mu_0 c^2 = 1$. These equations can be written in a covariant form in which we identify ρ and $\boldsymbol{j}$ as the components of a *four-current* vector j^μ such that

$$j^\mu := (c\rho, j_1, j_2, j_3) \tag{2.6.3}$$

It follows from (2.6.1) and (2.6.2) that the components of the four current satisfy a *continuity equation*

$$\frac{\partial \rho}{\partial t} + \mathrm{div}\boldsymbol{j} = 0, \quad \text{or} \quad \partial_\mu j^\mu = 0, \tag{2.6.4}$$

so that the *total charge is a relativistic invariant.* For justification of this assertion, consider an observer whose velocity four-vector is V^μ, so that $V^\mu V_\mu = c^2$. In the observer's rest frame, $V^\mu = (c, 0, 0, 0)$, and so the scalar $j^\mu V_\mu$ takes the invariant value $c^2\rho$. If the four-current j^μ is due to a stream of particles of charge q with four-velocity U^μ, then $j^\mu = NqU^\mu$. The observer sees $NU^\mu V_\mu / c^2$ particles per unit volume, and a charge density $j^\mu V_\mu = NqU^\mu V_\mu / c^2$, so that he then reckons that each particle carries the same charge q independent of its motion.

The *electromagnetic field tensor* is a second rank *covariant* antisymmetric tensor defined by

$$F_{\mu\nu} := \begin{pmatrix} 0 & E_1 & E_2 & E_3 \\ -E_1 & 0 & -cB_3 & cB_2 \\ -E_2 & cB_3 & 0 & -cB_1 \\ -E_3 & -cB_2 & cB_1 & 0 \end{pmatrix} \tag{2.6.5}$$

and we note the algebraic definitions

$$F_{0i} = E_i,\ i = 1, 2, 3, \quad F_{ij} = -\epsilon_{ijk} cB_k,\ i, j, k = 1, 2, 3, \tag{2.6.6}$$

with the antisymmetry conditions $F_{\mu\nu} = -F_{\nu\mu}$. It is easy to verify that Maxwell's equations are equivalent to

$$\partial_\mu F_{\nu\rho} + \partial_\nu F_{\rho\mu} + \partial_\rho F_{\mu\nu} = 0, \tag{2.6.7}$$

corresponding to (2.6.1) and

$$\partial_\mu F^{\mu\nu} = \frac{1}{\epsilon_0 c} j^\nu \tag{2.6.8}$$

corresponding to (2.6.2). These equations are clearly Lorentz covariant provided $F_{\mu\nu}$ behaves like a *covariant* second rank tensor under Lorentz transformations.

The equations (2.6.1) imply that there is a scalar ϕ and a 3-vector $\boldsymbol{A}$ such that

$$\boldsymbol{B} = \text{curl}\, \boldsymbol{A}, \quad \boldsymbol{E} = -\frac{\partial \boldsymbol{A}}{\partial t} - \text{grad}\, \phi \tag{2.6.9}$$

This does not define the potentials ϕ and $\boldsymbol{A}$ uniquely because *gauge transformations* of the form

$$\phi \to \phi + \frac{\partial \Lambda}{\partial t}, \quad \boldsymbol{A} \to \boldsymbol{A} - \text{grad}\, \Lambda \tag{2.6.10}$$

give the same field vectors. The potentials can be regarded as components of the *covariant four-potential*

$$a_\mu := (\phi, -c\boldsymbol{A}) \tag{2.6.11}$$

so that

$$F_{\mu\nu} = \partial_\mu a_\nu - \partial_\nu a_\mu \tag{2.6.12}$$

When this is substituted into (2.6.8), we obtain

$$\Box a^\mu - \partial^\mu(\partial_\nu a^\nu) = \frac{1}{\epsilon_0 c} j^\mu . \tag{2.6.13}$$

We define the *dual electromagnetic field tensor* as the tensor $\widetilde{F}$ with components

$$\widetilde{F}_{\mu\nu} := \frac{1}{2}\, \epsilon_{\mu\nu\rho\sigma} F^{\rho\sigma}, \tag{2.6.14}$$

where $\epsilon_{\mu\nu\rho\sigma}$ is the *alternating tensor*, taking the value +1 if $\mu\nu\rho\sigma$ is an *even* permutation of 0123, -1 if it is *odd*, and zero otherwise, so that

$$\left(\widetilde{F}_{\mu\nu}\right) = \begin{pmatrix} 0 & -cB_1 & -cB_2 & -cB_3 \\ cB_1 & 0 & -E_3 & E_2 \\ cB_2 & E_3 & 0 & -E_1 \\ cB_3 & -E_2 & E_1 & 0 \end{pmatrix} .$$

The contractions

$$F_{\mu\nu} \widetilde{F}^{\mu\nu} = 4c\boldsymbol{E} \cdot \boldsymbol{B} \tag{2.6.15}$$

and

$$F_{\mu\nu} F^{\mu\nu} = 2(c^2 \boldsymbol{B} \cdot \boldsymbol{B} - \boldsymbol{E} \cdot \boldsymbol{E}) \tag{2.6.16}$$

are both Lorentz invariant. Clearly if $\boldsymbol{E} \cdot \boldsymbol{B}$ is null in some frame, it is null in all frames. We can then say that the tensor F is *purely magnetic* if $2(c^2 \boldsymbol{B} \cdot \boldsymbol{B} - \boldsymbol{E} \cdot \boldsymbol{E}) > 0$, as it must then be possible to find an inertial frame in which $\boldsymbol{E} = 0$. We say that F is *purely electric* if the inequality has the opposite sign.

2.6.2 ∗ Lagrangian formulation

The field equations can be derived variationally along the lines of Appendix B.9.1 by requiring that the Lagrangian action

$$S := \int_D \mathcal{L}_{em}\left(a_\mu, \partial_\nu a_\mu\right) d^4x, \tag{2.6.17}$$

be stationary,

$$\delta S = 0, \tag{2.6.18}$$

with respect to variations in the functional form of a_μ and its partial derivatives $\partial_\nu a_\mu$. The theory of Appendix B.9.1 asssumes that the system has a finite number of degrees of freedom. Here we need to remember that the number of degrees of freedom may be infinite.

The integration is over some domain $D \subset \mathbb{R}^4$ with 3-boundary ∂D. We require that $\mathcal{L}$ be constructed from Lorentz invariant terms; a suitable choice for the Maxwell field is

$$\mathcal{L}_{em} = -\frac{1}{4}\epsilon_0 F_{\mu\nu}F^{\mu\nu} - \frac{1}{c}j^\mu a_\mu. \tag{2.6.19}$$

where from (2.6.12)

$$F_{\mu\nu}F^{\mu\nu} = \left(\partial_\mu a_\nu - \partial_\nu a_\mu\right)\left(\partial^\mu a^\nu - \partial^\nu a^\mu\right).$$

In more familiar notation,

$$\mathcal{L}_{em} = \frac{1}{2}\epsilon_0(\boldsymbol{E}\cdot\boldsymbol{E} - c^2\boldsymbol{B}\cdot\boldsymbol{B}) - (\rho\phi - \boldsymbol{j}\cdot\boldsymbol{A}). \tag{2.6.20}$$

From (2.6.18) we see that the first order variation is

$$\delta S = \int_D \left\{\frac{\partial\mathcal{L}_{em}}{\partial a_\mu}\delta a_\mu - \frac{\partial\mathcal{L}_{em}}{\partial(\partial_\nu a_\mu)}\delta\left(\partial_\nu a_\mu\right)\right\} d^4x \tag{2.6.21}$$

where

$$\delta\left(\partial_\nu a_\mu\right) = \partial_\nu(a_\mu + \delta a_\mu) - \partial_\nu a_\mu = \partial_\nu(\delta a_\mu).$$

Now

$$\mathrm{d}\int_D \delta a_\mu \left\{\frac{\partial\mathcal{L}_{em}}{\partial a_\mu} - \partial_\nu\left(\frac{\partial\mathcal{L}_{em}}{\partial(\partial_\nu a_\mu)}\right)\right\} d^4x \tag{2.6.22}$$

differs from (2.6.21) by the integral of a four divergence,

$$\int_D \partial_\nu\left(\frac{\partial\mathcal{L}_{em}}{\partial(\partial_\nu a_\mu)}\delta a_\mu\right) d^4x$$

which can be converted into an integral over the boundary ∂D. If the δa_μ vanish on ∂D, but are otherwise arbitrary in D, then standard methods of the calculus of variations permit us to conclude that the field equations are

$$\frac{\delta S}{\delta(\partial_\nu a_\mu)} := \frac{\partial \mathcal{L}_{em}}{\partial a_\mu} - \partial_\nu \left(\frac{\partial \mathcal{L}_{em}}{\partial(\partial_\nu a_\mu)} \right) = 0 \tag{2.6.23}$$

Substituting

$$\frac{\partial \mathcal{L}_{em}}{\partial a_\mu} = -\frac{1}{c} j^\mu, \qquad \frac{\partial \mathcal{L}_{em}}{\partial(\partial_\nu a_\mu)} = \epsilon_0 (\partial^\mu a^\nu - \partial^\nu a^\mu) = \epsilon_0 F^{\mu\nu}$$

into (2.6.23), we recover (2.6.8) in the form

$$-\frac{1}{\epsilon_0 c} j^\mu + \partial_\nu F^{\nu\mu} = 0$$

The antisymmetry of $F^{\mu\nu}$ ensures current conservation:

$$\partial_\mu j^\mu = \epsilon_0 c \, \partial_\mu \partial_\nu F^{\nu\mu} = 0.$$

2.6.3 Gauge invariance

Gauge transformations of the form (2.6.10)

$$\phi \to \phi + \frac{\partial \Lambda}{\partial t}, \qquad -\boldsymbol{A} \to -\boldsymbol{A} + \operatorname{grad} \Lambda$$

or

$$a_\mu \to a_\mu + \partial_\mu \Lambda \tag{2.6.24}$$

in tensor notation, leave the field equations unchanged. However

$$\mathcal{L}_{em} \to -\frac{1}{4} \epsilon_0 F_{\mu\nu} F^{\mu\nu} - \frac{1}{c} j^\mu a_\mu - \frac{1}{c} j^\mu (\partial_\mu \Lambda).$$

The last term can be dropped because, after an integration by parts,

$$\int j^\mu (\partial_\mu \Lambda) d^4x = -\int (\partial_\mu j^\mu) \Lambda d^4x = 0, \tag{2.6.25}$$

when the current is conserved. The presence of conserved currents is both necessary and sufficient for *gauge invariance* of Maxwell's equations.

So far, we have not attempted to remove the arbitrariness of the four-potential. The final result for any physical observable should, of course, be independent of the choice of gauge. One way to do this is to impose a *gauge condition* to fix Λ, for example

$$\operatorname{div} \boldsymbol{A} = \partial_i A^i = 0 \qquad \text{(Coulomb gauge)} \tag{2.6.26}$$

or

$$\partial_\mu a^\mu = 0 \qquad \text{(Lorentz gauge)} \tag{2.6.27}$$

When we use the Coulomb gauge, we can eliminate a^0 from the Lagrangian, and reduce the number of independent field variables. From (2.6.13), we see that

$$\partial_\mu \partial^\mu \left(a^0\right) - \partial^0 (\partial_\mu \, a^\mu) = \frac{1}{\epsilon_0}\, \rho,$$

which reduces to Poisson's equation

$$\nabla^2 \phi = -\frac{\rho}{\epsilon_0},$$

from which we get the familiar Coulomb law

$$\phi(ct, \boldsymbol{x}) = \frac{1}{4\pi\epsilon_0} \int \frac{\rho(ct, \boldsymbol{x}')}{|\boldsymbol{x} - \boldsymbol{x}'|}\, d^3x'. \tag{2.6.28}$$

In the Lorentz gauge, $\partial_\mu a^\mu = 0$, (2.6.13) reduces to the wave equation

$$\Box \phi = \left(\frac{1}{c^2} \frac{\partial^2}{\partial t^2} - \nabla^2 \right) \phi = \frac{\rho}{\epsilon_0}.$$

Even if the driving term ρ vanishes everywhere, there is no necessity for ϕ to vanish, and it will in general be time-dependent. For this reason, the Coulomb gauge, which yields Coulomb's law directly, is often very convenient for formulating atomic and molecular problems, as we shall see later in this book. Unlike the Lorentz condition, it has the disadvantage that it is not manifestly Lorentz covariant. It is necessary to examine the gauge dependence of observables carefully when calculating electromagnetic properties of atoms and molecules.

It is possible to write (2.6.20) in the Coulomb gauge in a form in which only *transverse* fields appear. A vector field $\boldsymbol{V}$ is said to be transverse if $\mathrm{div}\boldsymbol{V} = 0$. Thus the electric field vector given by (2.6.9) in Coulomb gauge, $\mathrm{div}\boldsymbol{A} = 0$, consists of two parts, a transverse field $\boldsymbol{E}_\perp = -\partial \boldsymbol{A}/\partial t$ and a longitudinal field $\boldsymbol{E}_\| = -\mathrm{grad}\ \phi$. As $\boldsymbol{B} = \mathrm{curl}\ \boldsymbol{A}$, the $\boldsymbol{B}$ field is already transverse. Because

$$\mathrm{div}(\phi\, \mathrm{grad}\ \phi) = \phi\, \nabla^2 \phi + (\mathrm{grad}\ \phi)^2,$$

we see that

$$\frac{1}{2}\boldsymbol{E}.\boldsymbol{E} - \frac{1}{\epsilon_0}\rho\phi = \frac{1}{2}\boldsymbol{E}_\perp.\boldsymbol{E}_\perp + \frac{\partial \boldsymbol{A}}{\partial t}.\mathrm{grad}\ \phi - \frac{1}{2\epsilon_0}\rho\phi.$$

Now, using the Coulomb gauge condition $\mathrm{div}\boldsymbol{A} = 0$,

$$\frac{\partial \boldsymbol{A}}{\partial t}.\mathrm{grad}\ \phi + \boldsymbol{A}.\mathrm{grad}\ \frac{\partial \phi}{\partial t} = \frac{\partial}{\partial t}\left(\boldsymbol{A}.\mathrm{grad}\ \phi\right) = \frac{\partial}{\partial t}\,\mathrm{div}\left(\phi \boldsymbol{A}\right).$$

The term on the right can be dropped, because its contribution to the action can be converted to a surface integral which can be expected to vanish. The final result is that (2.6.20) can be written in the form

$$\mathcal{L}_{em} = \frac{1}{2}\epsilon_0(\boldsymbol{E}_\perp \cdot \boldsymbol{E}_\perp - c^2\boldsymbol{B}_\perp \cdot \boldsymbol{B}_\perp) - \frac{1}{2}\rho\phi + \boldsymbol{j}_\perp \cdot \boldsymbol{A}, \tag{2.6.29}$$

where we write $\boldsymbol{B}_\perp \equiv \boldsymbol{B}$ for emphasis, and

$$\boldsymbol{j}_\perp = \boldsymbol{j} - \epsilon_0 \,\mathrm{grad}\, \frac{\partial \phi}{\partial t}.$$

It is easy to confirm that $\boldsymbol{j}_\perp$ is transverse, because the divergence of the last equation reduces to the continuity equation, (2.6.4). Finally we can use (2.6.28) to write the Lagrangian density in the form

$$\begin{aligned}\mathcal{L}_{em} &= \frac{1}{2}\epsilon_0(\boldsymbol{E}_\perp \cdot \boldsymbol{E}_\perp - c^2\boldsymbol{B}_\perp \cdot \boldsymbol{B}_\perp) + \boldsymbol{j}_\perp \cdot \boldsymbol{A} \\ &\quad - \frac{1}{8\pi\epsilon_0}\int \frac{\rho(ct, \boldsymbol{x})\rho(ct, \boldsymbol{x}')}{|\boldsymbol{x} - \boldsymbol{x}'|} d^3x' \end{aligned} \tag{2.6.30}$$

in which the Coulomb interaction energy emerges naturally.

2.6.4 * Motion of a test charge

The Lagrangian of a free particle with 3-velocity $\boldsymbol{v}$ relative to a given frame is

$$L_{free} := -mc^2\sqrt{1 - \boldsymbol{v}^2/c^2}. \tag{2.6.31}$$

Each coordinate x^i has a *canonically conjugate momentum* p^i defined by

$$p^i := \partial L_{free}/\partial v_i, \quad i = 1, 2, 3, \tag{2.6.32}$$

from which we find

$$p^i = m\gamma(\boldsymbol{v})v^i, \quad \text{where} \quad \gamma(\boldsymbol{v}) = \left(1 - \boldsymbol{v}^2/c^2\right)^{-1/2}. \tag{2.6.33}$$

We can therefore construct a free particle Hamiltonian

$$H_{free}(\mathbf{p}) := \mathbf{p}.\boldsymbol{v} - L_{free} = mc^2\gamma(\boldsymbol{v}) = c\sqrt{m^2c^2 + \mathbf{p}^2}, \tag{2.6.34}$$

so that $H_{free}(\mathbf{p})$ is also the energy, as expected for a conservative system.

The Lagrangian for the Maxwell field derived from (2.6.19) is

$$L_{em} = -\frac{1}{4}\epsilon_0\int F_{\mu\nu}(x)F^{\mu\nu}(x)d^3x - \frac{1}{c}\int j^\mu(y)a_\mu(y)d^3x,$$

where the first term is the Lagrangian for the free Maxwell field, L_{field} and the second couples the field to the charged particle charge-current density. For a single point particle at space-time position x moving with velocity $\boldsymbol{v}$ we have $j^0(y) = cq\delta(x - y)$ and $\boldsymbol{j}(y) = q\boldsymbol{v}\delta(x - y)$ giving an interaction Lagrangian

$$L_{int}(x) = -q\phi(x) + q\boldsymbol{v} \cdot \boldsymbol{A}(x)$$

The total action is therefore

$$S = \int dt \, \{L_{field} - q\phi + q\boldsymbol{v} \cdot \boldsymbol{A} + L_{free}\}$$

The variation of the action with respect to the $\boldsymbol{A}$ field components yields Maxwell's equations as before. However, the momentum $\boldsymbol{p}$ (2.6.33) is modified by the interaction giving

$$\boldsymbol{p} = m\boldsymbol{v}\gamma(\boldsymbol{v}) + q\boldsymbol{A} \tag{2.6.35}$$

and the Lagrange equations of motion of the particle reduce to

$$\frac{d}{dt} m\boldsymbol{v}\gamma(\boldsymbol{v}) = q(\boldsymbol{E} + \boldsymbol{v} \times \boldsymbol{B}), \tag{2.6.36}$$

in which the *Lorentz force law* appears on the right-hand side. Also

$$\frac{d}{dt} mc^2\gamma(\boldsymbol{v}) = q\boldsymbol{E} \cdot \boldsymbol{v} \tag{2.6.37}$$

showing that only electric fields do work on the charges and currents. However, the Hamiltonian, which must be expresed in terms of the canonical momentum, now takes the form

$$H_{min}(\boldsymbol{p}) = mc^2\gamma(\boldsymbol{v}) = c\sqrt{m^2c^2 + (\boldsymbol{p} - q\boldsymbol{A})^2}. \tag{2.6.38}$$

where the replacement of the free electron canonical momentum $\boldsymbol{p}$ by $\boldsymbol{p} - q\boldsymbol{A}$ is termed *minimal substitution.*

It is instructive to compare this with the elementary derivation from Newton's second law. In the rest frame of a particle, the force is $q\boldsymbol{E}$, the four-velocity is $v = (c, 0)$ and the four-acceleration may be written $dv/d\tau = (0, \boldsymbol{a})$, so that $m\boldsymbol{a} = q\boldsymbol{E}$, where τ is the particle's proper time. Let G be the four-vector whose components are given by

$$G^\mu = \frac{q}{c} F^{\mu\nu} v_\nu$$

so that $G^\mu = (0, q\boldsymbol{E})$ in the particle's rest frame. So $mdv/d\tau = G$ in the instantaneous rest frame, and because we are equating two four-vectors, this must hold in every frame. The momentum is $p^\mu = mv^\mu$ where the rest mass, m, is constant and $dp^\mu/d\tau = q\, F^{\mu\nu}\, v_\nu/c$. The equations of motion (2.6.36), (2.6.37) follow after noting that $dt/d\tau = \gamma(\boldsymbol{v})$.

2.7 * Symmetries and local conservation laws

The derivation of Maxwell's equations from (2.6.17) combined with ideas from Appendix B.9.3 allow us to write down conservation equations for the Maxwell field, both free and interacting with charged particles. The argument is the same: we consider the effect of a change in the action due to infinitesimal translations, rotations or Lorentz transformations as well as variations in the field variables to deduce local conservation laws. We return to global conservation issues in Section 2.8 below.

In Section 2.6.2 we saw that the variation in the Lagrangian due to variations in the fields a_μ, (2.6.22), involved a space-time integration over

$$\delta\mathcal{L}_{em} = \delta a_\mu \left\{ \frac{\partial \mathcal{L}_{em}}{\partial a_\mu} - \partial_\nu \left(\frac{\partial \mathcal{L}_{em}}{\partial(\partial_\nu a_\mu)} \right) \right\} + \partial_\nu \left(\frac{\partial \mathcal{L}_{em}}{\partial(\partial_\nu a_\mu)} \delta a_\mu \right).$$

The first term vanishes if the fields satisfy Maxwell's equations, and when this is the case, the local variation in the Lagrangian reduces to

$$\delta\mathcal{L}_{em} = \partial_\nu \left(\frac{\partial \mathcal{L}_{em}}{\partial(\partial_\nu a_\mu)} \delta a_\mu \right). \tag{2.7.1}$$

Different choices for δa_μ then generate different local conservation laws.

Suppose first that the variation δa_μ is generated by an infinitesimal translation

$$x \to x' = x + \epsilon\, u$$

so that

$$\begin{aligned} a_\mu(x) \to a_\mu(x') &= a_\mu(x + \epsilon\, u) \\ &= a_\mu(x) + \delta a_\mu(x) \end{aligned} \tag{2.7.2}$$

so that, to first order in ϵ,

$$\delta a_\mu(x) = \epsilon \left(\frac{\partial a_\mu}{\partial \epsilon} \right)_{\epsilon=0} + O(\epsilon^2) = \epsilon\, u^\nu\, \partial_\nu a_\mu(x) + O(\epsilon^2). \tag{2.7.3}$$

Applying the same argument to $\mathcal{L}_{em}(a_\mu(x), \dot{a}_\mu(x))$ to order ϵ gives

$$\delta\mathcal{L}_{em} = \epsilon\, u^\nu\, \partial_\nu \mathcal{L}_{em}(a_\mu(x), \dot{a}_\mu(x)) \tag{2.7.4}$$

and equating this to (2.7.1), and using (2.7.3) we get

$$\epsilon\, u^\rho \left\{ \partial_\rho \mathcal{L}_{em} - \partial_\nu \left(\frac{\partial \mathcal{L}_{em}}{\partial(\partial_\nu a_\mu)} \partial_\rho a_\mu \right) \right\} = 0.$$

Because u^ρ is arbitrary, we see that

$$\partial_\nu \left(\frac{\partial \mathcal{L}_{em}}{\partial(\partial_\nu a_\mu)} \partial_\rho a_\mu - \delta^\nu{}_\rho \mathcal{L}_{em} \right) = 0 \tag{2.7.5}$$

which we can write as

$$\partial_\nu \widetilde{T}^{\nu\mu} = 0.$$

because $\partial\mathcal{L}_{em}/\partial(\partial_\mu a_\rho) = \epsilon_0 F^{\mu\rho}$, the *canonical energy-momentum tensor* is given by

$$\widetilde{T}^{\mu\nu} := \epsilon_0 F^{\mu\rho}\,\partial^\nu a_\rho - g^{\mu\nu}\mathcal{L}_{em}. \tag{2.7.6}$$

This definition of an energy-momentum tensor has several unpleasant features: in particular it is unsymmetric in the indices μ and ν, and it is also gauge dependent. A gauge transformation of the form $a_\mu \to a_\mu + \partial_\mu \Lambda$ gives

$$\widetilde{T}^{\mu\nu} \to \widetilde{T}^{\mu\nu} - \epsilon_0 F^{\mu\rho}\partial^\nu\partial_\rho\Lambda = \widetilde{T}^{\mu\nu} - \epsilon_0\partial_\rho(F^{\mu\rho}\partial^\nu\Lambda)$$

since $\partial_\rho F^{\mu\rho} = 0$. We have therefore to define an energy-momentum tensor that is gauge invariant if we are to use it to calculate observable quantities.

The Lagrangian density is not completely determined by the equations of motion; we can always add terms to the Lagrangian density, which, when integrated over the whole domain D, can be converted to integrals on the boundary ∂D that vanish when suitable boundary conditions apply. The canonical energy-momentum tensor can be modified in a rather similar way. Consider a conserved current $s^\mu(x)$ with $\partial_\mu s^\mu(x) = 0$, and suppose a transformation $s^\mu(x) \to s^\mu(x) + \Delta s^\mu(x)$. If the current is still locally conserved, then

$$\partial_\mu\,\Delta s^\mu(x) = 0.$$

At the same time, the total charge associated with $s^\mu(x)$ is $Q = \int s^0(x)d^3x$ and if this is to be unchanged, then

$$\int \Delta s^0(x)d^3x = 0.$$

Both conditions are satisfied if $\Delta s^\mu(x)$ is a 4-divergence

$$\Delta s^\mu(x) = \partial_\nu S^{\mu\nu}, \quad S^{\mu\nu} = -S^{\nu\mu},$$

where $S^{\mu\nu}$ depends locally on the fields. We can use this construction to add a term to $\widetilde{T}^{\mu\nu}$ of the form

$$\Delta T^{\mu\nu} = \partial_\rho T^{\mu\rho\nu}, \quad T^{\mu\rho\nu} = -T^{\rho\mu\nu}.$$

It is clear that the choice

$$T^{\mu\rho\nu} = -\epsilon_0 F^{\mu\rho}a^\nu$$

when added to the first term of (2.7.6) makes the whole expression gauge invariant, so we now adopt the expression

$$T_0^{\mu\nu} := \frac{1}{4}\epsilon_0 g^{\mu\nu}F_{\alpha\beta}F^{\alpha\beta} + \epsilon_0 F^{\mu\rho}F^\nu{}_\rho \tag{2.7.7}$$

as the energy momentum tensor of the free Maxwell field; this is symmetric, has zero trace, and is also conserved because $\partial_\mu T_0^{\mu\nu} = 0$.

The Lagrangian for a conserved external current distribution coupled to the Maxwell field, §2.6.2, is

$$\mathcal{L}_{em} = -\frac{1}{4}\epsilon_0 F_{\mu\nu}F^{\mu\nu} - \frac{1}{c}j^\mu a_\mu . \tag{2.7.8}$$

The canonical energy-momentum tensor is

$$\widetilde{T}^{\mu\nu} := \pi^{\mu\rho}\,\partial^\nu a_\rho - g^{\mu\nu}\mathcal{L}_{em} + \frac{1}{c}g^{\mu\nu}j^\rho a_\rho$$

and we can construct a gauge invariant energy-momentum tensor for the coupled system by writing

$$T^{\mu\nu} := T_0^{\mu\nu} + \frac{1}{c}\left(g^{\mu\nu}j^\rho a_\rho - j^\mu a^\nu\right). \tag{2.7.9}$$

When the current does not vanish,

$$\partial_\mu T^{\mu\nu} = \frac{1}{c}a_\rho \partial^\nu j^\rho, \tag{2.7.10}$$

so that

$$\partial_\mu T_0^{\mu\nu} = \partial_\mu\left\{\frac{1}{4}\epsilon_0\, g^{\mu\nu}F_{\alpha\beta}F^{\alpha\beta} + \epsilon_0 F^{\mu\rho}F_\rho{}^\nu\right\} = \frac{1}{c}j_\rho F^{\rho\nu}. \tag{2.7.11}$$

which exhibits $T_0^{\mu\nu}$ as the purely electromagnetic part of the energy momentum tensor. Thus

$$T_0^{00} = \frac{1}{2}\epsilon_0\left(\boldsymbol{E}\cdot\boldsymbol{E} + c^2\boldsymbol{B}\cdot\boldsymbol{B}\right) \tag{2.7.12}$$

can be identified as the energy density of the electromagnetic field, and

$$T_0^{i0} = \epsilon_0\, c^2 \epsilon_{ijk}E_j B_k, \quad i = 1, 2, 3, \tag{2.7.13}$$

known as the *Poynting vector* gives the energy flow.

Along with translations, it is also useful to consider infinitesimal Lorentz transformations of the form

$$x^\mu \to x'^\mu = x^\mu + \epsilon\,\omega^\mu{}_\nu x^\nu, \quad \omega_{\mu\nu} = -\omega_{\nu\mu}.$$

In this case,

$$a_\mu(x) \to a'_\mu(x') = \left(\delta_\mu{}^\nu + \frac{1}{2}\epsilon\, b_\mu{}^{\nu\sigma\tau}\omega_{\sigma\tau}\right)a_\nu(x), \qquad b_\mu{}^{\nu\sigma\tau} = -b_\mu{}^{\nu\tau\sigma},$$

and we obtain a canonical conserved rank three tensor, antisymmetric with respect to the indices ν and σ,

$$J^{\nu\sigma\tau} = L^{\nu\sigma\tau} + S^{\nu\sigma\tau}, \quad \partial_\tau J^{\nu\sigma\tau} = 0, \tag{2.7.14}$$

where

$$L^{\nu\sigma\tau} := x^\sigma \widetilde{T}^{\nu\tau} - x^\nu \widetilde{T}^{\sigma\tau},$$

$$S^{\nu\sigma\tau} := \epsilon_0 b_\mu{}^{\lambda\sigma\nu} a_\lambda F^{\tau\mu}$$

The term $S^{\nu\sigma\tau}$ can be interpreted as a spin angular momentum density, whilst $L^{\nu\sigma\tau}$ represents an orbital angular momentum density distribution. Note that whilst $J^{\nu\sigma\tau}$ is conserved, neither $S^{\nu\sigma\tau}$ nor $L^{\nu\sigma\tau}$ are separately conserved; this decomposition is not necessarily covariant or gauge invariant.

We can define a gauge invariant angular momentum tensor density from $T_0^{\mu\nu}$ by writing

$$J^{\nu\sigma\tau} = x^\sigma T_0^{\nu\tau} - x^\nu T_0^{\sigma\tau}. \tag{2.7.15}$$

The symmetry of $T_0^{\mu\nu}$ guarantees that this satisfies the conservation equation $\partial_\tau J^{\nu\sigma\tau} = 0$. Similar arguments show that gauge transformations, which are connected with conserved currents in interacting systems, for example the coupling of Maxwell and charged particle fields also lead to conservation laws.

2.8 * Global conservation laws

So far we have not specified the boundary ∂D of the region of integration envisaged in forming the action. Consider first a family of space-like hyperplanes

$$\sigma := \{x \,|\, n^\mu x_\mu = \tau, \quad n^\mu n_\mu = 1,\}$$

labelled by τ. The normal n points to the future if $n^0 > 0$ and to the past if $n^0 < 0$; for example in the simplest case, $n = (1, 0, 0, 0)$, $n^\mu x_\mu = x^0 = ct$, so that the hyperplane consists of all of $\mathbb{R}^3$ at time τ/c. Dynamics is a matter of relating field values on different hyperplanes. Suppose we calculate the action in (2.6.17) for a region of Minkowski space D bounded by hyperplanes σ_1 and σ_2. Assume that the fields vanish sufficiently fast at spatial infinity so that we can ignore any boundary effects. When the Maxwell fields satisfy the field equations, (2.7.5) holds at each space-time point, and integrating over the region D bounded by arbitrary hyperplanes σ_1 and σ_2 gives

$$\begin{aligned} 0 &= \int_D \partial_\nu \left(\frac{\partial \mathcal{L}_{em}}{\partial(\partial_\nu a_\mu)} \partial_\rho a_\mu - \delta^\nu{}_\rho \mathcal{L}_{em} \right) d^4x \\ &= F(\sigma_2) - F(\sigma_1) \end{aligned} \tag{2.8.1}$$

where

$$F(\sigma) := \int_\sigma \left(\frac{\partial \mathcal{L}_{em}}{\partial(\partial_\nu a_\mu)} \partial_\rho a_\mu - \delta^\nu{}_\rho \mathcal{L}_{em} \right) d\sigma_\nu = \int_\sigma \widetilde{T}^{\nu\mu} d\sigma_\nu$$

in which $d\sigma_\nu$ is the measure associated with the direction n^ν on σ. For example, if n is future-pointing along the x^0-axis, then the only non-zero component is $d\sigma_0 = dx^1 dx^2 dx^3$, the usual integration over 3-dimensional volume. Because σ_1 and σ_2 are arbitrary, we conclude that the four quantities

$$P^\mu = \int_\sigma \widetilde{T}^{\nu\mu}\, d\sigma_\nu, \quad \mu = 0, 1, 2, 3, \tag{2.8.2}$$

are constants of the motion for the free Maxwell field. Similarly,

$$J^{\mu\nu} = \int_\sigma J^{\lambda\mu\nu}\, d\sigma_\lambda, \quad J^{\mu\nu} = -J^{\nu\mu} \tag{2.8.3}$$

are also constants of the motion. We can replace $\widetilde{T}^{\nu\mu}$ by $T^{\nu\mu}$ in (2.8.2) without changing the result.

We can verify that (2.8.2) gives the same total momentum to the free field as we should expect on elementary grounds by taking $d\sigma_\nu = (d^3x, 0, 0, 0)$, so that

$$P^\mu = \int_\sigma T_0^{0\mu}\, d^3x$$

These are indeed constants of the motion, as from

$$\partial_\mu T_0^{\mu\nu} = 0$$

we see that

$$\partial_0 P^\mu = \int_\sigma \partial_0 T_0^{0\mu}\, d\sigma = -\int_\sigma \partial_i T_0^{i\mu}\, d\sigma$$

The 3-divergence on the right may be converted, using Green's theorem, to a surface integral whose contribution, according to our assumptions, vanishes in the absence of any charge-current sources. From (2.7.12) we see that

$$P^0 = \int_\sigma \frac{1}{2}\epsilon_0(\boldsymbol{E}\cdot\boldsymbol{E} + c^2\boldsymbol{B}\cdot\boldsymbol{B})\, d^3x$$

is the total energy associated with the field. When a current distribution is present, we integrate (2.7.10) over a finite volume V with boundary ∂V, giving

$$\frac{dP^0}{dt} + \int_{\partial V} \boldsymbol{P}\cdot d\boldsymbol{S} = \int_V \boldsymbol{j}\cdot\boldsymbol{E}\, d^3x$$

where $\boldsymbol{P} = \epsilon_0 c^2 \boldsymbol{E}\times\boldsymbol{B} = \mu_0^{-1}\boldsymbol{E}\times\boldsymbol{B} = \boldsymbol{E}\times\boldsymbol{H}$ is the Poynting vector and $d\boldsymbol{S}$ is the surface element on ∂V. The surface integral often does not vanish if V is finite. The right hand side represents the energy dissipated in Joule heating.

2.9 * Green's functions

The notion of a Green's function is used widely in physics and engineering [31]. Instead of dealing with an ordinary or partial differential equation with associated initial and/or boundary conditions, we focus on an integral equation whose kernel is the Green's function for the problem. This section deals with the construction of Green's functions and the corresponding field operators, or *propagators*, for the free Klein-Gordon, Maxwell, and Dirac equations.

2.9.1 Nonrelativistic Green's functions

For orientation, consider a simple nonrelativistic scattering problem with Hamiltonian $H(x) = H_0 + V(x)$, $H_0 = p^2/2m$, $x = (x^0, \boldsymbol{x})$ (for convenience we retain the relativistic notation $x^0 = ct$), and $V(x)$ is a smooth potential having a finite range. The Schrödinger equation is then

$$(ic\partial_0 - H)\,\psi(x) = 0. \tag{2.9.1}$$

Following Feynman [29, 32], we suppose that for times $x^0 > x'^{\,0}$ we can formally write the solution as some linear superposition

$$\theta(x^0 - x'^{\,0})\,\psi(x) = i\int G(x, x')\,\psi(x')\,d^3x' \tag{2.9.2}$$

in accordance with Huygens' principle, where the Heaviside step function

$$\theta(s) = \begin{cases} 1 & s > 0, \\ 1/2 & s = 0, \\ 0 & s < 0, \end{cases}$$

expresses our wish that there is no scattering at times earlier than $x'^{\,0}$. Thus the Green's function $G(x, x')$ propagates the solution from the space-time point x' to the point x. Using the relation $\delta(s) = \theta'(s)$, where prime denotes differentiation with respect to s, we see that

$$\begin{aligned}(ic\partial_0 - H)\,\theta(x^0 - x'^{\,0})\,\psi(x) &= i\int (ic\partial_0 - H)\,G(x, x')\,\psi(x')\,d^3x' \\ &= ic\delta(x^0 - x'^{\,0})\,\psi(x)\end{aligned}$$

from which we infer

$$(ic\partial_0 - H)\,G(x, x') = c\delta^{(4)}(x - x'), \tag{2.9.3}$$

with

$$G(x, x') = 0, \quad x^0 < x'^{\,0},$$

so that this is said to be a *retarded* Green's function. Similarly, we can set $V(x) = 0$ and express the solution of the free particle problem as

$$\theta(x^0 - x'^{\,0})\,\phi(x) = i\int G_0(x, x')\,\phi(x')\,d^3x' \tag{2.9.4}$$

where

$$\begin{aligned}(ic\partial_0 - H_0)\,G_0(x - x') &= c\delta^{(4)}(x - x'), \\ G_0(x - x') &= 0, \quad x^0 < x'^{\,0}.\end{aligned} \tag{2.9.5}$$

Here G_0 depends only on the relative separation $x - x'$ because the defining equation is invariant with respect to translation.

Now rewrite (2.9.1) in the form

$$(ic\partial_0 - H_0)\,\psi(x) = V(x)\,\psi(x).$$

Using the free particle Green's function, we replace this by the integral equation

$$\psi(x) = \phi(x) + i\int G_0(x-x')\,V(x')\,\psi(x')d^4x', \tag{2.9.6}$$

where $\phi(x)$ satisfies $(i\partial_0 - H_0)\phi(x) = 0$. A similar procedure applied to the differential equation

$$(ic\partial_0 - H_0)\,G(x,x') = c\delta^{(4)}(x-x') + V(x)\,G(x,x')$$

gives the Lippmann-Schwinger integral equation

$$G(x,x') = G_0(x-x') + \int G_0(x-x'')V(x'')\,G(x'',x')\,d^4x''. \tag{2.9.7}$$

when the interaction is subject to the appropriate boundary conditions as in (2.9.6). Equations (2.9.6) and (2.9.7) can formally be solved by iteration. Thus, if $G^{(k)}$ is the k-th approximation, with $G^{(0)} = G_0$, then we obtain a sequence of approximations of which the first two are

$$\begin{aligned}
G^{(1)}(x,x') &= G_0(x-x') + \int G_0(x-x_1)V(x_1)\,G_0(x_1-x')\,d^4x_1 \\
G^{(2)}(x,x') &= G_0(x-x') + \int G_0(x-x_1)V(x_1)\,G^{(1)}(x_1,x')\,d^4x_1 \\
&= G_0(x-x') + \int G_0(x-x_1)V(x_1)\,G_0(x_1,x')\,d^4x_1 \\
&+ \iint G_0(x-x_1)V(x_1)\,G_0(x_1-x_2)V(x_2)\,G_0(x_2-x')\,d^4x_1\,d^4x_2.
\end{aligned}$$

Thus $G^{(1)}(x,x')$ is the sum of two terms: in the first, particles propagate freely from x' to x without scattering; in the second, we sum over all possible events in which the particle propagates freely from x' to x_1 where it is scattered by the potential $V(x_1)$, and then the scattered particle propagates freely from x_1 to x. Similarly $G^{(2)}(x,x')$ contains beside these terms, a further contribution from double scattering. The k-th iterate will include terms with up to k-fold scattering by the potential, and we can regard the formal expansion, $G = G^{(0)} + G^{(1)} + G^{(2)} + \ldots$ as a perturbation series.

To construct the free particle Green's function for this problem consider

$$G_0(x-x') = \int \frac{d^4k}{(2\pi)^4}\mathrm{e}^{-ik\cdot(x-x')}\,G(k), \tag{2.9.8}$$

where $k = (z/c, \boldsymbol{k})$. The Hamiltonian is $H_0 = -\nabla^2/2m$ so that, provided differentiation under the integral sign is permissible,

$$(ic\partial_0 - H_0)G_0(x-x') = \int \frac{d^4k}{(2\pi)^4}\left(z - \boldsymbol{k}^2/2m\right) e^{-ik\cdot(x-x')}G(k)$$
$$= c\delta^{(4)}(x-x')$$

by (2.9.5). Because

$$\delta^{(4)}(x-x') = \int \frac{d^4k}{(2\pi)^4} e^{-ik\cdot(x-x')},$$

we conclude that

$$G(k) = \left(z - \boldsymbol{k}^2/2m\right)^{-1}.$$

It follows that

$$G_0(x-x') = \int \frac{d^4k}{(2\pi)^4} e^{-ik\cdot(x-x')}\left(z - \boldsymbol{k}^2/2m + i\epsilon\right)^{-1}$$

where the term $i\epsilon$, $\epsilon > 0$ has been introduced into the denominator to ensure that $G_0(x-x')$ is non-zero only when $x^0 > x'^0$, in line with the integral representation

$$\theta(s) = -\frac{1}{2\pi i}\lim_{\epsilon\to 0}\int_{-\infty}^{\infty}\frac{de^{-i\omega s}}{\omega + i\epsilon}\,d\omega.$$

Thus

$$G_0(x-x') = \int \frac{d^3k}{(2\pi)^3} e^{i\boldsymbol{k}\cdot(\boldsymbol{x}-\boldsymbol{x}')}\int \frac{dz}{2\pi} e^{-iz(t-t')}\left(z - \boldsymbol{k}^2/2m + i\epsilon\right)^{-1}$$
$$= -i\theta(t-t')\int \frac{d^3k}{(2\pi)^3}\exp\left(i\boldsymbol{k}\cdot(\boldsymbol{x}-\boldsymbol{x}') - i\boldsymbol{k}^2(t-t')/2m\right)$$

Because

$$\psi_{\boldsymbol{k}}(x) = (2\pi)^{-3/2}\exp(i\boldsymbol{k}\cdot\boldsymbol{x} - i\boldsymbol{k}^2t/2m), \qquad \int \psi_{\boldsymbol{k}}(x)\,\psi^*_{\boldsymbol{k}'}(x)d^3x = \delta^{(3)}(\boldsymbol{k}-\boldsymbol{k}')$$

is an eigenfunction of the Schrödinger equation,

$$(ic\partial_0 - H_0)\psi = 0,$$

we can conclude that

$$G_0(x-x') = -i\theta(t-t')\int d^3\boldsymbol{k}\,\psi_{\boldsymbol{k}}(x)\psi^*_{\boldsymbol{k}}(x'). \tag{2.9.9}$$

More generally, if a Hamiltonian H has normalized eigenstates $\psi_a(\boldsymbol{x},t)$ satisfying a completeness relation of the form

$$\sum_a \psi_a(\boldsymbol{x},t)\psi^*_a(\boldsymbol{x}',t) = \delta^{(3)}(\boldsymbol{x}-\boldsymbol{x}')$$

where the formal sum over a runs over all point and continuous spectra, then the Green's function

$$G(x, x') = -i\theta(t - t') \sum_a \psi_a(x)\, \psi_a^*(x'),$$

satisfies $(ic\partial_0 - H)G(x, x') = \delta^{(4)}(x - x')$. Using this representation, it follows that

$$i \int d^3x'\; G(x, x')\psi_b(x') = \theta(t - t')\psi_b(x)$$

so that $G(x, x')$ propagates the state ψ_b *forwards* in time. Equally

$$-i \int d^3x\; \psi_b^*(x) G(x, x') = \theta(t' - t)\psi_b^*(x')$$

so that the complex conjugate states are propagated *backwards* in time

2.9.2 Klein-Gordon operator

The Klein-Gordon equation provides a useful introduction to the construction of Green's functions for relativistic equations. We wish to solve the equation

$$\left(\Box + m^2c^2\right) \psi(x) = F(x) \tag{2.9.10}$$

where $F(x)$ is some source term. If, as in (2.9.8), we assume that there exists a Green's function of the form

$$\Delta(x - x') = \int \frac{d^4k}{(2\pi)^4}\, \mathrm{d}e^{-ik\cdot(x-x')}\, K(k), \tag{2.9.11}$$

an argument along the lines of the previous section leads to

$$\left(-k^2 + m^2c^2\right)\, K(k) = 1,$$

so that

$$K(k) = -\frac{1}{k^2 - m^2c^2}. \tag{2.9.12}$$

The denominator vanishes on the surface $k^2 = (k^0)^2 - \boldsymbol{k}^2 = m^2c^2$ in 4-space; for finite mass m this is a two-sheet hyperboloid, degenerating to the light cone $k^2 = 0$ for $m = 0$. It is useful to regard the expression (2.9.11) as an integral over the three degrees of freedom of the space-like components $\boldsymbol{k}$ and a contour integral over the time-like component k_0, choosing paths which avoid the poles at $k_0 = \pm\sqrt{\boldsymbol{k}^2 + m^2c^2}$. Thus

$$\Delta(z) = -\int \frac{d^3k}{(2\pi)^3} \exp\{i\boldsymbol{k}\cdot\boldsymbol{z}\} \int_C \frac{dk_0}{2\pi} \frac{\exp\{-ik_0z^0\}}{k_0^2 - \omega^2}$$

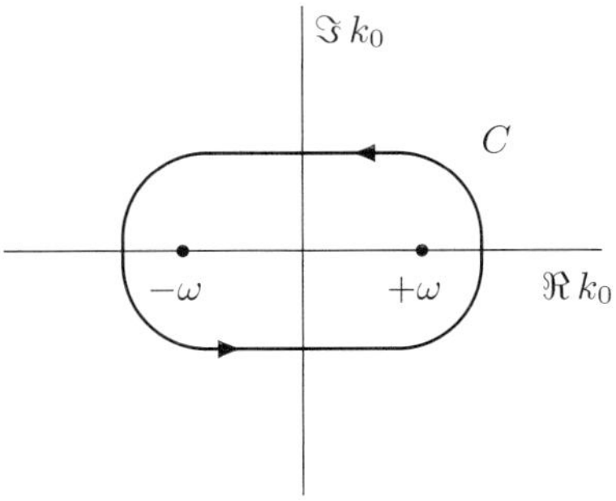

Fig. 2.1. Contour C for $\Delta(x)$, equation (2.9.13).

where $z = (z^0, \boldsymbol{z})$, $\omega = +\sqrt{\boldsymbol{k}^2 + m^2c^2}$ and C is some path in the k_0-plane. The boundary conditions which determine the way in which this kernel propagates solutions of the Klein-Gordon equation can be used to select the appropriate path C. We look first at the simple closed contour of Fig. 2.1, C, taken in the positive sense and surrounding both poles $k_0 = \pm\omega$. Then

$$\Delta(x) = -\frac{1}{(2\pi)^4}\int_C \frac{\mathrm{e}^{-ik\cdot x}}{k^2 - m^2c^2} \tag{2.9.13}$$

which is equivalent to

$$\Delta(x) = -\frac{i}{(2\pi)^3}\int d^4k\, \epsilon(k_0)\, \delta(k^2 - m^2c^2)\, \mathrm{e}^{-ik\cdot x} \tag{2.9.14}$$

$$= -\frac{1}{(2\pi)^3}\int_{\omega>0} \frac{d^3k}{\omega} \sin k_0x^0 \exp\{i\boldsymbol{k}\cdot\boldsymbol{x}\} \tag{2.9.15}$$

where $\epsilon(s)$ is the sign of s, and we have used

$$\epsilon(k_0)\, \delta(k^2 - m^2c^2) = \frac{\delta(k_0 - \omega) - \delta(k_0 + \omega)}{2|k_0|} \tag{2.9.16}$$

It is straightforward to show that $\Delta(x)$ is invariant under Lorentz transformations,

$$\Delta(\Lambda x) = \Delta(x)$$

and that it is an odd function of x,

$$\Delta(x) = -\Delta(-x) \tag{2.9.17}$$

as well as being a solution of the Klein-Gordon equation

$$\left(\Box + m^2c^2\right)\, \Delta(x) = 0. \tag{2.9.18}$$

Because $\Delta(x)$ is invariant, it must be a function of the single invariant argument x^2 when $x^2 < 0$ (x *space-like*), and of the invariant arguments x^2 and

$\epsilon(x^0)$ when $x^2 \geq 0$, (x *time-like*). Outside the light cone, $x^2 < 0$, there is some function f such that $\Delta(x) = f(x^2)$ then $\Delta(x) = \Delta(-x)$ which, with (2.9.17), implies $\Delta(x) = 0$. Inside the light cone, we can satisfy the conditions with an expression of the form $\Delta(x) = \epsilon(x^0)\, g(x^2)$ for some function g, and then

$$\Delta(0, \boldsymbol{x}) = 0. \tag{2.9.19}$$

Finally, if we differentiate (2.9.15) with respect to x^0 under the integral sign, we get

$$\left(\frac{\partial \Delta(x)}{\partial x^0}\right)_{x^0=0} = -\delta^{(3)}(\boldsymbol{x}) \tag{2.9.20}$$

The properties expressed by (2.9.18)–(2.9.20) enable us to use $\Delta(x)$ to solve the Cauchy problem for the Klein-Gordon equation. Let $F_\mu(x)$ be any function that vanishes as $|\boldsymbol{x}| \to \infty$. We apply Green's theorem

$$\int_\sigma F_\mu(x')\, d\sigma^\mu(x') - \int_{\sigma_0} F_\mu(x')\, d\sigma^\mu(x') = \int_\Omega \partial'_\mu\, F_\mu(x') d^4x'$$

to the space-time region Ω bounded by the space-like surfaces σ_0 and σ. Let $\Psi(x)$ be a wave-packet solution of the Klein-Gordon equation such that $\Psi(x) = \Psi_0(x)$ and $n^\mu \partial_\mu \Psi_0(x)$, where n^μ is the normal on σ_0, are given on σ_0. Now choose

$$F_\mu(x') = \Delta(x - x')\, \partial'_\mu \Psi(x') - \partial'_\mu \Delta(x - x')\, \Psi(x').$$

Because $\Delta(x - x')$ and $\Psi(x)$ are both solutions of the Klein-Gordon equation,

$$\partial^\mu F_\mu(x) = 0$$

in Ω. Take σ_0 to be a hyperplane $x^0 = x'^{\,0}$, $n^\mu(x) = (1, 0, 0, 0)$, and apply (2.9.19) and (2.9.20), giving

$$\Psi(x) = \int_{\sigma_0} \{\Delta(x - x')\, \partial'_\mu \Psi_0(x') - \partial'_\mu \Delta(x - x')\, \Psi_0(x')\}\, d\sigma^\mu(x'),$$

where $x^0 > x'^{\,0}$. Equation (2.9.16) shows that $\Delta(x)$ can be decomposed into positive and negative frequency parts by writing

$$\Delta(x) = \Delta^{(+)}(x) + \Delta^{(-)}(x) \tag{2.9.21}$$

where

$$\Delta^{(+)}(x) = \frac{-i}{(2\pi)^3} \int_{k_0>0} \mathrm{e}^{-ik\cdot x}\, \delta(k^2 - m^2c^2)\, d^4k. \tag{2.9.22}$$

$$\Delta^{(-)}(x) = \frac{+i}{(2\pi)^3} \int_{k_0<0} \mathrm{e}^{-ik\cdot x}\, \delta(k^2 - m^2c^2)\, d^4k. \tag{2.9.23}$$

in which we have assumed k_0 to be real. We can replace these expressions by

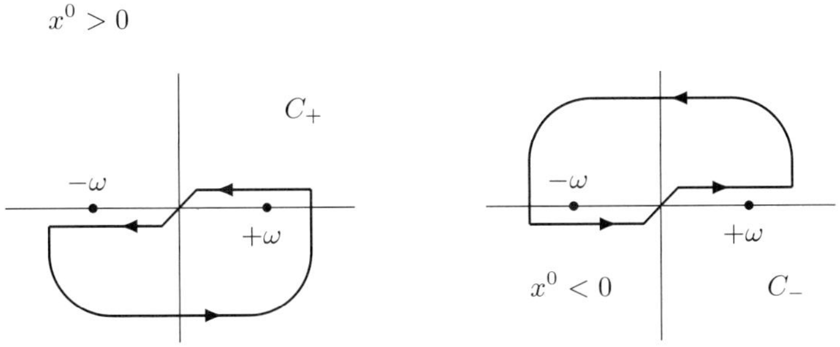

Fig. 2.2. Contours C_+, (2.9.24), and C_-, (2.9.25).

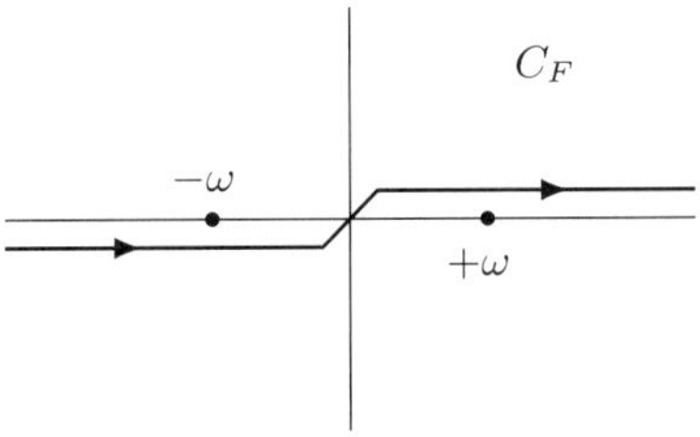

Fig. 2.3. Feynman contour for causal Green's function.

$$\Delta^{(+)}(x) = -\frac{1}{(2\pi)^4} \int_{C_+} \frac{\mathrm{e}^{-ik\cdot x}}{k^2 - m^2c^2}\, d^4k \tag{2.9.24}$$

valid only for $x^0 > 0$, and

$$\Delta^{(-)}(x) = -\frac{1}{(2\pi)^4} \int_{C_-} \frac{\mathrm{e}^{-ik\cdot x}}{k^2 - m^2c^2}\, d^4k \tag{2.9.25}$$

valid only for $x^0 < 0$, where C_+ encircles the single pole $+\omega$ and C_- encircles the single pole $-\omega$ once in the positive sense. The Feynman propagator, which propagates positive frequency amplitudes (associated with particles) forward in time and negative frequency amplitudes (associated with anti-particles) backward in time, is defined by

$$\Delta_F(x) = \begin{cases} +2i\Delta^{(+)}(x) \text{ for } x_0 > 0 \\ -2i\Delta^{(-)}(x) \text{ for } x_0 < 0 \end{cases} \tag{2.9.26}$$

Its contour integral representation is

$$\Delta_F(x) = \frac{2i}{(2\pi)^4} \int_{C_F} \frac{e^{-ik\cdot x}}{k^2 - m^2c^2}\, d^4k \tag{2.9.27}$$

where the path C_F passes below $-\omega$ along $k_0 < 0$ and above $+\omega$ along $k_0 > 0$. The path can be deformed onto the real k_0-axis if we displace the poles so that they occur at $\pm(\omega - i\epsilon)$, where $\epsilon > 0$ is infinitesimal.

2.9.3 Maxwell's equations: the zero-mass case

Our discussion of the covariant form of Maxwell's equations in §2.6.1 led to the equation

$$\Box a^\mu - \partial^\mu(\partial_\nu a^\nu) = \frac{1}{\epsilon_0 c} j^\mu. \tag{2.9.28}$$

for the four-potential (2.6.13) in terms of the generating currents, for which the associated Green's function $D_{\mu\nu}(x)$ must generate solutions of the form

$$a^\mu(x) = \frac{1}{c}\int D_{\mu\nu}(x - x')j^\nu(x')\, d^4x' \tag{2.9.29}$$

in coordinate space. It is convenient to write

$$D_{\mu\nu}(x) = \int \frac{d^4k}{(2\pi)^4} e^{-ik.x}\, D_{\mu\nu}(k)$$

The most general second rank 4-tensor $D_{\mu\nu}(k)$ that we may construct without imposing particular constraints will have the form

$$\epsilon_0 D_{\mu\nu}(k) = g_{\mu\nu}\, D(k^2) + k_\mu k_\nu D^{(1)}(k^2) \tag{2.9.30}$$

The requirement of current conservation, $\partial_\mu j^\mu = 0$, becomes

$$k_\mu j^\mu = 0$$

in momentum space, so that a replacement of the form

$$D_{\mu\nu}(k) \to D_{\mu\nu}(k) + k_\mu \chi_\nu + \chi_\mu k_\nu \tag{2.9.31}$$

in (2.9.30), where χ_μ is an arbitrary four-vector, leaves the four-potential unchanged. For the same reason, the form of $D^{(1)}(k^2)$ does not affect the four-potential, so that the choice of χ_μ and of $D^{(1)}(k^2)$ together fixes the gauge. From §2.9.2, the simplest choice, the so-called Feynman gauge, is to take $D^{(1)}(k^2) = 0$ and to take the zero-mass limit of (2.9.12), giving

$$D^F_{\mu\nu}(k) = -\frac{g_{\mu\nu}}{\epsilon_0}\frac{1}{k^2 + i\epsilon} \tag{2.9.32}$$

In this case, a^μ satisfies the Lorentz condition, because

$$\partial^\mu a_\mu(x) = \frac{1}{c}\int \partial^\mu D_{\mu\nu}(x-x')j^\nu(x')\,d^4x' = 0$$

so that $k^\mu D_{\mu\nu}(k)j^\nu(k) = 0$ for a conserved current. The Landau gauge

$$D^L_{\mu\nu}(k) = -\frac{1}{\epsilon_0 k^2}\left\{g_{\mu\nu} - \frac{k_\mu k_\nu}{k^2}\right\}, \tag{2.9.33}$$

for which $k^\mu D^L_{\mu\nu}(k) = 0$, is also useful. We shall make extensive use of the (non-covariant) Coulomb gauge, for which

$$\chi^\mu = k^\mu/2k^2\,|\boldsymbol{k}|^2, \tag{2.9.34}$$

giving

$$D^C_{00}(k) = \frac{1}{\epsilon_0|\boldsymbol{k}|^2}, \quad D^C_{ij}(k) = \frac{1}{\epsilon_0 k^2}\left\{\delta_{ij} - \frac{k_i k_j}{|\boldsymbol{k}|^2},\right\}, \quad i.j = 1,2,3, \tag{2.9.35}$$

whilst $D_{0i}(k) = D_{i0}(k) = 0$. Whilst the lack of covariance will be inconvenient if we need to compare solutions in different reference frames, most calculations of atomic and molecular properties fix the coordinate frame at the outset. We shall need to take this into account at a later stage. From (2.9.29) and (2.9.35), we see that the Coulomb gauge splits the four-potential into a scalar potential depending only on the $\nu = 0$ component of $j^\nu(x')$, the charge density, and a vector potential depending only on the 3-current vector. In coordinate representation, we see that

$$D^C_{00}(x) = \frac{1}{(2\pi)^4\epsilon_0}\int d^4k\,\frac{\mathrm{e}^{-ik\cdot x}}{|\boldsymbol{k}|^2} = \frac{\delta(x^0)}{4\pi\epsilon_0|\boldsymbol{x}|}, \tag{2.9.36}$$

which is just the Coulomb interaction kernel, acting instantaneously.

To calculate the kernels $D^C_{ij}(x)$, it is useful first to consider the Feynman kernel,

$$\begin{aligned} D^F(x) &= \frac{1}{(2\pi)^4\epsilon_0}\int \frac{d^4k}{k^2+i\epsilon}e^{-ik\cdot x} \\ &= \frac{1}{(2\pi)}\int dz\,\mathrm{e}^{-izx^0}.\frac{1}{(2\pi)^3\epsilon_0}\int d^3k\,\frac{\mathrm{e}^{i\boldsymbol{k}.\boldsymbol{x}}}{|\boldsymbol{k}|^2 - z^2 - i\epsilon} \end{aligned} \tag{2.9.37}$$

where we have written $k_0 = z$. Because the space-like integrand is spherically symmetric, apart from the exponential numerator, we can take $\boldsymbol{x}$ along the z-axis, and integrate over the azimuthal coordinate, giving

$$\begin{aligned} \int \frac{d^3k}{(2\pi)^3\epsilon_0}\frac{\mathrm{e}^{i\boldsymbol{k}.\boldsymbol{x}}}{k^2 - z^2 - i\epsilon} &= \int_0^\infty \frac{1}{k^2 - z^2 - i\epsilon}\int_{-1}^{+1}\mathrm{e}^{ikr\mu}d\mu\,\frac{k^2dk}{(2\pi)^2\epsilon_0} \\ &= \frac{-i}{(2\pi)^2\epsilon_0}\frac{1}{r}\int_{-\infty}^\infty \frac{\mathrm{e}^{ikr}k\,dk}{k^2 - z^2 - i\epsilon} \\ &= \frac{\mathrm{e}^{i|z|r}}{4\pi\epsilon_0 r}, \end{aligned} \tag{2.9.38}$$

where we have used symmetry to extend the range of integration to the whole real line, and only the pole at $k = (|z| + i\epsilon)$ contributes. Hence

$$D^F(x) = \int \frac{dz}{2\pi} \frac{e^{-izx^0 + i|z|r}}{4\pi\epsilon_0 r}. \tag{2.9.39}$$

This calculation also gives immediately the term proportional to δ_{ij} in (2.9.35). A similar construction, using

$$\partial_i\partial_j e^{i\boldsymbol{k}.\boldsymbol{x}} = -k_i k_j e^{i\boldsymbol{k}.\boldsymbol{x}},$$

leads to

$$D^C_{ij}(x) = -\int \frac{dz}{2\pi} e^{-izx^0} \left\{ \delta_{ij} \frac{e^{i|z|r}}{4\pi\epsilon_0 r} + \partial_i\partial_j \frac{e^{i|z|r} - 1}{4\pi\epsilon_0 r\,|z|^2} \right\} \tag{2.9.40}$$

This propagator introduces *retardation* effects arising from the finite speed of propagation of electromagnetic signals. For light elements, it is often a good approximation to assume that only long wavelength photons, in the sense that $|zr| \ll 1$, lead to measurable effects. In this case (2.9.40) gives

$$\begin{aligned} \lim_{z\to 0} D^C_{ij}(x) &= \frac{1}{4\pi\epsilon_0} \int \frac{dk_0}{2\pi} e^{-izx^0} . \left\{ \frac{\delta_{ij}}{r} - \frac{1}{2}\partial_i\partial_j\, r \right\} \\ &= -\frac{\delta(x^0)}{8\pi\epsilon_0 r} \left\{ \delta_{ij} + \frac{x^i x^j}{r^2} \right\} \end{aligned} \tag{2.9.41}$$

in which there is no time-dependence. This long wavelength approximation yields an interaction potential first derived by Breit [33].

2.9.4 Free-particle Dirac equation

Green's functions for the Dirac operator are matrix-valued and exist in similar forms to the Green's functions $\Delta, \Delta^{(+)}, \Delta^{(-)}$, and Δ_F already encountered. Corresponding to the causal propagator (2.9.27), for example, we have

$$\begin{aligned} S_F(x) &= -(\gamma^\mu p_\mu + mc)\, \Delta_F(x) \\ &= -\frac{2i}{(2\pi)^4} \int_{C_F} \frac{\gamma^\mu k_\mu + mc}{k^2 - m^2c^2} e^{-ik\cdot x}\, d^4k \end{aligned} \tag{2.9.42}$$

which satisfies the equation

$$(\gamma^\mu p_\mu - mc)\, S_F(x - x') = -2i\, \delta^{(4)}(x - x') \tag{2.9.43}$$

where $p_\mu = i\partial_\mu = i\partial/\partial x^\mu$. In many applications, it is useful to separate out the time dependence by writing

$$S_F(x - x') = \frac{i}{\pi} \int_{C_F} dz\, G(\boldsymbol{x}, \boldsymbol{x}'; z)\gamma_0\, e^{-iz(x-x')^0/c}. \tag{2.9.44}$$

Substituting into (2.9.43) and using

$$\gamma^\mu p_\mu - mc = \gamma^0(i\partial_0 - H/c), \quad H = c\,\boldsymbol{\alpha}\cdot\boldsymbol{p} + \beta mc^2$$

and

$$\gamma^\mu p_\mu + mc = (i\partial_0 + H/c)\,\gamma^0$$

we find, using (2.9.42), that

$$\begin{aligned} G(\boldsymbol{x}, \boldsymbol{x}'; z) &= \frac{z+H}{c} \int \frac{d^3k}{(2\pi)^3} \frac{e^{i\boldsymbol{k}.\boldsymbol{x}}}{k^2 - \boldsymbol{p}^2 - i\epsilon} \\ &= \frac{z+H}{c} \frac{e^{i|\boldsymbol{p}|R}}{4\pi R} \end{aligned} \tag{2.9.45}$$

where $R = |\boldsymbol{x} - \boldsymbol{x}'|$, $\boldsymbol{p}^2 = z^2 - m^2c^2$, and

$$(H - z)\,G(\boldsymbol{x}, \boldsymbol{x}'; z) = c\delta^{(3)}(\boldsymbol{x} - \boldsymbol{x}'). \tag{2.9.46}$$

As in the Klein-Gordon case, this is exactly what is required in QED to ensure that the theory incorporates the correct boundary conditions to describe the motions of both positrons and electrons.

References

1. Woodhouse N M J 1992 *Special relativity* (Berlin: Springer-Verlag).
2. Rindler W 1982 *Introduction to special relativity* (Oxford: Oxford University Press).
3. Misner C W, Thorne K S and Wheeler J A 1973 *Gravitation* (San Francisco: W H Freeman & Co.)
4. Drake G W F ed 2005 *Springer Handbook of Atomic, Molecular and Optical Physics* Second edn (New York: Springer-Verlag).
5. Wybourne B G 1974 *Classical Groups for Physicists* (New York: Wiley).
6. Elliott J P and Dawber P G 1979 *Symmetry in Physics* (Basingstoke: Macmillan Press).
7. Wigner E P 1939 *Ann. Math.* **40** 149.
8. Bargmann V and Wigner E P 1948 *Proc. Nat. Acad. Sci.* **34** 211.
9. Shirokov Yu M 1958 *Sov. Phys. JETP* **6** 664.
10. Foldy L L 1956 *Phys Rev* **102** 568.
11. Foldy L L and Wouthuysen S A 1950 *Phys. Rev.* **78** 29.
12. Dirac P A M 1928 *Proc Roy Soc (Lond) A* **117** 610..
13. Dirac P A M 1928 *Proc Roy Soc (Lond) A* **118** 351.
14. Anderson C D 1932 *Science* **76** 238.
15. Mohr P J and Taylor B N 2000 *Rev Mod Phys* **72** 351 (CODATA recommended values of the fundamental physical constants: 1998)
16. Schiff L I 1968 *Quantum Mechanics* (3rd edition) (New York: McGraw-Hill Book Company)
17. Pais A 1986 *Inward Bound* (Oxford: Clarendon Press)

18. Thaller B 1992 *The Dirac Equation* (New York: Springer-Verlag)
19. Greiner W 1990 *Relativistic Quantum Mechanics. Wave Equations* (Berlin: Springer-Verlag)
20. Newton T D and Wigner E P 1949 *Rev. Mod. Phys.* **21** 400.
21. Douglas M and Kroll N M 1974 *Ann Phys (NY)* **82** 89.
22. Hess B A 1986 *Phys Rev A* **33** 3742 – 3748.
23. Jansen G and Hess B A 1989 *Phys Rev A* **39** 6016.
24. Wolff A, Reiher M and Hess B A 2002 in *Relativistic Electronic Structure Theory. Part 1. Fundamentals* (ed. P Schwerdtfeger) pp. 622–663 (Amsterdam: Elsevier).
25. Bethe H A and Salpeter E E 1957 *Quantum Mechanics of One- and Two-Electron Atoms* (Berlin-Göttingen-Heidelberg: Springer-Verlag)
26. Bohr N 1913 *Nature* **73** 1109.
27. Dirac P A M 1930 *Proc Roy Soc A* **126** 360.
28. Anderson C D 1933 *Phys Rev* **43** 491.
29. Feynman R P 1949 *Phys. Rev.* **76** 749.
30. Stückelberg E C C 1942 *Helv. Phys. Acta* **15** 23.
31. Stakgold 1979 *Green's functions and Boundary Value Problems* (New York: John Wiley & Sons)
32. Feynman R P 1949 *Phys Rev* **76** 769.
33. Breit G 1929 *Phys Rev* **34** 353; — 1930 *Phys Rev* **36** 383; — 1932 *Phys Rev* **39** 616.

3

The Dirac Equation

The solutions of the Dirac equation for the free electron and for hydrogenic atoms with stationary nuclei are primary building blocks for calculations on more complex many-electron systems. Section 3.1 introduces many useful notions, including plane wave solutions, the bilinear covariant expressions representing physical quantities such as the electron charge-current density, along with energy and spin projection operators. Charge conjugation relates electron and positron solutions. The separation of angular and radial amplitudes of central field Dirac spinors is explained in detail, permitting solution of the radial equations for bound and continuum states. Applications include relativistic Coulomb scattering and relativistic quantum defect theory. We show how to construct partial wave Green's functions for the free and for hydrogenic Dirac electrons and sum the partial wave expansion for the free electron. Finally, we discuss the nonrelativistic limit and approximate relativistic Hamiltonians.

Supplementary material on relativistic notation (§A.1), Dirac matrices (§A.2), the properties of spherical Bessel functions (§A.3.1), confluent hypergeometric functions (§A.3.2), and frequently used properties of central field Dirac orbitals (§A.4) has been collected in the appendices.

3.1 Free particles

The form of Dirac's equation for free particles, (2.5.7), which is most convenient for demonstrating the relativistic covariance properties, is

$$\{\gamma^\mu p_\mu - mc\}\,\psi = 0, \tag{3.1.1}$$

where the 4×4 γ^μ matrices have the property (2.5.8)

$$\{\gamma^\mu, \gamma^\nu\} = 2g^{\mu\nu} I_4; \tag{3.1.2}$$

in particular

$$(\gamma^i)^2 = -I_4, \quad (\gamma^0)^2 = I_4; \tag{3.1.3}$$

The $\mu = 0$ matrix is therefore Hermitian, whilst the remainder are anti-Hermitian. We shall use the representation (2.5.11) in terms of Pauli matrices

$$\gamma^0 = \begin{pmatrix} I & 0 \\ 0 & -I \end{pmatrix}, \quad \gamma^i = \begin{pmatrix} 0 & \sigma^i \\ -\sigma^i & 0 \end{pmatrix}, \tag{3.1.4}$$

and define

$$\gamma^5 = \gamma_5 = i\gamma^0\gamma^1\gamma^2\gamma^3 = \begin{pmatrix} 0 & I \\ I & 0 \end{pmatrix}. \tag{3.1.5}$$

It follows that

$$(\gamma^5)^2 = I_4, \quad \{\gamma^5, \gamma^\mu\} = 0, \quad \mu = 0, 1, 2, 3. \tag{3.1.6}$$

3.1.1 Properties of Dirac matrices

Any 4×4 matrix can be represented as a linear combination of 16 linearly independent matrices generated by the four γ-matrices, forming what is known as a Clifford algebra. The 16 matrices, Γ^a, so generated are conventionally represented as follows

Γ^S: I_4	Scalar	1 matrix
Γ^V_μ : γ_μ	Vector	4 matrices
$\Gamma^T_{\mu\nu}$: $\sigma_{\mu\nu} = \frac{i}{2}[\gamma_\mu, \gamma_\nu]$	Antisymmetric tensor	6 matrices
Γ^A_μ: $\gamma_5\gamma_\mu$	Pseudovector	4 matrices
Γ^P: γ_5	Pseudoscalar	1 matrix

The reason for labelling these matrices as scalar, vector, tensor, pseudovector, and pseudoscalar is explained in §3.1.3. Their main properties are

1. $(\Gamma^a)^2 = \pm I_4$.
2. For any Γ^a ($\Gamma^a \neq \Gamma^S = I_4$) we can find Γ^b such that $\{\Gamma^a, \Gamma^b\} = 0$.
3. For any Γ^a, Γ^b with $a \neq b$, we can find a number $\eta = \pm 1, \pm i$ and a Γ^c such that $\Gamma^a\Gamma^b = \eta\Gamma^c$, where $\Gamma^c \neq \Gamma^S = I_4$.

We deduce first of all that the trace of each matrix $\Gamma^a \neq \Gamma^S = I_4$ vanishes. For $\operatorname{tr} \Gamma^a = \operatorname{tr}[\Gamma^a(\Gamma^b)^2] = -\operatorname{tr}[\Gamma^b\Gamma^a\Gamma^b]$, using property 2. Because the trace of a product of matrices is invariant under cyclic permutation, this gives $\operatorname{tr} \Gamma^a = -\operatorname{tr}[(\Gamma^b)^2\Gamma^a] = -\operatorname{tr} \Gamma^a$, so that $\operatorname{tr} \Gamma^a$ vanishes. The linear independence of the Γ^a now follows; the equation

$$\sum_a c_a\Gamma^a = 0$$

holds if and only if $c_a = 0$ for all a. (Multiply from the left by, say, Γ^b and take the trace to conclude that $c_b = 0$). A table of properties of Dirac matrices will be found in Appendix A.2.

3.1.2 Covariance properties

A Lorentz or Poincaré transformation from one inertial frame of reference to another preserves the form of Dirac's equation: in other words

$$\left(i\hbar\gamma^\mu \frac{\partial}{\partial x^\mu} - mc\right)\psi(x) = 0 \tag{3.1.7}$$

transforms into

$$\left(i\hbar\gamma^\mu \frac{\partial}{\partial x'^\mu} - mc\right)\psi'(x') = 0 \tag{3.1.8}$$

under the inhomogeneous Lorentz (Poincaré) transformation (2.3.1)

$$x \to x' = \Lambda x + a. \tag{3.1.9}$$

The invariance with respect to translation $x \to x' = x + a$ is obvious, so we can set $a = 0$ and restrict the discussion to pure Lorentz transformations. We assume a relation

$$\psi'(x') = S(\Lambda)\psi(x) \tag{3.1.10}$$

where $S(\Lambda)$ is a nonsingular 4×4 matrix. The relation between ψ and ψ' is then a local one so that an observer in the new frame can immediately construct ψ' when given ψ. Substituting in (3.1.8) gives

$$i\hbar\gamma^\mu \frac{\partial}{\partial x'^\mu} S(\Lambda)\psi(x) - mcS(\Lambda)\psi(x) = 0.$$

From (3.1.9),

$$\frac{\partial}{\partial x'^\mu} = \frac{\partial x^\nu}{\partial x'^\mu}\frac{\partial}{\partial x^\nu} = (\Lambda^{-1})^\nu{}_\mu \frac{\partial}{\partial x^\nu}$$

so that

$$i\hbar\gamma^\mu (\Lambda^{-1})^\nu{}_\mu \frac{\partial}{\partial x^\nu} S(\Lambda)\psi(x) - mcS(\Lambda)\psi(x) = 0,$$

and $S(\Lambda)$ is characterized by

$$S(\Lambda)\gamma^\mu S^{-1}(\Lambda) = (\Lambda^{-1})^\mu{}_\nu \gamma^\nu. \tag{3.1.11}$$

The next step is to construct $S(\Lambda)$; for an infinitesimal Lorentz transformation

$$\Lambda^\mu{}_\nu = g^\mu{}_\nu + \omega^\mu{}_\nu + \cdots, \qquad \omega_{\mu\nu} = -\omega_{\nu\mu}, \tag{3.1.12}$$

the corresponding transformation matrix is

$$S(\Lambda) = I - \frac{i}{4}\sigma_{\mu\nu}\,\omega^{\mu\nu} + \cdots, \quad S^{-1}(\Lambda) = I + \frac{i}{4}\sigma_{\mu\nu}\,\omega^{\mu\nu} + \cdots,$$

where $\sigma_{\mu\nu} = -\sigma_{\nu\mu}$. Substituting in (3.1.11), we find that

$$[\gamma^\mu, \sigma_{\nu\rho}] = 2i(g^\mu{}_\nu\,\gamma_\rho - g^\mu{}_\rho\,\gamma_\nu)$$

which is satisfied if we set

$$\sigma_{\nu\rho} = \frac{i}{2}[\gamma_\nu, \gamma_\rho].$$

From this, we can derive a finite transformation of the form

$$S(\Lambda) = e^{-(i/4)\omega_{\mu\nu}\sigma^{\mu\nu}} \tag{3.1.13}$$

where $\omega_{\mu\nu}$ is now finite, rather than infinitesimal, and antisymmetric in the indices μ and ν. Thus $S(\Lambda)$ is unitary for real rotations in $\mathbb{R}^3$, and Hermitian for Lorentz boosts. Equation (3.1.10) may also be written

$$\psi'(x) = S(\Lambda)\psi(\Lambda^{-1}x) \tag{3.1.14}$$

so that

$$\begin{aligned}
\psi'(x) &= \left(I - \frac{i}{4}\sigma_{\mu\nu}\,\omega^{\mu\nu}\right)\psi(x^\rho - \omega^\rho{}_\sigma x^\sigma) \\
&= \left(I - \frac{i}{4}\sigma_{\mu\nu}\,\omega^{\mu\nu} + x_\mu \omega^{\mu\nu}\partial_\nu\right)\psi(x) + o(\omega^{\mu\nu}) \\
&= \left(I - \frac{i}{2}\omega^{\mu\nu}\,J_{\mu\nu}\right)\psi(x) + o(\omega^{\mu\nu})
\end{aligned}$$

where the infinitesimal generator

$$J_{\mu\nu} = i\,(x_\mu\partial_\nu - x_\nu\partial_\mu) + \frac{1}{2}\sigma_{\mu\nu}. \tag{3.1.15}$$

can be recognised as the quantum mechanical total angular momentum operator.

The Poincaré algebra, §2.3, contains the infinitesimal generators

$$P_\mu = i\partial_\mu, \quad W_\mu = -\frac{1}{2}\epsilon_{\mu\nu\rho\sigma}J^{\nu\rho}P^\sigma = \frac{1}{4}\epsilon_{\mu\nu\rho\sigma}\sigma^{\nu\rho}P^\sigma, \tag{3.1.16}$$

which are important in determining the irreducible representations. Their scalar product vanishes,

$$W_\mu P^\mu = 0,$$

and

$$[W_\mu, P_\nu] = 0, \qquad [W_\mu, W_\nu] = -i\epsilon_{\mu\nu\rho\sigma}W^\rho P^\sigma.$$

Because the operators $P \cdot P$ and $W \cdot W$ commute with every operator of the group they can be used to label the irreducible representations. Now by operating on (2.5.7) from the left with $\gamma^\mu p_\mu + mc$, we see that

$$P \cdot P\psi = -\Box\psi = m^2c^2\psi$$

for any solution ψ of Dirac's equation, so that the rest mass m of the particle is one of the labels required. We can evaluate the other label in any convenient inertial frame, in particular the one in which P^0 takes the value $mc/\hbar$ and the space-like components P^i vanish. In this frame, equation (3.1.16) takes the values

$$W_\mu = mc(0, \mathbf{s})$$

where $\mathbf{s}$ is defined in terms of the Pauli matrices as

$$\mathbf{s}_i = \frac{1}{2}\sigma_i, \quad i = 1, 2, 3.$$

Thus

$$W \cdot W\psi = -m^2c^2\mathbf{s}^2\psi = -\frac{3}{4}m^2c^2\psi,$$

verifying that a Dirac particle has an intrinsic spin of $\frac{1}{2}$.

3.1.3 Bilinear covariants

We define the *Dirac adjoint* $\overline{\psi}(x)$ by

$$\overline{\psi}(x) = \psi^\dagger(x)\gamma^0, \tag{3.1.17}$$

where $\psi^\dagger(x)$ is the Hermitean conjugate of $\psi(x)$.[1] From (3.1.10) we see that, under a Lorentz transformation,

$$\begin{aligned}\overline{\psi}'(x') &= \overline{\psi}(x)\gamma^0 S(\Lambda)^\dagger\gamma^0 \\ &= \overline{\psi}(x)S^{-1}(\Lambda),\end{aligned} \tag{3.1.18}$$

where the last line follows by use of the explicit form (3.1.13). Then a bilinear expression of the form

$$\overline{\psi}(x)A\psi(x),$$

where A is any 4×4 matrix, transforms so that

$$\overline{\psi}'(x')A\psi'(x') = \overline{\psi}(x)S^{-1}(\Lambda)AS(\Lambda)\psi(x).$$

As a specific example, we defined the density (2.5.2) and the probability current (2.5.3) as the components of a conserved charge-current four-vector j^μ, with

$$j^\mu = (j^0, \boldsymbol{j}), \quad j^0/c = \rho = \psi^\dagger\psi, \quad \boldsymbol{j} = \psi^\dagger c\boldsymbol{\alpha}\psi;$$

in terms of the Dirac adjoint, this can be written

$$j^\mu = \overline{\psi}(x)c\gamma^\mu\psi(x). \tag{3.1.19}$$

Because (3.1.11)

$$S^{-1}(\Lambda)\gamma^\mu S(\Lambda) = \Lambda^\mu{}_\nu\gamma^\nu,$$

[1] We remind the reader that $\psi^\dagger(x)$ is a 4-column row vector whose elements are the complex conjugates of the elements of the 4-row column vector $\psi(x)$.

we see that

$$j'^{\mu}(x') = \overline{\psi}'(x')c\gamma^{\mu}\psi'(x') = \Lambda^{\mu}{}_{\nu}j^{\nu}(x),$$

verifying that $j^{\mu}(x)$ transforms like a four-vector as previously asserted. Similarly, by taking A as the identity matrix, we find that $\overline{\psi}(x)\psi(x)$ is a scalar density. These two quantities are examples of *bilinear covariants* that can be constructed from expressions of the form $\overline{\psi}(x)\,A\,\psi(x)$, where A is one of the Γ operators listed in Section 3.1.1. Thus the effect of a Lorentz transformation is represented by

$$\begin{aligned}
\text{S:}\quad \overline{\psi}'\,\Gamma^{S}\,(x')\,\psi'(x') &= \overline{\psi}(x)\Gamma^{S}\psi(x),\\
\text{V:}\quad \overline{\psi}'(x')\,(\Gamma^{V})^{\mu}\,\psi'(x') &= \Lambda^{\mu}{}_{\nu}\,\overline{\psi}(x)(\Gamma^{V})^{\nu}\,\psi(x),\\
\text{T:}\quad \overline{\psi}'(x')\,(\Gamma^{T})^{\mu\nu}\,\psi'(x') &= \Lambda^{\mu}{}_{\rho}\Lambda^{\nu}{}_{\sigma}\,\overline{\psi}(x)\,(\Gamma^{T})^{\rho\sigma}\psi(x),\\
\text{A:}\quad \overline{\psi}'(x')\,(\Gamma^{A})^{\mu}\,\psi'(x') &= \det(\Lambda)\,\Lambda^{\mu}{}_{\nu}\,\overline{\psi}(x)\,(\Gamma^{A})^{\nu}\,\psi(x),\\
\text{P:}\quad \overline{\psi}'(x')\,\Gamma^{P}\,\psi'(x') &= \det(\Lambda)\,\overline{\psi}(x)\,\Gamma^{P}\,\psi(x),
\end{aligned}$$

justifying identifying these combinations as scalar (S), vector (V), antisymmetric second rank tensor (T), axial- (or pseudo)-vector (A), and pseudoscalar (P). The last two change sign under inversions, for which $\det(\Lambda) = -1$. The bilinear covariants, in particular j^{μ}, represent observable quantities with well defined properties under Lorentz transformations, which are needed to construct coupling of the Dirac field to other fields. For example, the minimal coupling to the Maxwell field, used to study the motion of a test charge in Section 2.6.4, involves the invariant interaction $j^{\mu}A_{\mu}$, where A_{μ} is the 4-potential.

3.1.4 Plane wave solutions

The equation (3.1.1)

$$(i\hbar\gamma^{\mu}\partial_{\mu} - mc)\psi = 0$$

has plane wave solutions

$$\psi(x) = \mathrm{e}^{-ix\cdot p\,/\hbar}\,u(p); \tag{3.1.20}$$

the exponential has the invariant exponent

$$x\cdot p = x^{0}\,p_{0} - \boldsymbol{x}\cdot\boldsymbol{p} = t\,E(|\boldsymbol{p}|) - \boldsymbol{x}\cdot\boldsymbol{p}$$

where $E(|\boldsymbol{p}|) = cp_0$ is the energy, and the (4-component) *spinor* $u(p)$ satisfies

$$(\not{p} - mc)\,u(p) = 0, \quad \not{p} := \gamma_{\mu}p^{\mu}. \tag{3.1.21}$$

Multiplying from the left by $\not{p} + mc$ and using (2.5.8) we see that

$$(p^{2} - m^{2}c^{2})\,u(p) = 0 \tag{3.1.22}$$

verifying that p^{μ} is the particle's 4-momentum. The energy is $\pm E(|\boldsymbol{p}|)$, where

$$E(|\boldsymbol{p}|) = cp_0 = +\sqrt{m^2c^2 + \boldsymbol{p}^2}. \tag{3.1.23}$$

We therefore have (particle) solutions with *positive energy* and positive frequency $\omega = E(|\boldsymbol{p}|)/\hbar$ and (antiparticle) solutions with *negative energy* and negative frequency $\omega = -E(|\boldsymbol{p}|)/\hbar$.

The solutions in an arbitrary inertial reference frame can be obtained directly from the partitioned form of (3.1.21),

$$\begin{pmatrix} mc - p_0 & \boldsymbol{\sigma}\cdot\boldsymbol{p} \\ \boldsymbol{\sigma}\cdot\boldsymbol{p} & -mc - p_0 \end{pmatrix} u(p) = 0.$$

There are two linearly independent solutions for positive energies, $p^0 > 0$ and two for negative energies, $p^0 < 0$. Writing

$$\phi^1 = \begin{pmatrix} 1 \\ 0 \end{pmatrix}, \quad \phi^2 = \begin{pmatrix} 0 \\ 1 \end{pmatrix},$$

we have, when $p^0 > 0$,

$$u^{(r)}(p) = \left(\frac{E(|\boldsymbol{p}|) + mc^2}{2E(|\boldsymbol{p}|)}\right)^{1/2} \begin{pmatrix} \phi^r \\ \dfrac{c\boldsymbol{\sigma}\cdot\boldsymbol{p}}{E(|\boldsymbol{p}|) + mc^2}\,\phi^r \end{pmatrix} \quad r = 1, 2, \tag{3.1.24}$$

and, when $p^0 < 0$,

$$v^{(r)}(p) = \left(\frac{E(|\boldsymbol{p}|) + mc^2}{2E(|\boldsymbol{p}|)}\right)^{1/2} \begin{pmatrix} \dfrac{-c\boldsymbol{\sigma}\cdot\boldsymbol{p}}{E(|\boldsymbol{p}|) + mc^2}\,\phi^r \\ \phi^r \end{pmatrix} \quad r = 1, 2. \tag{3.1.25}$$

The Lorentz invariant normalization has been chosen so that the solutions form an orthonormal set,

$$\begin{aligned} \overline{u}^{(r)}(p)\,u^{(s)}(p) &= \overline{v}^{(r)}(p)\,v^{(s)}(p) = \delta_{rs} \\ \overline{u}^{(r)}(p)\,v^{(s)}(p) &= \overline{v}^{(r)}(p)\,u^{(s)}(p) = 0. \end{aligned}$$

We also see that if we write $p = (E(|\boldsymbol{p}|)/c,\ \boldsymbol{p})$, then $u^{(r)}(p), v^{(s)}(p)$ satisfy respectively

$$(\not{p} - mc)u^{(r)}(p) = 0, \quad (-\not{p} - mc)v^{(s)}(p) = 0 \tag{3.1.26}$$

The probability current-density 4-vector associated with the spinor wave function $\psi(x)$ is $j^\mu(x) = \overline{\psi}c\gamma^\mu\psi$. Writing

$$\psi^{(+)(r)} = \mathrm{e}^{-ix\cdot p/\hbar}\,u^{(r)}(p), \quad \psi^{(-)(s)} = \mathrm{e}^{+ix\cdot p/\hbar}\,v^{(s)}(p),$$

gives

$$\begin{aligned}\overline{\psi}^{(+)(r)} c\gamma^\mu \psi^{(+)(s)} &= \overline{u}^{(r)}(p) c\gamma^\mu u^{(s)}(p) = \overline{u}^{(r)}(p) \frac{\{\gamma^\mu, \not{p}\}}{2m} u^{(s)}(p) \\ &= \frac{p^\mu}{m} \delta_{rs} \end{aligned} \tag{3.1.27}$$

where the first step follows from (3.1.21) and the second from (2.5.8). From elementary relativistic mechanics, a particle with rest mass m and velocity $\boldsymbol{v}$ has velocity 4-vector

$$V^\mu = p^\mu/m = \gamma(v)(c, \boldsymbol{v}), \quad \gamma(v) = 1/\sqrt{1 - v^2/c^2}$$

so that (3.1.27) agrees with the classical result.

A similar argument for the negative energy (negative frequency) plane wave states gives

$$\overline{\psi}^{(-)(r)} c\gamma^\mu \psi^{(-)(s)} = -\frac{p^\mu}{m} \delta_{rs}. \tag{3.1.28}$$

This time, all the momentum components are reversed: the particles appear to be going backwards! However, because the energy is negative in (3.1.28), the probability density, $\rho = j^0/c$, remains positive for both positive and negative energy states.

3.1.5 Energy and spin projectors

The plane wave solutions of Dirac's equation can be divided into disjoint sets according to the sign of the energy. From (3.1.26), we can construct the orthogonal projection operators $\Lambda_+(p), \Lambda_-(p)$ onto the positive and negative energy manifolds:

$$\Lambda_+(p) = \frac{mc + \not{p}}{2mc} = \sum_\alpha u^{(\alpha)}(p) \otimes \overline{u}^{(\alpha)}(p) \tag{3.1.29}$$

$$\Lambda_-(p) = \frac{mc - \not{p}}{2mc} = \sum_\alpha v^{(\alpha)}(p) \otimes \overline{v}^{(\alpha)}(p), \tag{3.1.30}$$

where if a and $\overline{b}$ are respectively a 4-spinor with components a_i and an adjoint (row) 4-spinor with components b_j^*, then the 4×4 matrix $a \otimes \overline{b}$ has matrix elements $(a \otimes \overline{b})_{ij} = a_i\, b_j^*$. The projectors satisfy

$$\begin{aligned} \Lambda_\pm^2(p) &= \Lambda_\pm(p) \\ \Lambda_+(p) + \Lambda_-(p) &= I \\ \Lambda_+(p)\, \Lambda_-(p) = \Lambda_-(p)\, \Lambda_+(p) &= 0. \\ Tr \Lambda_\pm(p) &= 2 \end{aligned} \tag{3.1.31}$$

The irreducible representations of the Poincaré group can be characterized by the invariants $C_1 = P \cdot P$ and $C_2 = w \cdot w$, §2.3, which we can use to classify the plane wave solutions. Let n be any space-like four-vector satisfying $n \cdot p = 0$ and normalized so that $n \cdot n = -1$. From (3.1.16) we see that $w \cdot n = \frac{1}{4}\epsilon_{\mu\nu\rho\sigma} n^\mu \sigma^{\nu\rho} p^\sigma$, which reduces to $w \cdot n = -\frac{1}{2}\gamma_5 \not{n} \not{p}$. We have already seen that, in the rest frame,

$$w_\mu = \frac{mc}{2}(0, \boldsymbol{\Sigma})$$

where

$$\boldsymbol{\Sigma} = \begin{pmatrix} \boldsymbol{\sigma} & 0 \\ 0 & \boldsymbol{\sigma} \end{pmatrix}$$

so that, if $n = n^{(3)} = (0, 0, 0, 1)$ is directed along the z-axis,

$$w \cdot n^{(3)} = -\frac{mc}{2}\Sigma_3.$$

Thus, in the rest frame, solutions are eigenstates of $-w \cdot n^{(3)}/mc = w^3/mc$ corresponding to eigenvalues $+\frac{1}{2}$ (spin-up) for $u^{(1)}, v^{(1)}$ and $-\frac{1}{2}$ (spin-down) for $u^{(2)}, v^{(2)}$.

When n is a space-like unit vector, $n \cdot n = -1$,

$$P(n) = \frac{1}{2}(1 + \gamma_5 \not{n}), \quad P^2(n) = P(n), \quad P(n)P(-n) = 0. \tag{3.1.32}$$

is a projection operator. When $n = n^{(3)}$,

$$P(n^{(3)}) = \frac{1}{2}(1 + \gamma_5 \not{n}^{(3)}) = \frac{1}{2}\begin{bmatrix} 1 + \sigma_3 & 0 \\ 0 & 1 - \sigma_3 \end{bmatrix},$$

so that

$$\begin{aligned} P(n^3)u^{(1)}(mc, \mathbf{0}) &= u^{(1)}(mc, \mathbf{0}), \\ P(n^3)v^{(2)}(mc, \mathbf{0}) &= v^{(2)}(mc, \mathbf{0}), \\ P(-n^3)u^{(2)}(mc, \mathbf{0}) &= u^{(2)}(mc, \mathbf{0}), \\ P(-n^3)v^{(1)}(mc, \mathbf{0}) &= v^{(1)}(mc, \mathbf{0}), \end{aligned}$$

corresponding respectively to spin projections $+\frac{1}{2}$ for u^1 and v^1 and $-\frac{1}{2}$ for u^2 and v^2. Making a Lorentz transformation to a frame in which the particle has 3-momentum $\mathbf{p}$, the states $u^{(r)}(mc, \mathbf{0}), v^{(s)}(mc, \mathbf{0})$ become the states $u^{(r)}(p), v^{(s)}(p)$ which are also eigenstates of $-w \cdot n/mc$, where n is the Lorentz transform of $n^{(3)}$. Thus for any space-like vector n with $n \cdot p = 0$, $P(n)$ projects onto positive energy states that have spin $\boldsymbol{\Sigma} \cdot \boldsymbol{n} = +\frac{1}{2}$ and negative energy states that have spin $\boldsymbol{\Sigma} \cdot \boldsymbol{n} = -\frac{1}{2}$ in the rest frame.

We can also choose $n = (p^0, \mathbf{p})/mc$ so that its space part $\boldsymbol{n}$ is parallel to the 3-momentum vector $\boldsymbol{p}$; then

$$P(n)\Lambda_\pm(p) = \left(I \pm \frac{\boldsymbol{\Sigma} \cdot \boldsymbol{p}}{|\boldsymbol{p}|}\right)\Lambda_\pm(p),$$

defining the *helicity* representation.

It is usual to write the eigenvectors of $P(n)$ as $u(p, n)$ for positive energy and $v(p, n)$ for negative energy states, so that

$$P(n)u(p, n) = u(p, n), \qquad P(-n)v(p, n) = v(p, n).$$

The relations

$$[\Lambda_\pm(p), P(n)] = 0,$$
$$\Lambda_+(p)P(n) + \Lambda_-(p)P(n) + \Lambda_+(p)P(-n) + \Lambda_-(p)P(-n) = I,$$
$$Tr\Lambda_\pm(p)P(\pm n) = 1$$

enable us to classify the set of plane wave solutions in terms of spin polarization and 4-momentum together.

3.1.6 Charge conjugation

Minimal coupling to an external electromagnetic field, §2.5.5, replaces p_μ by Π_μ,

$$\{\gamma^\mu \, \Pi_\mu - mc\}\psi = 0,$$

which we can decompose into

$$\{i\gamma^\mu \, \partial_\mu - mc\}\psi = \frac{q}{c} \, \gamma^\mu a_\mu \, \psi, \tag{3.1.33}$$

in which the motion of the charged particle is driven by the interaction term on the right hand side. There exists a *charge conjugate* spinor, $\psi_c = C\overline{\psi}^t$, which satisfies (3.1.33) but with the sign of q reversed,

$$\{i\gamma^\mu \, \partial_\mu - mc\}\psi_c = -\frac{q}{c} \, \gamma^\mu a_\mu \, \psi_c. \tag{3.1.34}$$

Recall that $\overline{\psi} = \psi^\dagger\gamma^0$, where the dagger denotes Hermitian conjugation. The Dirac conjugate of (3.1.33) is

$$-i(\partial_\mu\overline{\psi})\gamma^\mu - mc\,\overline{\psi} = \frac{q}{c} \, \overline{\psi} \, \gamma^\mu a_\mu,$$

and taking the matrix transpose gives

$$(i\gamma^{\mu\, t}\partial_\mu + mc) \, \overline{\psi}^t = -\frac{q}{c} \, \gamma^{\mu\, t} \, a_\mu \, \overline{\psi}^t.$$

The *charge conjugation matrix*

$$C = i\gamma^2\gamma^0 = -i\sigma^1 \otimes \sigma^2 = \begin{pmatrix} 0 & -i\sigma^2 \\ -i\sigma^2 & 0 \end{pmatrix}, \tag{3.1.35}$$

has the property, Appendix A.2,

$$\gamma^{\mu\,t} = -C^{-1}\gamma^\mu C,$$

from which (3.1.34) follows, and

$$\psi_c = C\,\overline{\psi}^{\,t} = i\gamma^2\,\psi^* = \begin{pmatrix} 0 & -i\sigma^2 \\ +i\sigma^2 & 0 \end{pmatrix}\psi^*. \tag{3.1.36}$$

It is instructive to apply the charge conjugation transformation to the plane wave states (3.1.20). Complex conjugation maps $\exp(-ip\cdot x)$ into $\exp(+ip\cdot x)$. Effectively this transforms $p \to -p$; the 3-momentum is reversed, and positive frequency oscillations become negative frequency and *vice versa.* The spinor amplitudes (3.1.24) and (3.1.25) are both real, so we have only to operate on them with C. The result is simple in the particle's rest frame, in which

$$u^{(r)} = \begin{pmatrix} \phi^r \\ 0 \end{pmatrix}, \quad v^{(r)} = \begin{pmatrix} 0 \\ \phi^r \end{pmatrix}.$$

Since $i\sigma^2\,\phi^1 = -\phi^2$ and $i\sigma^2\,\phi^2 = +\phi^1$, we find

$$u^{(1)} \to v^{(2)}, \quad u^{(2)} \to -v^{(1)}, \quad v^{(1)} \to u^{(2)}, \quad v^{(2)} \to -u^{(1)}$$

so that when $p = (mc, 0, 0, 0)$,

$$\begin{aligned} \psi_c^{(+)(1)} &= \psi^{(-)(2)}, \quad \psi_c^{(+)(2)} = -\psi^{(-)(1)}, \\ \psi_c^{(-)(1)} &= \psi^{(+)(2)}, \quad \psi_c^{(-)(2)} = -\psi^{(+)(1)}. \end{aligned}$$

It follows that, for a particle at rest, charge conjugation maps positive energy states onto the corresponding negative energy states and flips the spin.

The expectation values of operators O are derived from

$$\langle\, O \,\rangle = \int \overline{\psi}\, O\, \psi\, d^3x \tag{3.1.37}$$

Substituting ψ_c for ψ and manipulating the resulting expression gives

$$\langle\, O \,\rangle_c = \int \overline{\psi}_c\, O\, \psi_c\, d^3x = -\langle \gamma^2\, O^*\, \gamma^2 \,\rangle^*, \tag{3.1.38}$$

from which we deduce the following correspondences:

$$\langle\, x^\mu \,\rangle_c = \langle\, x^\mu \,\rangle, \quad \langle\, p^\mu \,\rangle_c = -\langle\, p^\mu \,\rangle, \quad \langle\, j^\mu \,\rangle_c = \langle\, j^\mu \,\rangle, \quad \langle\, \beta \,\rangle_c = -\langle\, \beta \,\rangle,$$

$$\langle\, \boldsymbol{\Sigma} \,\rangle_c = -\langle\, \boldsymbol{\Sigma} \,\rangle \quad \langle\, \boldsymbol{L} \,\rangle_c = -\langle\, \boldsymbol{L} \,\rangle \quad \langle\, \boldsymbol{J} \,\rangle_c = -\langle\, \boldsymbol{J} \,\rangle$$

where $\boldsymbol{J} = \boldsymbol{L} + \frac{1}{2}\boldsymbol{\Sigma}$.

Equation (3.1.33) can be put into the Schrödinger form, (2.5.33),

$$i\frac{\partial\psi}{\partial t} = \left\{ c\boldsymbol{\alpha}\cdot(\boldsymbol{p} - q\boldsymbol{A}) + \beta mc^2 + q\Phi \right\}\psi,$$

which enables us to identify the corresponding Dirac Hamiltonian

$$H_D(q) := c\boldsymbol{\alpha} \cdot (\boldsymbol{p} - q\boldsymbol{A}) + \beta mc^2 + q\phi \tag{3.1.39}$$

The charge conjugation rules now give

$$\langle\, H_D(-q)\,\rangle_c = -\langle\, H_D(+q)\,\rangle, \tag{3.1.40}$$

connecting the negative energy states of a Dirac particle with the corresponding positive energy states of its anti-particle.

3.2 Spherical symmetry

Dirac's equation for a particle in a spherically symmetric scalar potential can be obtained from (3.1.39) by setting $\boldsymbol{A} = 0$ and writing $q\Phi = V(r)$, where $r = |\boldsymbol{x}|$ is the distance from the origin of coordinates in $\mathbb{R}^3$, so that

$$i\hbar\frac{\partial\psi}{\partial t} = \{c\boldsymbol{\alpha}\cdot\boldsymbol{p} + \beta mc^2 + V(r)\}\psi. \tag{3.2.1}$$

This equation is the starting point for much of the theory of atomic and molecular structure, so that we shall need a thorough understanding of its properties. We first look for stationary states of the form

$$\psi(x) = \mathrm{e}^{-iEt/\hbar}\,\phi(\boldsymbol{x})$$

so that E will be an eigenvalue of the Dirac Hamiltonian

$$H\,\phi(\boldsymbol{x}) = \{c\boldsymbol{\alpha}\cdot\boldsymbol{p} + \beta mc^2 + V(r)\}\phi(\boldsymbol{x}) = E\,\phi(\boldsymbol{x}). \tag{3.2.2}$$

The Hamiltonian is only invariant under rotations and space-like reflections. The infinitesimal generators of rotations are the components of $\boldsymbol{J} = \boldsymbol{L} + \boldsymbol{S}$, where $\boldsymbol{S} = \frac{1}{2}\,\boldsymbol{\Sigma}$, so that we expect to be able to characterize the eigenstates in the usual way by the eigenvalues of $\boldsymbol{J}^2$ and J_3. By elementary algebra, we find $\boldsymbol{S}^2 = \frac{3}{4}I_4$, so that $\boldsymbol{S}^2$ is a multiple of the identity and therefore commutes with H. Finally, the operator

$$\mathcal{P} = \beta P, \quad P\,f(\boldsymbol{x}) = f(-\boldsymbol{x}) \tag{3.2.3}$$

also commutes with H, and so its eigenvalues are also constants of the motion. We shall use lower case letters $\boldsymbol{j}, \boldsymbol{l}, \boldsymbol{s}$ to denote the angular momentum operators on a 2-spinor space.

In a spherical polar coordinate system, (r, θ, φ), eigenstates of (3.2.2) have the 4-spinor structure

$$\phi(\boldsymbol{x}) = \frac{1}{r}\begin{pmatrix} P(r)\chi_{\kappa m}(\theta,\varphi) \\ iQ(r)\chi_{-\kappa m}(\theta,\varphi) \end{pmatrix} \tag{3.2.4}$$

The spin-angle 2-spinors $\chi_{\kappa m}(\theta, \varphi)$ are eigenfunctions of $\boldsymbol{j}^2$ and j_3, belonging to a $(2j+1)$-dimensional irreducible representation of SO(3), so that

$$\boldsymbol{j}^2 \chi_{\kappa m}(\theta,\varphi) = j(j+1)\hbar^2\, \chi_{\kappa m}(\theta,\varphi), \quad j_3 \chi_{\kappa m}(\theta,\varphi) = m\hbar\, \chi_{\kappa m}(\theta,\varphi), \tag{3.2.5}$$

where $m = -j, -j+1, \ldots, j$, $j = \frac{1}{2}, \frac{3}{2}, \frac{5}{2}, \ldots$ $\chi_{\kappa m}$ is also an eigenfunction of $\boldsymbol{l}^2$ with eigenvalue $l(l+1)\hbar^2$ where l can take the values $j+1/2$ or $j-1/2$. The cleanest classification depends on the relation

$$K\chi_{\kappa m}(\theta,\varphi) = \kappa\chi_{\kappa m}(\theta,\varphi), \tag{3.2.6}$$

where

$$K = -[1 + \boldsymbol{j}^2 - \boldsymbol{l}^2 - \boldsymbol{s}^2] = -(1 + \boldsymbol{\sigma}\cdot\boldsymbol{l}), \tag{3.2.7}$$

so that the notation

$$\kappa = (j+1/2)\eta \;\text{ when } l = j + \frac{1}{2}\eta, \quad \eta = \pm 1,$$

enables us to track which coupling of $\boldsymbol{l}$ and $\boldsymbol{s}$ is involved. The conventions used in labelling Dirac 4-spinors in spherical symmetry are given in Table A.1.

3.2.1 Angular structure

We construct spin-angular functions by diagonalizing the product representation $\mathcal{D}^{(l)} \times \mathcal{D}^{(1/2)}$ in the usual way using the the Clebsch-Gordon decomposition

$$\mathcal{D}^{(l)} \times \mathcal{D}^{(1/2)} = \mathcal{D}^{(l+1/2)} \oplus \mathcal{D}^{(l-1/2)}$$

The representation $\mathcal{D}^{(1/2)}$ is two-dimensional; we can choose a basis of eigenvectors of $\boldsymbol{s}^2$ and s_3, where $\boldsymbol{s} = \frac{1}{2}\boldsymbol{\sigma}$, namely

$$\phi_{1/2} = \begin{pmatrix} 1 \\ 0 \end{pmatrix}, \quad \phi_{-1/2} = \begin{pmatrix} 0 \\ 1 \end{pmatrix},$$

so that

$$\boldsymbol{s}^2\,\phi_\sigma = \frac{3}{4}\phi_\sigma, \quad s_3\,\phi_\sigma = \sigma\phi_\sigma, \quad \sigma = \pm\frac{1}{2}.$$

The representation $\mathcal{D}^{(l)}$ is $(2l+1)$-dimensional; its basis vectors can be taken to be the spherical harmonics

$$\{Y_l^m(\theta,\varphi)\,|\, m = -l, -l+1, \ldots, l\},$$

so that

$$\begin{aligned}
\mathbf{l}^2\, Y_l^m(\theta,\varphi) &= l(l+1)\, Y_l^m(\theta,\varphi), \\
l_3\, Y_l^m(\theta,\varphi) &= m\, Y_l^m(\theta,\varphi) \\
l_\pm Y_l^m(\theta,\varphi) &= [l(l+1) - m(m\pm 1)]^{1/2} Y_l^{m\pm 1}(\theta,\varphi).
\end{aligned}$$

The spherical harmonics, defined in (B.3.23), with the required phase conventions are given by

$$\begin{aligned}
Y_l^m(\theta,\varphi) &= \left(\frac{2l+1}{4\pi}\right)^{1/2} C_l^m(\theta,\varphi)\\
C_l^m(\theta,\varphi) &= (-1)^m \left[\frac{(l-m)!}{(l+m)!}\right]^{1/2} P_l^m(\theta)e^{im\varphi} \text{ if } m \geq 0 \qquad (3.2.8)\\
C_l^{-m}(\theta,\varphi) &= (-1)^m C_l^m(\theta,\varphi)^*.
\end{aligned}$$

Basis functions for the representations $\mathcal{D}^{(j)}$ with $j = l \pm \frac{1}{2}$ can now be constructed from sums of products of the form[2]

$$\chi_{\kappa m}(\theta,\varphi) = \sum_\sigma (l, m-\sigma, 1/2, \sigma \,|\, l, 1/2, j, m)\; Y_l^{m-\sigma}(\theta,\varphi)\phi_\sigma \qquad (3.2.9)$$

where $l = j + \frac{1}{2}\eta$, $\eta = \operatorname{sgn}\kappa$. Inserting explicit expressions for the Clebsch-Gordon coefficients[3] gives

$$\chi_{+|\kappa|m}(\theta,\varphi) = \begin{pmatrix} -\left(\dfrac{j+1-m}{2j+2}\right)^{1/2} Y_{j+1/2}^{m-1/2}(\theta,\varphi) \\ \left(\dfrac{j+1+m}{2j+2}\right)^{1/2} Y_{j+1/2}^{m+1/2}(\theta,\varphi) \end{pmatrix} \qquad (3.2.10)$$

$$\chi_{-|\kappa|m}(\theta,\varphi) = \begin{pmatrix} \left(\dfrac{j+m}{2j}\right)^{1/2} Y_{j-1/2}^{m-1/2}(\theta,\varphi) \\ \left(\dfrac{j-m}{2j}\right)^{1/2} Y_{j-1/2}^{m+1/2}(\theta,\varphi) \end{pmatrix} \qquad (3.2.11)$$

The eigenfunctions $\chi_{\kappa m}(\theta,\varphi)$ are orthonormal on the unit sphere with respect to the inner product[4]

$$(\chi_{\kappa' m'} \,|\, \chi_{\kappa m}) = \int\!\!\int \chi^\dagger_{\kappa' m'}(\theta,\varphi)\, \chi_{\kappa m}(\theta,\varphi)\, \sin\theta\, d\theta\, d\varphi = \delta_{\kappa'\kappa}\, \delta_{m'm}. \qquad (3.2.12)$$

3.2.2 The operator σ_r

The operator $\sigma_r = \boldsymbol{\sigma}\cdot\boldsymbol{e}_r$ sets up an involution connecting basis functions with opposite signs of κ:

---

2 In this book we use the convention that the order of coupling is l, s, j. The same spin-angle basis functions are obtained if we use the order s, l, j but there is a phase difference $(-1)^{l-j+1/2}$. *It is vital not to mix formulae based on different conventions!*.

3 Compare [2, §2.12]; Louck uses $\mathcal{Y}^{(j\pm 1/2, 1/2)jm}$ for our $\chi_{\pm|\kappa|m}$

4 $\chi^\dagger$ is the Hermitian conjugate of χ, a row vector whose two elements are the complex conjugates of those of χ. The scalar product is therefore an ordinary function of the variables θ, φ.

$$\sigma_r\,\chi_{\kappa m}(\theta,\varphi) = -\chi_{-\kappa m}(\theta,\varphi). \tag{3.2.13}$$

Recall that $\chi_{\kappa m}(\theta,\varphi)$ and $\chi_{-\kappa m}(\theta,\varphi)$ are both eigenfunctions of $\boldsymbol{j}^2$ and j_3 belonging to the same eigenvalue pair j, m, but with opposite parity. The commutator relations

$$[\boldsymbol{l}, \sigma_r] = -i\hbar\,\frac{\boldsymbol{x}\times\boldsymbol{\sigma}}{r} = -\left[\boldsymbol{s}, \sigma_r\right],$$

can be verified by straightforward calculation, so that as $\boldsymbol{j} = \boldsymbol{l} + \boldsymbol{s}$,

$$[\boldsymbol{j}, \sigma_r] = 0.$$

Thus

$$\boldsymbol{j}\sigma_r\,\chi_{\kappa m}(\theta,\varphi) = \sigma_r\boldsymbol{j}\,\chi_{\kappa m}(\theta,\varphi),$$

verifying that both $\chi_{\kappa m}(\theta,\varphi)$ and $\sigma_r\,\chi_{\kappa m}(\theta,\varphi)$ are simultaneous eigenfunctions of $\boldsymbol{j}^2$ and j_3 belonging to the same pair j, m. Similarly, as $P\boldsymbol{x} = -\boldsymbol{x}$ and P does not act on spin matrices, we have

$$\{P, \sigma_r\} = 0$$

so that

$$P\sigma_r\,\chi_{\kappa m}(\theta,\varphi) = -\sigma_r P\,\chi_{\kappa m}(\theta,\varphi).$$

It follows that $\chi_{-\kappa m}(\theta,\varphi)$ is proportional to $\sigma_r\,\chi_{\kappa m}(\theta,\varphi)$ and it is only necessary to find the constant of proportionality. The simplest way to do this is to observe that, in spherical polar coordinates,

$$\boldsymbol{e}_r = (\sin\theta\,\cos\varphi, \sin\theta\,\sin\varphi, \cos\theta)$$

so that

$$\sigma_r = \boldsymbol{\sigma}\cdot\boldsymbol{e}_r = \begin{pmatrix} \cos\theta & \sin\theta e^{-i\varphi} \\ \sin\theta e^{+i\varphi} & -\cos\theta \end{pmatrix}$$

Now because the constant of proportionality must not depend on θ and φ, we can pick a particular direction, say $\theta = 0$, $\varphi = 0$, to verify equation (3.2.13). It is also instructive to observe that this result holds for general directions by direct computation: the relations for contiguous associated Legendre polynomials [3, p. 161, equations (15) – (18)] give, for example,

$$\sqrt{\frac{j+m}{2j}}Y^{m-1/2}_{j-1/2}(\theta,\varphi) =$$

$$\cos\theta\sqrt{\frac{j+1-m}{2j+2}}Y^{m-1/2}_{j+1/2}(\theta,\varphi) - \sin\theta\,\mathrm{e}^{-i\varphi}\sqrt{\frac{j+1+m}{2j+2}}Y^{m+1/2}_{j+1/2}(\theta,\varphi)$$

$$\sqrt{\frac{j-m}{2j}}Y_{j-1/2}^{m+1/2}(\theta,\varphi) =$$

$$\sin\theta e^{+i\varphi}\sqrt{\frac{j+1-m}{2j+2}}Y_{j+1/2}^{m-1/2}(\theta,\varphi) + \cos\theta\sqrt{\frac{j+1+m}{2j+2}}Y_{j+1/2}^{m+1/2}(\theta,\varphi)$$

reproduces the components of $\chi_{-|\kappa|m}(\theta,\varphi)$ from $\sigma_r\chi_{+|\kappa|m}(\theta,\varphi)$. The inverse transformation is left as an exercise for the reader.

3.2.3 The operator $c\boldsymbol{\sigma}\cdot\boldsymbol{p}$

We start by expanding $\boldsymbol{x}\times\boldsymbol{l} = \boldsymbol{x}\times(\boldsymbol{x}\times\boldsymbol{p})$ and rearranging to give

$$\boldsymbol{p} = \boldsymbol{e}_r p_r - \frac{1}{r}\boldsymbol{e}_r\times\boldsymbol{l}, \quad p_r = \boldsymbol{e}_r\cdot\boldsymbol{p} = -i\partial_r,$$

taking care not to change the order of the noncommuting quantum mechanical operators $\boldsymbol{x}$ and $\boldsymbol{p}$. Thus

$$c\boldsymbol{\sigma}\cdot\boldsymbol{p} = c\sigma_r p_r - \frac{c}{r}\boldsymbol{\sigma}\cdot(\boldsymbol{e}_r\times\boldsymbol{l}),$$

Because (A.2.16),

$$(\boldsymbol{\sigma}\cdot\boldsymbol{e}_r)(\boldsymbol{\sigma}\cdot\boldsymbol{l}) = (\boldsymbol{e}_r\cdot\boldsymbol{l}) + i\boldsymbol{\sigma}\cdot(\boldsymbol{e}_r\times\boldsymbol{l}),$$

$(\boldsymbol{e}_r\cdot\boldsymbol{l}) = 0$ and $K = -1-\boldsymbol{\sigma}\cdot\boldsymbol{l}$ we arrive at the important formula

$$c\boldsymbol{\sigma}\cdot\boldsymbol{p} = -ic\sigma_r\left(\partial_r + \frac{K+1}{r}\right), \tag{3.2.14}$$

3.2.4 Separation of radial and spin-angular parts

The Dirac central field Hamiltonian can be partitioned in conformity with (3.2.4) so that, with the corresponding representation of $\boldsymbol{\alpha}$ and β matrices and the formula (3.2.14),

$$\begin{aligned} H &= c\boldsymbol{\alpha}\cdot\boldsymbol{p} + \beta mc^2 + V(r) \qquad (3.2.15)\\ &= \begin{pmatrix} mc^2 + V(r) & -ic\sigma_r\left(\partial_r + \frac{K+1}{r}\right) \\ -ic\sigma_r\left(\partial_r + \frac{K+1}{r}\right) & -mc^2 + V(r) \end{pmatrix}, \end{aligned}$$

and

$$
\begin{aligned}
0 &= (H - E)\,\phi(\boldsymbol{x}) \\
&= (c\boldsymbol{\alpha}\cdot\boldsymbol{p} + \beta mc^2 + V(r) - E)\begin{pmatrix} \dfrac{P}{r}\chi_{\kappa m}(\theta,\varphi) \\ i\dfrac{Q}{r}\chi_{-\kappa m}(\theta,\varphi)\end{pmatrix} \\
&= \frac{1}{r}\begin{pmatrix} \left\{(mc^2 + V(r) - E)P + c\left(-\dfrac{dQ}{dr} + \dfrac{\kappa}{r}Q\right)\right\}\chi_{\kappa m}(\theta,\varphi) \\ i\left\{c\left(\dfrac{dP}{dr} + \dfrac{\kappa}{r}P\right) + (-mc^2 + V(r) - E)Q\right\}\chi_{-\kappa m}(\theta,\varphi)\end{pmatrix}.
\end{aligned}
\tag{3.2.16}
$$

The substitution

$$\left(\frac{d}{dr} + \frac{1}{r}\right)\frac{f(r)}{r} = \frac{1}{r}\frac{df(r)}{dr},$$

allows extraction of a common prefactor $1/r$. The linear independence of the spin-angle functions allows us to separate the radial parts to give the coupled equations

$$
\begin{aligned}
&(mc^2 + V(r) - E)\,P + \hbar c\left(-\frac{dQ}{dr} + \frac{\kappa}{r}Q\right) = 0 \\
&\hbar c\left(\frac{dP}{dr} + \frac{\kappa}{r}P\right) + (-mc^2 + V(r) - E)\,Q = 0.
\end{aligned}
\tag{3.2.17}
$$

The 4-spinors (3.2.4) with $m = -j, \dots, +j$ span a $(2j+1)$-dimensional irreducible representation of the rotation group. Their symmetries, exploited in [1], simplified the construction of the atomic relativistic self-consistent field equations and helped make Dirac-Hartree-Fock calculations feasible on the computers of the 1960s. Chapter 10 also exploits these properties for molecular calculations in a manner that is even more striking.

3.2.5 Angular density distributions

The Dirac particle density (2.5.2) for a stationary central field state

$$\psi_{E\kappa m}(x) = \phi_{E\kappa m}(\boldsymbol{x})\,\mathrm{e}^{-iEt}$$

is given by

$$\rho_{E\kappa m} = \psi^{\dagger}_{E\kappa m}(x)\,\psi_{E\kappa m}(x) = \phi^{\dagger}_{E\kappa m}(x)\phi_{E\kappa m}(x). \tag{3.2.18}$$

Substituting from (3.2.4) gives

$$\rho_{E\kappa m}(\boldsymbol{x}) = \left\{|P_{E\kappa}(r)|^2 A_{\kappa m}(\theta) + |Q_{E\kappa}(r)|^2 A_{-\kappa m}(\theta)\right\}/r^2. \tag{3.2.19}$$

where, making use of (3.2.13), we have

$$A_{\kappa m}(\theta) = \chi_{\kappa,m}(\theta,\varphi)^{\dagger}\chi_{\kappa,m}(\theta,\varphi) = A_{-\kappa m}(\theta), \tag{3.2.20}$$

independently of φ. Thus (3.2.19) factorizes into the product,

$$\rho_{E\kappa m}(\boldsymbol{x}) = \frac{D_{E\kappa}(r)}{r^2}.A_{|\kappa|m}(\theta), \tag{3.2.21}$$

of a radial density

$$D_{E\kappa}(r) = |P_{E\kappa}(r)|^2 + |Q_{E\kappa}(r)|^2 \tag{3.2.22}$$

with the common angular distribution $A_{|\kappa|m}(\theta)$. The result (3.2.20) was first derived by Hartree [4] as an identity in terms of associated Legendre polynomials. Explicit formulae for the angular densities $A_{|\kappa|,m}(\theta)$ and for the

Table 3.1. Angular density functions

A. Relativistic			B. Nonrelativistic		
$\|\kappa\|$	$\|m\|$	$4\pi.A_{\|\kappa\|,m}(\theta)$	l	$\|m\|$	$4\pi.A_{l,m}(\theta)_{nr}$
1	$\frac{1}{2}$	1	0	0	1
2	$\frac{3}{2}$	$\frac{3}{2}\sin^2\theta$	1	1	$\frac{3}{2}\sin^2\theta$
	$\frac{1}{2}$	$\frac{1}{2}(1+3\cos^2\theta)$		0	$3\cos^2\theta$
3	$\frac{5}{2}$	$\frac{15}{8}\sin^4\theta$	2	2	$\frac{15}{8}\sin^4\theta$
	$\frac{3}{2}$	$\frac{3}{8}\sin^2\theta(1+15\cos^2\theta)$		1	$\frac{15}{2}\sin^2\theta\cos^2\theta$
	$\frac{1}{2}$	$\frac{3}{4}(5\cos^4\theta - 2\cos^2\theta + 1)$		0	$\frac{5}{4}(3\cos^2\theta - 1)^2$

corresponding Schrödinger angular densities

$$A_{lm}(\theta)_{nr} = |Y_l^m(\theta,\varphi)|^2 = \frac{2l+1}{4\pi}\frac{(l-|m|)!}{(l+|m|)!}|P_l^{|m|}(\mu)|^2$$

for a selection of values of $|\kappa|$ and l are shown in Table 3.1. The relativistic angular distribution $A_{|\kappa|m}(\theta)$ is very similar to that of $A_{lm}(\theta)_{nr}$ for $l = |\kappa| - 1$. The main difference between the relativistic and nonrelativistic angular densities, shown in Figures 3.1 and 3.2, is that the nonrelativistic densities for $|m| < l$ have "wasp waists", whereas the corresponding relativistic densities for $|m| < j - 1/2$ have "middle-aged spreads". The zeros in the angular functions are washed out by relativity. The sum of the angular densities over the projection quantum numbers m is independent of the angles for both relativistic and nonrelativistic distributions, because

$$\sum_{m=-j}^{j} A_{\kappa m}(\theta) = \frac{2j+1}{4\pi}, \quad \sum_{m=-l}^{l} A_{lm}(\theta)_{nr} = \frac{2l+1}{4\pi}.$$

These results can also be derived in an elementary fashion from the properties of spherical harmonics [4]. For this reason, the set of functions $\{\psi_{E,\kappa,m} \,|\, m = -j \ldots j\}$ for *bound* states is said to define a *subshell* just as in the Schrödinger theory.

$$A_{1,1/2}(\theta,\phi), \ A_{0,0}(\theta,\phi)_{nr}$$

$$A_{2,3/2}(\theta,\phi), \ A_{1,1}(\theta,\phi)_{nr}$$

$$A_{2,1/2}(\theta,\phi) \qquad\qquad A_{1,0}(\theta,\phi)_{nr}$$

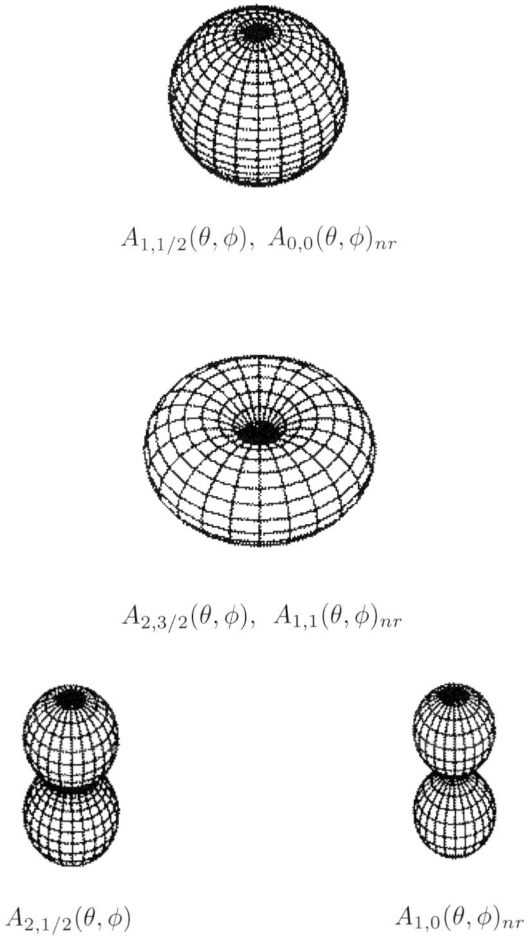

Fig. 3.1. Angular densities, $A_{|\kappa|,|m|}(\theta)$ and $A_{l,|m|}(\theta)_{nr}$.

$$A_{3,5/2}(\theta,\phi),\ A_{2,2}(\theta,\phi)_{nr}$$

$$A_{3,3/2}(\theta,\phi) \qquad\qquad A_{2,1}(\theta,\phi)_{nr}$$

$$A_{3,1/2}(\theta,\phi) \qquad\qquad A_{2,0}(\theta,\phi)_{nr}$$

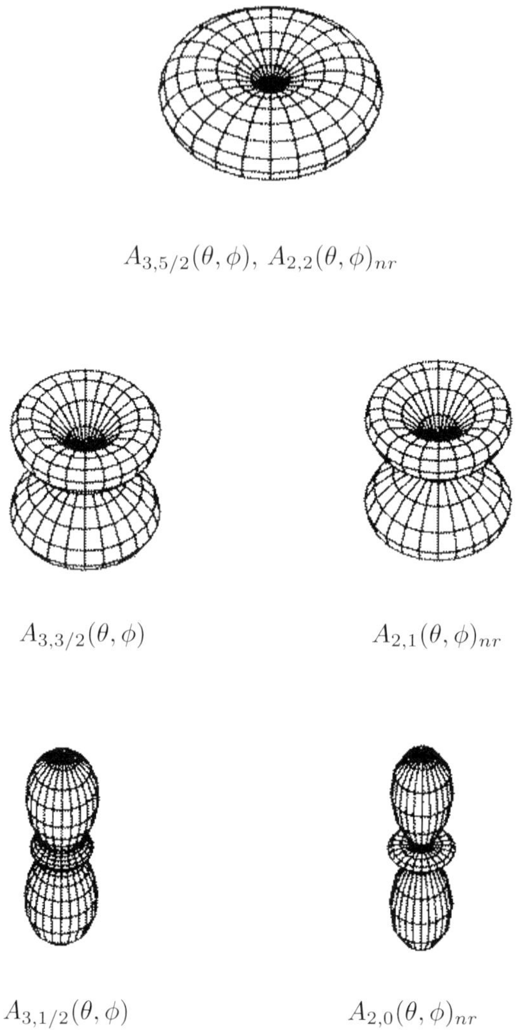

Fig. 3.2. Angular densities, $A_{|\kappa|,|m|}(\theta)$ and $A_{l,|m|}(\theta)_{nr}$.

3.2.6 Radial solutions for the free particle

The radial Dirac equations (3.2.17) for a free particle, for which $V(r) \equiv 0$, are

$$\begin{pmatrix} mc^2 - E & c\left(-\dfrac{d}{dr} + \frac{\kappa}{r}\right) \\ c\left(\dfrac{d}{dr} + \dfrac{\kappa}{r}\right) & -mc^2 - E \end{pmatrix} \begin{pmatrix} P_{E\kappa}(r) \\ Q_{E\kappa}(r) \end{pmatrix} = 0.$$

By eliminating one or other of the components, we see that both $P_{E\kappa}(r)$ and $Q_{E\kappa}(r)$ satisfy the second order equations

$$\begin{aligned} \frac{d^2 P_{E\kappa}}{dr^2} + \left\{p^2 - \frac{\kappa(\kappa+1)}{r^2}\right\} P_{E\kappa} = 0. \\ \frac{d^2 Q_{E\kappa}}{dr^2} + \left\{p^2 - \frac{\overline{\kappa}(\overline{\kappa}+1)}{r^2}\right\} Q_{E\kappa} = 0. \end{aligned} \tag{3.2.23}$$

where p^2 is the square of the relativistic 3-momentum

$$c^2 p^2 = E^2 - m^2 c^4$$

and

$$\overline{\kappa} = -\kappa.$$

These equations are the defining equations for the Riccati-Bessel functions (see, for example, [5, §10.3] and Appendix A.3). Because

$$\kappa(\kappa+1) = l(l+1), \quad \overline{\kappa}(\overline{\kappa}+1) = \overline{l}(\overline{l}+1),$$

we can express the solutions of (3.2.23) in the form

$$P_{E\kappa}(r) = Axf_l(x), \qquad Q_{E\kappa}(r) = Bxf_{\overline{l}}(x), \qquad x = pr, \tag{3.2.24}$$

where $f_l(x)$ is a spherical Bessel function of the first, second, or third kind [5, §10.1.1], Appendix A.3. The ratio A/B in (3.2.24) can be determined by using one of the two component equations of (3.2.17), say

$$(mc^2 + E)\, Q_{E\kappa}(r) = c\left(\frac{d}{dr} + \frac{\kappa}{r}\right) P_{E\kappa}(r), \tag{3.2.25}$$

so that

$$\frac{B}{A} = \eta\left(\frac{cp}{E + mc^2}\right) = \eta\left(\frac{E - mc^2}{E + mc^2}\right)^{1/2}$$

where $\eta = \operatorname{sgn}(\kappa)$. The relation (3.2.25) is closely related to equation (A.4.12) of the appendix. The solution of the radial reduced equations (3.2.17) is therefore

$$\boldsymbol{u}(r) = \begin{pmatrix} P_{E\kappa}(r) \\ Q_{E\kappa}(r) \end{pmatrix} = \begin{pmatrix} \mathcal{N}\left(\dfrac{E+mc^2}{\pi E}\right)^{1/2} x f_l(x) \\ \mathcal{N}\eta\left(\dfrac{E-mc^2}{\pi E}\right)^{1/2} x f_{\bar{l}}(x) \end{pmatrix} \tag{3.2.26}$$

where $\mathcal{N}$ is an overall normalizing constant. Selected properties of the spherical Bessel functions will be found in Appendix A.3. The equations (3.2.23) are of second order, so that linearly independent solutions come in pairs. These are either of standing wave type, (j_l, y_l), or of progressive wave type $(h_l^{(1)}, h_l^{(2)})$. The solutions of the first kind, j_l, are bounded everywhere, including the singular points $x = 0, \infty$, whereas the solutions of the second kind, y_l, are bounded at infinity but have poles at $x = 0$. The spherical Hankel functions, the solutions of the third kind, as linear combinations of the functions of the first and second kinds,

$$h_l^{(1)}(x) = j_l(x) + i y_l(x), \qquad h_l^{(2)}(x) = j_l(x) - i y_l(x),$$

have poles at the origin but are bounded at infinity. The structure of (3.2.25) implies that the two components of $\mathbf{u}(r)$ must contain Riccati-Bessel functions of the same kind. For a free particle $|E| \geq mc^2$ and $x = pr \geq 0$.

3.2.7 Partial wave normalization

The inner product of two free particle solutions is given by

$$\langle \psi_{E'\kappa'm'} \,|\, \psi_{E\kappa m} \rangle = \int \psi^{\dagger}_{E'\kappa'm'}(\boldsymbol{x}, t).\psi_{E\kappa m}(\boldsymbol{x}, t) d\boldsymbol{x} \tag{3.2.27}$$

where the integral is over all space. The orthonormality condition (3.2.12) ensures that this is diagonal in the angular quantum numbers κ, m and the integral reduces to

$$\langle \psi_{E'\kappa'm'} \,|\, \psi_{E\kappa m} \rangle = \delta_{\kappa',\kappa}.\delta_{m',m} \int_0^{\infty} \left\{ P^{\dagger}_{E'\kappa}(r).P_{E\kappa}(r) + Q^{\dagger}_{E'\kappa}(r).Q_{E\kappa}(r) \right\} dr.$$

For *standing waves*, the integral on the right-hand side can be evaluated in terms of the well-known integral [6, pages 90-91]

$$\int_0^{\infty} j_l(pr) j_l(p'r) r^2 dr = \frac{\pi}{2p^2} \delta(p - p')$$

so that

$$\int_0^{\infty} \left\{ P^{\dagger}_{E'\kappa}(r).P_{E\kappa}(r) + Q^{\dagger}_{E'\kappa}(r).Q_{E\kappa}(r) \right\} dr = \mathcal{N}^2 \delta(p - p').$$

Thus $\mathcal{N} = 1$ normalizes (3.2.26) with respect to *momentum*. Because

$$\delta(E-E') = \left|\frac{dp}{dE}\right| \delta(p-p') = \frac{c^2 p}{|E|}\delta(p-p'),$$

normalization with respect to *energy* requires

$$\mathcal{N} = \left(|E|/c^2 p\right)^{1/2}.$$

The spherical Hankel functions $h_l^{(1)}(x)$ represent outgoing *progressive waves*

$$\psi_E^{(1)}(r) = \mathcal{N}\begin{pmatrix} \left(\dfrac{E+mc^2}{E-mc^2}\right)^{1/4} & h_l^{(1)}(pr)\chi_{\kappa,m}(\theta,\phi) \\ i\eta\left(\dfrac{E-mc^2}{E+mc^2}\right)^{1/4} & h_{\bar{l}}^{(1)}(pr)\chi_{-\kappa,m}(\theta,\phi) \end{pmatrix} \tag{3.2.28}$$

$$\sim \frac{\mathcal{N}}{pr}\,\mathrm{e}^{i(pr-(l+1)\pi/2)}\begin{pmatrix} \left(\dfrac{E+mc^2}{E-mc^2}\right)^{1/4} & \chi_{\kappa,m}(\theta,\phi) \\ -\left(\dfrac{E-mc^2}{E+mc^2}\right)^{1/4} & \chi_{-\kappa,m}(\theta,\phi) \end{pmatrix}$$

(see Appendix A.3). The relation (3.2.13) enables us to write the asymptotic radial particle current density as

$$s(\boldsymbol{r}) = \left[\psi^\dagger\, c\boldsymbol{\alpha}\psi\right]\cdot \boldsymbol{e}_r = c\left(\frac{\mathcal{N}}{pr}\right)^2\left[\chi^\dagger_{\kappa,m}\cdot\chi_{\kappa,m} + \chi^\dagger_{-\kappa,m}\cdot\chi_{-\kappa,m}\right], \tag{3.2.29}$$

and integrating over the surface of a sphere of radius $r = R$ using (3.2.12) gives the outgoing particle flux

$$S(R) = \int_{r=R} s(\boldsymbol{r})\, r^2 \sin\theta\, d\theta\, d\phi = 2c\left(\frac{\mathcal{N}}{p}\right)^2, \tag{3.2.30}$$

so that

$$\mathcal{N}/p = (2c)^{-1/2}.$$

gies unit outgoing flux. Because $h_l^{(2)}(pr) = \left[h_l^{(1)}(pr)\right]^*$, the ingoing partial wave has the same normalization. Thus, in the formal nonrelativistic limit $c \to \infty$,

$$\psi_E^{(1)}(\boldsymbol{r}) \sim p^{-1/2}\,\frac{\mathrm{e}^{i(pr-(l+1)\pi/2)}}{r}\begin{pmatrix}\chi_{\kappa,m}(\theta,\phi)\\ 0\end{pmatrix},$$

as expected.

3.3 Hydrogenic atoms

For hydrogenic atoms, the potential energy in atomic units is $V(r) = -Z/r$, so that equations (3.2.17) take the form

$$\begin{pmatrix} (mc^2 - Z/r - E) & c\left(-\dfrac{d}{dr} + \dfrac{\kappa}{r}\right) \\ c\left(\dfrac{d}{dr} + \dfrac{\kappa}{r}\right) & (-mc^2 - Z/r - E) \end{pmatrix} \begin{pmatrix} P_{E\kappa}(r) \\ Q_{E\kappa}(r) \end{pmatrix} = 0. \tag{3.3.1}$$

We seek solutions of (3.3.1) on $0 \leq r < \infty$, whose end-points are singular points of the matrix differential operator. These boundary values determine the nature of the solutions.

We start with the limit $r \to \infty$; then (3.3.1) reduces asymptotically to

$$\begin{pmatrix} mc^2 - E & -c\dfrac{d}{dr} \\ c\dfrac{d}{dr} & -mc^2 - E \end{pmatrix} \begin{pmatrix} P_{E\kappa}(r) \\ Q_{E\kappa}(r) \end{pmatrix} \to 0,$$

so that

$$P_{E\kappa}(r),\, Q_{E\kappa}(r) \sim \exp(\pm\lambda r), \quad \lambda^2 = m^2c^2 - E^2/c^2. \tag{3.3.2}$$

There are two situations:

$-mc^2 \leq E \leq mc^2$: λ is *real.*

Either both components increase exponentially like $\exp(|\lambda| r)$ or both approach zero like $\exp(-|\lambda| r)$ as r increases. Thus the integral of the radial density (3.2.22) over the interval (R, ∞),

$$\int_R^\infty D_{E\kappa}(r) dr,$$

where $D_{E\kappa}(r) = \left\{|P_{E\kappa}(r)|^2 + |Q_{E\kappa}(r)|^2\right\}$, is only finite for the exponentially decreasing solution, which therefore represents a *bound* state.

$-\infty < E < -mc^2$, $mc^2 < E < \infty$: λ is *pure imaginary.*

When $E > 0$ we set $\lambda = -ip$ where

$$p^2 = E^2/c^2 - m^2c^2$$

in which we take $p = |\boldsymbol{p}|$ to be the *positive* root. The radial amplitudes $P_{E\kappa}(r)$ and $Q_{E\kappa}(r)$ either represent progressive waves, outgoing if they are proportional to $\exp(+ip\,r)$, incoming if proportional to $\exp(-ip\,r)$, or standing waves if proportional to some linear combination of $\sin p\,r$ and $\cos p\,r$ with real coefficients; in any case $\int_R^\infty D_{E\kappa}(r) dr$ diverges.

We can develop power series solutions near $r = 0$, with leading term

$$\left.\begin{matrix} P_{E\kappa}(r) \\ Q_{E\kappa}(r) \end{matrix}\right\} \approx r^{\pm\gamma}\{1 + O(r)\}, \quad \gamma^2 = \kappa^2 - \alpha^2 Z^2 \tag{3.3.3}$$

where the exponent γ depends only upon the angular parameter κ and the atomic number Z is independent of the energy parameter E. We want the probability density to be bounded near the origin:

$$0 < \int_0^{R'} D_{E\kappa}(r)dr < M_1 \qquad (3.3.4)$$

for some finite M_1. Because $D_{E\kappa}(r) \propto r^{\pm 2\gamma}$ near $r = 0$, (3.3.4) is satisfied if

$$\pm 2\gamma > -1.$$

Thus we must discard the solution with $\gamma < 0$ if $|\gamma| > \frac{1}{2}$, corresponding to $\alpha Z < \sqrt{\kappa^2 - 1/4}$. This condition is most restrictive for the case $|\kappa| = 1$ or $j = \frac{1}{2}$, which requires $Z < c\sqrt{3}/2 \approx 118.6$. Coddington and Levinson [7] classify the situation in which only *one* of the two solutions satisfies (3.3.4) as the *limit point case* (3.3.4); clearly this solution is unique. For $Z > c\sqrt{3/2}$ we have the *limit circle* case: both the solution proportional to $r^{-\gamma}$ and that proportional to $r^{+\gamma}$ satisfy (3.3.4). Any linear combination of both solutions is then admissible, and we need an additional constraint to make the solution unique. The potential energy operator $-Z/r$ must have a *finite* expectation value,

$$0 < \int_0^{R'} \frac{1}{r} D_{E\kappa}(r)dr \leq M_2 \qquad (3.3.5)$$

for some finite M_2, so that we must discard the solution with $\gamma < 0$. The solution with $\gamma > 0$ is therefore valid for all $Z < c\,|\kappa|$, extending the range of acceptable solutions to $Z = 137$ for $|\kappa| = 1$. The integral (3.3.5) for the $\gamma > 0$ solution is always finite for $|\kappa| > 1$, so that then (3.3.5) is redundant. Most discussions of the Dirac hydrogenic equation ignore (3.3.5), and avoid discussing the status of the analytic bound state solutions; however Greiner [8, §9.12] goes so far as to suggest that bound solutions for $Z > c\sqrt{3/2}$ have no physical significance! In fact, the same problem occurs for s states in the Schrödinger theory of the hydrogen atom. The indicial equation has solutions $s = l, -l-1$; the negative power solution is always (rightly) discarded, even for $l = 0$, where it is in fact square integrable. The fact that (3.3.5) is violated for $s = -1$ seems usually to have gone unnoticed.

For $Z > 137$, γ becomes pure imaginary for $|\kappa| = 1$ and there are no normalizable solutions near $r = 0$. Real nuclei have a finite size; this permits solutions for much higher values of Z. Applications to the physics of superheavy elements and to heavy ion collisions are treated in detail in [8, 9].

The discussion of charge conjugation in §3.1.6 and of negative energy states in §2.5.6 relates solutions for negative energy to scattering solutions of positive energy. The formalism of this subsection therefore remains valid for the negative energy continuum, $E < -mc^2$, if we make the substitutions

$$E \to -E, \quad Z \to -Z, \quad \kappa \to -\kappa, \quad P_{E,\kappa} \leftrightarrow Q_{E,-\kappa}.$$

We have therefore the machinery for treating positron processes as well as electron processes.

3.3.1 Solution of the radial equations

There have been many studies of the radial equations for hydrogenic atoms, for example, Darwin [10], Gordon [11], Mott [12], Hylleraas [13], and Johnson and Cheng [14] as well as Dirac himself [15, Chapter XI]. The asymptotic exponent, λ, (3.3.2), is real for bound states, and we can define an *apparent principal quantum number*, N, by the equation

$$\lambda = Z/N = +mc\sqrt{1 - E^2/m^2c^4}. \tag{3.3.6}$$

By rearranging, we obtain[5]

$$E = +c^2\sqrt{1 - Z^2/N^2c^2}. \tag{3.3.7}$$

We assume a solution incorporating (3.3.3) at the origin such that

$$\begin{aligned} P_{E\kappa}(r) &= \mathcal{N}_{E\kappa}\,(c + E/c)^{1/2}\,\rho^{\gamma}\mathrm{e}^{-\rho/2}[X(\rho) + Y(\rho)] \\ Q_{E\kappa}(r) &= \mathcal{N}_{E\kappa}\,(c - E/c)^{1/2}\,\rho^{\gamma}\mathrm{e}^{-\rho/2}[X(\rho) - Y(\rho)] \end{aligned} \tag{3.3.8}$$

where $\rho = 2\lambda r$, $\mathcal{N}_{E\kappa}$ is a normalization factor and $X(\rho)$ and $Y(\rho)$ are to be determined. Substituting (3.3.8) into (3.3.1) we find, after some algebra, that $Y(\rho)$ satisfies

$$\frac{d^2Y(\rho)}{d\rho^2} + (b - \rho)\frac{dY(\rho)}{d\rho} - aY(\rho) = 0 \tag{3.3.9}$$

which is Kummer's confluent hypergeometric equation [3, Chapter VI], [5, §13.1.1], Appendix A.3, where

$$a = \gamma - NE/c^2, \quad b = 2\gamma + 1, \tag{3.3.10}$$

whilst

$$X(\rho) = -\frac{1}{\kappa - N}\left(aY(\rho) + \rho\frac{dY(\rho)}{d\rho}\right). \tag{3.3.11}$$

Equation (3.3.9) has pairs of linearly independent solutions satisfying different boundary conditions at the regular singular point $\rho = 0$ and the irregular singular point at ∞. A solution bounded as $\rho \to 0$ is

$$Y_1(\rho) = M(a, b; \rho) = 1 + \frac{(a)_1}{(b)_1}\rho + \frac{(a)_2}{(b)_2}\frac{\rho^2}{2!} + \ldots + \frac{(a)_n}{(b)_n}\frac{\rho^n}{n!} + \ldots \tag{3.3.12}$$

where $(a)_0 = 1, \quad (a)_n = (a+n-1)(a)_{n-1} = a(a+1)\ldots(a+n-1), \quad n \geq 1.$ A solution which is singular at $\rho = 0$ and is linearly independent of Y_1 is

[5] The negative root would be inappropriate for a bound state.

$$Y_5(\rho) = U(a,b;\rho) \tag{3.3.13}$$
$$= \frac{\pi}{\sin \pi b} \left\{ \frac{M(a,b;\rho)}{\Gamma(1+a-b)\Gamma(b)} - \rho^{1-b} \frac{M(1+a-b,2-b;\rho)}{\Gamma(a)\Gamma(2-b)} \right\}.$$

This is a multi-function, its principal branch being defined by $-\pi < \arg \rho \leq \pi$. The general solution of (3.3.9) is a linear combination $Y = AY_1 + BY_5$ with arbitrary complex coefficients, A, B. We shall also need another pair of linearly independent solutions:

$$Y_2(\rho) = \rho^{1-b} M(1+a-b, 2-b;\rho), \quad Y_7(\rho) = \mathrm{e}^{\rho} U(b-a,b;-\rho). \tag{3.3.14}$$

The conventional labelling of solutions follows [5, §13.1.12 -§13.1.19]. Tbe pairs Y_1 and Y_5 have a simple behaviour at the origin, whilst Y_2 and Y_7 have simple exponential behaviour as $\rho \to \infty$. The functions X_i can be obtained from the Y_i using [5, equations 13.4.10, 13.4.23]. We now define u_i to be the (unnormalized) two component vector with elements P_i, Q_i as in (3.2.26), so that the first pair of solutions can be written

$$u_1 = \rho^{\gamma} \mathrm{e}^{-\rho/2} \tag{3.3.15}$$
$$\begin{pmatrix} (c+E/c)^{1/2} \left\{ aM(a+1,b;\rho) + (N-\kappa) M(a,b;\rho) \right\} \\ (c-E/c)^{1/2} \left\{ aM(a+1,b;\rho) - (N-\kappa) M(a,b;\rho) \right\} \end{pmatrix}$$

and

$$u_2 = \rho^{-\gamma} \mathrm{e}^{-\rho/2} \tag{3.3.16}$$
$$\begin{pmatrix} (c+E/c)^{1/2} \left\{ (1+a-b) M(2+a-b,2-b;\rho) \right. \\ \left. \qquad +(N-\kappa) M(1+a-b,2-b;\rho) \right\} \\ (c-E/c)^{1/2} \left\{ (1+a-b) M(2+a-b,2-b;\rho) \right. \\ \left. \qquad -(N-\kappa) M(1+a-b,2-b;\rho) \right\} \end{pmatrix},$$

whilst the other pair of solutions is

$$u_5 = \rho^{\gamma} \mathrm{e}^{-\rho/2} (N-\kappa) \tag{3.3.17}$$
$$\begin{pmatrix} (c+E/c)^{1/2} \left\{ (N+\kappa) U(a+1,b;\rho) + U(a,b;\rho) \right\} \\ (c-E/c)^{1/2} \left\{ (N+\kappa) U(a+1,b;\rho) - U(a,b;\rho) \right\} \end{pmatrix},$$

and

$$u_7 = \rho^{\gamma} \mathrm{e}^{+\rho/2} \tag{3.3.18}$$
$$\begin{pmatrix} (c+E/c)^{1/2} \left\{ -U(b-a-1,b;-\rho) + (N-\kappa) U(b-a,b;-\rho) \right\} \\ (c-E/c)^{1/2} \left\{ -U(b-a-1,b;-\rho) - (N-\kappa) U(b-a,b;-\rho) \right\} \end{pmatrix}.$$

There is at most one pair of linearly independent solutions, so that the two sets of solutions are related by

$$u_1 = Au_5 + Bu_7, \quad u_2 = Cu_5 + Du_7, \tag{3.3.19}$$

where

$$A = \mathrm{e}^{i\epsilon\pi a}\Gamma(b)/\Gamma(b-a), \qquad B = \mathrm{e}^{i\epsilon\pi(a-b)}\Gamma(b)/\Gamma(a),$$
$$C = \mathrm{e}^{i\epsilon\pi(a-b)}\Gamma(2-b)/\Gamma(1-a), \quad D = \mathrm{e}^{i\epsilon\pi(a-b)}\Gamma(2-b)/\Gamma(1+a-b).$$

3.3.2 The bound state solutions

The probability integral

$$\int_0^\infty D_{E\kappa}(r)dr. \tag{3.3.20}$$

must be finite for $D_{E\kappa}(r)$ to represent a bound state. The power series defining the confluent hypergeometric functions appearing in u_1 are finite at the origin and converge uniformly to functions that increase asymptotically like $\exp(\rho)$ as $\rho \to \infty$. Thus (3.3.20) will only be finite if the power series terminate, which requires a to be a *non-positive* integer. The solution u_2 has a non-normalizable radial density at the origin, and can be discarded. We therefore set

$$a = -n_r, \quad n_r = 0, 1, 2, \ldots;$$

n_r is known as the *inner quantum number*. Inserting this into (3.3.10), we see that

$$n_r = -a = -\gamma + NE/c^2.$$

Substituting for E from (3.3.7) and rearranging the result, gives the *apparent principal quantum number*

$$\begin{aligned} N &= \left[(n_r+\gamma)^2 + \alpha^2 Z^2\right]^{1/2} \\ &= \left[n^2 - 2n_r(|\kappa| - \gamma)\right]^{1/2} \end{aligned} \tag{3.3.21}$$

where

$$n = n_r + |\kappa|$$

is the *principal quantum number*. It follows that, for bound states, $(n_r + \gamma)/N \leq 1$, so that $0 < E = c^2(n_r+\gamma)/N \leq c^2$. In terms of n, n_r and N, the *normalized* bound state solutions of the Dirac hydrogenic problem are therefore

$$P_{E\kappa}(r) = \mathcal{N}_{E\kappa}\,(c + E/c)^{1/2}\,\rho^\gamma e^{-\rho/2} \tag{3.3.22}$$
$$[-n_r M(-n_r+1, 2\gamma+1; \rho) + (N-\kappa)M((-n_r, 2\gamma+1; \rho)]$$
$$Q_{E\kappa}(r) = \mathcal{N}_{E\kappa}\,(c - E/c)^{1/2}\,\rho^\gamma e^{-\rho/2} \tag{3.3.23}$$
$$[-n_r M(-n_r+1, 2\gamma+1; \rho) - (N-\kappa)M((-n_r, 2\gamma+1; \rho)]$$

where

$$\mathcal{N}_{E\kappa} = \left\{ \frac{\alpha Z}{2N^2(N-\kappa)} \cdot \frac{\Gamma(2\gamma+n_r+1)}{n_r[\Gamma(2\gamma+1)]^2} \right\}^{1/2}.$$

3.3.3 Charge distributions and energy levels in hydrogenic atoms

The notion that qualitative differences between relativistic and nonrelativistic energies and charge distributions are significant for understanding where to expect physical differences in the predictions for atoms, molecules, and solids was introduced in §1.2. These differences are entirely due to the dynamics of the electron in hydrogenic atoms. The way in which electromagnetic interactions between electrons modify the picture in many-electron systems was reviewed in §1.3 and will be investigated more deeply later in the book.

Dirac and Schrödinger angular distributions, §3.2.5, are somewhat different. Most obviously, the zeros of the nonrelativistic angular density, which occur at angles at which the associated Legendre polynomials $P_l^{|m|}(\cos\theta)$ vanish, become minima in the relativistic case. Relativity smears the sharp edges of the distribution. Following [16], we look at relativistic effects on the radial probability distribution and on the energy levels. The corresponding relativistic charge densities are more compact and the relativistic bound energies are more negative than their nonrelativistic counterparts. The Schrödinger eigenvalues in hydrogenlike atoms depend only on the principal quantum number n; the Dirac eigenvalues depend on n and $|\kappa|$ (or j) giving rise to *fine structure*. When Z is *sub-critical* (that is, $Z < c\,|\kappa|$) and $\gamma = +\sqrt{\kappa^2 - Z^2/c^2}$,

$$|\kappa| - \gamma = \frac{Z^2}{2c^2|\kappa|} + O\left(Z^4/c^4\right) > 0. \tag{3.3.24}$$

from which we see that

$$N = \left[n^2 - 2n_r(|\kappa| - \gamma)\right]^{1/2} \leq n.$$

Thus, relative to the usual nonrelativistic zero of energy, $E = mc^2$,

$$\epsilon_{n\kappa} = c^2\left\{\sqrt{1 - \frac{Z^2}{N^2c^2}} - 1\right\} < c^2\left\{\sqrt{1 - \frac{Z^2}{n^2c^2}} - 1\right\} < -\frac{Z^2}{2n^2} = \epsilon_{nl}. \tag{3.3.25}$$

A similar argument using (3.3.24) shows that the eigenvalue $\epsilon_{n\kappa}$ increases monotonically as $|\kappa|$ increases. In nonrelativistic terms, the centrifugal barrier increases with increasing angular momentum, reducing the probability for an electron to penetrate the high potential region near the nucleus. Its maximum orbital speed will therefore be less and its binding energy will be closer to the nonrelativistic value.

The relativistic hydrogenic electron is always more tightly bound than its nonrelativistic counterpart, depending on the size of the *relativistic coupling constant* $Z/c = \alpha Z$. Clearly the difference vanishes in the mathematical *nonrelativistic* limit $c \to \infty$ in which light is effectively regarded as propagating with infinite speed.

Similar inequalities can be constructed for the radial moments of the charge distribution $\langle (2Zr)^s \rangle_{n\kappa}$. Table 3.2 displays the formulae derived by Garstang

Table 3.2. Radial moments $\langle (2Zr)^s \rangle_{n\kappa}$ of hydrogenic bound states

s	Nonrelativistic	Relativistic
2	$2n^2[5n^2+1-3l(l+1)]$	$2\left\{N^2(5N^2-2\kappa^2)\left(1-Z^2/N^2c^2\right) +N^2(1-\gamma^2)-3\kappa N^2\sqrt{1-Z^2/N^2c^2}\right\}$
1	$3n^2-l(l+1)$	$(3N^2-\kappa^2)\sqrt{1-Z^2/N^2c^2}-\kappa$
0	1	1
-1	$1/2n^2$	$n\gamma+(\lvert\kappa\rvert-\gamma)\lvert\kappa\rvert/2\gamma N^3$
-2	$\dfrac{1}{2n^3(2l+1)}$	$\dfrac{\kappa^2\sqrt{1-Z^2/N^2c^2}}{2\gamma^2N^3(2\gamma-\mathrm{sgn}\kappa)}$
-3	$\dfrac{1}{4n^3l(l+1)(2l+1)}$	$\dfrac{-3N^2\kappa\sqrt{1-Z^2/N^2c^2}+N^2+2\gamma^2\kappa^2}{4N^5(\gamma-1)\gamma(\gamma+1)(2\gamma-1)(2\gamma+1)}$

and Mayers [17] for $s = 1, 2$, by Burke and Grant [16] for $s = -3, -2, -1, 0$ and, using a different approach, by Kobus *et al.* [18] for $s = -2$. A comparison of relativistic and nonrelativistic density profiles for the hydrogenic mercury ion ($Z = 80$) will be found in Figs. 1.1, 1.2, and 1.3.

3.3.4 * The continuum solutions

The parametrization must be changed for continuum solutions. The conventional choices for the positive energy continuum, $E > c^2$ (in atomic units), are

$$\lambda \to -ip \quad \text{where} \quad p = +\sqrt{\frac{E^2}{c^2} - c^2} \tag{3.3.26}$$

$$N \to i\nu' \quad \text{where} \quad \nu' = \frac{Z}{p} \tag{3.3.27}$$

$$\frac{NE}{c^2} \to i\nu \quad \text{where} \quad \nu = \frac{ZE}{c^2 p} \tag{3.3.28}$$

where, as usual, $p = |\boldsymbol{p}|$ is the relativistic momentum, so that

$$a \to \gamma - i\nu, \quad b = 2\gamma + 1 \tag{3.3.29}$$

$$\rho \to -2ipr, \quad (c^2 - E)^{1/2} \to -i(E - c^2)^{1/2}, \tag{3.3.30}$$

With these substitutions, the solutions regular at the origin can be written

$$P_{E\kappa}(r) = \mathcal{N}\left(\frac{E+c^2}{E-c^2}\right)^{1/4} (-2ipr)^{\gamma}\, \mathrm{e}^{ipr} \tag{3.3.31}$$

$$\times \left\{ -\frac{\gamma - i\nu}{\kappa - i\nu'} M(\gamma + 1 - i\nu, 2\gamma + 1, -2ipr) + M(\gamma - i\nu, 2\gamma + 1, -2ipr) \right\}$$

$$Q_{E\kappa}(r) = -i\mathcal{N}\left(\frac{E-c^2}{E+c^2}\right)^{1/4} (-2ip\,r)^{\gamma}\,\mathrm{e}^{ip\,r} \tag{3.3.32}$$
$$\times\left\{-\frac{\gamma-i\nu}{\kappa-i\nu'}\,M(\gamma+1-i\nu,2\gamma+1,-2ip\,r) - M(\gamma-i\nu,2\gamma+1,-2ip\,r)\right\}$$

The asymptotic behaviour when $r \to \infty$ can be found using [5, 13.5.1]:

$$(-2ip\,r)^{\gamma}\,\mathrm{e}^{ip\,r} M(\gamma+1-i\nu,2\gamma+1,-2ip\,r)$$
$$\sim \frac{\Gamma(2\gamma+1)\mathrm{e}^{-\nu\pi/2}}{\Gamma(\gamma+1-i\nu)}\mathrm{e}^{-i(p\,r+\nu\ln 2p\,r)},$$

$$(-2ip\,r)^{\gamma}\,\mathrm{e}^{ip\,r} M(\gamma-i\nu,2\gamma+1,-2ip\,r)$$
$$\sim \frac{\Gamma(2\gamma+1)\mathrm{e}^{-\nu\pi/2}}{\Gamma(\gamma+1+i\nu)}\mathrm{e}^{i(p\,r+\nu\ln 2p\,r-\gamma\pi)}.$$

Introducing

$$\sigma_\kappa = \arg\,\Gamma(\gamma+i\nu)), \quad e^{2i\rho_\kappa} = \frac{-\kappa+i\nu'}{\gamma+i\nu}, \tag{3.3.33}$$

where ρ_κ is *real*, gives

$$f_{E\kappa}(r) \sim \mathcal{N}'\begin{pmatrix} \left(\dfrac{E+c^2}{E-c^2}\right)^{1/4} \cos\phi(r) \\ -\left(\dfrac{E-c^2}{E+c^2}\right)^{1/4} \sin\phi(r) \end{pmatrix} \tag{3.3.34}$$

where

$$\phi(r) = p\,r + \nu\ln 2p\,r - (l+1)\pi/2 + \delta_\kappa. \tag{3.3.35}$$

All factors independent of r have been absorbed into the normalizing constant $\mathcal{N}'_{E\kappa}$ and the *Coulomb phase shift* is given by

$$\delta_\kappa = \rho_\kappa - \sigma_\kappa - \pi\gamma/2 + (l+1)\pi/2 \tag{3.3.36}$$

in agreement with [19, page 79], where the definition (3.3.36) ensures $\delta_\kappa \to 0$ as $Z \to 0$. A second standing wave solution, which will not be regular at the origin, will be

$$g_{E\kappa}(r) \sim \mathcal{N}'\begin{pmatrix} \left(\dfrac{E+c^2}{E-c^2}\right)^{1/4} \sin\phi(r) \\ \left(\dfrac{E-c^2}{E+c^2}\right)^{1/4} \cos\phi(r) \end{pmatrix}. \tag{3.3.37}$$

Progressive spherical partial wave solutions can be constructed as

$$H^{\pm}_{E\kappa}(r) = f_{E\kappa}(r) \pm i g_{E\kappa}(r) \qquad (3.3.38)$$
$$\sim \mathcal{N}' \exp(\pm i\phi(r)) \begin{pmatrix} (E + c^2/E - c^2)^{1/4} \\ \pm i\,(E - c^2/E + c^2)^{1/4} \end{pmatrix}.$$

As in §3.2.7 the normalization

$$\mathcal{N}' = (2c)^{-1/2} \qquad (3.3.39)$$

ensures that these solutions give *unit outgoing/incoming particle flux* consistent with nonrelativistic conventions. For *delta function* normalization,

$$\int_0^\infty f^\dagger(r).f'(r)dr = \int_0^\infty g^\dagger(r).g'(r)dr = \delta(E - E').$$

where $f'(r)$ and $g'(r)$ denote functions with the parameters E, p replaced by E', p', the appropriate choice is

$$\mathcal{N}' = (\pi c)^{-1/2} \qquad (3.3.40)$$

3.4 Scattering by a centre of force

In this section, we discuss potential scattering of electrons from a fixed centre of force. Quantum mechanics and classical mechanics give the same cross section for Rutherford scattering [20], where the force field is coulombic. This serves as a good introduction to Mott's relativistic version [12].

3.4.1 Nonrelativistic potential scattering

We idealize the problem in terms of the motion of a particle in the field of a potential $V(r)$, where r is the distance of the particle from the origin. For the present, we exclude the Coulomb case by assuming that $rV(r) \to 0$ as $r \to \infty$. A particle with momentum p is incident along the z-axis in the positive direction, and we require the probability of scattering through an angle θ as it passes the centre of force. As shown in textbooks [19, Chapter II], [21], the wave function far from the centre will have asymptotic form

$$\psi \sim \mathrm{e}^{ipz} + r^{-1}\mathrm{e}^{ipr} f(\theta), \qquad (3.4.1)$$

in which the first term represents the incident beam and the second describes the outgoing scattered particles. The number crossing an area element dS at the point (r, θ, φ) when r is large is then $v|f(\theta)|^2\, dS/r^2$ where $v = \hbar p/m$ is the particle speed. The incident wave, e^{ipz}, gives a density of one particle per unit volume, and v electrons per unit area per unit time are therefore incident along the z-axis. Thus, if $d\Omega = dS/r^2$ is the element of solid angle, the number of particles scattered into $d\Omega$ per unit time is $I(\theta) = |f(\theta)|^2$, and the number of particles scattered between θ and $\theta + d\theta$ is

$$|f(\theta)|^2 \, 2\pi \sin\theta \, d\theta.$$

The incident plane wave is the solution of a free particle Schrödinger equation

$$(\nabla^2 + p^2)\psi = 0,$$

whilst the wave equation satisfied by the electron moving in the central force field is

$$\left(\nabla^2 + [p^2 - U(r)]\right)\psi = 0 \qquad (3.4.2)$$

where $U(r) = 2mV(r)$. This has axially symmetric solutions of the (unnormalized) form

$$\psi_l = r^{-1} F_l(r) P_l(\cos\theta)$$

where $P_l(\cos\theta)$ is a Legendre polynmial and $F_l(r)$ satisfies

$$\mathrm{d}\frac{d^2F}{dr^2} + \left(p^2 - U(r) - \frac{l(l+1)}{r^2}\right) F = 0;$$

we require the solution which is well-behaved at the origin, and discard the other solution.

Because (3.4.2) is a linear homogeneous equation, the most general axisymmetrical solution is of the form

$$\psi = \sum_{l=0}^{\infty} A_l \, r^{-1} F_l(r) \, P_l(\cos\theta) \qquad (3.4.3)$$

with arbitrary coefficients A_l. The incident plane wave can also be expanded in a similar fashion [5, 10.1.47]

$$\mathrm{e}^{ipz} = \sum_{l=0}^{\infty} (2l+1) \, i^l \, j_l(pr) \, P_l(\cos\theta), \qquad (3.4.4)$$

where $j_l(pr)$ is a spherical Bessel function [5, §10.1], with asymptotic form

$$j_l(pr) \sim r^{-1} \sin\left(pr - \frac{1}{2} l\pi\right), \quad r \to \infty. \qquad (3.4.5)$$

It is not difficult to prove that the asymptotic form of $F_l(r)$ is

$$F_l(r) \sim \sin\left(pr - \frac{1}{2} l\pi + \eta_l\right), \quad r \to \infty \qquad (3.4.6)$$

so that η_l vanishes in the absence of the potential $U(r)$. If (3.4.3) is to have the correct asymptotic form (3.4.1) then the difference $\psi - \exp(ipz)$ must contain only outgoing waves. Inserting (3.4.5) and (3.4.6), we find the converging waves can be eliminated if

$$A_l = (2l+1)\, i^l\, \mathrm{e}^{i\eta_l},$$

so that the required solution is

$$\psi = \sum_{l=0}^{\infty} (2l+1)\, i^l\, \mathrm{e}^{i\eta_l}\, r^{-1} F_l(r)\, P_l(\cos\theta), \tag{3.4.7}$$

from which the scattering amplitude is

$$f(\theta) = \frac{1}{2ip} \sum_{l=0}^{\infty} (2l+1)\, (\mathrm{e}^{2i\eta_l} - 1)\, P_l(\cos\theta). \tag{3.4.8}$$

The total cross-section, Q, is obtained by integrating over the angles:

$$Q = 2\pi \int_0^{\pi} |f(\theta)|^2 \sin\theta\, d\theta = \frac{4\pi}{p^2} \sum_{l=0}^{\infty} (2l+1)\, \sin^2 \eta_l.$$

Whilst there may be practical problems associated with generating and summing such infinite series, the *phase shifts*, η_l, completely determine the cross-section.

The solutions for a Coulomb have a slightly more complex asymptotic form than (3.4.6), but can be treated using the same methodology. The potential energy of two charged particles $Z_1 e$, $Z_2 e$ is $V(r) = Z_1 Z_2 / r$ in Hartree atomic units, so that the radial equation satisfied by $F_l(r)$ now becomes

$$\left\{ \frac{d^2}{dr^2} + p^2 - \frac{l(l+1)}{r^2} - \frac{2Z_1 Z_2}{r} \right\} F_l(r) = 0 \tag{3.4.9}$$

The solution which is regular at the origin can be expressed in terms of confluent hypergeometric functions [5, Chapter 13]

$$F_l(r) = \text{const.}\ \mathrm{e}^{ip\,r}\, (p\,r)^{l+1}\, M(l+1-i\nu, 2l+2; -2ip\,r), \tag{3.4.10}$$

where $\nu = -Z_1 Z_2 / p$, with the asymptotic form

$$F_l(r) \sim \text{const.}\ \frac{(2l+1)!\, \mathrm{e}^{-\frac{1}{2}\pi\nu + i\sigma_l}}{2^l\, \Gamma(l+1-i\nu)} \sin\left(p\,r - \frac{1}{2} l\pi + \nu \ln 2p\,r + \sigma_l \right)$$

where

$$\sigma_l = \arg \Gamma(l+1-i\nu).$$

It follows that the Coulomb scattering amplitude can be written down using the preceding work provided we normalize $F_l(r)$ so that

$$F_l(r) \sim \sin\left(p\,r + \nu \ln 2p\,r - \frac{1}{2} l\pi + \sigma_l\right), \qquad (3.4.11)$$

giving the normalized expression

$$F_l(r) = \frac{2^l\, \mathrm{e}^{\frac{1}{2}\pi\nu}\, |\Gamma(l+1-i\nu)|}{(2l+1)!}\, e^{ip\,r}\, (p\,r)^{l+1}\, M(l+1-i\nu, 2l+2; -2ip\,r). \qquad (3.4.12)$$

The Schrödinger equation for this problem can also be solved in parabolic coordinates (ξ, η, ϕ), where $\xi = r - z, \eta = r + z, \tan\phi = y/x$. Asymptotic analysis of the solution reveals that (3.4.1) is not quite right for Coulomb waves as terms involving $\ln p\,r$ appear in the exponents. The equivalent Dirac equation does not appear to be separable in parabolic coordinates: the interested reader should consult [19, 21] for further details.

3.4.2 ∗ Relativistic Coulomb scattering

The treatment of relativistic potential scattering due to Mott [19, Chapter IV] follows the nonrelativistic scheme outlined above. The advantages of working with 2-spinor or 4-spinor structures were not yet recognized in 1928, and Mott's equations treat each of the four components as an independent entity. Here we use a more modern notation.

We choose the same coordinate frame as in the nonrelativistic calculation, starting with the same unnormalized incident positive energy plane wave, proportional to e^{-ipz}, from (3.1.20) and (3.1.24):

$$\Psi^r(x) = \mathrm{e}^{-ip\cdot x} \begin{pmatrix} \phi^r \\ \dfrac{c\,\boldsymbol{\sigma}\cdot\boldsymbol{p}}{E+c^2}\,\phi^r \end{pmatrix}, \quad r = 1, 2 \qquad (3.4.13)$$

where

$$\phi^1 = \alpha = \begin{pmatrix} 1 \\ 0 \end{pmatrix}, \quad \phi^2 = \beta = \begin{pmatrix} 0 \\ 1 \end{pmatrix}, \qquad (3.4.14)$$

Because we are concerned with elastic scattering, we can drop the time dependence and choose $\boldsymbol{p} = (0, 0, p)$ as in the nonrelativistic calculation; the time-independent spinors are

$$\psi^1(\boldsymbol{x}) = \mathrm{e}^{ip\,z} \begin{pmatrix} \alpha \\ C(p)\,\alpha \end{pmatrix}, \quad \psi^2(\boldsymbol{x}) = \mathrm{e}^{ip\,z} \begin{pmatrix} \beta \\ -C(p)\,\beta \end{pmatrix} \qquad (3.4.15)$$

for the two spin states, where $C(p) = cp/(E+c^2) = \sqrt{(E-c^2)/(E+c^2)}$. The next step is to expand e^{ipz} as in (3.4.4), replacing the Legendre polynomials $P_l(\cos\theta)$ by the equivalent spherical harmonics

$$\mathrm{e}^{ip\,z} = \sum_{l=0}^{\infty} [4\pi(2l+1)]^{1/2}\, i^l\, j_l(p\,r)\, Y_{l,0}(\theta, \varphi)$$

In order to make contact with our spherical wave conventions, we now need to express the uncoupled states in terms of the $\chi_{\kappa m}(\theta,\varphi)$, which we do by inverting (A.4.4), so that (omitting arguments θ,φ)

$$Y_{l,0}\otimes\alpha = \frac{1}{\sqrt{2l+1}}\left(\sqrt{l}\,\chi_{l,1/2} - \sqrt{l+1}\,\chi_{-l-1,1/2}\right)$$

and

$$Y_{l,0}\otimes\beta = \frac{1}{\sqrt{2l+1}}\left(\sqrt{l}\,\chi_{l,-1/2} + \sqrt{l+1}\,\chi_{-l-1,-1/2}\right)$$

Thus

$$\psi^1(\boldsymbol{x}) = \sqrt{4\pi}\sum_{l=0}^{\infty} i^l\, j_l(pr) \times \left(\sqrt{l}\,\chi_{l,1/2} - \sqrt{l+1}\,\chi_{-l-1,1/2}\right)\otimes\begin{pmatrix}1\\ C(p)\end{pmatrix}, \tag{3.4.16}$$

when the helicity is along the direction of motion and

$$\psi^2(\boldsymbol{x}) = \sqrt{4\pi}\sum_{l=0}^{\infty} i^l\, j_l(pr) \times \left(\sqrt{l}\,\chi_{l,-1/2} + \sqrt{l+1}\,\chi_{-l-1,-1/2}\right)\otimes\begin{pmatrix}1\\ C(p)\end{pmatrix}, \tag{3.4.17}$$

when it is anti-parallel. In both cases, the lower pair of components are constant multiples of the upper components. Equations (3.4.16) and (3.4.17) are linear combinations of the (unnormalized) spherical wave solutions (3.2.26)

$$\psi_{\kappa m}(\boldsymbol{x}) = \begin{pmatrix} j_l(pr)\chi_{\kappa m} \\ i\eta C(p) j_{\bar{l}}(pr)\chi_{-\kappa m}\end{pmatrix}$$

where $\eta = \operatorname{sgn}\kappa$, $l = j+\eta/2$, $\bar{l} = j-\eta/2 = l-\eta$ and $m = \pm 1/2$. With the notation

$$\widetilde{\chi}_{0,m} = 0, \quad \widetilde{\chi}_{\kappa,m}(\theta,\varphi) = \sqrt{4\pi k}\chi_{\kappa,m}(\theta,\varphi), \quad k = |\kappa| > 0, \tag{3.4.18}$$

we can write

$$\widetilde{\psi}_{+k,1/2} = \begin{pmatrix} j_k(pr)\widetilde{\chi}_{k,1/2} \\ iC(p) j_{k-1}(pr)\widetilde{\chi}_{-k,1/2}\end{pmatrix}, \tag{3.4.19}$$

$$\widetilde{\psi}_{-k,1/2} = \begin{pmatrix} j_{k-1}(pr)\widetilde{\chi}_{-k,1/2} \\ -iC(p) j_{k}(pr)\widetilde{\chi}_{k,1/2}\end{pmatrix}, \tag{3.4.20}$$

so that

$$\psi^1(\boldsymbol{x}) = \sum_{k=1}^{\infty}\left\{ i^k\widetilde{\psi}_{+k,1/2} - i^{k-1}\widetilde{\psi}_{-k,1/2}\right\} \tag{3.4.21}$$

and

$$\psi^2(\boldsymbol{x}) = \sum_{k=1}^{\infty} \left\{ i^k \widetilde{\psi}_{+k,-1/2} + i^{k-1} \widetilde{\psi}_{-k,-1/2} \right\}. \tag{3.4.22}$$

Similarly, the regular Coulomb solutions $\psi^C_{\kappa m}(\boldsymbol{x})$ have asymptotic amplitudes (3.3.34) and (3.3.37)

$$\widetilde{\psi}^C_{k,1/2}(\boldsymbol{x}) \sim \frac{1}{pr} \begin{pmatrix} \sin\left(pr + \nu \ln 2pr - \frac{1}{2}k\pi + \delta_k\right) \widetilde{\chi}_{k,1/2} \\ iC(p) \cos\left(pr + \nu \ln 2pr - \frac{1}{2}k\pi + \delta_k\right) \widetilde{\chi}_{-k,1/2} \end{pmatrix} \tag{3.4.23}$$

$$\widetilde{\psi}^C_{-k,1/2}(\boldsymbol{x}) \sim \frac{1}{pr} \begin{pmatrix} \cos(pr + \nu \ln 2pr - \frac{1}{2}k\pi + \delta_{-k}) \widetilde{\chi}_{-k,1/2} \\ -iC(p) \sin(pr + \nu \ln 2pr - \frac{1}{2}k\pi + \delta_{-k}) \widetilde{\chi}_{k,1/2} \end{pmatrix}.$$

so that the Coulomb scattering wavefunction is

$$\Psi \sim \sum_{k=1}^{\infty} \left\{ i^k A_k \widetilde{\psi}^C_{k,1/2} - i^{k-1} A_{-k} \widetilde{\psi}^C_{-k,1/2} \right\}. \tag{3.4.24}$$

The linear independence of the functions $\chi_{\kappa,m}$, allows us to choose the coefficients $A_{\pm k} = \mathrm{e}^{i\delta_{\pm k}}$ so that Ψ has an asymptotic form that is the sum of an incident plane wave propagating forward along Oz and outgoing spherical waves as in the nonrelativistic case. The asymptotic scattered wave function is

$$\Psi_{scatt} \sim \frac{\mathrm{e}^{ipr}}{r} \begin{pmatrix} F(\theta, \varphi) \\ C(p) F(\theta, \varphi) \end{pmatrix} \tag{3.4.25}$$

where

$$F(\theta, \varphi) = \frac{1}{2ip} \sum_{k=1}^{\infty} \left\{ \left(\mathrm{e}^{2i\delta_k} - 1\right) \widetilde{\chi}_{k,1/2}(\theta, \varphi) - \left(\mathrm{e}^{2i\delta_{-k}} - 1\right) \widetilde{\chi}_{-k,1/2}(\theta, \varphi) \right\}$$

substituting

$$\widetilde{\chi}_{k,1/2}(\theta, \varphi) = \begin{pmatrix} -kP_k(\cos\theta) \\ -P^1_k(\cos\theta) e^{i\varphi} \end{pmatrix}, \quad \widetilde{\chi}_{-k,1/2}(\theta, \varphi) = \begin{pmatrix} kP_{k-1}(\cos\theta) \\ -P^1_{k-1}(\cos\theta) e^{i\varphi}, \end{pmatrix},$$

and rearranging gives the well-known result

$$F(\theta, \varphi) = \begin{pmatrix} f(\theta) \\ g(\theta) \mathrm{e}^{i\varphi} \end{pmatrix} \tag{3.4.26}$$

where

$$f(\theta) = \frac{1}{2ip} \sum_{k=0}^{\infty} \left\{ (k+1) \left(\mathrm{e}^{2i\delta_{-k}} - 1\right) + k \left(\mathrm{e}^{2i\delta_k} - 1\right) \right\} P_k(\cos\theta),$$

$$g(\theta) = \frac{1}{2ip} \sum_{k=1}^{\infty} \left(\mathrm{e}^{2i\delta_k} - \mathrm{e}^{2i\delta_{-k-1}}\right) P_k^1(\cos\theta)$$

The differential cross-section is the ratio of the number of particles scattered into a solid angle element $d\omega$ in unit time to the number of particles in the incident beam crossing unit area in unit time. Because the (particle) current density vector is defined by $\boldsymbol{j} = \psi^\dagger c\boldsymbol{\alpha}\psi$, the contribution dN to the particle flux crossing an area element $\boldsymbol{n}dS$, where $|\boldsymbol{n}| = 1$, at the point $\boldsymbol{r}$ is

$$dN = \boldsymbol{j} \cdot \boldsymbol{n}\, dS = \psi^\dagger (c\,\boldsymbol{\alpha} \cdot \boldsymbol{n})\, \psi\, dS \tag{3.4.27}$$

For the incident plane wave $\psi^1(\boldsymbol{x})$ this gives the flux crossing unit area normal to the beam, $\boldsymbol{n} = \boldsymbol{e}_3$, is

$$N_{inc} = \psi^1(\boldsymbol{x})^\dagger (c\,\boldsymbol{\alpha} \cdot \boldsymbol{e}_3)\, \psi^1(\boldsymbol{x}) = 2c\, C(p) = \frac{2c^2 p}{E + c^2} \tag{3.4.28}$$

$N_{inc} \to p$ in the limit $E \to c^2$ in agreement with the nonrelativistic result $N_{inc}^{nr} = v = p/m$ because $m = 1$ for electrons The number of scattered particles crossing an area element $dS = r^2\, d\omega$, with normal $\boldsymbol{e}_r$ at $\boldsymbol{r}$ in unit time is

$$dN_{scatt} = \Psi^\dagger_{scatt}\, (c\,\boldsymbol{\alpha} \cdot \boldsymbol{e}_r)\, \Psi_{scatt} r^2\, d\omega. \tag{3.4.29}$$

To evaluate this we use (A.4.10), giving

$$\boldsymbol{\alpha} \cdot \boldsymbol{e}_r \Psi_{scatt} \sim \frac{e^{ipr}}{r} \begin{pmatrix} C(p)F(\theta,\varphi) \\ F(\theta,\varphi) \end{pmatrix}, \tag{3.4.30}$$

from which

$$dN_{scatt} = 2c\, C(p)\, F(\theta,\varphi)^\dagger F(\theta,\varphi) d\omega$$

so that the differential cross-section for relativistic Coulomb scattering with incident spin parallel to the motion is

$$\frac{d\sigma}{d\omega} = F(\theta,\varphi)^\dagger F(\theta,\varphi) = |f(\theta)|^2 + |g(\theta)|^2; \tag{3.4.31}$$

clearly this is independent of the helicity of the incident beam.

3.4.3 * Polarization effects in Coulomb scattering

By definition, the spin of the particle is defined as the total angular momentum in the particle's rest frame. The free particle spin projector $P(n)$, (3.1.32), for the direction $n = (0, \boldsymbol{n})$, $|\boldsymbol{n}| = 1$, becomes

$$P(\boldsymbol{n}) = \frac{1}{2} \begin{pmatrix} I + \boldsymbol{\sigma} \cdot \boldsymbol{n} & 0 \\ 0 & I - \boldsymbol{\sigma} \cdot \boldsymbol{n} \end{pmatrix} \tag{3.4.32}$$

For spin oriented along $\boldsymbol{n} = (\sin\alpha\cos\beta, \sin\alpha\sin\beta, \cos\alpha)$,

$$\boldsymbol{\sigma}\cdot\boldsymbol{n} = \begin{pmatrix} (1+\cos\alpha)/2 & \sin\alpha\,\mathrm{e}^{-i\beta}/2 \\ \sin\alpha\,\mathrm{e}^{i\beta}/2 & (1-\cos\alpha)/2 \end{pmatrix}$$

This has the real eigenvalues $\lambda = \pm 1$ with eigenvectors s_λ given by

$$s_1 = \begin{pmatrix} \cos\dfrac{\alpha}{2} \\ \sin\dfrac{\alpha}{2}\,\mathrm{e}^{i\beta} \end{pmatrix}, \quad s_{-1} = \begin{pmatrix} \sin\dfrac{\alpha}{2} \\ -\cos\dfrac{\alpha}{2}\,\mathrm{e}^{i\beta} \end{pmatrix}$$

each of which can be regarded as a coherent linear superposition

$$s_\lambda = A\begin{pmatrix}1\\0\end{pmatrix} + B\begin{pmatrix}0\\1\end{pmatrix}$$

of the spin states parallel and antiparallel to the original direction of quantization, Oz. The generalization of (3.4.25) for general spin orientation (α,β) becomes

$$\Psi_{scatt} \sim \frac{e^{ip\,r}}{2ip\,r}\begin{pmatrix} F'(\theta,\varphi) \\ iC(p)F'(\theta,\varphi) \end{pmatrix}, \quad F'(\theta,\varphi) = \begin{pmatrix} Af(\theta) - Bg(\theta)\mathrm{e}^{-i\varphi} \\ Bf(\theta) + Ag(\theta)\mathrm{e}^{i\varphi} \end{pmatrix},$$

with differential cross-section

$$\frac{d\sigma}{d\omega} = F'(\theta,\varphi)^\dagger F'(\theta,\varphi) = \left(|f(\theta)|^2 + |g(\theta)|^2\right)\{1 + S(\theta)\,h(\varphi)\}. \qquad (3.4.33)$$

The real function

$$S(\theta) = i\,\frac{f(\theta)\,g^*(\theta) - f^*(\theta)\,g(\theta)}{|f(\theta)|^2 + |g(\theta)|^2} \qquad (3.4.34)$$

is known as the Sherman function [22], and

$$h(\varphi) = i\,\frac{AB^*e^{i\varphi} - A^*Be^{-i\varphi}}{|A|^2 + |B|^2}$$

describes the departure from axial symmetry. When the incident spin is in a pure state with direction given by polar angles (α,β), we find a two-lobed angular dependence,

$$h(\varphi) = \pm\frac{1}{2}\sin\alpha\,\sin(\varphi - \beta),$$

with $+$ sign for the parallel and $-$ sign for the antiparallel case.

Further study of electron polarization processes would take us too far afield. The monograph [23] provides an introduction to the physics of polarized electrons and to relevant experimental techniques of this still developing field.

3.4.4 Historical note

The derivation by Mott [12] of the differential cross-section for Coulomb scattering was one of the early applications of Darwin's calculation of Coulomb wavefunctions in 1928 [10, 19]. This work has been summarized in books such as [24, §15] and has been a key feature of the interpretation of many experiments. However, it was written at an early stage before the development of the machinery for spherical Dirac spinors, and the notation used by different authors is confusing. This is particularly the case with Coulomb phase shifts, which, in this book, we label with the quantum number κ. Thus where we write δ_k and δ_{-k-1}, Mott writes respectively η_{-k-1} and η_k. Walker [25], who was one of the first to publish on relativistic effects on low-energy scattering from non-hydrogenic atoms, quotes Mott's formula using δ_k^- and δ_k^+.

Here we have used the interpretation of $c\boldsymbol{\alpha}$ as the Dirac current operator to re-derive Mott's formula. Mott argued that "asymptotically the scattered wave may be regarded as made up of a number of plane waves proceeding outwards from the centre in different directions". This argument allowed him to adopt the nonrelativistic definition for both the incident and the scattered particle current as the product of the particle density, $\|\psi\|^2$, and the speed v.. As (3.4.28) shows, this is only an approximation in Dirac theory. It gives the correct cross-section because the upper and lower 2-spinors making up the Dirac incident plane wave (3.4.13) differ only by a real multiplicative factor $C(p)$, as does Ψ_{scatt} in the asymptotic region, and because the leading terms in the asymptotic expansions of Ψ_{scatt} and $\alpha_r \Psi_{scatt}$ are the same.

Some light can be shed on this fortuitous agreement by Gordon's decomposition of the Dirac current [26], which divides the total current density into two parts

$$j^\mu = j_1^\mu + j_2^\mu,$$

a *convection current*

$$j_1^\mu = -\frac{i\hbar}{2m}\left(\tilde{\psi}.\partial^\mu\psi - \partial^\mu\tilde{\psi}.\psi\right) - \frac{q}{m}A^\mu\tilde{\psi}.\psi$$

similar to the nonrelativistic expression invoked by Mott and others in their treatment of Coulomb scattering, and a *spin current*

$$j_2^\mu = \frac{q}{2m}\partial_\nu\tilde{\psi}\sigma^{\mu\nu}\psi.$$

This can be proved by writing

$$j^\mu = \frac{1}{2}c\left(\tilde{\psi}\gamma^\mu\psi + \tilde{\psi}\gamma^\mu\psi\right)$$

and using Dirac's equation to substitute $\psi = \gamma^\nu(i\partial_\nu - qA_\nu)\psi$ for ψ in the first term and $\tilde{\psi}$ by its adjoint in the second, together with the formula $\gamma^\mu\gamma^\nu = g^{\mu\nu} - i\sigma^{\mu\nu}$ from Appendix A.2. In the absence of magnetic fields, the radial convection current reduces to $p_r/m.\tilde{\psi}\psi$, which reproduces the nonrelativistic value for *positive energy* states. Also the particle density for a positive energy particle is $[E/mc^2 - (q/2mc).\phi]\tilde{\psi}\psi$, which only reduces to the density $\tilde{\psi}\psi$ in the weak field limit with $E/mc^2 \to 1$. There is no nonrelativistic counterpart of the spin current in this picture.

3.5 * Relativistic quantum defect theory

The concept of a quantum defect seems to have appeared early in the history of atomic spectroscopy when Rydberg found in 1889 that the term values of series of lines in the spectrum of alkali atoms could be fitted to a formula

$$T_n = T_\infty - \frac{RZ^2}{(n-\mu)^2}, \tag{3.5.1}$$

where R is Rydberg's constant, Z is atomic number, and μ, which is nearly independent of the integer n, is called the *quantum defect.* Modern *quantum defect theory* (QDT) provides a framework for exploiting this notion in a variety of applications capable, for example, of extracting values of T_∞ to spectroscopic accuracy, and hence determining accurate ionization potentials. Quantum defects can be used to aid spectral line identifications and to locate unobserved members of a spectral series. But most importantly, a knowledge of quantum defects summarizes a great deal of information on a Rydberg series concisely and can be used to study series perturbations, autoionisation, photoionisation, and resonance structures in electron-atom collisions. A comprehensive review by Seaton [27] deals with the foundations of QDT, mainly from a nonrelativistic viewpoint, and sets out the formulae needed for most of these applications. Connerade [28] makes extensive use of QDT in his survey of the physics of highly excited atoms. Single channel QDT, focusing on one Rydberg series, must be supplemented by multi-channel QDT to deal with more complicated situations in many-electron systems where there is coupling between one or more spectral series of bound states as well as continuum states.

A relativistic version of QDT developed by Johnson and Cheng [14, 29], is needed for the study of spectra of highly stripped atomic ions with pronounced relativistic fine structure occurring in laboratory and astrophysical plasmas. It is appropriate to study the single channel RQDT at this point as a natural extension of the theory of Coulomb wavefunctions presented in previous sections. We begin by replacing the effective principal quantum number N of the hydrogenic atom by an *effective principal quantum number*

$$N^* = \sqrt{(n_r^* + \gamma)^2 + \alpha^2 Z^2}, \quad n_r^* = n_r - \mu_{n\kappa} \tag{3.5.2}$$

so that the relativistic term values are given by

$$T_{n\kappa} = T_{\infty\kappa} - Rc^2 \left[\left(1 - \frac{Z^2}{N^{*2}c^2} \right)^{1/2} - 1 \right]. \tag{3.5.3}$$

As usual, $n = n_r + |\kappa|$. Table 3.3 lists typical results from fitting (3.5.1) to the observed term values from [14] to give nonrelativistic quantum defects μ_{NR} and from fitting (3.5.3) to give the equivalent relativistic quantum defects μ_R. The two columns of nonrelativistic defects μ_{NR} are somewhat different, because they must account for the difference in relativistic dynamics as well as the non-Coulomb part of the atomic potential. μ_R has to account only for the latter so that the last two columns are essentially identical.

Table 3.3. Relativistic (μ_R) and nonrelativistic (μ_{NR}) quantum defects for $nd_{3/2}$, $nd_{5/2}$ series in the observed spectra of C IV and N V. Reprinted , with permission, from [30]

		μ_{NR}		μ_R	
Ion	n	$nd_{3/2}$	$nd_{5/2}$	$nd_{3/2}$	$nd_{5/2}$
C IV	3	0.001614	0.001533	0.001507	0.001498
	4	0.001912	0.001839	0.001779	0.001777
	5	0.002075	0.002015	0.001926	0.001936
	6	0.002211	0.002137	0.002051	0.002049
	7	0.002520	0.002461	0.002353	0.002365
N V	3	0.001490	0.001380	0.001324	0.001324
	4	0.001778	0.001668	0.001570	0.001571
	5	0.001876	0.001765	0.001643	0.001643
	6	0.001879	0.001768	0.001629	0.001630
	7	0.001817	0.001705	0.001556	0.001554
	8	0.001704	0.001592	0.001434	0.001433

RQDT aims to express the quantum defect below threshold, $E < mc^2$, as a holomorphic function of energy, $\mu_\kappa(E)$, in the complex E-plane, and to relate this to the short range non-Coulomb scattering phaseshift $\delta(E)$ for $E > mc^2$. Below threshold, $\mu_\kappa(E)$ takes the value $\mu_{n\kappa}$ when $E = E_{n\kappa}$. We shall only be concerned with one symmetry, κ, in this section, so we can omit the κ label for brevity from now on. As in §3.3.1 we use the notation $u(r) : \mathbb{R}_+ \to \mathbb{C} \times \mathbb{C}$ to denote the two-component radial function

$$u(r) = \begin{pmatrix} P(r) \\ Q(r) \end{pmatrix}.$$

We select two such linearly independent solutions of the Dirac Coulomb equation, u_R and u_I, which are entire functions[6] of the energy parameter E such that

$$u_1 = c_1 \, u_R, \quad u_2 = c_2 \, u_I$$

where

[6] A function $F(E, z)$ is said to be an entire function of E if, for all finite values of E, the power series $\sum_{n=0}^{\infty} E^n f_n(z)$ converges uniformly and absolutely to $F(E, z)$ for all values of z.

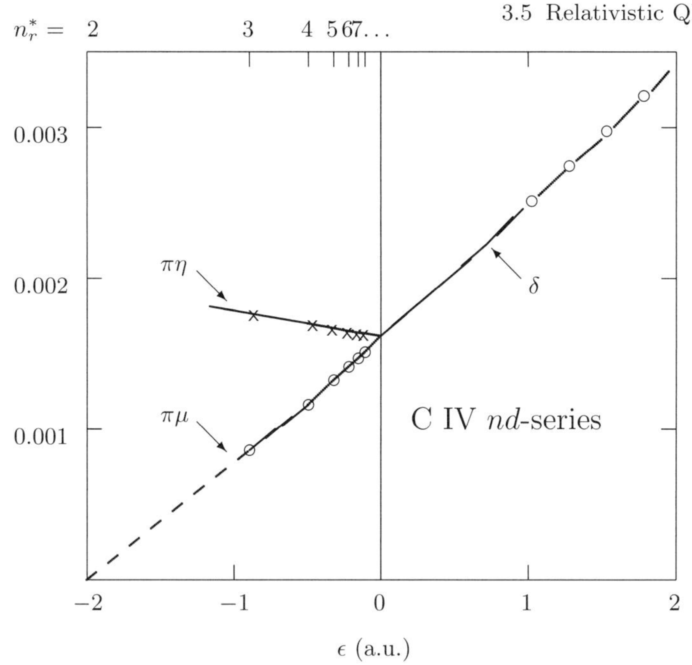

Fig. 3.3. Model potential calculations of quantum defects as a function of the term energy. The quantum defect functions $\delta(E)$, $\mu(E)$, and $\eta(E)$ are defined in the text. Reproduced, with permission, from [14].

$$c_1 = \left(\frac{E}{c} + c\right)^{1/2} (2\lambda)^\gamma(-\kappa + \gamma + N - NE/c^2),$$

$$c_2 = \left(\frac{E}{c} + c\right)^{1/2} (2\lambda)^\gamma(-\kappa - \gamma + N - NE/c^2).$$

These functions enable us to relate the solutions at threshold, $E = c^2$. Johnson and Cheng show that there are coefficients d_1, e_1, e_2 such that the two independent solutions $f(r)$, (3.3.34), and $g(r)$, (3.3.37), above threshold with delta function normalization can be written

$$f(r) = d_1\, u_R, \quad g(r) = e_1\, u_R + e_2\, u_I$$

where

$$d_1 = c_1/N_1, \quad N_1 = \frac{2p^{1/2}\Gamma(b)}{|\Gamma(\gamma + i\nu)|} \exp(-\pi\nu/2 - i\pi\gamma/2 + i\sigma_\kappa)$$

$$e_1 = -(\csc \pi b\, \mathrm{e}^{-2\pi\nu} + \cot \pi b)\, d_1, \quad e_2 = \frac{i\, \mathrm{e}^{-i\pi a}\, \Gamma^2(b)}{\gamma(\gamma + i\nu)|\Gamma(\gamma + i\nu)|^2} \frac{c_2}{N_1}$$

The coefficients d_1, e_1, e_2 are real and continuous at threshold.

The motion of a Rydberg electron in an atomic ion will asymptotically approach that of an electron in a Coulomb field due to the residual charge Z on the ion, perturbed by potentials which decay faster at infinity than $1/r$. The asymptotic wavefunction will therefore have the form

$$u(r) \sim a(E)\, u_R(r) + b(E)\, u_I(r)$$

in which the function

$$\beta(E) = -b(E)/a(E)$$

is a holomorphic function of E [14, §2.2]. Below threshold, we can express $u(r)$ as a linear combination of $u_1(r)$ (3.3.15) and $u_5(r)$ (3.3.17), where $u_1 \sim \mathrm{e}^{\lambda r}$ and $u_5 \sim \mathrm{e}^{-\lambda r}$ as $r \to \infty$,

$$u \sim \left(1 - \frac{c_1}{c_2} \frac{\Gamma(a)\Gamma(2-b)}{\Gamma(b)\Gamma(1+a-b)} \beta(E)\right) u_1 - \frac{c_1}{c_2} \frac{\Gamma(a)}{\Gamma(b-1)} u_5,$$

so that the coefficient of u_1 must vanish when $E \to E_n$, a bound state eigenvalue, enabling us to identify

$$\beta(E_n) = \frac{c_2}{c_1} \frac{\Gamma(b)\Gamma(1+a_n-b)}{\Gamma(a_n)\Gamma(2-b)}$$

where $a_n = -n_r^* = -(n_r - \mu_n)$, where μ_n is the quantum defect, and $b = 2\gamma + 1$. By using the relation

$$\Gamma(z)\Gamma(1-z) = \pi \csc \pi z$$

we can rewrite this as

$$\beta(E_n) = -\frac{c_2}{c_1} \frac{\Gamma^2(b)\, \Gamma(1+z_n-\gamma)}{2\pi\gamma\, \Gamma(1+z_n+\gamma)} \left(\cot \pi\mu_n - \cot \pi b\right)^{-1}$$

where $z_n = ZE_n/c^2\lambda_n$, $\lambda_n = \sqrt{c^2 - E_n^2/c^2}$. The Rydberg eigenvalues E_n form an increasing countable set converging to $E_n = mc^2$. We can therefore construct $\beta(E)$, $E \in \mathbb{C}$, as a unique holomorphic function from its values, $\beta(E_n)$, on this set. We can similarly define a quantum defect function $\mu(E)$ from its values $\mu(E_n)$ which is properly defined at energies above threshold. Set $z = ZE/c^2\lambda$, $z' = Z/\lambda$ below threshold, so that z, z' are both infinite at threshold. Because $\Gamma(z+\alpha)/\Gamma(z+\beta) \sim z^{\alpha-\beta}(1 + O(z^{-1}))$ as $|z| \to \infty$, we can define a new function

$$\mathcal{B}(z, \gamma) = \frac{\Gamma(z+\gamma)}{z^{2\gamma-1}\Gamma(1+z-\gamma)}$$

which, along with

$$R(z, z') = \left(\frac{\gamma + z}{z}\right) \left(\frac{-\kappa + \gamma + z' - z}{-\kappa - \gamma + z' - z}\right),$$

is finite and continuous at threshold. This means that we can take

$$\beta(E) = -\frac{\Gamma^2(b)\left(\cot\pi\mu(E) - \cot\pi b\right)^{-1}}{2\pi\gamma\,(2ZE/c^2)^{2\gamma}R(z,z')\mathcal{B}(z,\gamma)} \tag{3.5.4}$$

Note that $\mathcal{B}(z,\gamma)$ is real below threshold. Above threshold, we have to make the replacements $z \to i\nu$, $z' \to i\nu'$, where $\nu = ZE/c^2p$, $\nu' = Z/p$, so that its real and imaginary parts are

$$B(i\nu,\gamma) := \Re\mathcal{B}(i\nu,\gamma) = (1 + \cos\pi b\,\mathrm{e}^{-2\pi\nu})A(\nu,\gamma)$$

and

$$C(i\nu,\gamma) := \Im\mathcal{B}(i\nu,\gamma) = -\sin\pi b\,\mathrm{e}^{-2\pi\nu}A(\nu,\gamma),$$

with

$$A(\nu,\gamma) = |\Gamma(\gamma + i\nu)|^2\mathrm{e}^{\pi\nu}/2\pi\nu^{2\gamma-1}.$$

Above threshold, the asymptotic form is a linear combination of (3.3.34) and (3.3.37)

$$u(r) \sim \cos\delta(E)\,f(r) + \sin\delta(E)\,g(r)$$

where $\delta(E)$ is the phase shift due to short range non-Coulomb forces. Then above threshold,

$$\beta(E) = -\frac{\Gamma^2(b)\left(\cot\delta(E) - \csc\pi b\,\mathrm{e}^{-2\pi\nu} - \cot\pi b\right)^{-1}}{2\pi\gamma\,(2ZE/c^2)^{2\gamma}R(i\nu,i\nu')B(i\nu,\gamma)}$$

from which the relation between the quantum defect and the phase shift is given by

$$\cot\delta(E) = (1 + \cos\pi b\,\mathrm{e}^{-2\pi\nu})\cot\pi\mu(E) + \sin\pi b\,\mathrm{e}^{-2\pi\nu}. \tag{3.5.5}$$

In the nonrelativistic limit, $b \to 2|\kappa| + 1$ becomes an odd integer, so that (3.5.5) reduces to the formula

$$\cot\delta_{nr}(E) = (1 - \mathrm{e}^{-2\pi\nu})\cot\pi\mu_{nr}(E) \tag{3.5.6}$$

given by Seaton [31]. At threshold, $\nu \to \infty$ and $\cot\delta(E) = \cot\pi\mu(E)$, $E \to c^2$, and it is customary to choose the solution for which $\delta = \pi\mu$ at threshold in both relativistic and nonrelativistic versions of the theory.

There is an alternative form of quantum defect which is usually more slowly varying than $\mu(E)$, and hence is better suited for numerical interpolation and extrapolation in applications. This can be obtained by rewriting (3.5.4) in the form

$$\beta(E) = -\frac{\Gamma^2(b)}{2\pi\gamma\,(2Z)^{2\gamma}}\left(\frac{\kappa-\gamma}{\kappa+\gamma}\right)\frac{1}{\mathcal{A}(z,\kappa)\left(\cot\pi\mu(E) - \cot\pi b\right)}$$

where

$$\mathcal{A}(z,\kappa) = \frac{(z+z'+\kappa+\gamma)(z+z'-\kappa-\gamma)}{2z(z+z')}\left(\frac{z}{z'}\right)^{2\gamma} B(z,\gamma).$$

isolates the energy dependent terms. $\mathcal{A}(z,\kappa)$ vanishes when z takes any non-negative values $\gamma-1, \gamma-2, \ldots$ for $\kappa = -l-1$, or $\gamma, \gamma-1, \ldots$ when $\kappa = l$. The only way for $\beta(E)$ to remain holomorphic is if $\mu(E)$ tends to an integer at these energies. This constraint also appears in the nonrelativistic theory. By defining a new quantum defect variable $\eta(E)$ such that

$$\cot \pi\eta = \mathcal{A}(z,\kappa) \cot \pi\mu,$$

remains finite at the critical energies, we see that $\eta(E)$ has no integer values and is therefore likely to vary more slowly as a function of E as claimed. Moreover, $\mathcal{A}(z,\kappa) \to 1$ at threshold where η and μ agree. A typical model calculation for the relativistic $nd_{3/2,5/2}$ series in the C IV ion is shown in Figure 3.3.

3.6 Green's functions

In this section, we extend the construction of the Green's function for the free Dirac electron, §2.9.4, to the case of an electron in a spherically symmetric potential such as the electron-nucleus Coulomb interaction. The equations are no longer translationally invariant, so that the Green's function is a function of both x and x' separately rather than of $x - x'$. Thus (2.9.44) must be replaced by

$$S_F(x,x') = \frac{i}{\pi} \int_{C_F} dz\, G(\boldsymbol{x},\boldsymbol{x}';z)\gamma_0 \exp(-iz(x-x')^0/c). \tag{3.6.1}$$

where the resolvent kernel $G(\boldsymbol{x},\boldsymbol{x}';z)$ satisfies

$$(H-z)\, G(\boldsymbol{x},\boldsymbol{x}';z) = c\delta^{(3)}(\boldsymbol{x}-\boldsymbol{x}') \tag{3.6.2}$$

and, (3.2.15),

$$H := c\boldsymbol{\alpha}\cdot\boldsymbol{p} + \beta mc^2 + V(r) \tag{3.6.3}$$

with $V(r) = -Ze^2/4\pi\epsilon_0 r$ in the case of a hydrogenic ion with a point nuclear charge Ze; we shall normally use atomic (Hartree units) in which $e^2/4\pi\epsilon_0 = 1.0$. As H and the angular operator K commute, the Green's function can be written as a sum of independent radial reduced Green's functions

$$G(\boldsymbol{x},\boldsymbol{x}';z) = \sum_{\kappa} G_{\kappa}(\boldsymbol{x},\boldsymbol{x}';z). \tag{3.6.4}$$

3.6.1 * Partial wave Green's functions

We first construct the kernel for the inhomogeneous radial equation

$$(H_\kappa - z)\, u(r;z) = v(r) \tag{3.6.5}$$

where, from (3.2.15),

$$H_\kappa = \begin{pmatrix} V(r) + c^2 & c\left(-\dfrac{d}{dr} + \dfrac{\kappa}{r}\right) \\ c\left(\dfrac{d}{dr} + \dfrac{\kappa}{r}\right) & V(r) - c^2 \end{pmatrix}. \tag{3.6.6}$$

We wish to determine

$$u(r;z) = \begin{pmatrix} P(r;z) \\ Q(r;z) \end{pmatrix} \tag{3.6.7}$$

on $0 < r < \infty$ subject to suitable boundary conditions. In terms of the reduced Green's function we write the solution as

$$u(r;z) = w(r;z) + \int_0^\infty G_\kappa(r,s;z)\, v(s)ds \tag{3.6.8}$$

where $w(r;z)$ is a solution of the homogeneous equation

$$(H_\kappa - z)\, w(r;z) = 0.$$

When $V(r) = -Z/r$, the spectrum of H_κ consists of an infinite point set of bound eigenvalues in $(0, c^2)$ with an accumulation point at c^2, and continuous eigenvalues on the line segments $(-\infty, -c^2)$ and (c^2, ∞). Most of the potentials we shall encounter in atomic and molecular calculations have this kind of spectrum. We assume that z belongs to the *resolvent set*, consisting of the whole of the *complex plane* cut from $-\infty$ to $-mc^2$ and from mc^2 to ∞ with the point eigenvalues removed. We seek a solution that satisfies

1. $u(r;z)$ is bounded as $r \to 0$;
2. $u(r;z) \sim \exp(ip\,r)$ as $r \to \infty$.

Because $p := p\,(z)$ is a multi-function of z, we need to define which branch is to be used. It is convenient to take

$$\begin{aligned} p\,(z) &= q_1(z)\, q_2(z) \\ q_1(z) &= (z + c^2)^{1/2}, \quad \arg q_1(0) = 0 \\ q_2(z) &= (z - c^2)^{1/2}, \quad \arg q_2(0) = \pi/2. \end{aligned} \tag{3.6.9}$$

This is equivalent to the use of the Feynman path C_F, Fig. 2.3, in the contour integral representation (3.6.1).

The Green's function $G_\kappa(r,s;z)$ for this problem is matrix-valued on the quarter-plane $\mathbb{R}_+ \times \mathbb{R}_+$ at each $z \in \mathbb{C}$, and has a jump discontinuity on the line $r = s$. Thus

$$G_\kappa(r,s;z) = \begin{cases} G_\kappa^{(0)}(r,s;z), & \text{for } 0 \le r < s < \infty \\ G_\kappa^{(\infty)}(r,s;z), & \text{for } 0 \le s < r < \infty \end{cases} \tag{3.6.10}$$

so that (3.6.8) is equivalent to

$$u(r;z) = w(r;z) + \int_r^\infty G_\kappa^{(0)}(r,s;z)\,v(s)ds + \int_0^r G_\kappa^{(\infty)}(r,s;z)\,v(s)ds. \tag{3.6.11}$$

Substituting (3.6.11) into (3.6.5) gives

$$\begin{aligned} &\int_0^\infty (H_\kappa - z)\,G_\kappa(r,s;z)v(s)ds \\ &\quad + cJ\,\{G_\kappa^{(\infty)}(r,r;z) - G_\kappa^{(0)}(r,r;z)\}\,v(r) = v(r), \end{aligned} \tag{3.6.12}$$

where the second term on the left comes from differentiation with respect to the limits of integration and $J = -i\sigma^2 = \begin{pmatrix} 0 & -1 \\ 1 & 0 \end{pmatrix}$. For this equation to be satisfied for all values of r,

$$\begin{aligned} (H_\kappa - z)\,G_\kappa^{(0)}(r,s;z) &= 0, \quad 0 \le r < s, \\ (H_\kappa - z)\,G_\kappa^{(\infty)}(r,s;z) &= 0, \quad 0 \le s < r, \end{aligned} \tag{3.6.13}$$

for all values of s, whilst

$$[G_\kappa^{(\infty)}(r,r;z) - G_\kappa^{(0)}(r,r;z)] = (cJ)^{-1} = -J/c \tag{3.6.14}$$

for all values of r. If $u(r;z)$ is to satisfy the required boundary conditions, then

$$\begin{aligned} &\lim_{r\to 0} G_\kappa^{(0)}(r,s;z) \quad \text{is bounded} \\ &\lim_{r\to\infty} G_\kappa^{(\infty)}(r,s;z) \sim e^{ipr}, \quad z > c^2, \quad p > 0, \end{aligned} \tag{3.6.15}$$

so that we can take solutions of equations (3.6.13) and (3.6.14) having the form

$$\begin{aligned} G_\kappa^{(0)}(r,s;z) &= u^{(0)}(r;z) \otimes A^t(s), \quad r < s \\ G_\kappa^{(\infty)}(r,s;z) &= u^{(\infty)}(r;z) \otimes B^t(s), \quad s < r \end{aligned}$$

where $u^{(0)}(r;z)$ and $u^{(\infty)}(r;z)$ are linearly independent (unnormalized) solutions satisfying the boundary conditions respectively at 0 and ∞. The condition (3.6.14) can now be satisfied on the line $r = s$ if we choose

$$A(s) = \begin{pmatrix} P^{(\infty)}(s;z) \\ Q^{(\infty)}(s;z) \end{pmatrix}, \quad B(s) = \begin{pmatrix} P^{(0)}(s;z) \\ Q^{(0)}(s;z) \end{pmatrix},$$

making

$$G_\kappa^{(0)}(r,s;z) = \begin{pmatrix} P^{(0)}(r;z)\,P^{(\infty)}(s;z) & P^{(0)}(r;z)\,Q^{(\infty)}(s;z) \\ Q^{(0)}(r;z)\,P^{(\infty)}(s;z) & Q^{(0)}(r;z)\,Q^{(\infty)}(s;z) \end{pmatrix}, \quad r < s,$$

and

$$G_\kappa^{(\infty)}(r,s;z) = \begin{pmatrix} P^{(\infty)}(r;z)\,P^{(0)}(s;z) & P^{(\infty)}(r;z)\,Q^{(0)}(s;z) \\ Q^{(\infty)}(r;z)\,P^{(0)}(s;z) & Q^{(\infty)}(r;z)\,Q^{(0)}(s;z), \end{pmatrix}, \quad s < r,$$

together with the Wronskian normalization

$$\begin{aligned} &\Delta\{u^{(0)}(r;z), u^{(\infty)}(r;z)\} \\ &\qquad = P^{(\infty)}(r;z)\,Q^{(0)}(r;z) - P^{(0)}(r;z)\,Q^{(\infty)}(r;z) = 1/c \end{aligned} \tag{3.6.16}$$

for all values of r. The outcome is that

$$G_\kappa(r,s;z) = \begin{pmatrix} g_\kappa^{11}(r,s;z) & g_\kappa^{12}(r,s;z) \\ g_\kappa^{21}(r,s;z) & g_\kappa^{22}(r,s;z) \end{pmatrix} \tag{3.6.17}$$

where, if $r_< = \min(r,s),\ r_> = \max(r,s)$,

$$\begin{aligned} g_\kappa^{11}(r,s;z) &= P^{(0)}(r_<;z)\,P^{(\infty)}(r_>;z), \\ g_\kappa^{12}(r,s;z) &= P^{(0)}(r_<;z)\,Q^{(\infty)}(r_>;z), \\ g_\kappa^{21}(r,s;z) &= Q^{(0)}(r_<;z)\,P^{(\infty)}(r_>;z), \\ g_\kappa^{22}(r,s;z) &= Q^{(0)}(r_<;z)\,Q^{(\infty)}(r_>;z). \end{aligned}$$

To complete the construction we include the angular parts. Define

$$\Pi_{\kappa,\kappa'}(\hat{\boldsymbol{r}}, \hat{\boldsymbol{s}}) = \sum_m \chi_{\kappa m}(\hat{\boldsymbol{r}}).\chi^\dagger_{\kappa' m}(\hat{\boldsymbol{s}}).$$

where $\hat{\boldsymbol{r}}$ denotes a unit vector corresponding to the polar angles θ, φ. Then

$$\begin{aligned} &G_\kappa(\boldsymbol{r}, \boldsymbol{s}; z) \\ &= A \begin{pmatrix} g_\kappa^{11}(r,s;z)\,\Pi_{\kappa,\kappa}(\hat{\boldsymbol{r}}, \hat{\boldsymbol{s}}) & -ig_\kappa^{12}(r,s;z)\,\Pi_{\kappa,\bar\kappa}(\hat{\boldsymbol{r}}, \hat{\boldsymbol{s}}) \\ +ig_\kappa^{21}(r,s;z)\,\Pi_{\bar\kappa,\kappa}(\hat{\boldsymbol{r}}, \hat{\boldsymbol{s}}) & g_\kappa^{22(r,s;z)}\,\Pi_{\bar\kappa,\bar\kappa}(\hat{\boldsymbol{r}}, \hat{\boldsymbol{s}}) \end{pmatrix} \end{aligned} \tag{3.6.18}$$

where $A = 1/w\{u^{(0)}(r;z), u^{(\infty)}(r;z)\}$ if the normalization does not satisfy (3.6.16). The structure of (3.6.18) is similar to that of $\psi^{(0)}_{z\kappa}(\boldsymbol{r}_<) \otimes \psi^{(\infty)\dagger}_{z\kappa}(\boldsymbol{r}_>)$, although the radial parts are not conjugated in the second factor.

3.6.2 The partial wave Green's function for the free Dirac particle

In the free particle case, (3.2.26) gives

$$\boldsymbol{u}^{(\cdot)}(x) = x \begin{pmatrix} (z + c^2/cp)^{1/2} f_l^{(\cdot)}(x) \\ \eta\,(z - c^2/cp)^{1/2} f_{\bar{l}}^{(\cdot)}(x) \end{pmatrix} \tag{3.6.19}$$

after dropping unwanted normalization factors. Here $\eta = \mathrm{sgn}\,(\kappa)$ and

$$f_l^{(0)}(x) := j_l(x), \quad f_l^{(\infty)}(x) := h_l^{(1)}(x), \quad x = p\,r$$

are spherical Bessel functions. Substituting from (3.6.19) into (3.6.16) and noting that (Appendix A.3)

$$f_{\bar{l}}(x) = \eta \left(\frac{d}{dx} + \frac{\kappa}{x} \right) f_l(x), \quad \bar{l} = l - \eta \tag{3.6.20}$$

we find that (A.3.7)

$$\begin{aligned} \Delta\{u^{(0)}(r;z), u^{(\infty)}(r;z)\} &= (p\,r)^2 \eta\, \{h_l^{(1)}(p\,r)\, j_{\bar{l}}(p\,r) - h_{\bar{l}}^{(1)}(p\,r)\, j_l(p\,r)\} \\ &= (p\,r)^2\, W(h_l^{(1)}(p\,r), j_l(p\,r)) = -i \end{aligned}$$

so that

$$A = 1/c\Delta = i/c.$$

The fact that the Wronskian is not only independent of r but also of the parameter z means that there are no values of the energy at which it vanishes. Were this the case, we could infer that the two solutions $u^{(0)}(r;z)$ and $u^{(\infty)}(r;z)$ are linearly dependent, which is the condition that z is an eigenvalue. We deduce that the free Dirac particle has no point spectrum. Thus

$$\begin{aligned} g_\kappa^{11} &= \left(\frac{z + c^2}{cp} \right) j_l(x_<)\,.\, h_l^{(1)}(x_>), \\ g_\kappa^{12} &= \eta\, j_l(x_<)\,.\, h_{\bar{l}}^{(1)}(x_>), \\ g_\kappa^{21} &= \eta\, j_{\bar{l}}(x_<)\,.\, h_l^{(1)}(x_>), \\ g_\kappa^{22} &= \left(\frac{z - c^2}{cp} \right) j_{\bar{l}}(x_<)\,.\, h_{\bar{l}}^{(1)}(x_>), \end{aligned}$$

where $x_< = p\,r_<, x_> = p\,r_>$. Regarded as a function of the complex variable z, we see that this is holomorphic on the complex z-plane cut along the real axis from $-\infty < z < c^2$ and $c^2 < z < \infty$ when $p = p(z)$ is defined as in (3.6.9).

3.6.3 Summation over partial waves in the free electron case

A demonstration that for the free electron

$$G(\boldsymbol{R}, z) = \sum_\kappa G_\kappa(\boldsymbol{r}, \boldsymbol{s}; z),$$

where $\boldsymbol{R} = \boldsymbol{r} - \boldsymbol{s}$, verifies that we have indeed constructed the Green's function correctly. To see this, it is convenient to combine the angular parts $\Pi_{\kappa,\kappa}(\hat{\boldsymbol{r}}, \hat{\boldsymbol{s}})$ that have a common l value:

$$\Pi_{l,\,l}(\hat{\boldsymbol{r}}, \hat{\boldsymbol{s}}) + \Pi_{-l-1,-l-1}(\hat{\boldsymbol{r}}, \hat{\boldsymbol{s}}) = \frac{2l+1}{4\pi} \begin{pmatrix} P_l(\hat{\boldsymbol{r}} \cdot \hat{\boldsymbol{s}}) & 0 \\ 0 & P_l(\hat{\boldsymbol{r}} \cdot \hat{\boldsymbol{s}}) \end{pmatrix}$$

after using the addition theorem for spherical harmonics

$$\frac{2l+1}{4\pi} P_l(\hat{\boldsymbol{r}} \cdot \hat{\boldsymbol{s}}) = \sum_m Y_l^m(\hat{\boldsymbol{r}})\, Y_l^{m*}(\hat{\boldsymbol{s}}).$$

It is therefore straightforward to see that

$$\begin{aligned} \sum_\kappa G_\kappa^{11}(\boldsymbol{r}, \boldsymbol{s}; z) &= i \left(\frac{z + c^2}{c}\right) \sum_{l=0}^{\infty} \frac{2l+1}{4\pi} j_l(p\,r_<).\, h_l^{(1)}(p\,r_>)\, P_l(\hat{\boldsymbol{r}} \cdot \hat{\boldsymbol{s}}) \\ &= \left(\frac{z + c^2}{c}\right) \frac{\mathrm{e}^{ip\,R}}{4\pi R}, \qquad R = |\boldsymbol{r} - \boldsymbol{s}|, \end{aligned} \tag{3.6.21}$$

Similarly,

$$\sum_\kappa G_\kappa^{22}(\boldsymbol{r}, \boldsymbol{s}; z) = \left(\frac{z - c^2}{c}\right) \frac{\mathrm{e}^{ipR}}{4\pi R},$$

The off-diagonal blocks can be summed by noting first that

$$\boldsymbol{\sigma} \cdot \boldsymbol{p}\; [f_l(p\,r)\, \chi_{\kappa m}(\hat{\boldsymbol{r}})] = i\eta p\, f_{\bar{l}}(p\,r) \chi_{\bar{\kappa} m}(\hat{\boldsymbol{r}})$$

from which

$$\sum_\kappa G_\kappa^{12}(\boldsymbol{r}, \boldsymbol{s}; z) = \sum_\kappa G_\kappa^{21}(\boldsymbol{r}, \boldsymbol{s}; z) = \boldsymbol{\sigma} \cdot \boldsymbol{p}\, \frac{\mathrm{e}^{ipR}}{4\pi R},$$

and the final result is

$$\begin{aligned} \sum_\kappa G_\kappa(\boldsymbol{r}, \boldsymbol{s}; z) &= \frac{1}{c} \begin{pmatrix} z + c^2 & c\,\boldsymbol{\sigma} \cdot \boldsymbol{p} \\ c\,\boldsymbol{\sigma} \cdot \boldsymbol{p} & z - c^2 \end{pmatrix} \frac{\mathrm{e}^{ipR}}{4\pi R} \\ &= \frac{z + H}{c} \frac{\mathrm{e}^{ipR}}{4\pi R} \end{aligned} \tag{3.6.22}$$

as in (2.9.45).

3.6.4 ∗ Green's function for hydrogenic ions

The Green's function for the relativistic hydrogen ions was constructed along similar lines by Brown and Schaefer [32] and by Wichmann and Kroll [33] independently. The solution regular at the origin is (3.3.15)

$$\begin{pmatrix} P^{(0)}(r;z) \\ Q^{(0)}(r;z) \end{pmatrix} = \rho^{\gamma} \mathrm{e}^{-\rho/2} \times \begin{pmatrix} (c+z/c)^{1/2} \{aM(a+1,b;\rho) + (\nu'-\kappa)M(a,b;\rho)\} \\ (c-z/c)^{1/2} \{aM(a+1,b;\rho) - (\nu'-\kappa)M(a,b;\rho)\} \end{pmatrix} \tag{3.6.23}$$

and the solution giving outgoing waves at infinity is (3.3.17)

$$\begin{pmatrix} P^{(\infty)}(r;z) \\ Q^{(\infty)}(r;z) \end{pmatrix} = \rho^{\gamma} \mathrm{e}^{-\rho/2} \times \begin{pmatrix} (c+z/c)^{1/2} \{(\nu'+\kappa)U(a+1,b;\rho) + U(a,b;\rho)\} \\ (c-z/c)^{1/2} \{(\nu'+\kappa)U(a+1,b;\rho) - U(a,b;\rho)\} \end{pmatrix}, \tag{3.6.24}$$

in which

$$a = \gamma - \nu, \quad b = 2\gamma + 1, \quad \rho = 2\lambda r,$$

where γ and λ have the same meaning as before. The complex parameter z has been substituted for the energy E and we have suppressed an unimportant constant factor $\nu' - \kappa$. The bound state apparent principal quantum number N is replaced by

$$\nu' = Z/\lambda, \quad \lambda = c(1 - z^2/m^2c^4)^{1/2}$$

and we introduce another parameter

$$\nu = \nu'(1 - \lambda^2/c^2)^{1/2}$$

so that

$$\nu^2 - \gamma^2 = \nu'^2 - \kappa^2.$$

To complete the construction, we need the Wronskian (3.6.16). This is most easily found by using the limits as $\rho \to 0$, given by (A.3.15) and (A.3.16) respectively:

$$M(a,b,\rho) \to 1, \quad U(a,b,\rho) \to \frac{\Gamma(2\gamma)}{\Gamma(a)} \rho^{-2\gamma}$$

from which

$$\begin{pmatrix} P^{(0)}(r;z) \\ Q^{(0)}(r;z) \end{pmatrix} \to \rho^{\gamma} \begin{pmatrix} (c+z/c)^{1/2} ((\gamma-\nu) + (\nu'-\kappa)) \\ (c-z/c)^{1/2} ((\gamma-\nu) - (\nu'-\kappa)) \end{pmatrix}$$

and

$$\begin{pmatrix} P^{(\infty)}(r;z) \\ Q^{(\infty)}(r;z) \end{pmatrix} \to \rho^{-\gamma} (\nu'-\kappa) \frac{\Gamma(2\gamma)}{\Gamma(\gamma-\nu)} \begin{pmatrix} (c+z/c)^{1/2} \left(\dfrac{\nu'+\kappa}{\gamma-\nu} + 1 \right) \\ (c-z/c)^{1/2} \left(\dfrac{\nu'+\kappa}{\gamma-\nu} - 1 \right) \end{pmatrix}$$

so that

$$\Delta = 4\gamma\,\lambda\,\frac{\Gamma(2\gamma)}{\Gamma(\gamma-\nu)} \tag{3.6.25}$$

and

$$A = 1/c\Delta = \frac{\Gamma(\gamma-\nu)}{4c\,\gamma\,\lambda\,\Gamma(2\gamma)}. \tag{3.6.26}$$

As a function of z, the Coulomb Green's function $G_\kappa(\boldsymbol{r}, \boldsymbol{s}; z)$ has branch points at $z = \pm c^2$ as in the free particle case. However, the Wronskian Δ has simple zeros (so that the Green's function has simple poles) at the poles of $\Gamma(\gamma - \nu)$ when $\gamma - \nu = -n_r$, where n_r is a non-negative integer, as in Section 3.3.2.

An important check on the above results, left as a tedious exercise for the reader, is to verify that we recover the free particle formulas in the limit as the nuclear charge Z approaches zero. This entails the use of relations connecting confluent hypergeometric functions with Bessel functions [5, §13.4, §13.6]. Although the sum over partial waves can be performed analytically for the nonrelativistic hydrogenic Green's function, no similar result is known for the relativistic hydrogenic case.

3.7 The nonrelativistic limit: the Pauli approximation

The relativistic equations of motion should transform into nonrelativistic equations in the mathematical limit, $c \to \infty$, in which light signals appear to propagate instantaneously. This limit has been much studied for several reasons, some theoretical, others more practical. Analytic techniques were all that were available in the early days of relativistic quantum theory, and a perturbation expansion in powers of the fine structure constant, α, threw light on the size of "relativistic corrections" and on the origin of spin-dependent effects. The simple Pauli approximation of this section, despite its limitations, continues to be of major importance in quantum chemistry.

3.7.1 The Pauli approximation

The behaviour of the solutions of the Dirac equation in the nonrelativistic limit is a matter of perturbation theory, in which the strength of the perturbation is proportional to $\alpha = 1/c$. We start from the Hamiltonian form of (3.1.39)

$$(H_D - E)\psi = \left\{c\boldsymbol{\alpha}\cdot(\boldsymbol{p} - q\boldsymbol{A}) + \beta mc^2 + q\phi - E\right\}\psi = 0, \tag{3.7.1}$$

In order to pass to the nonrelativistic limit, we shift the energy zero so that

$$\epsilon = E - mc^2, \quad |\epsilon| \ll mc^2,$$

and partition (3.7.1) in the form

$$\begin{pmatrix} q\,\phi - \epsilon & c\,\boldsymbol{\sigma}\cdot\boldsymbol{\Pi} \\ c\,\boldsymbol{\sigma}\cdot\boldsymbol{\Pi} & -2mc^2 + q\,\phi - \epsilon \end{pmatrix} \begin{pmatrix} \psi^+(\boldsymbol{r}) \\ \psi^-(\boldsymbol{r}) \end{pmatrix} = 0$$

where the superscripts + and − indicate upper and lower two-component amplitudes respectively, and

$$\boldsymbol{\Pi} = \boldsymbol{p} - q\boldsymbol{A}.$$

This can be regarded as a pair of simultaneous linear differential equations, of which the lower can be rearranged to read

$$\psi^-(\boldsymbol{r}) = \frac{1}{2mc}\left\{1 - \frac{q\,\phi - \epsilon}{2mc^2}\right\}^{-1} \boldsymbol{\sigma}\cdot\boldsymbol{\Pi}\,\psi^+(\boldsymbol{r}).$$

Pauli's approximation results from replacing the bracketed term by unity:

$$\psi^-(\boldsymbol{r}) \approx \frac{1}{2mc}\,\boldsymbol{\sigma}\cdot\boldsymbol{\Pi}\,\psi^+(\boldsymbol{r})$$

and substituting this in the upper equation gives

$$\left\{\boldsymbol{\sigma}\cdot\boldsymbol{\Pi}\,\frac{1}{2m}\,\boldsymbol{\sigma}\cdot\boldsymbol{\Pi} + q\,\phi - \epsilon\right\}\psi^+(\boldsymbol{r}) \approx 0.$$

The well-known algebraic identity (A.2.16)

$$(\boldsymbol{\sigma}\cdot\boldsymbol{A})(\boldsymbol{\sigma}\cdot\boldsymbol{B}) = \boldsymbol{A}\cdot\boldsymbol{B} + i\boldsymbol{\sigma}\cdot\boldsymbol{A}\times\boldsymbol{B}$$

gives

$$(\boldsymbol{\sigma}\cdot\boldsymbol{\Pi}\,\boldsymbol{\sigma}\cdot\boldsymbol{\Pi})\,\psi^+(\boldsymbol{r}) = (\boldsymbol{\Pi}\cdot\boldsymbol{\Pi} - q\,\boldsymbol{\sigma}\cdot\mathrm{curl}\;\boldsymbol{A})\,\psi^+(\boldsymbol{r}).$$

so that $\psi^+(\boldsymbol{r})$ satisfies a modified Schrödinger equation

$$\left\{\frac{1}{2m}(\boldsymbol{p} - q\boldsymbol{A})^2 - \boldsymbol{\mu}\cdot\boldsymbol{B} + q\,\phi - \epsilon\right\}\psi^+(\boldsymbol{r}) \approx 0, \tag{3.7.2}$$

where $\boldsymbol{B} = \mathrm{curl}\;\boldsymbol{A}$ is the magnetic induction. The operator $(\boldsymbol{p} - q\boldsymbol{A})^2/2m$ describes the interaction of a charged particle with a magnetic field of classical physics described in Section 2.6.4. The new term $-\boldsymbol{\mu}\cdot\boldsymbol{B}$, which is not predicted by classical theory, is the potential energy of a magnetic dipole of magnitude

$$\boldsymbol{\mu} = \frac{q}{2m}\boldsymbol{\sigma} \tag{3.7.3}$$

in the magnetic field $\boldsymbol{B}$. The operator $(\boldsymbol{p} - q\boldsymbol{A})^2/2m$ embodies the usual minimal substitution of Section 2.6.4. Dirac's prediction that the electron possesses an intrinsic magnetic dipole was a major success of the theory.

The magnetic dipole moment of the electron has been measured to within an uncertainty of a few parts in 10^{12} [34, p. 375]. The magnetic moment of a lepton, the generic term for a spin-1/2 particle such as an electron or positron, can be written

$$\boldsymbol{\mu} = g\,\frac{q}{2m}\,\boldsymbol{s}$$

where in terms of the elementary charge e, $q = -e$ for electrons and $q = +e$ for positrons, m is the particle mass, and $\boldsymbol{s}$ is its spin. The g-factor thus has the same absolute value for electron and positron and its sign is that of the particle's charge. It is conventional to write the magnitude of the electron or positron moment as

$$\mu_e = \frac{g_e}{2}\,\mu_B$$

where $\mu_B = e\hbar/2m$ is the *Bohr magneton* in SI units. Because $\boldsymbol{s} = \hbar\boldsymbol{\sigma}/2$ for Dirac electrons, (3.7.2) predicts $|g_e| = 2$. The experimental value for the free electron or positron differs by the *electron anomaly*,

$$a_e = \frac{|g_e|}{2} - 1 = \frac{|\mu_e|}{\mu_B} - 1,$$

due to quantum electrodynamic effects. The recommended value [34, Table IX], $a_e = 1.159\,652\,1883(42) \times 10^{-3}$, is based on measurements for both electrons and positrons that agree at this level of accuracy. The theory, reviewed recently by Kinoshita [35], involves perturbative calculations including electromagnetic, hadronic and weak interaction contributions, achieving an accuracy comparable with the experimental uncertainties of order 10^{-12}.

3.7.2 The Foldy-Wouthuysen and related transformations

The Pauli approximation was the earliest attempt to devise an approximate treatment of the Dirac equation, and it is still a popular method of accounting approximately for relativistic effects. The Foldy-Wouthuysen transformation [36] for a free particle, §2.5.3 and §2.5.4, served to relate the Dirac to the canonical representation and to throw light on the meaning of position and spin in Dirac theory. However, the main motivation has been to derive simpler equations in which relativistic effects are represented by perturbations to a nonrelativistic model [37, p. 46]. Quantum chemists have made much use of schemes such as the Breit-Pauli approximation, §1.5.2; see [38] for a recent review.

The simplest approach is to partition the Dirac Hamiltonian into coupled equations for the upper and lower component as in the Pauli reduction, §3.7.1, and to try to eliminate the lower component from the equations. The technique, [38, §2.1], has been developed to give a sequence of "regular approximations" – ZORA (zero order), FORA (first order), etc. – which avoid the singular effective potentials generated by the Pauli approximation. Another approach seeks unitary transformations that block diagonalize the Dirac operator. Because off-diagonal blocks involve odd powers of $\boldsymbol{\alpha}$ matrices, the effect is to replace such terms with effective operators in which $\boldsymbol{\alpha}$ matrices only appear to even order. The earliest programme of this sort was due to Foldy and Wouthuysen [36]; currently the most popular version is based on the scheme of Douglas and Kroll [39], and developed by Hess [40] and his collaborators for applications to quantum chemistry [41]. Another version was proposed by Jørgensen [42], building on earlier work by Van Vleck. The effective Hamiltonian to $O(1/c^2)$ is the sum of several terms:

$$\begin{aligned} H_0 &= c^2 \\ H_2 &= \frac{1}{2}[\boldsymbol{\sigma}\cdot\boldsymbol{\Pi}]^2 - e\Phi = \frac{1}{2}(\boldsymbol{p}+e\boldsymbol{A})^2 - e\Phi + \frac{e}{2}\boldsymbol{\sigma}\cdot\boldsymbol{B} \\ H_4 &= -\frac{1}{8c^2}\left\{[\boldsymbol{\sigma}\cdot\boldsymbol{\Pi}]^4 + [\boldsymbol{\sigma}\cdot\boldsymbol{\Pi}, -e\Phi]^{(2)}\right\} \\ &= -\frac{1}{8c^2}\{[(\boldsymbol{p}+e\boldsymbol{A})^2 + e\boldsymbol{\sigma}\cdot\boldsymbol{B}]^2 - e\,\mathrm{div}\boldsymbol{E} \\ &\qquad\qquad - 2e\boldsymbol{\sigma}\cdot\boldsymbol{E}\times(\boldsymbol{p}+e\boldsymbol{A})\} \\ &\ \vdots \end{aligned} \tag{3.7.4}$$

where $\boldsymbol{\Pi} = \boldsymbol{p}+e\boldsymbol{A}$ and the operators act on the Hilbert space of 2-component functions.

After subtracting the rest energy term, we are left with the Pauli Hamiltonian H_2 and the $O(1/c^2)$ corrections H_4. In the absence of an external magnetic field, $\boldsymbol{A} = 0$, these simplify to

$$H_4 = -\frac{[(\boldsymbol{p})^2]^2}{8c^2} + \frac{e}{8c^2}\mathrm{div}\boldsymbol{E} + \frac{e}{4c^2}\boldsymbol{\sigma}\cdot\boldsymbol{E}\times\boldsymbol{p}.$$

If the external electric field $\boldsymbol{E}$ is due to a point nuclear charge Ze at the origin, then $-e\Phi = -Z/r$, $-e\boldsymbol{E} = -Z\boldsymbol{r}/r^3$ in atomic units, so that

$$H_4 = -\frac{[(\boldsymbol{p})^2]^2}{8c^2} - \frac{\pi Z}{2c^2}\delta^3(\boldsymbol{r}) - \frac{Z}{4c^2}\frac{1}{r^3}\boldsymbol{\sigma}\cdot\boldsymbol{L} \tag{3.7.5}$$

where $\boldsymbol{L} = \boldsymbol{r}\times\boldsymbol{p}$ is the orbital angular momentum of the electron.

The classical free particle Hamiltonian (2.6.34) can be expanded in powers of $\lambda = 1/c$ when $\boldsymbol{p}^2/c^2 \ll 1$ giving

$$\mathrm{d}H_{free}(\boldsymbol{p}) = c^2\sqrt{1+\boldsymbol{p}^2/c^2} = c^2 + \frac{\boldsymbol{p}^2}{2} - \frac{[\boldsymbol{p}^2]^2}{8c^2} + \ldots$$

The term of order $1/c^2$ in this expansion, the lowest order correction for the *relativistic variation of mass with velocity*, is the leading term in (3.7.5). The next term in (3.7.5), the *Darwin correction*, has an expectation proportional to the electron density at the origin and contributes only for s-states. The third, spin-orbit coupling, term contributes only for $p, d, \ldots$- states with $l > 0$. The Darwin term is said to provide heuristic evidence of the *Zitterbewegung* phenomenon, on the ground that for spherical distributions, the expectation of $V(\boldsymbol{r}+\delta\boldsymbol{r})-V(\boldsymbol{r})$ is proportional to $\langle(\delta\boldsymbol{r})^2\,\nabla^2 V(\boldsymbol{r})\rangle$, with $|\delta\boldsymbol{r}|$ of the order of the Compton wavelength $\hbar/mc$. However, this does little to promote belief in *Zitterbewegung* as anything more than an unfortunate feature of the procedure having little or no physical meaning.

It is instructive to use (3.7.5) to compute relativistic corrections to the Schrödinger energy for hydrogenic ions to lowest order. The necessary non-relativistic matrix elements of $r^{-s}, s = 1,2,3$ will be found in Section 3.3.3. Because $(\boldsymbol{p})^2\,\psi_{nl}(\boldsymbol{r}) = 2\,(\epsilon_{nl} + Z/r)\,\psi_{nl}(\boldsymbol{r})$, $\langle Z/r\rangle_{nl} = Z^2/n^2 = -2\epsilon_{nl}$ and $\langle Z^2/r^2\rangle_{nl} = Z^4/n^3(l+1/2)$, the lowest order mass-velocity term is

$$\begin{aligned}\left\langle -\frac{[(\boldsymbol{p})^2]^2}{8c^2}\right\rangle_{nl} &= -\frac{1}{2c^2}\left\langle\left(\epsilon_{nl} + \frac{Z}{r}\right)^2\right\rangle_{nl}\\ &= -\frac{1}{2c^2}\frac{Z^4}{n^4}\left(\frac{n}{l+\frac{1}{2}} - \frac{3}{4}\right).\end{aligned}$$

Because $\langle\delta^3(\boldsymbol{r})\rangle_{nl} = |\psi_{nl}(0)|^2 = \delta_{l,0}Z^3/\pi n^3$, the Darwin term gives

$$\left\langle\frac{\pi Z}{2m^2c^2}\delta^3(\boldsymbol{r})\right\rangle_{nl} = \frac{Z^4}{2c^2n^3}\delta_{l,0}.$$

The spin-orbit term has a j-dependence given by

$$\begin{aligned}\langle\boldsymbol{\sigma}\cdot\boldsymbol{L}\rangle_{lj} = \langle 2\boldsymbol{s}\cdot\boldsymbol{L}\rangle_{lj} &= \boldsymbol{j}^2 - \boldsymbol{l}^2 - \boldsymbol{s}^2\\ &= j(j+1) - l(l+1) - s(s+1)\end{aligned}$$

where $\boldsymbol{s} = \frac{1}{2}\hbar\boldsymbol{\sigma}$ so that the quantum number $s = 1/2$ and

$$\left\langle -\frac{Z}{4c^2}\frac{1}{r^3}\boldsymbol{\sigma}\cdot\boldsymbol{L}\right\rangle_{nlj} = -\frac{Z}{4c^2}\left\langle\frac{1}{r^3}\right\rangle_{nl}\langle 2\boldsymbol{s}\cdot\boldsymbol{L}\rangle_{lj}.$$

The hydrogenic spin-orbit energy for $l \geq 1$ is therefore

$$-\frac{1}{4c^2}\frac{Z^4}{n^3(l+1/2)(l+1)}\left[j(j+1) - l(l+1) - \frac{3}{4}\right]$$

Putting everything together, the total energy to this order is (once again remembering $m = 1$ in atomic units)

$$\epsilon_{nlj} = -\frac{Z^2}{2n^2} - \frac{Z^4}{2c^2n^4}\left(\frac{n}{j+\frac{1}{2}} - \frac{3}{4}\right) \tag{3.7.6}$$

in exact agreement with the Uhlenbeck and Goudsmit result, §2.4, and the expansion of the Sommerfeld fine structure formula (3.3.7) to $O(1/c^2)$.

3.8 Other aspects of Dirac theory

The subject matter of this chapter has been directed to understanding the construction and properties of solutions of the Dirac equation which are important for the treatment of the many-electron problems that form the core of this book. As such we have ignored many of the issues that have received attention from mathematicians and physicists over the last 70 years. The theoretical background expounded, for example, in Thaller's monograph [44], provides a rigorous mathematical description using linear functional analysis. Practical methods of numerical calculation must respect this analysis when relevant. The monograph of Bagrov and Gitman [45] surveys a number of exact solutions to a number of relativistic quantum mechanical problems involving Klein-Gordon and Dirac equations, especially involving charged particles in electromagnetic fields of various conformations. There has also been a good deal of research into the foundations of electron theory and electromagnetism whose flavour can be sampled in the volume edited by Dowling [46].

References

1. Grant I P 1961 *Proc Roy Soc (Lond) A* **262** 555; — 1965 *Proc Phys Soc* **86** 523.
2. Drake G W F ed 2005 *Springer Handbook of Atomic, Molecular and Optical Physics* Second edn (New York: Springer-Verlag)
3. Staff of the Bateman Manuscript Project 1953 *Higher Transcendental Functions* Vol. 1 (New York: McGraw-Hill Book Company)
4. Hartree D R 1929 *Proc Camb Phil Soc* **25** 225.
5. Abramowitz M and Stegun I A 1970 *Handbook of Mathematical Functions* (New York: Dover)
6. Gottfried K 1966 *Quantum Mechanics* (New York: W A Benjamin, Inc.)
7. Coddington E A and Levinson N 1955 *Theory of Ordinary Differential Equations* (New York: McGraw Hill)
8. Greiner W 1990 *Relativistic Quantum Mechanics. Wave Equations* (Berlin: Springer-Verlag)
9. Greiner W, Müller B and Rafelski J 1985 *Quantum Electrodynamics of Strong Fields* (Berlin: Springer-Verlag)
10. Darwin C G 1928 *Proc Roy Soc (Lond) A* **118** 654.
11. Gordon W 1928 *Z f Phys* **48** 11.
12. Mott N F 1929 *Proc. Roy. Soc. A* **124** 425.
13. Hylleraas E A 1955 *Z. Phys.* **140** 626.
14. Johnson W R and Cheng K T 1979 *J. Phys. B: Atom. molec. phys.* **12** 863.
15. Dirac P A M 1958 *The Principles of Quantum Mechanics* 4th ed. (Oxford: Clarendon Press).
16. Burke V M and Grant I P 1967 *Proc Phys Soc* **90** 297.
17. Garstang R H and Mayers D F 1966 *Proc. Camb. Phil. Soc.* **62** 777.
18. Kobus J, Karwowski J and Jaskolski W 1987 *J. Phys. A.: Math. Gen. phys* **20** 3347.
19. Mott N F and Massey H S W (1949) *The theory of atomic collisions* (2nd edition) (Oxford: Clarendon Press)

20. Rutherford E 1911 *Phil. Mag.* **21** 669.
21. Joachain C J (1983) *Quantum collision theory* (3rd edition) (Amsterdam: North-Holland Publishing Company)
22. Sherman N 1956 *Phys. Rev.* **103** 1601.
23. Kessler J 1985 *Polarized Electrons* (2nd edition) (Berlin: Springer-Verlag)
24. Bethe H A and Salpeter E E 1957 *Quantum Mechanics of One- and Two-Electron Atoms* (Berlin-Göttingen-Heidelberg: Springer-Verlag)
25. Walker D W 1971 *Adv. Phys.* **20** 257.
26. Gordon W 1928 *Z f Phys* **50** 630.
27. Seaton M J 1983 *Rep. Prog. Phys.* **46** 167.
28. Connerade J P 1998 *Highly Excited Atoms* (Cambridge: Cambridge University Press)
29. Lee C M and Johnson W R 1980 *Phys. Rev. A* **22** 272.
30. Bashkin S and Stoner J O 1975 *Atomic Energy Levels and Grotrian Diagrams* Vol 1. (Amsterdam : North-Holland Publishing Company)
31. Seaton M J 1958 *Mon. Not. R. Astr. Soc.* **118** 504.
32. Brown G E and Schaefer G W 1956 *Proc. Roy. Soc. A* **233** 527.
33. Wichmann E H and Kroll N M 1956 *Phys. Rev.* **101** 843.
34. Mohr P J and Taylor B N 2000 *Rev Mod Phys* **72** 351 (CODATA recommended values of the fundamental physical constants: 1998)
35. Kinoshita T 2001 in [47, pp. 155–175]
36. Foldy L L and Wouthuysen S A 1950 *Phys. Rev.* **78** 29.
37. Bjorken J D and Drell S D 1964 *Relativistic Quantum Mechanics* (New York: McGraw-Hill Book Co.)
38. Wolf A, Reiher M and Hess B A 2002 in [41, Chapter 11].
39. Douglas M and Kroll N M 1974 *Ann Phys (NY)* **82** 89.
40. Hess B A 1986 *Phys Rev A* **32** 756; — *Phys Rev A* **33** 3742.
41. Schwerdtfeger P (ed.) 2002 *Relativistic Electronic structure Theory, Part 1: Fundamentals.* (Theroetical and Computational Chemistry, Vol. 11) (Amsterdam: Elsevier Science B.V.)
42. Jørgensen F 1975 *Mol. Phys.* 1137.
43. Bellman R 1970 *Introduction to Matrix Analysis* (2e) (New York: Mc Graw-Hill Book Company)
44. Thaller B 1992 *The Dirac Equation* (New York: Springer-Verlag)
45. Bagrov B G and Gitman D M *Exact Solutions of Relativistic Wave Equations* (Dordrecht: Kluwer Academic Publishers)
46. Dowling J P (ed.) 1997 *Electron Theory and Quantum Electrodynamics: 100 years later* (New York: Plenum Press)
47. Karshenboim S G, Pavone F S, Bassani F, Inguscio M and Hänsch T W 2001 *The Hydrogen Atom. Precision Physics of Simple Atomic systems* (Berlin: Springer-Verlag)

4

Quantum electrodynamics

Quantum electrodynamics (QED), the study of the motion of electrically charged particles such as electrons, positrons, and charged nuclei, provides the formal framework for the relativistic theory of atoms, molecules, and other forms of matter. Quantum field theory [1, 2], of which QED is an example, was invented to model physical processes in which the number of particles is not necessarily fixed. The coupling of the electron-positron field with the Maxwell photon field in QED allows us to build a relativistic theory of atoms and molecules. Feynman diagrams serve to clarify the radiative and collision processes that contribute to atomic and molecular physics. A subset of these diagrams corresponds to the familiar self-consistent field theory, which is both the starting point for more accurate calculations as well as a popular model in its own right. Diagrams associated with "radiative corrections", which are not normally included in theories of atomic or molecular electronic structure, pose additional technical challenges. The interaction of a charged particle with the fluctuations of the Maxwell photon field leads to a correction to the particle's energy and to its magnetic moment, whilst the particle's charge modifies the electromagnetic field close by. These radiative corrections can be significant in some applications.

4.1 Second quantization

4.1.1 Quantization of the Schrödinger equation

The standard theory of quantized fields rests on the elementary theory of Lagrangian and Hamiltonian methods of Appendix B.9. We start by writing down a Lagrangian density for structureless particles of mass m moving in a potential $V(x)$:

$$\begin{aligned}\mathcal{L} := &\frac{1}{2}\psi^*(x).\, i\partial_0\psi(x) - \frac{1}{2}i\partial_0\psi^*(x).\psi(x)\\ &- \frac{1}{2m}\,\partial_i\psi^*(x)\cdot\partial_i\psi(x) + V(x)\psi^*(x)\psi(x).\end{aligned} \tag{4.1.1}$$

The complex quantities $\psi(x)$ and $\psi^*(x)$ are to be treated as independent classical fields. As usual, $\partial_\mu = \partial/\partial x^\mu$, $\mu = 0, 1, 2, 3$, but in this section $\partial_0 = \partial/\partial t$ (without the usual factor c). One of the Euler-Lagrange equations is

$$\partial_\mu \frac{\partial \mathcal{L}}{\partial(\partial_\mu \psi^*)} - \frac{\partial \mathcal{L}}{\partial \psi^*} = -\frac{i}{2}\partial_0 \psi - \frac{i}{2}\partial_0 \psi - \frac{1}{2m}\nabla^2 \psi(x) + V(x)\psi(x) = 0.$$

The first occurrence of $\partial_0\psi$ comes from $\partial\mathcal{L}/\partial(\partial_0\psi^*)$, the second from $\partial\mathcal{L}/\partial\psi^*$. We recover the usual Schrödinger equation

$$i\partial_0\psi = -\frac{1}{2m}\nabla^2\psi(x) + V(x)\psi(x); \tag{4.1.2}$$

the second Euler-Lagrange equation gives its complex conjugate. The momentum conjugate to ψ will be denoted[1]

$$\pi^*(x) = \frac{\partial \mathcal{L}}{\partial(\partial_0\psi)} = \frac{i}{2}\psi^*(x), \tag{4.1.3}$$

and that conjugate to ψ^* by

$$\pi(x) = \frac{\partial \mathcal{L}}{\partial(\partial_0\psi^*)} = -\frac{i}{2}\psi(x), \tag{4.1.4}$$

We can now construct a Hamiltonian density,

$$\begin{aligned}\mathcal{H} &= \pi^*(x)\,\partial_0\psi(x) + \partial_0\psi^*(x)\,\pi(x) - \mathcal{L} \\ &= \frac{1}{2m}\,\partial_i\psi^*(x)\cdot\partial_i\psi(x) + V(x)\psi^*(x)\psi(x)\end{aligned} \tag{4.1.5}$$

which is just the sum of kinetic and potential energy terms.

In ordinary quantum mechanics, we interpret the dynamical variables q_i and p_j as operators satisfying commutation relations

$$[q_i, p_j] = i\hbar\,\delta_{ij}, \quad [q_i, q_j] = 0, \quad [p_i, p_j] = 0,$$

for all pairs i, j. Similarly we now interpret ψ as an operator on some as yet undefined space in the Heisenberg picture and replace ψ^* by the operator adjoint $\psi^\dagger$. Then the Hamiltonian operator is

$$H = \int d^3x\,\mathcal{H} = \int d^3x \left\{\frac{1}{2m}\nabla\psi^\dagger(x)\cdot\nabla\psi(x) + V(x)\psi^\dagger(x)\psi(x)\right\} \tag{4.1.6}$$

For simplicity, we suppose that the real function $V(x)$ does not depend on time. Suppose also either that the field variables have *commutation* relations

---

[1] The labelling of π and π^* looks more intelligible in the context of spinor fields below.

$$\begin{aligned}\left[\psi(\boldsymbol{x}), \psi^\dagger(\boldsymbol{x}')\right] = \psi(\boldsymbol{x})\,\psi^\dagger(\boldsymbol{x}') - \psi^\dagger(\boldsymbol{x}')\,\psi(\boldsymbol{x}) &= \delta^3(\boldsymbol{x}-\boldsymbol{x}'),\\ \left[\psi(\boldsymbol{x}), \psi(\boldsymbol{x}')\right] = \left[\psi^\dagger(\boldsymbol{x}), \psi^\dagger(\boldsymbol{x}')\right] = 0,\end{aligned} \tag{4.1.7}$$

or *anti-commutation* relations

$$\begin{aligned}\left\{\psi(\boldsymbol{x}), \psi^\dagger(\boldsymbol{x}')\right\} = \psi(\boldsymbol{x})\,\psi^\dagger(\boldsymbol{x}') + \psi^\dagger(\boldsymbol{x}')\,\psi(\boldsymbol{x}) &= \delta^3(\boldsymbol{x}-\boldsymbol{x}'),\\ \left\{\psi(\boldsymbol{x}), \psi(\boldsymbol{x}')\right\} = \left\{\psi^\dagger(\boldsymbol{x}), \psi^\dagger(\boldsymbol{x}')\right\} = 0.\end{aligned} \tag{4.1.8}$$

The remaining results of this section will be derived for the set (4.1.7) but they also hold for the set (4.1.8).

We must first confirm that this quantization procedure leads to the usual equations of motion. For a conservative system, the Heisenberg equation of motion for the field ψ is

$$i\partial_0\psi = [\psi, H],$$

where

$$[\psi(\boldsymbol{x}), H] = \left[\psi(\boldsymbol{x}), \int d^3x' \left\{\frac{1}{2m}\nabla'\psi^\dagger(\boldsymbol{x}')\cdot\nabla'\psi(\boldsymbol{x}') + V(x)\psi^\dagger(x)\psi(x)\right\}\right].$$

Using (4.1.7) and (4.1.8) and the fact that V is independent of t for a conservative system, we find

$$\begin{aligned}[\psi(\boldsymbol{x}), H] &= \left[\psi(\boldsymbol{x}), \int d^3x'\,\psi^\dagger(\boldsymbol{x}')\left\{-\frac{1}{2m}\nabla'^2 + V(\boldsymbol{x}')\right\}\psi(\boldsymbol{x}')\right]\\ &= \int d^3x'\,\left[\psi(\boldsymbol{x}), \psi^\dagger(\boldsymbol{x}')\right]\left\{-\frac{1}{2m}\nabla'^2 + V(\boldsymbol{x}')\right\}\psi(\boldsymbol{x}')\\ &= \int d^3x'\,\delta^3(\boldsymbol{x}-\boldsymbol{x}')\left\{-\frac{1}{2m}\nabla'^2 + V(\boldsymbol{x}')\right\}\psi(\boldsymbol{x}')\\ &= \left\{-\frac{1}{2m}\nabla^2 + V(\boldsymbol{x})\right\}\psi(\boldsymbol{x})\end{aligned}$$

from which Schrödinger's equation,

$$i\partial_0\psi(\boldsymbol{x}) = \left\{-\frac{1}{2m}\nabla^2 + V(\boldsymbol{x})\right\}\psi(\boldsymbol{x}),$$

follows. The first line involves an integration by parts and assumes that the boundary conditions permit surface terms to be dropped. The adjoint equation,

$$-i\partial_0\psi^\dagger(\boldsymbol{x}) = \left\{-\frac{1}{2m}\nabla^2 + V(\boldsymbol{x})\right\}\psi^\dagger(\boldsymbol{x}),$$

can be derived in the samne way.

Because H is a conservative Hamiltonian in which time does not appear explicitly and $[H, H] = 0$, its equation of motion reduces to $i\partial_0 H = 0$, showing that H is a constant of the motion. We expect

$$N = \int d^3x \, \psi^\dagger \psi \tag{4.1.9}$$

to be the operator giving the number of particles in the system. Its equation of motion is $i\partial_0 N = [N, H]$, and a calculation of the commutator along similar lines show that $[N, H] = 0$ so that N is a constant of the motion. Next we show that

$$[N, \psi^\dagger(\boldsymbol{x})] = \psi^\dagger(\boldsymbol{x}), \quad [N, \psi(\boldsymbol{x})] = -\psi(\boldsymbol{x}). \tag{4.1.10}$$

Proof: It suffices to consider the first of these results; the second is obtained in the same way.

$$\begin{aligned} [N, \psi^\dagger(\boldsymbol{x})] &= \int d^3x' \, \{\psi^\dagger(\boldsymbol{x}')\psi(\boldsymbol{x}')\psi^\dagger(\boldsymbol{x}) - \psi^\dagger(\boldsymbol{x})\psi^\dagger(\boldsymbol{x}')\psi(\boldsymbol{x}')\} \\ &= \int d^3x' \, \psi^\dagger(\boldsymbol{x}') \, [\psi(\boldsymbol{x}'), \, \psi^\dagger(\boldsymbol{x})] \\ &= \int d^3x' \, \psi^\dagger(\boldsymbol{x}') \, \delta^3(\boldsymbol{x} - \boldsymbol{x}') \\ &= \psi^\dagger(\boldsymbol{x}). \end{aligned}$$

■

Suppose now that N has an eigenstate $|\,\alpha\,\rangle$ with eigenvalue α. Then from (4.1.10) we obtain immediately

$$\begin{aligned} N \, \psi^\dagger(\boldsymbol{x}) \,|\, \alpha \,\rangle &= (\alpha + 1)\psi^\dagger(\boldsymbol{x}) \,|\, \alpha \,\rangle, \\ N \, \psi(\boldsymbol{x}) \,|\, \alpha \,\rangle &= (\alpha - 1)\psi(\boldsymbol{x}) \,|\, \alpha \,\rangle, \end{aligned} \tag{4.1.11}$$

Thus $\psi^\dagger(\boldsymbol{x})$ operates on $|\,\alpha\,\rangle$ to give an (unnormalized) eigenstate of N belonging to the eigenvalue $\alpha+1$ and $\psi(\boldsymbol{x})$ operates on $|\,\alpha\,\rangle$ to give an (unnormalized) eigenstate of N belonging to the eigenvalue $\alpha - 1$. If the eigenvalues of N are to represent the number of particles in the system, α must be a non-negative integer $n = 0, 1, 2, \ldots$ Then (4.1.11) shows that (4.1.9) is a consistent definition of the number operator and that $\psi^\dagger(\boldsymbol{x})$ creates a particle at location $\boldsymbol{x}$ whilst $\psi(\boldsymbol{x})$ destroys one. We refer to the state with $n = 0$ as the particle *vacuum* state. 5

4.1.2 Identical particles: the symmetric case

A system of independent particles is said to be indistinguishable if the expectation value of a physical observable is unaltered by permutation of particle labels. We suppose that there exists a complete orthonormal set of eigenstates $\phi_i(\boldsymbol{x})$ and corresponding eigenvalues ϵ_i of the Schrödinger Hamiltonian on some suitable domain

$$\left\{ -\frac{1}{2m}\nabla^2 + V(\boldsymbol{x}) \right\} \phi_i(\boldsymbol{x}) = \epsilon_i \, \phi_i(\boldsymbol{x}).$$

We can represent the field operators by

$$\psi(\boldsymbol{x}) = \sum_i a_i\, \phi_i(\boldsymbol{x}), \quad \psi^\dagger(\boldsymbol{x}) = \sum_i {a_i}^\dagger\, \phi_i^\dagger(\boldsymbol{x}), \tag{4.1.12}$$

where if the field operators depend on time then so do a_i and $a_i^\dagger$. We can determine the a_i and $a_i^\dagger$ from (4.1.12) using the orthonormality of the eigenfunctions, giving

$$a_i = \int \phi_i^\dagger(\boldsymbol{x})\, \psi(\boldsymbol{x})\, d^3x, \quad {a_i}^\dagger = \int \psi^\dagger(\boldsymbol{x})\, \phi_i(\boldsymbol{x})\, d^3x. \tag{4.1.13}$$

Multiply the first equation of (4.1.8) by $\phi_i^\dagger(\boldsymbol{x})\phi_j(\boldsymbol{x}')$ and integrating, we see that

$$\left[a_i, {a_j}^\dagger\right] = \iint \phi_i^\dagger(\boldsymbol{x})\, \phi_j(\boldsymbol{x}')\, \delta^3(\boldsymbol{x}-\boldsymbol{x}')\, d^3x\, d^3x' = \delta_{ij}, \tag{4.1.14}$$

and, similarly,

$$[a_i, a_j] = \left[{a_i}^\dagger, {a_j}^\dagger\right] = 0.$$

Thus defining the number operator

$$N_i = {a_i}^\dagger a_i$$

and using the orthonormality of the eigenfunctions, we see that

$$N = \sum_i N_i, \quad H = \sum_i N_i \epsilon_i. \tag{4.1.15}$$

Because the operators N_i commute with each other, they can be diagonalized simultaneously with H.

Now we can choose a basis of kets $|n_1, n_2, \ldots\rangle$ labelled by the eigenvalues of the operators $N_1, N_2, \ldots$ Equation (4.1.14) implies that the operators $a_i, {a_i}^\dagger$ satisfy relations

$$\begin{aligned} a_i\, |n_1, n_2, \ldots, n_i, \ldots\rangle &= n_i^{1/2}\, |n_1, n_2, \ldots, n_i - 1, \ldots\rangle \\ {a_i}^\dagger\, |n_1, n_2, \ldots, n_i, \ldots\rangle &= (n_i + 1)^{1/2}\, |n_1, n_2, \ldots, n_i + 1, \ldots\rangle \end{aligned} \tag{4.1.16}$$

Notice that the n_i are non-negative integers, limited only by the fact that $\sum n_i = n$ for each fixed n. Although the total number of particles, N, is a constant of the motion, there will be cases in which H induces transitions between eigenstates of the Hamiltonian; then the N_i will change in time according to the equation $i\hbar\partial_0\, N_i = [N_i, H]$.

The wavefunctions of a system of independent indistinguishable particles will either be completely symmetric (the case of Bose-Einstein statistics) or completely ant-isymmetric (Fermi-Dirac statistics). Let $X = (x_1, x_2, \ldots x_n)$ denote the space-time coordinates of n indistinguishable particles. Let $|\Phi\rangle$,

$|\Psi\rangle$ be two states in the relevant Hilbert space $\mathcal{H}^{(n)}$. The configuration space amplitudes are given by the functions

$$\Phi(X) = \langle X \,|\, \Phi\rangle, \quad \Psi(X) = \langle X \,|\, \Psi\rangle,$$

with scalar product

$$\langle \Phi \,|\, \Psi\rangle = \int dX \Phi^\dagger(X)\Psi(X) \tag{4.1.17}$$

where the integration runs over all $X = (x_1, x_2, \ldots x_n)$. Let $X \to X' = PX$ denote a permutation of the particle labels $X = (x_1, x_2 \ldots x_n)$ into $X' = (x_{\alpha_1}, x_{\alpha_2}, \ldots x_{\alpha_n})$ so that $\Psi(X) = \Psi(x_1, x_2, \ldots x_n)$ maps into $\Psi(PX) = \Psi'(X)$. Because the linear transformation $X \to X'$ has unit Jacobian, the value of the scalar product (4.1.17) is preserved, and

$$\langle \Phi \,|\, \Psi\rangle = \langle \Phi' \,|\, \Psi'\rangle$$

so that for each permutation P there exists a unitary transformation $\mathcal{P} : \mathcal{H}^{(n)} \to \mathcal{H}^{(n)}$ for which

$$\Psi'(X) = \Psi(PX) = \mathcal{P}\Psi(X).$$

Let $\mathcal{O}$ be any observable. If the particles are indistinguishable, then under permutation of particle coordinates

$$\langle \Phi,\, \mathcal{O}\,\Psi\rangle = \langle \Phi',\, \mathcal{O}\,\Psi'\rangle = \langle \mathcal{P}\Phi,\, \mathcal{O}\,\mathcal{P}\Psi\rangle = \langle \Phi,\, \mathcal{P}^{-1}\mathcal{O}\mathcal{P}\,\Psi\rangle$$

so that

$$\mathcal{O} = \mathcal{P}^{-1}\mathcal{O}\mathcal{P}, \longrightarrow [\mathcal{O},\, \mathcal{P}] = 0.$$

Thus every such operator $\mathcal{O}$ must be a symmetric function of the particle coordinates $X = (x_1, x_2, \ldots x_n)$.

A transposition $\mathcal{T}_{ij}$ is a permutation in which only two indices, say i and j, are interchanged. A *symmetric* function is left unchanged by a transposition,

$$\mathcal{T}_{ij}\Psi(x_1, x_2, \ldots x_n) = \Psi(x_1, x_2, \ldots x_n), \tag{4.1.18}$$

whereas an *anti-symmetric* function changes sign:

$$\mathcal{T}_{ij}\Psi(x_1, x_2, \ldots x_n) = -\Psi(x_1, x_2, \ldots x_n). \tag{4.1.19}$$

It is elementary to prove that all symmetric states in $\mathcal{H}^{(n)}$ are orthogonal to all anti-symmetric states, and no observable on $\mathcal{H}^{(n)}$ can couple symmetric and anti-symmetric states.

Every permutation P can be written, usually in more than one way, as a product of transpositions. However the number of transpositions in any such representation is always either even (in which case P is said to have parity $\pi_P = +1$) or odd ($\pi_P = -1$). The set of all permutations on n objects forms a group: see, for example [3, Chap. 7] or [4, Chap. 17]. We can construct two projection operators, the *symmetrizer* $\mathcal{S}$ and *anti-symmetrizer* $\mathcal{A}$

$$\mathcal{S} = \frac{1}{n!} \sum_{P \in S_n} P, \quad \mathcal{A} = \frac{1}{n!} \sum_{P \in S_n} \pi_P P \tag{4.1.20}$$

which allow us to project any state onto the corresponding symmetric or anti-symmetric subspace of $\mathcal{H}^{(n)}$. We are concerned only with the symmetrizer $\mathcal{S}$ in this section. Suppose we label one-particle amplitudes with eigenvalues α_i so that the amplitude $\phi_{\alpha_i}(x)$ occurs n_i times in the n-particle wavefunction. The numbers $n_1, n_2, \ldots$ define what we may call a *configuration*. The set of vectors $|\, n; n_1, n_2 \ldots \rangle$ spans the Hilbert space of realizable states of n identical Bose particles so that the relations

$$\sum_{[n]} |\, n; n_1, n_2 \ldots \rangle \langle\, n; n_1, n_2 \ldots | = 1$$

$$\langle\, n; n_1, n_2 \ldots |\, n'; n_1', n_2' \ldots \rangle = \delta_{nn'} \delta_{n_1 n_1'} \delta_{n_2 n_2'} \cdots$$

express the completeness and orthonormality of the set. Consider now the amplitude

$$\begin{aligned} &\langle\, x_1, x_2, \ldots, x_n; n |\, n_1, n_2, \ldots \rangle \\ &= \sqrt{\frac{n!}{n_1! n_2! \ldots}} \mathcal{S}\, \phi_{\alpha_1}(x_1) \ldots \phi_{\alpha_n}(x_n) \end{aligned} \tag{4.1.21}$$

where the numerical factor in front normalizes the expression. In the next section, we study the case in which the projection operator $\mathcal{S}$ is replaced by the anti-symmetrizer $\mathcal{A}$; the right-hand side then becomes a determinant. Expanding in terms of states of the particle at x_1 gives

$$\begin{aligned} &\langle\, x_1, x_2, \ldots, x_n; n |\, n_1, n_2, \ldots, n_i, \ldots \rangle = \\ &\sum_{i=1}^{\infty} \sqrt{\frac{n_i}{n}} \phi_{\alpha_i}(x_1) \cdot \langle\, x_2, x_3, \ldots, x_n; n-1 |\, n_1, n_2, \ldots, n_i - 1, \ldots \rangle \end{aligned} \tag{4.1.22}$$

where the sum runs over all possible eigenstates, empty as well as occupied. Alternatively, we can write

$$\begin{aligned} \langle\, x_1, x_2, \ldots, x_n; n |\, n_1, n_2, \ldots, n_i, \ldots \rangle &= \sum_{j=1}^{n} \sqrt{\frac{1}{n n_i}} \phi_{\alpha_i}(x_j) \\ \times \langle\, x_1, x_2, \ldots, x_{j-1}, x_{j+1}, \ldots, x_n; n-1 |\, &n_1, n_2, \ldots, n_i - 1, \ldots \rangle. \end{aligned} \tag{4.1.23}$$

Fock space is defined as a direct sum of subspaces,

$$\mathcal{H} = \mathcal{H}^{(0)} \oplus \mathcal{H}^{(1)} \oplus \ldots, \tag{4.1.24}$$

each of which is characterized by a fixed value of n. The field operators $\psi(x)$ and $\psi^{\dagger}(x)$ acting on the elements of Fock space respectively create and annihilate particles:

$$\psi^\dagger(x)\,|\,x_1,\ldots,x_{n-1};n-1\,\rangle = n^{1/2}\,|\,x,x_1,\ldots,x_{n-1};n\,\rangle, \tag{4.1.25}$$

$$\begin{aligned}\psi(x)\,|\,x_1,\ldots,x_{n+1};n+1\,\rangle &= \\ (n+1)^{-1/2}\sum_{i=1}^{n}\delta(x_i-x)\,|\,x_1,\ldots,x_{i-1},x_{i+1},\ldots,x_{n+1};n\,\rangle,\end{aligned} \tag{4.1.26}$$

relating states in the subspace $\mathcal{H}^{(n)}$ to those of $\mathcal{H}^{(n\pm1)}$.

4.1.3 Identical particles: the antisymmetric case

Fermi particles such as electrons satisfy Pauli's exclusion principle: each non-degenerate state may be occupied by at most one particle, so that the numbers n_i can only take the values 0 or 1. This can be accommodated by replacing the symmetrizer $\mathcal{S}$ in (4.1.21) by the anti-symmetrizer $\mathcal{A}$, so that

$$\langle\, x_1,x_2,\ldots,x_n;n\,|\,n_1,n_2,\ldots\rangle = \sqrt{n!}\,\mathcal{A}\,\phi_{\alpha_1}(x_1)\ldots\phi_{\alpha_n}(x_n). \tag{4.1.27}$$

The anti-symmetrizer generates a determinantal product

$$\langle\, x_1,x_2,\ldots,x_n;n\,|\,n_1,n_2,\ldots\rangle = \frac{1}{\sqrt{n!}}\begin{vmatrix}\phi_{\alpha_{j_1}}(x_1) & \phi_{\alpha_{j_2}}(x_1) & \ldots & \phi_{\alpha_{j_n}}(x_1)\\ \phi_{\alpha_{j_1}}(x_2) & \phi_{\alpha_{j_2}}(x_2) & \ldots & \phi_{\alpha_{j_n}}(x_2)\\ \vdots & \vdots & & \vdots\\ \phi_{\alpha_{j_1}}(x_n) & \phi_{\alpha_{j_2}}(x_n) & \ldots & \phi_{\alpha_{j_n}}(x_n)\end{vmatrix}. \tag{4.1.28}$$

The determinant vanishes if any two rows or columns are identical, and changes sign under transposition of either rows or columns. Thus α_i can appear at most once if the determinant is not to vanish so that n_i can take only the values 0 or 1. The subscripts $(j_1,\ldots,j_n)$ distinguish the n occupied states of the full (infinite) list of eigenstates in some order. Expanding (4.1.28) by the first row gives

$$\begin{aligned}\langle\, x_1,x_2,\ldots,x_n;n\,|\,n_1,n_2,\ldots\rangle &= \frac{1}{\sqrt{n}}\sum_{i=1}^{n} n_{j_i}\,\phi_{\alpha_{j_i}}(x_1)\\ &\times(-1)^{s_i}\langle\, x_2,x_3,\ldots,x_n;n-1\,|\,n_1,n_2,\ldots,n_{j_i}-1,\ldots\rangle,\end{aligned} \tag{4.1.29}$$

where

$$s_i = \sum_{k=1}^{j_i-1} n_k$$

is the number of states occupied up to the j_i-th. Alternatively we can expand by the j_i-th column, so that

$$\begin{aligned}\langle\, x_1,x_2,\ldots,x_n;n\,|\,n_1,n_2,\ldots\rangle &= \frac{1}{\sqrt{n}}\sum_{k=1}^{n}(-1)^{k-1}\phi_{\alpha_{j_i}}(x_k)\\ \times\langle\, x_1,x_2,\ldots,x_{k-1},x_{k+1},\ldots,x_n;n-1\,|\,n_1,n_2,\ldots,n_{j_i}-1,\ldots\rangle.\end{aligned} \tag{4.1.30}$$

The creation and annihilation operators have the properties

$$\begin{aligned} a_i{}^{\dagger}\,|\,n_1, n_2, \ldots, n_i, \ldots\rangle &= (-1)^{s_i}(1-n_i)\,|\,n_1, n_2, \ldots, n_i+1, \ldots\rangle \\ a_i\,|\,n_1, n_2, \ldots, n_i, \ldots\rangle &= (-1)^{s_i} n_i\,|\,n_1, n_2, \ldots, n_i-1, \ldots\rangle, \end{aligned} \tag{4.1.31}$$

and satisfy the anti-commutation relations

$$\{a_i, a_j\} = \{a_i{}^{\dagger}, a_j{}^{\dagger}\} = 0, \quad \{a_i, a_j{}^{\dagger}\} = \delta_{ij}. \tag{4.1.32}$$

Because $\{a_i, a_i\} = 0$, we have $a_i a_i = 0$, and similarly $a_i{}^{\dagger} a_i{}^{\dagger} = 0$. Also, as $n_i = 0, 1$, $n_i^2 = n_i$ and therefore the number operator, $N_i = a_i{}^{\dagger} a_i$ is also a projection operator, as may easily be checked using the relations (4.1.32).

In Fock space, the relations corresponding to (4.1.25) and (4.1.26) are replaced by

$$\mathcal{A}\,|\,x_1, x_2, \ldots, x_n; n\,\rangle = \frac{1}{\sqrt{n!}}\psi^{\dagger}(x_n)\ldots\psi^{\dagger}(x_1)\,|\,0\,\rangle \tag{4.1.33}$$

This result is particularly convenient when constructing matrix elements for complex electron configurations for operators that act symmetrically on all the particles present so that

$$F(x_1, x_2, \ldots, x_n; n) = \sum_{j=1}^{n} f(x_j)$$

when they act only on a single particle, or

$$G(x_1, x_2, \ldots, x_n; n) = \sum_{i<j} g(x_i, x_j) = \frac{1}{2}\sum_{i \neq j} g(x_i, x_j)$$

when they involve the coordinates of two particles. The Fock space equivalent operators are

$$F = \int \psi^{\dagger}(x)\, f(x)\, \psi(x)\, dx \tag{4.1.34}$$

$$G = \frac{1}{2}\iint \psi^{\dagger}(x')\,\psi^{\dagger}(x)\, g(x, x')\,\psi(x)\,\psi(x')\, dx\, dx' \tag{4.1.35}$$

in which the order of the coordinate arguments in the field operators is significant. This can easily be extended to symmetric operators involving three or more particles simultaneously.

4.2 Quantization of the electron-positron field

4.2.1 The Furry picture

Furry [5] proposed a formalism that is widely used in atomic and molecular structure theory. Suppose, as a first approximation, that we can treat the electrons as moving independently in a conservative potential field $V(\boldsymbol{x})$. The unquantized Dirac Hamiltonian in configuration space is

$$h_D = c\boldsymbol{\alpha}\cdot\boldsymbol{p} + \beta mc^2 + V(\boldsymbol{x}), \tag{4.2.1}$$

where the electron rest mass $m = 1$ a.u.. When $V(\boldsymbol{x})$ is negative-definite (and satisfies other reasonable conditions which we shall discuss later), as for example in hydrogenic ions,

$$V(\boldsymbol{x}) = -\frac{Z}{r}, \quad r = |\boldsymbol{x}|,$$

we know [6, §7.4] that h_D has a spectrum consisting of two continua, $-\infty < E < -c^2$ and $c^2 < E < \infty$ together with a countably infinite set of bound states in the gap $(-c^2, +c^2)$ having a point of accumulation at $+c^2$.

We assume that the quantized Dirac field amplitudes can be expanded in terms of a complete set of eigenstates of h_D in the form

$$\psi(x) = \sum_r a_r\psi_r(x), \quad \psi^\dagger(x) = \sum_r {a_r}^\dagger\psi_r^\dagger(x)$$

where a_r and ${a_r}^\dagger$ are anti-commuting fermion annihilation and creation operators respectively, (4.1.32), and $\psi_r(x)$ is an eigenfunction of h_D with the energy E_r. Writing

$$N_r = {a_r}^\dagger a_r$$

for the number operator we see that, as in Section 4.1.1, not only is the total number of electrons defined as the expectation of the operator

$$N = \sum_r N_r,$$

but the Hamiltonian operator is

$$H = \sum_r N_r E_r$$

which, because E_r can have either positive or negative values whilst N_r is a non-negative operator, can be either positive or negative. Were the operators a_r to satisfy Bose commutation relations like (4.1.14), N_r could have arbitrary non-negative integer eigenvalues and the field could have an arbitrarily large negative energy. On the other hand, if the operators a_r satisfy Fermi anti-commutation relations like (4.1.32), there can be at most one particle in each nondegenerate eigenstate in accordance with the Pauli exclusion principle.

We split the energy spectrum $\mathcal{S}$ into two disjoint pieces: $\mathcal{S}^{(+)} = \{E_r \mid E_r > 0\}$, which includes the positive energy continuum and the bound states, and its complement $\mathcal{S}^{(-)} = \{E_r \mid E_r < 0\}$ so that the number operator is

$$N = \sum_r^{(+)} N_r^{(+)} + \sum_s^{(-)} N_s^{(-)},$$

and the Hamiltonian operator is

$$H = \sum_r^{(+)} N_r^{(+)} E_r^{(+)} - \sum_s^{(-)} N_s^{(-)} |E_s^{(-)}|, \tag{4.2.2}$$

Likewise, if we write $q = -e$ for the elementary charge on the electron, the total charge operator, which is also a constant of the motion, is

$$Q = -e \sum_r^{(+)} N_r^{(+)} - e \sum_s^{(-)} N_s^{(-)}. \tag{4.2.3}$$

Now according to Dirac's hole theory, Section 2.5.6, the vacuum state is one in which the positive energy states are all empty, $N_r^{(+)} = 0$, and the negative energy states are all occupied, $N_s^{(-)} = 1$. The vacuum is therefore characterized by the infinite unobservable quantities

$$Q_{vacuum} = -e \sum_s^{(-)} 1, \quad E_{vacuum} = - \sum_s^{(-)} |E_s^{(-)}|.$$

The hole theory therefore postulates that only the differences $Q' = Q - Q_{vacuum}$ and $E' = E - E_{vacuum}$ are observable, so that

$$Q' = -e \sum_r^{(+)} N_r^{(+)} + e \sum_s^{(-)} (1 - N_s^{(-)})$$

$$E' = \sum_r^{(+)} N_r^{(+)} E_r^{(+)} + \sum_s^{(-)} (1 - N_s^{(-)}) |E_s^{(-)}|.$$

A "hole" in a negative energy state ($N_s^{(-)} = 0$) therefore contributes a charge $+e$ to Q' and a positive contribution $|E_s^{(-)}|$ to the field energy. We therefore interpret such a hole as an "anti-particle", in this case a positron, carrying charge $+e$ and positive energy.

To emphasize the interpretation of "holes" in the filled vacuum as anti-particles, we introduce new operators, $b_s, {b_s}^\dagger$, which respectively destroy and create antiparticles, such that

$$b_s = {a_s}^\dagger, \quad {b_s}^\dagger = a_s, \quad s \in \mathcal{S}^{(-)} \tag{4.2.4}$$

so that we can redefine the field variables as

$$\psi(x) = \sum_{r}^{(+)} a_r \phi_r(x) + \sum_{s}^{(-)} {b_s}^{\dagger} \phi_s(x), \tag{4.2.5}$$

$$\psi^{\dagger}(x) = \sum_{r}^{(+)} {a_r}^{\dagger} \phi_r^{\dagger}(x) + \sum_{s}^{(-)} b_s \phi_s^{\dagger}(x). \tag{4.2.6}$$

The non-vanishing anti-commutators become

$$\left\{ {a_r}^{\dagger}, a_s \right\} = \delta_{rs}, \quad r, s \in \mathcal{S}^{(+)}, \qquad \left\{ {b_r}^{\dagger}, b_s \right\} = \delta_{rs}, \quad r, s \in \mathcal{S}^{(-)}, \tag{4.2.7}$$

whilst all other anti-commutator pairs vanish.

Any sequence of creation and annihilation operators is said to be in *normal order* if it is arranged so that *all* creation operators stand to the left of *all* annihilation operators in the sequence. Thus ${a_r}^{\dagger} b_s$ is in normal order, but $a_r {b_s}^{\dagger}$ is not. The latter can be put into normal order by transposing adjacent operators and changing signs as if all anti-commutators vanished. We denote such a *normal product* by writing $: a_r {b_s}^{\dagger} :$, sometimes written $N(a_r {b_s}^{\dagger})$, so that, for every pair of indices r, s,

$$\begin{aligned} &: a_r a_s : = +a_r a_s, \qquad &&: b_r b_s : = +b_r b_s, \\ &: a_r a_s^{\dagger} : = -{a_s}^{\dagger} a_r, \qquad &&: b_r b_s^{\dagger} : = -{b_s}^{\dagger} b_r, \\ &: a_r b_s^{\dagger} : = -{b_s}^{\dagger} a_r, \qquad &&: b_r a_s^{\dagger} : = -{a_s}^{\dagger} b_r. \end{aligned} \tag{4.2.8}$$

We suppose the distributive law holds. It is easy to see that (4.2.3) and (4.2.4) are equivalent to

$$H = : H : + E_{vacuum}, \qquad Q = : Q : + Q_{vacuum}, \tag{4.2.9}$$

so that the operators which must be used to give physical results are $: H :$ and $: Q :$. The infinite vacuum energy and charge can then be discarded. We shall later need the normally ordered charge-current density vector

$$j^{\mu}(x) = -ec\widetilde{\psi}(x)\gamma^{\mu}\psi(x) = \; -ec : \widetilde{\psi}(x)\gamma^{\mu}\psi(x) : + j^{\mu}_{vacuum}(x) \tag{4.2.10}$$

so that the total charge is $Q = \int d^3x j^0(x)/c$.

4.2.2 The free electron case

Textbooks of quantum electrodynamics, such as [1, 2, 7, 8], traditionally present the canonical quantization of the electron-positron field for *free* electrons for which the potential $V(\boldsymbol{x})$ vanishes. This has two advantages: in the first place, the absence of external forces ensures that all the space-time symmetries, in particular space-like translations, can be invoked. Secondly, one can define the field variables $\psi(x)$ and $\psi^{\dagger}(x)$ without reference to an eigenfunction expansion (4.2.5). If the electron is firmly confined in an atom or molecule, a perturbation treatment encounters difficulties, and the Furry picture is appropriate. Nevertheless an understanding of the elements of free particle QED helps to understand QED of electrons in all environments.

The free particle eigenfunctions, (3.1.20), are

$$\psi(x) = \mathrm{e}^{-ix\cdot p/\hbar}\, u(p)$$

where the spinors $u(p)$, (3.1.24), are

$$u^{(r)}(p) = \left(\frac{E(|\boldsymbol{p}|)+c^2}{2E(|\boldsymbol{p}|)}\right)^{1/2} \begin{pmatrix} \phi^r \\ \dfrac{c\,\boldsymbol{\sigma}\cdot\boldsymbol{p}}{E(|\boldsymbol{p}|)+c^2}\,\phi^r \end{pmatrix} \qquad r = 1, 2,$$

for positive frequency (energy), $p^0 > 0$, and by (3.1.25),

$$v^{(r)}(p) = \left(\frac{E(|\boldsymbol{p}|)+c^2}{2E(|\boldsymbol{p}|)}\right)^{1/2} \begin{pmatrix} \dfrac{-c\,\boldsymbol{\sigma}\cdot\boldsymbol{p}}{E(|\boldsymbol{p}|)+c^2}\,\phi^r \\ \phi^r \end{pmatrix} \qquad r = 1, 2$$

for negative frequency, $p^0 < 0$, with

$$\phi^1 = \begin{pmatrix} 1 \\ 0 \end{pmatrix}, \quad \phi^2 = \begin{pmatrix} 0 \\ 1 \end{pmatrix}.$$

The Heisenberg field operators are given by

$$\psi(x) = \psi^{(+)}(x) + \psi^{(-)}(x), \quad \overline{\psi}(x) = \overline{\psi}^{(+)}(x) + \overline{\psi}^{(-)}(x), \tag{4.2.11}$$

where, with $p^0 = +E(|\boldsymbol{p}|)/c$, the positive frequency components

$$\psi^{(+)}(x) = \int \frac{d^3p}{(2\pi)^{3/2}} \left(\frac{c}{p^0}\right)^{1/2} \sum_{r=1,2} a_r(\boldsymbol{p})\, u^{(r)}(\boldsymbol{p})\, \mathrm{e}^{-ix\cdot p/\hbar} \tag{4.2.12}$$

$$\overline{\psi}^{(+)}(x) = \int \frac{d^3p}{(2\pi)^{3/2}} \left(\frac{c}{p^0}\right)^{1/2} \sum_{r=1,2} b_r(\boldsymbol{p})\, \overline{v}^{(r)}(\boldsymbol{p})\, \mathrm{e}^{-ix\cdot p/\hbar} \tag{4.2.13}$$

respectively destroy an electron (4.2.12) or a positron (4.2.13), and the negative frequency components

$$\overline{\psi}^{(-)}(x) = \int \frac{d^3p}{(2\pi)^{3/2}} \left(\frac{c}{p^0}\right)^{1/2} \sum_{r=1,2} a_r{}^{\dagger}(\boldsymbol{p})\, \overline{u}^{(r)}(\boldsymbol{p})\, \mathrm{e}^{+ix\cdot p/\hbar} \tag{4.2.14}$$

$$\psi^{(-)}(x) = \int \frac{d^3p}{(2\pi)^{3/2}} \left(\frac{c}{p^0}\right)^{1/2} \sum_{r=1,2} b_r{}^{\dagger}(\boldsymbol{p})\, v^{(r)}(\boldsymbol{p})\, \mathrm{e}^{+ix\cdot p/\hbar} \tag{4.2.15}$$

respectively create an electron (4.2.14) or a positron (4.2.15). The integrals over momentum replace the sum over eigenfunctions of (4.2.5) and (4.2.6). Clearly,

$$\overline{\psi^{(+)}}(x) = \overline{\psi}^{(-)}(x), \quad \overline{\psi^{(-)}}(x) = \overline{\psi}^{(+)}(x)$$

and the vacuum state $|\,\Phi_0\rangle$ is characterized by

$$\psi^{(+)}(x)\,|\,\Phi_0\rangle = \overline{\psi}^{(+)}(x)\,|\,\Phi_0\rangle = 0. \tag{4.2.16}$$

The anti-commutators

$$\begin{aligned} \left\{\psi^{(+)}(x),\, \psi^{(-)}(x')\right\} &= 0, \quad \left\{\overline{\psi}^{(-)}(x),\, \overline{\psi}^{(+)}(x')\right\} = 0 \\ \left\{\overline{\psi}^{(+)}(x),\, \psi^{(+)}(x')\right\} &= 0, \quad \left\{\overline{\psi}^{(-)}(x),\, \psi^{(-)}(x')\right\} = 0, \end{aligned} \tag{4.2.17}$$

are a consequence of (4.2.7); however,

$$\begin{aligned} &\left\{\psi^{(+)}(x),\, \overline{\psi}^{(-)}(x')\right\} = \\ &\quad \frac{1}{(2\pi)^3}\int d^3p \int d^3p' \frac{c}{\sqrt{p^{\,0}\,p^{\prime\,0}}} \\ &\quad \times \sum_{r,s=1}^{2} \left\{a_r(\boldsymbol{p}),\, a_s^\dagger(\boldsymbol{p}')\right\} u^{(r)}(\boldsymbol{p}) \otimes \overline{u}^{(s)}(\boldsymbol{p}')\,\mathrm{e}^{-ix\cdot p + ix'\cdot p'} \\ &= \frac{1}{2(2\pi)^3}\int_{p^0>0} \frac{d^3p}{p^{\,0}}\,(\not p + c)\,\mathrm{e}^{-ix\cdot(p-p')} \\ &= \frac{1}{2(2\pi)^3}(i\gamma^\mu\partial_\mu + c)\int_{p^0>0}\frac{d^3p}{p^{\,0}}\,\mathrm{e}^{-ix\cdot(p-p')} \\ &= i\,(i\gamma^\mu\partial_\mu + c)\,\Delta^{(+)}(x-x'). \end{aligned} \tag{4.2.18}$$

Equation (3.1.29) has been used in the second line and (2.9.22) and (2.9.16) are needed for the final result, usually written

$$\left\{\psi^{(+)}(x),\, \overline{\psi}^{(-)}(x')\right\} = -iS^{(+)}(x-x'), \tag{4.2.19}$$

and similarly

$$\{\psi^{(-)}(x),\, \overline{\psi}^{(+)}(x')\} = -iS^{(-)}(x-x'). \tag{4.2.20}$$

Assembling the pieces, using (4.2.11), gives

$$\left\{\psi(x),\, \overline{\psi}(x')\right\} = -iS(x-x') \tag{4.2.21}$$

where

$$S(x-x') = S^{(+)}(x-x') + S^{(-)}(x-x').$$

The equal time anti-commutators are somewhat simpler: using (2.9.20), we see that

$$\left\{\psi(x),\, \overline{\psi}^{(+)}(x')\right\}\Big|_{x^0=x'^{\,0}} = -\gamma^0\partial_0\Delta(x-x')\Big|_{x^0=x'^{\,0}}$$
$$= +\gamma^0\delta^{(3)}(\boldsymbol{x}-\boldsymbol{x}'),$$

equivalent to

$$\left\{\psi(x),\, \psi^{\dagger}(x')\right\}\Big|_{x^0=x'^{\,0}} = \delta^{(3)}(\boldsymbol{x}-\boldsymbol{x}'),$$

verifying the canonical quantization rule (4.1.8).

The charge conjugation operator C, (3.1.36), interchanges particle and anti-particle states:

$$\psi_c(x) = C\,\overline{\psi}^{\,t}(x), \quad \psi(x) = C\,\left[\overline{\psi}_c(x)\right]^t.$$

The corresponding positive and negative frequency operators are

$$\psi^{(-)}(x) = C\,\left[\overline{\psi_c^{(+)}}(x)\right]^t, \quad \psi^{(+)}(x) = \left[C^{-1}\psi_c^{(+)}(x)\right]^t \tag{4.2.22}$$

so that

$$\psi_c^{(+)}(x) = \int \frac{d^3p}{(2\pi)^{3/2}}\left(\frac{c}{p^0}\right)^{1/2}\sum_{r=1,2} b_r(\boldsymbol{p})\,u_c^{(r)}(\boldsymbol{p})\,\mathrm{e}^{-ix\cdot p/\hbar} \tag{4.2.23}$$

$$\overline{\psi_c^{(+)}}(x) = \int \frac{d^3p}{(2\pi)^{3/2}}\left(\frac{c}{p^0}\right)^{1/2}\sum_{r=1,2} b_r^{\dagger}(\boldsymbol{p})\,\overline{u}_c^{(r)}(\boldsymbol{p})\,\mathrm{e}^{+ix\cdot p/\hbar} \tag{4.2.24}$$

where $u_c^{(r)}(\boldsymbol{p}) = C\overline{v^{(r)}}(\boldsymbol{p})^t$, reinforcing the interpretation of $\psi_c^{(+)}(x)$ as an anti-particle destruction operator and $\overline{\psi_c^{(+)}}(x)$ as a creation operator. Then, because $C^{-1}S^{(\pm)}(-x)C = -S^{(\pm)}(x)^t$,

$$\left\{\psi_c^{(+)}(x), \overline{\psi_c^{(+)}}(x')\right\} = -iS^{(+)}(x-x') \tag{4.2.25}$$

and

$$\left\{\psi_c(x), \overline{\psi_c}(x')\right\} = -iS(x-x'). \tag{4.2.26}$$

A covariant expression for the electron number operator N^e is

$$N^e = \int d\sigma^{\mu}x\,\overline{\psi^{(+)}}(x)\gamma_{\mu}\,\psi^{(+)}(x) = \int d^3x\,\overline{\psi^{(+)}}(x)\gamma_0\,\psi^{(+)}(x) \tag{4.2.27}$$

and similarly, the positron number operator is

$$N^p = \int d^3x\,\overline{\psi_c^{(+)}}(x)\gamma_0\,\psi_c^{(+)}(x) \tag{4.2.28}$$

The total charge is then

$$Q = \int d\sigma^\mu x\, j_\mu(x)/c = -e\,(N^e - N^p), \tag{4.2.29}$$

consistent with (4.2.10) if we drop the vacuum current, and define the current as a normal product

$$j_\mu(x) = -ec : \overline{\psi}(x)\,\gamma^\mu\,\psi(x) : = -ecN(\overline{\psi}(x)\,\gamma^\mu\,\psi(x)). \tag{4.2.30}$$

Because normal products are designed to have zero vacuum expectation, this gives us a convenient expression for the vacuum current $-e\,c\langle\overline{\psi}(x)\,\gamma^\mu\,\psi(x)\rangle_0$ in terms of propagators. Using (4.2.11), (4.2.16), (4.2.17), and (4.2.20) we find

$$\overline{\psi}(x)\psi(x') = N(\overline{\psi}(x)\psi(x')) - iS^{(-)}(x'-x)^t \tag{4.2.31}$$

so that

$$\begin{aligned} -ec\langle\overline{\psi}(x)\,\gamma^\mu\,\psi(x)\rangle_0 &= -ec \lim_{x\to x'} \langle\Phi_0, \overline{\psi}(x)\,\gamma^\mu\,\psi(x')\Phi_0\rangle \\ &= -iec \lim_{x\to x'} \mathrm{tr}\,(\gamma^\mu S^{(-)}(x'-x)) \\ &= -iec\,\mathrm{tr}\,(\gamma^\mu S^{(-)}(0)). \end{aligned} \tag{4.2.32}$$

This is infinite, corresponding to the charge of the "negative energy sea" in the language of Dirac hole theory.

4.3 Quantization of the Maxwell field

Canonical quantization of the Maxwell field is less straightforward because of of the gauge constraint needed to define the four-potential uniquely. We saw that Maxwell's equations can be derived from the classical Lagrangian, (2.6.19)

$$\mathcal{L}_{em} = -\frac{1}{4}\epsilon_0 F_{\mu\nu}F^{\mu\nu} - \frac{1}{c}j^\mu a_\mu, \tag{4.3.1}$$

and that charge-current four-vector conservation,

$$\partial_\mu j^\mu(x) = 0,$$

is satisfied if

$$\partial_\mu a^\mu(x) = 0. \tag{4.3.2}$$

The canonically conjugate momenta to the $a_\mu(x)$ are given by

$$\pi^\mu = \frac{\partial\mathcal{L}_{em}}{\partial(\partial_0 a_\mu)} = \epsilon_0(\partial^\mu a^0 - \partial^0 a^\mu) = \epsilon_0 F^{\mu 0} \tag{4.3.3}$$

so that π^0 is identically zero. This is inconsistent with the equal time rules for canonical quantization,

$$[a_\mu(x), \pi^\nu(x')]_{x^0=x'^0} = i\delta_\mu{}^\nu \delta^3(\boldsymbol{x} - \boldsymbol{x}'), \tag{4.3.4}$$

which cannot be satisfied when $\pi^0 = 0$. One way to preserve (4.3.4) is to incorporate the Lorentz condition in the Lagrangian density using a Lagrange multiplier, λ, for example

$$\mathcal{L} = \epsilon_0 \left\{ -\frac{1}{4}(\partial_\mu a_\nu - \partial_\nu a_\mu)(\partial^\mu a^\nu - \partial^\nu a^\mu) + \frac{1}{2}\lambda(\partial_\mu a^\mu)^2 \right\} - \frac{1}{c} j^\mu a_\mu. \tag{4.3.5}$$

This modifies the Maxwell equations so that

$$\Box a^\mu + (\lambda - 1)\partial^\mu(\partial_\nu a^\nu) = \frac{1}{\epsilon_0 c} j^\mu \tag{4.3.6}$$

The canonical momenta are now

$$\pi^\mu = \epsilon_0 \left(F^{\mu 0} - \lambda g^{\mu 0}(\partial_\nu a^\nu) \right) \tag{4.3.7}$$

so that π^0 no longer vanishes as long as $\lambda \neq 0$.

When the current is conserved, the four-divergence of (4.3.6) reduces to

$$\Box \chi = 0, \quad \chi = \partial_\nu a^\nu$$

whenever $\lambda \neq 0$, so that χ can be thought of as a scalar field. Classically, we could solve ths equation uniquely by imposing suitable initial conditions, for example $\chi = 0, \quad \partial_0 \chi = 0$ as $x^0 \to -\infty$; regrettably, this would make χ is identically zero. Fortunately, a consistent quantized Maxwell theory can be constructed using a weaker subsidiary condition. Assume that the field operators are Hermitean, with the equal time commutation rules

$$[\partial_0 a_\mu(x),\, a_\nu(x')]_{x^0=x'^0} = -icg_{\mu\nu}\delta^3(\boldsymbol{x} - \boldsymbol{x}')/\epsilon_0, \tag{4.3.8}$$

equivalent to (4.3.4), together with

$$[a_\mu(x),\, a_\nu(x')]_{x^0=x'^0} = 0, \quad [\partial_0 a_\mu(x),\, \partial_0 a_\nu(x')]_{x^0=x'^0} = 0. \tag{4.3.9}$$

Because $g_{00} = -g_{11} = -g_{22} = -g_{33} = +1$, the rule (4.3.8) for time-like components has a different sign from that for space-like components.

Now set $\lambda = 1$ for simplicity, so that the free field form of (4.3.6) is just the wave equation

$$\Box a_\mu(x) = 0.$$

As in the free particle case, it is convenient to expand the field operator components in plane waves so that

$$a_\mu(x) = \sqrt{\frac{c}{2(2\pi)^3\epsilon_0}} \int \frac{d^3k}{k_0} \times \sum_{\lambda=0}^{3} \left\{ c^{(\lambda)}(k)\, \epsilon_\mu^{(\lambda)}(k)\, \mathrm{e}^{-ik\cdot x} + c^{(\lambda)\dagger}(k)\, \epsilon_\mu^{(\lambda)}(k)\, \mathrm{e}^{+ik\cdot x} \right\}, \tag{4.3.10}$$

where the four-vector $k^\mu = (k_0, \boldsymbol{k})$ defines the direction of travel of the plane wave and, as the particles are massless, $k^0 = k_0 = |\boldsymbol{k}|$. The four linearly independent polarization vectors $\epsilon^{(\lambda)}(k)$, here taken as real, are chosen so that

$$\epsilon_\mu^{(\lambda)}(k) \cdot \epsilon^{(\lambda')\mu}(k) = g^{\lambda\lambda'} \tag{4.3.11}$$

where $g^{\lambda\lambda'}$ is the Minkowski metric tensor. We define

$$c_\mu(k) = \sum_{\lambda=0}^{3} \epsilon_\mu^{(\lambda)}(k) c^{(\lambda)}(k), \quad c_\mu^\dagger(k) = \sum_{\lambda=0}^{3} \epsilon_\mu^{(\lambda)}(k) c^{(\lambda)\dagger}(k). \tag{4.3.12}$$

Then the commutation relations (4.3.8) and (4.3.9) are satisfied if

$$\begin{aligned} \left[c_\mu(k), c_\nu^\dagger(k')\right] &= -g_{\mu\nu} k_0 \delta^3(\boldsymbol{k} - \boldsymbol{k}'), \quad k_0 = |\boldsymbol{k}| \\ \left[c_\mu(k), c_\nu(k')\right] &= \left[c_\mu^\dagger(k), c_\nu^\dagger(k)'\right] = 0 \end{aligned} \tag{4.3.13}$$

from which we get

$$[a_\mu(x), a_\nu(x')] = -icg_{\mu\nu} D(x - x') \tag{4.3.14}$$

where, as in Section 2.9.3, $\epsilon_0 D(x)$ is given by (2.9.13) with the mass m set zero.

However, all is not yet quite secure. Suppose we define the vacuum state $|\,0\rangle$ by

$$c_\mu(k)|\,0\rangle = 0,$$

for all k and μ. Ignoring polarization for the moment, consider a one-particle state

$$|\,1\rangle = \sqrt{\frac{c}{2(2\pi)^3\epsilon_0}} \int \frac{d^3k}{k_0} f(k) c_\mu^\dagger(k)|\,0\rangle$$

Then a short calculation gives

$$\langle 1\,|\,1\rangle = -g_{\mu\nu} k_0 \langle 0\,|\,0\rangle \frac{c}{2(2\pi)^3\epsilon_0} \int \frac{d^3k}{k_0} |f(k)|^2.$$

Although space-like components $\mu = 1, 2, 3$ make positive contributions to $\langle 1\,|\,1\rangle$, the time-like component $\mu = 0$ has a negative contribution. Thus $\langle 1\,|\,1\rangle$ is not positive definite for all non-trivial vectors $|\,1\rangle$ and so the span of all such vectors is not, as it stands, a Hilbert space.

Until 1950, the only way to avoid this inconsistency was to abandon manifest Lorentz covariance and to eliminate the scalar and longitudinal modes with the aid of the (noncovariant) Coulomb gauge condition, $\operatorname{div} \boldsymbol{A} = 0$ (2.6.26). The Hamiltonian density $\mathcal{H} = T^{00}$ follows from (2.7.9) and (2.7.12)

$$\mathcal{H} = T^{00} = \frac{1}{2}\epsilon_0(\boldsymbol{E}_\perp \cdot \boldsymbol{E}_\perp + c^2 \boldsymbol{B} \cdot \boldsymbol{B}) + \frac{1}{2}\rho\phi - \boldsymbol{j}_\perp . \boldsymbol{A} \tag{4.3.15}$$

where $\boldsymbol{j}_\perp = \boldsymbol{j} - \epsilon_0 \,\mathrm{grad}\, \partial\phi/\partial t$ satisfies the transversality condition $\mathrm{div}\, \boldsymbol{j}_\perp = 0$ and $E_\perp = -\partial \boldsymbol{A}/\partial t$ and $\boldsymbol{B}$ both have zero divergence. Then Gupta [9] found a manifestly covariant way to quantize the free Maxwell equations that treats all four components of a^μ on the same footing and Bleuler [10] extended his argument to include the coupling to the electron field. This Lorentz covariant and gauge invariant formalism gives physical results which are identical to those derived from the Hamiltonian density (4.3.15) by restricting the class of admissable vectors $|\, 1\rangle$ to a proper Hilbert space.

We need only a brief sketch of the Gupta-Bleuler construction. Were the Lorentz condition to hold only in the mean, we should expect that the relevant Hilbert space would contain only states $|\,\psi\rangle$ for which

$$\langle \psi \,|\, \partial_\mu a^\mu \,|\, \psi\rangle = 0;$$

However this is still too restrictive and it is sufficient to require only that the positive frequency part of (4.3.10) annihilates any state in the one-photon space $\mathcal{H}_1$:

$$\partial_\mu a^{(+)\mu} \,|\, \psi\rangle = 0, \quad |\, \psi\rangle \in \mathcal{H}_1. \tag{4.3.16}$$

From (4.3.10),

$$\partial_\mu a^{(+)\mu}(x) = -i\sqrt{\frac{c}{2(2\pi)^3\epsilon_0}} \int \frac{d^3k}{k_0} \sum_{\lambda=0}^{3} k^\mu c_\mu(k)\, \mathrm{e}^{-ik\cdot x},$$

so that (4.3.16) simplifies to

$$k^\mu c_\mu(k) \,|\, \psi\rangle = 0. \tag{4.3.17}$$

To make things specific, we now take $\epsilon^0 = n$, where n is a unit vector along the time axis, $\epsilon^3 = \boldsymbol{k}/k^0$ along the direction of propagation, and ϵ^1, ϵ^2 perpendicular to ϵ^3 so as to satisfy the orthogonality conditions (4.3.11). A general n-photon state can be constructed as a linear superposition of n-fold products of states in $\mathcal{H}_1$, in particular those of the form

$$|\, \psi\rangle = |\, \psi_{tr}\rangle \times |\, \phi\rangle \tag{4.3.18}$$

where $|\, \psi_{tr}\rangle$ involves only products of photon operators polarized in the transverse directions 1 and 2, and $|\, \phi\rangle$ involves only the 0 (scalar) and 3 (longitudinal) polarizations. Because (4.3.17) automatically annihilates states of the form $|\, \psi_{tr}\rangle$, we need only consider its effect on the $|\, \phi\rangle$ states. With our choice of coordinate directions, (4.3.17) reduces still further to

$$[c_0(k) - c_3(k)]|\, \phi\rangle = 0. \tag{4.3.19}$$

The number operator for scalar and longitudinal photons is

$$N' = \int \frac{d^3k}{k_0} \left[c_3{}^\dagger(k) c_3(k) - c_0{}^\dagger(k) c_0(k)\right]$$

which will have eigenstates $|\,\phi^{(n)}\rangle$ with eigenvalue n, so that in general $|\,\phi\rangle$ will be a linear superposition of states satisfying (4.3.19):

$$|\,\phi\rangle = \sum_n d_n |\,\phi^{(n)}\rangle, \quad |\,\phi^{(0)}\rangle = |\,0\rangle.$$

This entails

$$n\,\langle\,\phi^{(n)}\,|\,\phi^{(n)}\rangle = \delta_{n0}$$

so that all such states with $n \neq 0$ have zero norm. Only the term with $n = 0$ contributes so that $\|\phi\|^2 = |d_0|^2 \geq 0$. With this constraint, the elements of $\mathcal{H}_1$ form a proper Hilbert space.

Whilst a lot of arbitrariness about the states $|\,\phi\rangle$ remains, it does not affect the expectation values of observables. Thus the Hamiltonian for the free Maxwell field is

$$\begin{aligned} H &= -\int \frac{d^3k}{k_0}\, c_\mu^\dagger(k) c^\mu(k) \frac{\hbar k_0}{c} \\ &= \int \frac{d^3k}{k_0} \left\{ \sum_{i=1}^{3} c^{i\dagger}(k) c^i(k) - c^{0\dagger}(k) c^0(k) \right\} \hbar k_0, \end{aligned} \tag{4.3.20}$$

so that, taking account of (4.3.19), its expectation value for the state $|\,\psi\rangle$ is

$$\frac{\langle\psi\,|\,H\,|\,\psi\rangle}{\langle\psi\,|\,\psi\rangle} = \frac{\left\langle \psi_{tr} \,\middle|\, \int d^3k/k_0 \sum_{i=1,2} c^{i\dagger}(k) c^i(k)\, \hbar k_0 \,\middle|\, \psi_{tr} \right\rangle}{\langle \psi_{tr}\,|\,\psi_{tr}\rangle}. \tag{4.3.21}$$

We get the same result for all states of the form (4.3.18) whatever the choice of $|\,\phi\rangle$, so that the vacuum state, $|\,\phi\rangle = |\,0\rangle$, can be selected as the representative of the equivalence class satisfying (4.3.19). The same sort of calculation gives the total 4-momentum associated with the field as

$$\frac{\langle\psi\,|\,\boldsymbol{P}\,|\,\psi\rangle}{\langle\psi\,|\,\psi\rangle} = \frac{\left\langle \psi_{tr} \,\middle|\, \int d^3k/k_0 \sum_{i=1,2} c^{i\dagger}(k) c^i(k)\, \hbar \boldsymbol{k} \,\middle|\, \psi_{tr} \right\rangle}{\langle \psi_{tr}\,|\,\psi_{tr}\rangle}.$$

The Gupta-Bleuler indefinite metric formalism plays little part in the rest of ths book. However, the full Hilbert space is needed for sums over a complete set of intermediate states, and then states with more than one scalar/longitudinal photon must be present to preserve locality properties.

4.4 Interaction of photons and electrons

4.4.1 The equations of motion

Most applications of QED in atomic and molecular physics assume interaction of the quantized electron-positron field with the quantized photon field in the presence of classical electromagnetic fields due to the charged nuclei. Thus, high precision tests of QED probe mainly the electromagnetic interactions of atomic electrons with the internal structure of heavy element nuclei [11, 12]. For simplicity, we shall adopt the Born-Oppenheimer approximation, which fixes nuclear positions and neglects contributions due to nuclear recoil. These simplifications can be relaxed when necessary.

The Lagrangian density for the coupled fields is

$$\mathcal{L}(x) = \mathcal{L}_{em}^{(\lambda)}(x) + \mathcal{L}_D(x) + \mathcal{L}_{int}(x) \tag{4.4.1}$$

where

$$\mathcal{L}_{em}^{(\lambda)}(x) = -\frac{1}{4}\epsilon_0\, F_{\mu\nu}F^{\mu\nu} + \frac{1}{2}\epsilon_0 \lambda \chi^2, \quad \chi(x) = \partial_\mu a^\mu(x)$$

is the Maxwell field Lagrangian with some gauge parameter λ as in (4.3.5),

$$\begin{aligned}\mathcal{L}_D(x) := \frac{1}{2}\overline{\psi}\,(i\gamma^\mu \partial_\mu - e\gamma^\mu (a_{nuc})_\mu - c)\,\psi \\ +\frac{1}{2}\left(-i\partial_\mu \overline{\psi}\gamma^\mu - e\overline{\psi}(a_{nuc})_\mu \gamma^\mu - c\overline{\psi}\right)\psi\end{aligned}$$

is the Lagrangian for the electron-positron field moving in the classical electromagnetic 4-potential $(a_{nuc})_\mu(x)$ due to the nuclei (2.5.12), coupled to the photon field through

$$\mathcal{L}_{int}(x) = -\frac{1}{c} j^\mu(x) a_\mu(x).$$

The field equations deduced from this Lagrangian are

$$\begin{aligned}(i\gamma^\mu \partial_\mu - e\gamma^\mu (a_{nuc})_\mu - c)\,\psi &= ec\gamma^\mu a_\mu(x)\psi(x), \\ -i\partial_\mu \overline{\psi}\gamma^\mu - e\overline{\psi}(a_{nuc})_\mu \gamma^\mu - c\overline{\psi} &= ec\overline{\psi}(x)\gamma^\mu a_\mu(x), \\ \partial_\mu F^{\mu\nu}(x) &= j^\nu(x)/\epsilon_0 c.\end{aligned} \tag{4.4.2}$$

The 4-current density of electrons and positrons in the quantized theory is given by

$$j^\mu(x) = -\frac{1}{2} ec\left[\overline{\psi}(x)\,\gamma^\mu, \psi(x)\right], \tag{4.4.3}$$

which is formally the same as (4.2.30), though here the operators ψ and $\overline{\psi}$ are determined in the presence of the external field.

Equations (4.4.2) describe the evolution of the operators ψ, $\overline{\psi}$ and $a_\mu(x)$ in the Heisenberg picture, the state vector $|\Psi\rangle$ being kept fixed. We can write down equal time commutation rules

$$\begin{aligned}\left\{\psi(x), \overline{\psi}(x')\right\}_{x_0=x_0'} &= \gamma_0\, \delta^{(3)}(\boldsymbol{x} - \boldsymbol{x}') \\ \left\{\psi(x), \psi(x')\right\}_{x_0=x_0'} &= \left\{\overline{\psi}(x), \overline{\psi}(x')\right\}_{x_0=x_0'} = 0 \\ \left[a_\mu(x), \partial_0 a_\nu(x')\right]_{x_0=x_0'} &= -icg_{\mu\nu}\delta^{(3)}(\boldsymbol{x} - \boldsymbol{x}')/\epsilon_0 \\ \left[\psi(x), a_\mu(x)\right]_{x_0=x_0'} &= \left[\overline{\psi}(x), a_\mu(x)\right]_{x_0=x_0'} = 0.\end{aligned} \tag{4.4.4}$$

Commutation rules for the operators at different times require solution of the coupled equations of motion, and are therefore impossible to write down simply. However, the current is conserved,

$$\partial_\mu j^\mu(x) = 0,$$

and so the total charge operator

$$Q = -\frac{1}{c}\int_\sigma d\sigma^\mu(x)\, j_\mu(x) \tag{4.4.5}$$

taken over a space-like surface σ is independent of σ and is constant. It follows that

$$\Big[Q, \psi(x)\Big] = +e\psi(x), \quad \Big[Q, \overline{\psi}(x)\Big] = -e\overline{\psi}(x), \tag{4.4.6}$$

so that the Heisenberg operator $\psi(x)$ destroys an amount of charge $-e$ or creates an amount of charge $+e$. Similarly, $\overline{\psi}(x)$ destroys an amount of charge $+e$ or creates an amount of charge $-e$.

4.4.2 The Furry picture

The complete Hamiltonian for the system corresponding to the Lagrangian density (4.4.1) is

$$H = H_{em} + H_D + H_{int}. \tag{4.4.7}$$

After dropping the interaction with the charge-current terms, the Hamiltonian for the Maxwell field (4.3.15) in Coulomb gauge can be written

$$H_{em} = \int d^3x\, \frac{1}{2}\epsilon_0(\boldsymbol{E}_\perp \cdot \boldsymbol{E}_\perp + c^2 \boldsymbol{B}\cdot\boldsymbol{B}), \tag{4.4.8}$$

or, if quantized according to the Gupta-Bleuler scheme,

$$H_{em} = \int \frac{d^3k}{k_0} \sum_{i=1,2} c^{i\dagger}(k)c^i(k)\, \hbar k_0 \tag{4.4.9}$$

where only transverse modes appear.

So far we have assumed that the electrons interact only with bare nuclei, represented by the term $-e\,(a_{nuc})_\mu(x)\gamma^\mu$. Whilst this is a valid starting point for atomic and molecular calculations, it is usually better to add an interaction, $U(\boldsymbol{x})$, which represents the mean field screening of the nuclear charges by the ambient electrons. This may be a (Dirac-)Hartree-Fock potential, or else a simpler parametrized local model potential [13]. The Furry picture states will be determined from a normal ordered quantized Hamiltonian with a *local* potential, $V(\boldsymbol{x})$,

$$H_D = \int :\psi^\dagger(x)\left\{c\boldsymbol{\alpha}.\boldsymbol{p} + V(\boldsymbol{x}) + \beta c^2)\right\}\psi(x) :\, d^3x; \tag{4.4.10}$$

If the nucleus is treated as a point charge, then

$$V(\boldsymbol{x}) = -\frac{Z}{|\boldsymbol{x}|} + U(\boldsymbol{x}).$$

When the field operators are expanded in terms of the eigenstates of the unquantized Hamiltonian (4.2.1) using (4.2.5) and (4.2.7),

$$H_D = \sum_r^{(+)} a_r^\dagger a_r\, E_r - \sum_s^{(+)} b_s^\dagger b_s\, E_s. \tag{4.4.11}$$

The Hamiltonian operator (4.4.7) can be partitioned as

$$H = H_0 + H_1, \tag{4.4.12}$$

where

$$H_0 = H_{em} + H_D,$$

describes the uncoupled fields (including the mean field potential) and, (2.9.29), the perturbation

$$H_1 = \int \left\{ \frac{1}{c} : j^\mu(x) : a_\mu(x) - U(\boldsymbol{x}) \, : \psi^\dagger(x)\, \psi(x) : \right\} d^3x$$

contains a balancing counter-term $U(\boldsymbol{x})$.

4.4.3 The interaction picture

In the Heisenberg picture, the states are fixed but the operators evolve in time, whilst in the Schrödinger picture, the unperturbed states, $\Psi(t)$, evolve according to the equation

$$i\frac{\partial \Psi(t)}{\partial t} = H_0\, \Psi(t), \tag{4.4.13}$$

where, of course, H_0 is independent of time. The connection between the two pictures is provided by the formal canonical transformation

$$\Psi(t) = \mathrm{e}^{-iH_0 t}\, \Phi \tag{4.4.14}$$

where Φ is the state vector in the Heisenberg picture. Let O be some operator in the Schrödinger picture. Matrix elements must have the same value in both pictures for physical consistency, so that

$$\langle \Psi_a(t) |\, O \,| \Psi_b(t) \rangle = \langle \Phi_a |\, O(t) \,| \Phi_b \rangle, \tag{4.4.15}$$

and the Heisenberg operator O and the Schrödinger operator $O(t)$ must be related by

$$e^{+iH_0t}\,O\,e^{-iH_0t} = O(t). \tag{4.4.16}$$

Differentiating (4.4.16) with respect to t gives the Heisenberg equation of motion

$$i\frac{\partial O(t)}{\partial t} = [O(t), H_0]. \tag{4.4.17}$$

The stationary states Φ_a of H_0 will satisfy

$$H_0\,\Phi_a = E_a\,\Phi_a,$$

and it follows from (4.4.15) that

$$\langle\Phi_a|\,O(t)\,|\Phi_b\rangle = e^{iE_a - E_b)t}\langle\Phi_a|\,O\,|\Phi_b\rangle. \tag{4.4.18}$$

In the presence of the interaction H_1, the system evolves in the Schrödinger picture so that

$$i\partial_t\,\Psi(t) = (H_0 + H_1)\,\Psi(t). \tag{4.4.19}$$

Define the Dirac (or interaction) picture vector $\Psi_D(t)$ by

$$\Psi_D(t) = e^{iH_0t}\Psi(t), \tag{4.4.20}$$

with equation of motion

$$i\partial_t\,\Psi_D(t) = V(t)\,\Psi_D(t), \quad V(t) = e^{iH_0t}\,H_1\,e^{-iH_0t}. \tag{4.4.21}$$

In the absence of interaction, $V(t) = 0$, the state $\Psi_D(t)$ is independent of time and coincides with the state vector of the Heisenberg picture. The label D can now be dropped provided we stay in the interaction picture.

Equation (4.4.21) is the usual starting point for solving the interacting field problem. Define $U(t, t_0)$ to be the time-development operator connecting the interaction picture vectors at times t and t_0,

$$\Psi(t) = U(t, t_0)\,\Psi(t_0), \tag{4.4.22}$$

so that $U(t, t_0)$ satisfies the differential equation

$$i\partial_t\,U(t, t_0) = V(t)\,U(t, t_0), \tag{4.4.23}$$

with initial condition $U(t_0, t_0) = 1$. State normalization is preserved if

$$U(t_2, t_1)U(t_1, t_0) = U(t_2, t_0), \quad U(t, t_0) = U^{-1}(t_0, t) = U^\dagger(t_0, t) \tag{4.4.24}$$

Equation (4.4.23), with initial condition $U(t_0, t_0) = 1$, is equivalent to the Volterra integral equation

$$U(t, t_0) = 1 - i\int_{t_0}^{t} V(t')\,U(t', t_0)\,dt' \tag{4.4.25}$$

which, for continuous $V(t')$, can be solved iteratively using the sequence [14, p. 250]

$$U^{(0)}(t,t_0)=1,\quad U^{(n+1)}(t,t_0)=1-i\int_{t_0}^{t}V(t')\,U^{(n)}(t',t_0)\,dt',$$

for $n=0,1,2,\ldots$ This generates the Neumann-Liouville series solution

$$\begin{aligned}U(t,t_0)=&1-i\int_{t_0}^{t}dt_1\,V(t_1)\\&+(-i)^2\int_{t_0}^{t}dt_1\int_{t_0}^{t_1}V(t_1)\,V(t_2)\\&+(-i)^3\int_{t_0}^{t}dt_1\int_{t_0}^{t_1}dt_2\int_{t_0}^{t_2}dt_3\,V(t_1)\,V(t_2)\,V(t_3)\\&+\ldots\end{aligned}\tag{4.4.26}$$

Consider the n-th term of (4.4.26). Because $t_1\geq t_2\geq\ldots\geq t_n$, the product of interactions is in time-ordered form, so that we may as well write this as

$$(-i)^n\int_{t_0}^{t}dt_1\int_{t_0}^{t_1}dt_2\ldots\int_{t_0}^{t_{n-1}}dt_n\,T(V(t_1)\,V(t_2)\,\ldots\,V(t_n))$$

This expression is symmetric with respect to the interchange of the arguments $t_1,\ldots,t_n$ because each $V(t)$ contains an even number of fermion operators and a photon operator, so that we can now average over the $n!$ different permutations of $t_1,\ldots,t_n$ giving

$$\begin{aligned}U(t,t_0)&=1+\sum_{n=1}^{\infty}U^{(n)}(t,t_1)\\&=1+\sum_{n=1}^{\infty}\frac{(-i)^n}{n!}\int_{t_0}^{t}dt_1\int_{t_0}^{t}dt_2\ldots\int_{t_0}^{t}dt_n\,T(V(t_1)\,V(t_2)\,\ldots\,V(t_n))\end{aligned}\tag{4.4.27}$$

The symbol

$$T\ \exp\left\{-i\int_{t_0}^{t}dt_1V(t_1)\right\}$$

is often used for the infinite sum on the right-hand side of (4.4.27). With this notation, and with $V(t)$ defined in terms of the Furry picture Hamiltonian (4.4.12), we find that the system evolves according to

$$U(t,t_0)=T\ \exp\left\{-\frac{i}{c}\int_{t_0}^{t}dt_1\int d^3x_1:\tilde{\psi}(x_1)(-ec)\gamma^{\mu}\psi(x_1):A_{\mu}(x_1)\right\},\tag{4.4.28}$$

where

$$A_{\mu}(x)=a_{\mu}(x)-A^e_{\mu}(x),$$

in which $A^e_{\mu}(x)$ is a classical four potential equivalent to the counter term $U(\boldsymbol{x})$ of (4.4.12).

4.5 Wick's theorems

Wick's theorems [15, 16, 17] give a systematic way of reducing the collection of operator products which occur in (4.4.27). The first theorem expresses a product of field operators as a sum of *normal-ordered* products, the second expresses a product of field operators as a sum of *time-ordered* products.

In this section we denote any quantum field at the space-time point x by $\phi(x)$, be it scalar, spinor or vector in character. We have seen that we can always decompose a field in the manner of (4.2.5) and (4.3.10) into two parts, one with positive frequency time-dependence, the other with negative frequency time-dependence, so that

$$\phi(x) = \phi^{(+)}(x) + \phi^{(-)}(x) \tag{4.5.1}$$

The positive frequency parts are associated with annihilation operators and the negative frequency parts with creation operators. We can treat both commuting and anticommuting fields together by writing

$$[a, b]_\eta = ab + \eta ba,$$

where $\eta = +1$ for the Bose case and $\eta = -1$ for the Fermi case. This is, of course. a c-number. We next note that

$$\phi(x)\,\phi^\dagger(y) = N\left(\phi(x)\,\phi^\dagger(y)\right) + \left[\phi^{(+)}(x), \phi^{(-)\dagger}(y)\right]_\eta \tag{4.5.2}$$

where $N(\phi\psi)$ denotes the normal product of ϕ and ψ. We now define the *contraction* of two fields as their corresponding vacuum expectation value:

$$\underbrace{\phi(x)\phi^\dagger(y)} = \langle 0\,|\,\phi(x)\,\phi^\dagger(y)|\,0\rangle \tag{4.5.3}$$

By definition, a normal product has a null vacuum expectation value, so that

$$\phi(x)\,\phi^\dagger(y) = N\left(\phi(x)\,\phi^\dagger(y)\right) + \underbrace{\phi(x)\phi^\dagger(y)} \tag{4.5.4}$$

Now consider a more general product of field operators $\phi_j := \phi_j(x_j)$. We define

$$\begin{aligned} &N\left(\phi_1\cdots\underbrace{\phi_j\cdots\phi_k}\cdots\phi_n\right) \\ &:= (-1)^p\,\underbrace{\phi_j\phi_k}\;N\left(\phi_1\phi_2\cdots\phi_{j-1}\phi_{j+1}\cdots\phi_{k-1}\phi_{k+1}\cdots\phi_n\right) \end{aligned} \tag{4.5.5}$$

where p denotes the number of interchanges of *Fermi* operators needed to rearrange the ordered set to bring the contracted pair ϕ_j and ϕ_k to the extreme left. Movement of boson operators does not affect the value of p. Although this notation permits us to state Wick's theorem concisely, it is formally somewhat inconsistent as the vacuum expectation of a normal product is zero, so that contractions inside the normal product are not formally permitted. However if we accept (4.5.5) as a definition of the left-hand side, we can write

$$
\begin{aligned}
\phi_1 \cdots \phi_n = {} & N\left(\phi_1 \cdots \phi_n\right) \\
& + N\left(\underbrace{\phi_1\phi_2}\cdots\phi_n\right) + N\left(\underbrace{\phi_1 \;\; \phi_2 \;\; \phi_3}\cdots\phi_n\right) + \dots \\
& + N\left(\underbrace{\phi_1\phi_2}\;\underbrace{\phi_3\phi_4}\cdots\phi_n\right) + \dots \\
& + \dots
\end{aligned}
\tag{4.5.6}
$$

The number of non-zero terms depends on which pairs of operators have non-zero contractions. The maximum number of terms will appear if there are no zero contractions; then the first line of (4.5.6) has a single uncontracted term; the second has $\binom{n}{2}$ single contractions as defined by (4.5.4); the third has $\frac{1}{2}\binom{n}{2}\binom{n-2}{2}$ double contractions, and so on. The straightforward proof of (4.5.6) is inductive, and can be found in many of the standard textbooks such as [2] as well as in the original papers.

The second theorem is used for reducing time-ordered products such as appear in (4.4.27). We recall first that

$$
T\left(\phi(x)\phi^\dagger(y)\right) = \begin{cases} \phi(x)\phi^\dagger(y) & x^0 > y^0 \\ \eta\phi^\dagger(y)\phi(x) & x^0 < y^0 \end{cases}
$$

where again $\eta = +1$ for Bose fields and $\eta = -1$ for Fermi fields. Using (4.5.3),

$$
T\left(\phi(x)\phi^\dagger(y)\right) = \begin{cases} N\left(\phi(x)\,\phi^\dagger(y)\right) + \underbrace{\phi(x)\,\phi^\dagger}(y) & x^0 > y^0 \\ \eta N\left(\phi^\dagger(y)\,\phi(x)\right) + \eta\,\underbrace{\phi^\dagger(y)\,\phi}(x) & x^0 < y^0 \end{cases}
\tag{4.5.7}
$$

Now because $\eta N\left(\phi^\dagger(y)\,\phi(x)\right) = N\left(\phi(x)\,\phi^\dagger(y)\right)$ the normal products are unchanged, and we can define a *time-ordered contraction* such that

$$
\overbrace{\phi(x)\phi^\dagger}(y) = \begin{cases} \underbrace{\phi(x)\,\phi^\dagger}(y) & x^0 > y^0 \\ \eta\,\underbrace{\phi^\dagger(y)\,\phi}(x) & x^0 < y^0 \end{cases}
\tag{4.5.8}
$$

so that (4.5.7) reduces to

$$
T\left(\phi(x)\phi^\dagger(y)\right) = N\left(\phi(x)\,\phi^\dagger(y)\right) + \overbrace{\phi(x)\phi^\dagger}(y)
\tag{4.5.9}
$$

which allows us to express the time-ordered contraction as a vacuum expectation value

$$
\overbrace{\phi(x)\phi^\dagger}(y) = \langle 0 \,|\, T\left(\phi(x)\phi^\dagger(y)\right) \,|\, 0\rangle.
\tag{4.5.10}
$$

Thus for time-ordered products, we replace (4.5.6) by

$$
\begin{aligned}
T(\phi_1 \cdots \phi_n) = {} & N\,(\phi_1 \cdots \phi_n) \\
& + N\left(\overbrace{\phi_1\phi_2}\cdots\phi_n\right) + N\left(\overbrace{\phi_1\ \phi_2\ \phi_3}\cdots\phi_n\right) + \ldots \\
& + N\left(\overbrace{\phi_1\phi_2}\ \overbrace{\phi_3\phi_4}\cdots\phi_n\right) + \ldots \\
& + \ldots
\end{aligned}
\tag{4.5.11}
$$

4.6 Propagators

Time-ordered contractions can be regarded as propagators, proportional to the causal Green's functions defined in Chapter 2, which allow us to relate the fields at different space-time points.

4.6.1 Photon propagators

The Maxwell field variables in Feynman gauge, (4.3.10), are

$$a_\mu(x) = a_\mu^{(+)}(x) + a_\mu^{(-)}(x) \tag{4.6.1}$$

where

$$
\begin{aligned}
a_\mu^{(+)}(x) &= \sqrt{\frac{1}{2(2\pi)^3\epsilon_0}} \int \frac{d^3k}{k_0} c_\mu(k)\,\mathrm{e}^{-ik\cdot x}, \\
a_\mu^{(-)}(x) &= \sqrt{\frac{1}{2(2\pi)^3\epsilon_0}} \int \frac{d^3k}{k_0} c_\mu{}^\dagger(k)\,\mathrm{e}^{+ik\cdot x}
\end{aligned}
$$

are the positive and negative frequency parts. The contraction of a pair of these field components gives

$$
\begin{aligned}
& \langle 0|\, T(a_\mu(x)a_\nu(y))\,|0\rangle \\
& = \langle 0|\, a_\mu^{(+)}(x)\, a_\nu^{(-)}(y)\theta(x^0-y^0) + a_\nu^{(+)}(y)\, a_\mu^{(-)}(x)\theta(y^0-x^0)\,|0\rangle
\end{aligned}
\tag{4.6.2}
$$

Using (4.6.1) and the commutation relations (4.3.13) reduces this expression to

$$-\frac{g_{\mu\nu}}{(2\pi)^3\epsilon_0}\int\frac{d^3k}{2k_0}\left\{\mathrm{e}^{-ik\cdot(x-y)}\theta(x^0-y^0) + \mathrm{e}^{-ik\cdot(y-x)}\theta(y^0-x^0)\right\}$$

The dependence on the relative time $\tau = x^0 - y^0$ can be written

$$\mathrm{e}^{-ik_0\tau}\theta(\tau) = \frac{1}{2\pi i}\int_{-\infty}^{+\infty}\frac{\mathrm{e}^{-iz\tau}\,dz}{k_0 - z - i\epsilon}$$

where $\epsilon > 0$. By first relabelling so that $k_0 = |\boldsymbol{k}|$ and then identifying z with the time-like integration variable k_0, we get

$$ig_{\mu\nu}\int \frac{d^4k}{(2\pi)^4\epsilon_0}\,\frac{\mathrm{e}^{-ik.(x-y)}}{2|\boldsymbol{k}|}\left\{\frac{1}{|\boldsymbol{k}|-k_0-i\epsilon}+\frac{1}{|\boldsymbol{k}|+k_0-i\epsilon}\right\},$$

so that, compare (2.9.32), we have

$$\begin{aligned}\langle 0|\, T(a_\mu(x)a_\nu(y))\,|0\rangle &= -\frac{ig_{\mu\nu}}{\epsilon_0}\int \frac{d^4k}{(2\pi)^4}\,\frac{\mathrm{e}^{-ik.(x-y)}}{k^2+i\epsilon}\\ &= iD^F_{\mu\nu}(x-y). \qquad (4.6.3)\end{aligned}$$

Similar expression can be found for other gauges. For example, if we start from the Lagrangian (4.3.5) and the corresponding modified Maxwell equation (4.3.6), we find [2, equation (3-131)]

$$\begin{aligned}&\langle 0|\, T(a_\mu(x)a_\nu(y))\,|0\rangle \qquad (4.6.4)\\ &= -i\int \frac{d^4k}{(2\pi)^4\epsilon_0}\,\mathrm{e}^{-ik.(x-y)}\left\{\frac{g_{\mu\nu}}{k^2+i\epsilon}+\frac{1-\lambda}{\lambda}\,\frac{k_\mu k_\nu}{(k^2+i\epsilon)^2}\right\}\end{aligned}$$

which recovers the Feynman gauge propagator (4.6.3) when $\lambda = 1$ and the Landau gauge propagator when $\lambda \to \infty$. The corresponding result for the Coulomb gauge is given by (2.9.35) in momentum space or by (2.9.36) and (2.9.40) in coordinate space:

$$\langle 0|\, T(a_\mu(x)a_\nu(y))\,|0\rangle = iD^C_{\mu\nu}(x-y). \qquad (4.6.5)$$

where, setting $R = |\boldsymbol{R}|$, $\boldsymbol{R} = \boldsymbol{x} - \boldsymbol{y}$, $\partial_i = \partial/\partial R_i$, $i = 1, 2, 3$,

$$\begin{aligned}D^C_{00}(x-y) &= \frac{\delta(x^0-y^0)}{4\pi\epsilon_0 R},\\ D^C_{0i}(x-y) &= D^C_{i0}(x-y) = 0,\\ D^C_{ij}(x-y) &= -\int \frac{dz}{2\pi}\,\mathrm{e}^{-iz(x^0-y^0)}\left\{\delta_{ij}\frac{\mathrm{e}^{i|z|R}}{4\pi\epsilon_0 R}+\partial_i\partial_j\,\frac{\mathrm{e}^{i|z|R}-1}{4\pi\epsilon_0 R\,|z|^2}\right\}.\end{aligned}$$

It can be shown, using (4.6.4), that integrals like

$$\int \langle 0|\, T(a_\mu(x)a_\nu(y))\,|0\rangle\, j^\nu(y)d^4y$$

are independent of the value of λ when $j^\nu(y)$ is a smooth conserved current, so that the corresponding physical results are also gauge independent.

4.6.2 Electron-positron propagators

The contraction of two free electron field operators can be written in terms of the Green's function of (2.9.42)

$$\langle 0|\, T(\psi_\alpha(x)\,\overline{\psi}_\beta(y))\,|\,0\rangle = -\frac{1}{2}S_{F\alpha\beta}(x-y), \tag{4.6.6}$$

where the spinor indices label rows and columns of the propagator. The order of the factors is significant; it is easy to see that

$$\langle 0|\, T(\overline{\psi}_\beta(y)\,\psi_\alpha(x))\,|\,0\rangle = +\frac{1}{2}S_{F\alpha\beta}(x-y), \tag{4.6.7}$$

so that it is important to respect the order of the contracted operators in the case of electron field operators. In the Furry picture, the result, (3.6.1), is

$$S_F(x,y) = \frac{i}{\pi}\int_{C_F} dz\, G(\boldsymbol{x},\boldsymbol{y};z)\,\gamma_0\,\mathrm{e}^{-iz(x-y)^0/c}.$$

with

$$G(\boldsymbol{x},\boldsymbol{y};z) = \sum_{E_r>0}\frac{\phi_r(\boldsymbol{x})\,\phi_r^\dagger(\boldsymbol{y})}{z-E_r+i\epsilon} + \sum_{E_s<0}\frac{\phi_s(\boldsymbol{x})\,\phi_s^\dagger(\boldsymbol{y})}{z-E_s-i\epsilon}, \tag{4.6.8}$$

where the sign of ϵ ensures that positive energy states propagate forwards in time and the negative energy states backwards in time.

4.6.3 Feynman diagrams

The Hamiltonian density for interaction of the quantized Maxwell and electron-positron fields in the Furry picture is given by (4.4.11)

$$H_1(x^0) = \int\left\{\frac{1}{c}:j^\mu(x):a_\mu(x) - U(\boldsymbol{x})\;:\psi^\dagger(x)\,\psi(x):\right\}d^3x \tag{4.6.9}$$

where $j^\mu(x) = -ec\overline{\psi}(x)\gamma^\mu\psi(x)$ is the electron current operator and $a_\mu(x)$ the Maxwell field operator. The field operators can be decomposed into positive and negative frequency components,(4.2.5) and (4.2.7) (or for free electrons (4.2.11) and (4.2.12)), for which

$\psi^{(+)}(x)$ destroys an electron at x; $\psi^{(-)}(x)$ creates a positron at x
$\overline{\psi}^{(+)}(x)$ creates an electron at x; $\overline{\psi}^{(-)}(x)$ destroys a positron at x.

Feynman [18] represented operators by directed lines in a space-time diagram. In Fig. 4.1, time increases up the page: the operators $\overline{\psi}$ are represented by lines (up or down) directed away from the space-time point x and ψ operators by lines directed towards x. Positive frequency lines are below (earlier than) x and negative frequency lines above (later than) x.

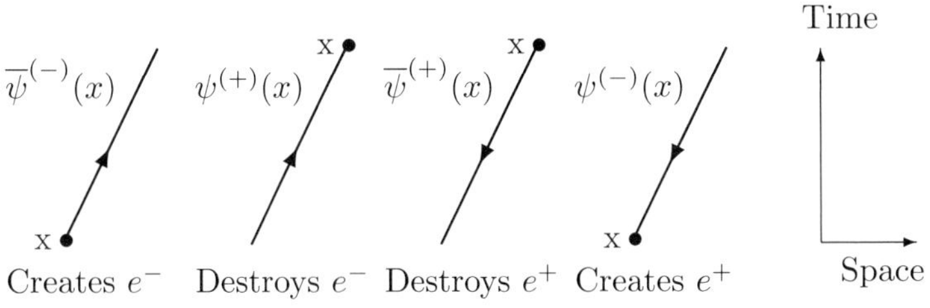

Fig. 4.1. Feynman representation of field operators.

We start with the case of an electron interacting with a classical static charge density:

$$j^\nu(x) = c\rho(\boldsymbol{x}).\delta^{\nu 0}.$$

According to (2.9.29) this generates a four-potential

$$A^e_\mu(x) = \frac{1}{c}\int D_{\mu\nu}(x - x')j^\nu(\boldsymbol{x}')d^4x'.$$

The configuration space propagator in Coulomb gauge is given by (2.9.36) and (2.9.40), so that the potential

$$A^e_\mu(x) = \frac{\delta_{\mu 0}}{4\pi\epsilon_0}\int \frac{\rho(\boldsymbol{x}')}{|\boldsymbol{x} - \boldsymbol{x}'|}d^3x'. \tag{4.6.10}$$

is purely electrostatic. For the specific case of a point nucleus of charge $+Ze$, for which $\rho(\boldsymbol{x}) = Ze\delta(\boldsymbol{x})$, we recover the familiar result

$$A^e_\mu(x) = \frac{\delta_{\mu 0}Ze}{4\pi\epsilon_0|\boldsymbol{x}|}.$$

We can therefore write the contribution to the interaction Hamiltonian as

$$\int :\psi^\dagger(x)V(\boldsymbol{x})\psi(x): d^3x. \tag{4.6.11}$$

where the potential energy is

$$V(\boldsymbol{x}) = -\frac{Ze^2}{4\pi\epsilon_0}\frac{1}{|\boldsymbol{x}|} = -\frac{Z}{|\boldsymbol{x}|}\ \text{a.u.}$$

The four normal ordered terms give diagrams that are shown in Fig. 4.2. Each diagram describes an interaction a space-time point x, represented by a vertex with two electron lines, one entering and one leaving, together with the local potential $V(\boldsymbol{x})$, represented by a horizontal dashed line terminating in a cross. From Fig. 4.1, we see that the four diagrams can be interpreted respectively as

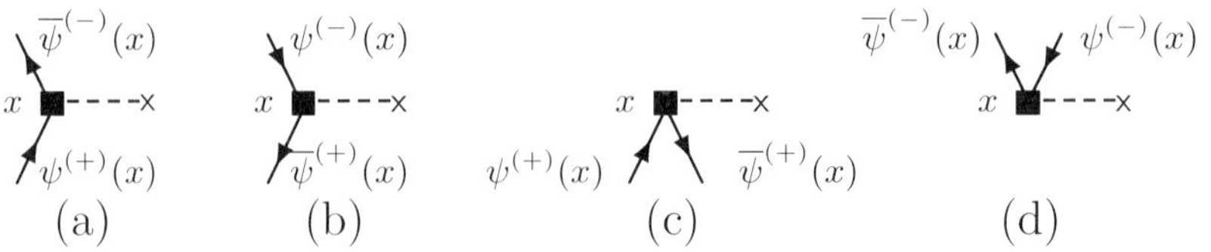

Fig. 4.2. First order processes in an external field: (a) represents electron scattering by the potential $V(\boldsymbol{x})$, (b) scattering of a positron; (c) represents pair destruction and (d) pair creation.

a scattering of an electron by the potential, scattering of a positron, creation of an electron-positron pair, and destruction of a pair at the point x. All these processes have a null expectation for the vacuum state because of the normal ordering. For a single particle state, we can suppose, for example, that the initial state is that of an electron in state α and the final state is also that of an electron in a state β, so that

$$|\,\Phi_\alpha\rangle = a_\alpha^\dagger\,|\,\Phi_0\rangle, \quad \langle\Phi_\beta\,| = \langle\Phi_0\,|\,a_\beta.$$

Only one term of (4.6.11), namely

$$\int \phi_\beta^\dagger(x) V(\boldsymbol{x}) \phi_\alpha(x) d^3x \tag{4.6.12}$$

survives, giving the expected result for scattering of an electron by the potential $V(\boldsymbol{x})$ from the state α to the state β.

4.6.4 Second order interaction: $U^{(2)}(t, t_0)$

Next consider the second-order QED contribution $U^{(2)}(t, t_0)$ to (4.4.26),

$$-\frac{e^2}{2!\hbar^2c^2}\iint T\left(:\overline{\psi}(x)\gamma^\mu\psi(x):A_\mu^e(x)\ :\overline{\psi}(y)\gamma^\nu\psi(y):A_\nu^e(y)\right)d^4x d^4y. \tag{4.6.13}$$

where we have retained the four-potential to ease the transition to the quantized field case. Wick's theorem (4.5.11) gives

$$\begin{aligned}
T\left(N(\overline{\psi}(x)\,\gamma^\mu A_\mu^e(x)\,\psi(x))\,N(\overline{\psi}(y)\,\gamma^\nu A_\nu^e(y)\,\psi(y))\right)& \\
= N\left(\overline{\psi}(x)\,\gamma^\mu A_\mu^e(x)\,\psi(x)\,\overline{\psi}(y)\,\gamma^\nu A_\nu^e(y)\,\psi(y)\right)& \\
-\frac{1}{2}N\left(\overline{\psi}(x)\,\gamma^\mu A_\mu^e(x)\,S_F(x,y)\,\gamma^\nu A_\nu^e(y)\,\psi(y)\right)& \\
-\frac{1}{2}N\left(\overline{\psi}(y)\,\gamma^\mu A_\mu^e(y)\,S_F(y,x)\,\gamma^\nu A_\nu^e(x)\,\psi(x)\right)& \\
-\frac{1}{4}\gamma^\mu A_\mu^e(x)\,S_F(x,y))\,\gamma^\nu A_\nu^e(y)\,S_F(y,x)&
\end{aligned} \tag{4.6.14}$$

We ignore the first term for the present. The next two terms generate the diagrams of Fig. 4.3 representing processes involving two interactions with the external field whose overall effect is simply electron scattering, positron scattering, pair destruction, and pair creation as in Fig. 4.2. The only new fea-

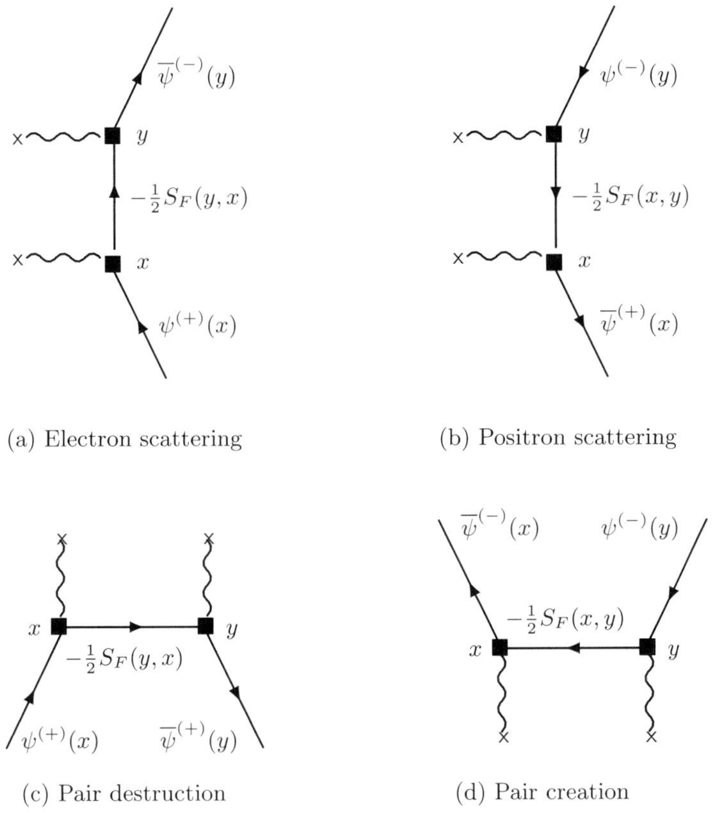

Fig. 4.3. Second order processes in an external field.

ture of these diagrams is the internal line connecting x and y, which indicates propagation of the electron/positron field amplitude represented in (4.6.13) by the propagator $-\frac{1}{2}S_F(x,y)$. Clearly permuting the labels x and y does not alter the value of the corresponding integral, and we can add the two integrals to cancel the pre-factor 2! in (4.6.13); the same thing happens in higher order diagrams. The last term of (4.6.13) has no free lines and represents a vacuum process: the corresponding diagram, Fig. 4.4, describes the creation of an electron-positron pair at y and its destruction at x leaving the vacuum as it was. Such diagrams are found to have an infinite negative imaginary part, independent of the field. It can be shown that they do not contribute

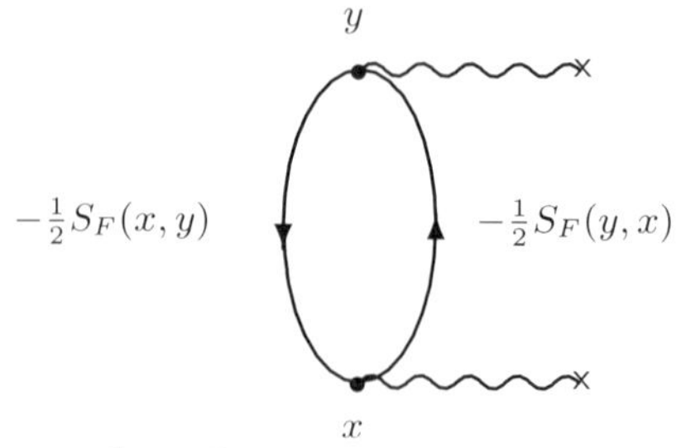

Fig. 4.4. Second order vacuum process in an external field.

to amplitudes for observable processes so that they can be dropped from the calculation.

The one-to-one correspondence of the terms of (4.6.14) with the Feynman diagrams of Figs. 4.3 and 4.4 illustrates an important general result. In this approach, graphs such as which are similar except for the labelling of vertices can be treated as distinct. One unlabelled diagram is equivalent to a class of labelled diagrams, as we saw in the case of the two middle terms of (4.6.14). We can therefore start with the Feynman diagrams and write down the matrix elements without performing the detailed manipulation required by Wick's theorems. However, the diagram approach requires more care to ensure that each term is prefixed by the correct sign.

Electronic structure calculations also require the calculation of matrix elements involving virtual photons of the quantized radiation field. Figure 4.5 shows the two disconnected diagrams corresponding to the first term of the Wick decomposition (4.6.13),

$$N\left(\overline{\psi}(x)\,\gamma^{\mu}a_{\mu}(x)\,\psi(x)\,\overline{\psi}(y)\,\gamma^{\nu}a_{\nu}(y)\,\psi(y)\right),$$

in which a photon is either emitted or absorbed at each vertex. These processes only give nonvanishing amplitudes for bound electrons, as it is not possible to satisfy all the conditions for momentum conservation at each vertex for a free particle.

The remaining diagrams of this order, Fig. 4.6, have internal photon lines. Diagrams 4.6(a) to 4.6(d) represent particle-particle scattering with a photon being emitted at one vertex and absorbed at the other:

$$N\left(\overline{\psi}(x)\,\gamma^{\mu}\psi(x)\,D_{\mu\nu}(x-y)\,\overline{\psi}(y)\,\gamma^{\nu}\psi(y)\right),$$

together with a similar term with x and y interchanged. Figure 4.6(d) represents annihilation of an electron-positron pair with emission of a virtual photon, followed by the creation of a pair of particles at a later time.

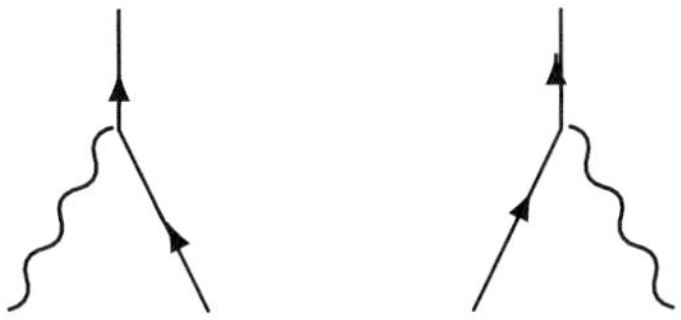

Fig. 4.5. Disconnected pair of first order diagrams.

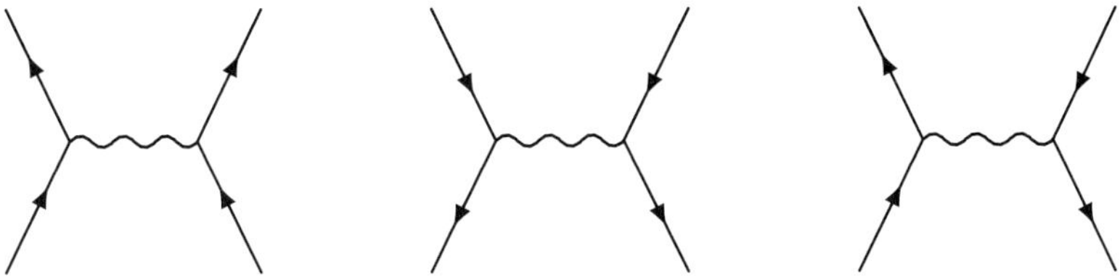

(a) $e^- - e^-$ (b) $e^+ - e^+$ (c) $e^- - e^+$

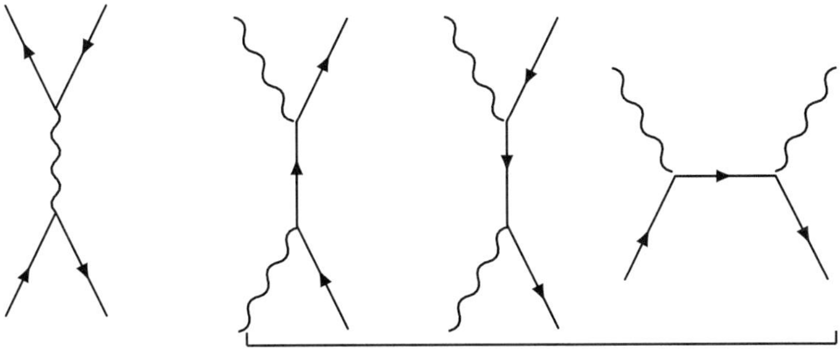

(d) $e^- - e^+$ (e) Compton scattering

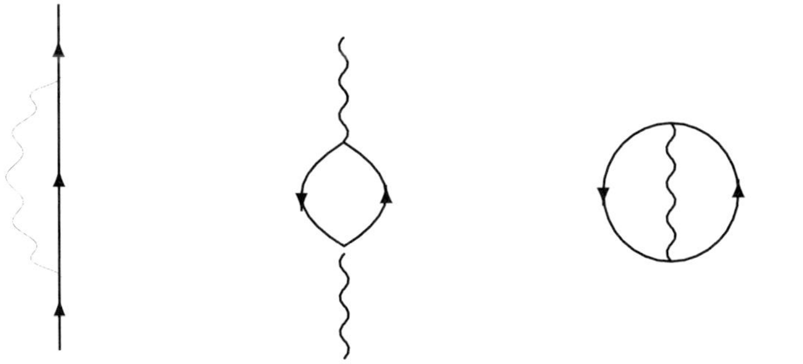

(f) Self-energy diagram (g) Vacuum polarization diagram (h) Vacuum-vacuum diagram

Fig. 4.6. Second order diagrams.

Figure 4.6(e) shows processes in which a pair of electron operators is contracted, corresponding to the kernel

$$N\left(\overline{\psi}(x)\,\gamma^{\mu}a_{\mu}(x)\,S_F(x,y)\,\gamma^{\nu}a_{\nu}(y)\,\psi(y)\right).$$

Here two of the diagrams represent Compton scattering of a photon from an electron or positron, whilst the others involve destruction of an electron-positron pair with two-photon emission. There is a corresponding diagram for the inverse process. Figures 4.6(f) and 4.6(g) describe respectively the lowest order contribution to the electron self-energy and the lowest order vacuum polarization. Their kernels

$$N\left(\overline{\psi}(x)\,\gamma^{\mu}S_F(x,y)\,\gamma^{\nu}\,\psi(y)\right)D_{\mu\nu}(x-y)$$

and

$$N\left(a_{\mu}(x)a_{\nu}(y)\right).\gamma^{\mu}S_F(x,y)\,\gamma^{\nu}\,S_F(y,x)$$

are examples of *loop diagrams*, both of which give an infinite result; a process of renormalization is needed to extract finite physical results. Finally, Figure 4.6(h), in which there are no free lines,

$$\gamma^{\mu}S_F(x,y)\,\gamma_{\nu}\,S_F(y,x)D_{\mu\nu}(x-y)$$

gives the second order amplitude for the vacuum to remain a vacuum. We shall examine such terms further in Section 4.7.

4.6.5 Feynman rules

By writing down the Feynman diagrams with n vertices, we can obtain all contributions to the n-th order U-matrix directly. We drop all vacuum bubbles and retain only topologically distinct diagrams, that is to say diagrams in which the vertices are not labelled and which cannot be deformed into one another without breakage. The expansion (4.4.28) (in covariant notation) is a sum of terms

$$\begin{aligned}\lim_{t'\to-\infty,t\to+\infty} U^{(n)}(t,t') &= \left(\frac{-i}{c}\right)^{n}\int d^4x_1\ldots\int d^4x_n\\ &\times T\left(:\overline{\psi}(x_1)(-e\gamma^{\mu_1})\psi(x_1):\ldots:\overline{\psi}(x_n)(-e\gamma^{\mu_n})\psi(x_n):\right)\\ &\times a_{\mu_1}(x_1)\ldots a_{\mu_n}(x_n).\end{aligned} \tag{4.6.15}$$

The expression for each Feynman diagram can be written down according to the following rules:

(a) A pre-factor $(-i/c)^n$ from the perturbation series.
(b) A factor $-e(\gamma^{\mu})_{\alpha\beta}$ from each bilinear expression $\sum_{\alpha\beta}:\overline{\psi}_{\alpha}(x_i)(-e\gamma^{\mu_i})_{\alpha\beta}\psi_{\beta}(x_i):$.

(c) A factor $iD_{F\mu\nu}(x-y)$ for each internal photon line connecting the vertices x and y, where μ labels the Dirac matrix associated with x and ν that associated with y, arising from $\langle 0|\, T(a_\mu(x)a_\nu(y))\,|\, 0\rangle$.
(d) A factor $-\frac{1}{2}S_{F\alpha\beta}(x,y)$ for each internal electron/positron line directed from y to x, arising from $\langle 0|\, T(\psi_\alpha(x)\overline{\psi}_\beta(y)\,|\, 0\rangle$.
(e) A creation/destruction operator $\psi^{(\pm)}(x)$ or $\overline{\psi}^{(\pm)}(x)$ or $a_\mu^{(\pm)}(x)$ for each external line entering/leaving x.
(f) A factor -1 for each internal fermion loop such as the vacuum polarization bubble, Figure 4.6(h).
(g) Integrate over all variables $x_1, \ldots, x_n$.

It is important to assemble each incoming fermion spinor, vertex operator, propagator and outgoing fermion spinor along the fermion line, following the arrows, in order to get a well-formed matrix element.

4.7 The S-matrix

The states of the interaction picture evolve in time according to (4.4.22) and remain constant when $V(t)$ vanishes. Suppose that the interaction is switched on at $t = -T/2$ and switched off again at $t = +T/2$. Initially, we suppose the system is in a (Heisenberg) state Φ_α,

$$\Psi(t) = \Phi_\alpha, \quad t < -T/2.$$

The system evolves according to (4.4.22) from $t = -T/2$ to $t = +T/2$ into a linear superposition of (Heisenberg) states Φ_β and subsequently

$$\Psi(t) = \Psi(T/2), \quad t > +T/2.$$

We define the S-matrix by writing

$$U(T/2, -T/2)|\,\Phi_\alpha) = N(T)\sum_\beta S_{\beta\alpha}(T/2, -T/2)|\,\Phi_\beta) \tag{4.7.1}$$

where $N(T)$ is a normalizing constant. If the states Φ_α and Φ_β are *orthonormal*, we see that

$$S_{\beta\alpha}(T/2, -T/2) = (\Phi_\beta\,|U(T/2, -T/2)|\,\Phi_\alpha)/N(T).$$

We define the elements of the S-matrix for the transition from Φ_α to Φ_β by

$$S_{\beta\alpha} = \lim_{T\to\infty} S_{\beta\alpha}(T/2, -T/2) \tag{4.7.2}$$

whenever the limit exists.

In QED the vacuum state, Φ_0, is one for which there are no free electrons or positrons and no photons, and we shall choose

$$N(\infty) = \lim_{T\to\infty} N(T) = (\Phi_0 \,|U(\infty,-\infty)|\, \Phi_0).$$

We can assume that Φ_0 is unique so that $(\Phi_\alpha \,|U(\infty,-\infty)|\, \Phi_0) = 0$ when $\alpha \neq 0$ in order that the vacuum be stable. It then follows that

$$\begin{aligned}
1 &= (\Phi_0 \,|\, \Phi_0) \\
&= (\Phi_0 \,|\, U^\dagger(\infty,-\infty) U(\infty,-\infty) \,|\, \Phi_0) \\
&= \sum_\alpha (\Phi_0 \,|\, U^\dagger(\infty,-\infty) \,|\, \Phi_\alpha)(\Phi_\alpha \,|\, U(\infty,-\infty) \,|\, \Phi_0) \\
&= (\Phi_0 \,|\, U^\dagger(\infty,-\infty) \,|\, \Phi_0)(\Phi_0 \,|\, U(\infty,-\infty) \,|\, \Phi_0) \\
&= |(\Phi_0 \,|\, U(\infty,-\infty) \,|\, \Phi_0)|^2,
\end{aligned}$$

so that there is some real c-number, say C, not necessarily finite, such that

$$(\Phi_0 \,|\, U(\infty,-\infty) \,|\, \Phi_0) = e^{iC}$$

This normalization ensures that $S_{\beta\alpha}$ is independent of overall c-number phases. From (4.4.27) we see that adding a c-number C to $V(t)$ means that

$$U(t,t_0) \to e^{iC(t-t_0)} U(t,t_0)$$

so that $U(t,t_0)$ is multiplied by an infinite phase e^{iCT} as $T = t - t_0 \to \infty$. The same phase factor appears in both numerator and denominator, so that $\lim_{T\to\infty} S_{\beta\alpha}(T/2, -T/2)$ is well-defined. This argument permits us to drop vacuum-vacuum diagrams such as Figure 4.6(h) from the calculation.

4.8 Bound states

4.8.1 A perturbation expansion

We can circumvent much of the delicate analysis of convergence of QED perturbation series by adopting the adiabatic switching approach. We replace the perturbation $V(t)$ (we use Hartree units) by the expression [19]

$$V_\epsilon(t) = V(t) e^{-\epsilon|t|}, \quad \epsilon > 0,$$

and take the limit $\epsilon \to 0$ after performing the integrations. We denote the n-th term of the resulting perturbation series (4.6.15) by $U_\epsilon^{(n)}(t,t_0)$ whose contribution is

$$\begin{aligned}
\langle \Phi_f \,|\, S_\epsilon^{(n)} |\Phi_i\rangle = \lim_{T'\to\infty} \lim_{T\to-\infty} \frac{(-i)^n}{n!} \int_T^{T'} dt_1 \int_T^{T'} dt_2 \ldots \int_T^{T'} dt_n \\
\times \langle \Phi_f \,|\, T(V_\epsilon(t_1)\, V_\epsilon(t_2) \,\ldots\, V_\epsilon(t_n)) |\Phi_i\rangle.
\end{aligned}$$

This is easy to evaluate if $V_\epsilon(t)$ is an unquantized perturbation in a nonrelativistic framework. From (4.4.18) we have

$$\langle \Phi_b | \, V_\epsilon(t) \, | \Phi_a \rangle = \mathrm{e}^{i(E_b - E_a)t - \epsilon|t|} \langle \Phi_b | \, H_1 \, | \Phi_a \rangle .$$

so that

$$\langle b \,|\, S_\epsilon^{(1)} | a \rangle = -2\pi i \delta(E_b - E_a) \langle b \,|\, H_1 \,|\, a \rangle \tag{4.8.1}$$

and

$$\begin{aligned}
\langle b \,|\, S_\epsilon^{(2)} \,|\, a \rangle &= (-i)^2 \int_{-\infty}^{\infty} dt_1 \int_{-\infty}^{t_1} dt_2 \, \langle b \,|\, V_\epsilon(t_1) \, V_\epsilon(t_2) \,|\, a \rangle \\
&= (-i)^2 \sum_k \int_{-\infty}^{\infty} dt_1 \int_{-\infty}^{t_1} dt_2 \\
&\quad \times \mathrm{e}^{i(E_b - E_k)t_1 - \epsilon|t_1|} \mathrm{e}^{i(E_k - E_a)t_2 - \epsilon|t_2|} \langle b \,|\, H_1 \,|\, k \rangle \langle k | \, H_1 \, | a \rangle \\
&= -2\pi i \delta(E_b - E_a) \sum_k \frac{\langle b \,|\, H_1 \,|\, k \rangle \langle k \,|\, H_1 \,|\, a \rangle}{E_a - E_k + i\epsilon}
\end{aligned} \tag{4.8.2}$$

which is what would have been obtained by simple quantum mechanical perturbation theory. More generally, one finds for $n \geq 2$

$$\begin{aligned}
&\langle b \,|\, S_\epsilon^{(n)} \,|\, a \rangle \\
&= -2\pi i \delta(E_b - E_a) \\
&\times \sum_{k_1, \ldots, k_{n-1}} \frac{\langle b \,|\, H_1 \,|\, k_1 \rangle \langle k_1 \,|\, H_1 \,|\, k_2 \rangle \ldots \langle k_{n-1} \,|\, H_1 \,|\, a \rangle}{(E_a - E_{k_1} + i\epsilon)(E_a - E_{k_2} + i\epsilon) \ldots (E_a - E_{k_{n-1}} + i\epsilon)}
\end{aligned}$$

4.8.2 Gell-Mann, Low, Sucher energy shift

Atomic and molecular structure calculations start from a zero-order scheme in which the independent electrons are described by the Furry picture. The state is modified by coupling to the quantized photon field, and the resulting energy shift may be calculated using the adiabatic S-matrix formalism [19, 20]. It is convenient to introduce a coupling parameter λ so that

$$H_\epsilon(\lambda) = H_0 + \lambda V_\epsilon(t), \quad H(\lambda) = H_0 + \lambda V(t), \tag{4.8.3}$$

making the n-th term of the perturbation series proportional to λ^n. Suppose that the eigenstates and eigenvalues of $H_\epsilon(\lambda)$ are continuous with respect to ϵ in a neighbourhood of $\epsilon = 0$ and of λ in a neighbourhood of $\lambda = 0$ and that the nondegenerate eigenstate $|\Phi_0\rangle$ of H_0 with energy E_0 evolves into the eigenstate $|\Phi_\epsilon(\lambda)\,\rangle$ of $H(\lambda)$ with energy $E(\lambda)$ at $t = 0$. Clearly

$$\lim_{\lambda \to 0} E(\lambda) = E_0, \quad \lim_{\lambda \to 0} |\Phi_\epsilon(\lambda)\,\rangle = |\Phi_0\rangle .$$

independent of ϵ. Gell-Mann and Low [19] and Sucher [20] give the energy shift for a non-degenerate state in the point spectrum as

$$\Delta E(\lambda) = E(\lambda) - E_0 = \lim_{\epsilon \to 0} \frac{1}{2} i\epsilon\lambda \frac{\partial}{\partial \lambda} \ln \langle \Phi_0 | S_\epsilon(\lambda) | \Phi_0 \rangle, \tag{4.8.4}$$

where

$$S_\epsilon(\lambda) = U_\epsilon(\infty, -\infty).$$

To derive (4.8.4), we start with the unnormalized state

$$|\Phi'_\epsilon(\lambda)\rangle = U_\epsilon(0, -\infty) |\Phi_0\rangle \tag{4.8.5}$$

where the adiabatic U matrix satisfies

$$i\frac{\partial}{\partial t} U_\epsilon(t, t_0) = \lambda V_\epsilon(t) U_\epsilon(t, t_0), \quad U_\epsilon(t_0, t_0) = I.$$

Next we note that

$$(H_0 - E_0) |\Phi'_\epsilon(\lambda)\rangle = [H_0, U_\epsilon(0, -\infty)] \, |\Phi_0\rangle. \tag{4.8.6}$$

From (4.4.27), we can express the commutator in the form

$$[H_0, U_\epsilon(0, -\infty)] = \sum_n \frac{(-i\lambda)^n}{n!} \left[H_0, \int_{-\infty}^0 dt_1 \ldots \int_{-\infty}^0 dt_n e^{\epsilon(t_1 + \ldots + t_n)} T(V(t_1) \ldots V(t_n)) \right],$$

and, because H_0 generates time displacements in the interaction picture,

$$[H_0, F(t)] = -i\partial F(t)/\partial t,$$

we see that

$$[H_0, U_\epsilon(0, -\infty)] = -i \sum_n \frac{(-i\lambda)^n}{n!} \left[H_0, \int_{-\infty}^0 dt_1 \ldots \int_{-\infty}^0 dt_n \, e^{\epsilon(t_1 + \ldots + t_n)} \sum_k \frac{\partial}{\partial t_k} T(V(t_1) \ldots V(t_n)) \right].$$

The symmetry of the integrand allows us to replace $\sum_k \partial/\partial t_k$ by $n\partial/\partial t_1$, and using

$$\frac{\partial}{\partial t_1} \left\{ e^{\epsilon(t_1 + \ldots + t_n)} T(V(t_1) \ldots V(t_n)) \right\} = e^{\epsilon(t_1 + \ldots + t_n)} \left\{ \epsilon T(V(t_1) \ldots V(t_n)) + \frac{\partial}{\partial t_1} T(V(t_1) \ldots V(t_n)) \right\}$$

gives eventually

$$[H_0, U_\epsilon(0, -\infty)] = \left[-\lambda V(0) + i\epsilon\lambda \frac{\partial}{\partial \lambda} \right] U_\epsilon(0, -\infty), \tag{4.8.7}$$

and inserting this in (4.8.6) gives

$$\left(H(\lambda) - E_0 - i\epsilon\lambda\frac{\partial}{\partial\lambda}\right) U_\epsilon(0,-\infty)\,|\Phi_0\rangle = 0. \tag{4.8.8}$$

Now define

$$|\Phi_\epsilon(\lambda)\rangle = \frac{|\Phi'_\epsilon(\lambda)\rangle}{\langle\Phi_0|\Phi'_\epsilon(\lambda)\rangle} = \frac{U_\epsilon(0,-\infty)\,|\Phi_0\rangle}{\langle\Phi_0|U_\epsilon(0,-\infty)\,|\Phi_0\rangle}.$$

Equation (4.8.8) gives

$$\left(H(\lambda) - E_0 - i\epsilon\lambda\frac{\partial}{\partial\lambda}\right)|\Phi_\epsilon(\lambda)\rangle = i\epsilon\left(\lambda\frac{\partial}{\partial\lambda}\ln\langle\Phi_0|U_\delta(0,-\infty)\,|\Phi_0\rangle\right)|\Phi_\delta(\lambda)\rangle,$$

provided the limit

$$|\Phi(\lambda)\rangle := \lim_{\epsilon\to 0}|\Phi_\epsilon(\lambda)\rangle$$

exists and then, provided $\lambda\partial(|\Phi_\epsilon(\lambda)\rangle)/\partial\lambda$ is bounded as $\epsilon \to 0$,

$$(H(\lambda) - E_0)|\Phi(\lambda)\rangle = \Delta E(\lambda)|\Phi(\lambda)\rangle,$$

where

$$\Delta E(\lambda) = \lim_{\epsilon\to 0} i\epsilon\left(\lambda\frac{\partial}{\partial\lambda}\ln\langle\Phi_0|U_\epsilon(0,-\infty)\,|\Phi_0\rangle\right). \tag{4.8.9}$$

Thus $|\Phi(\lambda)\rangle$ is an eigenfunction of $H(\lambda)$ with shifted energy $E_0 + \Delta E(\lambda)$.

The same calculation with the operator $U_\epsilon(0,+\infty)$ gives

$$|\Phi_\epsilon(\lambda)\rangle = \frac{U_\epsilon(0,+\infty)\,|\Phi_0\rangle}{\langle\Phi_0|U_\epsilon(0,+\infty)\,|\Phi_0\rangle}, \tag{4.8.10}$$

provided we are dealing with a discrete state. In the continuous spectrum, $U_\epsilon(0,\pm\infty)\,|\Phi_0\rangle$ give the in- and out-states, respectively $\Phi^\pm$, which satisfy different boundary conditions and are quite distinct eigensolutions of the Hamiltonian $H(\lambda)$.

The symmetrical form (4.8.4) is more convenient for practical calculation because only straightforward Feynman diagram techniques are needed. Equations (4.8.8) and (4.8.9) give

$$\begin{aligned}\Delta E_\epsilon(\lambda) &= \lim_{\epsilon\to 0} i\epsilon\lambda\frac{\langle\Phi_\epsilon(\lambda)|\partial U_\epsilon(0,-\infty)/\partial\lambda|\Phi_0\rangle}{\langle\Phi_\epsilon(\lambda)|U_\epsilon(0,-\infty)|\Phi_0\rangle}\\ &= \lim_{\epsilon\to 0} i\epsilon\lambda\frac{\langle\Phi_0|U_\epsilon^{-1}(0,+\infty)\partial U_\epsilon(0,+\infty)/\partial\lambda|\Phi_0\rangle}{\langle\Phi_0| - U_\epsilon^{-1}(0,+\infty)U_\epsilon(0,-\infty)|\Phi_0\rangle}\\ &= \lim_{\epsilon\to 0} i\epsilon\lambda\frac{\langle\Phi_0|U_\epsilon(+\infty,0)\partial U_\epsilon(0,-\infty)/\partial\lambda|\Phi_0\rangle}{\langle\Phi_0|U_\epsilon(\infty,0)U_\epsilon(0,-\infty)|\Phi_0\rangle}\end{aligned}$$

Combining this with the similar expression [20]

$$\Delta E_\epsilon(\lambda) = \lim_{\epsilon\to 0} i\epsilon\lambda \frac{\langle \Phi_0|\partial U_\epsilon(+\infty,0)/\partial\lambda U_\epsilon(0,-\infty)|\Phi_0\rangle}{\langle \Phi_0|U_\epsilon(\infty,0)U_\epsilon(0,-\infty)|\Phi_0\rangle}.$$

gives (4.8.4). The normalization of the Gell-Mann, Low, Sucher formula (4.8.4) is similar to that of the S-matrix as defined in Section 4.7 so that vacuum diagrams make no contribution to the energy shift and it is only necessary to evaluate connected Feynman diagrams. Inserting

$$S_\epsilon(\lambda) = 1 + \sum_{n=1}^{\infty} \lambda^n S_\epsilon^{(n)}(1)$$

in (4.8.4), we see that as $\lambda \to 1$, the energy shift becomes

$$\Delta E = \lim_{\epsilon\to 0} \frac{1}{2} i\epsilon \frac{\langle \Phi_0|\sum_{n=1}^{\infty} nS_\epsilon^{(n)}|\Phi_0\rangle}{\langle \Phi_0|1+\sum_{n=1}^{\infty} S_\epsilon^{(n)}|\Phi_0\rangle}, \tag{4.8.11}$$

where $S_\epsilon^{(n)} = \lim_{\lambda\to 1} S_\epsilon^{(n)}(\lambda)$. Thus the leading contributions to the perturbation series are

$$\begin{aligned}\Delta E = \lim_{\epsilon\to 0} \frac{1}{2} i\epsilon \Big\langle \Phi_0 \Big| & S_\epsilon^{(1)} + 2S_\epsilon^{(2)} - \left(S_\epsilon^{(1)}\right)^2 & (4.8.12)\\ & + 3S_\epsilon^{(3)} - 3S_\epsilon^{(1)}S_\epsilon^{(2)} + \left(S_\epsilon^{(1)}\right)^3 \\ & + 4S_\epsilon^{(4)} - 4S_\epsilon^{(1)}S_\epsilon^{(3)} - 2\left(S_\epsilon^{(2)}\right)^2 \\ & + 4\left(S_\epsilon^{(1)}\right)^2 S_\epsilon^{(2)} - \left(S_\epsilon^{(1)}\right)^4 \ldots \Big| \Phi_0 \Big\rangle & (4.8.13)\end{aligned}$$

to which only connected diagrams contribute.

4.9 Effective interactions

Effective interactions between electrons and positrons appear in second order of QED perturbation theory from diagrams such as Figure 4.6 (a)–(d) in which a virtual photon transfers energy from the particle passing through the space-time point x to a second particle passing through another point y. The treatment of this section is similar to that of Lindgren [21].

4.9.1 One-photon exchange: Feynman gauge

We start with electron-electron scattering; the corresponding calculations for electron-positron scattering or positron-positron scattering are very similar. We first express the electron field operators, (4.2.5) and (4.2.6), as a sum

$$\psi(x) = \sum_r^{(+)} a_r \psi_r(x) + \sum_s^{(-)} b_s^\dagger \psi_s(x),$$

Substituting

$$\psi_r(x) = \phi_r(\boldsymbol{x}) \mathrm{e}^{-iE_r t},$$

where E_r is the relativistic energy, we can write

$$\psi(x) = \sum_r^{(+)} a_r(t) \phi_r(\boldsymbol{x}) + \sum_s^{(-)} b_s^\dagger(t) \phi_s(\boldsymbol{x}) \tag{4.9.1}$$

where $a_r(t) = a_r \exp(-iE_r t)$. The lowest order diagrams contributing to scattering of one particle by another are Figures 4.6(a)–(d); Feynman's rules give

$$\begin{aligned} S^{(2)} = \left(-\frac{ie}{c}\right)^2 \int d^4x_2 \int d^4x_1 \, \mathrm{e}^{-\epsilon|t_1|-\epsilon|t_2|} \\ \times iD^F_{\mu\nu}(x_1 - x_2) T\left(: \overline{\psi}(x_1)\gamma^\mu \psi(x_1) :: \overline{\psi}(x_2)\gamma^\nu \psi(x_2) : \right), \end{aligned} \tag{4.9.2}$$

where we have adopted the Feynman gauge for the photon propagator and $\epsilon > 0$ is a small adiabatic parameter. From (4.9.1), the electron part of the current at the vertex x_1 is

$$\sum_{pq} a_q(t_1)^\dagger a_p(t_1) \, \overline{\phi}_q(\boldsymbol{x}_1) \, \phi_p(\boldsymbol{x}_1),$$

and we write down a similar expression at the vertex x_2. The Feynman gauge photon propagator (2.9.39) can be written

$$iD^F_{\mu\nu}(x_1 - x_2) = ig_{\mu\nu} \int \frac{dz}{2\pi} \frac{\mathrm{e}^{-iz(t_1-t_2)+i|z|R/c}}{4\pi\epsilon_0 R}, \tag{4.9.3}$$

where $R = |\boldsymbol{x}_1 - \boldsymbol{x}_2|$. We can perform the time integrations using

$$\int_{-\infty}^{\infty} dt_1 \, \mathrm{e}^{i(E_q - E_p - z)t_1 - \epsilon|t_1|} = 2\pi \Delta_\epsilon(E_q - E_p), \tag{4.9.4}$$

and

$$\int_{-\infty}^{\infty} dt_2 \, \mathrm{e}^{i(E_s - E_r + z)t_2 - \epsilon|t_2|} = 2\pi \Delta_\epsilon(z + E_s - E_r), \tag{4.9.5}$$

where

$$\lim_{\epsilon \to 0} \Delta_\epsilon(x) = \lim_{\epsilon \to 0} \frac{1}{\pi} \frac{\epsilon}{x^2 + \epsilon^2} = \delta(x). \tag{4.9.6}$$

defines the Dirac delta distribution. Because

$$\int_{-\infty}^{\infty} dz\, \Delta_\epsilon(z - E_q + E_p)\Delta_\epsilon(z + E_s - E_r) = \Delta_{2\epsilon}(E_q + E_s - E_p - E_r) \quad (4.9.7)$$

the sum of the incoming energies $E_p + E_r$ and the outgoing energies $E_q + E_s$ are equal, and we can set $z = \omega = E_q - E_p = E_r - E_s$ as $\epsilon \to 0$, so that

$$S^{(2)F} = -2\pi i \sum_{p,q,r,s} (a_q^\dagger a_p)(a_s^\dagger a_r)\Delta_{2\epsilon}(E_q + E_s - E_p - E_r) \times \int d^3x_2 \int d^3x_1\, \phi_q^\dagger(\boldsymbol{x}_1)\gamma^0\gamma^\mu\phi_p(\boldsymbol{x}_1)\, g_{\mu\nu}\frac{\mathrm{e}^{i\omega R/c}}{R}\, \phi_s^\dagger(\boldsymbol{x}_2)\gamma^0\gamma^\nu\phi_r(\boldsymbol{x}_2),$$

which can be put in the form

$$S^{(2)F} = -2\pi i \sum_{p,q,r,s} (a_q^\dagger a_p)(a_s^\dagger a_r) \times \Delta_{2\epsilon}(E_q + E_s - E_p - E_r)(qp\,|k^M(R;\omega)|\,sr), \quad (4.9.8)$$

where

$$(qp\,|\,k^M(R;\omega)|\,sr) = \int d^3x_2 \int d^3x_1\, j_{qp}(\boldsymbol{x}_1)^\mu\, \frac{\mathrm{e}^{i\omega R/c}}{c^2R}\, j_{sr}(\boldsymbol{x}_2)_\mu. \quad (4.9.9)$$

This definition of the matrix element emphasizes its interpretation as the interaction between two transition charge-current densities. However, tradition dictates a different expression,

$$\langle qs\,|\,g^M(R;\omega)\,|\,pr\rangle \equiv (qp\,|k(R;\omega)|\,sr) = \int d^3x_2 \int d^3x_1\, \phi_q^\dagger(\boldsymbol{x}_1)\phi_s^\dagger(\boldsymbol{x}_2)\, g^M(R;\omega)\, \phi_p(\boldsymbol{x}_1)\phi_r(\boldsymbol{x}_2), \quad (4.9.10)$$

where

$$g^M(R;\omega) = (1 - \boldsymbol{\alpha}_1 \cdot \boldsymbol{\alpha}_2)\, \frac{\mathrm{e}^{i\omega R/c}}{R} \quad (4.9.11)$$

is usually known as the Møller interaction [22]. Here the matrix element has been written to look like a nonrelativistic interaction of uncoupled initial and final product wavefunctions, with

$$(\gamma^0\gamma^\mu)_1(\gamma^0\gamma_\mu)_2 = (1 - \boldsymbol{\alpha}_1 \cdot \boldsymbol{\alpha}_2)$$

where the subscripts 1 and 2 label the two vertices of the Feynman diagram at which the Dirac matrices act. It is unfortunate that this notation obscures the physical meaning of the effective interaction.

The Møller interaction is usually regarded as an *effective potential* function in accordance with (4.9.10), although the dependence of $g^M(R;\omega)$ on ω makes it inconsistent with the usual interpretation of the word "potential". The derivation enforces overall energy conservation; this is what we should expect for two particles in isolation, but we shall encounter cases where this restriction is lifted in higher orders of perturbation theory. So for the present, we have only succeeded in defining the Møller interaction "on the energy shell": $E_q + E_s = E_p + E_r$.

The energies involved in light elements are such that $\omega R \ll 1$ over most of the region of interest. It is therefore often acceptable to set the exponential in (4.9.11) to unity so that

$$\lim_{\omega \to 0} g^M(R;\omega) = (1 - \boldsymbol{\alpha}_1 \cdot \boldsymbol{\alpha}_2)\,\frac{1}{R} \tag{4.9.12}$$

so that (4.9.9) becomes a $1/R$ interaction between two charge current densities,

$$\begin{aligned}(qp\,|k(R;0)|\,sr) &= \int d^3x_2 \int d^3x_1\, j_{qp}(\boldsymbol{x}_1)^{\mu}\,\frac{1}{c^2R}\, j_{sr}(\boldsymbol{x}_2)_{\mu} \qquad (4.9.13)\\ &= \int d^3x_2 \int d^3x_1 \frac{1}{R}\left\{\rho_{qp}(\boldsymbol{x}_1)\,\rho_{sr}(\boldsymbol{x}_2) - \frac{1}{c^2}\boldsymbol{J}_{qp}(\boldsymbol{x}_1)\cdot \boldsymbol{J}_{sr}(\boldsymbol{x}_2)\right\}.\end{aligned}$$

The first term is the nonrelativistic Coulomb interaction, the second is an interaction between currents (in $\mathbb{R}^3$) which, can be rewritten as the matrix element, (4.9.10), of the interaction proposed by Gaunt [23]:

$$g^G(R) = -\frac{\boldsymbol{\alpha}_1 \cdot \boldsymbol{\alpha}_2}{R}$$

4.9.2 One-photon exchange: Coulomb gauge

In Coulomb gauge, we have to replace the Feynman gauge propagator (4.9.3) with the expressions of (4.6.5). The part due to the Coulomb interaction alone can therefore be written in the familiar form

$$\begin{aligned}S^{(2)C} &= -2\pi i \sum_{p,q,r,s} (a_q^{\dagger}a_p)(a_s^{\dagger}a_r)\\ &\qquad \times \Delta_{2\epsilon}(E_q + E_s - E_p - E_r)(qp\,|1/R|\,sr) \qquad (4.9.14)\end{aligned}$$

where

$$(qp\,|1/R|\,sr) = \int d^3x_2 \int d^3x_1\, \rho_{qp}(\boldsymbol{x}_1)\,\frac{1}{R}\,\rho_{sr}(\boldsymbol{x}_2). \tag{4.9.15}$$

A corresponding calculation of the transverse photon interaction part gives

$$\begin{aligned}S^{(2)T} &= -2\pi i \sum_{p,q,r,s} (a_q^{\dagger}a_p)(a_s^{\dagger}a_r)\\ &\qquad \times \Delta_{2\epsilon}(E_q + E_s - E_p - E_r)\,(qp\,|k^T(R;\omega)|\,sr) \qquad (4.9.16)\end{aligned}$$

where

$$(qp\,|k^T(R;\omega)|\,sr) = \int d^3x_2 \int d^3x_1 \Big(\boldsymbol{J}_{qp}(\boldsymbol{x}_1)\cdot \boldsymbol{J}_{sr}(\boldsymbol{x}_2) \frac{e^{i\omega R/c}}{R} + (\boldsymbol{J}_{qp}(\boldsymbol{x}_1)\cdot \nabla_{\boldsymbol{R}})(\boldsymbol{J}_{sr}(\boldsymbol{x}_2)\cdot \nabla_{\boldsymbol{R}}) \frac{e^{i\omega R/c}-1}{\omega^2 R}\Big), \quad (4.9.17)$$

with $\omega = |E_q - E_p| = |E_r - E_s|$ as in (4.9.8).

The Breit interaction, which, when expanded in powers of $1/c^2$ gives the Breit-Pauli Hamiltonian [24, Sect. 39] used frequently to approximate relativistic effects in atomic and molecular physics, is obtained from (4.9.17), retaining only terms independent of ω,

$$(qp\,|k^B(R)|\,sr) = -\int d^3x_1 \int d^3x_2 \times \left(\frac{1}{2R}\right) \Big(\boldsymbol{J}_{qp}(\boldsymbol{x}_1)\cdot \boldsymbol{J}_{sr}(\boldsymbol{x}_2) + (\boldsymbol{J}_{qp}(\boldsymbol{x}_1)\cdot \widehat{\boldsymbol{R}})(\boldsymbol{J}_{sr}(\boldsymbol{x}_2)\cdot \widehat{\boldsymbol{R}})\Big), \quad (4.9.18)$$

where $\widehat{\boldsymbol{R}} = \boldsymbol{R}/R$, so that the kernel has an error $O(\omega R/c)$. As in the previous section, we can express these results in terms of an effective potential $g(R;\omega)$, such that

$$S^{(2)}_{qs,pr} = -2\pi i \Delta_{2\epsilon}(E_q + E_s - E_p - E_r)\langle qs\,|\,g(R;\omega)\,|\,pr\rangle, \quad (4.9.19)$$

where

$$\langle qs\,|\,g(R;\omega)\,|\,pr\rangle = \int d^3x_2 \int d^3x_1\, \phi^\dagger_q(\boldsymbol{x}_1)\phi^\dagger_s(\boldsymbol{x}_2)\, g_{12}(R;\omega)\, \phi_p(\boldsymbol{x}_1)\phi_r(\boldsymbol{x}_2),$$

so that

$$g(R;\omega) = 1/R + g^T(R;\omega), \quad (4.9.20)$$

where the transverse photon interaction kernel is

$$g^T(R;\omega) = -\boldsymbol{\alpha}_1\cdot\boldsymbol{\alpha}_2 \frac{e^{i\omega R/c}}{R} - (\boldsymbol{\alpha}_1\cdot \boldsymbol{\nabla}_{\boldsymbol{R}})(\boldsymbol{\alpha}_2\cdot \boldsymbol{\nabla}_{\boldsymbol{R}}) \frac{e^{i\omega R/c}, -1}{\omega^2 R/c^2}. \quad (4.9.21)$$

In the long wavelength limit, $\omega \to 0$, this reduces to the Breit interaction kernel corresponding to (4.9.18),

$$g^B(R) = \lim_{\omega\to 0} g^T(R;\omega) = -\frac{1}{2R}\Big(\boldsymbol{\alpha}_1\cdot\boldsymbol{\alpha}_2 + (\boldsymbol{\alpha}_1\cdot\widehat{\boldsymbol{R}})(\boldsymbol{\alpha}_2\cdot\widehat{\boldsymbol{R}})\Big). \quad (4.9.22)$$

4.9.3 * Off-shell potentials: heuristic argument

The effective potentials have been derived "on the energy shell", which forces $\omega_{sr} = -\omega_{qp} = \omega$, where $\omega_{sr} = (E_s - E_r)/c$ and $\omega_{qp} = (E_q - E_p)/c$. What are we to do when we encounter situations off the energy shell when $\omega_{sr} \neq -\omega_{qp}$? A simple heuristic argument gives part of the answer. After dropping time-dependent factors, we can write (4.9.2) in terms of the interaction matrix element

$$\langle qs \,|\, g(R) \,|\, pr \rangle = \frac{1}{c} \int d^3x \int d^3y j^{\mu}_{qp}(x)\, D_{\mu\nu}(x-y)\, j^{\nu}_{sr}(y),$$

where $D_{\mu\nu}(x-y)$ is the photon propagator in some unspecified gauge. The 4-potential generated by the charge-current density amplitude $j^{\nu}_{sr}(y)$ at the space-time vertex x is

$$A_{\mu}(x;\omega_{sr}) = \frac{1}{c} \int d^3y\, D_{\mu\nu}(x-y)\, j^{\nu}_{sr}(y),$$

where the frequency $\omega_{sr} = (E_s - E_r)/c$ is fixed by energy conservation at the vertex y. We can therefore interpret the interaction matrix element in terms of this 4-potential:

$$\langle qs \,|\, g(R;\omega_{sr}) \,|\, pr \rangle = \int d^3x\, j^{\mu}_{qp}(x)\, A_{\mu}(x;\omega_{sr}).$$

If we interchange the roles of the two vertices, we get an alternative expression

$$\langle qs \,|\, g(R;\omega_{qp}) \,|\, pr \rangle = \int d^3y\, j^{\nu}_{sr}(y)\, A_{\nu}(y;\omega_{qp}),$$

which suggests that the effective interaction *off the energy shell* should be defined by

$$\frac{1}{2} \langle qs \,|\, g(R;\omega_{sr}) + g(R;\omega_{qp}) \,|\, pr \rangle \tag{4.9.23}$$

Whilst this does not give the complete story, it throws some light on the more complex argument that follows in §4.10.

4.9.4 One-photon exchange: the first order energy shift

The one-photon exchange interaction contributes a single term to the Gell-Mann, Low, Sucher energy shift in the form (4.8.12), namely

$$\varDelta E = \lim_{\epsilon \to 0} i\epsilon\, S^{(2)}_{\epsilon},$$

where, from (4.9.8) or (4.9.14),

$$S^{(2)}_{\epsilon} = -2\pi i \sum_{p,q,r,s} (a^{\dagger}_q a_p)(a^{\dagger}_s a_r) \varDelta_{2\epsilon}(E_q + E_s - E_p - E_r)\, (qp \,|\, k(R;\omega) \,|\, sr).$$

Consider a two-electron state

$$|\Phi\rangle = \frac{1}{\sqrt{2!}}\, a_a^\dagger a_b^\dagger\, |\Phi_0\rangle,$$

where Φ_0 is the electron vacuum, and write

$$\Delta E(\Phi) = \lim_{\epsilon\to 0} \langle \Phi \,|\, i\epsilon S_\epsilon^{(2)} \,|\, \Phi\rangle. \tag{4.9.24}$$

Because the initial and final states are the same, the matrix elements must be evaluated on the energy shell so that, by (4.9.6), $\lim_{\epsilon\to 0} \epsilon\Delta_\epsilon(0) = 1/\pi$, and we obtain the overlap charge-current form

$$\Delta E(\Phi) = (aa \,|\, k(R;\omega) \,|\, bb) - (ba \,|\, k(R;\omega) \,|\, ab).$$

as in (4.9.9). Alternatively, using the traditional expression (4.9.10), we get

$$\Delta E(\Phi) = \langle ab \,|\, g(R;\omega) \,|\, ab\rangle - \langle ba \,|\, g(R;\omega) \,|\, ab\rangle.$$

4.10 * Off-shell potentials

The S-matrix approach of the last section is limited to processes that are diagonal in the interaction from which we can calculate energy shifts and transition rates for energy-conserving processes. A similar procedure that allows us to calculate matrix elements off the energy shell can be applied to the evolution operator $U(t, -\infty)$, where we can set $t = 0$ without loss of generality. We rework the calculations of the last section replacing the Feynman

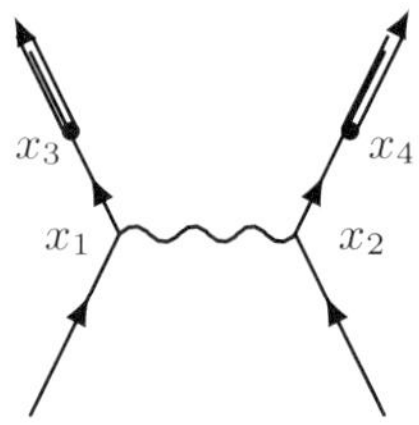

Fig. 4.7.

diagram of Figure 4.6(a) with Figure 4.7, which contains additional (double) lines representing electrons propagating in the external field from x_1 to x_3 and from x_2 to x_4, with $x_2^0 = x_4^0 = 0$. The amplitude for this process is given by

$$U_\epsilon^{(2)}(0,-\infty) = \left(-\frac{ie}{c}\right)^2 \int d^4x_4 \int d^4x_3 \int d^4x_2 \int d^4x_1 \, \mathrm{e}^{-\epsilon|t_1|-\epsilon|t_2|}$$
$$\times\delta(x_4^0)\,\psi^\dagger(x_4)\left(-\frac{1}{2}S_F(x_4,x_2)\right)\;\delta(x_3^0)\,\psi^\dagger(x_3)\left(-\frac{1}{2}S_F(x_3,x_1)\right)$$
$$\times iD^F_{\mu\nu}(x_1-x_2)\times T\left(:\tilde{\psi}(x_1)\gamma^\mu\psi(x_1)::\tilde{\psi}(x_2)\gamma^\nu\psi(x_2):\right). \qquad (4.10.1)$$

For simplicity, we shall start with the Feynman gauge photon propagator (4.6.3), and express the electron propagator in a spectral expansion as in (4.6.8),

$$-\frac{1}{2}S_F(x,y) = \frac{1}{2\pi i}\int_{-\infty}^{\infty} dz\; G(\boldsymbol{x},\boldsymbol{y};z)\,\gamma_0\,\mathrm{e}^{-iz(x-y)^0/c}, \qquad (4.10.2)$$

with

$$G(\boldsymbol{x},\boldsymbol{y};z) = \sum \frac{\phi_r(\boldsymbol{x})\,\phi_r^\dagger(\boldsymbol{y})}{z-E_r+i\eta_r\epsilon},$$

where η_r is +1 for electron states and -1 for positron (negative energy) states. It is also convenient to write $c_r = a_r$ when $\eta_r = +1$ and $c_r = b_r^\dagger$ when $\eta_r = -1$. After a trivial integration over x_4^0, the orthonormality of the eigenfunctions in (4.10.2) gives

$$\int d^3x_4\,\psi^\dagger(x_4)\left(-\frac{1}{2}S_F(x_4,x_2)\right) = \frac{1}{2\pi i}\int_{-\infty}^{\infty} dz_2 \frac{\mathrm{e}^{iz_2t_2}}{E_s-z_2-i\eta_s\epsilon}\,\psi_s^\dagger(x_2).$$

with a similar expression for the integration over x_3. Thus (4.10.1) becomes

$$U_\epsilon^{(2)F}(0,-\infty) = 2\pi i \sum_{p,q,r,s} :c_q^\dagger c_p::c_s^\dagger c_r: \qquad (4.10.3)$$
$$\times\int_{-\infty}^{\infty}\frac{dz_1}{2\pi i}\int_{-\infty}^{\infty}\frac{dz_2}{E_s-z_2-i\eta_s\epsilon}\int_{-\infty}^{\infty}\frac{dz_3}{E_q-z_3-i\eta_q\epsilon}$$
$$\times\,\Delta_\epsilon(z_2-E_r-z_1)\,\Delta_\epsilon(z_3-E_p+z_1)\,(qp\,|\,k^M(R;|z_1|)\,|\,sr)$$

where the two Δ_ϵ distributions come from the integrals over t_1 and t_2 and we have taken $e^2/4\pi\epsilon_0 = 1$. Before performing the integrations over z_2 and z_3, we rewrite the product of the z-dependent denominators in the form

$$\frac{1}{E_s-z_2-i\eta_s\epsilon}\frac{1}{E_q-z_3-i\eta_q\epsilon} \qquad (4.10.4)$$
$$=\left[\frac{1}{E_s-z_4-i\eta_s\epsilon}+\frac{1}{E_q-z_3-i\eta_q\epsilon}\right]\frac{1}{E_q+E_s-z_2-z_3-i(\eta_s+\eta_q)\epsilon}$$

and exploit the well-known relation that, as $\epsilon\to 0$,

$$\frac{1}{E-z\mp i\epsilon} = P\left(\frac{1}{E-z}\right)\pm i\pi\delta(E-z),$$

where $P(\ldots)$ denotes a Cauchy principal value. Equation (4.10.4) thus breaks up into a sum of terms

$$
\begin{aligned}
&(i\pi)^2 \left(\eta_s\delta(E_s - z_2) + \eta_q\delta(E_q - z_3)\right) (\eta_s + \eta_q)\delta(E_q + E_s - z_2 - z_3)\\
&+ i\pi \left(\eta_s\delta(E_s - z_2) + \eta_q\delta(E_q - z_3)\right) P\left(\frac{1}{E_q + E_s - z_2 - z_3}\right)\\
&+ i\pi \left[P\left(\frac{1}{E_s - z_2}\right) + P\left(\frac{1}{E_q - z_3}\right)\right] (\eta_s + \eta_q)\delta(E_q + E_s - z_2 - z_3)\\
&+ \left[P\left(\frac{1}{E_s - z_2}\right) + P\left(\frac{1}{E_q - z_3}\right)\right] P\left(\frac{1}{E_q + E_s - z_2 - z_3}\right).
\end{aligned}
$$

The calculation is now straightforward but tedious. The first line gives zero if the outgoing particle states are different, $\eta_s + \eta_q = 0$; when the outgoing particles are both electrons or both positrons we recover the *on-shell* amplitude

$$
-2\pi i \sum_{pqrs} : c_q^\dagger c_p :: c_s^\dagger c_r : \Delta_{2\delta}(\omega_{qp} + \omega_{sr})\,(qp \,|\, k^M(R; |\,\omega_{qp}|) \,|\, sr). \tag{4.10.5}
$$

The Δ factor enforces energy conservation, $\omega_{qp} + \omega_{sr} = 0$. The second line gives, as suggested by the heuristic argument,

$$
-2\pi i \sum_{pqrs} : c_q^\dagger c_p :: c_s^\dagger c_r : \frac{(qp \,|\, k^M(R; |\,\omega_{pq}|) \,|\, sr) + (qp \,|\, k^M(R; |\,\omega_{rs}|) \,|\, sr)}{2(E_s + E_q - E_r - E_p)} \tag{4.10.6}
$$

when the outgoing particles are both electrons, $\eta_q = \eta_s = 1$. The third line gives two terms that cancel, and the fourth line gives two principal value contributions

$$
\sum_{pqrs} : c_q^\dagger c_p :: c_s^\dagger c_r : \int_{-\infty}^{\infty} dz \left(\frac{P}{z - \omega_{sr}} - \frac{P}{z + \omega_{qp}}\right) \frac{(qp \,|\, k^M(R; |\,z|) \,|\, sr)}{E_s + E_q - E_r - E_p} \tag{4.10.7}
$$

which are not given by the heuristic argument. It can be transformed into a form that is more amenable to computation by first expressing each part of (4.10.7) as an integral on the positive real line:

$$
\int_{-\infty}^{\infty} \mathrm{e}^{i|z|R/c} \frac{P}{z - \omega}\, dz = \int_0^{\infty} \mathrm{e}^{izR/c} \left[\frac{P}{z - \omega} - \frac{P}{z + \omega}\right] dz.
$$

Now consider the contour integral

$$
I = \int_C \mathrm{e}^{izR/c} \frac{P}{z - \Omega}\, dz
$$

where C consists of the real axis $0 < z < a$, the quarter circle $z = a\mathrm{e}^{i\phi}$, $0 < \phi < \pi/2$, and the imaginary axis from $z = iy, a > y > 0$. If $\Omega > 0$, it is necessary to indent the contour to pass round the pole $z = \Omega$, which, in the limit $a \to \infty$, gives

$$I \to \int_0^\infty \mathrm{e}^{-yR/c} \frac{i\,dy}{iy - \Omega} + i\pi\theta(\Omega)\mathrm{e}^{i\Omega R/c},$$

where $\theta(\Omega) = +1$ if $\Omega > 0$, $\theta(\Omega) = 0$ otherwise. Collecting terms, we find that (4.10.7) is equal to

$$\begin{aligned} &-2\pi i \sum_{pqrs} : c_q^\dagger c_p :: c_s^\dagger c_r : \frac{1}{E_s + E_q - E_r - E_p} \qquad (4.10.8) \\ &\times \Bigg[-\frac{1}{\pi} \int_0^\infty dy \left(\frac{\omega_{rs}}{y^2 + \omega_{rs}^2} + \frac{\omega_{pq}}{y^2 + \omega_{pq}^2} \right) (qp \,|\, k^M(R; iy) \,|\, sr) \\ &+ i\pi\theta(\omega_{rs})(qp \,|\, k^M(R; \omega_{rs}) \,|\, sr) + i\pi\theta(\omega_{pq})(qp \,|\, k^M(R; \omega_{pq}) \,|\, sr) \Bigg] \end{aligned}$$

where

$$k^M(R; \omega) = \mathrm{e}^{i\omega R/c}/c^2 R, \quad k^M(R; iy) = \mathrm{e}^{-yR/c}/c^2 R.$$

The integral in the square brackets is real. However, the last two terms, which appear only if $E_r > E_s$ or $E_p > E_q$, are complex. So if p and r label ground state orbitals, these terms are absent. When they label excited orbitals, there may be decay channels and the imaginary part is related to the decay rate of the excited state. These can also be expressed [21] in terms of the sine and cosine integrals $Si(x)$ and $Ci(x)$ using [25, 3.354 (1)].

For electron-electron or positron-positron interactions, we can take the *low frequency limit* in which the kernels all reduce to $k^M(R; 0) = 1/R$, corresponding to the effective potential (4.9.12) and can be taken outside the sum of (4.10.5),(4.10.6), and (4.10.8). The result is that

$$\begin{aligned} &U_\epsilon^{(2)F}(0, -\infty) \to -2\pi i \sum_{p,q,r,s} : c_q^\dagger c_p :: c_s^\dagger c_r : (qp \,|\, k^M(R; 0) \,|\, sr) \\ &\times \Big\{ \Delta_{2\epsilon}(E_{qs,pr}) \qquad (4.10.9) \\ &\quad + \frac{\eta_q + \eta_s - \operatorname{sgn}(\omega_{rs}) - \operatorname{sgn}(\omega_{pq}) + i\pi[\theta(\omega_{rs}) + \theta(\omega_{pq})]}{2E_{qs,pr}} \Big\}, \end{aligned}$$

where $E_{qs,pr} = \omega_{qp} + \omega_{sr}$. This is not applicable to electron-positron interactions, where the photon energies are at least of order $2mc^2$.

In the Coulomb gauge, the results are more complicated. We replace $k^M(R; |\,z|)$ by $1/R + k^T(R; |\,z|)$. The Coulomb part is trivial. From (4.9.17) the transverse kernel has two parts

$$k^{(1)}(R; |\,z|) = \frac{\mathrm{e}^{i|\,z|R/c}}{R}, \quad k_{ij}^{(2)}(R; |\,z|) = \partial_i \partial_j \frac{\mathrm{e}^{i|\,z|R/c} - 1}{|\,z|^2 R}, \quad i, j = 1, 2, 3,$$

where i, j denote indices of Cartesian components and z is *real*. Clearly the scalar part, $k^{(1)}(R; |\,z|)$, gives the same singular and principal value terms as the Feynman kernel; the tensor part, $k^{(2)}(R; |\,z|)$, as it stands, has a singularity at $z = 0$. We can exhibit this as a removable singularity by evaluating the partial derivatives with respect to the Cartesian components of $\boldsymbol{R}$, giving

$$k^{(2)}_{ij}(R; |\,z|) = \left(\delta_{ij} - \frac{3X_iX_j}{R^2}\right)\frac{f(|\,z|)}{R^3} - \frac{X_iX_j}{R^2}\,k^{(1)}(R; |\,z|),$$

where

$$f(|\,z|) = \frac{1 - e^{i|\,z|R/c}(1 - i|\,z|R/c)}{z^2} = -\frac{R^2}{2c^2} + O(|\,z|), \quad |\,z| \to 0.$$

The principal value integral over z analogous to (4.10.7) can now be handled in much the same way as before, so that in Coulomb gauge we can replace $k^M(R;\omega)$ in (4.10.5), (4.10.6), and (4.10.8) with $k^T(R;\omega)$. In the same way, in the low frequency limit, we can replace $(qp\,|\,k^M(R;0)\,|\,sr)$ in (4.10.9) with the Breit potential $(qp\,|\,k^B(R)\,|\,sr)$ of (4.9.22).

4.11 Many-body perturbation theory

The effective interactions introduced in the last section enable us to set up a consistent scheme for studying the properties of many-electron systems modelled on nonrelativistic many-body perturbation theory. This is a widely used approach that is capable of giving results of high accuracy when relativistic effects are not relevant, mainly for light atoms and molecules. It is therefore not surprising that a similar theory based on the effective potentials of the last section, *relativistic many-body perturbation theory* (or RMBPT), is also very effective when relativistic effects cannot be ignored. We shall later discuss small effects, conveniently described as *radiative corrections*, whose diagrams are omitted in RMBPT.

MBPT was first introduced and applied in the theory of nuclear structure, which has some formal similarities with atomic and molecular structure. There is one important difference: nucleons experience a strong short range repulsive force that cannot be treated successfully by ordinary perturbation theory. The difficulty was overcome by Brueckner, who constructed an effective one-body potential U based on the two-body interaction g whose residual corrections to the nuclear energy were small, although the corrections to the wavefunction were large [26, 27, 28]. Brueckner [29] used this formalism to study *nuclear matter*, for which surface effects on the structure could be neglected. This established that, for a fixed nucleon density, the energy was proportional to the number of particles; the energy had a minimum at the nucleon density observed in large nuclei. Brueckner found terms in the perturbation expansion that were quadratic, rather than linear, in the nucleon number due to the presence of what he called "reducible" or "unlinked clusters" that could be expressed as products of energy contributions of lower order. These terms are clearly unphysical; Brueckner [30] showed that the perturbation theory could be recast so that offending terms of the first few orders cancelled but was unable to demonstrate this in full generality. Goldstone [31], who used a field theoretic formalism similar to ours, was able to show that the unphysical terms could always be represented in terms of "unlinked" Feynman diagrams that had two or more disconnected pieces. We have already encountered this situation in our discussion of the S-matrix, where such diagrams contribute only to the phase of the wavefunction but not to the energy of the system.

4.11.1 Nonrelativistic many-body theory

For application to atomic structure, the starting point of the Goldstone procedure is an unquantized nonrelativistic Hamiltonian for N orbital electrons,

$$H = \sum_{i=1}^{N} h_i + \sum_{i<j} g_{ij}, \tag{4.11.1}$$

where

$$h = \frac{\boldsymbol{p}^2}{2m} - \frac{Z(r)}{r}$$

is the Schrödinger Hamiltonian for a single electron subject to the Coulomb attraction of a fixed nucleus and

$$g_{ij} = \frac{1}{R_{ij}}, \quad R_{ij} = |\boldsymbol{R}_{ij}|, \quad \boldsymbol{R}_{ij} = \boldsymbol{x}_i - \boldsymbol{x}_j.$$

describes the Coulomb repulsion of an electrons at position $\boldsymbol{x}_i$ by an electron at position $\boldsymbol{x}_j$. The potential term $-Z(r)/r$, $r = |\boldsymbol{x}|$, allows mainly for the nuclear charge distribution to have a finite "radius" R_0, so that $Z(r) \to Z$ when $r > R_0$. We discuss commonly used nuclear models in the next chapter. We shall introduce an effective potential $u(r)$, and assume that we can construct a complete orthogonal set of *orbitals* spanning the relevant Hilbert space,

$$\left\{ \frac{\boldsymbol{p}^2}{2m} - \frac{Z(r)}{r} + u(r) \right\} \phi_i(\boldsymbol{r}) = \varepsilon_i \, \phi_i(\boldsymbol{r}), \tag{4.11.2}$$

where i orders the states in terms of increasing energy eigenvalue ε_i.

Goldstone used the Fock space formalism of §4.1.3. We shall do the same, although we shall here distinguish Fock space operators from their unquantized counterparts with a *hat* accent. Spectral resolution of the field operators $\hat{\psi}(x)$ and $\hat{\psi}^\dagger(x)$ allows us to write

$$\widehat{F} = \sum_{ij} a_i{}^\dagger a_j \, \langle i \,|\, f \,|\, j \rangle \tag{4.11.3}$$

for the Fock-space operator of equation (4.1.34) corresponding to the unquantized one-body operator $f(\boldsymbol{r})$ and

$$\widehat{G} = \frac{1}{2} \sum_{ijkl} a_i{}^\dagger a_j{}^\dagger a_l a_k \, \langle ij \,|\, g \,|\, kl \rangle \tag{4.11.4}$$

for the Fock space two-body operator of (4.1.35). The sums run over a complete set of indices of the orbital eigenfunctions, which we assume to be normalized as well as orthogonal. We now write

$$\widehat{H} = \widehat{H}_0 + \widehat{V}, \tag{4.11.5}$$

where

$$\widehat{H}_0 = \sum_{i=1}^{N} a_i{}^\dagger a_i \, \varepsilon_i, \quad \widehat{V} = -\widehat{U} + \widehat{G}, \quad \widehat{U} = \sum_{ij} a_i{}^\dagger a_j \, \langle i \,|\, u \,|\, j \rangle$$

and $\widehat{G}$ is given by (4.11.4).

Goldstone's analysis was similar to §4.8.1, the main difference being that he calculated the perturbed wavefunction Ψ at time $t = 0$ starting with the unperturbed *nondegenerate* wavefunction Φ_0 at $-\infty$, and used diagrams in which the vertices were time-ordered; all time-orderings are implied in a single Feynman diagram. He found, as in (4.8.8),

$$\Psi_\epsilon = \lim_{\epsilon \to 0} \frac{U_\epsilon(0, -\infty)\Phi_0}{\langle \Phi_0 \,|\, U_\epsilon(0, -\infty) \,|\, \Phi_0 \rangle}$$

where $U_\epsilon(0, -\infty)$ is now often called the *wave operator*. After carrying out the time integrations, this gave

$$\Psi_\epsilon = \lim_{\epsilon \to 0} \sum_L \frac{1}{E_0 - \widehat{H}_0 + ni\epsilon} \widehat{V} \ldots \frac{1}{E_0 - \widehat{H}_0 + 2i\epsilon} \widehat{V} \frac{1}{E_0 - \widehat{H}_0 + i\epsilon} \widehat{V} \Phi_0$$

where $\sum_L$ indicates that only linked diagrams are included and a sum over n from 1 to ∞ is to be understood. In the nonrelativistic case, where Φ_0 is the state of lowest energy E_0, there can be no zero denominators when taking the limit $\epsilon \to 0$, so that this equation becomes formally

$$\Psi = \sum_L \left(\frac{1}{E_0 - \widehat{H}_0} \widehat{V} \right)^n \Phi_0. \tag{4.11.6}$$

and the energy shift is

$$\begin{aligned} \Delta E = \langle \Phi_0 \,|\, \widehat{V} \,|\Psi \rangle &= \lim_{\epsilon \to 0} \frac{\langle \Phi_0 \,|\, \widehat{V} \widehat{U}_\epsilon(0, -\infty) \,|\, \Phi_0 \rangle}{\langle \Phi_0 \,|\, U_\epsilon(0, -\infty) \,|\, \Phi_0 \rangle} \\ &= \sum_L \left\langle \Phi_0 \,\middle|\, \widehat{V} \left(\frac{1}{E_0 - \widehat{H}_0} \widehat{V} \right)^n \,\middle|\, \Phi_0 \right\rangle. \end{aligned} \tag{4.11.7}$$

In the simplest model of an atom, we assume that the zero order ground state Φ_0 is one in which each of the N *non-degenerate* orbitals of lowest energy given by (4.11.2) accomodate one electron, all higher orbitals being unoccupied. We can think of this configuration as forming a *core*, and its wavefunction will be a Slater determinant. We divide the spectrum of (4.11.2) into two classes

- *Occupied/core* orbitals denoted by $a, b, c, \ldots$
- *Unoccupied/virtual* orbitals denoted by $r, s, t, \ldots$
- *Unspecified* orbitals which may belong to either of the above two classes are denoted by $i, j, k, \ldots$

From (4.1.27) or (4.1.33) we get

$$|\,\Phi_0\rangle = \frac{1}{\sqrt{N!}}\, a^\dagger_{a_1} a^\dagger_{a_2} \ldots a^\dagger_{a_N}\,|\,0\rangle \tag{4.11.8}$$

so that the zero order energy is given by

$$E_0 = \langle \Phi_0\,|\,\widehat{H}_0\,|\,\Phi_0\rangle = \sum_{p=1}^{N} \varepsilon_{a_p} \tag{4.11.9}$$

and the first-order correction to the energy due to the perturbation $\widehat{V}$ is

$$E_1 = \langle \Phi_0\,|\,\widehat{V}\,|\,\Phi_0\rangle = \langle \Phi_0\,|\,-\widehat{U}+\widehat{G}\,|\,\Phi_0\rangle \tag{4.11.10}$$

where

$$\langle \Phi_0\,|\,\widehat{G}\,|\,\Phi_0\rangle = \frac{1}{2}\sum_{ab}\left[\langle ab\,|\,g\,|\,ab\rangle - \langle ba\,|\,g\,|\,ab\rangle\right].$$

The *Hartree-Fock* effective one-body potential, u_{HF}, is defined for all orbitals by

$$\langle i\,|\,u_{HF}\,|\,j\rangle = \sum_{b}\left[\langle ib\,|\,g\,|\,jb\rangle - \langle bi\,|\,g\,|\,jb\rangle\right], \tag{4.11.11}$$

where the sum over b runs over core orbital indices. It follows that to this order of approximation, there is an *effective one-body potential* v such that

$$\langle i\,|\,v\,|\,j\rangle = \langle i\,|\,-u+u_{HF}\,|\,j\rangle. \tag{4.11.12}$$

Thus, from (4.11.10),

$$E_1 = \sum_a \langle a\,|\,v\,|\,a\rangle = -\sum_a \langle a\,|\,u\,|\,a\rangle + \frac{1}{2}\sum_a \langle a\,|\,u_{HF}\,|\,a\rangle$$

If we now identify $u(r)$ with the Hartree- Fock potential u_{HF}, then

$$E_1 = -\frac{1}{2}\sum_a \langle a\,|\,u_{HF}\,|\,a\rangle = -\frac{1}{2}\sum_{ab}\left[\langle ab\,|\,g\,|\,ab\rangle - \langle ba\,|\,g\,|\,ab\rangle\right]. \tag{4.11.13}$$

With u chosen in this way, then

$$\varepsilon_a = \langle a\,|\,h + u_{HF}\,|\,a\rangle,$$

so that the eigenvalue sum E_0 (4.11.9) counts all pair repulsion terms twice; the effect of E_1, (4.11.13) is therefore to correct for this double counting.

4.12 MBPT for atoms and molecules

The first person to apply Goldstone's formalism to atomic physics was Kelly [32, 33, 34, 35], who initially studied closed-shell atoms such as beryllium as well as atoms with nondegenerate unperturbed reference states. Extensions of the theory to handle atomic open shells followed; Brandow [36, 37] used the Brillouin-Wigner expansion, whilst Sandars [38] used the simpler Rayleigh-Schrödinger scheme. Sandars also exploited the topological similarity of Feynman diagrams to the angular momentum diagrams introduced by Yutsis *et al.* [39] to find representations for various effective operators, a technique that was further developed in [40]. Atoms with several open shells required an extension of the formalism made independently by Lindgren [41] and Kvasnička [42]. This extension is based on a *generalized Bloch equation* for the *wave operator* $\boldsymbol{\Omega}$, which generates the perturbed wavefunction Ψ from the unperturbed wavefunction Φ_0 belonging to a *multi-reference model space* in much the same way as the U-matrix of (4.8.5). Whilst this was a major advance, the perturbation series often converges quite slowly. So-called *all-order* methods, in which certain classes of diagrams are generated iteratively, and *coupled cluster* schemes have become popular in both atomic and molecular physics as well as in quantum chemistry.

4.12.1 Particle-hole formalism

A prominent feature of the quantization of the electron-positron field in §4.2 was the division of the energy spectrum into two disjoint pieces: the electron states of positive energy and the positron (or hole) states of negative energy. With a few modifications, this *particle-hole* formalism proves to be just as useful for nonrelativistic MBPT in the Furry picture. The Poincaré invariant, sgn E, no longer has a meaning. Instead of the QED vacuum state Φ_0, which has no particles or anti-particles, it is more convenient to start from a nearby *reference state* Φ representing a *closed-shell*, and then to generate an N-particle state by creating particles and holes with respect to Φ. The zero-order independent electron spectrum consists of the *core* states which are occupied in Φ and *particle* states, which comprise the rest of the spectrum.[2] The *particles* are associated with creation and destruction operators $a_i^\dagger, a_j$ whilst the *holes*, or excitations out of the core, are associated with creation and destruction operators $b_a^\dagger = a_a$ and $b_a = a_a^\dagger$. As in §4.10, it is convenient to write $c_i = a_i$ if i refers to a particle state and $c_i = a_i^\dagger$ if it refers to a hole state. A string of creation and destruction operators is in *normal order* if all particle-hole destruction operators c_i stand to the right of the particle-hole creation operators $c_i^\dagger$. We shall use curly brackets, $\{\ldots\}$, to denote normal ordering of a string of particle-hole operators.

In this scheme, the operator $\widehat{V}$ of (4.11.5) can be decomposed into three normally ordered parts

[2] There are, of course, no negative energy states in the nonrelativistic theory.

$$\widehat{V} = V_0 + \{\widehat{V}_1\} + \{\widehat{V}_2\},$$

where

$$V_0 = -\sum_a \langle a \,|\, u \,|\, a \rangle + \frac{1}{2}\sum_{ab} [\langle ab \,|\, g \,|\, ab \rangle - \langle ba \,|\, g \,|\, ab \rangle]$$

is a pure number (or *zero-body* operator),

$$\{\widehat{V}_1\} = \sum_{ij} \{c_i^\dagger c_j\} \langle i \,|\, v \,|\, j \rangle$$

is a *one-body* operator, and

$$\{\widehat{V}_2\} = \frac{1}{2}\sum_{ijkl} \{c_i^\dagger c_j^\dagger c_l c_k\} \langle ij \,|\, g \,|\, kl \rangle$$

is a *two-body* operator. Here the indices i, j, k, l run over the complete spectrum, whereas a, b run over core states only. The advantage of this formalism is that each term can be displayed in the manner of a Goldstone diagram [31] similar to the Feynman diagrams already encountered, a considerable help in taking MBPT to higher orders. Notice that $\{\widehat{V}_1\}$ vanishes when u is chosen to be the Hartree-Fock potential u_{HF}, which is one reason for the popularity of the Hartree-Fock approximation as a good starting point for more precise calculations. With this formalism, Wick's theorem and diagrammatic

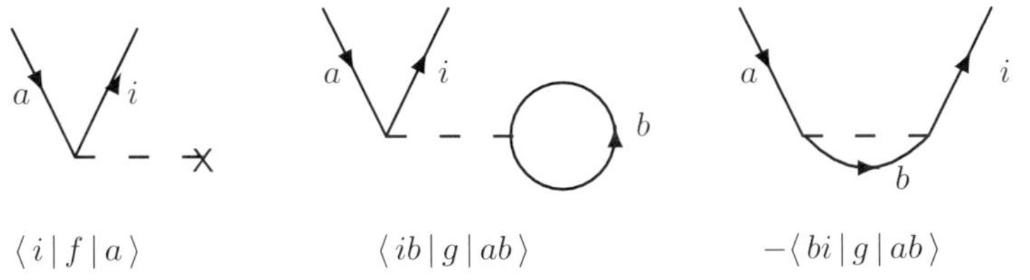

Fig. 4.8.

techniques can be used in much the same way as in QED perturbation theory. However, the Hamiltonian is time-independent in nonrelativistic quantum mechanics of atoms and molecules even when relativistic corrections are included at the Breit-Pauli level. It is therefore possible to formulate MBPT in a time-independent manner and it is customary to employ Goldstone diagrams, in which time is directed up the page as in Feynman diagrams, particle lines are directed upwards and hole lines directed downwards. One- and two-body interactions are instantaneous and appear as horizontal dashed lines as in Figure 4.8. However, one Feynman diagram incorporates all possible time-orderings, so the rules for expressing Goldstone diagrams in terms of contributions to the scattering amplitude are a little different from those for Feynman diagrams [43, Chapter 1].

4.12.2 Computational methods

The early calculations by Kelly and others used atomic Hartree-Fock wavefunctions constructed using finite difference numerical methods. The orbitals are of standard central-field form, in which

$$\phi(\boldsymbol{x}) = \text{const.}\,\frac{P(r)}{r}\,Y_{lm}(\theta,\varphi).$$

The integration over the angular variables in matrix element calculation can be done analytically [40] but the integration over the radial coordinates must be done numerically. Moreover, the spectral sums in perturbation formulas such as (4.8.2) involve summing in general over a countable infinity of bound states as well as an integration over positive energy continuum states. The technical challenges presented by these numerical procedures limited the accuracy of the final results.

Quantum chemists have taken a different route to numerical approximation. Although some attempts have been made at direct numerical solution of the differential equations for diatomic molecules, the majority of molecular calculations are based on expansion of the one-electron orbitals in terms of sets of analytic functions, generally known as basis sets. This replaces the infinite dimensional spectrum of the differential and integral operators of the original formulation with finite dimensional matrix approximations. The hope is that results good enough for practical purposes can be obtained with a relatively small basis set, and that in any case the results will converge if the basis set is enlarged sufficiently. The evaluation of perturbation expansions becomes a matter of repeated matrix multiplication, so that the use of finite dimensional approximations has become cheaper, easier and more attractive with continual increases in computer speed and memory. Lindgren [43, Chapter 1] and others therefore present the formalism of MBPT in terms of matrix algebra and Wilson [44] surveys applications to quantum chemistry. Although both books are now somewhat dated, they continue to give a good idea of the successes and limitations of such calculations.

4.13 Relativistic approaches to atomic and molecular structure

4.13.1 The no-virtual-pair approximation (NVPA)

The most direct way to introduce relativistic dynamics into approaches such as MBPT is to replace the one-electron Schrödinger operator, h, of (4.11.1) with the corresponding Dirac operator

$$h_D = c\boldsymbol{\alpha}\cdot\boldsymbol{p} + \beta mc^2 - \frac{Z(r)}{r} \tag{4.13.1}$$

and the electron-electron Coulomb repulsion operator g with one of the effective interactions derived from QED in §4.9 and §4.10. However, this is not quite straightforward, mainly because of the Dirac negative energy spectrum. The majority of writers on RMBPT, for example Lindgren [43, Chapter 1] and Johnson [43, Chapter 2], have chosen a different starting point, replacing (4.11.1) by the so-called *no-virtual-pair Hamiltonian* (NVP),

$$H = \Lambda_{++} \left\{ \sum_{i=1}^{N} h_{Di} + \sum_{i<j} \left(\frac{1}{R_{ij}} + g^B(R_{ij}) \right) \right\} \Lambda_{++}, \tag{4.13.2}$$

where Λ_{++} is a formal projection operator on to the *positive eigenstates* of h_D. This restricts the sums over orbital indices to those that belong to the positive energy states only. The operator $g^B(R)$ is the *instantaneous* Breit operator defined in (4.9.22).

The decision to adopt of the *no-virtual-pair approximation* (NVPA) is often motivated by the natural desire to ignore the positron states that are expected to contribute little to the physical processes of most interest in atoms and molecules. In practical calculations with finite matrix schemes, the positron states double (and, in some approaches, more than double) the dimension of the basis. The cost of calculations is dominated by the construction of atomic and molecular two-electron interaction integrals: $\langle pq \,|\, g \,|\, rs \rangle$, where p, q, r, s label basis functions. If there are M basis functions, then the work involved scales roughly like M^4, so that having to deal with the positron states might increase the work load dramatically.

4.13.2 The NVPA as an antidote to "continuum dissolution"

The NVP Hamiltonian is also supposed to prevent the alleged destructive role of "continuum dissolution", first highlighted by Brown and Ravenhall [45]. Breit [46] originally wrote down the two-electron wave equation that bears his name

$$\left(h_D^{(1)} + h_D^{(2)} + e^2/r_{12} \right) U(\boldsymbol{r}_1, \boldsymbol{r}_2) = E\, U(\boldsymbol{r}_1, \boldsymbol{r}_2)$$

to study the fine structure levels of helium. He derived relativistic corrections to the Schrödinger equation for helium as a series expansion in powers of α that disagreed with the the experimental fine structure splitting. The culprit appeared to be certain terms of order α^2 arising from $g^B(R)$ that could be associated with "free electron components of negative energy", which had to be omitted to resolve the disagreement. The critique of this work by Brown and Ravenhall [45], supported by a detailed account of the problem in Bethe and Salpeter [24] and later by Sucher [47]– [51] in a series of articles, has led to the almost universal adoption of the NVPA Hamiltonian (4.13.2) as the unquestioned starting point for many developments, in particular in relativistic molecular structure calculations.

Brown and Ravenhall's paper was written when QED was still at a formative stage and reflects a period in which opinion on several issues had still to crystallize. Their principal criticism was that the time-dependent form of the Breit equation,

$$\left\{ h_{D1} + h_{D2} + \frac{1}{R_{12}} \right\} \psi(\boldsymbol{x}_1, \boldsymbol{x}_2) = i\, \frac{\partial \psi(\boldsymbol{x}_1, \boldsymbol{x}_2)}{\partial t},$$

nowadays known as the Dirac-Coulomb (DC) model, was "meaningless" as a wave equation for the two-electron system. Without the Coulomb interaction, this equation has a stationary ground state that is a product of two $1s$ Dirac Coulomb functions. When the Coulomb repulsion term is "turned on slowly", the wavefunction evolves into a general sum of products of Dirac Coulomb functions some of which may be terms with one electron in the negative energy continuum and the other in the positive energy continuum. Such an expression would not be square integrable, from which Brown and Ravenhall inferred that the relativistic DC equation can have no bound state solutions and that the initial bound state must "dissolve" into the continuum. They concluded [45] that one could only obtain meaningful results by treating the relativistic parts using first order perturbation theory to calculate the energy shift from the nonrelativistic equation and that the DC equation should not be used to find the change in the wave function. This pessimistic conclusion denies any possibility of constructing a rigorous relativistic theory of the many-electron atom or molecule.

Sucher [47]– [51] put this argument even more vividly by claiming that the Dirac-Coulomb Hamiltonian is "sick". He suggested that Brown and Ravenhall's postulated mechanism of "continuum dissolution" (CD) was mathematically analogous to the physical process of autoionization; consequently the only way to obtain stable bound state solutions and to restore the unquantized DC equation to "health" is to use the NVP Hamiltonian (4.13.2). The projection operators Λ_{++} are intended to restrict the domain and range of the Hamiltonian to two-electron "states of positive energy". The Achilles' heel of this approach is that there is no way of constructing Λ_{++} without first solving the whole problem. Whilst the *free particle* positive energy projectors are given explicitly by (3.1.29), no such simple expression can be found in the presence of interactions. It is generally accepted that Furry picture QED, in which the Pauli exclusion principle is an integral part of the formalism, is the appropriate way to describe a system of interacting particles in atoms, molecules or solids; see, for example the formulation of Sapirstein [52, §IIIF]. The electron field operator in Furry QED is expanded in terms of a set of one-electron spinors, say $\{\, |\alpha(U)\rangle \,\}$, generated by solving the Dirac equation for some mean-field potential U. This is assumed to form a complete orthonormal set in some underlying Hilbert space $\mathcal{H}$. The basis vectors generated by one such potential, U, will be related to those generated by another potential, U', by a unitary transformation:

$$|\alpha(U)\rangle = \sum_{\alpha'} |\alpha'(U')\rangle . \langle \alpha'(U') \,|\, \alpha(U)\rangle$$

where, in general, the sum over states will run over the whole spectrum, including both positive and negative energy solutions. Since we cannot solve many-particle problems in an exact analytical form, we have to use an iterative scheme in which the interaction between particles depends on the trial solution at each step. Even if our initial trial solution includes only products of positive energy spinors, the next iteration changes the effective potential and introduces negative energy elements. One cannot define Λ_{++} *a priori*, as the only way to construct it uses the unknown positive energy orbitals. If these orbitals are constructable in some other way, then the projectors are not needed. Conversely, there is no way to construct the orbitals if the projectors are needed.

Rossky and Karplus [53] applied perturbation theory to a simple one-electron model, in which the nuclear charge of a hydrogenic ion is perturbed, $Z \to Z + z$. This problem has an analytic solution which serves to benchmark the perturbation theory results. In the nonrelativistic case, the energy of the *perturbed* Schrödinger ground state is

$$\varepsilon_{1s}(Z+z) = -(Z+z)^2/2 = -Z^2/2 - zZ - z^2/2. \tag{4.13.3}$$

A perturbation expansion of the energy in powers of z should give the exact result to second order, but terms of higher order contain diagrams that should sum to zero, order by order. Rossky and Karplus were able to demonstrate this cancellation at order z^3 and higher although their method for dealing with integration over the continuous spectrum was rather inaccurate. Better results can be obtained by using a finite matrix approximation [54] which replaces the continuum integrations by finite sums over (pseudo)-states.

When this method is applied to the relativistic hydrogenic problem, the results are even more instructive. One difference from the nonrelativistic problem is that the power series expansion of the analytic expression, (3.3.7), for the perturbed ground state eigenvalue,

$$\varepsilon_{1s}(Z+z) = \varepsilon_{1s}(Z) + z\delta_1(Z) + z^2\delta_2(Z) + \ldots, \tag{4.13.4}$$

does not terminate [55]. The coefficients, $\delta_k(Z)$, can be split into two parts:

$$\delta_k(Z) = \delta_k^+(Z) + \delta_k^-(Z).$$

$\delta_k^+(Z)$ includes only the positive energy intermediate states that would have appeared in the nonrelativistic calculation and $\delta_k^-(Z)$ includes all other terms of order k involving contributions from negative energy intermediate states. This reproduces the numerical values of the coeffficients from the analytic formula, (3.3.7), order by order *only if* the $\delta_k^-(Z)$ contributions are included. Unsurprisingly $\delta_2^+(Z)$ grows relatively slowly from a nearly nonrelativistic value -0.504 at $Z = 10$ to -1.61 at $Z = 100$, whilst $\delta_2^-(Z)$ grows like Z^3 from

$+10^{-4}$ at $Z = 10$ to $+0.047$ at $Z = 100$ [56, Table VI.]. As indicated earlier in this discussion, the perturbed ground state includes non-negligible negative energy components whose relative importance grows rapidly as Z increases.

Brown and Ravenhall believed that (4.13.2) was a "valid" relativistic two-electron wave equation. They gave orders of magnitude estimates for terms with negative energy components, and for their contributions in first order perturbation theory [24, §38], [45]. Their derivation of $g^B(R)$ from QED is essentially the same as §4.9 and §4.10; their conclusion that $g^B(R)$ should only be used in first order perturbation theory to predict fine structure splittings highlights the inconsistency of using an interaction which assumes that the exchanged virtual photon has low frequency to compute electron-positron interactions which involve energies of order $2mc^2$.

4.13.3 The NVPA and "variational collapse"

Sucher [57, pp. 1–54] also advocated the use of (4.13.2) as the starting point for DHF calculations, in the belief that positive energy projection operators would also eliminate "variational collapse". Belief in the existence of "variational collapse" [58] or "finite basis set disease" [59] became widespread when early attempts to use finite matrix methods to solve the Dirac problem for the hydrogen atom in spherical coordinates gave highly inaccurate results; for example [58, 60]. Spurious low energy "intruder" states appeared, especially for orbitals with $\kappa > 0$, along with the expected solutions resembling physical low-lying states. The results were sensitive to basis set size and, unlike the non-relativistic schemes on which the calculations were modelled, there was little sign of convergence with systematic enlargement of the basis set [60, 61, 62]. The lack of a global lower bound to the one-electron Dirac spectrum for atomic mean field potentials was naturally blamed for this, because the existence of a finite lower bound to the spectrum is conventionally used to deduce the existence of a lowest eigenvalue in variational methods. It is only recently [63] that a variational theory for Dirac Hamiltonians has been formulated, Chapter 5, in which the bound spectrum has a rigorous lower bound. "Variational collapse" appears in calculations in which the four components of the trial spinors are allowed to vary independently; the bound derived in [63] requires the use of spinor basis sets in which the components are related in a particular way. The components of unconstrained 4-spinors usually fail to reproduce the correct analytic behaviour close to the nuclei and the variational procedure has no way to correct this.

Whilst the use of variational methods for the one-electron Dirac problem is now on a sound footing, there is still controversy over the many-electron self-consistent field problem. The SCF equations are customarily derived variationally and, in practice, the problems encountered in solving them are much the same as in the nonrelativistic case. The solution of the SCF equations for orbital spinors is subject to the same constraints as the single particle equations. A perturbation analysis about the converged solution of the

SCF equations shows that the most important perturbations arising from the electron-electron interaction involve matrix elements linking the unperturbed state to low-lying excited orbitals associated with small energy denominators. Whilst the corresponding negative energy contributions are always present, their large energy denominators ensure that their effect is relatively small. In fact "variational collapse" never happens when the wavefunction is properly constructed.

4.13.4 Semirelativistic approaches

The perceived difficulty of making reliable calculations with Dirac wavefunctions led quantum chemists to fall back on more approximate schemes, reviewed briefly in §1.5.2. The Dirac Hamiltonian was replaced by quasi-relativistic Hamiltonians [64] such as the Breit-Pauli Hamiltonian [24, §39], often derived by a sequence of Foldy-Wouthuysen transformations [65]. Usually only terms of order $(\alpha Z)^2$ are presented, but it is easy enough to continue formally to write down terms of higher order as in §3.7.2. The additional terms introduce operators of order $(p^2/c^2)^n, n \geq 3$, which have infinite expectation values on the Hilbert space of nonrelativistic hydrogenic wavefunctions, making this approach effectively unusable. Whilst the Breit-Pauli method gives good results for light elements to order $(\alpha Z)^2$, its accuracy degrades in the third and lower rows of the Periodic Table. The 2-component effective Hamiltonian of Douglas and Kroll [66], modified by Hess [67], avoids introducing divergent operators into the expansion and has therefore become very popular. Similar ideas underly the *regular Hamiltonian* approach of Chang *et al.* [68] and the Amsterdam group [69, 70] whose ZORA (zero order regular approximation) Hamiltonian has been well used.

Relativistic effective core potentials (RECP) or pseudo-potentials (PP) have also been used to approximate the treatment of electron-electron interactions in atomic and molecular calculations. These relatively cheap models, briefly summarized in §1.5.2, exploit the idea that chemical properties are largely determined by the properties of atomic valence orbitals and their overlaps with valence orbitals on neighbouring atoms in a molecule. Calculations involving only valence electrons moving in a relativistic effective core potential are relatively cheap. The RECP are parametrized by benchmarking against atomic DHF calculations, although there is no way to assess the errors in subsequent applications. Nevertheless, much chemical information has been accumulated using these approximate schemes, which are currently the most popular amongst chemists.

4.14 A strategy for atomic and molecular calculations

Relativistic atomic structure literature comes in different flavours, depending on whether the focus is atoms with not more than two or three electrons or on general many-electron atoms. In hydrogen and helium, the focus is on experimental and theoretical advances in *precision physics* [71]. There is wide interest in experiments for determination of fundamental constants using simple atoms; precise tests of QED; search for violation of fundamental symmetries and for variation of fundamental constants; and the construction of new frequency standards. The required theoretical support therefore focuses on the calculation of QED corrections of increasing complexity.

Sapirstein [52] gives an excellent review of the extensive work on highly-charged ions and the relation between QED perturbation theory and RMBPT. There have been a limited number of studies of the effect of including negative energy states in the second order of RMBPT [72] for the ground states of helium-like systems and Sapirstein, Cheng and Chen [73] studied low-lying states of helium-like ions going up to third order in RMBPT. The former showed that the negative energy state energy contributions to the second order pair correlation energy in the ground state grow like Z^4 and reach barely detectable values only for $Z > 80$. The more recent and comprehensive study of Sapirstein *et al.* showed that the results depended slightly on the choice of potential used to define the orbital set – bare Coulomb, core-Hartree, Kohn-Sham density functional or modified core-Hartree – and that this could be more or less eliminated by including negative energy state contributions within a standard QED S-matrix scheme, corroborating the discussion of §4.13.

Most calculations for many-electron atoms and molecules start from the effective Coulomb gauge Hamiltonian

$$H = \sum_{i=1}^{N} h_{Di} + \sum_{i<j} \left(\frac{1}{R_{ij}} + g^B(R_{ij}) \right) \tag{4.14.1}$$

where $g^B(R)$ (4.9.22) is the instantaneous Breit interaction. If we ignore all coupling to the negative energy states, we have the full machinery of nonrelativistic MBPT, and the corresponding Goldstone diagram technology, at our disposal in RMBPT. Negative energy state corrections are mostly relatively straightforward to compute, apart from self-energy and vacuum polarization diagrams. There are two such one-photon exchange contributions in second order of QED perturbation theory that give energy shifts of similar order to the Breit correction. The electron self-energy diagram is given by Figure 4.6(f) and the vacuum polarization energy diagram by Figure 4.6(h). Corrections due to replacement of the Breit interaction by the more exact ω-dependent transverse photon interaction $g^T(R;\omega)$, (4.9.21), can usually be ignored except for X-ray and inner-shell processes in heavy atoms or molecules.

The most popular approximations within this scheme have been of the self-consistent field (SCF) type such as the Dirac-Hartree-Fock model (DHF). The Dirac-Hartree-Fock-Breit model (DHFB) is based on the full effective Hamiltonian (4.14.1). The contribution of $g^B(R_{ij})$ is quite small, and the large number of Breit interaction integrals for open shell systems has served ot inhibit their inclusion. This is not such a problem for closed shell systems, where the number of additional interaction terms is relatively small. The majority of calculations have treated the Breit or transverse interactions using first-order perturbation theory; it has been shown that the perturbed energies agree well with DHFB energies in the small number of cases in which comparisons have been made. However, the wavefunction in perturbation calculations is not corrected for the Breit interaction, and this will have to be taken into account in higher order correlation calculations.

Electron correlation is often more important than higher order relativistic effects for physical applications. A study of electron correlation in the alkali atoms [52] used a finite orbital basis set needed generated with the u_{DHF} one-body potential. The number of diagrams in the perturbation expansion increases rapidly as the order of perturbation increases and the cost escalates accordingly. Coupled cluster theory [74, 75], in which certain infinite classes of diagrams are summed completely, can be very effective. An overview of the theory as applied to (nonrelativistic) molecular calculations including several of its variants has been given by Paldus [76].

The CI method and the related technique of multi-configurational SCF (MCSCF) calculations are conceptually simpler. Here the idea is, as before, to start from a finite set of orbitals in some one-body potential. From these orbitals, one can construct a set of N-electron Slater determinants according to some scheme for generating 1-, 2- and many-particle excitations out of some reference set of determinantal wavefunctions. Diagonalizing the Hamiltonian in this N-particle basis gives a set of approximate wavefunctions and energy levels for comparison with experiment. The MCSCF technique requires also adjustment of the potential and the orbitals that it generates to make the whole system self-consistent. The implementation of this approach for molecules has been described, for example, by [77, Werner, pp. 1–62; Shepard, pp. 63–200]. In atoms, there have as yet been few calculations using finite matrix CI or MCSCF techniques. Fischer and colleagues [78] use finite difference solution of the coupled orbital equations to solve MCHF problems. Similar numerical techniques were used by Desclaux [79] to set up MCDHF calculations, and also by Grant *et al.* [80, 81, 82] in various versions of the GRASP package, which can be used either for relativistic atomic CI calculations or for MCDHF calculations. Radiative corrections for the lowest order electron self-energy and for vacuum polarization can also be included approximately. Relativistic molecular codes, for example DIRAC [83], MOLFDIR [84] and BERTHA [85, 86, 87, 88] are being developed actively.

4.15 Density functional theories

Although density functional theory (DFT) has been a vital tool in the study of condensed matter for many years, its extensive application to other fields such as quantum chemistry is more recent. The need for a relativistic extension of DFT (RDFT) to study systems containing heavy elements was recognized some 30 years ago [89, 90, 91] providing the basis for major progress in the last decade. The development of RDFT, as with other methods examined in this book, has been based on QED, and this section aims at a brief summary of its achievements. There is an extensive literature: Dreizler and Gross's monograph on DFT [92] includes a chapter on RDFT, and other expositions will be found in, for example [93, 94, 95, 96].

4.15.1 Basic ideas of RDFT

Relativistic density functional theory (RDFT), §4.15, is based on a generalization [92] of the Hohenberg-Kohn Theorem [98] in which the ground state energy is represented as a universal functional of the electron density. Whereas nonrelativistic DFT is based on the idea that all physical quantities pertaining to the ground state of a many-body system can be expressed as a unique functional of the particle density, RDFT requires the relativistic four-current, $j^\mu = (c\rho, \boldsymbol{j})$. The starting point is the usual QED Lagrangian density, (4.4.1), augmented by a time-independent external field interaction

$$\mathcal{L}_{ext} = -\frac{1}{c} j^\mu(x) v_\mu(x).$$

The requirement that the theory be locally gauge invariant ensures that the electron current, and therefore total charge, is conserved. An energy-momentum tensor, (2.7.10) can be defined that satisfies a conservation equation

$$\partial_\mu T^{\mu\nu}(x) = \frac{1}{c} j_\rho(x)\, \partial^\nu v^\rho(x), \;\Rightarrow\; \partial_\mu T^{\mu 0}(x) = 0. \tag{4.15.1}$$

when $v^\mu(x)$ is time-independent. Hence

$$\partial_0 \int d^3x\, T^{0\,0}(x) = 0, \tag{4.15.2}$$

so that energy is conserved in the rest frame of the sources. Hence there exists a QED Hamiltonian operator H

$$H = H_e + H_\gamma + H_{int} + H_{ext}, \tag{4.15.3}$$

where

$$H_e(x^0) = \frac{1}{2}\int d^3x \left[\psi^\dagger(x),\, (c\boldsymbol{\alpha}\cdot\boldsymbol{p} + \beta mc^2)\,\psi(x)\right], \tag{4.15.4}$$

is the energy of the noninteracting electron-positron field,

$$H_\gamma(x^0) = \frac{1}{2}\epsilon_0 \int d^3x(\boldsymbol{E}(x)\cdot\boldsymbol{E}(x) + c^2\boldsymbol{B}(x)\cdot\boldsymbol{B}(x)) \tag{4.15.5}$$

is the energy of the radiation field, and

$$H_{int}(x^0) = \frac{1}{c}\int d^3x\, j^\mu(x)A_\mu(x), \quad H_{ext}(x^0) = \frac{1}{c}\int d^3x\, j^\mu(x)v_\mu(x), \quad (4.15.6)$$

are the interaction energy of the electron-positron field with the photons and the external potential respectively. The four-potential $A_\mu(x)$ generates the electric field $\boldsymbol{E}(x)$ and the magnetic field $\boldsymbol{B}(x)$.

The expectation values of H and $j^\mu(x)$ diverge unless measures are taken to modify the model. For noninteracting electrons, the energy zero must be redefined to eliminate the contribution of the negative continuum to the vacuum state. The external field can create virtual electron-positron pairs, and this makes it necessary to renormalize the current $j^\mu(x)$. When we permit interaction with the photon field, the resulting perturbation theory generates classes of divergent terms that must be renormalized according to the standard procedures of QED.

Suppose that the ground state Φ of the system is in the Fock space sector with charge $Q = -Ne$ and is also nondegenerate. Then the renormalized total electron binding energy E_R is the difference between the energy of Φ and of the vacuum state $Q = 0$, so that

$$E_R \equiv E_{tot} = \langle \Phi \,|\, H \,|\, \Phi \rangle - \langle 0 \,|\, H \,|\, 0 \rangle + \Delta E_{tot}, \quad (4.15.7)$$

$$j^\mu_R(x) = \langle \Phi \,|\, j^\mu(x) \,|\, \Phi \rangle + \Delta j^\mu(x), \quad (4.15.8)$$

where ΔE_{tot} and $\Delta j^\mu(x)$ are renormalization counter-terms. The need for renormalization shows up in RDFT in the basic existence theorem, below, in the single-particle equations, as well as in the design of exchange-correlation functionals [92]–[96].

4.15.2 The relativistic Hohenberg-Kohn theorem

The relativistic Hohenberg-Kohn theorem (RHK) [97, p. 539] can be written

$$\{\, \Phi \,|\Phi \text{ from } A_\mu + \partial_\mu \Lambda \,\} \iff \{ j^\mu(x) \text{ implies } \Phi = \Phi[j] \} \quad (4.15.9)$$

where A_μ is the four-potential of the photon field and Λ is an arbitrary gauge potential. This statement asserts that the ground state is a unique functional, $\Phi[j]$, of the four-current j^μ once the gauge has been fixed. The lengthy proof [97], which relies on a perturbation expansion with respect to both the electron-electron interaction and the external potential, requires renormalization, order by order, to extract the required counter-terms. It implies that all ground state observables, in particular the total energy $E_{tot}[j]$, are unique functionals of the ground state four-current. In principle this energy functional accounts for the relativistic motion of the electrons and photons with which they interact and also all radiative corrections. The RDFT equations of motion are obtained from the equation

$$\frac{\delta}{\delta j^\mu}\left\{ E_{tot}[j] - \lambda \int d^3x\, \rho(x) \right\}_{j=j_0} = 0, \quad (4.15.10)$$

where λ is a Lagrange multiplier to incorporate a subsidiary condition expressing charge conservation and j_0 is the ground state four-current consistent with a four-potential $A[j_0]$ and, if present, an external field.

If there is no external magnetic field $\boldsymbol{B}_{ext}$ (4.15.9) simplifies [91] so that al ground state quantities can be defined in terms of the charge density $\rho = j^0/c$:

$$\{v_{ext}\,|v_{ext} + \text{ const. }\} \Longleftrightarrow \{\Phi\,|\,\Phi \text{ from } v_{ext} + \text{ const. }\} \Longleftrightarrow \rho(\boldsymbol{x}).$$

Thus we can now write $E_{tot}[\rho]$ rather than $E_{tot}[j]$, and we can regard the electron current density itself as a functional of ρ. The system may still have a magnetic moment. MacDonald and Vosko [91] have also developed a generalization of RDFT [91] to accomodate spin-polarized ground states.

4.15.3 The relativistic Kohn-Sham equations

The Kohn-Sham approach expresses the elementary variables of RDFT in terms of a set of auxiliary set of four spinors $\psi_k(x)$. Although the representation can describe all vacuum corrections to the ground state four-current and energy [93], they are usually irrelevant in practice. The four-current in the NVP approximation can be written in terms of the auxiliary spinors as [96]

$$j^\mu(x) = -ec \sum_k \Theta_k \overline{\psi}_k^\dagger(x)\, \gamma^\mu\, \psi_k(x) \tag{4.15.11}$$

where Θ_k projects onto occupied electron states:

$$\Theta_k = \begin{cases} 1 & \text{for } -2mc^2 < \epsilon_k \le \epsilon_F, \\ 0 & \text{otherwise,} \end{cases}$$

after shifting the energy zero to coincide with the nonrelativistic definition. The energy expression may be partitioned so that

$$E_{tot} = T_s + E_H + E_{ext} + E_{xc}, \tag{4.15.12}$$

where the first three terms have a simple representation in terms of the auxiliary spinor set and the last contains all the many-body effects. The free-particle Dirac Hamiltonian contributes

$$T_s = \int d^3x \sum_k \Theta_k \psi_k^\dagger(x)\, (c\boldsymbol{\alpha}\cdot\boldsymbol{p} + (\beta - 1)mc^2)\, \psi_k(x). \tag{4.15.13}$$

The term

$$E_H = E_H^C + E_H^T \tag{4.15.14}$$

is gauge-dependent, and in the long wavelength (Breit) approximation is given by (4.9.13), from which

$$E_H^C[j] = \frac{1}{2}\int d^3x \int d^3x' \frac{\rho(\boldsymbol{x})\rho(\boldsymbol{x}')}{|\boldsymbol{x}-\boldsymbol{x}'|},$$
$$E_H^T[j] = -\frac{1}{2c^2}\int d^3x \int d^3x' \frac{\boldsymbol{j}(\boldsymbol{x})\cdot\boldsymbol{j}(\boldsymbol{x}')}{|\boldsymbol{x}-\boldsymbol{x}'|}.$$

The interaction with the external field is given by (4.15.6), and E_{xc} is the sum of the exchange counterpart, E_x of the Hartree energy E_H and a correlation term E_c which accounts for all the many-body effects.

Requiring that the expression E_{tot} be stationary with respect to the spinors $\psi_k(x)$ leads to the (time-independent) relativistic Kohn-Sham equations [90, 91]

$$\left(c\boldsymbol{\alpha}\cdot\boldsymbol{p} + \beta mc^2 + \alpha^\mu\, a_{s,\mu}(\boldsymbol{x})\right)\psi_k(\boldsymbol{x}) = E_k\,\psi_k(\boldsymbol{x}), \tag{4.15.15}$$

in which $\alpha^\mu = \gamma^0\gamma^\mu$ and the local four-potential $a_{s,\mu}(\boldsymbol{x})$ is

$$a_{s,\mu}(\boldsymbol{x}) = v_\mu(\boldsymbol{x}) + v_{H,\mu}(\boldsymbol{x}) + v_{xc,\mu}(\boldsymbol{x})$$

with

$$v_{H,\mu}(\boldsymbol{x}) = \frac{1}{c}\int d^3x' \frac{j^\mu(\boldsymbol{x}')}{|\boldsymbol{x}-\boldsymbol{x}'|}, \quad v_{xc,\mu}(\boldsymbol{x}) = c\frac{\delta E_{xc}[j]}{\delta j^\mu(x)}. \tag{4.15.16}$$

These equations must be solved self-consistently in the same manner as the Dirac-Hartree-Fock equations. It is apparent that this system of equations has much in common with the DHFB equations studied in Chapters 7 and 9. Only the occupied orbitals are needed to define the electromagnetic fields which bind the electrons, and the local four-potential $a_{s,\mu}(\boldsymbol{x})$ implicitly couples the spinors $\psi_k(\boldsymbol{x})$. There is no explicit mention of virtual orbitals, and they play a purely passive role in the self-consistent solution of the Dirac-Kohn-Sham equations (4.15.15). Nevertheless, they are inevitably present in finite matrix approximations and can be used to go beyond the NVP approximation if desired.

4.15.4 Exchange and correlation functionals

The catch-all expression $E_{xc}[j]$ can be divided into two parts,

$$E_{xc}[j] = E_x[j] + E_c[j],$$

where $E_x[j]$ is the exchange counterpart of the Hartree energy $E_H[j]$, and $E_c[j]$ contains all the many-body correlation effects. In the long wavelength approximation

$$E_x = -\frac{1}{2}\sum_{k,l}\Theta_k\Theta_l \int d^3x \int d^3x' \frac{\rho_{kl}(\boldsymbol{x})\rho_{lk}(\boldsymbol{x}') - \boldsymbol{j}_{kl}(\boldsymbol{x})\cdot\boldsymbol{j}_{lk}(\boldsymbol{x}')/c^2}{|\boldsymbol{x}-\boldsymbol{x}'|}. \tag{4.15.17}$$

The unphysical self-interaction terms with $l = k$ cancel corresponding terms in E_H, one of the most important sources of error in DFT. Whilst the generalized Hohenberg-Kohn theorem asserts that E_x, like the other contributions to the energy, can be expressed as a functional of j, or of the density ρ, it does not explain how to construct it. The traditional response, going back to the development of Thomas-Fermi theory [99, 100], has been to approximate (4.15.17) by a *local density approximation* (LDA), using an explicit function of ρ based on the theory of a homogeneous electron gas (HEG). This has the virtue of simplicity, but has several failings, in particular an inability to cancel the self-interaction terms exactly.

The simplest LDA adopts the expression for the energy density of a nonrelativistic HEG:

$$E_x^{LDA}[\rho] = \int d^3x\, \epsilon_{xc}^{NRHEG}[\rho(\boldsymbol{x})] \tag{4.15.18}$$

where

$$\epsilon_x^{NRHEG}[\rho(\boldsymbol{x})] = -\frac{3}{4}\left(\frac{3}{\pi}\right)^{1/3} [\rho(\boldsymbol{x})]^{4/3}. \tag{4.15.19}$$

The effective potential $v_{xc}(\boldsymbol{x})$ is the functional derivative

$$v_x(\boldsymbol{x}) = \frac{\delta E_{xc}}{\delta \rho} = \frac{\partial \epsilon_{xc}^{NRHEG}[\rho(\boldsymbol{x})]}{\partial \rho} = -\left(\frac{3}{\pi}\right)^{1/3} [\rho(\boldsymbol{x})]^{1/3}. \tag{4.15.20}$$

A model of this sort was first suggested by Slater [101], whose expression is larger by a factor 3/2, and hence the nonrelativistic Kohn-Sham equations in which the exchange potential is identified by (4.15.20) are often referred to as the Hartree-Slater SCF model. The Xα model prefaces $v_x(\boldsymbol{x})$ with an adjustable parameter α which can be varied to make some theoretical observable agree with its measured value.

Attempts to improve on this recognize that the electron motions should be treated relativistically and that the electron distribution in an atom or molecule is by no means homogeneous. The energy density in a relativistic HEG is given by [102]

$$\epsilon_x^{RHEG}[\rho(\boldsymbol{x})] = \epsilon_x^{NRHEG}[\rho(\boldsymbol{x})]\, \Phi_{x,0}(\beta) \tag{4.15.21}$$

where

$$\beta = \frac{(3\pi^2\rho)^{1/3}}{mc}.$$

The relativistic correction factor

$$\Phi_{x,0}(\beta) = 1 - \frac{3}{2}\left(\frac{\eta}{\beta} - \frac{\sinh^{-1}\beta}{\beta^2}\right)^2,$$

where $\eta = \sqrt{1+\beta^2}$, can be split into corrections arising from the Coulomb interaction

$$\Phi^L_{x,0}(\beta) = \frac{5}{6} + \frac{1}{3\beta^2} + \frac{2\eta}{3\beta}\sinh^{-1}\beta - \frac{2\eta^4}{3\beta^4}\ln\eta - \frac{1}{2}\left(\frac{\eta}{\beta} - \frac{\sinh^{-1}\beta}{\beta^2}\right)^2 \tag{4.15.22}$$

and a transverse contribution

$$\Phi^T_{x,0}(\beta) = \frac{1}{6} - \frac{1}{3\beta^2} - \frac{2\eta}{3\beta}\sinh^{-1}\beta + \frac{2\eta^4}{3\beta^4}\ln\eta - \left(\frac{\eta}{\beta} - \frac{\sinh^{-1}\beta}{\beta^2}\right)^2. \tag{4.15.23}$$

This is still not enough; the four-current density of electrons in an atom or molecule is not homogeneous, so that the LDA can only be a first approximation. The relativistic generalized gradient approximations (RGGA) attempts to correct for this by adding terms dependent on $\boldsymbol{\nabla}\rho(\boldsymbol{x})$ so that, following the nonrelativistic GGA due to Becke [103] and Perdew *et al.* [104] the relativistic local exchange energy density can be written [105, 106]

$$\epsilon_x^{RGGA}[\rho(\boldsymbol{x})] = \epsilon_x^{NRHEG}[\rho(\boldsymbol{x})]\,[\Phi_{x,0}(\beta) + g(\xi)\,\Phi_{x,2}(\beta)] \tag{4.15.24}$$

where

$$\xi = \left[\frac{\boldsymbol{\nabla}\rho}{2(3\pi^2\rho)^{1/3}\rho}\right]^2$$

and the function $g(\xi)$ is the nonrelativistic GGA correction. This function is already quite complicated and although it is, in principle, possible to calculate $\Phi_{x,2}(\beta)$ from the first order response function of RHEG, a simpler semi-empirical expression has been used in practice. More details may be found in [96].

The construction of a local correlation potential on similar lines is more problematic. It seems that adding an LDA or GGA approximation of the correlation energy to E_x gives no consistent improvement over an exchange-only scheme [95, 107]. It has been suggested [108, 109] that a more promising scheme would apply perturbation theory starting from the auxiliary KS Hamiltonian. A relativistic version has been formulated by Engel *et al.* [96], which is potentially capable of including all transverse and vacuum corrections. The lowest order term in this scheme using the NVP approximation is [96, Equation (4.51)]

$$E_c^{(2)} = E_c^{MP2} + E_c^{\Delta HF}$$

where

$$E_c^{MP2} = \frac{1}{2}\sum_{ijkl}\Theta_i\Theta_j\frac{(ij\,|\,g|\,|\,kl)[(ki\,|\,g\,|\,lj) - (kj\,|\,g\,|\,li)]}{\epsilon_i + \epsilon_j - \epsilon_k - \epsilon_l}$$

$$E_c^{\Delta HF} = \sum_{il}\frac{\Theta_i}{\epsilon_i - \epsilon_l}\left|\frac{1}{c}\int d^3x\, j^{\mu}_{il}(x)v_{\mu}(x) + \sum_j \Theta_j(ij\,|g\,|\,jl)\right|^2$$

in which i, j are occupied orbitals and k, l are virtual states with $\epsilon_k, \epsilon_l > \epsilon_F$. The contribution of the transverse interaction is neglected in [96, Equation (4.51)]. The contribution E_c^{MP2} is second order in the electron-electron interaction g (or fourth order in QED). The contribution $E_c^{\Delta HF}$ reflects the difference between the exchange-only ROPM and DHF ground state energies. The extra computational labour needed for $E_c^{(2)}$ has limited application of this scheme to correlation energies in the helium isoelectronic sequence. The quality of the results is somewhat mixed, and there is evidence that higher order terms are needed for some applications. A proper discussion would take us too far afield; see [96] for more details.

4.15.5 The optimized potential method

A recent development [94] rests on the idea that if the relativistic KS spinors are unique functionals of j^μ then the functional derivatives of E_{xc} with respect to j^μ can be replaced in the evaluation of the local exchange-correlation potential v^μ_{xc} by functional derivatives with respect to the spinors ψ_k. This potential is the solution of an integral equation

$$\int d^3x' \chi_0^{\mu\nu}(\boldsymbol{x}, \boldsymbol{x}')\, v_{xc,\nu}(\boldsymbol{x}') = \Lambda^\mu_{xc}(\boldsymbol{x}), \tag{4.15.25}$$

where in the NVP approximation,

$$\chi_0^{\mu\nu}(\boldsymbol{x}, \boldsymbol{x}') = -\sum_k \Theta_k \left\{ \overline{\psi}_k^\dagger(\boldsymbol{x})\, \gamma^\mu\, G_k(\boldsymbol{x}, \boldsymbol{x}')\, \gamma^\nu\, \psi_k(\boldsymbol{x}') + \text{ c.c.} \right\},$$

which features the KS static response function

$$G_k(\boldsymbol{x}, \boldsymbol{x}') = \sum_{l \neq k} \frac{\psi_l(\boldsymbol{x})\, \overline{\psi}_l^\dagger(\boldsymbol{x}')}{\epsilon_l - \epsilon_k},$$

and

$$\Lambda^\mu_{xc}(\boldsymbol{x}) = -\sum_k \int d^3x' \left\{ \overline{\psi}_k^\dagger(\boldsymbol{x})\, \gamma^\mu\, G_k(\boldsymbol{x}, \boldsymbol{x}') \frac{\delta E_{xc}}{\delta \overline{\psi}_k(\boldsymbol{x}')} + \text{ c.c.} \right\} + \sum_k j_{kk}(\boldsymbol{x}) \frac{\partial E_{xc}}{\partial \epsilon_k}$$

The numerical evaluation of (4.15.25) is relatively complicated, but results for exchange-only ROPM calculations for the ground states of closed shell atoms agree well with DHF results [96, Table 4.1]. Results have also been obtained including transverse photon contributions perturbatively at the Breit and full retardation levels. The conclusions on the magnitude of retardation and relativistic effects are similar to those from DHF and DHFB calculations presented in Chapters 7 and 9.

References

1. Schweber S S 1961 *An Introduction to Relativistic Quantum Field Theory* (New York: Harper & Row).
2. Itzykson C and Zuber J-B 1980 *Quantum Field Theory* (New York: McGraw-Hill Book Company).
3. Hamermesh M, 1962 *Group Theory* (Reading, Mass.: Addison-Wesley Publishing Co. Inc.).
4. Elliott J P and Dawber P G 1979 *Symmetry in Physics* (Basingstoke: Macmillan Press).
5. Furry W H 1951 *Phys. Rev.* **81** 115.
6. Thaller B 1992 *The Dirac Equation* (Berlin: Springer-Verlag).
7. Schweber S S 1994 *QED and the men who made it: Dyson, Feynman, Schwinger and Tomonaga* (Princeton NJ: Princeton University Press).
8. Greiner W, Müller B and Rafelski J 1985 *Quantum Electrodynamics of Strong Fields* (Berlin: Springer-Verlag).
9. Gupta S N 1950 *Proc. Phys. Soc. (London)* **A63**, 681.
10. Bleuler K 1950 *Helv. Phys. Acta* **23**, 567.
11. Plunien G, Müller B, Greiner W and Soff G 1989 *Phys Rev A* **39** 5428.
12. Plunien G, Müller B, Greiner W and Soff G 1991 *Phys Rev A* **43** 5853.
13. Bar-Shalom A, Klapisch M and Oreg J 2001 *J. Quant. Spectrosc. Rad. Transf.* **71**, 169.
14. Stakgold 1979 *Green's functions and Boundary Value Problems* (New York: John Wiley & Sons).
15. Dyson, F J 1951 *Phys. Rev.* **82** 428.
16. Dyson, F J 1951*Phys. Rev.* **83** 608.
17. Wick G C 1950 *Phys. Rev.* **80** 268.
18. Feynman R P 1949 *Phys Rev* **76** 749, 769.
19. Gell-Mann M and Low F, 1951 *Phys. Rev.* **84** 350.
20. Sucher J 1957 *Phys. Rev.* **107** 1448.
21. Lindgren I 1989 in *Relativistic, Quantum Electrodynamic and Weak Interaction Effects in Atoms* (ed. W R Johnson, P J Mohr and J Sucher), pp. 371–392 (New York : American Institute of Physics).
22. Møller C 1932 *Ann. der Phys.* **14** 531.
23. Gaunt J A *Proc. Roy. Soc. A* **122** 513.
24. Bethe H A and Salpeter E E 1957 *Quantum Mechanics of One- and Two-Electron Atoms* (Berlin-Göttingen-Heidelberg: Springer-Verlag).
25. Gradshteyn I S and Ryzhik I M 1980 *Table of Integrals, Series and Products* (4th edition, ed. A Jeffrey) (London: Academic Press).
26. Brueckner K A and Levinson C A 1955 *Phys. Rev.* **97**, 1344.
27. Bethe H A 1956 *Phys. Rev.* **103**, 1353.
28. Eden R J 1956 *Proc. Roy. Soc. A* **235**, 408.
29. Brueckner K A 1955 *Phys. Rev.* **97**, 1353.
30. Brueckner K A 1955 *Phys. Rev.* **100**, 36.
31. Goldstone J 1957 *Proc. Roy. Soc. A* **239** 267.
32. Kelly H P 1963 *Phys. Rev.* **131**, 684.
33. Kelly H P and Sessler A M 1963 *Phys. Rev.* **132**, 2091.
34. Kelly H P 1964 *Phys. Rev.* **134**, A1450.
35. Kelly H P 1964 *Phys. Rev.* **136**, B896.

36. Brandow H B 1967 *Rev. Mod. Phys.* **39**, 771.
37. Brandow H B 1977 *Adv. Quantum Chem.* **10**,187.
38. Sandars P G H 1967 in *La Structure Hyperfine Magnétique des Atomes et des Molècules* (Paris: CNRS).
39. Yutsys AP Levinson I B and Vanagas V V 1962 *Mathematical Apparatus of the Theory of Anglar Momentum* (Jerusalem: Israel Program for Scientific Translations).
40. Lindgren I and Morrison J 1982 *Atomic Many-Body Theory* (Berlin: Springer-Verlag).
41. Lindgren I 1974 *J. Phys. B* **7**, 2441.
42. Kvasnička V 1974 *Czech. J. Phys. B* **24**, 605; see also Kvasnička V 1977 in *Adv. in Chem. Phys.* Vol. 36 (ed. Prigogine I and Rice S A), p.345. (New York: Interscience).
43. Boyle J J and Pindzola M S 1998 *Many-body atomic physics* (Cambridge University Press).
44. Wilson S 1984 *Electron correlation in molecules* (Oxford: Clarendon Press).
45. Brown G E and Ravenhall D G 1951 *Proc. Roy. Soc. A* **208**, 552.
46. Breit G 1929 *Phys Rev* **34** 353; — 1930 *Phys Rev* **36** 383; — 1932 *Phys Rev* **39** 616.
47. Sucher J 1980 *Phys. Rev. A* **22**, 348.
48. Sucher J 1980 in *Proceedings of the Argonne Workshop on the Relativistic Theory of Atomic Structure*, ed. H G Berry, K T Cheng, W R Johnson and Y-K Kim. ANL-80-116 (Argonne National Laboratory).
49. Sucher J 1982 *Relativistic Effects in Atoms, Molecules and Solids*, ed. G L Malli, pp. 1–53. (NATO ASI, Series B: Vol. 87) (New York, Plenum).
50. Sucher J 1984 *Int. J. Quant. Chem.* **25**, 3.
51. Sucher J 1985 *Atomic Theory Workshop on Relativistic and QED Effects in Heavy Atoms*, ed. H P Kelly and Y-K Kim. pp. 1–25 (AIP Conference Series No. 136) (New York, American Insitute of Physics).
52. Sapirstein J 1998 *Rev. Mod. Phys.* **70**, 55.
53. Rossky P and Karplus M 1977 *J. Chem. Phys.* **67**, 5419.
54. Quiney H M, Grant I P and Wilson S 1985 *J. Phys. B* **18**, 577.
55. Quiney H M, Grant I P and Wilson S 1985 *J. Phys. B* **18**, 2805.
56. Grant I P and Quiney H M 2000 *Phys. Rev. A* **62**, 022508.
57. Malli G L (ed.) 1983 *Relativistic Effects in Atoms, Molecules and Solids* (NATO Advanced Study Institutes, Series B., Physics, Vol. 87) (New York: Plenum Press).
58. Schwarz W H E and Wallmeier H 1982 *Mol. Phys.* **46**, 1045.
59. Schwarz W H E and Wechsel-Trakowski E 1982 *Chem. Phys. Lett.* **85**, 94.
60. Wallmeier H and Kutzelnigg W 1981 *Chem. Phys. Lett.* **78**, 341.
61. Mark F and Rosicky F 1980 *Chem. Phys. Lett.* **74**, 562.
62. Mark F and Schwarz W H E 1982 *Phys. Rev. Lett.* **48**, 673.
63. Grant I P and Quiney H M 2000 *Int. J. Quant. Chem¿* **80** 283.
64. Kutzelnigg W 1984 *Int. J. Quant. Chem.* **25**, 107.
65. Foldy L L and Wouthuysen S A 1950 *Phys. Rev.* **78** 29.
66. Douglas M and Kroll N M 1974 *Ann Phys (NY)* **82** 89.
67. Hess B A 1986 *Phys Rev A* **32** 756; – ; *Phys Rev A* **33** 3742.
68. Chang Ch., Pélissier M and Durand P 1986 *Phys. Scr.* **34**, 394.
69. van Lenthe E, Baerends E-J and Snijders J G 1993 *J. Chem. Phys.* **99**, 4597.

70. van Lenthe E, van Leeuven R, Baerends E-J and Snijders J G 1996 *Int. J. Quant. Chem.* **57**, 281.
71. Karshenboim S G, Pavone F S, Bassani F, Inguscio M and Hänsch T W 2001 *The Hydrogen Atom. Precision Physics of Simple Atomic systems* (Berlin: Springer-Verlag).
72. Quiney H M, Grant I P and Wilson S 1989, in Many-Body Methods in Quantum Chemistry (Lecture Notes in Chemistry 52) ed. U Kaldor *pp.* 307–344 (Berlin: Springer-Verlag).
73. Sapirstein J, Cheng K T and Chen M H 1999 *Phys. Rev. A* **59** 259.
74. Bishop R and Kummel H 1987 *Phys. Today* **40**, 42.
75. Bartlett R J 1991 *Theor. Chim. Acta* **80**, 71.
76. Paldus J 1992 in *Methods in Computational Molecular Physics* (ed. S Wilson and G H F Diercksen) pp. 99–194, (NATO ASI Series B: Physics Vol. 293) (New York: Plenum Press).
77. Lawley K P (ed.) 1987 *Adv. Chem. Phys.* **69** (New York: John Wiley).
78. Fischer C F, Brage T and Jönsson P 1997 *Computational Atomic Structure. An MCHF approach* (Bristol and Philadelphia: Institute of Physics).
79. Desclaux J-P 1975 *Comput. Phys. Commun.* **9**, 31.
80. Grant I P, McKenzie B J, Norrington P H, Mayers DF and Pyper N C 1980 *Comput. Phys. Commun.* **21**, 207.
81. Dyall K G, Grant I P, Johnson C T and Plummer E P 1989 *Comput. Phys. Commun.* **55**, 425.
82. Parpia F A, Fischer C F and Grant I P 1996 *Comput. Phys. Commun.* **94**, 249.
83. Jensen H J A, Dyall K G, Saue T, Fægri K 1996 *J. Chem. Phys.* **104**, 4083.
84. Saue T and Jensen H J A 1999 *J. Chem. Phys.* **111**, 6211.
85. Quiney H M, Skaane H and Grant I P 1998 *Chem. Phys. Lett.* **290**, 473.
86. Quiney H M, Skaane H and Grant I P 1999 *Adv. Quant. Chem.* **32**, 1.
87. Quiney H M 2002 in *Handbook of Molecular Physics and Quantum Chemistry* (Chichester: John Wiley).
88. Grant I P and Quiney H M 2002 in *Relativistic Quantum Chemistry* (ed. P Schwerdtfeger) pp. 107–202 (Amsterdam: Elsevier).
89. Rajagopal A K and Callaway J 1973 *Phys. Rev. B* **59**, 1912.
90. Rajagopal A K 1978 *J. Phys. C* **11**, L943.
91. Macdonald A H and Vosko S H 1979 *J. Phys. C* **12**, 2977.
92. Dreizler R M and Gross E K U 1990 *Density Functional Theory* (Berlin:: Springer-Verlag).
93. Engel E and Dreizler R M 1996 *Topics in Current Chem.* **181**, 1.
94. Engel E, Facco Bonetti A, Keller S, Andrejkocics I and Dreizler R M 1998 *Phys. Rev. A* **58**, 964.
95. Engel E and Dreizler R M 1999 *J. Comp. Chem.* **20**, 31.
96. Engel E, Dreizler R M, Varga S and Fricke B 2003 in *Relativistic Effects in Heavy Element Chemistry and Physics* (ed. B A Hess), Chapter 4. Relativistic Density Functional Theory. (Chichester: John Wiley and Sons, Ltd.).
97. Engel E 2002 in *Relativistic Quantum Chemistry* (ed. P Schwerdtfeger) pp. 523–621. (Amsterdam: Elsevier).
98. Hohenberg P and Kohn W 1964 *Phys. Rev.* **136 B**, 864.
99. Thomas L H 1927 *Proc. Camb. Phil. Soc.* **23**, 542.
100. Fermi E 1928 *Z. Physik* **48**, 73.
101. Slater J C 1951 *Phys. Rev.* **81**, 385.

102. Akhiezer I A and Peletminskii SV 1960 *Zh. Eksp. Teor. Fiz.* **38**, 1829 [*Sov. Phys. JETP* **11**, 1318.]
103. Becke A D 1988 *Phys. Rev. A* **38**, 3098.
104. Perdew J P, Chevary J A, Vosko S H, Jackson K A, Pederson M R, Singh D J and Fiolhais C 1992 *Phys. Rev. B* **46**, 6671.
105. Engel E, Keller S and Dreizler R M 1996 *Phys. Rev. A* **53**, 1367.
106. Engel E, Keller S and Dreizler R M 1998 in *Electronic Density Functional Theory: Recent Progress and New Directions* (ed J F Dobson, G Vignals and M P Das), p. 149 (New York: Plenum).
107. Kim Y-H, Städele M and Martin R M 1999 *Phys. Rev. A* **60**, 3633.
108. Sham L J 1985 *Phys. Rev. B* **32**, 3876.
109. Görling A and Levy M 1994 *Phys. Rev. A* **50**, 196.

Part III

Computational atomic and molecular structure

5

Analysis and approximation of Dirac Hamiltonians

Computational problems in atomic and molecular structure and processes require understanding of the mathematics of Dirac operators and of methods for constructing numerical solutions of the Dirac equation. Whilst simple problems such as the structure of hydrogenic atoms can be solved analytically, the majority of applications to many-electron systems can only be solved approximately. Some progress can be made using methods of functional analysis and symmetry properties, but the interpretation of experiments often needs high-precision numerical predictions. Meeting these demands requires cost-effective and reliable algorithms for constructing solutions of the Dirac equation.

Sections 5.1 to 5.3 present analytical properties of Dirac and Schrödinger operators, using the definitions and properties of linear operators on Hilbert spaces of Appendix B.1, on which the numerical analysis of the algorithms of Sections 5.5 to 5.12 used in relativistic electronic structure codes depend. Quantum mechanics relies on Hamiltonian self-adjointness; §5.1 examines the functional analysis of Schrödinger and Dirac Hamiltonians for free-particles and §5.2 considers the effect of adding a one-body potential. Limitations of the functional analysis approach for the Dirac operator are avoided in §5.3, which emphasizes the critical role of spinor boundary conditions in the study of the radial Dirac differential operator. The boundary conditions also play an important part in the analysis, Appendix B.7, of the convergence of eigenfunction expansions in two-point boundary value problems involving the radial Dirac differential operator.

Numerical methods for solving Dirac equations of atomic and molecular physics are of three types. Rayleigh-Ritz methods, whose theory is presented in §5.5 and §5.6, require the expansion of radial 2-spinor components in terms of "kinetically matched" pairs of analytic functions; popular choices are presented in §5.8 to §5.10. Finite difference methods are the most popular choice for solving atomic problems, §5.11. Iterative procedures are needed to construct eigenvalues and eigensolutions for bound states. Finite element methods, §5.12, have become popular more recently as an alternative to finite differences, but some technical problems with applications to Dirac operators remain. Finite element Rayleigh-Ritz methods lead to equations having some similarity to those occurring with basis sets, although there are some similarities with finite difference equations.

5.1 Self-adjointness of free particle Hamiltonians

5.1.1 Free particles: the Schrödinger case

The self-adjointness properties of the Schrödinger and Dirac operators for free particles can be discussed most conveniently in terms of the properties of Fourier transformations. According to the Fourier-Plancherel theorem [1, Chapter 3], each square-integrable function $u(x) \in L^2(E)$ has a Fourier transform $\hat{u}(k) \in L^2(\hat{E})$

$$\widehat{u}(k) = (2\pi)^{-n/2} \int_E \mathrm{e}^{-ik\cdot x} u(x) d^n x, \tag{5.1.1}$$

$$u(x) = (2\pi)^{-n/2} \int_{\widehat{E}} \mathrm{e}^{-ik\cdot x} \widehat{u}(k) d^n k, \tag{5.1.2}$$

where E and $\widehat{E}$ are both copies of R^n so that

$$k \cdot x = \sum_{j=1}^{n} k_j x_j, \ \forall\, x \in E, \ k \in \hat{E}.$$

It is convenient to think of $\mathcal{H} = L^2(E)$ and $\widehat{\mathcal{H}} = L^2(\widehat{E})$ as different spaces and the map $u \mapsto \widehat{u} = Uu$ as defining a unitary operator U on $\mathcal{H}$ into $\widehat{\mathcal{H}}$.

The free-particle Schrödinger operator on $\mathbb{R}^n$ is proportional to the Laplacian

$$\Delta = \frac{\partial^2}{\partial x_1^2} + \ldots + \frac{\partial^2}{\partial x_n^2}.$$

There are several ways in which we can define Δ as an operator on $\mathcal{H} = L^2(E)$. A function $\phi \in \mathcal{H}$ must be rather smooth for the expression $\Delta\phi$ to have a meaning. Suppose we first make the assumption that ϕ is a smooth function with *compact support* in $\mathbb{R}^n$: that is to say ϕ vanishes outside a compact subset of $\mathbb{R}^n$ that may depend on ϕ. In particular, we can think of the set C_0^∞ that contains all *infinitely differentiable functions with compact support*. The operator $\dot{T} = -\Delta$ with domain C_0^∞ is called the *minimal operator* constructed from the formal Laplacian $-\Delta$. In physical applications, we only need functions that have partial derivatives up to the second order, which contains C_0^∞ as a subset. The term *minimal* recognizes this restriction of the domain. Now the Fourier transform $U(T\phi)(k) = |k|^2\widehat{\phi}(k)$ where $|k|^2 = k_1^2 + \ldots + k_n^2$. Let K^2 be the *maximal operator* for multiplication by $|k|^2$ on $\widehat{\mathcal{H}}$, that is to say, the operator with the largest domain such that $|k|^2\widehat{\phi}(k) \in \widehat{\mathcal{H}}$. Then K^2 is self-adjoint, and its transform

$$H_0 = \frac{1}{2} U^{-1} K^2 U \tag{5.1.3}$$

is a self-adjoint operator on $\mathcal{H}$ into itself with domain $\mathcal{D}(H_0) = U^{-1}\mathcal{D}(K^2)$. Clearly $\mathcal{D}(H_0) \supset C_0^\infty$ so that H_0 is a self-adjoint extension of $\frac{1}{2}\dot{T}$. It can be shown [2, p. 300] that $\frac{1}{2}\dot{T}$ has H_0 as its closure: thus, in the language of §B.1.1, $\dot{T}$ is essentially self-adjoint and $\mathcal{D}(\dot{T}) = C_0^\infty$ is a core for H_0.

5.1.2 Free particles: the Dirac case

The construction of self-adjoint domains for the free-particle Dirac operator follows a similar path. Let

$$F = c\boldsymbol{\alpha} \cdot \boldsymbol{p} + mc^2\beta \tag{5.1.4}$$

where $\boldsymbol{\alpha}, \beta$ are the usual 4×4 Dirac matrices. This formal differential operator acts on 4-spinors constructed from complex-valued functions on $\mathbb{R}^3$. The underlying Hilbert space is therefore $\mathcal{H} = [L^2(\mathbb{R}^3)]^4$, consisting of all 4-component objects

$$u(\boldsymbol{r}) = \begin{bmatrix} u_1(\boldsymbol{r}) \\ u_2(\boldsymbol{r}) \\ u_3(\boldsymbol{r}) \\ u_4(\boldsymbol{r}) \end{bmatrix} \tag{5.1.5}$$

with the inner product

$$(u, v) = \int u(\boldsymbol{r})^\dagger v(\boldsymbol{r})\, d^3r,$$

where

$$u(\boldsymbol{r})^\dagger v(\boldsymbol{r}) = \sum_{j=1}^{4} u_j(\boldsymbol{r})^* v_j(\boldsymbol{r}),$$

and with the norm $\|u\|^2 = (u, u)$. We define a minimal operator $\dot{F}$ with domain $\mathcal{D}(\dot{F}) = [C_0^\infty(\mathbb{R}^3)]^4$ by requiring that $\dot{F}u = Fu$ for all 4-spinors $u \in \mathcal{D}(\dot{F})$. As in the Schrödinger case, the Fourier transform, applied component-wise to $u(\mathbf{r})$, induces a unitary map U on $\mathcal{H}$ into $\hat{\mathcal{H}}$ such that $u \mapsto Uu$. When $u \in \mathcal{D}(\dot{F})$, the Fourier transform of $v = \dot{F}u$ is given by

$$K\widehat{u}(\boldsymbol{k}) := \widehat{v}(\boldsymbol{k}) = [c\boldsymbol{\alpha} \cdot \boldsymbol{k} + mc^2\beta]\, \widehat{u}(\boldsymbol{k}), \tag{5.1.6}$$

so that the operator K is defined in terms of matrix multiplication of each 4-spinor $\widehat{u}(\boldsymbol{k})$ by the 4×4 matrix $[c\boldsymbol{\alpha} \cdot \boldsymbol{k} + mc^2\beta]$. As in the Schrödinger case, K is a self-adjoint operator, and its inverse Fourier representation

$$H_0 = U^{-1} K U \tag{5.1.7}$$

is a self-adjoint operator on $\mathcal{H}$ with domain $\mathcal{D}(H_0) = U^{-1}\mathcal{D}(K)$. Making use of the anticommutation properties of the Dirac matrices, we see that

$$[c\boldsymbol{\alpha} \cdot \boldsymbol{k} + mc^2\beta]^2 = c^2|\boldsymbol{k}|^2 + m^2c^4$$

so that the assertion that $\widehat{u} \in \mathcal{D}(K)$ implies that

$$\int (c^2|\boldsymbol{k}|^2 + m^2c^4)\widehat{u}(\boldsymbol{k})^\dagger \widehat{u}(\boldsymbol{k})\, d^3k = \|Ku\|^2 < \infty. \tag{5.1.8}$$

Thus $H_0 \supset \dot{F}$, and $\dot{F}$ is essentially self-adjoint if and only if its closure is H_0, which can be proved by the same sort of arguments as in the Schrödinger case.

5.2 Self-adjointness of Hamiltonians with a local potential

One-particle operators of the form $H = H_0 + V$, where V is some local one-body potential function, occur everywhere in atomic and molecular physics. As examples, we have the Coulomb energy $-Z/r$ a.u. of an electron relative to a point nucleus of charge Ze or the energy of an electron in some mean-field potential of an atom or molecule. We can study (essential) self-adjointness of the full operator H by considering V as a perturbation of the self-adjoint operator H_0.

Suppose first that A and B are densely defined operators on $\mathcal{H}$, that $\mathcal{D}(B) \supset \mathcal{D}(A)$ and that, for some real numbers a, b and for all $\phi \in \mathcal{D}(A)$,

$$\|B\phi\| \leq a\|A\phi\| + b\|\phi\|. \tag{5.2.1}$$

Then B is said to be A-*bounded*; the infimum of a is called the *relative bound* of B with respect to A. If the relative bound is zero, then B is said to be *infinitesimally small* relative to A, and we write $B \ll A$. Sometimes it is more convenient to replace (5.2.1) with a related inequality

$$\|B\phi\|^2 \leq \tilde{a}^2\|A\phi\|^2 + \tilde{b}^2\|\phi\|^2. \tag{5.2.2}$$

When (5.2.2) is satisfied, then (5.2.1) is also satisfied with $a = \tilde{a}, b = \tilde{b}$. However, if (5.2.1) is satisfied, then (5.2.2) holds, but with $\tilde{a}^2 = (1+\epsilon)a^2$ and $\tilde{b}^2 = (1+\epsilon^{-1})b^2$ for each $\epsilon > 0$. Thus the infimum of a in (5.2.1) and the infimum of $\tilde{a}$ in (5.2.2) are equal. It is also sufficient to prove estimates of either form on a core for A.

The basic result is the Kato-Rellich theorem [3, Vol. II, p. 162]:

Theorem 5.1. *Let A be self-adjoint and let B be symmetric and satisfy the inequality*

$$\|B\phi\| \leq a\|A\phi\| + b\|\phi\|$$

with relative bound $a < 1$. Then $A+B$ is self-adjoint on $\mathcal{D}(A)$ and essentially self-adjoint on any core of A.

Also if A is bounded below by M, then $A + B$ is also bounded below by $M - \max(b/(1-a), a|M|+b)$, where a, b are the numbers appearing in (5.2.1).

Wüst has extended this theorem to the case when the relative bound $a = 1$; then $A + B$ is only essentially self-adjoint on $\mathcal{D}(A)$ or on any core for A.

5.2.1 The Schrödinger case

We start [2] with the simple Schrödinger operator

$$H = H_0 + V(\boldsymbol{r})$$

where $H_0 = -\frac{1}{2}\Delta$ is the self-adjoint operator of (5.1.3) for the case of three space dimensions, $n = 3$. Suppose that the real-valued potential $V(\boldsymbol{r})$ can be split into two parts,

$$V(\boldsymbol{r}) = V_1(\boldsymbol{r}) + V_2(\boldsymbol{r}), \tag{5.2.3}$$

where $V_1(\boldsymbol{r}) \in L^2(\mathbb{R}^3)$ is square integrable and $V_2(\boldsymbol{r}) \in L^\infty(\mathbb{R}^3)$ is bounded. Because V is real-valued, the operator defined as multiplication by V on the domain

$$\mathcal{D}(V) = \{\phi \,|\, \phi \in L^2(R^3), V\phi \in L^\infty(R^3)\}$$

is self-adjoint. Using standard inequalities from integration theory, we have

$$\|V\phi\|_2 \leq \|V_1\|_2\|\phi\|_\infty + \|V_2\|_\infty\|\phi\|_2 \tag{5.2.4}$$

so that $\mathcal{D}(V)$ contains $C_0^\infty(\mathbb{R}^3)$ which is a core for H_0.

If $\phi \in L^2(\mathbb{R}^3)$ is also in $\mathcal{D}(H_0)$, then ϕ is a bounded and continuous function. So for any $a > 0$ there is a $b > 0$ such that

$$\|\phi\|_\infty \leq a\|H_0\,\phi\|_2 + b\|\phi\|_2 \tag{5.2.5}$$

[3, Vol. II, Theorem IX.28]. Combining (5.2.4) and (5.2.5) gives

$$\|V\phi\|_2 \leq a\|V_1\|_2\|H_0\|_2 + (b + \|V_2\|_\infty)\|\phi\|_2,$$

which holds for all $\phi \in C_0^\infty$. Thus all potentials $V(\boldsymbol{r})$ for which (5.2.3) holds are bounded by (H_0) on C_0^∞ with arbitrarily small relative bound a. Theorem 5.1 then allows us to infer that H is essentially self-adjoint on C_0^∞.

This analysis covers most of the cases of practical importance. In the case of the Coulomb potential,

$$V(\boldsymbol{r}) = -Z/r,$$

we can take

$$V_1(\boldsymbol{r}) = \begin{cases} -Z/r \text{ when } 0 < |\boldsymbol{r}| < R_1, \\ \quad 0 \text{ when } R_1 < |\boldsymbol{r}| < \infty, \end{cases}$$

and a similar partition works for the screened Coulomb and mean field potentials that are typically used in atomic structure. The extension to N-electron systems with pairwise Coulomb interactions used in nonrelativistic atomic and molecular structure theory is straightforward, and is thoroughly documented in [2, Chapter 5, §5] and [3, Vol. II, Theorem X.16].

5.2.2 The Dirac case

The Dirac operators require more delicate handling, though the underlying principles are the same. Consider the important case of the Coulomb potential $V(\boldsymbol{r}) = -Z/r$. The "uncertainty principle lemma" [2, p. 307], [3, Vol. II, p. 169] states that

$$\int \frac{1}{|\boldsymbol{r}|^2} u(\boldsymbol{r})^\dagger u(\boldsymbol{r})\, d^3x \leq 4 \int [\nabla u(\boldsymbol{r})]^\dagger \cdot [\nabla u(\boldsymbol{r})]\, d^3x = 4 \int |\boldsymbol{k}|^2 |\widetilde{u}(\boldsymbol{k})|^2\, d^3k.$$

In conjunction with (5.1.8) this yields

$$\|Vu\|^2 \leq a\|Ku\|^2, \quad a = \frac{4Z^2}{c^2}$$

when $u \in \mathcal{D}(\dot{F})$. To apply the Kato-Rellich theorem we need $a < 1$, giving $Z < c/2 \approx 68$, which makes the range of Z for which the Dirac operator

$$\dot{H} = H_0 + V \tag{5.2.6}$$

is self-adjoint on $\mathcal{D}(\dot{F})$ disappointingly small. A more subtle treatment is given by Kalf *et al.* [4]; their result coincides with that obtained in §3.3, namely $Z/c < \sqrt{3}/2$ or $Z < 118$, which covers almost all the interesting cases. Because $H_0 \supset \dot{F}$, these conclusions extend to the whole of $\mathcal{D}(H_0)$.

If we take the domain of H_0 instead to be

$$\mathcal{D}(H_0) = \{\phi \,|\, \phi \in \mathcal{H}, H_0\,\phi \in \mathcal{H}\}, \quad \mathcal{H} = [L^2(R^3)]^4$$

then it can be shown that the closure of the minimal operator $\dot{H}$ defined by (5.2.6) has the same domain when $Z/c < \sqrt{3}/2$. We can identify $\mathcal{D}(H_0)$ with the Sobolev space $[W^{(1)}(\mathbb{R}^3)]^4$ of 4-component spinors whose components and first partials are all quadratically integrable. We shall see in §5.4 that the Dirac Coulomb operator makes a transition from the limit point case at $r = 0$ to the limit circle case when $Z = \sqrt{3}c/2 \approx 118$ consistent with the theoretical restriction.

The failure of this analysis to tell us anything about the self-adjointness of the Dirac Coulomb operator for $Z > 118$ is disappointing. The discussion of boundary conditions in §3.3 showed that a further restriction of the domain to spinors u such that $\| r^{-1/2} u \|^2$ is bounded restores the problem to the limit point case at $r = 0$, and allows us to extend the range of Z for which computations can be done up to $Z = 137$. Kalf *et al.* [4] have also found a rather complicated way to extend the range of self-adjointness as far as $Z = 137$. However, real nuclei are not point charges, and the finite size of the nuclear charge density starts to produce noticeable physical effects for relatively modest values of Z. It is then desirable to use a model with a finite size charge distribution for determining V. A major change is the replacement of the $1/r$ singularity with a finite cut-off. This restores the situation and allows calculations to proceed to much higher values of Z. This behaviour has been exploited in theories of superheavy elements and of quasi-molecules with heavy element constituents [5] as well as the atomic and molecular calculations described in this book.

5.3 The radial Dirac differential operator

The properties of central field wavefunctions are of fundamental importance for atomic and molecular physics. The eigenfunctions of the Dirac equation in spherical polar coordinates have the form (3.2.4)

$$\phi(\boldsymbol{x}) = \frac{1}{r}\begin{pmatrix} P(r)\chi_{\kappa m}(\theta,\varphi) \\ iQ(r)\chi_{-\kappa m}(\theta,\varphi) \end{pmatrix}$$

where the radial amplitudes satisfy the *radially reduced* equations (3.2.17)

$$\Big(mc^2 + V(r) - E\Big)P(r) + \hbar c\left(-\frac{dQ(r)}{dr} + \frac{\kappa}{r}Q(r)\right) = 0$$

$$\hbar c\left(\frac{dP(r)}{dr} + \frac{\kappa}{r}P(r)\right) + \Big(-mc^2 + V(r) - E\Big)Q(r) = 0.$$

After shifting the energy to the usual nonrelativistic zero by writing $\varepsilon = E - mc^2$, this takes the form

$$T_\kappa\, u_{\varepsilon\kappa}(r) = \varepsilon\, u_{\varepsilon\kappa}(r), \quad u_{\varepsilon\kappa}(r) = \begin{pmatrix} P_{\varepsilon\kappa}(r) \\ Q_{\varepsilon\kappa}(r) \end{pmatrix}, \tag{5.3.1}$$

for all $r \in \mathbb{R}_+$, where

$$T_\kappa := \begin{pmatrix} V(r) & -c\left(\dfrac{d}{dr} - \dfrac{\kappa}{r}\right) \\ c\left(\dfrac{d}{dr} + \dfrac{\kappa}{r}\right) & -2mc^2 + V(r) \end{pmatrix}. \tag{5.3.2}$$

The matrix operator T_κ can be written as a self-adjoint differential operator

$$T_\kappa = cJ\frac{d}{dr} + W_\kappa(r), \tag{5.3.3}$$

where

$$J := -i\sigma^2 = \begin{pmatrix} 0 & -1 \\ 1 & 0 \end{pmatrix}, \quad W_\kappa(r) = \frac{1}{r}\begin{pmatrix} -Z(r) & c\kappa \\ c\kappa & -2r\,c^2 - Z(r) \end{pmatrix},$$

so that $Z(r) = -rV(r)$. The radial two-component spinor $u_{\varepsilon\kappa}$ has the norm

$$\|u_{\varepsilon\kappa}\|^2 = \int_{\mathbb{R}_+} u^{\dagger}_{\varepsilon\kappa}(r) u_{\varepsilon\kappa}(r) dr = \int_{\mathbb{R}_+} \left(P^2_{\varepsilon\kappa}(r) + Q^2_{\varepsilon\kappa}(r)\right) dr. \tag{5.3.4}$$

The integral is finite for bound states, but diverges for scattering states.

$Z(r)$ is a continuous function of r in applications to atomic and molecular physics. For a point nucleus $Z(r) = Z$ for all value of r. For a spherical nucleus in which the Z protons are confined to a ball of radius R_{nuc}, $V(r)$ has a finite well-depth so that as $r \to 0$, $V(r) = -Z(r)/r \approx -v_0 + v_2 r^2 + \ldots$, where the constants $v_0, v_2, \ldots$ depend on the model and as $r \to \infty$, $Z(r) \to z$, where z is the degree of ionization of the system. When these conditions apply, the operator T_κ is self-adjoint on any finite interval $[a, b]$ with domain $\mathcal{D}(T_\kappa)$ provided

$$u(r) \in (L^2[a,b])^2, \quad T_\kappa u(r) \in (L^2[a,b])^2, \tag{5.3.5}$$

provided the boundary condition

$$\left[cv^{\dagger}(r) J u(r)\right]_a^b = 0 \tag{5.3.6}$$

holds for every $u, v \in \mathcal{D}(T_\kappa)$, which corresponds, in the manner of the discussion of the Dirac operator in three dimension in §5.2.2, to the Sobolev space $\left[W^{(1)}(\mathbb{R}^3)\right]^2$. The crucial boundary condition (5.3.6) mixes the large and small components of u so that the handling of the boundary conditions is more involved than for the Schrödinger equation.

5.3.1 The boundary condition at a singular endpoint

The linear differential system $du/dr + A(r)u = 0$ is said to have a singular point at s if the square matrix $A(r)$ has a singularity at $r = s$. From (5.3.3) we see that $W_\kappa(r)$ has a singularity at $r = 0$, which is therefore a singular point of the radial Dirac equation. Similarly by writing $\rho = 1/r$ and examining the behaviour as $\rho \to 0$ we see that it also has a singular point at ∞. Thus the radial Dirac equation, like the radial Schrödinger equation, has two singular endpoints, and the theory can be developed along very similar lines [6, Chapter 10].

It is useful to start by looking at a result for second order differential equations of the form

$$T_0\phi := -\phi'' + V(x)\phi(x) = \lambda\phi(x) \tag{5.3.7}$$

on $[0, \infty)$ with $V(x)$ continuous is from [3, Vol. II, p. 151].

Theorem 5.2. *(i) If $\Im(\lambda) \neq 0$, then at least one solution of $T_0\phi = \lambda\phi$ is in L^2 near $x = 0$ and at least one solution is in L^2 near infinity.*

(ii) If, for one $\lambda \in \mathcal{C}$, both solutions of $T_0\phi = \lambda\phi$ are in L^2 near infinity (zero) then, for all *$\lambda \in \mathcal{C}$, both solutions of $T\phi = \lambda\phi$ are in L^2 near infinity (zero).*

Theorem B.3 shows that the existence and number of self-adjoint extensions of an operator T are determined by the deficiency indices n_+, n_-, which are respectively the number of linearly independent solutions of $T^*\phi = \pm i\,\phi$, where T^* is its adjoint operator. To specify a suitable minimal operator we really need a boundary condition at ∞, which is not yet available. We can avoid this by choosing $\mathcal{D}(T) = C_0^\infty(0,\infty)$, the space of infinitely differentiable functions with bounded support, such that $T\phi = T_0\phi$ for $\phi \in C_0^\infty$. Although this operator T is not closed. it is symmetric and T^* is the operator with $\mathcal{D}(T^*) = \{\phi \in L^2 \,:\, T_0\phi \in L^2\}$ such that $T^*\phi = T_0\phi$.

Because T is a second order differential operator, the differential equation $T_0\phi = \lambda\phi$ has two linearly independent solutions for any λ. A solution ϕ of this equation is in $\mathcal{D}(T^*)$ if and only if it is in $L^2(\mathbb{R}_+)$. According to Theorem 5.2, two linearly independent solutions of $T_0\phi = \lambda\phi$ may be in $L^2(\mathbb{R}_+)$, in which case all solutions are in $L^2(\mathbb{R}_+)$. It is known that if this happens for any choice of λ in either the upper or the lower half-plane, then it happens for every λ in either half-plane [6, Lemma, p. 153]. If $T_0\phi = \lambda\phi$ then $T_0^*\phi^* = \lambda^*\phi^*$ and as $\phi \in L^2(\mathbb{R}_+)$ then $\phi^* \in L^2(\mathbb{R}_+)$; what happens in one half-plane is mirrored in the other, and all solutions are in $L^2(\mathbb{R}_+)$ for all λ, real or complex. So the possible deficiency indices (n_+, n_-) are (0,0), (1,1) or (2,2). It is easy to show that there is always at least one square integrable solution in either half-plane, which rules out the (0,0) case.

The next step is to examine the behaviour of the solutions near a singular endpoint [6, p. 202]. For Sturm-Liouville operators such as T_0, there are two possibilities. Either for some complex λ with $\Im\lambda > 0$ there is one square-integrable solution in the neighbourhood of the singular endpoint, the so-called *limit-point case*, or there are two square-integrable solutions, the so-called *limit-circle case*, and every linear combination of them is a valid solution. In the limit-point case, the condition that a solution is square integrable near the endpoint selects a homogeneous boundary condition which fixes the solution uniquely. This is what is usually assumed without discussion in applications of nonrelativistic quantum mechanics. In the limit-circle case, an additional boundary condition, usually determined by physical considerations, is needed.

5.3.2 The Dirac radial operator with one singular endpoint

The Dirac equation cannot be transformed into a Sturm-Liouville problem to which the above argument applies. However, the analysis of the radial Dirac operator on $\mathbb{R}_+$ is remarkably similar to that of the radial Schrödinger equation and gives useful information on the self-adjointness properties which we shall need to construct numerical solutions.

Lemma 5.3. *Let $u_1(r)$ and $u_2(r)$ satisfy (5.3.1) in some finite interval $[a,b]$ with parameters (in general complex) ε_1 and ε_2 respectively. Define*

$$s_{12}(r) = c{u_1}^\dagger(r)\,J\,u_2(r) = c\,[\,Q_2^*(r)P_1(r) - P_2^*(r)Q_1(r)\,]\,. \tag{5.3.8}$$

so that $s_{12}(r) = -s_{21}^*(r)$. *Then*

$$s_{12}(b) - s_{12}(a) = (\varepsilon_1 - \varepsilon_2^*) \int_a^b u_2^\dagger(r)\, u_1(r)\, dr. \tag{5.3.9}$$

The proof is straightforward; just calculate ds_{12}/dr using (5.3.1) and integrate over (a, b). The lemma has a useful corollary:

Corollary 5.4. *In the case when* $\varepsilon_1 = \varepsilon_2 = \varepsilon$, *we can write* $u_1 = u_2 = u$. *Then*

$$s(b) - s(a) = 2i\Im\varepsilon \int_a^b u^\dagger(r)\, u(r)\, dr. \tag{5.3.10}$$

It follows that if ε is real then $s(r)$ is independent of r, and that with this notation (5.3.6) is just $s(a) = s(b)$.

Consider now the initial-value problem for the two-dimensional linear differential equation (5.3.1). Suppose that we can write, with arbitrary normalization,

$$u(a) = w_1 + m\, w_2, \tag{5.3.11}$$

where w_1 and w_2 are any two linearly independent 2-vectors, say

$$w_1 = \begin{pmatrix} 1 \\ 0 \end{pmatrix}, \quad w_2 = \begin{pmatrix} 0 \\ 1 \end{pmatrix},$$

and $m = Q(a)/P(a)$ is an arbitrary parameter. Let $u_1(r)$ and $u_2(r)$ be solutions of the initial value problem with respective initial values w_1 and w_2 so that

$$u(r) = u_1(r) + m\, u_2(r) \tag{5.3.12}$$

is the solution of the initial value problem with initial value $u(a)$ (5.3.11). This generates $s(r)$ as a quadratic complex expression in m,

$$s(r) = s_{11}(r) + m s_{12}(r) + m^* s_{21}(r) + |m|^2 s_{22}(r), \tag{5.3.13}$$

where $s_{ij}(r) = c u_i^\dagger(r) J u_j(r)$. The initial conditions are

$$s_{11}(a) = s_{22}(a) = 0, \quad s_{12}(a) = -c, \quad s_{21}(a) = +c, \tag{5.3.14}$$

so that

$$s(a) = -m + m^* = -2ic\Im m.$$

Inserting this in (5.3.10) and writing $\lambda = \varepsilon/c$ gives

$$s(b) = 2ic\left(-\Im m + \Im\lambda \int_a^b u^\dagger(r)\, u(r)\, dr\right). \tag{5.3.15}$$

If we replace $u(r)$ by $u_2(r)$ in (5.3.10) we find

$$s_{22}(b) = 2ic\Im\lambda \int_a^b u_2^\dagger(r)\, u_2(r)\, dr. \tag{5.3.16}$$

We focus on $s(b)$; it is convenient to set $s(b) = 2icF(b, m)$ and to write

$$F(b, m) := A\,|\,m|^2 + B\,\Re m + C\,\Im m + D \tag{5.3.17}$$

where A, B, C, D are all real numbers defined by

$$\begin{aligned} 2icA &:= s_{22}(b) = c(P_2Q_2^* - Q_2P_2^*), \\ 2icD &:= s_{11}(b) = c(P_1Q_1^* - Q_1P_1^*), \\ ic(B + iC) &:= s_{12}(b) = c(P_2Q_1^* - Q_2P_1^*), \\ ic(B - iC) &:= s_{21}(b) = c(P_1Q_2^* - Q_1P_2^*), \end{aligned}$$

in which we have omitted the arguments $r = b$ from the components of $u_1(r)$ and $u_2(r)$, For fixed b, the equation $F(b, m) = 0$ represents a circle in the complex m-plane, with centre $m(b) = -(B + iC)/2A$ and radius $R(b)$ given by

$$4A^2R(b)^2 = B^2 + C^2 - 4AD = |P_1Q_2 - Q_1P_2|^2 = |u_1^t(b)Ju_2(b)|^2.$$

Equation (5.3.16) shows that

$$A = \Im\lambda \int_a^b u_2^\dagger(r)u_2(r)\, dr > 0. \tag{5.3.18}$$

It is easy to show that $u_1^t(r)Ju_2(r)$ is independent of r, and using the initial conditions gives

$$R(b) = 1/2A.$$

Denote by D_b the interior of the circle $F(b, m) = 0$, where we take $\Im m > 0$; the other case is similar. Points of the m-plane will be outside D_b if $F(b, m) > 0$ and inside if $F(b, m) < 0$. Because the radius $R(b)$ of the disk is, by (5.3.17), strictly decreasing as a function of b, $D_{b'} \subset D_b$ whenever $b' > b$, so that the disks shrink as $b \to \infty$. The limit-point case occurs when $R(b) \to 0$, and there is a unique value $m = m(\infty)$ such that $u(r) = u_1(r) + m(\infty)u_2(r)$ for which the solution is square integrable. It follows from (5.3.15) that the solution is square integrable, with norm

$$\|u\|^2 = \int_a^\infty u^\dagger(r)u(r)\, dr < \Im m(\infty)/\Im\lambda.$$

The condition of square integrability is equivalent to a boundary condition. However, it is important to remember that this argument tells us nothing about the behaviour of the individual components of $u(r)$ as $r \to \infty$, and this must be recalled when constructing numerical solutions. In the limit-circle case, $R(b) \to R(\infty) > 0$; every value of m in D_∞ gives a square integrable solution, and we require a supplementary boundary condition to make the operator self-adjoint.

5.4 The radial Dirac equation for atoms

Atomic and molecular calculations often start from a solution of Dirac's equation for some potential $V(r)$ given by

$$V(r) = -Z(r)/r \tag{5.4.1}$$

where $Z(r)$, the effective central charge seen by an electron at a distance r from the nucleus, decreases as r increases. Thus we suppose

$$Z(r) \to z \geq 0, \quad r \to \infty$$

where $z < Z$ is the residual ionic charge seen by the electron, and also that

$$Z(r) = Z_0 + Z_1 r + Z_2 r^2 + \ldots \tag{5.4.2}$$

in the neighbourhood of $r = 0$. This property characterizes a number of commonly used nuclear models.

In §3.3, we saw that $u(r) \sim \exp(\lambda r)$ as $r \to \infty$, where $\lambda^2 = c^2 - E^2/c^2$ and E is real. There are two situations:

- $|E| < c^2$: All solutions in this part of the spectrum correspond to bound states. Only one solution with $\lambda < 0$ is square integrable at infinity, so this is a limit-point case.
- $|E| > c^2$: $\lambda = \pm ip$, where $p = +\sqrt{E^2/c^2 - c^2}$ is interpreted as the magnitude of the 3-momentum vector. Solutions are linear combinations of particular solutions with asymptotic form $\exp(\pm ip\, r)$ corresponding to the limit-circle case.

The nuclear model potentials determine the behaviour near $r = 0$, depending on the coefficients of (5.4.2):

- *Point nucleus:* $Z_0 \neq 0$; $Z_n = 0, \quad n > 0$.
  This is the normal choice in nonrelativistic calculations. Physical consequences of the finite size of the nuclear charge distribution become significant in the lower half of the Periodic Table, especially for inner-shell processes.
- *Uniform nuclear charge distribution:* A model which, though unphysical, is nevertheless very useful, spreads the nuclear charge uniformly over the interior of a sphere of radius R_N, so that

$$\rho_N(r) = \begin{cases} 3Z/4\pi R_N^3, & \text{when } 0 \leq r \leq R_N\,, \\ 0, & \text{when } r > R_N, \end{cases}$$

  gives

$$V_N(r) = \begin{cases} -\dfrac{3Z}{2R_N}\left(1 - \dfrac{r^2}{3R_N^2}\right), & \text{when } 0 \leq r \leq R_N\,, \\ -\dfrac{Z}{r}, & \text{when } r > R_N. \end{cases} \tag{5.4.3}$$

Thus when $r \leq R_N$, the non-zero coefficients are

$$Z_0 = 0, \quad Z_1 = +3Z/2R_N, \quad Z_2 = 0, \quad Z_3 = -Z/2R_N^3,$$

whilst for $r > R_N$ there is only one non-zero coefficient, $Z_0 = Z$, giving an unscreened Coulomb potential.

- *Spherical surface charge distribution:* Like the uniform charge distribution above, this is unphysical. It recognizes the fact that the mutual repulsion of the protons inside the nucleus makes it more probable that they will be located near the nuclear boundary, $r = R_N$. Real nuclear electrostatic potentials are likely to lie between the uniform charge model and the surface charge model inside the nucleus. The potential is constant for $r < R_N$, so that there is only one non-zero coefficient, $Z_1 = Z/R_N$ for $r < R_N$, and $Z_0 = Z$ for $r > R_N$. Whilst this is simple and has been used to understand the effect of changes in the nuclear parameters on electron binding energies and wavefunctions it has some technical disadvantages. Whilst the potential is continuous, the electric field is discontinuous at $r = R_N$ because all the nuclear charge resides on the surface; the potential and the field are continuous for the uniform charge density distribution, but it has a discontinuity in the field gradient. Numerical methods of solution must take this into account.
- *Gaussian nuclear density distribution:* The majority of molecular structure codes are based on wavefunctions written as a linear combination of Gaussian type functions (GTF). This is also the case for relativistic molecular structure codes, making it easy to incorporate a Gaussian density distribution to model nuclear charge distributions. The density distribution

$$\rho_N(\boldsymbol{r}) = Z\,(\lambda/\pi)^{3/2} \exp(-\lambda r^2)$$

gives the potential

$$V_N(\boldsymbol{r}) = -(Z/r)\,\mathrm{erf}\,(\sqrt{\lambda} r), \tag{5.4.4}$$

whose power series expansion about $r = 0$ has coefficients

$$Z_{2n} = 0, \quad n \geq 0; \quad Z_1 = 2\sqrt{\lambda/\pi},\ Z_3 = -2(\lambda/3)\sqrt{\lambda/\pi}, \ldots$$

where λ is related to the RMS radius of the distribution, $\bar{R}_N$ (in atomic units) by

$$\lambda = 1.50/\bar{R}_N^2.$$

A more realistic nuclear density distribution can be constructed by fitting to a sum of Gaussian densities.

- *Fermi distribution:* The nuclear charge density has the form

$$\rho_N(r) = \frac{\rho_0}{1 + \exp[(r - R_N)/d]},$$

where ρ_0 is chosen so that

$$Z = \int_0^\infty 4\pi r^2 \rho_N(r)\, dr.$$

This two-parameter model has a uniform core with a "skin" in which the density falls from ~90% to ~10% of its central value in a short distance typically $t \approx 10^{-5}$ a.u. Then $d = t/4\ln 3$.

All these models of the nuclear charge distribution have been chosen for their convenience in practical calculations rather than as faithful representations of actual nuclear charge distributions. The main reason for this is that the dominant shift in orbital energies, responsible for the *nuclear volume isotope effect*, is sensitive primarily to the RMS radius, $\bar{R}_N$, of the nuclear charge distribution [7, §8.3]. Measured values of $\bar{R}_N$, which is isotope dependent, can be obtained from the literature, but it is often sufficient to use one of several statistical expressions found in the literature. Thus for $A > 16$ [7, §8.3] suggest)

$$\bar{R}_N \approx (1.115\, A^{1/3} + 2.151\, A^{-1/3} - 1,742A^{-1}) \times 10^{-5}$$

atomic units; many similar formulae can be found in the literature. Real nuclei are mostly non-spherical, and generate both electric and magnetic multipole fields which are responsible for *hyperfine interactions*.

5.4.1 Power series solutions near $r = 0$

The radial amplitudes

$$u(r) = \begin{pmatrix} P(r) \\ Q(r) \end{pmatrix},$$

can be expanded in a power series

$$u(r) = r^\gamma [u_0 + u_1 r + u_2 r^2 + \ldots], \quad u_k = \begin{pmatrix} p_k \\ q_k \end{pmatrix} \tag{5.4.5}$$

near the origin, where the index γ, p_k, and q_k are constants that depend on the nuclear potential model.

Point nucleus model: For a Coulomb singularity, $Z_0 \neq 0$, the leading coefficients satisfy indicial equations

$$\begin{aligned} -Z_0 p_0 + c(\kappa - \gamma) q_0 &= 0\,, \\ c(\kappa + \gamma) p_0 - Z_0 q_0 &= 0\,, \end{aligned} \tag{5.4.6}$$

so that

$$\gamma = \pm\sqrt{\kappa^2 - Z_0^2/c^2}\,, \quad \frac{q_0}{p_0} = \frac{Z_0}{c(\kappa - \gamma)} = \frac{c(\kappa + \gamma)}{Z_0}\,. \tag{5.4.7}$$

- *Finite size nuclear models:* In this case $Z_0 = 0$, so that the potential is finite at $r = 0$. The indicial equation (5.4.7) reduces to $\gamma = \pm|\kappa|$, so that for $\kappa < 0$,

$$P(r) = p_0 r^{l+1} + O(r^{l+3}), \quad Q(r) = q_1 r^{l+2} + O(r^{l+4}), \tag{5.4.8}$$

with

$$q_1/p_0 = (E - mc^2 + Z_1)/[c(2l+3)], \quad q_0 = p_1 = 0,$$

and for $\kappa \geq 1$,

$$P(r) = p_1 r^{l+1} + O(r^{l+3}), \quad Q(r) = q_0 r^{l} + O(r^{l+2}), \tag{5.4.9}$$

with

$$p_1/q_0 = -(E - mc^2 + Z_1)/[c(2l+1)], \quad p_0 = q_1 = 0.$$

For finite size nuclei, the power series expansions consist of purely *even* powers for one component with purely *odd* powers of r for the other component. The power series of both components for point nuclei consist of all powers $\gamma + n$ for positive integer n.

5.4.2 Power series solutions in the nonrelativistic limit

Many of the pathologies that arise from a naive treatment of the numerical solution of Dirac problems can be understood after examining the behaviour in the nonrelativistic limit $c \to \infty$, where we expect

$$P(r) \to O(r^{l+1}).$$

Finite nuclear models.

The behavior is entirely regular as $c \to \infty$:

$$P(r) = O(r^{l+1}), \quad Q(r) = O(1/c) \to 0.$$

Point nuclear models.

The behaviour when $\kappa < 0$ is entirely regular. However things are different when κ is positive. Because

$$\gamma = |\kappa| - \frac{Z^2}{2c^2|\kappa|} + \dots,$$

(5.4.7) shows that the leading coefficient p_0 *vanishes* when $c \to \infty$, so that,

$$P(r) \approx p_1 r^{l+1}(1 + O(r^2)), \text{when } \kappa \geq 1, \quad l = \kappa \tag{5.4.10}$$

All higher powers of *odd* relative order vanish in the limit for both components.

5.4.3 The boundary condition at the origin

The requirement for radial spinors to be square integrable as $r \to 0$

$$\int_0^{R'} D_{E,\kappa}(r)dr < \infty, \quad R' > 0,$$

where

$$D_{E,\kappa}(r) = |P_{E,\kappa}(r)|^2 + |Q_{E,\kappa}(r)|^2,$$

is equivalent to a boundary condition. For a point nuclear model, $D_{E,\kappa}(r) \sim r^{\pm 2\gamma}$ as $r \to 0$, this condition holds when $\pm 2\gamma > -1$. Only the solution with $\gamma > 0$ is acceptable when $2|\gamma| > 1$, or $Z < \alpha^{-1}\sqrt{\kappa^2 - 1/4}$, and the second solution must be dropped. This corresponds to the *limit point case.* In the special case $|\kappa| = 1$ or $j = \frac{1}{2}$ this implies $Z < c\sqrt{3}/2 \approx 118.6$. For $Z > c\sqrt{3}/2$, both solutions are square integrable near the origin (the *limit circle case*) and we need an additional constraint to recover a self-adjoint operator. A physically acceptable solution must also have a finite value for the expectation of the Coulomb potential so that

$$\int_0^{R'} D_{E,\kappa}(r)\frac{dr}{r} < \infty, \quad R' > 0\,. \tag{5.4.11}$$

This is always satisfied by the solution with $\gamma > 0$ for all $|Z| < \alpha^{-1}|\kappa|$, but not by any solution with $\gamma < 0$. Imposing this condition restores essential self-adjointness (on a restricted domain) for $118 < Z \leq 137$.

In the case of a finite nucleus, the two leading exponents reduce to $l + 1$ and $-l$ respectively. When $l > 0$, it is clear that we have the limit point case and square integrability selects the exponent $l + 1$. However when $l = 0$, the second solution is bounded as $r \to 0$ and so we have the limit circle case again. We now have to restrict the domain of the operator to solutions for which (5.4.11) holds.

It is seldom remarked that the same thing happens with the Schrödinger equation; for example Pauling and Wilson [8, p. 122] simply state that the negative exponent "does not lead to an acceptable wavefunction". This is entirely correct, although neither Pauling and Wilson nor, as far as I am aware, the authors of other books on elementary quantum mechanics note that it is necessary to impose the condition (5.4.11) when $l = 0$, just as in the Dirac case. In contrast, Dirac [9, §38] argues that the second solution for $l = 0$ would not be acceptable because it would imply the presence of an additional delta function singularity on the right-hand side of the Schrödinger equation.

5.5 Variational methods in quantum mechanics

Variational methods are widely used in the quantum mechanics of atoms and molecules. The Ritz method seems to have been applied first to the ground state of helium by Kellner [10]; the work of Hylleraas [11, 12, 13] on the ionization potentials of the helium isoelectronic series, in particular his use of inter-particle coordinates, seems to be much better known. Most of the standard derivations of nonrelativistic [7, 14, 15] and of relativistic atomic self-consistent field theory [16, 17, 18] have depended on variational principles, as have many of the methods employed in scattering theory [19].

Most expositions of variational methods in nonrelativistic quantum mechanics make much of the fact that the Schrödinger operator is semi-bounded. Although Swirles [16] wrote down relativistic atomic Hartree-Fock equations using a variational principle, she carefully noted that the Dirac Hamiltonian was not semi-bounded so that her approach was "provisional". Nevertheless, the equations she derived were mathematically consistent and no unexpected problems were encountered in constructing numerical solutions using the finite difference methods developed by Hartree and his colleagues [14]. The provisional approach proved well-founded, because it relied only upon relations which make the energy functional stationary without reference to any lower bound.

Ritz methods, in which the orbitals are approximated by a linear combination of suitable functions, usually referred to as *basis sets*, provide the only practical route for solving quantum mechanical problems for molecules and solids. Roothaan's students Synek [20] and Kim [21, 22] in the mid-1960s were the first to devise relativistic self-consistent field equations for atoms in this way. Whilst Kim was able to generate a plausible solution for the ground state of the beryllium atom, he encountered some numerical instabilities that were not understood at the time. The method was not competitive with finite difference methods on the computers of the day, and although further work on the atomic problem was done by Kagawa and others [23, 24] some 10 years later, progress was slow. Real trouble was first encountered when quantum chemists tried to apply the customary nonrelativistic Ritz procedures to the full Dirac equation as well as to the radially reduced equation. Thus Schwarz and Wallmeier [25] and Wallmeier and Kutzelnigg [26] obtained unacceptable results even for the simplest case, the hydrogen atom. Spurious low-energy intruder states appeared along with solutions resembling the physical low-lying states. The pathological behaviour, discussed also in §4.13.3, was attributed to "variational collapse" [25] or "finite basis set disease" [27]. The results were sensitive to basis set dimension and failed to stabilize in the manner of nonrelativistic calculations as the dimension of the basis set was increased systematically [26, 28, 29]. In the absence of any consensus on remedies, attention switched by 1984 to semirelativistic approximations [30]. We now understand the importance of using spinor basis functions for Dirac calculations which incorporate the limiting behaviour of §5.4.1 and §5.4.3 at the singular endpoints of the Dirac radial operator. It is necessary for the basis functions to embody the correct connections between the radial components in the region in which relativistic motion is most probable.

To make maximal use of the Rayleigh-Ritz method in relativistic, as well as nonrelativistic quantum mechanics, we need criteria for convergence of energy eigenvalues, wavefunctions and expectation values for ground and excited states and also for transition amplitudes. We seek ways to characterize Rayleigh-Ritz approximations that work for both nonrelativistic and relativistic quantum mechanics.

5.5.1 Min-max theorems and the Ritz method

Let T be a self-adjoint operator with a domain $\mathcal{D}$ that is dense in the Hilbert space $\mathcal{H}$. Assume, for simplicity, that T has a simple point spectrum, ordered so that $-\infty < t_0 \leq t_1 \leq t_2 \leq \ldots$, with corresponding eigenfunctions $u_0, u_1, u_2, \ldots$ Let $\mathcal{M}_n = \{u_0, u_1, \ldots, u_{n-1}\}$ be the linear span of the eigenfunctions corresponding to the n lowest eigenvalues of T. Using Dirac bra-ket notation, we define the Rayleigh quotient

$$R(u) := \langle u \,|\, T \,|\, u \rangle \,/\, \langle u \,|\, u \rangle \tag{5.5.1}$$

for any nontrivial function $u \in \mathcal{D}$.

Theorem 5.5. *Let $\mathcal{M}_n^{\perp}$ be the orthogonal complement of $\mathcal{M}_n$ in $\mathcal{D}$. Then*

$$t_n = \min_{u \in \mathcal{M}_n^{\perp}} R(u).$$

Proof: For all $u \in \mathcal{D}$ we have

$$\langle u \,|\, T \,|\, u \rangle = \sum_k t_k |\langle u \,|\, u_k \rangle|^2.$$

If $u \in \mathcal{M}_n^{\perp}$, $\langle u \,|\, u_k \rangle = 0, \; k = 0, 1, \ldots, n-1$, so that

$$\langle u \,|\, T \,|\, u \rangle = \sum_{k \geq n} t_k |\langle u \,|\, u_k \rangle|^2 \geq t_n \sum_{k \geq n} |\langle u \,|\, u_k \rangle|^2 \geq t_n \langle u \,|\, u \rangle.$$

So $R(u) \geq t_n$. Since $u_n \in \mathcal{M}_n^{\perp}$, $R(u)$ attains its lower bound at $u = u_n$, which proves the theorem. ■

Theorem 5.5 is familiar from elementary quantum mechanics textbooks. It assumes throughout that the first n *exact* eigenfunctions are known, but this is rarely the case in practice. What we usually need is a set of functions, say $\mathcal{W}_n := \{w_0, w_1, \ldots, w_{n-1}\}$ which in some sense approximates the manifold $\mathcal{M}_n$.

Theorem 5.6 (Weyl-Courant). *Let*

$$\theta(\mathcal{W}_n) = \min_{u \in \mathcal{W}_n^{\perp}} R(u);$$

then

$$t_n = \max_{\forall \mathcal{W}_n} \theta(\mathcal{W}_n) = \max_{\mathcal{W}_n \in \mathcal{S}_n} \min_{w \in \mathcal{W}_n^\perp} R(w),$$

where $\mathcal{S}_n$ is the collection of all n-dimensional linear manifolds in $\mathcal{D}$.

Proof: From the previous theorem, $\theta(\mathcal{M}_n) = t_n$ so that $\max \theta(\mathcal{W}_n) \geq t_n$. The theorem is therefore proved if we can exhibit, for each choice of $\mathcal{W}_n$, an element $w \in \mathcal{W}_n^\perp$ such that $R(w) \leq t_n$. Suppose that

$$w = c_0 u_0 + c_1 u_1 + \ldots + c_n u_n,$$

where the coefficients $c_0, c_1, \ldots, c_n$ must be chosen so that, with respect to a basis $\{w_0, w_1, \ldots, w_{n-1}\}$ of $\mathcal{W}_n$, w is non-null and $(w, w_0) = (w, w_1) = \ldots, (w, w_{n-1}) = 0$. This is always possible, as we have to find $n+1$ unknown coefficients $c_0, c_1, \ldots, c_n$ subject only to n linear equations of constraint. If the eigenfunctions $\{u_0, u_1, \ldots, u_n\}$ are orthonormal, then

$$R(w) = \frac{\langle w \,|\, T \,|\, w \rangle}{\langle w \,|\, w \rangle} = \frac{\sum_{k=0}^{n} t_k |\, c_k|^2}{\sum_{k=0}^{n} |\, c_k|^2} \leq t_n$$

which completes the proof. ■

We usually require the set $\{w_k\}$ to be linearly independent and to be normalized. We shall assume, for the purposes of this section, that the set is also mutually orthogonal, although we shall later want to lift this restriction. So now we can write each $w \in \mathcal{W}_n$ in the form

$$w = c_0 w_0 + c_1 w_1 + \ldots + c_{n-1} w_{n-1};$$

so that

$$R(w) = \sum_{i,j=0}^{n-1} c_i^* T_{ij} c_j \Bigg/ \sum_{k=0}^{n-1} |c_k|^2$$

where the simple form of the denominator is a consequence of the assumed orthonormality, and where

$$T_{ij} = \langle w_i \,|\, T \,|\, w_j \rangle, \quad i, j = 0, 1, \ldots, n-1.$$

The effect of restricting w to the linear manifold $\mathcal{W}_n$ is therefore to replace the operator T, which is in general unbounded, by a bounded operator T_n on $\mathcal{W}_n$ defined by

$$T_n = P_n T P_n$$

where P_n is the orthogonal projector onto $\mathcal{W}_n$. The operator T_n is represented by the square matrix $\mathbf{T}_n$ whose elements are the numbers T_{ij} defined above. It is usually convenient to choose the basis so that the matrix $\mathbf{T}_n$ is real symmetric; its n eigenvalues $\tau_i^{(n)}, i = 0, 1, \ldots, n-1$ can be characterized exactly as in Theorem 5.5, with $\mathcal{W}_n$ replacing $\mathcal{M}_n$. We order the eigenvalues so that $\tau_0^{(n)} \leq \tau_1^{(n)} \leq \ldots \leq \tau_{n-1}^{(n)}$.

Theorem 5.7 (Poincaré). *For each integer $n > i$,*

$$\tau_i^{(n)} \geq t_i, \quad i = 0, 1, 2, \ldots$$

Proof: We prove $\tau_i^{(n)} \geq t_i$ for some fixed value of i. Pick a trial element from $\mathcal{W}_n$ in the form

$$w = d_0 w_0 + d_1 w_1 + \ldots + d_i w_i$$

such that w is non-null and orthogonal to $w_0, w_1, \ldots, w_{i-1}$. We can always find such an element since we have only to solve i linear equations for $i+1$ unknown coefficients $d_0, d_1, \ldots, d_i$. Theorem 5.5 gives us $R(w) \geq t_i$. However,

$$R(w) = \frac{\sum_{k=0}^{i} \tau_k^{(n)} |d_k|^2}{\sum_{k=0}^{i} |d_k|^2} \leq \tau_i^{(n)}$$

establishing the result. ■

This theorem has a useful corollary which enables us to characterize approximations to the eigenvalues of T without requiring a knowledge of the eigenvalues of lower index and without having to search for an extremum over the infinite subspace $\mathcal{W}_n^{\perp}$.

Theorem 5.8 (Poincaré). *Let $\mathcal{S}_n$ be the set of all n-dimensional linear manifolds of $\mathcal{D}$ and let $\mathcal{W}_n$ be a particular member of $\mathcal{S}_n$. Define*

$$\theta(\mathcal{W}_n) = \max_{w \in \mathcal{W}_n} R(w);$$

then

$$t_{n-1} = \min_{\mathcal{W}_n \in \mathcal{S}_n} \theta(\mathcal{W}_n).$$

Proof: We follow the same method as in the last theorem. This establishes that $\theta(\mathcal{W}_n)$ is the maximum eigenvalue of the restriction of T to $\mathcal{W}_n$, and $\theta(\mathcal{W}_n) \leq t_{n-1}$ for all $\mathcal{W}_n \in \mathcal{S}_n$. We observe that $\theta(\mathcal{M}_n) = t_{n-1}$ which proves the theorem. Theorem 5.8 gives the theoretical justification for the Galerkin method, in which we seek approximate solutions of the equation

$$(T - \tau)w = 0, \quad w \in \mathcal{W}_n$$

by requiring that the residual

$$\delta := (T - \tau)w$$

is orthogonal to each of the eigenfunctions w_i of the restriction of T to $\mathcal{W}_n$:

$$(\delta, w_i) = 0, \quad i = 0, 1, \ldots, n-1.$$

If $w \in \mathcal{W}_n$, we can write

$$w = c_0 w_0 + c_1 w_1 + \ldots + c_{n-1} w_{n-1}$$

from which the condition

$$\sum_{j=0}^{n-1} (T_{ij} - \tau\delta_{ij}) c_j = 0, \quad i = 0, 1, \ldots, n-1 \tag{5.5.2}$$

follows. Thus τ must be one of the eigenvalues of the matrix operator $\mathbf{T}_n$. ■

5.5.2 Convergence of the Rayleigh-Ritz eigenvalues in nonrelativistic quantum mechanics

The theorems of the last section tell us that we can approximate the eigenvalues and eigenfunctions of operators that are bounded below by diagonalizing finite matrices on suitable linear manifolds in the Hilbert space $\mathcal{H}$. We actually need rather more: it is essential to be able to use these approximations to estimate physical observables. This requires a deeper analysis of the convergence of Rayleigh-Ritz approximations in quantum mechanics. Klahn and Bingel [31, 32] first studied the convergence of basis set approximations of eigenvalues and eigenvectors, and Klahn and Morgan [33] extended this analysis to the convergence of expectation values and transition amplitudes of common quantum mechanical operators. The theory is sufficiently complete to provide some foundation for practical calculations in both nonrelativistic and relativistic quantum mechanics.

We consider functions defined on a suitable L^2 space $\mathcal{H}$ with inner product (u, v) and corresponding norm $\|u\| = (u, u)^{1/2}$. Lebesgue measure is implicit; Klahn and Bingel [31, 32] use that of $3N$-dimensional Euclidean space, but their results are not restricted to this choice. Suppose that B is a *positive-definite* and self-adjoint linear operator with domain $\mathcal{D}_B$, dense in $\mathcal{H}$. Then there exists a constant $\beta > 0$ such that

$$(u, Bu) \geq \beta(u, u), \quad \forall\, u \in \mathcal{D}_B.$$

Assume now that the lower part of the spectrum of B contains a finite or infinite number of isolated eigenvalues. Suppose we pick a basis set, $\Phi := \{\phi_m, m = 1, 2, \ldots\}$, which is complete in $\mathcal{H}$. As Φ is complete, we can choose a linear combination of the functions ϕ_m which approximates any element $\psi \in \mathcal{H}$ as closely as we please. Is this enough for *E-convergence*: can we guarantee that Rayleigh-Ritz approximations to eigenvalues and eigenstates converge as we enlarge the linear manifold of basis functions? Mikhlin [34] was able to prove E-convergence if the set Φ is complete in the *energy space* $\mathcal{H}_B$, defined as the closure of $\mathcal{D}_B$ with respect to the *B-norm*

$$\|u\|_B = (u, Bu)^{1/2}.$$

Bonitz [35] then extended this to excited states.

Theorem 5.9 (Bonitz). *Let the set Φ be complete in $\mathcal{H}_B$. Then the Rayleigh-Ritz method for the positive-definite operator B using trial functions of the form*

$$w^{(M)} = \sum_{m=1}^{M} c_m \phi_m$$

is equivalent to a matrix eigenvalue problem of the form (5.5.2). The eigenvalues $b_i^{(M)}$ of the matrix $\mathbf{B}_M$ converge as $M \to \infty$ to the exact eigenvalues $b_0, b_1, \ldots b_I$, where b_I is the lowest exact (isolated) degenerate eigenvalue of B.

We refer the reader to the original papers for the full proof. It rests on the expression

$$(w^{(M)}, Bw^{(M)}) - b_i \leq 2b_i\beta^{-1/2}\|w^{(M)} - w_i\|_B + \|w^{(M)} - w_i\|_B^2$$

for the difference between the estimate $(w^{(M)}, Bw^{(M)})$ and the exact eigenvalue b_i. This inequality is valid, in particular, for $w^{(M)} = w_i^{(M)}$, where $w_i^{(M)}$ is the i-th eigenvector of $\mathbf{B}_M$. The Ritz eigenvalues are characterized by the Weyl-Courant theorem 5.6 as

$$b_i^{(M)} = \min_{w^{(M)} \in \mathcal{D}_i^{(M)}} \frac{(w^{(M)}, Bw^{(M)})}{(w^{(M)}, w^{(M)})} \tag{5.5.3}$$

where

$$\mathcal{D}_i^{(M)} = \{w \in \mathcal{D}^{(M)} \,|\, (w, w_j^{(M)}) = 0,\, j < i\}.$$

and $\mathcal{D}^{(M)} \subset \Phi$ is the linear span of $\phi_0, \ldots, \phi_{M-1}$. Thus $b_0^{(M)}$ is the overall minimum of (5.5.3) on $\mathcal{D}^{(M)}$. If the Ritz eigenfunction $w_0^{(M)}$ does not converge to the ground state, then by (5.5.3), no sequence $w^{(M)} \in \mathcal{D}^{(M)}$ can be found such that $b_0^{(M)}$ converges to b_0, contradicting the assumed completeness of the set Φ in $\mathcal{H}_B$.

More specifically, the nonrelativistic molecular Hamiltonians describing atoms, molecules and solids have the form

$$H = T + V,$$

where

$$T = \frac{1}{2}\mathbf{p}^2 = \frac{1}{2}\sum_{k=1}^{N} \mathbf{p}_k^2.$$

with domain

$$\mathcal{D}_T = \{u(\mathbf{r}) \,|\, (1 + \mathbf{p}^2)\widehat{u}(\mathbf{p}) \in L^2\}$$

and $\widehat{u}(\mathbf{p})$ is the Fourier transform of $u(\mathbf{r})$, whilst

$$V = -\sum_A Z_A \sum_{i=1}^N \frac{1}{r_{iA}} + \sum_{i<j} \frac{1}{r_{ij}},$$

where Z_A is the charge on nucleus A, r_{iA} is the distance between A and electron i and r_{ij} is the separation of electrons i and j. It can be shown that V is T-bounded so that

$$\|Vu\| \le a\|Tu\| + b\|u\|, u \in \mathcal{D}_T, \tag{5.5.4}$$

where a can be taken arbitrarily small. It follows from Theorem 5.1 that H is self-adjoint with domain $\mathcal{D}_T$ and that all eigenfunctions of H are elements of $\mathcal{D}_T$. Using (5.2.2) we can deduce that V is also relatively bounded by T in the *quadratic form* sense,

$$|(u, Vu)| \le a'\|T^{1/2}u\| + b'\|u\|, \tag{5.5.5}$$

where the constants a' and $b' > 0$ are positive, and a' can be made arbitrarily small and $T^{1/2}$ is the square root of the positive-definite operator T, with domain

$$\mathcal{D}_{T^{1/2}} = \{u(\mathbf{r}) \,|\, (1 + |\mathbf{p}|)\hat{u}(\mathbf{p}) \in L^2\}$$

Since $T^{1/2}$ is a maximal multiplication operator, it is self-adjoint on $\mathcal{D}_{T^{1/2}}$ as is the operator $(1 + T)^{1/2}$ on the same domain. The condition (5.5.4) is more restrictive in this respect than (5.5.5) as the latter can still be valid for potentials that violate (5.5.4).

The Hamiltonians, H, are not positive definite although they are always bounded below. However, the operator

$$B = c + H$$

is positive definite for some positive values of c, and $\mathcal{D}_B = \mathcal{D}_H$. Now choose a basis set Φ complete in the energy space $\mathcal{H}_B$ with norm $\|u\|_B = (u, Bu)^{1/2}$. Then Theorem 5.9 assures us that the Rayleigh-Ritz procedure based on Φ converges.

Criteria for convergence in suitably defined Hilbert spaces of this sort which do not depend on the details of the potential V have been devised by Mikhlin, Kato and others. This is straightforward when V satisfies the condition (5.5.5): define a new energy space $\mathcal{H}_A$ that is the closure of $\mathcal{D}_T$ with respect to the norm $\|u\|_A = (u, Au)^{1/2}$ with $A = c + T$. The condition (5.5.5) is sufficient to make the A-norm and B-norm equivalent, in the sense that we can find positive constants k_1, k_2 such that

$$k_1\|u\|_A \le \|u\|_B \le k_2\|u\|_A, \forall\, u \in \mathcal{D}_T.$$

This leads to

Theorem 5.10 (Mikhlin's criterion). *Let the basis set Φ be complete in $\mathcal{H}_A$. Let H have eigenvalues $E_0 < E_1 < \ldots < E_I < \ldots$, where E_I is the first exact degenerate eigenvalue. Then the Rayleigh-Ritz method with respect to the basis set Φ converges.*

The proof follows the same lines as Theorem 5.9, being based on an estimate of the form

$$(w^{(M)}, Hw^{(M)}) - E_i \leq 2|E_i|c^{-1/2}\|w^{(M)} - w_i\|_A \\ +(2 + 2a' + b'c^{-1})\|w^{(M)} - w_i\|_A^2.$$

The conclusion does not depend upon the precise value of the constant c because T is itself positive definite.

The energy space $\mathcal{H}_A$ can also be defined as the closure of $\mathcal{D}_T$ with respect to the A'-norm $\|u\|_{A'} = (u, (c+T^{1/2})^2 u)^{1/2}$. This choice enables us to identify the domain as the Sobolev space $W_2^{(1)}$ consisting of all L^2 functions whose (generalized) first derivatives are also in L^2.

It is usually more convenient to express Mikhlin's criterion as a condition on sets of functions in L^2 rather than as completeness in $\mathcal{H}_A$. This can be done using the following lemma [36]:

Lemma 5.11. *Let L^2_ρ be the Hilbert space of all quadratically integrable functions with positive-definite weight function ρ. Then Φ is complete in L^2_ρ if and only if $\{\rho^{1/2}\phi_m\}$ is complete in L^2.*

This makes $\mathcal{H}_A$ an L^2_ρ space in the momentum representation with $\rho = c + T$ if we use closure with respect to the A-norm or $\rho = c + T^{1/2}$ if we use closure with respect to the A'-norm. With this interpretation, we can write the last theorem in the form

Theorem 5.12. *The set $\Phi = \{\phi_m\}$ is complete in $\mathcal{H}_A$ if and only if either $\{(c+T)^{1/2}\phi_m\}$ or $\{(c+T^{1/2})\phi_m\}$ with $c > 0$ is complete in L^2.*

This sort of consideration leads to

Theorem 5.13 (Kato's criterion). *Let the system $\{(c+T)\phi_m\}$, $c > 0$ be complete in L^2. Then the Rayleigh-Ritz method for the eigenvalues of H based on the set $\Phi = \{\phi_m\}$ converges for all states to the exact eigenvalues of H.*

The set $\{(c+T)\phi_m\}$, $c > 0$ is complete in L^2 if and only if $\Phi = \{\phi_m\}$ is complete in $\mathcal{H}_{A^2}$, defined as the closure of $\mathcal{D}_T$ with respect to the A^2-norm $\|u\|_{A^2} = (u, A^2 u)^{1/2} = \|(c+T)u\|$.

The first part of the theorem is based on a characterization of the Ritz eigenvalues using H itself rather than B, using an inequality

$$(w^{(M)}, Hw^{(M)}) - E_i \leq \left[\frac{a + |E_i|}{c} + b + 1\right] \|w^{(M)} - w_i\|_{A^2}.$$

The second part involves recognizing that $\mathcal{H}_{A^2}$ is an L^2_ρ space in the momentum representation with $\rho = (c+T)^2$ and $\mathcal{D}_T = \mathcal{H}_{A^2} = W_2^{(2)}$, where the Sobolev space $W_2^{(2)}$ consists of all L^2 functions whose (generalized) second derivatives are also in L^2.

The different convergence criteria are, of course, inequivalent though related. The spaces involved satisfy

$$L^2 \supset \mathcal{H}_A \supset \mathcal{H}_{A^2}$$

and we have

Theorem 5.14. *Let $\Phi = \{\phi_m\}$ be complete in $\mathcal{H}_{A^2}$: then Φ is complete in $\mathcal{H}_A$. Also let $\Phi = \{\phi_m\}$ be complete in $\mathcal{H}_A$: then Φ is complete in L^2.*

Note that assertions of the converses are false: there exist sequences which are complete in L^2 but which are not complete in $\mathcal{H}_A$, and so on.

5.5.3 Convergence of the Rayleigh-Ritz method in nonrelativistic quantum mechanics

So far we have criteria for convergence of Rayleigh-Ritz eigenvalues to the exact eignvalues of nonrelativistic Hamiltonians. In practice, we need to work within a finite M-dimensional subspace, and we should like to know how close our eigenvalue estimates are to the exact values, in what sense approximate eigenfunctions $w_i^{(M)} = P_M u_i$ defined on this subspace are close to the true eigenfunctions u_i, and whether the approximate expectation values and transition matrix elements derived from them are sufficiently close to the true values to be of practical use.

Rayleigh-Ritz eigenfunctions converge in the mean to the exact eigenfunction whenever the Ritz eigenvalue converges to the exact eigenvalue. This follows from the inequalities

$$\|w_0^{(M)} - u_0\|^2 \leq 2\frac{E_0^{(M)} - E_0}{E_1 - E_0}$$

[37] and

$$\|w_i^{(M)} - u_i\|^2 \leq 2\frac{E_i^{(M)} - E_i}{E_{i+1} - E_i}\left\{1 + G_i^2 \sum_{j=0}^{i-1}(E_{i+1} - E_j)\right\}$$

[38] where

$$G_i = \sum_{j=0}^{i-1}(E_i - E_j^{(M)})^{-1/2}.$$

Convergence of the wavefunction in the A-norm (Ψ-convergence) follows from

$$\|w_i^{(M)} - u_i\|_A^2 \leq (E_i^{(M)} - E_i)[(1 - a')^{-1} + 2(E_{i+1} - E_i)^{-1} H_i]$$

where

$$H_i = [c + (b' + |E_i|)(1 - a')^{-1}] \left\{ 1 + G_i^2 \sum_{j=0}^{i-1} (E_{i+1} - E_j) \right\}.$$

Thus Ψ-convergence (in the A-norm) is a necessary condition for E-convergence, and we have

Theorem 5.15. *A necessary and sufficient condition for the Rayleigh-Ritz method to converge to the lowest N exact non-degenerate eigenvalues is that the lowest N exact eigenfunctions of the Hamiltonian can be approximated in the A-norm with any desired accuracy in terms of the basis.*

This is about as far as we can go to characterize the convergence of eigenvalues and eigenfunctions. It seems to be impossible to make definite statements about Rayleigh-Ritz convergence in the A^2 norm, even when the method is E-convergent. We might hope to say something about convergence of the Schrödinger equation in the mean, that is to say of the mean square residual

$$\|(H - E_i^{(M)})u_i^{(M)}\|^2,$$

when the Rayleigh-Ritz method is E-convergent, but this seems not to be the case. It is known however that, when the Rayleigh-Ritz method converges, the mean square residual and $\|w_i^{(M)} - u_i\|_{A^2}^2$ converge or diverge together.

The outcome of this lengthy investigation, according to Klahn and Bingel [31, 32], can be summarized quite simply. As usual, we build many-electron wavefunctions from products of one-body wavefunctions for which the space A can be identified with the Sobolev space $W_2^{(1)}(\mathbb{R}^3) \subset L^2(\mathbb{R}^3)$ equipped with the norm

$$\|\psi\|_{W_2^{(1)}}^2 = \langle u|\, 1 + T\,|u\rangle = \|(1 + T)^{1/2}u\|_{L^2}^2. \tag{5.5.6}$$

Thus we have only to construct an M-dimensional basis set $\mathcal{W}^{(M)}$ that, as $M \to \infty$, is complete in $W_2^{(1)}(\mathbb{R}^3)$ to be certain of E-convergence both to eigenvalues and eigenfunctions of the Hamiltonian.

This analysis has been extended by Klahn and Morgan [33] to convergence of expectation values and transition matrix elements. They rely on the

Lemma 5.16. *For any trial function $w^{(M)} \in \mathcal{W}^{(M)}$, the sequence*

$$\langle A\rangle^{(M)} := \langle w^{(M)}|\, A\,|\, w^{(M)}\rangle$$

converges to $\langle A\rangle = \langle u|\, A\,|u\rangle$ if and only if $w^{(M)} \to u$ as $M \to \infty$ in the A-norm.

This is a direct consequence of two inequalities

$$|\langle A\rangle^{(M)} - \langle A\rangle| \le \|w^{(M)} - u\|_A^2 + 2\langle A\rangle^{1/2}\|w^{(M)} - u\|_A$$

and

$$\|w^{(M)} - u\|_A^2 \leq |\langle A\rangle^{(M)} - \langle A\rangle| + 2\|Au\| \cdot \|w^{(M)} - u\|.$$

Thus the set $\mathcal{W}$ spanning $W_2^{(1)}$ must be A-complete for the sequence of Rayleigh-Ritz eigenfunctions $\{w^{(M)}\}$ also to give a convergent sequence of approximations $\langle A\rangle^{(M)}$.

We can avoid having to deal with A-completeness if the operator A is *relatively form-bounded* by T: that is, there exists a pair of non-negative numbers a, b such that

$$|\langle u|\, A\,|u\rangle| < a\langle u\,|\,u\rangle + b\langle u|\,T\,|u\rangle, \quad \forall\, u \in \mathcal{D}(T). \tag{5.5.7}$$

This includes a wide range of operators: *bounded* operators, for which we can set $b = 0$; Coulomb potentials; T itself (with $a = 0$ and $b = 1$); components of the momentum operator $\boldsymbol{p}$; and nonrelativistic atomic and molecular Hamiltonians, H. Clearly, T can be relatively form bounded by $H + c$, where $c > 0$ is chosen large enough that $H + c$ has a purely positive spectrum. Then if $\langle u|\,A\,|u\rangle$ satisfies (5.5.7), we choose $c > 0$ so that $T + c$ is strictly positive, and the sequence $w^{(M)}$ is E-convergent to ψ in the $T + c$ norm, then

$$\|w^{(M)} - u\|_A^2 \leq \max(a/c, b)\|w^{(M)} - u\|_{T+c}^2$$

so that $\{w^{(M)}\}$ is also A-convergent to u. It is sufficient for $\mathcal{W}$ to be complete in the Sobolev space $\mathcal{W}_2^{(1)}$. Finally, it is straightforward to show that transition matrix elements of the form $\langle w_i^{(M)}|\,A\,|w_j^{(M)}\rangle$ converge to the desired limit $\langle u_i|\,A\,|u_j\rangle$ as $M \to \infty$ provided the sequences $\{w_i^{(M)}\}$ and $\{w_j^{(M)}\}$ are also A-convergent.

5.6 The Rayleigh-Ritz method in relativistic quantum mechanics

The Rayleigh-Ritz method can be applied to relativistic problems in atomic and molecular physics along similar lines by replacing nonrelativistic single particle functions by Dirac 4-spinors. The procedures of the last section are not applicable to such problems as they stand because the Dirac atomic Hamiltonian has no global lower bound. Following [39, 40, 41] we shall show that the nonrelativistic theory can be adapted for relativistic problems so that similar computational strategies can be used. We begin with the Galerkin equations for the one-electron problem, after which the way is open for applications to the structure of atoms, molecules and solids .

5.6.1 The finite matrix method for the Dirac equation

The finite matrix method approximates Dirac spinors in a problem with many atomic nuclear centres by writing down a trial solution

$$\psi_a(\mathbf{x}) = \begin{bmatrix} \sum_{\mu=1}^{N} c_{\mu a}^{+1} M[+1,\mu,\mathbf{x}] \\ i\sum_{\mu=1}^{N} c_{\mu a}^{-1} M[-1,\mu,\mathbf{x}] \end{bmatrix}, \tag{5.6.1}$$

where a labels each atomic or molecular spinor and $\beta = \pm 1$ labels the upper and lower components respectively. It is often sufficient just to use the sign of β as a label, and we shall do so whenever this causes no ambiguity.[1] The expansion coefficients $c_{\mu a}^{\beta}$ are in general complex numbers. The form of the 2-component functions $M[\beta,\mu,\mathbf{x}]$ will be discussed in more detail later. The multi-index μ completely specifies each basis spinor; in general it will have the form

$$\mu := \{\mathbf{A}, \kappa, j, m, \ldots\}$$

where $\mathbf{A}$ labels the atomic centre taken as the origin for $\mathbf{x}$ and the dots signify other parameters which may be required. When we wish to focus on one parameter belonging to a particular multi-index, we shall write $\mathbf{A}_\mu$, or j_μ, etc. For the time being, we shall consider just a single atomic centre, so that the label $\mathbf{A}$ is redundant. We write the Dirac Hamiltonian for an electron in the field of a static potential energy function $U(\boldsymbol{x})$, $\boldsymbol{x} \in \mathbb{R}^3$, in the usual form

$$\widehat{h}_D = c\,\boldsymbol{\alpha}\cdot\boldsymbol{p} + \beta\, mc^2 + U(\boldsymbol{x}). \tag{5.6.2}$$

The Rayleigh-Ritz method based on the trial function of (5.6.1) generates a $2N$-dimensional set of approximate eigenvalues and eigensolutions of the Dirac equation. When partitioned into 2×2 blocks in comformity with the spinor structure of (5.6.1), the Dirac Hamiltonian (5.6.2) becomes

$$\widehat{h}_D = c\boldsymbol{\alpha}\cdot\mathbf{p} + \beta mc^2 + U(\mathbf{x}) = \begin{pmatrix} mc^2 + U(\boldsymbol{x}) & c\,\boldsymbol{\sigma}\cdot\mathbf{p} \\ c\,\boldsymbol{\sigma}\cdot\mathbf{p} & -mc^2 + U(\boldsymbol{x}) \end{pmatrix}, \tag{5.6.3}$$

where a 2×2 identity matrix multiplying the scalar entries on the diagonal has been left implicit. The expectation of this operator with respect to the trial function (5.6.1) is a Hermitian form in the expansion coefficients

$$\mathbf{c} = \begin{pmatrix} \mathbf{c}^{+1} \\ \mathbf{c}^{-1} \end{pmatrix}, \tag{5.6.4}$$

where the $\mathbf{c}^{\beta}$, $\beta = \pm 1$ are N-rowed column vectors. The Rayleigh quotient corresponding to (5.5.1) can then be written

$$R(\mathbf{c}) = \mathbf{c}^{\dagger}\mathbf{H}\,\mathbf{c}\,/\,\mathbf{c}^{\dagger}\mathbf{S}\,\mathbf{c} \tag{5.6.5}$$

where $\mathbf{H}$ and $\mathbf{S}$ are $2N\times 2N$ Hermitean matrices which we partition into $N\times N$ blocks so that

---

[1] The labels $T = L$ for $\beta = +1$ and and $T = S$ for $\beta = -1$ were used in the original papers.

$$\mathbf{H} = \begin{bmatrix} mc^2\mathbf{S}^{++} + \mathbf{U}^{++} & c\boldsymbol{\Pi}^{+-} \\ c\boldsymbol{\Pi}^{-+} & -mc^2\mathbf{S}^{--} + \mathbf{U}^{--} \end{bmatrix},$$
$$\mathbf{S} = \begin{bmatrix} \mathbf{S}^{++} & 0 \\ 0 & \mathbf{S}^{--} \end{bmatrix}. \tag{5.6.6}$$

The Galerkin equations that result from (5.6.5) can then be written

$$\begin{bmatrix} (mc^2 - E)\mathbf{S}^{++} + \mathbf{U}^{++} & c\boldsymbol{\Pi}^{+-} \\ c\boldsymbol{\Pi}^{-+} & -(mc^2 + E)\mathbf{S}^{--} + \mathbf{U}^{--} \end{bmatrix} \begin{bmatrix} \mathbf{c}^+ \\ \mathbf{c}^- \end{bmatrix} = 0, \tag{5.6.7}$$

where E is one of the eigenvalues of this $2N$-dimensional algebraic system.

The original mathematical problem has now been approximated by an algebraic system which, we hope, exhibits many of the properties of the original. We discussed the convergence of Rayleigh-Ritz calculations in nonrelativistic quantum mechanics in the last section, and it is clear that the choice of 2-component basis spinor families $M[\beta, \mu, \mathbf{x}]$, which generate the matrix blocks of (5.6.7), needs some care. One consideration will be the ease with which we can generate the various matrices, especially when they involve basis spinors centred on different atomic nuclei. In the present case, the *Gram* (or *overlap*) matrices $\mathbf{S}^{\beta\beta'}$ have elements

$$S_{\mu\nu}^{\beta\beta'} = \delta_{\beta\beta'} \int M[\beta, \mu, \mathbf{x}]^\dagger \, M[\beta', \nu, \mathbf{x}] \, d\mathbf{x}, \tag{5.6.8}$$

and similarly the atomic mean field potential matrices $\mathbf{U}^{\beta\beta'}$ have elements

$$U_{\mu\nu}^{\beta\beta'} = \delta_{\beta\beta'} \int M[\beta, \mu, \mathbf{x}]^\dagger \, U(r) \, M[\beta', \nu, \mathbf{x}] \, d\mathbf{x}. \tag{5.6.9}$$

The kinetic matrices $\boldsymbol{\Pi}^{\beta\beta'} = (\boldsymbol{\Pi}^{\beta'\beta})^\dagger$ have elements

$$\boldsymbol{\Pi}_{\mu\nu}^{\beta\beta'} := \delta_{\beta',-\beta} \int M[\beta, \mu, \mathbf{x}]^\dagger \, \boldsymbol{\sigma} \cdot \mathbf{p} \, M[-\beta, \nu, \mathbf{x}] \, d\mathbf{x}, \tag{5.6.10}$$

However there is more at stake than computational convenience, and we shall see that it is essential to incorporate specific relations between the basis spinor components to make the method successful.

5.6.2 Convergence of Rayleigh-Ritz methods for Dirac Hamiltonians

The existence of a lower bound for nonrelativistic Hamiltonians of atomic and molecular physics plays a crucial role in Section 5.5. In particular, the eigenvalues $\tau_i^{(N)}, i = 1, 2, \ldots, N$ of the Galerkin equations (5.5.2) derived from a self-adjoint, non-negative compact operator T are, according to Theorem 5.7, upper bounds to the corresponding exact eigenvalues, t_i, of T itself:

$$0 < t_1 \leq \tau_1^{(N)} \leq \ldots \leq t_N \leq \tau_N^{(N)}. \tag{5.6.11}$$

However, this is not quite the end of the story, and Stakgold [42] points out a number of corollaries, in particular

1. If T is *non-positive* then -T is non-negative and (5.6.11) holds with the inequalities reversed.
2. If T is *indefinite* then (5.6.11) holds for the positive eigenvalues and also, with the inequalities reversed, for the negative ones.

When these results are combined with the next theorem, we have sufficient to extend the Raleigh-Ritz method to Dirac operators:

Theorem 5.17. *Let $\widehat{h}_D$ be the Dirac operator of (5.6.2) with potential function $U(\boldsymbol{x}) < 0$, and suppose that $0 > (\psi \,|\, U \,|\, \psi)/(\psi \,|\, \psi) > U_{min} > -2mc^2$ for all trial functions ψ in the dense domain, $\mathcal{D}(\widehat{h}_D)$, in the Hilbert space $\mathcal{H}$. Assume, as is usually the case, that the spectrum of $\widehat{h}_D$ consists of a point spectrum $E_1 < E_2 < \ldots$ in the interval $(-mc^2, +mc^2)$ with a point of accumulation at $+mc^2$, and a continuous spectrum with one branch $E > mc^2$ and the other in $E < -mc^2$. Then there exists a lower bound $\underline{E}$ to the point spectrum with $-mc^2 < \underline{E} < E_1$, and an upper bound $\overline{E} \leq -mc^2$ to the lower branch of the continuous spectrum.*

Proof: Consider the family of operators

$$\widehat{h}_D(\nu) = c\,\boldsymbol{\alpha} \cdot \boldsymbol{p} + \beta\, mc^2 + \nu\, U(\boldsymbol{x}), \quad 0 \leq \nu \leq 1$$

so that $\widehat{h}_D(0)$ is the free-particle Dirac Hamiltonian, and $\widehat{h}_D(1) = \widehat{h}_D$ The Rayleigh quotient

$$R_\psi(\nu) = \frac{(\,\psi \,|\, \widehat{h}_D(\nu) \,|\, \psi\,)}{(\,\psi \,|\, \psi\,)} = R_\psi(0) + \nu\,(\,\psi \,|U(\boldsymbol{x})\,|\, \psi\,)/(\,\psi \,|\, \psi\,).$$

represents an estimate of the eigenvalue corresponding to the trial function ψ. When $\nu = 0$, we know that the spectrum has two disjoint branches, one with $E \geq +mc^2$, the other $E \leq -mc^2$. Suppose we choose ψ such that $R_\psi(0) > +mc^2$; because $U_{min} \leq (\,\psi \,|U(\boldsymbol{x})\,|\, \psi\,)/(\,\psi \,|\, \psi\,) < 0$ by hypothesis, we have $R_\psi(\nu) > mc^2 + U_{min} > -mc^2$ so that, in particular, there exists an $\underline{E} > -mc^2$ so that each trial eigenvalue $R_\psi(1) > \underline{E}$. Because $\mathcal{D}(\widehat{h}_D)$ is dense in $\mathcal{H}$, we conclude that every point eigenvalue satisfies $E_i > \underline{E}$. Similarly, if we choose trial functions such that $R_\psi(0) < -mc^2$, we conclude that there exists $\overline{E}$ such that $R_\psi(1) < \overline{E} \leq -mc^2$. ■

Extension of the Rayleigh-Ritz method to the Dirac equation depends upon several conclusions that we can draw from these results:

1. For Dirac Hamiltonians appearing in atomic and molecular structure, the lower (negative energy) spectrum has an upper bound $\overline{E}$. The Galerkin

equations (5.6.7) have $2N$ eigenvalues which we may label $E_i^{(N)}$, $i = 1, 2, \ldots, 2N$ in increasing order. Then

$$E_1^{(N)} \leq E_2^{(N)} \leq \ldots \leq E_N^{(N)} < \overline{E} \leq -mc^2$$

provide N square integrable wavepacket approximations to scattering solutions in the lower Dirac continuum.

2. Similarly $\underline{E}$ furnishes a lower bound to the bound state eigenvalues and the upper (positive energy) continuum.

$$-mc^2 < \underline{E} < E_1 \leq E_{N+1}^{(N)} \leq E_2 \leq E_{N+2}^{(N)} \leq \ldots < E_{2N}^{(N)}.$$

The eigenvalues indexed $N+1, N+2, \ldots$ are upper bounds, hopefully good approximations, to the lowest bound state eigenvalues of $\widehat{h}_D$. As in similar nonrelativistic Rayleigh-Ritz calculations, some of the higher eigenvalues will be below the accumulation point $+mc^2$ and represent Rydberg states. Those above $+mc^2$ will represent square integrable wavepacket approximations to scattering states in the upper Dirac continuum.

3. The estimate $U_{min} > -2mc^2$, which ensures that eigenvalues of the upper set do not fall below $-mc^2$, provides a very slack lower bound $\underline{E}$. From the nonrelativistic virial theorem, for which the expectation value $\langle\, U\,\rangle = -2\langle\, T\,\rangle$, where T is the kinetic energy operator, we expect that the minimum bound eigenvalue $E_{N+1}^{(N)}$ is of order $U_{min}/2$. This is confirmed by numerical calculations [41].
4. Stakgold [42] comments that increasing the dimension, N, of the basis set generally improves the approximation. For this we need a basis set that can be completed in a suitable sense as $N \to \infty$. The notions of nonrelativistic E-convergence and A-convergence must be adapted to the relativistic problem so that $E_{N+i}^{(N)} \to E_i$, $i = 1, 2, \ldots$ for bound states in the point spectrum.
5. The Rayleigh-Ritz eigenvalues in the continuum do not converge as N increases. However for each fixed N the eigenvalues and eigenfunctions can be used in sum-over-states perturbation formulas such as (4.8.2) replacing the numerical integration over the energy parameter. The sum-over-states must then converge as $N \to \infty$. Spinor basis sets in relativistic calculations must have completeness properties that guarantee this convergence.

The domain of the operator $\widehat{\mathcal{K}} = \widehat{h}_D(0)$, $\mathcal{D}(\widehat{\mathcal{K}})$, can be written

$$\mathcal{D}(\widehat{\mathcal{K}}) = \mathcal{D}_+(\widehat{\mathcal{K}}) \cup \mathcal{D}_-(\widehat{\mathcal{K}}) \tag{5.6.12}$$

where

$$\mathcal{D}_+(\widehat{\mathcal{K}}) = \left\{\psi \,|\langle \psi \,|\, \mathcal{K} \,|\, \psi\rangle \geq mc^2\right\}, \quad \mathcal{D}_-(\widehat{\mathcal{K}}) = \left\{\psi \,|\langle \psi \,|\, \mathcal{K} \,|\, \psi\rangle \leq -mc^2\right\}$$

Clearly $\mathcal{D}_+(\widehat{\mathcal{K}}) \cap \mathcal{D}_-(\widehat{\mathcal{K}}) = \emptyset$; no Cauchy sequence in $\mathcal{D}_+(\widehat{\mathcal{K}})$ can have a limit in $\mathcal{D}_-(\widehat{\mathcal{K}})$ and *vice versa*.

For bound state calculations, we are interested in using the Rayleigh-Ritz method to approximate the point spectrum, which we have seen are connected to states ψ in $\mathcal{D}_+(\widehat{\mathcal{K}})$. This enables us to define a D-norm on $\mathcal{D}_+(\widehat{\mathcal{K}})$ by

$$\|\psi\|_D^2 = \langle \psi \,|\, \mathcal{K} \,|\, \psi \rangle, \quad \psi \in \mathcal{D}_+(\widehat{\mathcal{K}})$$

so that we only need to choose a basis which is complete with respect to the D-norm in $\mathcal{D}_+(\widehat{\mathcal{K}})$ to be sure that the Rayleigh-Ritz method will generate reliable estimates of point spectra.

Fortunately, most of the operators that occur in relativistic calculations in atomic and molecular physics are form bounded with respect to $\mathcal{K}$ in the sense of (5.5.5), namely

$$|\langle \psi \,|\, V \,|\, \psi \rangle| \le a \langle \psi \,|\, \mathcal{K} \,|\, \psi \rangle + b\|\psi\|^2, \quad \psi \in \mathcal{D}_+(\widehat{\mathcal{K}}),$$

so that we can define a form of A-convergence for such Dirac operators. The list includes

1. *Bounded operators*: Dirac $\boldsymbol{\alpha}$ and β matrices; operators such as $\boldsymbol{\alpha} \times \boldsymbol{x}$ and $\boldsymbol{\alpha} \cdot \boldsymbol{A}$, where $\boldsymbol{A}$ is the vector potential of some external field.
2. *Powers* r^λ with $\lambda \ge -1$ (including the Coulomb operator).
3. *Components* of the (3-)momentum operator $\boldsymbol{p}$, the position $\boldsymbol{x}$ and combinations such as $\boldsymbol{\alpha} \cdot \boldsymbol{p}$.
4. Other pieces of $\widehat{h}_D$.

5.7 Spinor basis sets

It is important to take the functional relations between the four components of a Dirac spinor into account when designing approximation schemes. The main factors influencing the design are:

1. The nuclear Coulomb field dominates the dynamics near each centre. It is therefore desirable that spinor basis functions behave asymptotically like central field spinors (3.2.4) near $r = 0$:
$$\Phi_{\kappa m}(\boldsymbol{x}) = \begin{pmatrix} \Phi^+_{\kappa m}(\boldsymbol{x}) \\ i\Phi^-_{\kappa m}(\boldsymbol{x}) \end{pmatrix} \approx \frac{1}{r} \begin{pmatrix} f^+_\kappa(r)\chi_{\kappa m}(\theta,\varphi) \\ i f^-_\kappa(r)\chi_{-\kappa m}(\theta,\varphi) \end{pmatrix}. \tag{5.7.1}$$
2. Basis spinors of the form (5.7.1) should, as far as practicable, be constructed so that the approximate Dirac spinors satisfy the boundary conditions, §5.3, as $r \to 0$ and $r \to \infty$.
3. We should the equations to exhibit nonrelativistic behaviour in the mathematical limit $c \to \infty$. When the dynamics is dominated by the nuclear Coulomb field, (3.2.2, basis spinors should satisfy approximately an equation of the form

$$\begin{pmatrix} U(r)-\epsilon & c\boldsymbol{\sigma}\cdot\boldsymbol{p} \\ c\boldsymbol{\sigma}\cdot\boldsymbol{p} & -2mc^2-\epsilon+U(r) \end{pmatrix} \begin{pmatrix} \Phi^{+}_{\kappa m}(\boldsymbol{x}) \\ i\Phi^{-}_{\kappa m}(\boldsymbol{x}) \end{pmatrix} = 0. \tag{5.7.2}$$

where $\epsilon = E - mc^2$ is the energy relative to the usual nonrelativistic zero. For c sufficiently large, $2mc^2 \gg \epsilon - U(r)$, this reduces to

$$\begin{pmatrix} U(r)-\epsilon & c\boldsymbol{\sigma}\cdot\boldsymbol{p} \\ c\boldsymbol{\sigma}\cdot\boldsymbol{p} & -2mc^2 \end{pmatrix} \begin{pmatrix} \Phi^{+}_{\kappa m}(\boldsymbol{x}) \\ i\Phi^{-}_{\kappa m}(\boldsymbol{x}) \end{pmatrix} \approx 0,$$

and the lower equation gives the Pauli approximation, (3.2.14) and (A.4.12),

$$i\Phi^{-}_{\kappa m}(\boldsymbol{x}) \to \frac{1}{2mc}\boldsymbol{\sigma}\cdot\boldsymbol{p}\,\Phi^{+}_{\kappa m}(\boldsymbol{x}). \tag{5.7.3}$$

Substituting this into the upper equation

$$(U(r)-\epsilon)\Phi^{+}_{\kappa m}(\boldsymbol{x}) + c\,\boldsymbol{\sigma}\cdot\boldsymbol{p}\; i\Phi^{-}_{\kappa m}(\boldsymbol{x}) = 0, \tag{5.7.4}$$

gives

$$\left\{\frac{1}{2m}(\boldsymbol{\sigma}\cdot\boldsymbol{p})^2 + (U(r)-\epsilon)\right\}\Phi^{+}_{\kappa m}(\boldsymbol{x}) = 0, \tag{5.7.5}$$

so that, with the aid of the formal operator identity,

$$(\boldsymbol{\sigma}\cdot\boldsymbol{p})^2 = \boldsymbol{p}^2, \tag{5.7.6}$$

we see that $\Phi^{+}_{\kappa m}(\boldsymbol{x})$ satisfies Schrödinger's equation in the limit. This suggests that the radial amplitudes, (5.7.1), should satisfy

$$f^{-}_{\kappa}(r) \to \frac{1}{2mc}\left(\frac{d}{dr}+\frac{\kappa}{r}\right) f^{+}_{\kappa}(r). \tag{5.7.7}$$

as $c \to \infty$. We shall say that basis spinors satisfying (5.7.3) and (5.7.7) *apart from a normalization constant* are *kinetically matched.*

4. Acceptable basis spinors must be complete in a suitable Hilbert space. In particular, matrix elements of components of the Dirac Hamiltonian, $U(r)$, $\boldsymbol{\alpha}\cdot\boldsymbol{p}$ and β must be finite.

Let us now apply similar reasoning to the matrix Dirac equation (5.6.7). The matrix analogue of (5.7.3) is

$$\mathbf{c}^{-} = \frac{1}{2mc}\left(\mathbf{S}^{--}\right)^{-1}\boldsymbol{\Pi}^{-+}\mathbf{c}^{+}$$

and that of (5.7.5) is

$$\left\{\frac{1}{2m}\boldsymbol{\Pi}^{+-}\left(\mathbf{S}^{--}\right)^{-1}\boldsymbol{\Pi}^{-+} + (\mathbf{U}^{++} - \epsilon\mathbf{S}^{++})\right\}\mathbf{c}^{+} = 0 \tag{5.7.8}$$

For this to be the matrix Schrödinger equation in the space spanned by the 2-component basis spinors $\Phi^{+}_{\kappa m}(\boldsymbol{x})$, we need to ensure that

$$\mathbf{T}^{++} = \frac{1}{2m} \boldsymbol{\Pi}^{+-} \left(\mathbf{S}^{--}\right)^{-1} \boldsymbol{\Pi}^{-+}, \tag{5.7.9}$$

where $\mathbf{T}^{++}$ is the matrix of the nonrelativistic kinetic energy, $\mathbf{p}^2/2m$.

This is by no means straightforward. Elementary quantum mechanical texts often make the point that it is not possible to reproduce the canonical commutation relations $[x, p] = i\hbar$ in a finite matrix representation. The presence of $\left(\mathbf{S}^{--}\right)^{-1}$, the inverse of the small component Gram matrix, indicates that we have an approximate resolution of the identity; in general, the Dirac matrix representation of $\mathbf{T}$ has too small an expectation value unless the $\beta = -1$ basis set is complete [39, 43, 44, 45]. The error due to incompleteness of the basis set lowers estimates of Dirac eigenvalues and predicts binding energies which are larger than they should be.

Fortunately, there is a simple way out: *kinetic matching* of the basis spinors [40]. In the notation of (5.6.1), we introduce 2-spinor basis elements $M[\beta, \mu, \mathbf{x}]$ satisfying

$$M[-1, \mu, \mathbf{x}] \propto \boldsymbol{\sigma} \cdot \boldsymbol{p}\, M[+1, \mu, \mathbf{x}]$$

(compare (5.6.8), (5.7.3)). Introduce normalization factors N_μ^β such that

$$M[\beta, \mu, \mathbf{x}] = N_\mu^\beta m[\beta, \mu, \mathbf{x}]. \tag{5.7.10}$$

Then

$$S_{\mu\nu}^{\beta\beta} = N_\mu^\beta N_\nu^\beta s_{\mu\nu}^{\beta\beta}, \quad s_{\mu\nu}^{\beta\beta} = \int m[\beta, \mu, \mathbf{x}]^\dagger m[\beta, \nu, \mathbf{x}] d\mathbf{x} \tag{5.7.11}$$

The normalization constants may be chosen for convenience, for example to make the diagonal elements of the Gram matrix unity:

$$S_{\mu\mu}^{\beta\beta} = \left(N_\mu^\beta\right)^2 s_{\mu\mu}^{\beta\beta} = 1$$

for all values of μ. From (5.6.10), this means that

$$\boldsymbol{\Pi}_{\mu\nu}^{-+} = \boldsymbol{\Pi}_{\nu\mu}^{+-*} = \int M[-1, \mu, \mathbf{x}]^\dagger \boldsymbol{\sigma} \cdot \mathbf{p}\, M[+1, \nu, \mathbf{x}]\, d\mathbf{x} = S_{\mu\nu}^{--}, \tag{5.7.12}$$

from (5.7.10) and

$$\boldsymbol{\Pi}^{+-} \left(\mathbf{S}^{--}\right)^{-1} \boldsymbol{\Pi}^{-+} = \mathbf{S}^{--} \tag{5.7.13}$$

which makes the calculation of the kinetic matrices a trivial matter.

To verify equivalence of $\mathbf{S}^{--}/2m$, (5.7.13), with $\mathbf{T}^{++}$, (5.7.9), we first integrate over angles to reduce the problem to a radial integration. We write

$$m[\beta, \mu, \mathbf{x}] = \frac{g_\mu^\beta(r)}{r} \chi_{\kappa m}(\theta, \varphi) \tag{5.7.14}$$

The radial kinetic energy T_r is diagonal with respect to the angular quantum number κ and its $\beta = +1$ matrix block has elements

$$t_{\mu\nu}^{++} = \frac{1}{2m} N_\mu^+ N_\nu^+ \int_0^\infty g_\mu^{+*}(r) \left(-\frac{d^2}{dr^2} + \frac{l(l+1)}{r^2} \right) g_\nu^+(r) dr$$

The differential operator can be factorized and an integration by parts gives

$$\begin{aligned} 2mt_{\mu\nu}^{++} &= \int_0^\infty \left[N_\mu^+ \left(\frac{d}{dr} + \frac{\kappa}{r} \right) g_\mu^+(r) \right]^* \left[N_\nu^+ \left(\frac{d}{dr} + \frac{\kappa}{r} \right) g_\nu^+(r) \right] dr \\ &= N_\mu^- N_\nu^- \int_0^\infty g_\mu^{-*}(r) g_\nu^-(r)\, dr \\ &= N_\mu^- N_\nu^- s_{\mu\nu}^{--} = S_{\mu\nu}^{--} \end{aligned}$$

as required. We could also apply a similar argument to negative energy states to give a nonrelativistic Schrödinger equation for low-lying positron states. This involves setting $\epsilon = E + mc^2$, interchanging the roles of the β components, replacing κ by $\bar{\kappa} = -\kappa$ and l by $\bar{l} = l \pm 1$, depending on the sign of $\bar{\kappa}$.

Kinetic matching is therefore central to basis set design for approximation of Dirac wavefunctions and to the explanation of the pathological "disorders" described in §4.13 and §5.5. The kinetic matching connection, (5.7.7), generates functions that do not belong to the $\beta = +1$ basis set; the disorders described in the literature stemmed from choosing a $\beta = -1$ basis set consisting of the $\beta = +1$ functions together with additional functions intended to "balance" the set kinetically [46]. This makes it impossible to satisfy the one-to-one kinetic matching relation (5.7.7). Moreover, the increase in the dimension of the $\beta = -1$ basis accentuates linear dependence problems, itself undesirable, and the algebraic equivalence of $\boldsymbol{\Pi}^{+-} \left(\mathbf{S}^{--}\right)^{-1} \boldsymbol{\Pi}^{-+}/2m$ and $\mathbf{T}^{++}$ is also lost. The unmatched small component basis functions lead to spurious states having no physical meaning. Kinetic matching, as defined here, eliminates the problem.

5.8 L-spinors

L-spinors were first mentioned in [47], although they were introduced with a different name in an earlier paper [49, eq. (71)] (see also [50, p. 240] and [51, §22.6.3] and [41]). The name differentiates them from the similar "relativistic Coulomb Sturmian functions" introduced by Szmytkowski [52] whose definition does not satisfy the full kinetic matching criteria.

Nonrelativistic Coulomb Sturmian functions [53, 54] constitute a countable basis of analytic functions which can be shown to be complete on suitable function spaces for the representation of both bound and scattering nonrelativistic radial wavefunctions. There have been many applications. The definition of L-spinors, which are relativistic analogues, ensures that L-spinor properties, in particular orthogonality and completeness, reduce smoothly to those of the Coulomb Sturmians in the limit $c \to \infty$.

The Dirac Coulomb eigenfunctions (3.3.22) and (3.3.23), from which L-spinors are derived, are best expressed in terms of Laguerre polynomials, Appendix A.3.3, rather than the corresponding confluent hypergeometric functions, Appendix A.3.2, used elsewhere in this book. See (A.3.28) for the connection. In the notation of (5.7.14), the unnormalized radial L-spinor amplitudes [51, 22.147 and 22.148] are

$$f^{+}_{n_r\kappa}(x) = x^{\gamma} e^{-x/2} \left\{ -(1-\delta_{n_r,0}) L^{2\gamma}_{n_r-1}(x) + \frac{N_{n_r\kappa}-\kappa}{n_r+2\gamma} L^{2\gamma}_{n_r}(x) \right\}, \tag{5.8.1}$$

and

$$f^{-}_{n_r\kappa}(x) = x^{\gamma} e^{-x/2} \left\{ -(1-\delta_{n_r,0}) L^{2\gamma}_{n_r-1}(x) - \frac{N_{n_r\kappa}-\kappa}{n_r+2\gamma} L^{2\gamma}_{n_r}(x) \right\}, \tag{5.8.2}$$

where, for some positive constant λ, $x = 2\lambda r$, n_r is a non-negative integer, and

$$\gamma = +\sqrt{\kappa^2 - Z^2/c^2}, \quad N_{n_r\kappa} = +\sqrt{n_r^2 + 2n_r\gamma + \kappa^2}, \tag{5.8.3}$$

are respectively the leading exponent of the power series expansion of the functions about $x = 0$ and the apparent principal quantum number. The L-spinors are solutions of the differential equation system

$$\begin{bmatrix} \dfrac{1}{2} - \dfrac{\alpha_{n_r\kappa} Z\mu^2}{dc}\dfrac{1}{x} & -\dfrac{d}{dx} + \dfrac{\kappa}{x} \\ \dfrac{d}{dx} + \dfrac{\kappa}{x} & -\dfrac{1}{2} - \dfrac{Z}{\alpha_{n_r\kappa}\mu^2 c}\dfrac{1}{x} \end{bmatrix} \begin{bmatrix} \mu^{-1} f^{+}_{n_r\kappa}(x) \\ \mu f^{-}_{n_r\kappa}(x) \end{bmatrix} = 0, \tag{5.8.4}$$

where c is the speed of light and μ^2 is a root of the equation

$$\mu^4 - \frac{2c}{\lambda}\mu^2 + 1 = 0. \tag{5.8.5}$$

We choose

$$\mu^2 = \frac{c}{\lambda}\left(1 + \sqrt{1 - \frac{\lambda^2}{c^2}}\right), \quad \mu^{-2} = \frac{c}{\lambda}\left(1 - \sqrt{1 - \frac{\lambda^2}{c^2}}\right).$$

which ensures that $f^{+}_{n_r\kappa}(x) \to \text{const.}\ S_{n_r l}(x)$ in the nonrelativistic limit $c \to \infty$ (see below). The analogue of the nonrelativistic energy parameter $E_0 = -\lambda^2/2$ is

$$E^R_0 = c^2\sqrt{1 - \lambda^2/c^2} = c^2 + E_0 + O(1/c^2). \tag{5.8.6}$$

The boundary conditions as $r \to 0$ and $r \to \infty$ are satisfied when

$$\alpha_{n_r\kappa} = N_{n_r\kappa}\lambda/Z$$

and, when $\alpha_{n_r\kappa} = 1$, the L-spinor amplitudes coincide with the Dirac-Coulomb eigenfunctions with principal quantum number $n = n_r + |\kappa|$ given by (3.3.22), (3.3.23).

5.8.1 Kinetic matching and the nonrelativistic limit

The definitions (5.8.1) and (5.8.2) and relations listed in Appendix A.3.3 can be used to verify that L-spinors satisfy the kinetic matching condition (5.7.7), with suitable normalization, in the nonrelativistic limit. because $\gamma \to |\kappa|$ and $N_{n_r\kappa} \to n$, we have

$$\begin{aligned} f^+(x) &\to \text{const. } x^{l+1} e^{-x/2} \left\{ -(1-\delta_{n_r 0}) L^{2l+2}_{n_r-1}(x) + L^{2l+2}_{n_r}(x) \right\} \\ &= \text{const. } x^{l+1} e^{-x/2} L^{2l+1}_{n_r}(x) \end{aligned} \tag{5.8.7}$$

for negative values of $\kappa = -l-1$, using [55, (22.7.30)] in the second line. Similarly for positive $\kappa = l$, for which $n_r \geq 1$, we have

$$\begin{aligned} f^+(x) &\to \text{const. } x^{l} e^{-x/2} \left\{ -(1-\delta_{n_r 0}) L^{2l}_{n_r-1}(x) + \frac{n_r}{n_r+2l} L^{2l}_{n_r}(x) \right\} \\ &= \text{const. } x^{l} e^{-x/2} \left\{ -(n_r+2l) L^{2l}_{n_r-1}(x) + n_r L^{2l}_{n_r}(x) \right\} \\ &= \text{const. } x^{l+1} e^{-x/2} L^{2l+1}_{n_r}(x), \end{aligned} \tag{5.8.8}$$

using [55, 22.7.31] in the third line. Because $n_r = n - l - 1$, we verify that as $c \to \infty$, for both signs of κ,

$$f^+(x) \to \text{const. } S_{nl}(x),$$

where $S_{nl}(x)$ is the Coulomb Sturmian of (B.5.12). A similar analysis shows that

$$f^-(x) \to \text{const.} \left(\frac{d}{dx} + \frac{\kappa}{x} \right) S_{nl}(x),$$

which verifies the kinetic matching property (5.7.7).

5.8.2 Orthogonality properties

The standard orthogonality properties of Laguerre polynomials can be used to write down L-spinor generalizations of Sturmian properties. A natural starting point is to mimic the derivation of orthogonality relations for nonrelativistic Sturmians starting from the defining equation (5.8.4). Multiplying from the left by the adjoint vector $[\mu^{-1} f^+_{n'_r\kappa}, \mu f^-_{n'_r\kappa}]$ and the weight function $1/x$, and subtracting the result from the corresponding equation with n_r and n'_r interchanged gives

$$\begin{aligned} (\alpha_{n_r\kappa} - \alpha_{n'_r\kappa}) \int_0^\infty & \left\{ f^+_{n'_r\kappa}(x) f^+_{n_r\kappa}(x) \right. \\ & \left. - (\alpha_{n_r\kappa}\alpha_{n'_r\kappa})^{-1} f^-_{n'_r\kappa}(x) f^-_{n_r\kappa}(x) \right\} \frac{dx}{x} = 0 \end{aligned}$$

Thus the integral vanishes if the eigenvalues $\alpha_{n_r\kappa}$ and $\alpha_{n'_r\kappa}$ are different. Although this reduces to the usual Sturmian orthogonality relation in the nonrelativistic limit, the integrand is not obviously positive definite and the result is therefore not very useful. A more profitable approach makes use of the unweighted Gram matrix. We choose the same normalization factor $\mathcal{N}_{n_r\kappa}$ for both $f^+_{n_r\kappa}(x)$ and $f^-_{n_r\kappa}(x)$:

$$\begin{aligned} g^{(\kappa)}_{n_r n_r} &= \mathcal{N}^2_{n_r\kappa}\left\{(1-\delta_{n_r 0})\frac{\Gamma(2\gamma+n_r)}{(n_r-1)!} + \left(\frac{N_{n_r\kappa}-\kappa}{2\gamma+n_r}\right)^2 \frac{\Gamma(2\gamma+n_r+1)}{(n_r)!}\right\} \\ &= \mathcal{N}^2_{n_r\kappa}\, 2N_{n_r\kappa}(N_{n_r\kappa}-\kappa)\frac{\Gamma(2\gamma+n_r)}{n_r!(2\gamma+n_r)} = 1, \end{aligned}$$

so that

$$\mathcal{N}_{n_r\kappa} = \left[\frac{n_r!\,(2\gamma+n_r)}{2N_{n_r\kappa}(N_{n_r\kappa}-\kappa)\,\Gamma(2\gamma+n_r)}\right]^{1/2} \tag{5.8.9}$$

In a similar fashion we can show that the off-diagonal elements of the L-spinor Gram matrix are

$$\begin{aligned} g^{(\kappa)}_{n_r,(n_r+1)} &= g^{(\kappa)}_{(n_r+1),n_r} \\ &= -\frac{1}{2}\beta\left[\frac{(n_r+1)(2\gamma+n_r+1)(N_{n_r\kappa}-\kappa)}{N_{n_r\kappa}N_{(n_r+1),\kappa}(N_{(n_r+1),\kappa}-\kappa)}\right]^{1/2} \end{aligned} \tag{5.8.10}$$

for both blocks $\beta = \pm 1$. This reduces to the Coulomb Sturmian Gram expression (B.5.19) (up to an unimportant sign difference) in the nonrelativistic limit.

5.8.3 Linear independence of L-spinors

The linear independence behaviour of L-spinors is very similar to that of the Coulomb Sturmians. Writing $g^{(N)} = G^{(N)} - I^{(N)}$, we see, by expanding with respect to the last row, that $f^{(N)}(\sigma) = \det(g^{(N)} - \sigma I^{(N)})$ satisfies

$$f^{(N)}(\sigma) = -\sigma f^{(N-1)}(\sigma) - g^2_{N,N-1} f^{(N-2)}(\sigma)$$

with $f^{(1)}(\sigma) = -\sigma$ and $f^{(2)}(\sigma) = \sigma^2 - g^2_{12}$. We conclude inductively that $f^{(2k)}(\sigma)$ and $f^{(2k+1)}(\sigma)/\sigma$ are polynomials in σ^2 of degree k, so that the eigenvalues of $G^{(N)}$ are in the interval $(1-\rho_N, 1+\rho_N)$, where

$$\rho_N = 1 - C/N^2 + O(1/N^3). \tag{5.8.11}$$

where C is a positive constant. The eigenvalues of $f^{(N)}$ are distributed symmetrically about $\sigma = 0$ when N is even, and there is an additional zero eigenvalue when N is odd. Thus $G^{(N)}$ has condition number $k_N = (1+\rho_N)/(1-\rho_N) \sim 2N^2/C$ when N is large. The numerical stability observed in L-spinor calculations supports this analysis [41].

5.8.4 Completeness of L-spinors

We can establish completeness of L-spinors in a variety of Hilbert spaces by exploiting the following [56, Lemma 5]:

Lemma 5.18 (Klahn). *Let $\{\varphi_n\}_{n=1}^{\infty}$ be a complete system in a Hilbert space H. Moreover, let $a_{n\mu}$, $(1 \le \mu \le n)$ be arbitrary complex numbers with $a_{nn} \neq 0$. Then the system*

$$\{\psi_n = \sum_{\mu=1}^{n} a_{n\mu}\varphi_\mu\}_{n=1}^{\infty}$$

is also complete in H.

To apply this to the L-spinors, we note that equations (5.8.1), (5.8.2) can be written

$$f^{\beta}_{n_r\kappa}(x) = a_{n_r,n_r-1}\, x^\gamma e^{-x/2} L^{2\gamma}_{n_r-1}(x) + \beta\, a_{n_r,n_r}\, x^\gamma e^{-x/2} L^{2\gamma}_{n_r}(x).$$

Because $a_{n_r,n_r-1} = -(1 - \delta_{n_r,0})$, only the second term contributes when $n_r = 0$ for both signs of κ. Also since $N_{0,\kappa} = |\kappa|$, the first non-vanishing L-spinor for $\kappa > 0$ has $n_r = 1$. Thus we can use the properties of Coulomb Sturmians, §B.5.1, to infer that the L-spinors are both complete and minimal on the Sobolev spaces $[W^{(p)}_2(\mathbb{R}^3)]^2$ for $p = 1, 2$. This is exactly what we need for constructing trial wavefunctions of the form (5.6.1) for Rayleigh-Ritz approximation of Dirac four-component wavefunctions.

5.8.5 Charge conjugation and L-spinors

Dirac four-spinors transform under charge conjugation, §3.1.6, so that

$$\psi \to \psi_c = C\,\overline{\psi}^t \tag{5.8.12}$$

where the superscript t denotes transposition and $\overline{\psi} = \psi^*\gamma^0$ is Dirac conjugation. The matrix C is given by (§A.2)

$$C = i\gamma^2\gamma^0 = \begin{bmatrix} 0 & -i\sigma^2 \\ -i\sigma^2 & 0 \end{bmatrix}.$$

When the radial amplitudes $P(r), Q(r)$ are real, it is easy to show that if

$$\psi = \frac{1}{r}\begin{bmatrix} P(r)\chi_{\kappa,m} \\ iQ(r)\chi_{-\kappa,m} \end{bmatrix}$$

then

$$\psi_c = -i(-1)^{m+1/2}\begin{bmatrix} Q(r)\chi_{-\kappa,-m} \\ iP(r)\chi_{\kappa,-m} \end{bmatrix}$$

Under this transformation, expectation values of the charge-current vector remain invariant, whilst those of spin, orbital, and total angular momentum change sign, as does the sign of the energy parameter E and the sign of Z coupling the electron to the external Coulomb potential. By making the corresponding changes

$$Z \leftrightarrow -Z, \quad f^{+}_{n_r\kappa}(x) \leftrightarrow f^{-}_{n_r\kappa}(x), \quad \kappa \leftrightarrow -\kappa, \quad \mu \leftrightarrow \mu^{-1}$$

in (5.8.4), we see that L-spinors retain the charge conjugation symmetries of the Dirac eigenfunctions on which they are modelled. Because the mapping $\mu \leftrightarrow \mu^{-1}$ is equivalent to changing the sign of the energy parameter $E_0^R = +\sqrt{1-\lambda^2/c^2}$, (5.8.6), so that L-spinor expansions will be able correctly to represent positron (negative energy electron) states as well as bound states and electron scattering states.

5.8.6 Construction of $\boldsymbol{\Pi}^{\beta\beta'}$, $\mathbf{S}^{\beta\beta'}$, and $\mathbf{U}^{\beta\beta'}$ matrices for hydrogenic atoms

Let $a := 2\gamma$, and define

$$G_0(a) = 1, \quad G_k(a) = \frac{a+k}{k} G_{k-1}(a), \quad k = 1, 2, \ldots \tag{5.8.13}$$

and

$$H_{mn}(a) = \sum_{k=0}^{\min\{m,n\}} G_k(a-1). \tag{5.8.14}$$

Then $\mathbf{U}^{\beta\beta'}$ has matrix elements

$$\begin{aligned} U^{\beta\beta'}_{mn} &= -\delta_{\beta\beta'} Z\lambda \left[\frac{m!n!(a+m)(a+n)}{N_{m\kappa}N_{n\kappa}(N_{m\kappa}-\kappa)(N_{n\kappa}-\kappa)(a)_m(a)_n} \right]^{1/2} \\ &\times \left\{ H_{m-1,n-1}(a) - \beta \frac{N_{n\kappa}-\kappa}{n+a} H_{m-1,n}(a) \right. \\ &\left. - \beta \frac{N_{m\kappa}-\kappa}{m+a} H_{m,n-1}(a) + \frac{N_{m\kappa}-\kappa}{m+a}\frac{N_{n\kappa}-\kappa}{n+a} H_{mn}(a) \right\}. \end{aligned} \tag{5.8.15}$$

The components of the kinetic matrices $\boldsymbol{\Pi}^{\beta,-\beta}$ are

$$
\begin{aligned}
\Pi^{-+}_{mn} &= \Pi^{+-}_{nm} \\
&= \frac{\lambda}{2}\left[\frac{m!n!(a+m)(a+n)}{N_{m\kappa}N_{n\kappa}(N_{m\kappa}-\kappa)(N_{n\kappa}-\kappa)(a)_m(a)_n}\right]^{1/2} \\
&\times\left\{(2N_{n\kappa}+2n-2+a)\left[H_{m-1,n-1}(a)+\frac{N_{m\kappa}-\kappa}{m+a}H_{m,n-1}(a)\right]\right. \\
&\quad -\frac{N_{n\kappa}-\kappa}{n+a}(2n+2\kappa+a)\left[H_{m-1,n}(a)+\frac{N_{m\kappa}-\kappa}{m+a}H_{m,n}(a)\right] \\
&\quad -2(n+a-1)\left[H_{m-1,n-2}(a)-\frac{N_{m\kappa}-\kappa}{m+a}H_{m,n-2}(a)\right] \\
&\quad -G_{m-1,n-1}(a)-\frac{N_{m\kappa}-\kappa}{m+a}G_{m,n-1}(a) \\
&\quad \left.+\frac{N_{n\kappa}-\kappa}{n+a}G_{m-1,n}(a)+\frac{N_{m\kappa}-\kappa}{m+a}\frac{N_{n\kappa}-\kappa}{n+a}G_{mn}(a)\right\}, \qquad (5.8.16)
\end{aligned}
$$

whilst the symmetric tridiagonal Gram matrices $\boldsymbol{S}^{\beta\beta}$ are related to the expressions (5.8.10) by

$$S^{\beta\beta}_{ij} = g^{\kappa}_{ij}/2\lambda,$$

where the scale factor 2λ arises from the change of independent variable from x to r. It is easy to assess the effect of variations in the parameter λ as it appears here only as a constant multiplier.

5.8.7 Numerical study of L-spinor performance in hydrogenic atoms

Equations (5.6.7) constitute a generalized eigenvalue problem for the *pseudo*-eigenvalues of of the Dirac equation for the potential $U(r)$. This system can be solved using standard numerical software from the EISPACK collection of algebraic eigensystem routines [57]. The generalized eigensystem $\mathbf{Ax} = \lambda\mathbf{Bx}$, in which $\mathbf{A}$ and $\mathbf{B}$ are both symmetric and $\mathbf{B}$ is also positive definite, is first transformed to an ordinary matrix eigenproblem, $\mathbf{Cx} = \lambda\mathbf{x}$ by Cholesky factorization of $\mathbf{B}$ followed by determination of all eigenvectors and eigenvalues of $\mathbf{Cx} = \lambda\mathbf{x}$ using the QL algorithm.[2] The final output of this procedure consists of the eigenvectors and eigenvalues of (5.6.7) with the latter arranged in increasing order. Thus eigenvalues $1,\ldots,N$ are in the negative energy region, whilst $N+1,\ldots,2N$ are in the bound state and positive energy region, so that E_{N+1} approximates the eigenvalue of the lowest bound state. For ease of interpretation, we shift the energy zero to $E = mc^2$ and write $\epsilon = E - mc^2$.

Table 5.1 gives the lowest bound state eigenvalues for the case of a hydrogenic atom of $Z = 50$ as a function of the dimension N of the L-spinor basis set [41]. The eigenvalues are given in atomic units ($E_h = 27.2$ eV) and the speed of light was taken to be $c = \alpha^{-1} = 137.0359895$ atomic units.

[2] A description of the algorithms can be found in [58, Chapter 11].

Table 5.1. Computed bound state eigenvalues of a hydrogenic atom with $Z = 50$.

N	ϵ_{1s}	ϵ_{2s}	ϵ_{3s}	ϵ_{4s}	ϵ_{5s}
			$\lambda = 50.0$		
20	-1294.62616	-326.494806	-143.829353	-79.573094	-35.139167
40	-1294.62616	-326.494806	-143.829802	-80.370331	-51.192342
60	-1294.62616	-326.494806	-143.829802	-80.370332	-51.197724
80	-1294.62616	-326.494806	-143.829802	-80.370332	-51.197724
	$\epsilon_{2p_{1/2}}$	$\epsilon_{3p_{1/2}}$	$\epsilon_{4p_{1/2}}$	$\epsilon_{5p_{1/2}}$	$\epsilon_{6p_{1/2}}$
			$\lambda = 25.0$		
20	-326.494806	-143.829807	-80.370337	-51.197247	-35.202715
40	-326.494806	-143.829803	-80.370333	-51.197725	-35.433571
60	-326.494806	-143.829802	-80.370332	-51.197725	-35.433571
80	-326.494806	-143.829802	-80.370332	-51.197725	-35.433571
100	-326.494806	-143.829802	-80.370332	-51.197725	-35.433571
	$\epsilon_{2p_{3/2}}$	$\epsilon_{3p_{3/2}}$	$\epsilon_{4p_{3/2}}$	$\epsilon_{5p_{3/2}}$	$\epsilon_{6p_{3/2}}$
			$\lambda = 25.0$		
20	-315.144355	-140.457874	-78.952058	-50.473186	-34.755474
40	-315.144355	-140.457874	-78.952058	-50.473867	-35.015794
60	-315.144355	-140.457874	-78.952058	-50.473867	-35.015794
	$\epsilon_{3d_{3/2}}$	$\epsilon_{4d_{3/2}}$	$\epsilon_{5d_{3/2}}$	$\epsilon_{6d_{3/2}}$	$\epsilon_{7d_{3/2}}$
			$\lambda = 15.0$		
20	-140.457874	-78.952058	-50.473867	-35.015794	-25.703485
40	-140.457874	-78.952058	-50.473867	-35.015794	-25.703739
60	-140.457874	-78.952058	-50.473867	-35.015794	-25.703739

The variation as N increases is much as would be expected from a comparable nonrelativistic bound state calculation with Coulomb Sturmians, and the eigenvalues quickly settle down to an asymptotic value (9 significant figures) for all the states listed. Calculations are given for a representative selection of values of the scaling parameter λ, to which the results are relatively insensitive. Note that the exact Dirac-Coulomb eigenvalue $\epsilon_{n\kappa}$ of the $n\kappa$ eigenstate is represented by a single L-spinor when $\lambda = Z/N_{n\kappa}$; all other states are a linear superposition. Since $\epsilon_{n\kappa}$ is independent of the sign of κ, we expect to find, for example, that $\epsilon_{2s} = \epsilon_{2p_{1/2}}$, $\epsilon_{3p_{3/2}} = \epsilon_{3d_{3/2}}$ and so on, and this is clearly displayed in Table 5.1. The eigenvalues $\epsilon_{n\kappa}$ converge to the exact values calculated using the Sommerfeld formula (3.3.25) as N increases. Table 5.2 shows the corresponding distribution of the three highest negative eigenvalues of each symmetry (numbered $N, N-1, N-2, \ldots$) relative to its boundary at $E = -mc^2$ or $\epsilon = -2mc^2$ as a function of the block matrix dimension N. As these eigenvalues represent the energies of continuum states, we expect

Table 5.2. Highest negative energy eigenvalues for $Z = 50$ (a.u.) relative to $-2mc^2$.

$N =$	20	40	60	80	100
λ			$\kappa = -1$		
30	-57.6	-26.6	-17.0	-12.4	-9.7
40	-80.3	-36.7	-23.4	-17.0	-13.3
50	-104.3	-47.4	-30.1	-21.8	-17.1
			$\kappa = +1$		
20	-46.6	-20.7	-13.0	-9.4	-7.3
25	-61.1	-27.0	-16.9	-12.1	-9.4
30	-76.4	-33.6	-20.9	-15.0	-11.6
			$\kappa = -2$		
20	-34.8	-16.6	-10.7	-7.9	-6.2
25	-44.9	-21.2	-13.7	-10.1	-7.9
30	-55.4	-26.0	-16.8	-12.3	-9.7
			$\kappa = +2$		
10	-19.6	-9.1	-5.9	-4.3	-3.3
15	-31.4	-14.6	-9.3	-6.7	-5.3
20	-44.2	-20.4	-12.9	-9.4	-7.3

no convergence and we see none. The theory predicts that all these states should have eigenvalues less than zero, and the numerical values confirm this. However the actual numerical values of the eigenvalues are sensitive both to λ and to N. Whilst these *pseudo*-states are of no concern for bound state calculations, they are required for completeness of the *pseudo*-spectrum and act as integration points, together with the higher positive energy states, for sum-over-states expressions in perturbation theory.

Table 5.3. Gram matrix condition numbers, k_N, for $N = 100$.

κ	$Z = 10$	$Z = 100$
-1	4134	9203
1	2061	4558
-2	2061	4568
2	1277	2815
-3	1277	2813
3	885	884
-4	885	884
4	658	143
-5	658	143

The calculation is very stable numerically, as can be seen by examining the condition numbers[3] in Table 5.3. Note that the condition number is the same for matrices with both values of β belonging to the same symmetry κ and is independent of the choice of λ. These numbers are relatively modest, and it is possible to increase N considerably without running into any numerical difficulties; calculations have been done for $N \leq 500$ without difficulty.

Table 5.4. U_{min} in atomic units. $N = 100$.

Z	$\kappa = -1$	$\kappa = +1$	$\kappa = -2$	$\kappa = +2$	$\kappa = -3$
10	-552	-201	-201	-107	-107
20	-1117	-406	-402	-215	-215
30	-1708	-618	-607	-324	-323
40	-2342	-843	-815	-434	-432
50	-3039	-1086	-1028	-547	-542
60	-3827	-1355	-1248	-664	-653
70	-4748	-1663	-1476	-785	-767
80	-5868	-2026	-1715	-910	-883
90	-7296	-2476	-1966	-1042	-1002
100	-9243	-3069	-2232	-1181	-1123
110	-12176	-3928	-2517	-1328	-1249
120	-17440	-5407	-2823	-1487	-1378
130	-31921	-9286	-3157	-1657	-1512

$-2mc^2 = -37557.7248$ a.u.

Table 5.2 verifies the assertion that the N lowest eigenvalues are bounded above by $\epsilon = -2mc^2$ and the argument of Theorem 5.17 postulated the existence of a lower bound $\underline{E}$ to the computed bound state eigenvalues provided that $U_{min} > -2mc^2$, where U_{min} is the lowest eigenvalue of the potential matrix. As L-spinors do not form orthonormal sets, we need to solve the generalized eigenvalue problems, $\mathbf{U}^{\beta\beta}\mathbf{c} = u\mathbf{S}^{\beta\beta}\mathbf{c}$, for $\beta = \pm 1$. The results for the case $N = 100$ shown in Table 5.4 show that the lower bound $\underline{E}$ is safely within the permitted range. Beyond $Z = 130$, it is necessary to increase N to get results of the accuracy of Table 5.1, but then calculations have been done successfully for the critical $\kappa = -1$ case with Z as large as 137.035 989, just below the assumed value of c!

[3] The condition number of the Gram matrix is defined by $k_N = \max(|\sigma|)/\min(|\sigma|)$, where σ runs over the N eigenvalues of the matrix.

5.9 S-spinors

S-spinors [59] are relativistic spinor analogues of the popular exponential-type (or Slater) functions used in nonrelativistic atomic calculations [65, §6.6]. Their functional form is derived from the corresponding L-spinor with the minimum value of n_r. In the notation of (5.7.14), the unnormalized radial components are

$$g^{\beta}(r) = A^{\beta} r^{\gamma} e^{-\lambda r} + B\lambda r^{\gamma+1} e^{-\lambda r}, \quad \beta = \pm 1, \tag{5.9.1}$$

where $A^+ = A^- = 1$, $B = 0$ for $\kappa < 0$ and

$$A^+ = \frac{(\kappa + 1 - N_{1,\kappa})(2\gamma + 1)}{2(N_{1,\kappa} - \kappa)},$$

$$A^- = \frac{(\kappa - 1 - N_{1,\kappa})(2\gamma + 1)}{2(N_{1,\kappa} - \kappa)},$$

and $B = 1$ for $\kappa > 0$, with

$$\gamma = \sqrt{\kappa^2 - Z^2/c^2}, \quad N_{1,\kappa} = \sqrt{\kappa^2 + 2\gamma + 1}. \tag{5.9.2}$$

A basis set is formed as a collection of S-spinors with positive real exponents $\{\lambda_\mu, \mu = 1, 2, \ldots, N\}$. Methods of generating suitable exponent sets which possess some of the desirable linear independence and completeness properties, Appendix B.5, are described in Appendix B.5.3.

S-spinors, like L-spinors, are designed to be used with point charge nuclear models. They have the kinetic matching property (5.7.7) and, with properly chosen exponent sets, $U_{min} > -2mc^2$, ensuring separation of the positive and negative *pseudo*-spectrum when $Z \leq 137$. To verify the kinetic matching property, we write down the nonrelativistic limit of (5.9.1) and (5.9.2). Consider first the s-states having $\kappa = -1$. Because $\gamma \to 1$ as $c \to \infty$ we get the limits

$$g^{\beta}_{\mu}(r) \stackrel{c\to\infty}{\longrightarrow} r e^{-\lambda_\mu r},$$

so that

$$\left(\frac{d}{dr} - \frac{1}{r}\right) g^{+}_{\mu}(r) \stackrel{c\to\infty}{\longrightarrow} -\lambda g^{-}_{\mu}(r)$$

up to a constant multiplier. Similarly we can show that for $\kappa = +1$, the $p_{1/2}$ states, we get the asymptotic forms

$$g^{+}_{\mu}(r) \to r^2 e^{-\lambda_\mu r}, \; g^{-}_{\mu}(r) \to -(3r - \lambda_\mu r^2) e^{-\lambda_\mu r}$$

so that

$$\left(\frac{d}{dr} + \frac{1}{r}\right) g^{+}_{\mu}(r) \to -g^{-}_{\mu}(r)$$

in the limit as required for kinetic matching. The $p_{1/2}$ and $s_{1/2}$ states have the same relativistic exponent, γ, governing their behaviour near $r = 0$, but

$g_\mu^+(r)$ has to have a p-type Slater function as its nonrelativistic limit. Because $N_{1,+1} \to 2$,

$$A^+ \to O(Z^2/c^2), \; A^- \to -3 + O(Z^2/c^2).$$

In the $s_{1/2}$ case A^+ remains finite, so that the two nonrelativistic radial functions have the correct cusp behaviour as $r \to 0$. L-spinors have the same cusp behaviour for all values of κ.

5.9.1 Construction of $\Pi^{\beta\beta'}$, $\mathrm{S}^{\beta\beta'}$, and $\mathrm{U}^{\beta\beta'}$ for hydrogenic atoms

The matrix elements of the Dirac Coulomb operator on the same nuclear centre can all be expressed in terms of Gamma functions through

$$\int_0^\infty r^{z-1} e^{-\lambda r} dr = \Gamma(z)/\lambda^z, \quad \Re z > 0. \tag{5.9.3}$$

Then, from (5.9.1), the Gram matrix for normalized S-spinors becomes

$$\begin{aligned} S_{\mu\nu}^{\beta\beta'} &= \delta_{\beta\beta'} N_\mu^\beta N_\nu^\beta \frac{\Gamma(2\gamma+1)}{\lambda_{\mu\nu}^{2\gamma+1}} \\ &\times \left\{ (A^\beta)^2 + A^\beta(2\gamma+1) + (2\gamma+1)(2\gamma+2)\frac{\lambda_\mu \lambda_\nu}{\lambda_{\mu\nu}^2} \right\}. \end{aligned}$$

where $\lambda_{\mu\nu} = \lambda_\mu + \lambda_\nu$. Write $\sigma_{\mu\nu} = 2\sqrt{\lambda_\mu \lambda_\nu}/\lambda_{\mu\nu}$ and normalize so that $S_{\mu\nu}^{\beta\beta'} = \delta_{\beta\beta'}$; then

$$S_{\mu\nu}^{\beta\beta'} = \delta_{\beta\beta'} \frac{\sigma_{\mu\nu}^{2\gamma+1}}{D^\beta(\gamma)} \left\{ (A^\beta)^2 + A^\beta(2\gamma+1) + (2\gamma+1)(2\gamma+2)\frac{\lambda_\mu \lambda_\nu}{\lambda_{\mu\nu}^2} \right\} \tag{5.9.4}$$

where $D^\beta(\gamma) = (A^\beta)^2 + (2\gamma+1)A^\beta + (2\gamma+1)(2\gamma+2)/4$,

$$\begin{aligned} N_\mu^\beta &= \left(\sqrt{2\lambda_\mu}\right)^{2\gamma+1} \Big/ \left[\Gamma(2\gamma+1) D^\beta(\gamma)\right]^{1/2}, \\ U_{\mu\nu}^{\beta\beta'} &= -\delta_{\beta\beta'} Z \frac{\lambda_{\mu\nu}\, \sigma_{\mu\nu}^{2\gamma+1}}{2\gamma\, D^\beta(\gamma)} \\ &\qquad \times \left\{ (A^\beta)^2 + 2\gamma\, A^\beta + 2\gamma(2\gamma+1)\, \frac{\lambda_\mu \lambda_\nu}{\lambda_{\mu\nu}^2} \right\}, \end{aligned} \tag{5.9.5}$$

$$\begin{aligned} \Pi_{\mu\nu}^{-+} = &\frac{\lambda_{\mu\nu}\, \sigma_{\mu\nu}^{2\gamma+1}}{2\gamma \sqrt{D^+(\gamma) D^-(\gamma)}} \quad \{(\gamma+\kappa) A^+ A^- \\ &+ \left[(\gamma+\kappa+1-A^+)\lambda_\nu A^- + (\gamma+\kappa)\lambda_\mu A^+\right] \frac{2\gamma}{\lambda_{\mu\nu}} \\ &+ \left[(\gamma+\kappa+1-A^+)\lambda_\mu - \lambda_\nu A^-\right] 2\gamma(2\gamma+1)\frac{\lambda_\nu}{\lambda_{\mu\nu}^2} \\ &\qquad -2\gamma(2\gamma+1)(2\gamma+2)\frac{\lambda_\mu \lambda_\nu^2}{\lambda_{\mu\nu}^3} \Big\}. \end{aligned} \tag{5.9.6}$$

Table 5.5. Eigenvalues of hydrogen-like uranium ion computed with an even-tempered S-spinor basis set with exponents $\lambda_n = \alpha\beta^n$, $n = 0 - 9$.

κ	nl_j	α	β	$\epsilon_{n\kappa}$: S-spinors	$\epsilon_{n\kappa}$: (3.3.7)
-1	$1s_{1/2}$	110.273	1.5	-4861.181 785 840	-4861.181 785 840
1	$2p_{1/2}$	27.7273	1.89453	-1257.406 533 822	-1257.390 676 718
-2	$2p_{3/2}$	21.0036	1.5	-1089.610 779 076	-1089.610 779 076
2	$3d_{3/2}$	19.8173	1.60004	-489.036 867 314	-489.036 704 004
-3	$3d_{5/2}$	20.6622	1.5	-476.261 476 646	-476.261 476 646
3	$4f_{5/2}$	23.3027	1.44015	-268.965 819 975	-268.965 970 051
-4	$4f_{7/2}$	15.4468	1.5	-266.389 410 530	-266.389 410 530
4	$5g_{7/2}$	18.5363	1.37100	-170.828 908 695	-170.828 906 977
-5	$5g_{9/2}$	12.3339	1.5	-170.049 919 428	-170.049 919 428

It is less easy to give a clear picture of the numerical performance of S-spinors than of L-spinors because the matrix properties depend upon the way in which the exponents λ_μ have been chosen. However, methods that rely on systematic sequences of basis sets constructed along the lines of Appendix B.5.3 appear to work well. Table 5.5 compares the eigenvalues of the nine lowest states of hydrogen-like uranium ($Z = 92$) computed using an even-tempered S-spinor basis [59] with eigenvalues calculated using the Sommerfeld formula (3.3.7). The speed of light has been taken as 137.0373 atomic units and the S-spinor exponents belong to a geometric sequence, (B.5.20), $\lambda = \alpha\beta^n, n = 0, 1, \ldots, N$ with $N = 9$. The parameters α and β for each symmetry are displayed alongside the numerical eigenvalues.

5.10 G-spinors

G-spinors are defined as the relativistic analogue of GTF so that, in the notation of (5.7.14), the unnormalized G-spinor components are

$$m[\beta, \mu, \mathbf{x}] = \frac{g_\mu^\beta(r)}{r} \chi_{\kappa m}(\theta, \varphi),$$

where

$$\begin{aligned} g_\mu^+(r) &= r^{l_\mu+1}\, e^{\mathrm{d}-\lambda_\mu r^2}, \qquad (5.10.1)\\ g_\mu^-(r) &= \left(\frac{d}{dr} + \frac{\kappa}{r}\right) g_\mu^+(r) = \left[t_\mu - 2\lambda_\mu r^2\right] r_\mu^l\, e^{\mathrm{d}-\lambda_\mu r^2}. \end{aligned}$$

where

$$t_\mu = \kappa_\mu + l_\mu + 1 = \begin{cases} 0 & \text{for } \kappa_\mu = -l_\mu - 1, \\ 2l_\mu + 1 & \text{for } \kappa_\mu = l_\mu \end{cases}$$

so that the kinetic matching condition (5.7.7) is satisfied (up to a constant multiplier) for all values of c. When we choose the normalization so that the diagonal elements of the Gram matrices $S_{\mu\mu}^{\beta\beta'} = \delta_{\beta\beta'}$, the normalization factors N_μ^β are given by

$$N^+ = \left[\frac{2(2\lambda_\mu)^{l_\mu+3/2}}{\Gamma(l_\mu+3/2)}\right]^{1/2}, \quad N^- = \left[\frac{2(2\lambda_\mu)^{l_\mu+1/2}}{\Gamma(l_\mu+5/2)}\right]^{1/2} \tag{5.10.2}$$

which depend only on $l_\mu = j_\mu + \eta_\mu/2$, where $\eta_\mu = \text{sgn}\ \kappa_\mu$. The Gram matrix elements on the same nuclear centre are particularly simple; they depend only on the ratio

$$\sigma_{\mu\nu} = \frac{2\sqrt{\lambda_\mu\lambda_\nu}}{\lambda_\mu+\lambda_\nu}$$

independent of the sign of κ_μ and κ_ν, so that

$$S_{\mu\nu}^{++} = \sigma_{\mu\nu}{}^{l_\mu+3/2}\delta_{l_\mu l_\nu}\delta_{m_\mu m_\nu}, \qquad S_{\mu\nu}^{--} = \sigma_{\mu\nu}{}^{l_\mu+1/2}\delta_{l_\mu l_\nu}\delta_{m_\mu m_\nu} \tag{5.10.3}$$

The kinetic matching construction (5.7.7) then gives the one-centre kinetic matrix elements

$$\Pi_{\mu\nu}^{-+} = \sqrt{(2l_\mu+3)\lambda_\mu} S_{\mu\nu}^{--}, \quad \Pi_{\mu\nu}^{+-} = \sqrt{(2l_\nu+3)\lambda_\mu} S_{\mu\nu}^{--}, \tag{5.10.4}$$

Whilst these matrices are particularly simple to calculate, the one-centre nuclear potential matrices $\mathbf{U}^{\beta\beta'}$ are rather more complicated and depend on the choice of nuclear model charge distribution. The uniform density model (5.4.3) gives perhaps the simplest expression in terms of the dimensionless variables $\sigma_{\mu\nu}$ and $x_{\mu\nu} = (\lambda_\mu+\lambda_\nu)R_N^2$ We have

$$U_{\mu\nu}^{++} = -\frac{Z}{R_N}\frac{(2\sigma_{\mu\nu})^{l+3/2}}{\sqrt{2\pi}(2l+1)!!}x_{\mu\nu}^{1/2}\Phi_l(x_{\mu\nu}). \tag{5.10.5}$$

where

$$\Phi_l(x) = l! - \gamma(l+1,x) + \frac{3}{2}x^{-1/2}\gamma(l+3/2,x) - \frac{1}{2}x^{-3/2}\gamma(l+5/2,x)$$

in which the incomplete gamma function is defined by

$$\gamma(a,x) = \int_0^x e^{-t}t^{a-1}\,dt, \quad x>0, \quad \Re a > 0.$$

Similarly

$$U_{\mu\nu}^{--} = -\frac{Z}{R_N}\frac{(2\sigma_{\mu\nu})^{l+5/2}}{\sqrt{2\pi}(2l+3)!!}x_{\mu\nu}^{1/2}\Phi_{l+1}(x_{\mu\nu}), \quad \kappa<0, \tag{5.10.6}$$

and

$$U_{\mu\nu}^{--} = -\frac{Z}{R_N}\frac{(2\sigma_{\mu\nu})^{l+5/2}}{\sqrt{2\pi}(2l+3)!!}x_{\mu\nu}^{1/2} \tag{5.10.7}$$
$$\times\left\{\Phi_{l+1}(x_{\mu\nu}) - \frac{2(2l+1)}{\sigma_{\mu\nu}}\Phi_l(x_{\mu\nu}) + \frac{(2l+1)^2}{\sigma_{\mu\nu}}\Phi_{l-1}(x_{\mu\nu})\right\}, \quad \kappa > 0.$$

Because $\Phi_l(x) \to l!$ as $x \to 0$, we recover the formula for a point nucleus when $R_N \to 0$. However, one should not use G-spinors with a point nuclear model, nor S-spinors with a finite size nuclear model. Even at relatively small atomic numbers, the relativistic cusp condition dominates the behaviour near $r = 0$ and the basis set mismatch degrades the numerical results if the wrong basis set is used.

5.11 Finite difference methods

The majority of calculations in relativistic atomic structure have been done using finite difference methods originating in the work of Hartree and his collaborators [14]. Finite difference methods generate bound radial wavefunctions $u_{n\kappa}(r)$ or continuum solutions $u_{\epsilon\kappa}(r)$ one at a time, so that the numerical procedures are quite different from those needed for calculations with finite basis sets. The most general form we shall encounter in DHF and MCDHF calculations is represented by inhomogeneous Dirac equations of the form

$$\left\{cJ\frac{d}{dr} + [W(r) - \epsilon]\right\} u(r) = X(r) \tag{5.11.1}$$

over a finite interval (R_0, R_1) where, as in (5.3.1) and (5.3.3),

$$u(r) = \begin{pmatrix} P(r) \\ Q(r) \end{pmatrix}, \quad J = \begin{pmatrix} 0 & -1 \\ 1 & 0 \end{pmatrix},$$
$$r\,W(r) = \begin{pmatrix} -Y(r) & c\kappa \\ c\kappa & -2rc^2 - Y(r) \end{pmatrix},$$

subject to boundary conditions at one or both ends of the interval. A power series expansion about $r = 0$ is normally used to calculate $u(R_0)$ provided R_0 is sufficiently small. For a bound state, we need a solution which decays sufficiently rapidly at large r to make the state normalizable. The inhomogeneous term $X(r)$ expresses the coupling between different classses of orbitals:

$$X(r) = x(r) + \int_0^\infty K(r,s)\,u(s)\,ds$$

where the integral includes all exchange interactions with the orbital $u(r)$ and $x(r)$ includes Lagrange multiplier terms expressing orthogonality of the orbital $u(r)$ to other orbitals of the same symmetry. Appendix B.6 presents the mathematical background to several finite difference methods for problems of the class (5.11.1). This section deals with their implementation.

The smooth mapping $t \to r = f(t)$ defines a new independent variable t which is discretized uniformly, $t_j = jh, \ j = 0, 1, \ldots, N$ so as to obtain conveniently spaced radial grid points $r_j = f(t_j)$. The mapping can be characterized by $k(t) = f'(t)/f(t)$; $k(t) = 1$ if we choose $f(t) = R_0 e^t$. We start with the initial value problem

$$J\frac{dw}{dt} = F(t; w), \quad w(0) = w_0.$$

where $w(t)$ is a continuously differentiable 2-vector. The constant matrix J is present in (5.11.1) in order to make the differential operator self-adjoint, and it is convenient to keep it explicit rather than absorbing it in the right-hand side. The differential equation with its initial conditions is equivalent to an integral equation

$$J(w(t) - w_0) = \int_0^t F(s, w(s))\, ds, \ 0 \leq t \leq T,$$

which we can discretize by replacing the integral on the right hand side by a numerical quadrature on a grid $t_j = jh, \ j = 0, 1, \ldots, N$ spanning the interval $(0, T = Nh)$. Choosing the *midpoint rule* for simplicity gives

$$J(v_{j+1} - v_j) = hF_{j+1/2} + h\,\tau_{j+1/2}, \quad v_0 = w_0, \quad j = 0, 1, \ldots, N-1 \quad (5.11.2)$$

connecting consecutive $v_j \approx w(t_j)$, where $F_{j+1/2}$ approximates the value of $F(t, w(t))$ at the midpoint $t_{j+1/2}$ of the interval (t_j, t_{j+1}). Theorem B.13 shows that if the *local truncation error* satisfies

$$\|\tau_{j+1/2}\| = O(h^2),$$

then the midpoint rule gives the error estimate

$$\| e_j \| = \| w(t_j) - v_j \| = O(h^2) \quad (5.11.3)$$

The error in the numerical solution therefore decreases only slowly as the step length h is reduced, and we need to look for a method in which the asymptotic error is proportional to a higher power of h.

However, there are advantages in retaining the simplicity of (5.11.2) but correcting the right-hand side iteratively to give a final result with the desired higher order accuracy. The solution of self-consistent field problems, which are inherently nonlinear because of the electromagnetic coupling between electrons, requires an iterative algorithm, and the *method of deferred correction* exploits this by retaining parts of $h\,\tau_{j+1/2}$ to $O(h^4)$ or $O(h^6)$ evaluated using values from previous iterations. Deferred correction schemes have long been used in both nonrelativistic [60, p. 241] as well as in relativistic self-consistent field calculations [61, 63]. At the $\nu+1$-st iteration, we determine $v_j \equiv v_j^{[\nu+1]}$ by solving equations

$$J(v_{j+1} - v_j) = hF^{[\nu]}_{j+1/2} + h\,\tau^{[\nu]}_{j+1/2}, \quad v_0 = w_0, \quad j = 0, 1, \ldots, N-1 \tag{5.11.4}$$

where

$$hF^{[\nu]}_{j+1/2} = -hW^{[\nu]}_{j+1/2} v_{j+1/2} + hR'_{j+1/2} \tag{5.11.5}$$

where

$$W^{[\nu]}_{j+1/2} = k_{j+1/2} \times \begin{pmatrix} -\left[Y(r) + \epsilon^{(\nu)} r\right]_{j+1/2}/c & \kappa \\ \kappa & -\left[Y(r) + \epsilon^{(\nu)} r + 2rc^2\right]_{j+1/2}/c \end{pmatrix}. \tag{5.11.6}$$

There a variety of ways to incorporate deferred difference corrections in (5.11.4) and to preserve the symmetry of the difference operator, we rewrite it as

$$J(v_{j+1} - v_j) + \frac{h}{2} W^{[\nu]}_{j+1/2}(v_{j+1} + v_j) = hR_{j+1/2} \tag{5.11.7}$$

with

$$hR_{j+1/2} = \frac{h}{2}\left(R'^{[\nu]}_{j+1} + R'^{[\nu]}_{j}\right) + T(\delta) hF^{[\nu]}_{j+1/2} \tag{5.11.8}$$

where

$$T(\delta) = -\frac{1}{12}\delta^2 + \frac{61}{2830}\delta^4 - \frac{563}{120960}\delta^6 + \ldots \tag{5.11.9}$$

is the central difference operator, (B.6.41), generating higher order contributions to the local truncation error $h\,\tau_{j+1/2}$ Setting $T(\delta)$ null is equivalent to using the trapezoidal rule for the original quadrature.

5.11.1 Methods of solution

The derivation of algorithms for solving the equations (5.11.7) is described in Appendix B.6.8. Whilst it is possible to treat them in principle as a system of simultaneous algebraic equations, there are advantages in using a shooting method to determine the estimates v_j by marching along the radial grid. Equations (5.11.7) are rearranged in the form

$$v_{j+1} = S_{j+1/2}\, v_j + T_{j+1/2}, \; j = 0, 1, \ldots, J-1, \tag{5.11.10}$$

where

$$S_{j+1/2} = \left(J + \frac{h}{2} W^{[\nu]}_{j+1/2}\right)^{-1} \left(J - \frac{h}{2} W^{[\nu]}_{j+1/2}\right)$$

and

$$T_{j+1/2} = \left(J + \frac{h}{2} W^{[\nu]}_{j+1/2} \right)^{-1} hR_{j+1/2}.$$

We can take advantage of the 2×2 structure of the matrices to simplify these formulae for practical computation. Thus, given v_0, we can generate successively $v_1, v_2, \ldots, v_J$. The initial value, v_0, can be obtained as in §5.4.1 by expansion in power series. The leading coefficient A (either p_0 or q_0) can be taken as arbitrary, and all coefficients of higher powers of the radius r can be computed as multiples of A. The lowest order coefficients depend only on the choice of ϵ and on the coefficients of $Y(r) \sim Y_0 + Y_1 r + \ldots$ near $r = 0$. The leading power of r in the expansion depends on the symmetry κ and on whether $Y_0 \neq 0$, as for a point charge nucleus, or $Y_0 = 0$ for one with a distributed nuclear charge. The marching process (5.11.10) is stable – that is to say errors introduced by rounding or the finite difference scheme are damped – out to the neighbourhood of the *classical turning point* of the effective potential. As r increases, square integrable solutions first reach a maximum and then oscillate out to the classical turning point.

Normalizable bound states decrease exponentially with j beyond the classical turning point. If the matrix W were constant, the solution of the linear difference system (5.11.10) could be written $v_j = \pi_j + A_1 {x_1}^j + A_2 {x_2}^j$, where π_j is a particular solution of the difference equation and x_1, x_2 are the eigenvalues of the (now constant) matrix $S_{j+1/2}$. Up to the turning point the two eigenvalues are of the form $x \sim \exp(\pm i\lambda)$ with *real*-valued λ, so that the numerical solution has a trigonometric dependence on j. Beyond the turning point, $\lambda = \pm i|\lambda|$; one root, say $x_1 \sim \exp(+|\lambda|)$, will dominate for j sufficiently large, no matter how small is A_1. This makes the marching process unstable beyond the turning point; powers of the dominant x_1 solution, introduced by tiny initial errors, soon overwhelm the decreasing solution. Although W is not independent of j, the qualitative behaviour is the same.

We need a different approach for the tail in bound state calculations. In principle we could still use (5.11.10) with $h \to -h$:

$$v_{j-1} = S_{j-1/2}\, v_j - T_{j-1/2},\; j = N, N-1, \ldots, J+1.$$

The solution that increases inwards dominates so that this process is stable. However, we usually have only a rough idea of what may be a suitable value of N so that a method that determines N automatically is to be preferred. The 2-component vector

$$v_j = \begin{pmatrix} P_j \\ Q_j \end{pmatrix}$$

is replaced by

$$w_{j+1/2} = \begin{bmatrix} -Q_j \\ P_{j+1} \end{bmatrix},\; j = J, \ldots, N-1 \qquad (5.11.11)$$

with the initial value

$$w_{J-1/2} = \begin{pmatrix} 0 \\ P_J \end{pmatrix}$$

at the join point. The original system (5.11.7) can simply be rearranged to give a block tridiagonal system of equations, Appendix B.6.10, for the vectors $w_{j+1/2}$:

$$-B_{j+1/2}w_{j-1/2} + C_{j+1/2}w_{j+1/2} - D_{j+1/2}w_{j+3/2} = hR_{j+1/2}$$

where $B_{j+1/2}, C_{j+1/2}, D_{j+1/2}$ are 2×2 matrices, modified appropriately at the outer region end-points $j = J$ and $j = N$ The system can be solved by Gauss elimination without pivoting, giving recurrence relations that are particularly simple because the matrices $B_{j+1/2}$ and $D_{j+1/2}$ are both rank 1.

Algorithm 5.1

(a) With the initial values

$$E_J := 0,\ x_{J-1/2} := w_{J-1/2},$$

form the matrices E_j and the vectors $x_{j+1/2}$ for $j = J, \ldots, N-1$ from the equations

$$E_{j+1} := (C_{j+1/2} - B_{j+1/2}E_j)^{-1}D_{j+1/2} \tag{5.11.12}$$

and

$$x_{j+1/2} := (C_{j+1/2} - B_{j+1/2}E_j)^{-1}(\widetilde{R}_{j+1/2} + B_{j+1/2}x_{j-1/2}). \tag{5.11.13}$$

(b) Starting from the boundary value $w_{N+1/2}$ form the solution for $j = N, N-1, \ldots, J+1$ from

$$w_{j-1/2} := E_j w_{j+1/2} + x_{j+1/2}. \tag{5.11.14}$$

■

The rank 1 character of $D_{j+1/2}$ in (5.11.12) means that E_j is also rank 1:

$$E_j = \begin{pmatrix} e_j^+ & 0 \\ e_j^- & 0 \end{pmatrix}.$$

Also

$$w_{N+1/2} = \begin{pmatrix} -Q_N \\ 0 \end{pmatrix} \to 0$$

for a bound state when N is sufficiently large. We can therefore determine the value of N from which to start the inward sweep (5.11.14) by checking when $\|x_{j+1/2}\|$ falls below some pre-assigned threshold for several steps.

5.11.2 Acceptable solutions

The result of this procedure is a solution v_j in two parts, $0 \leq j \leq J$ and $J \leq j \leq N$. We have made the component P_J of v_J the same in both sets, but the left and right values of Q_J will usually be different; we denote the difference by ΔQ_J. The solution will be said to be *acceptable* if

1. $|\Delta Q_J| < \delta$, where δ is a prescribed tolerance.
2. A has an appropriate sign
3. The large component P has the correct number of sign changes in $0 \leq j \leq J$. Because the zeros of P and Q interlace, this determines also the number of zeros of Q in $0 \leq j \leq J$.

We also want to choose A so that the solution is normalized. This is trivial for a single Dirac equation with a null right-hand side, $X(r) = 0$, because every term in the equation has A as a factor, but is less trivial for coupled DHF and MCDHF equations.

The number of zeros of the large component essentially depends on the value of ϵ. Appendix B.6.4 shows that the positions at which P has zeros move smoothly inwards as ϵ increases. Bound states, for which $-2mc^2 < \epsilon \leq 0$, have a finite number of zeros; continuum functions have an infinite number. The lowest bound state has just one zero, and extra ones appear as ϵ increases. A trial solution has the requisite n_r zeros when ϵ lies in an interval $(\underline{\epsilon}_{n_r}, \overline{\epsilon}_{n_r})$. Iteration strategies seek to reduce the width of this band until it determines ϵ to within the desired limits.

Hartree [14] proposed a scheme for improving A and ϵ together used in [62, 63]. The aim is to determine δA and $\delta\epsilon$ such that the first order correction to ΔQ_J

$$\Delta Q_J(A + \delta A, \epsilon + \delta\epsilon) = \Delta Q_J(A, \epsilon) + \frac{\partial \Delta Q_J}{\partial A}\delta A + \frac{\partial \Delta Q_J}{\partial \epsilon}\delta\epsilon$$

vanishes. As ΔQ_J is just the difference of two estimates of Q_J, we have

$$\Delta\frac{\partial Q_J}{\partial A}\delta A + \Delta\frac{\partial Q_J}{\partial \epsilon}\delta\epsilon = -\Delta Q_J(A, \epsilon) \tag{5.11.15}$$

Similarly, if we require the trial solution to be normalized to first order in the parameters,

$$\frac{\partial N}{\partial A}\delta A + \frac{\partial N}{\partial \epsilon}\delta\epsilon = 1 - N(A, \epsilon), \tag{5.11.16}$$

with

$$N(A, \epsilon) = \int_0^\infty (P^2(r) + Q^2(r))\, dr.$$

The normalization condition is not required for the homogeneous equation, so that we can then set $\delta A = 0$ in (5.11.15) and ignore (5.11.16). We require the partial derivatives with respect to A and ϵ which are easily obtained from (5.11.1):

$$\left\{ cJ\frac{d}{dr} + [W(r) - \epsilon] \right\} \frac{\partial u}{\partial \epsilon} = u(r)$$

whilst $\partial u/\partial A$ satisfies the homogeneous Dirac equation. These equations can be integrated at the same time as the original equation. Once we have obtained δA and $\delta\epsilon$ from (5.11.15) and (5.11.16), we can start a new iteration with these parameters. Unfortunately, this procedure can fail to give an acceptable trial solution. Several schemes, for example [64, §7.5–7.6], have been proposed to overcome the problem in codes that use the Hartree method.

A better strategy relies on the fact that we can always find an acceptable continuous solution to the inhomogeneous equation (5.11.4) satisfying the boundary conditions as long as ϵ is not an eigenvalue of the corresponding homogeneous equation (with $X(r) = 0$). Given values of the parameters A and ϵ, we can find a particular solution $v_j^{(P)}$ of the inhomogeneous problem and a corresponding solution $v_j^{(H)}$ of the homogeneous problem satisfying the boundary conditions. Unless the driving term $X(r)$ is very small, these will give different jumps $\Delta Q_J^{(P)}$ and $\Delta Q_J^{(H)}$ at the join, and the linear combination

$$v_j = v_j^{(P)} + \alpha v_j^{(H)}, \quad \alpha = -\Delta Q_J^{(P)}/\Delta Q_J^{(H)} \tag{5.11.17}$$

therefore gives a solution of (5.11.4) with $\Delta Q_J = 0$. The size of $\Delta Q_J^{(P)}$ in the next iteration is, however, a measure of the extent to which the DHF or MCDHF system has converged.

Appendix B.6.8 shows that the $2(N+1)$-vector $\boldsymbol{v}$, whose elements are the 2-vectors v_j obtained above, satisfy the matrix equation (B.6.4)

$$\boldsymbol{v}^\dagger \left(T^{[\nu]} - \epsilon^{[\nu]} \boldsymbol{S} \right) \boldsymbol{v} = \boldsymbol{v}^\dagger \, \boldsymbol{R}^{[\nu]} \tag{5.11.18}$$

where $T^{[\nu]}$ and $\boldsymbol{S}$ are $2N \times 2N$ symmetric block tri-diagonal matrices whose elements are 2×2 square matrices, and $\boldsymbol{R}^{[\nu]}$ and $\boldsymbol{v}$ are formed in the same way. This system is equivalent to the equations (5.11.10)–(5.11.14) of the shooting method. The matrix $T^{[\nu]}$, which is a representation of the Dirac operator, has non-zero blocks

$$\begin{aligned} T_{jj}^{[\nu]} &= \overline{V}_j^{[\nu]}, \\ T_{j,j+1}^{[\nu]} &= +J/h + \frac{1}{2} V_{j+1/2}^{[\nu]}, \\ T_{j+1,j}^{[\nu]} &= -J/h + \frac{1}{2} V_{j+1/2}^{[\nu]}, \end{aligned} \tag{5.11.19}$$

where

$$V_{j+1/2}^{[\nu]} = k_{j+1/2} \begin{pmatrix} -\dfrac{1}{c} Y_{j+1/2}^{(\nu)} & \kappa \\ \kappa & -\dfrac{1}{c} Y_{j+1/2}^{(\nu)} - 2cr_{j+1/2} \end{pmatrix},$$

and, setting $V_{-1/2}^{[\nu]} = V_{N+1/2}^{[\nu]} = 0$, we define

$$\overline{V}_j^{[\nu]} = \frac{1}{2}\left(V_{j+1/2}^{[\nu]} + V_{j-1/2}^{[\nu]}\right), \; j = 0, 1, \ldots, N.$$

The antisymmetry of the matrix J ensures that ${T_{j,j+1}}^T = T_{j+1,j}$, guaranteeing overall symmetry of the matrix $T^{[\nu]}$. Similarly the elements of $\boldsymbol{S}$ are defined in terms of 2×2 matrices $S_{j+1/2}$ by

$$S_{jj} = \overline{S}_j, \quad S_{j,j+1} = S_{j+1,j} = \frac{1}{2}S_{j+1/2}$$

with $S_{-1/2} = S_{N+1/2} = 0$, $\overline{S}_j = \left(S_{j+1/2} + S_{j-1/2}\right)/2$ and

$$S_{j+1/2} = \frac{1}{c}k_{j+1/2}r_{j+1/2}\begin{pmatrix} 1 & 0 \\ 0 & 1 \end{pmatrix}.$$

Appendix B.6.8 explains the origin of this averaging. The difference between $\overline{V}_j^{[\nu]}$ and $V_j^{[\nu]}$ is probably not all that significant in practice unless the variation with j is very nonlinear; the same will be true for other quantities appearing in (5.11.18).

This reformulation expresses the finite difference equations for the Dirac equation as a matrix problem similar in character to nonrelativistic atomic structure theory [7, §3.11.4] and motivates

Algorithm 5.2 *For* $\nu = 0, 1, 2, \ldots,$

(i) Given the acceptable solution $\boldsymbol{v}^{[\nu]}$, *use the generalized Rayleigh quotient to generate the estimate*

$$\epsilon^{[\nu]} = \boldsymbol{v}^{[\nu]\dagger}(\boldsymbol{T}^{[\nu]}\boldsymbol{v}^{[\nu]} - \boldsymbol{R}^{[\nu]})/\|\boldsymbol{v}^{[\nu]}\|^2 \qquad (5.11.20)$$

where $\|\boldsymbol{v}^{[\nu]}\|^2 = \boldsymbol{v}^{[\nu]\dagger}\boldsymbol{S}\boldsymbol{v}^{[\nu]}$ *is an expression for the integral* N *of (5.11.16). The matrix* $\boldsymbol{S}$ *accounts for the change of variable* $t \to r = f(t)$.

(ii) Generate a new acceptable trial solution $\boldsymbol{w}^{[\nu+1]}$ *from*

$$(\boldsymbol{T}^{[\nu]} - \epsilon^{[\nu]}\boldsymbol{S})\boldsymbol{w}^{[\nu+1]} = \boldsymbol{R}^{[\nu]} \qquad (5.11.21)$$

or its shooting equivalent (5.11.10)–(5.11.14) with (5.11.17).

(iii) Normalize to give the new trial solution

$$\boldsymbol{v}^{[\nu+1]} = \boldsymbol{w}^{[\nu+1]}/\|\boldsymbol{w}^{[\nu+1]}\|. \qquad (5.11.22)$$

■

This algorithm is expected to be satisfactory except when $X(r)$ is very small. In the limit $X(r) = 0$ this is a true eigenvalue problem, and the assumptions underlying Algorithm 5.2 are no longer valid. The inverse iteration step in the next algorithm overcomes this problem: :

Algorithm 5.3 *For* $\nu = 0, 1, 2, \ldots,$

(i) Given the acceptable solution $\boldsymbol{v}^{[\nu]}$, *use the generalized Rayleigh quotient to generate the estimate*

$$\epsilon^{[\nu]} = \boldsymbol{v}^{[\nu]\dagger}(\boldsymbol{T}^{[\nu]}\boldsymbol{v}^{[\nu]} - \boldsymbol{R}^{[\nu]})/\|\boldsymbol{v}^{[\nu]}\|^2$$

as before.

(ii) Generate a new acceptable trial solution $\boldsymbol{w}^{[\nu+1]}$ *by solving both*

$$(\boldsymbol{T}^{[\nu]} - \epsilon^{[\nu]}\boldsymbol{S})\boldsymbol{w}'^{[\nu+1]} = \boldsymbol{R}^{[\nu]}$$

and

$$(\boldsymbol{T}^{[\nu]} - \epsilon^{[\nu]}\boldsymbol{S})\boldsymbol{w}''^{[\nu+1]} = \boldsymbol{S}\boldsymbol{v}^{[\nu]} \qquad (5.11.23)$$

for $\boldsymbol{w}''^{[\nu+1]} \approx \partial\boldsymbol{v}/\partial\epsilon$ *as in Algorithm 5.2.*

(iii) Choose the new $\boldsymbol{v}^{[\nu+1]}$ *such that*

$$\boldsymbol{v}^{[\nu+1]} = \boldsymbol{w}'^{[\nu+1]} + \beta\boldsymbol{w}''^{[\nu+1]}$$

is normalized: $\|\boldsymbol{v}^{[\nu+1]}\| = 1$.

■

With any of these algorithms, the first few iterations may be atypical, and the Rayleigh quotient may fail to give a value of ϵ that is acceptable. Chapter 7 discusses the way in which this can be handled in practice.

5.12 Finite element methods

The routine use of finite element methods for solving partial differential equations in all forms of continuum mechanics and structural enginering owes much to the fact that computational elements can be placed where they are needed, which is particularly useful for dealing with irregular regions [66]. These problems usually lead to banded matrix equations that are well-suited to calculations on modern computers. The long-range electromagnetic interactions of atomic and molecular physics generate systems of equations with full matrices, making the advantages of finite elements less obvious. Shore [67] examined the use of cubic-splines in 1974 to solve one-dimensional Schrödinger equations. Altenberger-Siczek and Gilbert [68] examined both cardinal splines and B-splines for the helium atom in 1976, and ruled them out on cost grounds. Interest in the use of B-splines revived in the 1980s: for example, Bottcher and Strayer [69] considered time-dependent atomic and nuclear problems, Johnson *et al.* [70, 71, 72] used the approach for RMBPT, and Fischer *et al.* [73, 74, 75, 76] for Hartree-Fock and continuum calculations. Sapirstein and Johnson [77] have reviewed the use of B-splines in atomic physics. This section discusses the use of B-splines in solving single particle Schrödinger and Dirac equations.

5.12.1 B-splines

Following de Boor [78], we divide the interval $[0, R]$ into subintervals whose endpoints are defined by the *knot sequence* $\{t_i,\, i = 1, 2, \ldots, n+k\}$. The B-splines of order k are defined recursively. The first order B-splines (with $k = 1$) are *piecewise constant* between knots

$$B_i^1(r) = \begin{cases} 1, \; t_i \le r \le t_{i+1} \\ 0, \text{ otherwise.} \end{cases} \tag{5.12.1}$$

B-splines for $(k = 2, 3, \ldots)$ are defined by the recursion

$$B_i^k(r) = \frac{r - t_i}{t_{i+k-1} - t_i} B_i^{k-1}(t) + \frac{t_{i+k} - r}{t_{i+k} - t_{i+1}} B_{i+1}^{k-1}(t). \tag{5.12.2}$$

Thus $B_i^k(r)$ vanishes outside of the interval $t_i \le r \le t_{i+k}$ and

$$\sum_i B_i^k(r) = 1.$$

The set of B-splines of order k on the knot sequence t_1 forms a complete basis of piecewise polynomials of degree $k-1$ on the interval spanned by the knot sequence with all derivatives continuous up to order $k-2$. At the end-points, the knots have k-fold multiplicity:

$$t_1 = t_2 = \ldots = t_k = 0, \quad t_{n+1} = t_{n+2} = \ldots = t_{n+k} = R.$$

Limiting forms of the definitions must be used at multiple knots. Most B-splines vanish at the endpoints save for one that takes the value 1 at r=0, and one that takes the value 1 at $r = R$, for connecting to boundary values. For a mathematical presentation of approximation theory of splines in general, and B-splines in particular, see Powell [79, Chapters 18–24].

The user has considerable freedom to choose the knot sequence. Fischer *et al.* [74] introduced a general grid for atomic physics calculations with four parameters: the step size $h = 2^{-m}$ for some integer m, a maximum step size h_{max}, the maximum value R and the spline order k. The knot sequence is such that

$$\begin{aligned} t_i &= 0 && \text{for } i = 1, \ldots, k, \\ t_{i+1} &= t_i + h && \text{for } i = k, \ldots, k+m, \\ t_{i+1} &= t_i(1+h) && \text{for } 1 \le t_{i+1} - t_i < h_{max}, \\ t_{i+1} &= t_i + h_{max} && \text{for } t_i < ZR. \end{aligned}$$

and $r_i = t_i/Z$ at all knots. Sapirstein and Johnson [77] used an exponentially increasing set $t_{i+1} = t_i e^{\alpha h}$, $i = 1, 2, \ldots$.

5.12.2 Variational formulation of finite element schemes

Suppose that we wish to solve a (partial or ordinary) differential equation

$$\mathcal{L}y(r) = f(r) \tag{5.12.3}$$

where $\mathcal{L}$ is, say, an operator defined by

$$\mathcal{L}y(r) := y'' + (V(r) - \lambda)y(r)$$

on a suitable interval $[0, R]$. Although the finite element method is much more general [80], applicable to linear and nonlinear problems in several dimensions, we here restrict the discussion to approximation with a B-spline expansion

$$y(r) \approx Y(r) = \sum_{i=1}^{N} a_i\, B_i^k(r) \tag{5.12.4}$$

and determine the N coefficients a_i from the equations

$$\langle \psi_j,\ \mathcal{L}Y - f \rangle = 0, \quad j = 1, \ldots, N, \tag{5.12.5}$$

where

$$\langle u,\ v \rangle := \int_0^R u^*(r)\, v(r)\, dr$$

Several types of *test function* $\psi_j(r)$ have been used to generate finite element algorithms, of which the following have been most exploited in atomic physics:

Galerkin methods: $\psi_j = B_j^k,\ j = 1, \ldots, N$. Equation (5.12.5) gives an N-dimensional matrix equation of the form

$$\boldsymbol{La} = \boldsymbol{f} \tag{5.12.6}$$

where $\boldsymbol{a}$ is a vector of coefficients a_i in (5.12.4), $\boldsymbol{f}$ is a vector with elements $f_i = \langle B_i^k,\ f \rangle$ and $\boldsymbol{L}$ is a square matrix with elements

$$L_{ij} = \langle B_i^k,\ \mathcal{L}B_j^k \rangle. \tag{5.12.7}$$

Collocation methods: We can avoid having to evaluate integrals by choosing

$$\psi_j(r) = \delta(r - r_j), j = 1, \ldots, N,$$

where $r_j,\ j = 1, \ldots, N$ is some distribution of points on the interval $[0, R]$. This is equivalent to a set of equations for the coefficients a_i of the form

$$\mathcal{L}Y - f = 0 \text{ at } r = r_1, \ldots r_N.$$

The main problem with this approach is that the results are often sensitive to the distribution of the *collocation points* r_i and that the order k of the spline must be taken sufficiently high for $\mathcal{L}B_i^k(r)$ to have a meaning.

5.12.3 Schrödinger equations

Equation (5.12.3) becomes a Schrödinger equation with

$$\mathcal{L}y(r) := y'' - \left(\frac{l(l+1)}{r^2} + 2V(r) - \varepsilon\right) y = 0.$$

where, for a hydrogenic atom, $V(r) = -Z/r$. Following Fischer *et al.* [73], we choose $n+k-1$ basis splines $B_i^k(r),\ i = 0, 1, \ldots, n+k-2$ and write, as in (5.12.4),

$$y(r) \approx Y(r) = \sum_{i=1}^{n+k-3} c_i\, B_i^k(r).$$

The B-splines for $i = 0$ and $i = n+k-2$, which alone are nonvanishing at the endpoints, are dropped to ensure that $Y(0) = Y(R) = 0$. The Galerkin procedure then gives

$$\boldsymbol{F}\,\mathbf{c} = \widetilde{\varepsilon}\,\boldsymbol{S}\,\mathbf{c} \tag{5.12.8}$$

where $\boldsymbol{F}$ and $\boldsymbol{S}$ are banded square symmetric matrices of dimension $m = n+k-3$ with elements

$$F_{ij} = \int_0^\infty B_i^k(r)\left(\frac{d^2}{dr^2} - \frac{l(l+1)}{r^2} - 2V(r)\right) B_j^k(r)dr$$

$$S_{ij} = \int_0^\infty B_i^k(r)\, B_j^k(r)dr$$

and $\widetilde{\varepsilon}$ is an approximation to the eigenvalue. Gaussian quadrature of order k provides an economical way of evaluating the integrals because of the piecewise polynomial character of the B-splines.

For applications to HF and CI calculations, the main issue is the influence of the choice of knot sequence and the size R of the confining box upon the solutions.[4] Table 5.6 is based on a calculation in which the knot distribution was linear near the origin and exponential thereafter:

$$r_j = jh/Z,\ j = 0, \ldots, h^{-1}, \quad r_{j+1} = r_j(1+h), j > h^{-1}.$$

With this distribution, relations like

$$\langle B_i^k, r^p B_i^k \rangle = (1+h)^{p+1} \langle B_{i-1}^k, r^p B_{i-1}^k \rangle$$

hold when the B-splines are entirely in the outer region, minimizing the number of integrals to be calculated. Although the matrix is quite large, only a small number of eigenvalues and eigenvectors of (5.12.8) are required, so that this type of calculation has much in common with finite difference schemes for one-particle problems in which each eigensolution is determined on its own.

---

[4] We defer discussion of the search for an efficient method of evaluating the large number of Slater integrals needed in HF and CI calculations to Chapter 6.

Table 5.6. Accuracy of ns eigenvalues of hydrogenic Schrödinger equation as a function of knot distribution and box size using splines of order $k = 7$ (after Fischer, Guo and Shen [76]).

n	$h = 1/8$			$h = 3/32$
	R=50 $N = 48$	200 59	2000 79	625 91
1	0.33 (-12)	0.29 (-12)	0.27 (-12)	0.09 (-12)
2	0.64 (-12)	0.69 (-12)	0.69 (-12)	0.04 (-12)
3	0.11 (-8)	0.27 (-10)	0.12 (-10)	0.98 (-12)
4	0.20 (-4)	0.68 (-11)	0.68 (-11)	0.26 (-12)
5		0.21 (-10)	0.28 (-10)	0.77 (-12)
6		0.63 (-10)	0.19 (-9)	0.22 (-11)
7		0.30 (-9)	0.69 (-9)	0.23 (-10)
8		0.23 (-6)	0.13 (-10)	0.34 (-9)
9		0.30 (-4)	0.12 (-8)	
10			0.45 (-8)	

The importance of taking a sufficiently large box in order to get good results for higher values of n is obvious.

The procedure described by Sapirstein and Johnson [77] is essentially similar, although the Galerkin equations are derived from a hydrogenic functional

$$S[y] := \frac{1}{2}\int_0^R \left\{ \frac{1}{2}(y')^2 + \left(-\frac{Z}{r} + \frac{l(l+1)}{2r^2} \right) y^2\, dr \right\} - \frac{1}{2}\int_0^R y^2\, dr \qquad (5.12.9)$$

and they used a purely exponential radial knot distribution. The qualitative behaviour of the eigensolutions as a function of the knot distribution and the box size was similar to that reported by Fischer *et al.* However, Johnson *et al.* were more interested in calculations using many-body theory rather than HF theory and therefore solved for all eigenvalues and eigenvectors.

5.12.4 Dirac equations

Both collocation [69] and Galerkin methods [72, 77] have been used to solve the Dirac equation with B-splines. The Galerkin procedure presented in [72] is based on the functional

$$\begin{aligned} S = \frac{1}{2}\int_0^R \Big\{ & cP(r)\left(\frac{d}{dr} - \frac{\kappa}{r}\right) Q(r) - cQ(r)\left(\frac{d}{dr} + \frac{\kappa}{r}\right) P(r) \\ & + V(r)\left(P^2(r) + Q^2(r)\right) - 2mc^2 Q^2(r) \Big\} dr \\ & - \frac{1}{2}\epsilon \int_0^R \left(P^2(r) + Q^2(r)\right) dr, \qquad (5.12.10) \end{aligned}$$

having a solution normalized so that

$$\int_0^R \left(P^2(r) + Q^2(r)\right) dr = 1.$$

The radial amplitudes are expanded in terms of B-splines so that

$$P(r) \approx \sum_{i=1}^{n} p_i B_i^k(r), \quad Q(r) \approx \sum_{i=1}^{n} q_i B_i^k(r).$$

The first variation of S is given by

$$\begin{aligned}\delta S = \frac{1}{2}\int_0^R \Big\{ &\delta P(r) \left[c\left(\frac{d}{dr} - \frac{\kappa}{r}\right) Q(r) + V(r)P(r) \right] \\ &+\delta Q(r) \left[-c\left(\frac{d}{dr} + \frac{\kappa}{r}\right) P(r) + (V(r) - 2mc^2)Q(r) \right] \Big\}\, dr \\ &- \epsilon \int_0^R (\delta P(r)P(r) + \delta Q(r)Q(r))\, dr \\ &+ \frac{1}{2}\, c\left[P(r)\delta Q(r) - Q(r)\delta P(r)\right]_0^R . \end{aligned} \tag{5.12.11}$$

Setting $\delta S = 0$ leads to the usual radial Dirac equations, subject to

$$\frac{1}{2} c\left[P(r)\delta Q(r) - Q(r)\delta P(r)\right]_0^R . \tag{5.12.12}$$

The method chosen in [72] is to add another functional

$$S' = \begin{cases} \frac{1}{4}\, c[P^2(R) - Q^2(R)] + \frac{1}{2}\, c[P^2(0) - P(0)Q(0)], & \text{when } \kappa < 0, \\ \frac{1}{4}\, c[P^2(R) - Q^2(R)] + c^2 P^2(0) + \frac{1}{2}\, cP(0)Q(0), & \text{when } \kappa > 0 \end{cases} \tag{5.12.13}$$

so as to make $\delta(S+S') = 0$. When the Dirac equations are satisfied at internal points, the boundary contribution now looks like

$$0 = \delta(S + S') = \begin{cases} \frac{1}{2}\, c[P(R) - Q(R)][\delta P(R) + \delta Q(R)] \\ \qquad + cP(0)(\delta P(0) - \delta Q(0)) & \kappa < 0 \\ \frac{1}{2}\, c[P(R) - Q(R)][\delta P(R) + \delta Q(R)] \\ \qquad + 2c^2 P(0)\delta P(0) + cP(0)\delta Q(0) & \kappa > 0 \end{cases}$$

which vanishes identically for *unconstrained* $\delta P(0)$, $\delta Q(0)$, $\delta P(R)$, and $\delta Q(R)$ provided

$$P(0) = 0, \qquad P(R) = Q(R). \tag{5.12.14}$$

The condition at R is that applied in the MIT bag model [81]. The factor c^2 was introduced into the $\kappa > 0$ condition of (5.12.13) in order to prevent the appearance of spurious states [77, p. 5222]. These states have eigenvalues in the bound state spectrum lying *below* the physical lowest bound state; we have seen, §5.4.3, that this is mainly attributable to the use of incorrect boundary conditions at the origin. The spurious states are highly oscillatory with respect to r and usually contribute little to sum-over-states calculations; when such states occur, they are simply moved to the end of the spectrum.

This pragmatic solution to the problem disguises the fact that more work needs to be done on the boundary conditions, especially at the origin. Most relativistic calculations are done with finite size nuclear models, and it is usual to place several knots inside the nucleus much as in finite difference calculations. The use of a finite radial box discretizes the spectrum; as with analytic basis sets such as S- or G-spinors, the lowest eigenstates are little affected by the box size, but the higher positive energy states (and the negative energy states) have a discrete spectrum whose physical interpretation is not so obvious. It is often the case that sum-over-states perturbation formulae converge quite well in a B-spline representation: in effect, the discrete spectrum allows one to integrate over the higher bound states and the continuum in much the same way as with S- and G-spinors. The complete spectrum includes the negative energy states, and their small contribution is often useful for improving the accuracy of numerical results. See [77] for more details.

References

1. Titchmarsh E C 1948 *Introduction to the Theory of Fourier Integrals* (2nd. ed.) (Oxford: Clarendon Press).
2. Kato T 1976 *Perturbation Theory for Linear Operators* (Berlin: Springer-Verlag).
3. Reed M and Simon B 1972-1979 *Methods of Modern Mathematical Physics* (4 vols.) (New York: Academic Press).
4. Kalf, H, Schmincke, U-W, Walter, J and Wüst, R 1975 *On the spectral theory of Schrödinger and Dirac operators with strongly singular potentials*, in Spectral theory and differential equations, (Lecture Notes in Mathematics No. 448) ed. W N Everitt, pp. 182 – 226. (Berlin: Springer-Verlag).
5. Greiner W, Müller B and Rafelski J 1985 *Quantum Electrodynamics of Strong Fields* (Berlin: Springer-Verlag).
6. Richtmyer R D 1978 *Methods of Advanced Mathematical Physics* (2 vols) (New York: Springer-Verlag).
7. Fischer C F, Brage T and Jönsson P 1997 *Computational Atomic Structure. An MCHF approach* (Bristol and Philadelphia: Institute of Physics).
8. Pauling L and Wilson Jr E B 1985 *Introduction to Quantum Mechanics with applications in chemistry* (Reprint of the 1935 edition) (New York: Dover).
9. Dirac P A M 1958 *The Principles of Quantum Mechanics* 4th ed. (Oxford: Clarendon Press).
10. Kellner G W 1927 *Z. Phys.* **44** 91, 110.
11. Hylleraas E A 1928 *Z. Phys.* **48**, 469.
12. Hylleraas E A 1929 *Z. Phys.* **54**, 347.
13. Hylleraas E A 1930 *Z. Phys.* **65**, 209.
14. Hartree D R 1957 *The calculation of atomic structures* (New York: John Wiley).
15. Fischer C F 1977 *The Hartree-Fock Method for Atoms* (New York: John Wiley).
16. Swirles B 1935 *Proc Roy Soc (Lond) A* **152** 625; — 1936 *Proc Roy Soc (Lond) A* **157** 680.
17. Grant I P 1961 *Proc Roy Soc (Lond) A* **262** 555; — 1965 *Proc Phys Soc* **86** 523.
18. Grant I P 1970 *Adv. Phys.* **19** 747.
19. Joachain C J (1983) *Quantum collision theory* (3rd edition) (Amsterdam: North-Holland Publishing Company).
20. Synek M 1964 *Phys. Rev. A* **136** 1552.
21. Kim Y K 1967 *Phys. Rev. A* **154** 17.
22. Kim Y K 1967 *ibid.* **155** 190.
23. Kagawa T 1975 *Phys. Rev. A* **12**, 2245.
24. Kagawa T 1980 *Phys. Rev. A* **22**, 2340.
25. Schwarz W H E and Wallmeier H 1982 *Mol. Phys.* **46**, 1045.
26. Wallmeier H and Kutzelnigg W 1981 *Chem. Phys. Lett.* **78**, 341.
27. Schwarz W H E and Wechsel-Trakowski E 1982 *Chem. Phys. Lett.* **85**, 94.
28. Mark F and Rosicky F 1980 *Chem. Phys. Lett.* **74**, 562.
29. Mark F and Schwarz W H E 1982 *Phys. Rev. Lett.* **48**, 673.
30. Kutzelnigg W 1984 *Int. J. Quant. Chem.* **25**, 107.
31. Klahn B and Bingel W A 1977 *Theoret. Chim. Acta (Berl.)* **44** 9.
32. Klahn B and Bingel W A 1977 *Theoret. Chim. Acta (Berl.)* **44** 27.
33. Klahn B and Morgan III J D 1984 *J. Chem. Phys.* **81** 410.
34. Mikhlin S G 1964 *Variational Methods in Mathematical Physics* (transl. T. Boddington) (Oxford: Pergamon Press).

35. Bonitz G 1971 *Zum Ritzschen Verfahren.* (Berlin: VEB Deutscher Verlag der Wissenschaften).
36. Kato T 1951 *Trans. Am. Math. Soc.* **70**, 195.
37. Eckart C 1930 *Phys. Rev.* **36** 878.
38. Löwdin P O 1959 *Adv. Chem. Phys.* **2** 207.
39. Grant I P 1982 *Phys. Rev. A* **25**, 1230.
40. Grant I P 1986 *J. Phys. B: Atom. molec. phys.* **19**, 3187.
41. Grant I P and Quiney H M 2000 *Phys. Rev. A* **62**, 022508.
42. Stakgold 1979 *Green's functions and Boundary Value Problems* (New York: John Wiley & Sons).
43. Dyall K G, Grant I P and Wilson S 1984 *J. Phys. B* **17**, L45.
44. Dyall K G, Grant I P and Wilson S 1984 *J. Phys. B* **17**, 493.
45. Dyall K G, Grant I P and Wilson S 1984 *J. Phys. B* **17**, 1201.
46. Lee Y S and McLean A D 1982 *J. Chem. Phys.* **76**, 735.
47. Quiney H M, Grant I P and Wilson S 1989, in [48], p. 307.
48. Kaldor U (ed.) 1989 *Many-Body Methods in Quantum Chemistry* (Lecture Notes in Chemistry 52) (Berlin: Springer-Verlag).
49. Grant I P and Quiney H M 1988 *Adv. At. Mol. Phys.* **23**, 37.
50. Grant I P 1989 in *Relativistic, Quantum Electrodynamic and Weak Interaction Effects in Atoms* (ed. W R Johnson, P J Mohr and J Sucher), p. 235.
51. Drake G W F ed 2005 *Springer Handbook of Atomic, Molecular and Optical Physics* Second edn (New York: Springer-Verlag).
52. Szmytkowski R 1997 *J. Phys. B: At. Mol. Opt. Phys.* **30**, 825.
53. Rotenberg M 1962 *Ann. Phys. (N. Y.)* **19**, 262.
54. Rotenberg M 1970 *Adv. Atom. Molec. Phys.* **6**, 233.
55. Abramowitz M and Stegun I A 1970 *Handbook of Mathematical Functions* (New York: Dover).
56. Klahn B and Bingel W A 1977 *Theoret. Chim. Acta (Berl.)* **44**, 9, 27.
57. Smith B T *et al.* 1976 *Matrix Eigensystem Routines - EISPACK Guide*, (2e) (Lecture Notes in Computer Science, Vol. 6) (New York: Springer-Verlag).
58. Press W H, Teukolsky S A, Vetterling W T and Flannery B P 1992 *Numerical Recipes in FORTRAN* (2e) (Cambridge: University Press).
59. Quiney H M 1987 *Finite basis set studies of the Dirac equation* (Unpublished D Phil Dissertation, Oxford University).
60. Fischer C F 1972 *Atomic Data* **4** 302; — 1973 *At Data Nucl Data Tab* **12** 87.
61. Mayers D F 1957 *Proc. Roy. Soc. A* **241** 93.
62. Mayers D F and O'Brien F 1968 *J. Phys. B* **1**, 145.
63. Desclaux J P, Mayers D F and O'Brien F 1971 *J. Phys. B: At. Mol. Phys.* **4** 631.
64. Cowan R D 1981 *The Theory of Atomic Structure and Spectra* (Berkeley: University of California Press).
65. Wilson S 1984 *Electron correlation in molecules* (Oxford: Clarendon Press).
66. Gladwell I and Wait R 1979 *A survey of numerical methods for partial differential equations.* (Oxford: Clarendon Press).
67. Shore B 1974 *J. Phys. B: At. Mol. Phys.* **7**, 2502.
68. Altenberger-Siczek A and Gilbert T L 1976 *J. Chem. Phys.* **64**, 432.
69. Bottcher C and Strayer M R 1987 *Ann. Phys. (N.Y.)* **175**, 64.
70. Johnson W R and Sapirstein J 1986 *Phys. Rev. Lett.* **57**, 1126.
71. Johnson W R, Idrees M and Sapirstein J 1987 *Phys. Rev. A* **35**, 3218.

72. Johnson W R, Blundell S A and Sapirstein J 1988 *Phys. Rev. A* **37** 307.
73. Froese Fischer C and Idrees M 1989 *Comput. Phys.* **3**, 53.
74. Froese Fischer C and Idrees M 1990 *J. Phys. B: At. Mol. Phys.* **23**, 679.
75. Froese Fischer C and Guo W 1990 *J. Comput. Phys.* **90**, 486.
76. Froese Fischer C, Guo W and Shen Z 1992 *Int. J. Quant. Chem.* **42**, 849.
77. Sapirstein J and Johnson W R 1996 *J. Phys. B: At. Mol. Phys.* **29**, 5213.
78. de Boor C 1978 *A practical guide to splines* (New York: Springer Verlag).
79. Powell M J D 1981 *Approximation theory and methods* (Cambridge: Cambridge University Press).
80. Davies A J 1980 *The finite element method: a first approach* (Oxford: Clarendon Press).
81. Chodos A, Jaffe R L, Johnson K, Thorn C B and Weisskopf V F 1974 *Phys. Rev. D* **9** 3471.

6

Complex atoms

6.1 Dirac-Hartree-Fock theory

The effective Hamiltonian (4.14.1) for many-electron atoms and molecules is

$$H = \sum_{i=1}^{N} h_i + \sum_{i<j} g_{ij}, \tag{6.1.1}$$

where h_i is the one-electron Dirac Hamiltonian for a bare nucleus for electron i and $g_{ij} = 1/R_{ij} + g^B(R_{ij})$ represents the interaction energy of electrons i and j. In most applications, the central core of the atom is dominated by electrons in spherically symmetric closed shells, so that we expect an independent particle central field model to be a good starting point. This motivates the use of wavefunctions for atoms and molecules built from anti-symmetrized products of one-electron central field wavefunctions. The theory of electronic structure of isolated complex atoms is dominated by applications of angular momentum algebra, Appendix B.3, exploiting the fact that such systems can have no preferred orientation. Angular momentum theory can only play a subordinate role in molecules where, away from the nuclei, the electrons move in a nonspherical force-field shaped by the nuclear skeleton.

The simplest model of a closed subshell atom, configuration $\mathcal{C}$, is described by a single determinant wavefunction. We can adapt the nonrelativistic particle-hole formalism of §4.12.1, so that the energy of the system is given by the Fock space operator

$$\widehat{H} = \widehat{H}_0 + \widehat{V}, \quad \widehat{V} = V_0 + \{\widehat{V}_1\} + \{\widehat{V}_2\}, \tag{6.1.2}$$

where

$$\widehat{H}_0 = \sum_{i=1}^{N} a_i^\dagger a_i \, \varepsilon_i, \tag{6.1.3}$$

ε_i being the i-th eigenvalue of the Dirac equation,

$$(h + u)\psi_i = \varepsilon_i \psi_i, \qquad h = c\boldsymbol{\alpha} \cdot \boldsymbol{p} + (\beta - 1)c^2 + v_{nuc}(r),$$

where $v_{nuc}(r)$ is the electron-nucleus interaction energy, and u is some one-electron *mean field* potential approximating the overall electron-electron interaction. The energy zero has been shifted to coincide with nonrelativistic conventions. The effective Hamiltonian comprises the *zero-body* operator (a pure number),

$$V_0 = -\sum_a \langle a \,|\, u \,|\, a \rangle + \frac{1}{2} \sum_{ab} \left[\langle ab \,|\, g \,|\, ab \rangle - \langle ba \,|\, g \,|\, ab \rangle \right];$$

a *one-body* operator

$$\{\widehat{V}_1\} = \sum_{ij} \{a_i^\dagger a_j\} \langle i \,|\, v \,|\, j \rangle;$$

and a *two-body* operator

$$\{\widehat{V}_2\} = \frac{1}{2} \sum_{ijkl} \{a_i^\dagger a_j^\dagger a_l a_k\} \langle ij \,|\, g \,|\, kl \rangle.$$

The indices i, j, k, l run over the complete (electron) spectrum, whereas a, b run over occupied (core) states only, and g is the electron-electron interaction. The effective one-body potential v is defined so that

$$\langle i \,|\, v \,|\, j \rangle = \langle i \,|\, -u + u_{HF} \,|\, j \rangle. \tag{6.1.4}$$

where the Hartree-Fock effective one-body potential, u_{HF}, is defined for all orbitals by

$$\langle i \,|\, u_{HF} \,|\, j \rangle = \sum_b \left[\langle ib \,|\, g \,|\, jb \rangle - \langle bi \,|\, g \,|\, jb \rangle \right], \tag{6.1.5}$$

with the sum over b running over core orbital indices. When we identify u with u_{HF} so that v vanishes, then

$$\varepsilon_a = \langle a \,|\, h + u_{HF} \,|\, a \rangle.$$

for each core orbital. However, the pair repulsions are counted twice, so that the total energy is

$$\begin{aligned} E &= \sum_a \left\{ \varepsilon_a - \frac{1}{2} \langle a \,|\, u_{HF} \,|\, a \rangle \right\} \\ &= \sum_a \left\{ \langle a \,|\, h \,|\, a \rangle + \frac{1}{2} \sum_b \left[\langle ab \,|\, g \,|\, ab \rangle - \langle ba \,|\, g \,|\, ab \rangle \right] \right\}. \end{aligned} \tag{6.1.6}$$

6.2 One-electron matrix elements of tensor operators

The next step is to find expressions for the matrix elements of one- and two-electron operators appearing above with respect to Dirac central field orbital spinors, (5.6.1), which we write

$$\phi_{\gamma,\kappa,m}(\boldsymbol{x}) = \begin{pmatrix} M[+1;\gamma,\ \kappa,\boldsymbol{r}] \\ M[-1;\gamma,-\kappa,\boldsymbol{r}] \end{pmatrix}, \tag{6.2.1}$$

with 2-spinor components labelled by an index $\beta = \pm 1$ [1, Equation (5.2)],

$$M[\beta;\gamma,\beta\kappa,\boldsymbol{r}] = \frac{\omega_\beta}{r} R_{\gamma,\beta\kappa}(r)\, \chi_{\beta\kappa,m}(\theta,\varphi). \tag{6.2.2}$$

In the notation of (3.2.4) the radial amplitudes are

$$R_{\gamma,+\kappa}(r) = P_{\gamma,\kappa}(r), \quad R_{\gamma,-\kappa}(r) = Q_{\gamma,\kappa}(r),$$

$\omega_{+1} = 1$ and $\omega_{-1} = i$ and the angular 2-spinors, (3.2.9), are

$$\chi_{\beta\kappa,m}(\theta,\varphi) = \sum_\sigma (l_\beta, m-\sigma, s, \sigma \,|\, l_\beta, s, j, m)\; Y_{l_\beta}^{m-\sigma}(\theta,\varphi)\phi_\sigma \tag{6.2.3}$$

with $l_\beta = j + \frac{1}{2}\beta\eta$, $\eta = \text{sgn}\,\kappa$ and $s = 1/2$. Clearly either the triple $l_\beta\, s j$ or the index $\beta\kappa$ can be used to label the component uniquely.

Compatible quantum mechanical operators, $\mathcal{O}$, are therefore 4×4 matrices partitioned into 2×2 blocks

$$\mathcal{O} = \begin{pmatrix} \mathcal{O}_{1,1} & \mathcal{O}_{1,-1} \\ \mathcal{O}_{-1,1} & \mathcal{O}_{-1,-1} \end{pmatrix} \tag{6.2.4}$$

labelled by the appropriate values of β. It is convenient to divide operators of the form (6.2.4) into two classes [1, 2]:

$$\text{Even: } \begin{pmatrix} \mathcal{O}_{1,1} & 0 \\ 0 & \mathcal{O}_{-1,-1} \end{pmatrix}, \quad \text{Odd: } \begin{pmatrix} 0 & \mathcal{O}_{1,-1} \\ \mathcal{O}_{-1,1} & 0 \end{pmatrix} \tag{6.2.5}$$

Most even operators are either multiples of the 4×4 identity operator, so that $\mathcal{O}_{1,1} = \mathcal{O}_{-1,-1}$ or of the matrix β for which $\mathcal{O}_{1,1} = -\mathcal{O}_{-1,-1}$. Odd operators are due to the presence of Dirac $\boldsymbol{\alpha}$ matrices: the most common are the kinetic energy operator, $c\boldsymbol{\alpha}\cdot\boldsymbol{p}$, and the interaction, $\boldsymbol{\alpha}\cdot\boldsymbol{A}$, with an Hermitian external vector potential $\boldsymbol{A}$. In this case, $\mathcal{O}_{1,-1} = \mathcal{O}_{-1,1} = \boldsymbol{\sigma}\cdot\boldsymbol{p}$ or $\mathcal{O}_{1,-1} = \mathcal{O}_{-1,1} = \boldsymbol{\sigma}\cdot\boldsymbol{A}$ respectively. We can therefore build everything we need for even operators from 2-spinor matrix elements of tensor operators $S_q^k(\boldsymbol{r}) = v_k(r)\, C_q^k(\theta,\varphi)$, and from expressions of the form $X_q^{(1\nu)k}(\boldsymbol{r}) = [\boldsymbol{\sigma}\times\mathbf{T}^\nu]^k$ for odd operators, where $\mathbf{T}^\nu = F_\nu \mathbf{C}^\nu(\theta,\varphi)$ in which F_ν operates on functions of r. The matrix element expressions derived in this chapter were first set out in [2] (with corrections in [1], [3]), [4], [5], and [6]. Frequently used results from this chapter are collected for convenience in Appendix A.4.

6.2.1 2-spinor matrix elements of even operators

We start with the matrix element of an even tensor operator

$$\langle \gamma, \beta\kappa, m \,|\, S^k_q \,|\, \gamma', \beta\kappa', m' \rangle,$$

where $S^k_q(\boldsymbol{r}) = v_k(r)\, C^k_q(\theta, \varphi)$ with integer k and q. The integration over the coordinates is separable [2], so that we can extract the dependence on m, m', q using the Wigner-Eckart theorem (B.2.26),

$$\langle \gamma, \beta\kappa, m \,|\, S^k_q \,|\, \gamma', \beta\kappa', m' \rangle = \begin{pmatrix} m & k & j' \\ j & q & m' \end{pmatrix} \langle \beta\kappa \,\|\boldsymbol{S}^k\|\, \beta\kappa' \rangle \tag{6.2.6}$$

where, because $\omega^*_\beta \omega_\beta = 1$,

$$\langle \beta\kappa \,\|\boldsymbol{S}^k\|\, \beta\kappa' \rangle = V_k(\gamma, \beta\kappa; \gamma', \beta\kappa') \langle \beta\kappa \,\|\boldsymbol{C}^k\|\, \beta\kappa' \rangle,$$

the radial integral is

$$V_k(\gamma, \beta\kappa; \gamma', \beta\kappa') = \int_0^\infty R^*_{\gamma,\beta\kappa}(r)\, R_{\gamma',\beta\kappa'}(r)\, v_k(r)\, dr,$$

and the reduced matrix element is

$$\langle \beta\kappa \,\|\boldsymbol{C}^k\|\, \beta\kappa' \rangle = (-1)^{l_\beta + j' + k + 1/2} [j, j']^{1/2} \langle l_\beta \| \,\boldsymbol{C}^k\, \| l'_\beta \rangle \begin{pmatrix} j & j' & k \\ l'_\beta & l_\beta & 1/2 \end{pmatrix}, \tag{6.2.7}$$

with $l_\beta = j + \frac{1}{2}\beta\eta$, $l'_\beta = j' + \frac{1}{2}\beta\eta'$. The derivation of this result requires a number of standard pieces; first (B.3.136),

$$(l \,\|\, \boldsymbol{C}^k \,\|\, l') = (-1)^l [\, l, l'\,]^{1/2} \begin{pmatrix} l & k & l' \\ 0 & 0 & 0 \end{pmatrix} \tag{6.2.8}$$

where the *angular momentum selection rules* can be expressed in terms of the triangular delta $\{\, l\, k\, l'\,\}$, (B.3.43), and (B.3.51) gives rise to the *parity selection rule*: $l + k + l'$ is an *even* integer. Equation (6.2.7) can be further simplified by noting the identity (see, for example, [7, p. 519])

$$(-1)^{j+j'+k+1} \begin{pmatrix} j & k & j' \\ 1/2 & 0 & -1/2 \end{pmatrix} = [\, l, l'\,]^{1/2} \begin{pmatrix} l & k & l' \\ 0 & 0 & 0 \end{pmatrix} \begin{pmatrix} j & j' & k \\ l' & l & 1/2 \end{pmatrix}, \tag{6.2.9}$$

giving finally

$$\langle \beta\kappa \,\|\boldsymbol{C}^k\|\, \beta\kappa'' \rangle = (-1)^{j+1/2} [\, j, j'\,]^{1/2} \begin{pmatrix} j & k & j' \\ 1/2 & 0 & -1/2 \end{pmatrix}. \tag{6.2.10}$$

This is independent of the index β so that we can write

$$\langle\, \beta\kappa \,\|\boldsymbol{C}^k\|\, \beta\kappa'' \,\rangle \equiv \langle\, j \,\|\boldsymbol{C}^k\|\, j' \,\rangle.$$

Following [2], we can now write the matrix element of (6.2.6) in the form

$$\langle\, \gamma, \beta\kappa, m \,|\, S^k_q \,|\, \gamma', \beta\kappa', m' \,\rangle = d^k(jm, j'm')\, V_k(\gamma, \beta\kappa; \gamma', \beta\kappa') \qquad (6.2.11)$$

where

$$d^k(jm, j'm') = \begin{pmatrix} m & k & j' \\ j & q & m' \end{pmatrix} \varPi^e(\kappa\kappa' k)\, \langle\, j \,\|\, \boldsymbol{C}^k \,\|\, j' \,\rangle \qquad (6.2.12)$$

The factor $\varPi^e(\kappa\kappa' k)$ incorporates the parity selection rules. Parity conservation requires that $l + k + l' = j + k + j' + \beta(\eta + \eta')/2$ should be an even integer; because $(\eta + \eta')/2 = 0, \pm 1$ this is the same for both values $\beta = \pm 1$. The $3jm$-symbol requires that $\{\, j\, k\, j' \,\}$ should not vanish without reference to the choice of β and β'. In other words, [2, Equation (29)], $d^k(jm, j'm')$ is nonvanishing if

$$\{\, j\, k\, j' \,\} = 1, \quad j + k + j' \text{ is } \begin{cases} \text{even if } \eta = -\eta' \\ \text{odd if } \eta = \eta' \end{cases} \qquad (6.2.13)$$

for the 4-spinor as a whole. The parity factor can be written (cf. [6, Equation (59)]),

$$\varPi^e(\kappa\kappa' k) = \frac{1}{2}[1 - \eta\eta'(-1)^{j+j'+k}] \qquad (6.2.14)$$

The symmetry relations

$$\begin{aligned} d^k(jm, j'm') &= (-1)^{m-m'} d^k(jm', j'm) \\ &= (-1)^{j+j'+k+1} d^k(j, -m, j', -m') \end{aligned} \qquad (6.2.15)$$

have been used in Appendix A.4 to shorten the numerical table of $d^k(jm, j'm')$ for $j, j' = 1/2, 3/2, 5/2$.

Although he did not use this notation, the coefficients $d^k(jm, j'm')$ are identical (up to a phase factor) with coefficients calculated by Inglis [8] in an early investigation of weak electrostatic interaction in jj-coupling. It is interesting to compare our result (6.2.12) with the corresponding nonrelativistic formula, known as Gaunt's integral, [9, Equations $8^6(6)$, $8^6(11)$] which, using Racah's results [11, Equations (52), (50) and (16') of the 1942 paper], can be put in the equivalent form

$$c^k(lm, l'm') = \begin{pmatrix} m & k & l' \\ l & q & m' \end{pmatrix} (l \,\|\, \boldsymbol{C}^k \,\|\, l'). \qquad (6.2.16)$$

Gaunt [12] evaluated (6.2.16) as an integral of a product of three normalized associated Legendre polynomials [9, Equation $8^6(6)$]

$$c^k(lm, l'm') = \sqrt{\frac{2}{2k+1}} \int_0^\pi \varTheta(kq)\, \varTheta(lm)\, \varTheta(l'm') \sin\theta \, d\theta$$

where $q = m - m'$ and, in the notation of (B.3.23),

$$\Theta(lm) = (-1)^m \left[\frac{(2l+1)(l-m)!}{2(l+m)!}\right]^{1/2} P_l^m(\cos\theta), \quad m \geq 0$$

with $\Theta(l,-m) = (-1)^m \Theta(lm)$. The result was expressed algebraically in a manner similar to the expression (B.3.49) for Clebsch-Gordan coefficients. The relativistic coefficients corresponding to $d^k(jm, j'm')$ obtained by Swirles [13] and Jacobsohn [14] in the earliest studies of relativistic atomic structure reproduce the numerical values calculated from (6.2.12). However, they were compelled to express the matrix element $\langle\, \gamma, \beta\kappa, m \,|\, S_q^k \,|\, \gamma', \beta\kappa', m' \,\rangle$ in terms of Gaunt integrals, so that the symmetry and closure properties of the $d^k(jm, j'm')$ as well as the functional similarity of (6.2.12) and (6.2.16) were hard to resolve.

6.2.2 2-spinor matrix elements of odd operators

Similarly, the matrix elements of an odd operator are always of the form

$$\langle\, \gamma, -\lambda, m \,|\, T_q^k \,|\, \gamma', \lambda', m' \,\rangle = \begin{pmatrix} m & k & j' \\ j & q & m' \end{pmatrix} \langle\, \gamma, -\lambda \,\|\boldsymbol{T}^k\|\, \gamma', \lambda' \,\rangle, \qquad (6.2.17)$$

where $\lambda = \beta\kappa,\ \lambda' = \beta\kappa', \beta = \pm 1$,

$$\boldsymbol{T}^k = F_\nu \boldsymbol{X}^{(1\nu)k}(\boldsymbol{r}), \quad X_q^{(1\nu)k}(\boldsymbol{r}) = [\,\boldsymbol{\sigma} \times \boldsymbol{C}^\nu]_q^k,$$

F_ν operates only on functions of r, and $\boldsymbol{X}^{(1\nu)k}$ involves only spin-angular coordinates. We again separate out the integration over r so that

$$\begin{aligned} &\langle \gamma, -\lambda \,\|\, \boldsymbol{T}^k \,\|\, \gamma', \lambda' \rangle \qquad (6.2.18) \\ &= -i\beta \int_0^\infty R^*_{\gamma,-\lambda}(r)\; F_\nu\; R_{\gamma',\lambda'}(r)\; dr\, \langle\, -\lambda \| \boldsymbol{X}^{(1\nu)k} \|\, \lambda' \rangle; \end{aligned}$$

the prefactor results from $\omega^*_{-\beta}\omega_\beta = -i\beta$. As a composite tensor operator, $X_q^{(1\nu)k}(\boldsymbol{r})$ only has nonvanishing matrix elements when $\beta' = -\beta$ and, like the 2×2 matrix operator $\boldsymbol{\sigma}$, operates on the space of 2-spinors. We need only cases in which the coupled tensor operators of $X_q^{(1\nu)k}(\boldsymbol{r})$ commute, so that

$$X_q^{(1\nu)k}(\boldsymbol{r}) = [\,\boldsymbol{\sigma} \times \boldsymbol{C}^\nu]_q^k = (-1)^{\nu-1+k}[\,\boldsymbol{C}^\nu \times \boldsymbol{\sigma}\,]_q^k.$$

This class of tensor operator occurs in electromagnetic interactions, and is therefore of importance for electronic structure as well as for electromagnetic radiation processes.

Following [4, 15], we apply (B.3.163), (B.3.136), and (B.3.134), and after evaluating the $9j$-symbol explicitly, we find

$$\langle\, -\lambda \| \boldsymbol{X}^{(1\nu)k} \|\, \lambda' \,\rangle = \Pi^o(\kappa\kappa' k) \langle\, j \,\| \boldsymbol{C}^k \|\, j' \,\rangle E^\nu(-\lambda, +\lambda', k). \qquad (6.2.19)$$

Here the parity conditions are the opposite of those in (6.2.13),

$$\Pi^o(\kappa\kappa' k) = 1 - \Pi^e(\kappa\kappa' k), \tag{6.2.20}$$

where $\nu = k-1, k$ or $k+1$. The coefficients $E^\nu(\lambda, \lambda', k)$ are given by[1]

$$\begin{aligned} E^{k-1}(\lambda,\lambda',k) &= \frac{k+\lambda-\lambda'}{[k(2k-1)]^{1/2}}, \\ E^{k}(\lambda,\lambda',k) &= \frac{-\lambda-\lambda'}{[k(k+1)]^{1/2}}, \\ E^{k+1}(\lambda,\lambda',k) &= \frac{-k-1+\lambda-\lambda'}{[(k+1)(2k+3)]^{1/2}}. \end{aligned} \tag{6.2.21}$$

6.3 Angular reduction of the Dirac Hamiltonian for a central potential

As a first application of the previous section, we compute matrix elements of the Dirac Hamiltonian with a spherically symmetric scalar potential, (6.1.3).

$$h = c\boldsymbol{\alpha}\cdot\boldsymbol{p} + (\beta-1)mc^2 + V(r) \tag{6.3.1}$$

This can be partitioned as in (6.2.4) so that

$$\begin{aligned} &h_{1,1} = V(r), \quad h_{-1,-1} = -2mc^2 + V(r), \\ &h_{1,-1} = h_{-1,1} = c\boldsymbol{\sigma}\cdot\boldsymbol{p}. \end{aligned} \tag{6.3.2}$$

Then

$$\begin{aligned} \langle \gamma,\kappa,m\,|\,h\,|\,\gamma',\kappa',m\rangle &= \sum_{\beta,\beta'} \langle \gamma,\beta\kappa,m\,|\,h_{\beta,\beta'}\,|\,\beta'\gamma',\kappa',m\rangle \\ &= \langle \gamma,+\kappa,m\,|\,V(r)\,|\,\gamma',+\kappa,'m\rangle + \langle \gamma,-\kappa,m\,|\,-2mc^2+V(r)\,|\,\gamma',-\kappa',m\rangle \\ &\quad + \langle \gamma,+\kappa,m\,|\,c\boldsymbol{\sigma}\cdot\boldsymbol{p}\,|\,\gamma',-\kappa',m\rangle + \langle \gamma,-\kappa,m\,|\,c\boldsymbol{\sigma}\cdot\boldsymbol{p}\,|\,\gamma',+\kappa',m\rangle. \end{aligned} \tag{6.3.3}$$

The diagonal matrix elements can be obtained from (6.2.11) with $k = q = 0$, giving

$$\langle \gamma,\kappa,m\,|\,V(r)\,|\,\gamma',\kappa',m'\rangle = \delta_{\kappa,\kappa'}\delta_{mm'}\int_0^\infty P^*_{\gamma,\kappa}(r)V(r)P_{\gamma',\kappa}(r)\,dr, \tag{6.3.4}$$

and

[1] The present notation $E^\nu(\lambda,\lambda',k)$ seems clearer than the notation $E^\nu_\beta(\kappa,\kappa';k)$ used in [4, Equation (16)].

$$\langle \gamma, -\kappa, m \,|\, -2mc^2 + V(r) \,|\, \gamma', -\kappa', m\rangle$$
$$= \delta_{\kappa,\kappa'}\delta_{mm'} \int_0^\infty Q^*_{\gamma,\kappa}(r)(-2mc^2 + V(r))Q_{\gamma',\kappa}(r)\, dr.$$

For off-diagonal contributions, we start from (B.3.147),

$$\langle \gamma, \kappa, m \,|\, c\boldsymbol{\sigma}\cdot\boldsymbol{p} \,|\, \gamma', -\kappa', m\rangle = -\sqrt{3}\langle \gamma, \kappa, m \,|\, [c\boldsymbol{\sigma}\times\boldsymbol{p}]^0_0 \,|\, \gamma', -\kappa', m\rangle,$$

and apply (B.3.163) with $k = 0$, $k_1 = k_2 = 1$, $j_1 = l$, $j_1' = l'$, $j_2 = j_2' = 1/2$ so that

$$\langle \gamma, +\kappa, m \,|\, c\boldsymbol{\sigma}\cdot\boldsymbol{p} \,|\, \gamma', -\kappa', m\rangle = -\delta_{\kappa\kappa'}\delta_{mm'}\sqrt{3}\,[j]$$
$$\times \begin{Bmatrix} l & l' & 1 \\ 1/2 & 1/2 & 1 \\ j & j & 0 \end{Bmatrix} \langle\, \gamma, l \| \,\boldsymbol{p}\, \| \gamma', l'\,\rangle \,\langle\, 1/2 \| \,\boldsymbol{\sigma}\, \| 1/2\,\rangle. \quad (6.3.5)$$

The reduced matrix elements of $\boldsymbol{p}$ and $\boldsymbol{\sigma}$ are given respectively by (B.3.152) and (B.3.134), and with explicit values for the $9j$-symbol, we find

$$\langle \gamma, +\kappa, m \,|\, c\boldsymbol{\sigma}\cdot\boldsymbol{p} \,|\, \gamma', -\kappa', m'\rangle$$
$$= -\delta_{\kappa,\kappa'}\delta_{mm'}\, c \int_0^\infty R^*_{\gamma,+\kappa}(r)\left(\frac{d}{dr} - \frac{\kappa}{r}\right) R_{\gamma',-\kappa}(r)\, dr, \quad (6.3.6)$$

and similarly

$$\langle \gamma, -\kappa, m \,|\, c\boldsymbol{\sigma}\cdot\boldsymbol{p} \,|\, \gamma', +\kappa', m\rangle$$
$$= +\delta_{\kappa,\kappa'}\delta_{mm'}\, c \int_0^\infty R^*_{\gamma,-\kappa}(r)\left(\frac{d}{dr} + \frac{\kappa}{r}\right) R_{\gamma',+\kappa}(r)\, dr, \quad (6.3.7)$$

in agreement with results obtained by more elementary methods. Combining these results gives

$$\langle a \,|\, h \,|\, b\rangle = \delta_{\kappa_a,\kappa}\delta_{\kappa_b,\kappa}\delta_{m_a m_b}\, I(a,b) \quad (6.3.8)$$

where

$$I(a,b) = \int_0^\infty c\,Q^*_{\gamma_a,\kappa}(r)\left(\frac{d}{dr} + \frac{\kappa}{r}\right) P_{\gamma_b,\kappa}(r)$$
$$- c\,P^*_{\gamma_a,\kappa}(r)\left(\frac{d}{dr} - \frac{\kappa}{r}\right) Q_{\gamma_b,\kappa}(r) - 2mc^2 Q^*_{\gamma_a,\kappa}(r)Q_{\gamma_b,\kappa}(r)$$
$$+ v_{nuc}(r)\left[P^*_{\gamma_a,\kappa}(r)P_{\gamma_b,\kappa}(r) + Q^*_{\gamma_a,\kappa}(r)Q_{\gamma_b,\kappa}(r)\right]\, dr. \quad (6.3.9)$$

6.4 Matrix elements of 2-body operators

Scalar two-body operators such as the Coulomb and Breit interactions can typically be written as a sum of terms of the form

$$g(1,2) = \sum_K g_K(r_1, r_2)\, \boldsymbol{T}^K(1) \cdot \boldsymbol{T}^K(2) \tag{6.4.1}$$

where $g_K(r_1, r_2)$ carries the dependence on the radial coordinates and the tensor operators $\boldsymbol{T}^K(a)$ act on the other spatial coordinates and possibly on the spin variables of the particle labelled a. Using the Wigner-Eckart Theorem, matrix elements of g can be written in terms of *effective interaction strengths* $X^K(ab;cd)$ of order K such that

$$\langle ab \,|\, g \,|\, cd\rangle = \sum_k \sum_q \begin{pmatrix} m_a & K & j_c \\ j_a & Q & m_c \end{pmatrix} \begin{pmatrix} m_b & Q & j_d \\ j_b & K & m_d \end{pmatrix} X^K(abcd) \tag{6.4.2}$$

where, in the simplest case,

$$X^K(abcd) = (-1)^K \langle a \,\|\boldsymbol{T}^K\|\, c\rangle \langle b \,\|\boldsymbol{T}^K\|\, d\rangle\; R^K(abcd), \tag{6.4.3}$$

in which the integral $R^K(abcd)$ involves radial amplitudes only. These equations serve as a template for the more complicated expressions encountered in treating the Breit and related interactions.

6.4.1 The Coulomb interaction

The Coulomb interaction kernel is given by

$$g^C(R) = \frac{1}{R}, \qquad R = |\boldsymbol{r}_1 - \boldsymbol{r}_2|.$$

This can be identified with the generating function for Legendre polynomials [16, 22.9.12] so that

$$g^C(R) = \sum_{k=0}^{\infty} U_k(r_1, r_2) P_k(\cos\Theta) \tag{6.4.4}$$

where

$$U_k(r_1, r_2) = \begin{cases} r_1^k / r_2^{k+1} & \text{if } r_1 \le r_2 \\ r_2^k / r_1^{k+1} & \text{if } r_2 < r_1 \end{cases} \tag{6.4.5}$$

and from (B.3.143) and (B.3.147)

$$P_k(\cos\Theta) = \boldsymbol{C}^k(1) \cdot \boldsymbol{C}^k(2). \tag{6.4.6}$$

This expansion acts on space coordinates only so that the effective Coulomb interaction strength is

$$X_C^k(abcd) = \{j_a, j_c, k\}\{j_b, j_d, k\}\Pi^e(\kappa_a\kappa_c k)\Pi^e(\kappa_b\kappa_d k) \\ \times (-1)^k \langle j_a\|\boldsymbol{C}^k\|j_c\rangle\langle j_b\|\boldsymbol{C}^k\|j_d\rangle R_C^k(abcd), \qquad (6.4.7)$$

where the first line expresses angular momentum and parity conservation, $\langle j\|\boldsymbol{C}^k\|j'\rangle$ is given by (6.2.10), and the radial integral is

$$R_C^k(abcd) = \sum_{\beta,\beta'} R^k(\lambda_a, \lambda_b', \lambda_c, \lambda_d'), \qquad (6.4.8)$$

so that, with $\lambda_a = \beta\kappa_a$, $\lambda_b' = \beta'\kappa_b$, $\lambda_c = \beta\kappa_c$ and $\lambda_d' = \beta'\kappa_d$,

$$R^k(\lambda_a, \lambda_b', \lambda_c, \lambda_d') = \int_0^\infty \int_0^\infty \rho_{ac}(r)\, U_k(r,s)\, \rho_{bd}(s)\, drds. \qquad (6.4.9)$$

The Dirac radial *overlap charge densities* appearing here are defined by

$$\begin{aligned} \rho_{ac}(r) &= \textstyle\sum_\beta R^*_{\gamma_a,\lambda_a}(r)R_{\gamma_c,\lambda_c}(r) = P_a^*(r)P_c(r) + Q_a^*(r)Q_c(r), \\ \rho_{bd}(s) &= \textstyle\sum_{\beta'} R^*_{\gamma_b,\lambda_b'}(s)R_{\gamma_d,\lambda_d'}(s) = P_b^*(s)P_d(s) + Q_b^*(s)Q_d(s). \end{aligned} \qquad (6.4.10)$$

Alternatively [1, 2] we can write the matrix element using (6.2.12), giving

$$\langle ab \mid g^C(R) \mid cd\rangle \\ = \sum_k d^k(j_a m_a, j_c m_c)\, d^k(j_d m_d, j_b m_b)\, \delta_{m_a+m_b, m_c+m_d}\, R_C^k(abcd). \quad (6.4.11)$$

The parity selection rules embodied in (6.4.7) can conveniently be summarized by

$$j_a + j_c + k : \begin{cases} \text{even if } \eta_a \neq \eta_c \\ \text{odd if } \eta_a = \eta_c \end{cases}, \quad j_d + j_b + k : \begin{cases} \text{even if } \eta_d \neq \eta_b \\ \text{odd if } \eta_d = \eta_b \end{cases} \qquad (6.4.12)$$

which hold along with the restrictions imposed by the triangular deltas.

These Dirac results are very similar to those of the nonrelativistic theory given in classic texts such as that of Condon and Shortley [9, Chapter 6] and Judd [17, Chapter 4]. The nonrelativistic addition of a weak electrostatic perturbation in nonrelativistic jj-coupling theory was studied by Inglis [8] and his formulae, quoted in [9, Chapter 10], agree with (6.4.11).

6.4.2 Relativistic corrections to the Coulomb interaction

The investigation in §4.9 of effective interactions from QED identified several approximations which can be important in applications.

The simplest, the Gaunt interaction [12], is the *unretarded* current-current component of the Feynman gauge effective interaction in (4.9.13). This is usually presented in terms of an "effective potential"

$$g^G(R) = -\frac{\boldsymbol{\alpha}_1 \cdot \boldsymbol{\alpha}_2}{R}. \tag{6.4.13}$$

where $R = |\boldsymbol{r}_1 - \boldsymbol{r}_2|$. The assumption that $g^G(R)$ is unretarded implies that the wavelength of the exchanged virtual photon propagating the interaction is large compared with system dimensions.

In the Coulomb gauge, the photon propagator splits into two parts: the Coulomb potential arises from virtual photons that are polarized in the direction along which the photon propagates, and a part from virtual photons polarized perpendicular to this direction. In the long wavelength limit the latter, transverse, part gives the Breit interaction (4.9.22):

$$g^B(R) = \lim_{\omega \to 0} g^T(R;\omega) = -\frac{1}{2R}\left(\boldsymbol{\alpha}_1 \cdot \boldsymbol{\alpha}_2 + (\boldsymbol{\alpha}_1 \cdot \widehat{\boldsymbol{R}})(\boldsymbol{\alpha}_2 \cdot \widehat{\boldsymbol{R}})\right), \tag{6.4.14}$$

When the interaction involves electrons that are sufficiently strongly bound, the long wavelength assumption fails and we need to use configuration space kernels that allow for this. In the Feynman gauge we obtain the Møller interaction kernel (4.9.11) [18]

$$g^M(R;\omega) = (1 - \boldsymbol{\alpha}_1 \cdot \boldsymbol{\alpha}_2)\,\frac{\mathrm{e}^{i\omega R/c}}{R}, \tag{6.4.15}$$

This gives relativistic corrections to the retarded Coulomb interaction. In the Coulomb gauge (6.4.14) is replaced by the transverse photon kernel (4.9.21),

$$g^T(R;\omega) = -\boldsymbol{\alpha}_1 \cdot \boldsymbol{\alpha}_2\,\frac{e^{i\omega R/c}}{R} - (\boldsymbol{\alpha}_1 \cdot \boldsymbol{\nabla}_{\boldsymbol{R}})(\boldsymbol{\alpha}_2 \cdot \boldsymbol{\nabla}_{\boldsymbol{R}})\,\frac{\mathrm{e}^{i\omega R/c} - 1}{\omega^2 R/c^2}, \tag{6.4.16}$$

which reduces to (6.4.14) as $\omega \to 0$. We present expressions below for the matrix elements of these interactions [1, 4, 6, 15].

6.4.3 The Gaunt interaction

The Gaunt interaction has a multipole expansion

$$g^G(R) = \sum_{k,K} (-1)^{k+K} U_k(r_1, r_2)\, \mathsf{X}^{(1k)K}(1) \cdot \mathsf{X}^{(1k)K}(2), \tag{6.4.17}$$

of the form (6.4.1) where $U_k(r_1, r_2)$ is defined by (6.4.4),

$$\mathsf{X}^{(1k)K} = \left[\boldsymbol{\alpha} \times \boldsymbol{C}^k\right]^K = \begin{pmatrix} 0 & \boldsymbol{X}^{(1k)K} \\ \boldsymbol{X}^{(1k)K} & 0 \end{pmatrix},$$

and $\boldsymbol{X}^{(1k)K}$ is defined by (6.2.17). The derivation of the 2-body matrix elements of (6.4.17) is on similar lines to that of the Coulomb interaction with the help of results from §6.2.2. The effective interaction strength is

$$X^k_G(abcd) = -(-1)^k \Pi^o(\kappa_a, \kappa_c, k)\Pi^o(\kappa_b, \kappa_d, k)$$
$$\times \sum_{\nu=k-1}^{k+1} \sum_{\beta,\beta'} \langle -\lambda_a \| \boldsymbol{X}^{(1\nu)k} \| \lambda_c \rangle \langle -\lambda'_b \| \boldsymbol{X}^{(1\nu)k} \| \lambda'_d \rangle R^\nu(-\lambda_a, -\lambda'_b, \lambda_c, \lambda'_d), \tag{6.4.18}$$

where the Slater integral is defined by (6.4.9). Alternatively, we introduce Gaunt coefficients $g^\nu(\beta, \kappa m, \kappa' m', k)$ analogous to the $d^k(jm, j'm')$ Coulomb coefficients,

$$g^\nu(\beta, \kappa m, \kappa' m', k) = \begin{pmatrix} m & k & j' \\ j & q & m' \end{pmatrix} \langle -\lambda \| \boldsymbol{X}^{(1\nu)k} \| \lambda' \rangle, \tag{6.4.19}$$

and obtain

$$\langle ab \,|\, g^G(R) \,|\, cd \rangle$$
$$= -\sum_{k=0}^{\infty} (-1)^k \sum_{\nu=k-1}^{k+1} \sum_{\beta,\beta'} g^\nu(\beta, \kappa_a m_a, \kappa_c m_c, k)\, g^\nu(\beta', \kappa_d m_d, \kappa_b m_b, k)$$
$$\times R^\nu(-\lambda_a, -\lambda'_b, \lambda_c, \lambda'_d)\, \delta_{m_a+m_b, m_c+m_d}. \tag{6.4.20}$$

6.4.4 The Møller interaction

The Møller interaction, (6.4.15), is a sum of two terms

$$g^M(R; \omega) = g^{M1}(R; \omega) + g^{M2}(R; \omega), \tag{6.4.21}$$

where

$$g^{M1}(R; \omega) = \frac{e^{i\omega R/c}}{R} \overset{\omega\to 0}{\longrightarrow} g^C(R),$$
$$g^{M2}(R; \omega) = -\boldsymbol{\alpha}_1 \cdot \boldsymbol{\alpha}_2 \frac{e^{i\omega R/c}}{R} \overset{\omega\to 0}{\longrightarrow} g^G(R)$$

In place of the Coulomb multipole expansion (6.4.4) we have [16, 10.1.45, 10.1.46],

$$\frac{e^{i\omega R/c}}{R} = i\frac{\omega}{c} \sum_{k=0}^{\infty} [\,k\,]\, j_k(\omega r_</c)\, h_k^{(1)}(\omega r_>/c)\, P_k(\cos\Theta), \tag{6.4.22}$$

where $r_<$ is the smaller and $r_>$ the larger of r_1 and r_2. Thus the multipole Coulomb potential $U_k(r_1, r_2)$ is replaced by a product of ω-dependent spherical Bessel functions,

$$U_k(r_1, r_2; \omega) = i\,\frac{\omega}{c}\,[\,k\,]\, j_k(\omega r_</c)\, h_k^{(1)}(\omega r_>/c). \tag{6.4.23}$$

It is easy to verify from Appendix A.3 that this reduces to $U_k(r_1, r_2)$ as $\omega \to 0$. The reduced matrix elements are

$$X_{M1}(\omega; abcd) = \sum_{K=0}^{\infty} (-1)^K \langle j_a \| \boldsymbol{C}^K \| j_c \rangle \langle j_b \| \boldsymbol{C}^K \| j_d \rangle \, R^K(\omega; abcd), \quad (6.4.24)$$

and

$$X_{M2}(\omega; abcd) = -\sum_{K=0}^{\infty} (-1)^K \sum_{\beta,\beta'} \sum_{k=K-1}^{K+1}$$
$$\times \langle -\lambda_a \| \boldsymbol{X}^{(1k)K} \| \lambda_c \rangle \langle -\lambda'_b \| \boldsymbol{X}^{(1k)K} \| \lambda'_d \rangle R^k(\omega; -\lambda_a, -\lambda'_b, \lambda_c, \lambda'_d) \quad (6.4.25)$$

where the radial Slater integrals $R^K(\omega; abcd)$ and $R^k(\omega; -\lambda_a, -\lambda'_b, \lambda_c, \lambda'_d)$ are respectively given by (6.4.8) and (6.4.9) with $U_k(r_1, r_2)$ replaced by $U_k(r_1, r_2; \omega)$. The multipole potential $V_k(r_1, r_2; \omega)$ of [4, Equations (4), (6)] is the *real part* of $U_k(r_1, r_2; \omega)$.

6.4.5 The transverse photon interaction in Coulomb gauge

As in the case of the Møller interaction, the transverse photon interaction (6.4.13) can be divided into two parts,

$$g^T(R; \omega) = g^{M2}(R; \omega) + g^{Ret}(R; \omega) \quad (6.4.26)$$

where

$$g^{Ret}(R; \omega) = -(\boldsymbol{\alpha}_1 \cdot \boldsymbol{\nabla}_{\boldsymbol{R}})(\boldsymbol{\alpha}_2 \cdot \boldsymbol{\nabla}_{\boldsymbol{R}}) \frac{e^{i\omega R/c} - 1}{\omega^2 R/c^2}$$

is normally designated, somewhat misleadingly, as the *retardation* part of the operator. Matrix elements of the real part of this interaction were derived in [4, 5], and it requires only minor changes to write down the matrix elements of (6.4.26) as a whole. We require additional multipole functions

$$W_K(r_1, r_2; \omega) = [U_K(r_1, r_2; \omega) - U_K(r_1, r_2)] \frac{c^2}{\omega^2} \quad (6.4.27)$$
$$W_{kk'K}(r_1, r_2; \omega) = D_1^{kK} \, D_2^{k'K} \, W_K(r_1, r_2; \omega)$$

where

$$D_i^{kK} = \frac{\partial}{\partial r_i} - \frac{k(k+1) - K(K+1) - 2}{2r_i}, \qquad i = 1, 2.$$

The functions $W_{\nu\nu' k}(r_1, r_2; \omega)$ have the explicit form

$$W_{K-1,K-1,K}(r_1, r_2; \omega) = \frac{2K+1}{2K-1} U_{K-1}(r_1, r_2; \omega)$$
$$(6.4.28)$$
$$W_{K+1,K+1,k}(r_1, r_2; \omega) = \frac{2K+1}{2K+3} U_{K+1}(r_1, r_2; \omega)$$

$$W_{K-1,K+1,K}(r_1,r_2;\omega) \tag{6.4.29}$$
$$= \begin{cases} -i\,\frac{\omega}{c}[\,K\,]\, j_{K-1}(\omega r_1/c)\, h^{(1)}_{K+1}(\omega r_2/c) + \frac{[\,K\,]^2c^2}{\omega^2}\frac{r_1^{K-1}}{r_2^{K+2}} & \text{if } r_1 < r_2 \\ -i\,\frac{\omega}{c}[\,K\,]\, h^{(1)}_{K+1}(\omega r_1/c)\, j_{K-1}(\omega r_2/c) & \text{if } r_1 > r_2 \end{cases}$$

so that

$$W_{K+1,K-1,K}(r_1,r_2;\omega) = W_{K-1,K+1,K}(r_2,r_1;\omega).$$

We can now write the interaction in the form (6.4.1), [4, Equation (8)], as

$$\begin{aligned} g^T(R;\omega) \qquad & \qquad\qquad (6.4.30) \\ = \sum_{K=0}^{\infty}\Bigg\{ & \sum_{k=K-1}^{K+1} v_{kK} U_k(r_1,r_2;\omega)\, \mathsf{X}^{(1k)K}(1)\cdot\mathsf{X}^{(1k)K}(2) \\ & + w_K\Big[W_{K-1,K+1,K}(r_1,r_2;\omega)\,\mathsf{X}^{(1,K-1)K}(1)\cdot\mathsf{X}^{(1,K+1)K}(2) \\ & \quad + W_{K+1,K-1,K}(r_1,r_2;\omega)\,\mathsf{X}^{(1,K+1)K}(1)\cdot\mathsf{X}^{(1,K-1)K}(2)\Big]\Bigg\} \end{aligned}$$

where

$$v_{KK} = 1, \quad v_{K-1,K} = -\frac{K+1}{2K+1}, \quad v_{K+1,K} = -\frac{K}{2K+1},$$

and

$$w_K = -\frac{[K(K+1)(2K-1)(2K+3)]^{1/2}}{(2K+1)^2}.$$

The matrix elements can therefore be written

$$\begin{aligned} X_T(\omega;abcd) = & \sum_{K=0}^{\infty}(-1)^{K+1}\sum_{\beta\beta'} \\ \times\Bigg\{ & \sum_{k=K-1}^{K+1} \langle -\lambda_a\|\boldsymbol{X}^{(1k)K}\|\,\lambda_c\,\rangle\langle -\lambda'_b\|\boldsymbol{X}^{(1k)K}\|\,\lambda'_d\,\rangle\; v_{kK}\, R^k(\omega;-\lambda_a,-\lambda'_b,\lambda_c,\lambda'_d) \\ & + w_K\Big[\langle -\lambda_a\|\boldsymbol{X}^{(1K-1)K}\|\,\lambda_c\,\rangle\langle -\lambda'_b\|\boldsymbol{X}^{(1K+1)K}\|\,\lambda'_d\,\rangle \\ & \qquad \times T^{K-1,K+1,K}(\omega;-\lambda_a,-\lambda'_b,\lambda_c,\lambda'_d) \\ & \quad + \langle -\lambda_a\|\boldsymbol{X}^{(1K+1)K}\|\,\lambda_c\,\rangle\langle -\lambda'_b\|\boldsymbol{X}^{(1K-1)K}\|\,\lambda'_d\,\rangle \\ & \qquad \times T^{K-1,K+1,K}(\omega;-\lambda'_b,-\lambda_a,\lambda'_d,\lambda_c)\Big]\Bigg\}, \end{aligned} \tag{6.4.31}$$

where $T^{K-1,K+1,K}(\omega;-\lambda_a,-\lambda'_b,\lambda_c,\lambda'_d)$ is defined by (6.4.18) after replacing $U_k(r_1,r_2)$ by $W_{K\mp1,K\pm1,K}(r_1,r_2;\omega)$ and using integral symmetry. Equation (6.4.30) is equivalent to the expressions given in the original papers [4, 5, 6] but is now in a form that makes the dependence on the indices β and β', which label the "large" and "small" components of the Dirac spinors more obvious. Similar formulas were derived in [19, 20, 21] by slightly different methods.

We have now to decide appropriate values of ω. On the energy shell we have $\epsilon_a + \epsilon_b = \epsilon_c + \epsilon_d$, so that the frequency of the virtual photon is given by $\omega = \omega_{ac} = -\omega_{bd}$ where $\omega_{ac} = (\epsilon_a - \epsilon_c)/c$ and $\omega_{bd} = (\epsilon_b - \epsilon_d)/c$. The position off the energy shell when $\omega_{ac} \neq -\omega_{bd}$ is more complicated, as discussed in §4.10, but the averaging of matrix elements with $\omega = \omega_{ac}$ and $\omega = \omega_{db}$ suggested in (4.9.23) has proved quite effective in bound state calculations involving atomic inner shells.

6.4.6 The Breit interaction

The effective interaction strength for the Breit interaction, (6.4.14), can be obtained by taking the long wavelength limit of (6.4.31):

$$g^B(R) = \lim_{\omega \to 0} g^T(R; \omega) = -\frac{1}{2R}\left(\boldsymbol{\alpha}_1 \cdot \boldsymbol{\alpha}_2 + (\boldsymbol{\alpha}_1 \cdot \widehat{\boldsymbol{R}})(\boldsymbol{\alpha}_2 \cdot \widehat{\boldsymbol{R}})\right). \quad (6.4.32)$$

We need the limits [4]

$$\lim_{\omega \to 0} U_k(r_1, r_2; \omega) = U_k(r_1, r_2);$$

$$\lim_{\omega \to 0} W_{K-1,K-1,K}(r_1, r_2; \omega) = \frac{2K+1}{2K-1} U_{K-1}(r_1, r_2);$$

$$\lim_{\omega \to 0} W_{K+1,K+1,K}(r_1, r_2; \omega) = \frac{2K+1}{2K+3} U_{K+1}(r_1, r_2);$$

and

$$\lim_{\omega \to 0} W_{K-1,K+1,K}(r_1, r_2; \omega) = -\frac{2K+1}{2}\left(\overline{U}_{K-1}(r_1, r_2) - \overline{U}_{K+1}(r_1, r_2)\right)$$

$$\lim_{\omega \to 0} W_{K+1,K-1,K}(r_1, r_2; \omega) = -\frac{2K+1}{2}\left(\overline{U}_{K-1}(r_2, r_1) - \overline{U}_{K+1}(r_2, r_1)\right)$$

where

$$\overline{U}_k(r, s) = \begin{cases} r^k/s^{k+1} & \text{if } r < s\,; \\ 0 & \text{otherwise}\,. \end{cases}$$

so that, from (6.4.30),

$$g^B(R;\omega) \tag{6.4.33}$$
$$= \sum_{K=0}^{\infty}(-1)^K \left\{ \sum_{k=K-1}^{K+1} v_{kK}\, U_k(r_1,r_2)\, \mathsf{X}^{(1k)K}(1)\cdot \mathsf{X}^{(1k)K}(2) - \frac{2K+1}{2} w_K \right.$$
$$\times \Big[\left(\overline{U}_{K-1}(r_1,r_2) - \overline{U}_{K+1}(r_1,r_2)\right)\, \mathsf{X}^{(1,K-1)K}(1)\cdot \mathsf{X}^{(1,K+1)K}(2)$$
$$\left. + \left(\overline{U}_{K-1}(r_2,r_1) - \overline{U}_{K+1}(r_2,r_1)\right)\, \mathsf{X}^{(1,K+1)K}(1)\cdot \mathsf{X}^{(1,K-1)K}(2) \Big] \right\}.$$

It is convenient to introduce a one-sided Slater integral (compare [4, Equation (19)])

$$S^k(-\lambda_a, -\lambda'_b, \lambda_c, \lambda'_d) = \int_0^{\infty}\int_0^{r_2} R^*_{\gamma_a,-\lambda_a}(r_1) R_{\gamma_c,\lambda_c}(r_1)$$
$$\times\, \overline{U}_k(r_1,r_2)\, R^*_{\gamma_b,-\lambda'_b}(r_1) R_{\gamma_d,\lambda'_d}(r_1)\, dr_1 dr_2, \tag{6.4.34}$$

so that, from (6.4.18),

$$R^k(-\lambda_a, -\lambda'_b, \lambda_c, \lambda'_d) = S^k(-\lambda_a, -\lambda'_b, \lambda_c, \lambda'_d) + S^k(-\lambda'_b, -\lambda_a, \lambda'_d, \lambda_c) \tag{6.4.35}$$

We can therefore write the effective interaction strength in terms of the S^k integrals as in [4, Equation (18)]

$$X_B(abcd) = \sum_{K=0}^{\infty}(-1)^K \sum_{\beta\beta'}$$
$$\times \left\{ \sum_{k=K-1}^{K+1} \langle -\lambda_a \| \boldsymbol{X}^{(1k)K} \| \lambda_c \rangle \langle -\lambda'_b \| \boldsymbol{X}^{(1k)K} \| \lambda'_d \rangle\, v_{kK}\, R^k(-\lambda_a, -\lambda'_b, \lambda_c, \lambda'_d) \right.$$
$$- \frac{2K+1}{2} w_K \Big(\langle -\lambda_a \| \boldsymbol{X}^{(1,K-1)K} \| \lambda_c \rangle \langle -\lambda'_b \| \boldsymbol{X}^{(1,K+1)K} \| \lambda'_d \rangle$$
$$\times \left[S^{K-1}(-\lambda_a, -\lambda'_b, \lambda_c, \lambda'_d) - S^{K+1}(-\lambda_a, -\lambda'_b, \lambda_c, \lambda'_d) \right]$$
$$+ \langle -\lambda_a \| \boldsymbol{X}^{(1,K+1)K} \| \lambda_c \rangle \langle -\lambda'_b \| \boldsymbol{X}^{(1,K-1)K} \| \lambda'_d \rangle$$
$$\left. \times \left[S^{K-1}(-\lambda'_b, -\lambda_a, \lambda'_d, \lambda_c) - S^{K+1}(-\lambda'_b, -\lambda_a, \lambda'_d, \lambda_c) \right] \Big) \right\}. \tag{6.4.36}$$

Equation (6.4.36) is equivalent to [4, Equation (22)], after collecting the coefficients of radial integrals S^ν using (6.4.35).

6.5 Interaction strengths for the magnetic interactions

6.5.1 The transverse photon interaction

In this section, we follow [15] to put the effective interaction strengths for the transverse photon interaction into computable form. Define

$$R^\nu[\alpha\gamma\,|\,\beta\delta;\omega] = \int_0^\infty \int_0^\infty D^o_{\alpha\gamma}(r)\, V_\nu(r,s;\omega)\, D^o_{\beta\delta}(s)\, dr\, ds. \tag{6.5.1}$$

and

$$S^k[\alpha\gamma\,|\beta\delta;\omega] = \int_0^\infty \int_0^\infty D^o_{\alpha\gamma}(r)\, W_{k-1,k+1,k}(r,s;\omega)\, D^o_{\beta\delta}(s)\, dr\, ds. \tag{6.5.2}$$

where the overlap radial densities are defined in the manner of (6.4.9) by

$$D^o_{\alpha\gamma}(r) = P_\alpha(r) Q_\gamma(r) = \left[R^*_{\gamma_\alpha,-\lambda_\alpha}(r)\, R_{\gamma_\gamma,+\lambda_\gamma}(r)\right]_{\beta=-1},$$

where $\lambda_\alpha = \beta\kappa_\alpha$ and $\lambda_\gamma = \beta\kappa_\gamma$[2]. These basic integrals appear with different permutations of orbital labels in combinations

$$R^\nu_\mu(ABCD) = \frac{1}{2}\left[R^\nu[\alpha_\mu\gamma_\mu\,|\,\beta_\mu\delta_\mu;\omega_{\alpha_\mu\gamma_\mu}] + R^\nu[\alpha_\mu\gamma_\mu\,|\,\beta_\mu\delta_\mu;\omega_{\beta_\mu\delta_\mu}]\right], \tag{6.5.3}$$
$$S^k_\mu(ABCD) = \frac{1}{2}\left[S^k[\alpha_\mu\gamma_\mu\,|\,\beta_\mu\delta_\mu;\omega_{\alpha_\mu\gamma_\mu}] + S^k[\alpha_\mu\gamma_\mu\,|\,\beta_\mu\delta_\mu;\omega_{\beta_\mu\delta_\mu}]\right],$$

listed in Table 6.1 so that the effective interaction strength for multipole k is

$$\begin{aligned} X^k(ABCD) &= (-1)^k \langle j_A \,\|\, \boldsymbol{C}^k \,\|\, j_C\rangle \langle j_B \,\|\, \boldsymbol{C}^k \,\|\, j_D\rangle \{j_A, j_C, k\}\{j_B, j_D, k\} \\ &\times \left\{ \sum_{\nu=k-1}^{k+1} \Pi^o(\kappa_A,\kappa_C,\nu)\, \Pi^o(\kappa_B,\kappa_D,\nu) \sum_{\mu=1}^{4} r^{\nu k}_\mu(ABCD) R^\nu_\mu(ABCD) \right. \\ &\left. + \Pi^o(\kappa_A,\kappa_C,k-1)\, \Pi^o(\kappa_B,\kappa_D,k+1) \sum_{\mu=1}^{8} s^k_\mu(ABCD) S^k_\mu(ABCD) \right\}. \end{aligned} \tag{6.5.4}$$

The associated coefficients are given in Tables 6.2 and 6.3. In both tables we have $K = \kappa_C - \kappa_A, \quad K' = \kappa_D - \kappa_B$, with $Q = 1/(2k+1)^2$ and

$$P = \begin{cases} \dfrac{(k+1)}{k(2k-1)(2k+1)} & \text{for } \nu = k-1 \\ -\dfrac{(\kappa_A+\kappa_C)(\kappa_B+\kappa_D)}{k(k+1)} & \text{for } \nu = k \\ \dfrac{k}{(k+1)(2k+1)(2k+3)} & \text{for } \nu = k+1 \end{cases}$$

The general form of the effective interaction strength simplifies when two or more subshell labels are equal:

---

[2] The symmetries exploited in Table 6.1 assume that the radial amplitudes are *real*. The table will need modification if we have to deal with complex scattering solutions of outgoing or ingoing type.

Table 6.1. Permutation of labels in (6.5.4)

$R^{\nu}_{\mu}(ABCD)$			$S^{k}_{\mu}(ABCD)$					
μ	$\alpha\gamma$	$\beta\delta$	μ	$\alpha\gamma$	$\beta\delta$	μ	$\alpha\gamma$	$\beta\delta$
1	AC	BD	1	AC	BD	5	AC	DB
2	CA	DB	2	BD	AC	6	DB	AC
3	AC	DB	3	CA	DB	7	CA	DB
4	CA	BD	4	DB	CA	8	BC	CA

Table 6.2. Coefficients $r^{\nu k}_{\mu}(ABCD)$

μ	$\nu = k-1$	$\nu = k$	$\nu = k+1$
1	$P(K+k)(K'+k)$	P	$P(K-k-1)(K'-k-1)$
2	$P(K-k)(K'-k)$	P	$P(K+k-1)(K'+k-1)$
3	$P(K+k)(K'-k)$	P	$P(K-k-1)(K'+k+1)$
4	$P(K-k)(K'+k)$	P	$P(K+k+1)(K'-k-1)$

See text for symbols.

Table 6.3. Coefficients $s^{k}_{\mu}(ABCD)$

μ		μ	
1	$Q(K+k)(K'-k-1)$	5	$Q(K+k)(K'+k+1)$
2	$Q(K'+k)(K-k-1)$	6	$Q(K'-k)(K-k-1)$
3	$Q(K-k)(K'+k+1)$	7	$Q(K-k)(K'-k-1)$
4	$Q(K'-k)(K+k+1)$	5	$Q(K'+k)(K+k+1)$

See text for symbols.

- $A = C, B \neq D$ (or $A \neq C, B = D$)

$$X^k(ABAD) = \langle j_A \,\|\, \boldsymbol{C}^k \,\|\, j_A \rangle \, \langle j_B \,\|\, \boldsymbol{C}^k \,\|\, j_D \rangle \times \frac{4\kappa_A(\kappa_B + \kappa_D)}{k(k+1)} R^k(A; BD; \omega_{BD} \tag{6.5.5}$$

where

$$R^k(A; BD; \omega_{BD}) = \frac{1}{2} \int_0^{\infty} \int_0^{\infty} D^o_{AA}(r) \left[U_k(r,s;\omega) + U_k(r,s) \right] \overline{D}^o_{BD}(s) \, drds \tag{6.5.6}$$

Table 6.4. Coefficients $g_\gamma^{\nu k}(AB)$

γ	$\nu = k-1$	$\nu = k$	$\nu = k+1$
+1	$-P(K+k)^2$	P	$-P(K-k-1)^2$
0	$-2P(K^2-k^2)$	$2P$	$-2P[K^2-(k+1)^2]$
-1	$-P(K-k)^2$	P	$-P(K+k+1)^2$

with

$$\overline{D}^o_{BD}(s) = D^o_{BD}(s) + D^o_{DB}(s) = P_B(s)Q_D(s) + P_D(s)Q_B(s).$$

- $A = C, B = D\,(A \neq B)$

$$X^k(ABAB) = \langle j_A \,\|\, \boldsymbol{C}^k \,\|\, j_A\rangle \langle j_B \,\|\, \boldsymbol{C}^k \,\|\, j_B\rangle \times \frac{16\kappa_A(\kappa_B+\kappa_D)}{k(k+1)} F^k(A,B) \tag{6.5.7}$$

where

$$F^k(A,B) = \int_0^\infty \int_0^\infty D^o_{AA}(r)\, U_k(r,s)\, D^o_{BB}(s)\, drds.$$

- $A = D, B = C$

$$X^k(ABBA) = (-1)^{j_A-j_B+k} \langle j_A \,\|\, \boldsymbol{C}^k \,\|\, j_B\rangle \langle j_B \,\|\, \boldsymbol{C}^k \,\|\, j_A\rangle \{j_A, j_B, k\} \times \left\{ \sum_{\nu=k-1}^{k+1} \Pi^o(\kappa_A,\kappa_B,\nu) \sum_{\gamma=-1}^{+1} g_\gamma^{\nu k}(AB)\, G_\gamma^\nu(AB) + \Pi^o(\kappa_A,\kappa_B,k\pm 1) \sum_{\delta=1}^{4} h_\delta^k(AB)\, H_\delta^k(AB) \right\}. \tag{6.5.8}$$

where

$$G_\gamma^\nu(AB) = \begin{cases} 2R^\nu[AB\,|\,AB;\omega_{AB}] & \text{for } \gamma = 1 \\ 2R^\nu[AB\,|\,BA;\omega_{AB}] & \text{for } \gamma = 0 \\ 2R^\nu[BA\,|\,BA;\omega_{AB}] & \text{for } \gamma = -1 \end{cases} \tag{6.5.9}$$

and

$$H_\delta^k(ab) = \begin{cases} S^k[AB\,|\,AB;\omega_{AB}] & \text{for } \delta = 1 \\ S^k[BA\,|\,BA;\omega_{AB}] & \text{for } \delta = 2 \\ S^k[AB\,|\,BA;\omega_{AB}] & \text{for } \delta = 3 \\ S^k[BA\,|\,AB;\omega_{AB}] & \text{for } \delta = 4 \end{cases} \tag{6.5.10}$$

The corresponding coefficients are shown in Tables 6.4 and 6.5, where $K = \kappa_B - \kappa_A$, and P and Q are as in Tables 6.2 and 6.3. See §6.7 for the construction of the integrals listed above.

Table 6.5. Coefficients $h^k_\delta(AB)$

δ		δ	
1	$-2Q(K+k)^(K-k-1)$	3	$-2Q(K+k)(K+k+1)$
2	$-2Q(K-k)(K+k+1)$	4	$-2Q(K-k)(K-k-1)$

6.5.2 The Breit interaction

The corresponding results for the simpler Breit interaction can be taken from [4]. The effective interaction strength is

$$X^k(ABCD) = (-1)^k \langle j_A \,\|\, \boldsymbol{C}^k \,\|\, j_C\rangle \, \langle j_B \,\|\, \boldsymbol{C}^k \,\|\, j_D\rangle \times \sum_{\nu=k=1}^{k+1} \Pi^o(\kappa_A, \kappa_C, \nu)\Pi^o(\kappa_B, \kappa_D, \nu) \sum_{\mu=1}^{8} s^{\nu k}_\mu(ABCD)\, S^\nu_\mu(ABCD). \tag{6.5.11}$$

The eight radial integrals $S^\nu_\mu(ABCD)$ are all of the form

$$S^\nu(\alpha\beta\gamma\delta) = \int_0^\infty dr \int_r^\infty ds\, D^o_{\alpha\gamma}(r)\overline{U}_k(r,s)D^o_{\beta\delta}(s) \tag{6.5.12}$$

where the arguments $\alpha\gamma$ and $\beta\delta$ for each value of μ are given in Table 6.1. The coefficients $s^{kk}_\mu(ABCD)$ are independent of μ, and are given by

$$s^{kk}_\mu(ABCD) = -\frac{(\kappa_A+\kappa_C)(\kappa_B+\kappa_D)}{k(k+1)}, \tag{6.5.13}$$

subject to the parity selection rules $l_A + l_C + k$, $l_B + l_D + k$ both *odd*. When $\nu = k \pm 1$, the coefficients are given in Table 6.6. These expressions simplify in particular cases:

- $A = C$, $B \neq D$ (or $A \neq C$, $B = D$)

$$X^k(ABAC) = \frac{4\kappa_A(\kappa_B+\kappa_C)}{k(k+1)} \times \langle j_A \,\|\, \boldsymbol{C}^k \,\|\, j_C\rangle \, \langle j_B \,\|\, \boldsymbol{C}^k \,\|\, j_D\rangle \, R^k(A;BC) \tag{6.5.14}$$

where

$$R^k(A;BC) = \int_0^\infty \int_0^\infty D^o_{AA}(r)\, U_k(r,s)\, \overline{D}^o_{BC}(s)\, drds.$$

- $A = C$, $B = D$ $(A \neq B)$

$$X^k(ABAC) = \frac{16\kappa_A\kappa_B}{k(k+1)} \langle j_A \,\|\, \boldsymbol{C}^k \,\|\, j_C\rangle \, \langle j_B \,\|\, \boldsymbol{C}^k \,\|\, j_D\rangle \, F^k_{Breit}(AB) \tag{6.5.15}$$

Table 6.6. Coefficients $s_\mu^{\nu k}(ABCD)$, $\nu = k \pm 1$

μ	$\nu = k-1$	$\nu = k+1$
1	$(k+K)(b'+c'K')$	$(K'-k-1)(b+cK)$
2	$(k+K')(b'+c'K)$	$(K-k-1)(b+cK')$
3	$(k-K)(b'-c'K')$	$(-K'-k-1)(b-cK)$
4	$(k-K')(b'-c'K)$	$(-K-k-1)(b-cK')$
5	$-(k+K)(b'-c'K')$	$(K'+k+1)(b+cK)$
6	$-(k-K')(b'+c'K)$	$(-K+k+1)(b-cK')$
7	$-(k-K)(b'+c'K')$	$(-K'+k+1)(b-cK)$
8	$-(k+K')(b'-c'K)$	$(K+k+1)(b+cK')$

$K = \kappa_C - \kappa_A,\ K' = \kappa_D - \kappa_B$

$b' = (k+1)/2(2k-1),\ c' = -(k-2)/2k(2k-1),\ k \geq 1$

$b = k/2(2k+3),\ c = (k+3)/(2k+2)(2k+3),\ k \geq 0.$

Table 6.7. Coefficients $\xi_\gamma^{\nu k}(AB)$, $\nu = k \pm 1$

γ	$\nu = k-1$	$\nu = k+1$
1	$(-K-k)(b'+c'K)$	$(k+1-K)(b+cK)$
2	$2[b'k - c'K^2]$	$-2[b(k+1)+cK^2]$
3	$(K-k)(b'-c'K)$	$(k+1+K)(b-cK)$

b, c, b', c' as in Table 6.6

where

$$F^k_{Breit}(AB) = \int_0^\infty \int_0^\infty D^o_{AA}(r)\, U_k(r,s)\, D^o_{BB}(s)\, dr ds.$$

- $A = D,\ B = C$

(i) $\nu = k,\ l_A + l_B + k$ odd

$$X^k(ABBA) = (-1)^{k+j_A+j_B} \frac{[(\kappa_A + \kappa_B)\langle j_A \,\|\, \boldsymbol{C}^k \,\|\, j_B\rangle]^2}{k(k+1)} G^k_{Breit}(AB) \tag{6.5.16}$$

with

$$G^k_{Breit}(AB) = \int_0^\infty \int_0^\infty \overline{D}^o_{AB}(r)\, U_k(r,s)\, \overline{D}^o_{AB}(s)\, dr ds.$$

(ii) $\nu = k \pm 1,\ l_A + l_B + k$ even

$$X^k(ABBA) = (-1)^{k+j_A-j_B} [\langle j_A \,\|\, \boldsymbol{C}^k \,\|\, j_B\rangle]^2 \times \sum_{\gamma=-1}^{+1} \sum_{\nu=k\pm1} \xi_\gamma^{\nu k} G^{\nu\gamma}_{Breit}(AB) \tag{6.5.17}$$

where, in (6.5.16) and (6.5.17),

$$G^k_{Breit}(AB) = G^{k,+1}_{Breit}(AB) + 2G^{k,0}_{Breit}(AB) + G^{k,-1}_{Breit}(AB)$$

and

$$\begin{aligned}
G^{\nu,+1}_{Breit}(AB) &= \int_0^\infty \int_0^\infty D^o_{AB}(r)\, U_\nu(r,s)\, D^o_{AB}(s)\, drds,\\
G^{\nu,0}_{Breit}(AB) &= \int_0^\infty \int_0^\infty D^o_{AB}(r)\, U_\nu(r,s)\, D^o_{BA}(s)\, drds,\\
G^{\nu,-1}_{Breit}(AB) &= \int_0^\infty \int_0^\infty D^o_{BA}(r)\, U_\nu(r,s)\, D^o_{BA}(s)\, drds.
\end{aligned}$$

The coefficients $\xi^{\nu k}_\gamma$ are shown in Table 6.7.

6.6 Closed shells and configuration averages

6.6.1 The Dirac-Hartree-Fock model

The energy of a configuration of closed subshells with a wavefunction which is a single determinant, §6.1, has the form

$$E = \sum_a \left\{ \langle a \,|\, h \,|\, a \rangle + \frac{1}{2} \sum_b [\langle ab \,|\, g \,|\, ab \rangle - \langle ba \,|\, g \,|\, ab \rangle] \right\} \tag{6.6.1}$$

where the indices a, b run over all *occupied orbitals*. For simplicity, we begin with the Dirac-Coulomb model in which $g = g^C$. We adopt a notation in which jj-coupled subshells are labelled by capital letters, $A, B, \ldots$ and identify lower-case labels as members of the corresponding subshell, so that $a \in A$, $b \in B$ and so on. The subshell A can be associated with quantum numbers n_A, κ_A, where n_A is a *principal quantum number* defining an energy eigenstate in the spherically symmetric model potential u of (6.1.3), and κ_A is an *angular quantum number*. The members $a \in A$ are therefore the orbitals with quantum numbers $\{n_A, \kappa_A, m_A \,|\, m_A = -j_A, -j_A+1, \ldots, +j_A\}$; we denote the dimension of this set by $q^0_A = [j_A] = 2j_A + 1$.

When the orbitals are Dirac central field spinors, we can use the results of the last two sections to express each of the terms contributing to (6.6.1) in terms of radial integrals. Thus each orbital $a \in A$ contributes

$$\langle a \,|\, h \,|\, a \rangle = I(A, A), \tag{6.6.2}$$

where, from (6.3.8).

$$\begin{aligned}
I(A,A) = \int_0^\infty & c\, Q^*_{n_A,\kappa_A}(r) \left(\frac{d}{dr} + \frac{\kappa_A}{r} \right) P_{n_A,\kappa_A}(r) \\
& -c\, P^*_{n_A,\kappa_A}(r) \left(\frac{d}{dr} - \frac{\kappa_A}{r} \right) Q_{n_A,\kappa_A}(r) - 2mc^2 Q^*_{n_A,\kappa_A}(r) Q_{n_A,\kappa_A}(r) \\
& +v_{nuc}(r) \left[P^*_{n_A,\kappa_A}(r) P_{n_A,\kappa_A}(r) + Q^*_{n_A,\kappa}(r) Q_{n_A,\kappa_A}(r) \right] \, dr,
\end{aligned} \tag{6.6.3}$$

depends only on the radial amplitudes $P_{n_A,\kappa_A}(r), Q_{n_A,\kappa_A}(r)$. Summing over the q_A^0 states making up each of the *occupied* subshells, A, gives the one-electron contribution

$$E^{(1)} = \sum_a \langle a \,|\, h \,|\, a\rangle = \sum_A q_A^0 \, I(A,A) \tag{6.6.4}$$

to the total energy. The Coulomb interactions contribute [2]

$$E^{(2)} = \frac{1}{2} \sum_a \sum_b \left[\langle ab \,|\, g^C \,|\, ab\rangle - \langle ba \,|\, g^C \,|\, ab\rangle \right], \tag{6.6.5}$$

where, from (6.4.2),

$$\langle ab \,|\, g \,|\, cd\rangle = \sum_k \sum_q \begin{pmatrix} m_a & k & j_C \\ j_A & q & m_C \end{pmatrix} \begin{pmatrix} m_b & q & j_D \\ j_B & k & m_D \end{pmatrix} X_C^k(ABCD).$$

We denote the *direct* interaction terms by

$$\begin{aligned} J(a,b) &= \langle ab \,|\, g^C \,|\, ab\rangle \\ &= \sum_k \begin{pmatrix} m_a & k & j_A \\ j_A & 0 & m_a \end{pmatrix} \begin{pmatrix} m_b & 0 & j_B \\ j_B & k & m_b \end{pmatrix} X_C^k(ABAB), \end{aligned} \tag{6.6.6}$$

and the *exchange* terms by

$$\begin{aligned} K(a,b) &= \langle ba \,|\, g^C \,|\, ab\rangle \\ &= \sum_k \begin{pmatrix} m_a & k & j_B \\ j_A & q & m_b \end{pmatrix} \begin{pmatrix} m_b & q & j_A \\ j_B & k & m_a \end{pmatrix} X_C^k(BAAB). \end{aligned} \tag{6.6.7}$$

It is customary to express (6.4.7) in terms of Slater integrals $F_C^k(AB)$ and $G_C^k(AB)$:

$$X_C^k(ABAB) = (-1)^k \langle j_a \| \boldsymbol{C}^k \| j_a\rangle \langle j_b \| \boldsymbol{C}^k \| j_b\rangle F_C^k(AB), \tag{6.6.8}$$

and

$$X_C^k(BAAB) = (-1)^k \langle j_b \| \boldsymbol{C}^k \| j_a\rangle \langle j_a \| \boldsymbol{C}^k \| j_b\rangle G_C^k(AB), \tag{6.6.9}$$

where $F_C^k(AB) = R_C^k(ABAB)$ and $G_C^k(AB) = R_C^k(BAAB)$ depend only on the subshell labels and not on the magnetic quantum numbers.

The contribution to the total energy from an electron in orbital $a \in A$ interacting with all electrons in B is

$$\sum_{m_B} J(a,b) = \frac{[j_B]^{1/2}}{[j_A]^{1/2}} \, X_C^0(ABAB),$$

independent of the value of m_a. This makes use of the result

$$\begin{pmatrix} m_a & 0 & j_B \\ j_A & 0 & m_b \end{pmatrix} = [j_B]^{-1/2} \delta_{j_A, j_B} \delta_{m_a, m_b},$$

from which we can evaluate the sum

$$\begin{aligned} S &= \sum_{m_b} \begin{pmatrix} m_b & k & j_B \\ j_B & 0 & m_b \end{pmatrix} \\ &= [j_B]^{1/2} \sum_{m_b, m'_b} \begin{pmatrix} m_b & k & j_B \\ j_B & 0 & m'_b \end{pmatrix} \begin{pmatrix} m_b & 0 & j_B \\ j_B & 0 & m'_b \end{pmatrix} \\ &= [j_B]^{1/2} \delta_{k,0} \end{aligned}$$

by the unitarity relation (B.3.53). Setting $k = 0$ in (6.6.8) and substituting numerical values for the reduced matrix elements we arrive at

$$\sum_{m_b} J(a, b) = [j_B] \, F_C^0(AB), \tag{6.6.10}$$

demonstrating that an electron interacts with a full subshell, B, as if B were a classical spherical distribution with total charge $-e[j_B]$. Similarly,

$$\sum_{m_b, q} \begin{pmatrix} m_a & k & j_B \\ j_A & q & m_b \end{pmatrix} \begin{pmatrix} m_b & q & j_A \\ j_B & k & m_a \end{pmatrix} = (-1)^{k + j_B - j_A} \, [j_A]^{-1},$$

so that

$$\sum_{m_b} K(ab) = \sum_k (-1)^{k + j_A - j_B} [j_A]^{-1} \, X_C^k(BAAB),$$

which can be expressed as

$$\sum_{m_b} K(ab) = \frac{1}{2} \, [j_B] \sum_k \Gamma_{j_A k j_B} \, G_C^k(AB), \tag{6.6.11}$$

where

$$\Gamma_{j_A k j_B} = 2 \begin{pmatrix} 1/2 & k & j_B \\ j_A & 0 & 1/2 \end{pmatrix}^2 . \tag{6.6.12}$$

The closed shell exchange coefficients $\Gamma_{jkj'} = \Gamma_{j'kj}$ for $1/2 \le j' \le j \le 7/2$ are given in Table A.5; values for $j' > j$ can be obtained by swapping the roles of j and j'.

Some formulations of relativistic atomic structure use $d^k(jm, j'm')$, (6.2.12), and related coefficients. Thus (6.6.6) can be written

$$J(a, b) = \sum_k a^k(j_A m_a, j_B m_b) \, F^k(A, B), \tag{6.6.13}$$

where

$$a^k(j_A m_a, j_B m_b) = d^k(j_A m_a, j_A m_a) \, d^k(j_B m_b, j_B m_b);$$

similarly

$$K(a,b) = \sum_k b^k(j_A m_a, j_B m_b)\, G^k(A,B), \tag{6.6.14}$$

where

$$b^k(j_A m_a, j_B m_b) = \left[d^k(j_A m_a, j_B m_b)\right]^2.$$

When the electron a belongs to the same subshell as the other electrons, the Coulomb energy of the closed subshell A is

$$\frac{1}{2} q_A^0 (q_A^0 - 1)\, F^0(A,A) - \sum_{k>0} \frac{1}{4} {q_A^0}^2\, \Gamma_{j_A k j_A}\, G^k(A,A), \tag{6.6.15}$$

where we have used the fact that $G^0(A,A) = F^0(A,A)$. Each fully occupied subshell has just $\frac{1}{2} q_A^0 (q_A^0 - 1)$ interacting pairs, so that we can identify $F^0(A,A)$ as the average interaction energy of q_A^0 electrons with the same radial charge distribution. The $k = 0$ exchange terms eliminate spurious self-interaction contributions to the energy formula. The energy of two different interacting subshells, A and B, is

$$q_A^0 q_B^0 \left[F^0(A,B) - \sum_k \frac{1}{2}\, \Gamma_{j_A k j_B}\, G^k(A,B) \right]. \tag{6.6.16}$$

The first term is the classical interaction energy of q_A^0 spherically distributed electrons in subshell A and q_B^0 spherically distributed electrons in subshell B. These results motivate the assumptions of the crude Pyper-Grant model [22] used to discuss shell filling in §1.3.5.

The DHF energy of an electronic configuration consisting only of closed subshells can therefore be written [23]

$$\begin{aligned} E_{DHF} = {} & \sum_A q_A^0 \\ & \times \left\{ I(A,A) + \frac{1}{2} \sum_B \sum_k \left[C^0(ABk)\, F^k(A,B) + D^0(ABk)\, G^k(A,B) \right] \right\}, \end{aligned} \tag{6.6.17}$$

where

$$C^0(ABk) = q_B^0\, \delta_{k,0}, \quad D^0(ABk) = -\frac{1}{2}\, q_B^0\, \Gamma_{j_A k j_B}. \tag{6.6.18}$$

6.6.2 Inclusion of magnetic interactions

The closed subshell contribution of magnetic interactions can be treated in much the same way; here we give results in terms of the effective interaction strengths, starting from

$$\langle ab\,|g|\,cd\rangle = \sum_{K=0}^{\infty} \begin{pmatrix} m_a & K & j_C \\ j_A & q & m_c \end{pmatrix} \begin{pmatrix} m_b & K & j_D \\ j_B & -q & m_d \end{pmatrix} X^K(ABCD) \tag{6.6.19}$$

where we use (6.4.32) for the Breit interaction, for example, (6.4.20) for the Gaunt interaction, and the corresponding expressions extracted from (6.4.31), (6.4.25), or (6.4.24) for the transverse photon, and the Møller interactions respectively. The selection rules for the retarded Coulomb interaction (M1), (6.4.24), are the same as for the Coulomb interaction. The current-current terms (M2) require the magnetic parity conditions (6.2.18) and (6.2.19):

$$\{\,j\,K\,j'\,\} = 1, \quad j + K + j' \text{ is } \begin{cases} \text{odd if } \eta = -\eta' \\ \text{even if } \eta = \eta' \end{cases} \tag{6.6.20}$$

The direct magnetic matrix elements are non-vanishing only if $K = 0$; since (6.6.20) also requires that K takes *odd* values when $\kappa = \kappa'$, this makes the *direct* magnetic interaction of an electron $a \in A$ with the subshell B vanish identically. The *exchange* magnetic contribution does not vanish and we find that the magnetic energy of a single electron of subshell A with the closed subshell B is

$$\sum_{m_B} \langle ab\,|g|\,ba\rangle = [j_A]^{-1} \sum_{k=0}^{\infty} X^k(ABBA), \tag{6.6.21}$$

where $X^k(ABBA)$ is defined in (6.5.16) and (6.5.17).

6.6.3 Average of configuration models

We can write (6.6.17) in terms of one-electron mean energies

$$e^0_A = I(A,A) + \sum_B e^0_{AB} \tag{6.6.22}$$

where

$$e^0_{AB} = \sum_k \left[C^0(ABk)\,F^k(A,B) + D^0(ABk)\,G^k(A,B)\right]$$

represents the Coulomb repulsion energy of a single electron in subshell A with the closed B subshell. With this notation, (6.6.17) gives

$$E_{DHF} = \frac{1}{2} \sum_A q^0_A \,\left[I(A,A) + e^0_A\right]. \tag{6.6.23}$$

Whilst it cannot give all the detail, the *average of configuration* model [24, Chapter 14] has proved a useful first approximation for open shell systems. This is a classical scheme exploiting closed shell formulae in which the interaction of a single electron in subshell A with a single electron in subshell B is given by e^0_{AB}/q^0_B when the subshells are different and $e^0_{AA}/[\frac{1}{2}(q^0_A - 1)]$

when the two electrons belong to the same subshell A. The j–j average of configuration energy following (6.6.23) is

$$E_{jj} = \frac{1}{2}\sum_A q_A \left[I(A,A) + e_A\right]. \tag{6.6.24}$$

where

$$e_A = I(A,A) + \sum_B e_{AB},$$

$$e_{AB} = \sum_k \left[C(ABk)\, F^k(A,B) + D(ABk)\, G^k(A,B)\right]$$

q_A is the *actual* subshell occupation, and

$$C(ABk) = \frac{q_B - \delta_{AB}}{q_B^0 - \delta_{AB}}\, C^0(ABk), \quad D(ABk) = \frac{q_B - \delta_{AB}}{q_B^0 - \delta_{AB}}\, D^0(ABk).$$

This model can be derived in another way. We define a *configurational state function* (CSF) $\Psi^{\mathcal{C}}_{\Gamma JM}$ for a distribution $\mathcal{C}$ of N electrons over partially filled subshells as a linear combination of $N \times N$ Slater determinants $\Phi^{\mathcal{C}}_{\{\kappa_i m_i\}}$, where $\{\kappa_i m_i\}$ is an abbreviation of $\{\kappa_1 m_1, \ldots \kappa_N m_N\}$ specifying the set of κm values of Dirac electrons in *partially filled* subshells. Thus

$$\Psi^{\mathcal{C}}_{\Gamma JM} = \sum_{\{\kappa_1 m_i\}} \Phi^{\mathcal{C}}_{\{\kappa_i m_i\}} \langle \{\kappa_i m_i\} \,|\, \Gamma JM \rangle.$$

where $\langle \{\kappa_i m_i\} \,|\, \Gamma JM \rangle$ is a generalized Clebsch-Gordan coefficient. For a free atom, the total energy $E_{\Gamma J}$ is independent of the projection M so that

$$\begin{aligned}(2J+1)E_{\Gamma J} &= \sum_{m_i} \sum_{m'_j} \sum_{\Gamma JM} \langle \{\kappa_i m_i\} \,|\, \Gamma JM \rangle \langle \Gamma JM \,|\, \{\kappa'_j m'_j\} \rangle \\ &\qquad \times \langle \{\kappa'_j m'_j\} \,|\, H_{DHF} \,|\, \{\kappa_i m_i\} \rangle \\ &= \sum_{m_j} \langle \{\kappa_i m_i\} \,|\, H_{DHF} \,|\, \{\kappa_i m_i\} \rangle\end{aligned}$$

by GCG unitarity. The sums over projections are unrestricted, so that we can evaluate the Hamiltonian matrix elements using closed shell formulae. A simple counting argument shows that the number of Slater determinants and of CSFs is the same, so that we recover the j–j average of configuration formula (6.6.24).

Bound state Dirac central field orbitals are characterized by quantum numbers $n\kappa m$, where n depends on the radial potential and κ and m are angular quantum numbers. In one-electron systems, the orbital eigenvalues $\epsilon_{n\kappa}$ depend only on $|\kappa| = j + 1/2$, so that j and m are good quantum numbers. The interaction between electrons lifts the $\pm\kappa$ degeneracy in a manner that depends both on the principal quantum number n and on the atomic number.

Except in the second row of the Periodic Table, inner shells can usually be characterized by nj. Things are more complicated in outer subshells, where *intermediate coupling* is the rule, although the structure is often close to LS coupling. The LS average of configuration DHF model, [23, 25, 26] may then be useful. In this scheme, the two subshells nl with $j = l \pm 1/2$ are treated as degenerate and contain q_{nl} electrons, divided between the two Dirac subshells in the ratio of their statistical weights, $2j+1$. Each subshell, $n_A l_A$, can hold up to $q_A^0 = 4l_A + 2$ electrons so that

$$E_{LS} = \frac{1}{2} \sum_A q_A \left[I(A,A) + e'_A \right]. \tag{6.6.25}$$

Here q_A is the actual number of electrons in the LS subshell and

$$e'_A = I(A,A) + \sum_B e'_{AB},$$

where

$$e'_{AB} = \sum_k \sum_{B' \in B} \left[C(AB'k)\, F^k(A,B') + D(AB'k)\, G^k(A,B') \right].$$

includes contributions from both j–j subshells B' in B.

The *average pair energy* of a pair of Dirac s, p, d and f electrons, one in subshell A the other in subshell B, defined by

$$\sum_k \left[C^0(ABk)\, F^k(A,B) + D^0(ABk)\, G^k(A,B) \right] / (q_B^0 - \delta_{AB}), \tag{6.6.26}$$

has been calculated by Larkins [26] and appears in Table A.8. For comparison, the nonrelativistic LS coupling equivalents [24] appear in Table A.6. The effect of averaging over j–j coupling subshell energies with weights $2l/(4l+2)$ for $j = l - 1/2$ and $(2l+2)/(4l+2)$ for $j = l + 1/2$ can be represented by replacing the Slater integrals of Table A.6 with the weighted sums of relativistic Slater integrals.

6.7 DHF integro-differential equations

Following Swirles [13] and Grant [2], we can derive the integro-differential equations for DHF and average of configuration models by variational methods. The energy expression E_{DHF}, (6.6.17), has been derived assuming that the radial amplitudes are normalized. Following Fischer [27, p. 20] we shall work with *unnormalized* amplitudes, although overlap densities must then be normalized explicitly when calculating interaction intregrals. We first write

$$(A|B) = \int_0^\infty D_{AB}(r) dr = \int_0^\infty u_A^\dagger(r) u_B(r)\, dr \tag{6.7.1}$$

for the overlap of two radial spinors. When using unnormalized wavefunctions we must make the replacements $u_A \to u_A/(A\,|A)^{1/2}$ so that (6.6.17) becomes

$$E_{DHF} = \sum_A \frac{q_A^0}{(A|A)} \Bigg\{ I(A,A) + \frac{1}{2}\sum_B \sum_k \left[C^0(ABk)\,\frac{F^k(A,B)}{(B\,|B)} + D^0(ABk)\,\frac{G^k(A,B)}{(B\,|B)} \right] \Bigg\}, \quad (6.7.2)$$

Introducing a Lagrange multiplier, $\lambda_{AB} = \lambda_{BA}$, to enforce orthogonality of pairs of orbitals of the same symmetry, $\kappa_A = \kappa_B$,

$$(A|B) = 0,\ B \neq A,$$

we consider the functional

$$J[A,B,\ldots] = E_{DHF} - \sum_A \sum_B \lambda_{AB} \frac{(A\,|B)}{(A\,|A)^{1/2}(B\,|B)^{1/2}} \equiv \sum_A J[A]$$

where, from (6.6.22) and (6.6.23),

$$J[A] = q_A^0 \left\{ \frac{I(A,A)}{(A\,|A)} + \frac{1}{2}\sum_B \frac{e_{AB}^0}{(A\,|A)(B\,|B)} - \sum_{B\neq A} \frac{\lambda_{AB}\,(A\,|B)}{(A\,|A)^{1/2}(B\,|B)^{1/2}} \right\}, \quad (6.7.3)$$

with

$$e_{AB}^0 = \sum_k \left[C^0(ABk)\,F^k(A,B) + D^0(ABk)\,G^k(A,B) \right].$$

Consider the effect of a variation $u_A \to u_A + \delta u_A$. First, when $\kappa_B = \kappa_A$ and $B \neq A$, then $\delta(A|B) = 0$, whilst

$$\delta(A\,|A) = 2\int_0^\infty \delta u_A^\dagger(r) u_A(r)\,dr, \quad \delta\left[(A\,|A)^{-1}\right] = -\frac{\delta(A\,|A)}{(A\,|A)^2}. \quad (6.7.4)$$

Next, as T_{κ_A} is a symmetric differential operator, and

$$I(A,A) = \int_0^\infty u_A^\dagger(r) T_{\kappa_A} u_A(r) dr,$$

we have

$$\delta I(A,A) = 2\int_0^\infty \delta u_A^\dagger(r) T_{\kappa_A} u_A(r) dr \quad (6.7.5)$$

Because $F^k(A,B) = R^k(ABAB)$ and $G^k(A,B) = R^k(BAAB)$, where the general radial Slater integral is

$$R^k(ABCD) = \int_0^\infty D_{AC}(r)\,\frac{Y^k(BD;r)}{r}\,dr = \int_0^\infty \frac{Y^k(AC;r)}{r}\,D_{BD}(r)\,dr$$

where

$$D_{AC}(r) = u_A^\dagger(r) u_C(r), \qquad \frac{Y^k(AC;r)}{r} = \int_0^\infty U_k(r,s)\, D_{AC}(s)\, ds,$$

the effect of the variation $u_A \to u_A + \delta u_A$ is

$$\delta F^k(A,B) = 2(1+\delta_{AB}) \int_0^\infty \delta u_A^\dagger(r) u_A(r) \frac{Y^k(BB;r)}{r}\, dr \tag{6.7.6}$$

and similarly, using the fact that bound state amplitudes are real,

$$\delta G^k(A,B) = (1+\delta_{AB}) \int_0^\infty \delta u_A^\dagger(r) u_B(r) \frac{Y^k(AB;r)}{r}\, dr. \tag{6.7.7}$$

Thus

$$\delta e_{AB}^0 = 2(1+\delta_{AB}) \int_0^\infty \delta u_A^\dagger(r) \mathcal{F}_{AB}(r)\, dr \tag{6.7.8}$$

where

$$\mathcal{F}_{AB}(r) = \sum_k \left\{ C^0(ABk) \frac{Y^k(BB;r)}{r} u_A(r) + D^0(ABk) \frac{Y^k(BA;r)}{r} u_B(r) \right\}.$$

The first order variation $\delta J[A]$ can therefore be split into two pieces: one, $\delta J_1[A]$, coming from the variation of the common factor $(A|A)^{-1}$ in (6.7.3), the other, $\delta J_2[A]$, from the variation of the integrals inside the bracket

$$\delta J_1[A] = -\frac{2q_A^0}{(A|A)} \left\{ \frac{I(A,A)}{(A|A)} + \frac{1}{2} \sum_B \frac{(1+\delta_{AB})\, e_{AB}^0}{(A|A)(B|B)} \right\} \int_0^\infty \delta u_A^\dagger(r) u_A(r)\, dr,$$

and

$$\delta J_2[A] = \frac{2q_A^0}{(A|A)} \int_0^\infty \delta u_A^\dagger(r) \mathcal{F}_A(r)\, dr$$

where

$$\mathcal{F}_A(r) = T_{\kappa_A}\, u_A(r) + \frac{1}{2} \sum_B (1+\delta_{AB}) \mathcal{F}_{AB}(r) - \sum_{B \neq A} \frac{\lambda_{AB}}{2q_A^0} \frac{(A\,|A)^{1/2}}{(B\,|B)^{1/2}} u_B(r)$$

Setting $\delta J[A] = \delta J_1[A] + \delta J_2[A] = 0$ and using the fact that $\delta u_A^\dagger(r)$ is an arbitrary variation, we get the DHF equations

$$T_{\kappa_A}\, u_A(r) + \frac{1}{2} \sum_B (1+\delta_{AB}) \mathcal{F}_{AB}(r) - \sum_B \epsilon_{AB}\, u_B(r) = 0 \tag{6.7.9}$$

where

$$\epsilon_{AA} = \left[\frac{I(A,A)}{(A\,|A)} + \frac{1}{2} \sum_B \frac{(1+\delta_{AB})\, e_{AB}^0}{(A\,|A)(B\,|B)} \right] \tag{6.7.10}$$

and, when $B \neq A$ (and $\kappa_B = \kappa_A$),

$$\epsilon_{AB} = \frac{\lambda_{AB}}{2q_A^0} \frac{(A\,|A)^{1/2}}{(B\,|B)^{1/2}}. \tag{6.7.11}$$

Equation (6.7.9) is more usually written in a form resembling a single Dirac equation

$$\begin{aligned} \frac{-Z_{nuc}(r) + Y(A;r)}{r} P_A(r) + c\left(-\frac{d}{dr} + \frac{\kappa_A}{r}\right) Q_A(r) \\ - \sum_B \epsilon_{AB} P_B(r) = -X_{+1}(A;r), \\ c\left(\frac{d}{dr} + \frac{\kappa_A}{r}\right) P_A(r) + \left(-2mc^2 + \frac{-Z_{nuc}(r) + Y(A;r)}{r}\right) Q_A(r) \\ - \sum_B \epsilon_{AB} Q_B(r) = -X_{-1}(A;r), \end{aligned} \tag{6.7.12}$$

in which

$$Z_{nuc}(r) = -r\, v_{nuc}(r)$$

is the effective charge of the electron-nucleus potential at radius r, and $Y(A;r)/r$ is the classical interaction of a single electron in subshell A with the remainder of the charge density distribution, where

$$\begin{aligned} Y(A;r) = (q_A^0 - 1)\, \frac{Y^0(AA;r)}{(A\,|A)} \\ + \sum_{k>0} D^0(AA;k)\, \frac{Y^k(AA;r)}{(A\,|A)} + \sum_{B \neq A} q_B^0\, \frac{Y^0(BB;r)}{(B\,|B)}. \end{aligned} \tag{6.7.13}$$

Here we have used the fact that the terms with coefficients $C^0(AA0) = q_A^0$ and $D^0(AA;0) = -1$ sum to give the interaction of a single electron with the residual $q_A^0 - 1$ electrons of the subshell; the exchange term $D^0(AA;0)$ is necessary to get this right. The remaining terms, here placed on the right of the equation, are due to the use of an anti-symmetric trial wavefunction and couple subshell amplitudes of all symmetries:

$$X(A;r) = \frac{1}{r} \sum_{B \neq A} \sum_k D^0(AB;k)\, \frac{Y^k(AB;r)}{(B\,|B)} \begin{pmatrix} P_B(r) \\ Q_B(r) \end{pmatrix}. \tag{6.7.14}$$

It is these terms that are responsible for much of the practical difficulty of numerically solving the DHF equations by finite difference methods.

The relativistic corrections to the Coulomb interaction can be handled in the same way. Because of the large number of new integrals involved, these have not normally been included in relativistic SCF calculations. However, only exchange terms contribute to the energy of closed shell configurations, §6.6.2, and the number of new potentials to be included is relatively small. Experiments with basis set methods indicate that the first order Breit correction to the DHF energy is almost exactly equal to the corresponding difference between DHFB and DHF energies. Correlation calculations with DHFB wavefunctions therefore include all-order magnetic effects that may be important in correlation calculations. The DHF integro-differential equations are essentially the same for both normalized or unnormalized radial amplitudes. Provided we renormalize the trial solutions, $u_A(r) \to u_A(r)/(A\,|\,A)^{1/2}$, before using them to calculate new potentials the normalization of the converged solution takes care of itself.

6.7.1 Construction of electrostatic potentials

As suggested by Hartree [28, p. 50], the functions

$$Y_k(BD;r) = r\int_0^\infty U_k(r,s)\,D_{BD}(s)\,ds$$

can be evaluated economically by differential equation methods. Because

$$\frac{dY_k(BD;r)}{dr} + (k+1)Y_k(BD;r) = -\frac{2k+1}{r}Z_k(BD;r) \tag{6.7.15}$$

where

$$Z_k(BD;r) = \int_0^r \left(\frac{s}{r}\right)^k D_{BD}(s)\,ds$$

so that

$$\frac{dZ_k(BD;r)}{dr} + \frac{k}{r}Z_k(BD;r) = D_{BD}(r), \tag{6.7.16}$$

together with the conditions $Z_k(BD;0) = 0$ at the origin and $Y_k(BD;r) - Z_k(BD;r) \to 0$ as $r \to \infty$ for marching outwards and inwards respectively.

6.7.2 Construction of magnetic potentials

Although they do not appear in the DHF equations, it is convenient to consider the construction of magnetic interactions at this point. The transverse photon integrals can be handled in much the same way as those for Coulomb integrals, but the appearance of spherical Bessel functions in the kernels V_ν and $W_{k-1,k+1,k}$ introduces singular behaviour that complicates the numerical evaluation and necessitates some rearrangement of the kernels. Section §6.5 lists six types of radial integrals:

Type 1: $R^\nu[AC\,|\,BD]$, equation (6.5.1).

Set

$$R^\nu[AC\,|\,BD] = \frac{1}{2}\Big\{\overline{R}^\nu[AC\,|\,BD;\omega_{AC}] + \overline{R}^\nu[AC\,s|\,BD;\omega_{BD}]$$
$$+\overline{R}^\nu[AC\,|\,BD;\omega AC_] + \overline{R}^\nu[AC\,|\,BD;\omega_{BD}]\Big\}, \quad (6.7.17)$$

where

$$\overline{R}^\nu[AC\,|\,BD] = \int_0^\infty D^o_{BD}(r)\psi_\nu(r)\xi^\nu(AC;\omega,r)dr \quad (6.7.18)$$

with

$$\xi^\nu(AC;\omega,r) = r^{-\nu-1}\int_0^r D^o_{AC}(s)s^\nu\phi_\nu(\omega s)ds.$$

and the functions $\phi_\nu(z), \psi_\nu(z)$ are defined in terms of spherical Bessel functions, Appendix A.3,

$$\phi_\nu(z) = (2\nu+1)!!\,z^{-\nu}j_\nu(z), \quad \psi_\nu(z) = -z^{\nu+1}\left[(2\nu-1)!!\right]^{-1}y_\nu(z)$$

As both $\phi_\nu(z)$ and $\psi_\nu(z)$ are holomorphic in the neighbourhood of $z=0$, $\phi_\nu(1) = \psi_\nu(1) = 1$, the integrands are both smooth, and we can evaluate $\xi^\nu(AC;\omega,r)$ as the solution of the initial value problem

$$\frac{d\xi}{dr} + \frac{\nu+1}{r}\xi = \frac{1}{r}D^o_{AC}(r)\,\phi_\nu(\omega r) \quad (6.7.19)$$

where, assuming $D^o_{AC}(r)\sim r^\lambda$ as $r\to 0$, the initial condition is

$$\xi(r) \sim D^o_{AC}(r)\,\phi_\nu(\omega r)/(\nu+\lambda+1), \quad r\to 0.$$

Type 2: $S^k[\alpha\gamma\,|\beta\delta,]$, equation (6.5.2).
Write

$$S^k[\alpha\gamma\,|\beta\delta] = \frac{1}{2}\left\{S^k[\alpha\gamma\,|\beta\delta;\omega_{AC}] + S^k[\alpha\gamma\,|\beta\delta;\omega_{BD}]\right\}.$$

In order to avoid cancellations, we divide $S^k[\alpha\gamma\,|\beta\delta;\omega]$ into two pieces

$$S^k[\alpha\gamma\,|\beta\delta;\omega] = S^k_1[\alpha\gamma\,|\beta\delta;\omega] + S^k_2[\alpha\gamma\,|\beta\delta;\omega] \quad (6.7.20)$$

where

$$S^k_1[\alpha\gamma\,|\beta\delta;\omega] = \int_0^\infty dr\int_0^r ds.$$
$$\times\, D^o_{\alpha\gamma}(r)(2k+1)\omega\,y_{k-1}(\omega r)j_{k+1}(\omega s)D^o_{\beta\delta}(s),$$

and

$$S^k_2[\alpha\gamma\,|\beta\delta;\omega] = \int_0^\infty dr\int_r^\infty ds.$$
$$\times\, D^o_{\alpha\gamma}(r)\left[(2k+1)\omega\,y_{k-1}(\omega s)j_{k+1}(\omega r) + \frac{(2k+1)^2}{\omega^2}\frac{r^{k-1}}{s^{k+1}}\right]D^o_{\beta\delta}(s).$$

The first of these can be written

$$S_1^k[\alpha\gamma\,|\beta\delta;\omega] = -\int_0^\infty dr \int_0^r ds.\, D^o_{\alpha\gamma}(r)\frac{\omega^2 r^2 \psi_{k-1}(\omega r)}{(2k-1)2k+3)}\xi_1(\beta\delta;\omega,r)$$

where

$$\frac{d\xi_1}{dr} + \frac{k+2}{r}\xi_1 = \frac{1}{r}D^o_{\beta\delta}(r)\,\phi_{k+1}(\omega r);$$

assuming $D^o_{\beta\delta}(r) \sim r^\lambda$ as $r \to 0$, the initial condition is

$$\xi_1(\beta\delta;\omega,r) \sim D^o_{\beta\delta}(r)\,\phi_{k+1}(\omega r)/(k+\lambda), \quad r \to 0.$$

In $S_2^k[\alpha\gamma\,|\beta\delta;\omega]$, we first invert the order of integration over r and s and write

$$S_2^k[\alpha\gamma\,|\beta\delta;\omega] = \int_0^\infty ds\, D^o_{\beta\delta}(s)\left(\frac{2k+1}{\omega}\right)^2 s^{k-1}$$
$$\times \int_0^s dr\, r^{k+2}\,[1-\psi_{k+1}(\omega s)\phi_{k-1}(\omega r)]\; D^o_{\alpha\gamma}(r)$$

revealing cancellation in the expression in square brackets near $r = s = 0$. We can avoid numerical instabilities in the quadrature by writing this as

$$S_2^k[\alpha\gamma\,|\beta\delta;\omega] = \int_0^\infty ds\, D^o_{\beta\delta}(s)\left(\frac{2k+1}{\omega s}\right)^2$$
$$\times\,[\psi_{k+1}(\omega s)\xi_2(\alpha\gamma;\omega,s) + (1-\psi_{k+1}(\omega s))\,\xi_3(\alpha\gamma;\omega,s)]$$

where $\xi_2(\alpha\gamma;\omega,s)$ and $\xi_3(\beta\delta;\omega,s)$ are respectively solutions of the initial value problems

$$\frac{d\xi_2}{dr} + \frac{k}{r}\xi_2 = \frac{1}{r}D^o_{\alpha\gamma}(r)\,[1-\phi_{k-1}(\omega r)]$$

and

$$\frac{d\xi_3}{dr} + \frac{k}{r}\xi_3 = \frac{1}{r}D^o_{\alpha\gamma}(r)$$

with $\xi_2 \sim D^o_{\alpha\gamma}(r)\,[1-\phi_{k-1}(\omega r)]\,/(\lambda+k+2)$ and $\xi_3 \sim D^o_{\alpha\gamma}(r)/(\lambda+k)$ as $r \to 0$.

Type 3 $R^k(A;BD;\omega_{BD})$, equation (6.5.5). This is a special case of the Type 1 integral.

Type 4 $F^k(AB)$, equation (6.5.7). This is a special case of the Type 1 integral.

Type 5 $G^\nu_\gamma(AB)$, equation (6.5.9). This is a special case of the Type 1 integral.

Type 6 $H^k_\delta(AB)$, equation (6.5.10). This is a special case of the Type 2 integral.

6.7.3 Algorithms for potentials and Slater integrals

The numerical solution of radial equations by finite differences, §5.11, and finite elements, §5.12, leads to different schemes for evaluating the potentials and Slater integrals appearing in this section. We consider first determination of the functions $Y_k(BD:r)$ and $Z_k(BD:r)$ using the Hartree approach. For simplicity consider first a change of variable

$$s = \ln(r/r_0), \quad r > r_0. \tag{6.7.21}$$

Then equation (6.7.16) can be expressed as

$$\frac{dZ_k}{ds} + kZ_k = r_0 e^s D_{BD}(r_0 E^s), \tag{6.7.22}$$

together with the initial conditions $Z_k(0) = 0$. We can replace this by the integral relation

$$Z_k(s+h) = e^{-kh} Z_k(s) + e^{-k(s+h)} \int_s^{s+h} r_0 e^{kt} D_{BD}(r_0 e^t)\, dt \tag{6.7.23}$$

where s increases by a constant amount at each step. The potential Y_k is obtained in a rather similar manner from (6.7.15), which becomes

$$\frac{dY_k}{ds} - (k+1)Y_k = -(2k+1)Z_k \tag{6.7.24}$$

marching inwards with the intial condition $Y_k - Z_k \to 0$ as $s \to \infty$. The method requires choice of a suitable quadrature formula to approximate the integral interms of values of the integrand at the grid points, $s = ph,\ p = 0, 1, \ldots$. One possibility is to use a four-point central difference formula so that

$$\begin{aligned} Z_k(s+h) &= e^{-kh} Z_k(s) \\ &\quad + \frac{h}{24}\left[-e^{-2kh}\phi(s-h) + 13e^{-kh}\phi(s)\right. \\ &\qquad \left. +13\phi(s+h) - e^{kh}\phi(s+2h)\right] + O(h^5) \end{aligned} \tag{6.7.25}$$

where $\phi(s)$ is the integrand of (6.7.23). Another is to use Simpson's rule, which gives the same asymptotic error $O(h^5)$ [51] or a five-point formula with asymptotic error $O(h^7)$. Special care may need to be taken near the endpoints to ensure that all parts of the calculation have the same asymptotic error, consistent with the order of the deferred difference scheme being used in §5.11. Slater integrals, which can be expressed as a product of a radial density distribution $\phi(s)$ with a potential function $Y_k(s)$,

$$\int_0^{s_{max}} \phi(s) Y_k(s) ds$$

can be done using standard quadrature formulae, choosing one with the same asymptotic accuracy. It is necessary to add a correction for the interval $(0, r_0)$ to eliminate dependence on the choice of r_0.

Fischer and co-workers have examined several ways to evaluate Slater integrals and Y_k-potentials using spline approximation methods. One successful approach [52] notes that a Schrödinger radial amplitude can be written as a B-spline expansion

$$P_a(r) \approx \sum_i a_i B_i^k(r) \tag{6.7.26}$$

so that a Slater integral can be written as an integral over B-spline basis elements of the form

$$R^k(a,b,c,d) = \sum_{i,i',j,j'} a_i b_j c_{i'} d_{j'}\, R^k(i,j,i',j') \tag{6.7.27}$$

where

$$R^k(i,j,i',j') = \int_0^\infty \frac{1}{r} B_i^k(r) B_{i'}^k(r) Y_k(j,j';r)\, dr$$

with

$$\frac{Y_k(j,j';r)}{r} = \int_)^\infty B_j^k(s) B_{j'}^k(s) U_k(r,s) ds.$$

$U_k(r,s)$ is defined in (6.4.5). The differential equations (6.7.15) and (6.7.16) are combined to give a second order equation

$$\frac{d^2}{dr^2} Y_k(j,j';r) - \frac{k(k+1)}{r^2} Y_k(j,j';r) + \frac{2k+1}{r} B_j^k(r) B_{j'}^k(r) = 0. \tag{6.7.28}$$

with $Y_k(j,j';0) = 0$ and $dY_k(j,j';r)/dr \to -kY_k(j,j';r)/r$ as $r \to \infty$, which can be solved straightforwardly by using the spline-Galerkin method outlined in §5.12.

Another approach, a cell algorithm [53], significantly improves the performance of the B-spline approximation in many-electron calculations by exploiting the fact that $B_i^k(r)$ is a piecewise polynomial of degree $k-1$ which is non-zero only within the range $t_i \le r \le t_{i+k}$. The grid devized by Fischer and Idrees [54] described in §5.12 was used. The approach exploits scaling properties over individual cells in the region in which the knots increase exponentially. Because the splines are normalised so that $\sum_i B_i^k(r) = i$, B-splines defined by (5.12.1) and (5.12.2) have a displacement property:

Let the left-most knot of $B_i^k(r)$ be t_i and let $r = t_i + s$. Then

$$B_i^k(t_i + s) = B_{i+1}^k((1+h)(t_i+s)) = B_{i+1}^k(t_{i+1} + s(1+h)), \tag{6.7.29}$$

which induces several useful scaling laws [55], in particular

$$\langle B_{i+1}^k(r) \,|\, r^n \,|\, B_{j+1}^k(r)\rangle = (1+h)^{n+1} \langle B_i^k(r) \,|\, r^n \,|\, B_j^k(r)\rangle \tag{6.7.30}$$

$$R^k(i+1, j+1, i'+1, j'+1) = (1+h)\, R^k(i,j,i',j').$$

The cell method exploits scaling properties applying to individual cells. Write

$$\rho_{ij}(r) = B_i^k(r)B_j^k(r).$$

Then

$$\int_{r_{p+1}}^{r_{p+2}} \rho_{i+1,j+1}(r)r^n dr = (1+h)^{n+1} \int_{r_p}^{r_{p+1}} \rho_{i,j}(r)r^n dr \qquad (6.7.31)$$

and

$$\begin{aligned} &\int_{r_{p+1}}^{r_{p+2}} \int_{r_{q+1}}^{r_{q+2}} \rho_{i+1,j+1}(r)U_k(r,s)\rho_{i'+1,j'+1}(s)\, drds \qquad (6.7.32) \\ &= (1+h) \int_{r_p}^{r_{p+1}} \int_{r_q}^{r_{q+1}} \rho_{i,j}(r)U_k(r,s)\rho_{i',j'}(s)\, drds \end{aligned}$$

Cell contributions to $R^k(i,j,i',j')$ vanish when the splines are too far apart, so that $R^k(i,j,i',j') = 0$ when $|i-i'| \geq k$ or $|j-j'| \geq k$. Exploiting symmetry with respect to the indices i, i', j, j' reduces the number of individual contributions that need to be calculated; see [52, 53, 54, 55] for technical details.

6.8 Configurations with incomplete subshells

6.8.1 Atomic states with incomplete subshells

Closed shell configurations are often an important component of atomic models, but much of the technical complication of calculations that aim at some sort of precision originates from the presence of one or more open subshells. We shall proceed as in the NVPA by considering only electron states, so that the formulation is close to that of nonrelativistic atomic structure theory, with Dirac orbitals and relativistic operators replacing their nonrelativistic counterparts. Contributions to correlation energies from negative energy states are often small, but can be calculated when needed. On the other hand, self-energy and vacuum polarization diagrams require special treatment which we shall not discuss here. Positron states are always present in the background of relativistic atomic structure calculations implicitly even when they are not mentioned explicitly.

We write the electron field as

$$\psi(\boldsymbol{x}) = \sum_i a_i \psi_i(\boldsymbol{x}), \qquad (6.8.1)$$

where the orbital functions $\psi_i(\boldsymbol{x})$ are orthonormal eigenfunctions of a mean field Hamiltonian as in (6.1.3) and the a_i are annihilation operators for *electron states only.* Fock space one- and two-electron operators F (4.1.34) and G (4.1.35) are therefore given by

$$F = \int \psi^\dagger(x)\, f(x)\, \psi(x)\, dx$$
$$G = \frac{1}{2} \iint \psi^\dagger(x')\, \psi^\dagger(x)\, g(x,x')\, \psi(x)\, \psi(x')\, dx\, dx'$$

Substituting from (6.8.1) gives

$$F = \sum_{i.j} a_i{}^\dagger \langle i \,|\, f \,|\, j \rangle\, a_j \tag{6.8.2}$$

and

$$G = \frac{1}{2} \sum_{ijkl} a_i{}^\dagger a_j{}^\dagger \langle i, j \,|\, g \,|\, k, l \rangle\, a_l\, a_k \tag{6.8.3}$$

General normalized N-electron Fock space kets and bras can be written

$$|\, \Psi_\Gamma \rangle = \sum_{\{\alpha\}} c_{\{\alpha\}} |\, \Phi_{\{\alpha\}} \,\rangle, \tag{6.8.4}$$

and the corresponding bra is

$$\langle\, \Psi_\Gamma \,| = \sum_{\{\alpha\}} c_{\{\alpha\}}{}^* \langle\, \Phi_{\{\alpha\}} \,|, \tag{6.8.5}$$

where $\{\alpha\} = \{\alpha_1, \ldots, \alpha_N\}$ labels an N-electron determinantal state so that

$$|\, \Phi_{\{\alpha\}} \rangle = a_{\alpha_1}{}^\dagger \ldots a_{\alpha_N}{}^\dagger \,|\, 0 \,\rangle, \quad \langle \Phi_{\{\alpha\}} \,| = \langle\, 0 \,|\, a_{\alpha_N} \ldots a_{\alpha_1}$$

where $|\, 0 \rangle$ and $\langle\, 0 \,|$ refer to the vacuum state with no electrons (and no positrons). The coefficients $c_{\{\alpha\}}$ are normalized so that

$$\sum_{\{\alpha\}} |\, c_{\{\alpha\}} \,|^2 = 1.$$

It is convenient to build more general N-electron states from Dirac central field orbitals. An isolated atom has no preferred axis, so that we can classify atomic states as eigenstates of total angular momentum, $\boldsymbol{J}$, where $\boldsymbol{J} = \sum_{n=1}^{N} \boldsymbol{j}_n$. An N-electron configurational state (CSF) will be written $|\, \gamma, J^\pi, M \,\rangle$, where $\boldsymbol{J}^2$ has eigenvalues $J(J+1)$, J_z has eigenvalue M, π is the overall parity of the state, and γ represents all other data needed to complete the classification. In particular, γ represents the electron configuration (in terms of subshell occupation numbers N_A) and an angular momentum coupling scheme. The structure of such a state is a specific example of (6.8.4) in which the coefficients $c_{\{\alpha\}}$ are generalized Clebsch-Gordan coefficients (see

Appendix B.3). More generally, we can write for a general atomic state (ASF) a CI expansion of the form

$$| \Gamma, J^{\pi}, M \rangle = \sum_{\gamma} c_{\Gamma,\gamma} | \gamma, J^{\pi}, M \rangle. \tag{6.8.6}$$

The configuration interaction (CI) method simply diagonalizes the atomic Hamiltonian in a CSF basis with *fixed* orbitals, so that the *mixing coefficients* $c_{\Gamma,\gamma}$ can be regarded as the elements of the ASF eigenvector, $\boldsymbol{c}_{\Gamma}$, with energy eigenvalue E_{Γ}. The MCDHF and MCDHFB methods require a nonlinear iterative sequence of CI steps in which the orbitals used to generate the CSF set are also adjusted until the system of equations is self-consistent. The ASF $| \Gamma, J^{\pi}, M \rangle$, (6.8.6), will include subshells that are completely filled along with subshells with vacancies.

6.8.2 Partially filled subshells in jj-coupling

The set of central field Dirac spinors constituting a subshell in jj-coupling, $\{| n\kappa m\rangle \,|\, m = -j, \ldots, j\}$, forms a basis for the irreducible representation D^j of SO(3). Electrons in these states are often said to be *equivalent*. We shall use the notation $(n\kappa)^N$ (or κ^N for brevity when we are considering only a single subshell) to mean a set of N-electron states built from this subshell basis. The subshells labelled by κ and $-\kappa$ share a common value of j but are clearly inequivalent. There are

$$\mathcal{N}_{\kappa^N} = \binom{2j+1}{N} \quad 0 \leq N \leq 2j+1 \tag{6.8.7}$$

anti-symmetric states in the κ^N configuration. This section describes the construction of these states and the classification of open shell states with $0 < N < 2j+1$.

Denote by ${a_m}^{\dagger}, a_m,\ m = -j, \ldots, j$ a set of creation and annihilation operators for the subshell, so that

$$| n\kappa m\rangle = {a_m}^{\dagger} \,| 0 \rangle.$$

The $\mathcal{N}_{\kappa^N}$ anti-symmetric states

$${a_{m_1}}^{\dagger} \ldots {a_{m_N}}^{\dagger} \,| 0 \rangle, \quad m_1 > \ldots > m_N, \tag{6.8.8}$$

which, together span the states of the κ^N configuration are eigenstates of the component J_z with eigenvalue $M = \sum m_n$ of the total angular momentum $\boldsymbol{J}$ and belong to the reducible representation $D^j \times \ldots \times D^j$ (with N factors). We can make a preliminary classification of the Clebsch-Gordan decomposition in the usual way by listing all ordered sets $m_1, \ldots, m_N$ such that $M = \sum m_n$. The irreducible representation D^{J_1} appears in the Clebsch-Gordan series if J_1

is the highest value of $M = M_1$, say; we can construct a series of $(2J_1 + 1)$ states spanning this representation by starting with $|\,\kappa^N, J_1, M = J_1\rangle$ by repeated application of the step-down operator J_-. Remove one entry from the list at each step until $|\,\kappa^N, J_1, M = -J_1\rangle$ has been obtained. Now repeat the process with the next highest value of M and continue until everything is accounted for. The result is shown in Tables A.12 and A.13; this process works as far as $(7/2)^3$, but we see that for $(7/2)^4$ the values $J = 2$ and $J = 4$ both occur twice, and there are more such occurrences for $j \geq 9/2$. The algebras generated by the creation and annihilation operators provide the machinery for a more complete classification [29, 30].

6.8.3 Creation and annihilation operators as irreducible tensor operators. Quasispin.

Because the total angular momentum operator of a many-electron system, $\boldsymbol{J}$, is of Fock space type F, equation (6.8.2) gives its operator form

$$\boldsymbol{J} = \sum_{\kappa,\kappa'} \sum_{m,m'} a^{\kappa\,\dagger}_m \langle \kappa, m\,|\,\boldsymbol{j}\,|\,\kappa', m'\rangle\, a^{\kappa'}_{m'} \tag{6.8.9}$$

where the upper index labels the subshell subspace. Because

$$\langle \kappa, m\,|\,j_3\,|\,\kappa', m'\rangle = m\,\delta_{\kappa,\kappa'}\delta_{m,m'},$$

$$\langle \kappa, m\,|\,j_\pm\,|\,\kappa', m'\rangle = [j(j+1) - m'(m' \pm 1)]^{1/2}\delta_{\kappa,\kappa'}\delta_{m,m'\pm 1},$$

we can drop the κ label when we are considering a single subshell. A simple calculation gives the commutators

$$\begin{aligned} [J_3, a_m{}^\dagger] &= m\,a_m{}^\dagger, \\ [J_\pm, a_m{}^\dagger] &= [j(j+1) - m(m \pm 1)]^{1/2}\, a_{m\pm 1}{}^\dagger, \end{aligned} \tag{6.8.10}$$

which *define*, (B.3.130), the set

$$\{a_m{}^\dagger\,|\,m = j, j-1, \ldots, -j\}$$

as the $2j+1$ components of an irreducible tensor operator $\boldsymbol{a}^{\kappa\dagger}$ of rank j under SO(3). Unfortunately the members of the set $\{\,a_m\,\}$ do not satisfy the commutation rules (6.8.10), but the set

$$\{\tilde{a}_m := (-1)^{j-m}\,a_{-m}\,|\,m = j, j-1, \ldots, -j\}$$

does, yielding an acceptable definition of the adjoint tensor operator $\boldsymbol{a}^\kappa$.

Because both $\boldsymbol{a}^\kappa$ and $\boldsymbol{a}^{\kappa'}$ can be regarded as irreducible tensor operators of ranks j and j' respectively, we can use them to build composite tensor operators in the usual way, for example

$$\left[\boldsymbol{a}^{\kappa\dagger} \otimes \boldsymbol{a}^{\kappa'}\right]^k_q = \sum_{m,m'} a^{\kappa\,\dagger}_m \,\tilde{a}^{\kappa'}_{m'}\, \langle jm, j'm' \,|\, jj'kq\rangle \tag{6.8.11}$$

and the anti-commutation relations give

$$\left[\boldsymbol{a}^{\kappa} \otimes \boldsymbol{a}^{\kappa'\dagger}\right]^k_q + (-1)^{j+j'-k}\left[\boldsymbol{a}^{\kappa'\dagger} \otimes \boldsymbol{a}^{\kappa}\right]^k_q = [\,j\,]^{1/2}\delta_{jj'}\,\delta_{k,0}\,\delta_{q,0}. \tag{6.8.12}$$

When the operators act on the *same* subshell κ, this simplifies to

$$\left[\boldsymbol{a} \times \boldsymbol{a}^{\dagger}\right]^k_q - (-1)^k\left[\boldsymbol{a}^{\dagger} \times \boldsymbol{a}\right]^k_q = [\,j\,]\,\delta_{k,0}\,\delta_{q\,,0}. \tag{6.8.13}$$

We shall also need the commutators of coupled tensors such as

$$\begin{aligned}\left[\left[\boldsymbol{a} \times \boldsymbol{a}^{\dagger}\right]^{k_1}_{q_1}, \left[\boldsymbol{a} \times \boldsymbol{a}^{\dagger}\right]^{k_2}_{q_2}\right] &= \sum_{KQ}[K][k_1,k_2]^{1/2}\begin{Bmatrix} k_1 & k_2 & K \\ j & j & j\end{Bmatrix} \\ \times\,(-1)^{2j}\left[1-(-1)^{k_1+k_2+K}\right]&\langle k_1q_1, k_2q_2 \,|\, k_1k_2KQ\rangle \left[\boldsymbol{a} \times \boldsymbol{a}^{\dagger}\right]^K_Q.\end{aligned} \tag{6.8.14}$$

The three operators [32, 33],

$$\begin{aligned} Q_+ &= \frac{1}{2}[\,j\,]^{1/2}\left[\boldsymbol{a}^{\dagger} \times \boldsymbol{a}^{\dagger}\right]^0_0, \\ Q_- &= -\frac{1}{2}[\,j\,]^{1/2}\left[\boldsymbol{a} \times \boldsymbol{a}\right]^0_0, \\ Q_3 &= -\frac{1}{4}[\,j\,]^{1/2}\left\{\left[\boldsymbol{a}^{\dagger} \times \boldsymbol{a}\right]^0_0 + \left[\boldsymbol{a} \times \boldsymbol{a}^{\dagger}\right]^0_0\right\}, \end{aligned} \tag{6.8.15}$$

for which we have the closed set of commutators,

$$[Q_+, Q_-] = 2Q_3, \quad [Q_3, Q_\pm] = \pm Q_\pm, \tag{6.8.16}$$

define the generators of the spin group SU(2). The corresponding vector operator $\boldsymbol{Q}$ is therefore often called *quasispin*. Because the eigenvalues of $[\,j\,]^{1/2}\left[\boldsymbol{a}^{\dagger} \times \boldsymbol{a}\right]^0_0$ are $-N$ and, (6.8.13), the eigenvalues of $[\,j\,]^{1/2}\left[\boldsymbol{a} \times \boldsymbol{a}^{\dagger}\right]^0_0$ are $2j+1-N$, every state of κ^N is an eigenstate of Q_3 with eigenvalue

$$M_Q = (N - |\kappa|)/2, \quad 0 \leq N \leq 2|\kappa| = 2j+1, \tag{6.8.17}$$

whilst Q_+ and Q_- connect states of κ^N with those of κ^{N+2} and κ^{N-2} respectively. If $Q_\pm = Q_1 \pm Q_2$, then

$$\boldsymbol{Q}^2 = Q_1^2 + Q_2^2 + Q_3^2 = Q_+Q_- + Q_3(Q_3 - 1)$$

maps the space of states of κ^N onto itself. The components of $\boldsymbol{Q}$ are invariant under spatial rotations, and therefore commute with $\boldsymbol{J}^2$ and J_3. Thus we can classify the states of κ^N as simultaneous eigenstates of $\boldsymbol{Q}^2, Q_3, \boldsymbol{J}^2$ and J_3: $|Q, M_Q, J, M\rangle$.

An equivalent classification, due to Racah [11], involves the notion of the *seniority number* ν. In the seniority scheme, we write the states of κ^N as $|\kappa^N, \nu, J, M\rangle$, where ν is the *lowest value* of N at which a particular J appears. Because Q_- acts as a step-down operator with respect to N, we define ν by

$$Q_- \, |\kappa^N, \nu, J, M\rangle = 0, \quad \text{when } N = \nu.$$

So

$$\begin{aligned}\boldsymbol{Q}^2 |\kappa^\nu, \nu, J, M\rangle &= Q(Q+1)|\kappa^\nu, \nu, J, M\rangle \\ &= [Q_+Q_- + Q_3(Q_3 - 1)]\, |\kappa^\nu, \nu, J, M\rangle \\ &= M_Q(M_Q - 1)|\kappa^\nu, \nu, J, M\rangle\end{aligned}$$

with $M_Q = (\nu - |\kappa|)/2$, from which we deduce

$$Q = -M_Q = (|\kappa| - \nu)/2, \quad 0 \le \nu \le |\kappa|. \tag{6.8.18}$$

The classification in terms of seniority or quasispin is shown in Tables A.12 and A.13, from which it is clear that it fails only for the doubly degenerate states $(9/2)^4, J = 4$ and $(9/2)^6, J = 6$. These two cases can be classified by *ad hoc* methods; however the problem will be harder to deal with if we need to treat many-electron states with $j \ge 11/2$. A more systematic approach to classification uses the theory of Lie groups; a full treatment may be found in [29, 30] and [31, Chapter 11]. A brief account will be found in Appendix B.2.

6.8.4 Double tensor operators

The tensor operators $\boldsymbol{a}^\dagger$ and $\boldsymbol{a}$ are irreducible tensor operators of rank j under SO(3). The commutators of their components with respect to quasispin are

$$\left[\, Q_3, a^\dagger_m \,\right] = \frac{1}{2}\, a^\dagger_m, \quad \left[\, Q_+, a^\dagger_m \,\right] = 0, \quad \left[\, Q_-, a^\dagger_m \,\right] = \tilde{a}_m, \tag{6.8.19}$$

and

$$[\, Q_3, \tilde{a}_m \,] = -\frac{1}{2}\, \tilde{a}_m, \quad [\, Q_+, \tilde{a}_m \,] = a^\dagger_m, \quad [\, Q_-, \tilde{a}_m \,] = 0 \tag{6.8.20}$$

so that they can be regarded as the components of an irreducible tensor operator of rank $q = 1/2$ under the group $\mathrm{SU}^Q(2)$ whose infinitesimal generators are Q_+, Q_- and Q_3. We can therefore define a *double tensor operator* $\boldsymbol{A}^{q,j}$ with $4j+2$ components

$$A^{q,\,j}_{\frac{1}{2},m} = a^\dagger_m, \quad A^{q,\,j}_{-\frac{1}{2},m} = \tilde{a}_m, \tag{6.8.21}$$

which has rank $q = 1/2$ under $\mathrm{SU}^Q(2)$ and rank j under SO(3) [33]. The anti-commutation relations for the original creation and annihilation operators become

$$\left\{ A^{q,\,j}_{m_q,m},\ A^{q,\,j}_{m'_q,m'} \right\} = (-1)^{\epsilon} \delta_{m_q,-m'_q} \delta_{m,-m'}, \tag{6.8.22}$$

where $\epsilon = q + m_q + j + m + 1$. We can define coupled double tensor operators of the general form

$$\boldsymbol{X}^{Q,K} = \left[\boldsymbol{A}^{q,j} \times \boldsymbol{A}^{q,j} \right]^{Q,K}; \tag{6.8.23}$$

in particular,

$$\boldsymbol{Q} = -\left\{ \frac{2j+1}{2} \right\}^{1/2} \boldsymbol{X}^{1,0}, \quad \boldsymbol{J} = -\left\{ \frac{2j(j+1)(2j+1)}{3} \right\}^{1/2} \boldsymbol{X}^{0,1}. \tag{6.8.24}$$

The components of $\boldsymbol{X}^{Q,K}$, when written out, are

$$X^{Q,\,K}_{M_Q,M_K} = \sum_{m_q,m'_q} \sum_{m,m'} A^{q,\,j}_{m_q,m} A^{q,\,j}_{m'_q,m'} C^{1/2\ 1/2\ Q}_{m_q\ m'_q\ M_Q} C^{j\ \ j\ \ K}_{m\ m'\ M}$$

so that

$$X^{Q,K}_{0,M} = -\frac{1}{\sqrt{2}} \left[1 - (-1)^{Q+K} \right] U^K_M - \frac{(-1)^Q}{\sqrt{2}} \left[\, j \,\right]^{1/2} \delta_{K,0}\, \delta_{M,0} \tag{6.8.25}$$

where the U^K_M are the components of the unit tensor operator $\boldsymbol{U}^K$ of rank K. For a single particle, this becomes a single particle operator $\boldsymbol{u}^K$, whose reduced matrix element is given by

$$(\gamma j \| \boldsymbol{u}^K \| \gamma' j') = \delta_{\gamma,\gamma'}\, \delta j, j' \left[\, K \,\right]^{1/2} \tag{6.8.26}$$

so that

$$\boldsymbol{U}^K = -\left[\boldsymbol{a}^\dagger \times \boldsymbol{a} \right]^K \tag{6.8.27}$$

When $Q + K$ is even, equation (6.8.25) simplifies so that

$$X^{Q,K}_{0,M} = -\frac{1}{\sqrt{2}} \left[\, j \,\right]^{1/2} \delta_{Q,0}\, \delta_{K,0}\, \delta_{M,0} \tag{6.8.28}$$

which is the double tensor equivalent of (6.8.22).

6.8.5 Parentage

The construction of matrix elements of one- and two-particle operators for subshells of equivalent electrons forces us to consider the connection between states of N equivalent electrons and $N-1$ or $N-2$ electrons. Consider a configuration space wavefunction in the seniority scheme, which we can expand as a linear combination of determinants,

$$\langle \boldsymbol{x}_1, \ldots, \boldsymbol{x}_{N-1} \,|\, \kappa^{N-1}, \bar{\nu}, \bar{J}, \bar{M} \rangle = \sum_{\{m_i\}} C_{\{m_i\}} \{m_1, \ldots, m_{N-1}\} \tag{6.8.29}$$

where the sum runs over all possible sets of magnetic quantum numbers with $m_1 + \dots m_{N-1} = \bar{M}$ and

$$\{m_1, \dots, m_{N-1}\} = \langle \boldsymbol{x}_1, \dots, \boldsymbol{x}_{N-1} \,|\, a^\dagger_{m_1}, \dots, a^\dagger_{m_{N-1}} \,|\, 0 \rangle,$$

From this we can construct an anti-symmetric N-electron wavefunction

$$\langle \boldsymbol{x}_1, \boldsymbol{x}_2 \dots, \boldsymbol{x}_N \,|\, a^\dagger_m \,|\, \kappa^{N-1}, \bar{\nu}, \bar{J}, \bar{M} \rangle = \sum_{\{m_i\}} C_{\{m_i\}} \{m, m_1, \dots, m_{N-1}\}\,.$$

Each Slater determinant in this sum can be expanded by its first column as in (4.1.30) giving

$$\{m, m_1, \dots, m_{N-1}\} = N^{-1/2} \sum_{i=1}^{N} (-1)^{i-1} \{\, m \,\}_i \, \{m_1, \dots, m_{N-1}\}^{(i)}$$

where $\{\, m \,\}_i$ indicates that m is assigned coordinate $\boldsymbol{x}_i$ and $\{m_1, \dots, m_{N-1}\}^{(i)}$ that the $N-1$ columns of the cofactor are assigned the coordinates $\boldsymbol{x}_1, \dots, \boldsymbol{x}_N$ in order, omitting $\boldsymbol{x}_i$. This allows us to write

$$\begin{aligned} &\langle \boldsymbol{x}_1, \dots, \boldsymbol{x}_N \,|\, a^\dagger_m \,|\, \kappa^{N-1}, \bar{\nu}, \bar{J}, \bar{M} \rangle \\ &= N^{-1/2} \sum_{i=1}^{N} (-1)^{i-1} \langle \boldsymbol{x}_i \,|\, m \rangle \langle \boldsymbol{x}^{(i)} \,|\, j^{N-1}, \bar{\nu}, \bar{J}, \bar{M} \rangle. \end{aligned}$$

where $\boldsymbol{x}^{(i)}$ is the argument labelling the omitted column.

The matrix element of this expression with an anti-symmetric state of N electrons in the seniority scheme is then

$$\begin{aligned} \langle \kappa^N, \nu, J, M \,|\, a^\dagger_m \,|\, \kappa^{N-1}, \bar{\nu}, \bar{J}, \bar{M} \rangle &= N^{-1/2} \sum_{i=1}^{N} (-1)^{i-1} \\ \int d\boldsymbol{x}_1 \dots \int d\boldsymbol{x}_N \langle \kappa^N, \nu, J, M \,|\, \boldsymbol{x}_1, \dots, \boldsymbol{x}_N \rangle \langle \boldsymbol{x}_i \,|\, m \rangle \, &\langle \boldsymbol{x}^{(i)} \,|\, \kappa^{N-1}, \bar{\nu}, \bar{J}, \bar{M} \rangle. \end{aligned}$$

We now relabel the coordinates in the first factor of the integrand by interchanging $\boldsymbol{x}_i$ and $\boldsymbol{x}_1$; because this term is anti-symmetric, there is an overall sign change. The coordinate labels are just dummies, so we may as well interchange the labels i and 1 throughout, giving

$$\begin{aligned} \langle \kappa^N, \nu, J, M \,|\, a^\dagger_m \,|\, j^{N-1}, \bar{\nu}, \bar{J}, \bar{M} \rangle &= N^{-1/2} \sum_{i=1}^{N} (-1)^{i-1} \\ &\times \int d\boldsymbol{x}_1 \dots \int d\boldsymbol{x}_N \langle \kappa^N, \nu, J, M \,|\, \boldsymbol{x}_1, \dots, \boldsymbol{x}_N \rangle \\ &\qquad \times \langle \boldsymbol{x}_1 \,|\, m \rangle \, \langle \mathcal{P}_{i1} \boldsymbol{x}^{(i)} \,|\, \kappa^{N-1}, \bar{\nu}, \bar{J}, \bar{M} \rangle, \end{aligned}$$

where

$$\mathcal{P}_{i1}\boldsymbol{x}^{(i)} \equiv \boldsymbol{x}_i, \boldsymbol{x}_1, \ldots, \boldsymbol{x}_{i-1}, \boldsymbol{x}_{i+1}, \ldots, \boldsymbol{x}_N.$$

This can be put into natural order with $i-1$ interchanges, introducing a phase factor $(-1)^{i-1}$. Each term in the sum gives the same result so that we obtain

$$\begin{aligned} &\langle \kappa^N, \nu, J, M \,|\, a^\dagger_m \,|\, \kappa^{N-1}, \bar\nu, \bar J, \bar M\rangle \\ &\quad = N^{1/2}\langle \kappa^N, \nu, J, M \;\{|m\rangle \,|\, \kappa^{N-1}, \bar\nu, \bar J, \bar M\rangle\}^J_M, \end{aligned} \tag{6.8.30}$$

where the main term on the right is an abbreviation for the N-fold integral in the previous equation. Alternatively, we could have interchanged the arguments x_i and x_N, in which case (6.8.30) would have been replaced by

$$\begin{aligned} &\langle \kappa^N, \nu, J, M \,|\, a^\dagger_m \,|\, \kappa^{N-1}, \bar\nu, \bar J, \bar M\rangle \\ &\quad = (-1)^{N-1}\, N^{1/2}\langle \kappa^N, \nu, J, M \;\{|\, \kappa^{N-1}, \bar\nu, \bar J, \bar M\rangle\, |m\rangle\}^J_M. \end{aligned} \tag{6.8.31}$$

The additional phase factor $(-1)^{N-1}$ can be viewed as the result of $N-1$ anti-commutations of the creation operator of the active electron required to place it at the right instead of the left of the string.

The dependence on the magnetic quantum numbers can be eliminated by using the Wigner-Eckart theorem: on the left-hand side we have

$$\begin{aligned} &\langle \kappa^N, \nu, J, M \,|\, a^\dagger_m \,|\, \kappa^{N-1}, \bar\nu, \bar J, \bar M\rangle \\ &\qquad = \begin{pmatrix} M & j & \bar J \\ J & m & \bar M \end{pmatrix} \langle \kappa^N, \nu, J \|\, \boldsymbol{a}^\dagger \,\|\, \kappa^{N-1}, \bar\nu, \bar J\rangle, \end{aligned} \tag{6.8.32}$$

and on the right-hand side

$$\begin{aligned} &\langle \kappa^N, \nu, J, M \;\{|m\rangle \,|\, \kappa^{N-1}, \bar\nu, \bar J, \bar M\rangle\}^J_M \\ &\qquad = \langle J\, M \,|\, j\, m,\; \bar J\, \bar M\rangle\, \left(j^N, \nu, J\, \{|\, j^{N-1}, \bar\nu, \bar J\right) \\ &\quad + \begin{pmatrix} M & j & \bar J \\ J & m & \bar M \end{pmatrix} (-1)^{\bar J - j + J}[\,J\,]^{1/2}\left(j^N, \nu, J\, \{|\, j^{N-1}, \bar\nu, \bar J\right), \end{aligned} \tag{6.8.33}$$

using (B.3.93). The expression $\left(j^N, \nu, J\, \{|\, j^{N-1}, \bar\nu, \bar J\right)$, a (one-particle) *coefficient of fractional parentage* (cfp), is independent of the sign of κ. The state $\kappa^{N-1}, \bar\nu, \bar J, \bar M$ is said to be a *parent* of κ^N, ν, J, M whenever the right-hand side of (6.8.33) does not vanish and κ^N, ν, J, M is said to be a *daughter* of $\kappa^{N-1}, \bar\nu, \bar J, \bar M$.

This can be summarized conveniently in terms of relations between the reduced matrix elements of creation and annihilation operators with the cfp:

Theorem 6.1. *The reduced matrix elements of $a^\dagger$ and a are given in terms of cfps by*

$$\begin{aligned} &\langle j^N, \nu, J \|\boldsymbol{a}^\dagger\| j^{N-1}, \bar\nu, \bar J\rangle \\ &\qquad = N^{1/2}(-1)^{\bar J - j + J}[\,J\,]^{1/2}\left(j^N, \nu, J\, \{|\, j^{N-1}, \bar\nu, \bar J\right) \end{aligned} \tag{6.8.34}$$

where the cfp

$$(j^N, \nu, J\,\{|\, j^{N-1}, \bar{\nu}, \bar{J}) = (j^{N-1}, \bar{\nu}, \bar{J}|\}j^N, \nu, J)$$

can be taken as real. Also, the definition (B.3.142) ensures that

$$\langle j^{N-1}, \bar{\nu}, \bar{J}\|\boldsymbol{a}\|j^N, \nu, J\rangle = \langle j^N, \nu, J\|\boldsymbol{a}^\dagger\|j^{N-1}, \bar{\nu}, \bar{J}\rangle.$$

These relations hold for subshells $\kappa = \pm(j+1/2)$ *of either sign.*

Another way of expressing the conclusion of this section is in terms of the expansion

$$\langle \boldsymbol{x}_1, \ldots, \boldsymbol{x}_N \,|\, \kappa^N, \nu, J, M\rangle = \sum_{\bar{\nu}} \sum_{\bar{J}, \bar{M}, m} \langle j\, m\, \bar{J}\, \bar{M}|\, J\, M\rangle \times (j^{N-1}, \bar{\nu}, \bar{J}|\}j^N, \nu, J)\, \langle \boldsymbol{x}_1 \,|\, m\rangle\, \langle \boldsymbol{x}_2, \ldots, \boldsymbol{x}_N \,|\, \kappa^{N-1}, \bar{\nu}, \bar{J}, \bar{M}\rangle \quad (6.8.35)$$

The sum on the right-hand side may not look at first glance as if it is totally anti-symmetric, although our construction ensures that it is. The orthonormality of the anti-symmetrized states leads to the orthogonality relation

Theorem 6.2.

$$\sum_{\bar{\nu}\bar{J}} (j^N, \nu, J\,\{|\, j^{N-1}, \bar{\nu}, \bar{J})\,(j^{N-1}, \bar{\nu}, \bar{J}|\}j^N, \nu', J') = \delta_{\nu,\nu'}\delta_{J,J'}. \quad (6.8.36)$$

6.8.6 Coefficients of fractional parentage in the seniority scheme

Coefficients of fractional parentage in the jj-coupling scheme can be computed using a recursive scheme devised by Redmond [34, 35] based upon (6.8.13). Because the operators appearing in this equation are diagonal with respect to N, we have

$$\begin{aligned}
&\langle j^N, \nu, J \,\|\, \left[\boldsymbol{a}^\dagger \times \boldsymbol{a}\right]^K \|\, j^N, \nu', J'\rangle \qquad (6.8.37)\\
&= [\,K\,]^{1/2}(-1)^{J+K+J'} \sum_{\bar{J},\bar{\nu}} \begin{Bmatrix} j & K & j \\ J & \bar{J} & J' \end{Bmatrix} \\
&\qquad \times \langle j^N, \nu, J\,\|\boldsymbol{a}^\dagger\|j^{N-1}, \bar{\nu}, \bar{J}\rangle\, \langle j^{N-1}, \bar{\nu}, \bar{J}\|\boldsymbol{a}\|j^N, \nu' J'\rangle \\
&= -N[\,J, K, J'\,]^{1/2} \sum_{\bar{J},\bar{\nu}} (-1)^{J+\bar{J}+K+j} \begin{Bmatrix} j & K & j \\ J & \bar{J} & J' \end{Bmatrix} \\
&\qquad \times (j^N, \nu, J\,\{|\, j^{N-1}, \bar{\nu}, \bar{J})\,(j^{N-1}, \bar{\nu}, \bar{J}|\}j^N, \nu', J')v
\end{aligned}$$

where we have first used (B.3.166) and then Theorem 6.1. Combining this with a similar result for $\langle j^{N-1}, \bar{\nu}, \bar{J} \,\|\, \left[\boldsymbol{a} \times \boldsymbol{a}^\dagger\right]^K \|\, j^{N-1}, \bar{\nu}', \bar{J}'\rangle$ gives

$$N[K]^{1/2} \sum_{J'',\nu''} (-1)^{x}[J'']\begin{Bmatrix} j & K & j \\ \overline{J} & J'' & \overline{J}' \end{Bmatrix}$$
$$\times(j^{N-1},\overline{\nu},\overline{J}\,|\}\,j^{N},\nu'',J'')\,(j^{N},\nu'',J''\{|j^{N-1},\overline{\nu}',\overline{J}')$$
$$= \delta_{\overline{\nu},\overline{\nu}'}\,\delta_{\overline{J},\overline{J}'}\,\delta_{k,0}\,[j,\overline{J}]^{1/2} + (N-1)[K]^{1/2}\sum_{\widetilde{\nu},\widetilde{J}}(-1)^{y}\begin{Bmatrix} j & K & j \\ \overline{J} & \widetilde{J} & \overline{J}' \end{Bmatrix}$$
$$\times(j^{N-1},\overline{\nu},\overline{J}\,\{|\,j^{N-2},\widetilde{\nu},\widetilde{J})\,(j^{N-2},\widetilde{\nu},\widetilde{J}|\}j^{N-1},\overline{\nu}',\overline{J}'),$$

where $x = \overline{J}' + J'' + K + j$, $y = \overline{J} + \widetilde{J} - j$. This formula can then be simplified by multiplying through by

$$(-1)^{z}[K]^{1/2}\begin{Bmatrix} j & K & j \\ \overline{J} & J & \overline{J}' \end{Bmatrix}, \quad z = -\overline{J}' - J - K - j,$$

and summing over K, giving

$$N\sum_{\nu',J'}(j^{N-1},\overline{\nu},\overline{J}\,|\}\,j^{N},\nu',J')\,(j^{N},\nu',J'\{|j^{N-1},\overline{\nu}',\overline{J}')$$
$$= \delta_{\overline{\nu},\overline{\nu}'}\,\delta_{\overline{J},\overline{J}'} + (N-1)(-1)^{\overline{J}+\overline{J}'}[\overline{J},\overline{J}']^{1/2}$$
$$\times\sum_{\widetilde{\nu},\widetilde{J}}\begin{Bmatrix} \overline{J} & j & \widetilde{J} \\ \overline{J}' & j & J \end{Bmatrix}(j^{N-1},\overline{\nu},\overline{J}\,\{|\,j^{N-2},\widetilde{\nu},\widetilde{J})\,(j^{N-2},\widetilde{\nu},\widetilde{J}|\}j^{N-1},\overline{\nu}',\overline{J}').$$

We obtain Redmond's formula by observing that

$$\sum_{\nu',J'}|\,j^{N},\nu',J',M'\rangle\,(j^{N},\nu',J'\{|j^{N-1},\overline{\nu}',\overline{J}')$$

must be proportional to a seniority state $|\,j^{N},\nu,J,M'\rangle$:

Theorem 6.3. *Let $j^{N-1},\overline{\nu}',\overline{J}'$ be a fixed state of the κ^{N-1} configuration. Then the $N \to N-1$ cfps are obtainable from*

$$\mathcal{N}\,(j^{N-1},\overline{\nu},\overline{J}\,|\}j^{N},\nu,J) \tag{6.8.38}$$
$$= \delta_{\overline{\nu},\overline{\nu}'}\,\delta_{\overline{J},\overline{J}'} + (N-1)(-1)^{\overline{J}+\overline{J}'}[\overline{J},\overline{J}']^{1/2}\sum_{\widetilde{\nu},\widetilde{J}}\begin{Bmatrix} \overline{J} & j & \widetilde{J} \\ \overline{J}' & j & J \end{Bmatrix}$$
$$\times(j^{N-1},\overline{\nu},\overline{J}\,\{|\,j^{N-2},\widetilde{\nu},\widetilde{J})\,(j^{N-2},\widetilde{\nu},\widetilde{J}|\}j^{N-1},\overline{\nu}',\overline{J}')$$

where $\mathcal{N}$ is a real normalization factor chosen so that

$$\mathcal{N}^2 = N^2\sum_{\nu,J}(j^{N-1},\overline{\nu}',\overline{J}'|\}j^{N},\nu,J)^2$$

The Redmond formula provides a recursive method to construct cfps with the initial conditions

$$(j^0, 0, 0|\}j^1, 1, j) = 1, \quad (j^1, 1, j|\}j^2, \nu, J) = \frac{1}{2}[1 + (-1)^J] \tag{6.8.39}$$

where $\nu = 0$ when $J = 0$, and $\nu = 2$ otherwise (compare Table F-1). The right-hand side of (6.8.37) must be computed for each choice of *principal parent* state $j^{N-1}, \bar{\nu}', \bar{J}'$ which have j^N, ν, J as daughter. This ensures that the unnormalized cfps are obtained with consistent relative phases in each row of the table.

The equivalence of the quasispin representation with the seniority scheme can be exploited to give additional relations between cfps. Theorem 6.1 relates each cfp to the reduced matrix element of a creation or annihilation operator regarded as a tensor under SO(3). Because we can also regard each such operator as the component of a double tensor operator, (6.8.21), we can apply the Wigner-Eckart theorem again in quasispin space so that

$$\begin{aligned} &\langle Q, M_Q, J, M|A^{q,\,j}_{\frac{1}{2},m}|Q', M_Q', J', M'\rangle \\ &= \begin{pmatrix} M & j & J' \\ J & m & M' \end{pmatrix} \begin{pmatrix} M_Q & \frac{1}{2} & Q' \\ Q & \frac{1}{2} & M_Q' \end{pmatrix} \langle Q, J|||\boldsymbol{A}^{q,j}|||Q', J'\rangle \end{aligned} \tag{6.8.40}$$

Thus the cfp are proportional to the appropriate quasispin $3j$-symbol; this yields relations such as

Theorem 6.4.

$$(a) \quad \frac{(j^N, \nu, J\{|j^{N-1}, \nu - 1, \bar{J})}{(j^\nu, \nu, J\{|j^{\nu-1}, \nu - 1, \bar{J})} = \left\{\frac{\nu(2j + 3 - N - \nu)}{N(2j + 3 - 2\nu)}\right\}^{1/2}$$

$$(b) \quad \frac{(j^N, \nu, J\{|j^{N-1}, \nu + 1, \bar{J})}{(j^{\nu+2}, \nu, J\{|j^{\nu+1}, \nu + 1, \bar{J})} = \left\{\frac{(N - \nu)(\nu + 2)}{2N}\right\}^{1/2}$$

$$\begin{aligned} (c) \quad &\frac{(j^N, \nu, J\{|j^{N-1}, \bar{\nu}, \bar{J})}{(j^{N+2}, \nu, J\{|j^{N+1}, \bar{\nu}, \bar{J})} \\ &= \left\{\frac{(N + 2)(N + 1 - \bar{\nu})(2j + 2 - N - \bar{\nu})}{N(N + 2 - \nu)(2j + 1 - N - \nu)}\right\}^{1/2} \end{aligned}$$

Because $2j + 1 \geq N = \nu + 2(Q + M_Q) \geq \nu$, $N - \nu$ is an even non-negative integer in all these formulae. Similar relations can be found for reduced matrix elements of unit tensors (6.8.27):

Theorem 6.5. *(a) Matrix elements of the unit tensor* $\boldsymbol{U}^K$ *are diagonal in the seniority number when* K *is an odd integer and are independent of* N.

(b) When K is an even integer

$$\frac{\langle j^N, \nu, J\|\boldsymbol{U}^K\|j^N, \nu, J'\rangle}{\langle j^\nu, \nu, J\|\boldsymbol{U}^K\|j^\nu, \nu, J'\rangle} = \frac{2j+1-N}{2j+1-\nu}$$

and also

(c)

$$\frac{\langle j^N, \nu, J\|\boldsymbol{U}^K\|j^N, \nu-2, J''\rangle}{\langle j^\nu, \nu, J\|\boldsymbol{U}^K\|j^\nu, \nu-2, J''\rangle} = \left\{\frac{(N-\nu+2)(2j+3-N-\nu)}{2(2j+3-2\nu)}\right\}^{1/2}$$

These results are all deduced from (6.8.25). When $Q+K$ is odd, only the first term on the right contributes, and rearranging gives

$$U^K_{M_Q} = -\frac{1}{\sqrt{2}} X^{Q,K}_{0,M_Q}, \quad Q+K \text{ odd}.$$

Case (a) requires $Q = 0$, so the relevant $3j$-symbol is

$$\begin{pmatrix} M_Q & 0 & Q' \\ Q & 0 & M_Q \end{pmatrix} = \delta_{Q,Q'}[Q]^{-1/2}$$

independent of M_Q and therefore of N. Cases (b) and (c) require $Q = 1$, so that $Q' = Q \pm 1$. $Q = Q'$ in case (b), so that the ratio of the two cfps is

$$\begin{pmatrix} M_Q & 1 & Q \\ Q & 0 & M_Q \end{pmatrix} \Big/ \begin{pmatrix} \overline{M}_Q & 1 & Q \\ Q & 0 & \overline{M}_Q \end{pmatrix}$$

where $M_Q = (N-2j-1)/2$ and $\overline{M}_Q = (\nu-2j-1)/2$. Case (c) is similar but with $Q' = Q+1$.

The obvious symmetry of states with N electrons and N holes in the filled subshell can also be treated within the quasispin formalism. Define the conjugation operator C so that

$$C\, A^{q,\ j}_{m_q,m}\, C^{-1} = (-1)^{q-m_q}\, A^{q,\ j}_{-m_q,m} \tag{6.8.41}$$

replaces creation of particles by creation of holes; this can be extended to other double tensor operators

$$C\, X^{q,\ j}_{m_q,m}\, C^{-1} = (-1)^{q-m_q}\, X^{q,\ j}_{-m_q,m} \tag{6.8.42}$$

and to states,

$$C\,|Q, M_Q, J, M\rangle = (-1)^{Q-M_Q}\,|Q, -M_Q, J, M\rangle \tag{6.8.43}$$

provided $C^\dagger C = 1$ to maintain normalization. This relation follows by reinterpreting (6.8.41) as the pair of equations

$$C\boldsymbol{a}^{\dagger}C^{-1} = \boldsymbol{a}, \quad C\boldsymbol{a}C^{-1} = -\boldsymbol{a}^{\dagger}.$$

Then from the definitions, we get

$$CQ_{\pm}C^{-1} = -Q_{\mp}, \quad CQ_3C^{-1} = -Q_3.$$

so that the conjugation properties of quasispin are similar to those of time reversal. As well as being unitary, conjugation is antilinear:

$$C\,i\,C^{-1} = -i,$$

from which we infer that matrices of conjugate quasispin states satisfy

Lemma 6.6.

$$\begin{aligned}&\langle Q, M_Q, J, M | X^{q;\, j}_{m_q, m} | Q', M'_Q, J', M' \rangle \\ &\quad = (-1)^{Q-q-Q'} \langle Q, -M_Q, J, M | X^{q;\, j}_{-m_q, m} | Q', -M'_Q, J', M' \rangle.\end{aligned}$$

The proof uses straightforward calculation. When we apply this result to the operator $A^{q;\, j}_{m_q, m}$ we get a relation between cfps of particle states and hole states:

Theorem 6.7.

$$\frac{(j^N, \nu, J\{|j^{N-1}\bar{\nu}, \bar{J})}{(j^{2j+1-N}, \nu, J|\}j^{2j+2-N}\bar{\nu}, \bar{J})} = (-1)^x \left\{ \frac{2j+2-N}{N} \frac{[\bar{J}]}{[J]} \right\}^{1/2}$$

where $x = J - j - \bar{J} - \frac{1}{2}(\nu - \bar{\nu} - 1)$.

This important result not only halves the size of archived tables of cfps but fixes the relative phases of particle and hole states. A similar argument gives

$$\begin{aligned}&\langle j^N, \nu, J \| \boldsymbol{U}^K \| j^N, \nu', J' \rangle \\ &\quad = (-1)^y \langle j^{2j+1-N}, \nu, J \| \boldsymbol{U}^K \| j^{2j+1-N}, \nu', J' \rangle + \delta_{\nu,\nu'}\, \delta_{J,J'}\, \delta_{K,0}\, [J, j]^{1/2}\end{aligned}$$

where $y = K + 1 + \frac{1}{2}(\nu' + \nu)$. Many authors, for example [36, 37, 38], have given numerical tables of cfps in jj-coupling and a consistent set for the subshell states with $0 \le N \le j + 1/2$ appears in Table A.11 for $j = 3/2$, Table A.12 for $j = 5/2$, and Table A.13 for $j = 7/2$. The cfp for hole states with $N \ge j + 1/2$ can be obtained using Theorem 6.7. Some signs in the tables given by [36, 37] are inconsistent with the seniority scheme and have been corrected using [38]; see [39, 40].

6.8.7 Equivalent electrons in LS-coupling

Although jj-coupling is the natural choice for relativistic calculations, this is not the choice usually made in traditional nonrelativistic presentations of the theory of atomic spectra [9, 41, 42]. The electron orbital angular momentum $\boldsymbol{l}$

and spin $\boldsymbol{s}$ are independently constants of the motion in the absence of "spin-orbit coupling" – taking partial account of relativistic effects – and the open shell theory is thus more complicated than the relativistic theory we have studied so far.

As in the Dirac case, we define *subshells of equivalent electrons* by symbols $(nl)^N$, so that $0 \leq N \leq 2(2l+1)$ taking into account the degeneracy of each orbital nlm_l with respect to spin. Assuming these orbitals are states of some mean field central potential, we can construct electron configurational states (CSF) as determinantal products of the orbitals: for example an excited state of Ne I may have the configuration $1s^2\, 2s^2\, 2p^5\, 3d$ which will give rise to several CSFs that can be labelled by the $m_l m_s$ values assigned to the electrons in the two last (unfilled) shells. In this simple case, we can write down 6×10 independent degenerate CSFs that can be classified as states of the orbital angular momentum L and the total spin S, which can take values $L = 1, 2, 3$ and $S = 0, 1$, along with appropriate values of the projections M_L and M_S, accounting for all 60 CSFs. The Coulomb interaction between electrons is rotationally invariant, so that L, M_L, S, M_S will be good quantum numbers of the nonrelativistic atomic Hamiltonian

$$H_{nr} = \frac{1}{2}\boldsymbol{p}^2 - \frac{Z}{r} + \sum_{i<j} \frac{1}{R_{ij}}$$

whose eigenvalues will be of the form E_{LS}, independent of the projections [9, Chap. 7]. This is therefore referred to as the LS- (or Russell-Saunders) coupling scheme.

Spin-dependence is often modelled by a simple one-body *spin-orbit coupling* operator

$$H_{so} = \sum_i \xi(r_i) \boldsymbol{L}_i \cdot \boldsymbol{S}_i.$$

When this is a small perturbation, as is usual in low-Z atoms, this splits the Russell-Saunders *terms*, E_{LS}, into a number of *fine structure levels* E_{LSJ} depending also on the resultant angular momentum J. In this simple model, the levels are given by

$$E_{LSJ} = E_{LS} + \zeta(LSJ \,|\, \boldsymbol{L} \cdot \boldsymbol{S} \,|\, LSJ)$$

where $|\, LSJ)$ denotes a coupled angular momentum state and the *spin-orbit coupling constant* ζ measures the effective size of the potential $\xi(r)$. Whilst this works well in the simple spectra of low-Z elements, there are many violations of the simple Landé interval rules given by the above formula for which more elaborate models are needed [41, 42], some of which incorporate spin-dependent operators from the Breit-Pauli Hamiltonian. Configuration interaction effects can also play a part in complex spectra.

The underlying algebra for more complex configurations can be developed using second quantization methods as expounded by Judd [10]. Operators $a^{l\ s}_{m_l m_s}$ (s=1/2) (and its adjoint) replace the simpler operators a^j_m of

jj-coupling, with a consequent increase in complexity of the algebra. The classification of open shell states is complicated by the fact that certain LS combinations of the nl^N configuration are forbidden by the Pauli principle. For s^N and p^N configurations, the quantum numbers L and S, traditionally written in the form ${}^{2S+1}L$, suffice for a complete classification. The d^N configurations require the machinery of quasi-spin or seniority quantum number ν: the states can therefore be labelled $d^N(\alpha\nu LM_LSM_S)$ (Table A.14). More elaborate machinery still is required for the f^N configurations. The main problem of classification was famously solved by Racah [11] in his third paper, in which he made elaborate use of group theory; see Judd [43]. States of f^N are labelled $f^N(\alpha(UW)\nu LM_LSM_S)$, where U and W are certain group parameter labels. This still fails to classify the terms of f^N unambiguously for $4 \leq N \leq 10$, a problem which remains unsolved. The situation for $l > 3$ is even worse.

A full discussion of LS-coupling and its relation to the jj-coupling scheme, important though it is, would take us too far from our present theme. The first part of the chapter by Martin and Wiese [44] on atomic spectroscopy puts the present section into a wider context. The construction of fractional parentage coefficients in LS-coupling, following Racah [11], is described in [17], [42] (emphasizing the role of quasispin) and also, with tables of cfp, in [45, Chapters 5–7].

6.9 Atoms with complex configurations

In central field models, the electron configurations for atoms in their ground states can be constructed by filling vacancies in successive subshells, starting with the most tightly bound, until all N electrons have been allocated to a central field state. The net charge on the system is $(Z - N)e$; if $N < Z$, we have a positive ion, $N = Z$ corresponds to a neutral atom, and $N > Z$ to a positive ion. The simplest conceivable wave function takes the form of an $N \times N$ determinant from which we can derive the familiar (Dirac)-Hartree-Fock-(Breit) hierarchy of approximations for the ground state of the atom. (D)HF ground state calculations give a fair picture of ground state electron distributions and their properties and have made a big contribution to our understanding of atomic shell structure and dimensions. However, most of atomic and molecular physics is concerned with understanding the way the system responds to disturbance by photons or other particles so that a ground state calculation may be only the first step towards a suitable model. A single determinant is often a rather poor approximation for the ground state, especially when the outermost subshell is partially filled, and better (lower) ground state energies can only be obtained using a linear combination of determinants.

We write the *atomic state functions* (ASF) in the manner of (6.8.6) as linear combinations of multi-shell *configurational state functions* (CSF)

$$| \Gamma, J^{\pi}, M \rangle = \sum_{\gamma} c_{\Gamma, \gamma} | \gamma, J^{\pi}, M \rangle,$$

which, for isolated atoms, share the same overall total angular momentum and parity and which can represent excited states as well as the ground state.

The CSF will be constructed by first assigning the N atomic electrons to different subshells so that $N = \sum_i N_i$, where N_i is the number of electrons in subshell i and $0 \le N_i \le 2j_i + 1$. In this section, we focus on states of a single subshell; the construction of multi-shell CSF will be described in §6.9.2. The states of subshell i will be specified by a multi-index label $T_i = \kappa_i^{N_i}, \nu_i, w_i, J_i, M_i$ where ν_i is the seniority number, J_i the total angular momentum, M_i is its projection, and w_i represents any further labels needed when the simple seniority scheme fails to classify some states uniquely (for $j \ge 9/2$). It is often convenient to list some of these labels, particularly the projections, explicitly – for example $T_i M_i$ – when focussing on the angular momentum structure of CSFs and their matrix elements even though the presence of M_i is already implied by the use of the symbol T_i.

We write a normalized ket for the $n_i \kappa_i$ subshell as

$$|T_i M_i\rangle = (N_i!)^{-1/2} \underbrace{\left[a_i^{\dagger} \times \ldots \times a_i^{\dagger} \right]^{T_i M_i}}_{N_i\, factors} | 0 \rangle \qquad (6.9.1)$$

As usual, κ_i is equivalent to listing both j_i and l_i or, equivalently, j_i and parity $\pi_i = (-1)^{l_i}$, and

$$T_i \equiv (n_i \kappa_i)^{N_i} \nu_i, w_i, J_i \qquad (6.9.2)$$

so that for each subshell i, the list of labels T_i runs over a full set of orthonormal states of $(n_i \kappa_i)^{N_i}$. A ket for an n-particle state A with a coupling scheme X can be represented by the diagram, (B.3.107),

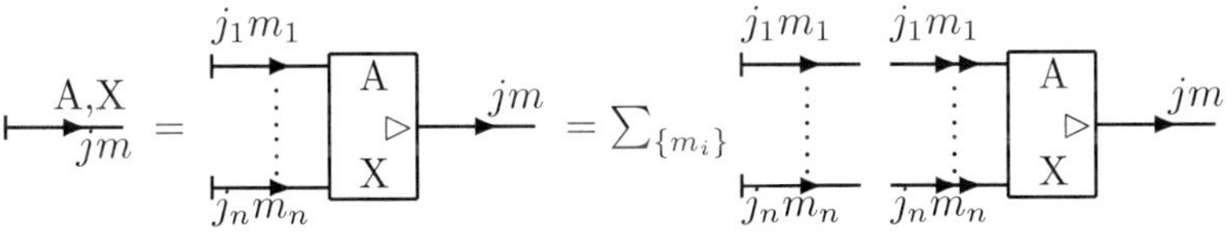

$$\begin{array}{c} \text{A,X} \\ \overline{jm} \end{array} = \begin{array}{c} j_1 m_1 \\ \vdots \\ j_n m_n \end{array}\!\!\boxed{\begin{array}{c} A \\ \triangleright \\ X \end{array}}\!\!jm = \sum_{\{m_i\}} \begin{array}{c} j_1 m_1 \\ \vdots \\ j_n m_n \end{array}\!\!\begin{array}{c} j_1 m_1 \\ \vdots \\ j_n m_n \end{array}\!\!\boxed{\begin{array}{c} A \\ \triangleright \\ X \end{array}}\!\!jm \qquad (6.9.3)$$

where the second block on the right-hand side is a generalized CGC. Each subshell state in the seniority scheme $|T_i M_i\rangle$ is constructed in this way, and can be represented by a diagram

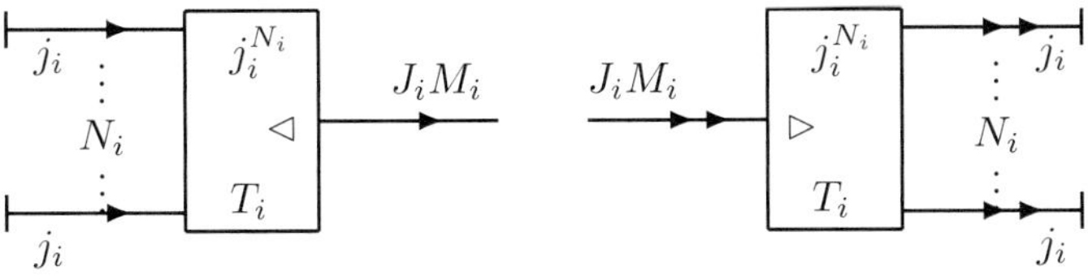

$$(6.9.4)$$

Subshell states in the seniority scheme with the *same* number of electrons, $(N_i = N_j)$, will be assumed to form an orthonormal set:

$$\langle T_i M_i \,|\, T_j M_j \rangle \equiv \big\langle (n_i\kappa_i)^{N_i}\nu_i, w_i, J_i M_i \,|\, (n_j\kappa_j)^{N_j}\nu_j, w_j, J_j M_j \big\rangle$$
$$= \delta_{N_i N_j}\,\delta_{n_i n_j}\,\delta_{\kappa_i\kappa_j}\,\delta_{\nu_i\nu_j}\,\delta_{w_i w_j}\,\delta_{J_i J_j}\,\delta_{M_i M_j} \qquad (6.9.5)$$

represented by the diagram

.

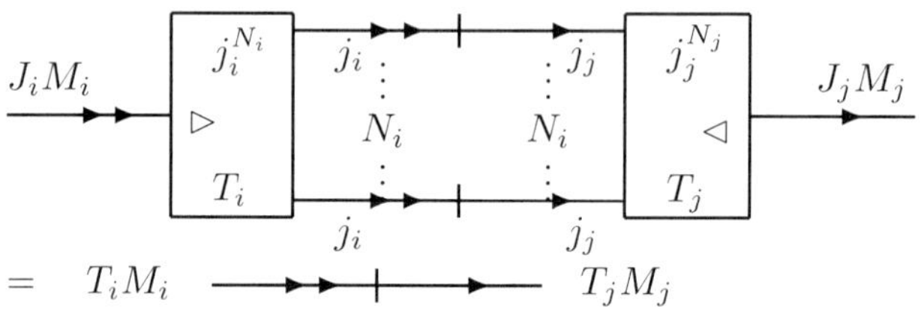

(6.9.6)

The calculation of matrix elements of one- and two-electron operators F (4.1.34) and G (4.1.35) requires detachment of up to two *active* electrons from each subshell using (6.8.30) or (6.8.33). The separation of a single electron is accomplished using

$$|\,T_i M_i\rangle = \sum_{\bar{J}_i,\bar{M}_i,\bar{\nu}_i,m_i} \langle j_i m_i, \bar{J}_i\bar{M}_i \,|\, J_i M_i\rangle \qquad (6.9.7)$$
$$\times\, (j_i^{\bar{N}_i}, \bar{\nu}_i, \bar{J}_i | \} j_i^{N_i}, \nu_i, J_i)\ \big\{ |j_i, m_i\rangle \otimes |\,\bar{T}_i, \bar{M}_i\rangle \big\}$$

or, diagrammatically,

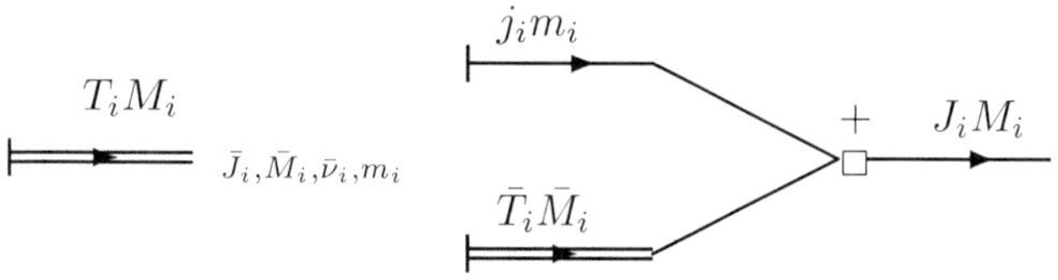

(6.9.8)

where the double lines represent subshell states for T_i and its parent $\bar{T}_i$ and we have made the projections explicit. The square on the node incorporates the fractional parentage coefficient so that

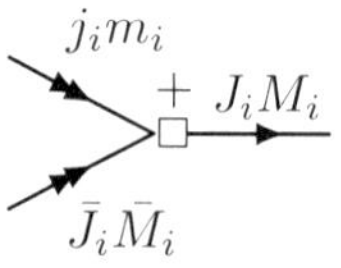

$= N_i^{1/2}(j_i^{\bar{N}_i}, \bar{\nu}_i, \bar{J}_i | \} j_i^{N_i}, \nu_i, J_i)$
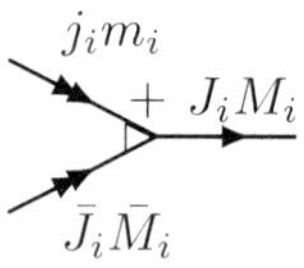

(6.9.9)

6.9.1 Recoupling coefficients

Consider an angular momentum state $|A; JM\rangle$ formed by coupling together states with angular momentum quantum numbers $J_1, \ldots J_n$ according to a coupling scheme A with $n - 2$ intermediate resultant angular momenta $X_1, \ldots, X_{n-2}$. Its scalar product with a state of the same total angular momentum but a different coupling scheme B is represented by the diagram

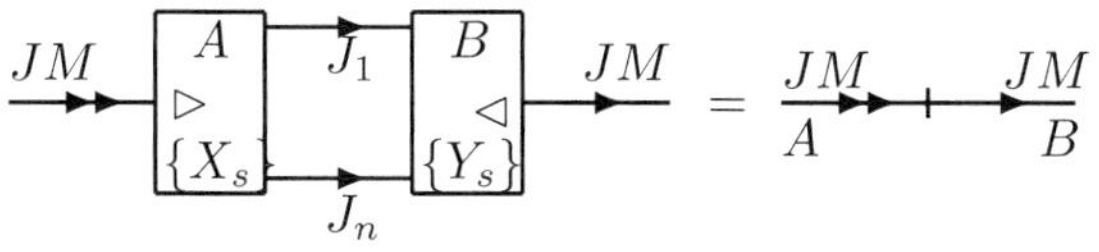

(6.9.10)

Because the value of this scalar product is independent of the projection M, we can define the angular momentum *recoupling coefficient* relating the two schemes A and B by averaging over M so that

$$\left\langle (J_1, \ldots, J_k)^B_{Y_1,\ldots,Y_{k-2}}; J \,|\, (J_1, \ldots, J_k)^A_{X_1,\ldots,X_{k-2}}; J \right\rangle$$

$$= [J]^{-1}$$

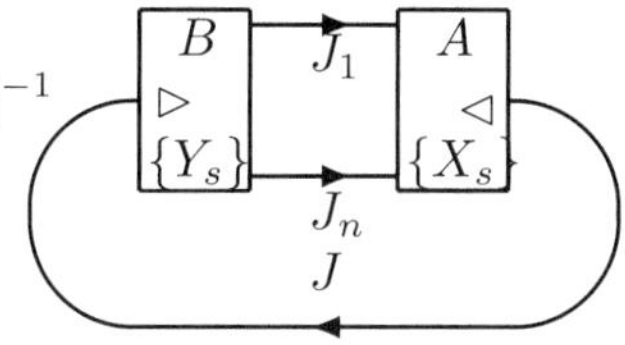

(6.9.11)

in which the diagram has no free lines.

6.9.2 Matrix elements between open shell states

The construction of matrix elements between open shell states can be simplified if we couple the angular momenta of the constituent subshells in a standard order. There is no loss of generality, as other CSF coupling schemes can be expressed as a linear combination of a standard set, in which a typical member T, is obtained by recursively coupling the subshells $T_1, T_2, \ldots$ in order:

$$\begin{aligned} T^{(1)} &= [T_1 \otimes T_2]^{X_1}; \\ T^{(k)} &= [T^{(k-1)} \otimes T_k]^{X_{k-1}}, \quad k = 2, 3, \ldots, n \end{aligned} \tag{6.9.12}$$

where $X_1, \ldots, X_{n-2}, X_{n-1} = J$ are the angular momenta chosen, in accordance with the selection rules, at each step, and the final member of the sequence, $T^{(n)} = T$, is the required CSF. The total number of electrons is given by $N = \sum_{i=1}^{n} N_i$ The corresponding diagrams are

$|T'M'\rangle =$

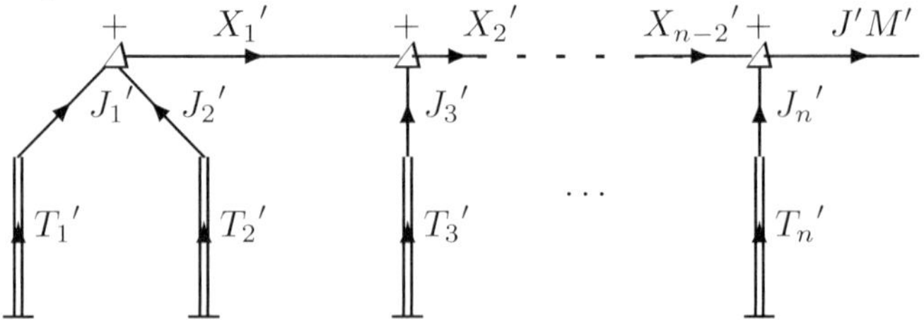

(6.9.13)

$\langle TM| =$

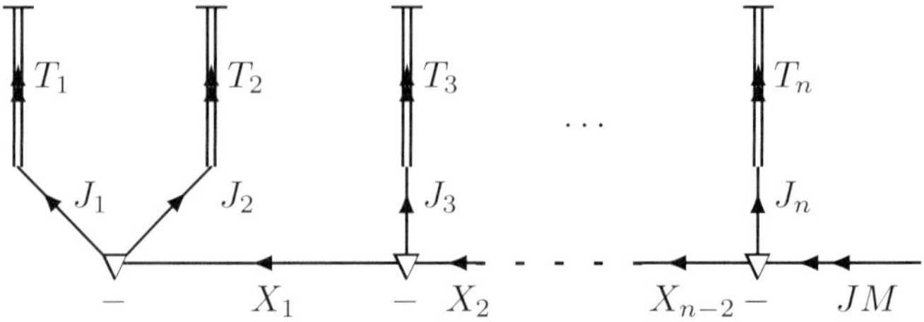

We start with the inner product $\langle TM|T'M'\rangle$. Provided the subshell states T_i from which the two CSFs are built are constructed from a common set of central field orbital spinors, we can exploit (6.9.5) and (6.9.6) to write

$\langle TM|T'M'\rangle$

$=$

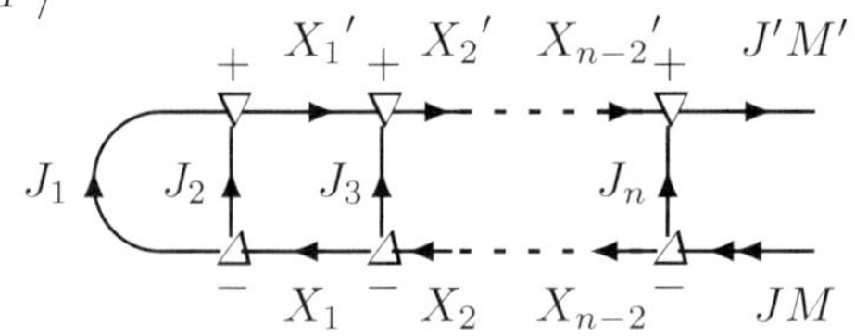

(6.9.14)

$= \delta_{JJ'}\delta_{MM'}[J]^{-1} \times \Big\{$

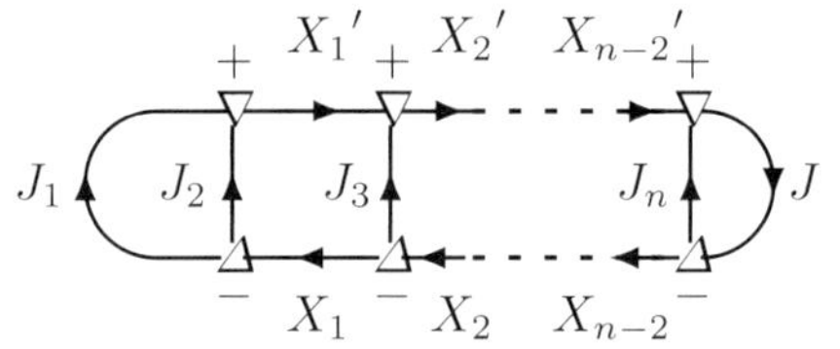

$\Big\}$

where the last step exploited (B.3.126). In general, the diagram in the braces is a recoupling coefficient, except when the coupling scheme is the same for both CSFs: $X_i = X_i{}'$, $i = 1, \dots, n-2$. We can then use (B.3.127) and (B.3.91) to separate the diagram on the two lines labelled X_1 and $X_1{}'$ giving a factor $[X_1]\{J_1 J_2 X_1\}\delta_{X_1 X_1{}'}$, and a closed diagram with an additional factor $[X_1]^{-1}$ with lines labelled J_1, J_2 replaced by one labelled X_1. Repeating this process eventually yields

$$\langle TM|T'M'\rangle = \delta_{TT'}\delta_{MM'}$$

where the delta function on the right indicates equality over all components of the multi-indices T and T'.

This example helps with the evaluation of matrix elements of the Fock space tensor operator

$$\boldsymbol{F}^k_q = \sum_{\alpha,\beta} a^\dagger_\alpha \langle \alpha \,|\, \boldsymbol{f}^k_q \,|\, \beta \rangle a_\beta \tag{6.9.15}$$

where $\boldsymbol{f}^k_q$ is an irreducible tensor operator acting on the coordinates of a single particle and α, β run over a complete set of states. When $\langle \alpha \,|\, \boldsymbol{f}^k_q \,|\, \beta \rangle \neq 0$, there will be a non-zero contribution to the sum obtained by contracting ${a_\alpha}^\dagger$ with one of the annihilation operators, $\boldsymbol{a}_i$ say, from $\langle TM \,|$ and a_β with ${\boldsymbol{a}_j}^\dagger$ from $|T'M'\rangle$. We say that α and β are *active subshells* for this matrix element. The triangle condition $\{j_\alpha k j_\beta\} = 1$ must hold and there will usually be a parity selection factor depending on the nature of $\boldsymbol{f}^k_q$; for example, if $\boldsymbol{f}^k_q \sim \boldsymbol{C}^k_q$, then $l_\alpha + k + l_\beta$ must be an even integer. The next step is to decouple one electron from the active subshell state T_α in the bra and another from the active subshell state ${T_\beta}'$ in the ket using (6.9.8). The argument used to evaluate the scalar product (6.9.14) reveals that ${T_\alpha}'$ must be the parent state of T_α and T_β must be the parent state of ${T_\beta}'$ if the CSF matrix element is not to vanish. Similarly $N_\alpha = {N_\alpha}' + 1$ and $N_\beta = {N_\beta}' - 1$.

The orthonormality of the subshell states can again be invoked to reduce the problem to evaluating a generalized CGC. The Wigner-Eckart theorem allows us to write the CSF matrix element as

$$\langle TM \,|\, \boldsymbol{F}^k_q \,|\, T'M' \rangle = \begin{pmatrix} M & k & J' \\ J & q & M' \end{pmatrix} \langle T \| \, \boldsymbol{F}^k \| \, T' \rangle \tag{6.9.16}$$

with

$$\begin{aligned} \langle T \| \, \boldsymbol{F}^k \| \, T' \rangle &= (-1)^{\Delta_\alpha + {\Delta_\beta}'} (N_\alpha {N_\beta}')^{1/2} (T_\alpha \{| {T_\alpha}') \, (T_\beta |\} {T_\beta}') \\ &\times \delta_{\widetilde{T}_\alpha, {T_\alpha}'} \delta_{\widetilde{T}'_\beta, T_\beta} \prod_{i \neq \alpha, \beta} \delta_{T_i, {T_i}'} \frac{[J]^{1/2}}{[j_\alpha]^{1/2}} \, \mathbb{R}_{\alpha,\beta}(T; T') \, \langle \alpha \| \, \boldsymbol{f}^k \| \, \beta \rangle, \end{aligned} \tag{6.9.17}$$

where we have used (6.9.7) to display the explicit dependence on the active subshell occupation numbers and CFPs. The phase factor $\Delta_{\alpha\beta}$ counts the number of anti-commutations needed to move the active electron operators a_α from the bra and ${a_\beta}^\dagger$ from the ket to contract with the corresponding operators in $\boldsymbol{F}^k_q$:

$$\Delta_\alpha = \sum_{i \geq \alpha} N_i, \quad {\Delta_\beta}' = \sum_{j \geq \beta} {N_j}' \tag{6.9.18}$$

The recoupling coefficient $\mathbb{R}_{\alpha,\beta}(T; T')$ is defined by

$\mathbb{R}_{\alpha,\beta}(T;T') = [J]^{-1}\times$

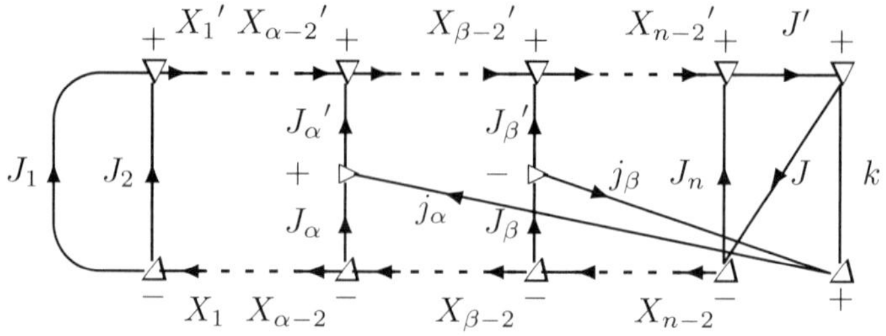

(6.9.19)

When T and T' are states of the same configuration so that the operator $\boldsymbol{f}^k_q$ can connect different orbital states in the same open subshell, the formula (6.9.17) reduces to

$$\langle T \| \boldsymbol{F}^k \| T' \rangle = \sum_\alpha \sum_{\widetilde{T}_\alpha} N_\alpha (T_\alpha \{ | \widetilde{T}_\alpha) \, (\widetilde{T}_\alpha \} | \} T_\alpha{}') \tag{6.9.20}$$
$$\times \prod_{i \neq \alpha} \delta_{T_i, T_i'} \frac{[J]^{1/2}}{[j_\alpha]^{1/2}} \mathbb{R}_{\alpha,\alpha}(T;T') \, \langle \alpha \| \boldsymbol{f}^k \| \alpha \rangle.$$

where T_α and $T_\alpha{}'$ are both daughter states of $\widetilde{T}_\alpha$. The additional summation over the active subshells α arises because several open subshells may contribute. In this case, the recoupling coefficient $\mathbb{R}_{\alpha,\alpha}(T;T')$ is defined by the diagram

$\mathbb{R}_{\alpha,\alpha}(T;T') = [J]^{-1}\times$

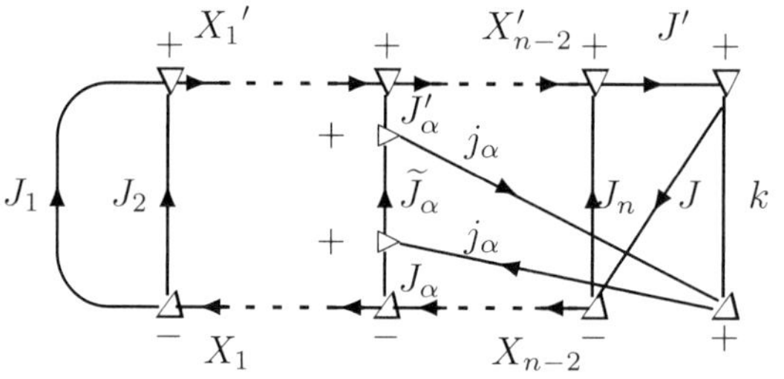

(6.9.21)

Examples of operators of type $\boldsymbol{f}^k_q$ are given in §6.2.

6.9.3 Matrix elements of two-electron operators of type *G*

The reduction of the CSF matrix elements of two-electron operators of type

$$\boldsymbol{G} = \sum_{\alpha,\beta,\gamma,\delta} a_\alpha{}^\dagger a_\beta{}^\dagger \langle \alpha\beta \,|\, g \,|\, \gamma\delta \rangle a_\delta a_\gamma \tag{6.9.22}$$

follows a similar pattern. For simplicity we consider only the Coulomb interaction, although the structure of the final result is much the same for all the interactions discussed in §6.4. The most general case involves four active subshells, say $\alpha, \beta, \gamma, \delta$; without loss of generality we can assume that, with our conventional ordering, $\alpha \leq \beta$ and $\gamma \leq \delta$. Then if $X^K(\alpha\beta\gamma\delta)$ is one of the effective interaction strengths defined in §6.4, we find

$$\begin{aligned}\langle TM \,|\, \boldsymbol{G} \,|\, T'M' \rangle &= \delta_{JJ'}\,\delta_{MM'} \\ &\times \sum_{\alpha\beta\gamma\delta} (-1)^{\Delta} [N_\alpha(N_\beta - \delta_{\alpha\beta}) N_\gamma{}'(N_\delta{}' - \delta_{\gamma\delta})]^{1/2} \\ &\times \prod_{i \neq \alpha,\beta,\gamma,\delta} \delta_{T_i,T_i{}'} \sum_k \sum_{\overline{T}} (T_\alpha\{|T_\alpha{}')\,(T_\beta\{|T_\beta{}')\,(T_\gamma|\}T_\gamma{}')(T_\delta|\}T_\delta{}') \\ &\times \Big\{ \mathbb{R}_d(T,T')\,(1+\delta_{\alpha\beta}\,\delta_{\gamma\delta})^{-1}\,[j_\alpha, j_\delta]^{-1/2}\,X^K(\alpha\beta\gamma\delta) \\ &\quad - \mathbb{R}_e(T,T')(1-\delta_{\alpha\beta})(1-\delta_{\gamma\delta})[j_\alpha, j_\gamma]^{-1/2}\,X^K(\alpha\beta\delta\gamma) \Big\}. \end{aligned} \tag{6.9.23}$$

The phase factor arising from anti-commutation of operators to extract the active subshell operators is

$$\Delta = \sum_{i=\alpha+1}^{\beta} N_i - \sum_{j=\gamma+1}^{\delta} N_j{}' \tag{6.9.24}$$

The *direct* recoupling coefficient is given by

$$\mathbb{R}_d(T;T') = [J]^{-1} \times \tag{6.9.25}$$

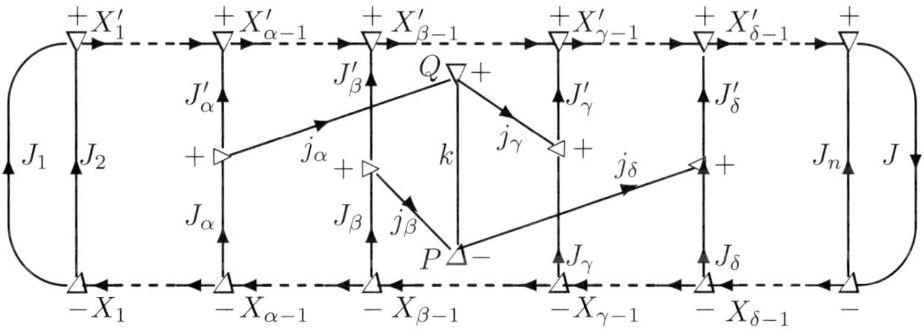

The *exchange* recoupling coefficient $\mathbb{R}_e(T,T')$ has a similar diagram, differing in that the the j_γ line is linked to the node P rather than Q, whilst the j_δ line is linked to Q rather than to P.

The expression (6.9.23) simplifies when, say, the ket $|T'M'\rangle$ has two active electrons in the same subshell. For example, setting $\gamma = \delta$ in (6.9.25) gives

$$
\begin{aligned}
\langle TM \,|\, \boldsymbol{G} \,|\, T'M' \rangle &= \delta_{JJ'}\, \delta_{MM'} \\
&\times \sum_{\alpha\beta\gamma} (-1)^{\Delta'} [N_\alpha (N_\beta - \delta_{\alpha\beta}) N_\gamma{}' (N_\gamma{}' - 1)]^{1/2} \qquad (6.9.26) \\
&\times \prod_{i \neq \alpha,\beta,\gamma} \delta_{T_i,T_i{}'} \sum_k \sum_{\widetilde{T}} (T_\alpha\{|T_\alpha{}')\,(T_\beta\{|T_\beta{}')\,(T_\gamma|\}\widetilde{T}_\gamma)(\widetilde{T}_\gamma|\}T_\gamma{}') \\
&\qquad \times \mathbb{R}_d(T,T')\,(1+\delta_{\alpha\beta})^{-1}\,[j_\alpha, j_\gamma]^{-1/2}\, X^K(\alpha\beta\gamma\gamma),
\end{aligned}
$$

where $\Delta' = \sum_{i=\alpha+1}^{\beta} N_i$ and the diagram representing $\mathbb{R}_d(T,T')$ is modified in the obvious way. Clearly, $\widetilde{T}_\gamma$ is a daughter of T_γ, and $T_\gamma{}'$ is a grand-daughter.

6.10 CI and MCDHF problems with large CSF sets

The preceding section exploits the quantum theory of angular momentum to simplify the calculation of matrix elements between CSFs. The energy of an atom in a state Γ given by an ASF of the form (6.8.6) is

$$
E_\Gamma = \boldsymbol{c}_\Gamma{}^\dagger \boldsymbol{H} \boldsymbol{c}_\Gamma \qquad (6.10.1)
$$

where $\boldsymbol{H}$ is the matrix of the Hamiltonian in the chosen CSF basis and $\boldsymbol{c}_\Gamma$ is a column vector of *mixing coefficients* with elements $c_{T,\Gamma}$ where T labels the N-electron CSFs involved. The formulae (6.9.16) and (6.9.17) for one-electron operators and (6.9.23) for two-electron operators, together with expressions for effective interaction strengths, $X^K(\alpha\beta\gamma\delta)$, in terms of radial (Slater) integrals by formulae with the generic form (6.4.3), enable us to write each matrix element of the Dirac-Coulomb Hamiltonian as a sum over radial integrals:

$$
H_{TT'} = \sum_{\alpha\beta} t_{TT'}(\alpha\beta)\, I(\alpha\beta) + \sum_K \sum_{\alpha\beta\gamma\delta} v^K_{TT'}(\alpha\beta\gamma\delta)\, R^K_C(\alpha\beta\gamma\delta) \qquad (6.10.2)
$$

where $I(\alpha\beta)$ is the radial part of the one-electron Hamiltonian as in (6.3.9) and $R^k_C(\alpha\beta\gamma\delta)$ is a radial Slater integral in the notation of (6.4.8). The coefficients $t_{TT'}(\alpha\beta)$ and $v^k_{TT'}(\alpha\beta\gamma\delta)$ depend upon the active subshells α, β, γ and δ) contributing to $H_{TT'}$ and on the given structure of the two CSFs T and T'. The matrix elements of the transverse photon interaction can be written

$$
H^{trans}_{TT'} = \sum_{ABCD} \sum_k \sum_{\tau=1}^{6} v^{k\tau}_{TT'}(ABCD)\, S^{k\tau}(ABCD) \qquad (6.10.3)
$$

where $S^{k\tau}(ABCD)$ denotes an entry of Type τ in the list of §6.7.2, with simplifications for the Breit and Gaunt interactions.

The energy expression (6.10.1) is the starting point for a number of self-consistent procedures. If we choose a basis of radial wavefunctions computed, say, using some model potential, then E_Γ is a quadratic function of the mixing coefficients $\boldsymbol{c}_\Gamma$. If we make this expression stationary with respect to weak variations of the components of $\boldsymbol{c}_\Gamma$ subject to the normalization condition ${\boldsymbol{c}_\Gamma}^\dagger \boldsymbol{c}_\Gamma = 1$, we get a simple algebraic eigenvalue problem,

$$\boldsymbol{H}\, \boldsymbol{c}_\Gamma = E_\Gamma\, \boldsymbol{c}_\Gamma,$$

whose eigenvalues and eigenvectors approximate the atomic energy levels and the ASF mixing coefficients.The procedure is usually referred to as the method of *configuration interaction* (CI). In practice, its numerical accuracy depends upon a number of factors, in particular whether the necessarily finite number of CSF trial functions is effectively complete for the states of interest. This is partly a question of choosing the most appropriate CSFs but also of the choice of the radial functions used to construct the interaction integrals.

Multi-configuration self-consistent field (MCSCF) schemes seek to make the energy expression (6.10.1) stationary with respect to variations in the radial functions as well as in the mixing coefficients. These lead to various generalizations of the Dirac-Hartree-Fock schemes considered earlier in this chapter, with the generic label MCDHF. We obtain generalizations of the integro-differential equations (6.7.7). The new complication is that the $Y^k(AB;r)$ interaction potentials in (6.7.4) now depend quadratically upon the mixing coefficient vectors $\boldsymbol{c}_\Gamma$. This necessitates the use of iterative methods of solution which will be discussed in Chapters 7 and 8.

The atomic MCSCF method was first proposed in [46] in the pre-computer era, and computer limitations on early relativistic MCDHF calculations, such as in [47], allowed the use of only a small number of CSFs. Today, CSF sets of dimension 10^2 to 10^4 are common, so that the calculation of the coefficients $t_{TT'}(\alpha\beta)$ and $v^k_{TT'}(\alpha\beta\gamma\delta)$ can become a major bottleneck. CI or MCSCF calculations require the user first to decide what CSFs are appropriate and then to calculate and store numerical values of these coefficients before proceeding further with the calculation. The algorithms are essentially scalar in character: the coefficients are calculated one by one so that the analysis of a given CSF coupling scheme has to be repeated for every Hamiltonian matrix element in which it appears, and storage of 10^8 or more coefficients is expensive.

6.10.1 Decoupling active electrons

Calculation of the recoupling coefficients $\mathbb{R}_d(T,T')$ and $\mathbb{R}_e(T,T')$ and their storage is a major burden in large-scale atomic calculations. There are now many programs available for evaluating recoupling coefficients. The NJGRAF utility program [48], for example, has two parts: the first analyses the topological structure of the angular momentum graph and sets up pointers that identify the arguments of the $6j$-symbols in each term of a Racah sum, along with $(2j+1)$ factors and phase factors. These tables are relatively short. The second stage of the program substitutes numerical values for each argument and evaluates the recoupling coefficient. In principle, there may be several equivalent Racah sums; there is no guarantee that the first stage analysis selects an optimal expression, but the program does well in practice.

A more efficient algorithm aimed at reducing the computational costs in large-scale calculations results from modifying (6.9.17) and (6.9.23) so that they consist of a sum of terms which are products of expressions which depend only on the *the internal structure* of each CSF T or T', together with a factor involving only the active electrons. Decoupling an active electron from a CSF is straightforward. Using (6.9.8) we can modify the ket of equation (6.9.13) so that

$|TM\rangle =$ 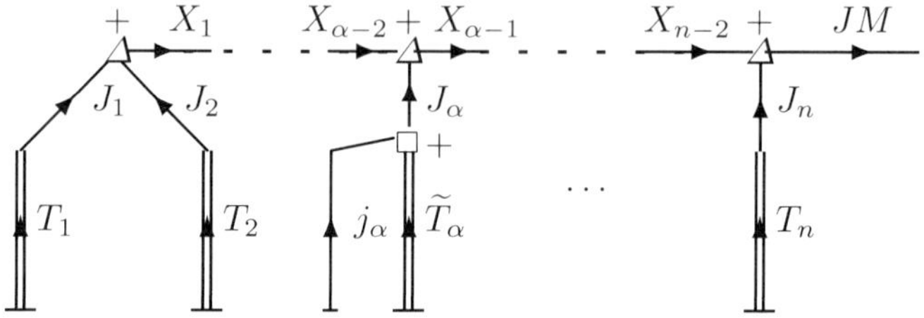

(6.10.4)

A standard parent state $|\widetilde{T}\widetilde{M}\rangle_\alpha$ has the same form as $|TM\rangle$ but with subshell states $\widetilde{T}_i,\ i = 1, \dots, n$ and coupling angular momenta $\widetilde{X}_i,\ i = 1, \dots, n-2$. We can therefore expand $|TM\rangle$ in the form

$$|TM\rangle = \sum_\alpha \prod_{i\neq\alpha} \delta_{T_i\widetilde{T}_i} \cdot \sum_{\widetilde{T}_\alpha, \widetilde{M}_\alpha m_\alpha} (-1)^{\Delta_\alpha} N_\alpha^{1/2}\, (\widetilde{T}_\alpha|\}\, T_\alpha)$$
$$\times\, C_\alpha(T, \widetilde{T}; JM). \left\{ |\widetilde{T}\widetilde{M}\rangle_\alpha \otimes |j_\alpha m_\alpha\rangle \right\}. \qquad (6.10.5)$$

The coefficient $C_\alpha(T, \widetilde{T}; JM)$ is a njm-symbol obtained from the scalar product $\langle \widetilde{T}\widetilde{M} \,|\, TM\rangle$, and we have moved the creation operator $a_{n_\alpha\kappa_\alpha m_\alpha}{}^\dagger$ to the extreme right, requiring $\Delta_\alpha = \sum_{i\geq\alpha} N_i - 1$ anti-commutations. The coupling is represented by the GCG coefficient

$C_\alpha(T, \widetilde{T}; JM) =$ 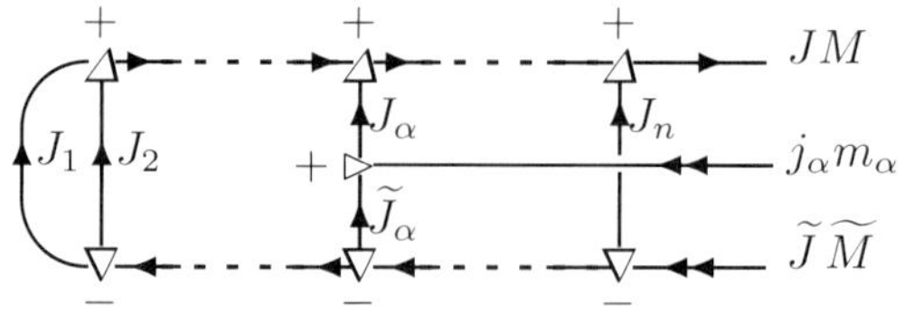

(6.10.6)

This can be written as the product of a recoupling coefficient $\mathbb{R}_\alpha(T, \widetilde{T})$ and a CG coefficient

$C_\alpha(T, \widetilde{T}; JM) = \mathbb{R}_\alpha(T, \widetilde{T})\, \langle \widetilde{J}\widetilde{M}, j_\alpha m_\alpha \,|\, JM\rangle$

$= [J]^{-1}$ 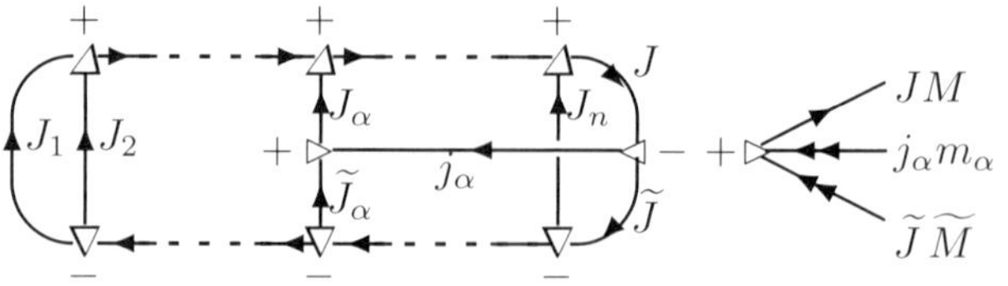

(6.10.7)

Thus (6.10.5) is equivalent to

$$|\,TM\,\rangle = \sum_{\alpha}\prod_{i\neq\alpha}\delta_{T_i\widetilde{T}_i}\cdot\sum_{\widetilde{T}_\alpha}(-1)^{\Delta_\alpha}N_\alpha{}^{1/2}\,(\widetilde{T}_\alpha|\}\,T_\alpha) \qquad (6.10.8)$$
$$\times\,\mathbb{R}_\alpha(T,\widetilde{T}).\sum_{\widetilde{M}m_\alpha}\Big\{|\,\widetilde{T}\widetilde{M}\rangle_\alpha\otimes|\,j_\alpha m_\alpha\rangle\Big\}\,\langle\widetilde{J}\widetilde{M},j_\alpha m_\alpha\,|\,JM\rangle.$$

There is a similar expression for the bra $\langle TM\,|$.

The shell recoupling coefficients $\mathbb{R}_\alpha(T,\widetilde{T})$ can be simplified. It is convenient to introduce a new expression

$$S(abc;def) = [a,b]^{1/2}(-1)^{a+b+c}\begin{Bmatrix} a & b & c\\ d & e & f\end{Bmatrix} \qquad (6.10.9)$$

This has the useful property that, when $f = 0$,

$$S(abc;de0) = \delta_{ae}\delta_{bd}\{a\,b\,c\},$$

which takes the value 1 when the triangular delta $\{a\,b\,c\}$ is non-zero, and vanishes otherwise. By cutting the diagram on successive pairs of lines $X_i, \widetilde{X}_i$ when $i < \alpha$ and similarly on three lines $X_i, j_\alpha, \widetilde{X}_i$ when $i > \alpha$, we reduce it to a product of factors

$$\mathbb{R}_\alpha(T,\widetilde{T}) = (-1)^{\Phi_\alpha(T,\widetilde{T})}.\frac{[J_\alpha]^{1/2}}{[J]^{1/2}}\;S(J_\alpha\widetilde{J}_\alpha j_\alpha;\widetilde{X}_{\alpha-1}X_{\alpha-1}X_{\alpha-2})$$
$$\times\prod_{j=1}^{\alpha-2}S(X_{j-1}X_jJ_{j+1};\widetilde{X}_j\widetilde{X}_{j-1}0)\prod_{j=\alpha+1}^{n}S(X_{j-2}\widetilde{X}_{j-2}j_\alpha;\widetilde{X}_{j-1}X_{j-1}J_j) \qquad (6.10.10)$$

where with $X_0 = \widetilde{X}_0 = J_1$, $\widetilde{X}_{n-1} = \widetilde{J}$, $X_{n-1} = J$, the phase factor has the exponent

$$\Phi_\alpha(T,\widetilde{T}) = (X_{\alpha-1} - \widetilde{X}_{\alpha-1} + j_\alpha) + (J_\alpha + \widetilde{J}_\alpha + j_\alpha)$$
$$+\sum_{j=0}^{\alpha-1}(X_{j-2} + X_{j-1} + J_j) + \sum_{j=\alpha+1}^{n}\Big(\widetilde{X}_{j-1} - \widetilde{X}_{j-2} - J_j\Big)$$

The first line of (6.10.10) involving the interaction with the active electron is always present and requires the evaluation of one $6j$-symbol. Those factors with index $j < \alpha$ contribute only to the phase $\Phi_\alpha(T,\widetilde{T})$ whilst those with index $j > \alpha$ may need more work. Both T and its parents have a simple structure, but each T may have several parents $\widetilde{T}$ with different values of $\widetilde{X}_j$ when $j > \alpha$. These will require non-trivial $6j$ evaluations whenever there is an open subshell with $J_j \neq 0$.

6.10.2 One-electron matrix elements

We can apply this immediately to construct matrix elements of one-electron irreducible tensor operators, $\langle TM \,|\, \boldsymbol{F}^k_q \,|\, T'M'\rangle$ in our standard CSF basis. The subshell CSF in the seniority scheme are designed to be orthonormal and those that appear in expansions of the bra and the ket of the form (6.10.8) must therefore be paired to give a non-zero result. The result is that

$$\langle TM|\, \boldsymbol{F}^k_q \,|T'M'\rangle = \sum_{\alpha,\beta}\sum_{\widetilde{T}} (-1)^{\Delta_\alpha}\, N_\alpha{}^{1/2}(T_\alpha\{|\,\widetilde{T}_\alpha)\; \mathbb{R}_\alpha(T,\widetilde{T})$$
$$\times\, (-1)^{\Delta_\beta}\, N'_\beta{}^{1/2}\,(T'{}_\beta\{|\,\widetilde{T'}_\beta)\; \mathbb{R}_\beta(T',\widetilde{T})\; \mathbb{F}^k_{\alpha\beta} \qquad (6.10.11)$$

where, after extracting a $3j$-symbol from the one-electron matrix element using the Wigner-Eckart theorem,

$$\mathbb{F}^k_{\alpha\beta} = \langle j_\alpha \|\, \boldsymbol{f}^k \,\| j_\beta\rangle.$$
$$\times \sum_{m_\alpha m_\beta \widetilde{M}} \langle JM\,|\,\widetilde{J}\widetilde{M}, j_\alpha m_\alpha\rangle\, \langle \widetilde{J}\widetilde{M}, j_\beta m_\beta\,|\,J'M'\rangle \begin{pmatrix} m_\alpha & k & j_\beta \\ j_\alpha & q & m_\beta \end{pmatrix}.$$

The sum over α, β runs over all subshells for which $\langle j_\alpha m_\alpha|\, \boldsymbol{f}^k_q |\, j_\beta m_\beta\rangle$ does not vanish, the sum over $\widetilde{T}$ runs over all common parent states, and we have used the symmetry of cfps. The Racah sum is given by the diagram

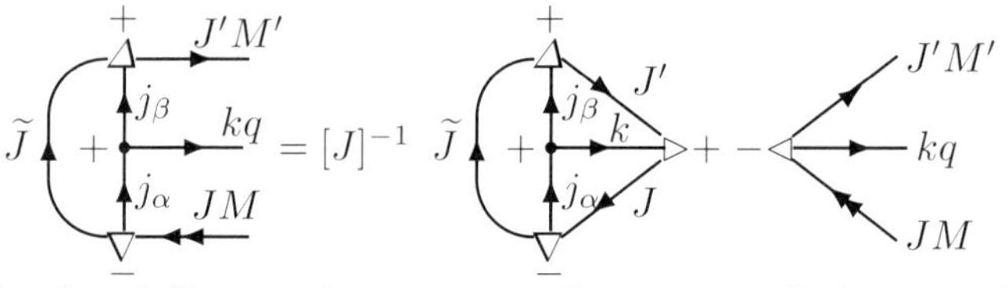

(6.10.12)

The closed diagram is proportional to a $6j$-symbol giving the final result

$$\langle T\|\, \boldsymbol{F}^k_q \,\| T'\rangle = \sum_{\alpha,\beta}\sum_{\widetilde{T}} (-1)^{\Delta_\alpha}\, N_\alpha^{1/2}(T_\alpha\{|\,\widetilde{T}_\alpha)\; \mathbb{R}_\alpha(T,\widetilde{T})$$
$$\times\, (-1)^{\Delta_\beta{}'}\, N_\beta{}'^{1/2}\,(T'{}_\beta\{|\,\widetilde{T}_\beta{}')\; \mathbb{R}_\beta(T',\widetilde{T})[J,J']^{1/2} \begin{Bmatrix} J & j_\alpha & \widetilde{J} \\ j_\beta & J' & k \end{Bmatrix} \langle j_\alpha \|\boldsymbol{f}^k\| j_\beta\rangle. \qquad (6.10.13)$$

The first line of this formula takes independent account of the structure of the bra, the second that of the ket and the last represents the matrix element of the active electrons.

6.10.3 Two-electron matrix elements

We can proceed in much the same way with the more complicated two-electron matrix elements of type $\boldsymbol{G}$. We have to decouple two active electrons on each side of the matrix element, which we can do by using (6.10.7) recursively. Denoting the parent state obtained by extracting an electron from the α subshell by $\widetilde{T}$ and the grandparent state obtained by extracting an electron from the β subshell by $\widetilde{\widetilde{T}}$, we have

$$
\begin{aligned}
|\,TM\,\rangle = &\sum_{\alpha}\prod_{i\neq\alpha}\delta_{T_i\widetilde{T}_i}\cdot\sum_{\widetilde{T}_\alpha}(-1)^{\Delta_\alpha}{N_\alpha}^{1/2}\,(\widetilde{T}_\alpha|\}\,T_\alpha)\mathbb{R}_\alpha(T,\widetilde{T}). \qquad (6.10.14)\\
&\times\sum_{\beta}\prod_{j\neq\beta}\delta_{\widetilde{T}_j\widetilde{\widetilde{T}}_j}\cdot\sum_{\widetilde{\widetilde{T}}_\beta}(-1)^{\widetilde{\Delta}_\beta}(\widetilde{N}_\beta-\delta_{\alpha\beta})^{1/2}\,(\widetilde{\widetilde{T}}_\beta|\}\,\widetilde{T}_\beta)\mathbb{R}_\beta(\widetilde{T},\widetilde{\widetilde{T}}).\\
&\times\sum_{\widetilde{M}\widetilde{\widetilde{M}}}\sum_{m_\alpha m_\beta}\left\{|\,\widetilde{\widetilde{T}}\,\widetilde{\widetilde{M}}\rangle_{\alpha\beta}\otimes|\,j_\beta m_\beta\rangle\otimes|\,j_\alpha m_\alpha\rangle\right\}\\
&\qquad\times\langle\widetilde{\widetilde{J}}\,\widetilde{\widetilde{M}},j_\beta m_\beta\,|\,\widetilde{J}\widetilde{M}\rangle\langle\widetilde{J}\,\widetilde{M},j_\alpha m_\alpha\,|\,JM\rangle.
\end{aligned}
$$

A similar expression with different active subshells can be written down for the bra state. According to (6.4.2), all interaction potentials can be expressed as a sum of terms of the type

$$
\langle ab\,|\,g\,|\,cd\rangle = \sum_K\sum_Q\begin{pmatrix} m_a & K & j_c\\ j_a & Q & m_c\end{pmatrix}\begin{pmatrix} m_b & Q & j_d\\ j_b & K & m_d\end{pmatrix}X^K(abcd)
$$

where $X^K(abcd)$ is an effective interaction strength for the term considered. The CSF matrix element is therefore

$$
\begin{aligned}
&\langle TM\,|\,g\,|\,T'M'\rangle = \qquad (6.10.15)\\
&\quad\sum_{\alpha}\prod_{i\neq\alpha}\delta_{T_i\widetilde{T}_i}\cdot\sum_{\widetilde{T}_\alpha}(-1)^{\Delta_\alpha}{N_\alpha}^{1/2}\,(\widetilde{T}_\alpha|\}\,T_\alpha)\mathbb{R}_\alpha(T,\widetilde{T}).\\
&\quad\times\sum_{\beta}\prod_{j\neq\beta}\delta_{\widetilde{T}_j\widetilde{\widetilde{T}}_j}\cdot\sum_{\widetilde{\widetilde{T}}_\beta}(-1)^{\widetilde{\Delta}_\beta}(\widetilde{N}_\beta-\delta_{\alpha\beta})^{1/2}\,(\widetilde{\widetilde{T}}_\beta|\}\,\widetilde{T}_\beta)\mathbb{R}_\beta(\widetilde{T},\widetilde{\widetilde{T}}).\\
&\quad\times\sum_{\gamma}\prod_{s\neq\gamma}\delta_{T'_s\widetilde{T'}_s}\cdot\sum_{\widetilde{T}_\gamma}(-1)^{\Delta'_\gamma}{N'_\gamma}^{1/2}\,(\widetilde{T'}_\gamma|\}\,T'_\gamma)\mathbb{R}_\gamma(T',\widetilde{T'}).\\
&\quad\times\sum_{\delta}\prod_{t\neq\beta}\delta_{\widetilde{T'}_t\widetilde{\widetilde{T'}}_t}\cdot\sum_{\widetilde{\widetilde{T'}}_\delta}(-1)^{\widetilde{\Delta'}_\delta}(\widetilde{N'}_\delta-\delta_{\gamma\delta})^{1/2}\,(\widetilde{\widetilde{T'}}_\delta|\}\,\widetilde{T'}_\delta)\mathbb{R}_\delta(\widetilde{T'},\widetilde{\widetilde{T}}).\\
&\quad\times\sum_K\left\{(1+\delta_{\alpha\beta}\delta_{\gamma\delta})^{-1}\mathbb{G}^K_{\alpha\gamma,\beta\delta}-(1-\delta_{\alpha\beta})(1-\delta_{\gamma\delta})\mathbb{G}^K_{\alpha\delta,\beta\gamma}\right\}
\end{aligned}
$$

$$
\mathbb{G}^K_{\alpha\gamma,\beta\delta} = X^K(\alpha\beta\gamma\delta)\;\mathbb{R}^K(\alpha\gamma,\beta\delta)_{JM,J'M'} \qquad (6.10.16)
$$

in which, after using rule (B.3.126), the Racah sum is

$$\mathbb{R}^K_{\alpha\beta\gamma\delta} = \delta_{JJ'}\delta_{MM'}[J]^{-1}$$

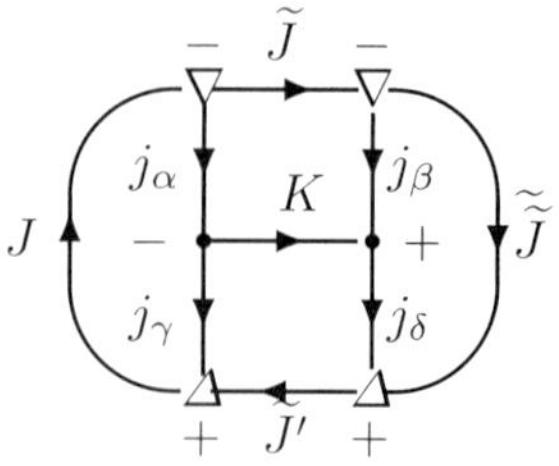

$$= \delta_{JJ'}\delta_{MM'}[J]^{-1}(-1)^{2(\widetilde{J}-\widetilde{J}')} \left\{ \begin{array}{ccc|c} J & j_\alpha & j_\gamma & \\ \widetilde{J} & K & \widetilde{J}' & 2 \\ \widetilde{\widetilde{J}} & j_\beta & j_\delta & \end{array} \right\}$$

$$= \delta_{JJ'}\delta_{MM'}[J]^{-1}(-1)^{2(\widetilde{J}-\widetilde{J}')}$$

$$\times \left\{ \begin{array}{ccc} \widetilde{J}' & \widetilde{J} & K \\ j_\alpha & j_\gamma & J \end{array} \right\} \left\{ \begin{array}{ccc} \widetilde{J}' & \widetilde{J} & K \\ j_\beta & j_\delta & \widetilde{\widetilde{J}} \end{array} \right\} \qquad (6.10.17)$$

which expresses the fact that the interaction is scalar so that the Fock matrix is diagonal with respect to J.

The exchange part of the two-electron interaction can be treated similarly. An exchange of the roles of j_δ and j_γ gives

$$\mathbb{R}^K_{\alpha\delta,\beta\gamma} = \delta_{JJ'}\delta_{MM'}[J]^{-1}$$

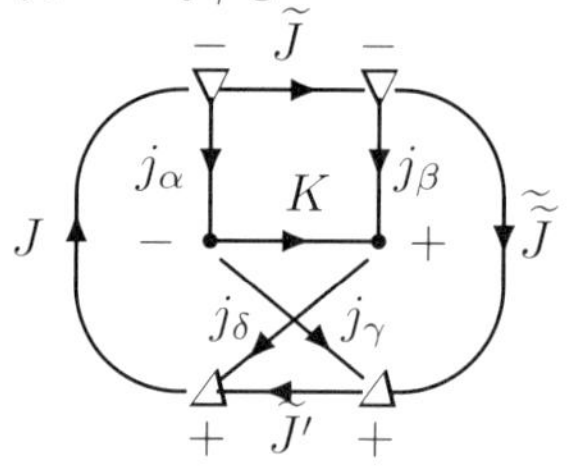

(6.10.18)

$$= \delta_{JJ'}\delta_{MM'}[J]^{-1}(-1)^{2(\widetilde{J}-\widetilde{J}')} \left\{ \begin{array}{ccc} j_\gamma & j_\alpha & K \\ \widetilde{J}' & J & j_\delta \\ \widetilde{\widetilde{J}} & \widetilde{J} & j_\beta \end{array} \right\}$$

The expression (6.10.15) divides up the calculation of matrix elements of two-electron interactions into several parts. The last line, which entangles the different states, only requires knowledge of the values of the total angular momentum of the two CSFs, their parent and grandparent states. The lines above it decouple one electron from each active subshell. The structure of the recoupling coefficients $\mathbb{R}_i(T, \widetilde{T})$ relating the CSF and its parents is simple, and it is only necessary to substitute the appropriate J values and to extract the appropriate CFPs once. The result can be recycled for each matrix element in which the CSF T participates, a considerable saving of effort in comparison with the older method based on (6.9.25). A similar construction has been given by Bar-Shalom *et al.* [56] in calculating electron-ion collisional excitation cross-sections in the relativistic distorted-wave approximation.

References

1. Grant I P 1970 *Adv. Phys.* **19** 747.
2. Grant I P 1961 *Proc Roy Soc (Lond) A* **262** 555.
3. Grant I P 1965 *Proc. Phys. Soc.* **86**, 523.
4. Grant I P and Pyper N C 1976 *J. Phys. B: Atom. Molec. Phys.* **9** 761.
5. Grant I P, McKenzie B J, Norrington P H, Mayers DF and Pyper N C 1980 *Comput. Phys. Commun.* **21**, 207.
6. Grant I P 1988 in *Methods in Computational Chemistry* Vol 2: Relativistic effects in atoms and molecules, ed. S WilsonS 1988 (New York: Plenum)[49], pp. 1.
7. de Shalit A and Talmi I 1963 *Nuclear Shell Theory* (New York: Academic Press).
8. Inglis D R 1931 *Phys. Rev.* **38**, 162.
9. Condon E U and Shortley G H 1953 *The Theory of Atomic Spectra* (Cambridge: University Press).
10. Judd B R 1967 *Second Quantization and Atomic Spectroscopy* (Baltimore: The Johns Hopkins Press).
11. Racah G 1942 *Phys Rev* **61** 186; — 1942 *Phys Rev* **62** 438; — 1943 *Phys Rev* **63** 367; — 1949 *Phys Rev* **76** 1352.
12. Gaunt J A *Proc. Roy. Soc. A* **122** 513.
13. Swirles B 1935 *Proc Roy Soc (Lond) A* **152** 625; — 1936 *Proc Roy Soc (Lond) A* **157** 680.
14. Jacobsohn B A 1947 Thesis, University of Chicago (unpublished).
15. Grant I P and McKenzie B J 1980 *J. Phys. B: Atom. Molec. Phys.* **13** 2671.
16. Abramowitz M and Stegun I A 1970 *Handbook of Mathematical Functions* (New York: Dover).
17. Judd B R 1963 *Operator Techniques in Atomic Spectroscopy* (New York: McGraw-Hill).
18. Møller C 1932 *Ann. der Phys.* **14** 531.
19. Doyle H T 1969 *Adv. Atom. Molec. Phys.***5**, 337.
20. Mann J B and Johnson W R 1971 *Phys. Rev. A* **4**, 41.
21. Huang K-N 1979 *Rev. Mod. Phys.* **51**, 215.
22. Pyper N C and Grant I P 1978 *Proc. Roy. Soc. A* **359**, 525.
23. Lindgren I and Rosén A 1974 *Case Studies in Atomic Physics* **4**, 93.
24. Slater J C 1960 *Quantum theory of atomic structure*, Vol. 1. (New York: McGraw-Hill).
25. Desclaux J P, Moser C M and Verhaegen G 1971 *J. Phys. B: Atom. Molec. Phys.* **4**, 296.
26. Larkins F P 1976 *J. Phys. B: Atom. Molec. Phys.* **9**, 37.
27. Fischer C F 1977 *The Hartree-Fock Method for Atoms* (New York: John Wiley).
28. Hartree D R 1957 *The calculation of atomic structures*, (New York: John Wiley).
29. Judd B R and Elliott J P 1970 *Topics in Atomic and Nuclear Theory* (University of Canterbury, New Zealand).
30. Wybourne B G 1974 *Classical Groups for Physicists* (New York: John Wiley).
31. Hamermesh M, 1962 *Group Theory* (Reading, Mass.: Addison-Wesley Publishing Co. Inc.).
32. Flowers B H and Szpikowski S 1964 *Proc. Phys. Soc. (Lond.)* **84**, 193.
33. Lawson R D and Macfarlane M H 1965 *Nuclear Phys.* **66**, 80.
34. Redmond P J 1954 *Proc Roy Soc A* **2**22, 84.
35. Hassitt A 1955 *Proc Roy Soc A* **2**29, 110.

36. Edmonds A R and Flowers B H 1952 *Proc Roy Soc A* **2**14, 515.
37. de-Shalit A and Talmi I 1963 *Nuclear Shell Theory* (New York: Academic Press).
38. Sivcev V I, Slepcov A A, Kičkin, Rudzikas Z B 1974 *Liet. fiz. rink., Lit. fiz. sb.* **14**, 189.
39. Grant I P 1972 *Comput Phys Commun* **4**, 377.
40. Grant I P 1972 *Comput Phys Commun* **14**, 311.
41. Cowan R D 1981 *The Theory of Atomic Structure and Spectra* (Berkeley: University of California Press).
42. Zenonas Rudzikas 1997 *Theoretical Atomic Spectroscopy* (Cambridge: University Press).
43. Judd B R 1996 in [50, Chapter 3].
44. Martin W C and Wiese W L 1996 in [50, Chapter 10].
45. Condon E U and Odabasi H 1980 *Atomic Structure* (Cambridge: University Press).
46. Hartree D R, Hartree W and Swirles B 1939 *Phil. Trans. Roy. Soc. Lond.* **238**, 229.
47. Grant I P, Mayers D F and Pyper N C 1976 *J. Phys. B* **9**, 2777.
48. Bar-Shalom A and Klapisch M 1988 *Comput. Phys. Commun.* **50**, 375.
49. Wilson S ed 1988 *Methods in Computational Chemistry* Vol 2: Relativistic effects in atoms and molecules (New York: Plenum).
50. Drake G W F ed 2005 *Springer Handbook of Atomic, Molecular and Optical Physics* Second edn (New York: Springer-Verlag).
51. Fischer C F 1986 *Computer Physics Reports* **3**, no. 5, 273.
52. Fischer C F, Guo W and Shen Z 1992 *Int. J. Quant. Chem.* **42**, 849.
53. Qiu Y and Fischer C F 1999 *J. Comput. Phys.* **156**, 257.
54. Froese Fischer C and Idrees M 1990 *J. Phys. B: At. Mol. Phys.* **23**, 679.
55. Fischer C F 1991 *Int. J. Supercom. Appl.* **5**, 5.
56. Bar-Shalom A, Klapisch M and Oreg J 1988 *Phys. Rev. A* **38**, 1773.

7

Computation of atomic structures

7.1 Atomic structure calculations with GRASP

GRASP [1] is a system for the calculation of relativistic atomic structure and properties. Table 7.1 lists the main modules with a brief description of their functions[1]. The software implements the finite difference numerical methods of Chapter 6. Having defined a basis of CSFs, $\{T\}$, the user invokes the MCP module to compute the angular momentum coefficients $t_{TT'}(\alpha\beta), v^{K}_{TT'}(\alpha\beta\gamma\delta)$ of (6.10.2). The MCDF code generates Dirac radial spinors, either with the user's choice of parametrized model potential or by solving the coupled DHF radial equations. It also generates the corresponding interaction integrals $I(\alpha\beta)$ and $R^K_C(\alpha\beta\gamma\delta)$ over the radial orbitals, and then assembles the Hamiltonian matrix $\boldsymbol{H}$ in the CSF basis using (6.10.2) and (6.10.3). The atomic state functions (ASF) are the eigenvectors of $\boldsymbol{H}$, and its eigenvalues represent the atomic energy levels. If the transverse photon interaction, self-energy, and vacuum polarization corrections are to be calculated, then MCBP must be called to compute the coefficients $v^{k\tau}_{TT'}(ABCD)$ of (6.10.3), after which BENA calculates the radial integrals $S^{k\tau}(ABCD)$ and QED corrections and assembles the perturbed Hamiltonian $\boldsymbol{H}$ in the ASF basis of the DC Hamiltonian before rediagonalizing. This corrects the total energy of the atom and the CSF mixing coefficients, but leaves the orbitals unperturbed. The outputs can be applied to bound state properties, radiative transition amplitudes, or target wavefunctions for scattering calculations as described in the following chapters. The structure of the original GRASP package reflects the limited memory and file storage of computers some 30 years ago. Free-standing modules pass data from one to another by way of formatted disc files (originally magnetic tapes). They are invoked noninteractively in the sequence in which data is presented by the input file. This arrangement survives in the most

[1] There are several versions of the code in circulation, mostly referenced in [1]. GRASP2, which had a limited circulation, is a precursor of GRASP92 [2]. Downloads are accompanied by documentation files.

Table 7.1. Main modules of GRASP

GRASP	DATAIN	Processes input batch file
	MCP	Constructs coefficients $v^K_{TT'}(abcd)$ (and $t_{TT'}(ab)$) of (6.10.2)
	MCBP	Constructs coefficients $v^{Kt}_{TT'}(abcd)$ for relativistic corrections to Coulomb interaction
	MCT	Constructs coefficients $d^L_{TT'}(ab)$ for radiative multipole transitions
	MCDF	Numerical integration of MCDHF equations, diagonalization of Hamiltonian in CSF basis, analysis and properties of wavefunctions
	BENA	Evaluates corrections to the Hamiltonian due to transverse photon (or Breit) interaction, vacuum polarization and electron self-energy, rediagonalizes Hamiltonian, analyses wavefunctions
	OSCL	Evaluates radiative transition rates

recent published version, GRASP92 [2], which can handle problems requiring of order 10^5 CSFs in multi-processor environments. For simplicity, the older structure is used to explain what needs to be done to carry out relativistic atomic structure calculations.

7.2 GRASP modules

We shall use calculations on the beryllium-like ion, Fe XXIII, to illustrate the use of GRASP as the low-lying states of this highly ionized atom can be approximated with fair accuracy using a small CSF basis, and the atomic number is large enough for relativistic effects to be of reasonable size. This ion has been observed, for example, in tokamak plasmas [3] and in an electron beam ion trap (EBIT) [4], and we shall compare outputs with experimental data where appropriate. Table 7.2 shows a batch input file for an MCDHF-EAL calculation[2] for the even parity states of Fe XXIII belonging to the $1s^2\{2s^2 + 2p^2\}$ manifold. The lines in Table 7.2 have been numbered only for the purposes of this discussion and are not part of the input file. The commands `ANG, MCP, MCT, MCBP, MCDF`, and so on, activate each program

[2] Different versions of GRASP use different conventions for the presentation of input data. In this chapter, we have used GRASP0 (part of Patrick Norrington's DARC98 package [5]) whose conventions are very similar to GRASP [1].

Table 7.2. A simple GRASP input file

```
1  Fe XXIII, even ! User's description of run
2 2 3 2 55.8      ! NMAN, NWM, IOP, ATW.
3  1s             ! Nonrelativistic
4  2s  2 0        ! subshell
5  2p  0 2        ! occupations.
6 ANG 1           ! Generate all possible CSFs
7 -1
8  MCP            ! Generate all angular coefficients
9  MCBP           ! Generate all angular coefficients
10  MCT 1         ! Dipole transition coefficients
11  MCDF          ! Generate wavefunctions,etc
12                !
13 26             ! Atomic number
14 EAL            ! Extended average energy functional
15 BENA           ! QED and relativistic corrections
16                !
17 OSCL           ! Radiative transitions
18                !
19 STOP
```

module in turn for specific parts of the calculation. Users may wish to carry out calculations step by step, in which case the unwanted modules can be dropped from the list. For example, if the user wants only the MCDHF wavefunctions and energy levels, then he or she can omit lines 15 to 18. Similarly, the user may wish to solve the radial equations in an MCDHF-OL fashion for individual levels using the same CSF data; in this case, the angular coefficients will already be available in a file and there will be no need to invoke `MCP` or `MCBP` to recalculate them.

Individual data items on a line are separated by blanks. Comments following the ! signs are not part of the data. `Typewriter` font indicates names used in the program. Thus we have

- Line 1 A title which identifies the calculation (*compulsory*).
- Line 2 The number of (nonrelativistic) configurations (`NMAN` = 2), the number of (nonrelativistic) orbitals (`NWM` = 3) are *compulsory*; data that follow include the data input mode (`IOP` = 2) to generate prints of the LSJ CSF and the $LS - jj$ transformation coefficients) and the atomic weight (`ATW` = 55.8).
- Lines 3–5 These `NWM` lines list the numbers of electrons in each of the `NMAN` configurations in order. The blanks in the $1s$ line indicate that it is a *core* shell that is filled (2 electrons) in all configurations.

- Line 6 `ANG` calls the angular coefficient modules. The coefficients are written to a binary file, and the user controls the amount of data to be output as text. Here the figure `1` sends output to a text file.
- Line 7 specifies what overall symmetries are to be used in calculation of angular coefficients by the `MCP`, `MCBP`, and `MCT` modules. If there is one non-negative number, then only CSFs with this value of J are considered; the tabular entry, `-1`, here indicates *all possible* J-values must be considered; an entry `-2` indicates that the following `NMAN` lines contain the limits on J for each nonrelativistic configuration. If there are 3 numbers, the first being negative, then the other two give limits on J for all CSF. If there are precisely `NMAN` numbers, these list the J-values of the CSF in order.
- Line 8 `MCP` invokes the module to generate angular coefficients $v^K_{TT'}(\alpha\beta\gamma\delta)$ (6.10.2) for the Coulomb interaction between electrons and one-electron coefficients $t_{TT'}(\alpha\beta)$.
- Line 9 `MCBP` invokes the module to generate the corresponding coefficients for the transverse photon (Breit) interaction: $v^{K\tau}_{TT'}(\alpha\beta\gamma\delta)$, (6.10.3).
- Line 10 `MCT KA` invokes the module to generate angular coefficients $d^k_{TT'}(\alpha\beta)$ for radiative transition matrix elements of tensor rank `KA`$= k$; this may be followed by an optional parameter, `IOPAR`, setting the parity: +1 for even parity (M1.E2. ...), -1 for odd parity (E1, M2, ...); coefficients for both parities are calculated if the parity is not specified.
- Line 11 `MCDF` invokes the module to solve the MCDHF integro-differential equations using finite difference methods, constructs and solves the eigenvalue problem for the $\boldsymbol{H}$ matrix, and controls the iterative process. The user can set a large number of options.
- Line 12 The blank line indicates that default settings should be used. The user can also supply data to instruct `MCDF` to use archived wavefunctions for input and to generate certain outputs.
- Line 13 Only the atomic number, `Z`=26, is compulsory. The user may also set the radial step size, `H` (default 0.05)and first grid point, `RNT` (default 10^{-5}), the precision (`ACCY`) for convergence of radial functions and mixing coefficients (default 10^{-8}), and two other parameters.
- Line 14 The commands `AL, EAL, OL, EOL` indicate which SCF mode is to be used.
- Line 15 `BENA` invokes the module to generate first order perturbation corrections to the Dirac-Coulomb Hamiltonian matrix for the transverse photon corrections to the Coulomb interaction, vacuum polarization and electron self-energy (Lamb shift). These are converted to the Dirac-Coulomb ASF basis before rediagonalizing the Hamiltonian. Many options are available.
- Line 16 The Breit interaction (the long wavelength limit of the full transverse photon interaction) can be used by setting the optional parameter `WFACT` (default 1.0) to, say, 10^{-3}. (This replaces each energy difference ω by $\omega\times$ `WFACT`.)

- Line 17 `OSCL` invokes the module to calculate radiative transition rates in both Babushkin ("relativistic length") and Coulomb ("relativistic velocity") gauges. There are a large number of options.
- Line 18 This lists initial and final levels between which transitions are to be calculated. All possible levels are considered if the line is empty.
- Line 19 `STOP` terminates data input for this calculation.

Table 7.3. Equivalent jj-coupled CSFs

```
  1s                ! Relativistic
  2s   2 0 0 0 0 0 ! subshell
  2p-  0 0 1 1 0 2 ! occupation
  2p   0 2 1 1 2 0 ! numbers.
ANG 1
       0 2 2 1 0 0 ! J-values
```

Table 7.4. jj- and $LSJ-$ coupled CSFs

jj-coupling	LSJ-coupling
1. $[1s^2_{1/2}2s^2_{1/2}]J = 0$	$^1S_0$
2. $[1s^2_{1/2}(2p_{3/2})^2]J = 2$	$^3P_2$
3. $[1s^2_{1/2}2p_{1/2}2p_{3/2}]J = 2$	$^1D_2$
4. $[1s^2_{1/2}2p_{1/2}2p_{3/2}]J = 1$	$^3P_1$
5. $[1s^2_{1/2}2p^2_{3/2}]J = 0$	$^3P_0$
6. $[1s^2_{1/2}2p^2_{1/2}]J = 0$	$^1S_0$

`DATAIN` first generates the jj-coupling CSFs used internally; for the even-parity states of Fe XXIII, the result is displayed in Table 7.3. In this case, the core $1s^2$ shell has angular momentum $J_1 = 0$ in all CSF, so that the overall angular momentum J depends only on the coupling of the angular momenta J_2, J_3 and J_4 of the relativistic open shell CSFs. Table 7.4 shows the same information in conventional spectroscopic notation.

The increased size of the CSF set in jj-coupling, even in this trivial example, is the most striking feature of Table 7.4. The number of CSFs increases, because of the greater number of jj-coupling subshell states and the various angular momentum coupling schemes that they permit. This number explodes when there are several open $d, f, \ldots$ shells such as in transition metals, lanthanides and actinides. For example, the ground state of neutral uranium is classified as [Rn] $5f^3(^4I^o_{9/2})6d7s^2$ $(9/2, 3/2)^o_6$ [6, Table 10.3] where [Rn] denotes the configuration: $1s^22s^22p^63s^23p^63d^{10}4s^24p^64d^{10}5s^25p^64f^{14}5d^{10}6s^26p^6$

of the Radon core. The representation of the ground state, $J = 6$, requires 44 jj-coupled CSF belonging to the $5f^3 6d 7s^2$ odd-parity manifold alone. Taking electron correlation into account involves adding manifolds such as $5f^3 6d^2 7s$ which generate even bigger CSF sets. Such calculations continue to pose major challenges for *ab initio* investigation. The structure of legacy codes such as GRASP was not designed to handle the huge amounts of data needed for a physically satisfactory model; GRASP92 aims to overcome this defect.

7.3 MCDHF integro-differential equations

The MCDHF integro-differential equations have much the same form as the DHF integro-differential equations of §6.7, but the multiconfiguration structure makes the coefficients more complicated. We first write the elements, H_{rs}, of the MCDHF Hamiltonian in a notation similar to that of §6.7. Using r, s (rather than T, T') for the CSF indices, the diagonal matrix elements are

$$H_{rr} = \sum_{A=1}^{nw} \Big\{ q_r(A)\, I(AA) \tag{7.3.1}$$

$$+ \sum_{B \geq A}^{nw} \left[\sum_{k=0,2,\ldots}^{k_0} f_r^k(AB)\, F^k(AB) + \sum_{k=k_1,k_1+2,\ldots}^{k_2} g_r^k(AB)\, G^k(AB) \right] \Big\},$$

where $I(AB), F^k(AB)$ and $G^k(AB)$ are the usual radial Slater integrals defined in (6.3.9), (6.6.8), and (6.6.9)[3] and $q_r(A)$ is the number of electrons in subshell A of CSF r. The angular coefficients of (6.6.15) and (6.6.16) are

$$f_r^0(AA) = \frac{1}{2} q_r(A)(q_r(A) - 1), \quad f_r^0(AB) = q_r(A) q_r(B), \ A \neq B. \tag{7.3.2}$$

and, for $k > 0$ and $q_r(A) = 2j_A + 1$ or $q_r(B) = 2j_B + 1$,

$$g_r^k(AA) = -\frac{1}{4}[q_r(A)]^2 \Gamma_{j_A k j_A}, \quad g_r^k(AB) = -\frac{1}{2} q_r(A) q_r(B) \Gamma_{j_A k j_B}, \tag{7.3.3}$$

whilst if neither subshell is fully occupied, $q_r(A) < 2j_A+1$ and $q_r(B) < 2j_B+1$, we need the coefficients defined in (6.10.2)

$$f_r^k(AB) = v_{rr}^k(ABAB), \quad g_r^k(AB) = v_{rr}^k(ABBA) \tag{7.3.4}$$

The summation limits in (7.3.1) are $k_0 = (2j_A - 1)\delta_{AB}$, $k_1 = |j_A - j_B|$ if $\kappa_A \kappa_B > 0$ or $|j_A - j_B| + 1$ if $\kappa_A \kappa_B < 0$ and $k_2 = j_A + j_B$ if $j_A + j_B + k$ is even and $j_A + j_B - 1$ otherwise.

Matrix elements that are not diagonal in CSF indices, $r \neq s$, are defined as in (6.10.3):

---

[3] We drop the subscripts C, which are not needed here.

$$H_{rs} = \sum_{AB} t_{rs}(AB)\, I(AB)\delta_{\kappa_A\kappa_B} + \sum_k \sum_{ABCD} v^k_{rs}(ABCD) R^k(ABCD). \quad (7.3.5)$$

The one-electron terms are relatively unimportant as they appear only if the two configurations r, s differ only by the excitation of a single electron from a subshell $n_A\kappa_A$ in one CSF to a subshell $n_B\kappa_B$ with $\kappa_A = \kappa_B$ in another. They were ignored in early versions of GRASP.

As in the DHF case, the MCDHF equations can be derived by making the energy of the atom stationary subject to orthogonality constraints on the orbitals and the CSF mixing coefficient vectors. Lagrange's method of undetermined multipliers leads to the functional

$$J\,[u_1, \dots u_{NW}, \boldsymbol{c}] = \sum_{r=1}^{NCF} \sum_{s=1}^{NCF} d_{rs} H_{rs} + \sum_A \sum_B (1 - \delta_{AB})\lambda_{AB}\, (A\,|B) \quad (7.3.6)$$

where the integer NCF is the dimension (in the notation used by GRASP) of the CSF set, NW is the number of orbitals, and u_A denotes the radial spinor for subshell A. The coefficients d_{rs} are generalized weights given by

$$d_{rs} = \mathrm{d} \sum_{i=1}^{n_L} (2J_i + 1)\, c_{ri}\, c_{si} / \mathrm{d} \sum_{i=1}^{n_L} (2J_i + 1), \quad (7.3.7)$$

where the (real) numbers c_{ri}, the *mixing coefficients*, are the components of the column vector $\boldsymbol{c}_r$ expressing the CSF composition of the ASF $\varGamma_r$, $(A\,|B)$ is the orthogonality integral (6.7.1), and $\lambda_{AB} = \lambda_{BA}$ are the Lagrange multipliers, which contribute only when $\kappa_A = \kappa_B$. The sum over i runs over a suitable set of n_L eigenvalues of $\boldsymbol{H}$. The ASFs are normalized so that

$$\boldsymbol{c}_r^\dagger \boldsymbol{c}_s = \delta_{rs}.$$

The diagonal energy parameter ϵ_{AA} is given by a generalization of (6.7.10) whilst

$$\lambda_{AB} = \overline{q}(A)\, \epsilon_{AB}, \quad \overline{q}(A) = \sum_{r=1}^{n_C} d_{rr} q_r(A), \quad (7.3.8)$$

where $q_r(A)$ is the number of electrons in subshell A in CSF r, so that we call $\overline{q}(A)$ a *generalized occupation number*. Because $\boldsymbol{H}$ is real symmetric in this type of calculation, its i-th eigenvalue is

$$E_i = \sum_{rs} c_{ri}\, c_{si}\, H_{rs},$$

so that we may interpret the first term on the right-hand side of (7.3.6) as a weighted sum over n_L eigenvalues:

$$\sum_{r=1}^{NCF} \sum_{s=1}^{NCF} d_{rs} H_{rs} = \frac{\sum_{i=1}^{n_L} (2J_i + 1)\, E_i}{\sum_{i=1}^{n_L} (2J_i + 1)}$$

The usual variational argument leads to the MCDHF integro-differential equations

$$-\frac{Z_{nuc}(r) - Y(A;r)}{r} P_A(r) + c\left(-\frac{d}{dr} + \frac{\kappa_A}{r}\right) Q_A(r) \\ -\epsilon_{AA}\, P_A(r) = -X_{+1}(A;r), \tag{7.3.9}$$

$$c\left(\frac{d}{dr} + \frac{\kappa_A}{r}\right) P_A(r) + \left(-2mc^2 - \frac{Z_{nuc}(r) - Y(A;r)}{r}\right) Q_A(r) \\ -\epsilon_{AA}\, Q_A(r) = -X_{-1}(A;r).$$

with new definitions of $Y(A;r)$ and $X_\beta(A;r)$ which reflect the multiconfigurational character of the model. The direct potential is

$$Y(A;r) = \sum_k \sum_{B=1}^{NW} \left\{ y^k(AB)\, Y^k(BB;r) \right. \\ \left. - \sum_{D=1}^{NW} y^k(ABAD) Y^k(BD;r) \right\}, \tag{7.3.10}$$

$$y^k(AB) = \frac{1+\delta_{AB}}{\overline{q}(A)} \sum_{r=1}^{NCF} d_{rr} f_r^k(AB),$$

$$y^k(ABAD) = \frac{1}{\overline{q}(A)} \sum_{r=1}^{NCF} \sum_{s=1}^{NCF} d_{rs} v_{rs}^k(ABAD)$$

whilst the right-hand sides are

$$X_\beta(A;r) = -\sum_{B\neq A} \delta_{\kappa_A \kappa_B}\, \epsilon_{AB}\, R_{\gamma_B,\beta\kappa_B}(r) \\ + \sum_k \left\{ \sum_{B\neq A} x^k(AB) \frac{Y^k(BA:r)}{r} R_{\gamma_B,\beta\kappa_B}(r) \right. \\ \left. - \sum_{BCD} (1-\delta_{AC}) x^k(ABCD) \frac{Y^k(BD;r)}{r} R_{\gamma_C,\beta\kappa_C}(r) \right\} \tag{7.3.11}$$

where $R_{\gamma_A,\beta\kappa_A}(r)$, $\beta = \pm 1$, has its usual meaning: (6.2.2), and

$$x^k(AB) = \frac{1}{\overline{q}(A)} \sum_{r=1}^{NCF} d_{rr} g_r^k(AB),$$

$$x^k(ABCD) = \frac{1}{\overline{q}(A)} \sum_{r=1}^{NCF} \sum_{s=1}^{n_C} d_{rs} v_{rs}^k(ABCD)$$

We can now give a more precise meaning to the different variants of MCDHF calculation [7]:

OL mode: The weights are chosen to solve for a single ASF I, so that $d_{rs} = c_{rI} c_{sI}$ equivalent to setting $n_L = 1$ in (7.3.7).

Each ASF must be determined from one such calculation. The orbitals which emerge from these calculations may differ significantly from one ASF to another.

EOL mode: The weights are given by (7.3.7) with $n_L > 1$.

The same set of orbitals is used to represent a selection of ASF.

AL mode: Here the weights are independent of the mixing coefficients and are given by $d_{rs} = \delta_{rs}(2J_r + 1)/\sum_{t=1}^{NCF}(2J_t + 1)$ This weights each CSF according to its J value. This choice greatly simplifies the iterative scheme by decoupling the diagonalization of the secular matrix from the solution of the radial equations. It also ensures that the limit $c \to \infty$ corresponds to a proper nonrelativistic problem. This is not always true for other MCDHF variants, which makes the comparison of relativistic and nonrelativistic calculations difficult.

EAL mode: The weights are independent of the mixing coefficients but are chosen by the user.

The AL, EAL, and EOL schemes average the potentials over several J values and the ASF are constructed from a common set of potentials. This is particularly useful for studying the highly ionized atomic species that are so common in laboratory and astrophysical plasmas. In any event, the secular matrix can be partitioned into a direct sum of secular matrices, one for each symmetry; each J matrix can then be diagonalized independently. This makes the AL, EAL, and EOL models cheaper than OL calculations because all the CSFs are constructed together from a common orbital set and only a single diagonalization is required for each symmetry. A common set of orbitals simplifies the calculation of radiative transition probabilities, but the results are not always accurate. An algorithm for calculation of transition probabilities taking account of orbital relaxation is described in §8.4 and §8.5 has recently been implemented in the latest version of grasp92.

7.4 Solving the integro-differential equations

The MCDHF system (7.3.9) can be written

$$c\mathbb{J}\,\frac{dU}{dt} + (\mathbb{W}[U] - \mathbb{E}\,r\,)\,U(t) = 0, \quad U(0) = U_0, \quad \lim_{t\to\infty} U(t) \to 0 \qquad (7.4.1)$$

in the manner of (5.11.1), where $U(t)$ is a `2*NW`-dimensional vector

$$U(t) = \begin{pmatrix} u_1(t) \\ \vdots \\ u_{NW}(t) \end{pmatrix}, \qquad (7.4.2)$$

indexed by the `NW` subshell labels. The `2*NW` matrix $\mathbb{E}$ has matrix elements ϵ_{AB}, and

$$\mathbf{\Lambda} = \mathbb{Q}\mathbb{E}, \quad \mathbb{Q} = \text{diag}\,\{\overline{q}_1\, I_2, \ldots, \overline{q}_{NW}\, I_2\}, \quad \mathbb{J} = \text{ diag }(J, \ldots, J) \qquad (7.4.3)$$

where I_2 denotes the 2×2 unit matrix, and J was defined in (5.3.3). The matrix $\mathbb{W}[U]$ has diagonal blocks (7.3.10)

$$W_{AA}[U] = \begin{pmatrix} -Z_{nuc}(r) + Y(A;r) & -c\kappa_A \\ -c\kappa_A & 2c^2 r - Z_{nuc}(r) + Y(A;r) \end{pmatrix}, \qquad (7.4.4)$$

and off-diagonal blocks (7.3.11)

$$W_{AB}[U] = -\sum_k \Big\{ x^k(AB)\, Y^k(AB;r) \\ -(1-\delta_{AB}) \sum_{CD} x^k(ACBD)\, Y^k(CD;r) \Big\}\, I_2 \qquad (7.4.5)$$

These equations must be solved together with the secular equations

$$[\mathbf{H} - E_\Gamma]\mathbf{c}_\Gamma = 0, \qquad (7.4.6)$$

subject to the usual orbital orthogonality constraints. In OL and EOL calculations, the CSF weights, d_{rs} defined by (7.3.7), depend on the mixing coefficient vectors $\mathbf{c}_i$. The d_{rs} are fixed in AL and EAL calculations, which decouples the orbital equations (7.4.1) from the secular equation (7.4.6).

There are many conceivable schemes for iterative solution of this nonlinear system. In practice, all of those implemented so far use a *single orbital improvement* scheme, in which we separate (7.4.1) into equations of the form (5.11.1)

$$\left\{ cJ\frac{d}{dt} + f(t)\,[W_{AA}[U] - \epsilon_{AA}] \right\} u_A(t) \\ = f(t) \sum_B \{\epsilon_{AB}\delta_{\kappa_A,\kappa_B} - W_{AB}[U]\}\, u_B(t) \qquad (7.4.7)$$

After ν iterations, we update orbital A using

$$\left\{ cJ\frac{d}{dt} + f(t)\left[W^{[\nu]}_{AA}[U] - \epsilon^{[\nu]}_{AA}\right] \right\} u^{[\nu+1]}_A(t) \\ = f(t) \sum_{B\neq A} \left\{\epsilon^{[\nu]}_{AB}\delta_{\kappa_A,\kappa_B} - W^{[\nu]}_{AB}[U]\right\} u^{[\nu]}_B(t). \qquad (7.4.8)$$

The algorithms used in GRASP to obtain the trial solution $u_A^{[\nu+1]}(t)$ have been described in §5.11 and Appendix B.6. The SOLV routine calls OUT to generate solutions for the inner region by marching outwards from r_0, near the origin, and then IN marches inwards from the outer boundary. The parameters ϵ and the normalizing coefficient A are adjusted using the Hartree perturbation equations (5.11.15) and (5.11.16) in order to obtain a *normalized* $u_A^{[\nu+1]}(t)$ with the correct number of zeros. This adjustment process may fail for several different reasons: the solution may not be normalizable, it may have the wrong sign, or the corrections may be so large that subsequent iterations diverge. SOLV includes strategies to encourage convergence when this happens..

7.5 Starting the calculation

The initial stages of the calculation for the model of Table 7.2 are clear from the extract of output in Table 7.5. The atomic number $Z = 26$ is specifed on line 13 of the input file; all other parameters C, RNT, H, ACCY, and CUTOFF have been given default values which can be modified by the user.

7.5.1 The radial grid

We introduced the continuous and differentiable mapping

$$r = f(t),\ k(t) = f'(t)/f(t). \tag{7.5.1}$$

in §5.11 in an attempt to get a smoother discrete representation of orbital wavefunctions on a radial grid using a uniform partition of the interval $[t_0, t_N]$

$$t_n = t_0 + n\mathrm{H},\ n = 0, 1, \ldots, N. \tag{7.5.2}$$

The most common mappings are

A. $r = f(t) = r_0\, e^t$, $k(t) = 1$.
B. $r = f(t) = r_0 + At^4$, where A is a positive constant such that $R = r_0 + At_N^4$.
C. r is defined implicitly by

$$ln(1 + r/r_0) + r/h_c = t,\ k(t) = \frac{r + r_0}{r}\frac{h_c}{r + r_0 + h_c}$$

.

Grid A is the default in GRASP as in nonrelativistic calculations [8, §3.11.1]. The radial step grows exponentially, roughly in line with the spacing of successive zeros of the wavefunction, so that each loop can be defined at approximately the same number of radial grid points. The default values of RNT$= r_0$ and H can be overridden by the user. Grid B has similar properties, but the interval between successive r values increases less rapidly. Grid C is exponential near $r = r_0$, since $k(0) \sim 2$; for $r_0 \ll r \ll h_c$, $k(t) \sim (1 + r/h_c)^{-1}$ so that intervals increase more slowly as t increases; finally when $r \gg h_c$ then $r \sim h_c t$ so that the grid becomes linear at large r. It is therefore useful in calculations that need to take weakly bound or continuum states into account.

Table 7.5. Starting the MCDHF calculation

```
 ***********************************************************
 routine INIT : start of MCDHF calculation
 ***********************************************************
 Title : Fe XXIII

 Run at : 23-Jun-03 12:12:58
 ***********************************************************

 atomic units are used except where indicated
 the following conversion factors are set in routine DATAIN
 using atomic weight = 55.80000
 1 a.u. = 2.1947247301E+05 cm-1
 1 a.u. = 2.7211128392E+01 eV
Z      (atomic number)                     = 26.00
C      (speed of light)                    = 1.3703598950E+02
RNT    (first grid point)                  = 1.000E-05
H      (grid step-size)                    = 5.000E-02
ACCY   (accuracy)                          = 1.000E-08
CUTOFF (wave-function cut-off parameter) = 1.000E-10
N      (maximum grid point at 200.0)       = 337

extended average level (EAL) calculation

orbital N  K 2J L

1 1S    1 -1  1 0
2 2S    2 -1  1 0
3 2P-   2  1  1 1
4 2P    2 -2  3 1
>>>> routine NUCPOT called : evaluate nuclear potential

point nucleus case


>>>> routine TFPOT called to evaluate the Thomas-Fermi
>>>> potential for degree of ionization = 22

*** Thomas-Fermi estimate *** 1S E = 3.35162E+02 241 points
*** Thomas-Fermi estimate *** 2S E = 8.10642E+01 257 points
*** Thomas-Fermi estimate *** 2P- E = 8.08238E+01 257 points
*** Thomas-Fermi estimate *** 2P E = 8.00979E+01 257 points

number of w.f. points = 1012 (max= 10000)
number of grid points = 257 (max= 400)
first grid point = 1.00000E-05
last grid point = 3.62217E+00
```

7.5.2 The nuclear mass

The finite mass and size of the nucleus must both be included in the model to obtain quantitative agreement with spectroscopic data. It is straightforward to separate the centre of mass motion of the atom as a whole in nonrelativistic quantum mechanics [8, §8.2]. The main effect on the internal motion is to replace the mass of the electron by a reduced mass which effectively replaces the Rydberg constant for infinite mass, R_∞, by

$$R_M = \frac{M}{M+m} R_\infty, \tag{7.5.3}$$

where m is the electron mass and M the nuclear mass. The energy conversion factors for the atomic unit reported at the beginning of Table 7.5 make this correction. This is often known as the *normal mass effect.* Additional terms in the Hamiltonian are due to the (nonrelativistic) coupling of the electron and nuclear motions:

$$H_{SMS} = \frac{1}{M} \sum_{i<j} \boldsymbol{p}_i \cdot \boldsymbol{p}_j \tag{7.5.4}$$

where the sum runs over all distinct pairs of electron indices. These give small corrections to atomic energy levels described as the *specific mass shift.* So far, nobody has succeeded in devising a rigorous and tractable relativistic scheme to take nuclear motion into account in atomic structure calculations. The nonrelativistic method of calculation described in [8, §8.2.1] has been implemented in GRASP92 [2].

7.5.3 The nuclear size

Real nuclei are not point charges, and the finite size of the nuclear charge distribution mu st be taken into account in relativistic calculations. Section 5.4.1 revealed the sensitivity of the relativistic wavefunctions near the nucleus to the shape of the nuclear charge density. The amplitudes of both components are proportional to r^γ for a point nucleus, whereas for a finite size nucleus the large component is proportional to r^{l+1} as $r \to 0$, whilst the small component is proportional to r^l when $\kappa > 0$ or r^{l+2} when $\kappa < 0$. The functions are therefore bounded at the nucleus and have finite derivatives, whereas for a point nucleus, the radial density of an s- or $p_{1/2}$-orbital, for which $0 < \gamma < 1$, has infinite slope at the origin, although the radial density is still normalizable. This major difference in behaviour makes it necessary at the outset of a calculation with GRASP to specify a model of the nuclear charge distribution, usually one of the simple models of §5.4 although more elaborate models are sometimes used. The critical parameter has been the RMS radius of the nuclear charge density, for which experimentally determined values should be used when available [9]. Many nuclei have nonspherical charge distributions, and the effect on the RMS radius should be taken into account for $Z > 90$ [10, 11].

The finite size of the nucleus makes an important contribution to X-ray transition energies.

The nuclear potential $-Z_{nuc}(r)/r$ is generated by the NUCPOT routine. The unscreened Coulomb potential is the default: $Z_{nuc}(r) = Z$; in our example $Z = 26$. Otherwise, an input line starting VNUC, inserted between lines 13 and 14 of Table 7.2 , permits the user to specify either the potential of a uniform distribution of nuclear charge (NUCTYP=1), (5.4.3), or a Fermi distribution (NUCTYP=2), both described in §5.4. The program recognizes a user supplied model if NUCTYP$>$ 2.

7.5.4 Initial estimates for radial wavefunctions

Initial estimates for SCF calculations can be obtained in several ways. The default model in GRASP0 is a Thomas-Fermi (TF) potential [12, 13], [14, §2^{1}4]. Thomas-Fermi theory applies (nonrelativistic) Fermi-Dirac statistics to the electron density distribution giving a universal local potential that approximates $Y(A;r)$. Let $z = Z + 1 - Q$, where $Q = \sum_A \overline{q}(A)$, be the residual charge on the atom after removing one electron to infinity. Then

$$Y(A;r) \approx Y^{TF}(r) = z - (z - Z_{nuc}(r))\,[\,f(x)\,]^2 \tag{7.5.5}$$

where $x = [\,(Q-1)^{1/3} r\,/\,0.8853\,]^{1/2}$ and

$$f(x) = \frac{0.60112x^2 + 1.81061x + 1}{0.04793x^5 + 0.21465x^4 + 0.77112x^3 + 1.39515x^2 + 1.81061x + 1}$$

The Dirac eigenvalues for a TF potential with $Z = 26$ and $Q = 4$ for Fe XXIII are shown in Table 7.5. In this example, they are more tightly bound than the converged MCDHF orbitals.

GRASP0 provides an alternative method of generating initial estimates in terms of Dirac Coulomb orbitals with an effective charge $Z_{n\kappa} = Z - \sigma_{n\kappa}$, where $\sigma_{n\kappa}$ is a suitably chosen screening parameter. Klapisch [15, 16] devised a simple model potential for use in relativistic CI calculations for highly ionized atoms. The nonrelativistic SUPERSTRUCTURE [17] code used scaled TF potentials in much the same way.

The orbital properties appear at the bottom of Table 7.6. E is the orbital eigenvalue in atomic units. SC is a *screening number*, $\sigma_{n\kappa} = Z - Z_{n\kappa}$ The effective nuclear charge $Z_{n\kappa}$ has been chosen such that the computed mean radius, $\langle\, r\,\rangle_{n\kappa}$, agrees with the Dirac hydrogenic expression[4], Table 3.3.3:

$$(3N_{n\kappa}^2 - \kappa^2)\sqrt{1 - Z_{n\kappa}^2/N_{n\kappa}^2 c^2} - \kappa.$$

The number of radial points needed to represent the orbital and the mean values of powers of the radius, `1/r, r` and `r*r`, together with the Lagrange multiplier $\epsilon_{1s,2s}$ and the orthogonality integral N(1s,2s) also are printed out.

[4] Note that $N_{n\kappa}$, (3.3.21), is itself a function of $Z_{n\kappa}$.

Table 7.6. EAL calculation for Fe XXIII $1s^2\{2s^2+2p^2\}$

```
*******************************************************
generalised orbital occupation numbers
2.000000E+00 1.250000E-01 6.250000E-01 1.250000E+00
CSF mixing coefficients
2.500000E-01 5.590170E-01 5.590170E-01 4.330127E-01
2.500000E-01 2.500000E-01
>>>> MCDF dump written
-------------------------------------------------------
>>>> routine SCF called
          ,,
          ,,
          ,,
  orbital properties
  ------------------
            E (a.u.)          SC
  1  1S  3.134905800718E+02 4.5307700E-01 249
  2  2S  7.147617671804E+01 1.5088053E+00 257
  3  2P- 6.919112481927E+01 2.1006627E+00 257
  4  2P  6.860433203665E+01 2.1228120E+00 257
           1/r                r             r*r
  1  1S  2.6099750E+01      5.8029267E-02 4.5174252E-03
  2  2S  6.1959696E+00      2.4211607E-01 6.8573220E-02
  3  2P- 6.1231129E+00      2.0641104E-01 5.1445883E-02
  4  2P  6.0210226E+00      2.0876794E-01 5.2510505E-02
  Lagrange multipliers
  1S 2S  5.9478353E-02
  orthogonality integrals
  1S 2S  5.3244318E-12
```

7.6 An EAL calculation

The `CSF mixing coefficients`, d_{rr}, are fixed in advance in AL and EAL calculations, which simplifies the SCF iterative calculation. The first line of the output text file, Table 7.6, gives `generalized occupation numbers`: $\overline{q}(A)$, and the second line shows the d_{rr}. The message `MCDF dump written` signals the output of a workspace image to a binary file; this is updated regularly as the calculation proceeds to facilitate restarts and to minimize the loss of information if the calculation halts prematurely. A call to `SCF` solves the orbital MCDHF equations (7.4.8); the first, low accuracy, sequence cycles through the orbital list from the most tightly to the least tightly bound, after which the next orbital to be improved is the one showing the biggest fractional change at the last iteration. Subsequent improvement cycles increase the precision until the orbital changes are below the final threshold.

We can solve the secular equations once the system has converged. Table 7.7 summarizes the predicted energy levels in all the common units: a.u., Rydbergs, cm.$^{-1}$, eV. The table is self-explanatory. One eigenstate, 3, is pure $J = 1^e$ because CSF 4 does not interact with others in the 6-dimensional basis. Levels 1,2 and 6 are linear combinations of 3 CSF with $J = 0^e$: the table gives the dominant CSF in each case, i say, and the absolute value of its coefficient c_{ri}. Next follows the predicted excitation energies relative to the ground state that are perhaps more informative. Before the matrix is diagonalized, the average energy is subtracted to improve the conditioning of the matrix and hence the accuracy of the eigenvalues and eigenvectors. Once the eigenvectors have been calculated the `NEWBAS` utility expresses them in the LSJ basis and the level scheme is printed out in this basis.[5]

7.7 Diagonal and off-diagonal energy parameters

The diagonal energy parameter in the DHF case is given by (6.7.10), and comparison with (6.6.22) shows that

$$\epsilon_{AA} = e^0_A, \tag{7.7.1}$$

which we have identified as the contribution of a single electron to the total energy of the atom. The inference that ϵ_{AA} can be identified as an estimate of the energy needed to remove an electron from subshell A without disturbing the remaining electrons is usually known as Koopmans' theorem [18]. Koopmans' theorem holds also for MCDHF models although the expression for ϵ_{AA} is now more complicated and the physical interpretation is quite different.

The general expression for the diagonal parameter can be written down by multiplying equation (7.3.9) from the left by $u^\dagger_A$ and integrating over r:

$$\begin{aligned}\epsilon_{AA} = I(AA) + \sum_k \Big\{ &\sum_B \left[y^k(AB)\, F^k(AB) + x^k(AB)\, G^k(AB)\right] \\ &- \sum_{BD} y^k(ABAD)\, R^k(ABAD) \\ &- \sum_{BCD} (1-\delta_{AC}) x^k(ABCD)\, R^k(ABCD) \Big\}.\end{aligned} \tag{7.7.2}$$

In practice, the Slater integrals must be replaced by finite difference quadratures, and (7.7.2) is only valid when used with the converged orbitals. The

[5] The Fortran code does not permit sub/superscripts to be printed, so that, for example, level 1 is labelled `1 S 0` rather than $^1S_0$.

Table 7.7. Predicted energy levels for Fe XXIII $1s^2\{2s^2+2p^2\}$

```
 atomic level properties
  -----------------------
  conversion of units using atomic weight = 55.80000
  1 a.u. = 2.1947247301E+05 cm-1
  1 a.u. = 2.7211128392E+01 eV
  eigenenergies
  -------------
             dominant
level J parity CSF mix       a.u.            Ryd.
  1   0 even 1 0.978    -8.1278670E+02 -1.62557340E+03
  2   0 even 6 0.930   -8.08413931E+02 -1.61682786E+03
  3   1 even 4 1.000   -8.08082170E+02 -1.61616434E+03
  4   2 even 3 0.911   -8.07855238E+02 -1.61571048E+03
  5   2 even 2 0.911   -8.07226228E+02 -1.61445246E+03
  6   0 even 5 0.920   -8.06224175E+02 -1.61244835E+03
 eigenenergies relative to the lowest
 ------------------------------------
             dominant
level J parity CSF mix      a.u.           Ryd.
  1   0 even 1 0.978   0.00000000E+00 0.00000000E+00
  2   0 even 6 0.930   4.37276677E+00 8.74553355E+00
  3   1 even 4 1.000   4.70452803E+00 9.40905606E+00
  4   2 even 3 0.911   4.93145991E+00 9.86291982E+00
  5   2 even 2 0.911   5.56046999E+00 1.11209400E+01
  6   0 even 5 0.920   6.56252241E+00 1.31250448E+01
>>>> MCDF dump written
>>>> routine NEWBAS called
>>>> eigenvectors transformed from jj to LS CSF basis
eigenenergies (absolute for groundstate, relative for others)
-------------
                  dominant
level state parity CSF    mix         Ryd.
  1     1   S 0 even 1   0.978 -1.625573395393E+03
  2     3   P 0 even 5   0.967   8.745533547791E+00
  3     3   P 1 even 4   1.000   9.409056057170E+00
  4     3   P 2 even 2   0.863   9.862919824754E+00
        1   D 2 even 3  -0.505
  5     1   D 2 even 3   0.863   1.112093998057E+01
        3   P 2 even 2   0.505
  6     1   S 0 even 6   0.948   1.312504482304E+01
```

generalized Rayleigh quotient (5.11.20) used in Algorithms 5.2 and 5.3 provides a more economical and appropriate estimate within the finite difference iterative scheme:

$$\epsilon_{AA}^{[\nu]} \approx \frac{\boldsymbol{v}_A^{[\nu]\dagger}\left(\boldsymbol{T}_A^{[\nu]}\boldsymbol{v}_A^{[\nu]} - \boldsymbol{R}_A^{[\nu]}\right)}{\boldsymbol{v}_A^{[\nu]\dagger}\boldsymbol{S}\boldsymbol{v}_A^{[\nu]}}. \tag{7.7.3}$$

The notation emphasizes that we are using the equations for orbital A. Because the exchange terms in $\boldsymbol{R}_A^{[\nu]}$ are proportional to $\boldsymbol{v}_A^{[\nu]}$, (7.7.3) does not require $\boldsymbol{v}_A^{[\nu]}$ to be normalized. A similar argument gives expressions for the off-diagonal parameters. Multiplying from the left by $u_B^\dagger$ and integrating over r gives the expression

$$\begin{aligned}\epsilon_{AB} &= I(B\,A)\\ &+\sum_k\Big\{\sum_B\left[y^k(AB)\,R^k(BCAC) + x^k(AB)\,R^k(BCCA)\right]\\ &\qquad -\sum_{CD} x^k(ACAD)\,R^k(BCAD)\\ &\qquad -\sum_{CDE}(1-\delta_{CD})x^k(ABDE)\,R^k(BCDE)\Big\}.\end{aligned} \tag{7.7.4}$$

Interchanging the roles of A and B gives

$$\begin{aligned}\epsilon_{BA} &= I(A\,B)\\ &+\sum_k\Big\{\sum_B\left[y^k(BA)\,R^k(ACBC) + x^k(BA)\,R^k(ACCB)\right]\\ &\qquad -\sum_{CD} x^k(BCBD)\,R^k(ACBD)\\ &\qquad -\sum_{CDE}(1-\delta_{CD})x^k(BADE)\,R^k(ACDE)\Big\}.\end{aligned} \tag{7.7.5}$$

Equations (7.7.4) and (7.7.5) can be used along with (7.4.3),

$$\overline{q}_A\epsilon_{AB} = \overline{q}_B\epsilon_{BA} = \lambda_{AB} \longrightarrow \lambda_{AB} = \frac{\overline{q}_A\overline{q}_B}{\overline{q}_A - \overline{q}_B}(\epsilon_{AB} - \epsilon_{BA}).$$

This is generally simpler than (7.7.4) and (7.7.5) unless $\overline{q}_A = \overline{q}_B$ because the expression $(\epsilon_{AB} - \epsilon_{BA})$ requires a much smaller number of Slater integrals. We can also obtain estimates in the spirit of (7.7.1) by using

$$\epsilon_{AB}^{[\nu+1]} \approx \frac{\boldsymbol{v}_B^{[\nu]\dagger}\left(\boldsymbol{T}_A^{[\nu]}\boldsymbol{v}_A^{[\nu]} - \boldsymbol{R}_A^{[\nu]}\right)}{\boldsymbol{v}_B^{[\nu]\dagger}\boldsymbol{S}\boldsymbol{v}_B^{[\nu]}}, \quad \epsilon_{BA}^{[\nu+1]} \approx \frac{\boldsymbol{v}_A^{[\nu]\dagger}\left(\boldsymbol{T}_B^{[\nu]}\boldsymbol{v}_B^{[\nu]} - \boldsymbol{R}_B^{[\nu]}\right)}{\boldsymbol{v}_A^{[\nu]\dagger}\boldsymbol{S}\boldsymbol{v}_A^{[\nu]}}. \tag{7.7.6}$$

The Dirac operators $\boldsymbol{T}_A^{[\nu]}$ and $\boldsymbol{T}_B^{[\nu]}$ have the same symmetry κ_A but different potentials and right hand sides. The relation (7.4.3) can be used in the same way to simplify the calculation of the off-diagonal parameters.

7.8 Koopmans' theorem and Brillouin's theorem

Koopmans [18] actually proved rather more than the result (7.7.1). He studied the effect of orthogonal transformations amongst the orbitals on the many-electron wavefunction and, in particular, the class of wavefunctions which remain invariant under these transformations. When this happens, further constraints are needed for the Hartree-Fock equations to have a unique solution.

Consider a group of equivalent electrons $\{n\kappa,\ n=1,2,\ldots\}$. It is sufficient to study infinitesimal rotations of the form

$$|n\kappa m\rangle \to |n\kappa m\rangle + \eta|n'\kappa m\rangle, \quad |n'\kappa m\rangle \to |n'\kappa m\rangle - \eta|n\kappa m\rangle, \tag{7.8.1}$$

which are normalized to $O(\eta)^2$. Suppose that Φ is a CSF and consider the effect of (7.8.1) on the DHF functional

$$J[\Phi] = \frac{\langle\, \Phi \,|\, H \,|\, \Phi \,\rangle}{\langle\, \Phi \,|\, \Phi \,\rangle}.$$

The transformation (7.8.1) induces the transformation

$$|\,\Phi\,\rangle \to |\,\Phi\,\rangle + \eta|\,\Phi^*\,\rangle + O(\eta^2),$$

where we can take $|\,\Phi^*\,\rangle$ to be orthogonal to $|\,\Phi\,\rangle$. In general,$|\,\Phi^*\,\rangle$ will be a linear combination of several CSF with the same overall symmetry, but it can happen that some substitutions are not allowed by the Pauli principle. The change this induces in $J[\Phi]$ is therefore

$$\delta J[\Phi] = 2\eta\langle\, \Phi \,|\, H \,|\, \Phi^*\,\rangle = 0,$$

if Φ is the DHF wavefunction. The result has been attributed to Brillouin [19, 20], although he did not himself express his conclusions so concisely. He observed that when the system has a ground state Φ that is represented by a single CSF with orbitals determined by the Hartree-Fock equations, the Hamiltonian matrix element linking Φ to any "singly-excited" CSF vanishes identically. This really applies only to configurations with completely closed shells of states that have just one valence electron outside a closed shell. Bauche and Klapisch [21] listed LS coupling cases covered by Brillouin's argument:

- $\gamma(nl) \to \gamma(n'l)$.
- $\gamma(nl)^{4l+2} \to \gamma(nl)^{4l+1}(n'l)$.
- $\gamma(nl)^q LS \to \gamma(nl)^{q-1}(n'l)LS$, when $\gamma(nl)^q LS$ has only one parent in $\gamma(nl)^{q-1}$.

and Labarthe [22] added a further case

- $\gamma(nl)^q\gamma(n'l)^{q'} \to (nl)^{q\pm1}\gamma(n'l)^{q'\mp1}$

for which the matrix elements vanish under certain conditions.

It is easy to give similar results for Dirac spinors in jj-coupling.

- $\gamma(n\kappa) \to \gamma(n'\kappa)$.
- $\gamma(n\kappa)^{2j+1} \to \gamma(n\kappa)^{2j}(n'\kappa),\ J=0$.
- $\gamma(n\kappa)^q \nu J \to \gamma\left[(n\kappa)^{q-1}\nu' J'(n'\kappa)\right] J$ when $(n\kappa)^{q-1}\nu' J'$ is the sole parent of $\gamma(n\kappa)^q \nu J$.

Examples of this last case include

- All configurations $(n\kappa)^q$ with $q = 1, 2$ or $2j+1$.
- $(5/2)^4,\ \nu = 0,\ J = 0$.
- $(7/2)^{4,6},\ \nu = 0,\ J = 0$.
- $(9/2)^4,\ \nu = 0, 4,\ J = 0$.

There is also a Dirac analogue of Labarthe's case.

The (D)HF method gives very good results for the ground state energy when Brillouin's theorem holds. When building a wavefunction that takes account of correlation effects, we can then omit singly substituted CSF from the trial wavefunction; the most important CSF needed to improve the wavefunction are then those with two or more excitations from the reference state. For example, consider a reference CSF of the form $(n\kappa)(n'\kappa), J$. The total angular momentum can have the values $J = 0, 1, 2, \ldots, 2j$. An orbital rotation will transform the ground state into some linear combination

$$(n\kappa)(n'\kappa), J \to c_0\,(n\kappa)(n'\kappa), J + c_1(n\kappa)^2, J + c_2(n'\kappa)^2, J.$$

Suppose first that J is an odd integer. Then the Pauli exclusion principle eliminates both $(n\kappa)^2$ states from the sum, so that the reference CSF, $(n\kappa)(n'\kappa), J$ can mix only with similar CSF $(n\kappa)(n''\kappa), J$ such that $n'' \neq n, n'$. When J is even there is no way of excluding $(n\kappa)^2$ staes from the sum, and in general the matrix elements with the reference CSF do not vanish.

7.8.1 Froese Fischer's analysis

The fact that the multi-configurational description of an atomic state may not be unique, in the sense that a representation in terms of a set of CSF may not be closed under orbital rotations is clearly important for both CI and MC(D)HF studies. The complex situations that can arise have been carefully analysed by Fischer [23, 24], and her papers are essential reading for anyone needing to understand the subject.

Let A, A' be a pair of (radial 2-component) orbitals with the same symmetry κ_A and consider a rotation

$$\begin{pmatrix} A \\ A' \end{pmatrix} \to \frac{1}{\sqrt{1+\eta^2}} \cdot \begin{pmatrix} 1 & \eta \\ -\eta & 1 \end{pmatrix} \begin{pmatrix} A \\ A' \end{pmatrix} \tag{7.8.2}$$

The induced changes in the atomic Hamiltonian $\mathbf{H}$, the eigenvalue E and the mixing coefficient vector $\mathbf{c}$ of (7.4.6) can be expanded as a regular perturbation series, for example

$$E \to E(\eta) = E + \eta E_1 + \eta^2 E_2 + \dots \tag{7.8.3}$$

There are two possibilities: either the energy functional from which the secular equations are derived is invariant, in which case $E_1 = E_2 = \dots = 0$; more usually, this is not the case. Then, we want E to be stationary with respect to small variations in η; setting $dE/d\eta = 0$ gives

$$0 = E_1 + 2\eta E_2 + \dots$$

so that, to lowest order, we should choose

$$\eta = -E_1/2E_2. \tag{7.8.4}$$

We therefore need both the first and second variation of the energy in order to determine the appropriate value of η

Let $\mathbf{H}(\eta)$ and $\mathbf{c}(\eta)$ be the perturbed Hamiltonian so that the perturbed secular equation is

$$\begin{aligned}(\mathbf{H} + \eta\mathbf{H}_1 + \eta^2\mathbf{H}_2 + \dots)(\mathbf{c} + \eta\mathbf{c}_1 + \eta^2\mathbf{c}_2 + \dots)\\ = (E + \eta E_1 + \eta^2 E_2 + \dots)(\mathbf{c} + \eta\mathbf{c}_1 + \eta^2\mathbf{c}_2 + \dots).\end{aligned}$$

Equating coefficients of successive powers of η gives

$$\begin{aligned}(\mathbf{H} - E)\mathbf{c} &= 0,\\ (\mathbf{H} - E)\,\mathbf{c}_1 &= (E_1 - \mathbf{H}_1)\,\mathbf{c} =: \mathbf{b}\\ (\mathbf{H} - E)\,\mathbf{c}_2 &= (E_1 - \mathbf{H}_1)\,\mathbf{c}_1 + (E_2 - \mathbf{H}_2)\,\mathbf{c}\end{aligned} \tag{7.8.5}$$

Denote the inner product of two vectors by

$$(\mathbf{u}, \mathbf{v}) = \mathbf{u}^\dagger \mathbf{v}.$$

Because $\mathbf{H}$ is a symmetric matrix and $\mathbf{c}$ is an eigenvalue of $\mathbf{H}$ belonging to the eigenvalue E,

$$(\mathbf{c}, \mathbf{b}) = (\mathbf{c}, (\mathbf{H} - E)\,\mathbf{c}_1) = ((\mathbf{H} - E)\,\mathbf{c}, \mathbf{c}_1) = (0, \mathbf{c}_1) = 0.$$

The fact that $(\mathbf{c}, \mathbf{b}) = 0$ gives the usual first order result

$$E_1 = (\mathbf{c}, \mathbf{H}_1\,\mathbf{c}). \tag{7.8.6}$$

The normalization condition $(\mathbf{c}, \mathbf{c}) = 1$ is satisfied to first order if

$$(\mathbf{c}, \mathbf{c}_1) = 0, \tag{7.8.7}$$

so that $\mathbf{c}$ and $\mathbf{c}_1$ are orthogonal. The same sort of reasoning applied to the second order terms gives

$$E_2 = (\mathbf{c}, \mathbf{H}_2\, \mathbf{c}) - (\mathbf{c}_1, \mathbf{b}). \tag{7.8.8}$$

To complete the analysis, we need to construct the perturbation matrices $\mathbf{H}_1$ and $\mathbf{H}_2$. Equations (7.3.1) and (7.3.5) express the Dirac-Coulomb matrix elements in the CSF basis in terms of radial integrals $I(AA')$ over the Dirac central field operator (including the nuclear potential) involving orbitals of the same symmetry κ_A, and Slater integrals of the generic form $R^k(ABCD)$ over the Coulomb interaction. We need consider only pairwise rotations of the form (7.8.2); expanding in powers of the parameter η, we see that

$$A \to A + \eta A' - \frac{1}{2}\eta^2 A + O(\eta^3), \quad A' \to A' - \eta A - \frac{1}{2}\eta^2 A' + O(\eta^3). \tag{7.8.9}$$

As an example, consider the case $A = 1s$, $A' = 2s$; it is easiest to write radial integrals $R^k(ABCD)$ in overlap charge notation as $(AC\|BD)^k$. The tensor order is irrelevant for present purposes, so to see what happens to $F^0(A, A)$ we just look at $T \equiv (AA\|AA)$ so that

$$\begin{aligned}
&(AA\|AA)\\
&\to (\, A + \eta A' - \frac{1}{2}\eta^2 A,\; A + \eta A' - \frac{1}{2}\eta^2 A \,\|\\
&\qquad\qquad A + \eta A' - \frac{1}{2}\eta^2 A,\; A + \eta A' - \frac{1}{2}\eta^2 A\,)
\end{aligned}$$

Because a Slater integral is linear in each of its arguments, it is easy to pick out the coefficients of powers of of η. The first order coefficient is

$$T_1 = 4\,(AA\|AA')^0 \equiv 4\,R^0(1s, 1s, 1s, 2s).$$

The second order coefficient T_2 is a little more complicated since it involves paired contributions of order η as well as terms of order η^2:

$$\begin{aligned}
T_2 &= 2\,(AA\|A'A')^0 + 4\,(AA'\|AA')^0 - 2\,(AA\|AA)^0\\
&\equiv 2\,F^0(1s, 2s) + 4\,G^0(1s, 2s) - 2\,F^0(1s, 1s),
\end{aligned}$$

in agreement with [24, p. 287].

Table 7.8. Approach to self-consistency, Fe XXIII

```
------------------------------------------------------

>>>> routine SCF called

         eigenvalue     Z mix XCA   RMAX     DMAX NPOINT

 1 1S  3.1261284E+02 26  0  1.0 2.44E-02 -4.75E-02 243
 2 2S  7.1400118E+01 26  0  1.0 1.34E-01  1.47E-01 251
 3 2P- 6.9074972E+01 26  0  1.0 8.96E-02 -1.76E-01 251
 4 2P  6.8549204E+01 26  0  1.0 9.90E-02 -1.80E-01 251
 4 2P  6.8606385E+01 26  0  1.0 4.44E-01 -7.69E-04 252
 3 2P- 6.9192333E+01 26  0  1.0 4.01E-01 -1.53E-03 252
 2 2S  7.1477879E+01 26  0  1.0 1.63E-01 -1.28E-03 252
 1 1S  3.1349638E+02 26  0  1.0 2.57E-02  2.09E-03 245
 1 1S  3.1349027E+02 26  0  1.0 2.21E-02 -3.11E-05 245
 3 2P- 6.9191056E+01 26  0  1.0 7.71E-02 -4.63E-05 252
 2 2S  7.1476119E+01 26  0  1.0 1.21E-01  5.85E-05 252
 4 2P  6.8604316E+01 26  0  1.0 8.96E-02 -6.98E-05 252
 4 2P  6.8604335E+01 26  0  1.0 4.44E-01 -2.48E-07 252
 2 2S  7.1476152E+01 26  0  1.0 1.40E-01 -7.79E-07 252
 3 2P- 6.9191097E+01 26  0  1.0 3.82E-01 -6.36E-07 252
 1 1S  3.1349059E+02 26  0  1.0 2.44E-02  8.86E-07 245

------------------------------------------------------

>>>> MCDF dump written

------------------------------------------------------

>>>> routine SCF called
          ''
          ''
          ''

>>>> routine SCF called
         eigenvalue     Z mix XCA   RMAX     DMAX NPOINT

 1 1S  3.1349058E+02 26  0  1.0 2.10E-01  3.45E-11 245
 2 2S  7.1476177E+01 26  0  1.0 4.02E-02  7.98E-10 252
 3 2P- 6.9191125E+01 26  0  1.0 3.63E-01 -2.17E-10 252
 4 2P  6.8604332E+01 26  0  1.0 5.70E-01  6.18E-11 252
```

It is clear that the calculation of the perturbation matrices $\mathbf{H}_1$ and $\mathbf{H}_2$ is a time-consuming business to be avoided except where it is absolutely essential.

The classic example is the HF calculation of He $1s2s\ ^1S$, where it is essential to rotate the basis at each iteration to ensure convergence [8, §3.5]. In that case, Fischer[6] suggests that it is better to use an MCHF approach with a minimal ^1S CSF basis of $\{1s^2, 1s2s\}$ for which $1s2s\ ^1S$ is the lowest excited state. MC(D)HF calculations with a large number of CSF are less likely to suffer from degeneracy with respect to orbital rotations, and rotations are relatively unimportant, except possibly for an initial HF calculation to obtain occupied core orbitals. Off-diagonal parameters will still be required, but only for pairs of subshells of the same symmetry of which one or both are partially filled.

7.9 Control of MCSCF iterations

Table 7.8 illustrates the approach to self-consistency leading to the energy levels of Table 7.7. The first SCF cycle starts by taking all the orbitals in order, starting with the most tightly bound. It then chooses the orbital showing the most change (measured by `DMAX` $= \max_i\{|\delta P(r_i)| + |\delta Q(r_i)|\}$) as the next to be improved. `RMAX` is the corresponding value of r_i. The table shows the first cycle of improvements during which `DMAX` is reduced to below 10^{-6} in all orbitals. Further cycles at higher accuracy 10^{-8} and 10^{-10} follow; higher order difference terms of order $O(h^6)$ are included at the final stage. `NPOINT`, which gives the number of radial points for each orbital, grows slightly as the calculation progresses.

This simple example shows rapid convergence to a self-consistent solution, but things are not always as straightforward, and the code has a number of devices to overcome convergence difficulties. The most difficult is frequently the initial construction of a normalized orbital. It is quite possible for the iterative scheme to produce a new trial solution with leading coefficient $A < 0$, which is unacceptable. The programme then halts. An orbital may have the wrong number of zeros: if too few, the diagonal parameter is too large, its magnitude must be reduced. The *join point* of the solutions of the `OUT` and `IN` modules can run off the end of the radial grid; in this case the calculation must be restarted with an extended radial grid.

The stability of positive ions can often be put to good use; an initial calculation for an ionic closed shell core is usually easy to converge and provides a stable platform on which to build the more sensitive outer orbitals. Increasing the nuclear charge Z to stabilize the calculation is sometimes an alternative; once an acceptable solution has been obtained, a slow reduction of Z to its physical value can overcome the difficulty. It may also be desirable to damp the change from one iteration to the next by taking as the next trial solution

$$u^{[\nu]} \to \alpha u^{[\nu]} + (1-\alpha)u^{[\nu-1]}, \tag{7.9.1}$$

[6] I am indebted to Charlotte Fischer for her advice on this topic.

where α is printed as `mix` in the table. The changes in the eigenvalue and A may also be damped in this way. Similarly, the parameter `XCA` has been used to reduce the size of the exchange terms in the hope that this will improve the rate of convergence.

The pseudo-code of Algorithm 7.1 outlines a procedure more in the spirit of the MCHF code [8]. This is based on the derivation of the integro-differential equations in §6.7 that does not require *a priori* orbital normalization. The potentials must, of course, be calculated with normalized orbitals, and trial orbitals of the same symmetry should be orthogonalized before calculating estimates of the diagonal parameters ϵ_{AA} and the off-diagonal parameters $\epsilon_{AA'}$.

Algorithm 7.1

1. ***Initialization:***
   - *Input basic physical parameters*
   - *Input or construct initial orbitals (Table 7.1)*
   - *Input CSF data and construct angular coefficients*
   - *Orthogonalize orbitals with same κ.*
   - *Estimate diagonal parameters ϵ_{AA} using Rayleigh quotient*
   - *Estimate off-diagonal parameters $\epsilon_{AA'}$ which are not to be set to zero.*
2. ***SCF cycle:***
   a) ***If*** *first cycle*
   then *take orbitals in natural order starting with the most tightly bound*
   else *choose orbital showing largest* `DMAX`*.*
   Next orbital is A.
   b) *Solve orbital equations for A using Algorithm 5.1.*
   c) ***Either*** *proceed as in Algorithm 5.2 by adding a multiple of solution of homogeneous equations as in (5.11.17) to obtain a solution with $\Delta Q_A = 0$ at the join point and normalize;*
   or*, if the right hand side is small, proceed as in Algorithm 5.3 by adding a multiple of the perturbed Dirac equation to give a solution normalized to first order.*
   d) ***If*** *number of nodes is wrong*
   then *adjust ϵ_{AA} to correct this; return to 2(b) with normalized estimate $u^{[\nu]}$ as in (7.9.1).*
   e) ***If*** $|\Delta Q_A^{(P)}|$ *is above threshold*
   then *return to 2(b)* ***else*** *exit.*
3. *Find orbital B with largest* `DMAX`*.*
   If $|$`DMAX`$|$ *is above threshold*
   then *set $A = B$ and return to 2(b)*
   else *exit from cycle.*
4. *Further cycles at higher precision if wanted.*

7.10 Corrections to the Coulomb interaction: Breit and other approximations

The `BENA` command activates the calculation of relativistic corrections to the Coulomb interaction given in §6.4.2 as a first order perturbation (MCDHF+B) of a preceding MCDHF or CI calculation. Experiments using the matrix SCF method that treat the Breit interaction self-consistently (MCDHFB) have shown that the energies predicted by the two approaches generally agree well. The perturbation calculation modifies the eigenvectors $\boldsymbol{c}_i$ and the corresponding eigenvalues E_i, but the orbitals are unaffected. The self-consistent MCDHFB calculation gives slightly different orbitals, having potentially important consequences for electron correlation calculations.

The BENA package was designed to implement the full transverse photon interaction of §6.4.5 in Coulomb gauge, but can also be used to calculate the Breit interaction. The angular coefficient packages list all angular momentum coefficients in one file: Table 7.9 summarizes this data for our test problem. There is just one *core* subshell, $1s^2, J = 0$; the remaining subshells listed in Tables 7.3 and 7.4 are said to form the *peel*. Table 7.10 gives an extract from the list of angular coefficients. The label `ITYPE` indicates the class to which the integral belongs: `ITYPE` $= 1, 2, \ldots, 6$ correspond to the classification of transverse photon radial integrals in §6.7.1. `ITYPE`$=7$ are one-electron integrals $I(A, B)$; `ITYPE`$= 8$ are direct Coulomb interaction integrals and `ITYPE`$=$ 9 are exchange integrals. Entries which involve only the core subshells are indicated by setting `ITYPE` negative. Core/core and core/peel interactions for the Coulomb interaction are automatically calculated by the MCDF module from the closed shell formulae of §6.6. The core/core coefficients that are common to all diagonal matrix elements are listed only for the first matrix element H_{RS}, where $(R, S) = (1, 1)$. The table is otherwise self-explanatory, save for the column labelled `ISTORE`, which codes for the labels A, B, C, D, K, where K denotes the tensor rank of the term. `ISTORE` is set negative for core/core and core/peel subshell entries.

The number of coefficients output by MCBP for this simple example is more than twice the number of the more important Coulomb coefficients, and the integrals are more complex to construct because of the presence of spherical Bessel functions. This makes the calculations relatively expensive, although the integrals need be evaluated only once in a first order perturbation calculation. Table 7.11 is a partial extract of the output from the `BENA` module for the Breit energy corrections in atomic units. This shows the ground state energy level and level excitation energies obtained from the MCDHF calculation along with the Breit energy corrections; the QED corrections, to be discussed below, and the level energies taking account of both transverse and QED corrections are not included in this table. The `BENA` command in Table 7.2 is followed by a blank line; this means that the wavelengths of the mediating photons responsible for the electron-electron interaction have been taken at their physical values. The long-wave limit, the Breit interaction, can be

Table 7.9. Summary of angular coefficient data for Fe XXIII model

		ITYPE
17 one-electron	(1 in core)	7
	(17 in core/core or core/peel)	-7
33 direct Coulomb	(1 in core)	8
	(19 in core/core or core/peel)	-8
23 exchange Coulomb	(0 in core)	9
	(16 in core/core or core/peel)	-9
73 total Coulomb	(2 in core)	
	(52 in core/core or core/peel)	
19 Breit type 1	(0 in core)	1
17 Breit type 2	(0 in core)	2
2 Breit type 3	(0 in core)	3
13 Breit type 4	(1 in core)	±4
83 Breit type 5	(0 in core)	5
34 Breit type 6	(0 in core)	6
168 total Breit	(1 in core)	

obtained by replacing the blank with a number WFACT, so that $\omega \to \omega$*WFACT; the value WFACT $=10^{-3}$ usually makes ω small enough. The effect is negligible in the present example.

The absolute contribution of the transverse photon interaction to the ground state of Be-like ions [25] is quite small for Fe XXIII, although it increases roughly like Z^4 for most of the Periodic Table, destabilizing the ground state by about 0.4 % of the MCDHF energy around $Z = 100$. Small though it is, the effect on the fine structure of atomic term values is much more significant, although the size of the shift is rarely sufficient to upset the level ordering predicted by the MCDHF calculation.

7.11 QED corrections

So far, we have considered only the NVP approximation in which negative energy states, although present in the model, are passive. This is not to say that processes to which they contribute actively can be ignored, and the BENA module allows some of them to be estimated. The self-energy of an electron in the quantized electromagnetic field is described in lowest order by the Feynman diagram of Figure 4.6(g) and the lowest order radiative correction to the nuclear Coulomb field due to polarization of the vacuum by the nuclear charge distribution is described by Figure 4.6(h). Like the electron-electron interaction, both diagrams are formally second order in QED perturbation theory, and the corresponding energy shifts are a similar order of magnitude to those of the relativistic corrections to the electron-electron interaction.

Table 7.10. Extract from angular coefficient file for Fe XXIII model

R	S	A	B	C	D	K		ISTORE	ITYPE
1	1	1S	1S	1S	1S	1	2.666666667E+00	781	-4
1	1	1S	1S	1S	1S	0	1.000000000E+00	-156	-8
1	1	2S	1S	2S	1S	1	1.333333333E+00	911	5
1	1	1S	2S	2S	1S	1	2.666666667E+00	811	5
1	1	1S	2S	1S	2S	1	1.333333333E+00	807	5
1	1	1S	2S	1S	2S	0	4.000000000E+00	-182	8
1	1	1S	2S	2S	1S	0	-2.000000000E+00	-186	9
1	1	2S	2S	2S	2S	1	2.666666667E+00	937	4
1	1	2S	2S	2S	2S	0	1.000000000E+00	-312	8
1	1	**	**	1S	1S		2.000000000E+00	-6	-7
1	1	**	**	2S	2S		2.000000000E+00	-12	7
1	9	2S	2S	2P	2P	1	-4.714045208E-01	949	8
1	9	2P	2S	2P	2S	0	-1.257078722E+00	572	1
1	9	2P	2S	2P	2S	1	1.047565602E-01	1197	2
1	9	2P	2S	2P	2S	1	1.047565602E-01	1197	2
1	9	2P	2S	2S	2P	1	-3.142696805E-01	1189	2
1	9	2P	2S	2S	2P	1	-3.142696805E-01	1189	2
1	9	2S	2P	2S	2P	2	-5.656854249E-01	1614	1
1	9	2P	2S	2P	2S	2	-4.399775527E-01	1822	1
1	9	2S	2P	2P	2S	2	-3.771236166E-01	1622	1
1	9	2P	2S	2S	2P	2	-3.771236166E-01	1814	1

The self-energy correction, Figure 4.6(g), is modelled in GRASP by the expression

$$H_{rr}^{SE} = \sum_{A=1}^{n_W} q_r(A)\, E_A^{SE} \tag{7.11.1}$$

where E_A^{SE} is the one-electron self-energy of an electron in subshell A. This would be expensive to evaluate from first principles in a many-electron atom. However, tables of the self-energy for a number of low-lying levels of hydrogenic systems as a function of atomic number, Z, are available, for example [6, Table 28.3], [26], and an effective, though nonrigorous, way of estimating the correction is to write

$$E_A^{SE} = \frac{\left(Z_A^{eff}\right)^4}{\pi c^3 n_A^3} \begin{cases} F_{n_A\kappa_A}(Z_A^{eff}/c) & \text{for } 1s, 2s, 2\bar{p}, 2p \text{ orbitals,} \\ F_{2\kappa_A}(Z_A^{eff}/c) & \text{for } n > 2 \\ 0 & \text{otherwise.} \end{cases} \tag{7.11.2}$$

where $F_{n\kappa}(y)$ is a smooth function of y and Z_A^{eff} is an effective point charge.

This approximation was first introduced in order to approximate energies of X-ray levels in heavy atoms, but has since been used also for valence orbitals. The innermost electrons were thought to be nearly hydrogenic, so that Z_A^{eff} would probably be insensitive to the method of estimation. GRASP equates the mean radius of the MCDHF orbital to that of a hydrogenic $(n_A\kappa_A)$ orbital with nuclear charge Z_A^{eff}. Welton [27] supposes that the self-energy is due to perturbation of the classical trajectory by fluctuations in the electromagnetic field of the vacuum. The fluctuations cause the electron to probe the potential at a displaced point $V_{nuc}(\boldsymbol{r}+\delta\boldsymbol{r})$ rather than $V_{nuc}(\boldsymbol{r})$. This gives a perturbing potential

$$\begin{aligned}\delta V_{nuc} &= \langle V_{nuc}(\boldsymbol{r}+\delta\boldsymbol{r}) - V_{nuc}(\boldsymbol{r})\rangle_{vacuum}\\ &\approx \left\langle \boldsymbol{\nabla} V_{nuc}(\boldsymbol{r}).\delta\boldsymbol{r} + \Delta V_{nuc}(\boldsymbol{r})(\delta\boldsymbol{r})^2 + \ldots\right\rangle_{vacuum}\\ &\approx \left\langle \Delta V_{nuc}(\boldsymbol{r})(\delta\boldsymbol{r})^2\right\rangle_{vacuum}\end{aligned}$$

where the first term vanishes because the vacuum fields average to zero. The nonvanishing second order term must be renormalized, and gives the hydrogenic formula of (7.11.2). Welton therefore argues that, at least for s-orbitals, the *ansatz*

$$E_A^{SE} = \frac{\langle A|V_{nuc}|A\rangle_{DHF}}{\langle A|V_{nuc}|A\rangle_{hydrogenic}} \cdot E_A^{SE}{}_{hydrogenic} \tag{7.11.3}$$

should be more reliable than (7.11.2), a view that has received some support in a number of recent investigations [9, 28].

The vacuum polarization correction Figure 4.6(h) can be treated similarly (see [6, Chapter 28]). GRASP evaluates an expression of the form

$$H_{rr}^{VP} = \sum_{A=1}^{nw} q_r(A) \int_0^\infty V^{VP}(r) D_{AA}(r)\, dr \tag{7.11.4}$$

where $V^{VP}(r)$, [29], includes vacuum polarization potentials of both second- and fourth-order in QED perturbation theory. BENA calculates these corrections along with the relativistic corrections to the electron-electron interaction. The MCDHF matrix has already been diagonalized to give an ASF basis, and the first stage is to express the perturbation Hamiltonian in this basis. After subtracting the average level energy, the perturbed Hamiltonian is rediagonalized and the perturbed energy levels can be re-ordered if necessary.

Table 7.11. Transverse interaction energy for Fe XXIII $1s^2\{2s^2+2p^2\}$

```
>>>> routine READA called : read angular coefficients

>>>> routine COUMAT called

>>>> routine COUMT2 called : diagonalise the Coulomb matrix

average energy (a.u.) -8.0843140651E+02

diagonalisation using the DSYEV routine

>>>> BENA dump has been written with the current data

>>>> routine BREMAT called

94 MCBP (Breit) angular coefficients read
1 core (Breit) angular coefficients read

50 radial integrals evaluated
50 radial integrals stored

>>>> routine BREMT2 called : diagonalise the Breit matrix

average energy (a.u.) -8.0814868669E+02

diagonalisation using the DSYEV routine

The ground state energy is absolute. Excited level energies are
relative to the ground state. Only CSF mixing coefficients whose
magnitude is larger than .1 are shown.

WFACT (factor used to multiply frequency) = 1.000E+00
```

Summary of contributions to energy levels in a.u.

level	CSF		zero-order	Breit
1 0	1	0.98	-8.128887E+02	2.70E-01
	5	0.15		
	6	0.14		
2 0	6	0.93	4.372767E+00	3.52E-02
	5	-0.36		
3 1	4	1.00	4.704528E+00	2.15E-02
4 2	3	0.91	4.931460E+00	6.46E-03
	2	0.41		
5 2	2	0.91	5.560470E+00	-1.06E-03
	3	-0.41		
6 0	5	0.92	6.562522E+00	1.68E-02
	1	-0.19		
	6	0.34		

Table 7.12. Transverse and QED corrections for Fe XXIII

```
The ground state energy is absolute. Excited level energies are
relative to the ground state. Only CSF mixing coefficients whose
magnitude is larger than .1 are shown.

Summary of contributions to energy levels in a.u.
```

level	CSF		Breit	QED	total energy
1 0	1	0.98	2.70E-01	3.01E-01	-8.122161E+02
	5	0.15			
	6	0.14			
2 0	6	0.93	3.52E-02	-3.11E-02	4.376820E+00
	5	-0.36			
3 1	4	1.00	2.15-02	-3.04E-02	4.695636E+00
4 2	3	0.91	6.46-03	-3.02E-02	4.907689E+00
	2	0.41			
5 2	2	0.91	-1.06-03	-2.95E-02	5.529946E+00
	3	-0.41			
6 0	5	0.92	1.68E-02	-2.84E-02	6.550916E+00
	1	-0.19			
	6	0.34			

7.12 Towards higher quality atomic models

Although we have the main ingredients for high-quality atomic calculations, the examples so far have used the MCDHF-EAL approach with a minimal CSF set. A single model potential generates a common set of orbital spinors from which all CSFs and atomic states (ASF) are built. This scheme can be used very effectively in highly ionized atoms such as Be-like iron, Fe XXIII, but it is not always appropriate. Column (b) of Table 7.13 gives the level spectra from an MCDHF-EAL calculation on Fe XXIII using ten jj-coupled CSF belonging to $1s^2(2s^2 + 2s2p + 2p^2)$. A comparison with the even parity levels in Table 7.6 shows that adding the four odd parity CSFs belonging to $1s^2 2s2p$ to the energy functional has made only a tiny difference to the orbital properties. The ground state energy is slightly raised in the sixth decimal place relative to Table 7.7, and there are similar small adjustments to the excited level energies. The MCDHF model potential is little changed because the electron-electron interaction has no matrix elements connecting even and odd parity levels. Level energies including the the Breit and QED perturbations, appear in column (c). Agreement with the recommended energies, column (a), taken from the NIST Atomic Spectra Database [33] is slightly improved.

Table 7.13. Levels of Fe XXIII $1s^2\{2s^2 + 1s2p + 2p^2\}$ with Breit and QED effects

Config.		Level(Ryd) (a)	(b)	(c)	Leading percentages		
$1s^22s^2$	$^1S_0$	0.	0.	0.	96	4	$1s^22p^2$
$1s^22s2p$	$^3P^o_0$	3.1728	3.1719	3.1841	100		
	$^3P^o_1$	3.4548	3.4732	3.4679	81	19	$1s^22s2p$
	$^3P^o_2$	4.2991	4.3444	4.3079	100		
	$^1P^0_1$	6.8603	7.0094	6.9866	81	19	$1s^22s2p$
$1s^22p^2$	$^3P_0$	8.712	8.3450	8.7538	86	13	$1s^22p^2$
	$^3P_1$	9.3605	9.4083	9.3912	100		
	$^3P_2$	9.7660	9.8625	9.8155	74	26	$1s^22p^2$
	$^1D_2$	10.973	11.121	11.061	74	26	$1s^22p^2$
	$^1S_0$	12.96	13.13	13.10	96	4	$1s^22s^2$

(a) NIST online Atomic Spectra Database [33].
(b) MCDHF. Ground state energy: -1.62557282E+03.
(c) MCDHF with Breit and QED.
Ground state energy: -1.62443139E+03.
Leading CSF percentage weights from GRASP0.

7.12.1 CSF sets for electron correlation: active space methods

The quality of the predictions in similar calculations for less highly ionized members of the isoelectronic sequence declines as Z decreases. A minimal CSF wavefunction is not good enough for such problems, which are similar to those encountered in nonrelativistic Hartree-Fock calculations [34], [8, Chapter 4]. The Hartree-Fock wave-function is an anti-symmetrized product of independent one-electron orbitals generated by some atomic mean-field electrostatic potential. *Electron correlation*, expressing the notion that interactions between the electrons destroy orbital independence, is essential to improve the quality of nonrelativistic Hartree-Fock calculations, along with relaxation, relativistic effects, and corrections due to the nuclear size and mass. For an MCDHF trial function

$$\Psi(\Gamma J^\pi) = \sum_i c_i\, \psi(\gamma_i J^\pi),$$

for which the CSF $\psi(\gamma_0 J^\pi)$ is a first approximation to $\Psi(\Gamma J^\pi)$, then first order perturbation theory for an interaction V gives the expansion coefficient

$$c_i = \frac{\langle \psi(\gamma_i J^\pi) \,|\, V \,|\, \psi(\gamma_0 J^\pi)\rangle}{E_0 - E_i} \tag{7.12.1}$$

for $i \neq 0$, suggesting that CSFs whose energy E_i is close to E_0 and that have non-negligible interaction matrix elements with the reference CSF,

$\langle\psi(\gamma_i J^\pi)\,|\,V\,|\,\psi(\gamma_0 J^\pi)\rangle$, are likely to be important. We need a convenient way to characterize the CSFs to be included in this expansion. One approach is to divide the orbital set O into two groups $\mathsf{O} = \mathsf{O}_1 \cup \mathsf{O}_2$, where O_1 is the set of *inactive* electrons and O_2 is the set of *active* electrons. The inactive subshells are always filled, whereas the active subshells may have any distribution of the $n_2 = N - n_1$ active electrons. The *complete active space* (CAS) then consists of all CSFs with n_1 inactive electrons and all possible CSFs allowed by the Pauli principle and angular coupling schemes. The number of CSFs can easily get out of control when n_2 is large, and it is clear that a slightly more flexible strategy is desirable.

Most commonly used CI expansions can be classed as restricted active set (RAS) schemes [35]. Here we subdivide the active set O_2 into three parts: $\mathsf{O}_2 = \mathsf{O}_{21} \cup \mathsf{O}_{22} \cup \mathsf{O}_{23}$, distributed so that $n_2 = n_{21} + n_{22} + n_{23}$. The set O_{21} consists of subshells with a *minimum* variable occupation $\bar{n}_1 : n_{21} \geq \bar{n}_1$. The set O_{22} consists of subshells O_{22} in which the distribution of the n_{22} electrons among the subshells is unconstrained. The set O_{23} consists of subshells in which no more than $\bar{n}_3$ electrons are permitted in any one subshell. For correlation of ground and low-lying excited atomic states, the inactive set O_1 normally comprises all deep core positive energy states and all negative energy states for which the energy denominators in (7.12.1) are so large they are unlikely to contribute significantly to the correlation. The subset O_{21} usually contains valence or deep core orbitals that are to be correlated or polarized. The restriction $n_{21} \geq \bar{n}_1$ then limits the number of electrons that can be excited from this group of subshells. We might identify O_{22} as the subshells that are expected to contribute most to the correlation and O_{23} as less important.

This partitioning has three main advantages:

(i) It encompasses very general types of CI expansion and of systematic study of the error incurred by truncation of the expansion.
(ii) RAS expansions have the important property of *closure under deexcitation*: the CSFs generated by removing an electron from subshell $n\kappa$ and placing it in $n'\kappa$ with $n' < n$ must be in the original CSF set.
(iii) The RAS expansions support orbital transformations that allow the use of *biorthonormal* orbitals: see §8.4.

The original version of GRASP [1] was not designed with this sort of calculation in mind. GRASP92 [2] kept much of the structure of the earlier versions of the code but introduced a user interface with interactive input along the lines of the MCHF package [8], capable of handling the large number of files generated in modern calculations. Early versions of GRASP made no attempt to partition the Hamiltonian matrix into independent J^π blocks (although users could do so manually); computer memories were too small for partitioning to be a major issue. Modern OL and EOL calculations require the extraction of only a small number of eigenvalues and eigenvectors of a very large partitioned Hamiltonian matrix. This can be done very efficiently using the popular Davidson/Liu algorithm [42, 43, 44, 45] which has been implemented by Stathopoulos and Froese Fischer [46] in GRASP92. Other improvements to take advantage of modern multi-processing environments are envisaged.

7.12.2 Example: intercombination transitions in Be-like ions

Ynnerman and Froese Fischer [36] used the RAS approach to study correlation effects on the spin-forbidden transitions $2s^2\ ^1S - 2s2p\ ^3P$ in ions belonging to the Be-like isoelectronic sequence. The 1909Å intercombination line in C III has provided an accurate diagnostic probe for stellar atmospheres and has been identified in the spectra of a wide variety of astronomical objects. Iron is present as an impurity in laboratory plasmas, and the corresponding Be-like resonance and intercombination lines in Fe XXIII have also been used for plasma diganostic purposes. The spin-forbidden transition provides a challenge to *ab initio* methods as the transition energies and transition rates are sensitive both to relativistic effects and to the way in which correlation has been taken into account.

The lowest levels of the Be-like ions associated with the $1s^22s^2$, $1s^22s2p$ and $1s^22p^2$ configurations can be found in the NIST on-line Atomic Spectra Database [33]. Relative to the $2s^2\ ^1S_0$ ground state, the centroid of the even parity $2p^2\ ^3P$ LS term of C III is at 137,478 cm^{-1}, $2p^2\ ^1D$ at 145,876 cm^{-1}, and $2p^2\ ^1S$ at 182,520 cm^{-1}. The odd-parity terms are $2s2p\ ^3P^o$ with centroid 52,420 cm^{-1} and $2s2p\ ^1P^o$ at 102,352 cm^{-1}. The multiplet splittings in $2s2p\ ^3P^o$ are 23.69 and 56.36 cm^{-1} whilst those of the $2p^2\ ^3P^e$ are 28.70 and 47.61 cm^{-1}. In nonrelativistic LS coupling, the terms are independent but L and S are no longer good quantum numbers in the presence of relativistic effects which mix singlet and triplet states with the same value of J. The Breit-Pauli theory is expected to work well in the spectra of low-Z ions like C^{++} as long as this mixing is small. The strength of the electric dipole (E1) intercombination line transition $^1S_0$–$^3P_1$ depends strongly on the mixing of the $^3P_1^o$ and $^1P_1^o$ states and also on the mixing of $2p^2\ ^3P_0$ with the ground state $1s^2\ ^1S_0$ [37, 38]. A technical problem is that when the two states are determined in different OL calculations, the $2p$ orbital for $^1P_1^o$ is much more diffuse than that for $^3P_1^o$, so that correlation orbitals introduced to improve the description of the triplet term are too compact to help improve $^1P_1^o$. An additional orbital of each symmetry is needed to capture the required diffuse character.

Similar tricks are needed to get the best results from MCDHF calculations. The simplest representation of the $^{1,3}P^o$ terms requires the four CSFs, $2s2p_{1/2}\ J = 0, 1$ and $2s2p_{3/2}\ J = 1, 2$. The CSFs with $J = 0, 2$ represent the two states $^3P_0^o$ and $^3P_2^o$ whose energies differ experimentally by 80.05 cm^{-1}. In this minimal CSF basis the splitting is entirely due to the difference in the radial wavefunctions for $2p_{1/2}$ and $2p_{3/2}$ orbitals; it vanishes in the nonrelativistic limit. On the other hand the two CSFs with $J = 1$ are strongly coupled and the secular equation gives two ASFs, one representing the lower $^3P_1^o$ state, the other the higher $^1P_1^o$ state. The lower state has to fall in the right place between the $^3P_0^o$ and $^3P_2^o$ levels, whose energy difference is three orders of magnitude smaller than the $^3P_1^o$ – $^1P_1^o$ splitting. Clearly, balancing the correlation contributions so that all the levels fall in the right place is likely to be a delicate matter.

The RAS solution adopted in [36] was to perform a sequence of EOL calculations adding correlation orbitals in successive layers characterized by the principal quantum number n:

- $n = 2,3$: An EOL calculation on the $2s2p\ ^3P^o_{0,1,2}$ levels, optimizing the $1s, 2s, 2p_{1/2}, 2p_{3/2}, 3s, 3p_{1/2}, 3p_{3/2}, 3d_{3/2}$ and $3d_{5/2}$ orbitals;
- $n = 4$: An OL calculation on the $^1P_1^o$ level, freezing all orbitals with $n \leq 3$ and optimizing with respect to $4s, 4p_{1/2}, \ldots 4f_{7/2}$ orbitals;
- $n = 5$: An OL calculation on the $2s^2\ ^1S_0$ ground state, freezing all orbitals with $n \leq 4$ and optimizing with respect to $5s, 5p_{1/2}, \ldots 5f_{9/2}$ orbitals;
- $n = 6, 7, \ldots$: continue the cycle until the system converges.

Two types of AS expansion were used:

- AS1: Substitutions using CSFs of the form $1s^2\, n\kappa\, n'\kappa'$ (*outer correlation*);
- AS2: The substitutions of AS1 together with CSFs in which one $1s$ orbital also has been excited to take some account of *core polarization.*

Angular momentum substitutions were limited to s, p, d, and f symmetries. The initial EOL calculation aims to get a good approximation to the $^3P^o$ multiplet structure and later stages of the calculation must avoid disturbing this. AS2 calculations aim to treat outer correlation and core polarization on the same footing so as to balance the correlation effect on both initial and final states of a transition.

The transition energies obtained by Ynnerman and Froese Fischer [36] for C III are given in Table 7.14. The transverse photon (Breit) interaction, which was incorporated at each stage as a perturbation as described previously, almost halves the fine structure splittings in C III [36, Table II]. The limiting values of the transition energies taking account of outer correlation only in C III are within 1.5% of the experimental values for both the resonance line and the intercombination line. The core polarization contribution improves this to 0.3% for the resonance line and 0.01 % for the intercombination line. Triple and quadruple excitations make the agreement with experimental transition energies slightly worse. The resonance line oscillator strengths of Table 7.15 are dominated essentially by nonrelativistic effects and are insensitive to the Breit interaction. The Coulomb (velocity) and Babushkin (length) gauge values show signs of coming together as n increases. The calculated intercombination line strengths for the two gauges do not agree well, probably because of large cancellations in the terms that contribute to the matrix elements, and convergence with n is likewise rather irregular.

Table 7.14. Transition energies (cm^{-1}) in C III. From [36, Table III] with permission. Copyright 1995 by the American Physical Society.

n	$^1S_0 - {}^3P_1^o$	$^3P_0^o - {}^3P_1^o$	$^3P_1^o - {}^3P_2^o$	$^1S_0 - {}^1P_1^o$	$^3P_1^o - {}^1P_1^o$
Outer correl.					
2	53452.62	22.20	54.21	115285.5	61832.9
3	53443.83	22.44	54.34	114219.5	60775.7
4	52683.27	22.65	54.26	103818.0	51134.7
5	52723.98	22.68	54.26	103693.3	50969.3
6	52729.80	22.70	54.26	103657.8	50928.0
7	52732.91	22.70	54.26	103688.3	50955.4
8	52717.66	22.70	54.27	103633.1	50915.4
Core polarn.					
2	53507.85	22.17	54.15	115215.0	61707.2
3	52630.20	22.70	56.24	111727.5	59097.3
4	52120.35	21.46	56.87	102948.4	50828.1
5	52291.09	21.00	57.29	102926.1	50635.0
6	52349.00	21.20	57.83	102870.8	50521.8
7	52382.41	21.03	57.64	102724.3	50341.9
8	52370.09	21.07	57.62	102765.0	50394.9
9	52398.97	21.02	57.67	102732.2	50333.2
Selected T,Q.					
6	52327.96	21.654	57.216	102867.0	50539.0
7	52360.61	21.483	57.036	102723.1	50362.5
8	52372.19	21.490	57.067	102725.6	50353.4
Experiment: NIST Atomic Spectra Database [33].					
	52390.75	23.69	56.36	102352.0	49961.3

Ynnerman and Froese Fischer [36] have performed similar calculations for several members of the isoelectronic sequence: C III, N IV, O V, Si XI, Fe XXIII, and Mo XXXIX. The CI expansions converge more rapidly at the high-Z end of the sequence. Results were compared with other calculations where available and with the very limited experimental data on transition rates for the resonance line and the intercombination line.

7.13 X-ray transition energies

Whilst the relativistic corrections to MCDHF predictions in the C III example are small, they cannot be ignored for processes involving inner shells and for heavy atoms. Table 7.16 summarizes contributions to the energy of the $K\alpha_1$

Table 7.15. Transition rates for the intercombination line and oscillator strengths for the resonance line of C III. From [36, Table III] with permission. Copyright 1995 by the American Physical Society.

n	Intercombination Transition rates (s^{-1}) $^3P_1^o - {}^1S_0$		Resonance Oscillator strengths $^1P_1^o - {}^1S_0$	
	Calculation			
Outer correl.	Babushkin	Coulomb	Babushkin	Coulomb
2	46.7	0.5	0.681	0.686
3	46.6	6.7	0.683	0.686
4	90.7	128.2	0.767	0.846
5	91.7	130.3	0.768	0.847
6	91.8	132.2	0.768	0.846
7	91.7	133.3	0.769	0.846
8	92.3	138.3	0.769	0.847
Core polarn.				
2	46.0	60.1	0.679	0.657
3	59.3	60.2	0.675	0.605
4	99.9	188.2	0.758	0.768
5	108.4	184.4	0.757	0.761
6	114.7	233.3	0.757	0.755
7	104.8	173.7	0.760	0.755
8	103.4	170.2	0.760	0.755
9	103.8	170.5	0.760	0.755
Selected T,Q.				
6	111.2	236.4	0.754	0.754
7	101.7	178.1	0.757	0.753
8	100.3	174.3	0.757	0.753
	Experiment			
	120.9 ± 7.0^a		0.754 ± 0.014^b	

[a]Kwong *et al.* [39]; [b]Reistad and Martinson [40].

X-ray line in neutral mercury from a DHF calculation [25]. The $K\alpha_1$ transition is one in which the upper state has a single vacancy in the $1s_{1/2}$ shell, which is then transferred to the $2p_{3/2}$ shell.

Table 7.16. The KL3 ($K\alpha_1$) line energy (eV) in mercury [25]

	E(1s)	E(2p)	E($K\alpha_1$)
DHF: point nucleus	-451,193.9	-532,479.0	71,285.2
DHF: finite nucleus correction	+77.8	+131.2	-53.4
Breit correction ($\omega = 0$)	+313.5	+581.3	-267.8
Transverse photon correction ($\omega \neq 0$)	-7.5	-10.9	+3.4
QED: self energy	+282.6	+481.1	-198.5
QED vacuum polarization	-64.7	-108.7	+44.1
Total	-450,592.2	-521,405.1	70,812.9

An obsolete value, α= 137.0373, was used in [25].

Experimental line energy [32] 70,819.5

Table 7.17. The KL2, KL3 line energies (eV) in mercury [28]

	Theoretical	Experimental
KL2	68 894.3 (23)	68 895.1 (17)
KL3	70 819.0 (22)	70 819.5 (18)

The difference of the two DHF atom energies predicts most of the energy of the X-ray line. The Breit energy, although only about 0.1% of the total, is the next biggest contributor. The correction for finite wavelength is less than 1% of the Breit energy but is needed if we aim for high accuracy. The estimated self-energy and vacuum polarization corrections are slightly smaller than the Breit contribution and have opposite signs. The vacuum polarization contribution is the expectation of the QED second and fourth order potentials of Fullerton and Rinker [29]. The estimate of the self-energy derived from (7.11.1) was based on Z^{eff} values derived from the DHF orbital mean radii, followed by interpolation in hydrogenic tables [30, 31]. Table 7.17 shows similar results for the KL2 and KL3 lines. The results would be slightly changed by adopting current recommended physical constants and slightly different approaches to the calculation of QED corrections [28].

References

1. Dyall K G, Grant I P, Johnson C T and Plummer E P 1989 *Comput. Phys. Commun.* **55**, 425.
2. Parpia F A, Fischer C F and Grant I P 1996 *Comput. Phys. Commun.* **94**, 249.
3. Davé J H, Feldman U, Seely J F, Wouters A, Suckewer S, Hinnov E and Schwob J L 1987 *J. Opt. Soc. Am. B* **4**, 635.
4. Levine M A, Marrs R E, Henderson J R, Knapp D A and Schneider M B 1988 *Physica Scripta* **T22**, 157.
5. Norrington P H 1998, private communication.
6. Drake G W F ed 2005 *Springer Handbook of Atomic, Molecular and Optical Physics* Second edn (New York: Springer-Verlag).
7. Grant I P, Mayers D F and Pyper N C 1976 *J. Phys. B* **9**, 2777.
8. Fischer C F, Brage T and Jönsson P 1997 *Computational Atomic Structure. An MCHF approach* (Bristol and Philadelphia: Institute of Physics).
9. Deslattes R D, Kessler,Jr. E G, Indelicato P, de Billy L, Lindroth E and Anton J 2003 *Rev. Mod. Phys.* **75**, 35.
10. Indelicato P and Desclaux J P 1990 *Phys. Rev. A* **42**, 5139.
11. Blundell S A, Johnson, W R and Sapirstein J 1990 *Phys. Rev. A* **42**, 3751.
12. Thomas L H 1927 *Proc. Camb. Phil. Soc.* **23**, 542.
13. Fermi E 1928 *Z. Physik* **48**, 73.
14. Condon E U and Shortley G H 1953 *The Theory of Atomic Spectra* (Cambridge: University Press).
15. Klapisch M 1971 *Comput Phys Commun* **2**, 239.
16. Bar-Shalom A, Klapisch M and Oreg J 2001 *J. Quant. Spectrosc. Rad. Transf.* **71**, 169.
17. Eissner W, Jones M and Nussbaumer H 1974 *Comput Phys Commun* **8**, 270.
18. Koopmans T 1933 *Physica* **1**, 104.
19. Brillouin L 1932 *J. Phys. Paris* **3**, 373.
20. Brillouin L 1933 *Actualités Scientifiques et Industrielles* no. 71 (Paris: Hermann).
21. Bauche J and Klapisch M 1972 *J. Phys. B: Atom. Molec. Phys.* **5**, 29.
22. Labarthe J J 1972 *J. Phys. B: Atom. Molec. Phys.* **5**,181.
23. Fischer C F 1977 *The Hartree-Fock Method for Atoms* (New York: John Wiley).
24. Fischer C F 1986 *Computer Physics Reports* **3**, no. 5, 273.
25. Grant I P and McKenzie B J 1980 *J. Phys. B: Atom. Molec. Phys.* **13**, 2671.
26. Johnson W R and Soff G 1985 *At. Data Nucl. Data Tables* **33**, 485.
27. Welton T A 1948 *Phys. Rev.* **74**, 1174.
28. Deslattes R D, Kessler,Jr. E G, Indelicato P, de Billy L, Lindroth E, Anton J, Coursey J, Schwab D J, Olsen K and Dragoset R A 2003 *X-ray Transition Energies* (version 1.0) [Online] Available http://physics.nist.gov/XrayTrans [2003, October 2] National Institute of Standards and Technology, Gaithersburg, MD.
29. Fullerton L W and Rinker, Jr., G A 1976 *Phys. Rev. A* **13**, 1283.
30. Mohr P J 1983 *At. Data Nucl. Data Tables* **29**, 453.
31. Mohr P J and Kim Y-K 1992 *Phys. Rev. A* **45** 2727.
32. Bearden J A and Burr A F 1967 *Rev. Mod. Phys.* **39**, 125.
33. On-line NIST Atomic Spectra Database URL: http://physics.nist.gov/cgi-bin/AtData/main_asd.
34. Froese Fischer C 2005 in [Chapter 21, §21.5.3]handbook.

35. Olsen J, Roos B O, Jørgensen and Jensen H J Aa 1988 *J. Chem. Phys,* **89** 2186.
36. Ynnerman A and Froese Fischer C 1995 *Phys. Rev. A* **51**, 2020.
37. Froese Fischer C 1994 *Physica Scripta* **49**, 323.
38. Fleming J, Hibbert A and Stafford R P *Physica Scripta* **49**, 316.
39. Kwong H S *et al.* 1993 *Astrophys. J.* **411**, 431.
40. Reistad N and Martinson I 1986 *Phys. Rev. A* **34**, 2632.
41. Brewer L 1971 *J. Opt. Soc. Am.* **61**, 1101.
42. Davidson E R 1975 *J. Comput. Phys.* **17**, 87.
43. Davidson E R 1989 *Comput. Phys. Commun.* **53**, 49.
44. Liu B 1978 in *Numerical Algorithms in Chemistry: Algebraic Methods.* (ed. Moler C and Shavitt I) (Lawrence Berkeley Laboratory).
45. Murray C W, Racine S C and Davidson E R 1992 *J. Comput. Phys.* **103**, 382.
46. Stathopoulos A and Froese Fischer C 1994 *Comput. Phys. Commun.* **79**, 268.

8

Computation of atomic properties

The calculation of radiative transition data, hyperfine interaction parameters, and isotope shifts are representative applications of relativistic atomic structure methods considered in this chapter. Line transitions and single channel photo-ionization are the main focus of attention.

8.1 Relativistic radiative transition theory

8.1.1 Line transitions

The huge range of radiative data needed for applications can be gauged by the experimental and theoretical effort portrayed, for example, in [1]. An atom in an excited state Γ can decay spontaneously to a lower state Γ' emitting a photon of energy $h\nu = E_\Gamma - E_{\Gamma'}$ with wavelength $\lambda = c/\nu$. The total power radiated per unit source volume into a solid angle $d\Omega$ is

$$P\, d\Omega = h\nu\, N_\Gamma\, A_{\Gamma\Gamma'}\, \frac{d\Omega}{4\pi} \tag{8.1.1}$$

where N_Γ is the number of atoms in state Γ per unit volume, and $A_{\Gamma\Gamma'}$ is the number of spontaneous transitions per unit time interval. Standard time-dependent perturbation theory gives the corresponding decay rate (in atomic units) as

$$A_{\Gamma\Gamma'} = 2\pi\, |\, \langle\Gamma'|\, H_{int}\, |\Gamma\rangle\, |^2 \tag{8.1.2}$$

where

$$H_{int} = \frac{1}{c}\int j^\mu(x) a_\mu(x)\, d^4x \tag{8.1.3}$$

is the electron-photon interaction Hamiltonian.

8.1.2 Multipole expansion of the radiation field

It is convenient to expand the electromagnetic potentials as a sum of multipole operators [2]. Derivations of these expressions (using different conventions for the relativistic metric and for electromagnetic units) may be found in [3, 4] and [5]. The four-potential for the free radiation field, $a_\mu(x) = (\Phi, c\boldsymbol{A})$,[1] satisfies

$$\Box\, a_\mu(x) = 0, \quad \partial_\mu\, a_\mu(x) = 0,$$

in Lorenz gauge. The wave equation $\Box\, a_\mu(x) = 0$ has scalar plane wave solutions

$$\Phi(x) \to \mathrm{e}^{-i(\omega t - \boldsymbol{k}\cdot\boldsymbol{x})}, \quad \omega^2 = |\boldsymbol{k}|^2.$$

Using the standard expansion

$$\mathrm{e}^{i\boldsymbol{k}\cdot\boldsymbol{x}} = \sum_{L=0}^{\infty} i^L\,(2L+1)\,j_L(\omega\, r/c)\,P_L(\cos\Theta) \tag{8.1.4}$$

gives a complete set of multipole solutions

$$\Phi_{LM}(\boldsymbol{x};\omega) = i^L\,(2L+1)\,j_L(\omega\, r/c)\,C_{LM}(\theta,\varphi), \tag{8.1.5}$$

satisfying

$$(\nabla^2 + \omega^2)\Phi_{LM}(\boldsymbol{x};\omega) = 0,$$

which can be used to develop general solutions of the wave equation.

We start with the scalar potential

$$a^0(x) = \Phi_{LM}(\boldsymbol{x};\omega)\,\mathrm{e}^{-i\omega t}. \tag{8.1.6}$$

The Lorenz condition

$$\mathrm{div}\boldsymbol{A} + \frac{1}{c^2}\frac{\partial\Phi}{\partial t} = 0$$

can be satisfied with a solution $\boldsymbol{A}^l_{LM}(\boldsymbol{x};\omega)\,\mathrm{e}^{-i\omega t}$ where

$$c\boldsymbol{A}^l_{LM}(\boldsymbol{x};\omega) = \frac{-ic}{\omega}\,\nabla\,\Phi_{LM}(\boldsymbol{x};\omega). \tag{8.1.7}$$

This potential is said to be *irrotational* because curl $\boldsymbol{A}^l_{LM} = 0$. Denoting the orbital angular momentum operator by $\boldsymbol{L} = -i\boldsymbol{x}\times\boldsymbol{\nabla}$, we have two more solutions

$$c\boldsymbol{A}^m_{LM}(\boldsymbol{x};\omega) = \frac{\boldsymbol{L}}{\sqrt{L(L+1)}}\,\Phi_{LM}(\boldsymbol{x};\omega), \tag{8.1.8}$$

and

---

[1] The notation differs from that of [2] where the components of the vector potential are written A_i rather than cA_i following [6, Equation 18-28], which uses the same electromagnetic units but writes the time component as x^4 rather than x^0.

$$c\boldsymbol{A}^{e}_{LM}(\boldsymbol{x};\omega) = \frac{c\boldsymbol{\nabla}\times\boldsymbol{L}}{\omega\sqrt{L(L+1)}}\,\Phi_{LM}(\boldsymbol{x};\omega). \tag{8.1.9}$$

The superscripts l, m, e identify the potentials as longitudinal, magnetic, and electric, type respectively. The methods of Appendix B.3.17 enable us to write

$$\begin{aligned}
c\mathbf{A}^{e}_{LM}(\mathbf{x}) &= -i^{L+1}\sqrt{L(2L+3)}\left(\mathbf{e}\times\mathbf{C}^{L+1}\right)_{LM} j_{L+1}(\rho)\\
&\quad -i^{L-1}\sqrt{(L+1)(2L-1)}\left(\mathbf{e}\times\mathbf{C}^{L-1}\right)_{LM} j_{L-1}(\rho)\\
c\mathbf{A}^{l}_{LM}(\mathbf{x}) &= -i^{L+1}\sqrt{L(2L+3)}\left(\mathbf{e}\times\mathbf{C}^{L+1}\right)_{LM} j_{L+1}(\rho)\\
&\quad +i^{L-1}\sqrt{(L+1)(2L-1)}\left(\mathbf{e}\times\mathbf{C}^{L-1}\right)_{LM} j_{L-1}(\rho)\\
c\mathbf{A}^{m}_{LM}(\mathbf{x}) &= -i^{L}(2L+1)\left(\mathbf{e}\times\mathbf{C}^{L}\right)_{LM} j_{L}(\rho)
\end{aligned} \tag{8.1.10}$$

where $\mathbf{e}$ is a unit vector. $\mathbf{A}^{e}_{LM}$ and $\mathbf{A}^{l}_{LM}$ have the same parity, $(-1)^{L+1}$, whilst $\mathbf{A}^{m}_{LM}$ and Φ_{LM} have the opposite parity $(-1)^{L}$. These multipole expressions are mutually orthogonal with respect to integration over the unit sphere.

A radiation plane wave polarized in the direction $\mathbf{e}$ propagating along the z-axis can be expressed in terms of these multipole fields. Writing the spherical vector components $\mathbf{e}_0 = \mathbf{e}_z$, $\mathbf{e}_{\pm 1} = \mp(\mathbf{e}_x \pm i\mathbf{e}_y)/\sqrt{2}$ so that $\mathbf{k} = (\omega/c)\mathbf{e}_0$, we find

$$\nabla\left(\mathbf{e}_0 \mathrm{e}^{i\mathbf{k}\cdot\mathbf{x}}\right) = \mathbf{e}_0\frac{d}{dz}\mathrm{e}^{i\omega z/c} = i\frac{\omega}{c}\mathbf{e}_0\mathrm{e}^{i\mathbf{k}\cdot\mathbf{x}},$$

so that

$$\mathbf{e}_0\mathrm{e}^{i\mathbf{k}\cdot\mathbf{x}} = -\frac{ic}{\omega}\nabla\mathrm{e}^{i\omega z/c} = \sum_L c\mathbf{A}^{l}_{L0}(\mathbf{x};\omega)$$

by (8.1.7), confirming our assertion that these fields are longitudinally polarized. Transversely polarized vector solutions are obtained by replacing $\mathbf{e}_0$ by $\mathbf{e}_{\pm 1}$, and we can write, for $q = \pm 1$,

$$\mathbf{e}_q\frac{d}{dz}\mathrm{e}^{i\omega z/c} = \sum_L a^{l}_{L}\mathbf{A}^{l}_{Lq}(\mathbf{x};\omega) + a^{e}_{L}\mathbf{A}^{e}_{Lq}(\mathbf{x};\omega) + a^{m}_{L}\mathbf{A}^{m}_{Lq}(\mathbf{x};\omega).$$

The divergence of the left-hand side vanishes, so that $a^{l}_{L} = 0$ for all values of L. To determine the remaining coefficients, we first take the curl of both sides, giving $a^{m}_{L} = qa^{e}_{L}$, and by evaluating $\mathbf{L}\cdot\mathbf{e}_q e^{di\omega z/c}$ we find $a^{e}_{L} = -1/\sqrt{2}$. Thus for propagation along the z-axis, the vector potential with polarization q is

$$A_q = \mathbf{e}_q\mathrm{e}^{i\omega z/c} = -\frac{1}{\sqrt{2}}\sum_L \mathbf{A}^{e}_{Lq}(\mathbf{x};\omega) + q\mathbf{A}^{m}_{Lq}(\mathbf{x};\omega),$$

and, more generally,

$$A_q = \mathbf{e}_q\mathrm{e}^{i\mathbf{k}\cdot\mathbf{x}} = -\frac{1}{\sqrt{2}}\sum_{LM}\left(\mathbf{A}^{e}_{LM}(\mathbf{x};\omega) + q\mathbf{A}^{m}_{LM}(\mathbf{x};\omega)\right)\mathcal{D}^{L}_{Mq}(R), \tag{8.1.11}$$

where R is the rotation that carries the z-axis into the direction of $\mathbf{k}$.

Physical predictions should be unaffected by gauge transformations of the form (2.6.24):

$$a_\mu(x) \to a_\mu(x) + \partial_\mu \Lambda(x), \quad \Box \Lambda(x) = 0.$$

A spherical tensor solution is $\Lambda(x) = G_L \Phi_{LM}(\mathbf{x};\omega)\exp(-i\omega t)$, where G_L is an arbitrary constant. By (8.1.7), $\partial_\mu \Lambda(x)$ is longitudinally polarized. The tensorial properties are such that the most general form of electric multipole potential is

$$\begin{aligned}\Phi_{LM}(\mathbf{x};\omega) &\to \left(1 - i\frac{\omega}{c}G_L\right)\Phi_{LM}(\mathbf{x};\omega)\\ \mathbf{A}^e_{LM}(\mathbf{x};\omega) &\to \mathbf{A}^e_{LM}(\mathbf{x};\omega) + G_L\mathbf{A}^l_{LM}(\mathbf{x};\omega);\end{aligned} \tag{8.1.12}$$

$\mathbf{A}^m_{LM}(\mathbf{x};\omega)$ does not depend on G_L. The Hamiltonian for interaction of a Dirac electron with the radiation field is

$$H_{int}(t) = \frac{1}{c}\int j^\mu(x) a_\mu(x)\, d^3x. \tag{8.1.13}$$

so that a gauge transformation adds $a_\mu \to a_\mu + \partial_\mu \Lambda$ giving an additional gauge dependent term

$$H_{int} \to H_{int} + \frac{G_L}{c}\int j^\mu(x)\partial_\mu \Lambda(x)\, d^4x.$$

Now

$$j^\mu(x)\partial_\mu \Lambda(x) = \partial_\mu\left(\Lambda(x) j^\mu(x)\right) - \Lambda(x)\partial_\mu j^\mu(x);$$

the space-time integral of the first term on the right can be converted to a surface integral that vanishes at infinity, and the gauge dependent term vanishes if, as in (2.6.25), *there is local conservation of charge*:

$$\partial_\mu j^\mu(x) = \frac{\partial \rho(x)}{\partial t} + \operatorname{div}\mathbf{j}(x) = 0.$$

Local charge conservation is built into all particle wave equations, §2.5, and also into the structure of QED. However, SCF equations are not constrained automatically by local charge conservation, and it is well-known that HF/DHF and MCHF/MCDHF wavefunctions give radiative transition probabilities that are sensitive to the choice of gauge parameter. The reason is that *individual SCF orbitals* are computed with nonlocal potentials coupled to other orbitals, so that charge conservation in SCF theories involves something much more complicated [2, §7]. It is now generally accepted that a weak dependence of radiation matrix elements on G_L is one indicator of wavefunction quality in many-electron systems, although it is insufficient on its own to guarantee the physical correctness of numerical radiative transition rates.

8.2 Emission and absorption by one-electron atoms

The quantized electron current density[2] can be written

$$j^\mu(x) = -ec \sum_{\alpha\beta} : q_\alpha^\dagger q_\beta : \overline{\psi}_\alpha(x)\gamma^\mu \psi_\beta(x)$$

whilst the monochromatic quantized radiation field becomes a combination of positive and negative frequencies,

$$a_\mu(x;\omega)\mathrm{e}^{-i\omega t} + a_\mu^\dagger(x;\omega)\mathrm{e}^{i\omega t}.$$

The four-potentials can be expanded in terms of multipoles

$$\begin{aligned} a_\mu(x;\omega) &= \sum_{LM} c_{LM}(\omega)\,(\omega/\pi c)^{1/2} \mathcal{A}_{LM\mu}(x;\omega), \\ a_\mu^\dagger(x;\omega) &= \sum_{LM} c_{LM}^\dagger(\omega)\,(\omega/\pi c)^{1/2} \mathcal{A}_{LM\mu}^\dagger(x;\omega) \end{aligned} \tag{8.2.1}$$

where $c_{LM}^\dagger(\omega)$ creates and $c_{LM}(\omega)$ annihilates a photon of frequency ω, and the components of the electric and magnetic multipole (covariant) four-potentials are given by

$$\mathcal{A}_{LM\mu}^{e,m}(x;\omega) = (\Phi_{LM}^{e,m}(\mathbf{x};\omega,G_L),\ -c\mathbf{A}_{LM}^{e,m}(\mathbf{x};\omega,G_L))\,. \tag{8.2.2}$$

From (8.1.12), the most general form for an electric multipole field is

$$\begin{aligned} \Phi_{LM}^e(\mathbf{x};\omega;G_L) &= (1 - i\omega G_L/c)\,\Phi_{LM}(\mathbf{x};\omega), \\ \mathbf{A}_{LM}^e(\mathbf{x};\omega,G_L) &= \mathbf{A}_{LM}^e(\mathbf{x};\omega) + G_L \mathbf{A}_{LM}^l(\mathbf{x};\omega)) \end{aligned} \tag{8.2.3}$$

and for magnetic multipoles

$$\Phi_{LM}^m(\mathbf{x};\omega;G_L) = 0, \quad \mathbf{A}_{LM}^m(\mathbf{x};\omega,G_L) = \mathbf{A}_{LM}^m(\mathbf{x};\omega) \tag{8.2.4}$$

The interaction Hamiltonian can therefore be written in terms of emission and absorption operators

$$H_{int} = \sum_{LM} \{[\mathcal{H}_{LM}^e]_{em} + [\mathcal{H}_{LM}^m]_{em} + [\mathcal{H}_{LM}^e]_{abs} + [\mathcal{H}_{LM}^m]_{abs}\}\,,$$

where

---

[2] We use $q, q^\dagger$ for electron annihilation and creation operators to avoid confusion with the four potentials.

$$
\begin{aligned}
[\mathcal{H}^{e,m}_{LM}]_{em} &= \sum_{\alpha\beta} q^\dagger_\alpha q_\beta \, c^\dagger_{LM}(\omega)\, [\mathcal{M}^{e,m}_{LM}(t)]^*_{\alpha\beta}, \\
[\mathcal{H}^{e,m}_{LM}]_{abs} &= \sum_{\alpha\beta} q^\dagger_\alpha q_\beta \, c_{LM}(\omega)\, [\mathcal{M}^{e,m}_{LM}(t)]_{\alpha\beta}.
\end{aligned}
\tag{8.2.5}
$$

with

$$
[\mathcal{M}^{e,m}_{LM}(t)]_{\alpha\beta} = \left(\frac{\omega}{\pi c}\right)^{1/2} \int \tilde{\psi}_\alpha(x)\gamma^\mu \psi_\beta(x)\, \mathcal{A}^{e,m}_{LM\mu}(x;\omega)\, d^3x\, \mathrm{e}^{-i\omega t}.
$$

If we now write $\psi_\alpha(x) = \phi_\alpha(\mathbf{x})\exp(-iE_\alpha t)$ and define the tensor operator

$$
\mathcal{O}^{e,m}_{LM} = [\Phi^{e,m}_{LM}(\mathbf{x};\omega) - c\alpha\cdot \mathbf{A}^{e,m}_{LM}], \tag{8.2.6}
$$

then

$$
[\mathcal{M}^{e,m}_{LM}(t)]_{\alpha\beta} = \left(\frac{\omega}{\pi c}\right)^{1/2} \mathrm{e}^{i(E_\alpha - E_\beta - \omega)t} \langle \alpha \,|\, \mathcal{O}^{e,m}_{LM} \,|\, \beta\rangle \tag{8.2.7}
$$

where, in 3-dimensional notation, the transition amplitude is

$$
\langle \alpha \,|\, \mathcal{O}^{e,m}_{LM} \,|\, \beta\rangle = \int \phi^\dagger_\alpha(\mathbf{x})\mathcal{O}^{e,m}_{LM}\phi_\beta(\mathbf{x})\, d^3x.
$$

This can be further reduced by applying the Wigner-Eckart theorem in the form

$$
\langle \alpha \,|\, \mathcal{O}^{e,m}_{LM} \,|\, \beta\rangle = \begin{pmatrix} m_\alpha & L & j_\beta \\ j_\alpha & M & m_\beta \end{pmatrix} \langle \alpha \,\|\, \mathbf{o}^L \,\|\, \beta\rangle^{e,m}, \tag{8.2.8}
$$

so that the rate, (8.1.2) for spontaneous emission of photons from the upper state β to a lower state α involving one active electron is

$$
A_{\alpha\beta} = \sum_{m_\alpha} \frac{1}{2j_\beta + 1} \sum_{m_\beta} 2\pi \left|\mathcal{M}^e_{\alpha\beta}\right|^2
$$

where we have summed over the $2j_\alpha + 1$ degenerate initial states and averaged over the $2j_\beta + 1$ degenerate final states. Substituting from (8.2.7) and (8.2.8) and summing over the magnetic quantum numbers gives

$$
A_{\alpha\beta} = \frac{2\omega}{c}\frac{1}{(2L+1)(2j_\beta+1)} |\langle \alpha \,\|\, \mathbf{o}^L \,\|\beta\rangle|^2 \tag{8.2.9}
$$

It is convenient also to define the *oscillator strength for emission* as

$$
f_{\alpha\beta} = -\frac{c}{2\omega^2}\frac{2j_\beta+1}{2j_\alpha+1}A_{\alpha\beta} = -\frac{1}{\omega}\frac{|\langle \alpha \,\|\, \mathbf{o}^L \,\|\beta\rangle|^2}{(2L+1)(2j_\alpha+1)}
$$

By convention, this is negative; the numerator on the right is symmetric with respect to the labels α, β and we define the (positive) *oscillator strength for absorption* by

$$f_{\beta\alpha} = -\frac{2j_\alpha + 1}{2j_\beta + 1} f_{\alpha\beta}$$

The *weighted oscillator strength* or gf-value, defined as the positive symmetric quantity

$$(gf)_{\alpha\beta} = g_\beta\, f_{\beta\alpha} = -g_\alpha\, f_{\alpha\beta}$$

where $g_\alpha = 2j_\alpha + 1$, is proportional to the intensity of the spectral line.

Table 8.1. Selection rules for radiative transitions

Matrix elements for multipole L are nonvanishing if		
$\lvert j_\alpha - j_\beta \rvert \le L \le j_\alpha + j_\beta$,		
$j_\alpha + j_\beta + L$ is even or odd:		
	Odd	Even
Electric multipole	$\kappa_\alpha \kappa_\beta > 0$	$\kappa_\alpha \kappa_\beta < 0$
Magnetic multipole	$\kappa_\alpha \kappa_\beta < 0$	$\kappa_\alpha \kappa_\beta > 0$

8.2.1 Evaluation of one-electron transition amplitudes

The one-particle reduced matrix elements of (8.2.8) can be written

$$\langle \alpha \,\|\, \mathbf{o}^L \,\|\, \beta \rangle^{e,m} = \langle j_\alpha \,\|\mathbf{C}^L\|\, j_\beta \rangle \,.\, \mathbf{M}^{e,m}_{\alpha\beta}(\omega; G_L) \tag{8.2.10}$$

where $\mathbf{M}^{e,m}_{\alpha\beta}(\omega; G_L)$ is a radial integral that can be expressed [2] as a linear combination of the three integrals

$$\begin{aligned} J_L &= \int_0^\infty \left(P^*_\alpha(r)P_\beta(r) + Q^*_\alpha(r)Q_\beta(r)\right)\, j_L(\omega r/c)\, dr \\ I^\pm_L &= \int_0^\infty \left(P^*_\alpha(r)Q_\beta(r) \pm Q^*_\alpha(r)P_\beta(r)\right)\, j_L(\omega r/c)\, dr \end{aligned} \tag{8.2.11}$$

The magnetic multipole amplitude

$$\mathbf{M}^m_{\alpha\beta}(\omega; G_L) = -i^{L+1}\frac{2L+1}{\sqrt{L(L+1)}}\,(\kappa_\alpha + \kappa_\beta)\, I^+_L \tag{8.2.12}$$

is independent of G_L. For electric multipoles,

$$\mathbf{M}^e_{\alpha\beta}(\omega; G_L) = \mathbf{M}^e_{\alpha\beta}(\omega; 0) + G_L \mathbf{M}^l_{\alpha\beta}(\omega), \tag{8.2.13}$$

where the Coulomb gauge integral is

$$\mathbf{M}^{e}_{\alpha\beta}(\omega;0) = -i^{L}\left\{\left(\frac{L}{L+1}\right)^{1/2}\left[((\kappa_\alpha-\kappa_\beta)I^{+}_{L+1}+(L+1)I^{-}_{L+1}\right]\right.$$
$$\left.-\left(\frac{L+1}{L}\right)^{1/2}\left[((\kappa_\alpha-\kappa_\beta)I^{+}_{L-1}-LI^{-}_{L-1}\right]\right\}$$

and the longitudinal part is

$$\mathbf{M}^{l}_{\alpha\beta}(\omega) = -i^{L}\left\{\left[((\kappa_\alpha-\kappa_\beta)I^{+}_{L+1}+(L+1)I^{-}_{L+1}\right]\right.$$
$$\left.+\left[((\kappa_\alpha-\kappa_\beta)I^{+}_{L-1}-LI^{-}_{L-1}\right]-(2L+1)J_L\right\}$$

8.2.2 The nonrelativistic limit: Pauli approximation

The expression, §3.7.1,

$$Q \approx \frac{1}{2c}\left(\frac{dP}{dr}+\frac{\kappa P}{r}\right)\{1+O(1/c^2)\} \tag{8.2.14}$$

gives a good approximation to the small component in the nonrelativistic limit when P satisfies the Schrödinger radial equation

$$-\frac{1}{2}\left(\frac{d^2P}{dr^2}-\frac{l(l+1)}{r^2}P\right)-(V+\epsilon)P = O(1/c^2).$$

The Dirac wavefunction is normalized so that

$$\int_0^\infty (|P|^2+|Q|^2)\,dr = 1,$$

giving

$$\int_0^\infty \left\{|P|^2+\frac{1}{4c^2}\left(\frac{dP^*}{dr}+\frac{\kappa P^*}{r}\right)\left(\frac{dP}{dr}+\frac{\kappa P}{r}\right)\right\}dr = \left\{1+O(1/c^4)\right\},$$

which can be written in terms of a normalized Schrödinger radial amplitude, $R(r)$, such that

$$\int_0^\infty |R|^2\,dr = 1, \quad R(r) = \left\{1+T_l/4c^2+O(1/c^4)\right\}P(r),$$

where T_l is the Schrödinger radial kinetic energy operator

$$T_l = -\frac{1}{2}\frac{d^2}{dr^2}+\frac{l(l+1)}{2r^2}.$$

Consider first the magnetic transition amplitude

$$\mathbf{M}^m_{\alpha\beta}(\omega) = -i^{L+1}\frac{2L+1}{\sqrt{L(L+1)}}\,\frac{\kappa_\alpha+\kappa_\beta}{2c} \tag{8.2.15}$$
$$\times \int_0^\infty \left(\frac{d}{dr}(P^*_\alpha P_\beta) + \frac{\kappa_\alpha+\kappa_\beta}{r}P^*_\alpha P_\beta\right) j_L(\omega r/c)\,dr(1+O(1/c^2)).$$

This can be simplified by integrating the derivative term by parts and using the spherical Bessel function relation

$$(2L+1)\,\frac{d\,j_L(x)}{dx} = L\,j_{L-1}(x) - (L+1)\,j_{L+1}(x),$$

giving

$$\mathbf{M}^m_{\alpha\beta}(\omega) = i^{L+1}\frac{1}{\sqrt{L(L+1)}}\,\frac{\kappa_\alpha+\kappa_\beta}{2c} \tag{8.2.16}$$
$$\times \int_0^\infty P^*_\alpha P_\beta \left\{\left[L\frac{\omega}{c}\,j_{L-1}(\omega r/c) - (2L+1)\frac{\kappa_\alpha+\kappa_\beta}{r}j_L(\omega r/c)\right]\right.$$
$$\left. -\,(L+1)\,\frac{\omega}{c}\,j_{L+1}(\omega r/c)\right\} dr\,(1+O(1/c^2)).$$

We can assume that the transition wavelength is large compared with atomic dimensions, so that we need retain only the leading terms of the power series expansion[3] of the Bessel function,

$$j_L(x) = \frac{x^L}{(2L+1)!!} - \frac{x^{L+2}}{2.(2L+3)!!} + \ldots$$

The square bracket in (8.2.16) provides the leading term in the long wavelength approximation,

$$\mathbf{M}^m_{\alpha\beta}(\omega) = \frac{i^{L+1}}{2c}\,\frac{\kappa_\alpha+\kappa_\beta}{\sqrt{L(L+1)}} \tag{8.2.17}$$
$$\times \frac{L-\kappa_\alpha-\kappa_\beta}{(2L-1)!!}\left(\frac{\omega}{c}\right)^L \int_0^\infty R^*_\alpha R_\beta r^{L-1}\,dr\,(1+O(1/c^2)).$$

The most important magnetic multipole transitions are the dipole, $L = 1$, for which $j_\alpha = j_\beta, j_\beta \pm 1$ depending on the relative signs of κ_α and κ_β, given in Table 8.1. In this case, the radial integral (8.2.14) reduces to the overlap of the two radial amplitudes, which is zero when $\alpha \neq \beta$ have the same angular symmetry; a rather tedious calculation involving higher order terms in $O(1/c^2)$ is then needed. A particularly interesting case is the magnetic dipole transition $1s^2\,{}^1S$ – $1s2s\,{}^3S$ in helium-like ions. Drake [7, 8] has shown that this

[3] The full expression may still be needed in calculations involving hard X-rays.

cancellation makes the dominant contribution to the transition rate scale like Z^{10} so that it is much brighter than the alternative two-photon electric dipole transition mechanism for modest values of the atomic number Z, making this line important in the spectra of both laboratory and astrophysical sources such as the solar corona.

Similarly, the Coulomb gauge electric multipole matrix element gives

$$\begin{aligned}\mathbf{M}^{e}_{\alpha\beta}(\omega;0) &= -i^{L}\frac{\sqrt{L(L+1)}}{(2L-1)!!}\frac{\omega^{L-1}}{c^{L}}\left(1+O(1/c^{2})\right)\\ &\times\int_{0}^{\infty}\left(r^{(L-1)/2}R^{*}_{\alpha}\right)\\ &\qquad\times\left(\frac{d}{dr}-\frac{(l_{\alpha}-l_{\beta})(l_{\alpha}+l_{\beta}+1)-L+1}{2r}\right)\left(r^{(L-1)/2}R_{\beta}\right)dr.\end{aligned} \tag{8.2.18}$$

The most important electric dipole case, $L=1$ has $l_{\alpha}=l_{\beta}\pm 1$, so that

$$\begin{aligned}\mathbf{M}^{e}_{\alpha\beta}(\omega;0) &= \frac{\sqrt{2}}{c}\left(1+O(1/c^{2})\right)\\ &\times\int_{0}^{\infty}R^{*}_{\alpha}(-i)\left(\frac{d}{dr}-\frac{(l_{\alpha}-l_{\beta})(l_{\alpha}+l_{\beta}+1)}{2r}\right)R_{\beta}\,dr,\end{aligned} \tag{8.2.19}$$

which is just the standard nonrelativistic *dipole velocity* matrix element [9, 10, 11, 12].

The choice $G_{L}=\sqrt{(L+1)/L}$, introduced by Babushkin [13, 14, 15], gives an expression that is dominated in the long wavelength limit by the J_{L} integral, replacing (8.2.19) by

$$\begin{aligned}\mathbf{M}^{e}_{\alpha\beta}(\omega;\sqrt{(L+1)/L}) &= -\left(\frac{L+1}{L}\right)^{1/2}\frac{i^{L}}{(2L-1)!!}\\ &\times\left(\frac{\omega}{c}\right)^{L}\left(1+O(1/c^{2})\right)\int_{0}^{\infty}R^{*}_{\alpha}r^{L}R_{\beta}\,dr.\end{aligned} \tag{8.2.20}$$

In the dipole case, this reduces to the *dipole length* matrix element,

$$\mathbf{M}^{e}_{\alpha\beta}(\omega;0) = -i\omega\frac{\sqrt{2}}{c}\left(1+O(1/c^{2})\right)\int_{0}^{\infty}R^{*}_{\alpha}\,r\,R_{\beta}\,dr. \tag{8.2.21}$$

The relation between (8.2.19) and (8.2.21) is usually established in textbooks by noting the nonrelativistic operator relation

$$[\mathbf{x},H] = i\hbar\mathbf{p}/m = i\hbar\mathbf{v},$$

where H is a Schrödinger Hamiltonian with a local potential. The matrix element between two eigenstates of H is then

$$\langle\alpha\,|\,[\mathbf{x},H]\,|\,\beta\rangle = (E_{\alpha}-E_{\beta})\langle\alpha\,|\,\mathbf{x}\,|\,\beta\rangle = i\langle\alpha\,|\,\mathbf{v}\,|\,\beta\rangle$$

which confirms our relation above. This derivation assumes that $H = T + V$, where T is the usual nonrelativistic kinetic energy $T = \mathbf{p}^2/2m$, and the operator V is multiplication by a function $V(\mathbf{x})$ so that the operators V and $\mathbf{x}]$ commute.. Such a local potential is needed for pointwise charge conservation, which is destroyed in self-consistent field models, along with gauge independence of the transition matrix elements.

8.3 Radiative transitions in many-electron atoms

The theory of §6.9.2 gives the matrix elements of quantized one-particle operators between CSFs T and T'. Because the multipole interaction operators making up H_{int}, (8.2.5), are of the type $\mathbf{F}$ of (6.9.15), we can use the decomposition (6.9.17) or (6.10.13) to express the reduced matrix elements in terms of the one electron reduced matrix elements (8.2.9):

$$\langle T \| \, \mathbf{O}^L \| \, T' \rangle = \sum_{\alpha\beta} (-1)^{\Delta_\alpha + \Delta_\beta{}'} \, (N_\alpha N_\beta{}')^{1/2} (T_\alpha \{ | T_\alpha{}') \, (T_\beta | \} T_\beta{}')$$

$$\times \, \delta_{\widetilde{T}_\alpha, T_\alpha{}'} \delta_{\widetilde{T}'_\beta, T_\beta} \prod_{i \neq \alpha, \beta} \delta_{T_i, T_i{}'} \, \frac{[J]^{1/2}}{[j_\alpha]^{1/2}} \, \mathbb{R}_{\alpha,\beta}(T; T') \, \langle \alpha \| \, \mathbf{o}^L \| \, \beta \rangle, \tag{8.3.1}$$

This procedure is followed in the GRASP `OSCL` program, where the above equation is written [16, Equation (60)]

$$\langle T \| \, \mathbf{O}^k \| \, T' \rangle = \sum_{\alpha\beta} d^L_{\alpha\beta}(TT') \langle \alpha \| \, \mathbf{o}^L \| \, \beta \rangle. \tag{8.3.2}$$

The CSFs T and T' are essentially linear combinations of products of Dirac spinor amplitudes that have been constructed either as eigenstates of a model potential or as SCF orbitals. The expression (8.3.1) therefore takes no account of orbital relaxation on the transition amplitudes. Labzowsky, Klimchitskaya, and Dmitriev [17, Chapter 3] go some way to remedy this using S-matrix methods, mainly for two-electron ions.

8.3.1 Transitions in highly ionized atoms: Fe XXIII

Results for selected transitions from the EAL calculation reported in Table 7.13 are presented in Table 8.2. The computed wavelengths are within about 1 Å of the NIST values [18] at the shorter wavelengths, but a more elaborate model would be needed to improve the agreement at longer wavelengths. The electric and magnetic dipole transition rates, computed with theoretical wavelengths in the Coulomb gauge, agree remarkably well with the rates given in the NIST tables. Better quality wavefunctions are needed to make the ratio, R, of Coulomb to Babushkin gauge closer to unity for most of the transitions.

Table 8.2. Selected line transitions in Fe XXIII

Electric dipole

$i-k$			$\lambda(\mathring{A})$ (a)	(b)	A_{ki} (10^8 s^{-1}) (a)	(b)	R
1 – 2	$^1S_0$ –	$^3P_1^o$	262.78	263.76	4.72 E –01	4.8 E –01	0.72
1 – 5	$^1S_0$ –	$^1P_1^o$	130.43	132.83	2.05 E+02	1.95 E+02	0.95
2 – 7	$^3P_0^o$ –	$^3P_1$	146.52	147.27	6.64 E+01	6.59 E+01	0.83
3 – 6	$^3P_1^o$ –	$^3P_0$	172.39	173.32	1.25 E+02	1.23 E+02	0.74
3 – 7	$^3P_1^o$ –	$^3P_1$	153.84	154.3	4.21 E+01	4.18 E+01	0.82
3 – 8	$^3P_1^o$ –	$^3P_2$	143.56	144.39	5.49 E+01	5.43 E+01	0.93
3 – 9	$^3P_1^o$ –	$^1D_2$	120.02	121.20	4.33 E+00	4.4 E+00	0.38
4 – 7	$^3P_2^o$ –	$^3P_1$	179.27	180.1	4.49 E+01	4.46 E+01	0.78
4 – 8	$^3P_2^o$ –	$^3P_2$	165.46	166.69	7.81 E+01	7.58 E+01	0.81
4 – 9	$^3P_2^o$ –	$^1D_2$	134.95	136.53	4.75 E+01	4.83 E+01	0.85
5 – 8	$^1P_1^o$ –	$^3P_2$	322.12	313.62	3.30 E+00	3.7 E+00	2.6
5 – 9	$^1P_1^o$ –	$^1D_2$	223.68	221.33	4.52 E+00	4.61 E+00	1.8
5 –10	$^1P_1^o$ –	$^1S_0$	148.97	149.22	3.26 E+02	3.27 E+02	0.2

Magnetic dipole

$i-k$			$\lambda(\mathring{A})$ (a)	(b)	A_{ki} (10^8 s^{-1}) (a)	(b)
1 – 7	$^1S_0$ –	$^3P_1$	97.03	97.35	1.06 E-04	1.3 E-04
2 – 3	$^3P_0^o$ –	$^3P_1^o$	3211.7	3230.1	5.29 E-06	4.73 E-06
2 – 5	$^3P_0^o$ –	$^1P_1^o$	239.66	247.12	3.10 E-04	2.9 E-04
3 – 4	$^3P_1^o$ –	$^3P_2^o$	1084.8	1079.3	1.03 E-04	9.98 E-05
3 – 5	$^3P_1^o$ –	$^1P_1^o$	258.98	267.59	1.85 E-04	1.2 E-04
4 – 5	$^3P_2^o$ –	$^1P_1^o$	340.20	355.80	1.37 E-04	1.2 E-04
6 – 7	$^3P_0$ –	$^3P_1$	1429.7	1406.5	5.75 E-05	7.2 E-05
7 – 8	$^3P_1$ –	$^3P_2$	2147.7	2246.5	1.03 E-05	9.1 E-06
7 – 9	$^3P_1$ –	$^1D_2$	545.9	565.0	2.02 E-04	2.1 E-04
7 –10	$^3P_1$ –	$^1S_0$	245.46	252.7	2.22 E-03	2.3 E-03
8 – 9	$^3P_2$ –	$^1D_2$	731.93	754.7	1.90 E-04	1.9 E-04

(a) MCDHF with Breit and QED.
(b) NIST online Atomic Spectra Database [18].

8.4 Orbital relaxation

Experience with nonrelativistic RAS calculations shows that calculated line strengths are more reliable when orbital relaxation is allowed, requiring an independent OL calculation for both initial and final ASFs. Each OL calculation generates in its own orthonormal orbital basis; it is essential to take into account the fact that their Gram (or overlap) matrix is nontrivial to get accurate transition probabilities.. The method presented in [19, 21] involves transforming the two orbital sets into two new biorthonormal sets, respectively $\{\phi_i^A\}$ and $\{\phi_i^B\}$, so that

$$\langle \phi_i^A \mid \phi_j^B \rangle = \delta_{ij}$$

for all pairs i, j. To start with, let ϕ^F and ϕ^I be m-dimensional row vectors whose components are the respective orbitals. Then there exist linear transformations

$$\phi^A = \phi^F \mathbf{C}^{FA}, \quad \phi^B = \phi^I \mathbf{C}^{IB}, \tag{8.4.1}$$

such that

$$\mathbf{C}^{IB} \mathbf{C}^{FA\,\dagger} = (\mathbf{S}^{FI})^{-1} \tag{8.4.2}$$

where $\mathbf{S}^{FI}$ is the overlap matrix with elements

$$S_{ij}^{FI} = \langle \phi_i^F \mid \phi_j^I \rangle. \tag{8.4.3}$$

We shall see that the matrices $\mathbf{C}^{FA}$ and $\mathbf{C}^{IB}$ can be constructed to be *upper triangular*.

Any ASF $|\Psi\rangle$ can be expanded in terms of CSFs formed either from I basis orbitals or the corresponding B basis orbitals:

$$|\Psi\rangle = \sum_{\mu} c_\mu^I \, |\Phi_\mu^I\rangle = \sum_{\mu} c_\nu^B \, |\Phi_\nu^B\rangle. \tag{8.4.4}$$

Similarly, we may expand $|\Psi\rangle$ un terms of the F and A basis orbitals. The coefficients c_ν^B and c_μ^I can be be calculated through the transformations (8.4.1) or their creation operator equivalents

$$\mathbf{a}^{A\dagger} = \mathbf{a}^{F\dagger} \mathbf{C}^{FA}, \quad \mathbf{a}^{B\dagger} = \mathbf{a}^{I\dagger} \mathbf{C}^{IB}, \tag{8.4.5}$$

where each $\mathbf{a}^\dagger$ is a row vector of creation operators. We can generate these transformations on the initial basis sequentially by writing

$$a_k^{I\dagger} = \sum_{l \le k} a_l^{B\dagger} t_{lk} + \sum_{l > k} a_l^{I\dagger} t_{lk}, \quad k = 1, 2, \ldots$$

or

$$\mathbf{a}^{I\dagger} = \mathbf{a}^{B\dagger} \mathbf{t}^U + \mathbf{a}^{I\dagger} \mathbf{t}^L \tag{8.4.6}$$

where $\mathbf{t}^U$ is an *upper triangular* matrix and $\mathbf{t}^L$ is *strictly lower triangular* matrix. Substituting for $\mathbf{a}^{B\dagger}$ from (8.4.5) and noting that $\mathbf{a}^{I\dagger}$ can be arbitrary, gives the matrix equation

$$\mathbf{C}^{IB} \mathbf{t}^U + \mathbf{t}^L = \mathbf{I}, \quad \mathbf{C}^{IB} = (\mathbf{I} - \mathbf{t}^L)(\mathbf{t}^U)^{-1}$$

which expresses $\mathbf{C}^{IB}$ as a product of a lower triangular matrix $(\mathbf{I} - \mathbf{t}^L)$ and an upper triangular matrix $(\mathbf{t}^U)^{-1}$. Setting $\mathbf{t}^L = 0$ gives

$$\mathbf{C}^{IB} = (\mathbf{t}^U)^{-1}. \tag{8.4.7}$$

which is a pure triangular matrix. There is a similar equation for $\mathbf{C}^{FA}$. . It follows from (8.4.2) that these upper triangular matrices can be written $\mathbf{C}^{FA} = \mathbf{L}^{FI\dagger}$ and $\mathbf{C}^{IB} = \mathbf{U}^{FI}$ so that

$$(\mathbf{S}^{FI})^{-1} = \mathbf{U}^{FI}\mathbf{L}^{FI} = \mathbf{C}^{IB}\mathbf{C}^{FA\dagger}. \tag{8.4.8}$$

A particularly useful attribute of this construction is that it works also when $\mathbf{C}^{FA}$ and $\mathbf{C}^{IB}$ are *block upper triangular* matrices, and is consistent with the partitioning of the RAS described at the end of §7.12. It also works [21, Appendix B] when the sets F, A and I, B have different dimensions n and m respectively, where $n < m$. Then $\mathbf{S}^{FI}$ can be written

$$\mathbf{S}^{FI} = (\mathbf{T} \vdots \mathbf{Z}) \tag{8.4.9}$$

where $\mathbf{T}$ is $n \times n$ and $\mathbf{Z}$ is $n \times (m-n)$. Choose the transformation matrices so that

$$\mathbf{C}^{FA\,\dagger}\mathbf{S}^{FI}\mathbf{C}^{IB} = (\mathbf{I}_n \vdots \mathbf{O}) \tag{8.4.10}$$

where $\mathbf{I}_n$ is the $n \times n$ identity and $\mathbf{O}$ is a null $n \times (m-n)$ matrix. It follows that

$$\mathbf{C}^{IB} = \begin{pmatrix} \overline{\mathbf{C}^{IB}} & \overline{\overline{\mathbf{C}^{IB}}} \\ 0 & \mathbf{I}_{m-n} \end{pmatrix}, \tag{8.4.11}$$

where, as in (8.4.8), we have the UL factorization

$$\mathbf{T}^{-1} = \overline{\mathbf{C}^{IB}}\mathbf{C}^{FA\,\dagger}, \tag{8.4.12}$$

and, as $\mathbf{C}^{FA\,\dagger}$ is nonsingular,

$$\overline{\overline{\mathbf{C}^{IB}}} = -\mathbf{T}^{-1}\mathbf{Z} = -\overline{\mathbf{C}^{IB}}\mathbf{C}^{FA\,\dagger}\mathbf{Z}. \tag{8.4.13}$$

We can now use (8.4.6) recursively to write, for each k, $|\Phi^{(k-1)}\rangle$ as a sum of CSFs in which all ϕ_i^I with $i \le k$ have been expressed in terms of the ϕ_j^B with $j \le k$ using

$$|\Phi^{(k-1)}\rangle = \left(\sum_{l<k} t_{lk}\, e_{k\to l} + t_{kk}\right) |\Phi^{(k)}\rangle, \quad k = 2, 3, \dots \tag{8.4.14}$$

where t_{lk}, $l \le k$, are the elements of $\mathbf{t}^U$ is defined by (8.4.7). If $\widehat{a}_k$ is the dual annihilation operator to $a_k^\dagger$, $e_{k\to l} = a_l^\dagger \hat{a}_k$ replaces the orbital ϕ_k^I by the orbital ϕ_l^B. Symmetry-breaking intermediates can be avoided by grouping the orbitals in $n\kappa$ subshells with common radial amplitudes. Let $a_m^\dagger$, $m = -j, \dots, j$ be a set of creation operators for this subshell; then

$$\prod_{m=-j}^{j} \left(\sum_{n' \le n} t_{n'n}\, e_{nm \to n'm} \right) = \prod_{m=-j}^{j} \left[\left(\widehat{1} + \sum_{n'<n} s_{n'}\, e_{nm \to n'm} \right) t_{nn} \right]$$

where $\widehat{1}$ is the identity on the $n\kappa$ subshell and $s_{n'} = t_{n'n}/t_{nn}$. One factor t_{nn} survives for each appearance of an $n\kappa$ orbital. Now $(e_{nm \to n'm})$ is nilpotent when $n' < n$, so that

$$\ln \left(1 + \sum_{n'<n} s_{n'}\, e_{nm \to n'm} \right) = \sum_{n'<n} s_{n'}\, e_{nm \to n'm}$$

and the product is equivalent to $\exp(\widehat{s})\ (t_{nn})^{\widehat{N}_n}$ where $\widehat{N}_{n\kappa}$ is the number of electrons occupying the $n\kappa$ subshell and

$$\widehat{s} = \sum_{n'<n} s_{n'} \sum_{m=-j}^{j} e_{nm \to n'm}.$$

Because $e_{nm \to n'm}$ is nilpotent, the powers $\widehat{s}^i$ vanish for $i > 2j+1$ so that the operator expansion terminates and

$$\exp(\widehat{s})\ (t_{nn})^{\widehat{N}_n} = \sum_{i=0}^{2j+1} \frac{\widehat{s}^i}{i!}\ (t_{nn})^{\widehat{N}_n}. \tag{8.4.15}$$

The exchange operator $e_{nm \to n'm} = a^{\dagger}_{n'm} \widehat{a}_{nm}$, $n' < n$, is a degenerate, rank 0, example of the operator appearing in (6.9.15) for which the method for constructing matrix elements between open-shell CSFs of §6.9.2 can be used. In the notation used in §7.3, the effect of $e_{nm \to n'm}$ on a CSF $|T'\rangle$ can be expressed as

$$e_{nm \to n'm}\, |T'\rangle = \sum_{T} t_{T'T}(n'\kappa, n\kappa)\, |T\rangle, \quad n' < n, \tag{8.4.16}$$

where the coefficients $t_{T'T}(n'\kappa, n\kappa)$ are given in (6.10.2) and (7.3.5).

The algorithm incorporated in the nonrelativistic MCHF-ASP program [20] and also in GRASP92, evaluates matrix elements of the form

$$\left\langle \Psi^F_f \left| \mathbf{O} \right| \Psi^I_i \right\rangle$$

where $\mathbf{O}$ is any electric or magnetic transition multipole Fock space operator.

Algorithm 8.1

1. *Check that the CI expansions for Ψ^F_f and Ψ^I_i satisfy the property of closure under de-excitation.*
2. *Evaluate the coefficients $t_{TT'}(n'\kappa, n\kappa)$ of (8.4.16) for each CSF appearing in both expansions Ψ^F_f and Ψ^I_i.*

3. *Evaluate the matrix* $\mathbf{S}^{FI}$ *and decompose it to obtain the block triangular matrices* $\mathbf{C}^{IB}$ *and* $\mathbf{C}^{FA}$ *which generate the biorthonormal orbital sets.*
4. *Use the sequence of single orbital orbital replacements described above to generate the CI expansion coefficients for the transformed states* Ψ_f^A *and* Ψ_i^B.
5. *Use the standard formulae, (8.3.2), to obtain the one-electron coefficients,* $d^L_{\alpha\beta}(TT')$, *and evaluate the associated radial integrals* $\langle \alpha^A \parallel \mathbf{o}^L \parallel \beta^B \rangle$ *using the biorthonormal radial functions to complete the calculation.*

Full details are given in [21].

Table 8.3. Excitation energies (cm^{-1}) relative to the C III $1s^22s^2\ ^1S_0$ ground state From [19] with permission. Copyright 1998 by the American Physical Society.

n	$1s^22s2p\ ^3P_1^o$		$1s^22s2p\ ^3P_2^o$		$^3P_2^o - ^3P_0^o$		$1s^22s2p\ ^1P_1^o$	
	A	B	A	B	A	B	A	B
3	52553	52393	52607	52446	78.17	77.32	104513	104362
4	52511	52223	52566	52278	80.03	79.46	103308	102982
5	52461	52298	52517	52353	80.32	79.59	102818	102656
6	52448	52357	52503	52412	80.44	79.74	102623	102528
7	52448	52373	52503	52428	80.48	79.81	102565	102496
8	52449	52382	52504	52437	80.50	79.84	102540	102480
a		52394		52449	80.52	79.86		102464
b		52370		52425		80.05		102440
c		52391		52447		80.05		102352

a: Extrapolated values (see text).
b: Corrected for QED and finite mass effects.
c: NIST online database [18].

8.5 Application to atomic transition calculations

The problems of modelling the intercombination lines of the C III spectrum were discussed in §7.12.2 following the paper of Ynnerman and Fischer [22]. Then Jönsson and Fischer [19] calculated transition probabilities for these lines using the biorthogonal representation of Algorithm 8.1. Separate MCDHF calculations were performed for the ground state $2s^2\ ^1S_0$ and each of the excited states $2s2p\ ^1P_1^o$ and $2s2p\ ^3P^o_{0,1,2}$. The $1s$ orbitals were determined from a minimal OL calculation and were frozen in all subsequent calculations. The ground state ASF included all CSFs generated by single and double substitutions from the active reference set $\{2s^2,\ 2p^2_{1/2},\ 2p^2_{3/2}\}$, allowing at most one excitation from the $1s^2$ core. The active set was increased systematically, as in [22], to include orbitals up to $n = 8$ with $l \leq 6$. The states of the $2s2p\ ^3P^o$ term were determined from EOL calculations in which the three CSF energies were weighted according to their $2J + 1$ degeneracies. The procedure was the same as for the ground state, with the CSF expansion comprising SD excitations from the reference set $\{2s2p_{1/2}, J = 0, 1,\ 2s2p_{3/2}, J = 1, 2\}$. The OL calculations for $2s2p\ ^1P_1^o$ followed the same pattern. As before, the Breit interaction (long wavelength, $\omega = 0$) energies were added perturbatively at each step.

Table 8.4. gf-values for the E1 resonance line in C III. From [19] with permission. Copyright 1998 by the American Physical Society.

	$^1S_0$–$^1P_1^o$			
	E1(Babushkin)		E1(Coulomb)	
n	A	B	A	B
3	0.7699	0.7850	0.7747	0.7401
4	0.7606	0.7627	0.7542	0.7414
5	0.7584	0.7586	0.7566	0.7537
6	0.7579	0.7591	0.7582	0.7559
7	0.7585	0.7587	0.7589	0.7568
8	0.7589	0.7588	0.7585	0.7571
a	0.7575	0.7579	0.7599	0.7581
b		0.7579		0.7576
c	Experimental: 0.7586			

a: Normalized to observed energy.
b: Corrected for QED and finite mass.
c: NIST online database [18].

The results are shown in the A columns of Table 8.3. The dependence on n is much smoother than in [22] and there are now clearer indications of convergence. The columns headed B in the table are intended to include the effect of correlation in a more balanced way. For example, the first stage of modelling the intercombination lines, $2s^2\ ^1S$ – $2s2p\ ^3P$ involves an EOL calculation for the states $2s^2\ ^1S_0$ and $2s2p\ ^3P_{0,1,2}$. The next step is to carry out OL calculations for each state with frozen $n = 2$ orbitals including CSFs with all possible excitations to $n = 3$. Subsequent steps including higher values of n augmented this wavefunction with core-valence excitations followed by a perturbation calculation of the Breit energy. The smooth behaviour with n is now good enough to extrapolate each sequence of results to a plausible limiting value. Let $\Delta x_n = x_n - x_{n-1}$ be successive increments down a column, and write $r_n = \Delta x_n / \Delta x_{n-1}$. Then if r_n is roughly constant, we can estimate the residual error at the n-th step to be $\Delta x_n r_n/(1 - r_n)$, giving corrected values in line a of the table. Corrections for QED and finite mass effects added in line b of the table make a small but significant improvement. The effect of the full transverse photon interaction, $\omega \neq 0$, on the energies and transition rates was investigated and found to be unimportant. The fine structure splitting was slightly decreased by about 0.32 cm^{-1} in case A and 0.40 cm^{-1} in case B. Table 8.4 shows that the gf values for the resonance line ${}^1S_0 - {}^1P_1$ are rel-

Table 8.5. Transition rates for C III intercombination E1 line (s^{-1}) and M2 line (10^{-3}s^{-1}). From [19] with permission. Copyright 1998 by the American Physical Society.

	${}^1S_0$–${}^3P_1^o$				${}^1S_0$–${}^3P_2^o$	
	E1(Babushkin)		E1(Coulomb)		M2	
n	A	B	A	B	A	B
3	77.36	66.12	91.35	3.14	5.210	5.172
4	89.49	81.45	89.34	80.99	5.187	5.089
5	98.58	95.95	127.90	136.78	5.163	5.089
6	101.47	100.92	132.68	157.55	5.159	5.119
7	102.46	102.38	137.17	158.86	5.160	5.128
8	102.91	102.72	137.22	160.14	5.160	5.134
a	102.57	102.77		160.17	5.132	5.139
b	102.94	102.85				
c		102.87				5.139
d		114			5.19	
e		102.94(14)				

a, b, c, d have the same meaning as in Table 8.4.
e: Ref. [23].

atively insensitive to correlation and converge smoothly as n increases. There is much better agreement between the Babushkin and Coulomb gauge results than in the earlier calculation [22], and the result is reasonably close to the value in the NIST on-line database. Table 8.5 displays similar results for the transition rates for the E1 and M2 intercombination lines. The Babushkin (nonrelativistic length) gauge values are much more stable than the Coulomb (nonrelativistic velocity) gauge ones and the extrapolated results agree well with a recent measurement [23]. The importance of correlation in obtaining a good result is obvious. The M2 line is relatively insensitive to correlation and the converged result is close to that in the NIST database.

8.5.1 Large-scale calculations of energies and transition rates

An example that requires the capacity of GRASP92 to handle very large orbital and CSF sets is that of the $3s^2\ {}^1S_0\ -\ 3s3p\ {}^{1,3}P_1^o$ transitions in the Mg-like isoelectronic sequence [24]. The low-lying spectrum of the heavy neutral Lu atom is even more demanding. The lowest levels of Lu (Z=71) are $[\mathrm{Xe}]4f^{14}6s^25d\ {}^2D_{3/2}$ and $[\mathrm{Xe}]4f^{14}6s^25p\ {}^2P^o_{1/2,3/2}$, and Yu Zou and Froese Fischer [25] have studied the transitions between them using methods similar to those of the last section. The upper and lower states in each transition were optimized independently, extending the CSF sets in layers including orbital angular momenta from f to h. The interaction of g orbitals with the core $4f$ orbitals was essential to get good energies; h orbitals had much smaller effects. The Breit interaction and QED corrections were included in the usual way.

The initial valence-valence (VV) correlation involved SD excitations from the $6s^25d$ and $6s^25p$ orbitals. Core-valence (CV) correlation was approximated with additional CSFs in which one electron was excited from the $4f$ core and another from a valence orbital. At this stage, the calculated $5d_{3/2} - 6p_{1/2}$ energy difference of 3989 cm^{-1} compared quite well with the observed value 4136 cm^{-1} Similar excitations from $6s$ were found to be unimportant and were therefore ignored. Unfortunately, CV excitations out of the $5p$ shell spoilt this promising result by doubling the predicted energy, demonstrating the need to investigate core correlation (CC).

By this time, the dimensions of the CSF sets for the ${}^2D_{3/2}$, ${}^2P^o_{1/2}$ and ${}^2P^o_{3/2}$ calculations had grown modestly to 5600, 2764, and 5073, respectively. The core correlation enlarged these sets enormously. At the final stage, which involved SD excitations from the $4d$, $5s$, and $5p$ orbitals, the respective dimensions were 305717, 87241, and 236554 CSF. The predicted $5d_{3/2} - 6p_{1/2}$ interval improved to 4186 cm^{-1} and that for $5d_{3/2} - 6p_{3/2}$ became 7462 cm^{-1} compared with the experimental values 4136 cm^{-1} and 7476 cm^{-1}. Once again, the Babushkin gauge gf values change more smoothly as the basis set is enlarged; core correlation increases the oscillator strength by a factor of order 3. The incomplete treatment of correlation seems likely to be responsible for the instability of the Coulomb gauge prediction.

The success of these calculations suggested a strategy for a similar calculation aimed to identify the ground state of Lr (Z=103) [25]. The low-lying levels are expected to have the structure $[\mathrm{Rn}]5f^{14}7s^26d\ {}^2D_{3/2}$ and $[\mathrm{Rn}]5f^{14}7s^26p\ {}^2P^o_{1/2,3/2}$. There are more orbitals than in the Lu calculation, but the active sets needed are much the same size. The $6d_{3/2}-7p_{1/2}$ transition in Lr had previously been studied using semi-empirical methods [26, 27, 28], MCDHF methods [29, 30] and relativistic coupled cluster methods [31] with reported transition energies ranging from 0.0 to -8000 cm^{-1} leaving considerable uncertainty as to the character of the ground state. As yet there are no experimental data. The computed interval $E({}^2D_{3/2} - {}^2P^o_{1/2})$ was -1298 cm^{-1} when only $5f$ VV and CV correlation was included; $6p$ CV correlation changed this to +1399 cm^{-1}; the final result, after including $5d, 6s$ correlations, was

-1127 cm^{-1}. The final value of the $E(^2D_{3/2} - ^2P^o_{3/2})$ interval was 7010 cm^{-1}, establishing the $^2P^o_{1/2}$ level as the ground state. The results are in reasonable agreement with the relativistic coupled cluster calculations [31].

8.6 Relativistic atomic photo-ionization theory

The theory of atomic photo-ionization reviewed, for example, by Starace [32, 33] and Amusia [35], has been elaborated to investigate many-body effects, especially near thresholds, revealed by autoionizing resonances, giant shape resonances, polarization and double ionization. These developments have gone hand-in-hand with new experimental techniques involving high-intensity laser sources and synchrotron radiation sources. Photoelectron spectroscopy has been used to partition total cross sections into contributions from different channels, and measurements of the angular distribution asymmetry parameter β and of spin-polarization serve to provide more detail on the mechanisms involved. The theoretical accounts are primarily nonrelativistic and any mention of relativistic effects in [32, 33, 35] is confined to the influence of spin-orbit coupling.

Formulae for the photo-ionization cross sections of the ground states of Dirac hydrogenic atoms were first presented in 1935 [36] with application to inner shells of heavy elements such as Pb and Hg. Spin-orbit effects at high atomic numbers motivated development of a more general relativistic theory of atomic photo-ionization by Walker and Waber [38]. This section presents a modernized version of their jj-coupling one-electron theory. The initial state of the target can be written

$$|\Theta_i\rangle = \sum_{f,a} (-1)^{\Delta_a} N_a^{1/2} (T_a\{|\widetilde{T}_a)\, \mathbb{R}_a(\Theta_i, \Phi_f) \\ \times \sum_{\widetilde{M},m_a} |\Phi_f\rangle \otimes |\psi_a(\boldsymbol{r})\rangle\, \langle \widetilde{J}\widetilde{M} j_a m_a \,|\, J_i M_i\rangle. \quad (8.6.1)$$

The photoelectron, angular quantum numbers j_a, m_a, is initially one of the N_a electrons in the subshell state T_a and its removal leaves $N_a - 1$ electrons in the subshell state $\widetilde{T}_a$. The ionized atom is left in the state Φ_f, with angular momentum quantum numbers $\widetilde{J}, \widetilde{M}$, and (8.6.1) expresses the channel state $|\Theta_i\rangle$ as a linear combination of final channel states using the parentage scheme. The first line of the formula takes care of the anti-symmetrization with with respect to all $N+1$ electrons and $\mathbb{R}_a(\Theta_i, \Phi_f)$ deals the angular recoupling involved when extracting a single electron from the many-electron state $|\Theta_i\rangle$. As in nonrelativistic photo-ionization theory [32, 33] the final scattering state in which the atom is left in a particular state Φ_f can be written

$$|\Psi_f^-\rangle = (\sqrt{2\pi}\sum_{km}[j]^{1/2}\,\mathcal{D}^{j\,*}_{m,\sigma}(\Omega)$$
$$\times\left\{i^l \mathrm{e}^{-i\delta_k}|\Theta_{f,k,m}\rangle + 2\sigma i^{\,l+1}\mathrm{e}^{-i\delta_{-k}}|\Theta_{f,-k,m}\rangle\right\}. \quad (8.6.2)$$

This is a linear combination of channel functions $\Theta_{f,\kappa,m}$,

$$\langle \boldsymbol{r}\,|\Theta_{f,\kappa,m}\rangle = |\Phi_f\rangle \otimes \psi_{\varepsilon,k,m}(\boldsymbol{r}),$$

which are products of final N-electron states with Dirac partial waves

$$\psi_{\varepsilon,\kappa,m}(\boldsymbol{r}) = \begin{pmatrix} P_{\varepsilon,\kappa}(r)\,\chi_{\kappa,m}(\theta,\varphi) \\ iC(p)Q_{\varepsilon,\kappa}(r)\,\chi_{-\kappa,m}(\theta,\varphi)\end{pmatrix},$$

As in (3.2.26), the radial amplitudes must be normalized with respect to energy, and have the asymptotic form

$$P_{\varepsilon,\kappa}(r) \sim \mathcal{N}\,\sin\left(p\,r + \nu\ln 2p\,r - \frac{1}{2}\,l\,\pi + \delta_\kappa\right),$$

$$Q_{\varepsilon,\kappa}(r) \sim \mathcal{N}\,\sin\left(p\,r + \nu\ln 2p\,r - \frac{1}{2}\,\bar{l}\,\pi + \delta_\kappa\right)$$

where δ_κ is given by (3.3.36), $\eta = \operatorname{sgn}\kappa$, and

$$\mathcal{N} = (\pi c C(p))^{-1/2}, \quad C(p) = \frac{cp}{|E| + c^2}, \quad l = j + \eta/2, \quad \bar{l} = j - \eta/2.$$

In the nonrelativistic limit, $C(p) \sim p/2c$, and $\mathcal{N} \sim (2/\pi p)^{1/2}$ in agreement with [32, Equation (4.24)]. The notation Ψ_f^- indicates that the coefficients of (8.6.2) have been chosen so that the electron scattering state represents asymptotically an outgoing (Coulomb modified) plane wave in the direction $\Omega = \boldsymbol{p}/p$ with helicity σ together with incoming spherical waves: a time-reversed version of (3.4.24). The scattering phases $\mathrm{e}^{-i\delta_\kappa}$, which must take account of both the Coulomb interaction and atomic effects, suppress outgoing spherical wave components. For both signs of κ, the Dirac partial waves transform under the same irreducible representation $\mathcal{D}^j$, and the rotation matrices $\mathcal{D}^j_{m,\sigma}(\Omega)$ ensure that the outgoing electron emerges in the direction Ω instead of along the axis of quantization as in the unrotated expression (3.4.24).

This closely resembles the wavefunction adopted by Walker and Waber [38, Equation (2.5)] from Akhiezer and Berestetskii [39, Equation (14.5')]. Akhiezer and Berestetskii used a four-component starting point [39, Equations (10.9), (10.9')] but, for reasons explained in §3.4.2 in connection with relativistic scattering by a point charge, they only retain two components in their equation (14.5'). The four-component form is preferable to avoid ambiguity.

8.6.1 The differential cross-section for photo-ionization

Using the methods of §8.1 and §8.2, the transition matrix element derived from (8.6.1) and (8.6.2) is

$$\langle \Psi_f^- \,|\, A_q \,|\Theta_i\rangle = \sqrt{2\pi} \sum_{\kappa m} \sum_{\widetilde{M}, m_a} d_a(\Theta_i, \Phi_f)\, \langle \widetilde{J}\widetilde{M} j_a m_a \,|\, J_i M_i\rangle \qquad (8.6.3)$$

$$\times i^{-l} [j]^{1/2}\, \mathcal{D}^j_{m,\sigma}(\Omega) \begin{pmatrix} m & L & j_a \\ j & q & m_a \end{pmatrix} (j \,\|\, \boldsymbol{C}^L \,\|\, j_a)\, R^L_\epsilon(\kappa, \kappa_a, \sigma)$$

where $q = 0, \pm 1$, defined by (8.1.11), specifies the polarization of the incident photon and

$$R^L_\epsilon(\kappa, \kappa_a, \sigma) = \frac{1}{2}[(1+\eta) - 2i\,\sigma(1-\eta)] e^{i\delta_\kappa} \boldsymbol{M}^L_\epsilon(\kappa, \kappa_a), \qquad (8.6.4)$$

where $\eta = \operatorname{sgn} \kappa == \pm 1$. The radial integrals $\boldsymbol{M}^L_\epsilon(\kappa, \kappa_a)$ are defined respectively by (8.2.12) for magnetic multipoles and (8.2.13) for electric multipoles (in an arbitrary gauge). For many applications we need only consider electric dipole (E1) transitions, but in general several multipoles may contribute for high energy processes, and in this case it is necessary to remember that, for example, M1 and E2 transitions have similar selection rules and the amplitudes will interfere. The factor

$$d_a(\Theta_i, \Phi_f) = (-1)^{\Delta_a} N_a^{1/2} (T_a \{|\widetilde{T}_a)\, \mathbb{R}_a(\Theta_i, \Phi_f).$$

contains information about the angular recoupling in going from the initial state of the atom to the final state as well as about the parentage.

For a single channel, the differential cross-section for photo-ionization is proportional to the squared transition amplitude

$$\begin{aligned} &|\langle \Psi_f^- \,|\, A_q \,|\Theta_i\rangle|^2 \\ &\quad = 2\pi \sum_{LL'} \sum_{\kappa\kappa'} \sum_{mm'} \sum_{m_a m'_a} \sum_{\widetilde{M}\widetilde{M}'} \langle \widetilde{J}\widetilde{M} j_a m_a \,|\, J_i M_i\rangle \langle J_i M_i \,|\, \widetilde{J}\widetilde{M}' j_a m'_a\rangle \\ &\qquad \times \begin{pmatrix} j & q & m_a \\ m & L & j_a \end{pmatrix} \begin{pmatrix} m' & L' & j_a \\ j' & q & m_a \end{pmatrix} \mathcal{D}^j_{m,\sigma}(\Omega) \mathcal{D}^{j'\,*}_{m',\sigma}(\Omega) \\ &\qquad \times |d_a(\Theta_i, \Phi_f)|^2\, i^{l'-l} [j, j']^{1/2}\, (j \,\|\boldsymbol{C}^L\|\, j_a)(j' \,\|\boldsymbol{C}^{L'}\|\, j_a) \\ &\qquad\qquad \times R^L_\epsilon(\kappa, \kappa_a, \sigma)\, R^{L'\,*}_\epsilon(\kappa', \kappa_a, \sigma) \qquad (8.6.5) \end{aligned}$$

This can be put in a form in which the dependence of the differential cross section on the polarization, $q = 0, \pm 1$, of the incident photon and on the helicity, $\sigma = \pm 1/2$ of the photoelectron is more transparent by first writing

$$\begin{aligned} &\sum_{mm'} \begin{pmatrix} j & q & m_a \\ m & L & j_a \end{pmatrix} \begin{pmatrix} m' & L' & j_a \\ j' & q & m_a \end{pmatrix} \mathcal{D}^j_{m,\sigma}(\Omega) \mathcal{D}^{j'\,*}_{m',\sigma}(\Omega) = \\ &\qquad \sum_\nu (-1)^q \begin{pmatrix} L & L' & \nu \\ -q & q & 0 \end{pmatrix} (-1)^{j-\sigma} \begin{pmatrix} j & j' & \nu \\ -\sigma & \sigma & 0 \end{pmatrix} \\ &\qquad\qquad \times (-1)^{j_a - j}\, [\nu] \begin{Bmatrix} L' & j' & j_a \\ j & L & \nu \end{Bmatrix} P_\nu(\cos\Theta) \qquad (8.6.6) \end{aligned}$$

where Θ is the angle of scattering and $P_\nu(x)$ is a Legendre polynomial. Using (A.4.21) for the reduced matrix elements of $\boldsymbol{C}^L$, averaging over initial projections and summing over final projections gives the differential cross-section

$$\begin{aligned}\frac{d\sigma_q(\boldsymbol{p})}{d\Omega} &= \frac{2\pi\alpha}{\omega}\frac{(2\widetilde{J}+1)(2j_a+1)}{2J_i+1}|d_a(\Theta_i,\Phi_f)|^2 \\ &\times \sum_{LL'}\sum_{\kappa\kappa'}\sum_{\nu} i^{l'-l}[j,j',j_a,\nu]\begin{Bmatrix} L' & j' & j_a \\ j & L & \nu \end{Bmatrix} P_\nu(\cos\Theta) \\ &\times (-1)^q \begin{pmatrix} L & L' & \nu \\ -q & q & 0 \end{pmatrix} \sum_\sigma (-1)^{j-\sigma}\begin{pmatrix} j & j' & \nu \\ -\sigma & \sigma & 0 \end{pmatrix} \\ &\times \begin{pmatrix} j & L & j_a \\ 1/2 & 0 & -1/2 \end{pmatrix}\begin{pmatrix} j_a & L' & j' \\ 1/2 & 0 & -1/2 \end{pmatrix} \\ &\times R^L_\epsilon(\kappa,\kappa_a,\sigma)\, R^{L'\,*}_\epsilon(\kappa',\kappa_a,\sigma). \qquad (8.6.7)\end{aligned}$$

We recover the one-electron model considered by Walker and Waber [38] by setting $\widetilde{J}=0$, $J_i=j_a$ and omitting the factor $|d_a(\Theta_i,\Phi_f)|^2$. As they noted, the angular terms with $\nu>0$ will involve interference between multipoles. The dependence on the photon spin q and the helicity of the outgoing electron is entirely contained in the $3j$-symbols on the third line of the formula.

Table 8.6. Electric dipole selection rules.

$R^1_\epsilon(\kappa,\kappa_a,\sigma)\neq 0$ when $j = j_a - 1\, j_a,\ j_a+1$				
κ_a	l_a	κ	j	l
$-(j_a+\frac{1}{2})$	$j_a-\frac{1}{2}$	$-(j_a-\frac{1}{2})$	j_a-1	l_a-1
		$j_a+\frac{1}{2}$	j_a	l_a+1
		$-(j_a+\frac{1}{2})$	j_a+1	l_a+1
$j_a+\frac{1}{2}$	$j_a+\frac{1}{2}$	$j_a-\frac{1}{2}$	j_a-1	l_a-1
		$-(j_a+\frac{1}{2})$	j_a	l_a-1
		$j_a+\frac{1}{2}$	j_a+1	l_a+1

$j=j_a-1$ only if $j_a>\frac{1}{2}$.

8.6.2 Low energies: the electric dipole case

The selection rules for electric dipole radiation, derived from Table 8.1, will be found in Table 8.6. Only $L=L'=1$, so that the triad $\{L,L',\nu\}$ in the $3j$-symbol in (8.6.7) restricts the values of ν to 0,1,2 only. The total cross-section requires $\nu=0$ only, and then $j=j'$, so that integrating over solid angles gives.

$$\sigma_q(\boldsymbol{p}) = \frac{8\pi^2\alpha}{3\omega}\frac{(2\widetilde{J}+1)(2j_a+1)}{2J_i+1}|d_a(\Theta_i,\Phi_f)|^2$$
$$\times (2j_a+1)\sum_{\kappa}(2j+1)\begin{pmatrix} j & 1 & j_a \\ 1/2 & 0 & -1/2 \end{pmatrix}^2 \left|R^1_\epsilon(\kappa,\kappa_a,1/2)\right|^2, \quad (8.6.8)$$

because $j = j'$ and $k = k'$ when $\nu = 0$. Inserting algebraic values for the $3j$-coefficients gives

$$\sigma_q(\boldsymbol{p}\,) = \frac{8\pi^2\alpha}{\omega}\frac{(2\widetilde{J}+1)(2j_a+1)}{2J_i+1}|d_a(\Theta_i,\Phi_f)|^2$$
$$\times (2j_a+1)\left\{\frac{(2j_a-1)}{12j_a}|R^1_\epsilon(\eta_a(|\kappa_a|-1),\kappa_a,1/2)|^2\right.$$
$$+\frac{1}{12j_a(j_a+1)}|R^1_\epsilon(-\eta_a|\kappa_a|,\kappa_a,1/2)|^2$$
$$\left.+\frac{(2j_a+3)}{12(j_a+1)}|R^1_\epsilon(\eta_a(|\kappa_a|+1),\kappa_a,1/2)|^2\right\}, \quad (8.6.9)$$

in agreement with Walker and Waber [38]. In the long wavelength limit $\omega \to 0$, it is only necessary in most cases to retain the leading term in the expansion of the spherical Bessel functions when evaluating radial integrals, neglecting powers $O(\omega r/c)^2$ and higher [2]. In the Coulomb (velocity) gauge (with parameter $G_1 = 0$) this gives

$$R^1_\epsilon(\kappa,\kappa_a,\sigma) = \frac{1}{2}[(1+\eta) - 2i\,\sigma(1-\eta)]\,\mathrm{e}^{i\delta_\kappa}$$
$$\times\, ic\int_0^\infty [(\kappa-\kappa_a-1)P^*_{\varepsilon\kappa}(r)Q_a(r) + (\kappa-\kappa_a+1)Q^*_{\varepsilon\kappa}(r)P_a(r)]\,dr. \quad (8.6.10)$$

In the Babushkin (length) gauge (with parameter $G_1 = \sqrt{2}$) we find

$$R^1_\epsilon(\kappa,\kappa_a,\sigma) = \frac{1}{2}[(1+\eta) - 2i\,\sigma(1-\eta)]\,\mathrm{e}^{i\delta_\kappa}$$
$$\times\, i\,\omega\int_0^\infty (P^*_{\varepsilon\kappa}(r)P_a(r) + Q^*_{\varepsilon\kappa}(r)Q_a(r))\,r\,dr, \quad (8.6.11)$$

so that

$$|R^1_\epsilon(\kappa,\kappa_a,\sigma)|^2 = \omega^2\langle r\rangle^2_{\epsilon\kappa,a}, \quad (8.6.12)$$

where

$$\langle r\rangle_{\epsilon\kappa,a} = \int_0^\infty (P^*_{\varepsilon\kappa}(r)P_a(r) + Q^*_{\varepsilon\kappa}(r)Q_a(r))\,r\,dr.$$

differs from the nonrelativistic dipole length matrix element by terms of order $1/c^2$ whilst (8.6.10) stands in a similar relationship to the nonrelativistic dipole velocity matrix element. It is common to use the notation R_j for the matrix elements of (8.6.10) or (8.6.11) [38]. In the one-electron approximation, the nonrelativistic limit of (8.6.9) is

$$\sigma_q(\boldsymbol{p}) \to \frac{8\pi^2\alpha}{3\omega}(2j_a+1)\left\{\frac{l_a R^2_{l_a-1} + (l_a+1)R^2_{l_a+1}}{2l_a+1}\right\} \tag{8.6.13}$$

where, following [32, Equation (9.15)], we write R_l for the nonrelativistic dipole radial matrix element

8.6.3 Angular distributions and polarization parameters for a single channel

The differential cross-section (8.6.7) when averaged over initial polarizations gives no contribution for $\nu = 1$ so that all asymmetry is due to the term with $\nu = 2$ for which

$$(-1)^q \begin{pmatrix} 1 & 1 & 2 \\ -q & q & 0 \end{pmatrix} = \begin{cases} 2/\sqrt{30} & \text{for } q = 0, \\ -1/\sqrt{30} & \text{for } q = \pm 1. \end{cases}$$

The differential cross-section for light polarized along Oz ($q = 0$) is therefore

$$\frac{d\sigma_q(\boldsymbol{p})}{d\Omega} = \frac{\sigma_q(\boldsymbol{p})}{4\pi}\big(1 + \beta\, P_2(\cos\Theta)\big), \tag{8.6.14}$$

and, if the light is circularly polarized ($q = \pm 1$), then

$$\frac{d\sigma_q(\boldsymbol{p})}{d\hat{\Omega}} = \frac{\sigma_q(\boldsymbol{p})}{4\pi}\left(1 - \frac{1}{2}\beta\, P_2(\cos\Theta\right). \tag{8.6.15}$$

Substituting algebraic values and making use of the notation of (8.6.12) gives [38, Equation (4.2)]

$$\begin{aligned}\beta = &\left\{\frac{(2j_a-3)(2j_a-1)}{48j_a^2}|R_{j_a-1}|^2 - \frac{(2j_a-1)(2j_a+3)}{48j_a^2(j_a+1)^2}|R_{j_a}|^2\right.\\ &+ \frac{(2j_a+3)(2j_a+5)}{48(j_a+1)^2}|R_{j_a+1}|^2 - \frac{(2j_a-1)(2j_a+3)}{8j_a(j_a+1)}[\Re(R_{j_a-1}R^*_{j_a+1})]\\ &\left.+ \frac{(2j_a-1)}{8j_a^2(j_a+1)}[\Re(R_{j_a}R^*_{j_a-1})] + \frac{(2j_a+3)}{8j_a(j_a+1)^2}[\Re(R_{j_a}R^*_{j_a+1})]\right\}\\ &\div\left\{\frac{2j_a-1}{12j_a}|R_{j_a-1}|^2 + \frac{1}{12j_a(j_a+1)}|R_{j_a}|^2 + \frac{2j_a+3}{12(j_a+1)}|R_{j_a+1}|^2\right\}.\end{aligned} \tag{8.6.16}$$

Modern experimental work involving photo-ionization aims to measure more than just total or differential cross-sections. The ability to prepare incident photons or electrons in almost pure initial states (monochromatic beams of collision partners) must be matched by detailed analyses of the final states. So far, there has been no complete analysis published within the jj-coupling relativistic framework used in this section, in which the differential cross-section for photo-emission is defined for given photoelectron helicity $\sigma = \pm 1/2$ in the direction $\boldsymbol{p}$. In the nonrelativistic theory, the channels are specified by labels

$$\alpha = n_a J_a s K l J M,$$

where J_a is the total angular momentum of the residual ion coupled to the photoelectron spin, s, to give a resultant K that in turn is coupled to the orbital angular momentum l of the photoelectron to give the final channel angular momentum labels JM, whilst n_a represents any additional quantum numbers that may be required. Jacobs [40] gives general expressions for both angular distribution and polarization parameters. Bartschat [41] considers a wide range of experiments in which either the projectile or the target are observed alone or are detected in coincidence in terms of a general density matrix formalism. His treatment of photo-ionization is based on the same coupling scheme as is used by Jacobs.

8.6.4 Other aspects of photo-ionization

The treatment of this section assumes that there is a single outgoing electron in a specific state. Channel coupling effects, which we shall discuss in Chapter 9, are therfore ignored. Nevertheless, there are still many physical problems to which the theory can be applied. Gauge dependence remains a problem for electric multipole transitions, just as it does with line transitions, particularly as the present formulation ignores orbital relaxation. Practical calculations require the ability to generate Dirac radial wavefunctions in the continuum numerically; see §9.1.

The relativistic random phase approximation (RRPA), which is discussed in §9.10, is formally gauge independent, at least when the first order wave function is expanded in a complete set of states. Practical calculations must have a finite termination, so that numerical results are inevitably gauge dependent to some extent. The RRPA is basically a single particle theory, but allows for coupling between different outgoing channels. Its relative simplicity has made it the most popular approach to photo-excitation and -ionization. A recent multi-channel version (MCRRPA), which has been worked out in detail for two active electrons, has not yet been much used.

The relativistic R-matrix method can also be applied to photo-ionization calculations, §9.5, though it is less well developed than the RRPA. The R-matrix formalism takes full account of channel coupling, and allows description of a wide range of continuum processes. The computational procedures have a good deal in common with bound state calculations, and the DARC relativistic R-matrix code is able to make use of many GRASP modules. The different approachs are compared in Chapter 9.

8.7 Hyperfine interactions

The nuclear charge density distribution is often nonspherical, and its charge-current density distribution will then generate nonspherical electromagnetic perturbations to the dominant Coulomb field. These interact with the orbital electrons to give the spectrum its *hyperfine structure*. The nonrelativistic theory is outlined, for example, by Fischer [42, Chapter 8] who describes how to apply the MCHF method to estimate the splittings. A relativistic formulation for one-electron atoms was given by Schwartz [43], later extended to many-electron atoms by Armstrong [44] and Lindgren and Rosén [45, 46], both of whom present a reduction to nonrelativistic effective operators; we shall not discuss this reduction here.

The contribution of hyperfine interactions to the Hamiltonian may be written in the generic form

$$H_{hfs} = \sum_{k \geq 1} \mathbf{T}^k \cdot \mathbf{M}^k \tag{8.7.1}$$

where $\mathbf{T}^k$ and $\mathbf{M}^k$ are spherical tensor operators of rank k in the spaces of electronic wavefunctions, $|\gamma J M_J\rangle$ and nuclear wavefunctions $|\nu I M_I\rangle$ respectively. This effective Hamiltonian is obtained fom a multipole expansion of the nuclear electromagnetic field [44, Chapter IV], [45, §II.2] [46]. Outside the nucleus, the nuclear charge distribution generates a scalar potential

$$\Phi(\mathbf{r}) = \frac{1}{4\pi\epsilon_0} \sum_k r^{-k-1} \mathbf{C}^k \cdot Q^k \tag{8.7.2}$$

where

$$Q^k = \langle \nu I I \,|\, \sum q r'^k \mathbf{C}_0^k(\theta', \varphi') \,|\, \nu I I \rangle$$

is the *generalized electric multipole moment* of the nucleus. The field inside the nucleus is usually ignored. The nucleus has no electric dipole moment in the absence of parity-violating weak interactions, §1.5.4, and the first non-zero nuclear moment is a quadrupole Q:

$$\frac{1}{2} Q = \langle \nu I I \,|\, r^2 \mathbf{C}_0^2(\theta, \varphi) \,|\, \nu I I \rangle. \tag{8.7.3}$$

This is associated with the quantized electron operator

$$\mathbf{T}^2 = \sum_{\alpha,\beta} a_\alpha{}^\dagger \, \langle \alpha \,|\, \mathbf{t}^2 \,|\, \beta \rangle \, a_\beta, \quad \mathbf{t}^2 = -\mathbf{C}^2(\theta, \varphi)/r^3 \tag{8.7.4}$$

The magnetic vector potential is derived from the nuclear internal currents. The magnetic multipole tensors are

$$\mathbf{M}^k = \mu_N \, \langle \nu I I \,|\, \sum \boldsymbol{\nabla} \left(r^k \mathbf{C}^k(\theta, \varphi) \right) \cdot \left(\frac{2}{k+1} g_l \mathbf{l} + g_s \mathbf{s} \right) \,|\, \nu I I \rangle, \tag{8.7.5}$$

where g_l and g_s are respectively the orbital and spin g-factors of the nucleons and $\mu_N = \mu_B/M$ is the nuclear magneton. The most important case is the magnetic dipole, for which

$$\mathbf{M}^1 = \boldsymbol{\mu} \tag{8.7.6}$$

and

$$\mathbf{T}^1 = \sum_{\alpha,\beta} a_\alpha{}^\dagger \, \langle \alpha \,|\, \mathbf{t}^1 \,|\, \beta \rangle \, a_\beta, \quad \mathbf{t}^1 = -\frac{i}{c} \boldsymbol{\alpha} \cdot (\mathbf{1}\, \mathbf{C}^1)(\theta, \varphi)/r^2. \tag{8.7.7}$$

All interactions are in atomic units.

The complete atomic Hamiltonian, $H = H_0 + H_{hfs}$, commutes with the total angular momentum $\mathbf{F} = \mathbf{J} + \mathbf{I}$ but not with the electronic angular momentum $\mathbf{J}$ and the nuclear spin $\mathbf{I}$ separately, so that the eigenstates of H can be classified in terms of the quantum numbers J, I, F, M_F:

$$|\gamma\nu JIFM_F\rangle = \sum_{M_I, M_F} |\gamma JM_J\rangle \otimes |\nu IM_I\rangle \, \langle JM_J, IM_I \,|\, JIFM_F\rangle \tag{8.7.8}$$

The hyperfine interaction (8.7.1) is very weak, so that a perturbation treatment starting with free atom eigenfunctions is a good first approximation. The complete Hamiltonian is diagonal with respect to F, M_F, so that each term of (8.7.1) will give a contribution

$$\langle \gamma\nu JIFM_F \,|\, \mathbf{T}^k \cdot \mathbf{M}^k \,|\, \gamma\nu JIFM_F\rangle.$$

Using (B.3.164), this reduces to

$$E_k(JIF) = (-1)^{J+I+F} \begin{Bmatrix} I & J & F \\ J & I & k \end{Bmatrix} \langle \gamma J \| \, \mathbf{T}^k \, \| \gamma J\rangle \, \langle \nu I \| \, \mathbf{M}^k \, \| \nu I\rangle. \tag{8.7.9}$$

This can be reduced to simple algebraic expressions in terms of the parameters I, J and $C = F(F+1) - J(J+1) - I(I+1)$ for the two cases of most interest:

$$E_{M1}(JIF) = \frac{1}{2} A(J,J)\, C, \tag{8.7.10}$$

and

$$E_{E2}(JIF) = B(J,J)\, \frac{\frac{3}{4}C(C+1) - I(I+1)J(J+1)}{2I(2I-1)J(2J-1)}. \tag{8.7.11}$$

The *hyperfine interaction constants* A, B are given by

$$A(J,J) = \frac{\mu_I}{I} \frac{1}{[J(J+1)(2J+1)]^{1/2}} \langle \gamma J \| \, \mathbf{T}^1 \, \| \gamma J\rangle, \tag{8.7.12}$$

and

$$B(J,J) = 2Q\left[\frac{J(2J-1)}{(J+1)(2J+1)(2J+3)}\right]^{1/2} \langle \gamma J \| \, \mathbf{T}^2 \, \| \gamma J \rangle. \qquad (8.7.13)$$

The electronic reduced matrix elements in many-electron systems can be evaluated using (6.10.17); for the ASF ΓJ^π, we have

$$\langle \Gamma J^\pi \| \, \mathbf{T}^k \, \| \Gamma J^\pi \rangle = \sum_{rs} c_{r\Gamma}{}^\dagger c_{s\Gamma} \langle \gamma_r J^\pi \| \, \mathbf{T}^k \, \| \gamma_s J^\pi \rangle \qquad (8.7.14)$$

where γ_r labels the r-th CSF, and the CSF matrix elements are

$$\langle \gamma_r J^\pi \| \, \mathbf{T}^k \, \| \gamma_s J^\pi \rangle = \sum_{\alpha,\beta} d^k_{\alpha,\beta}(rs) \, \langle n_\alpha \kappa_\alpha \| \, \mathbf{t}^k \, \| n_\beta \kappa_\beta \rangle. \qquad (8.7.15)$$

In the M1 case, the levels predicted by (8.7.10) are uniformly spaced with adjacent levels split by the Landé interval

$$\Delta E_{M1}(JIF) = A(J,J)\, F.$$

Hyperfine splittings are normally very small in relation to intervals between fine structure and other levels. For formulas relating to the M3 (magnetic octupole) and E4 (electric hexadecapole) interaction, see [43, 44, 45, 46].

8.7.1 Hyperfine interactions in the many-electron atom

The hyperfine interaction is a single particle operator, and the first approximation, treated above, assumes that the wavefunction remains unperturbed. However, it is well-known [44, Chapter V] that it is necessary to go beyond this to make proper comparisons with experiment. Certainly, an accurate representation of the electronic part of the interaction is essential to extract nuclear moments from the measured hyperfine splittings. Alternatively, we can test our understanding of atomic structure theory when reliable values of nuclear moments are available. The nonrelativistic theory is often adequate in light elements where relativistic effects scale like Z^2, but it is now generally accepted that it is better to use a fully relativistic theory. This is particularly relevant in hyperfine structure calculations because of the sensitivity of the results to the behaviour of the wavefunction near the nucleus. To first order in the hyperfine interaction, the perturbed wavefunctions are

$$|\gamma\nu JIFM_F\rangle^{(1)} = |\gamma\nu JIFM_F\rangle + \sum_{\gamma' J'} c_{\gamma' J'} \, |\gamma'\nu J'IFM_F\rangle, \qquad (8.7.16)$$

where

$$c_{\gamma' J'} = \frac{\langle \gamma'\nu J'IFM_F \,|\, \mathcal{H}_{hfs} \,|\, \gamma\nu JIFM_F\rangle}{E_{\gamma J} - E_{\gamma' J'}}.$$

The sum is formally over all electronic states $\gamma' J'$ with energy $E_{\gamma' J'} \neq E_{\gamma J}$, but the biggest contributions come from fine structure levels in the same multiplet. The denominators corresponding to excitations from other atomic states are usually unimportant and can be neglected. In this approximation, we require the M1 off-diagonal elements

$$E_{M1}(JIF, J'IF) = \langle \gamma\nu JIFM_F \,|\, \mathbf{T}^1 \cdot \mathbf{M}^1 \,|\, \gamma\nu J'IFM_F \rangle, \quad J' = J \pm 1,$$

from which we get

$$\begin{aligned} E_{M1}(JIF, (J-1)IF) &= \frac{1}{2} A(J, J-1) \\ \times \, [(I+J+F+1)(I+J-F)&(-I+J+F)(I-J+F+1)]^{1/2}, \end{aligned} \tag{8.7.17}$$

where

$$A(J, J-1) = \frac{\mu_I}{I} \frac{1}{[J(J+1)(2J+1)]^{1/2}} \langle \gamma J \| \,\mathbf{T}^1\, \| \gamma\,(J-1) \rangle.$$

The E2 off-diagonal elements are similarly given by

$$E_{E2}(JIF, J'IF) = \langle \gamma\nu JIFM_F \,|\, \mathbf{T}^2 \cdot \mathbf{M}^2 \,|\, \gamma\nu J'IFM_F \rangle,$$

where $J' = J \pm 1, J \pm 2,$, and

$$\begin{aligned} E_{E2}(JIF, (J-1)IF) &= B(J, J-1)\, [(F+I-1)(F-I) - J^2 + 1] \\ &\times \frac{[3(I+J+F+1)(I+J-F)(-I+J+F)(I-J+F+1)]^{1/2}}{2I(2I-1)J(J-1)}, \\ E_{E2}(JIF, (J-2)IF) &= B(J, J-2) \\ &\times [6(I+J+F)(I+J+F+1)(-I+J+F-1)(I+J-F)]^{1/2} \\ &\times \frac{[(I+J-F-1)(-I+J+F)(I-J+F+1)(I-J+F+2)]^{1/2}}{2I(2I-1)J(J-1)(2J-1)} \end{aligned} \tag{8.7.18}$$

The perturbed wavefunctions for hyperfine interaction require only CSFs generated by single excitations from the reference state either from valence subshells or from the core. The latter give rise to what is known as *core polarization.* A closed shell is spherically symmetric and contributes nothing to magnetic or electric hyperfien structure. However, the open subshells polarize the core through the Coulomb interaction, and this can have a big effect on the hyperfine structure.

The restricted active space MCDHF approach has now been used on several systems with considerable success. Bieroń *et al.* [47] studied the convergence of the M1 and E2 hyperfine coupling constants in the lithium atom, and compared the final results with other methods of calculation and with experimental data for the three lowest states. SDT (single, double, triple) substitutions from the reference DHF states were allowed to all orbitals with $n = 2, 3, 4, 5$; s, p, d, f, g symmetries were permitted for $n = 6$ and h orbitals were excluded. SD substitutions were permitted for $n \geq 7$ so that the final AS included $15s, 12p, 9d, 6f$, and $3g$ orbital layers. This kept the total number of CSFs below 4500, a convenient maximum for the computers available. Additional CI calculations were used to assess the error in this restricted calculation. The small effect of the Breit interaction and of the normal mass shift (NMS) and specific mass shift (SMS) were included in the total, along with a post-SCF estimate of the second-order vacuum polarization and the electron self-energy, neither of which contributed signifcantly to the hyperfine constants. Table 8.7 summarizes some of the results from this paper, including

Table 8.7. Diagonal hyperfine constants for the $1s^2 2s\ ^2S_{1/2}$, $1s^2 2p\ ^2P_{1/2}$ and $1s^2 2p\ ^2P_{3/2}$ states of ${}^7_3$Li. From [47, Table V] with permission. Copyright 1996 by the American Physical Society.

	A (Mhz)			B (Mhz)	
Method	$^2S_{1/2}$	$^2P_{1/2}$	$^2P_{3/2}$	$^2P_{3/2}$	Ref.
MCDHF	401.71	45.99	-3.106	-0.2190	[47]
MCHF	401.71	45.94	-3.098	-0.2148	[48]
CCSD	400.903	45.789	-2.879	-0.2160	[49]
RMBPT	402.47	45.96	-3.03	-0.2162	[50]
Expt.	401.752043				[51]
Expt.		45.914(25)	-3.055(14)	-0.221(29)	[52]

experimental values for $^2S_{1/2}$ by the atomic-beam magnetic resonance technique [51] and also for $^2P_{1/2}$ and $^2P_{3/2}$ states by the optical double-resonance method [52]. The agreement with the experimental A for the ground state is at the 0.01% level, and this improves to 0.003% when l-extrapolation is used to correct for higher angular momenta, indicating that the calculation is close to the MCDHF limit.

More recent calculations have been more ambitious, Bieroń *et al.* [53], used the measured electric quadrupole hyperfine constants for the $4d5s^2\ ^2D_{3/2}$ and $^2D_{5/2}$ levels to extract the nuclear quadrupole moment of the unstable 64-h ${}^{90}_{39}$Y isotope. A sequence of EOL calculations was performed in which the active set was enlarged systematically as previously described. The single CSF spectroscopic calculation gave estimates $Q = -0.1368$ b and -0.1384 b for the quadrupole moment for the 3/2 and 5/2 levels respectively. The first two stages of refinement changed these estimates by up to about 50% before converging to -0.1233 b for $J = 3/2$ (1580 CSFs) and -0.1257 b for $J = 5/2$ (1991 CSFs) after 5 or 6 stages in which only single substitutions from the reference CSF were allowed. Adding double substitutions from $4pd5s$ shells reduced these estimates to -0.1130 b (5008 CSFs) and -0.1181 b (5799 CSFs) respectively, and allowing further double substitutions from the $n = 3$ shell (13519 and 15782 CSFs) gave - 0.1118 b and -0.1160 b. Using these last results as a way of estimating the error of the calculation gave the final result $Q = -0.125(11)$ b, about 20% smaller than a previously computed semi-empirical value $Q = -0.155(3)$ b. Bieroń *et al.* [54] performed a similar calculation for the $3d^2 4s\ ^2D_{3/2}$ and $^2D_{5/2}$ and the $3d^3\ ^2G_{7/2,9/2}$ and $^2P_{3/2}$ levels of $^{49}_{22}$Ti. The average of these five calculations, each requiring around 5000 to 6000 CSFs, was 0.247(11) b, in good agreement with most other recent estimates using different methods to model the electronic contribution.

8.8 Isotope shifts

The spherically symmetric part of the nuclear charge distribution and the nuclear motion are responsible for the *isotope shifts* of atomic energy levels and transition energies. Whilst the effect of isotope dependence of the nuclear charge distribution is easy to model, motion of the nucleus is much more problematical. The difficulties of constructing a classical relativistic model of even a two-body system [55] are considerable, and the situation in relativistic quantum mechanics is even less satisfactory [56]. The nuclear motion in complex atoms is essentially nonrelativistic, and all practical calculations which model nuclear recoil have been modelled on the standard nonrelativistic treatment [42, Chapter 8].

8.8.1 Nuclear motion

In nonrelativistic quantum mechanics, nuclear motion contributes to level energies through the *normal mass shift* and the *specific mass shift.* Referred to the centre of mass (CMS) frame, the kinetic energy be expressed as [42, pp. 152–153]

$$T_{cms} = \frac{1}{2\mu} \sum_{i=1}^{N} \mathbf{p}_i^2 + \frac{1}{A} \sum_{i<j}^{N} \mathbf{p}_i \cdot \mathbf{p}_j. \tag{8.8.1}$$

where A is the nuclear mass in atomic units, $\mu = mA/(m + A)$ is the reduced mass, and $\mathbf{p}_i$ the momentum of the i-th electron relative to the centre of mass. The second term gives the *specific mass shift*, which we consider below. The replacement of the rest mass, m, of the electron by the reduced mass, μ, in (8.8.1) has two effects: first the energy of the atom with mass A, E_A, is scaled so that

$$E_A = \frac{\mu}{A} E_\infty = \frac{A}{A+m} E_\infty$$

where E_∞ is the energy for infinite mass, and internal lengths are scaled so that $r \to \rho = \mu r/m$. The *normal mass shift* is then defined as the difference

$$E_A^{nms} = E_A - E_\infty = -\frac{m}{A+m} E_\infty$$

and is easily accounted for by introducing the *finite mass Rydberg constant*

$$R_A = \frac{A}{A+m} R_\infty$$

to convert from atomic units to laboratory units (cm^{-1}). The correction to the kinetic energy can be written to lowest order in m/A as

$$T_{nms} = -\frac{1}{A} S_{nms}, \quad S_{nms} = \sum_{i=1} \frac{1}{2} \mathbf{p}_i^2 \tag{8.8.2}$$

The SMS92 module [57] of GRASP92 computes this by replacing each term $\mathbf{p}_i^2/2$ by the corresponding Dirac "kinetic energy", given by (6.3.9) with the potential set to zero:

$$I_{nms}(a,b) = \delta_{\kappa_a \kappa_b} \int_0^\infty c\, Q^*_{\gamma_a,\kappa_a}(r) \left(\frac{d}{dr} + \frac{\kappa_b}{r}\right) P_{\gamma_b,\kappa}(r) \tag{8.8.3}$$
$$-c\, P^*_{\gamma_a,\kappa_a}(r) \left(\frac{d}{dr} - \frac{\kappa_b}{r}\right) Q_{\gamma_b,\kappa_b}(r) - 2mc^2 Q^*_{\gamma_a,\kappa_a}(r) Q_{\gamma_b,\kappa_b}(r)\, dr.$$

The matrix element of S_{nms} in the CSF basis, in the notation of (6.10.2), can then be written [57, Equation (36)]

$$\langle TM\, | S_{nms} |\, T'M' \rangle = \sum_{\alpha\beta} t_{TT'}(\alpha\beta)\, I_{nms}(\alpha,\beta) \tag{8.8.4}$$

The rest of (8.8.1), the *specific mass* (or *mass polarization*) correction is

$$T_{sms} = \frac{1}{A} S_{sms}, \quad S_{sms} = \sum_{i<j}^{N} \mathbf{p}_i \cdot \mathbf{p}_j. \tag{8.8.5}$$

The orbital angular momentum reduced matrix elements of $\mathbf{p}$, (B.3.152), are

$$(l \,\|\, \mathbf{p} \,\|\, l') = -i \left(\frac{d}{dr} - \frac{l(l+1) - 2 - l'(l'+1)}{2r}\right) (l \,\|\, \mathbf{C}^1 \,\|\, l'). \tag{8.8.6}$$

From a relativistic viewpoint, this is an *even* operator, in the sense of (6.2.5), so that the Dirac matrix elements will be, as in (6.2.11),

$$(\beta\kappa \,\|\, \mathbf{p} \,\|\, \beta\kappa') = -i\mathcal{V}(\gamma\, \beta\kappa, \gamma'\, \beta\kappa')(\beta\kappa \,\|\, \mathbf{C}^1 \,\|\, \beta\kappa'), \tag{8.8.7}$$

where $\beta = \pm 1$ labels the large and small components, and $\mathcal{V}(\gamma\, \beta\kappa, \gamma'\, \beta\kappa')$ is the *Vinti integral*

$$\mathcal{V}(\gamma\, \beta\kappa, \gamma'\, \beta\kappa') = \tag{8.8.8}$$
$$\int_0^\infty R^*_{\gamma,\beta\kappa}(r) \left(\frac{d}{dr} - \frac{\beta\kappa(\beta\kappa+1) - 2 - \beta\kappa'(\beta\kappa'+1)}{2r}\right) R_{\gamma',\beta\kappa'}(r)\, dr$$

The selection rules (6.2.13) require that $j = j', j' \pm 1$, with $j + j'$ *odd* if $\eta = -\eta'$ and *even* otherwise. S_{sms} is a scalar operator with a similar structure to the $k = 1$ multipole of the Coulomb interaction, (6.4.4). Thus, in the notation of (6.10.2),

$$\langle TM \, | S_{sms} | \, T'M' \rangle = \sum_{abcd} \delta_{JJ'} \delta_{MM'} \, v^1_{TT'}(abcd) \tag{8.8.9}$$
$$\times \sum_{\beta,\beta'=\pm 1} \mathcal{V}(n_a \, \beta\kappa_a, n_c \, \beta\kappa_c) \mathcal{V}(n_b \, \beta' \kappa_b, n_d \, \beta' \kappa_d).$$

8.8.2 Nuclear volume effect

The *nuclear volume effect* or *field effect* isotope shift is attributed to variation of the spherically symmetric part of the nuclear charge density between different isotopes of an element. The classic theory [58, 59] perturbs the Coulomb potential of a point nucleus to give the expression

$$E_{fs} = \frac{2}{3} \pi Z \rho_e(0) \, \langle R^2 \rangle \tag{8.8.10}$$

where $\rho_e(0)$ is the electron density at the origin, and $\langle R^2 \rangle$ is the mean square radius of the nuclear charge density. It was later realized that E_{fs} is sensitive to relativistic effects, leading directly to a shift from isotope A to isotope A' of

$$\Delta E_{fs} = c \left[Q^A_{n\kappa} \, P^{A'}_{n\kappa} - P^A_{n\kappa} \, Q^{A'}_{n\kappa} \right]_{r=R_{nuc}} \tag{8.8.11}$$

where the Dirac radial components are evaluated just outside the nucleus at radius R_{nuc} [60, 61, 62, 63]. Only s orbitals have a (nonrelativistic) finite density at the origin and the s orbital density is all that is needed in this approximation. In relativistic calculations, both s and $p_{1/2}$ electrons have a density that is unbounded as $r \to 0$ in a hydrogenic model. The relativistic electron density is finite for extended nuclei, and the SMS92 code therefore evaluates CSF matrix elements [57, Equation (35),(38)]

$$\langle TM \, | \sum_{i=1}^{N} \delta(\mathbf{r}_i) | \, T'M' \rangle = \sum_{TT'} t_{TT'}(ab) \, I_{fs}(a,b) \tag{8.8.12}$$

where

$$I_{fs}(a,b) = \lim_{r \to 0} \rho_{ab}(r), \quad \rho_{ab}(r) = (P_a(r) P_b(r) + Q_a(r) Q_b(r))/4\pi r^2.$$

The above theories are essentially one-electron in character, but there has been relatively little investigation of many-electron effects. A perturbation analysis of the field effect in the DHF model [64] revealed that the change in the DHF eigenvalue could be split into three parts:

$$\Delta\epsilon = \Delta\epsilon_B + \Delta\epsilon_D + \Delta\epsilon_E$$

where $\Delta\epsilon_B$ the Broch term, is given by (8.8.11), $\Delta\epsilon_D$ is a correction due to the relaxation in the electron distribution leading to a change in the direct SCF potential. The corresponding exchange correction, $\Delta\epsilon_E$, is tiny. This was compared with the change in the eigenvalue calculated directly from DHF calculations for the 144 and 150 isotopes of europium (Z=63). The results in

Table 8.8. Field effect eigenvalue shifts (cm^{-1}) in DHF calculations for ^{144}Eu - ^{150}Eu. From [64] with permission.

	$\Delta\epsilon$	$\Delta\epsilon_B$	$\Delta\epsilon_D$	Sum
$1s$	1257	1288	-30	1258
$2s$	162	168	-7	161
$2p_{1/2}$	-2	7	-9	-2
$2p_{3/2}$	-8		-8	-8
$3s$	35	36	-2	34
$3p_{1/2}$	-1	2	-2	-1
$3p_{3/2}$	-2		-2	-2
$3d_{3/2}$	-2		-2	-2
$3d_{5/2}$	-2		-2	-2
$4s$	8	8	-0	8
$5s$	1	1	-0	1

Table 8.8 have been rounded to the nearest wavenumber. The second column shows the change in each orbital eigenvalue, which is dominated by the s contribution and verifies the Broch approximation in the third column. Orbitals with angular momentum greater than 3/2 give a negligible Broch shift. However, the next column reveals that relaxation of the electron distribution is non-negligible, even for the higher angular momentum orbitals, and the total of these columns agrees well with the DHF orbital shift in the second column. The electron relaxation has never been taken into account in applications. A study of the numerical errors shows that provided sufficient care is taken with the numerical algorithms and the arithmetic is performed with sufficient precision, as it is on most computer platforms today, the direct shifts calculated from such models are sufficiently accurate for practical purposes. A survey of practical applications can be found in [34, §16.3].

References

1. Mohr P J and Wiese W L 1998 *Atomic and Molecular Data and their Applications* (AIP Conference Proceedings 434). (Woodbury NY: American Institute

of Physics).
2. Grant I P 1974 *J. Phys. B*, **7**, 1458.
3. Akhiezer A I and Berestetskii V B 1965 *Quantum Electrodynamics* (New York: Interscience).
4. Berestetskii V B, Lifshitz E M and Pitaevskii L P 1971 *Relativistic Quantum Theory*, Vol. 1 (Oxford: Pergamon Press).
5. Brink D M and Satchler G R 1968 *Angular Momentum* (Oxford: Clarendon Press).
6. Panofsky W K H and Phillips M 1962 *Classical Electricity and Magnetism* (2e) (Reading, MA: Addison-Wesley Publishing Co. Inc.)
7. Drake G W F 1971 *Phys. Rev. A* **3**, 908.
8. Drake G W F 1972 *Phys. Rev. A* **5**, 1979.
9. Cohen M and Kelly P S 1967 *Can. J. Phys.* **45**, 1661.
10. Weiss A W 1963 *Astrophys. J.* **138**, 1261.
11. Weiss A W 1967 *Phys. Rev.* **162**, 71.
12. Stewart A L 1954 *Proc. Phys. Soc. A* **67**, 917.
13. Babushkin F A 1962 *Opt. Spectr.* **13**, 77.
14. Babushkin F A 1964 *Acta Phys. Polon.* **25**, 749.
15. Babushkin F A 1965 *Opt. Spectr.* **19**, 1.
16. Dyall K G, Grant I P, Johnson C T and Plummer E P 1989 *Comput. Phys. Commun.* **55**, 425.
17. Labzowsky L N, Klimchitskaya G L and Dmitriev Yu Yu 1993 *Relativistic effects in the spectra of atomic systems* (Bristol and Philadephia: IOP Publishing).
18. On-line NIST Atomic Spectra Database URL: http://physics.nist.gov/cgi-bin/AtData/main_asd
19. Jönsson P and Froese Fischer C 1998 *Phys. Rev. A* **57**, 4967.
20. Froese Fischer C 1991 *Comput. Phys. Commun.* **64**, 369.
21. Olsen J, Godefroid M R, Jönsson P, Malmqvist P A and Froese Fischer C 1995 *Phys. Rev. E* **52**, 4499.
22. Ynnerman A and Froese Fischer C 1995 *Phys. Rev. A* **51**, 2020.
23. Doerfert J, Träbert E, Wolf A, Schwalm D and Uwira O 1997 *Phys. Rev. Lett.* **78**, 4355.
24. Yu Zou and Froese Fischer C, 2000, *Phys. Rev. A* **62**, 062505.
25. Yu Zou and Froese Fischer C, 2002, *Phys. Rev. Lett.* **88**, 183001.
26. Brewer L 1971 *J. Opt. Soc. Am.* **61**, 1101.
27. Vander Sluis K L and Nugent L J 1972 *Phys. Rev. A* **6**, 86.
28. Nugent L J, Vander Sluis K L, Fricke B and Mann J B 1974 *Phys. Rev. A* **9**, 2270.
29. Desclaux J P and Fricke B 1980 *J. Phys.* **41**, 943.
30. Wijesundera W P, Vosko S H and Parpia F A 1995 *Phys. Rev. A* **51**, 278.
31. Eliav E, Kaldor U and Ishikawa Y 1995 *Phys. Rev. A* **52**, 291.
32. Starace A F 1982 in *Handbuch der Physik*, (ed. W. Mehlhorn), Vol.**31**, 1 – 121. (Berlin: Springer-Verlag).
33. Starace A F 2005 in [34, Chapter 24].
34. Drake G W F ed *Springer Handbook of Atomic, Molecular and Optical Physics* (New York: Springer-Verlag).
35. Amusia M Ya 1990 *Atomic Photoeffect* (New York: Plenum Press).
36. Hulme H R, McDougall J, Buckingham R A and Fowler R H 1935 *Proc. Roy. Soc.* **149**, 131; for an erratum see [37].

37. Grant I P 1957 *Phys. Rev.* **106**, 754.
38. Walker T E H and Waber J M 1973 *J. Phys. B: At. Mol. Phys.* **6**, 1165.
39. Akhiezer A I and Berestetskii V B 1965 *Quantum Electrodynamics* (New York: Interscience Publishers).
40. Jacobs V L 1972 *J. Phys. B: At. Mol. Phys.* **5**, 2257.
41. Bartschat K 1989 *Physics Reports* **180**, 1.
42. Fischer C F, Brage T and Jönsson P 1997 *Computational Atomic Structure. An MCHF approach* (Bristol and Philadelphia: Institute of Physics).
43. Schwartz C 1955 *Phys. Rev.* **97**, 380.
44. Armstrong, jr., L 1971 *Theory of the Hyperfine Structure of Free Atoms* (New York: Wiley-Interscience).
45. Lindgren I and Rosén A 1974 *Case Studies in Atomic Physics* **4**, 93.
46. Lindgren I and Rosén A 1974 *Case Studies in Atomic Physics* **4**, 197.
47. Bieroń J, Jönsson P and Froese Fischer C 1996 *Phys. Rev. A***53**, 2181.
48. Carlsson J, Jönsson P and Froese Fischer C 1992 *Phys. Rev. A***46**, 2420.
49. Mårtensson-Pendrill and Ynnerman A 1990 *Physica Scripta* **41** 329.
50. Blundell S A, Johnson W R, Liu Z W and Sapirstein J 1989 *Phys. Rev. A***40**, 2233.
51. Beckmann A, Böklen K D and Elke D, 1974 *Z. Phys. A* **270**, 173.
52. Orth H, Ackermann H and Otten E W 1975 *Z. Phys. A* **273**, 221.
53. Bieroń J, Grant I P and Froese Fischer C 1998 *Phys. Rev. A***58**, 4401.
54. Bieroń J, Froese Fischer C and Grant I P 1999 *Phys. Rev. A***59**, 4295.
55. Yennie D R 1989 in *Relativistic, Quantum Electrodynamic and Weak Interaction Effects in Atoms* (ed. W R Johnson, P J Mohr and J Sucher), pp. 93 (New York : American Institute of Physics)
56. Karshenboim S G, Pavone F S, Bassani F, Inguscio M and Hänsch (Eds).2001 *The Hydrogen Atom: Precision Physics of Simple Atomic Systems* (Berlin, Heidelberg: Springer-Verlag).
57. Jönsson P and Froese Fischer C 1997 *Comput. Phys. Commun.* **100**, 81.
58. Rosenthal J E and Breit G 1932 *Phys. Rev.* **41**, 459.
59. Racah G *Nature (Lond.)* 1932 **129**, 723.
60. Broch E K 1945 *Arch. Math. Naturvidensk.* **48**, 25.
61. Bodmer A R 1953 *Proc. Phys. Soc. A* **66**, 1041.
62. Bodmer A R 1959 *Nucl. Phys.* **9**, 371.
63. Seltzer E 1969 *Phys. Rev.* **188**, 1916.
64. Grant I P 1980 *Physica Scripta* **21**, 443.

9

Continuum processes in many-electron atoms

A detailed understanding of the physics of electron-atom and photon-atom collisions has been of use in many fields of application such as astrophysics, plasma physics and controlled thermonuclear fusion, laser physics, isotope separation, and surface science. Experiments have sought to reveal details of the dynamics of a myriad of processes at a fundamental level. In this chapter, we examine several topics in which a relativistic formulation is desirable. Low-energy elastic scattering of electrons by heavy atoms was one of the earliest of these to be investigated. At slightly higher energies, multi-channel close-coupling methods have been popular, and we focus on relativistic generalizations of the popular R-matrix method, both for electron-atom scattering and for photo-excitation and ionization, as well as the relativistic random phase approximation.

9.1 Relativistic elastic electron-atom scattering

Relativistic treatments of elastic scattering of electrons by heavy atoms and ions attracted attention [1] at about the same time as relativistic self-consistent field calculations [2]. Motz *et al.* [3] had surveyed a wide range of approximate cross-section formulae and had discussed electron polarization data at energies above 10 keV, whilst Kessler [4] had reviewed experimental low-energy data on electron polarization. The use of Dirac wavefunctions was arguably the most direct way to take account of spin-orbit effects in theoretical models [1]. It is now possible to make experiments to measure observable quantities such as angular distributions and spin polarizations simultaneously, posing new challenges to the theorist.

The basic Dirac formalism for potential scattering of electrons by atoms was presented in §3.4.2. The asymptotic scattering wavefunction is given by (3.4.25), so that the angular distribution of both upper and lower 2-spinors has the form (3.4.26)

$$F(\theta, \varphi) = \begin{pmatrix} f(\theta) \\ g(\theta) e^{i\varphi} \end{pmatrix}$$

where[1]

$$\begin{aligned} f(\theta) &= \frac{1}{2ip} \sum_{k=0}^{\infty} \left\{ (k+1) \left(e^{2i\delta_{-k-1}} - 1 \right) + k \left(e^{2i\delta_k} - 1 \right) \right\} P_k(\cos\theta), \\ g(\theta) &= \frac{1}{2ip} \sum_{k=1}^{\infty} \left(e^{2i\delta_k} - e^{2i\delta_{-k-1}} \right) P_k^1(\cos\theta) \end{aligned} \tag{9.1.1}$$

If $\boldsymbol{P}$ is the polarization of the incident electron beam, the polarization of the outgoing electron is [5, Equation (3.75)]

$$\boldsymbol{P}' = \{ [P_n + S(\theta)]\, \boldsymbol{n} + T(\theta)\, \boldsymbol{P}_p + U(\theta)\, (\boldsymbol{n} \times \boldsymbol{P}_p \} / [1 + P_n\, S(\theta)] \tag{9.1.2}$$

where $\boldsymbol{n}$ is a unit vector perpendicular to the scattering plane, $P_n = \boldsymbol{P}.\boldsymbol{n}$ is the component of $\boldsymbol{P}$ in the direction of $\boldsymbol{n}$, and $\boldsymbol{P}_p = \boldsymbol{P} - \boldsymbol{n} P_n$ is the component of $\boldsymbol{P}$ in the scattering plane and θ is the angle of scattering. Kessler [5, Chapter 3] gives a detailed analysis of the implications of this formula (see also Walker [1, §2.3]). The angular amplitudes $I(\theta), S(\theta), T(\theta), U(\theta)$, are defined by

$$I(\theta) = |f(\theta)|^2 + |g(\theta)|^2, \tag{9.1.3}$$

$$S(\theta) = i\, [f(\theta) g^*(\theta) - f^*(\theta) g(\theta)] \,/\, I(\theta), \tag{9.1.4}$$

$$T(\theta) = [\, |f(\theta)|^2 - |g(\theta)|^2] \,/\, I(\theta), \tag{9.1.5}$$

$$U(\theta) = [f(\theta) g^*(\theta) + f^*(\theta) g(\theta)] \,/\, I(\theta). \tag{9.1.6}$$

$I(\theta)$ gives the angular distribution for scattering of a longitudinally polarized electron beam [5, p. 36]. The Sherman function $S(\theta)$, (3.4.33), is required for scattering of a transversely polarized incident beam. When the incident beam is unpolarized, the Sherman function gives the distribution of polarization perpendicular to the scattering plane [5, p. 42]. The quantities $T(\theta)$ and $U(\theta)$ are needed to describe the scattering of an incident electron beam with arbitrary incident polarization $\boldsymbol{P}$ [5, §3.3.3].

Because $U^2 + S^2 + T^2 = 1$, the vector

$$\boldsymbol{n} = U\boldsymbol{e}_1 + S\boldsymbol{e}_2 + T\boldsymbol{e}_3$$

lies on the unit sphere. The stereographic projection of $\boldsymbol{n}$ on to the equatorial (x, y)-plane is the point associated with the complex number $x_- = x - iy$ where

$$U = x\zeta, \quad S = y\zeta, \quad T = 1 - \zeta$$

and $\zeta = 1 - T = 2/(x^2 + y^2 + 1)$. Baylis and Sienkiewicz [6] have suggested that the trajectories, $x_-(\theta)$, may be more useful for comparing experimental and theoretical results than a direct comparison of the parameters $U(\theta), S(\theta)$, and $T(\theta)$.

[1] Walker [1] and Kessler [5] use the notation δ_k^+ and δ_k^- respectively for our δ_{-k-1} and δ_k.

9.1.1 Model potentials

Following Walker [1], most calculations of elastic electron-atom scattering have used relatively simple atomic models. Bound states have generally been constructed at the (Dirac-)Hartree-Fock level, and the incident electron is perceived to scatter off (D)HF direct and/or exchange potentials. This frozen-core model is not sufficient: the scattering electron polarizes the atom, and this in turn perturbs the motion of the scattered electron. There is extensive literature dealing with studies of relativistic, exchange and polarization effects at various levels of sophistication.

This is exemplified in a paper of Sienkiewicz [7], who used a relatively simple DHF model to describe the dynamics of the continuum electron. He writes equations (7.3.9) for the continuum radial orbital in the non-self-adjoint form

$$\left(\frac{d}{dr}+\frac{\kappa}{r}\right)P_\kappa(r)=\left[2c+\frac{1}{c}\left(\epsilon-V_{fc}(r)-V_{pol}(r)\right)\right]Q_\kappa(r)+X_Q(r), \tag{9.1.7}$$

$$\left(\frac{d}{dr}-\frac{\kappa}{r}\right)Q_\kappa(r)=-\frac{1}{c}\left(\epsilon-V_{fc}(r)-V_{pol}(r)\right)P_\kappa(r)-X_P(r),$$

where the frozen core direct potential $V_{fc}(r)$ and the exchange potentials $X_P(r)$ and $X_Q(r)$ are constructed as in §7.3. The polarization of the target by the scattering electron is approximated by $V_{pol}(r)$, which constitutes the main correction to the frozen-core potentials. The model has two adjustable parameters:

$$V_{pol}(r)=-\frac{1}{2}\frac{\alpha_d\, r^2}{(r^3+r_0^3)^2}, \tag{9.1.8}$$

where α_d is the static dipole polarizability, and r_0 is a cut-off radius. In the mercury calculation $r_0=3\,a_0$ was chosen to match the p-wave shape resonance at an energy of 0.4 eV [8], and the value $\alpha_d=44.9\,a_0^3$ was taken from [9].

9.1.2 Computational issues

Linear multistep methods, some of which are described in Appendix B.6.3, can be used to solve equations (9.1.7) by marching outwards from the nucleus. The variant used in the mercury example, due to Sienkiewicz and Baylis [10, Appendix 1], generates a solution out to some suitably large radius. For a neutral atom, the asymptotic form of the large component is[2]

$$P_\kappa(r)/r \sim j_l(kr)\cos\delta_\kappa - y_l(kr)\sin\delta_\kappa \tag{9.1.9}$$

where $k=(2\epsilon+\epsilon^2/c^2)^{1/2}$ is the relativistic momentum and $l=j+1/2\,\eta$ as in §3.2.6. The relativistic effects on phaseshifts are most marked for low values

[2] The notation $n_l(z)$ is used for $y_l(z)$ in [10].

of $l \leq l_0$ (for suitable l_0) and it is enough to use approximate *nonrelativistic* phaseshifts to estimate the residual sum in equations (9.1.1). In the mercury example, relativistic phasehifts were calculated to $l_0 = 6$, and those for $6 < l \leq 50$ used a nonrelativistic effective range formula due to Ali and Fraser [11],

$$\tan \delta_l = a_l\, \alpha_d\, k^2 + (b_l\, \alpha_d + c_l\, \alpha'_q)\, k^4$$

based on a more elaborate polarization potential

$$V_{pol}(r) = -\frac{1}{2}\frac{\alpha_d\, r^2}{(r^3 + r_0^3)^2} - \frac{1}{2}\frac{\alpha'_q\, r^4}{(r^5 + r_0^5)^2}$$

where $\alpha'_q = \alpha_q - 6\beta$ includes a dynamical correction to α_d along with the static quadrupole polarizability α_q. The coefficients a_l, b_l, c_l are given in [11]. The k^4 terms were ignored in [7].

This relatively simple scheme gave a very good account of the differential cross-section at intermediate scattering angles from 50° to 150° in comparison with absolute data from [12] at incident energies from 9.0 eV to 25.0 eV. It improved considerably on the early calculations made by Walker [1], which were unable to reproduce the observed angular structure, and also on those of Sin Fai Lam [13], who used a second-order relativistic perturbation potential and nonrelativistic Hartree-Fock target wavefunctions. Sienkiewicz's polarization potential was probably responsible for much of the improvement. The calculated angular distribution was also good at scattering angles down to 10°.

9.1.3 Other approaches

Several attempts have been made to elaborate this simple model with a view to improving its performance. The polarized orbital approach pioneered by Temkin [14, 15] has been applied, for example, by Szmytkowski and Sienkiewicz [16], to elastic scattering from strontium and barium atoms at energies up to 100 eV. Szmytkowski [17, 18] developed a relativistic version of polarized orbital theory in which the continuum electron satisfies an equation [16, Equation 1] of the form (9.1.7). The presence of the continuum electron perturbs the target orbitals, and both direct and exchange polarization potentials can be obtained by solving coupled equations for the orbital perturbations [17, 18]. Only the direct polarization potential was considered by Szmytkowski and Sienkiewicz in [16].

More recently, Sienkiewicz, Fritzsche and Grant [19] proposed a multiconfiguration version based on GRASP. In this case, the $(N + 1)$-electron scattering wavefunction can be written [20]

$$\Psi_k(N+1, JM\Pi) = \mathcal{A} \sum_{a=1}^{m_a} c_{ak} \Phi_a(N, J_a M_a \Pi_a)\, u_{\kappa_a m_a} + \sum_{j=1}^{m_d} d_{jk}\, \phi_j(N+1, JM\Pi), \quad (9.1.10)$$

where $\Phi_a(N, J_a M_a \Pi_a)$ denotes a target CSF coupled to the wavefunction $u_{\kappa_a m_a}$ to form a state with total angular momentum quantum numbers J, M and parity Π and the pseudo-state CSFs $\phi_j(N+1, JM\Pi)$ are intended to account for polarization. This was applied in [20] to study the spin polarization in elastic scattering from krypton at energies from 5 to 20 eV. Although these calculations had mixed success, the method has the advantage of being an *ab initio*, parameter-free, approach applicable to any closed or open-shell atom.

A package for elastic scattering of electrons and positrons by atoms, positive ions and molecules has recently been published by Salvat, Jablonski, and Powell [21].

9.1.4 Determination of phase-shifts

The determination of phase-shifts is in principle straightforward provided the numerical solution of the radial equations is carried out to a sufficiently large radius, say R, at which only one or two terms of the asymptotic solution are required. In the case of a free particle, this means that (9.1.9) can be approximated by the leading term

$$\begin{aligned} P_\kappa(r) &\sim r\Big(j_l(kr)\cos\delta_\kappa - y_l(kr)\sin\delta_\kappa\Big) \\ &\sim k^{-1}\left\{\sin\left(kr - \frac{1}{2}l\pi\right)\cos\delta_\kappa + \cos\left(kr - \frac{1}{2}l\pi\right)\sin\delta_\kappa\right\} \\ &= k^{-1}\sin\left(kr - \frac{1}{2}l\pi + \delta_\kappa\right), \end{aligned} \quad (9.1.11)$$

whilst

$$\begin{aligned} Q_\kappa(r) &\sim \eta C(k).r\Big(j_{\bar{l}}(kr)\cos\delta_\kappa - y_{\bar{l}}(kr)\sin\delta_\kappa\Big) \\ &\sim \eta C(k).k^{-1}\sin\left(kr - \frac{1}{2}\bar{l}\pi + \delta_\kappa\right). \end{aligned} \quad (9.1.12)$$

where $\eta = \operatorname{sgn}\kappa$ and $C(k) = ck/(\epsilon + 2c^2) = \sqrt{\epsilon/(\epsilon + 2c^2)}$ as in (3.4.23) and (3.4.24). When the potential has finite range, the nonrelativistic Schrödinger amplitude is given by a formula of the form (9.1.11) and, denoting differentiation with respect to r by a prime, the logarithmic derivative

$$\gamma = P'(r)/P(r)|_{r=R} = k\,\frac{j_l{}'(kR)\cos\delta - y_l{}'(kR)\sin\delta}{j_l(kR)\cos\delta - y_l(kR)\sin\delta}$$

expresses the continuity of $P(r)$ and its derivative at $r = R$. For an ion of residual charge z, the Riccati-Bessel functions must be replaced by Coulomb functions. By inverting this equation, we obtain

$$\tan \delta = \frac{k\, j'_l(kR) - \gamma\, j_l(kR)}{k\, y'_l(kR) - \gamma\, y_l(kR)}. \tag{9.1.13}$$

In the relativistic case, a similar argument gives

$$\tan \delta_\kappa = \frac{k\, \eta C(k)\, j_{\bar{l}}(kR) - \gamma'\, j_l(kR)}{k\, \eta C(k)\, y_{\bar{l}}(kR) - \gamma'\, y_{\bar{l}}(kR)}, \qquad \gamma' = \frac{Q_\kappa(R)}{P_\kappa(R)}. \tag{9.1.14}$$

Because for any Riccati-Bessel function,

$$k\, \eta C(k)\, f_{\bar{l}}(kr) = \left(\frac{d}{dr} + \frac{\kappa}{r}\right) f_l(kr),$$

when $\bar{l} = l - \eta$, $\eta = \pm 1$, we can rewrite (9.1.14) in the form (9.1.13) by substituting

$$\gamma = \gamma' - \kappa/R. \tag{9.1.15}$$

The accuracy of this result may be poor when the target atom or ion has long-range potentials in $r > R$ due to residual ionic charge or target polarization, making it necessary to integrate out a very long way before reaching the asymptotic region. A number of schemes for extracting phase-shifts when this applies are discussed in Walker's review [1, §3.1]. Szmytkowski and Sienkiewicz [16, 18] suggest applying a relativistic version of the variable phase method. One such approach in which, (B.6.14),

$$P(r) = A(r)\, r^{-\kappa} \sin \phi(r), \quad cQ(r) = A(r)\, r^{\kappa} \cos \phi(r),$$

gives a pair of first order differential equations for the amplitude $A(r)$ and the phase angle $\phi(r)$, which proved useful for studying the nodal structure of Dirac radial wavefunctions in §B.6.4. Szmytkowski [18] suggests using

$$\begin{aligned} P_\kappa(r) &= \qquad x\Big(j_l(x)\, \cos \delta_\kappa(x) - y_l(x)\, \sin \delta_\kappa(x)\Big), \\ Q_\kappa(r) &= x\eta C(k)\Big(j_{\bar{l}}(x)\, \cos \delta_\kappa(x) - y_{\bar{l}}(x)\, \sin \delta_\kappa(x)\Big), \end{aligned} \tag{9.1.16}$$

where $x = kr$, from which

$$\begin{aligned} \frac{d\delta_\kappa(x)}{dx} &= -[cC(k)]^{-1} V(x)\Big(j_l(x)\, \cos \delta_\kappa(x) - y_l(x)\, \sin \delta_\kappa(x)\Big)^2 \\ &\quad -c^{-1} C(k) V(x)\Big(j_{\bar{l}}(x)\, \cos \delta_\kappa(x) - y_{\bar{l}}(x)\, \sin \delta_\kappa(x)\Big)^2. \end{aligned} \tag{9.1.17}$$

With the initial value of $\delta_\kappa(x)$ obtained from (9.1.14) at $x = kR$, the solution of (9.1.17) can be constructed by marching outwards until the value of the phaseshift $\delta_\kappa(x)$ has stabilized within the desired tolerance. Szmytkowski tested this scheme on positron scattering from mercury [18], using a sixth-order Adams predictor-corrector procedure, and found that it was necessary to integrate out several hundred Bohr radii in the presence of long-range polarization potentials before reaching the asymptotic region in which $\delta_\kappa(x)$ settles down. Szmytkowski has also presented a multi-channel variable-phase algorithm [22], which generalizes the method of this section.

9.1.5 Summation of the partial wave expansion

In view of the slow convergence of partial wave expansions (9.1.1), Walker [1, §3.4, §3.5] suggested employing some sort of acceleration scheme together with simpler approximations to the phase-shift at large values of l. In studying electron-atom scattering from mercury with a Hsrtree potential, he found that it was necessary to use some 77 terms when determining $f(\theta)$ at an incident energy of 2 keV. The spin-flip amplitude $g(\theta)$ required only 12 terms. At 100 eV only 24 and 8 terms respectively were needed. An elegant algorithm to sum Legendre series due to Clenshaw [23] is described and analysed in [24, §5.5].

9.2 Electron-atom scattering: the close-coupling method

9.2.1 Low-energy elastic and inelastic collisions

So far we have used a quasi-single particle model: the many-electron atomic target has been treated as though it were an inert charge distribution whose potential modifies the trajectory of the scattering particle. This basic model was improved first to take account of the Pauli principle by using anti-symmetric wavefunctions and then polarization potentials were added to model the static distortion of the target charge distribution in the presence of the scattering electron. At higher energies, new channels open in which atomic electrons have been promoted to excited states. More elaborate models are clearly essential.

The close-coupling method uses a multi-configurational CSF expansion of the wavefunction of an $(N+1)$-electron system of the form (6.8.6):

$$|\,\Gamma, J^\pi, M\,\rangle = \sum_\gamma c_{\Gamma,\gamma} |\,\gamma, J^\pi, M\,\rangle.$$

In order to apply this idea to scattering calculations [25, 47], we write the wavefunction as

$$\Psi^{(JM\Pi)} = \sum_{ij} c_{ij}\, \mathcal{A}[\Phi_i \otimes \psi_j]^{(JM\Pi)} + \sum_m d_m \Theta_m^{(JM\Pi)} \tag{9.2.1}$$

where Φ_i is an N-electron target ASF; $\psi_j(\boldsymbol{r})$ is an orbital wavefunction for the scattering electron; $\mathcal{A}$ is the antisymmetrizer; and Θ_m represents localized states of the $(N+1)$-electron system not included in the first sum. The terms are coupled so that each has the same overall angular momentum and parity quantum numbers J, M, Π. The index i labels the target states and j stands for a complete set of quantum numbers of the continuum electron. If E is the total energy of the system, then in each channel

$$E = E_i + e_j, \tag{9.2.2}$$

where E_i is the energy of the target state Φ_i, e_j is the kinetic energy of the scattering electron, whose squared momentum is

$$p_j^2 = 2e_j(1 + e_j/2c^2).$$

The one electron amplitudes ψ_j are of the usual central field type (3.2.4)

$$\psi_j(\boldsymbol{x}) = \frac{1}{r}\begin{pmatrix} P_j(r)\chi_{\kappa_j m}(\theta,\varphi) \\ iQ_j(r)\chi_{-\kappa_j m}(\theta,\varphi) \end{pmatrix}$$

For *open channels* ($e_i > 0$, the continuum solutions of §3.3.4 can be used. We can write the amplitude in terms of the two standing wave solutions (3.3.34) and (3.3.37)

$$\mathsf{s}_j(r) \sim \mathcal{N}'\begin{pmatrix} \omega_j \sin\zeta_j(r) \\ \omega_j^{-1}\cos\zeta_j(r) \end{pmatrix}, \tag{9.2.3}$$

and

$$\mathsf{c}_j(r) \sim \mathcal{N}'\begin{pmatrix} \omega_j \cos\zeta_j(r) \\ -\omega_j^{-1}\sin\zeta_j(r) \end{pmatrix} \tag{9.2.4}$$

respectively, where $\omega_j = \left[(e_j + 2c^2)/e_j\right]^{1/4}$, and using the conventions of [48],

$$\zeta_j(r) = p_j r + \nu_j \ln 2p_j r - l_j\pi/2 + \delta_{\kappa_j}, \tag{9.2.5}$$

with the Coulomb phase shift (3.3.36)

$$\delta_{\kappa_j} = \rho_{\kappa_j} - \sigma_{\kappa_j} - \pi\gamma_j/2 + l_j\pi/2. \tag{9.2.6}$$

Progressive spherical partial waves, (3.3.38), can be constructed from

$$\begin{aligned} \mathsf{e}_j^{\pm}(r) &= \mathsf{c}_j(r) \pm i\mathsf{s}_j(r) \\ &\sim \mathcal{N}'\exp(\pm i\zeta_j(r))\begin{pmatrix} \omega_j \\ \pm i\omega_j^{-1} \end{pmatrix}. \end{aligned} \tag{9.2.7}$$

For *closed channels* ($e_j < 0$) the asymptotic form is

$$\mathsf{b}_j(r) \sim \exp -\{\lambda_j r + \nu_j \ln(2|\lambda_j|r)\}, \tag{9.2.8}$$

where $\lambda_j = +\sqrt{-2e_j(1+e_j/2c^2)}$, so that $\mathsf{b}_j(r)$ decreases exponentially as $r \to \infty$.

The scattering amplitude and cross section can be obtained with a knowledge of the $\boldsymbol{K}$-matrix, whose elements are defined by the asymptotic radial scattering amplitudes as $r \to \infty$,

$$\mathsf{u}_{ij}(r) \sim \mathsf{s}_i(r)\,\delta_{ij} + \mathsf{c}_i(r) K_{ij}, \tag{9.2.9}$$

when channel i is one of the n_O *open* channels, and

$$\mathsf{u}_{ij}(r) \sim \mathsf{b}_i(r) \widetilde{K}_{ij}, \tag{9.2.10}$$

when channel i is one of the n_C *closed* channels. When all channels are open, then $\boldsymbol{K} := [K_{ij}]$ is a square matrix of dimension n_O, and the *scattering matrix* $\boldsymbol{S}$ is defined by [25, Equation 45.17]

$$\boldsymbol{S} = \frac{\boldsymbol{I} + i\boldsymbol{K}}{\boldsymbol{I} - i\boldsymbol{K}} \tag{9.2.11}$$

where $\boldsymbol{I}$ is the $n_O \times n_O$ identity matrix. The *transition matrix* $\boldsymbol{T}$ is defined by

$$\boldsymbol{T} = \boldsymbol{S} - \boldsymbol{I} = \frac{2i\boldsymbol{K}}{\boldsymbol{I} - i\boldsymbol{K}}. \tag{9.2.12}$$

The labels J, Π are not given explicitly; the channels with different J, Π do not interact unless an external field, for example, singles out some preferred orientation for the system. The relation between the $\boldsymbol{S}$ and $\boldsymbol{K}$ matrices has to be modified when some channels are closed; the nonrelativistic case has been treated by Seaton [26].

The cross-section for excitation of a target state, Γ_i, J_i to some final state Γ_f, J_f by an electron of energy E can be written [27],

$$\sigma_{i\to f}(E) = \frac{\pi a_0^2}{2p^2(2J_i+1)} \Omega_{if}(E) \tag{9.2.13}$$

where the *collision strength*

$$\Omega_{if}(E) = \frac{1}{2}(2J+1)\left|T_{if}^{J\Pi}\right|^2 \tag{9.2.14}$$

is a dimensionless quantity which is symmetric with respect to the interchange of the initial state i with the final state f. The total collision strength can be decomposed into partial collision strengths

$$\Omega_{if}(E) = \sum_J \Omega_{if}^J(E) \tag{9.2.15}$$

where the sum is over all values of the total angular momentum J of the combined target and scattering electron. The partial collision strength includes contributions from both even and odd parities. It is often necessary to go to high values of J to converge the sum, and acceleration algorithms will then be needed. The partial collision strengths are related to the $\boldsymbol{T}$ matrix elements by

$$\Omega^J_{if}(E) = \frac{1}{2}\sum_{\kappa\kappa'}(2J+1)\left|T^J(\Gamma_i J_i \kappa,\, \Gamma_f J_f \kappa')\right|^2 \tag{9.2.16}$$

where κ, κ' refer to the continuum orbital, and Γ_i, Γ_f refer to suppressed quantum numbers of the initial and final atomic states respectively. The sum runs over all open channels (states which can be excited by scattering electrons of energy E), and over all couplings of the electron angular momentum with N-electron states of the target.

9.2.2 The distorted wave approximation

In distorted wave calculations, the $\boldsymbol{T}$ matrix is approximated by writing

$$T^J(\Gamma_i J_i \kappa,\, \Gamma_f J_f \kappa') = 4i\,\langle \Gamma_i J_i \kappa; J)\,|\, g \,|\, \Gamma_f J_f \kappa'; J\rangle \tag{9.2.17}$$

where g denotes the electron-electron interaction potential. Most applications of the relativistic distorted wave method to inelastic collisions have involved highly stripped ions where channel interactions are unimportant and the collision strengths are small enough to make the approximation acceptable. As usual, the initial and final states can be expanded in terms of CSFs as in (6.8.6),

$$|\Gamma J\,\kappa; J\rangle = \sum_{\gamma} c_{\Gamma\gamma}|\gamma J\,\kappa; J\rangle. \tag{9.2.18}$$

The $\boldsymbol{T}$ matrix elements can therefore be reduced to a sum of matrix elements of the general form (6.9.23) and can be evaluated in the same way using the relevant modules of GRASP or similar programs. In most applications, considerable time can be saved by neglecting CSFs with very small expansion coefficients. Whilst some highly forbidden transitions are sensitive to this approximation, they do not play much of a role in the calculation of level populations in the hot plasmas in which these ions are found, so that this simplification is usually justified. See [27] for further information on the use of electron scattering data in plasma modelling.

9.3 The relativistic R-matrix method

The (nonrelativistic) R-matrix method was first introduced by Wigner [28] and Wigner and Eisenbud [29] to describe nuclear resonance reactions. Reviews of subsequent developments in this field have been written by Lane and Thomas [30], Breit [31], and Barrett *et al.* [32]. The same approach has proved remarkably effective for studying a wide range of processes in atomic and molecular physics, particularly in the hands of Burke and co-workers [33, 34]. Recent reviews of applications [34, 35] list electron-atom and electron-ion excitation and ionization processes, electron-molecule scattering, positron-atom and positron-molecule scattering, atomic and molecular photo-ionization processes, atomic and molecular bound state energies and oscillator strengths, atomic polarizabilities, atom-molecule reactive scattering, dissociative attachment and recombination processes, and atomic multiphoton processes. R-matrix calculations have also played a major role in the preparation of atomic data for the Opacity Project Collaboration [36] led by M. J. Seaton, whose objective was to compute Rosseland Mean Opacities which determine the flow of thermal radiation in stellar envelopes and interiors.

The bulk of this work has dealt with low-Z atoms so that, when relativistic effects have been needed, a version of R-matrix theory using the Breit-Pauli Hamiltonian (one-body mass-correction, Darwin and spin-orbit terms) [37, 38] has sufficed. A fully relativistic scheme is required for atoms in lower rows of the Periodic Table. A Dirac R-matrix scheme was first formulated by Goertzel [39] soon after the original paper by Wigner and Eisenbud [29] and later by Chang [40, 41], Norrington and Grant [42], and Thumm and Norcross [43, 44].

The R-matrix method for atomic processes assumes that the volume occupied by the $N+1$ electrons can be divided into two regions separated by the sphere $r = a$, where r is the coordinate of the scattering electron relative to the target nucleus. The *internal* region $r < a$ is assumed just large enough to enclose the charge distribution of the target states. The wavefunction in $r < a$ must describe electron exchange and correlation effects, and will therefore be of the usual close-coupling type. The *external* region $r > a$ is one in which exchange effects can be neglected so that the scattered electron can be considered to move in the asymptotic electric multipole potential field of the target.

9.3.1 The radial Dirac equation on a finite interval

We begin by solving the eigenvalue problem on the internal region, specifcally on the closed interval $[r_0, a]$. The radial Dirac equation is put in the form

$$(T - \varepsilon)\, u(r) = 0, \qquad u(r) = \begin{pmatrix} P(r) \\ Q(r) \end{pmatrix}, \tag{9.3.1}$$

where

$$T = cJ\frac{d}{dr} + W(r), \quad r_0 < r < a,$$

and

$$J = \begin{pmatrix} 0 & -1 \\ 1 & 0 \end{pmatrix}, \quad W(r) = \begin{pmatrix} V(r) & c\kappa/r \\ c\kappa/r & -2c^2 + V(r) \end{pmatrix}.$$

Appendix B.7 analyses the radial Dirac equation on a finite interval $0 < r_0 \leq r \leq a < \infty$. The elementary theory of differential systems shows that the initial value problem for equation (9.3.1) with a given $u(r_0)$ has a unique continuous solution subject only to mild restrictions on the matrix $W(r)$. The eigenvalue problem supplements (9.3.1) with two-point boundary conditions

$$u(r_0) = Mv, \quad u(a) = Nv, \tag{9.3.2}$$

where M and N are 2×2 matrices such that if $Mv = 0$ and $Nv = 0$, then $v = 0$. We use $\widehat{T}$ to denote the operator T together with the boundary conditions (9.3.2).

The operator $\widehat{T}$ is said to be Hermitian if

$$(u, \widehat{T}w) = (\widehat{T}u, w), \quad (u, w) = \int_{r_0}^{a} u^{\dagger}(r)\, w(r)\, dr \tag{9.3.3}$$

Theorem B.19 shows that the eigenvalues of $\widehat{T}$ are real, so that (9.3.3) holds if, by Lemma 5.3, $s(r) = cu^{\dagger}(r)\, J\, w(r)$ is independent of r and in particular if

$$s(r_0) = s(a). \tag{9.3.4}$$

By (9.3.2), this is equivalent to the condition $M^{\dagger}JM = N^{\dagger}JN$, for arbitrary nonvanishing v. Because $J^{\dagger} = -J$, then $(M^{\dagger}JM)^{\dagger} = -M^{\dagger}JM$, so that when M, N are real,

$$M^{\dagger}JM = N^{\dagger}JN = 0.$$

We can therefore decouple the boundary conditions at each end of the range by choosing *real* rank-1 matrices M and N

$$M = \begin{pmatrix} 0 & m \\ 0 & 1 \end{pmatrix}, \quad N = \begin{pmatrix} 1 & 0 \\ n & 0 \end{pmatrix}, \tag{9.3.5}$$

where M operates only on the lower component of v and N only on the upper component. The eigenvalues are determined by (B.7.16):

$$(N - U_{\varepsilon}(R_2)\, M)\, v_{\varepsilon} = 0,$$

where $U_{\varepsilon}(r)$ is the fundamental matrix solution of the initial value problem (9.3.1) for parameter ε with the initial condition $U_{\varepsilon}(r_0) = I$, where I is the identity matrix, and v_{ε} is determined by the eigenvalue equation up to an overall normalization constant. Thus the eigenvalue depends on the ratio of the two components of v but not on their absolute values and

$$u_{\varepsilon}(r_0) = \begin{pmatrix} P_{\varepsilon}(r_0) \\ Q_{\varepsilon}(r_0) \end{pmatrix}, \quad P_{\varepsilon}(r_0) = mv_{\varepsilon}^{(2)}, \quad v_{\varepsilon}^{(2)} = \mathcal{N} Q_{\varepsilon}(r_0),$$

$$u_{\varepsilon}(a) = \begin{pmatrix} P_{\varepsilon}(a) \\ Q_{\varepsilon}(a) \end{pmatrix}, \quad Q_{\varepsilon}(a) = nv_{\varepsilon}^{(1)}, \quad v_{\varepsilon}^{(1)} = \mathcal{N} P_{\varepsilon}(a).$$

where $\mathcal{N}$ is the arbitrary normalizing constant. The eigenvalues are discrete and nondegenerate (Theorem B.19). Theorem B.30, the main result of Appendix B.7, states that, when $f(r)$ is a measurable (2-component) function, the solution almost everywhere of the inhomogeneous equation

$$(T - \varepsilon)u(r) = -f(r), \quad r_0 \leq r \leq a, \tag{9.3.6}$$

subject to the boundary conditions (9.3.2) can be expanded in terms of the eigenfunctions of (9.3.1) so that

$$u(r) \sim \sum_k c_k\, u_k(r), \quad c_k = \int_{r_0}^{a} u_k^\dagger(r)\, u(r)\, dr$$

converges uniformly and absolutely to $u(r)$ on $r_0 \leq r \leq a$.

9.3.2 Bloch operators

In the last section, we denoted the radial Dirac operator by T and used the notation $\widetilde{T}$ when it was augmented by boundary conditions at $r = r_0$ and $r = a$ to make it self-adjoint. An alternative, suggested by Bloch [45] is to incorporate the boundary condition explicitly by adding operators with support only on the boundary. For the 3-dimensional Dirac Hamiltonian (3.1.39) for an electron, charge $q = -e$,

$$H_D := c\boldsymbol{\alpha} \cdot (\boldsymbol{p} + e\boldsymbol{A}) + \beta mc^2 - e\phi,$$

we find that

$$\begin{aligned} &\int_{\mathcal{V}} \left[\phi^\dagger\,(H_D\,\psi) - (H_D\,\phi)^\dagger\,\psi\right]\, d\boldsymbol{r} \\ &= -ic\int_{\mathcal{V}} \boldsymbol{\nabla} \cdot \left[\phi^\dagger \boldsymbol{\alpha}\psi\right]\, d\boldsymbol{r} = \int_{\partial\mathcal{V}} \left[\phi^\dagger \boldsymbol{\alpha}\psi\right] \cdot \boldsymbol{n} dS \end{aligned} \tag{9.3.7}$$

where $\boldsymbol{n}$ and dS are the outward normal and surface element on the bounding surface $\partial\mathcal{V}$ to the simple connected volume V. We restrict ourselves to the case when $\mathcal{V}$ is a spherical shell bounded by surfaces at $r = r_0$ and $r = a$. Following Bloch's lead, we define the Hermitian operator

$$\widehat{H}_D = H_D + L(a) - L(r_0), \tag{9.3.8}$$

where $L(r)$ is a 4-component analogue of Bloch's surface operator,

$$L(r_0) = \delta(r - r_0)\,\frac{c}{2}\,\left[\boldsymbol{b}(r_0) + i\,\boldsymbol{\alpha} \cdot \boldsymbol{e}_r\right],$$

$$L(a) = \delta(r - a)\,\frac{c}{2}\,\left[\boldsymbol{b}(a) + i\,\boldsymbol{\alpha} \cdot \boldsymbol{e}_r\right].$$

The outward normal on $r = r_0$ and $r = a$ is $\boldsymbol{e}_r = \boldsymbol{r}/r$ and $\boldsymbol{b}(r_0)$ and $\boldsymbol{b}(a)$ are arbitrary 4×4 Hermitian matrices. It is straightforward to verify that

$$\int_{\mathcal{V}} \left[\phi^\dagger\,(\widehat{H}_D\,\Psi) - (\widehat{H}_D\,\phi)^\dagger\,\psi\right]\, d\boldsymbol{r} = 0.$$

This can be reduced to a more familiar form involving only radial operators when $\boldsymbol{A} = 0$ and $-e\phi = V(r)$. Using the methods of §3.2, assuming that $\psi(\boldsymbol{r})$ is written in the form (3.2.4), then

$$H_D\,\psi(\boldsymbol{r}) = \frac{1}{r}\begin{pmatrix} \left\{(mc^2 + V(r) - E)P(r) + \hbar c\left(-\frac{dQ}{dr} + \frac{\kappa}{r}Q(r)\right)\right\}\chi_{\kappa m}(\theta,\varphi) \\ i\left\{\hbar c\left(\frac{dP}{dr} + \frac{\kappa}{r}P(r)\right) + (-mc^2 + V(r) - E)Q(r)\right\}\chi_{-\kappa m}(\theta,\varphi) \end{pmatrix} \tag{9.3.9}$$

Similarly, (3.2.13) gives

$$i\,\boldsymbol{\alpha}\cdot\boldsymbol{e}_r\,\psi(\boldsymbol{r}) = \frac{1}{r}\begin{pmatrix} Q(r)\,\chi_{\kappa m}(\theta,\varphi) \\ -iP(r)\,\chi_{-\kappa m}(\theta,\varphi)\end{pmatrix} \tag{9.3.10}$$

Without loss of generality, we can choose $\boldsymbol{b}(r_0)$ and $\boldsymbol{b}(a)$ to be block diagonal. The off-diagonal blocks mix components of opposite parity, and can therefore be discarded, whilst the diagonal blocks must be multiples of the identity in order to preserve the angular structure. Thus the 2-component version of (9.3.8) generalizing [45, Equation (7)] is

$$\widetilde{T} = T + l(a) - l(r_0), \quad l(s) := \frac{c}{2}\,(b(s) - J)\,\delta(r - s), \tag{9.3.11}$$

where

$$J = \begin{pmatrix} 0 & -1 \\ 1 & 0 \end{pmatrix}, \quad b(s) = \begin{pmatrix} b^+(s) & 0 \\ 0 & b^-(s) \end{pmatrix},$$

is the most general form of a radially reduced Hermitian Dirac operator on a finite interval.

Following Bloch, we now note that the result of operating on a 2-component trial function with $\widetilde{T} - \varepsilon$ is to generate a function of the form $f(r) + A\delta(r-a) - A'\delta(r-r_0)$ where $f(r)$ is a smooth function of r; the singular terms only contribute on the boundary and can be treated as independent of values on the interior of the interval. Thus we have

$$\begin{aligned} &(T - \varepsilon)\,u(r) = 0, \quad r_0 < r < a, \\ &l(s)u(r) = \frac{c}{2}\,(b(s) - J)\,\delta(r - s)u(r) = 0 \text{ at } s = r_0,\ a. \end{aligned} \tag{9.3.12}$$

The boundary equations have nontrivial solutions if $\det\,(b(s) - J) = 0$ that is if $b^+(s)b^-(s) + 1 = 0$. Comparing this with the boundary conditions (9.3.5), we see that

$$b^-(r_0) = -1/b^+(r_0) = m, \quad b^+(a) = -1/b^-(a) = -n.$$

The number $m = P(r_0)/Q(r_0)$ is determined by the power series expansion of the solution near the origin. We shall only deal with functions $u(r)$ which satisfy this condition, so that we shall not need to consider $l(r_0)$ again. On the other hand, the value of $n = Q(a)/P(a)$ is needed for matching to solutions outside the R-matrix boundary $r > a$, and $l(a)$ therefore plays an important part in what follows.

9.3.3 The inner region, $r \leq a$

Suppose that H_{N+1} is the Hamiltonian for the e-atom system,

$$H_{N+1} = \sum_{n=1}^{N+1} H_{Dn} + \sum_{m<n} g_{mn} \tag{9.3.13}$$

where H_{Dn} is the Dirac operator for the n-th electron as in (9.3.9) with the usual energy zero and g_{mn} is the electron-electron interaction. The scattering wavefunction, $\Psi(E)$, for energy E in channel $\Gamma \equiv \gamma JM$, where JM are the total angular momentum quantum numbers associated with the channel and γ denotes other identifying parameters, satisfies

$$(H_{N+1} - E)\,\Psi(E) = 0. \tag{9.3.14}$$

on $r_0 < r < a$. The boundary conditions at $r = r_0$ are set by the power series expansion about the origin. The boundary conditions at $r = a$ are imposed with the Bloch operator

$$L_{N+1} = \sum_{n=1}^{N+1} L^{(n)}(a), \quad L^{(n)}(a) = \delta(r_n - a)\,\frac{c}{2}\,\left[\boldsymbol{b}_n(a) + i\,\boldsymbol{\alpha}_n \cdot \boldsymbol{e}_r\right]. \tag{9.3.15}$$

so that $\widehat{H} = H_{N+1} + L_{N+1}$ is Hermitian on $r_0 \leq r \leq a$. We rewrite (9.3.14) formally as

$$\Psi_E = (\widehat{H} - E)^{-1} L_{N+1} \Psi_E \tag{9.3.16}$$

so that when E is not an eigenvalue of $\widehat{H}$, the formal solution is obtained by acting on the boundary values $L_{N+1}\Psi_E$ with the Hermitian resolvant operator $(\widehat{H} - E)^{-1}$. Introduce a set of orthonormal target eigenstates Φ_i and eigenvalues E_i^N so that

$$H_N\,\Phi_i = E_i^N\,\Phi_i, \quad \langle \Phi_i \,|\, \Phi_j \rangle = \delta_{ij}. \tag{9.3.17}$$

These will normally have been constructed in a self-consistent field calculation either using GRASP, or else in some form of CI calculation with Dirac orbitals generated with a model potential. We adopt a close-coupling expansion of the type (9.2.1) for the scattering wavefunction Ψ_E of the $(N+1)$-electron system,

$$\Psi_E = \sum_k \Psi_k A_{kE}, \tag{9.3.18}$$

where Ψ_k are eigensolutions of (9.3.14) satisfying the R-matrix boundary conditions, so that

$$\langle \Psi_k \mid \widetilde{H} \mid \Psi_{k'} \rangle = E_k \delta_{kk'} \tag{9.3.19}$$

where E_k is real. The diagonalization (9.3.19) produces eigensolutions of the form

$$\Psi_k = \sum_{ij} c_{ijk}\, \Theta_{ij} + \sum_{j} d_{mk} \theta_m. \tag{9.3.20}$$

We couple N-electron target states Φ with continuum Dirac orbitals ψ to form the channel functions

$$\Theta = \mathcal{A}[\Phi \otimes \psi]^{J\Pi}_{M} = \sum_{M^t m} \mathcal{A}[\Phi \otimes \psi]\, \langle J^t M^t, jm \mid JM \rangle. \tag{9.3.21}$$

The channel index, i, corresponds to the coupling $(J^t j)J$, with parity $\Pi = \Pi^t \pi$, where $\pi = (-1)^l$. The R-matrix algorithm defines an orbital basis, ψ_{ij}, for the i channel satisfying the boundary conditions (9.3.15) on $r = a$ with energies ε_{ij}. Those with $\varepsilon_{ij} > 0$ correspond to open channels whilst those with $\varepsilon_{ij} < 0$ generate normalizable "capture" states. The "continuum orbitals" (it is convenient to use this term for all ψ_{ij} which are non-vanishing on $r = a$ whatever the sign of ε_{ij}) must be orthogonal to the bound orbitals of the same symmetry used to construct the target states. The θ_m are capture states constructed in the same way, but with the continuum orbital replaced by a true bound or pseudo-orbital. It will later be convenient to use the compact notation of [46, Equation (7)])

$$\Psi_k = \sum_{\lambda} \Theta_\lambda V_{\lambda k}$$

for the sum (9.3.20), so that the scattering wave function (9.3.18) can be written in matrix notation

$$\Psi_E = \boldsymbol{\Theta} \boldsymbol{V} \boldsymbol{A}_E, \tag{9.3.22}$$

where $\boldsymbol{\Theta}$ is a row vector and $\boldsymbol{A}_E$ a column vector.

The radial parts of the continuum orbital spinors ψ_{ij} may be written

$$v_{ij}(r) = \begin{pmatrix} P_{ij}(r) \\ Q_{ij}(r) \end{pmatrix}, \tag{9.3.23}$$

The radial spinors of symmetry κ are orthonormalized in the sense

$$\int_{r_0}^{a} v^{\dagger}_{ij}(r) v_{ij'}(r)\, dr = \delta_{jj'}. \tag{9.3.24}$$

The bound target orbitals are similarly denoted $\underline{v}(r)$; in this case $\underline{v}(r)$ is effectively zero when $r > a$ so that their normalization is unaffected by the finite value of a. We construct the continuum orbitals by solving a set of equations of the form (9.3.6),

$$(T^0 - \epsilon_i) u_i(r) = \sum_l \lambda_{il} \, \underline{v}_l(r), \tag{9.3.25}$$

where the Lagrange multipliers λ_{il} enforce orthogonality of $u_i(r)$ to the bound orbitals $\underline{v}_l(r)$ of the same symmetry and the superscript on T^0 indicates that a local static model potential, $V^0(r)$, has been used. There is some freedom to choose the model potential most suited to the application.

We can project out the states of the $(N+1)$-st electron by defining

$$\phi_{ik}(\boldsymbol{r}) = \langle \Phi_i \,|\, \Psi_k \rangle_I = \sum_j C_{ijk} \psi_{ij}(\boldsymbol{r}) \tag{9.3.26}$$

where the subsecript I denotes integration over the inner region and, from (9.3.21),

$$C_{ijk} = c_{ijk} \, \langle J_i M_i, jm \,|\, JM \rangle$$

if the uncoupled electron is not equivalent to a bound orbital. If, as in photo-ionization processes, ionization takes place out of the α subshell in the initial $(N+1)$-electron state, equation (6.10.8) gives an additional factor

$$(-1)^{\Delta_\alpha} (N_\alpha + 1)^{1/2} (\widetilde{T}_\alpha \,|\}\, T_\alpha),$$

where N_α and $\widetilde{T}_\alpha$ characterize the α subshell in the target state. Denote the two-component radial functions corresponding to $\phi_{ik}(\boldsymbol{r})$ by $w_{ik}(r)$, with components $p_{ik}(r), q_{ik}(r)$. Then from (9.3.15)

$$\langle \Psi_k | \, L \, | \Psi_E \rangle_I = \sum_i w^\dagger_{ik}(a) \, l(a) \, w_{iE}(a) \tag{9.3.27}$$

where $l(a)$ is defined by (9.3.12) with

$$w_{iE}(a) = \sum_k w_{ik}(a) A_{kE}, \tag{9.3.28}$$

so that, (9.3.16),

$$A_{kE} = \frac{\langle \Psi_k | \, L \, | \Psi_E \rangle_I}{E_k - E}. \tag{9.3.29}$$

If we write $q_{jk}(a)/p_{jk}(a) = n_j$, then (9.3.12) gives $b^+(a) = -n_j$, $b^-(a) = 1/n_j$ so that

$$w^\dagger_{jk}(a) \, l(a) \, w_{jE}(a) = p_{jk}(a) \left[c q_{jE}(a) - c n_j \, p_{jE}(a) \right], \tag{9.3.30}$$

where $p_{jE}(a), q_{jE}(a)$ are the components of $w_{jE}(a)$. For most applications of R-matrix theory it is sufficient [42] to set

$$n_j = (b + \kappa_j)/2ac, \tag{9.3.31}$$

where b is an arbitrary constant. This boundary condition serves to define the solutions uniquely in $r_0 < r < a$ whatever the energy of the "continuum" electron. When this is well below $mc^2 \approx 0.5$MeV, it is consistent with the Pauli approximation, (B.7.13), which makes it easy to match to nonrelativistic amplitudes in $r > a$. It does not impose any upper limit on the energy as we can still use relativistic dynamics in $r > a$ if necessary. Collecting the results (9.3.27) to (9.3.31) gives the two equations

$$\begin{aligned} p_{iE}(a) &= \sum p_{ik}(a) A_{kE} \\ &= \sum_{jk} \frac{p_{ik}(a)\, p_{jk}(a)}{2(E_k - E)} \left[2c\, q_{jE}(a) - \frac{b + \kappa_j}{a}\, p_{jE}(a) \right], \end{aligned} \tag{9.3.32}$$

and

$$\begin{aligned} q_{iE}(a) &= \sum q_{ik}(a) A_{kE} \\ &= \sum_{jk} \frac{q_{ik}(a)\, p_{jk}(a)}{2(E_k - E)} \left[2c\, q_{jE}(a) - \frac{b + \kappa_j}{a}\, p_{jE}(a) \right]. \end{aligned}$$

Because of the boundary condition (9.3.12) this last equation is a multiple $(b + \kappa_i)/2ac$ of (9.3.32). It therefore plays no independent part in the rest of the calculation. As in the nonrelativistic theory, we define[3] the R-matrix (or resolvent) by

$$R_{ij}(E) := \sum_k \frac{p_{ik}(a)\, p_{jk}(a)}{2(E_k - E)}, \tag{9.3.33}$$

which enables us to express (9.3.32) as

$$p_{iE}(a) = \sum_j R_{ij}(E)\, Q_{jE} \tag{9.3.34}$$

where, following Seaton [49, Equation (4.34)], we introduce

$$Q_{jE} = 2c\, q_{jE}(a) - \frac{b + \kappa_j}{a}\, p_{jE}(a).$$

In the nonrelativistic limit

$$2c\, q_{jE}(r) \approx \left(\frac{dp_{jE}}{dr} + \frac{\kappa_j p_{jE}}{r} \right),$$

so that

$$\lim_{c \to \infty} Q_{jE} = \left(\frac{dp_{jE}}{dr} - \frac{b}{r}\, p_{jE} \right)_{r=a},$$

in agreement with the standard nonrelativistic expression [35, Equation (13.11)].

[3] Here, for later convenience, we have departed from the convention of Chang [40] and others by absorbing the factor a into the square bracket of (9.3.27).

9.3.4 The outer region, $r > a$

The R-matrix boundary a is normally chosen so that the target wavefunctions Φ_i have negligible overlaps with functions in the outer region so that exchange potentials $X_\beta(A; r)$ and the Lagrange multiplier terms can be dropped from (6.7.12). With a change of notation to conform to the conventions of this section, the functions in the outer region satisfy

$$\left(-\frac{z}{r} - \varepsilon_i\right) p_i(r) + c\left(-\frac{d}{dr} + \frac{\kappa_i}{r}\right) q_i(r) = -\sum_{i'} V_{ii'}(r) p_{i'}(r),$$
$$(9.3.35)$$
$$c\left(\frac{d}{dr} + \frac{\kappa_i}{r}\right) p_i(r) + \left(-2mc^2 - \frac{z}{r} - \varepsilon_i\right) q_i(r) = -\sum_{i'} V_{ii'}(r) q_{i'}(r),$$

(cf. [48, Equation (7.1)]). Here z is the effective charge on the atom in $r > a$. The potentials $V_{ii'}(r)$ couple the channels and are commonly expanded in an asymptotic multipole series [48, Equation (7.2)]

$$V_{ii'}(r) \sim \sum_\lambda C^\lambda_{ii'}/r^{\lambda+1}. \qquad (9.3.36)$$

It is normal to choose the radius a significantly larger than the effective outer radius of the target orbitals, r_{max}. Because $C^\lambda_{ii'} = O|(r^\lambda)_{ii'}| \le r^\lambda_{max}$ and $r \ge a$, the terms of (9.3.36) decrease geometrically like powers of r_{max}/r, so that the right-hand side of (9.3.35) can be treated as a small perturbation [49, §4.4].

The equations (9.3.35) must be integrated numerically out to some large radius, $R_{max} \gg a$, at which the solutions can be fitted to asymptotic values to deduce the K-matrix, the S-matrix and cross sections. Because it is usual to assume that the Pauli approximation can be applied to solutions in $r > a$, it is possible to use nonrelativistic packages for this problem, and most current implementations take advantage of this. The radial range that must be covered to reach an "asymptotic" distances at which the channel coupling can be neglected can be very large, and a variety of techniques have been developed to deal with the consequences (see §9.6).

9.3.5 Matching inner and outer solutions

It is convenient to express the connection between inner and outer regions in terms of matrix algebra. Suppose that there are n equations (9.3.28) of which n_O refer to open channels and $n_C = n - n_O$ to closed channels. Then $\boldsymbol{R}(E)$ is an $n \times n$ matrix with elements given by (9.3.33). Let $\boldsymbol{p}(r)$ and $\boldsymbol{q}(r)$ be $n \times n$ matrices with elements $p_{ik}(r), q_{ik}(r)$. Then (9.3.33) can be written

$$\boldsymbol{R}(E) = \boldsymbol{p}(a)[2(\boldsymbol{e} - E)]^{-1}\boldsymbol{p}^T(a) \qquad (9.3.37)$$

where $\boldsymbol{e}$ is the diagonal matrix of energies E_k. Let $\mathsf{p}_E(r)$ and $\mathsf{q}_E(r)$ be the vectors with elements $p_{jE}(r), q_{jE}(r)$, so that if $\boldsymbol{A}_E$ is the n-vector with elements A_{kE}, then

$$\mathsf{p}_E(r) = \boldsymbol{p}(r)\,\boldsymbol{A}_E, \quad \mathsf{q}_E(r) = \boldsymbol{q}(r)\,\boldsymbol{A}_E, \tag{9.3.38}$$

and the matching condition on $r = a$, (9.3.27), becomes

$$\mathsf{p}_E(a) = \boldsymbol{p}(a)\,\boldsymbol{A}_E = \boldsymbol{R}(E)\,\mathsf{Q}_E. \tag{9.3.39}$$

If $\boldsymbol{\beta}$ is the diagonal matrix with elements $(b + \kappa_j)/a$, then

$$\mathsf{Q}_E = [2c\mathsf{q}_E(a) - \boldsymbol{\beta}\mathsf{p}_E(a)] = \boldsymbol{R}^{-1}(E)\,\mathsf{p}_E(a),$$

whenever E does not coincide with a pole of the R-matrix. Substituting from (9.3.40) gives the expansion coefficients

$$\boldsymbol{A}_E = [2(\boldsymbol{e} - E)]^{-1}\boldsymbol{p}^T\,\mathsf{Q}_E \tag{9.3.40}$$

consistent with the nonrelativistic result [49, Equation (3.21)]. Provided the channel functions Ψ_k are orthonormal on the inner region, $(\Psi_k \,|\, \Psi'_k)_I = \delta_{kk'}$, this gives the norm

$$\begin{aligned} \|\boldsymbol{A}_E\|^2 &= \boldsymbol{A}_E^T.\boldsymbol{A}_E \\ &= \mathsf{Q}_E^T.\boldsymbol{p}\,[2(\boldsymbol{e} - E)]^{-2}\boldsymbol{p}^T\,.\mathsf{Q}_E \\ &= \mathsf{Q}_E^T\,\dot{\boldsymbol{R}}(E)\,\mathsf{Q}_E \end{aligned} \tag{9.3.41}$$

where $\dot{\boldsymbol{R}}(E)$ is the derivative of $\boldsymbol{R}(E)$ with respect to E. Thus we can eliminate all reference to the lower components in applications provided the $\boldsymbol{R}(E)$ has been computed relativistically.

In the outer region, the solutions of (9.3.35) can be either of open type ($\varepsilon_i > 0$) or closed type ($\varepsilon_i < 0$) depending on their asymptotic behaviour as $r \to \infty$. For the n_O open channels, we can construct solutions that are linear combinations of the standing wave vectors $\mathsf{s}_j(r)$, (9.2.3), and $\mathsf{c}_j(r)$, (9.2.4), and we write

$$\mathsf{w}_{jk}(r) \sim \mathsf{s}_j(r)\,\delta_{jk} + \mathsf{c}_j(r)\,K_{jk}, \quad r \to \infty, \tag{9.3.42}$$

where $1 \le j \le n_O, 1 \le k \le n_O$. Similarly for closed channels

$$\mathsf{w}_{jk}(r) \sim \mathsf{b}_j(r)\,K_{jk}, \quad r \to \infty, \tag{9.3.43}$$

where $n_O + 1 \le j \le n, 1 \le k \le n_O$. Whilst it is possible to construct the $\boldsymbol{K}$ matrix by solving (9.3.35) directly, the Pauli approximation allows us to work directly with the nonrelativistic equations as in [50, §8]. If all channels are open, then define functions

$$S_j(r) \sim \sin\zeta_j(r), \quad C_j(r) \sim \cos\zeta_j(r) \tag{9.3.44}$$

where $\zeta_j(r)$ is defined by (9.2.5). Then in the outer region, the upper components $P_{jk}(r)$[4] can be assembled into a $n_O \times n_O$ matrix $\boldsymbol{P}(r)$ so that

$$\boldsymbol{P}(r) = \boldsymbol{S}(r) + \boldsymbol{C}(r)\boldsymbol{K}, \quad r \geq a, \tag{9.3.45}$$

with $\boldsymbol{S}$ and $\boldsymbol{C}$ diagonal. Because solutions and their derivatives must be continuous at $r = a$, we can substitute from (9.3.45) into the matching condition (9.3.39) and rearrange to give [48, Equation (8.6)]

$$\boldsymbol{K} = \boldsymbol{G}^{-1}\boldsymbol{F} \tag{9.3.46}$$

where

$$\boldsymbol{F} = -\boldsymbol{S} + \boldsymbol{R}(E)(\boldsymbol{S}' - b\boldsymbol{S}/a), \quad \boldsymbol{G} = \boldsymbol{C} - \boldsymbol{R}(E)(\boldsymbol{C}' - b\boldsymbol{C}/a).$$

This result can easily be generalized to the case when some channels are open and some closed by extending $\boldsymbol{C}$ to a $n \times n$ matrix with the block structure

$$\widetilde{\boldsymbol{C}} = \begin{pmatrix} \boldsymbol{C}_{OO} & \boldsymbol{C}_{OC} \\ \boldsymbol{C}_{CO} & \boldsymbol{C}_{CC} \end{pmatrix}$$

where $\boldsymbol{C}_{OC}$ and $\boldsymbol{C}_{CO}$ are null, and $\boldsymbol{C}_{CC} = \boldsymbol{b}$, a diagonal matrix whose elements are the upper components of the vectors $\mathsf{b}_j(r)$. Also $\boldsymbol{S}$ and $\boldsymbol{K}$ now have dimension $n \times n_O$ so that

$$\widetilde{\boldsymbol{K}} = \begin{pmatrix} \boldsymbol{K}_{OO} \\ \boldsymbol{K}_{CO} \end{pmatrix}, \quad \widetilde{\boldsymbol{S}} = \begin{pmatrix} \boldsymbol{S}_{OO} \\ \boldsymbol{S}_{CO} \end{pmatrix},$$

where $\boldsymbol{S}_{CO}$ is null and $\boldsymbol{S}_{OO} = \boldsymbol{S}$ and we find [48, Equation (8.10)] in place of (9.3.44)

$$\widetilde{\boldsymbol{K}} = -\widetilde{\boldsymbol{G}}^{-1}\widetilde{\boldsymbol{F}} \tag{9.3.47}$$

with $\widetilde{\boldsymbol{C}}$ replacing $\boldsymbol{C}$ in the definition of $\widetilde{\boldsymbol{G}}$ and $\widetilde{\boldsymbol{S}}$ replacing $\boldsymbol{S}$ in the definition of $\widetilde{\boldsymbol{F}}$. The formalism of [48] is more general than we need here, because it was developed to account for the situation in which the matrices $\boldsymbol{C}(r)$ and $\boldsymbol{S}(r)$ are pure Coulomb, and the multipole terms of (9.3.32) are included in first order perturbation theory. The result is that $\widetilde{\boldsymbol{C}}$ and $\widetilde{\boldsymbol{S}}$ are now full matrices in which the sub-matrices are corrected for the perturbation potentials. We refer to [48] for more detail.

Seaton [49] has noted that it can be useful to use the R-matrix method to calculate bound states of $(N+1)$-electron systems, arguing that it generates energies and wavefunctions with an accuracy comparable to that which can be obtained with the best alternative methods. The same expansions can be used for close-coupling collision calculations and for 'frozen core' approximations, and the CC wavefunction is very convenient for calculating both radiative line strengths and photo-ionization cross-sections. Finally, it provides a very

[4] Capital letters distinguish outer region solutions.

efficient way to generate the large amounts of radiative data required in astrophysical applications. The matching condition (9.3.39) for bound states is

$$\mathtt{p}_E(a) = \boldsymbol{P}(a)\boldsymbol{A}_E = \boldsymbol{R}(E)(\boldsymbol{P}'(a) - b\boldsymbol{P}(a)/a)\boldsymbol{A}_E, \tag{9.3.48}$$

where the $n_C \times n_C$ bound state amplitudes (9.3.45) in the outer region generate the matrix $\boldsymbol{P}(r)$. Rearranging, we get

$$\boldsymbol{B}(E)\,\boldsymbol{A}_E = 0, \quad \boldsymbol{B}(E) = \boldsymbol{P}(a) - \boldsymbol{R}(E)(\boldsymbol{P}'(a) - b\boldsymbol{P}(a)/a)$$

so that the eigenvalues appear as solutions of the nonlinear characteristic equation $\det \boldsymbol{B}(E) = 0$. Practical solution is not quite straightforward, as the eigenvalues are often quite close to the poles of the R-matrix, and special methods are needed to overcome the resulting numerical instability; see [49, 51] for algorithmic details.

It is desirable, when calculating transition data, to normalize the bound state wavefunctions including contributions from the outer region. Setting $(\boldsymbol{P}\,|\,\boldsymbol{P})_O = \boldsymbol{I}$ gives the normalization integral

$$\boldsymbol{A}_E^T\left(\boldsymbol{I} + \mathtt{Q}^T\,\dot{\boldsymbol{R}}(E)\,\mathtt{Q}\right)\boldsymbol{A}_E = 1, \tag{9.3.49}$$

where the first term is the contribution of the outer region and the second is the contribution from the inner region from (9.3.41) [48, cf. Equation (8.4)].

9.4 The Buttle correction

Bloch's nonrelativistic R-matrix formalism and our relativistic adaptation of it have been presented as mathematically "formal" derivations. That is to say the conditions under which the various operators and functions appearing in the equations have a meaning are taken for granted, and it is assumed that the eigenstates of the $(N+1)$-electron Hamiltonian, Ψ_k, are complete in some undefined Hilbert space so that the R-matrix of (9.3.33) exists. Justification of these assumptions is reasonably straightforward in the single channel case by appealing to Theorems B.26, B.29, and B.30, although more work is needed in the general case. In practice, the basis is always finite; the error due to basis set truncation is often the major cause of discrepancies with experiment. An early and widely used attempt to deal with this problem in the R-matrix method is due to Buttle [52]. He exploited the observation that higher excited states contribute mainly to the diagonal elements of the R-matrix and are relatively insensitive to channel coupling. First solve equation (9.3.25)

$$(T^0 - \epsilon_i)u_i^0(r) = \sum_l \lambda_{il}\,\underline{v}_l(r).$$

at the fixed continuum energy $\epsilon_i = E - E_i^N$ without imposing the R-matrix boundary condition. Then if $p_{ik}^0(a)$ is the upper component of $u_{ik}^0(a)$ for the continuum energy ε_{ik} of the k-th eigensolution of (9.3.19) and $p_i^0(a)$ is the upper component of $u_i^0(a)$, the missing diagonal contribution is

$$\sum_{k=K+1}^{\infty} \frac{|p_{ik}(a)|^2}{2(\varepsilon_{ik} - \varepsilon_i)}$$

which we can approximate with

$$R_{ii}^K(\varepsilon_i) = \frac{p_i^0(a)}{Q_i^0} - \sum_{k=1}^{K} \frac{|p_{ik}^0(a)|^2}{2(\varepsilon_{ik} - \varepsilon_i)}, \tag{9.4.1}$$

where Q_i^0 is defined as in (9.3.39). The corrected R-matrix is therefore

$$R_{ij}(E) = \sum_k \frac{p_{ik}(a)\, p_{jk}(a)}{2(E_k - E)} + \delta_{ij}\, R_{ii}^K(\varepsilon_i). \tag{9.4.2}$$

The correction $R_{ii}^K(\varepsilon_i)$ is a continuous and monotonic function of ε_i when $\varepsilon_i < \varepsilon_{ik}$ and is often required for a large number of energy values. Seaton [53] has developed an efficient fitting procedure for $R_{ii}^K(\varepsilon_i)$ as a function of ε_i which is incorporated in many R-matrix packages.

9.5 R-matrix theory of photo-ionization

The R-matrix method expresses the initial state Ψ_i and final state Ψ_f^- in terms of R-matrix basis functions for the $(N+1)$ electron and N electron problems respectively. For the former, we can use (9.3.22),

$$\Psi_i = \boldsymbol{\Theta} \boldsymbol{V} \boldsymbol{A}_i \tag{9.5.1}$$

where $\boldsymbol{\Theta}$ is a row vector of $(N+1)$-electron CSFs, $\boldsymbol{V}$ diagonalizes the $(N+1)$-electron Hamiltonian, and $\boldsymbol{A}_i$ is obtained by solving the outer region equations for the initial bound state energy E_i,

$$\boldsymbol{A}_i = [2(\boldsymbol{e} - E_i)]^{-1} \boldsymbol{p}_i{}^T \boldsymbol{R}^{-1}(E_i)\, \mathsf{p}_i, \tag{9.5.2}$$

as in (9.3.42). Similarly, the final state, energy $E_f = E_i + \omega$, is given by

$$\boldsymbol{A}_f^- = [2(\boldsymbol{e} - E_f)]^{-1} \boldsymbol{p}_f{}^T \boldsymbol{R}^{-1}(E_f)\, \mathsf{p}_f^-, \tag{9.5.3}$$

where the final state energy is $E_f = E_i + \omega$ and p_f^- are partial waves whose upper components have the asymptotic form

$$\mathsf{p}_f^- = -i\, \boldsymbol{P}(a)(1 - i\boldsymbol{K})^{-1}, \tag{9.5.4}$$

with $\boldsymbol{P}(a)$ defined by the K-matrix boundary condition (9.3.47) so that the incoming spherical wave boundary condition of (8.6.2) holds. Thus the transition matrix element is

$$\left(\Psi_f^- \middle| A_q \middle| \Psi_i\right) = \boldsymbol{c}_f{}^\dagger \, \mathcal{A}_q \, \boldsymbol{c}_i \tag{9.5.5}$$

where $\mathcal{A}_q$ is the matrix with elements

$$\mathcal{A}_{q\lambda\lambda'} = \langle \Theta_\lambda \,|\, A_q \,|\, \Theta_{\lambda'} \rangle.$$

of the form of (8.3.1), where

$$\boldsymbol{c}_f{}^\dagger = \mathrm{p}_f^{-T} \boldsymbol{R}^{-1}(E_f)\boldsymbol{p}_f[2(\boldsymbol{e} - E_f)]^{-1}\boldsymbol{V}^T$$

and

$$\boldsymbol{c}_i = \boldsymbol{V}[2(\boldsymbol{e} - E_i)]^{-1}\boldsymbol{p}_i{}^T\boldsymbol{R}^{-1}(E_i)\mathrm{p}_i$$

are respectively row and column vectors.

9.6 The DARC relativistic R-matrix package

The DARC suite of programs [54] is closely modelled on the nonrelativistic RMATRX1 package [55] and uses many of the same program modules. The use of Dirac wavefunctions and jj-coupling means that the inner region modules are different from those of RMATRX1. The asymptotic wavefunction uses the Pauli approximation so that the time-consuming part of the calculation, especially when there are many channels, is virtually identical with that of RMATRX1.

A typical electron-atom collision calculation involves a number of steps, each carried out with one of DARC's modules:

- **grasp**: the calculation begins with the generation of states of the N-electron target using GRASP (or a similar relativistic atomic structure code). The output file may be based on the Dirac-Coulomb Hamiltonian but results taking account of the Breit interaction and QED effects can also be used.
- **dstg0**: creates a formatted file of *bound* orbital wavefunctions from **grasp**, possibly augmented by bound pseudo-orbitals[5] that are used to construct the target states Φ_k.
- **dstg1-orbs**: generates solutions of the R-matrix equations (9.3.19) for the inner region, $r < a$. The energies ε_{ik} and the corresponding boundary values of the amplitudes $p_{ik}(a)$ are written to a formatted file ORBS.OUT whilst the orbitals themselves are written to direct access files used by **dstg1-ints**. The Buttle corrections are calculated and stored at this stage.
- **dstg1-ints**: generates radial integrals needed in later stages of DARC. Different program modules are used according to whether the integrals are of bound-bound, bound-continuum, or continuum-continuum types.

[5] The program documentation refers to all orbitals whose surface amplitudes on $r = a$ are negligible as bound orbitals, and the remainder as continuum orbitals.

  - Radial matrix elements of the Dirac one-electron Hamiltonian
  - Coulomb interaction integrals
  - Radial moments needed for the coefficients $C^{\lambda}_{ii'}$ of (9.3.38)
  - Radiative transition integrals (E1) in the long wavelength limit in both length and velocity forms.
- **dstg2**: defines the target states and channels, evaluates the angular integrals, asymptotic coefficients and continuum Hamiltonian $\widetilde{H}$ and the dipole radiative matrix elements needed for photo-ionization if requested.
  - Target CSFs generate a target (N-electron) Hamiltonian to give the target states Φ_k and mixing coefficients to be used in the scattering calculation. The radial integrals are available from **dstg1-ints** and the angular integrals are generated at this stage.
  - Channel wavefunctions for each symmetry $J\Pi$ are generated as linear combinations of the form (9.3.23) and are then used to generate all required angular integrals needed for the $(N+1)$-electron Hamiltonian.
- **dstgh**: Diagonalizes $\widetilde{H}$ and generates the eigenvalues E_k (9.3.23) (R-matrix poles) and the surface amplitudes $p_{ik}(a)$ from (9.3.20) that generate input to the asymptotic modules **dstg3, pdstg3**, and **stgfjj**. Eigenvectors can also be generated if $(N+1)$-electron bound states and photo-ionization cross-sections are needed.
- **dstg3** and **pdstg3**: The module **dstg3** generates the outer region solutions, §9.3.4, from which the K-matrix, the eigenphases[6], the collision strengths, and collision cross-sections can be found. The module **pdstg3** evaluates photo-ionization cross-sections using the formalism of §8.6. There are several packages available to solve the external region equations (9.3.32):
  - `DCOUL` neglects channel coupling and uses nonrelativistic Coulomb solutions. The `COUL` program [56] is used for closed channels ($\varepsilon_i < 0$) and a version [57] of the `COULFG` program of [58] is used for open channels ($\varepsilon_i > 0$).
  - `DASYPCK` is based on the nonrelativistic `ASYPCK2` package [59, 60].
  - `DVPM` is a nonrelativistic program for neutral targets using the variable phase method adapted from `VPM` [61].
  - `DASYM` is a relativistic program [62].
- **stgfjj**: this is adapted from the Opacity Project STGF code [48], for calculating collision cross-sections for ionic targets. It uses the perturbation method of [51] to stabilize the calculation of cross-sections in the neighbourhood of R-matrix poles.

The current version, DARC-OXQUB [54], is dimensioned with at most 200 channels, 100 continuum κ-symmetries with 40 continuum orbitals per κ, and 300 target levels or correlation functions. The code was originally written for serial computers, and the use of a single processor to calculate every

[6] The eigenphases, δ_i, are obtained from the eigenvalues, $\tan\delta_i$, of the K-matrix.

partial wave and diagonalize the resulting Hamiltonian for every eigenvalue and eigenvector is very time-consuming and limits the problems that can be studied. This has become a major bottleneck for the study of heavy element spectral signatures in fusion plasma research [63]. In response to this challenge, a recent calculation on the Ni-like ion Xe^{26+} [64] made use of a new interface to adapt the output of **dstg2** modules for use with a suite of programs, **pstg3r** and **pstgf**, developed originally to run the Breit-Pauli R-matrix program [65] on parallel computer systems. This calculation to establish the feasibility of modelling the spectral characteristics of heavy ions used 129 target levels arising from the $3d^{10}$ and $3d^9nl$ configurations of Xe^{26+} with $l = 0$ to $n-1$ and $n = 4, 5$. The collision calculation involved up to 821 channels, and used 21 continuum basis orbitals for $1/2 \leq j \leq 43/2$ and 16 orbitals for $45/2 \leq j \leq 71/2$, resulting in maximum Hamiltonian dimensions of 17,356 and 13,136 respectively. The calculation is performed 'concurrently' on a multiple processor, ideally assigning one partial wave to each processor. Handling of such large quantities of data by serial programs is impossible, and this development opens the way to make calculations with more target states with a better representation of their structure, the inclusion of more resonance structure in the spectra and higher incident energies.

9.7 Truncation of the close-coupling expansion. The nonrelativistic CCC method

The Buttle correction goes some way to treat the consequences of truncation. However, the finite dimension of the close-coupling expansion stops us from taking full account of the flux scattered into excited discrete and continuum channels. It is likely that this is responsible for much of the disagreement between theory and experiment. Thus Castillejo, Percival, and Seaton [66] showed that 18.6% of the static dipole polarizability of hydrogen was due to excited states in the continuum and, more recently, McCarthy and Shang [67] and Odgers *et al.* [68] showed that the omission of continuum terms can lead to ∼15% overestimate of the of the $1s-2s$ and $1s-2p$ excitation cross sections of atomic hydrogen at intermediate energies.

Although no relativistic version of the convergent close-coupling (CCC) method of Bray and Stelbovics [69] has yet been devised, an understanding of the reasons for its success in systematically dealing with issues of truncation illuminates the failings of other methods such as the R-matrix. The CCC method aims to solve the coupled Lippman-Schwinger equations for the elements of the distorted-wave K and T matrices. The initial formulation for treating electron scattering from hydrogen was extended to quasi-one-electron targets [70]. The starting point is a Hartree-Fock calculation for the target ground state T:

$$(K + V^{HF} - \varepsilon_j)\,\psi_j(\boldsymbol{r}) = 0, \qquad \psi_j \in T \tag{9.7.1}$$

where for core states, $\psi_j \in C$,

$$V^{HF}\,\psi_j(\boldsymbol{r}) = \left(-\frac{Z}{r} + 2\sum_{\psi_{j'}\in C}\int d^3r'\,\frac{|\psi_{j'}(\boldsymbol{r}')|^2}{|\boldsymbol{r}-\boldsymbol{r}'|}\right)\psi_j(\boldsymbol{r}) - \sum_{\psi_{j'}\in C}\int d^3r'\,\frac{\psi_{j'}(\boldsymbol{r}')\psi_j(\boldsymbol{r}')}{|\boldsymbol{r}-\boldsymbol{r}'|}\,\psi_{j'}(\boldsymbol{r}).$$

This is used to define a *frozen core* potential

$$V^{FC}\,\phi_j(\boldsymbol{r}) = \left(-\frac{Z}{r} + 2\sum_{\psi_{j'}\in C}\int d^3r'\,\frac{|\psi_{j'}(\boldsymbol{r}')|^2}{|\boldsymbol{r}-\boldsymbol{r}'|}\right)\phi_j(\boldsymbol{r}) - \sum_{\psi_{j'}\in C}\int d^3r'\,\frac{\psi_{j'}(\boldsymbol{r}')\phi_j(\boldsymbol{r}')}{|\boldsymbol{r}-\boldsymbol{r}'|}\,\psi_{j'}(\boldsymbol{r}). \quad (9.7.2)$$

The (time-independent) Schrödinger equation for the scattering problem is

$$\left(E^{(+)} - H\right)\left|\Psi^{S(+)}_{i_0 k_0}\right\rangle = 0, \quad (9.7.3)$$

where the superscript $(+)$ denotes incoming plane- or Coulomb-wave and outgoing spherical wave boundary conditions, E is the total energy, $E^{(+)} = E + i0$, S is total spin and i_0, k_0 denote the initial target state and projectile momentum. The scattering Hamiltonian is

$$H = H_1 + H_2 + V_{12}, \qquad H_\alpha = K_\alpha + V_\alpha,$$

in which $\alpha = 1$ denotes the projectile and $\alpha = 2$ the target electron, V_{12} is the electron-electron interaction, K_α is the kinetic energy operator for particle α and

$$V_\alpha = V^{FC} + V^{pol}.$$

The polarization potential used in [70] is

$$V^{pol}(r) = -\frac{\alpha_d}{2r^4}\left(1 - \exp(-\rho^6/r^6)\right)$$

where α_d is the static dipole polarizability and the value of ρ is chosen to fit one-electron ionization energies of the target.

The next step is to expand the scattering wave function $|\Psi^{S(+)}_{i_0 k_0}\rangle$ in a set of square-integrable states whose configuration space representation is

$$\left\langle \boldsymbol{r}\,\middle|\,i^N_{nlm}\right\rangle = \frac{\phi^{N_l}_{nl}(r)}{r}\,Y_{lm}(\widehat{\boldsymbol{r}}), \quad \phi^{N_l}_{nl}(r) = \sum_{k=1}^{N_l} C^l_{nk}\xi_{kl}(r) \quad (9.7.4)$$

where the Laguerre functions,

$$\xi_{kl}(r) = \left[\frac{(k-1)!\,\lambda_l}{(2l+k+1)!}\right]^{1/2} (\lambda_l r)^{l+1} L_{k-1}^{2l+2}(\lambda_l r), \qquad k = 1, \ldots, N_l \tag{9.7.5}$$

are orthonormal and complete in $L^2(0,\infty)$. If $\boldsymbol{c}_n^{N_l}$ denotes the vector with elements C_{nk}^l, $k = 1, \ldots, N_l$, then

$$\boldsymbol{H}^{N_l} \boldsymbol{c}_n^{N_l} = \varepsilon_n^{N_l} \boldsymbol{c}_n^{N_l}, \tag{9.7.6}$$

so that $\boldsymbol{c}_n^{N_l}$ is the n-th eigenvector of the matrix $\boldsymbol{H}^{N_l}$, which represents the radial part of the operator H_2 with respect to the basis (9.3.41). As in Appendix B.5.1, the solutions depend on two parameters λ_l and N_l that are typically different for each l and that can be chosen to optimize the computational effort. The idea of the CCC method is to project the scattering states onto the finite basis,

$$\left|\Psi_{i_0 k_0}^{S(+)}\right\rangle \approx \sum_n \left|i_n^N\right\rangle \left\langle i_n^N \middle| \Psi_{i_0 k_0}^{S(+)}\right\rangle. \tag{9.7.7}$$

The completeness of the Laguerre polynomials (9.3.41) is expected to ensure that

$$\lim_{N\to\infty} \sum_n \left|i_n^N\right\rangle \left\langle i_n^N \middle| \Psi_{i_0 k_0}^{S(+)}\right\rangle = \left(\sum + \int\right)_{i\in I} \left|i\right\rangle\left\langle i\middle|\Psi_{i_0 k_0}^{S(+)}\right\rangle = \left|\Psi_{i_0 k_0}^{S(+)}\right\rangle$$

where i runs over the complete set I of exact discrete and continuous eigenstates of H_2 [71, 72], so that it should be possible to demonstrate numerical convergence by taking N_l sufficiently large.

The scattering wave function satisfies the symmetry condition

$$\left\langle \boldsymbol{r}_1 \boldsymbol{r}_2 \middle| \Psi_{i_0 k_0}^{S(+)}\right\rangle = (-1)^S \left\langle \boldsymbol{r}_2 \boldsymbol{r}_1 \middle| \Psi_{i_0 k_0}^{S(+)}\right\rangle \tag{9.7.8}$$

and when the projectile radial coordinate, r_1, is large, the boundary condition is

$$\left\langle \boldsymbol{r}_1 \boldsymbol{r}_2 \middle| \Psi_{i_0 k_0}^{S(+)}\right\rangle \sim \left\langle \boldsymbol{r}_1 \middle| \boldsymbol{k}_0^{(+)}\right\rangle \left\langle \boldsymbol{r}_1 \middle| i_0\right\rangle \tag{9.7.9}$$

where

$$\langle \boldsymbol{r}|\boldsymbol{k}^{(\pm)}\rangle = \sqrt{\frac{2}{\pi}\frac{1}{kr}} \sum_{LM} i^L \mathrm{e}^{\pm(\sigma_L+\delta_L)}\, u_L(kr) Y_{LM}(\widehat{\boldsymbol{r}})\, Y_{LM}^*(\widehat{\boldsymbol{k}})$$

is a distorted wave, σ_L is the Coulomb phase shift, δ_L the phase shift due to the scattering potential and the radial amplitude has the asymptotic form

$$u_L(kr) \sim F_L(kr)\cos\delta_L + G_L(kr)\sin\delta_L$$

where $F_L(kr)$, $G_L(kr)$ are the regular and irregular Coulomb functions respectively. Bray [70] goes on to show how the Lippmann-Schwinger equations can be written in terms of matrix elements of product states $|i_n^N \boldsymbol{k}_n^{(\pm)}\rangle$. The angular variables can be integrated out leaving a set of coupled equations for real reduced matrix elements, which he writes

$$\langle Lk_{nl}ln \| K^{SN}_{J\Pi} \| n_0l_0k_0L_0\rangle = \langle Lk_{nl}ln \| V^{SN}_{J\Pi} \| n_0l_0k_0L_0\rangle$$
$$+\sum_{l'L'}\sum_{n'=1}^{N_{l'}} \mathcal{P}\left(\sum_{k'}+\int\right) \frac{\langle Lk_{nl}ln \| V^{SN}_{J\Pi} \| n'l'k'L'\rangle}{E-\varepsilon_{n'l'}-\varepsilon_{k'}}\langle L'k'l'n' \| K^{SN}_{J\Pi} \| n_0l_0k_0L_0\rangle \tag{9.7.10}$$

where $\mathcal{P}$ indicates a principal value integration over k' with $\varepsilon_{k'} = k'^2/2$ and $V^{SN}_{J\Pi}$ is an effective potential. These equations are approximated by replacing the energy integration by a quadrature rule, splitting the range into several intervals, one of which is symmetric about the singular point of the denominator to take care of the principal value. The resulting linear equations for the K matrix elements can be written (using an abbreviated notation for the state labels)

$$V^{SN}_{n'i} = \sum_{n}\left(\delta_{n'n} - w_n V^{SN}_{n'n}\right) K^{SN}_{ni} \tag{9.7.11}$$

with weights w_n incorporating the denominators. The T matrix is obtained finally from the equations

$$\langle Lk_{nl}ln \| K^{SN}_{J\Pi} \| n_0l_0k_0L_0\rangle = \sum_{l'L'}\sum_{n'=1}^{N^o_{l'}} \langle Lk_{nl}ln \| T^{SN}_{J\Pi} \| n'l'k'L'\rangle \tag{9.7.12}$$
$$\times\left(\delta_{l'l_0}\delta_{L'L_0}\delta_{n'n_0} + i\pi k_{n'l'}\langle L'k'l'n' \| K^{SN}_{J\Pi} \| n_0l_0k_0L_0\rangle\right).$$

The sum over n' in (9.3.46) can often be truncated to use fewer than the $N_{l'}$ states of the Laguerre basis, and above the threshold it is often sufficient to use only the $N^o_{l'}$ states generating open channels.

This brief account of the CCC method neglects many technical details [70] that are needed to make the method work well. As summarized here, it applies only to atoms with one active electron outside a closed core; electron scattering from sodium at energies from 1 to 54.4 eV was used to illustrate the quality of results in [70] and other examples are reviewed in [73]. The same scheme has more recently been applied to quasi-two-electron systems [74], typified by the alkaline earths beryllium, magnesium, etc.; the main additional complication is the heavier algebra needed to treat anti-symmetric three-electron wave functions. The generalized formalism was applied in [74] to calculate cross sections and other quantities for scattering from the Be ground state to $n = 4$ levels. At the time of writing, the method has not been extended to deal with scattering from more complex atoms.

A major advantage of the CCC scheme is that the Laguerre functions form a complete orthonormal set. This makes mathematical and practical convergence of the solution easier to demonstrate. A relativistic version based on L-spinors, described in §5.8, seems technically feasible but has not yet been attempted.

9.8 The R-matrix method at intermediate energies

The use of a countable complete orthonormal radial basis on $(0, \infty)$ to describe both discrete and continuum states is not readily compatible with the R-matrix approach. A less rigorous, physically intuitive, scheme with motivation similar to that of the CCC method is the (nonrelativistic) R-matrix method with pseudo-states (RMPS) [75]. As usual, the normal starting point is a Hartree-Fock calculation, which provides one-electron orbitals whose radial parts are denoted $P_{nl}(r)$. These *physical* orbitals can be augmented with nonphysical *pseudo-orbitals* $P_{\overline{nl}}(r)$, generally having the form

$$P_{\overline{nl}}(r) = \sum_{i=l+1}^{n_{max}} a_{il}^{n} \, \mathrm{e}^{-\alpha r} \tag{9.8.1}$$

where α is a range parameter, similar to the λ_l parameters of (9.7.5). They are often included to lower the energy of a target eigenstate or to improve agreement with experiment for the oscillator strengths connecting particular target states. Pseudo-states are also used to improve estimates of the polarizability of the ground or excited states of the target atom or ion, or to act as surrogates for highly excited discrete or continuum states that cannot be included explicitly in the calculation for reasons of cost.

The continuum orbitals are obtained by solving equations of the form

$$\left(\frac{d^2}{dr^2} - \frac{l(l+1)}{r^2} + V(r) + k_{il}^2 \right) u_{il}(r) = \sum_n \lambda_{inl} \, P_{nl}(r) \tag{9.8.2}$$

where

$$u_{il}(0) = 0, \quad \left(a\frac{d}{dr} - \frac{b}{r} \right) u_{il}(r) \Big|_{r=a} = 0, \quad \int_0^a u_{il}(r) \, u_{i'l}(r) dr = \delta_{ii'},$$

analogous to the relativistic equations (9.3.25). In this form, the Lagrange multipliers on the right-hand side of (9.8.2) ensure that the continuum orbitals are orthogonal to the physical orbitals $P_{nl}(r)$ but not to the non-physical pseudo-orbitals. An obvious way to remedy this defect would be to include the $P_{\overline{nl}}(r)$ on the right-hand side, but Bartschat *et al.* [75] reject this solution. They observe [75, p. 117] that as well as ensuring orthogonality of the continuum orbitals to the bound states, the right-hand side simulates an exchange potential, and argue that this simulation would be compromised by including the pseudo-orbitals. A more practical justification is that "...experience shows that their (inappropriate) inclusion would slow down the convergence of the R-matrix expansion considerably."

Bartschat *et al.* [75] avoid these problems by Schmidt orthogonalizing the continuum orbitals to the nonphysical pseudo-orbitals, which simplifies the evaluation of the one-electron matrix elements of the Hamiltonian. As a particular example, consider e-H atom scattering using physical orbitals P_{1s}, P_{2s}, a pseudo-orbital $P_{\overline{3s}}$, and continuum orbitals $u_1, \ldots, u_n$. The latter are replaced by a new set of Schmidt orthogonalized orbitals $v_1, \ldots, v_n$ generated recursively through

$$v_i = \mathcal{N}_i \left(u_i - \sum_{j=1}^{i-1} \langle u_i \,|\, v_j \rangle v_j - \sum_{\overline{n}} \langle u_i \,|\, P_{\overline{ns}} \rangle P_{\overline{ns}} - \sum_{n} \langle u_i \,|\, P_{ns} \rangle P_{ns} \right),$$

where $\mathcal{N}_i$ is a normalization constant. The sum over bound orbitals, although theoretically unnecessary, is included to compensate for numerical inaccuracies introduced by the recursion.

Although the RMPS scheme offers a systematic way to improve the basis, it requires some skill to extract good results. There are problems if many target states must be considered and at energies where there are several open channels. The intermediate energy R-matrix method (IERM) [76] was developed to deal with these eventualities. Highly excited orbitals may extend way beyond the core, so that a better description of the scattering process would allow two interacting electrons outside the core. In this case, the inner region is divided into two parts, with radii $a_1 < a_2$. Suppose first that one of the two electrons $r_{N+1} \leq a_1$ whilst $r_{N+2} \geq a_1$. Then the total wavefunction can be expanded in the form [76, Equation (36)]

$$\theta_{kl} = \sum_{ij} \overline{\psi}_i(1, \ldots, N+1; \widehat{\boldsymbol{r}}_{N+2})\, u_{kj}(r_{N+2})\, \gamma^k_{ijl}$$

as in the usual R-matrix approach, where $u_{kj}(r_{N+2})$ is a radial amplitude for the outermost electron, and where the angular parts have been incorporated in $\overline{\psi}_i$. The coefficients γ^k_{ijl} are obtained by diagonalizing the $N+2$ electron Hamiltonian H_{N+2} on the two-dimensional interval $(0, a_1) \times (a_1, a_2)$ so that

$$\langle \theta_{kl} |\, H_{N+2} + L_{N+2} \,|\, \theta_{kl'} \rangle = E^{N+2}_{kl} \delta_{ll'}$$

where the Bloch operator is $L_{N+2} = l_{N+1}(a_1) - l_{N+1}(0) + l_{N+2}(a_2) - l_{N+2}(a_1)$. More generally, one can partition the range (a_1, a_2) into a sequence of intervals, and [76] defines a sequence of two-dimensional intervals of which the k-th has $c_k \leq r_{N+1} \leq c_{k+1}$ and $d_k \leq r_{N+2} \leq d_{k+1}$ on which solutions can be obtained in the same way. When both $r_{N+1}, r_{N+2} \geq a_1$ then [76, Equation (37)]

$$\theta_{kl} = \sum_{imn} \overline{\overline{\phi}}_i(1, \ldots, N; \widehat{\boldsymbol{r}}_{N+1})\, u_{km}(r_{N+1}, \widehat{\boldsymbol{r}}_{N+2})\, u_{kn}(r_{N+2})\, \gamma^k_{imnl}$$

The radial amplitudes $u_{kn}(r)$ satisfy (nonrelativistic) continuum equations similar to (9.3.35). The method of solution advocated in [76] is derived from the R-matrix propagator method of Light and Walker [77] implemented in the FARM package [78] (cf. Baluja *et al.* [79]). This divides the radial range into intervals, of which the p-th is (r^p_L, r^p_R), and gives equations for determining

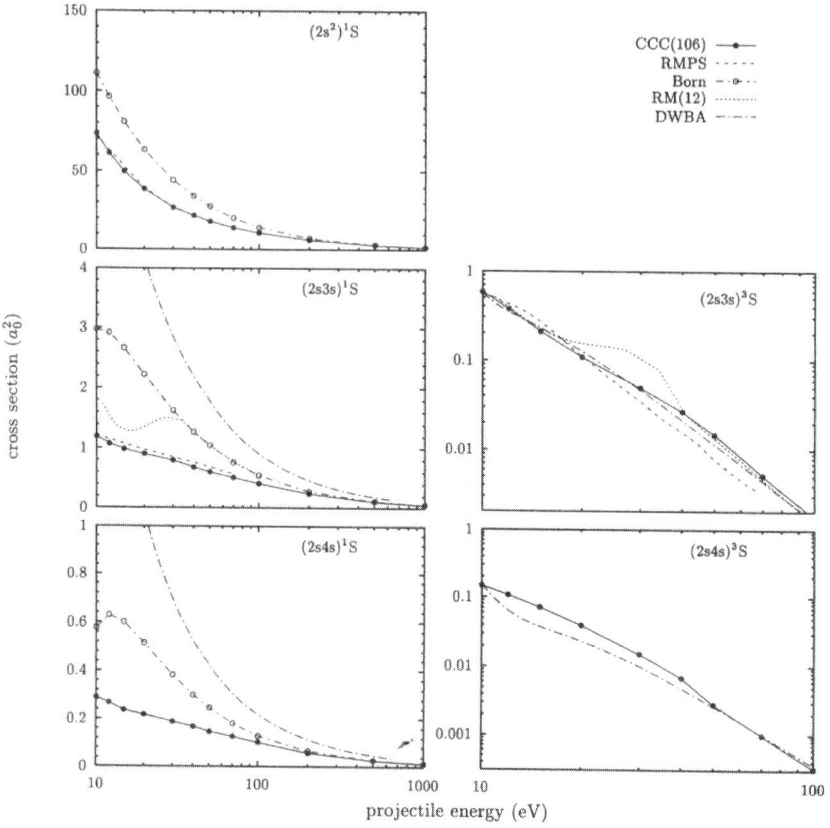

Fig. 9.1. Integrated cross sections for excitation of ^{1,3}S states by electron impact on the ground state of Be using different approximations [74]: CCC(106) from [74]; RMPS from [80]; RM(12) from [81]; DWBA from [82]. Reproduced, with permission, from [74].

$\boldsymbol{R}(r_R^p)$ given $\boldsymbol{R}(r_L^p)$. In this way, the R-matrix can be propagated economically outwards from a_1 to a point at which the solution can be matched to the appropriate asymptotic boundary conditions.

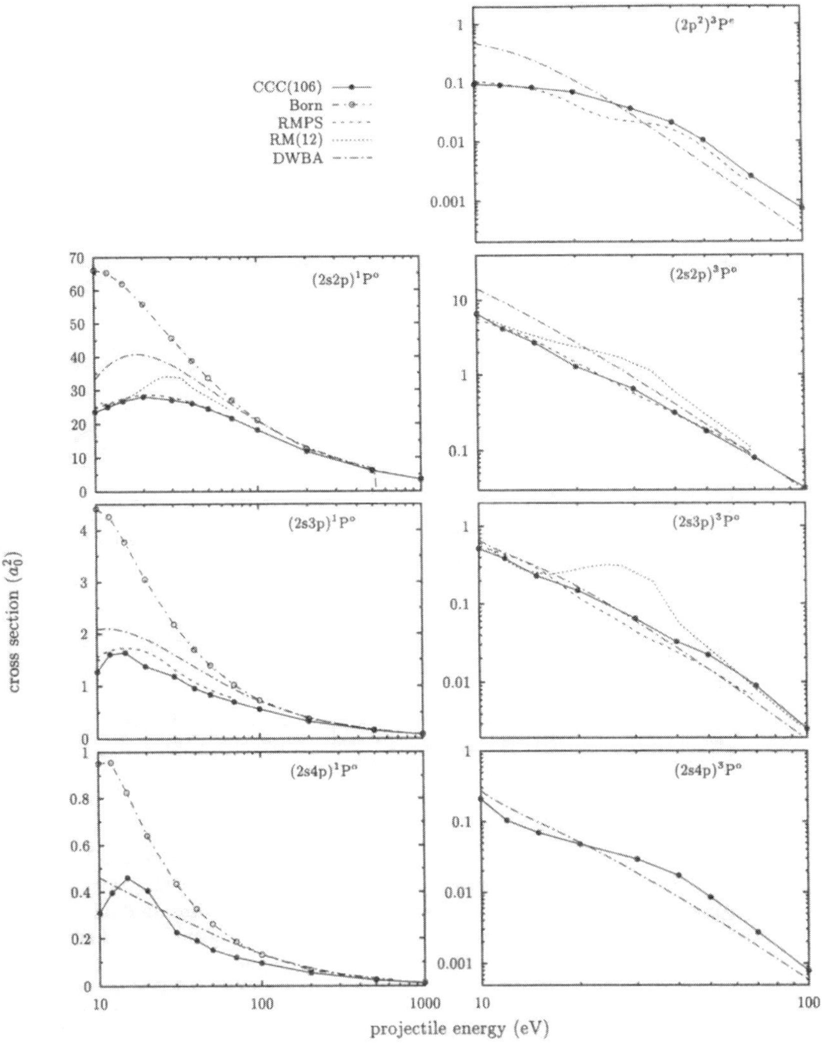

Fig. 9.2. Integrated cross sections for excitation of ^{1,3}P states by electron impact on the ground state of Be; details as in Figure 9.1. Reproduced, with permission, from [74].

Figures 9.1 and 9.2, taken from [74], display calculated total cross sections for electron scattering from beryllium in the range 10–1000 eV by several different methods. The CCC calculations used an orbital basis comprised of $2s-15s$, $2p-15p$, $3d-12d$, and $4f-11f$ orbitals, and solved the Lippmann-Schwinger equations for 106 states: 16 ^{1}S, 18 ^{1}P^o, 15 ^{1}D^e, 10 ^{1}F^o, 10 ^{3}S, 12 ^{3}P^o,10 ^{3}D^e, 7 ^{3}F^o, with two each of ^{1,3}P^e and ^{1,3}D^o. The figures also display results from first-order distorted wave (DWBA) calculations by Clark and Abdallah [82], R-matrix calculations involving only states of the discrete spectrum by Fon *et al.* [81] and RMPS calculations by Bartschat *et al.* [80] which include pseudo-states taking account of the continuum channels. The elastic cross-sections predicted by the CCC method generally agree well with those predicted by RMPS. The R-matrix excitation cross-sections for 2s2p $^{1,3}P^o$ states of Fon *et al.* [81] predict cross sections that are in general too large; by allowing for the flux going into continuum channels, both CCC and RMPS yield smaller cross-sections that are in good agreement. The less sophisticated DWBA results are poor except at the highest energies. More details will be found in [74].

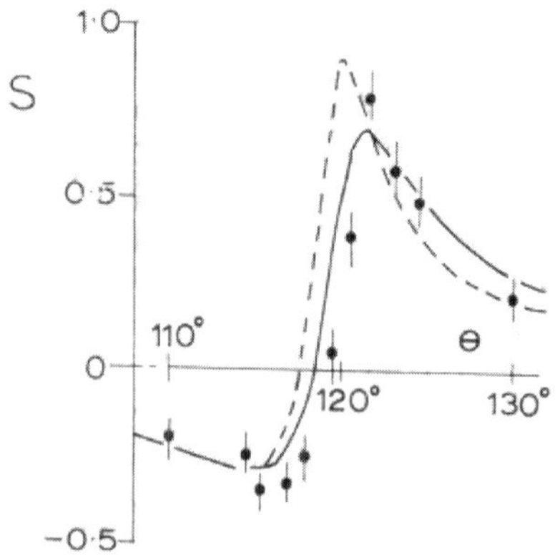

Fig. 9.3. The main polarization peak for elastic electron scattering from mercury at 300 eV. Dashed lines indicate a DH (nonexchange) potential and full lines a DHF potential. Reprinted with permission from [1].

9.9 Electron scattering from heavy atoms and ions

9.9.1 Early work

The vast range of electron-atom scattering experiments and calculations with a variety of theoretical models makes it almost impossible to give an adequate overview of the comparison of theory with experiment. Walker's 1971 review [1] covered high-energy elastic scattering at incident energies above 50 keV using pure and screened Coulomb potential models, examining both differential cross-sections, $I(\theta)$,(9.1.3), and spin polarizations, $S(\theta)$, (9.1.4). He described an extensive series of calculations of $I(\theta)$ and $S(\theta)$ for mercury

(Z=80) at energies up to 10 keV. Comparing results using both DH and DHF potentials, he found that the effect of exchange was small at energies greater than 100 eV: $S(\theta)$ was affected more strongly than $I(\theta)$, and the differential cross-section was reduced in the forward direction. The changes can largely be correlated with changes in the phase shifts of low order partial waves. Further calculations incorporated polarization potentials; although the results changed, the limited experimental data available permitted Walker to draw few conclusions. Some studies were also made for bismuth, gold, and inert gas atoms using nonexchange (DH) potentials for the target. Walker summarized his conclusions by saying that relativistic effects in electron-atom scattering are important if either the incident energy is high enough for the electron speed to be an appreciable fraction of the speed of light or at energies below 100 eV in heavy atoms. The main effects are interpreted largely in terms of spin-orbit coupling, which affects both the details of differential cross-sections and especially spin polarization.

9.9.2 Electron scattering from the mercury atom

Electron scattering from the mercury atom attracts a lot of attention, partly because of its widespread use in fluorescent and high-intensity discharge lamps and also because its atomic number is high enough to expect unambiguous signs of relativistic effects. A recent study by Fursa, Bray, and Lister [83] of the elastic and excitation cross-sections from the ground state of mercury was undertaken because the cross sections for excitation and ionization of mercury obtained from swarm data [87], which had previously been used extensively for modelling the characteristics of plasmas containing mercury, disagreed srongly with more recent measurements [88, 89].

Fursa *et al.* modelled the mercury atom as if it had two valence electrons above a Hg^{2+} inert frozen core: [Xe]$4f^{14}5d^{10}$. The Hg^{+} Hamiltonian was diagonalized in a large Laguerre basis for symmetries $l = 0, 1, 2$ using a phenomenological core-polarization potential similar to that used in [74] with $\alpha_c = 8.4$ and cut-off radius at $1.8a_0$. The orbitals so generated were then used to construct target Hg states using a standard CI procedure. Two types of scattering calculations were performed for incident electron energies in the range 4-500 eV. The first was a 21 state CC model comprising discrete Hg target states: four ^{1}S and ^{1,3}P^o and three ^{3}S and ^{1,3}D^e. The second was a 54 state CCC calculation that included the target continuum along the lines of [74]: nine ^{1}S, eight ^{3}S, ^{1,3}D^e and ^{1,3}P^o , two ^{3}P^e, one ^{1}P^c and one ^{1,3}D^o. The difference between the results of these two models gives an indication of the role of continuum channels. This basically nonrelativistic calculation was modified by adding a one-body spin-orbit potential and diagonalizing the two-electron Hamiltonian in LSJ coupling. Also an additional short range potential for $l = 0$ of the form $-A\exp(-Br)$, with coefficients $A = 3000$ and $B = 80$ chosen to fit the ground state energy of mercury, simulated the relativistic contraction of the 6s orbital.

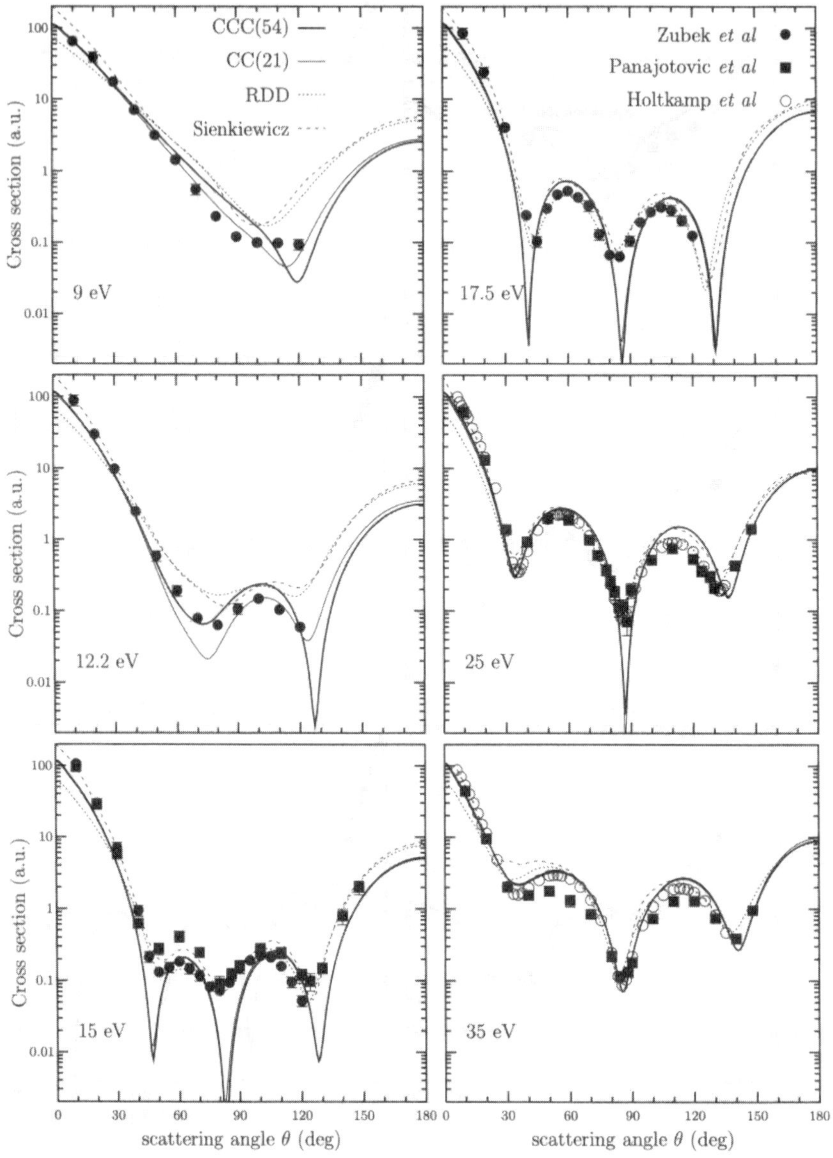

Fig. 9.4. Differential cross-sections for elastic scattering from the ground state of mercury at energies up to 35 eV comparing CC(21) and CCC(54) results [83] with those of Sienkiewicz [8, 7] and McEachran and Elford [84]. Experimental values from Zubek *et al.* [12], Panajotović *et al.* [85], and Holtkamp *et al.* [86]. Reprinted with permission from [83].

Differential cross-sections for elastic scattering from mercury for several energies up to 500 eV, taken from [83], are shown in Figures 9.4 and 9.5. This compares nonrelativistic CC(21) and CCC(54) results with recent relativistic calculations [8, 7, 84] and with various experiments. Sienkiewicz's calculations [8, 7] used a model core polarization potential $V_p(r) = -0.5\alpha_d r^2/(r^3 + r_0^3)^2$ with $\alpha_d = 44.9$ and $r_0 = 3$. McEachran and Elford's results [84], labelled RDD in the figures, used a more elaborate polarization potential together with a dynamic distortion potential. Earlier calculations such as those of Walker [1] gave differential cross-sections with the same general shape but with significant differences in the forward direction. It seems that coupling to the continuum has little effect on the differential cross-section; the main differences between different models stem from different choices of static polarization potential.

Fursa *et al.* [83] also studied inelastic scattering to the singly excited states 6s6p $^1P_1$, 6s6p $^3P_{0,1,2}$, 6s7s $^1S_0$, and 6s7s $^3S_1$. The CC(21) and CCC(54) results, which generally fit the experimental data quite well, were compared with nonrelativistic [91] and relativistic [92] distorted wave results. See [83] for a more detailed comparison.

Elastic scattering from the 6s^2 $^1S_0$ ground state and inelastic scattering into the 6s6p $^1P_1$, 6s6p $^3P_{0,1,2}$ excited states has also been studied using the DARC relativistic R-matrix package [90] at incident energies up to 8 eV. A preliminary 5-state DHF-AL GRASP calculation with a minimal CSF set [Xe]4f^{14}5d^{10} (6s^2 +6s6p) was used to obtain the target orbitals. The 7p orbitals were obtained from a DHF-OL calculation for the $J = 0$ ground state including the additional even parity configurations 5d^{10}6p7p and 5d^{10}7p^2 in the energy expression and keeping the remaining orbitals frozen. The five target states used in the R-matrix calculation were obtained from a CI calculation allowing for some valence correlation and a part of the core polarization involving 98 CSFs constructed from the configurations 5d^{10} 6s^x 6p^y 7p^z with $x+y+z = 2$ and 5d^{9}6s^x 6p^y 7p^z with $x+y+z = 3$. The improved agreement, Table 9.1, between calculated and observed level energies seemed sufficient for the subsequent R-matrix calculation. Whilst this construction goes some way to account for the effects of higher states and the continuum, the fact that there are almost 250 levels below the first, 6s, ionization threshold at 10.44 eV highlights the need to include many more states in the basis as in RMPS and CCC methods.

The R-matrix radius was fixed at $a = 24.156a_0$ and b was taken to be zero. Basis orbitals were determined for all $|\kappa| \leq 10$. The Hamiltonian was diagonalized for all channels with symmetry J^π in the range $\frac{1}{2}^{\pm}, \ldots, \frac{15}{2}^{\pm}$. The outer region was treated nonrelativistically and the Buttle correction was used. The calculation was done using both the theoretical CI energies from Table 9.1 and the observed energies; the cross sections agreed to about 5%. The resonance energies, corresponding to states of the negative ion Hg$^-$, listed in Table 9.2, were calculated using observed target level energies, which agree better with experiment than those based on theoretical target energies [90,

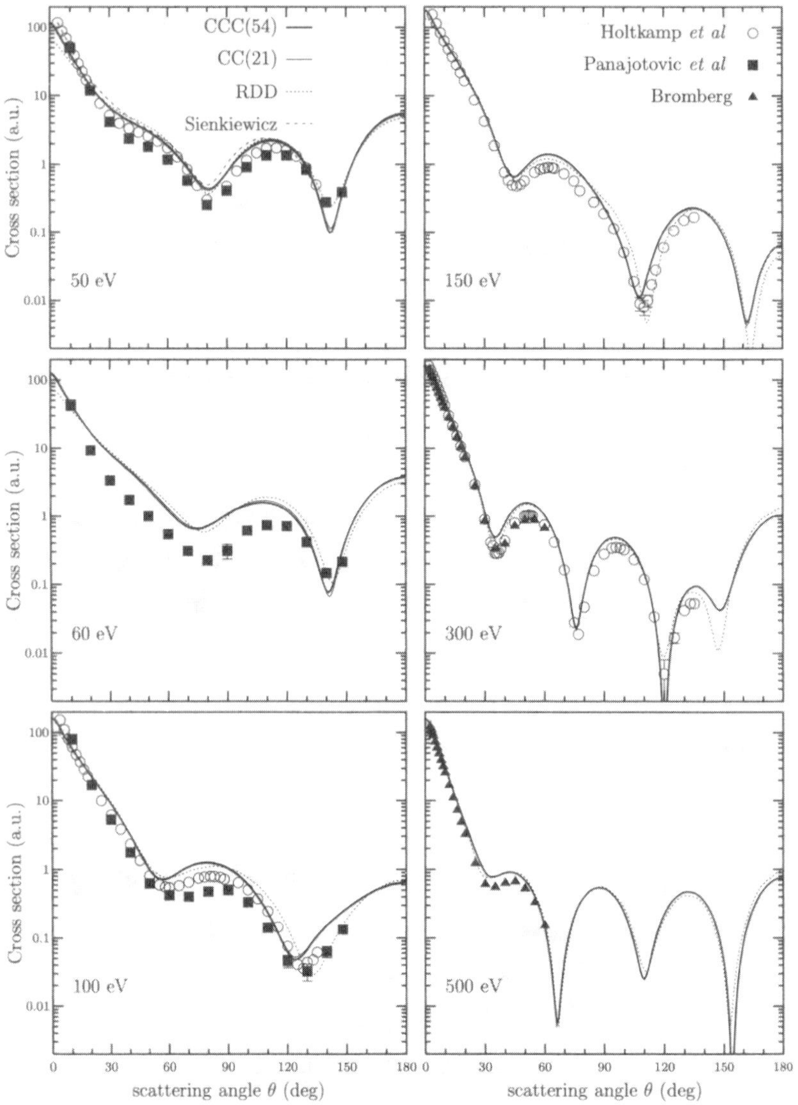

Fig. 9.5. Differential cross-sections for elastic scattering from the ground state of mercury at energies up to 500 eV. See caption to Figure 9.4 for further details. Reprinted with permission from [83].

Table 3]. Table 9.2 compares GRASP calculations of the low-lying Hg^- levels, the DARC resonance positions, and equivalent values taken from R-matrix calculations by Bartschat and Burke [93] who used three different approxima-

Table 9.1. Measured levels (eV) in the Hg I spectrum from NIST Atomic Spectra Database Version 3.0 (http://physics.nist.gov). AL,CI: GRASP calculations [90] – see text for details.

LSJ			Obs.	Minimal	Target
1	$6s^2$	$^1S_0$	0.	0.	0.
2	6s6p	$^3P^o_0$	4.67	3.20	4.32
3	6s6p	$^3P^o_1$	4.89	3.45	4.55
4	6s6p	$^3P^o_2$	5.46	3.97	5.06
5	6s6p	$^1P^o_1$	6.70	6.09	6.81

Table 9.2. Low-lying resonances in electron scattering (eV) from neutral mercury [90]. See text for details.

LSJ	GRASP	DARC	RM(5)	RMB(5)	RMB(11)
1 $6s^26p\ ^2P^o_{1/2}$	0.19	0.72			
2 $6s^26p\ ^2P^o_{3/2}$	0.21	0.96			
3 $6s\ 6p^2\ ^4P^e_{1/2}$	4.63		4.7	4.59	4.49
4 $6s\ 6p^2\ ^4P^e_{3/2}$	4.70	4.58	4.7	4.66	4.57
5 $6s\ 6p^2\ ^4P^e_{5/2}$	5.00	4.81	4.9	4.86	4.76
6 $6s\ 6p^2\ ^2D^e_{3/2}$	6.07	5.07	5.5	5.5	5.0
7 $6s\ 6p^2\ ^2D^e_{5/2}$	6.20	5.46	5.5	5.5	5.4
8 $6s\ 6p^2\ ^2P^e_{1/2}$	6.86	5.95			
9 $6s\ 6p^2\ ^2S^e_{1/2}$	7.26	5.83			

tions. The results labelled RM(5) used a nonrelativistic Hamiltonian with a core model potential together with a spin-orbit coupling potential and were based on the five target states of Table 9.1 with the same values of a, b as in the DARC calculation. The one-body relativistic correction and the Darwin mass correction were included in the RMB(5) calculations. The RMB(11) calculations, which took in six more target states, $6p^2\ ^3P_0$, $6p^2\ ^1S_0$, 6p6d $^3P_{0,1,2}$ and 6p6d $^1P_1$, had only a small effect on the resonance positions.

9.9.3 Scattering of polarized electrons from polarized atoms

Experiments in which polarized electrons are scattered from polarized atom targets, which have become practicable comparatively recently, provide the most detailed tests of theoretical models. CCC predictions of spin asymmetries for light alkali atoms sodium [94, 95] and lithium [96] demonstrated excellent agreement with experimental data. The heaviest alkali atom, cesium, has been extensively studied both theoretically and experimentally. Recent theoretical work includes nonrelativistic CCC [97], semirelativistic [98], as well as fully relativistic approaches [43, 44, 99]. A benchmark experiment measuring the

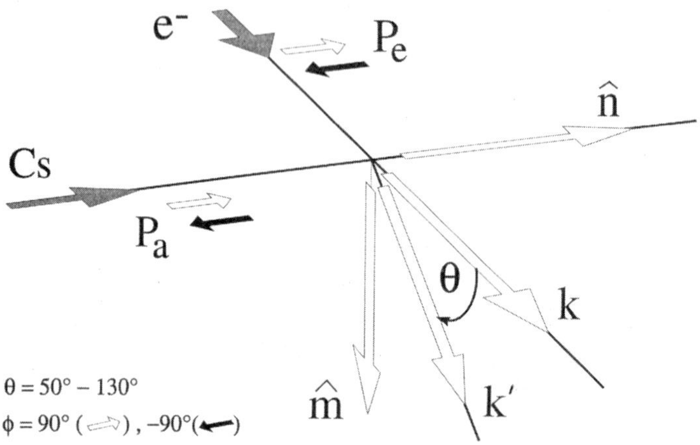

Fig. 9.6. Geometry of spin polarization experiment. Reprinted with permission from [100]. Copyright 1999 by the American Physical Society.

differential cross-section and spin asymmetries for elastic scattering from the ground state of cesium at 3 eV [100] emphasized the need to take account both of relativistic effects as well as the coupling to higher excited states.

Figure 9.6 is a schematic diagram of the experimental setup. A beam of cesium atoms enters from the left along the direction $\boldsymbol{n}$, which is perpendicular to the scattering plane. The incident electrons, linear momentum $\mathbf{k}$, have linear momentum $\mathbf{k}'$ after scattering and can be detected at an angle θ in the range $50^o - 130^o$. The experimental arrangement generates incident electrons with a polarization vector $\boldsymbol{P}_e$ parallel or antiparallel to $\boldsymbol{n}$ and cesium atoms with a polarization vector $\boldsymbol{P}_a$ parallel or anti-parallel to $\boldsymbol{n}$. Measured polarizations $P_e = |\boldsymbol{P}_e|$ of 0.65 and $P_a = |\boldsymbol{P}_a|$ of 0.9 have been achieved; see [100] for the full details.

A detailed analysis of spin-polarization in elastic scattering of electrons by one-electron atoms has been presented by Burke and Mitchell [101] using properties of the scattering amplitude under time reversal. space rotation and space reflection. They write the differential cross-section

$$\sigma(\theta) = \sigma_0(\theta)\,[1 + A_1(\theta)\,\boldsymbol{P}_a \cdot \boldsymbol{n} + A_2(\theta)\,\boldsymbol{P}_e \cdot \boldsymbol{n} - A_{nn}(\theta)\,\boldsymbol{P}_a \cdot \boldsymbol{n}\,\boldsymbol{P}_e \cdot \boldsymbol{n}\,], \quad (9.9.1)$$

where $\sigma_0(\theta)$ is the differential cross-section for unpolarized beams and $A_1(\theta)$, $A_2(\theta)$, and $A_{nn}(\theta)$ are termed asymmetry functions. Burke and Mitchell show [101, Equation (16)] that the 4×4 scattering amplitude $\boldsymbol{M}(\theta, \phi)$, whose rows and columns are labelled by the 4 pairs of spin projections $m_a \,/\, m_e$ for the final and initial atom/electron states respectively, can be written in terms of six complex amplitudes $a_1, \ldots, a_6$ such that

$$\begin{aligned} \boldsymbol{M} &= a_1 + a_2(\boldsymbol{\sigma}_a \cdot \boldsymbol{n}) + a_3(\boldsymbol{\sigma}_e \cdot \boldsymbol{n}) \\ &+ a_4(\boldsymbol{\sigma}_a \cdot \boldsymbol{n})(\boldsymbol{\sigma}_e \cdot \boldsymbol{n}) + a_5(\boldsymbol{\sigma}_a \cdot \boldsymbol{p})(\boldsymbol{\sigma}_e \cdot \boldsymbol{p}) + a_6(\boldsymbol{\sigma}_a \cdot \boldsymbol{q})(\boldsymbol{\sigma}_e \cdot \boldsymbol{q}) \end{aligned} \tag{9.9.2}$$

where $\boldsymbol{p}$ and $\boldsymbol{q}$ are unit vectors in the directions $\boldsymbol{k}+\boldsymbol{k}'$ and $\boldsymbol{k}-\boldsymbol{k}'$ respectively. In terms of these amplitudes [101, Equation (40)],

$$\sigma_0(\theta) = |a_1|^2 + |a_2|^2 + |a_3|^2 + |a_4|^2 + |a_5|^2 + |a_6|^2, \tag{9.9.3}$$

$$\sigma_0(\theta)A_1(\theta) = 2\Re(a_1\, a_2^* + a_3\, a_4^*), \tag{9.9.4}$$

$$\sigma_0(\theta)A_2(\theta) = 2\Re(a_1\, a_3^* + a_2\, a_4^*), \tag{9.9.5}$$

$$\sigma_0(\theta)A_{nn}(\theta) = 2\Re(a_1\, a_4^* + a_2\, a_3^* - a_5\, a_6^*). \tag{9.9.6}$$

The experiment measures event yields for four different settings of the electron and target spin projections N^{m_a, m_e} relative to the scattering plane, abbreviated $N^{\uparrow\uparrow}$, $N^{\downarrow\downarrow}$, $N^{\uparrow\downarrow}$ and $N^{\downarrow\uparrow}$. Then

$$\sigma_0(\theta) \propto \frac{1}{4}(N^{\uparrow\uparrow} + N^{\downarrow\downarrow} + N^{\uparrow\downarrow} + N^{\downarrow\uparrow}), \tag{9.9.7}$$

$$P_a A_1(\theta) = \frac{(N^{\uparrow\downarrow} + N^{\uparrow\uparrow}) - (N^{\downarrow\downarrow} + N^{\downarrow\uparrow})}{(N^{\downarrow\uparrow} + N^{\uparrow\downarrow}) + (N^{\uparrow\uparrow} + N^{\downarrow\downarrow})} \tag{9.9.8}$$

$$P_e A_2(\theta) = \frac{(N^{\uparrow\uparrow} + N^{\downarrow\uparrow}) - (N^{\uparrow\downarrow} + N^{\downarrow\downarrow})}{(N^{\downarrow\uparrow} + N^{\uparrow\downarrow}) + (N^{\uparrow\uparrow} + N^{\downarrow\downarrow})} \tag{9.9.9}$$

$$P_a P_e A_{nn}(\theta) = \frac{(N^{\downarrow\uparrow} + N^{\uparrow\downarrow}) - (N^{\uparrow\uparrow} + N^{\downarrow\downarrow})}{(N^{\downarrow\uparrow} + N^{\uparrow\downarrow}) + (N^{\uparrow\uparrow} + N^{\downarrow\downarrow})} \tag{9.9.10}$$

The angular distribution of the polarization of outgoing electrons due to the scattering of initially *unpolarized* electrons from *polarized* atoms is given by $A_1(\theta)$. It involves both exchange and spin-dependent interactions [102]. Polarization due to the scattering of *polarized* electrons from *unpolarized* atoms is given by $A_2(\theta)$, the Sherman function. The amplitude $A_{nn}(\theta)$, the "exchange asymmetry" parameter, involves both the relative orientation of electron and target spins as well as their orientation relative to the reaction plane. The "interference asymmetry" parameter, $A_1(\theta)$, is so called because [102] it requires both spin-orbit and exchange effects to be non-zero. Unpolarized atomic beams are simulated by averaging $N^{\uparrow\uparrow}$ and $N^{\downarrow\uparrow}$ for electrons with spin up and $N^{\downarrow\downarrow}$ and $N^{\uparrow\downarrow}$ for electrons with spin down. Whilst this experiment cannot measure all the 11 independent coefficients in (9.9.2), and so does not merit the label of a "perfect scattering experiment" [103], it nevertheless provides a severe test of our ability to interpret electron-atom scattering experiments.

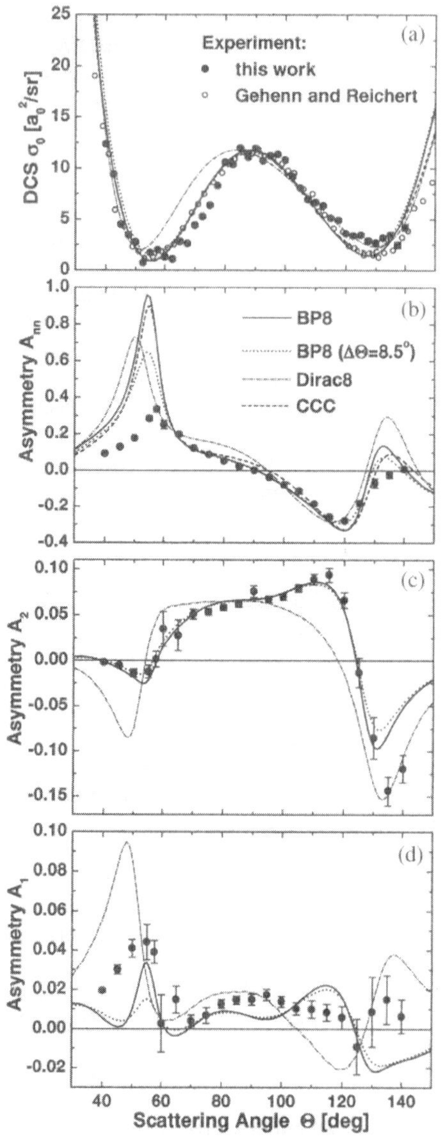

Fig. 9.7. Elastic scattering of polarized electrons from polarized Cs atoms at incident energy 3 eV [100]. (a) Differential cross-section $\sigma(\theta)$; asymmetry functions (b,c,d). See text for details. Reprinted with permission from [100]. Copyright 1999 by the American Physical Society.

Figure 9.7 compares the results of benchmark experiments presented in [100] with predictions from a semirelativistic 8 state Breit-Pauli R-matrix model (BP8), a relativistic Dirac R-matrix calculation (Dirac8), and nonrelativistic CCC calculations. Figure 9.7(a) shows the differential cross section $\sigma(\theta)$; the experimental benchmark results of [100], which have been normalized to the BP8 results at $\theta = 90^o$, were taken with an energy resolution of 150 meV and an angular resolution of 8.5^o (FWHM). The finite angular opening of the detector was modelled by convoluting the BP8 results with the experimental angluar resolution. The corresponding CCC results were scaled with a factor 0.82 and the Dirac8 results with a factor 1.12 in order to compare the shapes. The experimental data of Gehenn and Reichert [104] are also shown on this figure. Clearly, relativistic effects on $\sigma(\theta)$ are rather small at this energy and the main difference between BP8 and Dirac8 curves can be ascribed to the incomplete allowance for core polarization in the latter. The BP8 calculations have probably taken sufficient states into account to be regarded as converged. Figure 9.7(b,c) are evidence for similar conclusions, whilst the prominant disagreement of the Dirac 8 predictions from both BP8 prediction and experiment in Figure 9.7(d) suggests that the DARC calculations should be repeated with a suitable polarization potential.

Similar 5-state and 8-state DARC calculations [99] have been made at higher energies, 7 eV and 13.5 eV. As in the calculations at 3 eV, continuum states were constructed for symmetries $|\kappa| \leq 43$ and Hamiltonians for the scattering system were constructed for symmetries $J^\pi = 0^\pm, \ldots, 40^\pm$. Allowance was made also for higher partial waves using effective range theory. The asymmetry functions at 7 eV agree semi-quantitatively with quoted experimental data, and there was little difference between 5-state and 8-state results. There was evidence at 13.5 eV that more partial waves need to be taken into account explicitly. A comparison with CCC again suggested that a model potential may represent core polarization effects better than a first principles approach relying on large CSF expansions.

9.10 The relativistic random phase approximation

9.10.1 The RRPA equations

Many studies of photo-excitation and -ionization in atoms and ions of high nuclear charge have been performed using the relativistic random phase approximation (RRPA) [105], which has been reviewed in [106, 107]. A multi-configuration version (MCRRPA) has been given in [108].

The RRPA starts from a simple DHF model in which the wavefunction is built from N orbitals $u_i(\boldsymbol{r})$ satisfying DHF equations

$$(h_0 + V)\, u_i = \epsilon_i\, u_i, \quad i = 1, \ldots, N \tag{9.10.1}$$

where, in relativistic units ($\hbar = c = 1, e^2 = \alpha$),

$$h_0 = \boldsymbol{\alpha} \cdot \boldsymbol{p} + \beta m - Ze^2/r,$$

and the DHF potential V is given by

$$V\,u(\boldsymbol{r}) = \sum_{j=1}^{N} e^2 \int \frac{d^3\boldsymbol{r}'}{|\boldsymbol{r}-\boldsymbol{r}'|}\,[(u_j^\dagger u_j)'\,u - (u_j^\dagger u)'\,u_j].$$

The ground state energy of this model is taken in the many-body formalism as the Fermi level of a closed-shell ion with N electrons. This system is subjected to a time-dependent external perturbation

$$v_+\mathrm{e}^{-i\omega t} + v_-\mathrm{e}^{i\omega t}$$

so that

$$u_i(\boldsymbol{r}) \to u_i(\boldsymbol{r}) + w_{i+}(\boldsymbol{r})\mathrm{e}^{-i\omega t} + w_{i+}(\boldsymbol{r})\mathrm{e}^{i\omega t} + \ldots \tag{9.10.2}$$

ignoring terms involving higher powers of $\exp(\pm i\omega t)$. The linearized (TDHF) equations for the spinor amplitudes $w_{i\pm}(\boldsymbol{r})$ are

$$(h_0 + V - \epsilon_i \mp \omega)\,w_{i\pm} = (v_\pm - V_\pm^{(1)})\,u_i + \sum_j \lambda_{ij}\,u_j \tag{9.10.3}$$

where the Lagrange multipliers λ_{ij} ensure orthogonality of the $w_{i\pm}$ to the occupied orbitals u_j and the first order perturbation $V_\pm^{(1)}$ induces electron-electron correlations:

$$V_\pm^{(1)}\,u(\boldsymbol{r}) = \sum_{j=1}^{N} e^2 \int \frac{d^3\boldsymbol{r}'}{|\boldsymbol{r}-\boldsymbol{r}'|}\,[(u_j^\dagger w_{j\pm})'\,u_i + (w_{j\mp}^\dagger u_j)'\,u_i - (w_{j\mp}^\dagger u_i)'\,u_j - (u_j^\dagger u_i)' w_{j\pm}]$$

Solutions to (9.10.3) can be expanded in terms of a basis of solutions of the homogeneous equations

$$(h_0 + V - \epsilon_i)\,w_{i\pm} + V_\pm^{(1)})\,u_i - \sum_j \lambda_{ij}\,u_j = \pm\omega\,w_{i\pm}. \tag{9.10.4}$$

The eigenvalues ω_A and the corresponding spinors $w_{i\pm}^A$ approximate the excitation spectrum of the atom, and will have continuum as well as discrete basis elements subject to the orthonormality constraints

$$\int d^3\boldsymbol{r}\,w_{i\pm}^\dagger u_j = 0, \quad i,j = 1,\ldots,N, \tag{9.10.5}$$

and

$$\sum_{i=1}^{N} \int d^3\boldsymbol{r}\,[w_{i+}^{A\dagger} w_{i+}^B - w_{i-}^{A\dagger} w_{i-}^B] = \delta(\omega_A, \omega_B). \tag{9.10.6}$$

Figure 9.8 shows the lowest order Feynman diagrams contributing to the RRPA transition matrix elements. The electron-electron interaction, represented as horizontal broken lines, acts instantaneously. In the literature, this is always taken to be the Coulomb interaction; however, there is no reason (other than the extra computational cost) to ignore the instantaneous frequency-independent Breit interaction if this were to lead to useful improvements in the quality of the solution. It would be necessary to start from the DHFB equations rather than the DHF equations in this case.

The radiative transition vertex, Figure 9.8(a), implies the usual minimal coupling of the electron and photon fields. In the Coulomb gauge, the interaction is

$$v_+ = \boldsymbol{\alpha} \cdot \boldsymbol{A} = \quad v_-^* \tag{9.10.7}$$

where, if $\boldsymbol{e}$ is the polarization vector,

$$\boldsymbol{A} = \boldsymbol{e}\, \mathrm{e}^{i\boldsymbol{k}\cdot\boldsymbol{r}},$$

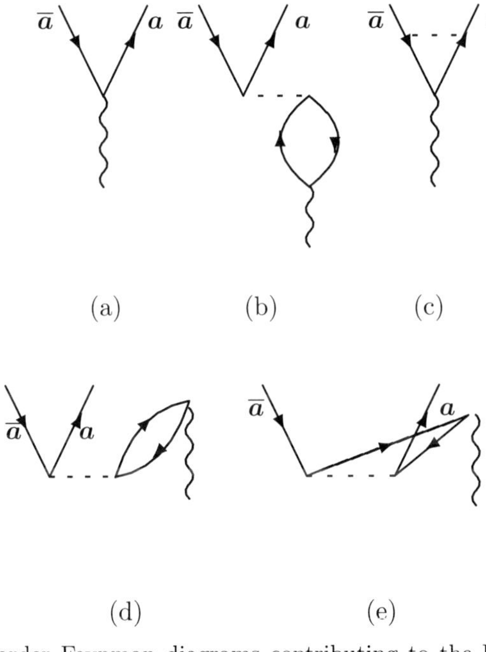

Fig. 9.8. Lowest order Feynman diagrams contributing to the RRPA transition matrix element: (a) uncorrelated matrix element, (b,c) positive frequency final state correlations, (d,e) ground state correlations. Time increases up the page. Solid lines: electrons/holes in the DHF basis. Dashed lines: instantaneous electron-electron interaction. Wiggly lines: photons. $a/\overline{a}$: particle/hole labels.

with $|\boldsymbol{k}| = \omega$ in relativistic units. The transition amplitude out of the ground state is then

$$T = \sum_i \int d^3\boldsymbol{r}\, [w^\dagger_{i+}\, \boldsymbol{\alpha} \cdot \boldsymbol{A}\, u_i + u^\dagger_i\, \boldsymbol{\alpha} \cdot \boldsymbol{A}\, w_{i-}] \tag{9.10.8}$$

One important advantage of the RRPA is that, with some qualifications, the amplitude T is gauge independent.

9.10.2 Radial equations

In this section, we shall adopt the notation used in [106], which is somewhat different from that used in the rest of this book, mainly in order to avoid transcription errors. The main reason for this is that, as well as using relativistic units rather than the SI and Hartree atomic units adopted in this book, Johnson *et al.* [106] write radial 4-spinors in the form

$$u(r) = \frac{1}{r} \begin{pmatrix} i\, G_{n\kappa}(r)\, \Omega_{\kappa m}(\widehat{\boldsymbol{r}}) \\ F_{n\kappa}(r)\, \Omega_{-\kappa m}(\widehat{\boldsymbol{r}}) \end{pmatrix} \tag{9.10.9}$$

where $\Omega_{\kappa m}(\widehat{\boldsymbol{r}})$ is the same as our $\chi_{\kappa m}(\widehat{\boldsymbol{r}})$, equation (3.2.9).[7] Comparing (9.10.9) with our definition (3.2.4), and extracting an overall factor i, we see that most of the expressions in this book can be transcribed into the notation of [106] with $P_{n\kappa}(r) \to G_{n\kappa}(r)$ and $Q_{n\kappa}(r) \to -F_{n\kappa}(r)$. Radial matrix elements may also require an additional overall phase factor. Using latin letters a for the subshell quantum numbers $n\kappa$, we write the radial 2-spinor

$$\mathsf{F}_a(r) = \begin{pmatrix} G_a(r) \\ F_a(r) \end{pmatrix} \tag{9.10.10}$$

so that the closed shell radial DHF equations become

$$(H_a - \epsilon_a)\mathsf{F}_a = 0 \tag{9.10.11}$$

where, for each occupied subshell,

$$H_a = \begin{pmatrix} m - \dfrac{Ze^2}{r} + V & \dfrac{d}{dr} - \dfrac{\kappa}{r} \\ -\dfrac{d}{dr} - \dfrac{\kappa}{r} & -m - \dfrac{Ze^2}{r} + V \end{pmatrix}$$

with the DHF potential defined by

$$V\, \mathsf{F}_a = \sum_b (2j_b + 1) \left\{ \frac{e^2}{r} Y_0(bb;r)\mathsf{F}_a - \sum_l \Lambda_l(ab) \frac{e^2}{r} Y_l(ba;r)\mathsf{F}_b \right\}.$$

[7] However, Huang and Johnson [108] use the phase convention of this book for radial decomposition but retain the notation G/F for large/small components respectively.

These equations are equivalent to (6.7.12). The exchange coefficients are given by

$$\Lambda_l(a,b) = \pi(l_a, l_b, l) \begin{pmatrix} j_a & j_b & l \\ -1/2 & 1/2 & 0 \end{pmatrix}^2,$$

where $\pi(l_a, l_b, l) = \frac{1}{2}\left[1 + (-1)^{l_a+l_b+l}\right]$ is the usual Coulomb parity factor. The perturbation induces transitions from orbitals $a = n\kappa$ into excitation channels $\bar{a}$ with symmetry $\bar{\kappa}\,\bar{m}$. The spinor amplitudes $w_{n\kappa m\pm}(\boldsymbol{r})$ describing the excitation can be expanded in terms of auxiliary functions $y_{\bar{\kappa}\,\bar{m}\pm}(\boldsymbol{r})$ so that

$$\begin{aligned} w_{n\kappa m+}(\boldsymbol{r}) &= \sum_{\bar{\kappa}\,\bar{m}} (-1)^{j-m} \langle j, -m, \bar{j}, \bar{m} \,|\, j\,\bar{j}\,J\,M \rangle \\ &\qquad \times \pi(l, \bar{l}, J+\lambda-1)\, y_{\bar{\kappa}\,\bar{m}+}(\boldsymbol{r}) \\ w_{n\kappa m-}(\boldsymbol{r}) &= \sum_{\bar{\kappa}\,\bar{m}} (-1)^{j-m+M} \langle j, -m, \bar{j}, \bar{m} \,|\, j\,\bar{j}\,J\,-M \rangle \\ &\qquad \times \pi(l, \bar{l}, J+\lambda-1)\, y_{\bar{\kappa}\,\bar{m}-}(\boldsymbol{r}) \end{aligned} \tag{9.10.12}$$

$\lambda = 0, 1$ depends on the parity of the excited state: $\lambda = 1$ corresponds to an electric 2^J-pole excitation with parity $(-1)^J$, $\lambda = 0$ corresponds to a magnetic 2^J-pole excitation with parity $(-1)^{J+1}$. In terms of the radial 2-component spinors

$$\mathsf{y}_{\bar{a}\pm}(r) = \begin{pmatrix} S_{\bar{a}\pm}(r) \\ T_{\bar{a}\pm}(r) \end{pmatrix}, \tag{9.10.13}$$

associated with $y_{\bar{a}\pm}(\boldsymbol{r})$, the radial RRPA equations are

$$\begin{aligned} [H_{\bar{a}} - (\epsilon_a \pm \omega)]\mathsf{y}_{\bar{a}\pm}(r) &= -C_J(a,\bar{a})\, V_J^{(1)}(r) \mathsf{F}_a(r) \\ &\quad + \sum_{b\bar{b}l} \Big[A(a\,b\,\bar{a}\,\bar{b}\,l\,J) \frac{e^2}{r} Y_l(a\,b; r)\, \mathsf{y}_{\bar{b}\pm}(r) \\ &\qquad + (-1)^{j_b - j_{\bar{b}}} A(a\,\bar{b}\,\bar{a}\,b\,l\,J) \frac{e^2}{r} Y_l(a\,\bar{b}\mp; r)\, \mathsf{F}_b(r) \Big] \\ &\qquad\qquad + \sum_b \delta_{\kappa_b \bar{\kappa}_a}\, \lambda_{\bar{a}b\pm} \mathsf{F}_b(r) \end{aligned} \tag{9.10.14}$$

for *electric multipole* perturbations, with $\pi = (-1)^J$, and

$$[H_{\bar a} - (\epsilon_a \pm \omega)]\mathsf{y}_{\bar a\pm}(r) =$$

$$+\sum_{b\bar b l}\Bigg[A(-a\ -b\,\bar a\,\bar b\, l\, J)\frac{e^2}{r}Y_l(a\,b;r)\,\mathsf{y}_{\bar b\pm}(r)$$

$$+(-1)^{j_b-j_{\bar b}}A(-a\ -\bar b\,\bar a\, b\, l\, J)\frac{e^2}{r}Y_l(a\,\bar b\mp;r)\,\mathsf{F}_b(r)\Bigg]$$

$$+\sum_b \delta_{\kappa_b\bar\kappa_a}\,\lambda_{\bar a b\pm}\mathsf{F}_b(r) \quad (9.10.15)$$

for *magnetic multipole* perturbations, with $\pi = (-1)^{J+1}$. The notation $-a$ in (9.10.15) denotes a change of sign $\kappa \to -\kappa$, which modifies the parity selection rule. The angular coefficients are

$$A(a\,b\,c\,d\,l\,J) = (-1)^{l+J-j_b-j_c}C_l(a\,c)C_l(b\,d)\begin{Bmatrix} j_a & j_b & l \\ j_d & j_c & J\end{Bmatrix}$$

where

$$C_J(a\,b) = (-1)^{j_a+1/2}[j_a, j_b]\begin{pmatrix} j_a & j_b & J \\ -1/2 & 1/2 & 0\end{pmatrix}\pi(l_a\,l_b\,J)$$

The potential $V_J^{(1)}(r)$ in (9.10.14) is

$$V_J^{(1)}(r) = \sum_{b\bar b}[J]^{-2}\,C_J(b\,\bar b)\,\frac{e^2}{r}\,[Y_J(b\,\bar b+;r) + Y_J(b\,\bar b-;r)]$$

where the sum ranges over all subshell perturbations $\bar b$ of subshells b in the atom. The notation $Y_l(b\,\bar b\mp;r)$ implies that the argument $\bar b$ refers to the radial functions $\mathsf{y}_{\bar b\mp}(r)$, whilst b refers to $\mathsf{F}_b(r)$. The Lagrange multipliers $\lambda_{\bar a b\pm}$ ensure that the functions $\mathsf{y}_{\bar a\pm}(r)$ are orthogonal to unperturbed orbitals of the same angular symmetry.

9.10.3 Multipole transition amplitudes

The vector potential $\boldsymbol{A}$, (9.10.7), has a multipole decomposition [106, Equation (25)], [109, §4],

$$\boldsymbol{A} = 4\pi\sum_{JM\lambda} i^{J-\lambda}\boldsymbol{Y}^{(\lambda)}_{JM}(\hat k)\cdot \boldsymbol{e}\,\boldsymbol{a}^{(\lambda)}_{JM} \quad (9.10.16)$$

where the Coulomb gauge radial potentials are

$$\boldsymbol{a}^{(0)}_{JM} = j_J(x)\,\boldsymbol{Y}^{(0)}_{JM}(\hat r)$$

$$(9.10.17)$$

$$\boldsymbol{a}^{(1)}_{JM} = \left(\frac{d}{dx}+\frac{1}{x}\right)j_J(x)\boldsymbol{Y}^{(1)}_{JM}(\hat r) + \sqrt{J(J+1)}j_J(\omega r)/\omega r\,\boldsymbol{Y}^{(-1)}_{JM}(\hat r)$$

with $x = \omega r$, and $\boldsymbol{Y}^{(\lambda)}_{JM}(\hat{k})$ is a *vector spherical harmonic* in the notation used in [109]. This can be related to the notation of §8.1 by noting that

$$\boldsymbol{Y}^{(0)}_{JM} = \frac{1}{\sqrt{J(J+1)}} \boldsymbol{L} Y_{JM}, \quad \boldsymbol{Y}^{(-1)}_{JM} = \boldsymbol{n} Y_{JM} \tag{9.10.18}$$

where $\boldsymbol{n}$ is a unit vector along the radius, whilst $\boldsymbol{Y}^{(1)}_{JM}$ is defined so that the set $\boldsymbol{Y}^{(\lambda)}_{JM}(\hat{k})$ are mutually orthogonal, with $\lambda = -1$ corresponding to longitudinal polarization. The orthonormality relations are

$$\int \boldsymbol{Y}^{(\lambda)}_{JM}(\hat{k})^* \cdot \boldsymbol{Y}^{(\lambda')}_{J'M'}(\hat{k}) d\hat{k} = \delta_{JJ'} \delta_{MM'} \delta_{\lambda\lambda'}. \tag{9.10.19}$$

Electromagnetic potentials in the Coulomb gauge vector potential can be related to potentials in other gauges by introducing a gauge function χ such that

$$\boldsymbol{A}' = \boldsymbol{A} + \nabla \chi, \quad \phi' = i\omega\, \chi$$

so the the corresponding interaction (9.10.7) is

$$v'_+ = v_+ + (\boldsymbol{\alpha} \cdot \boldsymbol{\nabla}\, \chi - i\omega\chi).$$

The Coulomb gauge potentials (9.10.17) reduce to the nonrelativistic velocity gauge in the limit $\alpha \to 0$ and the choice

$$\begin{aligned}
\phi'_{JM} &= -i\sqrt{\frac{J+1}{J}}\, j_J(\omega r)\, Y_{JM}(\hat{r}) \\
\boldsymbol{a}^{(0)'}_{JM} &= j_J(\omega r)\, \boldsymbol{Y}^{(0)}_{JM}(\hat{r}) \\
\boldsymbol{a}^{(1)'}_{JM} &= -j_{J+1}(\omega r) \left(\boldsymbol{Y}^{(1)}_{JM}(\hat{r}) - \sqrt{\frac{J+1}{J}} \boldsymbol{Y}^{(-1)}_{JM}(\hat{r}) \right)
\end{aligned} \tag{9.10.20}$$

gives the nonrelativistic length gauge. The magnetic multipoles are unaffected by this sort of gauge transformation. The Coulomb gauge multipole potentials give transition amplitudes

$$T^{(\lambda)}_{JM} = \sum_i \int d^3\boldsymbol{r}\, [w^\dagger_{i+}\, \boldsymbol{\alpha} \cdot \boldsymbol{a}^{(\lambda)}_{JM}\, u_i + u^\dagger_i\, \boldsymbol{\alpha} \cdot \boldsymbol{a}^{(\lambda)}_{JM}\, w_{i-}] \tag{9.10.21}$$

and $\boldsymbol{\alpha} \cdot \boldsymbol{a}^{(\lambda)}_{JM} \to \boldsymbol{\alpha} \cdot \boldsymbol{a}^{(\lambda)'}_{JM} - \phi'_{JM}$ in the length gauge. Notice that the only non-vanishing conribution to $T^{(\lambda)}_{JM}$ comes from solutions to the RRPA equations with the driving term $\boldsymbol{a}^{(\lambda)}_{JM}$.

Gauge transformations modify the transition amplitude so that

$$T \to T' = T + \sum_i \int d^3 r' [w^\dagger_{i+}(\boldsymbol{\alpha} \cdot \boldsymbol{\nabla}\, \chi - i\omega\chi) u_i + u^\dagger_i (\boldsymbol{\alpha} \cdot \boldsymbol{\nabla}\, \chi - i\omega\chi) w_{i-}].$$

It can be shown, by using equation (9.10.3), that the additional sum vanishes, so that $T' = T$. This assumes that the sum is taken over a complete set of states; in practice, the sum is always truncated, so that it will never vanish completely. The difference of T and T' can then be used as an estimate of the truncation error.

9.11 RRPA rates for photo-excitation and photo-ionization

9.11.1 Photo-excitation

Photoexcitation rate calculations use the multipole transition formula

$$A_k = 8\pi\alpha\omega_k \left| T^{(\lambda)}_{JM} \right|^2 , \tag{9.11.1}$$

where ω_k is the energy of the absorbed photon. $T^{(\lambda)}_{JM}$ is the amplitude (9.10.19), which can be expressed in the usual way in terms of radial integrals involving $\mathsf{F}_b(r)$ and $\mathsf{y}_{\bar{a}\pm}(r)$. For bound states, it is necessary to solve the coupled equations (9.10.12) and/or (9.10.13) subject to bound state boundary conditions. The resulting eigenvalue problem has much in common with the problem of numerical solution of the DHF equations. It requires solving the coupled equations, here abbreviated

$$\left[H^{(N-1)}_{\bar{a}} - (\epsilon_{\bar{a}} \pm \omega) \right] \mathsf{y}_{\bar{a}\pm} = R_{\bar{a}\pm}. \tag{9.11.2}$$

The Hamiltonian $H^{(N-1)}_{\bar{a}}$ differs from $H_{\bar{a}}$ of equations (9.10.12) and (9.10.13) by using the spherically averaged DHF potential for the $\bar{a}$ hole state; terms omitted from $H_{\bar{a}}$ are transferred to the term $R_{\bar{a}\pm}$ which also includes all interchannel coupling terms. A brief account of the iterative method of solution with references to more detailed explanation is given in [106].

9.11.2 Photo-ionization

The differential cross section for photo-ionization can be written in terms of the transition amplitude of (9.10.19) as [106, Equation (36)]

$$\frac{d\sigma}{d\Omega} = \frac{\alpha E p}{2\pi\omega} |T|^2 \tag{9.11.3}$$

The total cross section for photo-ionization from the closed subshell $a = n\kappa$, summed over photon polarizations can be written

$$\sigma_a(\omega) = \frac{2\pi^2\alpha}{\omega} \sum_{J\kappa\lambda} \frac{J+1}{J(2J+1)} \frac{\omega^{2J}}{[(2J-1)!!]^2} |\langle \bar{a} \,\| Q^{(\lambda)}_J \|\, a \rangle_{RRPA}|^2 \tag{9.11.4}$$

10.2 Dirac-Hartree-Fock-Breit equations for closed shell atoms

10.2.1 DHFB energy of a closed shell atom

The DHF energy expression for a configuration containing only closed subshells is given in (6.6.17)

$$E_{DHF} = \sum_A (2j_A + 1)\Big\{ I(A, A) + \frac{1}{2}\sum_B (2j_B + 1)\big[F^0(A, B) - \sum_k b^{\,k}_{\kappa_A,\kappa_B} G^k(A, B)\big]\Big\}, \quad (10.2.1)$$

where

$$b^{\,k}_{\kappa,\kappa'} = \begin{pmatrix} j & k & j' \\ -1/2 & 0 & 1/2 \end{pmatrix}^2,$$

The magnetic interaction,(6.6.21), between a pair of closed subshells in the Breit approximation ($\omega \to 0$), [6] is

$$E_{Breit} = \frac{1}{2}\sum_{AB}\sum_k (-1)^{j_A - j_B + k}\, X^k(ABBA). \quad (10.2.2)$$

When $l_A + l_b + k$ is *odd*,

$$(-1)^{j_A - j_B + k}\, X^k(ABBA) = -(2j_A + 1)(2j_B + 1)\, b^{\,k}_{\kappa_A,\kappa_B} \frac{(\kappa_A + \kappa_B)^2}{k(k+1)} \sum_{\gamma=-1}^{+1} G^k_\gamma(AB), \quad (10.2.3)$$

and when $l_A + l_b + k$ is *even*,

$$(-1)^{j_A - j_B + k}\, X^k(ABBA) = (2j_A + 1)(2j_B + 1)\, b^{\,k}_{\kappa_A,\kappa_B} \sum_{k'=k\pm 1}\sum_{\gamma=-1}^{+1} x^{k'k}_\gamma G^{k'}_\gamma(AB). \quad (10.2.4)$$

Only the exchange type terms contribute; the direct terms all vanish. The Slater integrals were defined in §6.5.

10.2.2 Spinor basis function representation

Each orbital a in subshell A, with symmetry κ, can be written in the form (5.6.1):

$$\psi_A(\boldsymbol{x}) = \begin{bmatrix} \sum_{\mu=1}^{N_\kappa} c^{+1}_{\mu A} M[+1, \mu, \boldsymbol{x}] \\ i\sum_{\mu=1}^{N_\kappa} c^{-1}_{\mu A} M[-1, \mu, \boldsymbol{x}] \end{bmatrix}, \quad (10.2.5)$$

$$
\begin{aligned}
&x^{k-1,k}_{\pm 1}(AB) \\
&\quad = -b^{\;k}_{\kappa_A,\kappa_B} \frac{[k \pm (\kappa_B - \kappa_A)][(k+1)(2k+1) \mp 2(\kappa_B - \kappa_A)(k-3)]}{2(2k-1)(2k+1)}, \\
&x^{k-1,k}_{0} \\
&\quad = b^{\;k}_{\kappa_A,\kappa_B} \frac{k(k+1)(2k+1) \mp 2(\kappa_B - \kappa_A)^2(k-3)}{(2k-1)(2k+1)}, \\
&x^{k+1,k}_{\pm 1} \\
&\quad = \mp b^{\;k}_{\kappa_A,\kappa_B} \frac{[\kappa_B - \kappa_A \mp (k+1)][k(k+1) \pm (\kappa_B - \kappa_A)(k+3)]}{(2k+1)(2k+3)}, \\
&x^{k+1,k}_{0} \\
&\quad = -2b^{\;k}_{\kappa_A,\kappa_B} \frac{k(k+1)^2 + (\kappa_B - \kappa_A)^2(k+3)}{(2k+1)(2k+3)},
\end{aligned}
\qquad (10.2.16)
$$

and

$$
x^{k,k}_{\pm 1} = -b^{\;k}_{\kappa_A,\kappa_B} \frac{(\kappa_A + \kappa_B)^2}{k(k+1)}, \quad x^{k,k}_{0} = -2b^{\;k}_{\kappa_A,\kappa_B} \frac{(\kappa_A + \kappa_B)^2}{k(k+1)}.
$$

The angular momentum and parity selection rules limit k' and k so that the coefficients are non-zero only if $l_A + l_b + k$ is *odd* for $k' = k$ and *even* for $k' = k \pm 1$.

10.2.6 The DHFB Fock matrix

The DHFB energy (6.6.17) can be written as a relativistic generalization of an equation familiar to quantum chemists (e.g. [7, Equation (6.2.5)]),

$$
E_{DHFB} = \sum_{\kappa} \sum_{\beta\beta'} \mathrm{tr} \left\{ \mathbf{h}^{\beta\beta'}_{\kappa} + \frac{1}{2}(\mathbf{J}^{\beta\beta'}_{\kappa} - \mathbf{K}^{\beta\beta'}_{\kappa} + \mathbf{B}^{\beta\beta'}_{\kappa}) \right\} (\boldsymbol{D}^{\beta\beta'}_{\kappa})^{\dagger}, \quad (10.2.17)
$$

emphasizing the close links between relativistic and nonrelativistic SCF theories. Equation (10.2.17) features *density matrices*

$$
D^{\;\beta\beta'}_{\kappa\mu\nu} = \sum_{A} (2j_A + 1) c^{\beta}_{\mu A} \, c^{\beta'}_{\nu A} \, \delta_{\kappa_A,\kappa} \qquad (10.2.18)
$$

where the sum runs over all N_κ occupied subshells A, and $N_\kappa \times N_\kappa$ matrices $\mathbf{J}^{\beta\beta'}_{\kappa}$, $\mathbf{K}^{\beta\beta'}_{\kappa}$, and $\mathbf{B}^{\beta\beta'}_{\kappa}$, which are components of the Fock matrix for symmetry κ [8, Equation (165)],

$$
\boldsymbol{F}_{\kappa} = \boldsymbol{h}_{\kappa} + \boldsymbol{J}_{\kappa} - \boldsymbol{K}_{\kappa} + \boldsymbol{B}_{\kappa} \qquad (10.2.19)
$$

with

$$
\boldsymbol{h}_{\kappa} = \begin{pmatrix} \boldsymbol{V}^{++}_{\kappa} & c\boldsymbol{\Pi}^{+-}_{\kappa} \\ c\boldsymbol{\Pi}^{-+}_{\kappa} & -2mc^2\boldsymbol{S}^{--}_{\kappa} + \boldsymbol{V}^{--}_{\kappa} \end{pmatrix},
$$

$$C^{\eta}_{lm} = s_{lm}\left(\frac{l+1/2+\eta m}{2l+1}\right)^{1/2}. \tag{10.6.5}$$

Then, if $\eta_\mu = \text{sgn}\,\kappa_\mu = \pm 1$, $l_\mu = j_\mu + \frac{1}{2}\eta_\mu$, and $j_\mu = |\kappa_\mu| - \frac{1}{2}$,

$$M[+1,\mu,\mathbf{r}_{\mathbf{A}_\mu}] = N^{+}_{\mu}\begin{bmatrix} -\eta_\mu\, C^{-\eta_\mu}_{l_\mu m_\mu}\; S[a_\mu, \mathbf{r}_{\mathbf{A}_\mu}\,;\, 0, l_\mu, m_\mu - 1/2] \\ C^{+\eta_\mu}_{l_\mu m_\mu}\; S[a_\mu, \mathbf{r}_{\mathbf{A}_\mu}\,;\, 0, l_\mu, m_\mu + 1/2] \end{bmatrix}, \tag{10.6.6}$$

and

$$M[-1,\mu,\mathbf{r}_{\mathbf{A}_\mu}] = N^{-}_{\mu}\begin{bmatrix} \eta_\mu C^{\eta_\mu}_{\bar{l}_\mu, m_\mu} \;\{t_\mu\, S[a_\mu, \mathbf{r}_{\mathbf{A}_\mu}\,;\, 0, \bar{l}_\mu, m_\mu - 1/2] \\ \qquad -2a_\mu\, S[a_\mu, \mathbf{r}_{\mathbf{A}_\mu}\,;\, 1, \bar{l}_\mu, m_\mu - 1/2]\,\} \\ C^{-\eta_\mu}_{\bar{l}_\mu, m_\mu} \;\{t_\mu\, S[a_\mu, \mathbf{r}_{\mathbf{A}_\mu}\,;\, 0, \bar{l}_\mu, m_\mu + 1/2] \\ \qquad -2a_\mu\, S[a_\mu, \mathbf{r}_{\mathbf{A}_\mu}\,;\, 1, \bar{l}_\mu, m_\mu + 1/2]\,\} \end{bmatrix}, \tag{10.6.7}$$

where $\bar{l}_\mu = l_\mu - \eta_\mu$, $t_\mu = \kappa_\mu + l_\mu + 1$ and $\mathbf{r}_{\mathbf{A}_\mu} = \mathbf{r} - \mathbf{A}_\mu$. All coefficients are defined in terms of the "nonrelativistic" (large component) value of $l_\mu = j_\mu + \frac{1}{2}\eta_\mu$. The normalization factors, which do not depend upon the sign of κ_μ, are given by

$$N^{+}_{\mu} = \sqrt{\frac{2(2a_\mu)^{l_\mu+3/2}}{\Gamma(l_\mu+3/2)}}, \quad N^{-}_{\mu} = \sqrt{\frac{2(2a_\mu)^{l_\mu+1/2}}{\Gamma(l_\mu+5/2)}}. \tag{10.6.8}$$

Notice that $t_\mu = 0$ when $\kappa_\mu < 0$ and $t_\mu = 2l_\mu + 1$ when $\kappa_\mu > 0$, so that the G-spinor small components for $\kappa_\mu < 0$ are less expensive to construct. The kinetic matching construction, §5.7, is responsible for the more complex structure of (10.6.7).

10.7 The charge-current density

All interaction integrals are built from the components of the relativistic charge-current density resulting from the overlap of two radial spinors. In nonrelativistic notation, $j = (c\varrho_{\mu\nu}, \boldsymbol{j}_{\mu\nu})$,

$$\begin{aligned} \varrho_{\mu\nu}(\boldsymbol{x}) &= \varrho^{++}_{\mu\nu}(\boldsymbol{x}) + \varrho^{--}_{\mu\nu}(\boldsymbol{x}), \quad \text{where} \\ \varrho^{\beta\beta}_{\mu\nu}(\boldsymbol{x}) &= -eM^{\dagger}(\beta,\mu,\boldsymbol{x}).\, M(\beta,\nu,\boldsymbol{x}), \quad \text{and} \\ \boldsymbol{j}^{+-}_{\mu\nu}(\boldsymbol{x}) &= \left(\boldsymbol{j}^{-+}_{\nu\mu}\right)^{*} = -i\,ecM^{\dagger}(+1,\mu,\boldsymbol{x})\,\boldsymbol{\sigma}M(-1,\nu,\boldsymbol{x}). \end{aligned} \tag{10.7.1}$$

Using the abbreviated notation

$$M[\beta,\mu,\mathbf{r}_{\mathbf{A}_\mu}] \to \begin{bmatrix} M(\beta,\mu)_1 \\ M(\beta,\mu)_2 \end{bmatrix}$$

The multi-index $\boldsymbol{k} = (\rho, \sigma, \tau)$ runs over the i_Λ triples in the set

$$\mathcal{T}_\Lambda = \{(\rho, \sigma, \tau) \,|\, 0 \leq \rho + \sigma + \tau \leq \Lambda\} \tag{10.8.4}$$

where $i_\Lambda = \dim \mathcal{T}_\Lambda = (\Lambda + 1)(\Lambda + 2)(\Lambda + 3)/6$ and $\Lambda = 2n + 2n' + l + l'$. The differential operator $D^P_{\boldsymbol{k}}$ of (10.8.3) can then be written

$$D^P_{\boldsymbol{k}} = \left(\frac{\partial}{\partial P_x}\right)^\rho \left(\frac{\partial}{\partial P_y}\right)^\sigma \left(\frac{\partial}{\partial P_z}\right)^\tau.$$

The expansion (10.8.2), which depends on writing $\boldsymbol{r}_A = \boldsymbol{r} - \boldsymbol{A} = \boldsymbol{r}_P + \boldsymbol{P}_A$, was first suggested by McMurchie and Davidson [20]. A relativistic generalization of this scheme proposed by Quiney, Skaane, and Grant [21] expresses the charge density in the analogous form

$$\varrho^{\beta\beta}_{\mu\nu}(\boldsymbol{r}) = -e \sum_{\boldsymbol{k} \in \mathcal{T}_\Lambda} E^{\beta\beta}_0(\mu, \nu; \boldsymbol{k})\, H(p, \boldsymbol{r}_P; \boldsymbol{k}) \tag{10.8.5}$$

and the components, $q = 0, \pm 1$, of the current density vector as

$$\left(\boldsymbol{j}^{\beta,-\beta}_{\mu\nu}(\boldsymbol{r})\right)_q = -iec \sum_{\boldsymbol{k} \in \mathcal{T}_\Lambda} E^{\beta,-\beta}_q(\mu, \nu; \boldsymbol{k})\, H(p, \boldsymbol{r}_P; \boldsymbol{k}). \tag{10.8.6}$$

10.8.1 Relativistic expansion coefficients

The $\beta\beta$ case with $\beta = +1$ illustrates the general strategy for constructing the relativistic expansion coefficients. To obtain

$$\varrho^{++}_{\mu\nu}(\boldsymbol{r}) = -e \sum_{\boldsymbol{k} \in \mathcal{T}_\Lambda} E^{++}_0(\mu; \nu; \mathbf{k})\, H(p_{\mu\nu}, \boldsymbol{r}_{P_{\mu\nu}}; \boldsymbol{k}), \tag{10.8.7}$$

where $p_{\mu\nu} = a_\mu + a_\nu$ and $\boldsymbol{P}_{\mu\nu} = (a_\mu \boldsymbol{A}_\mu + a_\nu \boldsymbol{A}_\nu)/p_{\mu\nu}$, we write

$$\begin{aligned} &M^\dagger[+1, \mu; \boldsymbol{r}]\, M[+1, \nu; \boldsymbol{r}] \\ &= N^+_\mu N^+_\nu \Big\{ \eta_\mu \eta_\nu\, C^{-\eta_\mu}_{l_\mu m_\mu}\, C^{-\eta_\nu}_{l_\nu m_\nu}\, \mathsf{S}^*[\mu;\, 0, l_\mu, m_\mu - 1/2]\, \mathsf{S}[\nu;\, 0, l_\nu, m_\nu - 1/2] \\ &\qquad + C^{\eta_\mu}_{l_\mu m_\mu}\, C^{\eta_\nu}_{l_\nu m_\nu}\, \mathsf{S}^*[\mu;\, 0, l_\mu, m_\mu + 1/2]\, \mathsf{S}[\nu;\, 0, l_\nu, m_\nu + 1/2] \Big\}. \end{aligned}$$

Applying (10.8.2) and the relation $Y^{m*}_l = (-1)^m Y^{-m}_l$ gives the relativistic E_0 coefficient

$$\begin{aligned} E^{++}_0[\mu; \nu; \boldsymbol{k}] = N^+_\mu N^+_\nu \Big\{ &\eta_\mu \eta_\nu\, C^{-\eta_\mu}_{l_\mu m_\mu}\, C^{-\eta_\nu}_{l_\nu m_\nu} \\ &\times (-1)^{m_\mu - 1/2}\, E[0, l_\mu, -m_\mu + 1/2\,;\, 0, l_\nu, m_\nu - 1/2\,;\, \boldsymbol{k}] \\ + C^{\eta_\mu}_{l_\mu m_\mu}\, C^{\eta_\nu}_{l_\nu m_\nu}\, &(-1)^{m_\mu + 1/2}\, E[0, l_\mu, -m_\mu - 1/2\,;\, 0, l_\nu, m_\nu + 1/2\,;\, \boldsymbol{k}] \Big\} \end{aligned} \tag{10.8.8}$$

where $Q_J^{(\lambda)}$ is the multipole moment operator. It is sufficient to consider only the electric dipole contribution $J = \lambda = 1$ for low-energy photons, so that

$$\sigma_a(\omega) = \frac{4\pi^2\alpha}{3}\,\omega\,\left(|D_{j\to j-1}|^2 + |D_{j\to j}|^2 + |D_{j\to j+1}|^2\right) \tag{9.11.5}$$

where

$$D_{j\to\bar{j}} = i^{l-\bar{l}} \mathrm{e}^{i\delta_{\bar{\kappa}}} \langle\bar{\kappa}\,\|Q_J^{(\lambda)}\,\|\kappa\rangle_{RRPA}.$$

The angular distribution for electric dipole photoelectrons has the form of (9.10.16). See [106, Equation (41)] for the asymmetry parameter $\beta_a(\omega)$ which has a similar form to that given by Walker and Waber quoted in (9.10.17).

The RRPA reduced matrix element is a sum of single particle terms

$$\langle\bar{a}\pm\,\|Q_J^{(\lambda)}\,\|a\rangle_{RRPA} = \sum_{b\bar{b}} \left[\langle\bar{b}+\,\|Q_1^{(1)}\,\|b\rangle + \langle\bar{b}-\,\|Q_1^{(1)}\,\|b\rangle\right] \tag{9.11.6}$$

where, in *length gauge*,

$$\langle\bar{b}\pm\,\|Q_1^{(1)}\,\|b\rangle = C_1(b\,\bar{b}) \int_0^\infty [S_{\bar{b}\pm}G_b + T_{\bar{b}\pm}F_b]\, r\, dr \tag{9.11.7}$$

and, in *velocity gauge*,

$$\langle\bar{b}\pm\,\|Q_1^{(1)}\,\|b\rangle = \tag{9.11.8}$$
$$\pm C_1(b\,\bar{b})\omega^{-1} \int_0^\infty [(\kappa_b - \bar{\kappa}_b + 1)S_{\bar{b}\pm}F_b + (\kappa_b - \bar{\kappa}_b - 1)T_{\bar{b}\pm}G_b]\, dr$$

where G_b, F_b are the components of F_b, (9.10.10), and $S_{\bar{b}}, F_{\bar{b}}$ are the components of $\mathsf{y}_{\bar{b}}$, (9.10.13).

Let N_P be the number of open channels $(a, \bar{a})$ with $\epsilon_a + \omega > m$ having angular momentum J^π. Photo-ionizing solutions of the RRPA equations have spherical outgoing waves in channel $(a, \bar{a})$ and incoming waves in all other channels. There are N_P independent solutions to the RRPA equations, labelled with the channel index $i = 1, 2, \ldots, N_P$. Starting from approximate single channel solutions to the homogeneous equations

$$[H_{\bar{a}}^{(N-1)} - \epsilon_{\bar{a}}]\chi_{\bar{a}} = 0,$$

solutions to the coupled system (9.11.2) are constructed which are regular at the origin and which have the asymptotic form

$$\mathsf{y}_j^{(i)}(r) \to \mathsf{f}_j(r)\,\delta_{ij} + \mathsf{g}_j(r)\,K_{ij} \tag{9.11.9}$$

as in (9.2.9). The N_P eigenvalues λ_α of the $\boldsymbol{K}$-matrix can be expressed in terms of the eigenphases δ_α by

$$\lambda_\alpha = \tan\delta_\alpha, \quad \alpha = 1, \ldots N_P, \tag{9.11.10}$$

with corresponding eigenvectors $U_{i\alpha}$. Define standing wave eigenchannel solutions by

$$\mathsf{Z}_j^{(\alpha)}(r) = \sum_i \mathsf{y}_j^{(l)}\, U_{l\alpha}\, \cos\delta_\alpha \quad j, \alpha = 1, \ldots N_P \tag{9.11.11}$$

The required S-matrix (physical) solutions with a normalized outgoing wave in one channel and incoming waves in all other channels are

$$\mathsf{w}_j^{(k)}(r) = \sum_\lambda \mathsf{Z}_j^{(\lambda)}(r)\, \mathrm{e}^{-i\delta_\lambda}\, U_{k\lambda} \tag{9.11.12}$$

having the asymptotic form

$$\mathsf{w}_j^{(k)}(r) \sim \frac{1}{2}\mathsf{h}_j^+(r)\,\delta_{jk} + \frac{1}{2}\mathsf{h}_j^-(r)\,S_{jk}^* \tag{9.11.13}$$

where[8]

$$\mathsf{h}_j^\pm(r) \sim \begin{pmatrix} \sqrt{(E+m)/\pi p} \\ \mp i\sqrt{(E-m)/\pi p} \end{pmatrix} \mathrm{e}^{\pm i\zeta_j(r)}.$$

From (9.2.11), the S-matrix is defined by

$$\boldsymbol{S} = \boldsymbol{U}\boldsymbol{\Lambda}\boldsymbol{U}^\dagger, \quad \boldsymbol{\Lambda} = \mathrm{diag}\; \mathrm{e}^{2i\delta_\alpha}. \tag{9.11.14}$$

Finally, we can complete the cross-section calculation by expressing the "physical" reduced matrix elements of the multipole operator, $Q_i^{(in)}$, in terms of eigenchannel amplitudes $Q^{(\alpha)}$:

$$Q_i^{(in)} = \sum_\alpha e^{\mathrm{d}i\delta_\alpha} U_{i\alpha} Q^{(\alpha)}, \quad Q^{(\alpha)} = Q_i U_{i\alpha} \cos\delta_\alpha \tag{9.11.15}$$

where Q_i is calculated using the standing wave solutions (9.11.9), so that (9.11.4) becomes

$$\sigma_a(\omega) = \frac{2\pi^2\alpha}{\omega} \sum_{J\kappa\lambda} \frac{J+1}{J(2J+1)} \frac{\omega^{2J}}{[(2J-1)!!]^2} \sum_i \left| Q_i^{(in)} \right|^2, \tag{9.11.16}$$

in which the sum over i comprises all those channels for which the final ion has a hole in subshell $a = n\kappa$. The asymmetry parameter is again given by (8.6.16).

---

[8] The functions $\mathsf{h}_j^\pm$ are the same as $\mathsf{e}_j^\pm$, (9.2.7), apart from the conventional differences in units and relative phases of the components.

9.12 Comparison with experiment

Space permits only a somewhat selective account of the confrontation of relativistic photo-excitation and photo-ionization theory with experiment. Most theoretical work has utilized the semi-relativistic Breit-Pauli R-matrix method, single channel calculations with DHF wavefunctions following Walker and Waber [110] and the RRPA scheme. A Dirac R-matrix photo-ionization package has been available comparatively recently. BP R-matrix theory has been quite successful for outer shells in the near-threshold region for atoms and ions [111, 112]. One problem which affects both BP and Dirac R-matrix theory is the rapidly increasing number of channels that must be taken into account in intermediate angular momentum coupling, although recent massive advances in computing power go some way to alleviate it. Because the BP approximation uses Schrödinger wavefunctions, it has no way to account for orbital expansion and contraction induced by relativistic dynamics, which is most easily taken into account using Dirac wavefunctions. The RRPA method has been very successful in modelling closed shell atoms and ions, but it is hard to generalize to more complex systems. It includes certain correlation diagrams to all orders but there is no simple way to account for those that have been neglected. Nor can it treat channels in which excitation accompanies ionization, for which the most promising approach remains the Dirac R-matrix.

9.12.1 Photo-ionization of outer atomic subshells at high Z

A prominent feature of the energy dependence of the cross section for photo-ionization of outer subshells is the existence of *Cooper minima* [113][9], due to the presence of a zero in the dominant transition matrix element at a certain photon energy. These minima occur in atoms throughout the Periodic Table and can strongly influence the shape of the cross section curves. As a general rule, the minimum of the cross section for a given subshell moves first to higher energies as Z increases and then back (though not monotonically) towards the threshold and down into the discrete spectrum [114]. This is easy to explain in terms of the relative shift of continuum and bound wavefunction nodes as Z increases.

Besides their effect on energy dependence of the cross section, Cooper minima are also associated with interesting effects on photoelectron angular distributions and subshell branching ratios. Kim *et al.* [114] suggested that relativistic effects on photo-ionization should be detectable even for outer electrons and at low energies. Figure 9.9 demonstrates dramatically the expected features using HS and DHS calculations. Inner shell contraction and

[9] The attribution to Cooper recognizes his explanation of the cross section minimum as a general phenomenon and not, as previously thought, confined to photo-ionization of s-shells. Kim *et al.* [114, Ref. 3] give reasons for associating it also with the names of Ditchburn, Bates, and Seaton.

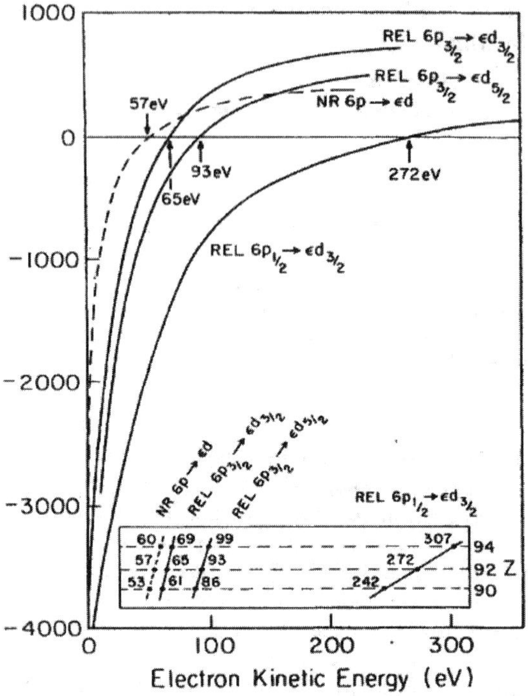

Fig. 9.9. HS/DHS (broken/solid lines) are radiative matrix elements (R_l/R_j of §8.6.2) for $6p \to \epsilon d$ transitions of uranium. The HS ionization energy is 26.1 eV and the DHS ionization energies are 34.1 eV for $6p_{1/2}$ and 24.6 eV for $6p_{3/2}$ respectively. Reprinted with permission from [114, Figure 1]. Copyright 1981 by the American Physical Society.

the consequent adjustments in outer shell charge distributions along with relativistic j-dependence (*alias* spin-orbit splitting) mean that one nonrelativistic $6p \to \epsilon d$ radial matrix element, with a Cooper zero at 57 eV, is replaced by $6p_{3/2} \to \epsilon d_{3/2}$, $6p_{3/2} \to \epsilon d_{5/2}$ and $6p_{1/2} \to \epsilon d_{3/2}$ matrix elements with Cooper zeros at 65, 93, and 373 eV respectively. Some of the differences in the radial matrix elements are due to the way in which the Dirac radial components depend on energy, and some to the replacement of the nonrelativistic $\boldsymbol{p} \cdot \boldsymbol{\epsilon}$ transition operator by $\boldsymbol{\alpha} \cdot \boldsymbol{\epsilon}$. The insert to Figure 9.9 shows how the relativistic Cooper zeros move as a function of Z in the neighbourhood of $Z = 92$.

A recent study of the angular distribution of Xenon $4d$ photoelectrons near the Cooper minimum [115] provided the first experimental verification of these theoretical predictions. In the nonrelativistic case, the asymmetry parameter is given by

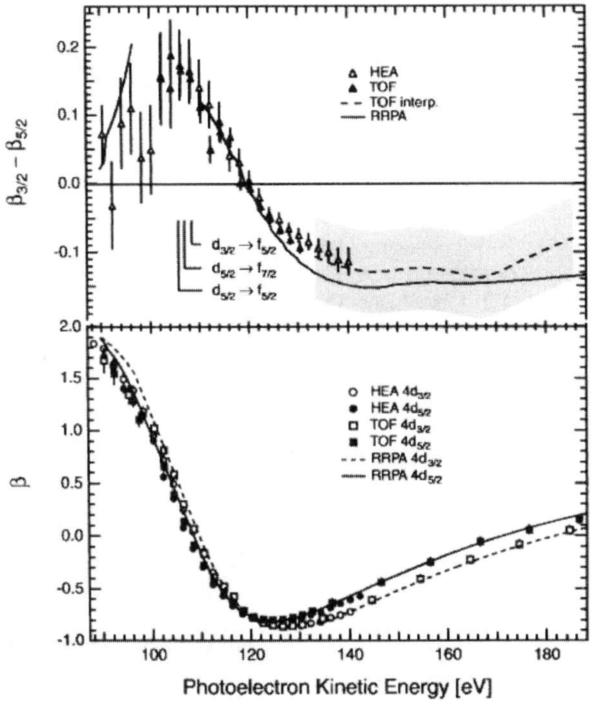

Fig. 9.10. Energy dependence of experimental asymmetry parameters $\beta_{3/2}$ and $\beta_{5/2}$ measured at the same photoelectron energy and corresponding theoretical results from 20 channel RRPA calculations for Xe $4d$ photo-ionization. The upper panel shows the difference $\beta_{3/2} - \beta_{5/2}$ derived from the results in the lower panel. The theoretical difference curve is not shown in the region in which the values are perturbed by $4p \rightarrow ns, nd$ resonances, and the shaded area above 130 eV represents error bars. The theoretical positions of Cooper minima are also shown in the upper panel. Reprinted, with permission, from [115]. Copyright 2001 by the American Physical Society.

$$\mathrm{d}\beta = \frac{2R_p^2 + 12R_f^2 - 36R_pR_f\cos\Delta_{pf}}{10R_p^2 + 15R_f^2}$$

where R_p, R_f are the two radial partial wave amplitudes in the $4d \rightarrow \epsilon p, \epsilon f$ channels and Δ_{pf} is their phase shift difference. Thus $\beta = 0.2$ when there is a Cooper minimum in the $d \rightarrow f$ channel. In the relativistic case, there are six channels: $4d_{5/2} \rightarrow \epsilon p_{3/2}, \epsilon f_{5/2}, \epsilon f_{7/2}$ and $4d_{3/2} \rightarrow \epsilon p_{3/2}, \epsilon p_{1/2}, \epsilon f_{5/2}$ giving different and more complex expressions for $\beta_{3/2}$ and $\beta_{5/2}$ with the same nonrelativistic limit. RRPA calculations taking 20 coupled channels with single excitations from $4s, 4p, 4d, 5s$ and $5d$ subshells into account were made for these asymmetries. The lower panel of Figure 9.10 shows that the energy dependence of the two asymmetries is different. The calculation showed that relativistic dynamical effects on the ϵp channels were insignificant, and that $\beta_{3/2} \approx 0.2$ at the Cooper minimum in the $4d_{3/2} \to \epsilon f_{5/2}$. The behaviour of $\beta_{5/2}$ was quite similar, taking the value 0.2 at a slightly lower energy. The energy variation of $\beta_{3/2} - \beta_{5/2}$ shown in the upper panel provides direct experimental evidence of relativistic effects. The experiments, described in more detail in [115], were done independently using hemispherical (HEA) and time of flight (TOF) electron spectrometers on different beam lines at the Advanced Light Source at Lawrence Berkeley National Laboratory. The authors point out that there is no reason to think that Xe $4d$ is a special case, and that the behaviour found in this experiment is likely to be quite general.

A 20 channel RRPA calculation for Ra 7s photo-ionization [116] included all relativistic dipole-allowed channels arising from $7s, 6p, 6s, 5d$ and $5p$ subshells. Figure 9.11(a) reveals cross section minima at about 0.4, 1.8 and 5.2 E_h. Agreement between the L and V forms worsens with increasing photon energy as the effects of channels neglected by the model become more important. The cross section below 1 E_h is due to interchannel coupling with the $6p$ channels [116] and the minimum just below 2 E_h and the maximum above it are due to coupling with the $5d$ channels, consistent with the opening of the $6p$ channels at about 1 E_h and the $5d$ channels at about 3 E_h and with the fact that these channels have their maxima near threshold.

The identification of the minima of Figure 9.11(a) as Cooper minima, is supported by the data for the angular asymmetry parameter

$$\beta = \frac{2R_{3/2}^2 + 4R_{1/2}R_{3/2}\cos(\delta_{1/2} - \delta_{3/2})}{R_{1/2}^2 + 2R_{3/2}^2},$$

Figure 9.11(b), where R_j is the dipole matrix element to ϵp_j and δ_j is the associated phase shift. In practice, the cosine is nearly one, and if the R_j are close in value then $\beta \approx 2$, independent of photon energy. The calculated values of β are far from this over much of the energy range and are close to -1 at each of the cross section minima. Although interchannel coupling modifies the picture slightly because the real and imaginary parts of the matrix elements vanish at slightly different energies, the notion of a Cooper minimum is cllearly relevant. Electron scattering from outer s subshells of high-Z elements can be expected to show similar behaviour.

9.12.2 Beyond RRPA

Like its nonrelativistic counterpart, RPAE, the RRPA method neglects important features of the photoionization process such as relaxation of the atomic electron distribution, polarization and accompanying excitation of one or more electrons. Theoretical improvements in nonrelativistic photo-absorption theory, including double photo-ionization processes have been described by Amusia [118]; no relativistic generalizations have appeared so far. The RRPAR

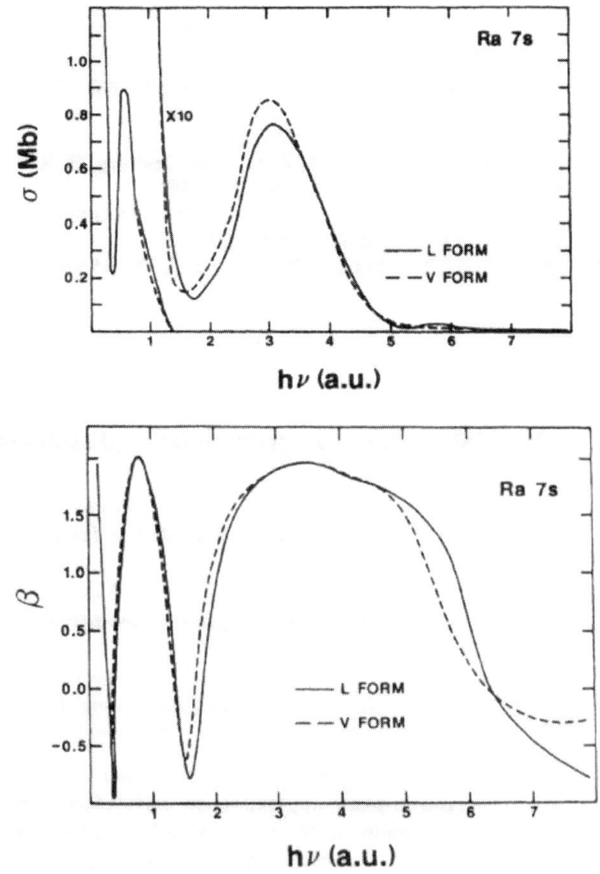

Fig. 9.11. (a) RRPA cross section, (b) RRPA angular distribution parameter β for Ra 7s photo-ionization as a function of photon energy. Length (L) and velocity (V) forms are shown in both graphs. Reprinted with permission from [116]. Copyright 1986 by the American Physical Society.

method provides a simple way to model some of these effects by calculating the perturbed orbitals using the DHF V^{N-1} potential of the relaxed ion rather than a frozen core V^N potential. In this approximation, the dipole matrix elements entering into the various expressions needed are of the form $\langle \Phi_n \,|\, \mathcal{D} \,|\, \Phi_0 \rangle = \gamma \, \langle \phi'_\epsilon \,|\, D \,|\, \phi_i \rangle$ where Φ_0, Φ_n are respectively the ground state and excited state CSFs, the many-particle dipole operator is a sum of single particle dipole operators $\mathcal{D} = \sum_{k=1}^{N} D_k$ and γ is the overlap of the $(N-1)$-electron CSFs using the original and relaxed orbitals. The RRPAR method neglects relaxation in the one-electron matrix elements; it is taken into ac-

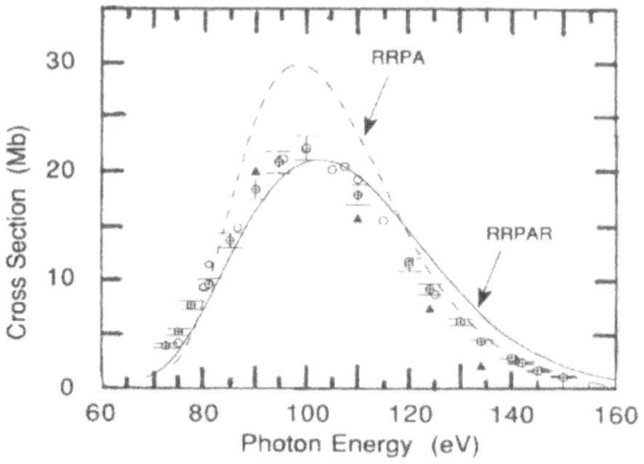

Fig. 9.12. Theoretical and experimental partial cross section for photo-ionization of $4d$ electrons from Xenon. Reprinted with permission from [117]. Copyright 1989 by the American Physical Society.

count in the RRPARA modification [119]. A comparison of the RRPA and RRPAR partial cross sections for the photo-ionization of $4d$ electrons is shown in Figure 9.12. The RRPAR cross sections fit the experiments well for photon energies up to the peak, but the unrelaxed RRPA result is rather better at energies above 120 keV. Intuitively, one expects the relaxation of the core to occur before a low energy photoelectron escapes in the manner envisaged by the RRPAR scheme, but this model is clearly less appropriate at higher photoelectron energies, an interpretation supported by the data of Figure 9.12.

The multichannel multiconfiguration DHF (MMCDF) model [120] has been used, for example, to study multiple excitation near the $5s$ excitation threshold in Xenon [121]. Discrete and continuum orbitals for the final ionic state were optimized with respect to the $5s$-hole configuration keeping previously optimized orbitals frozen. Orthogonality of core and excited orbitals was enforced by using Lagrange multipliers. The final state wave function involved first a CI calculation using 6 CSFs with $J = 1/2$ belonging to the $[5s]$ and $[5p^2]5d$ configurations[10]. The initial- and final-state orbitals and the ionic eigenvectors generated with the aid of the GRASP code provided input to a K-matrix code. Altogether 23 channels resulted when the ionic states were coupled to outgoing $\epsilon p_{1/2,3/2}$ orbitals. The cross sections and angular asymmetry parameters agreed with corresponding RRPA results [122, 123] to within a few per cent. A recent review by Huang *et al.* [124] of the MCRRPA approach [108] considers applications to atomic photoionization.

[10] Square brackets denote hole configurations; e.g. $[5s]$ means a state with one electron removed from the $5s$ subshell

References

1. Walker D W 1971 *Adv. Phys.* **20**, 257. On-line at `http://www.tandf.co.uk/journals`.
2. Grant I P 1970 *Adv. Phys.* **19**, 747.
3. Motz J W, Olsen H and Koch H W 1964 *Rev. Mod. Phys.* **36**, 881.
4. Kessler J 1969 *Rev. Mod. Phys.* **41**, 3.
5. Kessler J 1985 *Polarized Electrons* (2nd edition) (Berlin: Springer-Verlag).
6. Baylis W E and Sienkiewicz J E 1995 *J. Phys. B: At. Mol. Opt. Phys.* **28**, L549.
7. Sienkiewicz J E 1997 *J. Phys. B: At. Mol. Opt. Phys.* **30**, 1261.
8. Sienkiewicz J E 1990 *J. Phys. B: At. Mol. Opt. Phys.* **23**, 1869.
9. Kolb D, Johnson W R and Shorer P 1982 *Phys. Rev A* **26**, 19.
10. Sienkiewicz J E and Baylis W E 1987 *J. Phys. B: At. Mol. Phys.* **20**, 5145.
11. Ali M A and Fraser P A 1977 *J. Phys. B: At. Mol. Phys.* **10**, 3091.
12. Zubek M, Danjo A and King G C 1995 *J. Phys. B: At. Mol. Opt. Phys.* **28** 4117.
13. Sin Fai Lam L T 1980 *Aust. J. Phys.* **33**, 261.
14. Temkin A 1957 *Phys. Rev.* **107**, 1004.
15. Drachman R J and Temkin A 1972 *Case Studies in Atomic Collision Physics* vol. 2, ed. E W McDaniel and M R C Mc Dowell, p. 399 (Amsterdam: North Holland).
16. Szmytkowski R and Sienkiewicz J E 1994 *Phys. Rev A* **50**,4007.
17. Szmytkowski R 1991 *J. Phys. B: At. Mol. Opt. Phys.* **24**, 3895.
18. Szmytkowski R 1993 *J. Phys. B: At. Mol. Opt. Phys.* **26**, 535.
19. Sienkiewicz J E, Fritzsche S and Grant I P 1995 *J. Phys. B: At. Mol. Opt. Phys.* **28**, L633.
20. Sienkiewicz J E and Baylis W E 1997 *Phys. Rev A* **55**, 1108.
21. Salvat F, Jablonski A and Powell C J 2005 *Comput. Phys. Commun.* **165**, 157.
22. Szmytkowski R 1995 *Comput. Phys. Commun.* **90**, 244.
23. Clenshaw C W 1962 in *Mathematical Tables*, Vol. 5. (London: HMSO).
24. Press W H, Teukolsky S A, Vetterling W T and Flannery B P 1992 *Numerical Recipes in FORTRAN* (2e) (Cambridge: Cambridge University Press).
25. Burke P G 1996 in *Atomic, Molecular and Optical Physics Handbook* (ed. G W F Drake) Chapter 45. (Woodbury NY: American Institute of Physics)
26. Seaton M J 1983 *Rep. Prog. Phys.* **46**, 167.
27. Hagelstein P L and Jung R K 1986 *At. Data. Nucl. Data Tables* **37**, 121.
28. Wigner E P 1946 *Phys. Rev* **70**,15; Wigner E P 1946 *Phys. Rev* **70**, 606.
29. Wigner E P and Eisenbud L 1947 *Phys. Rev* **72**, 29.
30. Lane A M and Thomas R G 1958 *Rev. Mod. Phys.* **30**, 252.
31. Breit G 1959 in *Encyclopedia of Physics* Vol 41/1 (ed. S Flugge), p. 107 (Berlin: Springer-Verlag)
32. Barrett R F, Biedenharn L C, Danos M, Delsanto R P, Greiner W and Wahsweiler H G 1973 *Rev. Mod. Phys.* **45**, 44.
33. Burke P G and Robb W D 1975 *Adv. Atom. Molec. Phys.* **11**, 143.
34. Burke P G and Berrington K A 1993 *Atomic and Molecular Processes: an R-matrix approach* (Bristol: Institute of Physics).
35. Burke P G 1998 in *Many-body atomic physics* (ed. J J Boyle and M S Pindzola) pp. 305 – 324 (Cambridge University Press).
36. Compiled by the Opacity Project Team 1995 *The Opacity Project* Vol. 1. (Bristol: Institute of Physics Publishing)

37. Scott N S and Burke P G 1980 *J. Phys. B: At. Mol. Phys.* **13**, 4299.
38. Scott N S and Taylor K T 1982 *Comput. Phys. Commun.* **25**, 347.
39. Goertzel G 1948 *Phys. Rev.* **73**, 1463.
40. Chang J J 1975 *J. Phys. B: At. Mol. Phys.* **8**, 2327.
41. Chang J J 1977 *J. Phys. B: At. Mol. Phys.* **10**, 3335.
42. Norrington P H and Grant I P 1981 *J. Phys. B: At. Mol. Phys.* **14**, L261.
43. Thumm U and Norcross V W 1992 *Phys. Rev. A* **45**, 6349.
44. Thumm U and Norcross V W 1993 *Phys. Rev. A* **47**, 305.
45. Bloch C 1957 *Nucl. Phys.* **4**, 503. (Reprinted in [34], pp.113.)
46. Burke P G and Taylor K T 1975 *J. Phys. B: At. Mol. Phys.* **8**, 2620.
47. Wijesundera W J, Grant I P and Norrington P H 1998 in *Many-body atomic physics* (ed. J J Boyle and M S Pindzola) pp. 325 – 358 (Cambridge University Press).
48. Berrington K A, Burke P G, Butler K, Seaton M J, Storey P J and Yu Yan 1987 *J. Phys. B: At. Mol. Phys.* **20**, 6379; reprinted in [36, pp. 97 – 115].
49. Seaton M J 1985 *J. Phys. B: At. Mol. Phys.* **18**, 2111.
50. Seaton M J 1987 *J. Phys. B: At. Mol. Phys.* **20**, 6363.
51. Burke P G and Seaton M J 1984 *J. Phys. B: At. Mol. Phys.* **17**, L683.
52. Buttle P J A 1967 *Phys. Rev.* **160**, 712.
53. Seaton M J 1987 *J. Phys. B: At. Mol. Phys.* **20**, L69.
54. Norrington P H maintains the website `http://web.am.qub.ac.uk/DARC/`
55. Berrington K A, Eissner W and Norrington P H 1995 *Comput. Phys. Commun.* **92**, 290.
56. Bell K L and Scott N S 1980 *Comput. Phys. Commun.* **20**, 447.
57. Noble C J and Nesbet R K, 1984 *Comput. Phys. Commun.* **33**, 399.
58. Barnett A R, 1982 *Comput. Phys. Commun.* **27**, 147.
59. Crees M A, 1980 *Comput. Phys. Commun.* **19**, 103.
60. Crees M A, 1981 *Comput. Phys. Commun.* **23**, 181.
61. Croskery J P, Scott N S, Bell K L and Berrington K A, 1982 *Comput. Phys. Commun.* **27** 385.
62. Young I G and Norrington P H, 1994 *Comput. Phys. Commun.* **83**, 215.
63. ITER Physics Basis 1999 *Nucl. Fusion* **39**, 2137.
64. Badnell N R, Berrington K A, Summers H P, O'Mullane M G, Whiteford A D and Ballance C P 2004 *J. Phys. B: At. Mol. Opt. Phys.* **37**, 4589.
65. Mitnik D M, Griffin D C, Ballance C P and Badnell N R 2003 *J. Phys. B: At. Mol. Opt. Phys.* **36**, 717.
66. Castillejo L, Percival I C and Seaton M J 1960 *Proc. Roy. Soc. A* **254**, 259.
67. McCarthy I E and Shang B 1992 *Phys. Rev.* **46**, 3958.
68. Odgers B R, Scott M P and Burke P G 1994 *J. Phys. B: At. Mol. Opt. Phys.* **28**, 1973.
69. Bray I and Stelbovics A T 1992 *Phys. Rev. A* **46**, 6995.
70. Bray I 1994 *Phys. Rev. A* **49**, 1066.
71. Yamani H A and Reinhhart W P 1975 *Phys. Rev. A* **11**, 1144.
72. Stelbovics A T 1990 *J. Phys. B: At. Mol. Opt. Phys.* **22**, L159.
73. Bray I and Stelbovics A T 1995 *Adv. At. Mol. Phys.* **35**, 209.
74. Fursa D V and Bray I 1997 *J. Phys. B: At. Mol. Opt. Phys.* **30**, 5895.
75. Bartschat K, Hudson E T, Scott M P, Burke P G and Burke V M 1996 *J. Phys. B: At. Mol. Opt. Phys.* **29**, 115.
76. Burke P G, Noble C J and Scott P 1987 *Proc. Roy. Soc. A* **410**, 289.

77. Light J C and Walker R B 1976 *J. Chem. Phys.* **65**. 4272.
78. Burke V M and Noble C J 1995 *Comput. Phys. Commun.* **85**, 471.
79. Baluja K L, Burke P G and Morgan L A 1982 *Comput. Phys. Commun.* **27**, 299.
80. Bartschat K, Burke P G and Scott M P 1997 *J. Phys. B: At. Mol. Opt. Phys.* **30**, 5915.
81. Fon W C, Berrington K A, Burke P G, Burke V M and Hibbert A 1992 *J. Phys. B: At. Mol. Opt. Phys.* **25**, 507.
82. Clark R E H and Abdallah J 1996 *Phys. Scr. T* **62**, 7.
83. Fursa D V, Bray I and Lister G 2003 *J. Phys. B: At. Mol. Opt. Phys.* **36**, 4255.
84. McEachran R P and Elford M T 2003 *J. Phys. B: At. Mol. Opt. Phys.* **36**, 427.
85. Panajotović R, Pejčev V, Konstatinović M, Filipović D, Bocčvarski V and Marinković B 1993 *J. Phys. B: At. Mol. Opt. Phys.* **26**, 1005.
86. Holtkamp G, Jost K, Peitzmann F J and Kessler J 1987 *J. Phys. B: At. Mol. Opt. Phys.* **20**, 4543.
87. Rockwood S D 1973 *Phys. Rev. A* **8**, 2348.
88. Panajotović R, Pejčev V, Konstantinović M, Filipović D, Bočvarski V and Marinković B 1993 *J. Phys. B: At. Mol. Opt. Phys.* **26**, 1005.
89. Zubek M, Gulley N, Danjo A and King G C 1996 *J. Phys. B: At. Mol. Opt. Phys.* **29**, 5927.
90. Wijesundera W J, Grant I P and Norrington P H 1992 *J. Phys. B: At. Mol. Opt. Phys.* **25**, 2143.
91. Bartschat K and Madison D H 1987 *J. Phys. B: At. Mol. Phys.* **20**, 1609.
92. Srivastava R, Zuo T, McEachran R P and Stauffer A D 1993 *J. Phys. B: At. Mol. Opt. Phys.* **26**, 1025.
93. Bartschat K and Burke P G 1986 *J. Phys. B: At. Mol. Phys.* **19**, 1231.
94. Lorentz S R, Scholten R E, McClelland J J, Kelley M H and Celotta R J 1991 *Phys. Rev. Lett.* **67**, 3761.
95. McClelland J J, Lorentz S R, Scholten R E, Kelley M H and Celotta R J 1991 *Phys. Rev.A* **46**, 6079.
96. Baum G, Moede M, Raith W and Sillmen U 1986 *Phys. Rev. Lett.* **57**, 1855.
97. Bartschat K and Bray I 1996 *Phys. Rev. A* **54**, 1723.
98. Bartschat K 1993 *J. Phys. B: At. Mol. Opt. Phys.* **26**, 3695.
99. Ait-Tahar S, Grant I P and Norrington P H 1997 *Phys. Rev. Lett.* **79**, 2955.
100. Baum G, Raith W, Roth B, Tondera M, Bartschat K, Bray I, Ait-Tahar S, Grant I P and Norrington P H 1999 *Phys. Rev. Lett.* **82**, 1128.
101. Burke P G and Mitchell J F B 1974 *J. Phys. B: At. Mol. Phys.* **7**, 214.
102. Farago P S 1974 *J. Phys. B: At. Mol. Phys.* **7**, L28.
103. Bederson B 1969 *Comments At. Mol. Phys.* **1**, 41, 65.
104. Gehenn W and Reichert F 1977 *J. Phys. B: At. Mol. Phys.* **10**, 3105.
105. Johnson W R and Lin C D 1977 *J. Phys. B: At. Mol. Phys.* **10**, 331.
106. Johnson W R, Lin C D, Cheng K T and Lee C M 1980 *Physics Scripta* **21**, 409.
107. Johnson W R 1988 *Adv. Atom. Molec. Phys.* **25**, 375.
108. Huang K-N and Johnson W R 1982 *Phys. Rev. A* **25**, 634.
109. Akhiezer A I and Berestetskii V B 1965 *Quantum Electrodynamics* (New York: Interscience Publishers).
110. Walker T E H and Waber J M 1973 *J. Phys. B: At. Mol. Phys.* **6**, 1165.

111. Müller A, Phaneuf R A, Aguilar A, Gharaibeh M F, Schlachter A S, Alvarez I, Cisneros C, Hinojosa G and McLaughlin B M 2002 *J. Phys. B: At. Mol. Opt. Phys.* **35**, L137.
112. Schippers S, Müller A, McLaughlin B M, Aguilar A, Cisneros C, Emmons E D, Gharaibeh M F and Phaneuf R A 2003 *J. Phys. B: At. Mol. Opt. Phys.* **36**, 3371.
113. Cooper J W 1962 *Phys. Rev.* **128**, 681.
114. Kim Y S, Ron A, Pratt R H, Tambe B R and Manson S T 1981 *Phys. Rev. Lett.* **46**, 1326.
115. Wang H, Snell G, Hemmers O, Sant'Anna M M, Sellin I, Berrah N, Lindle D W, Desmukh P C, Haque N amd Manson S T 2001 *Phys. Rev. Lett.* **87**, 123004.
116. Desmukh P C, Radojevic V and Manson S T 1986 *Phys. Rev. A* **34**, 5162.
117. Kutzner M, Radojević V and Kelly H P 1989 *Phys. Rev. A* **40**, 5052.
118. Amusia M Ya 1990 *Atomic Photoeffect* (New York: Plenum Press).
119. Kutzner M, Shamblin Q, Vance S E and Winn D 1997 *Phys. Rev. A* **55**, 248.
120. Tulkki J and Aberg T 1985 *J. Phys B: At. Mol. Phys.* **18**, L489.
121. Tulkki J 1989 *Phys. Rev. Lett.* **62**, 2817.
122. Johnson W R and Cheng K T 1978 *Phys. Rev. Lett.* **40**, 1167.
123. Desmukh P C and Manson S T 1985 *Phys. Rev. A* **32**, 3109.
124. Huang K-N, Chi H-C and Chou H-S 1995 *Chinese J. Phys.* **33**, 565.

10

Molecular structure methods

10.1 Molecular and atomic structure methods

The nuclei in a molecule provide a skeleton that has usually no dominant centre of spherical symmetry, so that much of the elaborate technology of atomic theory developed in earlier chapters loses its relevance.. The history of the relativistic theory of molecular electronic structure over the past 20 years has been both turbulent and confusing. This chapter presents an account of the Dirac theory of molecular (and atomic) structure in terms of finite matrix methods. Just as nonrelativistic quantum chemistry is mostly built on the use of atomic and molecular orbitals constructed from Gaussian-type orbitals (GTOs), relativistic quantum chemistry makes use of their relativistic analogues: G-spinors. The straightforward similarity of the formulation may surprise those familiar with the conventional treatment of relativistic molecular structure in terms of a confusing plethora of relativistic corrections. The perturbation operators generating these corrections can be quite difficult to handle, and their complexity can obscure rather than illuminate the underlying physics.

Nonrelativistic molecular structure calculations start from a quantum mechanical Hamiltonian for an assembly of N electrons and a number of massive charged structureless nuclei, charges $Z_A, Z_B, \ldots$ at positions $\boldsymbol{A}, \boldsymbol{B}, \ldots$:

$$H = H_N + h_e + V_{eN}$$

where, with atomic units,

$$H_N = \sum_A \frac{1}{2M_A}\boldsymbol{p}_A^2 + \sum_{A<B} \frac{Z_A Z_B}{|\boldsymbol{A}-\boldsymbol{B}|}$$

is the nonrelativistic nuclear Hamiltonian,

$$h_e = \sum_1^N \frac{1}{2m}\boldsymbol{p}_i^2 + \sum_{i<j} \frac{1}{|\boldsymbol{r}_i - \boldsymbol{r}_j|}$$

is the nonrelativistic Hamiltonian for the electrons, including their Coulomb repulsion energy, and

$$V_{eN} = -\sum_A \sum_i \frac{1}{|\boldsymbol{r}_i - \boldsymbol{A}|}$$

is the electron-nuclear interaction energy. This translates into a Schrödinger equation in the configuration space of $N_A + N_e$ particles.

The difficulty of formulating a consistent relativistic theory of the atom including nuclear motion was discussed in §8.8, and this becomes even more intractable for molecules and condensed matter. Each nucleus has an internal structure whose effect on electronic motions is small but is not entirely negligible. How are we to model this? A "first principles" approach might use nuclear structure theory to describe each nucleus as a collection of nucleons whose states are determined by strong nuclear forces. This presents a formidable problem that we may try to avoid by treating the nuclei as relativistic particles with mass, momentum, charge, and electromagnetic moments. Whilst this has been attempted for simple systems like the hydrogen atom [3], it seems far too complicated for diatomic and polyatomic molecules, for which the nonrelativistic theory of nuclear motion used in molecular spectroscopy is already complicated enough [4]. In practice, there is little alternative but to treat the nuclear dynamics in the conventional nonrelativistic manner. We shall therefore focus first on the Born-Oppenheimer model: the nuclei have fixed positions, and we make no attempt to deal with nuclear motion. The nuclei will be treated as having finite spherical charge densities. Electric and magnetic multipole moments generate hyperfine interactions that we may incorporate as perturbations. As in §4.14, we treat electron dynamics using the Dirac operator, taking the relativistic electron-electron interaction in the Coulomb gauge. This means that the *unquantized* electronic Hamiltonian is, (4.14.1),

$$H = \sum_i h_{Di} + \sum_{i<j} \left(\frac{1}{|\boldsymbol{r}_i - \boldsymbol{r}_j|} + g^B(|\boldsymbol{r}_i - \boldsymbol{r}_j|) \right) + \frac{1}{2} \sum_{A \neq B} \frac{Z_A Z_B}{|\boldsymbol{A} - \boldsymbol{B}|} \tag{10.1.1}$$

where

$$h_{Di} = c\,\boldsymbol{\alpha}_i \cdot \boldsymbol{p}_i + \beta_i c^2 - \sum_A V_A(|\boldsymbol{r}_i - \boldsymbol{A}|)$$

is the Dirac Hamiltonian for the i-th electron in the presence of the bare nuclei and the second summation is the electron-electron interaction potential in Coulomb gauge. The last term is the Coulomb repulsion energy of the nuclei, which is constant in the Born-Oppenheimer approximation, but which must be included if we wish to calculate the molecular potential energy surface, which will enable us to model the nuclear motion adiabatically.

The electron-nuclear interaction, V_A, depends on the choice of nuclear model. We shall use a Gaussian density distribution for convenience; a wide range of more elaborate nuclear models for quantum chemistry is discussed, for example, by Andrae [5].

where the coefficients $c^{\beta}_{\mu A}$ can be assumed *real*. The normalized 2-spinors $M[\beta, \mu, \boldsymbol{x}]$, where $\beta = +1$ for upper and $\beta = -1$ for lower components[1], can be written as in (5.7.10)

$$M[\beta, \mu, \boldsymbol{x}] = N^{\beta}_{\mu}\, m[\beta, \mu, \boldsymbol{x}] = \frac{\mathrm{g}^{\beta}_{\mu}(r)}{r}\, \chi_{\kappa,m}(\theta, \varphi) \tag{10.2.6}$$

where $\mathrm{g}^{\beta}_{\mu}(r)$ is the *normalized radial basis function*

$$\mathrm{g}^{\beta}_{\mu}(r) = N^{\beta}_{\mu} g^{\beta}_{\mu}(r).$$

We postpone consideration of specific functional forms for $\mathrm{g}^{\beta}_{\mu}(r)$ to §10.3.

10.2.3 Matrix of the radial Dirac operator

The radial Dirac operator commutes with the components of the angular momentum operator and parity, and its matrix is therefore block-diagonal with respect to κ. Each such $2N_\kappa \times 2N_\kappa$ block can in turn be decomposed into four $N_\kappa \times N_\kappa$ blocks labelled by the component indices β and β'. Substituting (10.2.5) in (6.3.9) gives

$$I(A, B) = \sum_{\beta,\beta'} \sum_{\mu\nu} c^{\beta}_{\mu A}\, h^{\beta,\beta'}_{\mu\nu}\, c^{\beta'}_{\nu B} = \sum_{\beta,\beta'} \mathrm{tr}\, \left(\boldsymbol{c}^{\beta\dagger}_{A}\, \boldsymbol{h}^{\beta,\beta'}_{\kappa}\, \boldsymbol{c}^{\beta'}_{B} \right) \tag{10.2.7}$$

where the trace is over the indices μ, ν, and $\kappa_A = \kappa_B$. The Hermitian matrix $\boldsymbol{h}^{\beta,\beta'}_{\kappa}$ has four $N_\kappa \times N_\kappa$ blocks

$$h^{\beta,\beta}_{\mu\nu} = V^{\beta,\beta}_{\mu\nu} + (\beta - 1)mc^2 S^{\beta,\beta}_{\mu\nu}, \quad h^{\beta,-\beta}_{\mu\nu} = c\Pi^{\beta,-\beta}_{\mu\nu}, \quad \beta = \pm 1, \tag{10.2.8}$$

with matrix elements

$$\begin{aligned} V^{\beta,\beta}_{\mu\nu} &= \int_0^\infty \mathrm{g}^{\beta *}_{\mu}(r) v_{nuc}(r) \mathrm{g}^{\beta}_{\nu}(r)\, dr \\ S^{\beta,\beta}_{\mu\nu} &= \int_0^\infty \mathrm{g}^{\beta *}_{\mu}(r) \mathrm{g}^{\beta}_{\nu}(r)\, dr \\ \Pi^{\beta,-\beta}_{\mu\nu} &= \int_0^\infty \mathrm{g}^{\beta}_{\mu}(r) \left(-\beta \frac{d}{dr} + \frac{\kappa}{r} \right) \mathrm{g}^{-\beta}_{\nu}(r)\, dr \end{aligned} \tag{10.2.9}$$

as in the one-electron case.

10.2.4 Coulomb Slater integrals

The two-electron integrals in (10.2.1) are special cases of the general Slater integral, (6.4.8), which can be expanded as

[1] The indices $\beta = +1$ and $\beta = -1$ are equivalent to the labels L and S respectively.

$$R^k_C(ABCD) = \sum_{\beta,\beta'} \sum_{\mu\nu\sigma\tau} c^{\beta}_{\mu A}\, c^{\beta}_{\nu C} \cdot c^{\beta'}_{\sigma B}\, c^{\beta'}_{\tau D} \left(\mu^{\beta}\nu^{\beta} \,\middle|\, \sigma^{\beta'}\tau^{\beta'}\right)^k \tag{10.2.10}$$

where

$$\left(\mu^{\beta}\nu^{\beta'} \,\middle|\, \sigma^{\beta''}\tau^{\beta'''}\right)^k = \int_0^\infty \int_0^\infty \mathrm{g}^{\beta}_{\mu}(r)\mathrm{g}^{\beta'}_{\nu}(r)\, U_k(r,s)\, \mathrm{g}^{\beta''}_{\sigma}(s)\mathrm{g}^{\beta'''}_{\tau}(s)\, dr ds \tag{10.2.11}$$

from which

$$F^0(A,B) = \sum_{\beta\beta'} \sum_{\mu\nu\sigma\tau} c^{\beta}_{\mu A}\, c^{\beta}_{\nu A} \cdot c^{\beta'}_{\sigma B}\, c^{\beta'}_{\tau B} \left(\mu^{\beta}\nu^{\beta} \,\middle|\, \sigma^{\beta'}\tau^{\beta'}\right)^0 \tag{10.2.12}$$

and

$$G^k(A,B) = \sum_{\beta\beta'} \sum_{\mu\nu\sigma\tau} c^{\beta}_{\mu A}\, c^{\beta}_{\nu A} \cdot c^{\beta'}_{\sigma B}\, c^{\beta'}_{\tau B} \left(\mu^{\beta}\sigma^{\beta'} \,\middle|\, \nu^{\beta}\tau^{\beta'}\right)^k . \tag{10.2.13}$$

10.2.5 Breit integrals for closed shells

The magnetic interaction matrix $\mathbf{B}_\kappa$ can be treated in the same way. The Breit interaction (10.2.2) between a pair of closed subshells is [6, Equations (28), (29)]

$$(-1)^{j_A - j_B + k} X^k(ABBA) = [j_A, j_B] \sum_{k'=k-1}^{k+1} \sum_{\gamma=-1}^{+1} x^{k'k}_{\gamma}(AB)\, G^{k'}_{\gamma}(AB), \tag{10.2.14}$$

where

$$\begin{aligned} G^{k'}_{+1}(AB) &= \sum_{\mu\nu\sigma\tau} c^{+}_{\mu A}\, c^{-}_{\nu B} \cdot c^{+}_{\sigma A}\, c^{-}_{\tau B} \left(\mu^{+}_{A}\nu^{-}_{B} \,\middle|\, \sigma^{+}_{A}\tau^{-}_{B}\right)^{k'} \\ G^{k'}_{0}(AB) &= \sum_{\mu\nu\sigma\tau} c^{+}_{\mu A}\, c^{-}_{\nu B} \cdot c^{+}_{\sigma B}\, c^{-}_{\tau A} \left(\mu^{+}_{A}\nu^{-}_{B} \,\middle|\, \sigma^{+}_{B}\tau^{-}_{A}\right)^{k'} \\ G^{k'}_{-1}(AB) &= \sum_{\mu\nu\sigma\tau} c^{+}_{\mu B}\, c^{-}_{\nu A} \cdot c^{+}_{\sigma B}\, c^{-}_{\tau A} \left(\mu^{+}_{B}\nu^{-}_{A} \,\middle|\, \sigma^{+}_{B}\tau^{-}_{A}\right)^{k'} \end{aligned} \tag{10.2.15}$$

with coefficients

as in (10.2.7) and (10.2.8), and

$$\boldsymbol{J}_\kappa = \begin{pmatrix} \boldsymbol{J}_\kappa^{++} & 0 \\ 0 & \boldsymbol{J}_\kappa^{--} \end{pmatrix}, \quad \boldsymbol{K}_\kappa = \begin{pmatrix} \boldsymbol{K}_\kappa^{++} & \boldsymbol{K}_\kappa^{+-} \\ \boldsymbol{K}_\kappa^{-+} & \boldsymbol{K}_\kappa^{--} \end{pmatrix},$$

The Breit matrix $\boldsymbol{B}_\kappa$ has a similar structure to the exchange Coulomb matrix $\boldsymbol{K}_\kappa$. The DHFB equations for each symmetry,

$$\boldsymbol{F}_\kappa \boldsymbol{c}_K = \boldsymbol{\epsilon}_\kappa \boldsymbol{S}_\kappa \boldsymbol{c}_K, \tag{10.2.20}$$

are coupled to other symmetries through the matrices $\boldsymbol{J}_\kappa$, $\boldsymbol{K}_\kappa$, and $\boldsymbol{B}_\kappa$. We shall discuss methods of solution below. The *direct* matrices $\mathbf{J}_\kappa^{\beta\beta}$ and the *exchange* matrices $\mathbf{K}_\kappa^{\beta\beta'}$ are built respectively from the integrals

$$J^{0,\,\beta\beta,\,\beta'\beta'}_{\kappa\mu\nu,\kappa'\sigma\tau} = \left(\mu^\beta\nu^\beta \,\middle|\, \sigma^{\beta'}\tau^{\beta'}\right)^0, \quad K^{k,\,\beta\beta',\,\beta\beta'}_{\kappa\mu\sigma,\kappa'\nu\tau} = \left(\mu^\beta\sigma^{\beta'} \,\middle|\, \nu^\beta\tau^{\beta'}\right)^k.$$

Their matrix elements are

$$J^{\beta\beta}_{\kappa\mu\nu} = \sum_{\kappa'\sigma\tau}\sum_{\beta'} J^{0,\beta\beta,\beta'\beta'}_{\kappa\mu\nu,\kappa'\sigma\tau} D^{\beta'\beta'}_{\kappa'\sigma\tau}, \tag{10.2.21}$$

and

$$K^{\beta\beta'}_{\kappa\nu\sigma} = \sum_k b^k_{\kappa,\kappa'} \sum_{\kappa'\sigma\tau} K^{k,\,\beta\beta',\,\beta\beta'}_{\kappa\mu\nu,\kappa'\sigma\tau} D^{\beta\beta'}_{\kappa'\sigma\tau}. \tag{10.2.22}$$

The Breit matrix elements derived from (10.2.14) can be written[2]

$$B^{\beta\beta}_{\kappa\mu\nu} = \sum_k \sum_{\kappa'\sigma\tau} e^{\beta\beta}_k(\kappa,\kappa')\, D^{-\beta\,-\beta}_{\kappa'\sigma\tau}\, K^{k,\beta\beta,-\beta\,-\beta}_{\kappa\mu\nu,\,\kappa'\sigma\tau}, \tag{10.2.23}$$

$$B^{\beta\,-\beta}_{\kappa\mu\nu} = \sum_k \sum_{\kappa'\sigma\tau} f_k(\kappa,\kappa')\, D^{-\beta\,\beta}_{\kappa'\sigma\tau}\, K^{k,\beta\,-\beta,-\beta\,\beta}_{\kappa\mu\nu,\,\kappa'\sigma\tau}. \tag{10.2.24}$$

Numerical values of the coefficients $e^{\beta\beta}_k(\kappa,\kappa')$, $f_k(\kappa,\kappa')$ for s- and p-orbitals have been tabulated in [9, Table I].

10.3 One-centre interaction integrals

The construction of the atomic $\boldsymbol{h}_\kappa$ matrices with S-spinors and G-spinors was presented in §5.9 and §5.10 respectively. This subsection discusses the evaluation of the interaction integrals $(\mu^\beta, \nu^{\beta'} \,|\, \sigma^{\beta''}, \tau^{\beta'''})^k$ defined in (10.2.11). These can all be represented as linear combinations of integrals of the form

[2] Compare [9, Equations (27a-c)]; a factor $(2j'+1)$ has been absorbed here in the density matrix elements. The last term of [9, Equation (27b)], which vanishes identically, has here been omitted.

$$(\mu, \nu \,|\, \sigma, \tau)^k = \iint s_\mu(r) s_\nu(r)\, U_k(r,s)\, s_\sigma(s) s_\tau(s)\, dr ds,$$

where $s_\mu(r)$ denotes the monomial

$$s_\mu(r) = \begin{cases} r^{\gamma_\mu + 1}\, \mathrm{e}^{-\lambda_\mu r} & \text{(S-spinors)}, \\ r^{l_\mu + 1}\, \mathrm{e}^{\mathrm{d} - \lambda_\mu r^2} & \text{(G-spinors)}, \end{cases} \tag{10.3.1}$$

We first decompose these integrals into two parts,

$$(\mu, \nu \,|\, \sigma, \tau)^k = S^k(\mu\nu \,|\, \sigma\tau) + S^k(\sigma\tau \,|\, \mu\nu), \tag{10.3.2}$$

where

$$S^k(\mu\nu \,|\, \sigma\tau) = \int_0^\infty dr \int_r^\infty ds\; s_\mu(r) s_\nu(r)\, \frac{r^k}{s^{k+1}}\, s_\sigma(s) s_\tau(s).$$

In the S-spinor case, we put [10, p. 177]

$$a = \gamma_\sigma + \gamma_\tau + 2 - k, \quad b = \gamma_\mu + \gamma_\nu + 3 + k, \quad A = \lambda_\sigma + \lambda_\tau, \quad B = \lambda_\mu + \lambda_\nu$$

and transform the integral into

$$S^k(\mu\nu \,|\, \sigma\tau) = \Gamma(a)\, \Gamma(b)\, A^{-a}\, B^{-b}\, I_x(a,b) \quad (S) \tag{10.3.3}$$

where the label (S) signifies this equation refers to S-spinors,

$$I_x(a,b) = \frac{1}{B(a,b)} \int_0^x t^{a-1}(1-t)^{b-1} dt, \quad x = \frac{A}{A+B}, \quad 0 \le x \le 1,$$

is an incomplete Beta function defined in [11, (6.6.2)],

$$B(a,b) = \Gamma(a)\Gamma(b)/\Gamma(a+b),$$

and the Gamma functions $\Gamma(z)$ are defined in [11, (6.1.1)] or (5.9.3). Thus

$$(\mu, \nu \,|\, \sigma, \tau)^k = \Gamma(a)\, \Gamma(b)\, A^{-a}\, B^{-b}\, (I_x(a,b) + I_{1-x}(b,a)) \quad (S) \tag{10.3.4}$$

The G-spinor case is very similar; one first rewrites the integral in terms of $r' = r^2$ and $s' = s^2$ to put it into the required form, after which we find, [8, Equation (180)],

$$(\mu, \nu \,|\, \sigma, \tau)^k = \frac{1}{4} \Gamma(a)\, \Gamma(b)\, A^{-a}\, B^{-b}\, (I_x(a,b) + I_{1-x}(b,a)) \quad (G) \tag{10.3.5}$$

with the label (G) signifying G-spinors,

$$a = (l_\sigma + l_\tau + 2 - k)/2, \quad b = (l_\mu + l_\nu + 3 + k)/2,$$

and $A = \lambda_\sigma + \lambda_\tau$ and $B = \lambda_\mu + \lambda_\nu$ are the same as in the S-spinor case. These formulae allow us to evaluate all the closed shell interaction integrals we have encountered so far. An efficient recurrence scheme [10] for evaluating very large numbers of the normalized incomplete beta functions $I_x(a,b)$ will be found in Appendix A.3.6.

10.4 Numerical examples

Ishikawa, Quiney, and Malli [9] have studied the performance of G-spinor matrix DHF and DHFB models of a number of closed shell atoms: the rare gases He, Ne, Ar, Kr, Xe, and the alkaline-earth atoms Be, Mg, Ca, and Sr. All the calculations used even-tempered [12] or well-tempered exponents [13]; small- and medium-size exponent sets [14] were also used for Ne and Ar. A comparison of total energies for Ar obtained in this way as the basis was refined is shown in Table 10.1. The Breit interaction, here calculated as the difference of the two SCF energies, raises the energy by the amount shown in

Table 10.1. DHFB and DHF total energies (E_h) for Ar: dependence on basis set. Reprinted with permission from [9, Table IV]. Copyright 1991 American Physical Society.

Basis	E_{DHFB}	E_{DHF}	E_{Breit}
$10s\,7p$ G	-527.988 940 0	-528.120 319 3	+0.131 379 3
$14s\,10p$ G	-528.534 308 4	-528.666 604 0	+0.132 295 6
$16s\,11p$ G	-528.549 164 3	-528.681 481 6	+0.132 317 3
$17s\,13p$ G	-528.550 723 2	-528.683 044 8	+0.132 321 6
$17s\,14p$ G	-528.550 998 6	-528.683 321 3	+0.132 322 7
$17s\,15p$ G	-528.551 037 8	-528.683 360 6	+0.132 322 8
$27s\,22p$ G	-528.551 446 4	-528.683 769 4	+0.132 323 0
$28s\,23p$ G	-528.551 476 0	-528.683 799 0	+0.132 323 0
GRASP(finite)		-528.683 84	
$17s\,17p$ S	-528.552 124 9	-528.684 450 5	+0.132 365 3
GRASP(point)		-528.684 450 1	

the last column. This difference stabilizes more rapidly than the total energy. The line labelled GRASP(finite) gives the total energy from a finite difference DHF calculation for the same nuclear model.

The G-spinor calculations used a finite nucleus [9] and an old value, $c =$ 137.037 3, of the speed of light. Similar calculations [15] were earlier performed for a point nucleus, with a more recent value of the speed of light, c=137.035 989 5, and up to even-tempered $17s, 17p$ S-spinors. The results, at the bottom of the table, were of comparable quality. Both DHF and DHFB S-spinor energies with a point nucleus were about 0.65 mE_h lower than the corresponding G-spinor finite nucleus calculations, in good agreement with the difference in the corresponding GRASP calculations. The perturbation calculation of the Breit energy gave a total energy -528.552 085 19 E_h. This is some 40 μE_h higher than the variational value; the difference is likely to be due mainly to higher order effects not accounted for in the perturbation calculation.

The variational treatment of the Breit interaction reduces the size of the energy eigenvalues and dilates the orbitals, by about 1 part in 10^4 in the case of Ar [15]. This small effect is consistent with the close agreement of the variational and perturbative values of E_{Breit} in Ar. Quiney *et al.* [15] explored the second order correlation energy in Ar in the NVP scheme, ignoring the even smaller contribution of negative energy states. The total nonrelativistic second-order energy at the f-limit was -638 680 μE_h, to be compared with -639 424 μE_h and -646 206 μE_h when using DHF and DHFB orbitals respectively. The corresponding shifts were - 744 μE_h for DHF and -7526 μE_h for DHFB. Although much of the relativistic shift is due to the inner core electrons, in particular the $1s^2$ pair energy, the remainder comes from subtle interactions between the core and valence electrons. The surprisingly large shift due to the use of DHFB orbitals indicates that that there may be much to be gained by adopting the DHFB solution as a starting point for more accurate calculations.

The DHF and DHFB orbital eigenvalues of Xe (Z=54) using the 23s 21p 14d basis set of Huzinaga [13] are shown in Table 10.2. The Breit interaction destabilizes all orbitals by amounts ranging from 3 E_h in the case of $1s_{1/2}$ down to $\sim 10^{-4}$ E_h in the valence shell, $5p$. This is a measure of the self-consistent magnetic repulsion of the electron current distributions. The DHFB approximation provides an economical starting point for more accurate calculations using the open-shell coupled cluster method [16].

Table 10.2. DHF and DHFB orbital eigenvalues (E_h) for Xe. Reprinted with permission from [9, Table VIII] Copyright 1991 American Physical Society.

Orbital	DHF	DHFB
$1s_{1/2}$	-1277.258	-1274.292
$2s_{1/2}$	-202.465 0	-202.184 5
$2p_{1/2}$	-189.678 2	-189.198 8
$2p_{3/2}$	-177.704 5	-177.380 6
$3s_{1/2}$	-43.010 36	-42.969 91
$3p_{1/2}$	-37.659 54	-37.584 81
$3p_{3/2}$	-35.325 18	-35.280 06
$3d_{3/2}$	-26.023 29	-26.000 13
$3d_{5/2}$	-25.537 03	-25.526 86
$4s_{1/2}$	-8.429 814	-8.424 185
$4p_{1/2}$	-6.452 325	-6.440 767
$4p_{3/2}$	-5.982 693	-5.977 144
$4d_{3/2}$	-2.711 237	-2.711 006
$4d_{5/2}$	-2.633 670	-2.635 577
$5s_{1/2}$	-1.010 069	-1.009 779
$5p_{1/2}$	-0.492 489 3	-0.491 736 3
$5p_{3/2}$	-0.439 730 7	-0.439 636 7

10.5 The DHFB method for closed shell molecules

The construction of the DHFB equations for molecules with fixed nuclei can be done on similar lines. The equations differ in two ways. An isolated atom has no preferred orientation, so that the states can always be classified by the irreducible representation of the rotation group to which they belong, and the Fock matrix takes a block diagonal form. Similarly symmetric molecules have states that can be classified according to a (double) point group representation, and the Fock matrix can also be decomposed into block diagonal form. This is not possible in the general case. Secondly, molecular orbitals (MOs) are linear combinations of atomic orbitals (AOs) that are themselves linear combinations of spinor basis functions centred on the nuclei. The new technical challenge is to evaluate multi-centre integrals over the spinor basis set.

The energy of a closed shell molecule can be written in a form similar to (10.2.17)

$$E_{DHF} = \sum_{\beta\beta'} \mathrm{tr}\, \left\{ \mathbf{h}^{\beta\beta'} + \frac{1}{2}(\mathbf{G}^{\beta\beta'} + \mathbf{B}^{\beta\beta'}) \right\} (\boldsymbol{D}^{\beta\beta'})^{\dagger}, \qquad (10.5.1)$$

with a Fock matrix

$$\boldsymbol{F} = \boldsymbol{h} + \boldsymbol{G} + \boldsymbol{B}. \qquad (10.5.2)$$

where, as in (10.2.7), (10.2.8),

$$\boldsymbol{h} = \begin{pmatrix} \boldsymbol{V}^{++} & c\boldsymbol{\Pi}^{+-} \\ c\boldsymbol{\Pi}^{-+} & -2mc^2\boldsymbol{S}^{--} + \boldsymbol{V}^{--} \end{pmatrix},$$

and the two-electron Coulomb and Breit matrices are partitioned so that

$$\boldsymbol{G} = \begin{pmatrix} \boldsymbol{G}^{++} & \boldsymbol{G}^{+-} \\ \boldsymbol{G}^{-+} & \boldsymbol{G}^{--} \end{pmatrix}, \quad \boldsymbol{B} = \begin{pmatrix} \boldsymbol{B}^{++} & \boldsymbol{B}^{+-} \\ \boldsymbol{B}^{-+} & \boldsymbol{B}^{--} \end{pmatrix}.$$

When expanded in a multi-centre G-spinor basis set, the sub-matrix components of $\boldsymbol{h}$ are given formally by (10.2.8),

$$h^{\beta,\beta}_{\mu\nu} = V^{\beta,\beta}_{\mu\nu} + (\beta - 1)mc^2 S^{\beta,\beta}_{\mu\nu}, \quad h^{\beta,-\beta}_{\mu\nu} = c\Pi^{\beta,-\beta}_{\mu\nu}, \quad \beta = \pm 1,$$

where the indices μ, ν are no longer restricted to a single centre. The closed shell two-electron interactions can be expressed in terms of the density matrices

$$D^{\beta\beta'}_{\mu\nu} = \sum_{A} \left(c^{\beta}_{\mu A} \right)^{*} c^{\beta'}_{\nu A}. \qquad (10.5.3)$$

The Coulomb matrix components can be written in terms of primitive integrals over two-component basis functions as

$$G^{\beta\beta}_{\mu\nu} = \sum_{\sigma\tau} \Big\{ \left[(\mu\beta, \nu\beta \,|\, \sigma\beta, \tau\beta) - (\mu\beta, \tau\beta \,|\, \sigma\beta, \nu\beta) \right] D^{\beta\beta}_{\sigma\tau}$$
$$+ (\mu\beta, \nu\beta \,|\, \sigma - \beta, \tau - \beta)\, D^{-\beta\,-\beta}_{\sigma\tau} \Big\}. \quad (10.5.4)$$

and

$$G^{\beta,-\beta}_{\mu\nu} = -\sum_{\sigma\tau} (\mu\beta, \tau\beta \,|\, \sigma - \beta, \nu - \beta)\, D^{-\beta\beta}_{\sigma\tau}. \quad (10.5.5)$$

The Breit interaction has a similar form but the integrals have a different kernel. Explicit formulae for G-spinor basis functions appear below. This matrix structure is the same as for the closed shell atom, although new integrals in which the basis functions refer to different centres are needed in the molecular case.

10.6 G-spinor basis functions

The molecular orbitals (MO) of the matrix method will be linear combinations of G-spinors on the different nuclei labelled by multi-indices

$$\mu \equiv \{\boldsymbol{A}_\mu, a_\mu, \kappa_\mu, m_\mu\} \quad (10.6.1)$$

consisting of the nuclear position vector $\boldsymbol{A}_\mu$, the Gaussian exponent a_μ, and the usual angular momentum quantum numbers. The G-spinors are two-component objects defined as in (10.2.6),

$$M[\beta, \mu, \boldsymbol{x}] = N^\beta_\mu\, m[\beta, \mu, \boldsymbol{x}] = \frac{\mathrm{g}^\beta_\mu(r)}{r}\, \chi_{\kappa,m}(\theta, \varphi), \quad \beta = \pm 1,$$

with

$$\mathrm{g}^\beta_\mu(r) = N^\beta_\mu\, g^\beta_\mu(r).$$

For polyatomic problems it is convenient to write G-spinor components in terms of unnormalized scalar spherical Gaussian-type functions (SGTF) [17]

$$S(a, \mathbf{r}_A; n, l, m) = r_A^{2n}\, \mathcal{Y}_{lm}(\mathbf{r}_A) \exp(-a r_A^2), \quad (10.6.2)$$

in which $\mathcal{Y}_{lm}(\mathbf{r}_A)$ denotes the *normalized solid harmonic* defined by [19]

$$\mathcal{Y}_{lm}(\mathbf{r}_A) = s_{lm} r^l\, P_l^{|m|}(\cos\theta_A)\, \mathrm{e}^{im\varphi_A}, \quad (10.6.3)$$

with standard normalization and phase

$$s_{lm} = (-1)^{(m+|m|)/2} \left[\frac{2l+1}{4\pi} \frac{(l-|m|)!}{(l+|m|)!} \right]^{1/2}. \quad (10.6.4)$$

It is convenient to combine s_{lm} with the CG-symbols in each component of the four-spinor by writing

the charge density matrix elements are

$$\varrho^{\beta\beta}_{\mu\nu}(\boldsymbol{x}) = -e\left\{M^*(\beta,\mu)_1 . M(\beta,\nu)_1 + M^*(\beta,\mu)_2 . M(\beta,\nu)_2\right\} \tag{10.7.2}$$

and the spherical components[3] of $\boldsymbol{j}^{+-}_{\mu\nu}$ are

$$\begin{aligned} \left(\boldsymbol{j}^{+-}_{\mu\nu}\right)_{+1} &= +i\,ec\sqrt{2}M^*(+1,\mu)_1 M(-1,\nu)_2 \\ \left(\boldsymbol{j}^{+-}_{\mu\nu}\right)_{-1} &= -i\,ec\sqrt{2}M^*(+1,\mu)_2 M(-1,\nu)_1 \\ \left(\boldsymbol{j}^{+-}_{\mu\nu}\right)_{0} &= -i\,ec\left[M^*(+1,\mu)_1 M(-1,\nu)_1 - M^*(+1,\mu)_2 M(-1,\nu)_2\right] \end{aligned} \tag{10.7.3}$$

in which we have used the standard representation of the Pauli $\boldsymbol{\sigma}$ matrices.

10.8 Two-centre overlaps

Bilinear products of spinor components, (10.7.2) and (10.7.3), are themselves linear combinations of products of the SGTF defined in (10.6.2). The popularity of Gaussian type functions in quantum chemistry rests on the fact that the product of two "simple" GTF on different centres can be written as a multiple of a "simple" GTF on a third centre [17, p. 2]:

$$\exp(-a\boldsymbol{r}_A{}^2) . \exp(-b\boldsymbol{r}_B{}^2) = K_{AB}\exp(-p\,\boldsymbol{r}_P{}^2) \tag{10.8.1}$$

where $\boldsymbol{r}_A = \boldsymbol{r} - \boldsymbol{A}$ etc., and

$$p = a + b, \quad \boldsymbol{P} = (a\boldsymbol{A} + b\boldsymbol{B})/p, \quad K_{AB} = \exp\left(-ab|\boldsymbol{A} - \boldsymbol{B}|^2/p\right).$$

Since SGTF are products of "simple" GTF with homogeneous polynomials in (x_A, y_A, z_A) of degree $2n + l$, (10.6.2), each SGTF product can be expanded in the form [17]

$$\begin{aligned} &S(a,\mathbf{r}_A;n,l,m)\,S(b,\mathbf{r}_B;n',l',m') \\ &= \sum_{\boldsymbol{k}\in\mathcal{T}_\Lambda} E[n,l,m\,;\,n',l',m'\,;\boldsymbol{k}].H(p,\boldsymbol{r}_P;\boldsymbol{k}) \end{aligned} \tag{10.8.2}$$

where $E[n,l,m\,;\,n',l',m'\,;\boldsymbol{k}]$ is a numerical coefficient depending on the position vectors $\boldsymbol{A}$ and $\boldsymbol{B}$. The common factor K_{AB} multiplying all terms on the right hand side of (10.8.2) has been absorbed into the E-coefficients; the cost involved is negligible as the factor can be incorporated into the initial condition for the linear recurrence scheme used to generate them (see Appendix B.10). $H(p,\boldsymbol{k};\boldsymbol{r}_P)$ is an Hermite Gaussian function (HGTF) defined by

$$H(p,\boldsymbol{r}_P;\boldsymbol{k}) = D^P_{\boldsymbol{k}}\exp(-p\,\boldsymbol{r}^2_P). \tag{10.8.3}$$

---

[3] Recall that the spherical components are $j_{\pm 1} = \mp(j_x \pm i j_y)/\sqrt{2}$, $j_0 = j_z$.

The multi-index $\boldsymbol{k} = (\rho, \sigma, \tau)$ runs over the i_Λ triples in the set

$$\mathcal{T}_\Lambda = \{(\rho, \sigma, \tau) \,|\, 0 \le \rho + \sigma + \tau \le \Lambda\} \tag{10.8.4}$$

where $i_\Lambda = \dim \mathcal{T}_\Lambda = (\Lambda+1)(\Lambda+2)(\Lambda+3)/6$ and $\Lambda = 2n + 2n' + l + l'$. The differential operator $D^P_{\boldsymbol{k}}$ of (10.8.3) can then be written

$$D^P_{\boldsymbol{k}} = \left(\frac{\partial}{\partial P_x}\right)^\rho \left(\frac{\partial}{\partial P_y}\right)^\sigma \left(\frac{\partial}{\partial P_z}\right)^\tau .$$

The expansion (10.8.2), which depends on writing $\boldsymbol{r}_A = \boldsymbol{r} - \boldsymbol{A} = \boldsymbol{r}_P + \boldsymbol{P}_A$, was first suggested by McMurchie and Davidson [20]. A relativistic generalization of this scheme proposed by Quiney, Skaane, and Grant [21] expresses the charge density in the analogous form

$$\varrho^{\beta\beta}_{\mu\nu}(\boldsymbol{r}) = -e \sum_{\boldsymbol{k}\in\mathcal{T}_\Lambda} E^{\beta\beta}_0(\mu, \nu; \boldsymbol{k})\, H(p, \boldsymbol{r}_P; \boldsymbol{k}) \tag{10.8.5}$$

and the components, $q = 0, \pm 1$, of the current density vector as

$$\left(\boldsymbol{j}^{\beta,-\beta}_{\mu\nu}(\boldsymbol{r})\right)_q = -iec \sum_{\boldsymbol{k}\in\mathcal{T}_\Lambda} E^{\beta,-\beta}_q(\mu, \nu; \boldsymbol{k})\, H(p, \boldsymbol{r}_P; \boldsymbol{k}). \tag{10.8.6}$$

10.8.1 Relativistic expansion coefficients

The $\beta\beta$ case with $\beta = +1$ illustrates the general strategy for constructing the relativistic expansion coefficients. To obtain

$$\varrho^{++}_{\mu\nu}(\boldsymbol{r}) = -e \sum_{\boldsymbol{k}\in\mathcal{T}_\Lambda} E^{++}_0(\mu; \nu; \mathbf{k})\, H(p_{\mu\nu}, \boldsymbol{r}_{P_{\mu\nu}}; \boldsymbol{k}), \tag{10.8.7}$$

where $p_{\mu\nu} = a_\mu + a_\nu$ and $\boldsymbol{P}_{\mu\nu} = (a_\mu \boldsymbol{A}_\mu + a_\nu \boldsymbol{A}_\nu)/p_{\mu\nu}$, we write

$$\begin{aligned} &M^\dagger[+1, \mu; \boldsymbol{r}]\; M[+1, \nu; \boldsymbol{r}] \\ &\quad = N^+_\mu N^+_\nu \Big\{ \eta_\mu \eta_\nu\, C^{-\eta_\mu}_{l_\mu m_\mu}\, C^{-\eta_\nu}_{l_\nu m_\nu}\, \mathsf{S}^*[\mu;\, 0, l_\mu, m_\mu - 1/2]\; \mathsf{S}[\nu;\, 0, l_\nu, m_\nu - 1/2] \\ &\qquad + C^{\eta_\mu}_{l_\mu m_\mu}\, C^{\eta_\nu}_{l_\nu m_\nu}\, \mathsf{S}^*[\mu;\, 0, l_\mu, m_\mu + 1/2]\; \mathsf{S}[\nu;\, 0, l_\nu, m_\nu + 1/2] \Big\}. \end{aligned}$$

Applying (10.8.2) and the relation $Y^{m*}_l = (-1)^m Y^{-m}_l$ gives the relativistic E_0 coefficient

$$\begin{aligned} E^{++}_0[\mu; \nu; \boldsymbol{k}] = N^+_\mu N^+_\nu \Big\{ &\eta_\mu \eta_\nu\, C^{-\eta_\mu}_{l_\mu m_\mu}\, C^{-\eta_\nu}_{l_\nu m_\nu} \\ &\times (-1)^{m_\mu - 1/2}\, E[0, l_\mu, -m_\mu + 1/2\,;\, 0, l_\nu, m_\nu - 1/2\,;\, \boldsymbol{k}] \\ +C^{\eta_\mu}_{l_\mu m_\mu}\, C^{\eta_\nu}_{l_\nu m_\nu}\, &(-1)^{m_\mu + 1/2}\, E[0, l_\mu, -m_\mu - 1/2\,;\, 0, l_\nu, m_\nu + 1/2\,;\, \boldsymbol{k}] \Big\} \end{aligned} \tag{10.8.8}$$

where $\boldsymbol{k} = (\rho, \sigma, \tau)$. Similarly

$$\begin{aligned}
E_0^{--}[\mu;\nu;\boldsymbol{k}] = N_\mu^- N_\nu^- \Big\{ & \eta_\mu \eta_\nu C_{\bar{l}_\mu m_\mu}^{\eta_\mu} C_{\bar{l}_\nu m_\nu}^{\eta_\nu} (-1)^{m_\mu - 1/2} \\
& \times \{ t_\mu t_\nu \, E[0, \bar{l}_\mu, -m_\mu + 1/2; 0, \bar{l}_\nu, m_\nu - 1/2; \boldsymbol{k}] \\
& \quad - 2 t_\mu a_\nu \, E[0, \bar{l}_\mu, -m_\mu + 1/2; 1, \bar{l}_\nu, m_\nu - 1/2; \boldsymbol{k}] \\
& \quad - 2 t_\nu a_\mu \, E[1, \bar{l}_\mu, -m_\mu + 1/2; 0, \bar{l}_\nu, m_\nu - 1/2; \boldsymbol{k}] \\
& \quad + 4 a_\mu a_\nu \, E[1, \bar{l}_\mu, -m_\mu + 1/2; 1, \bar{l}_\nu, m_\nu - 1/2; \boldsymbol{k}] \} \\
& + C_{\bar{l}_\mu m_\mu}^{-\eta_\mu} C_{\bar{l}_\nu m_\nu}^{-\eta_\nu} (-1)^{m_\mu + 1/2} \\
& \times \{ t_\mu t_\nu \, E[0, \bar{l}_\mu, -m_\mu - 1/2; 0, \bar{l}_\nu, m_\nu + 1/2; \boldsymbol{k}] \\
& \quad - 2 t_\mu a_\nu \, E[0, \bar{l}_\mu, -m_\mu - 1/2; 1, \bar{l}_\nu, m_\nu + 1/2; \boldsymbol{k}] \\
& \quad - 2 a_\mu t_\nu \, E[1, \bar{l}_\mu, -m_\mu - 1/2; 0, \bar{l}_\nu, m_\nu + 1/2; \boldsymbol{k}] \\
& \quad + 4 a_\mu a_\nu \, E[1, \bar{l}_\mu, -m_\mu - 1/2; 1, \bar{l}_\nu, m_\nu + 1/2; \boldsymbol{k}] \} \Big\}.
\end{aligned} \tag{10.8.9}$$

For the current density components,

$$\begin{aligned}
E_{+1}^{+-}[\mu,\nu;\boldsymbol{k}] = -\sqrt{2}\, & \eta_\nu N_\mu^+ N_\nu^- C_{l_\mu, m_\mu}^{-\eta_\mu} C_{\bar{l}_\nu, m_\nu}^{-\eta_\nu} (-1)^{m_\mu - 1/2} \\
& \times \{ t_\nu E[0, l_\mu, -m_\mu + 1/2; 0, l_\nu, m_\nu + 1/2; \boldsymbol{k}] \\
& \quad - 2 a_\nu E[0, l_\mu, -m_\mu + 1/2; 1, l_\nu, m_\nu + 1/2; \boldsymbol{k}] \}
\end{aligned} \tag{10.8.10}$$

$$\begin{aligned}
E_{-1}^{+-}[\mu,\nu;\boldsymbol{k}] = +\sqrt{2}\, & \eta_\nu N_\mu^+ N_\nu^- C_{l_\mu, m_\mu}^{\eta_\mu} C_{\bar{l}_\nu, m_\nu}^{\eta_\nu} (-1)^{m_\mu + 1/2} \\
& \times \{ t_\nu E[0, l_\mu, -m_\mu - 1/2; 0, l_\nu, m_\nu - 1/2; \boldsymbol{k}] \\
& \quad - 2 a_\nu E[0, l_\mu, -m_\mu - 1/2; 1, l_\nu, m_\nu - 1/2; \boldsymbol{k}] \}
\end{aligned} \tag{10.8.11}$$

$$\begin{aligned}
E_0^{+-}[\mu,\nu;\boldsymbol{k}] = -N_\mu^+ N_\nu^- \Big\{ & -\eta_\mu \eta_\nu C_{l_\mu, m_\mu}^{-\eta_\mu} C_{\bar{l}_\nu, m_\nu}^{+\eta_\nu} (-1)^{m_\mu - 1/2} \\
& \times \{ t_\nu E[0, l_\mu, -m_\mu + 1/2; 0, l_\nu, m_\nu - 1/2; \boldsymbol{k}] \\
& \quad - 2 a_\nu E[0, l_\mu, -m_\mu + 1/2; 1, l_\nu, m_\nu - 1/2; \boldsymbol{k}] \} \\
& \qquad - (-1)^{m_\mu + 1/2} C_{l_\mu, m_\mu}^{+\eta_\mu} C_{\bar{l}_\nu, m_\nu}^{-\eta_\nu} (-1)^{m_\mu + 1/2} \\
& \times \{ t_\nu E[0, l_\mu, -m_\mu - 1/2; 0, l_\nu, m_\nu + 1/2; \boldsymbol{k}] \\
& \quad - 2 a_\nu E[0, l_\mu, -m_\mu - 1/2; 1, l_\nu, m_\nu + 1/2; \boldsymbol{k}] \} \Big\}.
\end{aligned} \tag{10.8.12}$$

The E_q coefficients are therefore simple linear combinations of the nonrelativistic McMurchie-Davidson E coefficients; see Appendix B.10.

10.8.2 Symmetry properties of E_q coefficients

The symmetry properties of G-spinors are the same as those of solutions of Dirac's equation in a central potential. The E_q coefficients thus inherit a structure that can be utilized both to check the correctness of algorithms and to eliminate unnecessary computation.

G-spinors labelled by μ (with a fixed value of κ_μ) and angular projections $m_\mu = -j_\mu, \ldots, j_\mu$ span an irreducible representation $\mathcal{D}^{(j_\mu)}$ of the rotation group. It follows that they can also be used to construct bases for the irreducible representations of the double point groups [22, 23, 24]. G-spinors also have well-defined properties under spatial and time inversions. The spatial inversion operator for Dirac 4-spinors can be written $\mathcal{P} = \beta\mathcal{P}'$, where $\mathcal{P}'f(\boldsymbol{x}) = f(-\boldsymbol{x})$ is the ordinary parity operator and β is the usual 4-component Dirac matrix. It follows that for any 4-spinor constructed as in (10.2.5), we have

$$\mathcal{P}\psi_A(\boldsymbol{x}) = (-1)^l \psi_A(-\boldsymbol{x}) \qquad (10.8.13)$$

where $l = j + \eta/2$ is the value of the orbital angular momentum associated with the large component 2-spinor. For time reversal, $\psi(t, \boldsymbol{x}) \to \psi(-t, \boldsymbol{x})$, the operator $\mathcal{T}$ takes the form

$$\mathcal{T} = -\begin{pmatrix} \sigma_y & 0 \\ 0 & \sigma_y \end{pmatrix} \mathcal{K} \qquad (10.8.14)$$

where σ_y is a Pauli matrix and $\mathcal{K}$ denotes complex conjugation. The way in which this works is illuminated by considering the effect of time reversal on a 2-component G-spinor $M[\beta, \mu, \boldsymbol{x}]$ of (10.2.6)

$$\begin{aligned} -\sigma_y \mathcal{K} M[\beta, \mu, \boldsymbol{x}] &= -\sigma_y \mathcal{K} \left[\frac{\mathrm{g}_\mu^\beta(r)}{\mathrm{d}r} \chi_{\kappa,m}(\theta, \varphi) \right] \\ &= (-1)^{l-j+m} \frac{(\mathcal{K}\mathrm{g}_\mu^\beta)(r)}{\mathrm{d}r} \chi_{\kappa,-m}(\theta, \varphi). \end{aligned} \qquad (10.8.15)$$

It is this symmetry that is responsible for the additional factor i preceding the lower component of (10.2.5), which ensures that the radial amplitudes can be assumed real. Dirac spinors ψ and $\mathcal{T}\psi$ are said to constitute a *Kramers' pair*. If ψ is an eigensolution of Dirac's equation with energy E, then so is $\mathcal{T}\psi$ in the absence of an interaction that is not invariant under time reversal. These considerations give the following symmetry relations (in which we refer explicitly to the magnetic quantum numbers only):

$$\begin{aligned} E_q^{\beta'\beta}[-m_\mu, -m_\nu; \boldsymbol{k}] &= \eta_\mu \eta_\nu (-1)^{m_\mu - m_\nu} E_q^{\beta\beta'}[m_\mu, m_\nu; \boldsymbol{k}] \\ E_q^{\beta\beta'}[-m_\mu, -m_\nu; \boldsymbol{k}] &= \eta_\mu \eta_\nu (-1)^{m_\mu - m_\nu} E_q^{\beta\beta'}[m_\mu, m_\nu; \boldsymbol{k}]^* \\ E_q^{\beta'\beta}[m_\nu, m_\mu; \boldsymbol{k}] &= E_q^{\beta'\beta}[m_\mu, m_\nu; \boldsymbol{k}]^* \end{aligned} \qquad (10.8.16)$$

10.9 Multi-centre interaction integrals

Interaction integrals of §10.5 require evaluation of expressions involving overlaps of G-spinors on different nuclear centres [21, 25]. The general expressions can be quite complicated, and the relative cheapness of the one-centre formulae of §10.2 means that they should be used preferentially whenever they are appropriate.

10.9.1 Auxiliary integrals involving HGTFs

The simplest auxiliary integral over HGTF is

$$\int H(a, \boldsymbol{x} - \boldsymbol{A}; \boldsymbol{k})\, d\boldsymbol{x} = \delta_{\boldsymbol{k},0} \left(\frac{\pi}{a}\right)^{3/2}. \tag{10.9.1}$$

where $\boldsymbol{k} = (\rho, \sigma, \tau)$ as defined by (10.8.3). The electron-electron interaction involves

$$(a, \boldsymbol{A}; \boldsymbol{k} \,|\, b, \boldsymbol{B}; \boldsymbol{k}') = \iint \frac{H(a, \boldsymbol{x} - \boldsymbol{A}; \boldsymbol{k}) H(b, \boldsymbol{x}' - \boldsymbol{B}; \boldsymbol{k}')}{|\boldsymbol{x} - \boldsymbol{x}'|}\, d\boldsymbol{x}\, d\boldsymbol{x}', \tag{10.9.2}$$

which can be evaluated using the identity

$$\frac{1}{R} = \frac{1}{\sqrt{\pi}} \int_{-\infty}^{\infty} \exp(-w^2 R^2) dw. \tag{10.9.3}$$

Using (10.7.3) and a double application of (10.7.1) gives

$$(a, \boldsymbol{A}; \boldsymbol{k} \,|\, b, \boldsymbol{B}; \boldsymbol{k}') = D^A_{\boldsymbol{k}}\, D^B_{\boldsymbol{k}'} \frac{2\pi^{5/2}}{ab\sqrt{a+b}} F_0 \left(\frac{ab}{a+b} |\boldsymbol{A} - \boldsymbol{B}|^2\right), \tag{10.9.4}$$

where

$$F_m(x) = \int_0^1 u^{2m} \exp(-xu^2)\, du, \; m = 0, 1, \ldots$$

The number of integrals that have to be computed is greatly reduced by using the relation

$$(a, \boldsymbol{A}; \boldsymbol{k} \,|\, b, \boldsymbol{B}; \boldsymbol{k}') = (-1)^{\rho' + \sigma' + \tau'} (a, \boldsymbol{A}; \boldsymbol{k} + \boldsymbol{k}' \,|\, b, \boldsymbol{B}; 0). \tag{10.9.5}$$

This result, which follows from (10.9.4), can be understood by writing $s = |\boldsymbol{A} - \boldsymbol{B}|^2$, so that $\partial s / \partial A_i = (A_i - B_i)/s = -\partial s / \partial B_i$ for $i = 1, 2, 3$ and

$$\frac{\partial}{\partial A_i} f(s) = f'(s) \frac{\partial s}{\partial A_i} = -f'(s) \frac{\partial s}{\partial B_i} = -\frac{\partial}{\partial B_i} f(s).$$

The internal electromagnetic fields can be calculated by exploiting the well-known representation of the Dirac δ-distribution [26, Equation (2.6-3)]

$$\lim_{b \to \infty} \left(\frac{b}{\pi}\right)^{3/2} H(b, \boldsymbol{x}; \boldsymbol{0}) = \delta(\boldsymbol{x}).$$

It is then easy to justify the limit

$$[a, \boldsymbol{A}; \boldsymbol{k}|\boldsymbol{x} - \boldsymbol{A}] = \lim_{b\to\infty} \left(\frac{b}{\pi}\right)^{3/2} [a, \boldsymbol{A}; \boldsymbol{k}|b, \boldsymbol{x}; \boldsymbol{0}]$$
$$= \int \frac{H(a, \boldsymbol{r} - \boldsymbol{A}; \boldsymbol{k})}{|\boldsymbol{r} - \boldsymbol{x}|}\, d\boldsymbol{r} \tag{10.9.6}$$

from which we can construct potentials for internal electromagnetic fields. The notation emphasizes that this expression is a continuous and differentiable function of $\boldsymbol{x} - \boldsymbol{A}$. The partial derivatives with respect to the coordinates x_r, $r = 1, 2, 3$ are

$$(\partial/\partial x_r)[a, \boldsymbol{A}; \boldsymbol{k}|\boldsymbol{x} - \boldsymbol{A}] = -(\partial/\partial A_r)[a, \boldsymbol{A}; \boldsymbol{k}|\boldsymbol{x} - \boldsymbol{A}]$$
$$= -[a, \boldsymbol{A}; \boldsymbol{k} + \boldsymbol{e}_r|\boldsymbol{x} - \boldsymbol{A}]. \tag{10.9.7}$$

where the first step uses the functional dependence on $\boldsymbol{x} - \boldsymbol{A}$. The second step takes the operator $(\partial/\partial A_r)$ inside the integral before using (10.7.3).

10.9.2 Multi-centre one-electron integrals

All one-electron integrals needed have the generic form

$$(\mu, \beta\,|\,\sigma_q\,|\,\nu, \beta') = \int M^\dagger[\beta, \mu, \boldsymbol{x} - \boldsymbol{A}_\mu]\; \sigma_q\; M[\beta', \nu, \boldsymbol{x} - \boldsymbol{A}_\nu]\, d\boldsymbol{x}, \tag{10.9.8}$$

where σ_q are the usual Pauli matrices for $q = 1, 2, 3$ and σ_0 is the 2×2 identity matrix. From (10.9.1) we find

$$(\mu, \beta\,|\,\sigma_q\,|\,\nu, \beta') = \left(\frac{\pi}{p_{\mu\nu}}\right)^{3/2} E_q^{\beta\beta'}[\mu, \nu; \boldsymbol{0}], \tag{10.9.9}$$

where $p_{\mu\nu} = a_\mu + a_\nu$ and $\boldsymbol{0} = (0, 0, 0)$.

2-centre Gram (overlap) matrix elements

The special case $\beta = \beta'$ with $q = 0$ gives the 2-centre Gram matrix elements

$$S_{\mu\nu}^{\beta\beta} = \int M^\dagger[\beta, \mu; \boldsymbol{x}]\; M[\beta', \nu; \boldsymbol{x}]\, d\boldsymbol{x}$$
$$= \left(\frac{\pi}{p_{\mu\nu}}\right)^{3/2} E_0^{\beta\beta}[\mu, \nu; \boldsymbol{0}]. \tag{10.9.10}$$

2-centre kinetic matrices: $\boldsymbol{\sigma} \cdot \boldsymbol{p}$

The kinetic matching condition

$$M[-1, \mu; \boldsymbol{x}] = \frac{N_\mu^-}{N_\mu^+}\, \boldsymbol{\sigma} \cdot \boldsymbol{p}\, M[+1, \mu; \boldsymbol{x}]$$

gives the particularly simple kinetic matrix

$$\Pi^{-+}_{\mu\nu} = \int M^\dagger[-1,\mu;\boldsymbol{x}]\, \boldsymbol{\sigma}\cdot\boldsymbol{p}\, M[+1,\nu;\boldsymbol{x}]\, d\boldsymbol{x} = S^{--}_{\mu\nu}\frac{N^-_\nu}{N^+_\nu} \tag{10.9.11}$$

along with its conjugate

$$\Pi^{+-}_{\mu\nu} = \left(\Pi^{-+}_{\nu\mu}\right)^*.$$

Non-relativistic kinetic energy matrix

We have seen that the kinetic matching condition ensures that the non-relativistic kinetic energy matrix is related to the matrices $\boldsymbol{\Pi}^{-+}$ and $\boldsymbol{\Pi}^{+-}$

$$T^{++}_{\mu\nu} = \frac{1}{2}\left(\boldsymbol{\Pi}^{+-}(\boldsymbol{S}^{--})^{-1}\boldsymbol{\Pi}^{-+}\right)_{\mu\nu} \tag{10.9.12}$$

This result can also been obtained more directly [17, §4.3.2] by noting that $\boldsymbol{\nabla}^2 = p_r^2 + \boldsymbol{L}^2/r^2$, where the radial momentum operator can be written $p_r = \partial/\partial r + 1/r$. Then if $\boldsymbol{e}_i$ is a unit vector in direction i,

$$\begin{aligned} T^{++}_{\mu\nu} &= -\frac{1}{2}\int M^\dagger[+1,\mu;\boldsymbol{x}]\,\boldsymbol{\nabla}^2\, M[+1,\nu;\boldsymbol{x}]\, d\boldsymbol{x} \qquad (10.9.13)\\ &= \left(\frac{\pi}{p_{\mu\nu}}\right)^{3/2}\Big\{\left((2l_\nu+3)a_\nu - 2a_\nu^2(\boldsymbol{P}_{\mu\nu}-\boldsymbol{A}_\nu)^2 + \frac{3}{2p_{\mu\nu}}\right)E_0^{++}[\mu,\nu;\boldsymbol{0}]\\ &\qquad - 4a_\nu^2\sum_{i=x,y,z}\left[(\boldsymbol{P}_{\mu\nu}-\boldsymbol{A}_\nu)_i\, E_0^{++}[\mu,\nu;\boldsymbol{e}_i] + E_0^{++}[\mu,\nu;2\boldsymbol{e}_i]\right]\Big\} \end{aligned}$$

where $(\boldsymbol{P}_{\mu\nu}-\boldsymbol{A}_\nu)_i = (\boldsymbol{P}_{\mu\nu}-\boldsymbol{A}_\nu).\boldsymbol{e}_i$.

Nuclear attraction potential matrix

The nuclear attraction integrals give the electrostatic energy between an electron charge density overlap (10.8.2) and a nuclear charge density, one for each atom in the molecule. It is convenient to use a simple Gaussian model

$$\rho_A(\boldsymbol{x}) = \frac{Z_A e}{4\pi\epsilon_0}\left(\frac{\zeta_A}{\pi}\right)^{3/2}\exp(-\zeta_A\left|\boldsymbol{x}-\boldsymbol{A}^2\right|) \tag{10.9.14}$$

which is itself a simple HGTF. Other useful models are listed by Andrae [5] and can often be approximated as a linear combination of simple Gaussians. It follows that the electron-nucleus potential energy is

$$\begin{aligned} [V_{nuc}]^{\beta\beta}_{\mu\nu} &= \sum_A \iint \frac{-e\left(M^\dagger[\beta,\mu;\boldsymbol{x}].M[\beta,\nu;\boldsymbol{x}]\right)\rho_A(\boldsymbol{x}')}{|\boldsymbol{x}-\boldsymbol{x}'|}\, d\boldsymbol{x}\, d\boldsymbol{x}'\\ &= -\sum_A \frac{Z_A e^2}{4\pi\epsilon_0}\left(\frac{\zeta_A}{\pi^{1/2}}\right)^{3/2} \qquad (10.9.15)\\ &\qquad \times\sum_{\boldsymbol{k}} E_0^{\beta\beta}[\mu,\nu;\boldsymbol{k}]\,(p_{\mu\nu},\boldsymbol{P}_{\mu\nu};\boldsymbol{k}\,|\,\zeta_A,\boldsymbol{A};\boldsymbol{0}). \end{aligned}$$

We shall see that this has the same form as a standard electrostatic repulsion integral. This passes smoothly to the expression for a point nucleus in the limit $\zeta_A \to \infty$.[4]

Moment intcgrals

Electromagnetic interactions involve expressions such as

$$(\mu, \beta \mid \sigma_j \,(\boldsymbol{x} - \boldsymbol{C})_k \mid \nu, \beta') = \tag{10.9.16}$$
$$\left(\frac{\pi}{p_{\mu\nu}}\right)^{3/2} \left\{ E_j^{\beta\beta'}[\mu, \nu; \boldsymbol{e}_k] + E_j^{\beta\beta'}[\mu, \nu; \boldsymbol{0}]\,(\boldsymbol{P}_{\mu\nu} - \boldsymbol{C})_k \right\}$$

where $j, k = x, y, z$. Scalar moments, with $j = 0$, occur when evaluating the electrostatic dipole moment. We shall also need matrix elements of $\boldsymbol{\alpha} \times (\boldsymbol{x} - \boldsymbol{C})$ in, for example, the relativistic theory of nuclear magnetic resonance chemical shifts. In Cartesian coordinates a vector product can be written

$$(\boldsymbol{a} \times \boldsymbol{b})_i = \sum_{j,k} \epsilon_{ijk} \boldsymbol{a}_j \boldsymbol{b}_k, \quad i = 1, 2, 3,$$

where ϵ_{ijk} is the Kronecker alternating symbol, equal to +1 if ijk is an even permutation of 123, -1 if it is an odd permutation, and zero otherwise. The resulting G-spinor matrix element is then obtained by multiplying (10.9.14) by ϵ_{ijk} and summing over j, k for each value of i. It is sometimes useful to use a decomposition of vectors into spherical components $a_o = a_z$, $a_{\pm 1} = \mp(a_x \pm a_y)/\sqrt{2}$. In this case we can use (B.3.149) to write the vector product as a tensor operator of rank 1:

$$(\boldsymbol{a} \times \boldsymbol{b})_q = -i\sqrt{2}[\boldsymbol{ab}]^1_q = -i\sqrt{2} \sum_{q_1, q_2} C^{1\ 1\ 1}_{q_1\, q_2\, q}\, a_{q_1} b_{q_2}$$

which simplifies to

$$(\boldsymbol{a} \times \boldsymbol{b})_{\pm 1} = \mp i(a_{\pm 1} b_0 - a_0 b_{\pm 1}), \quad (\boldsymbol{a} \times \boldsymbol{b})_0 = -i(a_1 b_{-1} - a_{-1} b_1).$$

Matrix elements between orbitals defined as in (10.2.5) involve combinations of G-spinor integrals. Operators which have no spin dependence give

$$\langle \psi_A \mid (\boldsymbol{x} - \boldsymbol{C}) \mid \psi_B \rangle = \sum_{\mu\nu} \Big\{ (c^{+}_{\mu A})^* c^{+}_{\nu B}\, (\mu, +1 \mid (\boldsymbol{x} - \boldsymbol{C}) \mid \nu, +1)$$
$$+ (c^{-}_{\mu A})^* c^{-}_{\nu B}\, (\mu, -1 \mid (\boldsymbol{x} - \boldsymbol{C}) \mid \nu, -1) \Big\} \tag{10.9.17}$$

whereas, when spin-dependent operators are involved, we get

[4] Compare (10.9.6); recall that $e^2/4\pi\epsilon_0 = 1$ in atomic units.

$$\langle\psi_A\,|\,\boldsymbol{\alpha}\times(\boldsymbol{x}-\boldsymbol{C})\,|\,\psi_B\rangle = \sum_{rst} \boldsymbol{e}_r \epsilon_{rst}$$

$$\times \sum_{\mu\nu} \Big\{ +i(c^+_{\mu A})^* c^-_{\nu B}\,(\mu,\,+1\,|\,\sigma_s(\boldsymbol{x}-\boldsymbol{C})_t\,|\,\nu,\,-1) \tag{10.9.18}$$

$$-i(c^-_{\mu A})^* c^+_{\nu B}\,(\mu,\,-1\,|\,\sigma_s(\boldsymbol{x}-\boldsymbol{C})_t\,|\,\nu,\,+1)\Big\}$$

so that care must be taken with the i factors associated with the lower components.

Internal electromagnetic fields

For simplicity, consider the long wavelength approximation, so that the electron-electron interaction can be treated as quasi-stationary. Then the scalar and vector potentials generated by the internal electron charge-current density in an atom or molecule are

$$V^{\beta\beta}_{\mu\nu}(\boldsymbol{x}) = \frac{1}{4\pi\epsilon_0}\int \frac{\rho^{\beta\beta}_{\mu\nu}(\boldsymbol{s})}{|\boldsymbol{x}-\boldsymbol{s}|}\,d\boldsymbol{s} \tag{10.9.19}$$

and

$$\boldsymbol{A}^{\beta,-\beta}_{\mu\nu}(\boldsymbol{x}) = \frac{\mu_0}{4\pi}\int \frac{\boldsymbol{j}^{\beta,-\beta}_{\mu\nu}(\boldsymbol{s})}{|\boldsymbol{x}-\boldsymbol{s}|}\,d\boldsymbol{s}, \tag{10.9.20}$$

where $\epsilon_0\mu_0 = 1/c^2$. These generate internal $\boldsymbol{E}$ and $\boldsymbol{B}$ fields from

$$\boldsymbol{E}^{\beta\beta}_{\mu\nu}(\boldsymbol{x}) = -\boldsymbol{\nabla}\,V^{\beta\beta}_{\mu\nu}(\boldsymbol{x}), \quad \boldsymbol{B}^{\beta,-\beta}_{\mu\nu}(\boldsymbol{x}) = \boldsymbol{\nabla}\times\boldsymbol{A}^{\beta,-\beta}_{\mu\nu}(\boldsymbol{x}). \tag{10.9.21}$$

The generalized McMurchie-Davidson expansions (10.8.2) and (10.8.3) along with (10.9.6) gives

$$V^{\beta\beta}_{\mu\nu}(\boldsymbol{x}) = -\frac{e}{4\pi\epsilon_0}\sum_{\boldsymbol{k}\in\mathcal{T}_\Lambda} E^{\beta\beta}_0(\mu,\nu;\boldsymbol{k})\,[p_{\mu\nu},\boldsymbol{P}_{\mu\nu};\boldsymbol{k}|\boldsymbol{x}-\boldsymbol{P}_{\mu\nu}], \tag{10.9.22}$$

$$\boldsymbol{A}^{\beta,-\beta}_{\mu\nu}(\boldsymbol{x}) = -i\frac{ec\mu_0}{4\pi}\sum_{\boldsymbol{k}\in\mathcal{T}_\Lambda} E^{\beta,-\beta}_q(\mu,\nu;\boldsymbol{k})\,[p_{\mu\nu},\boldsymbol{P}_{\mu\nu};\boldsymbol{k}|\boldsymbol{x}-\boldsymbol{P}_{\mu\nu}], \tag{10.9.23}$$

and, with the help of (10.9.7) and (10.9.21), the Cartesian components of the electric field are

$$\Big[\boldsymbol{E}^{\beta\beta}_{\mu\nu}(\boldsymbol{x})\Big]_r = \frac{e}{4\pi\epsilon_0}$$

$$\times \sum_{\boldsymbol{k}\in\mathcal{T}_\Lambda} E^{\beta\beta}_0(\mu,\nu;\boldsymbol{k})\,[p_{\mu\nu},\boldsymbol{P}_{\mu\nu};\boldsymbol{k}+\boldsymbol{e}_r|\boldsymbol{x}-\boldsymbol{P}_{\mu\nu}], \tag{10.9.24}$$

where $r = 1, 2, 3$. Similarly the components of the magnetic field are

$$\left[\boldsymbol{B}^{\beta,-\beta}_{\mu\nu}(\boldsymbol{x})\right]_r = +i\,\frac{ec\mu_0}{4\pi}\,\epsilon_{rst}$$
$$\times \sum_{\boldsymbol{k}\in\mathcal{T}_\Lambda} \{E^{\beta,-\beta}_s(\mu,\nu;\boldsymbol{k})\,[p_{\mu\nu},\boldsymbol{P}_{\mu\nu};\boldsymbol{k}+\boldsymbol{e}_t|\boldsymbol{x}-\boldsymbol{P}_{\mu\nu}],$$
$$-E^{\beta,-\beta}_t(\mu,\nu;\boldsymbol{k})\,[p_{\mu\nu},\boldsymbol{P}_{\mu\nu};\boldsymbol{k}+\boldsymbol{e}_s|\boldsymbol{x}-\boldsymbol{P}_{\mu\nu}]\Big\}, \quad (10.9.25)$$

Radiative transition amplitudes

The interaction between a charge-current density, $j^\mu(x)$ and the electromagnetic field is given by

$$H_{int} = \frac{1}{c}\int j^\mu(x)A_\mu(x)d^3x$$

where the quantized four-potential is given by (4.3.10),

$$A_\mu(x) = \sqrt{\frac{c}{2(2\pi)^3\epsilon_0}}\int\frac{d^3k}{k_0}$$
$$\times\sum_{\lambda=0}^{3}\left\{\mathrm{d}c^{(\lambda)}(k)\,\epsilon^{(\lambda)}_\mu(k)\,\mathrm{e}^{-ik\cdot x} + c^{(\lambda)\dagger}(k)\,\epsilon^{(\lambda)}_\mu(k)\,\mathrm{e}^{+ik\cdot x}\right\},$$

in which the four-vector $k^\mu = (k_0,\boldsymbol{k})$ defines the direction of travel of the plane wave and, as the particles are massless, $k^0 = k_0 = |\boldsymbol{k}|$. The four linearly independent polarization vectors $\epsilon^{(\lambda)}(k)$, here taken as real, are chosen so that

$$\epsilon^{(\lambda)}_\mu(k)\cdot\epsilon^{(\lambda')\mu}(k) = g^{\lambda\lambda'}$$

where $g^{\lambda\lambda'}$ is the Minkowski metric tensor. The components of $j^\mu(x)$ are built from the expressions (10.6.6) so that the G-spinor radiative matrix elements are of the form

$$\mathcal{M}^{(\lambda)\beta}_{\mu\nu}(\boldsymbol{K}) = (\mu,\beta\,|\sum_q \sigma_q\epsilon^{(\lambda)q}(K)\,\exp(i\boldsymbol{K}\cdot(\boldsymbol{x}-\boldsymbol{G}))\,|\,\nu,-\beta), \quad (10.9.26)$$

where $\beta = \pm 1$ and the point $\boldsymbol{G}$ determines the gauge potential. The polarization index λ takes the value 0 for scalar photons, 1 or 2 for transversely polarizazed photons and 3 for longitudinally polarized photons. The expansions (10.8.2) and (10.8.3) give the current components as functions of HGTF, and the integrals can be evaluated explicitly using (10.7.3):

$$\mathcal{M}^{(\lambda)\beta}_{\mu\nu}(\boldsymbol{K}) = \quad (10.9.27)$$
$$\left(\frac{\pi}{p_{\mu\nu}}\right)^{3/2}\exp\left((i\boldsymbol{K}\cdot(\boldsymbol{P}_{\mu\nu}-\boldsymbol{G}) - \boldsymbol{K}^2/\mathrm{d}4p_{\mu\nu})\right)$$
$$\times\sum_q\sum_{\boldsymbol{k}\in\mathcal{T}_\Lambda}\sigma_q\epsilon^{(\lambda)q}(K)\,E^{\beta,-\beta}_q(\mu,\nu;\boldsymbol{k})\,(iK_x)^r(iK_y)^s(iK_z)^t,$$

where $\boldsymbol{k} = (r, s, t)$. In the transverse photon gauge, the longitudinal and scalar fields do not contribute and we are left with the components perpendicular to the propagation vector $\boldsymbol{K}$.

10.9.3 Multi-centre two-electron integrals

Coulomb integrals

The electrostatic interaction energy of relativistic overlap charge distributions can be expanded in terms of similar integrals over basis spinors as

$$(AB\,|\,CD) = \sum_{\mu\nu\sigma\tau}\sum_{\beta\beta'} (c^{\beta}_{\mu A})^* c^{\beta}_{\nu B}\,(c^{\beta'}_{\sigma C})^* c^{\beta'}_{\tau D}\,(\mu\beta, \nu\beta\,|\,\sigma\beta', \tau\beta') \tag{10.9.28}$$

where

$$(\mu\beta, \nu\beta\,|\,\sigma\beta', \tau\beta') = \frac{1}{4\pi\epsilon_0}\iint \frac{\rho^{\beta\beta}_{\mu\nu}(\boldsymbol{x})\rho^{\beta'\beta'}_{\sigma\tau}(\boldsymbol{x}')}{|\boldsymbol{x}-\boldsymbol{x}'|}. \tag{10.9.29}$$

The charge densities are expressed in terms of HGTF by (10.8.5). It is now straightforward to derive

$$(\mu\beta, \nu\beta,\,|\,\sigma\beta', \tau\beta') = \tag{10.9.30}$$
$$e^2 \sum_{\boldsymbol{k}\in\mathcal{T}_{\Lambda_{\mu\nu}}}\ \sum_{\boldsymbol{k}'\in\mathcal{T}_{\Lambda_{\sigma\tau}}} E^{\beta\beta}_0\,[\mu,\nu;\boldsymbol{k}]\ E^{\beta'\beta'}_0\,[\sigma,\tau;\boldsymbol{k}']\,[p_{\mu\nu}, \mathbf{P}_{\mu\nu};\boldsymbol{k}|p_{\sigma\tau}, \mathbf{P}_{\sigma\tau};\boldsymbol{k}'],$$

where the HGTF interaction integral $[p_{\mu\nu}, \mathbf{P}_{\mu\nu};\boldsymbol{k}|p_{\sigma\tau}, \mathbf{P}_{\sigma\tau};\boldsymbol{k}']$ is given by (10.9.2). This has the same formal structure as the corresponding nonrelativistic formula, although the ranges of the HGTF indices are different.

Breit integrals

The matrix elements of the Breit interaction kernel (4.9.22), in terms of the electron current density overlaps, (4.9.18), are

$$(AB\,|k^B(R)|\,CD) = \tag{10.9.31}$$
$$-\iint \frac{\boldsymbol{j}_{AB}(\boldsymbol{x})\cdot\boldsymbol{j}_{CD}(\boldsymbol{x}') + (\boldsymbol{j}_{AB}(\boldsymbol{x})\cdot\widehat{\boldsymbol{R}})(\boldsymbol{j}_{CD}(\boldsymbol{x}')\cdot\widehat{\boldsymbol{R}})}{8\pi\epsilon_0 c^2 R}\,d\boldsymbol{x}\,d\boldsymbol{x}'.$$

where $\boldsymbol{R} = \boldsymbol{x} - \boldsymbol{x}'$, $R = |\boldsymbol{R}|$ and $\widehat{\boldsymbol{R}} = \boldsymbol{R}/R$. Expanding the current in terms of G-spinors as in (10.2.5) leads to

$$(AB\,|k^B(R)|\,CD) = \tag{10.9.32}$$
$$\sum_{\mu\nu\sigma\tau}\sum_{\beta\beta'} (c^{\beta}_{\mu A})^* c^{-\beta}_{\nu B}\,(c^{\beta'}_{\sigma C})^* c^{-\beta'}_{\tau D}$$
$$\times \iint \frac{\boldsymbol{j}^{\beta,-\beta}_{\mu\nu}(\boldsymbol{x})\cdot\boldsymbol{j}^{\beta',-\beta'}_{\sigma\tau}(\boldsymbol{x}') + (\boldsymbol{j}^{\beta,-\beta}_{\mu\nu}(\boldsymbol{x})\cdot\widehat{\boldsymbol{R}})(\boldsymbol{j}^{\beta',-\beta'}_{\sigma\tau}(\boldsymbol{x}')\cdot\widehat{\boldsymbol{R}})}{8\pi\epsilon_0 c^2 R}\,d\boldsymbol{x}\,d\boldsymbol{x}'.$$

Using (10.8.6) to express the overlap current density in terms of HGTF, and noting that $e^2/4\pi\epsilon_0$ is unity in atomic units, the double integral can be written

$$(\mu, \beta; \nu, -\beta\,|k^B(R)|\,\sigma, \beta', \tau, -\beta') = \tag{10.9.33}$$
$$\frac{1}{2}\sum_{q,q'}\sum_{\boldsymbol{k}\in\mathcal{T}_{\Lambda_{\mu\nu}}}\sum_{\boldsymbol{k}'\in\mathcal{T}_{\Lambda_{\sigma\tau}}} E_q^{\beta,-\beta}[\mu\nu;\boldsymbol{k}]\, E_{q'}^{\beta',-\beta'}[\sigma\tau;\boldsymbol{k}']$$
$$\times \iint H(p,\boldsymbol{x}-\boldsymbol{P}_{\mu\nu};\boldsymbol{k})\, H(q,\boldsymbol{x}'-\boldsymbol{P}_{\sigma\tau};\boldsymbol{k}')\; g_{i,i'}(\boldsymbol{x},\boldsymbol{x}')\, d\boldsymbol{x}d\boldsymbol{x}',$$

with kernel

$$g_{q,q'}(\boldsymbol{x},\boldsymbol{x}') = \frac{1}{R}\left(\delta_{q,q'} + \frac{(\boldsymbol{x}-\boldsymbol{x}')_q(\boldsymbol{x}-\boldsymbol{x}')_{q'}}{R^2}\right).$$

where q, q' are Cartesian labels. This interaction integral can be expressed in terms of the integrals defined by (10.9.2). The first term, involving $\delta_{i,i'}/R$, is a standard Coulomb integral:

$$[p, \boldsymbol{P}_{\mu\nu}; \boldsymbol{k}\,|\, p_{\sigma\tau}, \boldsymbol{P}_{\sigma\tau}; \boldsymbol{k}']_{\mu\nu}. \tag{10.9.34}$$

The term containing $(\boldsymbol{x}-\boldsymbol{x}')_q(\boldsymbol{x}-\boldsymbol{x}')_{q'}/R^3$ requires more effort. First of all,

$$-\frac{\partial}{\partial x_q}\frac{1}{R} = \frac{\partial}{\partial x'_q}\frac{1}{R} = \frac{(\boldsymbol{x}-\boldsymbol{x}')_q}{R^3}.$$

Secondly

$$(\boldsymbol{x}-\boldsymbol{x}')_{q'} = (\boldsymbol{x}-\boldsymbol{P}_{\mu\nu})_{q'} - (\boldsymbol{x}'-\boldsymbol{P}_{\sigma\tau})_{q'} + (\boldsymbol{P}_{\mu\nu}-\boldsymbol{P}_{\sigma\tau})_{q'}$$

so that we can use the (one-dimensional) HGTF properties

$$x\,H(a,x;i) = \frac{1}{2a}\,H(a,x;i+1) + i\,H(a,x;i-1)$$

and

$$\frac{d}{dA}H(a,x-A;i) = -\frac{d}{dx}H(a,x-A;i) = H(a,x-A;i+1)$$

to write

$$\begin{aligned}
&(\boldsymbol{x}-\boldsymbol{x}')_{q'}H(p_{\mu\nu},\boldsymbol{x}-\boldsymbol{P}_{\mu\nu};\boldsymbol{k})\,H(p_{\sigma\tau},\boldsymbol{x}'-\boldsymbol{P}_{\sigma\tau};\boldsymbol{k}') = \\
&\left(\frac{1}{2p_{\mu\nu}}H(p_{\mu\nu},\boldsymbol{x}-\boldsymbol{P}_{\mu\nu};\boldsymbol{k}+\boldsymbol{e}_{q'}) + \boldsymbol{k}_{q'}H(p_{\mu\nu},\boldsymbol{x}-\boldsymbol{P}_{\mu\nu};\boldsymbol{k}-\boldsymbol{e}_{q'})\right) \\
&\qquad\times H(p_{\sigma\tau},\boldsymbol{x}'-\boldsymbol{P}_{\sigma\tau};\boldsymbol{k}') \\
&\quad - H(p_{\mu\nu},\boldsymbol{x}-\boldsymbol{P}_{\mu\nu};\boldsymbol{k})\Big(\frac{1}{2p_{\sigma\tau}}H(p_{\sigma\tau},\boldsymbol{x}'-\boldsymbol{P}_{\sigma\tau};\boldsymbol{k}'+\boldsymbol{e}_{q'}) \\
&\qquad + \boldsymbol{k}'_{q'}H(p_{\sigma\tau},\boldsymbol{x}'-\boldsymbol{P}_{\sigma\tau};\boldsymbol{k}'-\boldsymbol{e}_{q'})\Big) \\
&\quad + (\boldsymbol{P}_{\mu\nu}-\boldsymbol{Q})_{q'}\,H(p_{\mu\nu},\boldsymbol{x}-\boldsymbol{P}_{\mu\nu};\boldsymbol{k})\,H(p_{\sigma\tau},\boldsymbol{x}'-\boldsymbol{P}_{\sigma\tau};\boldsymbol{k}').
\end{aligned} \tag{10.9.35}$$

This reduces (10.9.34) to a sum of integrals of the form

$$
\begin{aligned}
&-\iint H(p,\boldsymbol{x}-\boldsymbol{P};\boldsymbol{k})\,H(q,\boldsymbol{x}'-\boldsymbol{Q};\boldsymbol{k}')\frac{\partial}{\partial x_i}\left(\frac{1}{R}\right)d\boldsymbol{x}d\boldsymbol{x}'\\
&= -D^P_{\boldsymbol{k}}\,D^Q_{\boldsymbol{k}'}\int d\boldsymbol{x}\ \exp(-p(\boldsymbol{x}-\boldsymbol{P})^2)\left\{\frac{\partial}{\partial x_i}\int d\boldsymbol{x}'\ \exp(-q(\boldsymbol{x}'-\boldsymbol{Q})^2)\left(\frac{1}{R}\right)\right\}\\
&= +D^P_{\boldsymbol{k}}\,D^Q_{\boldsymbol{k}'}\int d\boldsymbol{x}\frac{\partial}{\partial x_i}\left[\exp(-p(\boldsymbol{x}-\boldsymbol{P})^2)\right]\int d\boldsymbol{x}'\ \exp(-q(\boldsymbol{x}'-\boldsymbol{Q})^2)\left(\frac{1}{R}\right)\\
&= -D^P_{\boldsymbol{k}+\boldsymbol{e}_i}\,D^Q_{\boldsymbol{k}'}\int d\boldsymbol{x}\ \exp(-p(\boldsymbol{x}-\boldsymbol{P})^2)\int d\boldsymbol{x}'\ \exp(-q(\boldsymbol{x}'-\boldsymbol{Q})^2)\left(\frac{1}{R}\right)\\
&= -(p,\boldsymbol{P};\boldsymbol{k}+\boldsymbol{e}_i\,|\,q,\boldsymbol{Q};\boldsymbol{k}') = +(p,\boldsymbol{P};\boldsymbol{k}\,|\,q,\boldsymbol{Q};\boldsymbol{k}'+\boldsymbol{e}_i)
\end{aligned}
\tag{10.9.36}
$$

The first step uses the HGTF definition (10.8.3). The second is an integration by parts, the third uses the fact that the partial derivative of the exponential by $\partial/\partial x_q$ can be replaced by that due to $-\partial/\partial P_q$ corresponding to the translation $\boldsymbol{k}\to\boldsymbol{k}+\boldsymbol{e}_q$, from which the last line follows. Collecting terms, and exploiting the symmetry of the two-electron HGTF integrals, we find

$$
\begin{aligned}
&(\mu,\beta;\nu,-\beta\,|k^B(R)|\,\sigma,\beta',\tau,-\beta') = \\
&\frac{1}{2}\sum_{q,q'}\sum_{\boldsymbol{k}\in\mathcal{T}_{\Lambda_{\mu\nu}}}\sum_{\boldsymbol{k}'\in\mathcal{T}_{\Lambda_{\sigma\tau}}} E^{\beta,-\beta}_q[\mu\nu;\boldsymbol{k}]\,E^{\beta',-\beta'}_{q'}[\sigma\tau;\boldsymbol{k}']\,M_{qq'}(\mu,\nu,\boldsymbol{k};\sigma,\tau,\boldsymbol{k}')
\end{aligned}
\tag{10.9.37}
$$

where

$$
\begin{aligned}
M_{qq'}(\mu,\nu,\boldsymbol{k};\sigma,\tau,\boldsymbol{k}') &= (p_{\mu\nu},\boldsymbol{P}_{\mu\nu};\boldsymbol{k}\,|\,p_{\sigma\tau},\boldsymbol{P}_{\sigma\tau};\boldsymbol{k}')\delta_{qq'}\\
&+\frac{1}{2}\left(\frac{1}{p_{\mu\nu}}+\frac{1}{p_{\sigma\tau}}\right)(p_{\mu\nu},\boldsymbol{P}_{\mu\nu};\boldsymbol{k}+\boldsymbol{e}_q\,|\,p_{\sigma\tau},\boldsymbol{P}_{\sigma\tau};\boldsymbol{k}'+\boldsymbol{e}_{q'})\\
&+(\boldsymbol{k}+\boldsymbol{k}')_{q'}(p_{\mu\nu},\boldsymbol{P}_{\mu\nu};\boldsymbol{k}+\boldsymbol{e}_q\,|\,p_{\sigma\tau},\boldsymbol{P}_{\sigma\tau};\boldsymbol{k}'-\boldsymbol{e}_{q'})\\
&+(\boldsymbol{P}_{\mu\nu}-\boldsymbol{P}_{\sigma\tau})_{q'}(p_{\mu\nu},\boldsymbol{P}_{\mu\nu};\boldsymbol{k}+\boldsymbol{e}_q\,|\,p_{\sigma\tau},\boldsymbol{P}_{\sigma\tau};\boldsymbol{k}').
\end{aligned}
$$

Equation (10.9.37) generalizes results of Rosicky [27] and Mohanty [28] to all symmetry types.

10.10 Fock matrix in terms of G-spinors

The construction of the $\boldsymbol{h}$ matrix is described in §10.5. In the closed shell case, the Coulomb, $\boldsymbol{G}$, and Breit, $\boldsymbol{B}$, matrices can be split into direct and exchange parts, for example

$$
\boldsymbol{G}^{\beta\beta'} = \boldsymbol{J}^{\beta\beta'} - \boldsymbol{K}^{\beta\beta'},
\tag{10.10.1}
$$

where from (10.9.4) the direct, J-matrix, has components

$$
\boldsymbol{J}^{\beta\beta'}_{\mu\nu} = \delta_{\beta\beta'}\sum_{\beta''}\sum_{\sigma\tau}(\mu\beta,\nu\beta\,|\,\sigma\beta'',\tau\beta'')\,D^{\beta''\beta''}_{\sigma\tau}
$$

and the exchange, K-matrix, has components

$$\boldsymbol{K}_{\mu\nu}^{\beta\beta'} = \sum_{\sigma\tau} (\mu\beta, \tau\beta \,|\, \sigma\beta', \nu\beta')\, D_{\sigma\tau}^{\beta'\beta}.$$

Substituting from (10.9.31) gives the G-spinor formulas

$$\boldsymbol{J}_{\mu\nu}^{\beta\beta'} = \delta_{\beta\beta'} \sum_{\boldsymbol{k},\boldsymbol{k}''} E_0^{\beta\beta}\,[\mu,\nu;\boldsymbol{k}] \sum_{\sigma\tau} [p_{\mu\nu},\mathbf{P}_{\mu\nu};\boldsymbol{k}|p_{\sigma\tau},\mathbf{P}_{\sigma\tau};\boldsymbol{k}''] \times \sum_{\beta''} E_0^{\beta''\beta''}\,[\sigma,\tau;\boldsymbol{k}'']\, D_{\sigma\tau}^{\beta''\beta''} \qquad (10.10.2)$$

and

$$\boldsymbol{K}_{\mu\nu}^{\beta\beta'} = \sum_{\boldsymbol{k},\boldsymbol{k}'} E_0^{\beta\beta}\,[\mu,\tau;\boldsymbol{k}] \sum_{\sigma\tau} [p_{\mu\tau},\mathbf{P}_{\mu\tau};\boldsymbol{k}|p_{\sigma\nu},\mathbf{P}_{\sigma\nu};\boldsymbol{k}'] \times E_0^{\beta'\beta'}\,[\sigma,\nu;\boldsymbol{k}']\, D_{\sigma\tau}^{\beta'\beta}. \qquad (10.10.3)$$

Almlöf [29, 30] noticed that it is possible to reduce the computational labour involved in J-matrix construction by making use of the fact that the two centre integrals like $[p_{\mu\nu},\mathbf{P}_{\mu\nu};\boldsymbol{k}|p_{\sigma\tau},\mathbf{P}_{\sigma\tau};\boldsymbol{k}'']$ depend only on the location of the nuclear centres, the exponents and the HGTF indices, but not on the angular symmetries of the individual basis functions. In order to exploit this, we note that the basis set label μ defined by (10.6.1) identifies a unique basis spinor with parameters

$$\mu = \{\boldsymbol{A}_\mu, a_\mu, \kappa_\mu, m_\mu\}$$

so that there will be a collection of basis spinors, say $\mu^{(i)}$, $i = 1,\ldots,n_\mu$ with common values of $\boldsymbol{A}_\mu$ and a_μ. We denote this set by

$$\{\mu\} = \{\mu^{(1)},\ldots,\mu^{(n_\mu)}\}. \qquad (10.10.4)$$

Define the Hermite density[5]

$$D_{\{\sigma\}\{\tau\}}(\boldsymbol{k}) = \sum_{\{\sigma\},\{\tau\}} \sum_{\beta} E_0^{\beta\beta}\,[\sigma,\tau;\boldsymbol{k}]\, D_{\sigma\tau}^{\beta\beta}, \qquad (10.10.5)$$

representing, together with the associated HGTF, the contribution to the total electron density from all spinor charges belonging to the classes $\{\sigma\}$ and $\{\tau\}$ with HGTF index $\boldsymbol{k}$. Combining this with the two-electron interaction integrals gives an effective one-electron potential energy matrix

$$U_{\{\mu\}\{\nu\}}(\boldsymbol{k}) = \sum_{\boldsymbol{k}'} \sum_{\{\sigma\}\{\tau\}} [p_{\mu\nu},\mathbf{P}_{\mu\nu};\boldsymbol{k}|p_{\sigma\tau},\mathbf{P}_{\sigma\tau};\boldsymbol{k}']\, D_{\{\sigma\}\{\tau\}}(\boldsymbol{k}') \qquad (10.10.6)$$

[5] This corresponds to the quantity denoted $\mathcal{H}[\alpha\beta; ijk]$ in the notation of Quiney [31], based on the paper by Almlöf [29]; $\boldsymbol{k} = (i,j,k)$ and the labels $\alpha\beta$ are equivalent to the classes $\{\sigma\}\{\tau\}$.

so that

$$J^{\beta\beta'}_{\mu\nu} = \delta_{\beta\beta'} \sum_{\boldsymbol{k}} E^{\beta\beta}_{0}\,[\mu,\nu;\boldsymbol{k}]\; U_{\{\mu\}\{\nu\}}(\boldsymbol{k}). \tag{10.10.7}$$

This construction involves a cost similar to that of a nuclear attraction integral, dramatically reducing the effort needed to construct the Coulomb direct repulsion integrals, but it is clear that the entanglement of indices in the exchange case frustrates any similar simplification of (10.9.36). This problem disappears when the energy is rewritten in terms of the energy of the internal electromagnetic fields in the manner of §10.11.

The Breit interaction can be treated in much the same way, though the Almlöf scheme is of less relevance as it can be shown that the direct Breit contribution for a relativistic closed shell system vanishes identically. Thus the closed shell Breit matrix is entirely due to exchange:

$$\begin{aligned} \boldsymbol{B}^{\beta,\beta}_{\mu\nu} &= -\frac{1}{2}\sum_{q,q'}\sum_{\boldsymbol{k},\boldsymbol{k}'}\sum_{\sigma\tau} E^{\beta,-\beta}_{q}[\mu\tau;\boldsymbol{k}] \\ &\qquad \times M_{qq'}(\mu,\tau,\boldsymbol{k};\sigma,\nu,\boldsymbol{k}')E^{-\beta,\beta}_{q'}[\sigma\nu;\boldsymbol{k}']\,D^{-\beta,-\beta}_{\sigma,\tau}. \end{aligned} \tag{10.10.8}$$

and

$$\begin{aligned} \boldsymbol{B}^{\beta,-\beta}_{\mu\nu} &= -\frac{1}{2}\sum_{q,q'}\sum_{\boldsymbol{k},\boldsymbol{k}'}\sum_{\sigma\tau} E^{\beta,-\beta}_{q}[\mu\tau;\boldsymbol{k}] \\ &\qquad \times M_{qq'}(\mu,\tau,\boldsymbol{k};\sigma,\nu,\boldsymbol{k}')E^{\beta,-\beta}_{q'}[\sigma\nu;\boldsymbol{k}']\,D^{\beta,-\beta}_{\sigma,\tau}. \end{aligned} \tag{10.10.9}$$

Magnetic interactions between open shells in general have both direct and exchange contributions.

10.10.1 The BERTHA integral package

In this section, we review the organisation needed to generate the Coulomb integrals $(\,ra\,|\,sb\,)$ over atomic or molecular spinors from the corresponding G-spinor integrals $(\,\mu\beta,\nu\beta\,|\,\sigma\beta',\tau\beta'\,)$. Similar methods can be used for generating Breit integrals and other quantities. As with all software, a process of incremental improvements is needed to make the fullest possible use of fast memory and of quantities generated at intermediate stages of the calculation. Each G-spinor is labelled by a multi-index pointer μ identifying a list of $\boldsymbol{A}_\mu, \kappa_\mu, m_\mu, a_\mu$ along with spinor component labels $\beta = \pm 1$, where $\boldsymbol{A}_\mu$ identifies the nuclear position, κ_μ, m_μ the spinor angular momentum quantum numbers and a_μ is one of a list of Gaussian exponents. The E_q-coefficients are best constructed in batches labelled by the orbital angular momentum indices $l_\mu, l_\nu, \ldots$. For example, p-type 4-spinors ($l = 1$) have relativistic angular quantum numbers $\kappa = +1$ ($j = 1/2$) and $\kappa = -2$ ($j = 3/2$). In constructing two-electron integrals over these p functions it makes sense to ensure that intermediate quantities contributing to more than one integral are used as efficiently as possible. Thus if all four orbitals are of p type, the G-spinor integrals will be characterized by 16 combinations of $\kappa_\mu, \kappa_\nu, \kappa_\sigma, \kappa_\tau$ with $l = 1$, namely the quadruples $(1,1,1,1), (-2,1,1,1), (1,-2,1,1) \ldots, (-2,-2,-2,-2)$; each charge-current pair $(\mu\nu)$, $(\sigma\tau)$ has 2 component indices $\beta, \beta' = \pm 1$, making 64 parameter sets in all. A single set of Gaussian exponents $a_\mu, a_\nu, a_\sigma, a_\tau$ will contribute the same set of HGTF two electron integrals in (10.9.30) for all 64 parameter combinations, and the calculation should be organized to exploit this fact.

G-spinor two-electron Coulomb integrals $(\,\mu\beta, \nu\beta \,|\, \sigma\beta', \tau\beta'\,)$ are therefore generated in batches of charge-current μ, ν overlaps, parametrized by a_μ, a_ν; $\boldsymbol{A}_\mu, \kappa_\mu, |m_\mu|$; $\boldsymbol{A}_\nu, \kappa_\nu, |m_\nu|; \beta$. Matrix elements for the different sign combinations $\pm|m_\mu|$; $\pm|m_\nu|$ can be obtained using the symmetry relations (10.8.16). A similar batch of four overlap charge distributions is obtained by replacing the labels $\mu\nu$ by $\sigma\tau$. If n_σ and n_τ denote the number of exponents a_σ and a_τ respectively, then there are $4n_\sigma n_\tau$ functions associated with the labels σ, τ, and BERTHA processes lists of $16n_\sigma n_\tau$ integrals, with a typical length of order 10,000.

The calculation of integrals over molecular spinors in terms of the G-spinor integrals (10.9.30) requires a four-index transformation (10.9.28). As in nonrelativistic integral generation codes, this can be done as a sequence of one-index transformations, the first being

$$
\begin{aligned}
(\,\mu\beta, \nu\beta \,|\, \sigma\beta', b\beta'\,) &= \textstyle\sum_\tau (\,\mu\beta, \nu\beta \,|\, \sigma\beta', \tau\beta'\,) c^{\beta'}_{\tau b}, \\
(\,\mu\beta, \nu\beta \,|\, \tau\beta', b\beta'\,) &= \textstyle\sum_\tau (\,\mu\beta, \nu\beta \,|\, \tau\beta', \sigma\beta'\,) c^{\beta'}_{\sigma b}
\end{aligned}
\qquad (10.10.10)
$$

where b denotes an occupied MS. This requires an array of length $4MN$, where M is the number of occupied MS and N the total number of basis functions (for both values of β). The transformation in the second line of (10.10.10) is needed only for integrals off-diagonal in σ, τ and uses the symmetries (10.8.16), to eliminate unnecessary calls to evaluate G-spinor integrals. The operational cost scales as MN^4. A second transformation step generates

$$
(\,\mu\beta, \nu\beta \,|\, sb\,) = \sum_{\beta'} \sum_{\bar\sigma} (\,\mu\beta, \nu\beta \,|\, \bar\sigma\beta', b\beta'\,) c^{*\beta'}_{\bar\sigma s}, \qquad (10.10.11)
$$

where $\bar\sigma$ runs over all possible basis spinor labels associated with the running value of β'. The operational cost is of order $MM'N^3$, where $2(M+M') = N$, so that M' is the number of virtual MS. The memory requirement is $4n_\mu n_\nu MM'$ numbers, each of which contributes to the interaction integrals involving every charge-current pair ra.

Similarly the next one-index transformation for each fixed a contributes to every charge-current pair ra and each β component:

$$
\begin{aligned}
(\,\mu\beta, a\beta \,|\, sb\,) &\overset{(+)}{=} \textstyle\sum_\nu (\,\mu\beta, \nu\beta \,|\, sb\,) c^{\beta}_{\nu a} \\
(\,\nu\beta, a\beta \,|\, sb\,) &\overset{(+)}{=} \textstyle\sum_\mu (\,\nu\beta, \mu\beta \,|\, sb\,) c^{\beta}_{\mu a}.
\end{aligned}
\qquad (10.10.12)
$$

This step involves $M^2M'N^2$ operations and $(n_\mu+n_\nu)M^2M'$ numbers in memory. The final one-index transformation gives the MS integrals

$$\begin{aligned}(\,ra\,|\,sb\,) &\overset{(+)}{=} \textstyle\sum_\beta \sum_\mu (\,\mu\beta, a\beta\,|\,sb\,) c^{*\beta}_{\mu r}\\ (\,ra\,|\,sb\,) &\overset{(+)}{=} \textstyle\sum_\beta \sum_\nu (\,\nu\beta, a\beta\,|\,sb\,) c^{\beta}_{\nu r}\end{aligned} \tag{10.10.13}$$

costing $m^2M'^2N$ operations and storage of $M^2M'^2/2$ numbers, ignoring savings due to symmetry relations. At each stage, the memory requirements are relatively modest, and the algorithm is implemented so that memory can be released once the final integrals have been generated.

10.11 Electromagnetic field energy

If N is the size of the basis set, then the complete set of integrals needed to build the Fock matrix from primitive integrals over pairs of charge-current overlap densities has dimension of order N^4. Can we devize a method of generating the Fock matrix (10.5.1) that grows more slowly with N? In this section we explore a promising method suggested by Quiney [32], which is based on the equivalence of the interaction energy of a charge-current distribution with the energy of the interacting fields to which it gives rise. The result is a factorization algorithm for computing Fock matrix elements based on integrating products of $O(N^2)$ fields generated by the charge-current overlaps over the space occupied by the molecule. Economies result from replacing calculation of $O(N^4)$ expensive interaction integrals with $O(N^2)$ less expensive field components at suitable integration points covering the whole volume.

10.11.1 Interaction energy in terms of internal fields

We start by writing the Coulomb interaction energy of two static charge distributions $\rho_{ac}(\boldsymbol{r})$ and $\rho_{bd}(\boldsymbol{r})$ as

$$(ac\,|\,bd) = \frac{1}{4\pi\epsilon_0} \iint d\boldsymbol{r}\, d\boldsymbol{s}\, \rho_{ac}(\boldsymbol{r})\, \frac{1}{R}\, \rho_{bd}(\boldsymbol{s}) \tag{10.11.1}$$

where $R = |\boldsymbol{r} - \boldsymbol{s}|$ and

$$\rho_{ac}(\boldsymbol{r}) = -e\psi^\dagger_a(\boldsymbol{r}\,)\psi_c(\boldsymbol{r}), \quad \rho_{bd}(\boldsymbol{s}) = -e\psi^\dagger_b(\boldsymbol{s}\,)\psi_d(\boldsymbol{s}).$$

We recall that the factor $e^2/4\pi\epsilon_0$ is unity in atomic units. The electron density distribution ρ_{bd} and the electrostatic potential V_{bd} are related by

$$\boldsymbol{\nabla}^2 V_{bd}(\boldsymbol{r}) = -\frac{1}{\epsilon_0}\rho_{bd}(\boldsymbol{r}), \quad V_{bd}(\boldsymbol{r}) = -\frac{1}{4\pi\epsilon_0}\int \frac{\rho_{bd}(\boldsymbol{s})}{|\boldsymbol{r}-\boldsymbol{s}|}\, d\boldsymbol{s}, \tag{10.11.2}$$

from which we get the electrostatic field

$$\boldsymbol{E}_{bd}(\mathbf{r}) = -\boldsymbol{\nabla}\, V_{bd}(\boldsymbol{r}). \tag{10.11.3}$$

We can therefore write (10.11.1) as

$$(ac\,|\,bd) = \int \rho_{ac}(\boldsymbol{r})\, V_{bd}(\boldsymbol{r})\, d\boldsymbol{r} = -\epsilon_0 \int \boldsymbol{\nabla}^2 V_{ac}(\boldsymbol{r})\, V_{bd}(\boldsymbol{r})\, d\boldsymbol{r}.$$

If ρ_{ac} and ρ_{bd} are everywhere continuous, then V_{ac} and V_{bd} have continuous second partial derivatives, and

$$\begin{aligned} &-\int \left[\nabla^2 V_{ac}(\mathbf{r})\right]\, V_{bd}(\mathbf{r})\, d\mathbf{r} \\ &\qquad = \int \nabla V_{ac}(\mathbf{r}) \cdot \nabla V_{bd}(\mathbf{r})\, d\mathbf{r} - \int \mathrm{div}\left[V_{bd}(\mathbf{r}) \nabla V_{ac}(\mathbf{r})\right]\, d\mathbf{r}. \end{aligned}$$

The last integral can be transformed into a surface integral over the boundary of the region of integration, and the smoothness and the asymptotic behaviour of the potentials normally suffice to make the boundary contribution vanish. When this applies, the interaction integral (10.11.1) is equivalent to

$$\langle ac\,|\,bd\rangle = \epsilon_0 \int \mathbf{E}_{ac}(\mathbf{r}) \cdot \mathbf{E}_{bd}(\mathbf{r})\, d\mathbf{r}, \tag{10.11.4}$$

which is clearly proportional to the mutual potential energy of the two electric fields.

Proposition 10.1. *Let $\mathbf{E}_{ac}(\mathbf{r})$ be the electric field strength due to the (Dirac or Schrödinger) electron overlap density $\rho_{ac}(\mathbf{r})$. Then*

(1) $\rho_{ac}(\mathbf{r}) = \rho^{\dagger}_{ca}(\mathbf{r})$, so that $\mathbf{E}_{ac}(\mathbf{r}) = \mathbf{E}^{\dagger}_{ca}(\mathbf{r})$ and, in particular, $\mathbf{E}_{aa}(\mathbf{r})$ is real and positive.
(2) The exchange integral $(ab\,|\,ba) = \epsilon_0 \int \mathbf{E}_{ab}(\mathbf{r}) \cdot \mathbf{E}_{ba}(\mathbf{r})\, d\mathbf{r}$ is therefore strictly positive [33, Appendix 19].

Proposition 10.1(2) reprises Slater's demonstration [33, Appendix 19] of the positive character of the exchange integral $(ab\,|\,ba)$. Feynman [34, Chapter 19] discussed the equvalence of the field and interaction formulations in connection with the principle of least action, and this result is implicit in his presentation. This equivalence is also cited as the foundation of schemes to evaluate the Coulomb energy by fitting the total electron density to an auxiliary Gaussian basis set of atom-centred functions [35], although the computation is performed in terms of conventional interaction integrals. Beebe and Linderberg [36] noted that the use of atom-centred basis sets made the two-electron Fock matrix linearly dependent, suggesting that economies could result from expressing it in terms of finite matrix representation of lower dimension of either the electron density matrix or the electrostatic interaction matrix. This led to the development of schemes for Cholesky factorization of the two-electron interaction matrix, to block-diagonal representations of the density matrix and also to several highly accurate variational schemes for the approximation of the Coulomb interaction energy. Applications to non-relativistic quantum chemistry have been reviewed recently in [37, 38]; see also Manby and Knowles [39]. The closed shell Hartree-Fock energy for the molecular Hamiltonian (10.1.1) is

$$E = E_1 + E_2 + \frac{e^2}{4\pi\epsilon_0}\sum_{A<B}\frac{Z_A Z_B}{|\boldsymbol{A}-\boldsymbol{B}|}, \quad E_2 = E_{dir} - E_{exch} \tag{10.11.5}$$

where

$$E_1 = \sum_a \langle a \,|\, h \,|\, a\rangle \tag{10.11.6}$$

is the one-electron energy, and the two-electron part, E_2, is the difference of *direct* and *exchange* contributions:

$$E_{dir} = \frac{1}{2}\epsilon_0 \int \sum_a \sum_b \mathbf{E}_{aa}(\mathbf{r})\cdot\mathbf{E}_{bb}(\mathbf{r})d\boldsymbol{r}, \tag{10.11.7}$$

$$E_{exch} = \frac{1}{2}\epsilon_0 \int \sum_a \sum_b \mathbf{E}_{ab}(\mathbf{r})\cdot\mathbf{E}_{ba}(\mathbf{r})d\boldsymbol{r}. \tag{10.11.8}$$

The indices a, b run over all occupied closed shells. Both E_{dir} and E_{exch} contain self-interaction terms in which $b = a$,

$$E_s = \frac{1}{2}\epsilon_0 \int \sum_a \mathbf{E}_{aa}(\mathbf{r})\cdot\mathbf{E}_{aa}(\mathbf{r})d\boldsymbol{r}$$

and their contribution to the total energy E cancels. Thus we can remove these terms from both direct and exchange sums, leaving

$$E'_{dir} = \epsilon_0 \int \sum_{a<b} \mathbf{E}_{aa}(\mathbf{r})\cdot\mathbf{E}_{bb}(\mathbf{r})d\boldsymbol{r}, \tag{10.11.9}$$

$$E'_{exch} = \epsilon_0 \int \sum_{a<b} \mathbf{E}_{ab}(\mathbf{r})\cdot\mathbf{E}_{ba}(\mathbf{r})d\boldsymbol{r}. \tag{10.11.10}$$

where the primes indicate that the unphysical self-interaction terms have been dropped. We can put E_1 into a similar form. Write

$$E_1 = T + V_{eN} \tag{10.11.11}$$

where

$$T = \begin{cases} \sum_a \langle a \,|\, \boldsymbol{p}^2/2m \,|\, a\rangle & \text{(Nonrelativistic)} \\ \sum_a \langle a \,|\, c\boldsymbol{\alpha}\cdot\boldsymbol{p} + (\beta - I)mc^2 \,|\, a\rangle & \text{(Relativistic)} \end{cases}$$

and

$$V_{eN} = \int \sum_a \rho_{aa}(\boldsymbol{r}). \sum_A V_A(\boldsymbol{r})\, d\boldsymbol{r} = \epsilon_0 \int \boldsymbol{E}_e(\boldsymbol{r}) \cdot \boldsymbol{E}_N(\boldsymbol{r})\, d\boldsymbol{r} \qquad (10.11.12)$$

where $\boldsymbol{E}_e(\boldsymbol{r}) = \sum_a \boldsymbol{E}_{aa}(\boldsymbol{r})$ and $\boldsymbol{E}_N(\boldsymbol{r}) = \sum_A \boldsymbol{E}_A(\boldsymbol{r})$ are the $\boldsymbol{E}$-fields generated by the occupied electrons and the nuclei respectively. The nuclear fields are classical; for the point nucleus

$$\boldsymbol{E}_A(\boldsymbol{r}) = \frac{Z_A e}{4\pi\epsilon_0} \frac{\boldsymbol{r} - \boldsymbol{A}}{|\boldsymbol{r} - \boldsymbol{A}|^3},$$

which needs modifying close to the nuclei when using a distributed charge model.

10.11.2 The nonrelativistic Fock matrix

The elements of the closed shell nonrelativistic Fock matrix (omitting the Coulomb replsion of the nuclei) are given by

$$F_{\mu\nu} = h_{\mu\nu} + \sum_{\sigma\tau} \left[(\mu\nu \,|\, \sigma\tau) - (\mu\tau \,|\, \sigma\nu) \right] D_{\sigma\tau} \qquad (10.11.13)$$

Equations (10.11.4) and (10.11.12) allow us to write this

$$\begin{aligned} F_{\mu\nu} &= T_{\mu\nu} \qquad (10.11.14) \\ &+ \epsilon_0 \int \left\{ (\boldsymbol{E}(\boldsymbol{r}) + \boldsymbol{E}_N(\boldsymbol{r})) \cdot \boldsymbol{E}_{\mu\nu}(\boldsymbol{r}) - \sum_{\sigma\tau} D_{\sigma\tau} \boldsymbol{E}_{\mu\tau}(\boldsymbol{r}) \cdot \boldsymbol{E}_{\sigma\nu}(\boldsymbol{r}) \right\} d\boldsymbol{r}, \end{aligned}$$

where

$$\boldsymbol{E}(\boldsymbol{r}) = \sum_a \boldsymbol{E}_{aa}(\boldsymbol{r}) = \sum_{\sigma\tau} D_{\sigma\tau} \boldsymbol{E}_{\sigma\tau}(\boldsymbol{r}).$$

10.11.3 The relativistic Fock matrix

The structure of the relativistic Fock matrix is very similar to (10.11.14), although we have now to take the block structure of the Fock matrix into account and to incorporate the magnetic energy terms. Thus the $\beta\beta'$ block of the Fock matrix is given by

$$F^{\beta\beta'}_{\mu\nu} = h^{\beta\beta'}_{\mu\nu} + G^{\beta\beta'}_{\mu\nu} + B^{\beta\beta'}_{\mu\nu} \qquad (10.11.15)$$

The one-electron Hamiltonian matrix is given, as in (10.5.2), by

$$h^{+-}_{\mu\nu} = \left[h^{-+}_{\nu\mu} \right]^* = c\boldsymbol{\Pi}^{+-}, \qquad (10.11.16)$$

whilst its diagonal blocks are

$$
\begin{aligned}
h^{++}_{\mu\nu} &= \epsilon_0 \int \boldsymbol{E}_N(\boldsymbol{r}) \cdot \boldsymbol{E}^{++}_{\mu\nu}(\boldsymbol{r})\, d\boldsymbol{r}, \\
h^{--}_{\mu\nu} &= -2mc^2 S^{--}_{\mu\nu} + \epsilon_0 \int \boldsymbol{E}_N(\boldsymbol{r}) \cdot \boldsymbol{E}^{--}_{\mu\nu}(\boldsymbol{r})\, d\boldsymbol{r}
\end{aligned}
\tag{10.11.17}
$$

The diagonal blocks of the $\boldsymbol{G}$ matrix are very like their nonrelativistic counterparts:

$$
G^{\beta\beta}_{\mu\nu} = \epsilon_0 \int \left\{ \boldsymbol{E}(\boldsymbol{r}) \cdot \boldsymbol{E}^{\beta\beta}_{\mu\nu}(\boldsymbol{r}) - \sum_{\sigma\tau} \boldsymbol{E}^{\beta\beta}_{\mu\tau}(\boldsymbol{r}) \cdot \boldsymbol{E}^{\beta,\beta}_{\sigma\nu}(\boldsymbol{r}) D^{\beta,\beta}_{\sigma\tau} \right\} d\boldsymbol{r} \tag{10.11.18}
$$

where now the total internal $\boldsymbol{E}$-field is

$$
\boldsymbol{E}(\boldsymbol{r}) = \sum_{\sigma\tau} \left[D^{++}_{\sigma\tau} \boldsymbol{E}^{++}_{\sigma\tau}(\boldsymbol{r}) + D^{--}_{\sigma\tau} \boldsymbol{E}^{--}_{\sigma\tau}(\boldsymbol{r}) \right],
$$

and

$$
G^{\beta,-\beta}_{\mu\nu} = -\epsilon_0 \int \sum_{\sigma\tau} \boldsymbol{E}^{\beta\beta}_{\mu\tau}(\boldsymbol{r}) \cdot \boldsymbol{E}^{-\beta,-\beta}_{\sigma\nu}(\boldsymbol{r}) D^{-\beta,\beta}_{\sigma\tau} d\boldsymbol{r} \tag{10.11.19}
$$

The magnetic interaction matrices have the structure

$$
B^{\beta\beta}_{\mu\nu} = -\frac{1}{\mu_0} \int \sum_{\sigma\tau} \boldsymbol{B}^{\beta,-\beta}_{\mu\tau}(\boldsymbol{r}) \cdot \boldsymbol{B}^{-\beta,\beta}_{\sigma\nu}(\boldsymbol{r}) D^{-\beta,-\beta}_{\sigma\tau} d\boldsymbol{r} \tag{10.11.20}
$$

and

$$
B^{\beta,-\beta}_{\mu\nu} = \frac{1}{\mu_0} \int \left\{ \boldsymbol{B}(\boldsymbol{r}) \cdot \boldsymbol{B}^{\beta,-\beta}_{\mu\nu}(\boldsymbol{r}) - \sum_{\sigma\tau} \boldsymbol{B}^{\beta,-\beta}_{\mu\tau}(\boldsymbol{r}) \cdot \boldsymbol{B}^{\beta,-\beta}_{\sigma\nu}(\boldsymbol{r}) D^{\beta,-\beta}_{\sigma\tau} \right\}, \tag{10.11.21}
$$

where the total $\boldsymbol{B}$-field is given by

$$
\boldsymbol{B}(\boldsymbol{r}) = \sum_{\sigma\tau} \left[D^{+-}_{\sigma\tau} \boldsymbol{B}^{+-}_{\sigma\tau}(\boldsymbol{r}) + D^{-+}_{\sigma\tau} \boldsymbol{B}^{-+}_{\sigma\tau}(\boldsymbol{r}) \right]
$$

The contribution to the total $\boldsymbol{B}$-field from closed relativistic subshells vanishes as in the conventional formulation. Provided we can take the internal fields in a molecule as quasi-stationary and thus neglect the time-dependence, the fields in a G-spinor representation can be constructed using the standard primitive elements from (10.9.25) and (10.9.26). However, new integrals appear when the finite wavelength of the exchanged photon is taken into account.

10.11.4 Implementation of the field formulation

The value of this approach becomes more obvious in the context of a recent method of numerical integration over the space occupied by the system that was developed for density functional calculations by Becke [40] (see §10.12). The assembly of the integrands at each integration point can be done from (10.11.15)–(10.11.21). The different parts of the calculation are independent so that there is much scope for parallel computation. We can exploit the fact that HGTF $H(p, \boldsymbol{x}; \boldsymbol{k})$ are essentially zero when $r > R(p) = C/\sqrt{p}$, where the constant C is typically of order 3–4. We can therefore make a multipole expansion of the function in (10.9.6), so that

$$[a, \boldsymbol{A}; \boldsymbol{k} \,|\, \boldsymbol{r}] = \sum_{l=0}^{\infty} \frac{M_l(a; \boldsymbol{k})}{r^{l+1}}, \quad r > C/\sqrt{a} \tag{10.11.22}$$

where

$$M_l(a; \boldsymbol{k}) = \int H(a, \boldsymbol{s}; \boldsymbol{k}) \, |\boldsymbol{s}|^l P_l(\cos\Theta) d\boldsymbol{s}$$

and $\cos\Theta = \boldsymbol{s} \cdot \boldsymbol{r}/rs$. The numbers $M_l(a; \boldsymbol{k})$ can be pre-tabulated. Only a few terms of ths expansion will be needed at many integration points, especially for contributions from other centres B, making the calculation relatively cheap. A simple DHF test calculation on the CO molecule was described in [41]. Table 10.3 shows the energies of the molecular orbital spinors for three different grids

Table 10.3. DHF energies (a.u.) for the CO molecule. [41, Table 2]

Eigenvalue	Grid 1	Grid 2	Grid 3	'Exact'
$1\sigma_{1/2}$	-20.70182	-20.70122	-20.70195	-20.70188
$2\sigma_{1/2}$	-11.39955	-11.39924	-11.39930	-11.39924
$3\sigma_{1/2}$	-1.56497	-1.56404	-1.56494	-1.56494
$4\sigma_{1/2}$	-0.80039	-0.80002	-0.80026	-0.80024
$1\pi_{1/2}$	-0.64624	-0.64694	-0.64703	-0.64701
$1\pi_{3/2}$	-0.64529	-0.64599	-0.64608	-0.64607
$5\sigma_{1/2}$	-0.55899	-0.55905	-0.55888	-0.55889
E_{DHF}	-112.7837	-112.7828	-112.7828	-112.7828

of increasing quality compared with the results of the conventional action-at-a-distance method of calculating electron repulsion integrals. The total energy of the molecule for each calculation appears at the bottom of the table. The calculation revealed unexpected features of the spatial distribution of electrostatic energy that are, nevertheless, easy to interpret in physical terms. The energy density near the boundary of each atom-centred region is asymptotically proportional to r^{-4}, so that the fields can then be generated very cheaply. Becke's scheme, §10.12, partitions the space occupied by the system into overlapping cells centred on each nuclear centre, A, and distributes the integrand over the cells, using a set of weights $w_A(\boldsymbol{r}_i)$ so that, at each integration point $\boldsymbol{r}_i$, we have $\sum_A w_A(\boldsymbol{r}_i) = 1$. The weight function w_A is close to unity near the nucleus A and vanishes at some point on the line joining A to each of the other nuclei. This partitioning necessitates representation by high order spherical harmonics near cell boundaries, which makes it necessary to use many angular integration points. To some extent this offsets the low cost of the multipole approximation of the fields. This means that some experiment is necessary, as in similar calculations in density functional theory, to select the most effective combination of integration points [42].

Visscher's method [43] for approximating long-range interactions with small component electron densities inexpensively uses an expansion on the lines of (10.11.22). The interaction energy of small component charge clouds on different centres $\boldsymbol{A}$ and $\boldsymbol{B}$ can be written as a classical multipole expansion whose leading term is

$$\frac{e^2}{4\pi\epsilon_0}\frac{q_A^- \, q_B^-}{|\boldsymbol{A}-\boldsymbol{B}|}$$

where

$$q_A^- = \int \sum_{\mu\nu} c_{\mu A}^{-1\,\dagger} c_{\nu A}^{-1} \rho_{\mu\nu}^{--}(\boldsymbol{r}) d\boldsymbol{r}.$$

is the total small component charge on nucleus $\boldsymbol{A}$. Assuming the exchange contributions can be neglected, the total energy arising from *multi-centre* small-small interactions is

$$E_2^{--} = \frac{1}{2}\sum_{k\neq l}\frac{q_{A_k}^- \, q_{A_l}^-}{|\boldsymbol{A}_k - \boldsymbol{A}_l|}$$

The corresponding contribution to the direct Coulomb matrix $\boldsymbol{G}^{--}$, (10.5.4) is

$$G_{\mu\nu}^{--} = \sum_k \int \rho_{\mu\nu}^{--}(\boldsymbol{r})\frac{q_{A_k}^-}{|\boldsymbol{r}-\boldsymbol{A}_k|}\delta_{A_\mu,A_\nu}(1-\delta_{A_\mu,A_k})(1-\delta_{A_\nu,A_k})d\boldsymbol{r}$$

where the sum runs over all nuclear centres, k, and the delta functions select one-centre overlaps on centres other than $\boldsymbol{A}_k$. The main small-small contributions are from single-centre interactions which can be evaluated inexpensively using the methods of §10.2.

10.12 Relativistic density functional calculations

The recent review of RDFT by Engel *et al.* [44] gives a comprehensive description of a wide range of density functional schemes and the results obtained from them. These authors concluded that even after 30 years of work, no applications of the most rigorous form of RDFT have yet been reported in the literature, which they attribute to the lack of practical and reliable approximations for the exchange-correlation density functional. Semi-empirical j-dependent exchange-correlation functionals have been proposed, but their performance is poor for both atoms and molecules. First principles functionals require calculation of the complete eigenvalue spectrum and interaction integrals between them, with a consequential dramatic increase in the computational effort compared with simple LDA schemes (cf. §4.15). The relativistic optimum potential method (ROPM) is capable of giving good agreement with exact cxchange at the DHF level, but attempts to deal with correlation in this way have not been too encouraging. Engel *et al.* conclude that the next step must be to derive an accurate, universally applicable and efficient correlation functional that can be used with exact exchange.

A framework for performing RDFT calculations using the G-spinor methods of this chapter has recently been proposed by Quiney and Belanzoni [31]. The simplest form of the relativistic Kohn-Sham equations has been given in (4.15.15) and (4.15.16):

$$\left(c\boldsymbol{\alpha}\cdot\boldsymbol{p}+\beta mc^2+\alpha^{\mu}\,a_{s,\mu}(\boldsymbol{x})\right)\psi_k(\boldsymbol{x})=E_k\,\psi_k(\boldsymbol{x}), \tag{10.12.1}$$

in which $\alpha^{\mu}=\gamma^0\gamma^{\mu}$ and the local four-potential $a_{s,\mu}(\boldsymbol{x})$ is

$$a_{s,\mu}(\boldsymbol{x})=v_{\mu}(\boldsymbol{x})+v_{H,\mu}(\boldsymbol{x})+v_{xc,\mu}(\boldsymbol{x}).$$

In practice, the space-like components are often neglected, so that only the Coulomb interaction, for which $\mu=0$, is taken into account, and the potentials are then functionals of the total electron density $\rho(\boldsymbol{x})$. In this case, the external potential $v_{\mu}(\boldsymbol{x})$ reduces to the electron-nuclear interaction, with matrix elements $[V_{nuc}]^{\beta\beta}_{\mu\nu}$ given by (10.9.15). The Hartree energy is a functional of $\rho(\boldsymbol{x})$

$$E_H=\frac{1}{2}\iint\frac{\rho(\boldsymbol{x})\rho(\boldsymbol{y})}{|\boldsymbol{x}-\boldsymbol{y}|}\,d\boldsymbol{x}d\boldsymbol{y}=\frac{1}{2}\int\rho(\boldsymbol{x})\,v_{H,0}[\rho](\boldsymbol{x})\,d\boldsymbol{x},$$

and, from (10.10.5),

$$\rho(\boldsymbol{x})=\sum_{\sigma\tau}H(p_{\sigma\tau},\boldsymbol{x}-\boldsymbol{P}_{\sigma\tau};\boldsymbol{k})D_{\{\sigma\}\{\tau\}}(\boldsymbol{k}) \tag{10.12.2}$$

so that the Almlöf prescription for the $\boldsymbol{J}^{\beta\beta}$ matrices (10.10.7) actually generates the direct interactions in the form required for RDFT. The explicit LDA energy functionals (including formal GGA terms) can be written

$$E_{xc,0}=\int F(\rho,\boldsymbol{\nabla}\rho)\,d\boldsymbol{x} \tag{10.12.3}$$

so that the exchange potential is

$$v_{xc,0}(\boldsymbol{x})=\frac{\delta E_{xc}[j]}{\delta\rho}=\frac{\partial F}{\partial\rho}-\boldsymbol{\nabla}\cdot\frac{\partial F}{\partial\boldsymbol{\nabla}\rho}. \tag{10.12.4}$$

The matrices of $v_{xc,0}$ can be calculated in the same way as the electron-nucleus interaction (10.9.15), and the kinetic and overlap matrices are given respectively by (10.9.11) and (10.9.10). The relativistic Kohn-Sham equations must be solved iteratively to practical self-consistency for the occupied spinors $\psi_k(x)$ and for the electron density in much as the same way as DHF.

More elaborate formulations of RDFT involve other components of the four-current density, their magnetic effects, and exchange-correlation schemes. The simplest form, presented here, illustrates the main features of RDFT calculations. Quiney and Belanzoni [31] follow DFT precedent by performing integrations in configuration space using Becke's numerical integration method [40], which was adapted to calculate electromagnetic energies in §10.11.4. In DFT calculations, each of the N_A nuclei act as the centre of a three-dimensional cell, and the contribution to a cubature at each integration point $\boldsymbol{x}$ is divided between the cells using weight functions $w_A(\boldsymbol{x})$, so that

$$\sum_{A=1}^{N_A} w_A(\boldsymbol{x}) = 1.$$

Then, every integral over the space occupied by the molecule can be partitioned as a sum

$$\int f(\mathbf{x})\, d\mathbf{x} = \sum_A \int w_A(\mathbf{x}) f(\mathbf{x})\, d\mathbf{x}. \tag{10.12.5}$$

The linearity of (10.12.5) allows each cell to be extended over all space and treated as independent of its neighbours.

Much of the relativistic effect is due to an accumulation of electron density in the neighbourhood of each nucleus. Becke [40] noted that the integrands in molecular structures are dominated by one-centre contributions and by a pairwise sum over two-centre contributions. The cell weights, $w_A(\boldsymbol{x})$, which distribute integrands over the different cells, must therefore be chosen so that sharp peaks in the electron density near each nucleus, A, are strongly weighted in the cell centred on $\boldsymbol{A}$, and suppressed in cells centred on other nuclei, say $\boldsymbol{B}$. The choice advocated by Becke, adopted in [31], uses the sequence of polynomials

$$f^{(1)}(t) = \frac{3}{2}t - \frac{1}{2}t^3, \quad f^{(k)}(t) = f\left(f^{(k-1)}(t)\right), \quad k = 1, 2, \ldots$$

together with the polynomials

$$P_A(\boldsymbol{x}) = \prod_{B \neq A} s(\mu_{AB}),$$

where, if $r_A = |\boldsymbol{x} - \boldsymbol{A}|$, $r_B = |\boldsymbol{x} - \boldsymbol{B}|$, $R_{AB} = |\boldsymbol{A} - \boldsymbol{B}|$, then

$$s(t) = \frac{1}{2}\left[1 - f^{(k)}(t)\right], \quad \mu_{AB} = (r_A - r_B)/R_{AB}.$$

The choice $k = 3$ is said to work well in practice. At each cubature point, Becke's weights are defined by

$$w_A(\boldsymbol{x}) = P_A(\boldsymbol{x}) / \sum_B P_B(\boldsymbol{x}). \tag{10.12.6}$$

The cell cubatures in [31] use a local spherical coordinate system $\boldsymbol{x}_{A,j}$ so that, at the j-th integration point, $\boldsymbol{x}_j = \boldsymbol{A} + \boldsymbol{x}_{A,j}$, and

$$\boldsymbol{x}_{A,j} \equiv (r_{A,j}, \theta_{A,j}, \varphi_{A,j}), \quad j = 1, \ldots, N_j.$$

The integrands are almost spherically symetrical near the nucleus, but the weight functions $w_A(\boldsymbol{x})$ at large radii are angular dependent. It has been found advisable to divide each cell into three concentric spherical shells, each with its own radial and angular mapping following the prescriptions of Treutler and Ahlrichs [42]. Radial integration is performed using Gauss-Chebychev quadrature on an auxiliary variable x,

$$\int_{-1}^{+1} G(x)\, dx \approx \sum_{j=1}^{N_j} w_j\, G(x_j), \quad w_j = \frac{\pi}{N_j}, \quad x_j = \cos\left[\frac{\pi(j-1/2)}{N_j}\right].$$

Suitable mappings $x \to r$ are discussed in [31]; the one used to generate Table 10.4 below was

$$x \to r = \frac{\xi}{\ln 2}(x+1)^\alpha \ln\left(\frac{2}{1-x}\right), \quad -1 \le x < 1, \ \alpha = 0.6.$$

The scale factor ξ ($r = \xi$ when $x = 0$) was chosen as the Bragg-Slater radius for the corresponding element. This empirical parameter includes any effect on the effective atomic radius due to relativistic effects. The integration over polar angles used an integration scheme

$$\int_{\theta=0}^{\pi} \int_{\varphi=0}^{2\pi} f(\theta, \varphi) \sin\theta d\theta\, d\varphi \approx 4\pi \sum_{i=1}^{N_L} s_i\, f(\theta_i, \varphi_i),$$

due to Lebedev and Laikov [45, 46]. A computer program is available [47] to generate the Lebedev-Laikov weights and integration points.

This approach has been tested [31] on atoms by comparing the G-spinor results with comparable finite difference calculations and also on a selection of small molecules. The values shown in Table 10.4 were computed using even tempered basis sets; for each value of the orbital quantum number l, the G-spinor exponents were determined from

$$a_k = \alpha\, \beta^{k-1}, \quad k = 1, 2, \ldots, N_l.$$

Values of the parameters α and β and the corresponding dimensions N_l can be found in the Appendix, Table A.15. The DHF values have comparable accuracy to those from finite difference calculations, and the RDFT results using G-spinors agree well with comparable calculations by Engel and Dreizler [48] (for the column RLDA^{x0}) and Engel, Keller, and Dreizler [49] (for the columns RLDA^{xL} and RLDA^{xT}), which used completely different numerical methods.

Table 10.4. Energies (a.u.) of atomic ground states in various RDFT local density schemes for a selection of closed shell atoms. [31]. The column labelled RLDA^{x0} uses the NRHEG exchange energy (4.15.19); RLDA^{xL} the RHEG exchange energy (4.15.21) using the relativistic Coulomb correction (4.15.22) only; and RLDA^{xT} uses the full relativistic correction including transverse contributions. Reproduced, with permission, from [31].

Element	DHF	RLDA^{x0}	RLDA^{xL}	RLDA^{xT}
He	-2.862	-2.724	-2.724	-2.724
Be	-14.576	-14.226	-14.226	-14.224
Ne	-128.692	-127.635	-127.628	-127.593
Mg	-199.935	-198.569	-198.556	-198.492
Ar	-528.684	-526.387	-526.337	-526.088
Ca	-679.710	-677.118	-677.047	-676.694
Zn	-1794.612	-1790.719	-1790.456	-1789.136
Kr	-2788.859	-2783.754	-2783.278	-2780.883
Sr	-3178.078	-3172.633	-3172.066	-3169.205
Pd	-5045.252	-5039.130	-5038.174	-5032.728
Cd	-5593.311	-5586.284	-5585.071	-5578.925
Xe	-7446.880	-7438.833	-7437.050	-7427.986
Ba	-8135.614	-8127.302	-8125.294	-8115.068
Yb	-14067.518	-14058.350	-14054.172	-14032.687
Hg	-19648.564	-19637.852	-19631.280	-19597.206
Rn	-23601.629	-23590.333	-23581.865	-23537.726
Ra	-25027.533	-25016.197	-25007.003	-24958.997

The extensive investigation of radial and angular grids for nonrelativistic DFT calculations carried out by Treutler and Ahlrichs [42] provided a good starting point. The effective potential is nearly spherical near the cell origin, but whilst the partitioning of each integrand between cells using (10.12.6) makes it possible to treat the cell integrations independently, it introduces strong angular variation in the inter-nuclear region. Thus fewer angular points are needed for the two inner concentric spherical shells, with more in the outer shell beyond the Bragg-Slater radius. A small increase in the number of angular points is necessary to accomodate the fact that when the symmetry number η_μ, is negative, the spherical harmonic in the the small component, (10.6.7), is one order higher, $\bar{l}_\mu = l_\mu + 1$ than in the associated large component (10.6.6).

Quiney and Belanzoni [31] also considered a selection of molecules at the Dirac-Hartree-Slater level of approximation (*alias* RLDA^{x0} with the nonrelativistic LDA exchange potential (4.15.19)) using a series of cell grids displayed in Table 10.5. This presents the number of angular points n_θ in the *outermost radial shell* ($r \geq n_r/2$, where n_r is the number of radial grid points). The number of Lebedev-Laikov grid points, n_θ, is set to 26 for the innermost shell, 86 for the middle shell in Grids 1-4, and 26 and 110 in Grid 5. The basis sets were of nonrelativistic triple zeta quality.

Table 10.5. Dimensions of Lebedev-Laikov grids used in relativistic molecular calculations of Table 10.6. Reproduced, with permission, from [31, Table IV].

Grid:		1	2	3	4	5
H–He	n_r	21	25	31	35	45
	n_θ	86	146	194	302	434
Li–Ne	n_r	25	31	35	41	51
	n_θ	246	239	302	434	770
Na–Ar	n_r	31	35	41	45	55
	n_θ	146	230	392	434	770
K–Kr	n_r	35	41	45	51	61
	n_θ	146	230	302	434	770

Table 10.6. Molecular geometries Reproduced, with permission, from [31, Table V].

	Symmetry	Bond (a.u.)	Angle
H_2O	C_{2v}	O–H 1.8104	H–O–H 104.5°
NH_3	C_{3v}	N–H 1.9124	H–N–H 106.7°
C_2H_4	D_{2h}	C–C 2.5303	H–C–H 117.8°
		C–H 2.0504	
P_4	T_d	P–P 4.1763	
CH_4	T_d	C–H 2.0635	
SiH_4	T_d	Si–H 2.7987	
$TiCl_4$	T_d	Ti–Cl 4.1399	

The molecular geometries are presented in Table 10.6. The integral of the electron density over the molecule, Table 10.7, serves as an indication of the

Table 10.7. Error in the computed total electron density in the DHS approximation Reproduced, with permission, from [31, Table V].

	Grid 1	Grid 2	Grid 3	Grid 4	Grid 5
H_2O	2.1(-4)	4.5(-5)	2.2(-6)	2.0(-7)	2.0(-8)
NH_3	2.3(-4)	1.0(-5)	1.1(-5)	2.0(-6)	1.0(-7)
C_2H_4	6.3(-4)	6.1(-5)	2.8(-6)	1.0(-7)	3.0(-7)
P_4	1.8(-4)	1.4(-4)	8.2(-6)	1.2(-6)	9.0(-7)
CH_4	3.7(-4)	4.8(-6)	3.9(-5)	3.7(-6)	1.3(-7)
SiH_4	6.2(-5)	4.0(-5)	8.0(-6)	4.3(-6)	1.0(-7)
$TiCl_4$	1.3(-3)	6.5(-5)	4.2(-5)	1.0(-5)	2.7(-6)

Table 10.8. Total electronic energy (a.u.) in the DHS approximation Entries for Grids 1–4 are energies relative to the last column (Grid 5). Reproduced, with permission, from [31, Table V].

	Grid 1	Grid 2	Grid 3	Grid 4	Grid 5
H_2O	-1.66(-5)	+1.43(-5)	+4.8(-6)	-1.8(-6)	-75.275 893 9
NH_3	-9.3(-6)	-1.13(-5)	-4.2(-6)	-1.8(-6)	-55.494 869 3
C_2H_4	-2.637(-4)	-1.13(-5)	-7.1(-6)	-2.0 (-6)	-76.895 526 6
P_4	-1.26(-3)	-1.65(-3)	-0.4(-4)	-0.2(-4)	-1359.162 88
CH_4	-1.760(-4)	-2.4(-6)	-9.5(-6)	+1.0(-7)	-39.539 952 3
SiH_4	-1.04(-4)	-3.97(-4)	2.4(-5)	-0.4(-6)	-290.031 578
$TiCl_4$	-6.12(-3)	1.2(-4)	1.9(-4)	-1.4(-4)	-2685.870 67

accuracy achieved with each of the 5 integration grids. The corresponding total energies in Table 10.8 suggest that the accuracy achieved in this calculation is comparable with that found in nonrelativistic practice. The only difference is the need to use slightly more integration points when calculating integrals involving small components in order to reach a given accuracy.

10.13 Computational strategies

The spinor structure of relativistic molecular structure calculations means that the potential for generating huge amounts of data is even larger than in equivalent nonrelativistic calculation. The need to find algorithms that minimize the cost of computation without degrading the model must never be forgotten when designing computational strategies. The cheapness of RDFT methods is one of their attractions.

The generalized eigenvalue equation $\boldsymbol{FC} = \boldsymbol{ESC}$ in all relativistic SCF calculations is most often solved iteratively according to the scheme

$$\boldsymbol{F}^n \boldsymbol{C}^{n+1} = \boldsymbol{E}^{n+1} \boldsymbol{S} \boldsymbol{C}^{n+1}, \quad n = 0, 1, 2, \dots \tag{10.13.1}$$

where the Fock matrix, $\boldsymbol{F}^n$, is calculated using the n-th iterate, $\boldsymbol{D}^n$, of the density matrix. In this section, we discuss a number of techniques for accelerating the process.

10.13.1 The Roothaan bound

As we have seen, the Fock matrix elements are calculated from sums of products of interaction integrals and density matrix elements. Thus a first step must be to find a way to avoid the calculation of integrals whose contributions to the Fock matrix are negligible. In the case of the Coulomb $\boldsymbol{G}^{\beta,\beta'}$ matrices of (10.5.4) and (10.5.5), each interaction integral $(\mu\beta,\nu\beta\,|\,\sigma\beta',\tau\beta')$ contributes

$$\delta G^{\beta,\beta}_{\mu\nu} = (\mu\beta,\nu\beta\,|\,\sigma\beta',\tau\beta')\, D^{\beta'\beta'}_{\sigma\tau} \tag{10.13.2}$$

to the *direct* matrix element $G^{\beta,\beta}_{\mu\nu}$ and

$$\delta G^{\beta,\beta'}_{\mu\nu} = -(\mu\beta,\tau\beta\,|\,\sigma\beta',\nu\beta')\, D^{\beta'\beta}_{\sigma\tau} \tag{10.13.3}$$

to the *exchange* matrix elements $G^{\beta,\beta'}_{\mu\nu}$ with $\beta' = \pm\beta$.

Roothaan [50] proved a version of the Schwartz inequality, here generalized to two-component spinor integrals, such that

$$\begin{aligned} 0 &\le |\,(\mu\beta,\nu\beta'\,|\,\sigma\beta'',\tau\beta''')| \\ &\le \sqrt{|\,(\mu\beta,\nu\beta'\,|\,\mu\beta,\nu\beta')|\,|\,(\sigma\beta'',\tau\beta'''\,|\,\sigma\beta'',\tau\beta''')|} \end{aligned} \tag{10.13.4}$$

For fixed β,β' there are N^2 diagonal integrals $(\mu\beta,\nu\beta'\,|\,\mu\beta,\nu\beta')$ for an N-dimensional basis set that can be computed and stored at the start of the iteration. By combining the estimate (10.13.4) with the density matrix components in (10.13.2) and (10.13.3), we can determine an upper bound on the contributions $\delta G^{\beta,\beta'}_{\mu\nu}$ and avoid calculating interaction integrals whose contributions are below threshold. The test is not without cost, but this is generally insignificant compared with the cost of evaluating unnecessary multi-centre integrals, especially if they involve $d, f, g, \ldots$ orbitals. The cost can become significant in very large systems, and it is necessary to construct program loop structures carefully to keep costs down.

10.13.2 Integral-direct Fock matrix evaluation

Whatever the details of the iterative scheme, the change in the Fock matrix at each iteration can be written formally

$$\delta\boldsymbol{F}^n \leftarrow \boldsymbol{I}\,\delta\boldsymbol{D}^n, \qquad (\delta\boldsymbol{F}^n = \boldsymbol{F}^{n+1} - \boldsymbol{F}^n) \tag{10.13.5}$$

where $\boldsymbol{I}$ represents a list of molecular integrals and $\delta\boldsymbol{D}^n$ is the change in the density matrix. The complete list is needed to construct the initial Fock matrix $\boldsymbol{F}^0$, but subsequent iterations need only those elements of $\delta\boldsymbol{F}^n$ above the chosen threshold. The cost of each cycle therefore decreases as the iteration approaches convergence.

10.13.3 Symmetry properties of interaction matrix elements

Symmetry properties of charge-current components (10.7.1) induce symmetries of interaction matrix elements that can be exploited to reduce the length of the list $\boldsymbol{I}$ to a minimum. Suppressing all quantum numbers other than angular momentum projections on the axis of quantization, we have

$$\left[j^{\beta',\beta}_{-m_\nu,-m_\mu}\right]_q = \eta_\mu\eta_\nu(-1)^{m_\mu-m_\nu}\left[j^{\beta,\beta'}_{m_\mu,m_\nu}\right]_q \tag{10.13.6}$$

$$\left[j^{\beta,\beta'}_{-m_\nu,-m_\mu}\right]_q = \eta_\mu\eta_\nu(-1)^{m_\mu-m_\nu}\left[j^{\beta,\beta'}_{m_\mu,m_\nu}\right]^*_q \tag{10.13.7}$$

$$\left[j^{\beta,\beta'}_{m_\mu,m_\nu}\right]^*_q = \left[j^{\beta',\beta}_{m_\nu,m_\mu}\right]_q. \tag{10.13.8}$$

Equation (10.13.7) is often identified with symmetry under time reversal, and the last equation follows from the Hermitian character of Pauli matrices. The two-electron matrix element symmetries that follow are

$$\begin{aligned}
&(m_\mu, m_\nu, q\,|\,m_\sigma, m_\tau, q') \\
&= \eta_\sigma\eta_\tau(-1)^{m_\sigma-m_\tau}(m_\mu, m_\nu, q\,|\,-m_\sigma, -m_\tau, q') && (10.13.9)\\
&= \eta_\mu\eta_\nu(-1)^{m_\mu-m_\nu}(-m_\nu, -m_\mu, q\,|\,m_\sigma, m_\tau, q') && (10.13.10)\\
&= (m_\sigma, m_\tau, q'\,|\,m_\mu, m_\nu, q) && (10.13.11)
\end{aligned}$$

The Hermitian symmetry of the Fock matrix means that it is only necessary to construct half of its matrix elements explicitly for production calculations. Verification of the symmetry relations can be used to check the correctness of computer programs.

10.13.4 Stepwise refinement

The LCAO method [7] often provides a very good first approximation to the molecular wavefunction. Atomic wavefunctions are needed to start the calculation; §10.2 enables us obtain them immediately. The contribution to the electron density due to basis functions centred on different nuclei is generally very small, so that a DHF scheme that starts by excluding all but the most important terms captures most of the physically important effects at the outset and reduces the computational cost overall.

- **Stage 0**: perform DHF calculations on atomic cores for all nuclei present and form the initial density matrix using LCAO combinations of atomic spinors.
- **Stage 1**: include one-centre integrals of all classes but only those two-electron interactions, $(\mu+, \nu+\,|\,\sigma+, \tau+)$, that would appear in the nonrelativistic Fock matrix.
- **Stage 2**: include the two-electron integrals $(\mu\beta, \nu\beta\,|\,\sigma-\beta, \tau-\beta)$ with $\beta = +, -$ which are next in importance.

- **Stage 3**: include the integrals $(\mu-, \nu- \,|\, \sigma-, \tau-)$.

Experience has shown that it is advisable to retain all classes of one-centre interaction integrals throughout the calculation. The one-centre integrals are relatively cheap to compute and are in any case needed to capture the dominant atomic-centred relativistic effects. The small component electron density in the neighbourhood of the nucleus is significant even for a moderately heavy atom and therefore shields the valence electrons from the full nuclear charge. These terms strongly perturb the solution if they are not introduced until Stage 2 or Stage 3, making it difficult, or even impossible, to make the calculation converge.

Roothaan's bounds allow us to drop classes of interaction integrals that make no significant contribution, which is particularly useful with multi-centre integrals over small components. One-centre Breit integrals dominate in closed shell molecules and are cheap to construct. The cost of computing multi-centre Coulomb and Breit integrals involving small components also is less than is commonly supposed, despite their complexity, because the charge-current components are only appreciable near the nuclei.

10.13.5 Level-shifting

Iterative SCF schemes construct perturbation operators that rotate the eigenvector basis. The total energy is invariant under rotations in the occupied space alone, so that energy improvement comes from the coupling with virtual levels. Perturbation of the Fock matrix from one iteration to the next may involve strong coupling between occupied and low-lying virtual states, and it is often desirable to damp the relatively large changes that may result in order to improve the convergence characteristics of the iterative scheme. *Level-shifting* modifies the Fock matrix by writing

$$\boldsymbol{F}_{shift} = \boldsymbol{F} + \sum_r w_r \boldsymbol{P}_r$$

where $\boldsymbol{P}_r$ projects onto the r-th eigenvector of $\boldsymbol{F}$ and the coefficients w_r are so-called *level shift parameters*. In effect, this performs the shift $E_r \to E_r + w_r$.

The simplest way to exploit this idea shifts *all virtual states*, say $r \in \mathcal{V}$, by the same amount $w_r = w > 0$. A change $\Delta\boldsymbol{F}$ in the Fock matrix introduces a first order correction to the unperturbed mixing coefficient vector $\boldsymbol{C}_0$ proportional to $\langle r \,|\, \Delta\boldsymbol{F} \,|\, 0 \rangle / (E_r - E_0)$. This may strongly perturb $\boldsymbol{C}_0$ when the denominator is small, causing large oscillations from one iteration to the next. The rate of convergence is often poor, and the amplitude may be so large that the next step is outside the basin of attraction to the wanted solution. Adding a positive w to each small denominator can therefore damp the change from one step to the next, leading to a more straightforward pattern of convergence. More sophisticated use of level-shifting can greatly improve the convergence characteristics in open-shell problems, where selective application of level shifts can help to stabilize oscillations between closed-shell, open-shell, and virtual state spinors. This can be particularly useful in compounds containing lanthanide or actinide elements, where relativistic effects are clearly of interest. There are often large numbers of electrons with relatively small binding energy having high angular momentum. These often produce a high density of states around the nonrelativistic energy zero, and level-shfting can be used to identify spinors that are nominally occupied in the reference configuration and distinguish them from those regarded as virtual states. Level-shifting can often stabilize an otherwise unstable iteration sequence in such applications. The negative energy orbitals, which are always present, play a largely passive role in these calculations. Their first order correction to the change in the Fock matrix has a large denominator of order $2mc^2$ so that their influence on the progress of SCF iterations is negligible in comparison with that from low-lying virtual states.

10.14 Multiconfigurational Dirac-Hartree-Fock theory

Multiconfiguration self-consistent field (MCSCF) methods play an important role in nonrelativistic molecular electronic structure calculations. The iterative solution of MCSCF equations now utilizes so-called second-order (or approximate second-order) procedures, and the introduction of "direct" procedures that avoid the explicit construction of the Hamiltonian and Hessian matrices have made it possible to fully optimize long CI expansions. Reliable and efficient MCSCF wavefunctions are often the starting point of more extensive multi-reference CI calculations. The subject has an extensive literature, summarized by Werner and Knowles [51], who describe an MCSCF approach that can be adapted to relativistic calculations. It involves rotations in the orbital space, which, in the relativistic context, includes the negative energy orbitals that are essential for a proper optimization [52]. The necessity for including the negative energy states has been highly controversial for the past 20 years; the resolution of the controversy, §4.13, is therefore essential to the success of the method. The Werner and Knowles scheme, like its precursors discussed in their paper [51], relies on simultaneous adjustment of the orbitals as well as the CSF mixing coefficients. There have been many such schemes, and Werner and Knowles describe many of the technical difficulties that they attempt to resolve in their paper. The relativistic version presented below provides a simple and convenient way to proceed.

10.14.1 Orbital optimization

The multi-configurational wavefunction of an atom or molecule may be written

$$\boldsymbol{\Psi} = \sum_{I=1}^{K} C_I \boldsymbol{\Phi}_I, \quad \langle \boldsymbol{\Phi}_I \,|\, \boldsymbol{\Phi}_J \rangle = \delta_{I,J}, \quad \sum_{I=1}^{K} |C_I|^2 = 1 \tag{10.14.1}$$

where $\{\boldsymbol{\Phi}_I | I = 1, 2, \ldots, K\}$ is a set of orthonormal CSFs and the C_I are expansion coefficients. In general, the $\boldsymbol{\Phi}_I$ are linear combinations of N-particle determinants so that we could also think of (10.14.1) as a linear combination of determinants. It is not necessary to be specific at this point.

An isolated atom can be characterized by its angular momentum and parity quantum numbers J, M, and Π. In this case, $\boldsymbol{\Psi}$ and all the $\boldsymbol{\Phi}_I$ can be labelled with the same values of J, M, and Π, and each CSF will itself be expressible as a linear combination of determinants. Similarly, if a molecule has point group symmetry, then $\boldsymbol{\Psi}$ and all the $\boldsymbol{\Phi}_I$ will be built from CSFs belonging to some common representation space for that group. The CSFs $\{\boldsymbol{\Phi}_I\}$ are constructed from 4-spinors $\{\psi_i\}$ for *occupied electron subshells* defined by the mean field used to construct them. The energy of the atom or molecule for the state (10.14.1) is

$$E = \langle \boldsymbol{\Psi} \,|\, \widehat{H} \,|\, \boldsymbol{\Psi} \rangle = \sum_{I=1}^{K} \sum_{J=1}^{K} C_I^* C_J H_{IJ}, \tag{10.14.2}$$

where H_{IJ} is the matrix element of the N-electron Hamiltonian $\widehat{H}$ between the N-electron CSFs $\boldsymbol{\Phi}_I$ and $\boldsymbol{\Phi}_J$, which can be written in the general form

$$H_{IJ} = \sum_{ij} (\,ij\,) \cdot \gamma_{IJ}^{ij} + \frac{1}{2} \sum_{ij} \sum_{kl} (\,ij \,|\, kl\,) \cdot \Gamma_{IJ}^{ijkl}, \tag{10.14.3}$$

where i, j, k, l denote occupied *electron* orbitals. Inserting this in (10.14.2) gives the energy expression

$$E = \sum_{ij} (\,ij\,) \cdot \gamma^{ij} + \frac{1}{2} \sum_{ij} \sum_{kl} (\,ij \,|\, kl\,) \cdot \Gamma^{ijkl}, \tag{10.14.4}$$

where the one- and two-electron density matrices have components

$$\gamma^{ij} = \sum_{I=1}^{K} \sum_{J=1}^{K} C_I^* C_J \, \gamma_{IJ}^{ij}, \quad \Gamma^{ijkl} = \sum_{I=1}^{K} \sum_{J=1}^{K} C_I^* C_J \, \Gamma_{IJ}^{ijkl}. \tag{10.14.5}$$

Integrals over orbital spinors

The notation $(\,ij\,)$ and $(ij \,|\, kl)$ for one- and two-electron interaction integrals reflects their construction from charge-current densities, labelled ij and kl, over a primitive G-spinor basis $\rho_{\mu\nu}^{\beta\beta}$ and $\boldsymbol{j}_{\mu\nu}^{\beta,-\beta}$ whose symmetries are defined by (10.7.1):

$$\rho_{\mu\nu}^{\beta\beta} = \left(\rho_{\nu\mu}^{\beta\beta}\right)^*, \quad \boldsymbol{j}_{\mu\nu}^{\beta,-\beta} = \left(\boldsymbol{j}_{\nu\mu}^{-\beta,\beta}\right)^* \tag{10.14.6}$$

or, in terms of the space-time charge-current vector,

$$(j^\sigma)^{\beta\beta'}_{\mu\nu} = \left[(j^\sigma)^{\beta'\beta}_{\nu\mu}\right]^*, \quad \sigma = 0, 1, 2, 3$$

The definition of the one-electron Hamiltonian, (10.2.7) – (10.2.9), applies as it stands to the molecular case, provided the basis set indices μ, ν range over all nuclear centres. It follows that the one-electron interaction energy can be written

$$(\, ij\,) = \sum_{\beta\beta'} \left(\boldsymbol{c}_i^{\beta}\right)^\dagger \boldsymbol{h}^{\beta\beta'} \boldsymbol{c}_j^{\beta'} \tag{10.14.7}$$

where $\boldsymbol{c}_i^{\beta}$ is a column matrix with entries $c_{\mu i}^{\beta}$ running over the G-spinor basis labels μ, with the spinor component index $\beta = \pm 1$, and the $N \times N$ block $\boldsymbol{h}^{\beta\beta'}$ of the Hamiltonian matrix, with elements $h_{\mu\nu}^{\beta\beta'}$, satisfies

$$\left(h_{\mu\nu}^{\beta\beta'}\right)^\dagger = h_{\nu\mu}^{\beta'\beta}$$

to make the full $2N \times 2N$ matrix $\boldsymbol{h}$ Hermitian as in (10.2.8). We can rewrite (10.14.7) in a density matrix notation as

$$(\, ij\,) = \mathrm{tr}\, \boldsymbol{h}\, (\boldsymbol{d}_{ji})^\dagger\,, \tag{10.14.8}$$

where $\boldsymbol{d}_{ji}$ has matrix elements

$$d_{ji}^{\beta'\beta} = \left(\boldsymbol{c}_j^{\beta'}\right)^\dagger \boldsymbol{c}_i^{\beta}. \tag{10.14.9}$$

We can proceed in the same way with two-electron interaction integrals:

$$(\, ij \,|\, kl\,) = \sum_{\beta\beta'} \sum_{\mu\nu} (\mu\beta, \nu\beta' \,|\, kl\,)\, d_{ij}^{\beta\beta'}(\mu\nu), \tag{10.14.10}$$

$$(\mu\beta, \nu\beta' \,|\, kl\,) = \sum_{\tilde{\beta}\tilde{\beta}'} \sum_{\rho\sigma} (\mu\beta, \nu\beta' \,|\, \rho\tilde{\beta}, \sigma\tilde{\beta}')\, d_{kl}^{\tilde{\beta}\tilde{\beta}'}(\rho\sigma). \tag{10.14.11}$$

(The trace notation is less helpful here.) Integrals over the Coulomb interaction involve the charge densities only and so $\beta' = \beta$ and $\tilde{\beta}' = \tilde{\beta}$. The Breit interaction involves the current components, and so $\beta' = -\beta$ and $\tilde{\beta}' = -\tilde{\beta}$. We can proceed as in (10.2.21) – (10.2.24) to write these in terms of the familiar J and K operators.

We have already seen that the symmetry of the E-coefficients can be used to reduce the computational cost of generating the basic one- and two-electron integrals. For present purposes, it is useful to note the consequences of the symmetries (10.14.6), which when combined with the Hermitian symmetry of the one-electron Hamiltonian and the symmetry of the two-electron kernels (Coulomb or Breit) give

$$(\, ij\,) = (\, ji\,)^*, \quad (\, ij \,|\, kl\,) = (\, kl \,|\, ij\,) = (\, ji \,|\, lk\,)^* \tag{10.14.12}$$

In fact, the interaction integrals are *real* so that the complex conjugation appearing in (10.14.12) can be ignored. The density matrices of (10.14.5) can be assumed real and symmetrized so that

$$\gamma^{ij} = \gamma^{ji}, \quad \Gamma^{ijkl} = \Gamma^{jikl} = \Gamma^{klij} \tag{10.14.13}$$

as in [51, Equation (9)].

Orbital rotations

After these preliminaries, the development is close to that of Werner and Knowles [51]. These authors distinguish between *internal* (or occupied) orbitals, denoted i, j, k, l, and all other orbitals $r, s, \ldots$. In relativistic terms, the internal set includes only *electron* ("positive energy") orbitals, whilst the complete set includes both electron and positron ("negative energy") orbitals. As we have seen the negative energy orbitals must be present to preserve normalization in orbital rotations that change the mean field potential. A new set of orthonormal MO spinors is obtained by an orthogonal transformation

$$\tilde{\psi}_i = \sum_r \psi_r \, U_{ri} \tag{10.14.14}$$

where the $2N \times 2N$ matrix $\boldsymbol{U}$ can be written

$$\boldsymbol{U} = \exp \boldsymbol{R} = \boldsymbol{I} + \boldsymbol{T}, \quad \boldsymbol{R}^{\dagger} = -\boldsymbol{R}. \tag{10.14.15}$$

We shall determine the self-consistent solution in terms of the $2N(2N-1)$ elements of the *real* anti-symmetric matrix $\boldsymbol{R}$. Introduce the operators $\boldsymbol{h}$, $\boldsymbol{J}^{kl}$ and $\boldsymbol{K}^{kl}$ by

$$(\, r \,|\, \boldsymbol{h} \,|\, s \,) = (\, rs \,), \quad \left(r \,\middle|\, \boldsymbol{J}^{kl} \,\middle|\, s \right) = (\, rs \,|\, kl \,), \quad \left(r \,\middle|\, \boldsymbol{K}^{kl} \,\middle|\, s \right) = (\, rk \,|\, ls \,), \tag{10.14.16}$$

the Fock matrix $\boldsymbol{F}^{ij}$ and the matrix $\boldsymbol{G}^{ij}$,

$$\boldsymbol{F}^{ij} = \gamma^{ij}\, \boldsymbol{h} + \sum_{kl} \Gamma^{ijkl} \boldsymbol{J}^{kl}, \quad \boldsymbol{G}^{ij} = \boldsymbol{F}^{ij} + 2 \sum_{kl} \Gamma^{ijkl} \boldsymbol{K}^{kl}, \tag{10.14.17}$$

and the matrices $\boldsymbol{A}$ and $\boldsymbol{B}$ with elements

$$A^{rs} = 2 \sum_j (\, r \,|\, \boldsymbol{F}^{sj} \,|\, j \,), \quad B^{rs} = A^{rs} + \sum_{kl} \left(r \,\middle|\, \boldsymbol{G}^{sk} \,\middle|\, l \right) T_{lk}, \tag{10.14.18}$$

where columns in which the label s refers to the external set vanish. Then we can expand the energy in powers of $\boldsymbol{T}$ to second order, giving the expression

$$E^{(2)}(\boldsymbol{T}) = E^{(2)}(\boldsymbol{0}) + \sum_{ri} T_{ri} B^{ri} \tag{10.14.19}$$

We shall seek a minimum of $E^{(2)}(\boldsymbol{T})$ with respect to the elements of $\boldsymbol{T}$ subject to the orthonormality constraints $\boldsymbol{U}^\dagger \boldsymbol{U} = \boldsymbol{I}$. Introducing the Lagrange multipliers $\epsilon_{ij} = \epsilon_{ji}$, gives us the equations

$$\frac{\partial}{\partial T_{rs}} \left(E^{(2)}(\boldsymbol{T}) - \sum_{ij} \epsilon_{ij} \left[(\boldsymbol{U}^\dagger \boldsymbol{U}) - \boldsymbol{I} \right]_{ij} \right)$$

for all r, s, which reduce to the matrix equation

$$\boldsymbol{B} - 2\boldsymbol{U}\boldsymbol{\epsilon} = 0 \longrightarrow \boldsymbol{\epsilon} = \frac{1}{2}\boldsymbol{U}^\dagger \boldsymbol{B} \tag{10.14.20}$$

after using the orthonormality constraint. The symmetry condition $\boldsymbol{\epsilon} = \boldsymbol{\epsilon}^\dagger$ allows us to eliminate the Lagrange multipliers to give the nonlinear equation [53]

$$\boldsymbol{U}^\dagger \boldsymbol{B} - \boldsymbol{B}^\dagger \boldsymbol{U} = 0. \tag{10.14.21}$$

Werner and Knowles [51] modified this approach by linearizing the equations about some arbitrary trial matrix $\boldsymbol{U}$. We can regard $\boldsymbol{U}$ as a function of the rotation parameters $\boldsymbol{R}$; we seek a small perturbation $\boldsymbol{U} \to \boldsymbol{U}(\boldsymbol{R}) \exp \boldsymbol{r}$ so that

$$\boldsymbol{T} \to \boldsymbol{T} + \Delta\boldsymbol{T}, \quad \Delta\boldsymbol{T} = \boldsymbol{U}(\boldsymbol{R}) \left(\boldsymbol{r} + \frac{1}{2}\boldsymbol{r}^2 + \dots \right), \quad \boldsymbol{r}^\dagger = -\boldsymbol{r}. \tag{10.14.22}$$

Inserting this in (10.14.19) gives

$$\begin{aligned} &E^{(2)}(\boldsymbol{T} + \Delta\boldsymbol{T}) \\ &= E^{(2)}(\boldsymbol{T}) + \operatorname{tr} \left[2\boldsymbol{U}^\dagger \boldsymbol{B} \left(\boldsymbol{r} + \frac{1}{2}\boldsymbol{r}^2 \right) + \boldsymbol{r}^\dagger \boldsymbol{U}^\dagger \boldsymbol{G}^{ij} \boldsymbol{U}\boldsymbol{r} \right] \end{aligned} \tag{10.14.23}$$

The stationary condition (10.14.21) is now replaced by

$$\boldsymbol{U}^\dagger \widetilde{\boldsymbol{B}} - \widetilde{\boldsymbol{B}}^\dagger \boldsymbol{U} - (\boldsymbol{E}\boldsymbol{r} + \boldsymbol{r}\boldsymbol{E}) = 0, \tag{10.14.24}$$

where

$$\boldsymbol{E} = \frac{1}{2}(\boldsymbol{U}^\dagger \boldsymbol{B} + \boldsymbol{B}^\dagger \boldsymbol{U}).$$

and

$$\widetilde{B}_{ri} = B_{ri} + \sum_j (\boldsymbol{G}^{ij} \boldsymbol{U}\boldsymbol{r})_{rj}$$

Equation (10.14.24) is linear in the rotation parameters $\boldsymbol{r}$. When $\boldsymbol{U} = \boldsymbol{I}$, this reduces to the corresponding equation of the Newton-Raphson method. Werner and Knowles [51] recommend that it is better to use the augmented Hessian (AH) method with step restriction to determine $\boldsymbol{r}$, on the grounds that it always determines a reasonable solution even if the Hessian matrix at the expansion point is not positive definite. This introduces a level shift into (10.14.24) giving an eigenvalue equation

$$\boldsymbol{U}^{\dagger}\widetilde{\boldsymbol{B}} - \widetilde{\boldsymbol{B}}^{\dagger}\boldsymbol{U} - (\boldsymbol{E}\boldsymbol{r} + \boldsymbol{r}\boldsymbol{E}) = \lambda\epsilon\,\boldsymbol{r}. \tag{10.14.25}$$

The number damping parameter $\lambda \geq 1$ can be adjusted automatically to restrict the "norm" of $\boldsymbol{r}$ so that

$$\sum_{ri} \boldsymbol{r}_{ri}^2 \leq s^2.$$

Davidson's method [54] may be used to solve (10.14.25). The original papers should be consulted for further implementation details.

Simultaneous optimization of CI coefficients and orbitals

Relaxation of the CSF mixing coefficients interacts with the orbital rotations discussed above, and it is desirable to find an iterative scheme that allows for this interaction. The formulae ((10.14.21) and (10.14.24) both make use of the one index transformed quantities

$$\widetilde{h}_{rj} = \sum_{s} (\, rs \,) U_{sj}, \quad \widetilde{\boldsymbol{J}}^{kl} = \boldsymbol{J}^{kl}\boldsymbol{U}, \quad \widetilde{\boldsymbol{K}}^{kl} = \boldsymbol{K}^{kl}\boldsymbol{T} \tag{10.14.26}$$

which are relatively cheap to construct, requiring only about $\frac{3}{2}(2N)^2.M^3$ operations, where M the number of orbitals in the internal set and $2N$ the total number of orbitals. The AH method above also requires one- and two-electron integrals that are correct to second order in $\varDelta\boldsymbol{T}$:

$$h_{ij}^{(2)} = \sum_{r} U_{ri}\widetilde{h}_{rj} = (\boldsymbol{U}^{\dagger}\widetilde{\boldsymbol{h}})_{ij} \tag{10.14.27}$$

and

$$\begin{aligned}(\, ij \,|\, kl \,)^{(2)} = -(\, ij \,|\, kl \,) + (\boldsymbol{U}^{\dagger}\boldsymbol{J}^{kl}\boldsymbol{U})_{ij} + (\boldsymbol{U}^{\dagger}\boldsymbol{J}^{ij}\boldsymbol{U})_{kl} \\ + (1 + \tau_{ij})(1 + \tau_{kl})(\boldsymbol{T}^{\dagger}\boldsymbol{K}^{ik}\boldsymbol{T})_{jl}\end{aligned} \tag{10.14.28}$$

where the permutation operator τ_{ij} exchanges the labels i and j. These relatively cheap integrals may be used in a direct CI step to improve the CI coefficients. When one configuration dominates, first order perturbation theory gives an update

$$\widetilde{C}_I = C_I - \frac{g_I^{(2)} - E^{(2)}\,C_I}{H_{II}^{(2)} - E^{(2)}}$$

for all except the dominant configuration, where $H_{II}^{(2)}$ is the diagonal element of the Hamiltonian matrix correct to second order in $\boldsymbol{T}$, $E^{(2)} = \sum_I g_I^{(2)} C_I \Big/ \sum_I C_I^2$, and $g_I^{(2)} = \sum_J C_J \left[h_{ij}^{(2)} \gamma_{IJ}^{ij} + \frac{1}{2}\sum_{ij}\sum_{kl}(\, ij \,|\, kl \,)^{(2)} \varGamma_{IJ}^{ijkl} \right]$. Werner and Knowles [51] carefully discuss the advantages and disadvantages of these methods. MCSCF calculations, relativistic or nonrelativistic, molecular or atomic, can still spring nasty surprises.

References

1. Born M and Oppenheimer J R 1927 *Ann der Phys* **84** 457.
2. Sutcliffe B 1992 in *Methods in Computational Molecular Physics* (ed. S Wilson and G H F Diercksen) pp. 19–46, (NATO ASI Series B: Physics Vol. 293) (New York: Plenum Press).
3. Karshenboim S G, Pavone F S, Bassani F, Inguscio M and Hänsch T W 2001 *The Hydrogen Atom. Precision Physics of Simple Atomic Systems* (Berlin: Springer-Verlag).
4. Sutcliffe B 1992 in *Methods in Computational Molecular Physics* (ed. S Wilson and G H F Diercksen) pp. 19–46, (NATO ASI Series B: Physics Vol. 293) (New York: Plenum Press).
5. Andrae D 2002 in *Relativistic Quantum Chemistry* (ed. P Schwerdtfeger) pp. 203–258 (Amsterdam: Elsevier).
6. Grant I P and Pyper N C 1976 *J. Phys. B: Atom. Molec. Phys.***9** 761.
7. McWeeny R 1989 *Methods of Molecular Quantum Mechanics* Second edn (London: Academic Press).
8. Grant I P and Quiney H M 2002 in *Relativistic Quantum Chemistry* (ed. P Schwerdtfeger) pp. 107–202 (Amsterdam: Elsevier).
9. Ishikawa Y, Quiney H M and Malli G L 1991 *Phys. Rev. A* **43** 3270.
10. Quiney H M 1990 in *Supercomputational Science* (ed. R G Evans and S Wilson), 159, 185.
11. Abramowitz M and Stegun I A 1970 *Handbook of Mathematical Functions* (New York: Dover).
12. Schmidt M W and Ruedenberg K 1979 *J. Chem. Phys.* **71**, 3951.
13. Huzinaga S and Klobukowski M 1985 *J. Mol. Struct. (Theochem.)* **63**, 1812.
14. Van Duijneveldt F B 1971 *IBM Research Report No.* RJ945.
15. Quiney H M, Grant I P and Wilson S 1990 *J. Phys. B: At. Mol. Opt. Phys.* **23**, L271.
16. Eliav E, Kaldor U and Ishikawa Y 1994 *Phys. Rev. A* **49**, 1724.
17. Saunders V R 1983 in *Methods of Computational Molecular Physics*, Vol. 1, p. 1. ed. G H F Diercksen and S Wilson (Dordrecht: Reidel).
18. Drake G W F ed 2005 *Springer Handbook of Atomic, Molecular and Optical Physics* Second edn (New York: Springer-Verlag).
19. Louck J D in [18, Chapter 2].
20. McMurchie L E and Davidson E R 1978 *J. Comput. Phys.* **26**, 218.
21. Quiney H M, Skaane H and Grant I P 1997 *J. Phys B: Ar. Mol. Opt. Phys.* **30**, L829.
22. Meyer J 1988 *Int. J. Quant. Chem.* **33**, 445.
23. Meyer J, Sepp W-D, Fricke B and Rosen A 1989 *Comput. Phys. Commun.* **54**, 55.
24. Meyer J, Sepp W-D, Fricke B and Rosen A 1996 *Comput. Phys. Commun.* **96**, 263.
25. Skaane H 1998, unpublished D. Phil. Thesis, University of Oxford.
26. Richtmyer R D 1978 *Methods of Advanced Mathematical Physics* (2 vols) (New York: Springer-Verlag).
27. Rosicky F 1982 *Chem. Phys. Lett.* **85**, 195.
28. Mohanty A K 1992 *Int. J. Quant. Chem.* **42**, 627.
29. Almlöf A 1966 *J. Chem. Phys.* **104**, 4685.

30. Ahmadi G R and Almlöf A 1966 *Chem. Phys. Lett.* **246**, 364.
31. Quiney H M and Belanzoni P 2002 *J. Chem. Phys.* **117**, 5550.
32. Quiney H M and Grant I P 2004 *Int. J. Quant. Chem.* **99**, 198.
33. Slater J C 1960 *Quantum theory of atomic structure*, Vol. 1. (New York: McGraw-Hill).
34. Feynman R P, Leighton R B and Sands M 1964 *The Feynman Lectures on Physics*, Vol. II. (Reading, MA: Addison-Wesley Publishing Co. Inc.).
35. Dunlap B I, Connolly J W D and Sabin J R 1979 *J. Chem. Phys.* **71**, 3396.
36. Beebe N H F and Linderberg J 1977 *Int. J. Quant. Chem.* **12**, 683.
37. Skylaris C-K, Gagliardi L, Handy N C, Ioannou A, Spencer S and Willetts A 2000 *J. Mol. Struct. (THEOCHEM)* **501–502**, 229.
38. Willetts A, Gagliardi L, Ioannou A, Simper A M, Skylaris C-K, Spencer S and Handy N C 2000 *Int. Rev. Phys. Chem.* **19**, 327.
39. Manby F R and Knowles P J 2001 *Phys. Rev. Lett.* **87**, 163001.
40. Becke A D 1988 *J. Chem. Phys.* **88**, 2547.
41. Quiney H M and Grant I P 2004 *Int. J. Quant. Chem.* **99**, 198.
42. Treutler O and Ahlrichs R 1995 *J. Chem. Phys.* **102**, 346.
43. Visscher L 1997 *Theor. Chim. Acta* **98**, 68.
44. Engel E, Dreizler R M, Varga S and Fricke B 2003 in *Relativistic Effects in Heavy Element Chemistry and Physics* (ed. B A Hess), Chapter 4. Relativistic Density Functional Theory. (Chichester: John Wiley and Sons, Ltd.).
45. Lebedev V I 1975 *Vychisl. Mat. Fiz.* **15**, 48.
46. Lebedev V I and Laikov D N 1999 *Dokl. Math.* **59**, 477.
47. Van Wüllen C. His program (FORTRAN) for generating Lebedev-Laikov grids [46] is available from Computational Chemistry List Ltd. (URL: http://www.ccl.net).
48. Engel E and Dreizler R M 1996 *Topics in Current Chem.* **181**, 1.
49. Engel E, Keller S and Dreizler R M 1996 *Phys. Rev. A* **53**, 1367.
50. Roothaan C C C 1951 *Rev. Mod. Phys.* **23**, 69.
51. Werner H-J and Knowles P J 1985 *J. Chem. Phys.* **82**, 5053.
52. Quiney H M 2002 in *Handbook of Molecular Physics and Quantum Chemistry* (Chichester: John Wiley).
53. Werner H-J and Meyer W 1980 *J. Chem. Phys.* **73**, 2341.
54. Davidson E R 1975 *J. Comput. Phys.* **17**, 87.

11

Relativistic calculation of molecular properties

This chapter applies the methods of Chapter 10 to the calculation of molecular properties. Section 11.1 discusses the way in which the BERTHA program handles molecular symmetry involving relativistic double point groups. Section 11.2 examines the Born-Oppenheimer potential energy surfaces of water and similar small molecules where the relativistic effects, although small, unexpectedly improved the calculated rotation-vibration spectra and made it possible to identify new lines in the solar spectrum. Sections 11.3–11.6 concern electromagnetic properties of atoms and molecules: Zeeman effect, hyperfine interactions, NMR shielding constants. Section 11.7 presents results for a few molecules containing high-Z atoms.

11.1 Molecular symmetry

Murrell *et al.* [1, Chapter 7] have remarked that the existence of molecular symmetry is a redeeming feature that makes it possible to reach conclusions about molecular wavefunctions without having to solve the quantum mechanical equations exactly. Whilst the number of highly symmetrical molecules is relatively small, their study provides insight into molecular properties. Well-known examples include the physics and chemistry of hydrocarbons and the octahedral and tetrahedral complexes that are common in much of transition metal chemistry.

So far we have exploited the theory of the Lorentz and Poincaré groups to construct powerful tools for approximate solution of the relativistic electronic structure of atoms and molecules. We have made no use of the relativistic point group symmetry of molecules although, as with isolated atoms, this can be used to reduce the Hamiltonian matrix to a block diagonal form in which each block is labelled by a representation of the symmetry group. In the relativistic LCAO approach, we construct molecular orbitals as linear combinations of 4-component G-spinors on each of the atoms. These G-spinors have well-defined transformation properties with respect to spatial rotations, parity and time reversal, which are set out in Appendix B.4. Four-component molecular symmetry orbitals (MS), $|\,\tau i\mu\rangle$, forming a space for the representation $D^{(i)}$ of the molecular symmetry group can be constructed as linear combinations of atomic four-spinors (AS) [2],

$$|\,\tau i\mu\rangle = |\,\rho a l j m \nu i \mu\rangle = \sum_{a'm'} C^{l j \nu i \mu}_{am,a'm'}\,|\,\rho a' l j m'\rangle, \tag{11.1.1}$$

where $|\,\rho a l j m\rangle$ is an AS on a typical centre a and the label τ distinguishes different representations of symmetry i. The sum over $a' = Sa$ ranges over all equivalent atom centres related by the group operations, S. If h is the order of the group and n_i the dimension of the representation,

$$C^{l j \nu i \mu}_{am,a'm'} = \frac{n_i}{h}\sum_{S} \delta_{a',Sa}\, D^{(i)\,*}_{\mu\nu}(S)\,(-1)^{l\tau_S}\, D^{j}_{m'm}(\alpha\beta\gamma). \tag{11.1.2}$$

where $D^{(i)\,*}_{\mu\nu}(S)$ is the representation matrix of the operation S, $D^{j}_{m'm}(\alpha\beta\gamma)$ is a rotation group matrix element with Euler angles α, β, γ for the group operation S, $\tau_S = 1$ if S contains an inversion and $\tau_S = 0$ otherwise.

In the BERTHA package, the AS are linear combinations of G-spinors with common rotational symmetry labels l, j, m so that it is sufficient to consider a single G-spinor component, (A.4.3),

$$|\,\rho a l j m \nu i \mu\rangle \to M[\beta; \gamma, \beta\kappa, \boldsymbol{r}] = \frac{1}{r}\, R_{\gamma,\beta\kappa}(r)\, \chi_{\beta\kappa,m}(\theta,\varphi),$$

where $\beta = \pm 1$ distinguishes the upper and lower 2-spinors, and the radial functions, denoted by

$$\rho_{nlj}(r) \to R_{\gamma,\beta\kappa}(r)$$

are supposed real, $\beta\kappa = \beta(j + 1/2)\eta, \quad \eta = \operatorname{sgn}\kappa = \pm 1, \quad l = j + \beta\eta/2$, and $\chi_{\beta\kappa,m}(\theta,\varphi)$ is a two-component spin-orbit function. A more extended description of this construction will be found in Appendix B.4, which also describes the software package `TSYM` [2] for generating the symmetry coefficients of (11.1.2).

Because j is always an odd multiple of $1/2$, each spatial transformation S corresponds to *two* group operations S and $\overline{S}$ differing by a rotation 2π, doubling the group dimension. However some classes of the *double group* are related by similarity transformations, so that not every classe is doubled. We use lower case letters to label MOs and upper-case letters to label total electronic states. The calculations that follow were done with small basis sets, Appendix B.5, which are not of the quality needed for state of the art calculations. They are only intended to give qualitative illustrations of the way in which molecular symmetry can be used in relativistic molecular structure. In particular, although no use has been made of symmetry to partition the Hamiltonian matrix into blocks corresponding to point group irreps, the symmetry classification of molecular spinors generated by BERTHA provides a strict test of correctness of the code's correctness.

11.1.1 Diatomic molecules

Diatomic molecules have an axis of rotation joining the atoms, making the classification of molecular states particularly simple. In a *nonrelativistic* model the orbital motion and spin can be treated independently. The rotational symmetry conserves the projection, m_l of the electron orbital angular momentum and the projection $m_s = \pm 1/2$ of the spin independently. Because m_l is always a signed integer, we can use $|m_l|$ to classify the orbitals; it is usual to label each MO as $\sigma, \pi, \delta, \ldots$ when $|m_l| = 0, 1, 2, \ldots$ respectively. The projection of the total electronic orbital angular momentum of an N-electron diatomic molecule on the symmetry axis is $M_l = \sum_{n=1}^{N} m_l(n)$, and its total spin projection is $M_s = \sum_{n=1}^{N} m_s(n)$. The electronic states can therefore be classified by the number $\Lambda = |M_l|$, so that $\Lambda = \Sigma, \Pi, \Delta, \ldots$ when $|M_l| = 0, 1, 2, \ldots$. Σ states are therefore nondegenerate, whereas those for which $\Lambda \neq 0$ are double degenerate: $\Lambda = \pm M_l$. Every plane containing the symmetry axis is a symmetry plane for the molecule, in which a reflection either changes the sign of the state (e.g Σ^-) or leaves it unchanged (e.g Σ^+). In the absence of spin-orbit coupling, the energy is independent of total spin, so that each state acquires a multiplicity $2S + 1$; the usual notation is, for example $^3\Sigma^+$. The states of molecules with an inversion centre can also be classified as $g$gerade, even) or u (*ungerade*, odd) with respect to inversion.

In Dirac theory, spin and orbital motion no longer provide independent constants of the motion. Each MO can be classified by the projection of its *total angular momentum*, $\omega = \pm m_j$, on the symmetry axis, so that ω is a signed odd multiple of 1/2. Molecular states can now be labelled by $\Omega = \sum_{n=1}^{N} \omega(n)$ in place of Λ, Σ. Before presenting examples of illustrative calculations on small molecules using BERTHA we need to examine the way in which the coupling between spin and orbital motion breaks down in low-Z molecules. Consider a single MS with $\omega = 1/2$, so that its $m_j = 1/2$ wavefunction is a linear combination of atomic spinors (AS). We can ignore the small effect of the $\beta = -1$ (small) component of the MS as we are dealing with low-Z molecules. The $\beta = +1$ (large) component is

$$\psi^+ = c_{-1}^+ \frac{R_{-1}^+(r)}{r} \chi_{-1,1/2}(\theta, \varphi) \tag{11.1.3}$$
$$+ c_{+1}^+ \frac{R_{+1}^+(r)}{r} \chi_{+1,1/2}(\theta, \varphi) + c_{-2}^+ \frac{R_{-2}^+(r)}{r} \chi_{-2,1/2}(\theta, \varphi)$$

where the subscripts -1, +1, -2 are the AS κ values and the c_κ^β are some coefficients. The spin-orbit functions are

$$\chi_{-1,1/2}(\theta, \varphi) = \begin{pmatrix} Y_0^0(\theta, \varphi) \\ 0 \end{pmatrix}$$

and

$$\chi_{+1,1/2}(\theta,\varphi) = \begin{pmatrix} -\sqrt{\frac{1}{3}}\, Y_1^0(\theta,\varphi) \\ \sqrt{\frac{2}{3}}\, Y_1^1(\theta,\varphi) \end{pmatrix}, \quad \chi_{-2,1/2}(\theta,\varphi) = \begin{pmatrix} \sqrt{\frac{2}{3}}\, Y_1^0(\theta,\varphi) \\ \sqrt{\frac{1}{3}}\, Y_1^1(\theta,\varphi) \end{pmatrix},$$

where the upper row gives the amplitude of α-spin ($m_s = 1/2$) and the lower gives the amplitude of β-spin ($m_s = -1/2$).

A molecular structure calculation will give numerical values to the mixing coefficients. If $c^+_{+1}/c^+_{-2} = -1/\sqrt{2}$ the β-spin component vanishes, and we get

$$\psi \approx \left[c^+_{-1} \frac{R^+_{-1}(r)}{r} Y_0^0(\theta,\varphi) + c_p \frac{R_p(r)}{r} Y_1^0(\theta,\varphi) \right] \alpha, \quad \alpha = \begin{pmatrix} 1 \\ 0 \end{pmatrix}$$

which is a product of a nonrelativistic σ orbital and an α spin ($m_s = 1/2$) wavefunction. Similarly if $c^+_{+1}/c^+_{-2} = \sqrt{2}$, we get a nonrelativistic π orbital and β spin $m_s = -1/2$). So if several MS have the same energy, we may expect to find a basis in which the states are products of a space part and a spin part as in this example. In practice, this separation can only be done approximately, and there will be situations, especially in molecules with high-Z atoms, for which no satisfactory factorization exists.

Carbon monoxide: $\mathbf{C}_{\infty v}$

Table 11.1. Carbon monoxide: orbital classification Reprinted, with permission, from Skaane [3].

	Relativistic		Nonrelativistic	
State	ϵ/E_h	ω	ϵ/E_h	Λ
1,2	-20.7323	±1/2	-20.7146	σ
3,4	-11.3969	±1/2	-11.3917	σ
5,6	-1.5470	±1/2	-1.5458	σ
7,8	-0.7932	±1/2	-0.7927	σ
9,10	-0.6474	±1/2	-0.6473	π
11,12	-0.6465	±3/2	-0.6473	π
13,14	-0.5584	±1/2	- 0.5583	σ
Total:	-112.6742	0^+	-112.6044	$^1\Sigma^+$

Table 11.1 (after [3]) shows results of an illustrative calculation with BERTHA for the $^1\Sigma^+$ ground state of the CO molecule using an uncontracted basis set (C:7s,4p ; O: 6s,4p), Table A.16, with an internuclear separation $d_{CO} = 3.132a_0$. No symmetry assumptions were made in setting up the Hamiltonian, so that a symmetry analysis of the resulting MS tests the

correctness with which the Hamiltonian has been constructed. The relativistic MS occur as Kramers' pairs with the same energy. The nonrelativistic results were obtained by repeating the calculation, increasing the speed of light by a factor of 100. The MS can be labelled in terms of ω in the relativistic case and of Λ, for both α and β spin, in the nonrelativistic case. Notice that although the nonrelativistic π MOs are degenerate, the relativistic MS show a small spin-orbit effect, consistent with results such as (11.1.3).

Nitrogen: $\mathbf{D}_{\infty h}$

Table 11.2. Nitrogen: orbital classification. Reprinted, with permission, from Skaane [3].

	Relativistic		Nonrelativistic	
State	ϵ/E_h	ω	ϵ/E_h	Λ
1,2	-15.6997	$\pm 1/2$	-15.6901	σ_g
3,4	-15.6963	$\pm 1/2$	-15.6866	σ_u
5,6	-1.5232	$\pm 1/2$	-1.5224	σ_g
7,8	-0.7705	$\pm 1/2$	-0.7696	σ_u
9,10	-0.6141	$\pm 1/2$	-0.6143	σ_g
11,12	-0.6102	$\pm 1/2$	-0.6103	π_u
13,14	0.6098	$\pm 3/2$	-0.6103	π_u
Total:	-108.7495	0_g^+	-108.6895	${}^1\Sigma_g^+$

Table 11.2 presents similar information on the homonuclear diatomic molecule N_2 using a slightly smaller basis (N: 6s,3p), Table A.16, with internuclear distance $d_{NN} = 2.0675a_0$. Here again the degeneracy of the nonrelativistic π_u MOs is lifted in the relativistic calculation.

11.1.2 Polyatomic molecules

Polyatomic symmetric molecules have states that transform under irreducible representations of more interesting point groups requiring machinery described in Appendix B.4. This enables the construction of molecular symmetry orbitals from which the Hamiltonian can, in principle, be reduced to diagonal blocks, each of which can be labelled by a particular irrep. This has not so far been implemented in BERTHA, but the application of a double group projection operator

$$\widehat{P}^{(i)} = \frac{n_i}{h} \sum_{S} \chi^*(\widehat{S})\, \widehat{S}, \tag{11.1.4}$$

where the sum runs over all group operations, quickly identifies whether a given molecular spinor belongs to a given irrep. Double group character tables are printed, for example, in [4, Volume 1, Appendix 1], and can be extracted from the TSYM software package [2].

Water: C_{2v}

Table 11.3. Water: orbital classification. Reprinted, with permission, from Skaane [3].

	Relativistic		Nonrelativistic	
State	ϵ/E_h	Label	ϵ/E_h	Label
1,2	-20.5848	$e_{1/2}$	-20.5669	$1a_1$
3,4	-1.3336	$e_{1/2}$	-1.3322	$2a_1$
5,6	-0.7049	$e_{1/2}$	-0.7052	$1b_2$
7,8	-0.5557	$e_{1/2}$	-0.5557	$3a_1$
9,10	-0.4970	$e_{1/2}$	-0.4971	$2b_1$
Total:	-75.9847		-79.9311	

The BERTHA calculation was done with a small basis set (O:6s,4p; H:6s), Table A.16, with bond length $d_{OH} = 1.81\ a_0$ and bond angle $\angle(\text{HOH}) = 104.36°$. The relativistic dynamics makes only a very small difference to the energies of the MS. However all relativistic orbitals are assigned the same double group label, contrasting with the nonrelativistic description. As usual, BERTHA orbitals appear in degenerate Kramers' pairs.

Table 11.4. Ammonia: orbital classification. Reprinted, with permission, from Skaane [3].

	Relativistic		Nonrelativistic	
State	ϵ/E_h	Label	ϵ/E_h	Label
1,2	-15.5320	$e_{1/2}$	-15.5221	$1a_1$
3,4	-1.1456	$e_{1/2}$	-1.1447	$2a_1$
5,6	-0.6124	$e_{1/2}$	-0.6124	$1e$
7,8	-0.6121	$e_{3/2}$	-0.6124	
9,10	-0.4089	$e_{1/2}$	-0.4089	$3a_1$
Total:	-56.0950		-56.0653	

Ammonia : C_{3v}

The calculation for ammonia was performed with $d_{NH} = 1.917\ a_0$ and $\angle(\text{HNH}) = 106.47°$ using a small basis (N: 6s,3p; H:6s), Table A.16. Relativistic energy differences are very small. As in the case of water, the relativistic orbitals appear in pairs; however, the nonrelativistic doubly degenerate e orbitals split into an $e_{1/2}$ pair and an $e_{3/2}$ pair with slightly different energies.

Table 11.5. Benzene: orbital classification. Reprinted, with permission, from Skaane [3].

	Relativistic		Nonrelativistic	
State	ϵ/E_h	Label	ϵ/E_h	Label
1,2	-11.2476	$e_{1/2,g}$	-11.2423	a_{1g}
3,4	-11.2471	$e_{1/2,u}$	-11.2417	e_{1u}
5,6	-11.2471	$e_{3/2,u}$	-11.2417	
7,8	-11.2459	$e_{5/2,g}$	-11.2406	e_{2g}
9,10	-11.2459	$e_{3/2,g}$	-11.2406	
11,12	-11.2454	$e_{5/2,u}$	-11.2400	b_{1u}
13,14	-1.1603	$e_{1/2,g}$	-1.1598	a_{1g}
15,16	-1.0216	$e_{3/2,u}$	-1.0212	e_{1u}
17,18	-1.0216	$e_{1/2,u}$	-1.0212	
19,20	-0.8280	$e_{3/2,g}$	-0.8278	e_{2g}
21,22	-0.8280	$e_{5/2,g}$	-0.8278	
23,24	-0.7163	$e_{1/2,g}$	-0.7164	a_{1g}
25,26	-0.6430	$e_{5/2,u}$	-0.6428	b_{1u}
27,28	-0.6239	$e_{5/2,u}$	-0.6241	b_{2u}
29,30	-0.5927	$e_{3/2,u}$	-0.5928	e_{1u}
31,32	-0.5926	$e_{1/2,u}$	-0.5928	
33,34	-0.5046	$e_{1/2,u}$	-0.5048	a_{2u}
35,36	-0.4932	$e_{5/2,g}$	-0.4933	e_{2g}
37,38	-0.4930	$e_{3/2,g}$	-0.4933	
39,40	-0.3384	$e_{1/2,g}$	-0.3385	e_{1g}
41,42	-0.3384	$e_{3/2,g}$	-0.3385	
Total:	-230.6810		-230.5866	

Benzene : D_{6h}

The benzene molecule, C_6H_6, is planar having the six carbons at the vertices of a regular hexagon and D_{6h} symmetry. The double point group has six additional fermion irreps, three *gerade* and three *ungerade* with respect to inversion. The calculation used an uncontracted basis (C: 7s,4p; H: 6s), Table A.16, $d_{CC} = 2.63a_0$ and $d_{CH} = 2.04a_0$.

The lowest 12 states are essentially carbon core orbitals. The delocalized π orbitals are (39,40) and (33,34); the former are classified as e_{1g} nonrelativistically or $e_{1/2,g}$ and $e_{3/2,g}$ relativistically, the latter a_{2u} nonrelativistically or $e_{1/2,u}$ relativistically. In the usual elementary model they are linear combinations of p_z orbitals on each of the carbon atoms. The MS 35–38, classified e_{2g} nonrelativistically and $e_{5/2,g}$ and $e_{3/2,g}$ relativistically are symmetric with respect to σ_h inversion and represent the σ bonds.

11.2 Relativistic effects in light molecules

Many expositions of atomic and molecular physics proceed from a nonrelativistic base with relativistic corrections included, when relevant, by treating the terms of the Breit-Pauli Hamiltonian, §3.7.2, as a first order perturbation. This description is so traditional and works so well for materials composed of atoms in the first two rows of the Periodic Table that we adopt it unthinkingly, even when relativistic effects are too large to be treated perturbatively, for example, when modelling high-Z atoms. Nevertheless the relativistic formulation can still be useful in high precision studies for low-Z materials.

About 70% of the absorption of sunlight in the atmosphere – and the bulk of the greenhouse effect – is attributable to water vapour [5]. Much research effort has been devoted to understanding the spectra of water, but some features, notably the details of atmospherically important weak line absorptions or the spectrum of superheated water, are difficult to determine experimentally [6]. It is claimed [7] that the precision of nonrelativistic treatments of molecules containing light elements, for example in recent *ab initio* results for potential energy hypersurfaces of H_2O [8] and H_2S [8, 9], now approaches the level at which remaining errors in the electronic energies are negligible in comparison with other effects. The small terms neglected in most treatments of Born-Oppenheimer potential energy hypersurfaces, including adiabatic and nonadiabatic corrections for nuclear motion as well as relativistic effects, become significant at the level needed to identify rotation-vibration bands important for atmospheric studies.

11.2.1 Nonrelativistic Breit-Pauli model

The effective Hamiltonian, H_4 (3.7.5), for relativistic corrections to the nonrelativistic motion of a single electron, was derived in §3.7.2. The many-body extension is derived in the same way from the DHFB Hamiltonian (10.1.1) in, for example, the classic texts of Bethe and Salpeter [10, §39] or Berestetskiĭ *et al.* [11, §83]. For a molecule, the effective Hamiltonian takes the form

$$H_{eff} = H_N + H_0 + H_{rel} \tag{11.2.1}$$

where

$$H_N = \sum_A \frac{1}{2M_A} \boldsymbol{p}_A^2 + \sum_{A<B} \frac{Z_A Z_B}{|\boldsymbol{A} - \boldsymbol{B}|}$$

is a Schrödinger operator describing the nuclear motion including the mutual Coulomb repulsion of the nuclei,

$$H_0 = \sum_i \frac{1}{2m} \boldsymbol{p}_i^2 + V \tag{11.2.2}$$

is the corresponding Schrödinger operator for the motion of the electrons in which

$$V = -\sum_A \sum_i \frac{Z_A}{|\boldsymbol{A} - \boldsymbol{r}_i|} + \sum_{i<j} \frac{1}{|\boldsymbol{r}_i - \boldsymbol{r}_j|}$$

is the potential energy of the electrons including nuclear attraction and their mutual repulsion. In the Born-Oppenheimer approximation, the nuclei have fixed positions, so that the only part of H_N contributing is the mutual repulsion of the nuclei.

The lowest order relativistic correction, H_{rel}, can be divided into two parts: $H_{rel}^{(1)}$ consists of a sum of one-body operators arising from the Dirac-Coulomb Hamiltonian and $H_{rel}^{(2)}$ contains additional terms derived from relativistic modifications to the Coulomb interaction between electrons. In the notation used by [7, Equations (5a-g)], which records an extensive investigation of relativistic corrections to properties of a large number of light molecules,

$$H_{rel}^{(1)} = H^{MV} + H^{D} + H^{SO} \tag{11.2.3}$$

$$H_{rel}^{(2)} = H^{SoO} + H^{SS} + H^{OO}. \tag{11.2.4}$$

The indivdual terms are

1. **Mass-velocity**

$$H^{MV} = -\frac{1}{8c^2} \sum_i \boldsymbol{p}_i^4$$

2. **Darwin, 1- and 2-body**: $H^D = H^{D1} + H^{D2}$

$$H^{D1} = \frac{\pi}{2c^2} \sum_A \sum_i Z_A \, \delta(|\boldsymbol{r}_i - \boldsymbol{A}|), \quad H^{D2} = -\frac{\pi}{2c^2} \sum_{i<j} \delta(|\boldsymbol{r}_i - \boldsymbol{r}_j|).$$

3. **Spin-orbit, 1- and 2-body**: $H^{SO} = H^{SO1} + H^{SO2}$

$$H^{SO1} = \frac{1}{2c^2} \sum_A \sum_i Z_A \frac{\boldsymbol{\sigma}_i \cdot (\boldsymbol{r}_i - \boldsymbol{A}) \times \boldsymbol{p}_i}{|\boldsymbol{r}_i - \boldsymbol{A}|^3}$$

$$H^{SO2} = -\frac{1}{2c^2} \sum_{i<j} \frac{\boldsymbol{\sigma}_i \cdot (\boldsymbol{r}_i - \boldsymbol{r}_j) \times \boldsymbol{p}_i}{|\boldsymbol{r}_i - \boldsymbol{r}_j|^3}$$

4. **Spin-other-orbit**:

$$H^{SoO} = -\frac{1}{c^2}\sum_{i<j}\frac{\boldsymbol{\sigma}_i\cdot(\boldsymbol{r}_i-\boldsymbol{r}_j)\times\boldsymbol{p}_j}{|\boldsymbol{r}_i-\boldsymbol{r}_j|^3}$$

5. **Spin-spin**: $H^{SS} = H^{FC} + H^{DP}$

$$H^{FC} = -\frac{1}{2c^2}\sum_{i<j}\boldsymbol{\sigma}_i\cdot\boldsymbol{\sigma}_j\,\frac{8\pi}{3}\delta(|\boldsymbol{r}_i-\boldsymbol{r}_j|) \quad \text{(Fermi contact)}$$

$$H^{DP} = \frac{1}{2c^2}\sum_{i<j}\left(\frac{\boldsymbol{\sigma}_i\cdot\boldsymbol{\sigma}_j}{|\boldsymbol{r}_i-\boldsymbol{r}_j|^3} - 3\frac{\boldsymbol{\sigma}_i\cdot(\boldsymbol{r}_i-\boldsymbol{r}_j)\,(\boldsymbol{r}_i-\boldsymbol{r}_j)\cdot\boldsymbol{\sigma}_j}{|\boldsymbol{r}_i-\boldsymbol{r}_j|^3}\right)$$

(spin-spin dipolar)

6. **Orbit-orbit**:

$$H^{OO} = \frac{1}{4c^2}\sum_{i<j}\left(\frac{\boldsymbol{p}_i\cdot\boldsymbol{p}_j}{|\boldsymbol{r}_i-\boldsymbol{r}_j|^3} + \frac{\boldsymbol{p}_i\cdot(\boldsymbol{r}_i-\boldsymbol{r}_j)\,(\boldsymbol{r}_i-\boldsymbol{r}_j)\cdot\boldsymbol{p}_j}{|\boldsymbol{r}_i-\boldsymbol{r}_j|^3}\right)$$

The survey [7] confirms that in light molecules such as C_2H_6, NH_3, H_2O, [H,C,N], HNCO, HCOOH, SiH_3^-, SiC_2, and H_2S the dominant relativistic energy corrections are associated with atomic core orbitals at DHF level, together with Breit and Lamb-shift corrections. Relativistic energy corrections must be included in high-accuracy theoretical treatments of geometrical dependence, for example inversion barriers (NH_3 and SiH_3^-), rotational barriers (C_2H_6), barriers to linearity (H_2O, H_2S and HNCO), conformational energy differences (HCOOH and SiC_2) and isomerization barriers ([H,C,N] system). In terms of the Pauli Hamiltonian, the dominant contributions are due to H^{MV} and H^{D1}. The two-electron Darwin term H^{D2} is often neglected, as it contributes only about 0.1 % of the molecular one-electron Darwin term; however its size is similar to the correlation contributions to H^{MV} and H^{D1} in these light molecules and should be included along with them. The spin-orbit H^{SO} effects are unimportant for the closed shell ground states of the light molecules surveyed, and become visible only when electrons are excited to higher states. However this approach breaks down, as we have seen in atomic spectra, as jj-coupling takes over with increasing Z.

The Breit interaction is the origin of the operators H^{SS}, H^{OO}; the simpler Gaunt interaction gives rise only to the H^{FC} term. This latter is spherically symmetric, and is therefore insensitive to changes in molecular geometry. Thus any geometry dependent corrections to potential energy hypersurfaces from the Breit interaction can only come from the terms H^{DP}, H^{OO}. Although necessary for getting very high precision, the Breit two-electron relativistic corrections are relatively small.

11.2.2 DHF and DHFB calculations for water using BERTHA

Table 11.6 displays energies of the H_2O molecule (units E_h) [12] for a succession of DHF and DHFB calculations with correlation consistent basis sets derived by Dunning *et al.* [13, 14]: see Appendix B.5, Tables A.17, A.22 and A.23. These were used [12] without any contraction, so that the nonrelativistic Hartree-Fock energies of the third column, obtained by increasing the speed of light by a factor 100, are slightly below the corresponding results due to Helgaker *et al.* [15] (-76.026799 for cc-pVDZ, -76.057168 for cc-pVTZ) in which the same basis sets were used in a contracted form. Table 11.7 shows the orbital energies from these correlation consistent basis sets [3, Table 8.2], which we may compare with the minimal basis set results of Table 11.3.

Table 11.6. DHF and DHFB calculations for water. Reprinted from [12, Table 2] with permission from Elsevier.

Basis set	E_{DHF}	E_{HF}	ΔE	ΔE_{Breit} ΔE_{Gaunt}
cc-pVDZ	-76.085331	-76.030375	-0.054956	+0.007565 +0.007864
aug-cc-pVDZ	-76.097556	-76.042604	-0.054952	+0.007557 +0.007857
cc-pVTZ	-76.112293	-76.057203	-0.055090	+0.007579 +0.007882
aug-cc-pVTZ	-76.115805	-76.060715	-0.055090	+0.007577 +0.007800

Table 11.7. DHF orbital energies (E_h) for water. Reprinted from [12] with permission from Elsevier.

	cc-pVDZ	aug-cc-pVDZ	cc-pVTZ	aug-cc-pVTZ
1,2	-20.565154	-20.595167	-20.572704	-20.584667
3,4	-1.339780	-1.358195	-1.347165	-1.355055
5,6	-0.699985	-0.718526	-0.708936	-0.716940
7,8	-0.567444	-0.585894	-0.577977	-0.585164
9,10	-0.493888	-0.508816	-0.504356	-0.510166

As expected, the relativistic corrections to Hartree-Fock total energies are relatively insensitive to basis sets that have been enhanced to improve valence correlation. The last column of Table 11.6 shows the corrections due to the two-electron magnetic interactions, Breit (upper entry) (6.4.14) or Gaunt (lower) (6.4.13); they also are relatively insensitive to basis set enlargement.

The total magnetic energy obtained self-consistently is about 20 μE_h smaller than that obtained by calculating the effect perturbatively [12, Table 3]. The simpler Gaunt interaction contributes the dominant two-electron magnetic effect, and the more complicated remainder of the Breit interaction (often called the retardation term) reduces the correction by about 4%.

11.2.3 Second-order many-body corrections

The BERTHA code uses an adaptation of the Head-Gordon and Pople [16] procedure in the no-virtual pair approximation [17] in which the negative-energy states are treated as inert. The second-order correlation energy of two electrons in occupied states a and b is

$$E_2(a,b) = \frac{1}{2}\sum_{rs}\frac{(\,ra\,|\,sb\,)(\,ar\,|\,bs\,)}{\epsilon_a+\epsilon_b-\epsilon_r-\epsilon_s} - \frac{1}{2}\sum_{rs}\frac{(\,ra\,|\,sb\,)(\,as\,|\,br\,)}{\epsilon_a+\epsilon_b-\epsilon_r-\epsilon_s} \tag{11.2.5}$$

where r, s label positive energy MS. The calculation [3] extends the White and Head-Gordon algorithm [18] to the relativistic case. The total second order energy summing over all occupied pairs a, b using the same basis sets as in §11.2.2 is shown in Table 11.8 along with the total energies resulting from DHF and HF calculations from Table 11.6. The contribution of relativistic and nonrelativistic correlation to the total energy is of the same order of magnitude, and the dependence on basis set is much the same in both calculations. Whereas relativity reduces the total HF energy by about 55 mE_h

Table 11.8. Second order correlation energy E_2 of water. Reprinted from [12, Table 5] with permission from Elsevier.

Basis set		$E_{(D)HF}/E_h$	E_2/E_h
cc-pVDZ	Relativistic	-76.085331	-0.257904
	Non-relativistic	-76.030375	-0.257797
	Difference	-0.054956	-0.000108
aug-cc-pVDZ	Relativistic	-76.097556	-0.273948
	Non-relativistic	-76.042604	-0.273838
	Difference	-0.054952	-0.000110
cc-pVTZ	Relativistic	-76.112293	-0.312813
	Non-relativistic	-76.057203	-0.312680
	Difference	-0.055090	-0.000133
aug-cc-pVTZ	Relativistic	-76.115805	-0.319271
	Non-relativistic	-76.060715	-0.319140
	Difference	-0.055090	-0.000131

and the two-electron magnetic interaction raises it by about 7.6 mE_h (Table 11.6), the relativistic correction to the correlation energy, -0.13 mE_h is tiny by comparison. However, the picture looks quite different at the orbital level where typical pair correlation energies can be as large as -30 mE_h, with relativistic corrections around -0.02 mE_h. In some cases, the relativistic effect is much larger, and the cancellation comes from summing terms of the form

$$E_2(a,b) + E_2(a,\bar{b})$$

where b and $\bar{b}$ are Kramers' pairs related by time reversal. The most striking example involves the pair (3,4), corresponding to the $2a_1$ orbitals in the nonrelativistic limit (Table 11.3). Here the relativistic corrections between the pairs (mE_h) are much larger,

$$\varDelta E_2(3,3) = \varDelta E_2(4,4) = -2.12, \quad \varDelta E_2(3,4) = 4.23,$$

compared with the correlation energies

$$E_2(3,3) = E_2(4,4) = -16.14, \quad E_2(3,4) = -33.35.$$

The total relativistic correction to the $2a_1$ correlation energy reduces to a tiny -2 μE_h. Similar, but much less surprising results were found for the other Kramers' pairs (1,2), (5,6), (7,8) and (9,10). Although spin and orbital motions are not separable in the relativistic calculation for this element, the outcome is hardly distinguishable in this case from nonrelativistic theory. For systems with higher Z elements, the cancellation can be expected to be less complete, and the picture will be much more complicated.

11.2.4 Relativistic study of the potential energy surface and vibration-rotation levels of water

Quiney *et al.* [19] report a detailed study of two-electron relativistic effects on the potential energy surface and vibration-rotation levels of water at over 300 points on the potential energy surface [19], ranging over $1.47a_0 \leq R_{OH} \leq 2.79a_0$ and $41° \leq \angle(HOH) \leq 172°$. This followed a preliminary exploration [12] that had examined the relativistic effects on the bond length R_{OH} and bond angle $\angle(HOH)$ near the equilibrium geometry, which revealed a small, significant and rather unexpected geometrical variation. The BERTHA calculations were compared with results from DIRCCR [20] for the two-electron Darwin term H^{D2} and MOLFDIR [21] for the Gaunt correction. Figure 11.1 shows the results for the two-body operators.

The main conclusion from this work was that the two electron relativistic corrections are significant for the calculated behaviour of the vibrational and rotational states of water. The one body corrections, $H^{MV} + H^{D1}$ dominate, and are due largely to atomic core contributions. The H^{D2} term is relatively easy to evaluate; most of the effect is recovered at the HF level and correlation

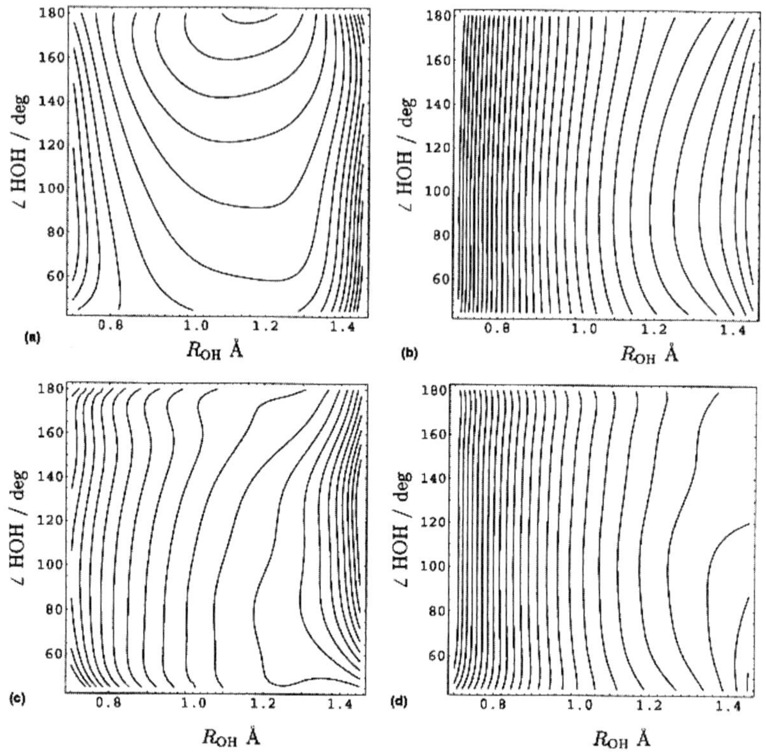

Fig. 11.1. Contour plots of components of the two-electron relativistic corrections to the potential energy surface of the water molecule. (a) Two-body Darwin term H^{D2}; contours are separated by 1 cm^{-1} with the maximum at the top of the plot. (b) Gaunt interaction; contours are separated by 2.5 cm^{-1} decreasing to the right. (c) Breit-Gaunt differences; contours are separated by 1 cm^{-1} increasing to the right. (d) Breit interaction; contours are separated by 2.5 cm^{-1} decreasing to the right. Reprinted from [19] with permission from Elsevier.

has only a small effect. The remaining Breit-Pauli two electron terms proved more difficult to calculate nonrelativistically because of slow convergence of the perturbation series with respect to angular momentum. The corresponding DHFB estimates of two-electron effects presented no such problems. The results gave confidence in the use of the Breit-Pauli Hamiltonian in these light molecular systems. The DHF and DHFB calculations were not supplemented by correlation calculations to the level used in the Breit-Pauli work and it would be interesting to explore this area further.

A similar set of calculations has been performed for H_2S [22]. The absolute value of the relativistic correction is an order of magnitude larger in H_2O, but the dependence on geometry is almost the same. The size of the relativistic corrections is a reflection of the higher binding energy of the core electrons, but the valence structure, which determines the variation of the relativistic effects across the PES, is similar to that of H_2O.

11.3 Electromagnetic properties of atoms and molecules

11.3.1 Gauge transformations in electromagnetic processes

The need for the physical predictions of the theory to be invariant to gauge transformations has been a recurrent theme of this book. The gauge dependence of computed electric multipole radiative transition probabilities was discussed at length in §8.1. Whilst the traditional choice of the relativistic Babushkin gauge (nonrelativistic length gauge) seems to give a stable approximation to measured transition rates, there is no *a priori* justification for choosing this particular gauge so that our theoretical result could be any real number from zero to infinity! The gauge dependence can be guaranteed to vanish if the transition moment of the atomic or molecular charge-current density for longitudinal photons vanishes. Whilst this can never be guaranteed, except for simple one-electron models using a local potential [23], there are many examples, §7.12, in atomic transitions in which the gauge dependence is small enough to be tolerable.

Similar gauge problems occur in the calculation of electromagnetic properties of molecules, even in quite simple situations. Consider a simple one-electron hydrogenic atom with a stationary vector potential $\boldsymbol{A}$,

$$\widehat{h}_D(\boldsymbol{A}) = c\boldsymbol{\alpha}\cdot(\boldsymbol{p}+\boldsymbol{A}) + \beta mc^2 - Z/r. \tag{11.3.1}$$

Under a gauge transformation

$$\boldsymbol{A} \rightarrow \boldsymbol{A}' = \boldsymbol{A} + \mathrm{grad}\,\Lambda,$$

the transformed Hamiltonian and state vector are

$$\begin{aligned} \widehat{h}_D(\boldsymbol{A}) \rightarrow \widehat{h}_D(\boldsymbol{A}') &= \exp[-i\Lambda]\,\widehat{h}_D(\boldsymbol{A})\exp[i\Lambda], \\ \psi \rightarrow \psi' &= \exp[-i\Lambda]\,\psi, \end{aligned} \tag{11.3.2}$$

where, the subsidiary condition $\mathrm{div}\boldsymbol{A} = 0 = \mathrm{div}\boldsymbol{A}'$ entails $\nabla^2\Lambda = 0$. Then this construction ensures

$$(\psi\,|\,\widehat{h}_D(\boldsymbol{A})\,|\,\psi) = (\psi'\,|\,\widehat{h}_D(\boldsymbol{A}')\,|\,\psi'); \tag{11.3.3}$$

the energy is gauge invariant.

However it is by no means obvious that if we try to solve the matrix Dirac equation (or the corresponding Schrödinger equation) for the ground state of the system in a fixed basis, the respective eigenvalues ϵ and ϵ' will be the same, even though we might expect $\epsilon = \epsilon'$ if the basis set is complete. Skaane [3, Table 9.16] studied the simple example in which $\boldsymbol{A} = 0, \Lambda = z$ for hydrogen-like neon ($Z = 10$). He used the same G-spinor basis set with exponents $a_i = \alpha\beta^{i-1}$, as in (B.5.20), setting $\alpha = 0.01$ and $\beta = 1.9$. The eigenvalues all converged to 9 decimal places within 10 iterations, giving $\epsilon = -12.504162870E_h$ compared with $\epsilon' = -11.938559855E_h$ for a 30s/30p basis set. Adding 30d G-spinors to the basis gave the improved value $\epsilon' = -12.478913786E_h$ and agreement to 4 decimals, $\epsilon' = -12.504151181E_h$ came only after adding 30f and 30g G-spinors to the basis. A similar test with the same geometric G-spinor basis set for H_2^+ gave the energies $\epsilon = -0.360731558E_h$ and $\epsilon' = -0.306137830E_h$ for a 20s/20p basis set and $\epsilon = -0.360871033E_h$, $\epsilon' = -0.362852656E_h$ when 20d, 20f and 20g G-spinors were added to the basis set.

This rather unsatisfactory situation poses a major problem in calculation of magnetic effects where, as we shall see below, a shift of the geometrical origin corresponds to precisely such a gauge transformation of the vector potential. Skaane pointed out that by trying to expand ψ and ψ' in terms of the same basis set, we are trying to match two expansions

$$\psi' \approx \sum_{\mu} c'^{\beta}_{\mu} M[\beta, \mu, \boldsymbol{x}]$$

and

$$\psi \approx \exp[-i\Lambda] \sum_{\mu} c^{\beta}_{\mu} M[\beta, \mu, \boldsymbol{x}]$$

so that it is not surprising that it is necessary to add many new functions to the basis in order to make the two approximate solutions match. We can avoid this mismatch by defining a new basis, *B-spinors*,

$$\omega^{\beta}_{\mu}(\boldsymbol{x}) = \exp[-i\Lambda(\boldsymbol{x})] \, M[\beta, \mu, \boldsymbol{x}] \tag{11.3.4}$$

a relativistic generalization of *London orbitals* [24], which have been used to remove dependence on the choice of origin in calculations of electromagnetic properties of nonrelativistic molecular calculations.

11.3.2 B-spinors

An external static homogeneous magnetic field $\boldsymbol{B}$ is associated with a time-independent vector potential

$$\boldsymbol{A} = \frac{1}{2}[\boldsymbol{B} \times (\boldsymbol{x} - \boldsymbol{G})] \tag{11.3.5}$$

where $\boldsymbol{G}$ is an arbitrary origin. A shift $\boldsymbol{G} \to \boldsymbol{G}'$ is equivalent to a gauge term

$$\Lambda(\boldsymbol{x}) = \frac{1}{2}[\boldsymbol{B} \times (\boldsymbol{G} - \boldsymbol{G}'] \cdot (\boldsymbol{x} - \boldsymbol{Q})$$

where $\boldsymbol{Q}$ is an arbitrary fixed point. If we take $\boldsymbol{G}$ as the (fixed) gauge origin for the whole system, and identify $\boldsymbol{G}'$ with a nuclear centre $\boldsymbol{A}_\mu$, we can define the *B-spinors*, or relativistic *London atomic orbitals* (LAO)[1], (also known as *gauge invariant atomic orbitals*, GIAO) by

$$B[\beta, \mu, \boldsymbol{x}; \boldsymbol{w}] = \exp\left[-i(\boldsymbol{B} \cdot \boldsymbol{w}_\mu)/2\right]\, M[\beta, \mu, \boldsymbol{x}]. \tag{11.3.6}$$

where

$$\boldsymbol{w}_\mu = [\boldsymbol{G} - \boldsymbol{A}_\mu] \times [\boldsymbol{x} - \boldsymbol{Q}]$$

This ensures that electromagnetic properties calculated using a basis of B-spinors will be independent of the choice of the point $\boldsymbol{G}$. It does not imply that the result will be gauge invariant in a more general sense.

In practice, matrix elements of a magnetic operator $\mathcal{O}$ with respect to B-spinors with indices μ, ν can be calculated as matrix elements of the transformed operator

$$\exp\left[+i(\boldsymbol{B} \cdot \boldsymbol{w}_\mu)/2\right]\, \mathcal{O}\, \exp\left[-i(\boldsymbol{B} \cdot \boldsymbol{w}_\nu)/2\right] \tag{11.3.7}$$

with respect to the original G-spinors. Ruud *et al.* [24] have highlighted several advantages of using the equivalent London orbitals in nonrelativistic calculations of magnetic properties. Perhaps the most important is the rapid convergence of basis set expansions in LAOs, noted above, along with the elimination of difficulties connected with the choice of the gauge origin $\boldsymbol{G}$. LAOs (and B-spinors) also preserve size extensivity properties of the wavefunction, important for correlation calculations. Ruud *et al.* deal with implementation issues, and many of their ideas have been incorporated in the relativistic BERTHA code [26].

Kutzelnigg *et al.* [27] introduced the IGLO (*individual gauge London orbitals*, which uses standard Gaussian basis functions. It uses local phase factors as in the London approach, but attaches them to localized MOs rather than to AOs. Completeness relations are used to simplify the calculation, and this makes it somewhat sensitive to basis set quality.

11.4 The Zeeman effect

The vector potential (11.3.5) is responsible for the splitting of atomic levels in a magnetic field known as the Zeeman effect [28, §13.2]. The nonrelativistic one-electron theory is based on adding a perturbation

$$W = \mu_B \boldsymbol{B} \cdot (\boldsymbol{l} + g_e \boldsymbol{s}) \tag{11.4.1}$$

to the free field Schrödinger Hamiltonian [10, Equation (45.9)], where $\mu_B = e\hbar/2m_e$ is the *Bohr magneton*, $\boldsymbol{l}$ and $\boldsymbol{s}$ are the electron's orbital and spin

[1] First introduced by London [25] in studies of π-electrons in aromatic hydrocarbons.

angular momenta, and $g_e \approx 2$ (the value derived from Dirac theory) is the *gyromagnetic ratio* or g-factor, which has now been determined experimentally to a few parts in 10^{12} [29]. The deviation of g_e from the Dirac value 2.0 is attributed to QED corrections, and its experimental determination in hydrogenic ions as a test of QED continues to be an active area of research [30].

If $\boldsymbol{B}$ is in the z-direction and the electron has quantum numbers $lsjm_j$, then the level splitting is [28, Equation (13.18)]

$$\Delta E_{m_j} = g\mu_B B m_j \tag{11.4.2}$$

and the *Landé splitting factor* g is given by

$$g = g_l \frac{j(j+1) - s(s+1) + l(l+1)}{2j(j+1)} + g_e \frac{j(j+1) + s(s+1) - l(l+1)}{2j(j+1)}, \tag{11.4.3}$$

where $g_l \approx 1 - m_e/M$, where m_e is the electron and M the nuclear mass. Substituting $g_l = 1$ and $g_e = 2$ in (11.4.3) gives the well-known approximation

$$g = (j + 1/2)/(l + 1/2). \tag{11.4.4}$$

The Pauli Hamiltonian from which (11.4.1) is derived also includes a term, neglected in elementary theory, which is proportional to $\boldsymbol{A}^2$, but which is important when $\boldsymbol{B}$ is sufficiently large. Experimental and theoretical studies on its effects in high Rydberg states have revealed complex patterns of energy levels and phenomena of "quantum chaos". Many references to this field may be found in [31]. The relativistic theory is somewhat simpler. The interaction of a single electron with an external static magnetic field is given by the usual interaction Hamiltonian, which in the present context reduces to

$$H_{int} = \frac{c}{2} \boldsymbol{\alpha} \cdot (\boldsymbol{B} \times \boldsymbol{x}). \tag{11.4.5}$$

This is *linear* in $\boldsymbol{B}$; there is nothing corresponding to the nonrelativistic $\boldsymbol{A}^2$ (or "diamagnetic") term. In perturbation theory such terms can only appear in second or higher orders. Equation (11.4.5) gives a formula for the magnetic moment operator

$$\boldsymbol{\mu} = -\frac{c}{2} \boldsymbol{\alpha} \times \boldsymbol{x} \tag{11.4.6}$$

The matrix elements in a G-spinor (or B-spinor) expansion are given by (10.9.18).

In the atomic case, the one-electron g-factor for an electron in the orbital $nljm$ can be written [32]

$$g = -\frac{\kappa c}{j} (jj \,|\, \boldsymbol{\alpha} \times \boldsymbol{x} \,|\, jj) = \frac{\kappa c}{j(j+1)} \int_0^\infty r(PQ + QP)\, dr. \tag{11.4.7}$$

This reduces correctly to the nonrelativistic result in the Pauli limit [32]. The corresponding result for a multiconfigurational atomic state T with total angular momentum J [32] in the notation of (6.10.17) is

Table 11.9. Electronic g-tensor components, $g_{\|}$ and $g_{\perp}$, and components of spin projection for selected diatomic molecules. Reprinted from [35] with permission of Elsevier.

	$10^6(1-\Sigma)\mu_B$		$10^4(g_{\|}-2)$		$10^4(g_{\perp}-2)$	
	$\|$	$\perp$	DHF	Exp. [36]	DHF	Exp. [36]
BeH	14	18	23	22(2)	21	21(3)
MgH	19	21	23	2(4)	8	20(4)
CaH	36	28	23	13(4)	-24	-34(4)
SrH	241	133	21	4(4)	-102	-135(4)
ZnH	88	63	2	3(3)	5	-148(6)
CdH	3111	182	20	30(4)	-149	-476(4)
PdH	4382	32323	-623	-350(20)	5973	2932(4)
YbH	1410	731	9	-47(4)	-281	-598(2)
BeF	17	20	23	14(5)	17	14(5)
MgF	22	24	22	20(5)	9	10(5)
CaF	27	24	22	20(5)	-1	0(10)
SrF	140	81	22	20(5)	-35	-30(5)
ZnF	300	208	19	20(10)	-414	-40(10)
YbF	415	224	19	-25(5)	5	-46(5)
BO	31	40	22	15(3)	9	12(3)
ScO	29	29	23	0(10)	18	0(10)
YO	143	87	21	30(20)	5	30(20)
LaO	226	123	21	100(100)	6	100(100)
BS	63	59	22	16(1)	-52	-58(1)

$$g_J = -c \sum_{ij} c^*_{iT} c_{jT} \sum_{\alpha\beta} d^1_{\alpha\beta}(T,T) \frac{\kappa_\alpha + \kappa_\beta}{\sqrt{J(J+1)}} \times [j_\alpha]^{-1/2} \langle j_\alpha \| \boldsymbol{C}^1 \| j_\beta \rangle \int_0^\infty r(P_\alpha Q_\beta + Q_\alpha P_\beta)\, dr, \quad (11.4.8)$$

where

$$d^1_{\alpha\beta}(T,T) = (-1)^{\Delta_\alpha + \Delta_\beta{}'} (N_\alpha N_\beta{}')^{1/2} (T_\alpha\{|T_\alpha{}')\,(T_\beta|\}T_\beta{}') \times \delta_{\widetilde{T}_\alpha, T_\alpha{}'} \delta_{\widetilde{T}'_\beta, T_\beta} \prod_{i \neq \alpha,\beta} \delta_{T_i, T_i{}'} \frac{[J]^{1/2}}{[j_\alpha]^{1/2}} \mathbb{R}_{\alpha,\beta}(T;T).$$

More generally, we can follow Abragam and Bleaney [33] to generalize (11.4.1) to give an effective Hamiltonian in the form

$$W = \frac{1}{4c} \sum_{kl} B_k\, g_{kl}\, \sigma_l, \quad (11.4.9)$$

where k, l label Cartesian coordinate directions σ_l, $l = 1, 2, 3$, are Pauli matrices and g_{kl} is customarily referred to as the *electronic g-tensor*.[2] In the four-component formalism used in this book, the g-tensor can be computed as the coefficients of the linear relation

$$\sum_l g_{kl} \, \Sigma_l = (\boldsymbol{\alpha} \times \boldsymbol{x})_k \tag{11.4.10}$$

where $\Sigma_l = \sigma_l \times I_2$. The G-spinor integrals required are available from (10.9.9), (10.9.16) and (10.9.18). The two-component representation devised by van Lenthe *et al.* [34] gives the lowest order relativistic approximation to the g-tensor; the four-component expression includes relativistic effects to all orders.

Quiney and Belanzoni [35] have made a study of electronic g-tensors of diatomic molecules in the DHF scheme using the BERTHA code [26]. The Cartesian axes are aligned with the principal axes of the molecule. The columns of Table 11.9 labelled with $\Sigma_{\|}$ and $\Sigma_{\perp}$ show values of the expectation of the spin operator along and perpendicular to the molecular axis respectively; the g-tensor is diagonal and we label the non-zero components g_{kk} as $g_{\|}$ and $g_{\perp}$. The comparison with experimental values for low-Z molecules, taken from [36], reveals that the deviation of the DHF value of $g_{\|}$ from the free-electron value $g = 2.00232$ is very small. This just confirms that these molecules are well described by the LS-coupling scheme, for which $\Sigma_{\|} = 1$. There are signs that the DHF single determinant model is not so satisfactory for molecules with high-Z constituents. The calculation of $g_{\perp}$ is a more sensitive test of the quality of the model as it depends on the anisotropy of the spin-current, which may not be so well represented at the single determinant DHF level. The value of $g_{\perp}$ for PdH is particularly poor, probably because of the neglect of configuration mixing in the Pd atom. However, the calculation reproduces qualitatively the huge anomalous increase in $g_{\perp}$ and the decrease in $g_{\|}$.

11.5 Hyperfine interactions

The fine structure levels predicted by the theory developed so far reveal further structure at high resolution. This *hyperfine structure* can be attributed to weak interaction of the electrons with electric and magnetic fields generated by the nuclear charge-current distribution. The theory of hyperfine structure for free atoms was surveyed by Armstrong [37] and also by Lindgren and Rosén [38]. The hyperfine interaction can be expanded in a product of multipoles of order k

$$H_{hfs} = \sum_{k \geq 1} \boldsymbol{T}^k \cdot \boldsymbol{M}^k \tag{11.5.1}$$

[2] This is a misnomer, because g_{kl} does not have the right transformation properties. However, the product $G_{ij} = \sum_k g_{ik} g_{jk}$ is a true Cartesian tensor.

Table 11.10. Components of the nuclear magnetic hyperfine tensor, $A_{\parallel}$ and $A_{\perp}$ (in Mhz), for selected diatomic molecules. Asterisks denote isotropic values. Reprinted from [35] with permission of Elsevier.

	$A_{\parallel}$		$A_{\perp}$	
	DHF	Exp. [36]	DHF	Exp. [36]
^{9}Be H	177	208(1)	164	194.8(3)
Be ^{1}H	80	201(1)	62	190.8
^{25}Mg H	217	226(3)	206	218(1)
Mg ^{1}H	93	298(1)	98	264(1)
Ca ^{1}H	45	138(1)	42	134(1)
Sr ^{1}H	38	123(2)	34	121(2)
Zn ^{1}H	162	485(2)	159	487(2)
^{111}Cd H	1370	4358(25)	1265	3966(3)
Cd ^{1}H	190	515(1)	187	515(1)
^{105}Pd H	538	867(10)	560	801(2)
Pd ^{1}H	68	103(6)	6	106(1)
^{171}Yb H	4656	5724(20)	4401	5266(5)
Yb ^{1}H	79	226(2)	76	224(2)
^{9}Be F	236	303	226	297
Be ^{19}F	190	241	136	227
Mg ^{19}F	170	331(3)	101	143(3)
Ca ^{19}F	85	149(3)	65	106(3)
^{87}Sr F	604	591(3)	587	570(3)
Sr ^{19}F	68	126(3)	47	95(3)
Zn ^{19}F	655	673(2)	19	143(2)
Cd ^{19}F	519	670(7)	195	64(2)
^{171}Yb F	6257	7822(5)	6167	7513(5)
Yb ^{19}F	215	220(2)	191	134(2)
^{11}B O	1051	1018(1)	964	1034(1)
^{45}Sc O	1561	2063(1)	1515	1990(10)
^{89}Y O	617		602	*803(3)
^{139}La O	4268	3724	4181	2639
^{11}B S	855	853.3(3)	751	766.7(3)
B ^{33}S	39		13	16(2)

where $\boldsymbol{T}^k$, which operates on electronic wavefunctions, and $\boldsymbol{M}^k$, which operates on the nuclear wavefunctions, are tensor operators of rank k. If $\boldsymbol{I}$ is the angular momentum operator for the nucleus and $\boldsymbol{J}$ the total angular momentum operator of the electrons, the angular momentum of the combined system is

$$\boldsymbol{F} = \boldsymbol{J} + \boldsymbol{I}.$$

The splitting of atomic energy levels for the interaction of rank k as a function of F can be written [28, Equation (16.6)]

$$\Delta E_k(JIF, JIF) = (-1)^{J+I+F} \begin{Bmatrix} J & J & k \\ I & I & F \end{Bmatrix} \times \left[\begin{pmatrix} J & k & J \\ J & 0 & -J \end{pmatrix} \begin{pmatrix} I & k & I \\ I & 0 & -I \end{pmatrix} \right]^{-1} A_k \quad (11.5.2)$$

where for $k \geq 1$,

$$A_k = (JJ \,|\, \boldsymbol{T}^k \,|\, JJ) \cdot (II \,|\, \boldsymbol{M}^k \,|\, II). \quad (11.5.3)$$

The commonly used *hyperfine shift parameters* $A, B \ldots$ are defined by

$$A = A_1/IJ, \quad B = 4A_2, \quad \ldots \quad (11.5.4)$$

We shall consider only the magnetic dipole interaction, $k = 1$, for which we can write

$$\Delta E_1(JIF, JIF) = A \, \langle \boldsymbol{I} \cdot \boldsymbol{J} \rangle \quad (11.5.5)$$

where

$$\langle \boldsymbol{I} \cdot \boldsymbol{J} \rangle = \frac{1}{2}[F(F+1) - J(J+1) - I(I+1)].$$

In this case,

$$\boldsymbol{T}^1 = -\frac{1}{c} \frac{\boldsymbol{\alpha} \times \boldsymbol{x}}{|\boldsymbol{x}|^3} \quad (11.5.6)$$

where $\boldsymbol{x}$ is the position of the electron relative to the relevant nucleus and the *nuclear magnetic dipole moment* is defined by

$$\mu_I = (II \,|\, \boldsymbol{M}^1 \,|\, II)$$

In molecules, the angular momentum classification used in atoms is no longer relevant, and we must evaluate the expectation value of

$$H_{hfs} = \frac{1}{c} \boldsymbol{\mu} \cdot \frac{\boldsymbol{\alpha} \times (\boldsymbol{x} - \boldsymbol{A}_n)}{|\boldsymbol{x} - \boldsymbol{A}_n|^3} \quad (11.5.7)$$

giving the interaction of the electron current with the nuclear magnetic dipole moment $\boldsymbol{\mu}$ of the nucleus at $\boldsymbol{A}_n$, where

$$\boldsymbol{\mu} = \mu_N \, g_I \, \boldsymbol{I},$$

with *nuclear moment* $g_I \, \boldsymbol{I}$ (g_I being the nuclear g-factor) and $\mu_N = 1/2m_p$ is the nuclear magneton, in terms of the proton mass m_p. The first-order energy shift can be written as the expectation of a spin Hamiltonian

$$H'_{hfs} = \frac{1}{2} \sum_{ij} \Sigma_i \, A_{ij} \, I_j \quad (11.5.8)$$

in the same manner as (11.4.9). A_{ij} is known as the *magnetic hyperfine tensor*; it is diagonal if the Cartesian axes are aligned with the principal axes of a diatomic molecule, and we write $A_{\|}$ for the diagonal tensor component along with the internuclear axis and $A_{\perp}$ for any diagonal component in a perpendicular direction. The necessary G-spinor integrals can be obtained by the methods described in the previous section.

Results for a number of diatomic molecules are shown in Table 11.10, selected according to the availability of experimental data and of suitable basis sets. As with similar nonrelativistic calculations, the hyperfine proton coupling is poorly represented by the DHF model because of the importance of electron correlation. Hyperfine parameters for molecules containing ^{19}F are usually underestimated, probably through neglect of core polarization. The electronic structure of the single determinant DHF model tends to emphasize the stable closed shell ion F^{-}, for which the atomic hyperfine tensor vanishes. This defect could be remedied by taking electron correlation into account.

11.6 NMR shielding in small molecules

Attempts to extract nuclear moments from hyperfine splittings are complicated by shielding of the atomic electrons. A simple relativistic theory [26, §6.2] starts from the combined interaction

$$H_{nmr} = c\boldsymbol{\alpha} \cdot \left\{ \frac{1}{2} \boldsymbol{B} \times \boldsymbol{x} + \frac{1}{c^2} \frac{\boldsymbol{\mu} \times (\boldsymbol{x} - \boldsymbol{A}_n)}{|\boldsymbol{x} - \boldsymbol{A}_n|^3} \right\}. \tag{11.6.1}$$

where $\boldsymbol{A}_n$ is the position vector of nucleus n. The lowest order energy correction due to the nuclear moment $\boldsymbol{\mu}$ is given by the spin Hamiltonian

$$H'_{nmr} = \sum_{ij} B_i \, \mu_j \, \sigma_{ij} \tag{11.6.2}$$

where i, j label Cartesian coordinate directions and σ_{ij} is the *shielding tensor*. The matrix elements of (11.6.1) in a B-spinor basis (11.3.6) are equivalent to matrix elements of the operator

$$\exp\left[+i\Lambda_{\mu}(\boldsymbol{x})\right] H_{nmr} \exp\left[-i\Lambda_{\nu}(\boldsymbol{x})\right],$$

in a G-spinor basis where, after putting the arbitrary position vector $\boldsymbol{Q} = 0$,

$$\boldsymbol{\Lambda}_{\mu}(\boldsymbol{x}) = \boldsymbol{B} \cdot [\boldsymbol{G} - \boldsymbol{A}_{\mu}] \times \boldsymbol{x}.$$

as in (11.3.5). The shielding tensor in lowest order is the sum of two terms

$$\sigma_{jk} = \sigma^{(1)}_{jk} + \sigma^{(2)}_{jk}. \tag{11.6.3}$$

The first is the result of linearizing

$$\exp\left[+i\boldsymbol{\Lambda}_\mu\right] \frac{\boldsymbol{\alpha}}{c} \cdot \frac{\boldsymbol{\mu} \times (\boldsymbol{x} - \boldsymbol{A}_n)}{|\boldsymbol{x} - \boldsymbol{A}_n|^3} \exp\left[-i\boldsymbol{\Lambda}_\nu\right]$$

with respect to $\boldsymbol{B}$, giving

$$\sigma_{jk}^{(1)} = \frac{i}{2} \sum_a \sum_{\mu\nu} c_\mu^{\beta\,*} c_\nu^{-\beta} \times \left\langle \mu, \beta \,\middle|\, [(\boldsymbol{A}_\mu - \boldsymbol{A}_\nu) \times \boldsymbol{x}]_j \, \frac{(\boldsymbol{\sigma} \times (\boldsymbol{x} - \boldsymbol{A}_n))_k}{|\boldsymbol{x} - \boldsymbol{A}_n|^3} \,\middle|\, \nu, -\beta \right\rangle \tag{11.6.4}$$

where n refers to the nucleus of interest. As expected, this term only contributes when there is more than one nuclear centre. The term $\sigma_{jk}^{(2)}$ is the first order interference between the two terms of (11.6.1),

$$\sigma_{jk}^{(2)} = \sum_{ar} \left\{ \frac{\langle a \,|\, [\boldsymbol{\alpha} \times (\boldsymbol{x} - \boldsymbol{G})]_j \,|\, r \rangle \, \langle r \,|\, [\boldsymbol{\alpha} \times (\boldsymbol{x} - \boldsymbol{A}_n)]_k / |\boldsymbol{x} - \boldsymbol{A}_n|^3 \,|\, a \rangle}{E_a - E_r} + \frac{\langle a \,|\, [\boldsymbol{\alpha} \times (\boldsymbol{x} - \boldsymbol{A}_n)]_k / |\boldsymbol{x} - \boldsymbol{A}_n|^3 \,|\, r \rangle \, \langle r \,|\, [\boldsymbol{\alpha} \times (\boldsymbol{x} - \boldsymbol{G})]_j \,|\, a \rangle}{E_a - E_r} \right\} \tag{11.6.5}$$

where a runs over occupied orbitals and r ranges over the whole spectrum *including states of negative energy*. We shall write

$$\sigma_{jk}^{(2)} = \sigma_{jk+}^{(2)} + \sigma_{jk-}^{(2)} \tag{11.6.6}$$

where $\sigma_{jk+}^{(2)}$ contains all terms of (11.6.5) with $E_r > -2mc^2$ (*positive energy* part) and $\sigma_{jk-}^{(2)}$ all terms with $E_r \leq -2mc^2$(*negative energy* part). Expressions for the G-spinor matrix elements can be found in §10.9.

The nonrelativistic theory of hyperfine shielding is due to Ramsey [39]. The expressions are somewhat more complicated, partly because the nonrelativistic Hamiltonian is quadratic in the vector potential, but also because of the need to include relativistic effects in the Pauli approximation. We refer the reader to [39] and to more recent work [40, 41] for details. Pyykkö's analysis [42] of the nonrelativistic theory associates the negative energy part of (11.6.5) with the *diamagnetic* contribution to the shielding constant and the positive energy part with the *paramagnetic* contribution. It has often been assumed that the negative energy states generated in 4-component basis set calculations are devoid of physical meaning and cannot be used to calculate the diamagnetic contribution. Authors such as [42, 43] have therefore tried to avoid explicit use of negative energy virtual states by approximating all denominators in terms where they occur by $2mc^2$ and applying closure relations to give a rough estimate. As we have seen, the division between positive and negative energy states depends upon the choice of mean field, so that the division of $\sigma_{jk}^{(2)}$ into two parts as in (11.6.6) is model dependent.

11.6.1 NMR shielding constants for ^{17}O in water

Table 11.11 reports estimates [12] of the spherically averaged second order hyperfine shielding tensor

$$\overline{\sigma^{(2)}} = (\sigma^{(2)}_{xx} + \sigma^{(2)}_{yy} + \sigma^{(2)}_{zz}),$$

defined by equation (11.6.5) for the DHF calculations on water reported in §11.2.2. The columns headed $\sigma^{(2)}_{+}$ and $\sigma^{(2)}_{-}$ record separately the sums over positive and negative virtual states; the latter dominate the calculation. Whilst there is considerable improvement as the basis set is refined, it is clear that even the CVTZ basis, Tables A.18 and A.24, does not saturate the Hilbert space completely. Table 11.12 displays the complete calculation with the CVTZ basis. Relativistic effects are small in the water molecule, and

Table 11.11. Second order magnetic hyperfine shielding constants (ppm) for ^{17}O in water. Reprinted from [12, Table 1] with the permission of Elsevier.

Basis set	$\sigma^{(2)}_{-}$	$\sigma^{(2)}_{+}$	$\overline{\sigma^{(2)}}$
cc-pVDZ	295.62	-45.20	257.14
aug-cc-pVDZ	310.52	-30.65	284.83
cc-pVTZ	323.24	-36.74	294.01
aug-cc-pVTZ	332.60	-31.83	300.77
CVTZ	354.94	-31.91	323.03

Table 11.12. Magnetic hyperfine shielding constants (ppm) for ^{17}O in water using the CVTZ basis. (The water molecule lies in the xz-plane with the origin on the O nucleus.) Reprinted from [12] with the permission of Elsevier.

	σ_{xx}	σ_{yy}	σ_{zz}	$\overline{\sigma}$
$\sigma^{(1)}$	3.54	11.73	8.52	7.93
$\sigma^{(2)}_{+}$	-12.11	-48.01	-35.60	-31.91
$\sigma^{(2)}_{-}$	335.94	354.70	354.18	354.94
Total	347.37	318.41	327.10	330.96

the results of our method should agree closely with nonrelativistic estimates using basis sets of comparable quality. A large number of nonrelativistic calculations are available for comparison with, say, our value $\overline{\sigma}$=330.96. Thus

Fukui *et al.* [44] obtained 295.8 using an SCF-CI method, 266.2 using a CHF approach and 267.3 using SOS-CI methods [45]. Lamanna *et al.* [46] gave 324.8 using the EOM-RPA method, Lazzaretti *et al.* [47] found 327.1 using the RPA and Chesnut and Foley gave 332.0 using a GIAO-FPT method [48]. Ruud *et al.* [24] went to great trouble to develop good nonrelativistic basis sets and numerical techniques, and their results can be taken as representing the state of the art at the time of writing. Their H II basis set consisted of H: (5s1p) $\rightarrow$ (3s1p), O:(9s5p1d) $\rightarrow$ (5s4p1d); H III was H:(6s2p) $\rightarrow$ (4s2p), O: (11s7p2d) $\rightarrow$ (7s6p2d); and H IV was H:(5s3p1d), O:(8s7p3d1f). They used two methods: with IGLO, they obtained 297.1, 314.6 and 321.4 using H II to H IV respectively, and using London orbitals, they got 328.2, 320.8 and 320.5 respectively.

Lee *et al.* [49] who made similar calculations using DFT noted that this sort of calculation requires diffuse basis functions on all nuclei as well as a good representation of the wavefunction near each nucleus, so that choosing a good basis set needs a lot of care.

Table 11.13. Magnetic hyperfine shielding constants (ppm) for ^{15}N in ammonia using the CVTZ basis. (The ammonia molecule has the nitrogen nucleus at the origin and two of its hydrogens in the xy-plane.) Reprinted from [12] with the permission of Elsevier.

	σ_{xx}	σ_{yy}	σ_{zz}	$\overline{\sigma}$
$\sigma^{(1)}$	-1.90	-1.90	-1.88	-1.89
$\sigma_+^{(2)}$	-34.08	-34.08	-61.44	-43.20
$\sigma_-^{(2)}$	313.66	313.66	314.65	313.99
Total:	277.68	277.68	251.33	268.90

11.6.2 NMR shielding constants for ^{15}N in ammonia

Another example is provided by the magnetic hyperfine shielding tensor of ^{15}N in ammonia, for which results are presented in Table 11.13. The geometry is the same as Table 11.4 and the calculation was done with the CVTZ basis sets of Tables A.18 and A.21 for hydrogen and nitrogen respectively. The value $\overline{\sigma} = 268.9$ ppm is similar to the nonrelativistic results of 265.2 using GIAO/FPT [48], 265.4 using the IGLO [27] and 244.1 using SOS-CI [45]. The behaviour as the basis set is refined through a sequence of basis sets based on the correlation consistent series of Dunning *et al.* [13, 14] shows much the same behaviour as in the case of water.

11.7 Molecules with high-Z constituents

The small number of calculations for molecules containing one or more high-Z nuclei at the DHF level and beyond is evidence that such 4-component calculations are not yet a matter of routine. Whilst the speed and memory of modern computers are now sufficient to make such calculations feasible, and the software is improving rapidly, there are only a limited number of state of the art examples in which the motivation to use a fully relativistic formulation has prevailed over other considerations.

Sandars [50] realized that the spin-rotational states of polar molecules containing heavy elements, such as the diatomics YbF and TlF, present an opportunity to detect the existence of an elementary or nuclear electric dipole moment (*edm*) predicted by some particle physics theories. Although such *edm* do not appear in the standard electroweak model of Glashow [51], Weinberg [52], and Salam [53], they appear in several particle physics theories [54] which invoke PT-odd nuclear forces (violating both parity and time inversion symmetries). Sandars suggested that the electronic structure of these polar molecules would permit PT-induced admixtures of energetically similar spin-rotational states of opposite parity. This would strongly amplify the PT-odd signal which, in atoms, is proportional to a high power of the atomic number.

Here we focus on the atomic and molecular electronic structure calculations required to make these estimates; an extensive discussion of the way in which PT-odd interactions perturb atomic and molecular structures may be found in, for example, [55] and its bibliography. The theories hypothesize effective interactions of the form [55, Equation (1)]

$$H_{eff} = -d\,\boldsymbol{\sigma}_N \cdot \boldsymbol{\lambda} \tag{11.7.1}$$

where $\boldsymbol{\sigma}_N$ is the nuclear spin operator, $\boldsymbol{\lambda}$ is a unit vector along the molecular axis and d is a coupling constant that depends upon the edm under investigation: an intrinsic proton edm, a weak neutral current interaction or a nuclear edm induced by PT-odd nuclear forces. The experiments aim to set limits on the coupling constants d. Because the direct interaction is confined to the nucleus, the calculation of PT-odd effects has much in common with the calculation of hyperfine interactions, and the basis sets employed must reflect both the necessity for high-quality representation of the electron density around the nucleus as well as the usual quantum chemical considerations relating to molecular electronic structure.

More generally, the chemistry of compounds containing heavy elements is of importance for its scientific, industrial, pharmaceutical, and environmental applications. Attention has naturally focussed on actinides, and the molecule UF_6 has been studied using a number of relativistic approaches, for example [56]. We consider the diatomics TlF and YbF at the DHF level below, followed by UF_6 using average of configuration DHF and CI methods.

11.7.1 Electronic structure of TlF

The need for wavefunction accuracy in the neighbourhood of the nuclei implies that more attention must be paid than is usual in quantum chemistry to the basis functions with large exponents, whereas it is those with the smallest exponents that are needed to represent the valence and subvalence electrons involved in chemical bonding. The molecular orbitals involved in bonding are rather insensitive to small inaccuracies in the representation of the inner shells although this does affect the total energy. This means that the basis set must give a balanced representation to both inner and outer shells. Thus the physical problem necessarily uses large basis sets, making calculations relatively expensive.

These issues were explored in depth by Quiney *et al.* [55]. The Tl nucleus was modelled by a normalized spherically symmetric Gaussian charge distribution, (5.4.4), with exponent

$$\lambda = 1.5 \times 10^{10}\,[\,0.529177249\,/(0.836A^{1/3} + 0.57)\,]^2 \qquad (11.7.2)$$

where A is the nuclear mass number. For ${}^{205}_{81}\mathrm{Tl}$ this gives $\lambda = 1.3888925203 \times 10^8$ and a root-mean-square nuclear radius $1.0392{\times}10^{-4}\,a_0$. The potential energy of an electron (in atomic units) due to this nuclear distribution, (5.4.4), can be expanded near the nucleus in a power series

$$V(r) = V_0 + V_1\,r + V_2\,r^2 + \ldots \qquad (11.7.3)$$

where

$$V_0 = -2Z\sqrt{\lambda/\pi}, V_1 = 0, V_2 = -\pi V_0^3/12Z^2 + \ldots$$

The model gives $V_0 = -1.0771 \times 10^6 E_h$. More elaborate models will change the numerical coefficients but we shall see that the predicted signals are rather insensitive to these changes, and greater accuracy is not yet warranted. Hyperfine and PT-odd parameters are sensitive to the amplitudes of large and small components near the origin. The ratio of the leading power series coefficients, $(q_0/p_0)_{-1}$ for $\kappa = -1$ or $(p_0/q_0)_{+1}$ for $\kappa = +1$, are given by (5.4.8) and (5.4.9), and dominate the behaviour of the wavefunction for small values of r. The values obtained from a GRASP DHF calculation, [55], were used as as numerical standard: from (5.4.8)

$$(q_0/p_0)_{-1} = \frac{\epsilon + V_0}{3c} = -2611.28 \qquad (11.7.4)$$

with $l = 0$ and $\epsilon \to \epsilon_{1s} = -3620.447145E_h$, and

$$(p_0/q_0)_{+1} = -\frac{\epsilon + V_0}{3c} = +2709.20 \qquad (11.7.5)$$

with $l = 1$ and $\epsilon \to \epsilon_{2p_{1/2}} = -930.962616E_h$. The large size of V_0 makes these ratios rather insensitive to the choice of ϵ.

Table 11.14. Ratios p_0/q_0 for s-orbitals and q_0/p_0 for $p_{1/2}$ orbitals for atomic Tl. Reprinted with permission from [55, Table IV]. Copyright 1998 by the American Physical Society.

	GRASP	Tl_{erg}	Tl_4
	s-states : p_0/q_0		
1s	-2611.14	-2603.56	-2614.94
2s	-2617.40	-2609.92	-2620.65
3s	-2618.43	-2611.02	-2621.02
4s	-2618.68	-2611.14	-2618.42
5s	-2618.74	-2611.54	-2616.96
6s	-2618.75	-2611.67	-2616.99
	$p_{1/2}$-states : q_0/p_0		
$2p_{1/2}$	2708.66	2049.27	2706.72
$3p_{1/2}$	2709.64	2050.17	2707.65
$4p_{1/2}$	2709.87	2050.39	2708.01
$5p_{1/2}$	2709.93	2050.47	2708.03
$6p_{1/2}$	2710.10	2050.49	2708.01

A calculation for hydrogen-like Tl using an even-tempered radial basis set with exponents

$$\lambda_i = \alpha \beta_N^{i-1}, \; i = 1, 2, \ldots, N, \tag{11.7.6}$$

and $\alpha = 0.04$ and $\lambda_N = 5 \times 10^6$ kept fixed, was first carried out to explore the quality of the numerical wavefunction near the nucleus; this demonstrated smooth convergence of $(q_0/p_0)_{-1}$ and $(p_0/q_0)_{+1}$ to the numerical values of (11.7.4) and (11.7.5) as N increased [55, Figures 2,3]. A more extensive series of calculations were then carried out for atomic Tl using both energy-optimized sets [57], Table A.28, and systematic sequences of even-tempered functions, Table A.29. The total energy, the orbital eigenvalues, and the ratios $(q_0/p_0)_{-1}$ and $(p_0/q_0)_{+1}$ were used to measure performance. The even-tempered exponents of Table A.29 were taken from a single list defined as in (11.7.6); the table shows the range of exponents selected for each l symmetry. The energy-optimized exponents of Table A.28 from [57] used two such lists, one for s, d symmetries, the other for p, f symmetries. The four smallest s and p exponents given by [57] were replaced respectively by 6 and 7 even tempered exponents respectively in order to get a better representation of the valence region and polarization of the thallium atom. Table 11.15 suggests that there is not a lot to choose between these basis sets if all one is interested is the total energy of the atom. However Table 11.14 shows that the energy-optimized basis set Tl_{erg} gives a relatively poor representation of the $p_{1/2}$ states near the origin; its p basis needs more large exponents to get a good value of q_0/p_0. The even-tempered set, Tl_4 seems to provide a better balance between the need

Table 11.15. Comparison of orbital eigenvalues and total energy (E_h) for atomic Tl. Reprinted with permission from [55, Table IV]. Copyright 1998 by the American Physical Society.

	GRASP	Tl_{erg}	Tl_4
1s	-3164.17970314	-3164.17850776	-3164.17985686
2s	-568.84401919	-568.84341313	-568.84386747
$2p_{1/2}$	-544.94952340	-544.94817872	-544.94939410
$2p_{3/2}$	-468.91691771	-468.91652351	-468.91678633
3s	-138.36335938	-138.36283869	-138.36279618
$3p_{1/2}$	-127.65241482	-127.65168577	-127.65192681
$3p_{3/2}$	-110.52805013	-110.52756958	-110.52751715
$3d_{3/2}$	-93.08378078	-93.08328155	-93.08269012
$3d_{5/2}$	-89.45995059	-89.45953160	-89.45944846
4s	-32.29256514	-32.29197689	-32.29176488
$4p_{1/2}$	-27.64426692	-27.64362415	-27.64350982
$4p_{3/2}$	-23.42745261	-23.42687563	-23.42668120
$4d_{3/2}$	-15.84356055	-15.84299121	-15.84269412
$4d_{5/2}$	-15.04648692	-15.04594324	-15.04578954
5s	-5.61908482	-5.61848196	-5.61859189
$4f_{5/2}$	-5.19080289	-5.19022259	-5.18981854
$4f_{7/2}$	-5.01478763	-5.01424273	-5.01396312
$5p_{1/2}$	-3.98513893	-3.98453012	-3.98468101
$5p_{3/2}$	-3.21732684	-3.21675075	-3.21690403
$5d_{3/2}$	-0.89449436	-0.89395643	-0.89409714
$5d_{5/2}$	-0.80617261	-0.80568678	-0.80580900
6s	-0.44919249	-0.44886131	-0.44905092
$6p_{1/2}$	-0.21135573	-0.21105758	-0.21125532
$6p_{3/2}$	-0.17654479	-0.17625310	-0.17644424
E_{tot}	-20274.85064428	-20274.83985165	-20274.83871260

for a good orbital representation near the origin along with enough flexibility for a satisfactory representation of molecular bonding.

The electronic structure near the fluorine nucleus is essentially nonrelativistic. Table A.25 lists exponents of a nonrelativistic 9s6p basis, augmented by two *d*-type functions to accommodate polarization involved in the formation of molecular orbitals, optimized for the negative fluorine ion [55]. Table 11.16 compares results for the equilibrium bond length r_{eq}, the harmonic force constant k_0 and frequency ν_0 for the vibrational ground state of TlF $^1\Sigma^+$ in HF and DHF calculations using the same Tl_{3b} basis set and the fluorine basis of Table A.25 with experimental values taken from the compilation of Huber and Herzberg [58]. A number of relativistic calculations were done with the BERTHA and DIRAC [59] codes, giving virtually the same results at several

Table 11.16. Comparison of HF and DHF estimates of equilibrium bond length r_{eq}, force constant k_0 and harmonic vibrational frequency for the ground state, $^1\Sigma^+$ of the thallium fluoride molecule. Reprinted with permisssion from [55, Table V] Copyright 1998 by the American Physical Society.

	$r_{eq}(\mathring{A})$	$k_0(\mathrm{Nm}^{-1})$	$\nu_0(\mathrm{cm}^{-1})$
HF	2.085	265	509
DHF	2.092	227	470
Expt. [58]	2.084	233	477

internuclear separations surrounding the experimental bond length. Nonrelativistic calculations with the same parameters were done using the DALTON code [60] on which DIRAC is based. The spectroscopic constants were then extracted from a quadratic fit to the molecular energy curve. The HF bond-length agrees rather well with the experimental value; correlation tends to stabilize the molecule so that both HF and DHF values are likely to decrease when this is done. However, the shape of the DHF energy curve already gives a rather better fit to the force constant and vibrational frequency.

These results were used to determine PT-odd parameters relevant to current experiments [55], settting bounds on the size of the electric dipole moment of the proton, the tensor coupling constant and the Schiff moment of the ^{205}Tl nucleus. A full discussion is outside the scope of this book.

11.7.2 Electronic structure of YbF

Similar calculations were performed for the $^2\Sigma$ ground state of the YbF radical in connection with the experimental search for a permanent dipole moment of the electron predicted by theories that give rise to PT-odd effects. The reasons for choosing YbF are similar to those of the previous section, the main difference being that this molecule has one unpaired electron in the ground state. The calculation closely resembles the calculation in TlF. The ytterbium atom was modelled [12] with a 26s26p15d8f basis set shown in Table A.27. The nonrelativistic 10s6p fluorine basis set of Table A.26 was taken from [61].

Table 11.17 compares the DHF eigenvalues and total energy of a BERTHA calculation for the $[\mathrm{Xe}]4f^{14}6s^2$ $J^\pi = 0^+$ ground state of ytterbium obtained with the basis set of Table A.27 with the finite difference GRASP [62] results, whilst Table 11.18 makes a similar comparison for the $[\mathrm{He}]2s^22p^5$ ground state of fluorine, the GRASP results being taken from [63]. Table 11.19 lists the molecular orbitals in a calculation of YbF using the $N-1$ potential method. This assumes that the unpaired outer electron does not perturb the YbF^+ positive ion core and can be represented to a first approximation by the first excited virtual state of the ion. The total energy of the YbF^+ ion core in this

Table 11.17. Comparison of orbital eigenvalues and total energy (E_h) for atomic Yb. From [3, Table 10.3] with the author's permission.

	GRASP	BERTHA
1s	-2267.65237	-2267.67392
2s	-388.89269	-388.91156
$2p_{1/2}$	-370.05522	-370.07438
$2p_{3/2}$	-331.48739	-331.50644
3s	-89.70956	-89.72108
$3p_{1/2}$	-81.42221	-81.43446
$3p_{3/2}$	-73.09396	-73.10549
$3d_{3/2}$	-59.19193	-59.20184
$3d_{5/2}$	-57.39060	-57.40247
4s	-18.67246	-18.67026
$4p_{1/2}$	-15.27510	-15.27284
$4p_{3/2}$	-13.37358	-13.37036
$4d_{3/2}$	-7.77796	-7.77437
$4d_{5/2}$	-7.42207	-7.41878
5s	-2.43951	-2.42930
$5p_{1/2}$	-1.41916	-1.40863
$5p_{3/2}$	-1.18279	-1.17207
$4f_{5/2}$	-0.53899	-0.52370
$4f_{7/2}$	-0.48019	-0.46668
6s	-0.19652	-0.19110
E_{tot}	-14067.67726	-14067.50185

DHF calculation was -14166.954285 E_h at an internuclear separation $d_{Yb-F} = 3.88\ a_0$.

Whilst the eigenvalue pattern shows that the innermost MOs are purely atomic, the most surprising feature that emerges is the strong overlap of the subvalence Yb $4f$ and F $2p$ AOs. The role of the $4f$ electrons is strongly influenced by relativity; in view of the high occupancy of the shell such hybridization is likely to play an important role in the structure of compounds containing lanthanide elements. This clearly needs further investigation.

The wavefunctions generated in this calculation have been used [64] also to calculate hyperfine interaction constants for ^{171}Yb both with and without allowance for core polarization. Table 11.20 compares the result with other calculations: Titov *et al.* [65] used relativistic effective core potentials (RECP)

Table 11.18. Comparison of orbital eigenvalues and total energy (E_h) for atomic F. From [3, Table 10.4] with the author's permission.

	GRASP	BERTHA
1s	-26.41175	-26.41192
2s	-1.57598	-1.57477
$2p_{1/2}$	-0.73133	-0.73004
$2p_{3/2}$	-0.72866	-0.72742
E_{tot}	-99.50162	-99.49711

Table 11.19. Molecular orbital (MO) eigenvalues (E_h) for YbF. [3, Table 10.5]. The highest occupied MO (*) is generated using the $N-1$ potential method. From [3, Table 10.5] with the author's permission.

	$\|m\|$	Eigenvalue		$\|m\|$	Eigenvalue
1,2	1/2	-2267.88934	41,42	1/2	-8.01456
3,4	1/2	-389.13489	43,44	5/2	-7.66032
5,6	1/2	-370.29437	45,46	3/2	-7.65904
7,8	3/2	-331.73138	47,48	1/2	-7.65837
9,10	1/2	-331.73131	49,50	1/2	-2.66967
11,12	1/2	-89.94945	51,52	1/2	-1.66770
13,14	1/2	-81.66247	53,54	1/2	-1.56455
15,16	3/2	-73.33541	55,56	3/2	-1.42266
17,18	1/2	-73.33522	57,58	1/2	-1.39396
19,20	3/2	-59.43363	59,60	5/2	-0.77675
21,22	1/2	-59.43360	61,62	1/2	-0.77646
23,24	5/2	-57.63243	63,64	3/2	-0.77544
25,26	3/2	-57.63233	65,66	1/2	-0.72313
27,28	1/2	-57.63230	67,68	3/2	-0.71927
29,30	1/2	-26.32166	69,70	7/2	-0.71874
31,32	1/2	-18.90876	71,72	5/2	-0.71699
33,34	1/2	-15.51239	73,74	1/2	-0.67964
35,36	3/2	-13.61162	75,76	1/2	-0.67381
37,38	1/2	-13.61027	77,78	3/2	-0.67180
39,40	3/2	-8.01577	(*)79	1/2	-0.18322

and restricted active space self-consistent field theory (RASSSCF) and Kozlov and Labzowsky [66] employed semi-empirical methods. The numerical DHF+CP values are much closer than the others to the experimental values, but $A_{\|}$ is 173 Mhz too large and the anisotropy, $A_{\|} - A_{\perp} = 180$ Mhz, is 129 MHz too small. The partial wave decomposition of the hyperfine tensor by atomic symmetry types of Table 11.21 shows that the magnitudes of $A_{\|}$ and

Table 11.20. Components, $A_{\|}$ and $A_{\perp}$ (Mhz), of the nuclear magnetic hyperfine tensor of ^{171}YbF [64, Table 2].

	$A_{\|}$	$A_{\perp}$
RECP [65]	5049	4873
RASSSCF [65]	4975	4794
DHF [64]	5987	5883
DHF+CP [64]	7905	7805
Experiment [67]	7822	7513

$A_{\perp}$ are dominated by (-1,-1) matrix elements, but the anisotropy is dominated by the (1,1) contributions which are proportional to the density of $p_{1/2}$ electrons near the ^{171}Yb nucleus. The sensitivity of the $p_{1/2}$ density to correlation makes it necessary to go beyond first order core polarization to make improvements.

Table 11.21. Decomposition, by atomic orbital symmetry, of the nuclear magnetic hyperfine tensor comonents, $A_{\|}$ and $A_{\perp}$ (Mhz), of ^{171}YbF [64, Table 2].

κ κ'	$A_{\|}$	$A_{\perp}$
-1 -1	5898	5898
1 1	51	-51
1 -2	17	8
-2 -2	13	26
2 2	3	-6
2 -3	2	1
-3 -3	2	6

11.7.3 DHF+CI study of uranium hexafluoride

The electronic structure of uranium hexafluoride, UF_6, has been extensively studied both experimentally and theoretically, largely because of its role in uranium isotope enrichment. The calculation is a challenging task mainly because of the large number of open shell electrons that contribute to the physical and chemical properties of the molecule. The complexity of the atomic structure is itself daunting. The ground state of the neutral atom has been classified as [Rn]$5f^36d7s^2$ with $J^\pi = 6^o$, but the low-lying [Rn]$5f^36d^27s$ configuration is very close in energy and this should not be forgotten in any attempt to model even the lowest states of the spectrum *ab initio* [68].

Models of the electronic structure of UF_6 have not attempted to scale such heights. Boring *et al.* [69] started with nonrelativistic $X\alpha$ calculations. Koelling *et al.* [70] carried out Dirac-Slater calculations using the discrete variational method (DVM). This calculation made it obvious that relativistic methods were needed and Rosén [71] showed that the relativistic DS method reproduced the experimental data better than nonrelativistic methods. Hay *et al.* [72, 73] devised a relativistic effective core potential (RECP) scheme, in which an essentially nonrelativistic approach incorporated the more important Breit-Paul operators into the model.

The first *ab initio* four-component study of UF_6 was performed by de Jong and Nieuwpoort [56] using the MOLFDIR package. The first stage was an average of configuration calculation, allowing for at most two open shell structures. The Gaunt interaction was available as a perturbation. A full-CI calculation on the open-shell states was used to project out the different open-shell states, in a manner reminiscent of that used in atomic calculations.

The conflict between the desirability of a large basis to adequately represent the spectrum and the practical limits on basis set size set by the available hardware and software is particularly severe for UF_6. The conflict is partially resolved by using contracted basis sets, in which a smaller basis set is constructed from linear combinations of primitive Gaussians with fixed coefficients. The coefficients are chosen to match the results of an uncontracted atomic DHF calculation. A nonrelativistically optimized (24s18p14d12f) primitive Gaussian basis for the uranium large components has been augmented with some tight functions to improve the representation of the p basis as well as diffuse functions to improve the representation of the d and f basis functions in the valence region. The DHF eigenvalues [56, Table II] for an average of configuration calculation for the atomic $5f^3 6d7s^2$ manifold using a contracted basis set given by (24s 21p 16d 13f) $\rightarrow$ (10s 13p 11d 8f) for the large components and (21s 24p 21d 16f 13g) $\rightarrow$ (8s 17p 18d 11f 8g) for the small components agree reasonably well with those from a comparable GRASP calculation. However, as found in earlier calculations [68], whilst the basis set results match the corresponding GRASP ones reasonably well, the low-lying level spectrum is poorly reproduced [56, Table III].

MOLFDIR does not used kinetically matched basis sets for the small and large components as advocated in this book, so that the small component basis functions have to be constructed explicitly by applying the kinetic balance relation. Because the large component contracted basis has 42 members, whilst the small component basis has 62 members, the Fock matrix has 20 spurious eigensolutions that are discarded after diagonalization. The remaining physical eigenspinors are expected to be very similar to those obtained using a matched G-spinor basis.

The closed shell ground state of the UF_6 calculation assumed the double group symmetry of O_h^*, at the experimental bond length $d_{U-F}= 1.999\ \overset{o}{A}$. Figure 11.2 shows the HF and DHF MO eigenvalue spectrum. It is notable that the $9a_{1g}$ Hartree-Fock highest occupied molecular orbital (HOMO) is stabilized in the relativistic calculation because of the increased binding of its large $6s$ component. In contrast, the 12 outermost MOs in the DHF calculation are mainly linear combinations of F $2p$ with U $5f, 6d$, and $7s$ orbitals. There are considerable shifts in the electron population between the two calculations, consistent with the usual dynamical contraction of the valence $s, p_{1/2}$ orbitals and the indirect expansion of orbitals with higher angular momentum. The DHF model predicts a stronger U-F bond than the HF model and one that is more ionic: the effective charge on U increases from 2.22 to 2.72 whilst that on each F centre decreases from -0.37 to -0.45 [56, Table VI].

Table 11.22. Koopmans' theorem estimates of photoelectron spectrum in UF_6. The predicted spectrum has been shifted by 3.14 eV to make the computed DHF+Gaunt energies of the $12\gamma_{8u}$ spinor agree with experiment [74]. (From [56, Table IX] with permission.)

Spinor type O_h^*	O_h	Eigenvalue (eV)	Experiment (eV)
$12\gamma_{8u}$	$4t_{1u}$	14.14	14.14
$10\gamma_{6u}$	$4t_{1u}$	15.31	
$10\gamma_{6g}$	$1t_{1g}$	15.31	15.30
$11\gamma_{8g}$		15.37	
$9\gamma_{6g}$	$3a_{1g}$	16.10	
$11\gamma_{8u}$	$1t_{2u}$	16.36	16.20
$3\gamma_{7u}$		16.39	
$10\gamma_{8u}$	$3t_{1u}$	16.78	16.71
$9\gamma_{6u}$		16.84	
$4\gamma_{7g}$	$1t_{2g}$	17.50	
$10\gamma_{8g}$		17.53	17.36
$9\gamma_{8g}$	$2e_g$	18.02	

The use of eigenvalues of molecular spinors to predict the photoelectron spectrum of UF_6 (Koopmans' theorem) overestimates the level energies [56, Table IX]; the DHF energy of the $12\gamma_{8u}$ HOMO is predicted as 17.31 eV, reducing to 17.28 eV when the Gaunt interaction is included as a perturbation, whilst the first photoelectron peak is at 14.14 eV [74]. A shift of the eigenvalue DHF+Gaunt spectrum by the difference, 3.14 eV produces a reasonable correspondence with experiment, shown in Table 11.22. Allowing for orbital relaxation by calculating the difference in energy of the neutral UF_6 and the ion UF_6^+ having a hole in the HOMO $12\gamma_{8u}$ gives an ionization energy 16.6 eV, going some way to improve the agreement with experiment. de Jong and Nieuwpoort examine several other features, including the equilibrium bond

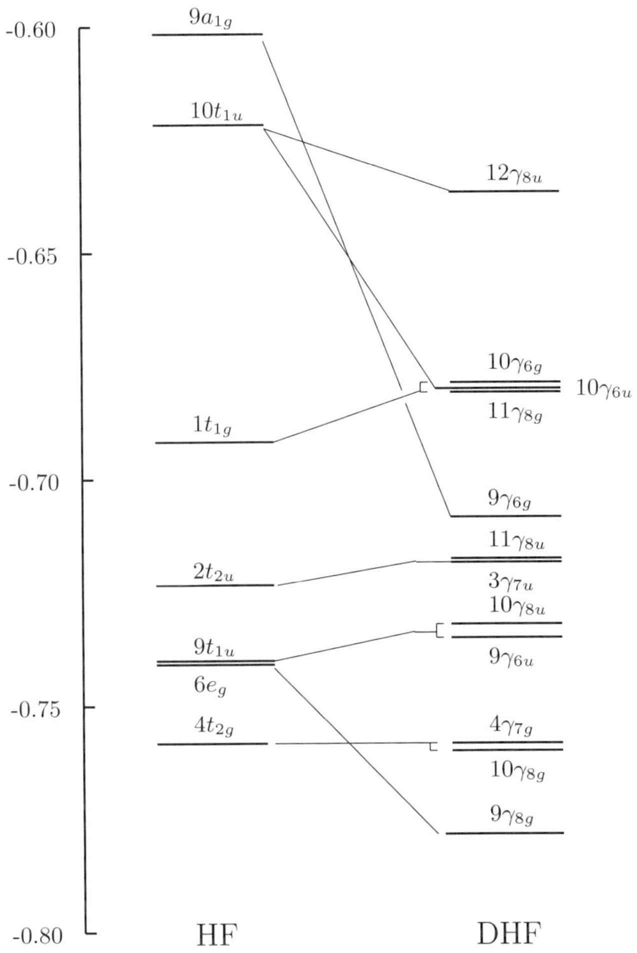

Fig. 11.2. Eigenvalues of molecular orbitals (E_h) for the UF_6 molecule. HF and DHF calculations were performed using O_h point group symmetry. The levels have been classified in terms of single and double group representations respectively. The nonrelativistic levels are split and shifted by relativity; the sloping lines indicate these movements. (After [56] with permission.)

length, the $5f$ spectrum of the negative ion UF_6^-, the electron affinity, and the atomization energy. The computed bond length, obtained in a 7-point fit, was 1.995 Å for HF and 1.994 Å for DHF, compared with the observed value

1.999 (±0.003) Å. The electron affinity [56, Table XIII], predicted to be 8.9 eV by the HF method, reduces to 6.4 eV with the use of DHF, rising to 6.5 eV when the Gaunt interaction is included, and falls to 5.3 eV when correlation is added in CI calculations for both the neutral molecule and the negative ion with single an double replacements. Experimental values quoted range from 4.9 eV to 8.1 eV, depending on the experiment.

Ab initio electronic structure calculations for molecules containing actinide atoms are likely to pose a major challenge to theorists for some time to come. Calculations on high-Z isolated atom spectra emphasize the importance of correlation, and it is clear that much more work will be needed to make the predictions more reliable.

References

1. Murrell J N, Kettle S F A and Tedder J M 1985 *The Chemical Bond* Second edn (New York: John Wiley)
2. Meyer J, Sepp W-D, Fricke B and Rosen A 1996 *Comput. Phys. Commun.* **96**, 263.
3. Skaane H 1998, unpublished D. Phil. Thesis, University of Oxford.
4. Elliott J P and Dawber P G 1979 *Symmetry in Physics* (Basingstoke: Macmillan Press).
5. Wayne R P 2000 *Chemistry of Atmospheres* 3 ed., pp. 50–58 (Oxford: Oxford University Press).
6. Polyansky O L, Császár A G, Shirin S V, Zobov N F, Barletta P, Tennyson J, Schwenke D W and Knowles P J 2003 *Science* **299**, 539.
7. Tarczay G, Császár A G, Klopper W and Quiney H M 2001 *Mole. Phys.* **99**, 1769.
8. Császár A G, Tarczay G, Leininger M L, Polyansky O L, Tennyson J and Allen W D 2001 *Spectroscopy from Space*, ed. J Demaison and K Sarka (Dordrecht: Klower).
9. Tarczay G, Császár A G, Polyansky O L and Tennyson J 2001 *J. Chem. Phys.* **115**, 1229.
10. Bethe H A and Salpeter E E 1957 *Quantum Mechanics of One- and Two-Electron Atoms* (Berlin-Göttingen-Heidelberg: Springer-Verlag).
11. Berestetskiĭ V B, Lifshitz E M and Pitaevskiĭ L P 1971 *Relativistic Quantum Theory*, Vol. 1. (Oxford: Pergamon Press).
12. Quiney H M, Skaane H and Grant I P 1998 *Chem. Phys. Lett.* **290**, 473.
13. Dunning T H Jr. 1989 *J. Chem. Phys.* **90**, 1007.
14. Kendall R A, Dunning T H Jr. and Harrison R J 1992 *J. Chem. Phys.* **96**, 6796.
15. Helgaker T U, Klopper W, Koch H amd Noga J 1997 *J. Chem. Phys.* **106**, 9639.
16. Head-Gordon M, Pople J A and Frisch M J 1988 *Chem. Phys. Lett.* **153**, 503.
17. Quiney H M, Grant I P and Wilson S 1988 in *Many-Body Methods in Quantum Chemistry* (Lecture Notes in Chemistry, Vol. 52 , ed. U Kaldor) pp. 307–344 (Berlin: Springer-Verlag).
18. White C A and Head-Gordon M 1996 *J. Chem. Phys.* **105**, 5061.
19. Quiney H M, Barletta P, Tarczay G, Cs'asz'ar A G, Polyansky O L and Tennyson J 2001 *Chem. Phys. Lett.* **344**, 413.

20. Noga J and Klopper W 1995 DIRCCR12-95, an integral-direct explicitly correlated coupled-cluster program.
21. Visscher L, Visser O, Aerts P J C, Merenga H and Nieuwpoort W C 1994 *Comput. Phys. Commun.* **81**, 120.
22. Barletta P, Császár A G, Quiney H M and Tennyson J 2002 *Chem. Phys. Lett.* **361**, 121.
23. Grant I P 1974 *J. Phys. B*, **7**, 1458.
24. Ruud K, Helgaker T, Kobayashi R, Jørgensen P, Bak K J and Jensen H J Aa 1994 *J. Chem. Phys.* **100**, 8178.
25. London F 1937 *J. Phys. Radium* **8**, 397.
26. Quiney H M, Skaane H and Grant I P 1999 *Adv. Quant. Chem.* **32**, 1.
27. Schindler M and Kutzelnigg W 1982 *J. Chem. Phys.* **76**, 1919.
28. Drake G W F ed 2005 *Springer Handbook of Atomic, Molecular and Optical Physics* Second edn (New York: Springer-Verlag).
29. Mohr P J and Taylor B N 2000 *Rev. Mod. Phys.* **72** 351 (CODATA recommended values of the fundamental physical constants: 1998).
30. Werth G, Häffner H, Hermanspahn N, Kluge H-J, Quint W and Verdú J 2001 in [31, pp. 204–220].
31. Karshenboim S G, Pavone F S, Bassani F, Inguscio M and Hänsch T W 2001 *The Hydrogen Atom. Precision Physics of Simple Atomic Systems* (Berlin: Springer-Verlag).
32. Pyper N C, Grant I P and Beatham N 1978 *Comput. Phys. Commun.* **15**, 387.
33. Abragam A and Bleaney B 1986 *Electron Paramagnetic Resonance of Transition Ions* (New York: Dover).
34. van Lenthe E, van der Avoird A and Wormer P E S 1998 *J. Chem. Phys.* **108**, 4783.
35. Quiney H M and Belanzoni P 2002 *Chem. Phys. Lett.* **353**, 253.
36. Weltner W. Jr 1983 *Magnetic Atoms and Molecules* (New York: Dover).
37. Armstrong, L jr., 1971 *Theory of the Hyperfine Structure of Free Atoms* (New York: Wiley-Interscience).
38. Lindgren I and Rosén A 1974 *Case Studies in Atomic Physics* **4**, 93, 197.
39. Ramsey N F 1950 *Phys. Rev.* **77**, 567.
40. Kowalewski J and Laaksonen A 1991 in *Theoretical Models of Chemical Bonding* (ed. Z B Maksić) (New York: Springer-Verlag).
41. Ruud K 1993 Master's thesis (University of Oslo, unpublished).
42. Pyykko P 1983 *Chem. Phys.* **74**, 1.
43. Pyper N C 1983 *Chem. Phys. Lett.* **96**, 204.
44. Fukui H, Miura K and Tada F 1983 *J. Chem. Phys.* **79**, 6112.
45. Fukui H, Yoshida H, Miura K 1981 *J. Chem. Phys.* **74**, 6988.
46. Lamanna U T, Guidotti C and Arrighini G P 1977 *J. Chem. Phys.* **67**, 604.
47. Lazzaretti, Rossi E and Zanasi R 1983 *Phys. Rev. A* **27**, 1301.
48. Chesnut D B and Foley C K 1986 *J. Chem. Phys.* **84**, 852.
49. Lee A M, Handy N C and Colwell S M 1995 *J. Chem. Phys.* **103**, 1095.
50. Sandars P G H 1967 *Phys. Rev. Lett.* **19**, 1396.
51. Glashow S L 1961 *Nucl. Phys.* **22**, 579.
52. Weinberg S 1967 *Phys. Rev. Lett.* **19**,1264.
53. Salam A in *Nobel Symposium No. 8* ed N Svartholm (Stockholm: Almquist and Wiksell).
54. Commins E D and Bucksmaum P H 1983 *Weak interactions of leptons and quarks* (Cambridge: Cambridge University Press).

55. Quiney H M , Lærdahl J K, Fægri K jr. and Saue T 1998 *Phys. Rev. A* **57** 920.
56. de Jong W A and Nieuwpoort W C 1996 *Int. J. Quant. Chem.* **58**, 203.
57. Dyall K G and Fægri K jr.1996 *Theor. Chim. Acta.* **94**, 39.
58. Huber G R and Herzberg G 1979 *Molecular Spectra and Molecular Structure, IV. Constants of diatomic molecules.* (New York: Litton).
59. URL: `http://dirac.chem.sdu.dk/doc/reference.shtml`
60. URL: `http://www.kjemi.uio.no/software/dalton/dalton.html`
61. Poirier R, Kari R, Csizmadia I G 1985 *Handbook of Gaussian Basis Sets* (Amsterdam: Elsevier).
62. Parpia F A 1998 *J. Phys. B: At. Mol. Opt. Phys.* **31**, 1409.
63. Visscher L and Dyall K G 1997 *At. Data Nucl. Data Tab.* **67**, 207.
64. Quiney H M, Skaane H and Grant I P 1998 *J. Phys. B: At. Mol. Opt. Phys.* **31**, L85.
65. Titov A V, Mosyagin N S and Ezhov V F 1996 *Phys. Rev. Lett.* **77**, 5346.
66. Kozlov M G and Labzowsky L N 1996 *J. Phys. B: At. Mol. Opt. Phys.* **28**, 1993.
67. van Zee R J, Seely M L, de Vore T C and Weltner W Jr 1978 *J. Phys. Chem.* **82**, 1192.
68. Pyper N C and Grant I P 1978 *J. Chem. Soc., Faraday Trans. II* **74**, 1885.
69. Boring M, Wood J H and Moskowitz J W 1974 *J. Chem. Phys.* **61**, 3800; Maylotte D H, St. Peters R L and Messmer R P 1976 *Chem. Phys. Lett.* **38**, 181; Boring M and Moskowitz J W 1974 *Chem. Phys. Lett.* **38**, 185.
70. Koelling D D, Ellis D E and Bartlett R J 1976 *J. Chem. Phys.* **65**, 3331.
71. Rosén A 1978 *Chem. Phys. Lett.* **55**, 311.
72. Hay P J, Wadt W R, Kahn L R, Raffenetti R C and Phillips D H 1979 *J. Chem. Phys.* **71**, 1767.
73. Hay P J 1983 *J. Chem. Phys.* **79**, 5469.
74. Mårtensson N, Malquist P-Å, Svensson A and Johansson B 1984 *J. Chem. Phys.* **80**, 5458.

A

Frequently used formulae and data

A.1 Relativistic notation

Vectors in Minkowski space-time	$(x^\mu) := (x^0, x^1, x^2, x^3) \equiv (x^0, \boldsymbol{x})$ where $x^0 = ct$, $\boldsymbol{x} = (x^1, x^2, x^3)$
Lorentz transformations[1]	$(x'^\mu) \to (x^\mu) = \Lambda^\mu{}_\nu x'^\nu + a^\mu$ $g_{\mu\nu} = \Lambda^\rho{}_\mu \, g_{\rho\sigma} \, \Lambda^\sigma{}_\nu$ or $g = \Lambda^t g \Lambda$
Metric tensor	$g_{\mu\nu} = g^{\mu\nu} = \mathrm{diag}(1, -1, -1, -1)$
Scalar products	$v \cdot w = g_{\mu\nu} v^\mu w^\nu \equiv v^0 w^0 - \boldsymbol{v}.\boldsymbol{w}$
Invariant length	$x' \cdot x' = x \cdot x$ if and only if $g = \Lambda^t g \Lambda$
Covectors	$(w_\mu) := g_{\mu\nu} w^\nu = (w^0, -\boldsymbol{w})$
Gradients	$(\partial_\mu) := \left(\dfrac{\partial}{\partial x^\mu}\right) = \left(\dfrac{1}{c}\dfrac{\partial}{\partial t}, \boldsymbol{\nabla}\right)$ $(\partial^\mu) := \left(\dfrac{\partial}{\partial x_\mu}\right) = \left(\dfrac{1}{c}\dfrac{\partial}{\partial t}, -\boldsymbol{\nabla}\right)$ $\boldsymbol{\nabla} := (\partial_1, \partial_2, \partial_3)$
d'Alembertian operator	$\Box := g^{\mu\nu}\partial_\mu\partial_\nu$ $= (\partial_0)^2 - (\partial_1)^2 - (\partial_2)^2 - (\partial_3)^2$ $= \dfrac{1}{c^2}\dfrac{\partial^2}{\partial t^2} - \boldsymbol{\nabla}^2$
Four-momentum operator	$p^\mu := (i\partial^\mu) = (i\partial^0, -i\boldsymbol{\nabla})$

[1] Summation over repeated Lorentz (Greek) indices (0,1,2,3) or space (Latin) indices (1,2,3) is to be understood unless the contrary is stated.

$$\text{Levi-Civita tensor in } \mathbb{R}^3 \quad \epsilon^{ijk} = \epsilon_{ijk} = \begin{cases} +1 \text{ if } \begin{pmatrix} i\ j\ k \\ 1\ 2\ 3 \end{pmatrix} \text{ is even;} \\ -1 \text{ if } \begin{pmatrix} i\ j\ k \\ 1\ 2\ 3 \end{pmatrix} \text{ is odd :} \\ 0 \text{ otherwise} \end{cases}$$

$$\text{Levi-Civita tensor in } \mathbb{R}^4 \quad \epsilon^{\mu\nu\rho\sigma} = -\epsilon_{\mu\nu\rho\sigma} = \begin{cases} +1 \text{ if } \begin{pmatrix} \mu\ \nu\ \rho\ \sigma \\ 1\ 2\ 3\ 4 \end{pmatrix} \text{ is even;} \\ -1 \text{ if } \begin{pmatrix} \mu\ \nu\ \rho\ \sigma \\ 1\ 2\ 3\ 4 \end{pmatrix} \text{ is odd :} \\ 0 \text{ otherwise} \end{cases}$$

$$\begin{aligned}
\epsilon^{\mu\nu\rho\sigma}\epsilon^{\mu'\nu'\rho'\sigma'} &= -\det(g^{\lambda\lambda'}), \quad \lambda = \mu,\nu,\rho,\sigma;\ \lambda = \mu',\nu',\rho',\sigma'. \\
\epsilon^{\mu\nu\rho\sigma}\epsilon_{\mu}{}^{\nu'\rho'\sigma'} &= -\det(g^{\lambda\lambda'}), \quad \lambda = \nu,\rho,\sigma;\ \lambda = \nu',\rho',\sigma'. \\
\epsilon^{\mu\nu\rho\sigma}\epsilon_{\mu\nu}{}^{\rho'\sigma'} &= -2(g^{\rho\rho'}g^{\sigma\sigma'} - g^{\rho\sigma'}g^{\sigma\rho'}) \\
\epsilon^{\mu\nu\rho\sigma}\epsilon_{\mu\nu\rho}{}^{\sigma'} &= -6g^{\sigma\sigma'} \\
\epsilon^{\mu\nu\rho\sigma}\epsilon_{\mu\nu\rho\sigma} &= -24 \\
\epsilon^{ijk}\epsilon_{ilm} &= \delta_{jl}\delta_{km} - \delta_{jm}\delta_{kl} \\
\epsilon^{ijk}\epsilon_{ijl} &= 2\delta_{kl} \\
\epsilon^{ijk}\epsilon_{ijk} &= 6
\end{aligned}$$

A.2 Dirac matrices

Basic anti-commutation relation:

$$\{\gamma^\mu, \gamma^\nu\} = 2g^{\mu\nu} \tag{A.2.1}$$

where γ^0 is Hermitian, γ^i anti-Hermitian.

$$\gamma^5 = \gamma_5 = i\gamma^0\gamma^1\gamma^2\gamma^3 = i\gamma^3\gamma^2\gamma^1\gamma^0 \tag{A.2.2}$$

$$= -\frac{i}{4!}\epsilon_{\mu\nu\rho\sigma}\gamma^\mu\gamma^\nu\gamma^\rho\gamma^\sigma = -i\gamma_0\gamma_1\gamma_2\gamma_3 \tag{A.2.3}$$

$$\gamma^5 = \gamma_5^\dagger, \quad (\gamma^5)^2 = I_4, \quad \{\gamma^5, \gamma^\mu\} = 0. \tag{A.2.4}$$

Connection with Dirac α and β matrices:

$$\gamma^0 = \beta, \quad \gamma^i = \beta\alpha^i, \quad i = 1,2,3. \tag{A.2.5}$$

σ matrices:

$$\sigma^{\mu\nu} = \frac{i}{2}[\gamma^\mu, \gamma^\nu], \quad \Sigma^i = \frac{1}{2}\epsilon_{ijk}\sigma^{jk}, \quad i = 1,2,3. \tag{A.2.6}$$

$$\gamma^\mu\gamma^\nu = g^{\mu\nu} - i\sigma^{\mu\nu}, \quad \left[\gamma^5, \sigma^{\mu\nu}\right] = 0, \quad \gamma^5\sigma^{\mu\nu} = \frac{i}{2}\epsilon^{\mu\nu\rho\sigma}\sigma_{\rho\sigma} \tag{A.2.7}$$

$$\gamma^5\gamma^0\gamma^i = \Sigma^1, \quad i = 1, 2, 3. \tag{A.2.8}$$

Hermitian conjugates:

$$\gamma^0\gamma^\mu\gamma^0 = \gamma^{\mu\dagger}, \quad \gamma^0\gamma^5\gamma^0 = -\gamma^{5\dagger} = -\gamma^5, \tag{A.2.9}$$

$$\gamma^0(\gamma^5\gamma^\mu)\gamma^0 = (\gamma^5\gamma^\mu)^\dagger, \quad \gamma^0\sigma^{\mu\nu}\gamma^0 = \sigma^{\mu\nu\dagger} \tag{A.2.10}$$

Charge conjugation:

$$C\gamma_\mu C^{-1} = -\gamma_\mu^t, \quad C\gamma_5 C^{-1} = \gamma_5^t, \tag{A.2.11}$$

$$C\sigma_{\mu\nu}C^{-1} = -\sigma_{\mu\nu}^t, \quad C(\gamma^5\gamma^\mu)C^{-1} = (\gamma^5\gamma^\mu)^t. \tag{A.2.12}$$

Pauli matrices:

$$\sigma^1 = \begin{pmatrix} 0 & 1 \\ 1 & 0 \end{pmatrix}, \quad \sigma^2 = \begin{pmatrix} 0 & -i \\ i & 0 \end{pmatrix}, \quad \sigma^3 = \begin{pmatrix} 1 & 0 \\ 0 & -1 \end{pmatrix}. \tag{A.2.13}$$

$$\boldsymbol{\sigma} = \left(\sigma^1, \sigma^2, \sigma^3\right). \tag{A.2.14}$$

$$\sigma^i\sigma^j = \delta_{ij} + i\,\epsilon_{ijk}\sigma^k \tag{A.2.15}$$

If $\boldsymbol{a}$ and $\boldsymbol{b}$ are vectors in $\mathbb{R}^3$ that commute with $\boldsymbol{\sigma}$

$$(\boldsymbol{\sigma}\cdot\boldsymbol{a})(\boldsymbol{\sigma}\cdot\boldsymbol{b}) = \boldsymbol{a}\cdot\boldsymbol{b} + i\,\boldsymbol{\sigma}\cdot(\boldsymbol{a}\times\boldsymbol{b}) \tag{A.2.16}$$

Dirac representation:

$$\beta = \gamma^0 = \sigma^3 \otimes I = \begin{pmatrix} I & 0 \\ 0 & -I \end{pmatrix}, \quad \boldsymbol{\alpha} = \sigma^1 \otimes \boldsymbol{\sigma} = \begin{pmatrix} 0 & \boldsymbol{\sigma} \\ \boldsymbol{\sigma} & 0 \end{pmatrix}, \tag{A.2.17}$$

$$\boldsymbol{\gamma} = \beta\boldsymbol{\alpha} = i\sigma^2 \otimes \boldsymbol{\sigma} = \begin{pmatrix} 0 & \boldsymbol{\sigma} \\ -\boldsymbol{\sigma} & 0 \end{pmatrix} \tag{A.2.18}$$

$$\gamma^5 = \gamma_5 = \sigma^1 \otimes I = \begin{pmatrix} 0 & I \\ I & 0 \end{pmatrix}, \tag{A.2.19}$$

$$\gamma^5\gamma^0 = -i\sigma^2 \otimes I = \begin{pmatrix} 0 & -I \\ I & 0 \end{pmatrix}, \tag{A.2.20}$$

$$\gamma^5\boldsymbol{\gamma} = -\sigma^3 \otimes \boldsymbol{\sigma} = \begin{pmatrix} -\boldsymbol{\sigma} & 0 \\ 0 & \boldsymbol{\sigma} \end{pmatrix}, \tag{A.2.21}$$

$$\gamma^5\gamma^0\boldsymbol{\gamma} = \boldsymbol{\Sigma} = I \otimes \boldsymbol{\sigma} = \begin{pmatrix} \boldsymbol{\sigma} & 0 \\ 0 & \boldsymbol{\sigma} \end{pmatrix} \tag{A.2.22}$$

$$\sigma^{0j} = i\sigma^1 \otimes \sigma^j = i\alpha^j = i\begin{pmatrix} 0 & \sigma^j \\ \sigma^j & 0 \end{pmatrix}, \tag{A.2.23}$$

$$\sigma^{ij} = \epsilon_{ijk} I \otimes \sigma^k = \epsilon_{ijk} \begin{pmatrix} \sigma^k & 0 \\ 0 & \sigma^k \end{pmatrix} \tag{A.2.24}$$

$$C = i\gamma^2\gamma^0 = -i\sigma^1 \otimes \sigma^2 = \begin{pmatrix} 0 & -i\sigma^2 \\ -i\sigma^2 & 0 \end{pmatrix}, \tag{A.2.25}$$

$$C^t = C^\dagger = -C, \quad CC^\dagger = C^\dagger C = I_4, \quad C^2 = -I_4 \tag{A.2.26}$$

In this book, we use the Dirac representation throughout. However, there are two other representations satisfying the defining anti-commutation relations that are sometimes used. They are related to the Dirac representation by a similarity transformation:

$$\gamma^\mu \to \gamma'^\mu = U\gamma^\mu_{Dirac} U^{-1}$$

Transformation to Majorana representation:

$$U = U^\dagger = \frac{1}{\sqrt{2}} \begin{pmatrix} I & \sigma^2 \\ \sigma^2 & -I \end{pmatrix}$$

Transformation to Chiral representation:

$$U = \frac{1}{\sqrt{2}}(1 - \gamma_5\gamma_0) = \frac{1}{\sqrt{2}} \begin{pmatrix} I & -I \\ I & I \end{pmatrix}$$

Useful identities:

$$\gamma_\mu a^\mu \gamma_\nu a^\nu = a \cdot b - i\sigma_{\mu\nu} a^\mu b^\nu \tag{A.2.27}$$

$$\gamma^\mu\gamma_\mu = 4, \quad \gamma^\mu\gamma^\nu\gamma_\mu = -2\gamma^\nu, \quad \gamma^\mu\gamma^\nu\gamma^\rho\gamma_\mu = 4g^{\nu\rho} \tag{A.2.28}$$

Traces:

$$\text{tr } I_4 = 4, \quad \text{tr } \gamma^\mu = \text{tr } \gamma^5 = 0 \tag{A.2.29}$$

The trace of any product of an odd number of γ^μ matrices vanishes.

$$\text{tr } \gamma^\mu\gamma^\nu = 4g^{\mu\nu}, \quad \text{tr } \sigma^{\mu\nu} = \text{tr } \gamma^5\gamma^\mu = 0. \tag{A.2.30}$$

A.3 Special functions

General references [1], [2]

A.3.1 Spherical Bessel functions

Differential equation for spherical Bessel functions (w)

$$w'' + \frac{2}{x}w' + \left[1 - \frac{l(l+1)}{x^2}\right] w = 0, \quad l = 0, \pm 1, \pm 2, \ldots \tag{A.3.1}$$

where $w' := dw/dx$ etc. and x is real.

Differential equation for Riccati-Bessel functions ($z = x\,w$)

$$z'' + \left[1 - \frac{l(l+1)}{x^2}\right] z = 0, \quad l = 0, \pm 1, \pm 2, \ldots, \tag{A.3.2}$$

Spherical Bessel functions of the first kind

$$j_l(x) = \sqrt{\frac{\pi}{2x}} J_{l+1/2}(x) \sim \frac{x^l}{(2l+1)!!} \text{ as } x \to 0 \tag{A.3.3}$$

Spherical Bessel functions of the second kind

$$y_l(x) = (-1)^{l+1} \sqrt{\frac{\pi}{2x}} J_{-l-1/2}(x) \sim -\frac{(2l-1)!!}{x^{l+1}} \text{ as } x \to 0, \tag{A.3.4}$$

where $l = 0, 1, 2, , \ldots$.

Spherical Bessel functions of the third kind

$$h_l^{(1)}(x) = j_l(x) + i y_l(x) = \sqrt{\frac{\pi}{2x}} H_{l+1/2}^{(1)}(x) \tag{A.3.5}$$

$$h_l^{(2)}(x) = j_l(x) - i y_l(x) = \sqrt{\frac{\pi}{2x}} H_{l+1/2}^{(2)}(x) \tag{A.3.6}$$

Wronskians

Let $W\{u(x), v(x)\} = u(x)v'(x) - u'(x)v(x)$. Then

$$W\{j_l(x), y_l(x)\} = \frac{1}{x^2}, \quad W\{h_l^{(1)}(x), h_l^{(2)}(x)\} = -\frac{2i}{x^2} \tag{A.3.7}$$

Spherical Bessel functions of orders $l = 0, 1, 2$

$$j_0(x) = \frac{\sin x}{x}, \qquad y_0(x) = -j_{-1}(x) = -\frac{\cos x}{x}$$

$$j_1(x) = \frac{\sin x}{x^2} - \frac{\cos x}{x} \qquad y_1(x) = j_{-2}(x) = -\frac{\cos x}{x^2} - \frac{\sin x}{x}$$

$$j_2(x) = \sin x\left(\frac{3}{x^3} - \frac{1}{x}\right) - \cos x\frac{3}{x^2}$$

$$\begin{aligned} y_2(x) &= -j_{-3}(x) \\ &= -\cos x\left(\frac{3}{x^3} - \frac{1}{x}\right) - \sin x\frac{3}{x^2} \end{aligned}$$

Recurrence relations

$u_l(x)$ can be any one of $j_l(x), y_l(x), h_l^{(1)}(x), h_l^{(2)}(x)$, $\quad l = 0, \pm 1. \pm 2, \ldots$.

$$u_{l-1}(x) + u_{l+1}(x) = \frac{2l+1}{x}u_l(x) \tag{A.3.8}$$

$$lu_{l-1}(x) - (l+1)u_{l+1}(x) = (2l+1)\frac{d}{dx}u_l(x) \tag{A.3.9}$$

$$\left(\frac{d}{dx} + \frac{l+1}{x}\right)u_l(x) = u_{l-1}(x) \tag{A.3.10}$$

$$\left(-\frac{d}{dx} + \frac{l}{x}\right)u_l(x) = u_{l+1}(x) \tag{A.3.11}$$

Finite descending series

$$h_l^{(1)}(x) = \frac{e^{i(x-\frac{1}{2}(l+1)\pi)}}{x}\sum_{k=0}^{l}\frac{(l+k)!}{k!(l-k)!}(-2ix)^{-k} \tag{A.3.12}$$

$$h_l^{(2)}(x) = \frac{e^{-i(x-\frac{1}{2}(l+1)\pi)}}{x}\sum_{k=0}^{l}\frac{(l+k)!}{k!(l-k)!}(2ix)^{-k} \tag{A.3.13}$$

A.3.2 Confluent hypergeometric functions

Kummer's differential equation

$$z\frac{d^2w}{dz^2} + (b-z)\frac{dw}{dz} - aw = 0 \tag{A.3.14}$$

This has a regular singularity at $z = 0$ and an irregular singularity at ∞.

Independent solutions

$$M(a,b,z) = 1 + \frac{az}{b} + \frac{(a)_2}{(b)_2}\frac{z^2}{2!} + \ldots + \frac{(a)_n}{(b)_n}\frac{z^n}{n!} + \ldots \tag{A.3.15}$$

where

$$(a)_0 = 1, (a)_n = (a+n-1)(a)_{n-1}, \quad n = 1,2,3,\ldots$$

$$U(a,b,z) = \frac{\pi}{\sin \pi b}\left\{\frac{M(a,b,z)}{\Gamma(1+a-b)\Gamma(b)} - z^{1-b}\frac{M(1+a-b,2-b,z)}{\Gamma(a)\Gamma(2-b)}\right\} \tag{A.3.16}$$

$U(a,b,z)$ is a many-valued function with principal branch given by $-\pi < \arg z \leq \pi$. It is defined even when $b \to \pm n$ with integer n.

Whittaker's differential equation

$$\frac{d^2w}{dz^2} + \left[-\frac{1}{4} + \frac{\kappa}{z} + \frac{\frac{1}{4} - \mu^2}{z^2}\right] w = 0 \tag{A.3.17}$$

where

$$\kappa = \frac{1}{2}b - a, \quad \mu = \frac{1}{2}b - \frac{1}{2}$$

Independent solutions

$$M_{\kappa,\mu}(z) = \mathrm{e}^{-z/2} z^{\mu+1/2} M\left(\frac{1}{2} + \mu - \kappa, 1 + 2\mu, z\right) \tag{A.3.18}$$

$$W_{\kappa,\mu}(z) = \mathrm{e}^{-z/2} z^{\mu+1/2} U\left(\frac{1}{2} + \mu - \kappa, 1 + 2\mu, z\right) \tag{A.3.19}$$

with $-\pi < \arg z \leq \pi$. From (A.3.16) it follows that

$$W_{\kappa,\mu}(z) = \frac{\Gamma(-2\mu)}{\Gamma(1/2 - \mu - \kappa)} M_{\kappa,\mu}(z) + \frac{\Gamma(2\mu)}{\Gamma(1/2 + \mu - \kappa)} M_{\kappa,-\mu}(z) \tag{A.3.20}$$

A.3.3 Generalized Laguerre polynomials

Rodrigues' formula

$$L_n^{(\alpha)}(x) = \frac{\mathrm{e}^x x^{-\alpha}}{n!}\frac{d^n}{dx^n}\left(\mathrm{e}^{-x} x^{n+\alpha}\right) \tag{A.3.21}$$

Differential equation

$$xy'' + (\alpha + 1 - x)y' + ny = 0, \quad y = L_n^{(\alpha)}(x) \tag{A.3.22}$$

Generating function

$$\sum_{n=0}^{\infty} L_n^{(\alpha)}(x)\, t^n = (1-t)^{-\alpha-1} \exp\left(\frac{tx}{t-1}\right), \quad |t| < 1. \tag{A.3.23}$$

Recurrence relation

$$(n+1)L_{n+1}^{(\alpha)}(x) = (2n+\alpha+1-x)L_n^{(\alpha)}(x) - (n+\alpha)L_{n-1}^{(\alpha)}(x) \qquad \text{(A.3.24)}$$

$$L_n^{(\alpha-1)}(x) = L_n^{(\alpha)}(x) - L_{n-1}^{(\alpha)}(x) \qquad \text{(A.3.25)}$$

$$x\,L_n^{(\alpha+1)}(x) = (n+\alpha+1)L_n^{(\alpha)}(x) - (n+1)L_{n+1}^{(\alpha)}(x) \qquad \text{(A.3.26)}$$

Orthonormalization

$$\int_0^\infty L_m^{(\alpha)}(x)\,L_n^{(\alpha)}(x)\,e^{-x}x^\alpha\,dx = \delta_{m,n}\,\frac{\Gamma(\alpha+n+1)}{n!}, \quad \alpha > -1. \qquad \text{(A.3.27)}$$

Relation to confluent hypergeometric functions

$$L_n^{(\alpha)}(x) = \binom{n+\alpha}{n} M(-n,\,\alpha+1,\,x) \qquad \text{(A.3.28)}$$

A.3.4 Hermite polynomials

Rodrigues' formula

$$H_n(x) = (-1)^n \mathrm{e}^{x^2} \frac{d^n}{dx^n}\mathrm{e}^{-x^2}, \qquad \text{(A.3.29)}$$

Differential equation

$$y'' - 2xy' + 2ny = 0, \quad y = H_n(x) \qquad \text{(A.3.30)}$$

Generating function

$$\sum_{n=0}^\infty H_n(x)\,\frac{t^n}{n!} = \mathrm{e}^{2xt-t^2} \qquad \text{(A.3.31)}$$

Recurrence relation

$$H_{n+1}(x) = 2xH_n(x) - 2nH_{n-1}(x) \qquad \text{(A.3.32)}$$

Orthonormalization

$$\int_{-\infty}^\infty H_m(x)H_n(x)\,\mathrm{e}^{-x^2}\,dx = \delta_{m,n}\,\pi^{1/2}2^n\,n! \qquad \text{(A.3.33)}$$

A.3.5 Incomplete gamma functions

Definitions

$$\gamma(a, x) = \int_0^x \mathrm{e}^{-t} t^{a-1}\, dt \to \Gamma(a) \text{ as } x \to \infty \tag{A.3.34}$$

$$\Gamma(a, x) = \Gamma(a) - \gamma(a, x) = \int_x^\infty \mathrm{e}^{-t} t^{a-1}\, dt \tag{A.3.35}$$

Relation to confluent hypergeometric functions

$$\begin{aligned}\gamma(a, x) &= a^{-1} x^a \mathrm{e}^{-x}\, M(1, 1+a, x)\\ &= a^{-1} x^a\, M(1, 1+a, -x),\\ &= x^a \mathrm{e}^{-x} \sum_{n=0}^{\infty} \frac{x^n}{(a)_{n+1}}\end{aligned} \tag{A.3.36}$$

Series converges rapidly for $0 < x < a + 1$

Continued fraction

$$\Gamma(a, x) = \mathrm{e}^{-x} x^a \left(\frac{1}{x+1-a-} \frac{1.(1-a)}{x+3-a-} \frac{2.(2-a)}{x+5-a-} \cdots \right). \tag{A.3.37}$$

Continued fraction converges rapidly for $x > a + 1$.

A.3.6 Incomplete beta functions

Definition

$$I_x(a, b) \equiv \frac{B_x(a, b)}{B(a, b)} = \frac{1}{B(a, b)} \int_0^x t^{a-1} (1-t)^{b-1}\, dt \tag{A.3.38}$$

where

$$B(a, b) = \frac{\Gamma(a)\Gamma(b)}{\Gamma(a+b)}, \quad 0 < x \leq 1, \quad , a > 0, \quad b > 0.$$

Symmetry relation

$$I_x(a, b) = 1 - I_{1-x}(b, a) \tag{A.3.39}$$

Continued fraction representation

$$I_x(a, b) = \frac{x^a (1-x)^b}{a\, B(a, b)} \left\{ \frac{1}{1+} \frac{d_1}{1+} \frac{d_2}{1+} \cdots \right\} \tag{A.3.40}$$

where

$$d_{2m+1} = -\frac{(a+m)(a+b+m)x}{(a+2m)(a+2m+1)}, \quad d_{2m} = \frac{m(b-m)x}{(a+2m-1)(a+2m)}$$

The continued fraction converges rapidly for $x < (a+1)/(a+b+2)$ taking at most $O(\sqrt{\max(a,b)})$ iterations. Use (A.3.40) to obtain an equivalent expression that also converges rapidly when $x > (a+1)/(a+b+2)$.

Recurrence relations

$$I_x(a,b) = xI_x(a-1,b) + (1-x)I_x(a,b-1) \tag{A.3.41}$$

$$(a+b)I_x(a,b) = aI_x(a+1,b) + bI_x(a,b+1) \tag{A.3.42}$$

A.3.7 Continued fraction evaluation

The general form of a continued fraction such as (A.3.37) or (A.3.40) is written

$$f(x) = b_0 + \frac{a_1}{b_1+}\,\frac{a_2}{b_2+}\,\frac{a_3}{b_3+}\cdots \tag{A.3.43}$$

where the coefficients a_i, b_i can themselves be functions of x, usually linear or quadratic monomials. Wallis's algorithm evaluates $f(x)$ as a sequence of convergents $f_n = A_n/B_n$ in which all coefficients following a_n, b_n are ignored.

Wallis's recurrence:

$$\begin{aligned} A_{-1} = 1,\ B_{-1} = 0, &\qquad A_0 = b_0,\ B_0 = 1, \\ A_j = b_jA_{j-1} + a_jA_{j-2}, &\qquad B_j = b_jB_{j-1} + a_jB_{j-2} \end{aligned} \tag{A.3.44}$$

for $j = 1, 2, \ldots, n$. This algorithm can generate very large or very small values for the partial numerators and denominators A_j, B_j. See [1, §5.2] for modified algorithms that can avoid these problems.

A.4 Central field Dirac spinors and their interactions

General references Chapter 3, Chapter 6.

A.4.1 Central field Dirac spinors

- **Standard notation, 4-spinors**

$$\phi(\boldsymbol{x}) = \frac{1}{r}\begin{pmatrix} P(r)\chi_{\kappa m}(\theta,\varphi) \\ iQ(r)\chi_{-\kappa m}(\theta,\varphi) \end{pmatrix} \tag{A.4.1}$$

- **Parametric notation for Dirac 4-spinor components**

  Partition central field 4-spinors [4, Equation (5.2)], (A.4.1)), :

$$\phi_{\gamma,\kappa,m}(\boldsymbol{x}) = \begin{pmatrix} \omega_{+1}M[+1;\gamma,\ \ \kappa,\boldsymbol{r}] \\ \omega_{-1}M[-1;\gamma,-\kappa,\boldsymbol{r}] \end{pmatrix}, \tag{A.4.2}$$

where

$$M[\beta;\gamma,\beta\kappa,\boldsymbol{r}] = \frac{1}{r}\, R_{\gamma,\beta\kappa}(r)\, \chi_{\beta\kappa,m}(\theta,\varphi), \quad \beta = \pm 1 \tag{A.4.3}$$

are 2-spinors and

$$\omega_\beta = \begin{cases} 1 \text{ if } \beta = +1, \\ i \text{ if } \beta = -1, \end{cases} \qquad R_{\gamma,\beta\kappa}(r) = \begin{cases} P_{\gamma,\kappa}(r) \text{ if } \beta = +1, \\ Q_{\gamma,\kappa}(r) \text{ if } \beta = -1. \end{cases}$$

- **Spin-angle 2-spinors**

$$\chi_{\kappa m}(\theta,\varphi) = \sum_\sigma (l, m-\sigma, 1/2, \sigma\,|\,l, 1/2, j, m)\; Y_l^{m-\sigma}(\theta,\varphi)\phi_\sigma \tag{A.4.4}$$

- **Spin-angle 2-spinors in detail**

$$\chi_{+|\kappa|m}(\theta,\varphi) = \begin{pmatrix} -\left(\dfrac{j+1-m}{2j+2}\right)^{1/2} Y_{j+1/2}^{m-1/2}(\theta,\varphi) \\ \left(\dfrac{j+1+m}{2j+2}\right)^{1/2} Y_{j+1/2}^{m+1/2}(\theta,\varphi) \end{pmatrix} \tag{A.4.5}$$

$$\chi_{-|\kappa|m}(\theta,\varphi) = \begin{pmatrix} \left(\dfrac{j+m}{2j}\right)^{1/2} Y_{j-1/2}^{m-1/2}(\theta,\varphi) \\ \left(\dfrac{j-m}{2j}\right)^{1/2} Y_{j-1/2}^{m+1/2}(\theta,\varphi) \end{pmatrix} \tag{A.4.6}$$

- **Spin-angle 2-spinors: compact form**

$$\chi_{\kappa m}(\theta,\varphi) = \begin{pmatrix} -\eta C_{lm}(-\eta)\, Y_l^{m-1/2}(\theta,\varphi) \\ C_{lm}(\eta)\, Y_l^{m+1/2}(\theta,\varphi) \end{pmatrix} \tag{A.4.7}$$

where [3, eq. (18)]

$$C_{lm}(\eta) = \left(\frac{l+1/2+\eta m}{2l+1}\right)^{1/2}.$$

- **2-spinor labels**

$$\boldsymbol{j}^2\chi_{\kappa m}(\theta,\varphi) = j(j+1)\hbar^2\chi_{\kappa m}(\theta,\varphi), \quad j_3\chi_{\kappa m}(\theta,\varphi) = m\hbar\chi_{\kappa m}(\theta,\varphi)$$

$$\boldsymbol{l}^2\chi_{\kappa m}(\theta,\varphi) = l(l+1)\hbar^2\chi_{\kappa m}(\theta,\varphi), \quad l = j + \frac{1}{2}\,\mathrm{sgn}\;\kappa$$

$$\boldsymbol{s}^2\chi_{\kappa m}(\theta,\varphi) = s(s+1)\hbar^2\chi_{\kappa m}(\theta,\varphi), \quad s = 1/2$$

$$P\,\chi_{\kappa m}(\theta,\varphi) = (-1)^l\,\chi_{\kappa m}(\pi-\theta,\pi+\varphi)$$

$$\boldsymbol{j} = \boldsymbol{l} + \boldsymbol{s}$$

- **The operator K**

$$K\chi_{\kappa m}(\theta,\varphi) = \kappa\chi_{\kappa m}(\theta,\varphi)$$

where

$$\hbar^2 (K+1) = -2\boldsymbol{s}\cdot\boldsymbol{l} = \boldsymbol{l}^2 + \boldsymbol{s}^2 - \boldsymbol{j}^2, \tag{A.4.8}$$

The operator K distinguishes the two possible $(2j+1)$-dimensional irreducible representations formed in the Kronecker products $\mathcal{D}^{(l)} \times \mathcal{D}^{(1/2)}$ with $l = j + 1/2$ or $l = j - 1/2$.

- **The operator σ_r**

$$\boldsymbol{\sigma}_r = \boldsymbol{\sigma}\cdot\boldsymbol{x}/r \tag{A.4.9}$$

$$\boldsymbol{x}/r = \boldsymbol{e}_r = (\sin\theta\,\cos\varphi, \sin\theta\,\sin\varphi, \cos\theta)$$

$$\sigma_r = \boldsymbol{\sigma}\cdot\boldsymbol{e}_r = \begin{pmatrix} \cos\theta & \sin\theta e^{-i\varphi} \\ \sin\theta e^{+i\varphi} & -\cos\theta \end{pmatrix}$$

$$\sigma_r\,\chi_{\kappa m}(\theta,\varphi) = -\chi_{-\kappa m}(\theta,\varphi). \tag{A.4.10}$$

$$[\boldsymbol{l},\sigma_r] = -i\hbar\,\frac{\boldsymbol{x}\times\boldsymbol{\sigma}}{r} = -[\boldsymbol{s},\sigma_r], \quad [\boldsymbol{j},\sigma_r] = 0,$$

- **The operator $\boldsymbol{\sigma}\cdot\boldsymbol{p}$**

$$\boldsymbol{\sigma}\cdot\boldsymbol{p} = -i\hbar\,\sigma_r\left(\partial_r + \frac{K+1}{r}\right), \tag{A.4.11}$$

$$\boldsymbol{\sigma}\cdot\boldsymbol{p}\,\frac{F(r)}{r}\chi_{\kappa m}(\theta,\varphi) = \frac{i\hbar}{r}\left(\frac{dF}{dr} + \frac{\kappa}{r}F\right)\chi_{-\kappa m}(\theta,\varphi) \tag{A.4.12}$$

- **Irreducible representations of SO(3)**

The set

$$\{\chi_{\kappa m}\,|\,m = -j,\dots,j\}$$

spans the irreducible representation $\mathcal{D}^{(j)}$ contained in the Clebsch-Gordon series of $\mathcal{D}^{(l)} \times \mathcal{D}^{(1/2)}$ with

$$l = j + \frac{1}{2}\eta, \quad \eta = \operatorname{sgn}\,\kappa, \quad \kappa = (j+1/2)\eta. \tag{A.4.13}$$

- **Orthonormality**

$$\begin{aligned}(\chi_{\kappa' m'}\,|\,\chi_{\kappa m}) &= \int\!\!\int \chi^{\dagger}_{\kappa' m'}(\theta,\varphi)\,\chi_{\kappa m}(\theta,\varphi)\,\sin\theta\,d\theta\,d\varphi \\ &= \delta_{\kappa'\kappa}\,\delta_{m' m}.\end{aligned} \tag{A.4.14}$$

- **Parity operator**

  Let $\widehat{\mathcal{P}}_0$ denote the parity operator: $\widehat{\mathcal{P}}_0 f(\boldsymbol{x}) \to f(-\boldsymbol{x})$. Then the four-component operator $\widehat{\mathcal{P}} = \beta\widehat{\mathcal{P}}_0$ operating on Dirac central-field four-spinors (A.4.1) gives

$$\widehat{\mathcal{P}}\phi(\boldsymbol{x}) = \phi(-\boldsymbol{x}) = (-1)^l \phi(\boldsymbol{x}) \tag{A.4.15}$$

  where $l = j + \eta/2$.
- **Time reversal operator**

  Time reversal refers to the mapping $f(t, \boldsymbol{x}) \to f(-t, \boldsymbol{x})$. Define the antilinear operator $\widehat{\mathcal{K}}$ mapping a function f to its complex conjugate: $\widehat{\mathcal{K}} f = f^*$, and let $\widehat{\mathcal{T}}_0 = -i\sigma_2\widehat{\mathcal{K}}$.
  - **2-spinor time reversal**

$$\widehat{\mathcal{T}}_0 \frac{F(r)}{r} \chi_{\kappa m}(\theta, \varphi) = (-1)^{l-j+m} \frac{\widehat{\mathcal{K}}F(r)}{r} \chi_{\kappa -m}(\theta, \varphi) \tag{A.4.16}$$

  - **4-spinor time reversal**
    This requires the four-component operator

$$\widehat{\mathcal{T}} = \widehat{\mathcal{T}}_0 \otimes I_2 = \begin{pmatrix} \widehat{\mathcal{T}}_0 & 0 \\ 0 & \widehat{\mathcal{T}}_0 \end{pmatrix}. \tag{A.4.17}$$

Table A.1. Dirac states in spherical coordinates

Spectroscopic label:	s	$\overline{p}$	p	$\overline{d}$	d	$\overline{f}$	f	$\overline{g}$ ...
κ:	-1	+1	-2	+2	-3	+3	-4	+4 ...
j:	$\frac{1}{2}$	$\frac{1}{2}$	$\frac{3}{2}$	$\frac{3}{2}$	$\frac{5}{2}$	$\frac{5}{2}$	$\frac{7}{2}$	$\frac{7}{2}$ ...
l:	0	1	1	2	2	3	3	4 ...
$\overline{l}$	1	0	2	1	3	2	4	3 ...

1. 4-spinors are labelled according to the value of κ for the *upper* 2-spinor component.
2. l and $\overline{l}$ are the orbital values associated with the upper and lower components respectively, so that if $\eta =$ sgn κ then $l - \overline{l} = \eta$.

A.4.2 Matrix elements of simple ITOs

- **Wigner-Eckart theorem for 2-spinors**
  For simple ITOs of the form

$$S^k_q(\boldsymbol{r}) = V_k(r)\, C^k_q(\theta, \varphi). \tag{A.4.18}$$

we write the matrix elements as

$$\langle \gamma lsjm \,|\, S_q^k \,|\, \gamma' l' s' j' m' \rangle = \begin{pmatrix} m & k & j' \\ j & q & m' \end{pmatrix} \langle \gamma lsj \| \boldsymbol{S}^k \| \gamma' l' s' j' \rangle \qquad \text{(A.4.19)}$$

where $s = s' = 1/2$ and $S_q^k(\boldsymbol{r}) = V_k(r)\, C_q^k(\theta, \varphi)$

- **Reduced matrix elements**

$$\langle \gamma lsj \,\| \boldsymbol{S}^k \|\, \gamma' l' s' j' \rangle = \langle j \,\| \boldsymbol{C}^k \|\, j' \rangle (\gamma l \,|\, V_k(r) \,|\, \gamma' l') \qquad \text{(A.4.20)}$$

where

$$\langle j \,\| \boldsymbol{C}^k \|\, j' \rangle = (-1)^{j+1/2} [\, j, j' \,]^{1/2} \begin{pmatrix} j & k & j' \\ 1/2 & 0 & -1/2 \end{pmatrix}. \qquad \text{(A.4.21)}$$

- **Alternative form**

$$\langle \gamma lsjm \,|\, S_q^k \,|\, \gamma' l' s' j' m' \rangle = d^k(jm, j'm')\, (\gamma l \,|\, V_k(r) \,|\, \gamma' l') \qquad \text{(A.4.22)}$$

where

$$d^k(jm, j'm') = \begin{pmatrix} m & k & j' \\ j & q & m' \end{pmatrix} (-1)^{j+1/2} [\, j, j' \,]^{1/2} \begin{pmatrix} j & k & j' \\ 1/2 & 0 & -1/2 \end{pmatrix}. \qquad \text{(A.4.23)}$$

Table A.2. Coefficients $d^k(jm, j'm')$, equation (6.2.12), for $j = \frac{1}{2}, \frac{3}{2}, \frac{5}{2}$ and $j' = \frac{1}{2}$. The numerical values are obtained by dividing the table entry by the common denominator, D_k, at the head of each column.

				$k = 0$	1	2	3
j	j'	m	m'	$D_k = 1$	3	5	7
$\frac{1}{2}$	$\frac{1}{2}$	$\frac{1}{2}$	$\frac{1}{2}$	1	-1		
		$-\frac{1}{2}$	$\frac{1}{2}$		$\sqrt{2}$		
$\frac{3}{2}$	$\frac{1}{2}$	$\frac{3}{2}$	$\frac{1}{2}$		$-\sqrt{3}$	1	
		$\frac{1}{2}$	$\frac{1}{2}$		$\sqrt{2}$	$-\sqrt{2}$	
		$-\frac{1}{2}$	$\frac{1}{2}$		-1	$\sqrt{3}$	
		$-\frac{3}{2}$	$\frac{1}{2}$		0	-2	
$\frac{5}{2}$	$\frac{1}{2}$	$\frac{1}{2}$	$\frac{1}{2}$			$\sqrt{5}$	-1
		$\frac{3}{2}$	$\frac{1}{2}$			-2	$\sqrt{2}$
		$\frac{1}{2}$	$\frac{1}{2}$			$\sqrt{3}$	$\sqrt{3}$
		$-\frac{1}{2}$	$\frac{1}{2}$			$-\sqrt{2}$	2
		$-\frac{3}{2}$	$\frac{1}{2}$			1	$-\sqrt{5}$

- **Closed subshell Coulomb coefficients**
  Direct interactions (6.6.13):

$$a^k(jm, j'm') = d^k(j\,|m|, j\,|m|)\, d^k(j'\,|m'|, j'\,|m'|) \tag{A.4.24}$$

for $k = 0, 2, \ldots \min\,(2j-1, 2j'-1)$.

Exchange interactions (6.6.14):

$$b^k(jm, j'm') = \left[d^k(jm, j'm')\right]^2 \tag{A.4.25}$$

for j, j', k satisfying (A.4.27).

Closed shell exchange coefficients (6.6.12):

$$\Gamma_{jkj'} = 2\begin{pmatrix} 1/2 & k & j' \\ j & 0 & 1/2 \end{pmatrix}^2 \tag{A.4.26}$$

for j, j', k satisfying (A.4.27).

- **Selection rules and symmetry relations**
  The following selection rules apply to (A.4.20)–(A.4.25):

$$\{\,j\,k\,j'\,\} = 1, \text{ or } j + j' \geq k \geq |j - j'|, \tag{A.4.27}$$

otherwise $\{\,j\,k\,j'\,\} = 0$, and with l, j, κ and l', j', κ' related by (A.4.13), the parity projector $\Pi^e(\kappa, \kappa', k)$ (which has been left implicit above) takes the value 1 if

$$j + k + j' \text{ is } \begin{cases} \text{even if } \eta = -\eta' \\ \text{odd if } \eta = \eta' \end{cases} \tag{A.4.28}$$

and is zero otherwise. The Tables A.2, A.3, and A.4 have been shortened by making use of the symmetry relations

$$\begin{aligned} d^k(jm, j'm') &= (-1)^{m-m'} d^k(jm', j'm) \\ &= (-1)^{j+j'+k+1} d^k(j\ -m, j'\ -m') \end{aligned} \tag{A.4.29}$$

Table A.3. Coefficients $d^k(jm, j'm')$, equation (6.2.12), for $j = \frac{3}{2}, \frac{5}{2}$ and $j' = \frac{3}{2}$. The numerical values are obtained by dividing the table entry by the common denominator, D_k, at the head of each column.

				$k = 0$	1	2	3	4
j	j'	m	m'	$D_k = 1$	15	35	35	21
$\frac{3}{2}$	$\frac{3}{2}$	$\frac{3}{2}$	$\frac{3}{2}$	1	-3	-7	3	
		$\frac{1}{2}$	$\frac{3}{2}$		$-\sqrt{6}$	$7\sqrt{2}$	-6	
		$-\frac{1}{2}$	$\frac{3}{2}$			$-7\sqrt{2}$	$3\sqrt{10}$	
		$-\frac{3}{2}$	$\frac{3}{2}$				$-6\sqrt{5}$	
$\frac{3}{2}$	$\frac{3}{2}$	$\frac{3}{2}$	$\frac{1}{2}$		$-\sqrt{6}$	$-7\sqrt{2}$	6	
		$\frac{1}{2}$	$\frac{1}{2}$	1	-1	7	-9	
		$-\frac{1}{2}$	$\frac{1}{2}$		$2\sqrt{2}$	0	$6\sqrt{3}$	
		$-\frac{3}{2}$	$\frac{1}{2}$			$-7\sqrt{2}$	$-3\sqrt{10}$	
$\frac{5}{2}$	$\frac{3}{2}$	$\frac{5}{2}$	$\frac{3}{2}$		$-3\sqrt{10}$	$\sqrt{30}$	$\sqrt{15}$	-1
		$\frac{3}{2}$	$\frac{3}{2}$		6	-6	-6	2
		$\frac{1}{2}$	$\frac{3}{2}$		-3	$3\sqrt{3}$	$3\sqrt{6}$	$-\sqrt{10}$
		$-\frac{1}{2}$	$\frac{3}{2}$			$-2\sqrt{3}$	$-2\sqrt{15}$	$2\sqrt{5}$
		$-\frac{3}{2}$	$\frac{3}{2}$				$3\sqrt{5}$	$-\sqrt{35}$
		$-\frac{5}{2}$	$\frac{3}{2}$					$2\sqrt{14}$
$\frac{5}{2}$	$\frac{3}{2}$	$\frac{5}{2}$	$\frac{1}{2}$			$2\sqrt{10}$	$5\sqrt{2}$	$-\sqrt{6}$
		$\frac{3}{2}$	$\frac{1}{2}$		$-3\sqrt{6}$	$-\sqrt{2}$	-7	$\sqrt{15}$
		$\frac{1}{2}$	$\frac{1}{2}$		$3\sqrt{6}$	$-\sqrt{6}$	$2\sqrt{6}$	$-2\sqrt{6}$
		$-\frac{1}{2}$	$\frac{1}{2}$		$-3\sqrt{3}$	5	$-\sqrt{2}$	$\sqrt{30}$
		$-\frac{3}{2}$	$\frac{1}{2}$			$-4\sqrt{2}$	$-\sqrt{10}$	$-\sqrt{30}$
		$-\frac{5}{2}$	$\frac{1}{2}$				$5\sqrt{3}$	$\sqrt{21}$

A.4.3 Magnetic interactions

- **Reduced matrix element with 2-spinors**

  Magnetic interactions involve the spin-dependent operator, (6.2.17),

$$X_q^{(1\nu)k}(\boldsymbol{r}) = [\boldsymbol{\sigma} \times \boldsymbol{C}^\nu]^k, \qquad (A.4.30)$$

$$\langle -\lambda \| \boldsymbol{X}^{(1\nu)k} \| \lambda' \rangle = \Pi^o(\kappa\kappa' k) \langle j \| \boldsymbol{C}^k \| j' \rangle E^\nu(-\lambda, +\lambda', k) \qquad (A.4.31)$$

  where $\lambda = \beta\kappa$, $\lambda' = \beta\kappa'$ and $\langle j \| \boldsymbol{C}^k \| j' \rangle$ is given by (A.4.21).

- **Selection rules**

  Because $\Pi^o(\kappa\kappa' k) = 1 - \Pi^e(\kappa\kappa' k)$

Table A.4. Coefficients $d^k(jm, j'm')$, equation (6.2.12), for $j = j' = \frac{5}{2}$. The numerical values are obtained by dividing the table entry by the common denominator, D_k, at the head of each column.

j	j'	m	m'	$k=0$	1	2	3	4	5
			D_k	$=1$	35	35	105	21	231
$\frac{5}{2}$	$\frac{5}{2}$	$\frac{5}{2}$	$\frac{5}{2}$	1	-5	-10	10	1	-5
		$\frac{3}{2}$	$\frac{5}{2}$		$\sqrt{10}$	$2\sqrt{30}$	$-4\sqrt{15}$	-2	$5\sqrt{6}$
		$\frac{1}{2}$	$\frac{5}{2}$			$-2\sqrt{15}$	$10\sqrt{3}$	3	$-5\sqrt{21}$
		$-\frac{1}{2}$	$\frac{5}{2}$				$-10\sqrt{2}$	$-\sqrt{14}$	$10\sqrt{7}$
		$-\frac{3}{2}$	$\frac{5}{2}$					$\sqrt{14}$	$-15\sqrt{14}$
		$-\frac{5}{2}$	$\frac{5}{2}$						$30\sqrt{7}$
$\frac{5}{2}$	$\frac{5}{2}$	$\frac{5}{2}$	$\frac{3}{2}$		$-\sqrt{10}$	$-2\sqrt{30}$	$4\sqrt{15}$	2	$-5\sqrt{6}$
		$\frac{3}{2}$	$\frac{3}{2}$	1	-3	2	-14	-3	25
		$\frac{1}{2}$	$\frac{3}{2}$		4	$4\sqrt{3}$	$2\sqrt{6}$	$\sqrt{10}$	$-10\sqrt{15}$
		$-\frac{1}{2}$	$\frac{3}{2}$			$-6\sqrt{3}$	$2\sqrt{15}$	$-\sqrt{5}$	$5\sqrt{105}$
		$-\frac{3}{2}$	$\frac{3}{2}$				$-8\sqrt{15}$	0	$-10\sqrt{35}$
		$-\frac{5}{2}$	$\frac{3}{2}$					$\sqrt{14}$	$15\sqrt{14}$
$\frac{5}{2}$	$\frac{5}{2}$	$\frac{5}{2}$	$\frac{1}{2}$			$-2\sqrt{15}$	$10\sqrt{3}$	3	$-5\sqrt{21}$
		$\frac{3}{2}$	$\frac{1}{2}$		-4	$-4\sqrt{3}$	$-2\sqrt{6}$	$-\sqrt{10}$	$10\sqrt{15}$
		$\frac{1}{2}$	$\frac{1}{2}$	1	-1	8	-8	2	-50
		$-\frac{1}{2}$	$\frac{1}{2}$		$3\sqrt{6}$	0	$8\sqrt{3}$	0	$10\sqrt{30}$
		$-\frac{3}{2}$	$\frac{1}{2}$			$-6\sqrt{3}$	$-2\sqrt{15}$	$-\sqrt{5}$	$-5\sqrt{105}$
		$-\frac{5}{2}$	$\frac{1}{2}$				$10\sqrt{2}$	$\sqrt{14}$	$10\sqrt{14}$

$$\{ j\, k\, j' \} = 1, \text{ or } j + j' \geq k \geq |k - k'| \tag{A.4.32}$$

(cf. (A.4.27)) with l, j, κ and l', j', κ' related by (A.4.13), and

$$j + k + j' \text{ is } \begin{cases} \text{odd if } \eta = -\eta' \\ \text{even if } \eta = \eta' \end{cases} \tag{A.4.33}$$

$E^\nu(\lambda, \lambda', k)$ coefficients

$$\begin{aligned} E^{k-1}(\lambda, \lambda', k) &= \frac{k + \lambda - \lambda'}{[k(2k-1)]^{1/2}}, \\ E^{k}(\lambda, \lambda', k) &= \frac{-\lambda - \lambda'}{[k(k+1)]^{1/2}}, \\ E^{k+1}(\lambda, \lambda', k) &= \frac{-k - 1 + \lambda - \lambda'}{[(k+1)(2k+3)]^{1/2}}. \end{aligned} \tag{A.4.34}$$

Table A.5. Closed shell exchange coefficients $\Gamma_{j_A k j_B}$, (6.6.12), for $1/2 \le j' \le j \le 7/2$.

$k =$	0	1	2	3	4	5	6
j				$j' = \frac{1}{2}$			
$\frac{1}{2}$	1	1/3					
$\frac{3}{2}$		1/3	1/5				
$\frac{5}{2}$			1/5	1/7			
$\frac{7}{2}$				1/7	1/9		
				$j' = \frac{3}{2}$			
$\frac{3}{2}$	1/2	1/30	1/10	9/70			
$\frac{5}{2}$		1/5	1/35	2/35	2/21		
$\frac{7}{2}$			9/70	1/42	5/126	5/66	
				$j' = \frac{5}{2}$			
$\frac{5}{2}$	1/3	1/105	8/105	8/315	2/63	50/693	
$\frac{7}{2}$		1/7	1/105	1/21	5/231	5/231	25/469
				$j' = \frac{7}{2}$			
$\frac{7}{2}$	1/4	1/252	5/84	3/308	9/308	75/4004	25/1716

- **2-spinor interaction strengths for odd one-body operators**

  Let
$$\boldsymbol{T}^k = F_\nu \boldsymbol{X}^{(1\nu)k}$$
  where F_ν is either a multiplicative function of r or a simple differential operator on functions of r. Then
$$\begin{aligned}&\langle \gamma, -\lambda \,\|\, \boldsymbol{T}^k \,\|\, \gamma', \lambda' \rangle \\ &= -i\beta \,\langle -\lambda \| \boldsymbol{X}^{(1\nu)k} \| \lambda' \rangle \int_0^\infty R^*_{\gamma,-\lambda}(r)\, F_\nu \, R_{\gamma',\lambda'}(r)\, dr.\end{aligned} \tag{A.4.35}$$

A.4.4 Effective interaction strengths for two-body operators

- **Coulomb interaction**

$$\begin{aligned}X^k_C(abcd) &= \{j_a, j_c, k\}\{j_b, j_d, k\}\Pi^e(\kappa_a\kappa_c k)\Pi^e(\kappa_b\kappa_d k) \\ &\times (-1)^k \langle j_a \| \boldsymbol{C}^k \| j_c \rangle \langle j_b \| \boldsymbol{C}^k \| j_d \rangle R^k_C(abcd)\end{aligned} \tag{A.4.36}$$

  Coulomb Slater integrals

$$R^k_C(abcd) = \sum_\beta \sum_{\beta'} R^k(\lambda_a, \lambda'_b, \lambda_c, \lambda'_d) \tag{A.4.37}$$

where, with $\lambda_a = \beta\kappa_a$, $\lambda'_b = \beta'\kappa_b$, $\lambda_c = \beta\kappa_c$ and $\lambda'_d = \beta'\kappa_d$,

$$R^k(\lambda_a, \lambda'_b, \lambda_c, \lambda'_d) = \int_0^\infty \int_0^\infty \left[R^*_{\gamma_a,\lambda_a}(r) R_{\gamma_c,\lambda_c}(r)\, U_k(r,s)\, R^*_{\gamma_b,\lambda'_b}(s) R_{\gamma_d,\lambda'_d}(s)\right] drds.$$

In terms of Dirac *overlap charge densities* such as

$$\rho_{ac}(r) = \sum_\beta R^*_{\gamma_a,\lambda_a}(r) R_{\gamma_c,\lambda_c}(r) = P^*_a(r) P_c(r) + Q^*_a(r) Q_c(r), \quad \text{(A.4.38)}$$

we have

$$R^k_C(abcd) = \int_0^\infty \int_0^\infty \rho_{ac}(r)\, U_k(r,s)\, \rho_{bd}(s)\, drds. \quad \text{(A.4.39)}$$

- **Gaunt interaction**

$$X_G(abcd) = \sum_k X^k_G(abcd) \quad \text{(A.4.40)}$$

where

$$X^k_G(abcd) = -(-1)^k \Pi^o(\kappa_a, \kappa_c, k) \Pi^o(\kappa_b, \kappa_d, k) \times \sum_{\nu=k-1}^{k+1} \sum_{\beta,\beta'} \langle -\lambda_a \| \boldsymbol{X}^{(1\nu)k} \| \lambda_c \rangle \langle -\lambda'_b \| \boldsymbol{X}^{(1\nu)k} \| \lambda'_d \rangle \times R^\nu(-\lambda_a, -\lambda'_b, \lambda_c, \lambda'_d) \quad \text{(A.4.41)}$$

Odd parity overlap densities

$$\rho^o_{ac}(\beta; r) = R^*_{\gamma_a,-\lambda_a}(r) R_{\gamma_c,\lambda_c}(r) = \begin{cases} Q^*_a(r) P_c(r) \text{ if } \beta = +1, \\ P^*_a(r) Q_c(r) \text{ if } \beta = -1, \end{cases} \quad \text{(A.4.42)}$$

Gaunt integrals

$$R^\nu(-\lambda_a, -\lambda'_b, \lambda_c, \lambda'_d) = \int_0^\infty \int_0^\infty \rho^o_{ac}(\beta; r)\, U_\nu(r,s)\, \rho^o_{bd}(\beta', s)\, drds. \quad \text{(A.4.43)}$$

- **Møller interaction**

$$X_M(\omega; abcd) = X_{M1}(\omega; abcd) + X_{M2}(\omega; abcd) \quad \text{(A.4.44)}$$

$$X_{M1}(\omega; abcd) = \sum_{k=0}^\infty X^k_{M1}(\omega; abcd)\, \Pi^e(\kappa_a, \kappa_c, k)\, \Pi^e(\kappa_b, \kappa_d, k) \quad \text{(A.4.45)}$$

where

$$X^{k}_{M1}(\omega; abcd) = (-1)^k \langle j_a \| \boldsymbol{C}^k \| j_c \rangle \langle j_b \| \boldsymbol{C}^k \| j_d \rangle \, R^k(\omega; abcd)$$

$$X_{M2}(\omega; abcd) = \sum_{k=0}^{\infty} \sum_{\nu=|k-1|}^{k+1} X^{\nu,k}_{M2}(\omega; abcd) \, \Pi^o(\kappa_a, \kappa_c, k) \, \Pi^o(\kappa_b, \kappa_d, k) \tag{A.4.46}$$

where

$$X^{\nu,k}_{M2}(\omega; abcd) = -(-1)^k \sum_{\beta,\beta'} \langle -\lambda_a \| \boldsymbol{X}^{(1\nu)k} \| \lambda_c \rangle \langle -\lambda'_b \| \boldsymbol{X}^{(1\nu)k} \| \lambda'_d \rangle \times R^\nu(\omega; -\lambda_a, -\lambda'_b, \lambda_c, \lambda'_d).$$

ω-dependent Slater integrals

$$R^k(\omega; abcd) = \int_0^\infty \int_0^\infty \rho_{ac}(r) \, U_k(r, s; \omega) \, \rho_{bd}(s); drds \tag{A.4.47}$$

has the same component structure as (A.4.39), and

$$R^\nu(\omega; -\lambda_a, -\lambda'_b, \lambda_c, \lambda'_d) = \int_0^\infty \int_0^\infty \rho^o_{ac}(\beta; r) \, U_\nu(r, s; \omega) \, \rho^o_{bd}(\beta', s) \, drds. \tag{A.4.48}$$

is the frequency-dependent analogue of (A.4.43).

- **Transverse photon interaction: Coulomb gauge**

The full Coulomb gauge interaction is a sum of an instantaneous Coulomb interaction with effective interaction strength given by (A.4.36) and the transverse photon interaction:

$$\begin{aligned} X_T(\omega; abcd) = \sum_{k=0}^{\infty} -(-1)^k \sum_{\beta\beta'} \\ \times \Bigg\{ \sum_{\nu=k-1}^{k+1} \langle -\lambda_a \| \boldsymbol{X}^{(1\nu)k} \| \lambda_c \rangle \langle -\lambda'_b \| \boldsymbol{X}^{(1\nu)k} \| \lambda'_d \rangle \\ \times \; v_{\nu k} \, R^\nu(\omega; -\lambda_a, -\lambda'_b, \lambda_c, \lambda'_d) \\ + w_k \Big[\langle -\lambda_a \| \boldsymbol{X}^{(1,k-1)k} \| \lambda_c \rangle \langle -\lambda'_b \| \boldsymbol{X}^{(1,k+1)k} \| \lambda'_d \rangle \\ \times \; T^{k-1,k+1,k}(\omega; -\lambda_a, -\lambda'_b, \lambda_c, \lambda'_d) \\ + \langle -\lambda_a \| \boldsymbol{X}^{(1,k+1)k} \| \lambda_c \rangle \langle -\lambda'_b \| \boldsymbol{X}^{(1,k-1)k} \| \lambda'_d \rangle \\ \times \; T^{k-1,k+1,k}(\omega; -\lambda'_b, -\lambda_a, \lambda'_d, \lambda_c) \Big] \Bigg\}, \end{aligned} \tag{A.4.49}$$

where

$$v_{kk} = 1, \quad v_{k-1,k} = -(k+1)/(2k+1), \quad v_{k+1,k} = -k/(2k+1),$$
$$w_k = -[k(k+1)(2k-1)(2k+3)]^{1/2}/(2k+1)^2,$$

$$T^{k-1,k+1,k}(\omega; -\lambda_a, -\lambda'_b, \lambda_c, \lambda'_d)$$
$$= \int_0^\infty \int_0^\infty \rho^o_{ac}(\beta; r)\, W_{k-1,k+1,k}(r, s; \omega)\, \rho^o_{bd}(\beta', s)\, drds. \quad \text{(A.4.50)}$$

with

$$W_{k-1,k+1,k}(r, s; \omega) = \begin{cases} -i\,\dfrac{\omega}{c}[\,k\,]\, j_{k-1}(\omega r/c)\, h^{(1)}_{k+1}(\omega s/c) \\ \qquad\qquad + \dfrac{[\,k\,]^2 c^2}{\omega^2} \dfrac{r^{k-1}}{s^{k+2}} & \text{if } r < s \\ -i\,\dfrac{\omega}{c}[\,k\,]\, h^{(1)}_{k+1}(\omega r/c)\, j_{k-1}(\omega s/c) \\ & \text{if } r > s \end{cases} \quad \text{(A.4.51)}$$

with $W_{k+1,k-1,k}(r, s; \omega) = W_{k-1,k+1,k}(s, r; \omega)$.

- **Breit interaction: long wavelength approximation** $\omega \to 0$.

$$X_B(abcd) = \sum_{k=0}^{\infty} (-1)^k \sum_{\beta\beta'} \quad \text{(A.4.52)}$$
$$\times \Bigg\{ \sum_{\nu=k-1}^{k+1} \langle -\lambda_a \| \boldsymbol{X}^{(1\nu)k} \| \lambda_c \rangle \langle -\lambda'_b \| \boldsymbol{X}^{(1\nu)k} \| \lambda'_d \rangle$$
$$\times\; v_{\nu k}\, R^\nu(-\lambda_a, -\lambda'_b, \lambda_c, \lambda'_d)$$
$$-\frac{2k+1}{2} w_k \Big(\langle -\lambda_a \| \boldsymbol{X}^{(1,k-1)k} \| \lambda_c \rangle \langle -\lambda'_b \| \boldsymbol{X}^{(1,k+1)k} \| \lambda'_d \rangle$$
$$\times\; \left[S^{k-1}(-\lambda_a, -\lambda'_b, \lambda_c, \lambda'_d) - S^{k+1}(-\lambda_a, -\lambda'_b, \lambda_c, \lambda'_d) \right]$$
$$+\; \langle -\lambda_a \| \boldsymbol{X}^{(1,k+1)k} \| \lambda_c \rangle \langle -\lambda'_b \| \boldsymbol{X}^{(1,k-1)k} \| \lambda'_d \rangle$$
$$\times\; \left[S^{k-1}(-\lambda'_b, -\lambda_a, \lambda'_d, \lambda_c) - S^{k+1}(-\lambda'_b, -\lambda_a, \lambda'_d, \lambda_c) \right] \Big) \Bigg\}$$

"Half-range" Coulomb integrals

$$S^\nu(-\lambda_a, -\lambda'_b, \lambda_c, \lambda'_d) = \mathrm{d} \int_0^\infty ds\, \rho^o_{bd}(\beta'; s) s^{-\nu-1} \int_0^s dr\, r^\nu\, \rho^o_{ac}(\beta; r). \quad \text{(A.4.53)}$$

so that

$$R^\nu(-\lambda_a, -\lambda'_b, \lambda_c, \lambda'_d) = S^\nu(-\lambda_a, -\lambda'_b, \lambda_c, \lambda'_d) + S^\nu(-\lambda'_b, -\lambda_a, \lambda'_d, \lambda_c)$$

Table A.6. Nonrelativistic average pair energies for equivalent electrons [6].

ss	$F^0(s,s)$
pp	$F^0(p,p) - \frac{2}{25}F^2(p,p)$
dd	$F^0(d,d) - \frac{2}{63}F^2(d,d) - \frac{2}{63}F^4(d,d)$
ff	$F^0(f,f) - \frac{4}{195}F^2(f,f) - \frac{2}{143}F^4(f,f) - \frac{100}{5577}F^6(f,f)$

Table A.7. Nonrelativistic average pair energies for inequivalent electrons [6].

ss'	$F^0(s,s') - \frac{1}{2}G^0(s,s')$
sp	$F^0(s,p) - \frac{1}{6}G^1(s,p)$
sd	$F^0(s,d) - \frac{1}{10}G^2(s,d)$
sf	$F^0(s,f) - \frac{1}{14}G^3(s,f)$
pp'	$F^0(p,p') - \frac{1}{6}G^0(p,p') - \frac{1}{15}G^2(p,p')$
pd	$F^0(p,d) - \frac{1}{15}G^1(p,d) - \frac{3}{70}G^3(p,d)$
pf	$F^0(p,f) - \frac{3}{70}G^2(p,f) - \frac{2}{63}G^4(p,f)$
dd'	$F^0(d,d') - \frac{1}{10}G^0(d,d') - \frac{1}{35}G^2(d,d') - \frac{1}{35}G^4(d,d')$
df	$F^0(d,f) - \frac{3}{70}G^1(d,f) - \frac{2}{105}G^3(d,f) - \frac{5}{231}G^5(d,f)$
ff'	$F^0(f,f') - \frac{1}{14}G^0(f,f') - \frac{2}{105}G^2(f,f') - \frac{1}{77}G^4(f,f') - \frac{50}{3003}G^6(f,f')$

Table A.8. Dirac average pair energies from (6.6.26) (after [5]). The spectroscopic labels l and $\bar{l}$ denote respectively states with $j = l + 1/2$ and $j = l - 1/2$. Continued in Table A.9.

	Equivalent electrons
ss	$F^0(s,s)$
$\bar{p}\bar{p}$	$F^0(\bar{p},\bar{p})$
pp	$F^0(p,p) - \frac{1}{15}F^2(p,p)$
$\bar{d}\bar{d}$	$F^0(\bar{d}\bar{d},) - \frac{1}{15}F^2(\bar{d},\bar{d})$
dd	$F^0(d,d) - \frac{24}{525}F^2(d,d) - \frac{10}{525}F^4(d,d)$
$\bar{f}\bar{f}$	$F^0(\bar{f},\bar{f}) - \frac{8}{175}F^2(\bar{f},\bar{f}) - \frac{2}{105}F^4(\bar{f},\bar{f})$
ff	$F^0(f,f) - \frac{5}{147}F^2(f,f) - \frac{9}{539}F^4(f,f) - \frac{25}{3003}F^6(f,f)$
	Inequivalent electrons
ss'	$F^0(s,s') - \frac{1}{2}G^0(s,s')$
$s\bar{p}$	$F^0(s,\bar{p}) - \frac{1}{6}G^1(s,\bar{p})$
sp	$F^0(s,p) - \frac{1}{6}G^1(s,p)$
$s\bar{d}$	$F^0(s,\bar{d}) - \frac{1}{10}G^2(s,\bar{d})$
sd	$F^0(s,d) - \frac{1}{10}G^2(s,d)$
$s\bar{f}$	$F^0(s,\bar{f}) - \frac{1}{14}G^3(s,\bar{f})$
sf	$F^0(s,f) - \frac{1}{14}G^3(s,f)$
$\bar{p}\bar{p}'$	$F^0(\bar{p},\bar{p}') - \frac{1}{2}G^0(\bar{p},\bar{p}')$
$\bar{p}p$	$F^0(\bar{p},p) - \frac{1}{10}G^2(\bar{p},p)$
$\bar{p}\bar{d}$	$F^0(\bar{p},\bar{d}) - \frac{1}{6}G^1(\bar{p},\bar{d})$
$\bar{p}d$	$F^0(\bar{p},d) - \frac{1}{14}G^3(\bar{p},d)$
$\bar{p}\bar{f}$	$F^0(\bar{p},\bar{f}) - \frac{1}{10}G^2(\bar{p},\bar{f})$
$\bar{p}f$	$F^0(\bar{p},f) - \frac{1}{18}G^4(\bar{p},f)$
pp'	$F^0(p,p') - \frac{1}{4}G^0(p,p') - \frac{1}{20}G^2(p,p')$
$p\bar{d}$	$F^0(p,\bar{d}) - \frac{1}{60}G^1(p,\bar{d}) - \frac{9}{140}G^3(p,\bar{d})$
pd	$F^0(p,d) - \frac{1}{10}G^1(p,d) - \frac{1}{35}G^3(p,d)$
$p\bar{f}$	$F^0(p,\bar{f}) - \frac{1}{70}G^2(p,\bar{f}) - \frac{1}{21}G^4(p,\bar{f})$
pf	$F^0(p,f) - \frac{9}{140}G^2(p,f) - \frac{5}{252}G^4(p,f)$

Table A.9. (Continuation of Table A.8.)

$\bar{d}\bar{d}'$	$F^0(\bar{d},\bar{d}') - \frac{1}{4}G^0(\bar{d},\bar{d}') - \frac{1}{20}G^2(\bar{d},\bar{d}')$
$\bar{d}d$	$F^0(\bar{d},d) - \frac{1}{70}G^2(\bar{d},d) - \frac{1}{21}G^4(\bar{d},d)$
$\bar{d}\bar{f}$	$F^0(\bar{d},\bar{f}) - \frac{1}{10}G^1(\bar{d},\bar{f}) - \frac{1}{35}G^3(\bar{d},\bar{f})$
$\bar{d}f$	$F^0(\bar{d},f) - \frac{1}{84}G^3(\bar{d},f) - \frac{5}{132}G^5(\bar{d},f)$
dd'	$F^0(d,d') - \frac{1}{6}G^0(d,d') - \frac{4}{105}G^2(d,d') - \frac{1}{63}G^4(\bar{d},d')$
$d\bar{f}$	$F^0(d,\bar{f}) - \frac{1}{210}G^1(d,\bar{f}) - \frac{4}{315}G^3(d,\bar{f}) - \frac{25}{693}G^5(d,\bar{f})$
df	$F^0(d,f) - \frac{1}{14}G^1(d,f) - \frac{1}{42}G^3(d,f) - \frac{5}{462}G^5(d,f)$
$\bar{f}\bar{f}'$	$F^0(\bar{f},\bar{f}') - \frac{1}{6}G^0(\bar{f},\bar{f}') - \frac{4}{105}G^2(\bar{f},\bar{f}') - \frac{1}{63}G^4(\bar{f},\bar{f}')$
$\bar{f}f$	$F^0(\bar{f},f) - \frac{1}{210}G^2(\bar{f},f) - \frac{5}{462}G^4(\bar{f},f) - \frac{25}{858}G^6(\bar{f},f)$
ff'	$F^0(f,f') - \frac{1}{8}G^0(f,f') - \frac{5}{168}G^2(f,f') - \frac{9}{616}G^4(f,f') - \frac{25}{3432}G^6(f,f')$

A.5 Open shells in jj-coupling

General reference Chapter 6.

Table A.10. States of j^N in jj-coupling: $j = 1/2, 3/2, 5/2, 7/2$.

Configuration	Q	$\|M_Q\|$	ν	J-values
$\left(\frac{1}{2}\right)^0, \left(\frac{1}{2}\right)^2$	$\frac{1}{2}$	$\frac{1}{2}$	0	0
$\left(\frac{1}{2}\right)^1$	0	0	1	$\frac{1}{2}$
$\left(\frac{3}{2}\right)^0, \left(\frac{3}{2}\right)^4$	1	1	0	0
$\left(\frac{3}{2}\right)^1, \left(\frac{3}{2}\right)^3$	$\frac{1}{2}$	$\frac{1}{2}$	1	$\frac{3}{2}$
$\left(\frac{3}{2}\right)^2$	1	0	0	0
	0	0	2	2
$\left(\frac{5}{2}\right)^0, \left(\frac{5}{2}\right)^6$	$\frac{3}{2}$	$\frac{3}{2}$	0	0
$\left(\frac{5}{2}\right)^1, \left(\frac{5}{2}\right)^5$	1	1	1	$\frac{5}{2}$
$\left(\frac{5}{2}\right)^2$ or $\left(\frac{5}{2}\right)^4$	$\frac{3}{2}$	$\frac{1}{2}$	0	0
	$\frac{1}{2}$	$\frac{1}{2}$	2	2,4
$\left(\frac{5}{2}\right)^3$	1	0	1	$\frac{5}{2}$
	0	0	3	$\frac{3}{2}, \frac{9}{2}$
$\left(\frac{7}{2}\right)^0, \left(\frac{7}{2}\right)^8$	2	2	0	0
$\left(\frac{7}{2}\right)^1, \left(\frac{7}{2}\right)^7$	$\frac{3}{2}$	$\frac{3}{2}$	1	$\frac{7}{2}$
$\left(\frac{7}{2}\right)^2, \left(\frac{7}{2}\right)^6$	2	1	0	0
	1	1	2	2,4,6
$\left(\frac{7}{2}\right)^3, \left(\frac{7}{2}\right)^5$	$\frac{3}{2}$	$\frac{1}{2}$	1	$\frac{7}{2}$
	$\frac{1}{2}$	$\frac{1}{2}$	3	$\frac{3}{2}, \frac{5}{2}, \frac{9}{2}, \frac{11}{2}, \frac{15}{2}$
$\left(\frac{7}{2}\right)^4$	2	0	0	0
	1	0	2	2,4,6
	0	0	4	2,4,5,8
$\left(\frac{9}{2}\right)^0, \left(\frac{9}{2}\right)^{10}$	$\frac{5}{2}$	$\frac{5}{2}$	0	0
$\left(\frac{9}{2}\right)^1, \left(\frac{9}{2}\right)^9$	2	2	1	$\frac{9}{2}$
$\left(\frac{9}{2}\right)^2, \left(\frac{9}{2}\right)^8$	$\frac{5}{2}$	$\frac{3}{2}$	0	0
	$\frac{3}{2}$	$\frac{3}{2}$	2	2,4,6,8
$\left(\frac{9}{2}\right)^3, \left(\frac{9}{2}\right)^7$	2	1	1	$\frac{9}{2}$
	1	1	3	$\frac{3}{2}, \frac{5}{2}, \frac{7}{2}, \frac{9}{2}, \frac{11}{2}, \frac{13}{2}, \frac{15}{2}, \frac{17}{2}, \frac{21}{2},$
$\left(\frac{9}{2}\right)^4, \left(\frac{9}{2}\right)^6$	$\frac{5}{2}$	$\frac{1}{2}$	0	0
	$\frac{3}{2}$	$\frac{1}{2}$	2	2,4,6,8
	$\frac{1}{2}$	$\frac{1}{2}$	4	$0,2,3,4^2,5,6^2,7,8,9,10,12$
$\left(\frac{9}{2}\right)^5$	2	0	1	$\frac{9}{2}$
	1	0	3	$\frac{3}{2}, \frac{5}{2}, \frac{7}{2}, \frac{9}{2}, \frac{11}{2}, \frac{13}{2}, \frac{15}{2}, \frac{17}{2}, \frac{21}{2}$
	0	0	5	$\frac{1}{2}, \frac{5}{2}, \frac{7}{2}, \frac{9}{2}, \frac{11}{2}, \frac{13}{2}, \frac{15}{2}, \frac{17}{2}, \frac{25}{2}$

Table A.11. Coefficients of fractional parentage for j^N configurations, $j = 3/2$. Values for $N = 1, 2$ from (6.8.39) and for $N = 3$ from [7]; values for $N > 3$ may be obtained using Theorem 6.7.

	$N = 3$	
$\overline{\nu}$:	0	2
$\overline{J}$:	0	2
ν J		
1 $\frac{3}{2}$	$\frac{1}{\sqrt{6}}$	$-\frac{\sqrt{5}}{\sqrt{6}}$

Table A.12. Coefficients of fractional parentage for j^N configurations, $j = 5/2$. Values for $N = 1, 2$ from (6.8.39) and for $N = 3, 4$ from [7]; values for $N > 4$ may be obtained using Theorem 6.7

		$N = 3$	
$\overline{\nu}$:	0	2	2
$\overline{J}$:	0	2	4
ν J			
1 $\frac{5}{2}$	$-\frac{\sqrt{2}}{3}$	$\frac{\sqrt{5}}{3\sqrt{2}}$	$\frac{1}{\sqrt{2}}$
3 $\frac{3}{2}$		$-\frac{\sqrt{5}}{\sqrt{7}}$	$\frac{\sqrt{2}}{\sqrt{7}}$
3 $\frac{9}{2}$		$\frac{\sqrt{3}}{\sqrt{14}}$	$-\frac{\sqrt{11}}{\sqrt{14}}$

		$N = 4$	
$\overline{\nu}$	1	3	3
$\overline{J}$	$\frac{5}{2}$	$\frac{3}{2}$	$\frac{9}{2}$
ν J			
0 0	1		
2 2	$-\frac{1}{2}$	$\frac{\sqrt{3}}{\sqrt{7}}$	$\frac{3}{2\sqrt{7}}$
2 4	$-\frac{1}{2}$	$-\frac{\sqrt{2}}{\sqrt{21}}$	$-\frac{\sqrt{55}}{2\sqrt{21}}$

Table A.13. Coefficients of fractional parentage for j^N configurations, $j = 7/2$. Values for $N = 1, 2$ from (6.8.39) and for $N = 3, 4$ from [7], with corrections; values for $N > 4$ may be obtained using Theorem 6.7

$N = 3$

	$\bar{\nu}$: 0	2		
	$\bar{J}$: 0	2	4	6
ν J				
1 $\frac{7}{2}$	$\frac{1}{2}$	$-\frac{\sqrt{5}}{6}$	$-\frac{1}{2}$	$-\frac{\sqrt{13}}{6}$
3 $\frac{3}{2}$		$\frac{\sqrt{3}}{\sqrt{14}}$	$-\frac{\sqrt{11}}{\sqrt{14}}$	
3 $\frac{5}{2}$		$\frac{\sqrt{11}}{3\sqrt{2}}$	$\frac{\sqrt{2}}{\sqrt{33}}$	$-\frac{\sqrt{65}}{3\sqrt{22}}$
3 $\frac{9}{2}$		$\frac{\sqrt{13}}{3\sqrt{14}}$	$-\frac{5\sqrt{2}}{\sqrt{77}}$	$\frac{7}{3\sqrt{22}}$
3 $\frac{11}{2}$		$-\frac{\sqrt{5}}{3\sqrt{2}}$	$\frac{\sqrt{13}}{\sqrt{66}}$	$\frac{2\sqrt{13}}{3\sqrt{11}}$
3 $\frac{15}{2}$			$\frac{\sqrt{5}}{\sqrt{22}}$	$-\frac{\sqrt{17}}{\sqrt{22}}$

$N = 4$

$\bar{\nu}$	1	3				
$\bar{J}$	$\frac{7}{2}$	$\frac{3}{2}$	$\frac{5}{2}$	$\frac{9}{2}$	$\frac{11}{2}$	$\frac{15}{2}$
ν J						
0 0	1					
2 2	$\frac{1}{\sqrt{3}}$	$\frac{3}{2\sqrt{35}}$	$-\frac{\sqrt{11}}{2\sqrt{10}}$	$-\frac{\sqrt{13}}{2\sqrt{42}}$	$-\frac{1}{2}$	
2 4	$\frac{1}{\sqrt{3}}$	$-\frac{\sqrt{11}}{2\sqrt{21}}$	$-\frac{1}{\sqrt{66}}$	$\frac{5\sqrt{5}}{\sqrt{462}}$	$\frac{\sqrt{13}}{2\sqrt{33}}$	$\frac{\sqrt{5}}{\sqrt{33}}$
2 6	$\frac{1}{\sqrt{3}}$		$\frac{\sqrt{5}}{2\sqrt{22}}$	$-\frac{7\sqrt{5}}{2\sqrt{858}}$	$\frac{\sqrt{2}}{\sqrt{11}}$	$-\frac{\sqrt{51}}{\sqrt{143}}$
4 2		$-\frac{\sqrt{11}}{\sqrt{35}}$	$-\frac{1}{2\sqrt{10}}$	$-\frac{3\sqrt{39}}{2\sqrt{154}}$	$\frac{1}{\sqrt{11}}$	
4 4		$\frac{\sqrt{13}}{2\sqrt{105}}$	$\frac{\sqrt{13}}{\sqrt{30}}$	$\frac{\sqrt{3}}{\sqrt{182}}$	$\frac{\sqrt{5}}{2\sqrt{3}}$	$-\frac{2}{\sqrt{139}}$
4 5		$\frac{3}{2\sqrt{5}}$	$-\frac{7}{2\sqrt{110}}$	$-\frac{3\sqrt{3}}{2\sqrt{22}}$	$\frac{\sqrt{65}}{2\sqrt{77}}$	$\frac{\sqrt{17}}{\sqrt{77}}$
4 8				$\frac{2\sqrt{5}}{\sqrt{143}}$	$\frac{3\sqrt{2}}{\sqrt{77}}$	$\frac{\sqrt{57}}{\sqrt{91}}$

Table A.14. States of l^N in LS-coupling: $l = 0, 1, 2$

Configuration	ν	Terms
s	1	^2S
s^2	0	^1S
p, p^5	1	^2P
p^2, p^4	0	^1S
	2	^3P, ^1D
p^3	1	^2P
	3	^4S, ^2D
d, d^9	1	^2D
d^2, d^8	0	^1S
	2	^3F, ^3P, ^1G, ^1D
d^3, d^7	1	^2D
	3	^4F, ^4P, ^2H, ^2G, ^2F, ^2P
d^4, d^6	0	^1S
	2	^3F, ^3P, ^1G, ^1D
	4	^5D, ^3H, ^3G, ^3F, ^3D, ^3P, ^1I, ^1G, ^1F, ^1D, ^1S
d^5	1	^2D
	3	^4F, ^4P, ^2H, ^2G, ^2F, ^2D, ^2P
	5	^6S, ^4G, ^4D, ^2I, ^2G, ^2F, ^2D, ^2S

A.6 Exponents for atomic and molecular G-spinors

Experience with relativistic calculations using G-spinors indicates that exponent sets derived from nonrelativistic collections are a good starting point. Table A.15 shows even-tempered exponents for a range of atoms following Quiney and Belanzoni [8, Table II]. and a nonrelativistic tabulation of Clementi [9]. The sequence of exponents, λ_n, has a length N_l corresponding to the usual orbital angular momentum quantum number, the same for both values $j = l \pm 1/2$, so that

$$\lambda_n = \alpha\beta^{n-1}, \quad n = 1, 2, \ldots, N_l.$$

When $\beta.1$, the value of $\lambda_1 = \alpha$ determines the most extended orbital that can be represented. The smaller its value, the larger the effective radius. Conversely, the value of λ_{N_l} fixes the most contracted orbital, which is needed to represent the behaviour in the neighbourhood of the nucleus.

Table A.15. Even-tempered exponent parameters for selected atoms. Reprinted with permission from [8]. Copyright 2002, American Institute of Physics. See text for explanation.

Element	α	β	N_s	N_p	N_d	N_f
He	0.048 506 70	2.147 745 20	20			
Be	0.020 310 60	2.089 324 30	23			
Ne	0.114 853 50	2.131 648 10	23	16		
Mg	0.012 158 50	2.121 836 20	26	20		
Ar	0.026 350 40	2.086 752 00	26	21		
Ca	0.016 465 70	2.014 656 50	29	22		
Zn	0.027 543 20	1.980 743 20	30	24	19	
Kr	0.029 604 30	2.019 604 30	31	24	19	
Sr	0.014 019 40	1.995 839 60	33	25	21	
Pd	0.017 625 00	1.976 728 50	32	26	21	
Cd	0.013 471 50	1.979 900 20	33	26	22	
Xe	0.014 301 30	1.977 844 50	35	27	22	
Ba	0.005 003 92	1.977 648 20	33	28	33	
Yb	0.013 127 70	1.976 443 60	32	26	22	18
Hg	0.021 778 60	1.980 451 90	32	26	22	21
Rn	0.024 292 20	1.972 942 40	33	27	22	19
Ra	0.024 292 20	1.972 942 40	33	27	22	19

Table A.16. Basis set exponents for selected first and second row atoms. Reprinted with permission from [10, Appendix D].

n	Hydrogen	Carbon	Nitrogen	Oxygen
		s-type		
1	82.6364	1267.1800	678.7187	1869.1275
2	12.4096	190.6040	102.2656	281.6302
3	2.8239	43.2477	22.9066	64.2526
4	0.7977	11.9649	6.0164	18.1915
5	0.2581	3.6631	0.8396	5.8769
6	0.0899	0.5392	0.2585	0.5685
7		0.1671		
		p-type		
1		4645.6	1280.17	7394.
2		2.0066	1.2613	3.8633
3		0.5469	0.2916	1.0486
4		0.1520		0.2761

Table A.17. Hydrogen basis sets. Exponents in parentheses are included in cc-aug-pVDZ and cc-aug-pVTZ sets.

cc-pVDZ			cc-pVTZ			
n	s	p	n	s	p	d
1	1.301 (1)	7.27 (-1)	1	3.387 (1)	1.407 (0)	1.057 (0)
2	1.962 (0)	{1.41 (-1)}	2	5.095 (0)	3.880 (-1)	{2.470 (-1)}
3	4.446 (-1)		3	1.159 (0)	{1.020 (-1)}	
4	1.220 (-1)		4	3.258 (-1)		
5	{2.974 (-2)		5	1.027 (-1)		
6			6	{2.526 (-2)}		

The value of N_l is therefore slightly larger than the value needed for non-relativistic calculations. See also [11] for similar methods of construction. Many other sets of basis functions for relativistic calculations with G-spinors have been proposed: a recent tabulation by Koga *et al.* [12] covers the atoms from H to Xe.

Table A.16 shows basis sets used for the examples of orbital classification in §11.1. The remaining tables display correlation consistent uncontracted basis sets of Dunning *et al.* [13, 14]. Exponents are entered as floating point numbers $a.b\,(c) \equiv a.b\,10^c$. Entries in paranthess are required for the augmented sets.

Table A.18. Hydrogen CVTZ basis set.

n	s	p
1	8.2636374 (1)	2.747748344 (0)
2	1.2409558 (1)	1.248976520 (0)
3	2.8238540 (0)	5.677166000 (-1)
4	7.9767000 (-1)	2.580530000 (-1)
5	2.5805300 (-1)	1.172968182 (-1)
6	8.9891000 (-2)	5.331673554 (-2)
7		2.423487979 (-2)
8		1.101585445 (-2)

Table A.19. Nitrogen cc-pVDZ basis set. Exponents in parentheses are included in the cc-aug-pVTZ basis set.

n	s	p	d
1	9.046 (3)	1.355 (1)	8.170 (-1)
2	1.357 (3)	2.917 (0)	{2.300 (-1)}
3	3.093 (2)	7.973 (-1)	
4	8.773 (1)	2.185 (-1)	
5	2.856 (1)	{5.611 (-2)}	
6	1.021 (1)		
7	3.838 (0)		
8	7.466 (-1)		
9	2.248 (-1)		
10	{6.124 (-2)}		

Table A.20. Nitrogen cc-pVTZ basis set. Exponents in parentheses are included in the cc-aug-pVTZ basis set.

n	s	p	d	f
1	1.142 (4)	2.663 (1)	1.654 (0)	1.093 (0)
2	1.712 (3)	5.948 (0)	4.690 (-1)}	{3.640 (-1)}
3	3.893 (2)	1.742 (0)	{1.510 (-1)}	
4	1.100 (2)	5.550 (-1)		
5	3.557 (1)	1.725 (-1)		
6	1.254 (1)	{4.910 (-2)}		
7	4.644 (0)			
8	1.293 (0)			
9	5.118 (-1)			
10	1.787 (-1)			
11	{5.760 (-2)}			

Table A.21. Nitrogen CVTZ basis set.

n	s	p	d
1	1.5882557 (4)	4.7926322 (3)	6.045046357 (0)
2	7.2193440 (3)	2.0837531 (3)	2.418018543 (0)
3	3.2815200 (3)	9.0597964 (2)	9.672074171 (-1)
4	1.4916000 (3)	3.9390419 (2)	3.868829668 (-1)
5	6.7871873 (2)	1.7126269 (2)	1.547531867 (-1)
6	1.0226555 (2)	7.4462040 (1)	6.190127470 (-2)
7	2.2906555 (1)	3.2374800 (1)	2.476050988 (-3)
8	6.0164040 (0)	1.4076000 (1)	9.904203951 (-3)
9	8.3954400 (-1)	6.1279700 (0)	
10	2.5853300 (-1)	1.2613100 (0)	
11	2.5853300 (-2)	2.9159000 (-1)	
12		2.9159000 (-2)	

Table A.22. Oxygen cc-pVDZ basis set. Exponents in parentheses are included in the cc-aug-pVDZ basis set.

n	s	p	d
1	1.172 (4)	1.770 (1)	1.185 (0)
2	1.759 (3)	3.854 (0)	{3.320 (-1)}
3	4.008 (2)	1.046 (0)	
4	1.137 (2)	2.753 (-1)	
5	3.703 (1)	{6.856 (-2)}	
6	1.327 (1)		
7	5.025 (0)		
8	1.013 (0)		
9	3.023 (-1)		
10	{7.896 (-2)}		

Table A.23. Oxygen cc-pVTZ basis set. Exponents in parentheses are included in the cc-aug-pVTZ basis set.

n	s	p	d	f
1	1.533 (4)	3.446 (1)	2.314 (0)	1.428 (0)
2	2.299 (3)	7.749 (0)	6.450 (-1)}	{5.000 (-1)}
3	5.224 (2)	2.280 (0)	{2.140 (-1)}	
4	1.473 (2)	7.156 (-1)		
5	4.755 (1)	2.140 (-1)		
6	1.676 (1)	{5.974 (-2)}		
7	6.207 (0)			
8	1.752 (0)			
9	6.882 (-1)			
10	2.384 (-1)			
11	{7.376 (-2)}			

Table A.24. Oxygen CVTZ basis set.

n	s	p	d
1	7.81654 (3)	3.51832 (1)	6.045046357 (0)
2	1.17582 (3)	7.90400 (0)	2.418018543 (0)
3	2.73188 (2)	2.30510 (0)	9.672074171 (-1)
4	8.11696 (1)	7.17100 (-1)	3.868829668 (-1)
5	2.71836 (1)	2.13700 (-1)	1.547531867 (-1)
6	9.53220 (0)	2.80000 (-2)	6.190127470 (-2)
7	3.41360 (0)		2.476050988 (-3)
8	9.39800 (-1)		9.904203951 (-3)
9	2.84600 (-1)		
10	3.20000 (-2)		

Table A.25. Fluorine 9s6p2d basis set. The 9s6p exponents are optimized for F^- and the basis set is intended for use in molecules with ionic character. Reprinted with permission from [15, Table 4]. Copyright by the American Physical Society.

n	s	p	d
1	1.38746194 (04)	6.30771751 (01)	2.0
2	2.08183460 (03)	1.44880120 (01)	0.8
3	4.74024179 (02)	4.38060140 (00)	
4	1.34275303 (02)	1.45214460 (00)	
5	4.37173394 (01)	4.61546462 (-01)	
6	1.57374155 (01)	1.26211456 (-01)	
7	6.04514678 (00)		
8	1.24305124 (00)		
9	3.39833174 (-01)		

Table A.26. Fluorine 10s6p basis set. Reprinted from [16, Table 9.38.1] with permission from Elsevier.

n	s	p
1	2.2451911 (04)	8.1849600 (01)
2	3.3661439 (03)	1.8934810 (01)
3	7.6743400 (02)	5.7716470 (00)
4	2.1736371 (02)	1.9765800 (00)
5	7.0048413 (01)	5.7872800 (-1)
6	2.4434934 (01)	2.1942700 (-1)
7	8.9649720 (00)	
8	2.9337530 (00)	
9	1.0398900 (00)	
10	3.3318300 (-1)	

Table A.27. Ytterbium relativistic 26s26p15d8f basis set. [17, Table I]

.

	s	p	d	f
1	1.5502155 (09)	1.2084686 (09)	3.9505578 (04)	2.2161936 (02)
2	6.2008625 (08)	4.8838743 (08)	1.5802231 (04)	7.6124071 (01)
3	2.4803450 (08)	1.9335497 (08)	6.3208925 (03)	3.1514318 (01)
4	9.9213800 (07)	7.7341989 (07)	2.5283570 (03)	1.4116035 (01)
5	3.9685520 (07)	3.0936796 (07)	7.6510381 (02)	6.4142517 (00)
6	1.5874209 (07)	1.2374718 (07)	2.9800187 (02)	2.8266169 (00)
7	2.3969203 (06)	4.9498873 (06)	1.3094908 (02)	1.1602936 (00)
8	5.5019050 (05)	1.9799549 (06)	6.1619422 (01)	4.1555735 (-1)
9	1.5713627 (05)	7.9198197 (05)	3.0071721 (01)	
10	5.1667659 (04)	3.1679279 (05)	1.4854616 (01)	
11	1.8782754 (04)	1.2671711 (05)	7.1662380 (00)	
12	7.3707771 (03)	5.0686846 (04)	3.3124755 (00)	
13	3.0714682 (03)	1.1956152 (04)	1.3811515 (00)	
14	1.3425378 (03)	3.8715950 (03)	5.5246060 (-1)	
15	6.0767024 (02)	1.4763972 (03)	2.2098424 (-1)	
16	2.7407132 (02)	6.2436731 (02)		
17	1.3358444 (02)	2.8307777 (02)		
18	6.4927758 (01)	1.3479853 (02)		
19	2.7274950 (01)	6.6072455 (01)		
20	1.4266406 (01)	3.2389585 (01)		
21	5.8267845 (00)	1.6283456 (01)		
22	2.8109416 (00)	7.5266633 (00)		
23	7.6402874 (-1)	3.5062020 (00)		
24	3.3868438 (-1)	1.4060090 (00)		
25	5.4530520 (-2)	6.0477513 (-1)		
26	2.2782330 (-2)	2.3518564 (-1)		

Table A.28. Thallium Tl_{erg}: relativistic dual-family 25s24p16d10f basis set. Reprinted with permission from [15, Table III]. Copyright 1998 by the American Physical Society.

λ	ℓ	λ	ℓ
50980195.85	s	18626562.61	p
11526467.50	s	3444971.589	p
3216581.844	s	800643.0337	p
984796.7808	s	215991.5182	p
326098.2911	s	65027.98447	p
114874.8526	s	21519.26032	p
42723.84963	s	7794.177919	p
16646.46727	s,d	3071.601225	p
6760.138769	s,d	1302.202583	p
2850.823938	s,d	585.0229722	p,f
1244.074062	s,d	274.7617212	p,f
558.4243869	s,d	133.3031257	p,f
256.5126586	s,d	64.77261541	p,f
122.9648384	s,d	32.73408726	p,f
60.83529156	s,d	16.41196338	p,f
30.79272153	s,d	8.065415659	p,f
15.31250874	s,d	3.882289818	p,f
7.516325476	s,d	1.866490000	p.f
3.622059784	s,d	0.897349000	p,f
1.741370000	s,d	0.431418000	p
0.837199000	s,d	0.207412000	p
0.402500000	s,d	0.099717500	p
0.193509000	s,d	0.047941100	p
0.093033400	s	0.023048600	p
0.044727600	s		

Table A.29. Even-tempered basis sets for Tl. The fourth column lists for each symmetry (in the order $s, p, d, \ldots$) the ranges $(n_1 : n_2)$ taken by the index i in (11.7.6). The fixed values $\alpha = 0.02$ and $\lambda_N = 5.0 \times 10^8$ for all N ensure a good description of both core and valence regions. Reprinted with permission from [15, Table II]. Copyright 1998 by the American Physical Society.

	β_N	Dimension	Ranges
Tl_1	2.606	25s 25p 12d 8f	(1:25) (2:26) (13:24) (15:22)
Tl_2	2.352	28s 28p 14d 8f	(1:28) (2:29) (14:27) (17:24)
$Tl_{3}a$	2.165	31s 31p 15d 8f	(1:31) (2:32) (16:30) (19:26)
$Tl_{3}b$	2.165	31s 31p 15d 8f 3g	(1:31) (2:32) (16:30) (19:26) (23:25)
Tl_4	2.022	34s 34p 16d 8f	(1:34) (2:35) (17:32) (21:29)

A.7 Software for relativistic molecular calculations

We give brief details of some software for relativistic *four-component* molecular calculations available mid-2004. All systems comprise a DHF module, one or more modules for treating correlation, modules to calculate properties and various utilities. The theoretical machinery and computational machinery of BERTHA is built on the G-spinor formalism of Chapter 10. The other packages assume each spinor component is a linear combination of Cartesian Gaussian functions. Group theoretical properties similar to those of G-spinors have to be built in subsequently. This means that the codes are very different at the lowest level, and their computational costs may be very different.

A.7.1 BERTHA

- **Hamiltonians**: Dirac-Coulomb, Dirac-Coulomb-Breit, Lévy-Leblond (non-relativistic limit).
- **Primitive basis functions**: G-spinors – §10.6
- **Nuclear models**: Point charge or simple Gaussian built in.
- **Integral evaluation**:
  - One-centre (atomic) – §10.2:
  - Multi-centre (molecular) – §10.9
- **Fock matrix construction**:
  - Conventional – §10.10.
  - Using $\boldsymbol{E}$ and $\boldsymbol{B}$ fields – §10.11
- **Economization strategies**: §10.13
- **Relativistic Density Functional Theory**: §10.12
- **Correlation**: MP2

A.7.2 DIRAC

This description is based on Dirac Version 3.2, released 6 October 2000.
Codes, manuals, etc. on-line: `http://dirac.chem.sdu.dk/`

- **Hamiltonians**: Dirac-Coulomb (default), Lévy-Leblond (non-relativistic limit), Dyall [18] spin-free Hamiltonian, ZORA [19].
- **Primitive basis functions**: CGTF, contracted and uncontracted. Small component basis set, derived using kinetic balance, need not have same dimension as large component basis set.
- **Nuclear models**: Point charge or simple Gaussian.
- **DHF module**: Direct SCF method
  - Closed shell; average of configuration open shell.
  - Point group symmetry: D_{2h} and subgroups.
  - Economization: screening scheme; time-reversal imposed using quaternion algebra.
  - Population analysis.

  - 1st order properties (dipole moment, electric field gradient, ...); 2nd order properties (polarizabilities, NMR parameters, ...)
- **Correlation**: Modules for MP2 energy; CCSD/CCSD(T) and CI (from MOLFDIR).
- **Geometry optimization**

A.7.3 MOLFDIR

This description is based on [20, 21, 22].
See `http://theochem.chem.rug.nl/~broer/Molfdir/Molfdir.html`

- **Hamiltonians**: Dirac-Coulomb -(Gaunt/Breit).
- **Basis sets** [20]: Kinetic balance (unmatched large and small component sets); atomic balance.

Components	Type	Usage
1	Primitive CGTF	Low-level integral evaluation.
1	General contracted CGTF	2-electron integral evaluation/storage
4	Non-symmetry-adapted	Fock matrix construction and storage of SCF vectors
4	Symmetry-adapted	Storage of 1-electron matrix elements and density matrices

- **Nuclear models**: 0 - Point charge; 1 - homogeneous sphere; 2- Fermi 2-parameter; 3- simple Gaussian.
- **MOLFDIR modules**:
  - MOLFDIR generates double group symmetry adapted basis functions
  - RELONEL, RELTWEL generate 1- and 2-electron integrals.
  - MFDSCF solves closed-shell and open-shell average-of-configuration equations, including Gaunt interaction if requested.
- **Correlation**:
  - Full CI (small spinor basis sets).
  - RASCI using direct method.
  - CCSD-T +T(T) coupled cluster calculations.
- **Analysis**
  - Electron density visualization, Mulliken population analysis.
  - Molecular properties
  - Linear response properties: RPA method

References

1. Press W H, Teukolsky S A, Vetterling W T and Flannery B P 1992 *Numerical Recipes* Second edn (Cambridge: Cambridge University Press)
2. Abramowitz M and Stegun I A 1970 *Handbook of Mathematical Functions* (New York: Dover)
3. Grant I P and Quiney H M 2000 *Int. J. Quant. Chem¿* **80** 283.
4. Grant I P 1970 Relativistic calculation of atomic structures, *Adv. Phys.* **19** 747-811.
5. Larkins F P 1976 *J. Phys. B: Atom. Molec. Phys.* **9**, 37.
6. Slater J C 1960 *Quantum theory of atomic structure*, Vol. 1. (New York: McGraw-Hill).
7. Edmonds A R and Flowers B H 1952 *Proc Roy Soc A* **2**14, 515.
8. Quiney H M and Belanzoni P 2002 *J. Chem. Phys.* **117**, 5550.
9. Clementi E 1990 *Modern Techniques in Computational Chemistry (MOTECC 90)*. (Leiden: ESCOM Science).
10. Skaane H 1998, unpublished D. Phil. Thesis, University of Oxford.
11. Tatewaki H, Mochizuki Y, Koga T and Karwowski J 2001 *J. Comp. Phys.* **115**, 9160.
12. Koga K, Tatewaki H and Matsuoka O 2001 *J. Comp. Phys.* **115**, 3561. See EPAPS document: E-JCPSA6-114-302124 for full details; retrievable from: http://www.aip.org/pubservs/epaps.html
13. Dunning T H Jr. 1989 *J. Chem. Phys.* **90**, 1007.
14. Kendall R A, Dunning T H Jr. and Harrison R J 1992 *J. Chem. Phys.* **96**, 6796.
15. Quiney H M, Lærdahl J K, Fægri K jr. and Saue T 1998 *Phys. Rev. A* **57** 920.
16. Poirier R, Kari R, Csizmadia I G 1985 *Handbook of Gaussian Basis Sets* (Amsterdam: Elsevier).
17. Quiney H M, Skaane H and Grant I P 1998 *J. Phys. B: At. Mol. Opt. Phys.* **31**, L85.
18. Dyall K G 1994 *J. Chem. Phys.* **100**, 2118.
19. van Lenthe E, Baerends E-J and Snijders J G 1993 *J. Chem. Phys.* **99**, 4597.
20. Visscher L, Visser O, Aerts P J C, Merenga H and Nieuwpoort W C 1994 *Comp. Phys. Commun.* **81**, 120.
21. Visscher L 1993 *Relativity and Electron Correlation in Chemistry.* Thesis: Groningen.
22. de Jong W A 1998 *Relativistic Quantum Chemistry Applied.* Thesis: Groningen.

B

Supplementary mathematics

B.1 Linear operators on Hilbert space

General references [1], [2], [3], [4]

B.1.1 Hilbert spaces

This appendix summarizes Hilbert space notions needed to work with Dirac operators in atomic and molecular physics and used in Chapters 3 and 5. Full details will be found in the general references above.

Elements

A Hilbert Space $\mathcal{H}$ is a *complete* linear vector space with elements here denoted $\{u, v, \ldots\}$. Any pair of vectors, $u, v \in \mathcal{H}$ has an *inner product* $(u, v) = (v, u)^*$ that is *anti-linear* in u and *linear* in v, where the asterisk denotes complex conjugation. Thus if a is any complex number, $(u, av) = a(u, v)$ and $(au, v) = a^*(u, v)$. $\mathcal{H}$ is endowed with a real and positive *norm* $\|u\|^2 = (u, u)$. The word 'complete' signifies that every sequence $\{u_n\}$ having the Cauchy property $\|u_n - u_m\| \to 0$ as $n, m \to \infty$ has a limit $u \in \mathcal{H}$.

B.1.2 Linear operators

A linear operator, T on $\mathcal{H}$ maps an element $u \in \mathcal{H}$ into the element $Tu \in \mathcal{H}$. T is said to have a *bound* M if $\|Tu\|/\|u\| \leq M$ for all $u \in \mathcal{H}$. In fact, most of the operators we have to deal with have no such bound: M is infinite. It is then necessary to restrict the operations of T to a subset, the *domain of definition* $\mathcal{D}(T) \subset \mathcal{H}$ on which T is bounded. The set of all vectors Tu with $u \in \mathcal{D}(T)$ is called the *range*, $\mathcal{R}(T)$. Then T has a norm, defined by

$$\|T\| = \sup_{u \in \mathcal{D}(T)} \frac{\|Tu\|}{\|u\|}.$$

The set of vectors $\mathcal{K}(T) = \{u \mid Tu = 0,\ u \in \mathcal{D}(T)\}$ is called the *kernel* of T. A subset $S \subset \mathcal{H}$ is said to be (*everywhere*) *dense* if its closure, $\widetilde{S}$ coincides with $\mathcal{H}$. If the set $\mathcal{D}(T)$ is dense in $\mathcal{H}$, then T is said to be *densely defined.* If $\mathcal{D}(T) = \mathcal{H}$ then T is said to be defined on $\mathcal{H}$.

If S and T are two operators with domains such that $\mathcal{D}(S) \subset \mathcal{D}(T)$ and $Su = Tu$ for all $u \in \mathcal{D}(S)$, then T is said to be an *extension* of S and S to be a *restriction* of T. These relations can be expressed as $S \subset T$ or $T \supset S$. The operator T is said to be T-*convergent* if any sequence $u_n \in \mathcal{D}(T)$ is such that both $\{u_n\}$ and $\{Tu_n\}$ are Cauchy sequences and $u_n \to u$: we use the notation

$$u_n \xrightarrow{T} u$$

to indicate this. Then T is said to be *closed* if $u_n \xrightarrow{T} u$ implies $u \in \mathcal{D}(T)$ and $Tu = \lim Tu_n$.

B.1.3 Spectrum and resolvent of linear operators

For operators on *finite dimensional* vector spaces, the *eigenvalues* of the linear operator T are the (generally complex) numbers λ such that $\det(\lambda I - T) = 0$, where I is the identity operator. The (finite) set of all such λ is called the *spectrum* of the operator T. If the vector space has dimension n, then there can be at most n such eigenvalues. Thus $\lambda I - T$ has an inverse when λ is not in the spectrum of T. More generally, when T is an operator on the infinite dimensional Hilbert space $\mathcal{H}$, the number λ is said to be in the *resolvent set* $\rho(T)$ if $\lambda I - T$ has a bounded inverse. The operator $R_\lambda(T) = (\lambda I - T)^{-1}$ is called the *resolvent operator* at the point λ. If $\lambda \notin \rho(T)$, then λ is said to be in the *spectrum* $\sigma(T)$ of T.

B.1.4 Self-adjoint operators

We associate with each linear operator T an *adjoint operator* T^* defined by the relation

$$(Tu, v) = (u, T^* v) \tag{B.1.1}$$

for every $u \in \mathcal{D}(T)$ and every $v \in \mathcal{D}(T^*)$ for which this relation makes sense. We say that T is *symmetric* if it is both densely defined and if

$$T^* \supset T. \tag{B.1.2}$$

A necessary and sufficient condition for T to be symmetric is that it be densely defined and that

$$(Tu, v) = (u, Tv)\ \forall\ u, v \in \mathcal{D}(T). \tag{B.1.3}$$

T is said to be *self-adjoint* if it is both symmetric and if also

$$T^* = T. \tag{B.1.4}$$

If $\mathcal{D}(T) = \mathcal{H}$ then (B.1.2) implies that $T^* = T$ as is the case in finite dimensional unitary spaces. The relation (B.1.3) shows that $(Tu, u) = (u, Tu) = (Tu, u)^*$ so that (Tu, u) must be *real.* If $(Tu, u) \geq 0$ then the symmetric operator T is said to be *non-negative.* According to (B.1.2), a symmetric operator T is always *closable* because T^* is closed and because its adjoint T^{**} is the closure $\widetilde{T}$ of T [3, p. 267]. The closure of a symmetric operator must itself be symmetric.

If T is symmetric and has a closure $\widetilde{T}$ which is self-adjoint then T is said to be *essentially self-adjoint.* If T is closed, then a subset $D \subset \mathcal{D}(T)$ is called a *core* for T if the closure of the restriction T_D of T to D is just T. Although there is a close connection between closed symmetric operators and self-adjoint operators, only the latter can be exponentiated to give the one-parameter unitary groups that generate dynamics in quantum mechanics. The distinction between the two classes is therefore important and we must be able to demonstrate that we can construct a suitable self-adjoint operator. This is not always easy. However, it is often possible to prove that a nonclosed symmetric operator T is *essentially self-adjoint.* Every such operator T has an unique self-adjoint closure $\widetilde{T} = T^{**}$ [2, Vol I, p.256]; this means that it is not necessary to define the whole domain of T but only a *core.* We can test for self-adjointness using the following:

Theorem B.1. *Let T be a symmetric operator on $\mathcal{H}$. Then the three statements*

1. *T is self-adjoint;*
2. *T is closed and $\mathcal{K}(T^* \pm i) = \{\emptyset\}$; and*
3. *$\mathcal{R}(T^* \pm i) = \mathcal{H}$*

are equivalent.

See [2, Vol I, p.256–7] for the proof. Note that if T is self-adjoint, and if there is a vector $\phi \in \mathcal{D}(T^*) = \mathcal{D}(T)$ such that $T^*\phi = i\phi$ then also $T\phi = i\phi$, and

$$-i(\phi, \phi) = (i\phi, \phi) = (T\phi, \phi) = (\phi, T^*\phi) = (\phi, T\phi) = i(\phi, \phi)$$

from which it follows that ϕ must be null. In the same way, we prove that the equation $T^*\phi = -i\phi$ has only the null solution.

Corollary B.2. *Let T be a symmetric operator on $\mathcal{H}$. Then the three statements*

1. *T is essentially self-adjoint;*
2. *$\mathcal{K}(T^* \pm i) = \{\emptyset\}$; and*
3. *$\mathcal{R}(T^* \pm i)$ are dense in $\mathcal{H}$*

are equivalent.

The discussion of self-adjoint extensions of symmetric operators is complicated by the fact that if such extensions exist, and in some cases there is no self-adjoint extension, there may be more than one. We first consider extensions of closed symmetric operators, as an operator and its closure have the same self-adjoint extensions.

Theorem B.3. *Let T be a closed symmetric operator on $\mathcal{H}$. Then*

1. *a) dim$[\mathcal{K}(\lambda I - T^*)]$ is constant throughout the open upper half λ-plane;*
   b) dim$[\mathcal{K}(\lambda I - T^*)]$ is constant throughout the open lower half λ-plane.
2. *The spectrum of T is one of the following:*
   a) The closed upper half λ-plane; or
   b) the closed lower half λ-plane; or
   c) the entire λ-plane; or
   d) a subset of the real axis.
3. *T is self-adjoint if and only if 2(d) holds.*
4. *T is self-adjoint if and only if the dimensions in both 1(a) and 1(b) are zero.*

For the proof see [2, Vol II, p. 136]. There is an important corollary: a closed symmetric operator that is semi-bounded, that is to say, an operator that satisfies an inequality of the form

$$(T\phi, \phi) \geq -M\|\phi\|^2,$$

has $dim[\mathcal{K}(\lambda I - T^*)]$ constant when $\lambda \in \mathcal{C}\backslash[-M, \infty]$. Part 3 of Theorem B.3 shows that a closed symmetric operator is self-adjoint if it has at least one real number in its resolvent set. For $\rho(T)$ is an open set; if it contains a point on the real axis it must contain points in both upper and lower half-planes.

The *deficiency subspaces* $\mathcal{K}_+(T)$ and $\mathcal{K}_-(T)$ of the symmetric operator T are defined by

$$\mathcal{K}_\pm(T) = \mathcal{K}(i \mp T^*) = \mathcal{R}(i \pm T)^\perp$$

where the superscript $\perp$ indicates the orthogonal subspace. The dimensions of these subspaces

$$n_\pm(T) = \dim\,[\mathcal{K}_\pm(T)]$$

are called the *deficiency indices* of T. T is self-adjoint if and only if $n_+ = n_- = 0$; T has self-adjoint extensions if and only if $n_+ = n_-$; there is a 1-1 correspondence between self-adjoint extensions of T and unitary maps from $\mathcal{K}_+(T)$ onto $\mathcal{K}_-(T)$; and finally if either $n_+ = 0$ or $n_- = 0$ and $n_+ \neq n_-$ then T has no non-trivial symmetric extensions [2, Vol II, §X.1]. In this last case, one says that T is *maximal symmetric.*

The theory of *deficiency indices* is due to Von Neumann, who has given a simple test for a symmetric operator to have self-adjoint extensions in terms of conjugation mappings. An antilinear map, $C : \mathcal{H} \to \mathcal{H}$, for which

$$C(\alpha u + \beta v) = \alpha^* u + \beta^* v$$

is called a *conjugation* if it is norm-preserving and if $C^2 = I$.

Theorem B.4. *Let T be a symmetric operator and suppose there exists a conjugation $C : \mathcal{D}(T) \to \mathcal{D}(T)$ with $TC = CT$. Then T has equal deficiency indices and therefore it has self-adjoint extensions.*

B.1.5 Observables and self-adjoint operators

Let $\mathcal{M}$ be a linear manifold in $\mathcal{H}$ and define the *projection operator* E such that

$$\begin{aligned} E\phi &= \phi \text{ when } \phi \in \mathcal{M} \\ &= 0 \text{ otherwise.} \end{aligned}$$

Then $E^2 = E = E^\dagger$, and the elements of $\mathcal{M}$ are projected on to the set

$$\{E\phi \mid \phi \in \mathcal{H}\}.$$

Projection operators E_1 and E_2 are *orthogonal* if their associated subspaces $\mathcal{M}_1$ and $\mathcal{M}_2$ are orthogonal; then either $E_1 E_2 = 0$ or $E_2 E_1 = 0$. When $\mathcal{M}_1 \subseteq \mathcal{M}_2$, then we write $E_1 \leq E_2$ or $E_2 \geq E_1$: either $E_1 E_2 = E_1$ or $E_2 E_1 = E_1$.

More generally, if E_1 projects onto $\mathcal{M}_1$ and E_2 projects onto $\mathcal{M}_2$ then

- If $E_1 E_2 = E_2 E_1$, then $E_1 E_2$ is projection operator onto the subspace $\mathcal{M}_1 \cap \mathcal{M}_2$, consisting of all vectors that are in both $\mathcal{M}_1$ and $\mathcal{M}_2$.
- If E_1 and E_2 are orthogonal, then $E_1 + E_2$ is the projection operator onto the subspace $\mathcal{M}_1 \oplus \mathcal{M}_2$, consisting of all vectors that are either in $\mathcal{M}_1$ or in $\mathcal{M}_2$.
- If $E_1 \leq E_2$, then $E_2 - E_1$ is a projection operator onto the orthogonal complement of $\mathcal{M}_1$ in $\mathcal{M}_2$, consisting of all vectors in $\mathcal{M}_2$ that are orthogonal to $\mathcal{M}_1$.

The set of projection operators $\{E_x \mid x \in \mathbb{R}\}$ is said to be a *spectral family* under the following conditions:

1. If $x \leq y$, then $E_x \leq E_y$ or $E_x E_y = E_x = E_y E_x$.
2. If $\varepsilon > 0$, then $E_{x+\varepsilon}\psi \to E_x\psi$ as $\varepsilon \to 0$ for any $\psi \in \mathcal{H}$ and any $x \in \mathbb{R}$.
3. $E_x\psi \to 0$ as $x \to -\infty$ and $E_x\psi \to \psi$ as $x \to +\infty$ for any $\psi \in \mathcal{H}$.

The notion of a spectral family enables us to formulate

Theorem B.5. *[1, p.42] Each self-adjoint operator A has a unique spectral family of projection operators E_x such that for every pair of vectors ϕ and ψ in $\mathcal{D}(A)$,*

$$(\phi, A\psi) = \int_{-\infty}^{+\infty} x \, d(\phi, E_x\psi). \tag{B.1.5}$$

When this is the case, we write

$$A = \int_{-\infty}^{+\infty} x \, dE_x. \tag{B.1.6}$$

This is the *spectral resolution* or *spectral decomposition* of the operator A. As x increases, we may encounter points at which E_x is constant, has a jump discontinuity, or increases smoothly. We define the *spectrum* of A to consist of all points on which E_x increases. A point x is therefore not in the spectrum if it lies in the interior of an interval in which E_x is constant. The set of *jump discontinuities* of E_x is called the *point spectrum*, and when x has a neighbourhood in which E_x increases continously, then x is said to be in the *continuous spectrum.*

Suppose now that we have a physical system in a state described by the vector $\psi \in \mathcal{H}$. Let A be a self-adjoint linear operator on $\mathcal{H}$ corresponding to some observable dynamical quantity (perhaps a component of the position or momentum, or the energy of the system) described quantum mechanically by a probability distribution. Then we require that A will be associated with a finite *expectation value*, denoted $\langle A\rangle$, with the following obvious properties:

- $\langle A\rangle$ is real if A is self-adjoint.
- $\langle A\rangle \geq 0$ if A is self-adjoint and positive.
- For any complex numbers α, β, $\langle \alpha A + \beta B\rangle = \alpha\langle A\rangle + \beta\langle B\rangle$.
- If I is the *identity* operator on $\mathcal{H}$, then $\langle I\rangle = 1$.
- If $E_1, E_2, \dots$ is a set of mutually orthogonal projectors, then

$$\left\langle \sum E_k \right\rangle = \sum \langle\, E_k\, \rangle.$$

The linear character of expectation can thus be extended to infinite sums.

For any state ψ in $\mathcal{H}$, a definition of the expectation value of a self-adjoint operator A that is in accordance with these conditions is

$$\langle A\rangle = (\psi, A\psi) = \int_{-\infty}^{+\infty} x\, d(\psi, E_x\psi) \tag{B.1.7}$$

Because $E_x^2 = E_x = E_x^\dagger$, we have that

$$(\psi, E_x\psi) = \|E_x\psi\|^2$$

so that

$$\langle A\rangle = (\psi, A\psi) = (A\psi, \psi) = \int_{-\infty}^{+\infty} x\, d\|E_x\psi\|^2. \tag{B.1.8}$$

By setting A equal to the identity, we see that

$$\int_{-\infty}^{+\infty} d\|E_x\psi\|^2 = 1,$$

which suggests the conventional probability interpretation

$$P[A \leq x] = \|E_x\psi\|^2 = (\psi, E_x\psi) = \langle E_x\rangle \tag{B.1.9}$$

Thus $\langle E_x\rangle$ is a strictly nondecreasing function of x and, as required of a probability distribution,

$$\lim_{x \to -\infty} \langle E_x \rangle = 0, \qquad \lim_{x \to +\infty} \langle E_x \rangle = 1.$$

If $f(x)$ is any function of x for which $\int_{-\infty}^{+\infty} f(x)\, d\|E_x\psi\|^2$ exists for every ψ in $\mathcal{D}(A)$, we can define

$$\langle f(A) \rangle = \int_{-\infty}^{+\infty} f(x)\, d\|E_x\psi\|^2 \tag{B.1.10}$$

Of course, there are functions $f(x)$ for which the integral does not exist and which therefore are inadmissible: for an example, suppose $\|E_x\psi\|^2 = 0$ for $x \le 0$ and $x > 1$, and $\|E_x\psi\|^2 = x$ for $0 < x \le 1$. Then $f(A) = A^{-1/2}$ is certainly admissible, but $f(A) = A^{-1}$ is not.

B.1.6 Commuting operators

Suppose that the self-adjoint operators A and B have a common domain and that

$$[A, B]\psi := (AB - BA)\psi = 0$$

for every ψ in the domain and we can assign a meaning to the operator identity $[A, B] = 0$. Then A and B have common eigenvectors, $A\psi = \alpha\psi$, $B\psi = \beta\psi$ and, if ψ is normalized, the expectation values are $\langle A \rangle = \alpha$, $\langle B \rangle = \beta$. We say that these observables are *simultaneously measurable.* A set of self-adjoint mutually commuting operators $A_1, \dots A_n$ is said to form a *complete commuting set* if and only if every bounded linear operator that commutes with every member of the set can be expressed as a function of $A_1, \dots A_n$. The states of the system can then be characterized by the eigenvalues of such a set so that (using Dirac notation)

$$A_i \,|\alpha_1, \dots, \alpha_i, \dots\rangle = \alpha_i \,|\alpha_1, \dots, \alpha_i, \dots\rangle, \qquad i = 1, \dots, n.$$

The choice of convenient sets of complete commuting operators to describe the states of atomic and molecular systems is an important component of the theory. A related notion is that of an *irreducible set* of self-adjoint operators, for which the only *bounded* operators that commute with every member of the set are multiples of the identity.

B.1.7 Unitary and anti-unitary operators

Let H be a self-adjoint operator with a well-defined spectral decomposition and define the *unitary* operator $U_t = \exp(-iHt)$, where t is a real parameter. Clearly

$$U_0 = I, \quad U_t U_{t'} = U_{t+t'} = U_{t'} U_t, \quad U_t^* = U_t^{-1} \tag{B.1.11}$$

so that the collection of operators $\{U_t\}$ generate a unitary group depending continuously on t. The converse is known as Stone's theorem.

Theorem B.6 (Stone [1]). *Let U_t satisfy (B.1.11). Then there exists a unique self-adjoint operator H such that $U_t = \exp(-iHt)$ for all t, and every bounded operator commuting with U_t also commutes with H.*

If

$$U_t = I - itH + o(t), \quad \lim_{t \to 0} \frac{U_t - I}{t} = -iH \tag{B.1.12}$$

where H is independent of t, then by (B.1.11)

$$i\frac{dU_t}{dt} = i \lim_{t' \to 0} \frac{U_{t+t'} - U_t}{t'} = iU_t \lim_{t' \to 0} \frac{U_{t'} - U_t}{t'} = HU_t \tag{B.1.13}$$

We see that U_t satisfies the differential equation (B.1.13) and that H can be interpeted as the *infinitesimal generator* of the group. Moreover, if $t = T/n$, and we consider the result of applying $U_{t/n}$ n times, we have

$$\left(U_{\frac{T}{n}}\right)^n = \left(I - iH\frac{T}{n} + o(t)\right)^n \to \exp(-iHT)$$

as $n \to \infty$ in accordance with our original definition of U_t. Thus if t is time, and H is the Hamiltonian operator of a dynamical system, then H is the infinitesimal generator of the group of time translations. If ψ represents the state of the system at time $t = 0$, then $U_t\psi$ represents the state of the system at time t. Similarly, if x is a space coordinate, then the linear momentum $p_x = -i\partial/\partial x$ is the infinitesimal generator of translation in the x-direction[1]. More generally, the momentum vector $\boldsymbol{p}$ generates translations in the direction of $\boldsymbol{p}$. The angular momentum operator $\boldsymbol{j} = \boldsymbol{r} \times \boldsymbol{p}$ generates rotations; an anticlockwise rotation through an angle θ about the unit vector $\boldsymbol{n}$ is generated by $\exp(-i\theta\boldsymbol{n} \cdot j)$.

Anti-linear operators involve complex conjugation:

$$A(a\psi + b\phi) = a^* A\psi + b^* A\phi \tag{B.1.14}$$

where a, b are complex numbers. A has an inverse, A^{-1} if and only if for each ψ there is one and only one ϕ such that $\psi = A\phi$. If A^{-1} exists and $\|A\psi\| = \|\psi\|$ for all ψ, then A is *anti-unitary*. If A is anti-linear and anti-unitary then $(A\phi, A\psi) = (\phi, \psi)^*$ for all vectors.

The probability for transition from a normalized state ψ to another normalized state ϕ is defined to be $|(\phi, \psi)|^2$. We say that a bijective map $\mathcal{O}$ from $\mathcal{H}$ into itself is a *symmetry transformation* if, for all ψ, ϕ, $(\mathcal{O}\psi, \mathcal{O}\phi) = (\phi, \psi)$ so that transition probabilities remain unchanged. Clearly $\mathcal{O}$ can be either *unitary* (like U_t) or *anti-unitary* (like A above).

Theorem B.7 (Wigner-Bargmann). *Every symmetry transformation in $\mathcal{H}$ is of the form $\mathcal{O}$, where $\mathcal{O}$ is either unitary or anti-unitary.*

[1] This is just Taylor's theorem of elementary calculus in disguise!

B.2 Lie groups and Lie algebras

General references [5], [6], [7]

B.2.1 Lie groups

Let $\{g(\boldsymbol{a})\}$ be a set of functions of the r-vector

$$\boldsymbol{a} = (a_1, \ldots, a_r)^t$$

where the a_i are a set of continuous real parameters. To be regarded as the elements of some group, we need to be able to exhibit a functional relation, $\boldsymbol{c}(\boldsymbol{a}, \boldsymbol{b})$, such that

$$g(\boldsymbol{a}).g(\boldsymbol{b}) = g(\boldsymbol{c})$$

for every pair of parameter vectors $\boldsymbol{a}, \boldsymbol{b}$ in the set. It is conventional to write the group identity $I = g(\mathbf{0})$. A simple 1-parameter example is the group of time translations, Appendix B.1.7.

Although such a group has an infinite number of elements, most of its properties can be deduced from its *infinitesimal operators*. Suppose that we have a representation of the group $\mathcal{G}$ in a space $\mathcal{H}$ denoted by $T(\boldsymbol{a})$, with $I = T(\mathbf{0})$. Suppose that $\boldsymbol{a}$ is small, so that

$$T(\boldsymbol{a}) = I + \sum_{i=1}^{r} a_i X_i + o(\boldsymbol{a}); \tag{B.2.1}$$

the operators X_i, $i = 1, \ldots, r$, are the infinitesimal operators of the group, for which the following must hold [5, 7]:

Theorem B.8. *A If two representations of the group $\mathcal{G}$ have the same infinitesimal operators, they are the same representation.*

B The infinitesimal operators defined in (B.2.1) must satisfy commutation relations

$$[X_i, X_j] = c_{ij}^k X_k \tag{B.2.2}$$

in which the structure constants c_{ij}^k are the same for all representations T of $\mathcal{G}$.

C Any set of operators X_i on a space $\mathcal{H}$ will be the infinitesimal operators of a representation of $\mathcal{G}$ if they satisfy the commutation relations (B.2.2).

Such a continuous group is known as an r-parameter *Lie group*, which we can think of as being generated by the operators X_i, the *infinitesimal generators* of the group.

The structure constants are not completely independent because of their definition (B.2.2) in terms of a commutator, leading to

$$c_{ij}^k = -c_{ji}^k \tag{B.2.3}$$

so that all infinitesimal generators self-commute,

$$[X_i, X_i] = 0.$$

Also any three such operators, X, Y, Z, satisfy the *Jacobi identity*

$$[X,[Y,Z]] + [Y,[Z,X]] + [Z,[X,Y]] = 0 \tag{B.2.4}$$

which imposes the set of constraints

$$c_{ij}^k c_{lk}^m + c_{jl}^k c_{ik}^m + c_{li}^k c_{jk}^m = 0.$$

An example, familiar from elementary quantum mechanics textbooks, is the group SO(3) of orthogonal transformations on three-dimensional Euclidean space. The infinitesimal generators of rotations about the coordinate axes satisfy

$$[X_1, X_2] = X_3, \quad [X_2, X_3] = X_1, \quad [X_3, X_1] = X_2, \tag{B.2.5}$$

so that the structure constants are

$$c_{12}^3 = c_{23}^1 = c_{31}^2 = 1 \tag{B.2.6}$$

The operators of the quantum theory of angular momentum are related to the X_k by $J_k = -iX_k$, so that we recover the standard angular momentum commutation relations

$$[J_1, J_2] = iJ_3, \quad [J_2, J_3] = iJ_1, \quad [J_3, J_1] = iJ_2, \tag{B.2.7}$$

B.2.2 Lie algebras

The infinitesimal operators that generate a Lie group can also be regarded as generating a *Lie algebra*. Formally, a Lie algebra is an r-dimensional vector space over a field $\mathbb{F}$ in which we associate with each pair of vectors X, Y a third vector

$$Z = [X, Y]$$

where the *commutator* $[\cdot,\cdot]$ is anti-symmetric and linear

$$[X,Y] + [Y,X] = 0, \quad [\alpha X + \beta Y, Z] = \alpha[X,Z] + \beta[Y,Z], \quad \alpha, \beta \in \mathbb{F}$$

along with the Jacobi identity (B.2.4). We say that the Lie algebra is *real* if $\mathbb{F} = \mathbb{R}$ and *complex* if $\mathbb{F} = \mathbb{C}$. It is usual to designate the Lie group with uppercase letters (SO(3) for example) and the corresponding Lie algebra with lowercase letters (so(3)). The infinitesimal generators are not determined uniquely by the commutation relations. Any nonsingular linear transformation on the X_i will satisfy the same commutation relations with transformed structure constants. The freedom conferred by this property has been exploited to define canonical forms (see §B.2.4) which can be used to classify the various algebraic structures.

We say that B is a *subalgebra* of A ($B \subset A$) if B is a linear subspace of A; if X, Y are both in B, then so is $[X, Y]$. An *Abelian* subalgebra is one for which all commutators $[X, Y]$ vanish. There is always a trivial Abelian subalgebra as $[X_i, X_i] = 0$. B is said to be an *ideal* or *invariant subalgebra* of A if it is a linear subspace of A, and if, for every $X \in B$ and $Y \in A$, the commutator $[X, Y] \in B$. If A contains vectors not in B, then B is a *proper ideal.* There are at least two *improper ideals*, namely $\{0\}$ (whose only element is the identity), and A itself.

A *simple* Lie algebra has no proper ideals. A *semisimple* Lie algebra no Abelian ideals save the trivial $\{0\}$. A simple algebra is clearly semisimple, but there is no reason to expect the converse to hold.

Theorem B.9. *A Lie algebra A is semisimple if and only if it can be written as a direct sum*

$$A = A_1 \oplus A_2 \oplus \ldots \oplus A_n.$$

Each of the A_i is both an ideal in A as well as being a simple Lie algebra in its own right.

The *metric tensor* (or *Killing form*) is defined by

$$g_{ij} = g_{ji} = c_{ik}^{l} c_{jl}^{k} \tag{B.2.8}$$

where summation over the indices k, l is understood.

Theorem B.10 (Cartan). *A Lie algebra is semi-simple if and only if its metric tensor has a non-vanishing determinant:* $\det[g_{ij}] \neq 0$.

The structure constants for so(3) were given in (B.2.6), from which it follows that $g_{ij} = -2\delta_{ij}$; it follows that $\det[g_{ij}]$ is non-singular so that so(3) is semi-simple. On the other hand, the group E_2 of linear transformations in the plane,

$$\boldsymbol{x}' = R\boldsymbol{x} + \boldsymbol{a}, \tag{B.2.9}$$

where R is a plane rotation and $\boldsymbol{a}$ is a translation, is not semi-simple. We can think of (B.2.9) as a linear transformation in a three-dimensional space with matrix

$$S = \begin{pmatrix} \cos\theta & -\sin\theta & a_1 \\ \sin\theta & \cos\theta & a_2 \\ 0 & 0 & 1 \end{pmatrix},$$

with infinitesimal generators

$$X_0 = \begin{pmatrix} 0 & -1 & 0 \\ 1 & 0 & 0 \\ 0 & 0 & 0 \end{pmatrix}, \quad X_1 = \begin{pmatrix} 0 & 0 & 1 \\ 0 & 0 & 0 \\ 0 & 0 & 0 \end{pmatrix}, \quad X_2 = \begin{pmatrix} 0 & 0 & 0 \\ 0 & 0 & 1 \\ 0 & 0 & 0 \end{pmatrix}.$$

There are two non-zero structure constants, $c^2_{01} = c^1_{20} = 1$ and the metric tensor has only one non-zero element $g_{00} = -2$. Thus E_2 does not have a semi-simple Lie algebra, but does contain a nontrivial Abelian subalgebra T_2 of translations in the plane generated by X_1, X_2 along with the one-dimensional algebra of plane rotations, so(2), generated by X_0.

A Lie group or Lie algebra is asid to be *compact* if its parameter space can be covered by a finite number of bounded parameter domains. A necessary and sufficient condition for compactness is that the Killing form (B.2.8) should be *negative definite*.

B.2.3 Representations of Lie groups and Lie algebras

The concept of a group representation is central to most of the applications of group theory in physics, and the notion of an irreducible representation is exploited in many parts of this book.

Suppose, then, we associate with each infinitesimal operator, X_i of a Lie algebra, an image D_i such that

$$[D_i, D_j] = c^k_{ij} D_k$$

where c^k_{ij} are the structure constants of the associated Lie algebra. If there is a one-to-one correspondence $X_i \leftrightarrow D_i$, then we say that we have a *faithful representation*. A representation on a linear vector space of N dimensions is said to be N-dimensional and the D_i will consist of $N \times N$ matrices. Two representations D_i and D'_i are said to be *equivalent* if there exists a constant non-singular linear transformation X such that

$$D'_i = X D_i X^{1}$$

for every X_i in the algebra. We then write $D_i \sim D'_i$.

Let R be the representation space of the representation D of the group G. A subspace $S \subset R$ is an *invariant subspace* if D maps S into itself. We say that R is *irreducible* if it has no proper invariant subspace; otherwise it is *reducible* and D_i can be partitioned in the form

$$D_i = \begin{pmatrix} D^S_i & B_i \\ 0 & D^{S^c}_i \end{pmatrix},$$

where $R = S \oplus S^c$. When B_i vanishes for every $A \in G$ then D_i is block diagonal and

$$D_i = D^S_i \oplus D^{S^c}_i .$$

The representation is said to be *fully reducible* if R can be expressed as the direct sum of irreducible invariant subspaces

$$R = S_1 \oplus \ldots \oplus S_k$$

and also

$$D = D^{(1)} \oplus \ldots \oplus D^{(k)}$$

in which each each representation $D^{(j)}$ acts on the invariant subspace S_j. A representation which is reducible but is not fully reducible in this way is said to be *indecomposable.*

B.2.4 The Cartan-Weyl classification

Cartan and Weyl have shown that any complex simple Lie algebra A can be put into a standard form that facilitates the classification of all such algebras. Suppose that A has r infinitesimal operators divisible into two classes. The *Cartan subalgebra*, $\Pi = \{H_i \,|\, i = 1, 2, \ldots, l\}$, contains a mutually commuting set of operators

$$[H_i, H_j] = 0, \quad i, j = 1, 2, \ldots, l \tag{B.2.10}$$

whilst the remainder $\{E_\alpha\}$ are simultaneously eigenvectors of the H_i:

$$[H_i, E_\alpha] = \alpha_i E_\alpha, \quad i = 1, \ldots, l \tag{B.2.11}$$

$$[E_\alpha, E_\beta] = c_{\alpha\beta}^{\alpha+\beta} E_{\alpha+\beta}, \quad \alpha + \beta \neq 0 \tag{B.2.12}$$

$$[E_\alpha, E_{-\alpha}] = \sum_{i=1}^{l} \alpha_i H_i \tag{B.2.13}$$

Thus the array $(\alpha_1, \ldots, \alpha_l)$ can be considered as a vector $\boldsymbol{\alpha}$ (the *root*) in an l-dimensional *weight space*. The collection of roots can be displayed in a *root figure* which can be used to classify the algebra.

The elements of the Cartan subalgebra possess simultaneous eigenvectors $|\Lambda\rangle := |\lambda_1, \ldots, \lambda_l\rangle$ such that

$$H_i|\Lambda\rangle = \lambda_i|\Lambda\rangle, \quad i = 1, \ldots, l$$

The eigenvalues $\lambda_1, \ldots, \lambda_l$ can be viewed as the components of a *weight vector* $|\Lambda\rangle$ of *weight* Λ in an l-dimensional *weight space* Δ associated with a representation D. Because

$$\begin{aligned} H_i E_\alpha|\Lambda\rangle &= \{[H_i, E_\alpha] + E_\alpha H_i\} |\Lambda\rangle \\ &= (\alpha_i + \lambda_i) E_\alpha|\Lambda\rangle \end{aligned}$$

by (B.2.11), we see that $E_\alpha|\Lambda\rangle$ is an eigenvector of weight $(\Lambda + \boldsymbol{\alpha})$ provided $(\Lambda + \boldsymbol{\alpha}) \in \Delta$; otherwise $E_\alpha|\Lambda\rangle = 0$. We can think of E_α as a *raising* operator on the eigenvalues of H_i if $\alpha > 0$ and a *lowering* operator if $\alpha < 0$. Also if H_i is taken to be Hermitian, as is usually the case, then $[H_i, E_\alpha]^\dagger = -[H_i, E_\alpha^\dagger]$ so that, from (B.2.11), we see that if E_α is a raising operator then $E_\alpha^\dagger$ is a lowering operator. The eigenvalues α generally appear in pairs of opposite sign.

These properties show that the representation space decomposes into a direct sum of weight subspaces R^Λ, each characterized by its weight Λ. We say that Λ is a *positive weight* if its first nonvanishing element is positive. We can then order the weights so that $\Lambda > \Lambda'$ if $\Lambda - \Lambda'$ is positive: the highest weight of a representation will then be higher than every other weight $\Lambda' \in \Delta$.

Cartan's classification identifies four general classes of semi-simple Lie algenras, A_l, B_l, C_l, D_l and five exceptions labelled G_2, F_4, E_6, E_7, E_8. For further details see [7] or [8, Chapter 3].

B.2.5 Casimir operators

A semisimple Lie algebra has, by Cartan's theorem, a nonsingular metric tensor $g = [g_{ij}]$. We denote the elements of the matrix inverse g^{-1} by g^{ij}, so that

$$g_{ik}g^{kj} = \delta_i^j.$$

We define the *Casimir operator* as the operator

$$C = g^{ij} X_i X_j. \tag{B.2.14}$$

It can be shown that the commutator $[C, X_i]$ vanishes for every element of the Lie algebra. The Casimir operator of the symmetry group of a quantum mechanical system is therefore an important tool for labelling the eigenstates of the system. Racah [9] has suggested a generalization. First define

$$X^i = g^{ij} X_j.$$

Then write

$$C_n = c^{j_2}_{i_1 j_1} c^{j_3}_{i+2 j_2} \cdots c^{j_1}_{i_n j_n} X^{i_1} \dots X^{i_n}, \tag{B.2.15}$$

each of which also commutes with all the elements of the Lie algebra. However, the generalized Casimir operators are not necessarily independent of C.

As a simple example, consider so(3) again, for which we found $g_{ij} = -2\delta_{ij}$ from which $g^{ij} = -\frac{1}{2}\delta_{ij}$, giving $C = -\frac{1}{2}(X_1^2 + X_2^2 + X_3^2)$ or, since $J_k = -iX_k$,

$$C = \frac{1}{2}(J_1^2 + J_2^2 + J_3^2) = \frac{1}{2}\boldsymbol{J}^2 \tag{B.2.16}$$

so that we recover the well-known result that the square of the total angular momentum operator commutes with all of its components: $[\boldsymbol{J}^2, J_i] = 0$ for all i.

This construction fails for Lie algebras that are not semisimple, although it may still be possible to construct invariants that commute with all elements of the Lie algebra.

B.2.6 Kronecker products of group representations

As well as being a mathematical tool for the classification and interpretation of many of the mathematical structures we meet in this book, the analysis of Kronecker products (direct products) of group representations provides us with a major computational tool.

The *Kronecker product of two vector spaces* U and V is defined as the linear space T spanned by all elements of the form $X \times Y$, where $X \in U$ and $Y \in V$, where $Z = X \times Y$ is linear in each argument X, Y. In the finite dimensional case, suppose that U, V have respective dimension $[U], [V]$. Then if $(X_1, \ldots, X_{[U]})$ is a basis for X and $(Y_1, \ldots, Y_{[V]})$ is a basis for Y, then the set $(X_i \times Y_j | i = 1, \ldots, [U], j = 1, \ldots, [V])$ is a basis for T. Similarly if $A : U \to U$ and $B : V \to V$ are linear transformations, with matrices having elements $a_i{}^j$ and $b_k{}^l$ with respect to the respective bases $(X_1, \ldots, X_{[U]})$ and $(Y_1, \ldots, Y_{[V]})$, then the linear transformation $C = A \times B$ relative to the Kronecker product basis has matrix elements $c_i{}^j{}_k{}^l = a_i{}^j b_k{}^l$. *Kronecker products of linear transformations* have the following properties:

1. $(A_1 A_2) \times (B_1 B_2) = (A_1 \times B_1)(A_2 \times B_2)$.
2. If A and B are invertible, then $(A \times B)^{-1} = A^{-1} \times B^{-1}$.
3. If I_U, I_V and I_T are the respective identities in U, V and T then $I_T = I_U \times I_V$.
4. If $U' \subset U$ is a subspace invariant under A and $V' \subset V$ is a subspace invariant under B then $U' \times V'$ is a subspace invariant under $C = A \times B$.

Kronecker products of group representations have an analogous definition. We use the notation (Λ) to label a unitary representation of a group G, so that $U_g(\Lambda)$ represents the group element g for every $g \in G$. In the Cartan-Weyl classification, we can take Λ to be the highest weight of a representation, and label the basis vectors of the representation by $|\Lambda\lambda\rangle$. where λ distinguishes vectors of the same multiplicity. An element $g \in G$ will have the matrix with elements

$$\mathcal{D}^{(\Lambda)}_{\lambda\lambda'}(g) = \langle \Lambda\lambda | U_g(\Lambda) | \Lambda\lambda' \rangle,$$

say and, in the Kronecker product $(\Lambda_1) \times (\Lambda_2)$ of two representations (Λ_1) and (Λ_2), it will be represented by the matrix

$$\mathcal{D}^{(\Lambda_1)}(g) \times \mathcal{D}^{(\Lambda_2)}(g).$$

This product matrix will, in general, be reducible. We recall that two representations (Λ_1) and (Λ_2) are equivalent, $(\Lambda_1) \sim (\Lambda_2)$ if their elements can be put into one-to-one correspondence by a change of basis. For Kronecker products, we have

1. If $(\Lambda_1) \sim (\Lambda_2)$ and $(\Lambda_1') \sim (\Lambda_2')$ then $(\Lambda_1) \times (\Lambda_1') \sim (\Lambda_2) \times (\Lambda_2')$.
2. $(\Lambda_1) \times (\Lambda_2) \sim (\Lambda_2) \times (\Lambda_1)$.
3. $[(\Lambda_1) \times (\Lambda_2)] \times (\Lambda_3) \sim (\Lambda_1) \times [(\Lambda_2) \times (\Lambda_3)]$.

4. $[(\Lambda_1) \oplus (\Lambda_2)] \times (\Lambda_3) \sim (\Lambda_1) \times (\Lambda_3) \oplus (\Lambda_2) \times (\Lambda_3)$.

It follows that a fully reducible product representation can be decomposed into a direct sum of irreducible representations

$$(\Lambda_1) \times (\Lambda_2) \sim \sum_{\oplus} \Gamma_{123}\,(\Lambda_3) \tag{B.2.17}$$

where Γ_{123} is the number of times the irreducible representation (Λ_3) appears in the decomposition. There will be a unitary mapping of the basis vectors $|\Lambda_1\lambda_1\rangle \times |\Lambda_2\lambda_2\rangle$ of the Kronecker product into vectors of the product basis, $|\Lambda_1\Lambda_2; a\Lambda_3\lambda_3\rangle$, of the form

$$|\Lambda_1\Lambda_2; a\Lambda_3\lambda_3\rangle = \sum_{\lambda_1\lambda_2} |\Lambda_1\lambda_1\rangle \times |\Lambda_2\lambda_2\rangle \, \langle \Lambda_1\lambda_1\Lambda_2\lambda_2|\Lambda_1\Lambda_2; a\Lambda_3\lambda_3\rangle, \tag{B.2.18}$$

with inverse

$$|\Lambda_1\lambda_1\rangle \times |\Lambda_2\lambda_2\rangle = \sum_{a\Lambda_3\lambda_3} |\Lambda_1\Lambda_2; a\Lambda_3\lambda_3\rangle \, \langle \Lambda_1\Lambda_2; a\Lambda_3\lambda_3|\Lambda_1\lambda_1\Lambda_2\lambda_2\rangle^*, \tag{B.2.19}$$

where the new label a serves to distinguish the different irreducible representations belonging to (Λ_3). We denote the *coupling coefficients* by

$$\langle \Lambda_1\lambda_1\Lambda_2\lambda_2|\Lambda_1\Lambda_2; a\Lambda_3\lambda_3\rangle,$$

often abbreviated to $\langle \lambda_1\lambda_2|a\Lambda_3\lambda_3\rangle$ when the meaning is clear, which are essential for practical computations. The unitary nature of the transformation leads directly to the relations

$$\sum_{\lambda_1\lambda_2} \langle a\Lambda_3\lambda_3|\lambda_1\lambda_2\rangle^* \, \langle \lambda_1\lambda_2|a'\Lambda_3'\lambda_3'\rangle = \delta_{aa'}\,\delta_{\Lambda_3\Lambda_3'}\,\delta_{\lambda_3\lambda_3'} \tag{B.2.20}$$

$$\sum_{a\Lambda_3\lambda_3} \langle \lambda_1\lambda_2|a\Lambda_3\lambda_3\rangle^* \langle a\Lambda_3\lambda_3|\lambda_1'\lambda_2'\rangle = \delta_{\lambda_1\lambda_1'}\,\delta_{\lambda_2\lambda_2'} \tag{B.2.21}$$

In general, (Λ_3) will reduce to the identity representation (0) if and only if $(\Lambda_1) \equiv (\Lambda_2)^*$ and $\lambda_1 = -\lambda_2$, where $(\Lambda)^*$ is the representation *contragredient*[2] to (Λ). Equation (B.2.21) then gives

$$\langle \Lambda\lambda\Lambda^* - \lambda|0\rangle^2 = [\Lambda]^{-1} \tag{B.2.22}$$

where $[\Lambda]$ is the dimension of the representation (Λ).

[2] The homomorphisms $A \to D(A)$ and $A \to D(A)^* = [D^\dagger(A)]^{-1}$ are said to define *contragredient* or *conjugate* representations.

B.2.7 Tensor operators and the Wigner-Eckart theorem

The $[\Lambda]$ linearly independent operators $T^{(\Lambda)}_{\lambda}$ are said to comprise a *tensor operator* under a group G, if the set transforms according to the representation (Λ) under the operations of the group. A tensor operator is said to be irreducible, reducible or equivalent to another tensor operator according as the representation (Λ) is irreducible, reducible or equivalent to another representation. In terms of the basis vectors $|\Lambda\lambda\rangle$, the defining relation is

$$U_g\, T^{(\Lambda)}_{\lambda}\, U_g^{-1} = \sum_{\lambda'} T^{(\Lambda)}_{\lambda'}\, \langle \Lambda\lambda' \,|\, U_g(\Lambda) \,|\, \Lambda\lambda\rangle. \tag{B.2.23}$$

For an infinitesimal representation

$$U_g = 1 + \sum_i \delta a^i\, X_i$$

the terms linear in δa^i give

$$\left[X_i,\, T^{(\Lambda)}_{\lambda}\right] = \sum_{\lambda'} T^{(\Lambda)}_{\lambda'}\, \langle \Lambda\lambda' \,|\, X_i \,|\, \Lambda\lambda\rangle \tag{B.2.24}$$

and

$$X_i\,|\Lambda\lambda\rangle = \sum_{\lambda'} |\,\Lambda\lambda'\rangle\langle\Lambda\lambda' \,|\, X_i \,|\, \Lambda\lambda\rangle. \tag{B.2.25}$$

Any operator acting on a linear vector space that can serve as the representation space for some representation (Λ) of G will be expressible as a linear combination of tensor operators under G. This emphasizes the importance of methods for constructing matrix elements of such tensor operators with respect to the basis vectors of an irreducible representation of G in practical calculations. For *compact* Lie groups, the principal result is the

Theorem B.11 (Wigner-Eckart).

$$\langle \Lambda_1\lambda_1 \,|\, T^{(\Lambda)}_{\lambda} \,|\, \Lambda_2\lambda_2\rangle = \sum_a \langle a\Lambda_1\lambda_1 \,|\, \lambda\lambda_2\rangle^{*}\, \langle a\Lambda_1 \| \boldsymbol{T}^{(\Lambda)} \| \Lambda_2\rangle \tag{B.2.26}$$

where $\langle a\Lambda_1 \| \boldsymbol{T}^{(\Lambda)} \| \Lambda_2\rangle$, *known as the reduced matrix element, is independent of the weights* λ_1 *and* λ_2 *of* (Λ_1) *and* (Λ_2) *respectively.*

In applications, the Wigner-Eckart theorem enables one to factor the symmetry properties of the tensor operator, expressed by the coupling coefficient, from other properties of the operator expressed by the reduced matrix element. In particular *all* tensor operators transforming under the same representation will satisfy (B.2.26) with the same coupling coefficients.

The implications of this theorem are best appreciated by examining its proof, for which we follow [10]. For a compact semi-simple Lie group, the mutually commuting operators H_i can be chosen real and simultaneously diagonal, so that $H_i^\dagger = H_i$. Similarly, we can choose the remaining operators so that $E_\alpha^\dagger = E_{-\alpha}$. Because the H_i are asociated with null roots, we can combine these choices by writing

$$X_\rho^\dagger = X_{-\rho}$$

for all the infinitesimal operators of the group. $\langle \Lambda_1\lambda_1 \,|\, T_\lambda^{(\Lambda)} \,|\, \Lambda_2\lambda_2\rangle$ transforms under group operations according to the Kronecker product representation $(\Lambda_1)^* \times (\Lambda) \times (\Lambda_2)$; if this is to have a well-defined invariant value, it must transform under the identity representation (0), so that

$$(\Lambda_1)^* \times (\Lambda) \times (\Lambda_2) \supset (0),$$

or

$$(\Lambda) \times (\Lambda_2) \supset (\Lambda_1)$$

or

$$(\Lambda_1)^* \times (\Lambda) \supset (\Lambda_2).$$

Picking a suitable vector, and using (B.2.24), (B.2.25) and (B.2.19), we find

$$\langle \Lambda_1\lambda_1 \,|\, T_\lambda^{(\Lambda)} \,|\, \Lambda_2\lambda_2\rangle = \sum_a \langle \Lambda_1\lambda_1 \,|\, T_\lambda^{(\Lambda)} \,|\, \Lambda_2; a\Lambda_1\lambda_1\rangle \, \langle a\Lambda_1\lambda_1 \,|\, \Lambda\lambda\Lambda_2\lambda_2\rangle.$$

Now the terms of the form $\langle \Lambda_1\lambda_1 \,|\, T_\lambda^{(\Lambda)} \,|\, \Lambda_2; a\Lambda_1\lambda_1\rangle$ can be thought of as some $\langle b\Lambda_1\lambda_1 \,|\, a\Lambda_1\lambda_1\rangle$. Such terms are independent of λ_1 and hence can be written formally using the reduced matrix element notation of (B.2.26). To prove this result, consider the result of operating on the identity

$$|a\Lambda_1\lambda_1\rangle = \sum_b |b\Lambda_1\lambda_1\rangle \, \langle b\Lambda_1\lambda_1 \,| \cdot |a\Lambda_1\lambda_1\rangle$$

with an arbitrary infinitesimal operator X_ρ. The result will be

$$|a\Lambda_1\lambda_1 + \mu\rangle = \sum_b |b\Lambda_1\lambda_1 + \mu\rangle \, \langle b\Lambda_1\lambda_1 \,|\, a\Lambda_1\lambda_1\rangle,$$

so that comparing these equations gives

$$\langle b\Lambda_1\lambda_1 \,|\, a\Lambda_1\lambda_1\rangle = \langle b\Lambda_1 + \mu\lambda_1 \,|\, a\Lambda_1\lambda_1 + \mu\rangle$$

for all non-zero μ, showing that this expression does not depend on the value of λ_1 This form of the Wigner-Eckart theorem is valid for finite or for compact continuous groups. We shall not have to consider non-compact groups, where its use needs some care. For our purposes, the most important case is rotation the group SO(3), Appendix B.3.

B.3 Quantum mechanical angular momentum theory

General references [8, Chapter 2], [11], [12], [13], [14], [15], [16], [17]

B.3.1 The rotation group

Let U be a 2×2 unitary matrix, $U^\dagger U = UU^\dagger = I_2$, where I_2 is the unit 2×2 matrix. U is unimodular when $\det U = 1$. If U_1 and U_2 are both unitary and unimodular, then so are U_1^{-1}, U_2^{-1}, and U_1U_2; the collection of all such matrices generates the group SU(2). Then if M is any 2×2 complex-valued matrix M

$$\det(UMU^\dagger) = \det M. \tag{B.3.1}$$

Let $\sigma_k,\ k = 1, 2, 3$, denote a Pauli matrix, defined in (2.2.3),

$$\sigma_1 := \begin{pmatrix} 0 & 1 \\ 1 & 0 \end{pmatrix}, \quad \sigma_2 := \begin{pmatrix} 0 & -i \\ i & 0 \end{pmatrix}, \quad \sigma_3 := \begin{pmatrix} 1 & 0 \\ 0 & -1 \end{pmatrix} \tag{B.3.2}$$

and let M be the traceless Hermitian matrix

$$M := x^k \sigma_k = \begin{pmatrix} x^3 & x^1 - ix_2 \\ x^1 + ix_2 & -x^3 \end{pmatrix}. \tag{B.3.3}$$

Then because of (B.3.1),

$$-\det M = (x^1)^2 + (x^2)^2 + (x^3)^2$$

is invariant under the operations of SU(2). If we interpret (x^1, x^2, x^3) as Cartesian coordinates in $\mathbb{R}^3$, then we see that $\det M$ is also an invariant for rotations in $\mathbb{R}^3$ and there exists a proper orthogonal transformation $R(U) : \mathbb{R}^3 \to \mathbb{R}^3$ such that

$$UMU^\dagger = R(U)\boldsymbol{x} \cdot \boldsymbol{\sigma}. \tag{B.3.4}$$

Let $\boldsymbol{n}$ be a unit vector in $\mathbb{R}^3$, and define

$$U_{\boldsymbol{n}}(\varphi) := \exp\{-i\boldsymbol{n} \cdot \boldsymbol{\sigma}\, \varphi/2\}. \tag{B.3.5}$$

In the simple case $\boldsymbol{n} = \boldsymbol{k} := (0, 0, 1)$, a unit vector along Ox^3, this reduces to

$$U_{\boldsymbol{k}}(\varphi) = \begin{pmatrix} \mathrm{e}^{-i\varphi/2} & 0 \\ 0 & \mathrm{e}^{+i\varphi/2} \end{pmatrix}$$

and then

$$R(U_{\boldsymbol{k}}(\varphi)) = \begin{pmatrix} \cos\varphi & -\sin\varphi & 0 \\ \sin\varphi & \cos\varphi & 0 \\ 0 & 0 & 1 \end{pmatrix}$$

which is the usual matrix for rotation through an angle φ about the Ox^3-axis, $\boldsymbol{k}$.

The rotations about a fixed direction $\boldsymbol{n}$ form a continuous 1-parameter subgroup of SU(2). Because

$$U_{\boldsymbol{k}}(2\pi) = -\begin{pmatrix} 1 & 0 \\ 0 & 1 \end{pmatrix}$$

whilst

$$R(U_{\boldsymbol{k}}(2\pi)) = \begin{pmatrix} 1 & 0 & 0 \\ 0 & 1 & 0 \\ 0 & 0 & 1 \end{pmatrix}$$

we see that the homomorphism SU(2) $\to$ SO(3) is double valued; SU(2) is therefore called the *universal covering group* of SO(3).

B.3.2 Abstract angular momentum

We have already seen that the abstract infinitesimal generators of SO(3) are the operators J_1, J_2, J_3, whose commutation relations are given in (B.2.7). The construction of the usual representation space is an example of the procedure of §B.2.4. $\boldsymbol{J}^2$ and J_3 (the conventional choice) may be taken as a complete commuting set of operators, so that we can generate an orthonormal basis

$$\{\, |jm\rangle \mid m = -j, -j+1, \ldots, j\}, \quad \langle jm'|jm\rangle = \delta_{m',m} \tag{B.3.6}$$

of their simultaneous eigenvectors

$$\boldsymbol{J}^2 \,|jm\rangle = j(j+1)\,|jm\rangle, \quad J_3\,|jm\rangle = m\,|jm\rangle \tag{B.3.7}$$

In terms of the operators

$$J_+ = J_1 + iJ_2, \quad J_- = J_1 - iJ_2,$$

the commutation relations (B.2.7) can be rewritten as

$$[J_3, J_+] = J_+, \quad [J_3, J_-] = -J_-, \quad [J_+, J_-] = 2J_3, \tag{B.3.8}$$

so that

$$\boldsymbol{J}^2 = J_1^2 + J_2^2 + J_3^2 = J_- J_+ + J_3(J_3 + 1).$$

$J_+, (J_-)$ are *stepping operators* connecting basis vectors of increasing (decreasing) weight:

$$J_+\,|jm\rangle = [(j-m)(j+m+1)]^{1/2}\,|jm+1\rangle, \tag{B.3.9}$$

$$J_-\,|jm\rangle = [(j+m)(j-m+1)]^{1/2}\,|jm-1\rangle \tag{B.3.10}$$

The highest weight vector is defined by

$$J_+\,|jj\rangle = 0, \quad J_3\,|jj\rangle = j\,|jj\rangle \tag{B.3.11}$$

from which the vectors of lower weight can be generated recursively so that

$$|jm\rangle = \left[\frac{(j+m)!}{(2j)!(j-m)!}\right]^{1/2} J_-{}^{j-m}|jj\rangle \tag{B.3.12}$$

This derivation makes it clear that the dimension of the representation is the non-negative integer $[j] = 2j+1$, so that in general j can only be a non-negative multiple of 1/2:

$$j = 0, \frac{1}{2}, 1, \frac{3}{2}, \ldots \quad [j] = 1, 2, 3, 4, \ldots$$

We denote the irreducible space spanned by this basis by $\mathcal{H}_j$.

B.3.3 Orbital angular momentum

When j is an integer, so that the representation has *odd* dimension, a representation can be found in terms of the eigenfunctions of the orbital angular momentum operator, $\boldsymbol{L} = -i\boldsymbol{x} \times \boldsymbol{\nabla}$ acting on the space of homogeneous polynomials in the Cartesian components x_1, x_2, x_3 of the position vector x. Its components satisfy the standard angular momentum commutation relations (B.3.8). We shall quote formulae in both Cartesian coordinates and in spherical polars

$$\begin{gathered}\boldsymbol{x} = (r\,\sin\theta\,\cos\varphi, r\,\sin\theta\,\sin\varphi, r\,\cos\theta) \\ 0 \le r < \infty, \quad 0 \le \varphi < 2\pi, \quad 0 \le \theta < \pi.\end{gathered} \tag{B.3.13}$$

We first note that

$$\boldsymbol{L}^2 = L_-L_+ + L_3(L_3+1) = -r^2\nabla^2 + (\boldsymbol{x}\cdot\boldsymbol{\nabla})^2 + (\boldsymbol{x}\cdot\boldsymbol{\nabla}). \tag{B.3.14}$$

The operators $\boldsymbol{L}^2$ and L_3 form a complete set of commuting Hermitian operators on this space, with eigenfunctions that are homogeneous polynomial solutions $\mathcal{Y}_{lm}(\boldsymbol{x})$ of Laplace's equation with the properties

$$\begin{aligned}\nabla^2\,\mathcal{Y}_{lm}(\boldsymbol{x}) &= 0, \\ \boldsymbol{x}\cdot\boldsymbol{\nabla}\,\mathcal{Y}_{lm}(\boldsymbol{x}) &= l\,\mathcal{Y}_{lm}(\boldsymbol{x}), \\ \mathcal{Y}_{lm}(\lambda\boldsymbol{x}) &= \lambda^l\,\mathcal{Y}_{lm}(\boldsymbol{x}).\end{aligned} \tag{B.3.15}$$

Equations (B.3.7), (B.3.9) and (B.3.10) become

$$\begin{aligned}L_3\mathcal{Y}_{lm}(\boldsymbol{x}) &= m\mathcal{Y}_{lm}(\boldsymbol{x}), \\ L_\pm\mathcal{Y}_{lm}(\boldsymbol{x}) &= [(l\mp m)(l\pm m+1)]^{1/2}\,\mathcal{Y}_{l,m\pm1}(\boldsymbol{x})\end{aligned} \tag{B.3.16}$$

and equation (B.3.15) yields

$$\boldsymbol{L}^2\,\mathcal{Y}_{lm}(\boldsymbol{x}) = l(l+1)\,\mathcal{Y}_{lm}(\boldsymbol{x})$$

as expected from (B.3.7).

The eigenfunctions are most conveniently expressed as homogeneous polynomials of degree l in the combinations $-x_1 - ix_2$, $x_1 - ix_2$ and x_3. The highest weight eigenfunction with $m = l$, for which $L_+\mathcal{Y}_{ll}(\boldsymbol{x}) = 0$, is given by

$$\mathcal{Y}_{ll}(\boldsymbol{x}) = \frac{1}{2^l l!}\left[\frac{(2l+1)!}{4\pi}\right]^{1/2}(-x_1 - ix_2)^l \tag{B.3.17}$$

from which those of lower weight can be generated as in (B.3.12),

$$\mathcal{Y}_{lm}(\boldsymbol{x}) = \left[\frac{(l+m)!}{(2l)!(l-m)!}\right]^{1/2} L_-{}^{l-m}\mathcal{Y}_{ll}(\boldsymbol{x}).$$

The explicit formula is

$$\mathcal{Y}_{lm}(\boldsymbol{x}) = \left[\frac{(2l+1)}{4\pi}(l+m)!(l-m)!\right]^{1/2} \times \sum_k \frac{(-x_1 - ix_2)^{k+m}(x_1 - ix_2)^k x_3^{l-m-2k}}{2^{2k+m}(k+m)!k!(l-m-2k)!} \tag{B.3.18}$$

where l is a non-negative integer, $m = -l, -l+1, \ldots, l$ and k runs over all non-negative integer values for which $0 \le k \le (l+m)/2$. Complex conjugation of (B.3.18) gives the important result

$$\mathcal{Y}^*_{lm}(\boldsymbol{x}) = (-1)^m \mathcal{Y}_{l,-m}(\boldsymbol{x}). \tag{B.3.19}$$

The same formulae are often required in spherical polar coordinates (B.3.13), for which we can write

$$\mathcal{Y}_{lm}(\boldsymbol{x}) = r^l \mathcal{Y}_{lm}\left(\frac{\boldsymbol{x}}{r}\right) \tag{B.3.20}$$

Because $\boldsymbol{x}/r$ is a function of the polar angles θ, φ only, we can define the *spherical harmonics*

$$Y_{lm}(\theta,\varphi) := \mathcal{Y}_{lm}\left(\frac{\boldsymbol{x}}{r}\right) \tag{B.3.21}$$

so that for $m \ge 0$, from (B.3.18),

$$Y_{lm}(\theta,\varphi) = (-1)^m \left[\frac{(2l+1)}{4\pi}(l+m)!(l-m)!\right]^{1/2} e^{im\varphi} \times \sum_k (-1)^k \frac{(\sin\theta)^{2k+m}(\cos\theta)^{l-m-2k}}{2^{2k+m}(k+m)!k!(l-m-2k)!}. \tag{B.3.22}$$

For $m < 0$, equation (B.3.19) gives

$$Y_{l,-m}(\theta,\varphi) = (-1)^m Y^*_{lm}(\theta,\varphi), \quad m > 0.$$

Equation (B.3.22) can be conveniently expressed in terms of associated Legendre polynomials, $P_l^m(\cos\theta)$, by

$$Y_{lm}(\theta,\varphi) = (-1)^m \left[\frac{(2l+1)(l-m)!}{4\pi(l+m)!}\right]^{1/2} P_l^m(\cos\theta)\, e^{im\varphi}, \tag{B.3.23}$$

where

$$P_l^m(x) = (1-x^2)^{m/2}\frac{d^m}{dx^m}P_l(x), \quad m \geq 0, \tag{B.3.24}$$

and $P_l(x)$ is the Legendre polynomial of degree l,

$$P_l(x) = \frac{1}{2^l l!}\frac{d^l}{dx^l}(x^2-1)^l, \quad l \geq 0.$$

The spherical harmonics are *orthonormal* on the unit sphere (radius $r = 1$) because

$$\int_0^{2\pi} d\varphi \int_0^{\pi} d\theta\, \sin\theta\, Y^*_{l'm'}(\theta,\varphi)\, Y_{lm}(\theta,\varphi) = \delta_{l'l}\delta_{m'm}. \tag{B.3.25}$$

For convenience, we note that the Cartesian components of $\boldsymbol{L}$ in spherical polar coordinates are

$$\begin{aligned} L_1 &= i\cos\varphi\cot\theta\frac{\partial}{\partial\varphi} + i\sin\varphi\frac{\partial}{\partial\theta} \\ L_2 &= i\sin\varphi\cot\theta\frac{\partial}{\partial\varphi} - i\cos\varphi\frac{\partial}{\partial\theta} \\ L_3 &= -i\frac{\partial}{\partial\varphi} \end{aligned} \tag{B.3.26}$$

so that

$$L_\pm = e^{\pm i\varphi}\left(\pm\frac{\partial}{\partial\theta} + i\cot\theta\frac{\partial}{\partial\varphi}\right)$$

and

$$\boldsymbol{L}^2 = -\frac{1}{\sin\theta}\frac{\partial}{\partial\theta}\left(\sin\theta\frac{\partial}{\partial\theta}\right) - \frac{1}{\sin^2\theta}\frac{\partial^2}{\partial\varphi^2}$$

B.3.4 Representation functions

The action of elements of SO(3) on solid harmonics is given by

$$(O_R\mathcal{Y}_{lm})(\boldsymbol{x}) = \mathcal{Y}_{lm}(R^{-1}\boldsymbol{x}) = \sum_{m'}\mathcal{D}^l_{m'm}(R)\,\mathcal{Y}_{lm'}(\boldsymbol{x}) \tag{B.3.27}$$

whilst the equivalent result for elements of SU(2) is

$$(T_U \mathcal{Y}_{lm})(\boldsymbol{x}) = \mathcal{Y}_{lm}(U^\dagger X U) = \sum_{m'} D^l_{m'm}(U)\, \mathcal{Y}_{lm'}(\boldsymbol{x}) \tag{B.3.28}$$

with

$$\mathcal{D}^l_{m'm}(R) = D^l_{m'm}(U(R)).$$

If $\boldsymbol{n}$ is a unit vector in $\mathbb{R}^3$, $|\boldsymbol{n}| = 1$, then a rotation through the angle ϕ about $\boldsymbol{n}$ is given by (B.3.5)

$$\begin{aligned} U_{\boldsymbol{n}}(\varphi) &= \exp\{-i\boldsymbol{n}\cdot\boldsymbol{\sigma}\,\varphi/2\} \\ &= \sigma_0 \cos(\varphi/2) - i\boldsymbol{n}\cdot\boldsymbol{\sigma}\,\sin(\varphi/2) \\ &= \begin{pmatrix} \cos(\varphi/2) - in_3 \sin(\varphi/2) & (-in_1 - n_2)\sin(\varphi/2) \\ (-in_1 + n_2)\sin(\varphi/2) & \cos(\varphi/2) + in_3 \sin(\varphi 2) \end{pmatrix}, \end{aligned} \tag{B.3.29}$$

where σ_0 denotes the 2×2 identity matrix and $0 \le \phi \le 2\pi$. On the representation space of (B.3.6) we have

$$T_U\,|jm\rangle = \sum_{m'} D^j_{m',m}(U)\,|jm'\rangle, \tag{B.3.30}$$

where the rotation functions are defined on the elements u_{ij} of the matrix U of (B.3.29) by the symmetric polynomial of degree $2j$

$$\begin{aligned} D^j_{m',m}(U) &= [(j+m)!(j-m)!(j+m')!(j-m')!]^{1/2} \\ &\times \sum_{abcd} \frac{u_{11}^a\, u_{12}^b\, u_{21}^c\, u_{22}^d}{a!\, b!\, c!\, d!} \end{aligned} \tag{B.3.31}$$

where the non-negative summation indices a, b, c, d satisfy the constraints

$$a + b = j + m', \quad c + d = j - m', \quad a + c = j + m, \quad b + d = j - m.$$

Equation (B.3.31) in this form is useful for displaying the symmetries of the representation functions. The constraints can be used to express any three of the summation indices in terms of the fourth, which is more convenient for numerical evaluation.

In order to use this abstract formula, we need a convenient parametrization of the group. Equation (B.3.29) shows that each U in SU(2) can be parametrized in terms of four real numbers $\alpha_0, \alpha_1, \alpha_2, \alpha_3$ such that

$$U(\alpha_0, \boldsymbol{\alpha}) = \alpha_0\sigma_0 - i\boldsymbol{\alpha}\cdot\boldsymbol{\sigma} = \begin{pmatrix} \alpha_0 - i\alpha_3 & -i\alpha_1 - \alpha_2 \\ -i\alpha_1 + \alpha_2 & \alpha_0 + i\alpha_3 \end{pmatrix} \tag{B.3.32}$$

where

$$\alpha_0^2 + \alpha_1^2 + \alpha_2^2 + \alpha_3^2 = 1.$$

These Euler-Rodrigues parameters therefore correspond to points on the unit sphere S^3 in $\mathbb{R}^4$. The corresponding element $R(\alpha_0, \boldsymbol{\alpha})$ of SO(3) is

$$R(\alpha_0, \boldsymbol{\alpha}) = \tag{B.3.33}$$
$$\begin{pmatrix} \alpha_0^2 + \alpha_1^2 - \alpha_2^2 - \alpha_3^2 & 2(\alpha_1\alpha_2 - \alpha_0\alpha_3) & 2(\alpha_1\alpha_3 + \alpha_0\alpha_2) \\ 2(\alpha_1\alpha_2 + \alpha_0\alpha_3) & \alpha_0^2 + \alpha_2^2 - \alpha_3^2 - \alpha_1^2 & 2(\alpha_2\alpha_3 - \alpha_0\alpha_1) \\ 2(\alpha_1\alpha_3 - \alpha_0\alpha_2) & 2(\alpha_2\alpha_3 + \alpha_0\alpha_1) & \alpha_0^2 + \alpha_3^2 - \alpha_1^2 - \alpha_2^2 \end{pmatrix}$$

Clearly

$$R(\alpha_0, -\boldsymbol{\alpha}) = R(\alpha_0, \boldsymbol{\alpha})$$

so that we need only half of S^3, say that portion with $\alpha_0 > 0$, to cover the parameter domain of SO(3). Rotations can be viewed either as *active* or as *passive*. For active rotations, we keep the same reference frame but map each vector $\boldsymbol{x} \in \mathbb{R}^3$ into a new vector $\boldsymbol{x}' = R\boldsymbol{x}$ or its equivalent Cartan form $X' = UXU^\dagger$. For passive rotations, we keep the vectors fixed, but move the reference frame, so that for a right orthonormal triad of vectors, $\boldsymbol{e}_1, \boldsymbol{e}_2, \boldsymbol{e}_3$,

$$\boldsymbol{e}'_j = \sum_i R_{ij}\boldsymbol{e}_i, \quad R_{ij} = \boldsymbol{e}_i \cdot \boldsymbol{e}'_j.$$

The coordinates (x'_1, x'_2, x'_3) of a point P in the new frame are therefore related to the original coordinates (x_1, x_2, x_3) in the unrotated frame by $\boldsymbol{x}' = R^t\boldsymbol{x}$, or

$$x_1\boldsymbol{e}_1 + x_2\boldsymbol{e}_2 + x_3\boldsymbol{e}_3 = x'_1\boldsymbol{e}'_1 + x'_2\boldsymbol{e}'_2 + x'_3\boldsymbol{e}'_3$$

for passive rotations. A convenient method of parametrization uses Euler angles. This involves three successive steps

A rotation about the axis $\boldsymbol{e}_3 = (0, 0, 1)$ through angle γ.
A rotation about the new axis $\boldsymbol{e}_2 = (0, 1, 0)$ through angle β.
A rotation about the new axis $\boldsymbol{e}_3$ through angle α.

For SU(2), this is effected by the operator

$$\begin{aligned} U(\alpha\beta\gamma) &= \mathrm{e}^{-i\alpha\sigma_3/2}\,\mathrm{e}^{-i\beta\sigma_2/2}\,\mathrm{e}^{-i\gamma\sigma_3/2} \\ &= \begin{pmatrix} \mathrm{e}^{-i\alpha/2}\cos\left(\frac{1}{2}\beta\right)\mathrm{e}^{-i\gamma/2} & -\mathrm{e}^{-i\alpha/2}\sin\left(\frac{1}{2}\beta\right)\mathrm{e}^{i\gamma/2} \\ \mathrm{e}^{i\alpha/2}\sin\left(\frac{1}{2}\beta\right)\mathrm{e}^{-i\gamma/2} & \mathrm{e}^{i\alpha/2}\cos\left(\frac{1}{2}\beta\right)\mathrm{e}^{i\gamma/2} \end{pmatrix} \end{aligned} \tag{B.3.34}$$

where either $0 \le \alpha < 2\pi, 0 \le \beta < \pi, 0 \le \gamma < 2\pi$ or $0 \le \alpha < 2\pi, 2\pi \le \beta < 3\pi, 0 \le \gamma < 2\pi$. Also

$$U(\alpha, \beta + 2\pi, \gamma) = -U(\alpha\beta\gamma).$$

In SO(3), the equivalent is

$$\begin{aligned} R(\alpha\beta\gamma) &= \begin{pmatrix} \cos\alpha & -\sin\alpha & 0 \\ \sin\alpha & \cos\alpha & 0 \\ 0 & 0 & 1 \end{pmatrix} \\ &\times \begin{pmatrix} \cos\beta & 0 & \sin\beta \\ 0 & 1 & 0 \\ -\sin\beta & 0 & \cos\beta \end{pmatrix} \begin{pmatrix} \cos\gamma & -\sin\gamma & 0 \\ \sin\gamma & \cos\gamma & 0 \\ 0 & 0 & 1 \end{pmatrix} \end{aligned} \tag{B.3.35}$$

which reduces to

$$\begin{pmatrix} \cos\alpha\cos\beta\cos\gamma-\sin\alpha\sin\gamma & -\cos\alpha\cos\beta\cos\gamma-\sin\alpha\cos\gamma & \cos\alpha\sin\beta \\ \sin\alpha\cos\beta\cos\gamma+\cos\alpha\sin\gamma & -\sin\alpha\cos\beta\cos\gamma+\cos\alpha\cos\gamma & \sin\alpha\sin\beta \\ -\sin\beta\cos\gamma & \sin\beta\sin\gamma & \cos\beta \end{pmatrix}$$

with $0 \le \alpha < 2\pi, \quad 0 \le \beta \le \pi, \quad 0 \le \gamma < 2\pi$.

The multiplication rule for representation functions (B.3.31) of two operators $U, U' \in \mathrm{SU}(2)$ is

$$D^j(U)D^j(U') = D^j(UU'), \quad (D^j(U))^\dagger = (D^j(U))^{-1} = D^j(U^\dagger)),$$

and this provides the most convenient route to write down formulae for the general representation functions (B.3.35) in terms of the Euler angles from (B.3.32) and (B.3.31). We note first that if U is characterized by the Euler-Rodrigues parameters $(\alpha_0, \boldsymbol{\alpha})$ and U' by $(\alpha_0', \boldsymbol{\alpha}')$, then $U'' = UU'$ is characterized by $(\alpha_0'', \boldsymbol{\alpha}'')$, where

$$\alpha_0'' = \alpha_0'\alpha_0 - \boldsymbol{\alpha}'\cdot\boldsymbol{\alpha}, \quad \boldsymbol{\alpha}'' = \alpha_0'\boldsymbol{\alpha} + \alpha_0\boldsymbol{\alpha}' + \boldsymbol{\alpha}'\times\boldsymbol{\alpha}.$$

For a rotation through an angle φ about the unit vector $\boldsymbol{n}$, we have (B.3.29)

$$\alpha_0 = \cos\frac{1}{2}\varphi, \quad \boldsymbol{\alpha} = \boldsymbol{n}\sin\frac{1}{2}\varphi,$$

from which we obtain

$$D^j_{m'm}(\alpha\beta\gamma) = \mathrm{e}^{-im'\alpha}\, d^j_{m'm}(\beta)\, \mathrm{e}^{-im\gamma} \tag{B.3.36}$$

with

$$\begin{aligned} d^j_{m',m}(\beta) &= \left\langle jm' \left| \mathrm{e}^{-i\beta J_2} \right| jm \right\rangle \\ &= [(j+m)!(j-m)!(j+m')!(j-m')!]^{1/2} \\ &\quad \times \sum_s \frac{(-1)^{m'-m+s}\left(\cos\frac{1}{2}\beta\right)^{2j+m-m'-2s}\left(\sin\frac{1}{2}\beta\right)^{m'-m+2s}}{(j+m-s)!\,s!\,(m'-m+s)!\,(j-m'-s)!}. \end{aligned} \tag{B.3.37}$$

We note the symmetry relations

$$\begin{aligned} d^j_{m',m}(\beta) &= (-1)^{m'-m} d^j_{-m',-m}(\beta) \\ &= (-1)^{m'-m} d^j_{m,m'}(\beta) \\ &= d^j_{m,m'}(-\beta). \end{aligned} \tag{B.3.38}$$

When j takes an integer value, l, the representation functions for $m' = 0$ are proportional to spherical harmonics,

$$\begin{aligned} d^l_{m,0}(\beta) &= (-1)^m \left[\frac{(l-m)!}{(l+m)!}\right]^{1/2} P_l^m(\cos\beta) \\ &= \left[\frac{(l-m)!}{(l+m)!}\right]^{1/2} P_l^{-m}(\cos\beta), \end{aligned} \tag{B.3.39}$$

so that

$$Y_{lm}(\beta\alpha) = \left(\frac{2l+1}{4\pi}\right)^{1/2} d^l_{m,0}(\beta)$$
$$= \left(\frac{2l+1}{4\pi}\right)^{1/2} D^{l*}_{m0}(\alpha\beta\gamma). \tag{B.3.40}$$

For further properties see [8, Chapter 3].

B.3.5 Kronecker products of irreducible representations

The Clebsch-Gordon series (B.2.17) for SO(3) is simply reducible, so that

$$D^{j_1} \times D^{j_2} = \sum_{\oplus} \{j_1 j_2 j_3\}\, D^{j_3} \tag{B.3.41}$$

where j_3 can take the values

$$|j_1 - j_2|, |j_1 - J_2| + 1, \ldots j_1 + j_2, \tag{B.3.42}$$

and

$$\{j_1 j_2 j_3\} = \begin{cases} +1 & \text{if } j_3 \text{ satisfies (B.3.42)}, \\ 0 & \text{otherwise.} \end{cases} \tag{B.3.43}$$

The coupling coefficients, also referred to as *Clebsch-Gordan coefficients* or *vector addition coefficients* may be written either as

$$\langle j_1 m_1 j_2 m_2 | j_3 m_3 \rangle \qquad \text{or} \qquad C^{j_1\ j_2\ j_3}_{m_1 m_2 m_3}$$

so that as described in §B.2.6, the expression

$$I = \sum_{m_1 m_2 m_3} C^{j_1\ j_2\ j_3}_{m_1 m_2 m_3} \langle j_3 m_3 | \left(|j_1 m_1\rangle \otimes |j_2 m_2\rangle \right)$$

is a rotational invariant. Equivalently, we can write the coupled kets as

$$|j_3 m_3\rangle = \sum_{m_1 m_2} |j_1 m_1\rangle |j_2 m_2\rangle \, \langle j_1 m_1 j_2 m_2 | j_3 m_3 \rangle \tag{B.3.44}$$

along with the inverse relation

$$|j_1 m_1\rangle |j_2 m_2\rangle = \sum_{j_3 m_3} |j_3 m_3\rangle \langle j_3 m_3 \,|\, j_1 m_1 j_2 m_2 \rangle, \tag{B.3.45}$$

with corresponding equations for the bras. It is sometimes convenient to denote the right-hand side of (B.3.44) by $|(j_1 j_2) j_3 m_3\rangle$. These notations are used freely throughout this book.

To obtain expressions for the coupling coefficients, choose a polynomial basis for $D^{j_i}, i = 1, 2, 3$ of the form

$$|j_i m_i\rangle = \frac{u_i^{j_i+m_i} v_i^{j_i-m_i}}{[(j_i+m_i)!(j_i-m_i)!]^{1/2}} \tag{B.3.46}$$

on which the infinitesimal operators of SO(3) take the form

$$(J_+)_i = u_i \frac{\partial}{\partial v_i}, \quad (J_-)_i = v_i \frac{\partial}{\partial u_i}, \quad (J_3)_i = \frac{1}{2}\left(u_i \frac{\partial}{\partial u_i} - v_i \frac{\partial}{\partial v_i}\right) \tag{B.3.47}$$

For the contragredient representation $D^{j_3 *}$, the basis vectors are

$$\langle j_3 m_3| = (-1)^{j_3-m_3} \frac{u_3^{j_3+m_3} v_3^{j_3-m_3}}{[(j_3+m_3)!(j_3-m_3)!]^{1/2}} \tag{B.3.48}$$

and inserting these expressions into the invariant I gives an explicit algebraic expression from which we can deduce the Clebsch-Gordon coefficients. The easiest way to do this is to use the observation that determinants like

$$\delta_1 = u_2 v_3 - u_3 v_2$$

are invariant under SU(2). Thus I must be expressible as a product of the form

$$I = \delta_1^{k_1} \delta_2^{k_2} \delta_3^{k_3}$$

and by comparing the two equivalent expressions, we find

$$\begin{aligned} C^{j_1\ j_2\ j_3}_{m_1 m_2 m_3} &= \delta_{m_1+m_2, m_3} \\ &\times \left[\frac{(2j_3+1)(j_1+j_2-j_3)!(j_1-j_2+j_3)!(-j_1+j_2+j_3)!}{(j_1+j_2+j_3+1)!}\right]^{1/2} \\ &\times [(j_1+m_1)!(j_1-m_1)!(j_2+m_2)!(j_2-m_2)!(j_3+m_3)!(j_3-m_3)!]^{1/2} \\ &\times \sum_r \left\{(-1)^r \frac{1}{r!(j_1-m_1-r)!(j_2+m_2-r)!(j_3-j_2+m_1+r)!} \right. \\ &\qquad \left. \times \frac{1}{(j_3-j_1-m_2+r)!(j_1+j_2-j_3-r)!}\right\} \end{aligned} \tag{B.3.49}$$

We shall usually give results in terms of the more symmetric $3jm$ coefficients, defined by

$$\begin{pmatrix} j_1 & j_2 & j_3 \\ m_1 & m_2 & -m_3 \end{pmatrix} = \frac{(-1)^{j_1-j_2+m_3}}{\sqrt{2j_3+1}} C^{j_1\ j_2\ j_3}_{m_1\, m_2\, m_3} \tag{B.3.50}$$

The most important symmetries are

$$\begin{pmatrix} j_1 & j_2 & j_3 \\ -m_1 & -m_2 & -m_3 \end{pmatrix} = (-1)^{j_1+j_2+j_3} \begin{pmatrix} j_1 & j_2 & j_3 \\ m_1 & m_2 & m_3 \end{pmatrix} \tag{B.3.51}$$

and

$$\begin{pmatrix} j_1 & j_2 & j_3 \\ m_1 & m_2 & m_3 \end{pmatrix} = \begin{pmatrix} j_2 & j_3 & j_1 \\ m_2 & m_3 & m_1 \end{pmatrix} = (-1)^{j_1+j_2+j_3} \begin{pmatrix} j_2 & j_1 & j_3 \\ m_2 & m_1 & m_3 \end{pmatrix}. \tag{B.3.52}$$

In all cases

$$m_1 + m_2 + m_3 = 0.$$

Because the transformation effected using Clebsch-Gordon coefficients or 3-j symbols is unitary, they satisfy the following conditions:

$$\begin{aligned} \sum_{m_1 m_2} (2j_3+1) \begin{pmatrix} j_1 & j_2 & j_3 \\ m_1 & m_2 & m_3 \end{pmatrix} \begin{pmatrix} j_1 & j_2 & j_3' \\ m_1 & m_2 & m_3' \end{pmatrix} &= \delta_{j_3 j_3'} \delta_{m_3 m_3'}, \\ \sum_{j_3 m_3} (2j_3+1) \begin{pmatrix} j_1 & j_2 & j_3 \\ m_1 & m_2 & m_3 \end{pmatrix} \begin{pmatrix} j_1 & j_2 & j_3 \\ m_1' & m_2' & m_3 \end{pmatrix} &= \delta_{m_1 m_1'} \delta_{m_2 m_2'}. \end{aligned} \tag{B.3.53}$$

B.3.6 Coupling of three or more angular momenta

The Clebsch-Gordan coefficients and the more symmetrical $3jm$-symbols arise from decomposing the direct product of *two* irreducible representations into a direct sum of irreducible representations in the Clebsch-Gordan series (B.3.41). When we reduce a direct product of three or more irreducible representations, we have two choices: we can first combine D^{j_1} and D^{j_2}, according to (B.3.41), form the direct product of each term with D^{j_3} and perform another Clebsch-Gordan reduction, or we can first combine D^{j_2} and D^{j_3} and then reduce the product of the direct sum with D^{j_1}. Whilst the output spaces are equivalent, the individual states are different, and are said to be related by a process of *recoupling*. The coefficients describing this linear transformation are called *recoupling coefficients*; in the case of three angular momenta they are the Racah or $6j$-coefficients. In general, such recoupling coefficients can be written in terms of multiples of three angular momenta. For example, the transformation from a scheme of LS-coupling of the orbital and spin angular momenta of two particles to the equivalent jj-coupling representation can be expressed as a $9j$-symbol connecting the states $|((l_1 l_2)L(s_1 s_2)S)JM\rangle$ with $|((l_1 s_1)j_1(l_2 s_2)j_2)JM\rangle$. The result is actually independent of the projection M, and is therefore independent of the choice of coordinate axes used to define the states.

The $3nj$-symbols are most conveniently described in terms of angular momentum diagrams of Appendix B.3.10. Below we list the properties and major relations of the $3j$, $6j$, and $9j$ symbols which are needed throughout this book. A $3nj$-symbol has $2n$ vertices, with exactly 3 lines starting or finishing at each vertex.

The *Wigner covariant notation* [13, pp. 295–6] for $3jm$-symbols exploits the fact that factors $(-1)^{j-m}$ in association with a projection $-m$ always appear in valid summation formulae. This is connected with the phase convention for representing bras in (B.3.73). Such combinations are called *covariant* and are indicated by interchanging the j and m symbols in a column of a $3jm$ symbol. The original j, m pairs are said, in contrast, to be *contravariant.*

B.3.7 The $3j$-symbol

The Einstein summation convention has been used below to sum over *repeated* pairs of contravariant and covariant m quantum numbers as, for example, in (B.3.55) below, where the implied summation is over all variables m_1, m_2 and m_3.

Graphical representation

$$\{j_1 j_2 j_3\} = -$$

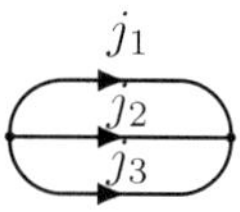

$$+ \tag{B.3.54}$$

Algebraic definition

$$\{j_1 j_2 j_3\} = \begin{pmatrix} j_1 & j_2 & j_3 \\ m_1 & m_2 & m_3 \end{pmatrix} \begin{pmatrix} m_1 & m_2 & m_3 \\ j_1 & j_2 & j_3 \end{pmatrix}. \tag{B.3.55}$$

Algebraic expression

$$\{j_1 j_2 j_3\} = \begin{cases} 1 \text{ if } j_3 = j_1 + j_2, j_1 + j_2 - 1, \ldots, |j_1 - j_2|, \\ 0 \text{ otherwise} \end{cases} \tag{B.3.56}$$

A triple abc for which $\{abc\} = 1$ is often said to satisfy the *triangle condition.*

Sum rules

$$\sum_j [j]\{j_1 j_2 j\} = [j_1 j_2] \tag{B.3.57}$$

$$\sum_j [j](-1)^j \{j_1 j_1 j\} = (-1)^{2j_1} [j_1] \tag{B.3.58}$$

B.3.8 The $6j$-symbol

For a fuller treatment see [8, §2.10].

Graphical representation

$$\left\{ \begin{matrix} j_1 & j_2 & j_3 \\ l_1 & l_2 & l_3 \end{matrix} \right\} = \tag{B.3.59}$$

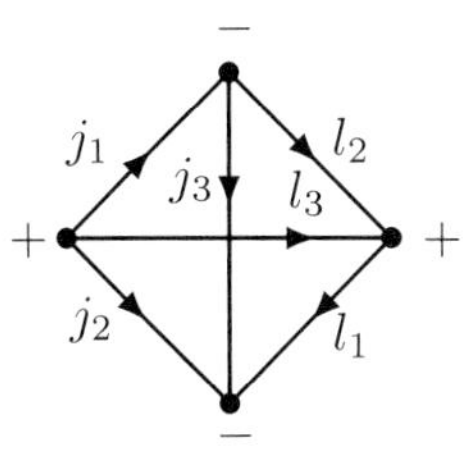

Algebraic definition (in terms of $3jm$-symbols)

$$\left\{ \begin{matrix} j_1 & j_2 & j_3 \\ l_1 & l_2 & l_3 \end{matrix} \right\} = \begin{pmatrix} j_1 & j_2 & m_3 \\ m_1 & m_2 & j_3 \end{pmatrix} \begin{pmatrix} m_1 & l_2 & l_3 \\ j_1 & n_2 & n_3 \end{pmatrix} \times \begin{pmatrix} l_1 & n_2 & j_3 \\ n_1 & l_2 & m_3 \end{pmatrix} \begin{pmatrix} n_1 & m_2 & n_3 \\ l_1 & j_2 & l_3 \end{pmatrix} \tag{B.3.60}$$

Algebraic expression

$$\left\{ \begin{matrix} a & b & e \\ d & c & f \end{matrix} \right\} = \Delta(abe)\,\Delta(cde)\,\Delta(acf)\,\Delta(bdf) \times \sum_k \frac{(-1)^k (k+1)!}{(k-a-b-e)!\,(k-c-d-e)!\,(k-a-c-f)!} \times \frac{1}{(k-b-d-f)!\,(a+b+c+d-k)!} \times \frac{1}{(a+d+e+f-k)!\,(b+c+e+f-k)!} \tag{B.3.61}$$

where the *triangle coefficient* $\Delta(abc)$ is defined for every triple abc whose sum is an integer and satisfies the *triangle condition* (B.3.56) by

$$\Delta(abc) = \left[\frac{(a+b-c)!\,(c+a-b)!\,(b+c-a)!}{(a+b+c+1)!} \right]^{1/2} \tag{B.3.62}$$

Symmetries: There are 144 symmetry relations among the $6j$ symbols. 24 of these were discovered by Racah: they consist of all $6j$ symbols which can be obtained (a) by permuting its columns:

$$\left\{ \begin{matrix} a & b & e \\ d & c & f \end{matrix} \right\} = \left\{ \begin{matrix} b & a & e \\ c & d & f \end{matrix} \right\} \tag{B.3.63}$$

and (b) by interchanging a pair of elements in the top row with the pair of elements below them:

$$\left\{ \begin{matrix} a & b & e \\ d & c & f \end{matrix} \right\} = \left\{ \begin{matrix} a & c & f \\ d & b & e \end{matrix} \right\} \tag{B.3.64}$$

For the full set of relations see, for example, [8, §2.10.5].

Racah sum rule

$$\sum_f (-1)^{e+g+f}[f]\begin{Bmatrix} a & b & e \\ d & c & f \end{Bmatrix}\begin{Bmatrix} a & d & g \\ b & c & f \end{Bmatrix} = \begin{Bmatrix} a & b & e \\ c & d & g \end{Bmatrix} \tag{B.3.65}$$

Biedenharn-Elliott identity

$$\begin{Bmatrix} a' & a & c' \\ b & b' & e \end{Bmatrix}\begin{Bmatrix} a' & e & b' \\ d & d' & c \end{Bmatrix} = \sum_f (-1)^{\phi}[f]\begin{Bmatrix} a & b & e \\ d & c & f \end{Bmatrix}\begin{Bmatrix} c' & b & b' \\ d & d' & f \end{Bmatrix}\begin{Bmatrix} a' & a & c' \\ f & d' & c \end{Bmatrix} \tag{B.3.66}$$

where $\phi = a + a' + b + b' + c + c' + d + d' + e + f$.
Relation to Racah W-coefficients

$$\begin{Bmatrix} a & b & e \\ d & c & f \end{Bmatrix} = (-1)^{a+b+c+d} W(abcd; ef) \tag{B.3.67}$$

B.3.9 The 9j-symbols

Graphical representation: 9j-symbol (of the first kind)

$$\begin{Bmatrix} j_1 & j_2 & j_3 \\ l_1 & l_2 & l_3 \\ k_1 & k_2 & k_3 \end{Bmatrix} = \tag{B.3.68}$$

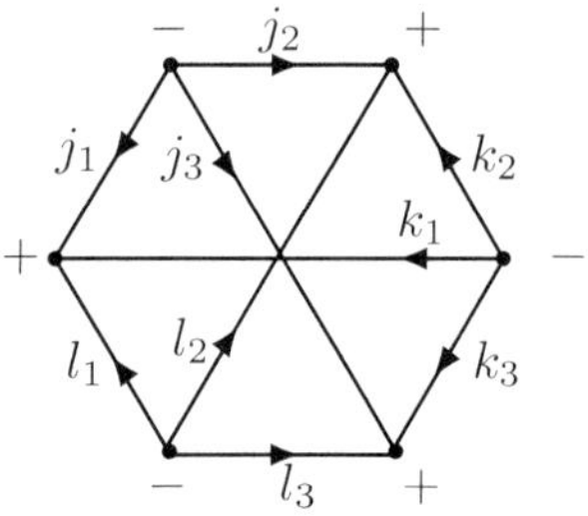

Algebraic definition (in terms of 3jm-symbols)

$$\begin{Bmatrix} j_1 & j_2 & j_3 \\ l_1 & l_2 & l_3 \\ k_1 & k_2 & k_3 \end{Bmatrix} = \begin{pmatrix} j_1 & j_2 & j_3 \\ m_1 & m_2 & m_3 \end{pmatrix}\begin{pmatrix} m_2 & n_2 & x_2 \\ j_2 & l_2 & k_2 \end{pmatrix} \times \begin{pmatrix} k_1 & k_2 & k_3 \\ x_1 & x_2 & x_3 \end{pmatrix}\begin{pmatrix} x_3 & m_3 & n_3 \\ k_3 & j_3 & l_3 \end{pmatrix} \times \begin{pmatrix} l_3 & l_1 & l_2 \\ n_3 & n_1 & n_2 \end{pmatrix}\begin{pmatrix} n_1 & x_1 & m_1 \\ l_1 & k_1 & j_1 \end{pmatrix} \tag{B.3.69}$$

The $9j$ symbol is nonvanishing only if the arguments of each row and column satisfy the triangle condition (B.3.56).

Symmetries: There are 72 symmetries arising from the 72 symmetries of the $3jm$-coefficients in (B.3.69).

A The $9j$-coefficient (B.3.68) is *invariant* under *even* permutation of its rows, *even* permutation of its columns and under interchange of rows and columns (matrix transposition).

B The $9j$-coefficient is multiplied by a factor $(-1)^{\phi}$ under odd permutation of its rows or columns, where $\phi = j_1+j_2+j_3+l_1+l_2+l_3+k_1+k_2+k_3$.

$9j$ *coefficients with one argument zero* reduce to $6j$ symbols.

$$\begin{Bmatrix} a & b & e \\ c & d & e \\ f & f & 0 \end{Bmatrix} = \frac{(-1)^{b+c+e+f}}{[e,f]^{1/2}} \begin{Bmatrix} a & b & e \\ d & c & f \end{Bmatrix} \tag{B.3.70}$$

Graphical representation: $9j$-symbol (of the second kind)

$$\left\{ \begin{matrix} j_1 & j_2 & j_3 \\ l_1 & l_2 & l_3 \\ k_1 & k_2 & k_3 \end{matrix} \middle| 2 \right\} =$$

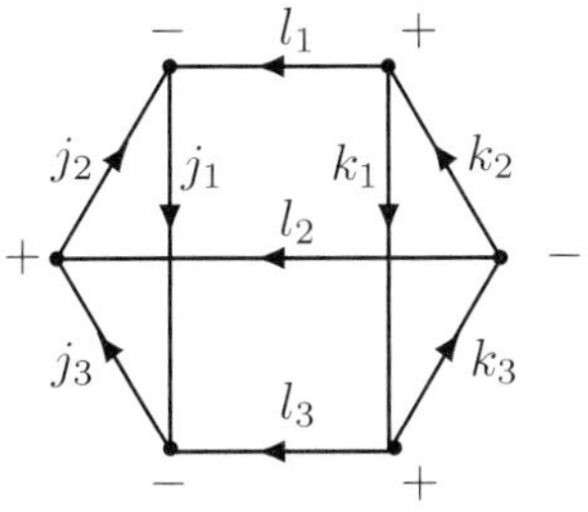

(B.3.71)

$$= (-1)^{2j_3+2k_3} \begin{Bmatrix} j_2 & j_3 & l_2 \\ l_3 & l_1 & j_1 \end{Bmatrix} \begin{Bmatrix} k_2 & k_3 & l_2 \\ l_3 & l_1 & k_1 \end{Bmatrix}$$

Further results may be found in [11, Chapter 5] for example

B.3.10 Graphical treatment of angular momentum algebra

Diagrammatic techniques techniques in angular momentum theory were pioneered by Yutsis (also transliterated as Jucys) and collaborators [12] and there are a number of variants and elaborations. One of the technical problems in evaluating expressions involving njm and nj symbols is the ease with which one can make sign errors. The diagrammatic version described by El Baz and Castel [11] noted that Wigner's covariant notation minimizes the number of additional phase factors that have to be listed explicitly; here we exploit their methods with one difference – we use $[j_1, j_2, \ldots]$ to denote the product $(2j_1+1)(2j_2+1)\ldots$ rather than its square root. Angular momentum states are represented as follows:

Kets:

$|\,j,m\rangle = a^{\dagger}_{jm}\,|\,0\rangle =$ 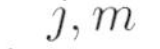

, $|\,j,-m\rangle = a^{\dagger}_{jm}\,|\,0\rangle =$ 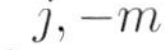

(B.3.72)

Bras:

$$\langle j,m\,| = \langle 0\,|\,\tilde{a}_j m = (-1)^{j-m}\langle 0\,|\,a_{j,-m} = (-1)^{j-m}$$ 

or (B.3.73)

$\langle j,m\,| :=$  $\qquad \langle j,-m\,| :=$ 

The double arrows signal the presence of the factors $(-1)^{j\mp m}$ in the bras $\langle j,\pm m\,|$ as required in summations over projection quantum numbers.

Scalar products: The scalar product of the ket $|\,u\rangle$ by the ket $|\,v\rangle$ is the number $\langle v\,|\,u\rangle$ that is linear in v, anti-linear in u. For angular momentum eigenstates we have

$\langle j,m\,|\,j',m'\rangle =$ 

$=$ 

$= \delta_{jj'}\delta_{mm'}$ (B.3.74)

Projection onto $\mathcal{D}^j$:

$P_j = \sum_m |\,j,m\rangle\langle j,m\,| = \sum_m$ 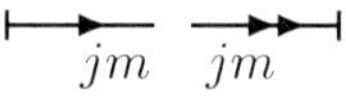

$=$ 

(B.3.75)

Because

$$\sum_m |\,j,m\rangle\langle j,m\,| = \sum_m |\,j,-m\rangle\langle j,-m\,|$$

it is unnecessary to fix the direction of the arrow in advance in a free projector:

 $=$  $=$  (B.3.76)

Because the basis vectors $\{|\,j,m\rangle,\ m = -j,\ldots,j\}$ for all possible values of j are both orthonormal and complete, we also have

$\sum_j \sum_m |\,j,m\rangle\langle j,m\,| = \sum_j$ 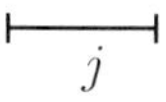

$= 1.$ (B.3.77)

Continuous variables: With minor changes, the diagrammatic representation can be used for continuous variables such as the position variable $\boldsymbol{r}$. For example, a Schrödinger wavefunction $\psi_a(\boldsymbol{r})$ can be represented as a scalar product of an abstract ket $|\,a\rangle$ (where a is one of a finite or countable set of state labels) with a bra $\langle\boldsymbol{r}|$:

$\psi_a(\boldsymbol{r}) = \langle\boldsymbol{r}\,|\,a\rangle =$ 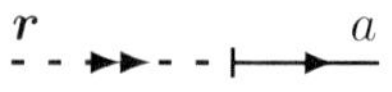

(B.3.78)

and

$\psi_a^*(\boldsymbol{r}) = \langle a\,|\,\boldsymbol{r}\rangle =$ 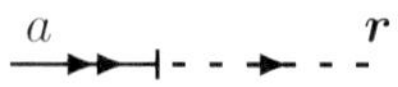

(B.3.79)

Thus

$\langle a \,|\, b\rangle = \int d\boldsymbol{r}$  $= \delta_{ab}$ (B.3.80)

An important example is the representation of the completeness and orthonormality of the spherical harmonics. We write as in (B.3.78), (B.3.79)

$Y_{lm}(\Omega) =$ 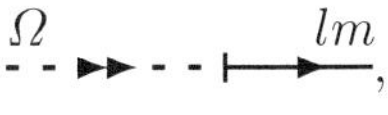 , $Y^*_{lm}(\Omega) =$ 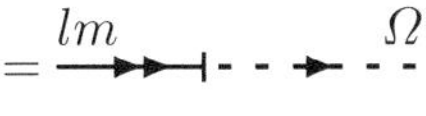 (B.3.81)

so that, summing first over m and then over l,

$\sum_{lm} Y^*_{lm}(\Omega) Y_{lm}(\Omega') = \sum_{lm}$ … $= \sum_l$ … $=$ … $= \delta(\Omega' - \Omega)$ (B.3.82)

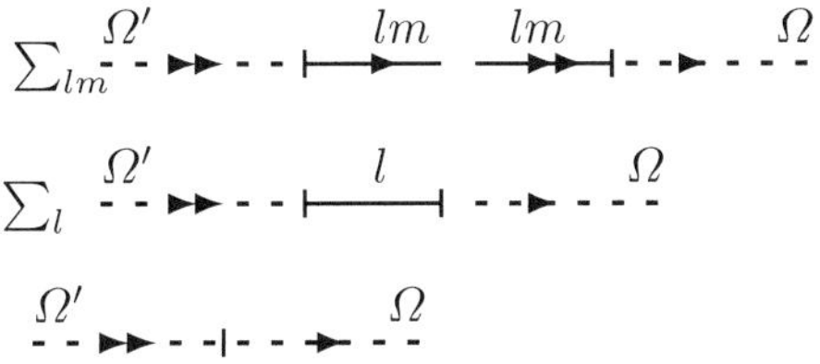

Similarly,

$\int Y^*_{lm}(\Omega) Y_{l'm'}(\Omega)\, d\Omega = \int d\Omega$ … $= \int d\Omega$ … $=$ … $= \delta_{ll'}\delta_{mm'}$ (B.3.83)

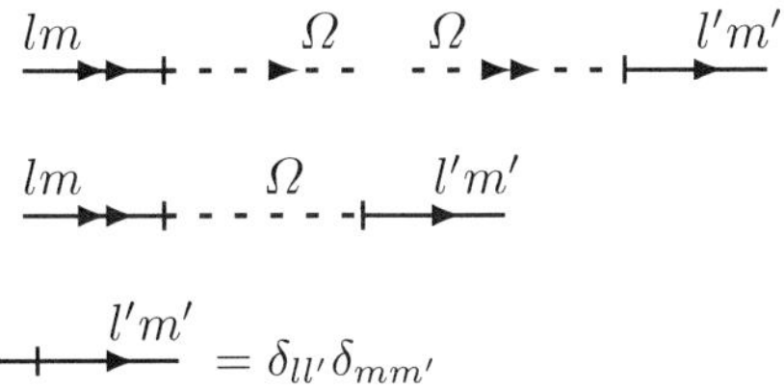

B.3.11 Diagrammatic treatment of Clebsch-Gordan coefficients

The Clebsch-Gordan coefficients can be represented by diagrams of the form

$\langle jm \,|\, j_1 m_1 j_2 m_2\rangle =$ 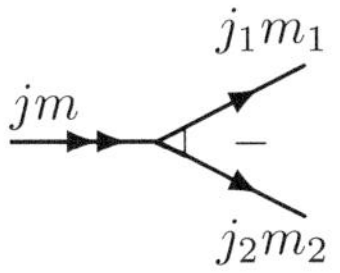 (B.3.84)

As before, single outgoing arrows indicate kets, double arrows indicate bras, labelled by the corresponding jm values. The order of coupling has to be shown explicitly; in this case, we label the vertex with +/- to indicate *anti-clockwise/clockwise* ordering. Thus the relation (B.3.45)

$$|j_1m_1\rangle|j_2m_2\rangle = \sum_{jm} |jm\rangle\langle jm\,|\,j_1m_1j_2m_2\rangle$$

may be represented diagrammatically by

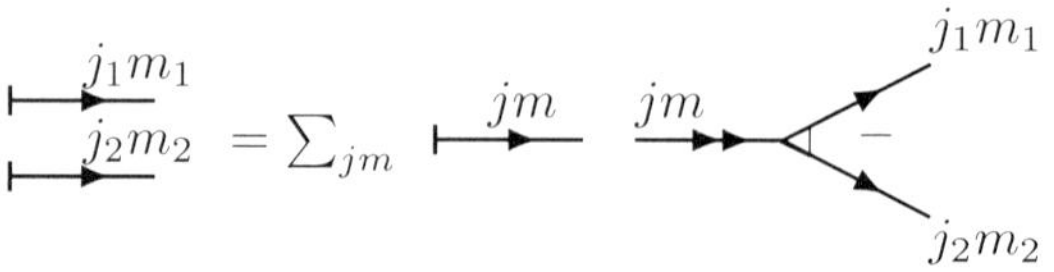

(B.3.85)

Similarly,

$\langle j_1m_1j_2m_2\,|\,jm\rangle =$

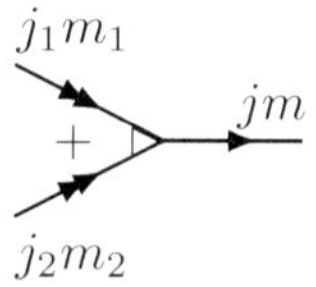

(B.3.86)

The two Clebsch-Gordan coefficients (B.3.84) and (B.3.86) are numerically equal; so that

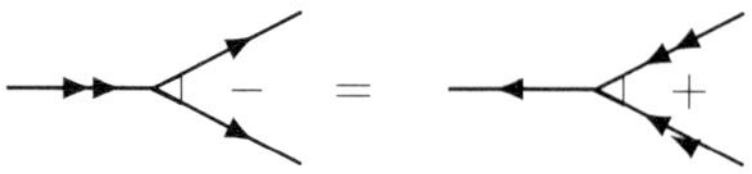

(B.3.87)

In terms of Clebsch-Gordan coefficients, the relations (B.3.53) can be written

$$\sum_{m_1m_2} \langle j_3m_3\,|\,j_1m_1j_2m_2\rangle\langle j_1m_1j_2m_2\,|\,j_3'm_3'\rangle = \delta_{j_3j_3'}\delta_{m_3m_3'} \qquad \text{(B.3.88)}$$

$$\sum_{j_3m_3} \langle j_1m_1j_2m_2\,|\,j_3m_3\rangle\langle j_3m_3\,|\,j_1m_1'j_2m_2'\rangle = \delta_{m_1m_1'}\delta_{m_2m_2'} \qquad \text{(B.3.89)}$$

The first of these is represented diagrammatically by

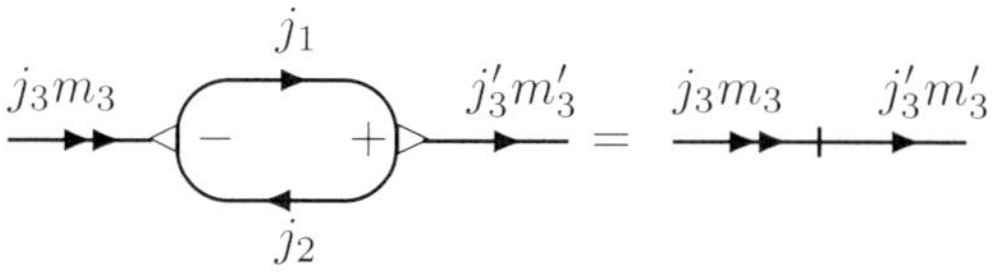

(B.3.90)

in which the j_1 and j_2 can be replaced by unity using (B.3.77) because both angular momenta have fixed values. If we also sum over m_3, each term of the sum gives a triangular delta, (B.3.54), so that this gives the dimension of the subspace $\mathcal{D}^{j_3}$, namely $[j_3] = 2j_3 + 1$. The diagrammatic form is

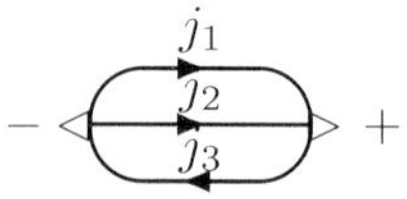

$= [j_3]\{j_1j_2j_3\}.$ (B.3.91)

Equation (B.3.89) is represented by

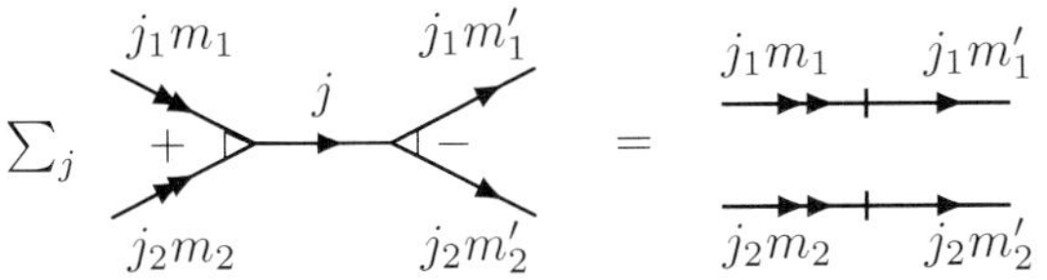

(B.3.92)

in similar fashion.

B.3.12 Diagrammatic treatment of $3jm$-symbols

The connection of Clebsch-Gordan and $3jm$-symbols is given by (B.3.50) which can be written

$$\langle j_3m_3 \,|\, j_1m_1j_2m_2\rangle = (-1)^{j_1-j_2+j_3}[\,j_3]^{1/2}(-1)^{j_3-m_3}\begin{pmatrix} j_1 & j_2 & j_3 \\ m_1 & m_2 & -m_3 \end{pmatrix}. \tag{B.3.93}$$

Using the equivalence of $\langle j_3m_3 \,|\, j_1m_1j_2m_2\rangle$ and $\langle j_1m_1j_2m_2 \,|\, j_3m_3\rangle$ together with $m_3 = m_1 + m_2$, we can rewrite (B.3.93) as

$$\begin{aligned}\langle j_1m_1j_2m_2 \,|\, j_3m_3\rangle &= (-1)^{j_1-j_2-j_3}[\,j_3]^{1/2} \\ &\quad \times (-1)^{j_1-m_1+j_2-m_2}\begin{pmatrix} j_1 & j_2 & j_3 \\ -m_1 & -m_2 & m_3 \end{pmatrix}.\end{aligned} \tag{B.3.94}$$

Note that while the numbers j_i, m_i can be either odd multiples of $1/2$ or integers, the combinations $j_i \pm m_i$ and $\pm j_1 \pm j_2 \pm j_3$ are always integers. In terms of Wigner's covariant notation, (B.3.93) and (B.3.94) become

$$\langle j_3m_3 \,|\, j_1m_1j_2m_2\rangle = (-1)^{j_1-j_2+j_3}[\,j_3]^{1/2}\begin{pmatrix} j_1 & j_2 & m_3 \\ m_1 & m_2 & j_3 \end{pmatrix} \tag{B.3.95}$$

and

$$\langle j_1m_1j_2m_2 \,|\, j_3m_3\rangle = (-1)^{j_1-j_2-j_3}[\,j_3]^{1/2}\begin{pmatrix} m_1 & m_2 & j_3 \\ j_1 & j_2 & m_3 \end{pmatrix}, \tag{B.3.96}$$

from which it follows that

$$\begin{pmatrix} j_1 & j_2 & m_3 \\ m_1 & m_2 & j_3 \end{pmatrix} = \begin{pmatrix} m_1 & m_2 & j_3 \\ j_1 & j_2 & m_3 \end{pmatrix}.$$

Using the Einstein-Wigner summation convention with (B.3.88) gives

$$\begin{pmatrix} j_1 & j_2 & m_3 \\ m_1 & m_2 & j_3 \end{pmatrix}\begin{pmatrix} m_1 & m_2 & m_3' \\ j_1 & j_2 & j_3' \end{pmatrix} = [\,j_3]^{-1}\delta_{j_3j_3'}\delta_{m_3m_3'}$$

with an implied sum over m_1 and m_2. and if we now set $j_3 = j_3'$ and $m_3 = m_3'$ and sum also over m_3 we get the triangular delta

$$\begin{pmatrix} j_1 & j_2 & m_3 \\ m_1 & m_2 & j_3 \end{pmatrix} \begin{pmatrix} m_1 & m_2 & m_3 \\ j_1 & j_2 & j_3 \end{pmatrix} = \{j_1 j_2 j_3\}.$$

This motivates the diagrammatic representation of $3jm$-symbols

$$\begin{pmatrix} j_1 & j_2 & j_3 \\ m_1 & m_2 & m_3 \end{pmatrix} =$$

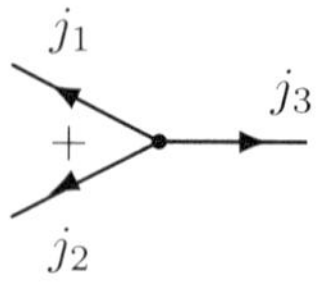

$$=$$

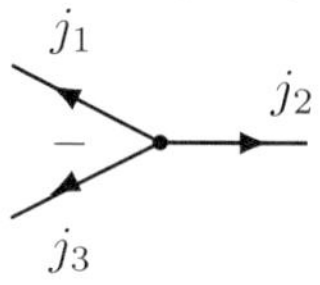

(B.3.97)

Here the single outgoing arrow on each *contravariant* line j_i indicates that the associated quantum number is $+m_i$. Similarly, if j_3 is a *covariant* line marked by a double ingoing arrow then

$$\begin{pmatrix} j_1 & j_2 & m_3 \\ m_1 & m_2 & j_3 \end{pmatrix} =$$

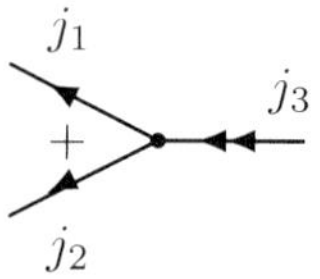

(B.3.98)

The $3jm$ symmetry relations (B.3.51) and (B.3.52) are expressed diagrammatically by

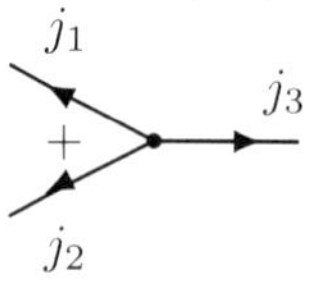

$$= (-1)^{j_1+j_2+j_3}$$

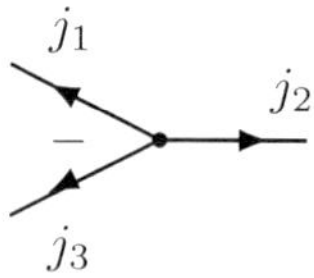

$$=$$

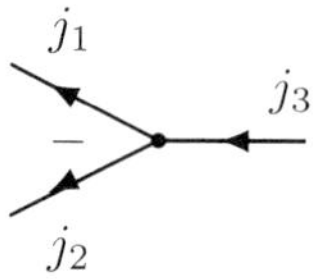

(B.3.99)

Another important symmetry is a change of orientation of the lines of a Clebsch-Gordan or $3jm$-symbol. Thus there is no change in phase when we reverse the sense of all arrows in a Clebsch-Gordan coefficient as in

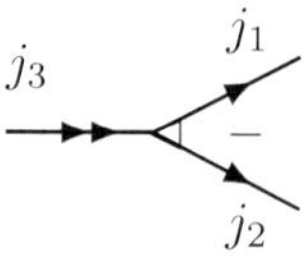

$$=$$

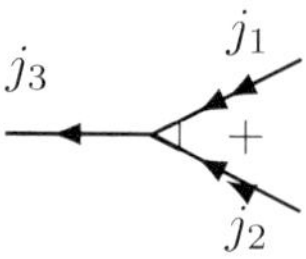

(B.3.100)

but the equivalent $3jm$ relation

$$\begin{pmatrix} j_1 & j_2 & m_3 \\ m_1 & m_2 & j_3 \end{pmatrix} = (-1)^{2j_1+2j_2} \begin{pmatrix} m_1 & m_2 & j_3 \\ j_1 & j_2 & m_3 \end{pmatrix}$$

is represented by

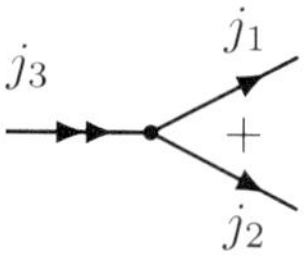

$$= (-1)^{2j_1+2j_2}$$

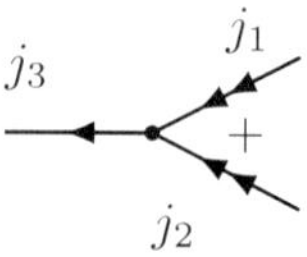

(B.3.101)

It follows that a reversal of the sense of all the lines in a $3jm$ diagram involves no change of sign:

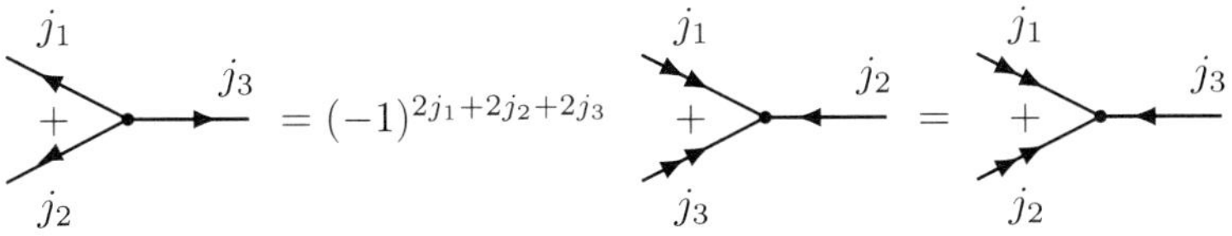

(B.3.102)

because the exponent of the phase factor $(-1)^{2j_1+2j_2+2j_3}$ is always an even integer.

B.3.13 Generalized angular momentum coupling schemes

We shall encounter many situations in which we wish to construct states of several independent identical particles (or subshells) of given angular momentum which form a basis for some representation $\mathcal{D}^j$ of $SO(3)$. Consider first the k-fold product states

$$|j_1 m_1\rangle\,|j_2 m_2\rangle \ldots |j_k m_k\rangle \tag{B.3.103}$$

We wish to choose linear combinations of these states that can partially be described as eigenstates of $\boldsymbol{j}_1^2$, $\boldsymbol{j}_2^2$, $\ldots$, $\boldsymbol{j}_k^2$, $\boldsymbol{j}^2$, and j_z. For a more complete description, we need to prescribe a *coupling scheme*. Thus if $k = 3$, we can first form

$$|j_{12} m_{12}\rangle = \sum_{m_1 m_2} |j_1 m_1\rangle\,|j_2 m_2\rangle\,\langle j_1 m_1 j_2 m_2\,|\,j_{12} m_{12}\rangle$$

followed by a final step

$$|jm\rangle = \sum_{m_{12} m_3} |j_{12} m_{12}\rangle\,|j_3 m_3\rangle\,\langle j_{12} m_{12} j_3 m_3\,|\,jm\rangle.$$

Because only terms with $m_{12} = m_1 + m_2$ contribute, the overall construction gives

$$|jm\rangle = \sum_{m_1 m_2 m_3} |j_1 m_1\rangle\,|j_2 m_2\rangle\,|j_3 m_3\rangle\,\langle j_1 m_1 j_2 m_2\,|\,j_{12} m_{12}\rangle\langle j_{12} m_{12} j_3 m_3\,|\,jm\rangle,$$

which we can write

$$|jm\rangle = \sum_{m_1 m_2 m_3} |j_1 m_1\rangle\,|j_2 m_2\rangle\,|j_3 m_3\rangle \times \langle j_1 m_1 j_2 m_2 j_3 m_3|(j_1 j_2 j_3)_A; jm\rangle, \tag{B.3.104}$$

where

$$\langle j_1 m_1 j_2 m_2 j_3 m_3|(j_1 j_2 j_3)_A; jm\rangle = \sum_{m_{12}} \langle j_1 m_1 j_2 m_2\,|\,j_{12} m_{12}\rangle\,\langle j_{12} m_{12} j_3 m_3\,|\,jm\rangle. \tag{B.3.105}$$

The coefficient $\langle j_1m_1j_2m_2j_3m_3|(j_1j_2j_3)_A;jm\rangle$ is called a *generalized Clebsch-Gordan coefficient* (GCG); the subscript A identifies the process described by (B.3.104). When $k = 3$ there are two alternative schemes in which, say $|j_2m_2\rangle$ and $|j_3m_3\rangle$ are coupled first or $|j_3m_3\rangle$ and $|j_1m_1\rangle$ are coupled first. In graphical form, (B.3.105) becomes

$\langle j_1m_1j_2m_2j_3m_3|(j_1j_2j_3)_A;jm\rangle =$

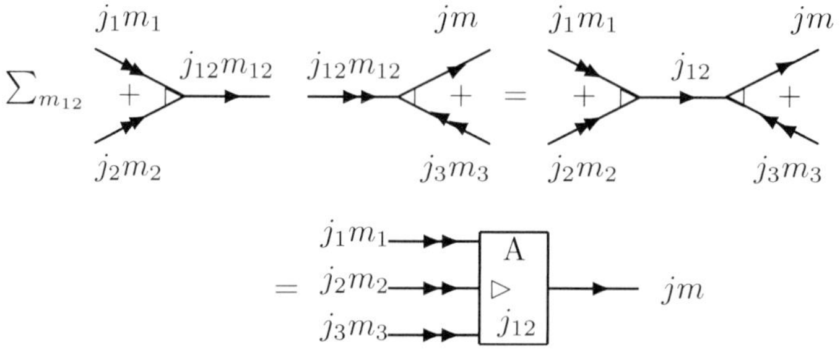

(B.3.106)

The number of possible coupling schemes grows as k increases, but the general structure is similar to the simplest case, $k = 3$. The GCG can be written $\langle j_1m_1 \ldots j_km_k \,|\, (j_1 \ldots j_k)_A; X, jm\rangle$ where the label A describes the manner in which the states are to be coupled and a set of intermediate angular momenta $X = \{X_1, \ldots, X_{k-2}\}$. We express this graphically by writing for the construction of the coupled state $|\,(j_1 \ldots j_k)_A; X, jm\rangle$

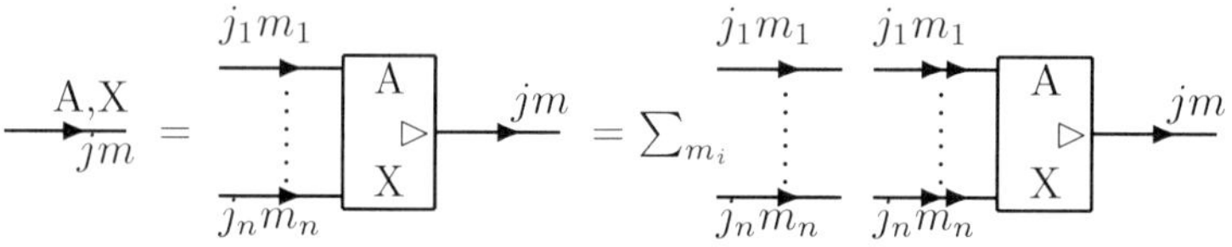

(B.3.107)

and conversely

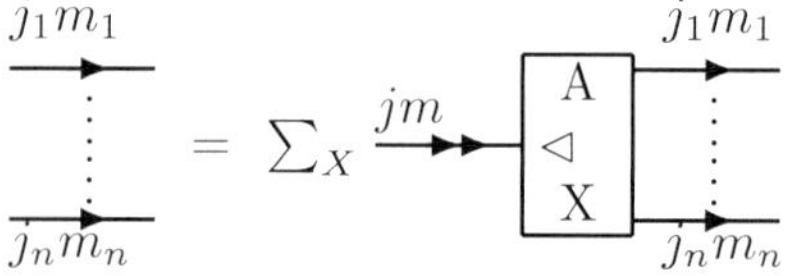

(B.3.108)

for expressing the product (B.3.103) in terms of the coupled states. As with simple Clebsch-Gordan coefficients, the GCG are real, and the m quantum numbers satisfy $m_1 + \ldots + m_k = m$. The rectangles represent tree graphs with $k-1$ vertices corresponding to the $k-1$ Clebsch-Gordan coefficients required to build the X coupling scheme.

The GCG satisfy orthogonality relations inherited from the simple Clebsch-Gordan coefficients. Thus, for two states with the same coupling scheme, A, but different intermediates X, X',

$$\sum_{m_1,\ldots,m_k} \langle (j_1 \ldots j_k)_A; X, jm \,|\, j_1 m_1 \ldots j_k m_k \rangle \tag{B.3.109}$$

$$\times \langle j_1 m_1 \ldots j_k m_k \,|\, (j_1 \ldots j_k)_A; X', j'm' \rangle = \delta_{jj'} \delta_{mm'} \prod_{i=1}^{k-2} \delta_{X_i X_i'}.$$

The graphical representation of this is

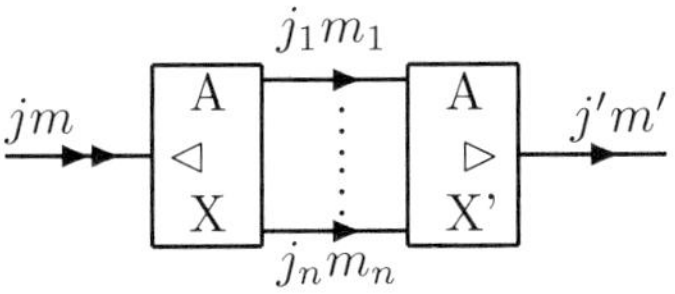

$$= \delta_{jj'} \delta_{mm'} \textstyle\prod_{i=1}^{k-2} \delta_{X_i X_i'}. \tag{B.3.110}$$

Similarly

$$\sum_{Xjm} \langle j_1 m_1 \ldots j_k m_k \,|\, (j_1 \ldots j_k)_A; X, jm \rangle \tag{B.3.111}$$

$$\times \langle\, (j_1 \ldots j_k)_A; X, jm \,|\, j_1 m_1' \ldots j_k m_k' \rangle = \prod_{i=1}^{k} \delta_{m_i m_i'},$$

with the graphical representation

$\sum_{\{j_i\},X}$

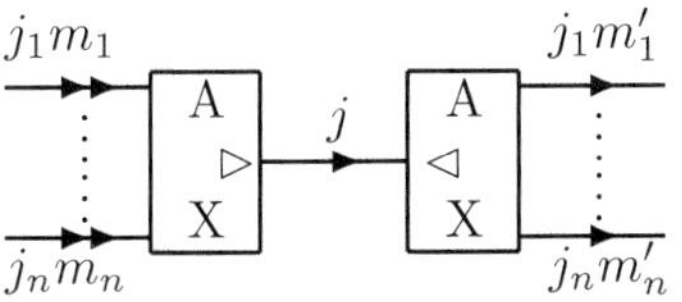

$$= \textstyle\prod_{i=1}^{k} \delta_{m_i m_i'}. \tag{B.3.112}$$

The meaning of these graphs is best clarified by looking at the simplest cases, $k = 1, 2, 3$. In the case $k = 1$, we have only one ket, $|j_1\rangle$, and no intermediates X_i so that (B.3.110) and (B.3.112) reduce respectively to (B.3.74) and (B.3.75). In the case $k = 2$, we have two kets and one final resultant and no intermediates so that (B.3.110) reduces to (B.3.90) and (B.3.112) to (B.3.92). The first non-trivial case is $k = 3$. Here, the bra and ket are represented by

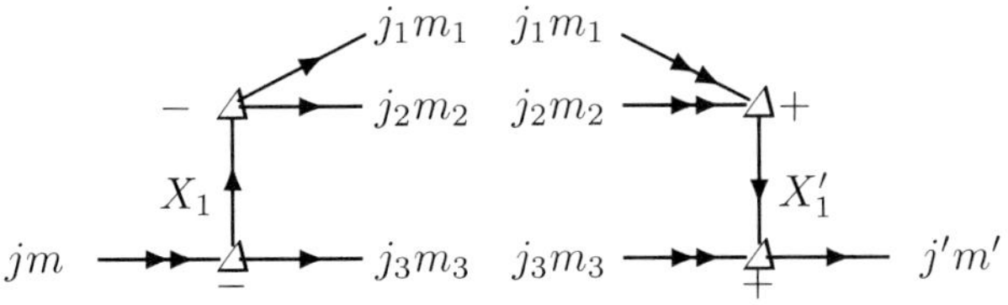

(B.3.113)

For (B.3.110) we sum over m_1, m_2 and m_3 by linking the corresponding lines so that

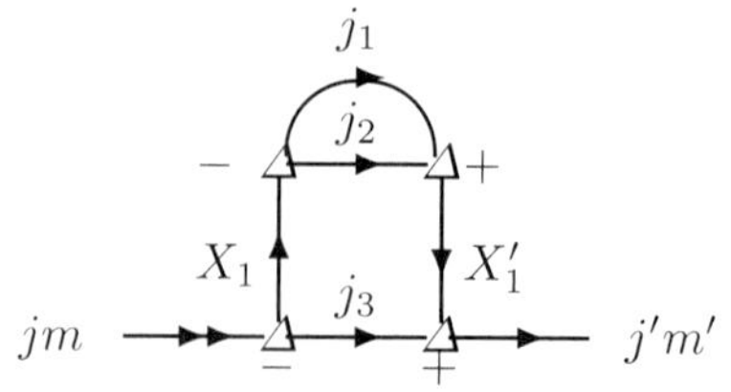

(B.3.114)

The lines labelled by j_1 and j_2 have given values, so their projectors are just unity and can be removed from the diagram; then we link the lines labelled by X_1 and X_1', giving a diagram of the form (B.3.90) with j_2 replaced by X_1, verifying (B.3.110). For (B.3.112) we have to link the jm lines giving

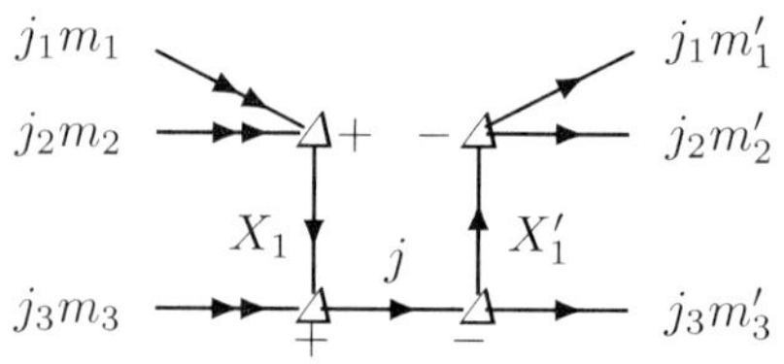

(B.3.115)

Next we apply (B.3.92) twice to give

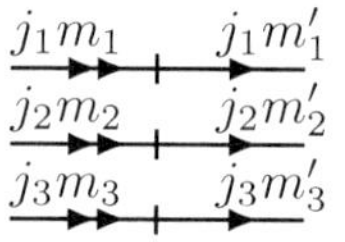

(B.3.116)

verifying (B.3.112). This analysis only deals with the case in which the same coupling scheme is used for both bra and ket. A change in the order of coupling leads to a different Clebsch-Gordan expansion; the linear transformation connecting two such schemes will be discussed below.

B.3.14 GCG and njm coefficients

The relations, (B.3.95) and (B.3.96), between Clebsch-Gordan and $3jm$ coefficients, can be represented by

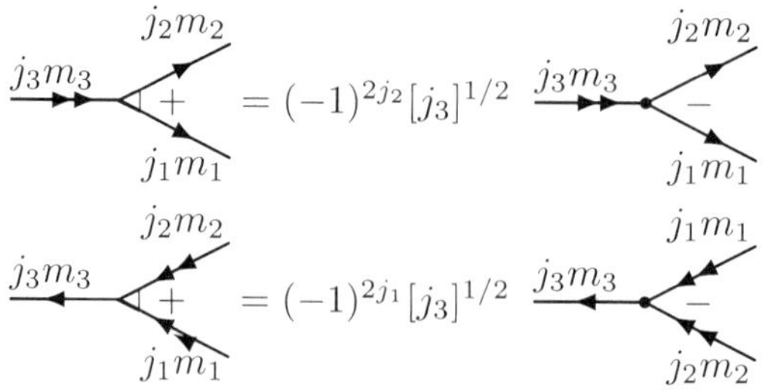

(B.3.117)

This is obtained by extracting a factor $(-1)^{\pm(j_1+j_2+j_3)}$ from both (B.3.95) and (B.3.96), represented by the sign change at the vertex, and the factor $[j_3]^{1/2}$ that is attached to the *single* line that is covariant in (B.3.95) and contravariant in (B.3.96). What is left is a phase $(-1)^{2j_2}$ in (B.3.95) and $(-1)^{2j_1}$ in (B.3.96); we can identify the line with which it is associated as the one that follows j_3 next in order at the vertex of the $3jm$-coefficient.

This generalizes straightforwardly to higher order GCG and njm coefficients. The GCG (B.3.107) is related to the corresponding njm coefficient by

$$\langle j_1 m_1 \ldots j_k m_k \,|\, (j_1 \ldots j_k)_A; X, jm \rangle = \tag{B.3.118}$$

$$\Phi\,[X_1, \ldots, X_{k-2}, j]^{1/2} \begin{pmatrix} m_1 & \cdots & m_k & j \\ j_1 & \cdots & j_k & m \end{pmatrix}_{A;X}$$

where the arguments $X_1, \ldots, X_{k-2}, j$ represent the single outgoing line at each vertex and Φ is the product of the corresponding phases "$(-1)^{2j_2}$" at each vertex. We distinguish the njm diagram from the GCG By omitting the CG "triangle" from the box, so that

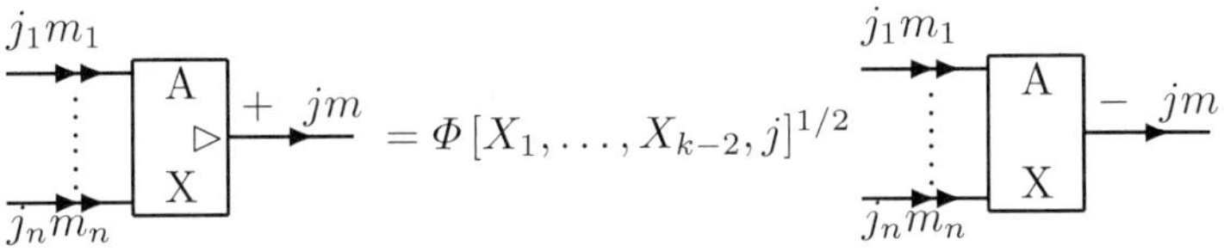

$$= \Phi\,[X_1, \ldots, X_{k-2}, j]^{1/2} \tag{B.3.119}$$

and changing the sign of all internal vertices. Similarly for (B.3.108) the diagram is

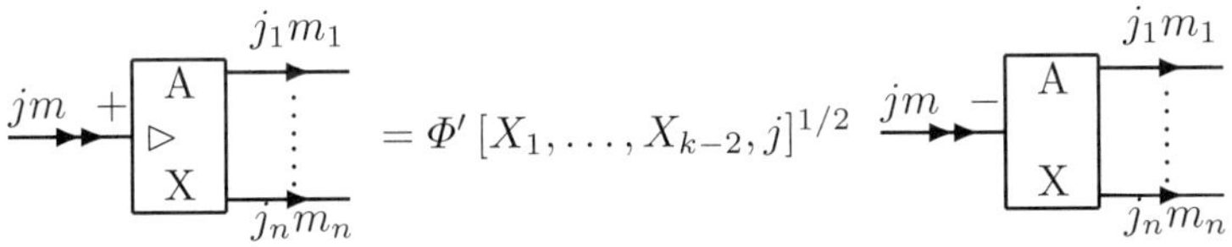

$$= \Phi'\,[X_1, \ldots, X_{k-2}, j]^{1/2} \tag{B.3.120}$$

We note that the njm symbols

$$\begin{pmatrix} j_1 & \cdots & j_k & j \\ m_1 & \cdots & m_k & m \end{pmatrix}_{A;X} \tag{B.3.121}$$

are real numbers; that $j_1 + \ldots + j_k + j$ is always an integer; that $m_1 + \ldots + m_k + m = 0$; and that reversing the signs on all m arguments must be accompanied by a phase factor $(-1)^{j_1 + \ldots + j_k + j}$. It follows that the completely contravariant njm symbol (B.3.121) and the corresponding completely covariant equvalent are equal:

$$\begin{pmatrix} j_1 & \cdots & j_k & j \\ m_1 & \cdots & m_k & m \end{pmatrix}_{A;X} = \begin{pmatrix} m_1 & \cdots & m_k & m \\ j_1 & \cdots & j_k & j \end{pmatrix}_{A;X}$$

B.3.15 Manipulating angular momentum diagrams

We have already seen that complex angular momentum diagrams can be obtained by coupling simpler diagrams in which internal lines come from summation over the appropriate m variables. Matrix elements between many-particle states lead to diagrams in which the free lines from different blocks are joined to give a more complex diagram. Where the graphical method shows its power is through the use of simple rules to decompose these diagrams into simpler, identifiable standard sub-diagrams whose algebraic form is relatively easy to evaluate numerically. We shall simply state these rules; proofs can be found in books such as [11, 12].

The assembly of diagrams involves units such as

$$\boxed{\alpha} \xrightarrow{\ jm\ } \qquad \xrightarrow{\ jm\ } \boxed{\beta} \tag{B.3.122}$$

where the labels α and β define coupling schemes and a set of intermediate angular momenta both coupled to give a resultant jm. Summation over the projections m is achieved by linking the two external lines

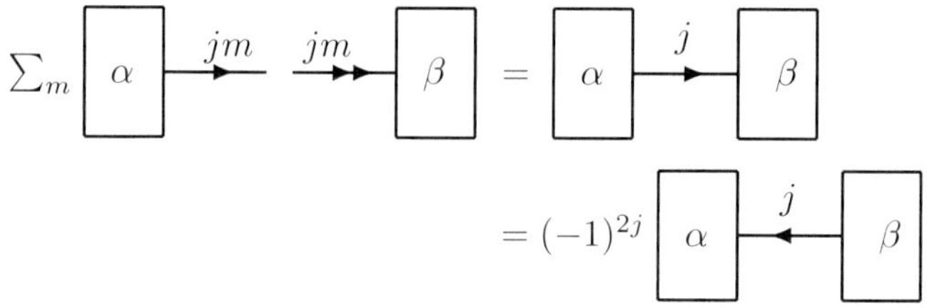

$$\sum_m \boxed{\alpha} \xrightarrow{jm}\ \xrightarrow{jm} \boxed{\beta} = \boxed{\alpha} \xrightarrow{\ j\ } \boxed{\beta}$$

$$= (-1)^{2j} \boxed{\alpha} \xleftarrow{\ j\ } \boxed{\beta} \tag{B.3.123}$$

Similarly,

$$\sum_j \boxed{\alpha} \underset{j_2}{\overset{j_1}{\lessgtr}} \xrightarrow{\ j\ }_{-} \underset{j_2}{\overset{j_1}{\gtrless}} \boxed{\beta} = \boxed{\alpha} \underset{j_2}{\overset{j_1}{}} \boxed{\beta} \tag{B.3.124}$$

after summing over the values of j.

We say that a diagram G is separable over n lines if it has the form

$$G: \quad \boxed{\overline{\alpha}} \overset{j_1}{\underset{j_n}{\vdots}} \boxed{\beta} \tag{B.3.125}$$

where the block labelled $\overline{\alpha}$ has no lines with free ends, so that any remaining lines with free ends are associated with the block β. The basic rule for simplifying such diagrams can be written

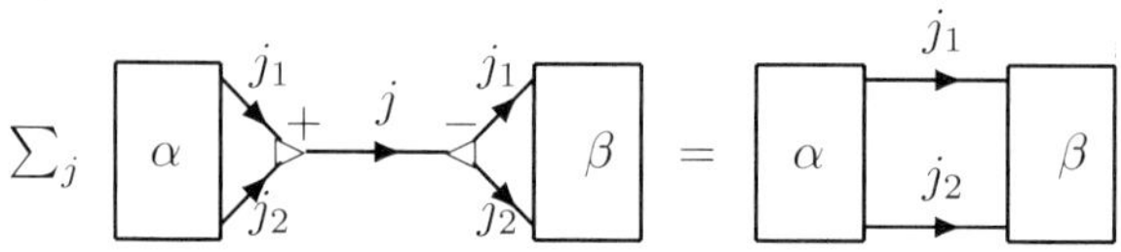

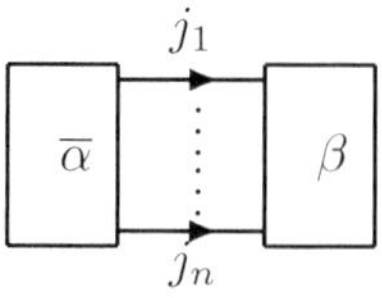

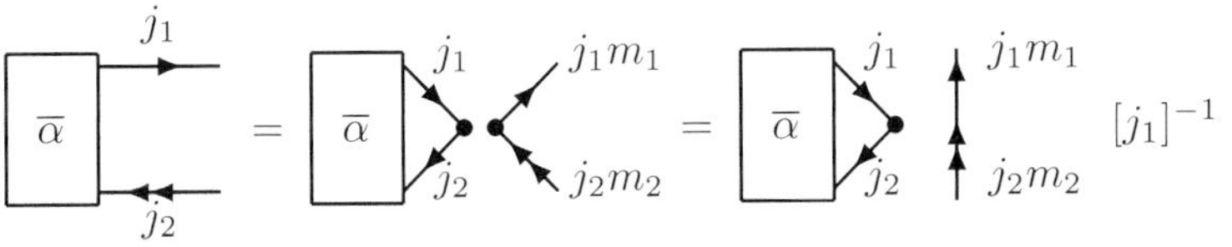

(B.3.126)

We merely link the two lines j_1 and j_2. For the proof see, for example [11, Chapter 4].

Separation on two lines with *opposite* orientations gives

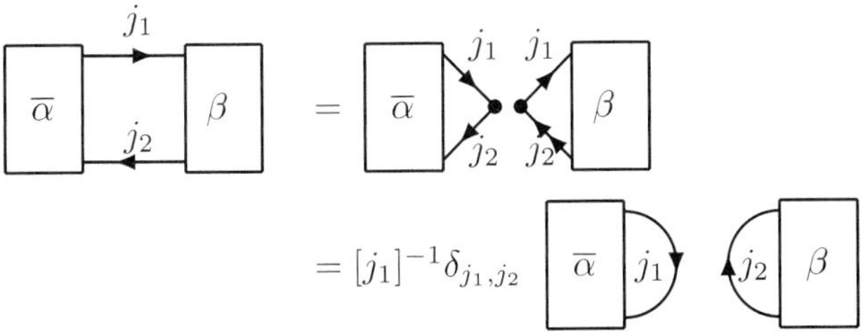

(B.3.127)

Separation on three lines with the *same* orientation is achieved with

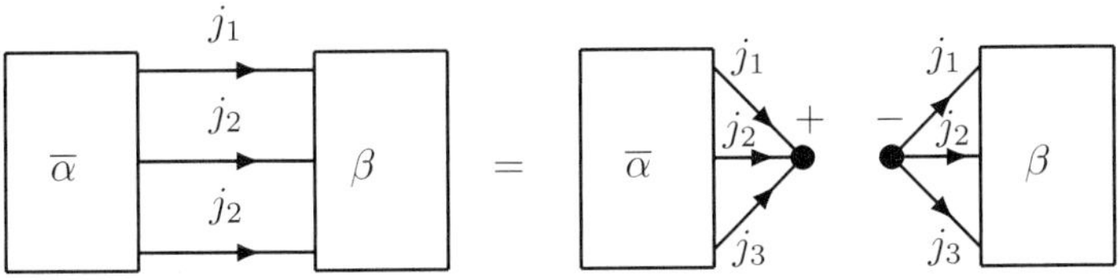

(B.3.128)

If one of the lines, say j_3, has a different orientation, we can change its direction, introducing an extra phase factor $(-1)^{2j_3}$, before using (B.3.128). The general case of separation over n lines with the *same* orientation requires an intermediate coupling mode, A, with intermediates $X_1, \ldots, X_{n-3}$:

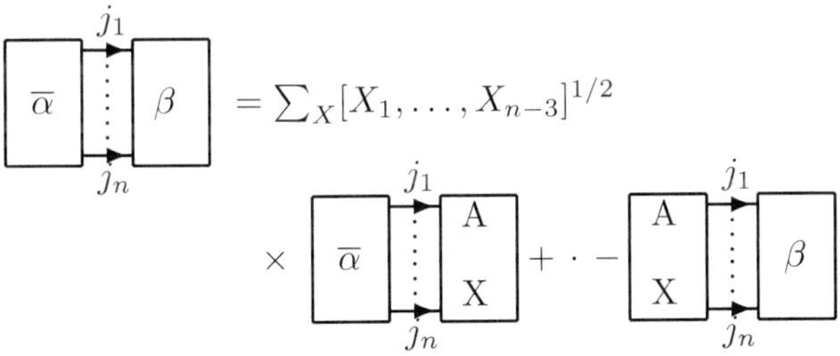

(B.3.129)

where the sum runs overr all available couplng schemes, X, and the $+$ and $-$ signs indicate that nodes appearing when cutting j_1 to j_n in succession have opposite senses in the two diagrams on the right-hand side. This preserves the ordering of the lines at each node as in the simplest case, (B.3.128).

B.3.16 Tensor operators and the Wigner-Eckart theorem

An irreducible tensor operator of rank k is a set of $2k + 1$ operators

$$\boldsymbol{T}^k = \{\, T^k_q \mid q = k, k-1, \ldots, -k \,\} \tag{B.3.130}$$

satisfying the relations (compare (B.2.24)

$$[j_3, a_m{}^\dagger] = m\, a_m{}^\dagger, \quad [j_\pm, a_m{}^\dagger] = [j(j+1) - m(m \pm 1)]^{1/2}\, a_{m\pm 1}{}^\dagger.$$

In practical calculations on atoms – and also on molecules – it is useful to decompose the operators encountered as a sum of terms, each classified by the irreducible representation of SO(3) to which it belongs. Once this has been done, we can invoke the Wigner-Eckart theorem, §B.2.7, to simplify the angular integrations needed to compute the matrix elements. In SO(3), this separates those parts of the matrix elements of irreducible tensor operators that depends upon the projections in terms of a 3-j symbol:

$$\langle\, jm \,|\, T^k_q \,|\, j'm' \,\rangle = (-1)^{j-m} \begin{pmatrix} j & k & j' \\ -m & q & m' \end{pmatrix} (\, j \,\|\, \boldsymbol{T}^k \,\|\, j'). \tag{B.3.131}$$

The reduced matrix elements $(\, j \,\|\, \boldsymbol{T}^k \,\|\, j')$ of simple operators are therefore essential building blocks, and we include formulae for reduced matrix elements for each of the operators listed below. We refer to the main text and to Appendix B.3.10 for the way in which these are used, in particular, the compound operators arising in relativistic atomic and molecular calculations

We follow initially the notation of Louck [8, §2.9], whose review contains many of the properties we shall exploit. Other results may be found in [11, 12, 14, 15, 16, 17]. The following reduced matrix elements are used frequently in this book:

- *Orbital angular momentum*: $\boldsymbol{L}$, which has rank 1. Its components are $L_0 = L_z$ and $L_{\pm 1} = \mp(L_x \pm iL_y)/\sqrt{2}$ and its reduced matrix elements are

$$(l\|\boldsymbol{L}\|l') = \{l(l+1)(2l+1)\}^{1/2}\, \delta_{ll'} \tag{B.3.132}$$

- *General angular momentum*: $\boldsymbol{J}$, also has rank 1. Its components are $J_0 = L_z$ and $J_{\pm 1} = \mp(J_x \pm iJ_y)/\sqrt{2}$ and its reduced matrix elements are

$$(j\|\boldsymbol{J}\|lj') = \{j(j+1)(2j+1)\}^{1/2}\, \delta_{jj'} \tag{B.3.133}$$

- *Spin* 1/2: $\boldsymbol{s} = \frac{1}{2}\boldsymbol{\sigma}$. Its components are $s_0 = s_z$ and $s_{\pm 1} = \mp(s_x \pm is_y)/\sqrt{2}$ and, setting $j = j' = 1/2$ in (B.3.133) gives

$$(1/2\|\boldsymbol{s}\|1/2) = \sqrt{3/2}, \quad (1/2\|\boldsymbol{\sigma}\|1/2) = \sqrt{6}. \tag{B.3.134}$$

- *Spherical harmonics*: We define tensor operators $\boldsymbol{C}^k$ of rank k in terms of the spherical harmonics defined by (B.3.23)

$$C_{kq}(\theta, \varphi) = \left(\frac{4\pi}{2k+1}\right)^{1/2} Y_{kq}(\theta, \varphi) \tag{B.3.135}$$

so that

$$(l \,\|\, \boldsymbol{C}^k \,\|\, l') = (-1)^l [l, l']^{1/2} \begin{pmatrix} l & k & l' \\ 0 & 0 & 0 \end{pmatrix} \tag{B.3.136}$$

Special cases:

$$\boldsymbol{C}^0 : \quad C_{00} = 1 \tag{B.3.137}$$

$$\boldsymbol{C}^1 = \frac{\boldsymbol{r}}{r} = \widehat{\boldsymbol{r}} : \quad C_{10} = \cos\theta = z/r, \tag{B.3.138}$$

$$C_{1\pm1} = \mp\sqrt{\frac{1}{2}} \sin\theta \mathrm{e}^{\pm i\varphi} = \mp\sqrt{\frac{1}{2}} \frac{(x \pm iy)}{r}$$

$$\boldsymbol{C}^2 : \quad C_{20} = \frac{1}{2}(3\cos^2\theta - 1) = \frac{1}{2}\frac{2z^2 - x^2 - y^2}{r^2} \tag{B.3.139}$$

$$C_{2\pm1} = \mp\sqrt{\frac{3}{2}} \cos\theta \sin\theta \mathrm{e}^{\pm i\varphi} = \mp\sqrt{\frac{3}{2}} \frac{z(x \pm iy)}{r^2}$$

$$C_{2\pm2} = \sqrt{\frac{3}{8}} \sin^2\theta \mathrm{e}^{\pm 2i\varphi} = \mp\sqrt{\frac{3}{8}} \frac{(x \pm iy)^2}{r^2}$$

B.3.17 Composite tensor operators

Composite tensor operators built from the simple operators considered above are needed in many applications.

(a) Multiplication of an irreducible tensor operator by a complex number or an invariant with respect to angular momentum $\boldsymbol{J}$ gives an irreducible tensor operator of the same rank.

(b) Addition of two irreducible tensor operators of the same rank k gives an irreducible tensor of rank k.

(c) In general, multiplication of irreducible tensor operators is not commutative; however multiple products of irreducible tensor operators are associative.

(d) Let the irreducible tensor operators $\boldsymbol{S}^{k_1}$ and $\boldsymbol{T}^{k_2}$ act on the same Hilbert space $\mathcal{H}$. Then their product $[\boldsymbol{S}^{k_1} \times \boldsymbol{T}^{k_2}]^k$ is an irreducible tensor operator of rank k with components

$$[\boldsymbol{S}^{k_1} \times \boldsymbol{T}^{k_2}]^k_q = \sum_{q_1, q_2} C^{k_1\,k_2\,k}_{q_1\,q_2\,q}\, S^{k_1}_{q_1} T^{k_2}_{q_2}, \quad q = k, k-1, \ldots -k, \tag{B.3.140}$$

where $k \in \{k_1 + k_2, k_1 + k_2 - 1, \ldots, |k_1 - k_2|\}$ (Clebsch-Gordan series).

(e) Similarly, the irreducible tensor operator $\boldsymbol{S}^{k_1}$ acting on the Hilbert space $\mathcal{H}$ and the irreducible tensor operator $\boldsymbol{T}^{k_2}$ acting on the Hilbert space $\mathcal{K}$ may first be multiplied by the tensor product rule so as to act on the tensor product space $\mathcal{H} \otimes \mathcal{K}$ and then coupled to obtain a new tensor operator $[\boldsymbol{S}^{k_1} \otimes \boldsymbol{T}^{k_2}]^k$ with components

$$[\boldsymbol{S}^{k_1} \otimes \boldsymbol{T}^{k_2}]^k_q = \sum_{q_1,q_2} C^{k_1\,k_2\,k}_{q_1\,q_2\,q}\, S^{k_1}_{q_1} \otimes T^{k_2}_{q_2}, \quad q = k, k-1, \dots -k, \tag{B.3.141}$$

acting on $\mathcal{H} \otimes \mathcal{K}$, where $k \in \{k_1 + k_2, k_1 + k_2 - 1, \dots, |k_1 - k_2|\}$. Common examples are operators like $\boldsymbol{S} \cdot \boldsymbol{L}$ and $\boldsymbol{\alpha} \cdot \boldsymbol{p}$ where the first operates on the spin space and the second on functions in $\mathbb{R}^3$.

(f) It is convenient (here we differ from the convention used by Louck) to define the tensor operator $\boldsymbol{T}^{k\dagger}$ *adjoint* to $\boldsymbol{T}^k$ as the operator with components

$$T^{k\dagger}_q = (-1)^{k-q} T^k_{-q} \tag{B.3.142}$$

See [8, §2.9.3] for other definitions and references to the literature.

We list a number of important composite operators below. For the sake of economy, we shall often ignore the distinction between $[\boldsymbol{S}^{k_1} \times \boldsymbol{T}^{k_2}]^k$ and $[\boldsymbol{S}^{k_1} \otimes \boldsymbol{T}^{k_2}]^k$. The context is always suffficient to make the meaning clear.

- When $\boldsymbol{C}^{k_1}$ and $\boldsymbol{C}^{k_2}$ act on the same coordinates (case (d) above) we have

$$(\boldsymbol{C}^{k_1} \times \boldsymbol{C}^{k_2})^k = (-1)^k [k]^{1/2} \begin{pmatrix} k_1 & k & k_2 \\ 0 & 0 & 0 \end{pmatrix} \boldsymbol{C}^k \tag{B.3.143}$$

In particular

$$\begin{aligned} (\boldsymbol{C}^1 \times \boldsymbol{C}^k)^{k+1} &= \sqrt{\frac{k+1}{2k+1}}\, \boldsymbol{C}^{k+1}, \\ (\boldsymbol{C}^1 \times \boldsymbol{C}^k)^k &= 0, \\ (\boldsymbol{C}^1 \times \boldsymbol{C}^k)^{k-1} &= -\sqrt{\frac{k}{2k+1}}\, \boldsymbol{C}^{k-1}. \end{aligned} \tag{B.3.144}$$

Also

$$\sqrt{(k+1)(2k+3)}(\boldsymbol{C}^{k+1} \times \boldsymbol{l})^k = \sqrt{k(2k-1)}(\boldsymbol{C}^{k-1} \times \boldsymbol{l})^k \tag{B.3.145}$$

and

$$(l \,\|(\boldsymbol{C}^k \times \boldsymbol{l})^k\|\, l') = \frac{l(l+1) - k(k+1) - l'(l'+1)}{\sqrt{2k(2k+2)}} (l \,\|(\boldsymbol{C}^k\|\, l'). \tag{B.3.146}$$

- Two tensors $\boldsymbol{R}^k$ and $\boldsymbol{S}^k$ having the same *integer* rank k have a *scalar product*, written $\boldsymbol{R}^k \cdot \boldsymbol{S}^k$, given by

$$\boldsymbol{R}^k \cdot \boldsymbol{S}^k = \sum_q (-1)^q R^k_q S^k_{-q} = (-1)^k [k]^{1/2} (\boldsymbol{R}^k \times \boldsymbol{S}^k)^0. \tag{B.3.147}$$

- A 3-vector $\boldsymbol{a}$ can also be regarded as a rank 1 tensor, such that $\boldsymbol{a} = |\boldsymbol{a}|\, \boldsymbol{C}^1$, with (spherical) components $a_\mu = |\boldsymbol{a}|\, C_{1\mu}(\theta\varphi)$. Then, setting $k = 1$ in (B.3.147), we get the scalar product of two 3-vectors

$$\boldsymbol{a} \cdot \boldsymbol{b} = -\sqrt{3}(\boldsymbol{ab})^0 \tag{B.3.148}$$

Similarly, the ordinary vector product of two 3-vectors is

$$\boldsymbol{a} \times \boldsymbol{b} = -i\sqrt{2}\,(\boldsymbol{ab})^1 \tag{B.3.149}$$

and the rank 2 tensor formed by $\boldsymbol{a}$ and $\boldsymbol{b}$ has components

$$\begin{aligned} (\boldsymbol{ab})^2_2 &= a_1 b_1, \\ (\boldsymbol{ab})^2_1 &= (a_0 b_1 + a_1 b_0)/\sqrt{2}, \\ (\boldsymbol{ab})^2_0 &= (3a_0 b_0 - \boldsymbol{a} \cdot \boldsymbol{b})/\sqrt{6} \end{aligned} \tag{B.3.150}$$

- Because $\boldsymbol{r} \times \boldsymbol{l} = \boldsymbol{r} \times (\boldsymbol{r} \times \boldsymbol{p}) = \boldsymbol{r}(\boldsymbol{r} \cdot \boldsymbol{p}) - r^2 \boldsymbol{p}$, the quantum mechanical momentum operator, $\boldsymbol{p} = -i\boldsymbol{\nabla}$ can be represented as a rank 1 tensor

$$\boldsymbol{p} = -i \left[\boldsymbol{C}^1 \frac{\partial}{\partial r} - \frac{\sqrt{2}}{r} (\boldsymbol{C}^1 \times \boldsymbol{l})^1 \right], \tag{B.3.151}$$

By using (B.3.146) with $k = 1$, we see that

$$\langle l \,\|\, \boldsymbol{\nabla} \,\|\, l' \rangle = (l \,\|(\boldsymbol{C}^1 \|\, l') \left[\frac{\partial}{\partial r} - \frac{(l - l')(l + l' + 1) - 2}{2r} \right]$$

so that

$$\langle l \,\|\, \boldsymbol{\nabla} \,\|\, l' \rangle = \begin{cases} l \left(\dfrac{\partial}{\partial r} - \dfrac{l'}{r} \right), & l = l' + 1 \\ 0, & l = l' \\ -\sqrt{l'} \left(\dfrac{\partial}{\partial r} + \dfrac{l' + 1}{r} \right), & l = l' - 1 \end{cases} \tag{B.3.152}$$

Equations (B.3.145), (B.3.146), (B.3.149), and (B.3.151) are due to Innes and Ufford [18].

B.3.18 Diagrammatic representation of tensor operators

An irreducible tensor operator $T^k_q : \mathcal{H} \to \mathcal{H}$ can be defined by

$$\langle u \,|\, T^k_q \,|u' \rangle = T^k_q(u)\, \delta(u - u'), \quad u \in \mathcal{H} \tag{B.3.153}$$

As in (B.3.78), we can represent this as a diagram

$T_q^k(u) =$ 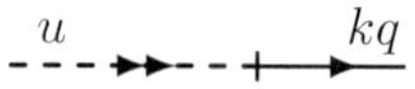 (B.3.154)

and, in accordance with its definition at the beginning of this section, we can represent the adjoint by

u

$T_q^{k\dagger}(u) =$ 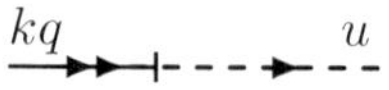 (B.3.155)

corresponding to (B.3.79). The matrix element $\langle\, jm \,|\, T_q^k \,|\, j'm' \,\rangle$ appearing in the statement (B.3.131) of the Wigner-Eckart theorem can be expressed as an integral

$$\langle\, jm \,|\, T_q^k \,|\, j'm' \,\rangle = \int du \,\langle\, jm \,|\, u\rangle\, T_q^k(u)\, \langle\, u \,|\, j'm' \rangle.$$

The integration over u is done diagrammatically by linking the ends of the u lines, so that

$\langle\, jm \,|\, T_q^k \,|\, j'm' \,\rangle =$ 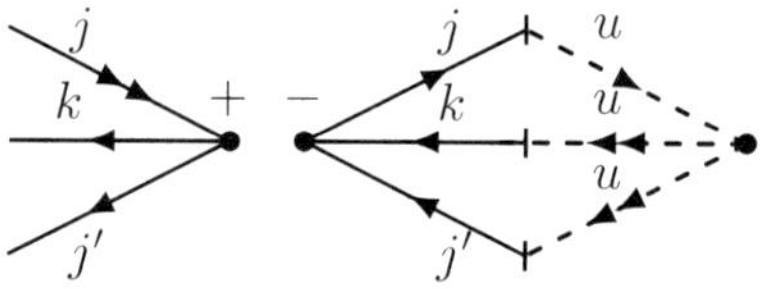 (B.3.156)

where we have used the rule (B.3.128) to put the diagram in the form required by the Wigner-Eckart theorem (B.3.131). The left-hand component of (B.3.158) is the $3jm$ symbol

$$\begin{pmatrix} m & k & j' \\ j & q & m' \end{pmatrix} = (-1)^{j-m} \begin{pmatrix} j & k & j' \\ -m & q & m' \end{pmatrix}$$

so that the rest corresponds to the reduced matrix element $(\, j \,\|\, \boldsymbol{T}^k \,\|\, j')$, which will be denoted by a *marking circle* on the vertex of the $3jm$ symbol

$\langle\, jm \,|\, T_q^k \,|\, j'm' \,\rangle =$ 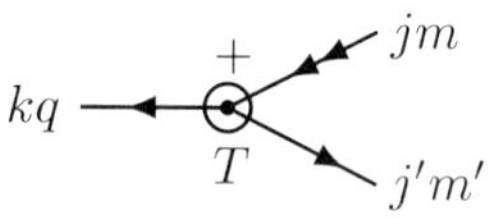 (B.3.157)

The composite tensor operator

$$\boldsymbol{R}^k = \left[\boldsymbol{S}^{k_1} \otimes \boldsymbol{T}^{k_2}\right]^k$$

which acts on the space $\mathcal{H}_1 \otimes \mathcal{H}_2$ is then represented by

$$R^k_q(u_1, u_2) = \sum_{q_1 q_2} \langle k_1 q_1, k_2 q_2 \,|\, kq \rangle \, S^{k_1}_{q_1}(u_1) \, T^{k_2}_{q_2}(u_2)$$

$$=$$ 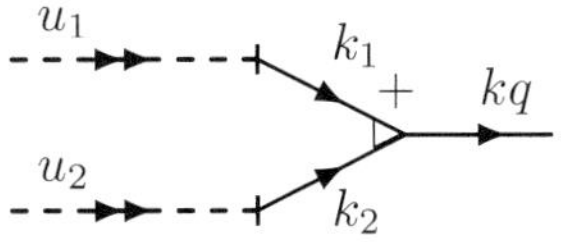

(B.3.158)

and its adjoint is represented by reversing the arrows,

$$R^{k\dagger}_q(u_1, u_2) =$$ 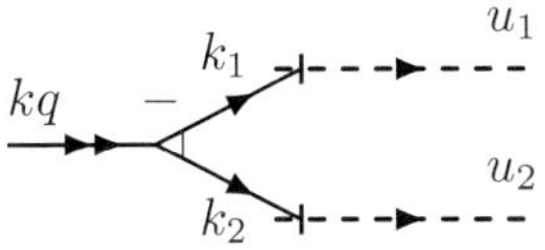

(B.3.159)

Coupled states on $\mathcal{H}_1 \otimes \mathcal{H}_2$ belonging to D^j will be written $|\gamma j_1 j_2 j m\rangle$, where γ symbolizes any other labels needed to make up a complete set of quantum numbers. An argument similar to (B.3.156) leads to the reduced matrix element

$$\langle \gamma j_1 j_2 j \,\|\, \boldsymbol{R}^k \,\|\, \gamma' j_1' j_2' j' \rangle = \langle \gamma j_1 j_2 j \,\|\, [\boldsymbol{S}^{k_1} \otimes \boldsymbol{T}^{k_2}]^k \,\|\, \gamma' j_1' j_2' j' \rangle$$

$$= \iint du_1 du_2$$ 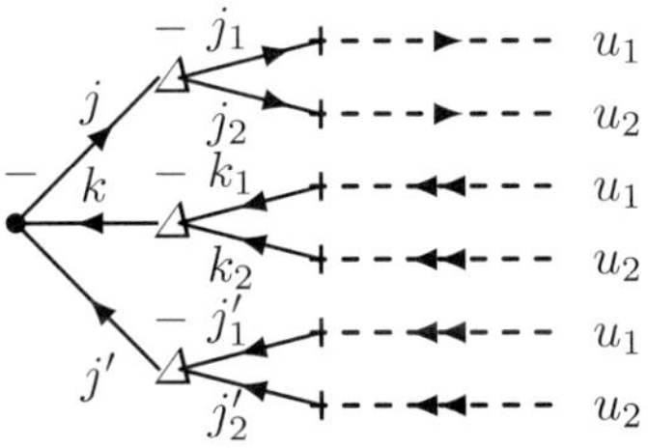

(B.3.160)

The integration over the variables u_1 (or u_2) is done diagrammatically by linking the three lines labelled by u_1 (or u_2) to form a new vertex. This introduces two reduced matrix elements, for $\boldsymbol{S}^{k_1}$ and $\boldsymbol{T}^{k_2}$, and a closed diagram:

$$\langle \gamma j_1 j_2 j \,\|\, \boldsymbol{R}^k \,\|\, \gamma' j_1' j_2' j' \rangle =$$ 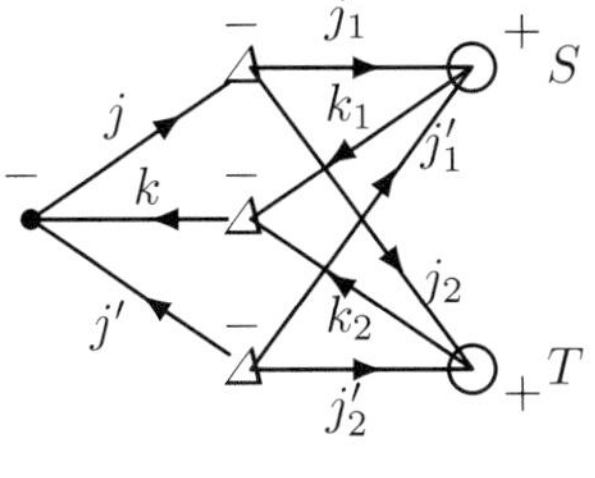

(B.3.161)

By pulling the right-hand vertex of the k line to the right, we can smoothly deform the diagram into the shape of (B.3.68) giving

$$(-1)^{j_1-j_1'+j_2-j_2'-k}$$ 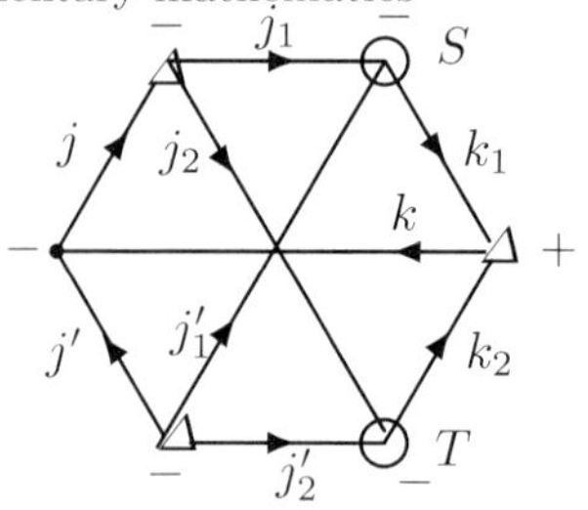 (B.3.162)

where the phase factor arises from change the sense of three of the nodes. Comparing this diagram with (B.3.68), we see that is necessary to change the sense of several nodes and the lines in order to put (B.3.162) into the standard form, with the final result being [14, Equation (3-35)]

$$\begin{aligned}&\langle\, \gamma j_1 j_2 j \,\|\, \boldsymbol{R}^k \,\|\, \gamma' j_1' j_2' j' \,\rangle = \\ &[jkj']^{1/2} \left\{ \begin{matrix} j_1 & j_1' & k_1 \\ j_2 & j_2' & k_2 \\ j & j' & k \end{matrix} \right\} \sum_{\gamma''} \langle\, \gamma j_1 \| \, \boldsymbol{S}^{k_1} \, \| \gamma'' j_1' \,\rangle \, \langle\, \gamma'' j_2 \| \, \boldsymbol{T}^{k_2} \, \| \gamma' j_2' \,\rangle. \end{aligned} \tag{B.3.163}$$

We shall need a number of results that are direct specializations of this equation:

- *Scalar product of two irreducible tensor operators*:
  In this case we set $k = 0$. Using (B.3.70) and then (B.3.64) we have

$$\left\{ \begin{matrix} j_1 & j_1' & k_1 \\ j_2 & j_2' & k_2 \\ j & j' & 0 \end{matrix} \right\} = \delta_{jj'} \delta_{k_1 K} \delta_{k_2 K} \frac{(-1)^{j_1'+j_2+K+j}}{[K, j]^{1/2}} \left\{ \begin{matrix} j_1' & j_2' & j \\ j_2 & j_1 & K \end{matrix} \right\}.$$

  Because $\boldsymbol{R}^0 = [\boldsymbol{S}^K \otimes \boldsymbol{T}^K]^0 = (-1)^K [K]^{-1/2} \boldsymbol{S}^K \cdot \boldsymbol{T}^K$, we have (replacing K by k)

$$\begin{aligned}\langle\, \gamma j_1 j_2 j \,\| \boldsymbol{S}^k \cdot \boldsymbol{T}^k \| \, \gamma' j_1' j_2' j' \,\rangle &= (-1)^{j_1'+j_2+j} \delta_{jj'} \\ \times \left\{ \begin{matrix} j_1' & j_2' & j \\ j_2 & j_1 & k \end{matrix} \right\} &\sum_{\gamma''} \langle\, \gamma j_1 \| \, \boldsymbol{S}^{k_1} \, \| \gamma'' j_1' \,\rangle \, \langle\, \gamma'' j_2 \| \, \boldsymbol{T}^{k_2} \, \| \gamma' j_2' \,\rangle. \end{aligned} \tag{B.3.164}$$

- $\boldsymbol{R}^k$ operates on $\mathcal{H}_1$ only; $k_2 = 0$:
  In this case, we get

$$\begin{aligned}\langle\, \gamma j_1 j_2 j \,\| \boldsymbol{S}^k \| \, \gamma' j_1' j_2' j' \,\rangle &= (-1)^{j_1+j_1'+k+j} \\ \times [j, j']^{1/2} \left\{ \begin{matrix} j & j' & k \\ j_1' & j_1 & j_2 \end{matrix} \right\} &\langle\, \gamma j_1 \| \, \boldsymbol{S}^k \, \| \gamma' j_1' \,\rangle. \end{aligned} \tag{B.3.165}$$

- The case in which $\boldsymbol{R}^k = \boldsymbol{S}^{k_1} \times \boldsymbol{T}^{k_2}$ gives

$$\begin{aligned}\langle\, \gamma J \,\| \boldsymbol{R}^k \| \, \gamma' J' \,\rangle &= [\, k\,]^{1/2} (-1)^{J+k+J'} \\ \times \sum_{\gamma'' J''} \left\{ \begin{matrix} k_2 & k & k_1 \\ J & J'' & J' \end{matrix} \right\} &\langle \gamma J \| \, \boldsymbol{S}^{k_1} \, \| \gamma'' J'' \rangle \langle \gamma'' J'' \| \, \boldsymbol{T}^{k_2} \, \| \gamma' J' \rangle. \end{aligned} \tag{B.3.166}$$

B.4 Relativistic symmetry orbitals for double point groups

General reference [19]

B.4.1 Construction of symmetry orbitals

The way in which the atomic nuclei are arranged in symmetric molecule is reflected in the symmetry properties of molecular orbitals [20, Chapters 5,9]. The LCAO method generates relativistic molecular orbitals as linear combinations of Dirac central field spinors (AS): see Appendix A.4 for definitions and properties. The symmetry properties of Dirac AS have been crucial to low-level techniques for evaluating matrix elements and efficient methods of computation for both atoms and molecules. On the other hand, the symmetry properties of molecular spinors (MS) have received less attention in this book. They can be generated conveniently by the `TSYM` program [19] to provide sets of MS that span an irreducible representation of a double point group in molecular structure calculations. In this book, `TSYM` has been used at the end of structure calculations to classify the orbital MS generated in molecular structure calculations on symmetric molecules. This can provide a powerful check on the numerical consistency of the computational scheme.

Following [19], let $D^{(i)}$ denote the i-th inequivalent irreducible representation of a molecular symmetry group. Its invariant subspaces constitute the *representation spaces* of $D^{(i)}$, which are spanned by the basis:

$$\{\,|\,\tau i\mu\rangle, \quad \mu = 1,\ldots,n_i\} \tag{B.4.1}$$

where μ indexes the n_i members of the basis and τ distinguishes different representation spaces of the same irrep $D^{(i)}$. Let S be a symmetry operation of the point group of the molecule. The effect of S on a basis vector can be represented by a linear operator $\widehat{S}$ given by

$$\widehat{S}\,|\,\tau,i,\mu\rangle = \sum_{\mu=1}^{n_i} D^{(i)}_{\nu\mu}(S)\,|\,\tau i\nu\rangle. \tag{B.4.2}$$

Assume throughout that the operators $\widehat{S}$ and the *representation matrices* $D^{(i)}(S)$ are unitary. Meyer *et al.* denote two-spinor AS centred on the a-th nucleus [19, Equation (2.4)] as

$$\langle \boldsymbol{r}\,|\,\rho a l j m\rangle = \rho_{nlj}(r_a)\,\mathcal{Y}_{ljm}(\theta_a,\varphi_a,1/2) \tag{B.4.3}$$

where ρ stands for a radial function. This employs the same structure and phase conventions as this book: for example we could interpret (B.4.3) by replacing it with G-spinors, (A.4.3), so that

$$\langle \boldsymbol{r}\,|\,\rho a l j m\rangle \to M[\beta;\gamma,\beta\kappa,\boldsymbol{r}_a] \equiv \frac{\omega_\beta}{r} R_{\gamma,\beta\kappa}(r)\,\chi_{\beta\kappa,m}(\theta_a,\varphi_a).$$

The angular parts are the same: $\mathcal{Y}_{ljm}(\theta_a, \varphi_a, 1/2) \equiv \chi_{\beta\kappa,m}(\theta_a, \varphi_a)$ and the notation of (A.4.3) identifies the two compoentns of a four-spinor with $\beta = \pm 1$. The index γ corresponds to the label nlj on the radial function in (B.4.3). We shall use the notation of Meyer *et al.* in what follows.

Operations S of the point group can be classified as *proper* (a pure rotation) or *improper* (a product of a pure rotation and a spatial inversion). Thus if α, β, γ are the Euler angles that rotate nucleus a to the position Sa, we can write

$$\widehat{S}\,|\,paljm\rangle = (-1)^{l\tau_S} \sum_{m'} D^j_{m'm}(\alpha\beta\gamma)\,|\,pSaljm'\rangle \tag{B.4.4}$$

where $D^j_{m'm}(\alpha\beta\gamma)$ is an element of the representation matrix of the three-dimensional rotation group SO(3) defined by(B.3.36),

$$\tau_S = \begin{cases} 1 \text{ if } S \text{ is an } \textit{improper} \text{ rotation,} \\ 0 \text{ if } S \text{ is a } \textit{proper} \text{ rotation.} \end{cases}$$

and $l = j + \eta/2$, where $\eta = \operatorname{sgn} \kappa$. We generate MS symmetry orbitals

$$|\,\tau i\mu\rangle := |\,paljm\nu i\mu\rangle$$

using projection operators

$$\widehat{P}^{(i)}_{\mu\nu} = \frac{n_i}{h} \sum_S D^{(i)\,*}_{\mu\nu}(S)\,\widehat{S} \tag{B.4.5}$$

acting on the AS so that

$$|\,\tau i\mu\rangle == \widehat{P}^{(i)}_{\mu\nu}\,|\,paljm\rangle. \tag{B.4.6}$$

When j is an odd multiple of $1/2$, each spatial transformation S is represented by *two* group operations, S and $\overline{S}$, differing by a rotation 2π so that

$$\widehat{\overline{S}}|\,paljm\rangle = \begin{cases} +\widehat{S}|\,paljm\rangle \text{ for integer } j, \\ -\widehat{S}|\,paljm\rangle \text{ for half-integer } j \end{cases}. \tag{B.4.7}$$

The group comprising all operations S and $\overline{S}$ is called a *double group*. All its irreps satisfy one or other of the relations

$$D^{(i)}(\overline{S}) = \pm D^{(i)}(S) \tag{B.4.8}$$

The Dirac 2- (or 4-) spinors are characterized by j values that are odd multiples of $1/2$ so that only the minus sign is relevant in the present context, and we shall assume from now on that this is the case. The MS symmetry orbitals can therefore be generated from the AS basis using

$$|\,\tau i\mu\rangle = |\,paljm\nu i\mu\rangle = \sum_{a'm'} C^{lj\nu i\mu}_{am,a'm'}\,|\,pa'ljm'\rangle \tag{B.4.9}$$

where $a' = Sa$ and the symmetry coefficients are given by

$$C^{lj\nu i\mu}_{am,a'm'} = \frac{n_i}{h} \sum_S \delta_{a',Sa}\, D^{(i)\,*}_{\mu\nu}(S)\,(-1)^{l\tau_S}\, D^j_{m'm}(\alpha\beta\gamma). \tag{B.4.10}$$

B.4.2 Linear independence of molecular symmetry orbitals

Suppose that we have A *equivalent* atoms whose positions are interchanged by the operations of the symmetry group. There are therefore $A(2j+1)$ vectors $\{|\,\rho aljm\rangle : m = -j,\ldots,j, a = 1,\ldots,A\}$ with fixed values of l, j and ρ. The application of (B.4.6) with varying values of i, μ, ν gives $Ah(2j+1)$ MS. However, because positions of the atoms are merely exchanged by the operations of the symmetry group, it suffices to consider only the $h(2j+1)$ MS generated from a single orbital on one atom centre, a. We conclude

1. If $h = A$, then the $h(2j+1)$ symmetry orbitals obtained by applying (B.4.6) to a single AS $|\,\rho aljm\rangle$ are linearly independent.
2. If $h > A$, then the number of representation spaces N_i of a given irrep $D^{(i)}$ contained in the space spanned by the $A(2j+1)$ AS is

$$N_i = \frac{1}{h}\sum_S \chi^{(i)\,*}(S) \sum_{a'm'} (-1)^{l\tau_S}\, D^j_{m'm'}(S)\, \delta_{a',Sa} \qquad \text{(B.4.11)}$$

where $\chi^{(i)}$ is a character of $D^{(i)}$. Then for each irrep $D^{(i)}$ we can find a basis of N_i linearly independent vectors

$$|\,\tau i\mu\rangle = |\,\rho aljm\nu i\mu\rangle, \quad \rho alji\mu \text{ fixed}, \qquad \text{(B.4.12)}$$

Then the $n_i N_i$ basis elements of the N_i basis systems

$$\{\,|\,\tau i\mu\rangle, \mu = 1,\ldots,n_i\}, \quad \tau = 1,\ldots,N_i \qquad \text{(B.4.13)}$$

of $D^{(i)}$ are linearly independent.

B.4.3 Reduction of operator matrices

The basis vectors constructed above satisfy the orthogonality relation

$$\langle \tau i\mu\,|\,\sigma k\lambda\rangle = \langle \tau i\|\,\sigma k\rangle \delta_{ik}\delta_{\mu\lambda} \qquad \text{(B.4.14)}$$

independent of the indices μ, λ. This is a special case of the Wigner-Eckart theorem (see B.2.26). The symmetry orbitals $|\,\tau i\mu\rangle$ are orthogonal with respect to i and μ but not, in general, with respect to τ. More generally, the matrix of any operator $\widehat{O}$ (such as the Hamiltonian or the overlap matrix) invariant under the symmetry group of a molecule satisfies

$$\langle \tau i\mu\,|\,\widehat{O}\,|\,\sigma k\lambda\rangle = \langle \tau i\|\,\widehat{O}\,|\,\sigma k\rangle \delta_{ik}\delta_{\mu\lambda} \qquad \text{(B.4.15)}$$

where $\langle \tau i\|\,\widehat{O}\,|\,\sigma k\rangle$ is the *reduced matrix element* of $\widehat{O}$, whose matrix decomposes into block diagonal form. The n_i diagonal blocks forming a representation of $D^{(i)}$ are equal. This can be exploited to reduce effective Hamiltonian matrix dimensions in molecular structure calculations.

B.4.4 Time reversal

The time reversal operator defined in (A.4.17) is

$$\widehat{\mathcal{T}} = \widehat{\mathcal{T}}_0 \otimes I_2 = \begin{pmatrix} \widehat{\mathcal{T}}_0 & 0 \\ 0 & \widehat{\mathcal{T}}_0 \end{pmatrix}. \tag{B.4.16}$$

where the operator $\widehat{\mathcal{T}}_0$ acts on two-spinor components so that

$$\widehat{\mathcal{T}}_0 \frac{F(r)}{r} \chi_{\kappa m}(\theta, \varphi) = (-1)^{l-j+m} \frac{\widehat{\mathcal{K}} F(r)}{r} \chi_{\kappa -m}(\theta, \varphi) \tag{B.4.17}$$

We assume from now on that the radial amplitudes $F(r)$ are *real* so that the space of orbital AS maps onto itself under time reversal. The spinors $\psi, \widehat{\mathcal{T}}\psi$ are said to constitute a *Kramers' pair* of states. If all interactions are invariant under time reversal then, in particular, $\widehat{\mathcal{T}}$ commutes with the Hamiltonian and both $\psi, \widehat{\mathcal{T}}\psi$ have the same energy eigenvalue E. This constitutes another useful practical check on the consistency and accuracy of our equations of motion.

We explore the connection between a representation $D^{(i)}$ and its image $D^{(i)*}$ under time-reversal. The symbol $\widehat{H}$ denotes any Hermitian operator which is invariant under both the symmetry group and time reversal.

A. $D^{(i)}$ and $D^{(i)*}$ are *equivalent to a real representation*; $D^{(i)}$ can therefore be taken as real, with $2n_i$ elements of the basis systems $\{\,|\,\tau i\mu\rangle\,\}$ and $\{\,|\,\overline{\tau} i\mu\rangle\,\}$ defined by

$$\begin{aligned} |\,\tau i\mu\rangle &= |\,\rho\, a\, l\, j\, m\, \nu\, i\, \mu\rangle, \\ |\,\overline{\tau} i\mu\rangle &= \widehat{\mathcal{T}}\,|\,\tau i\mu\rangle = (-1)^{l-j+m}|\,\rho\, a\, l\, j\, -m\, \nu\, i\, \mu\rangle. \end{aligned} \tag{B.4.18}$$

These are linearly independent, and

$$\begin{aligned} \langle \overline{\tau} i\,\|\widehat{H}\|\sigma i\rangle &= -\langle \tau i\,\|\widehat{H}\|\overline{\sigma} i\rangle^*, \\ \langle \overline{\tau} i\,\|\widehat{H}\|\overline{\sigma} i\rangle &= +\langle \tau i\,\|\widehat{H}\|\sigma i\rangle^* \end{aligned} \tag{B.4.19}$$

B. $D^{(k)} := D^{(i)*}$ and $D^{(i)}$ are *inequivalent.* In this case the basis systems $\{\,|\,\tau i\mu\rangle\,\}$ and $\{\,|\,\overline{\tau} k\mu\rangle\,\}$ with

$$\begin{aligned} |\,\tau i\mu\rangle &= |\,\rho\, a\, l\, j\, m\, \nu\, i\, \mu\rangle, \\ |\,\overline{\tau} k\mu\rangle &= \widehat{\mathcal{T}}\,|\,\tau i\mu\rangle = (-1)^{l-j+m}|\,\rho\, a\, l\, j\, -m\, \nu\, k\, \mu\rangle. \end{aligned} \tag{B.4.20}$$

span orthogonal subspaces, and matrix elements in the two subspaces satisfy

$$\langle \overline{\tau} k\,\|\widehat{H}\|\overline{\sigma} k\rangle = \langle \tau i\,\|\widehat{H}\|\sigma i\rangle^* \tag{B.4.21}$$

C. $D^{(i)}$ and $D^{(i)*}$ are *equivalent* ao that we can find a unitary matrix B, the same for all elements of the group, such that

$$D^{(i)*}(S) = B^{-1} D^{(i)} B \tag{B.4.22}$$

but it is not possible to transform $D^{(i)}$ into a real representation as in case A. It has been shown that the elements of B can be written

$$B_{\mu\lambda} = i\, f_{\lambda}^{(i)}\, \delta_{\overline{\mu},\lambda}, \quad \overline{\mu} = n_i - \mu + 1,\; f_{\lambda}^{(i)} = \pm 1,\; f_{\overline{\lambda}}^{(i)} = -f_{\lambda}^{(i)}. \tag{B.4.23}$$

The unitarity of B now shows that

$$D_{\mu\nu}^{(i)*}(S) = f_{\mu}^{(i)} f_{\nu}^{(i)} D_{\overline{\mu}\overline{\nu}}^{(i)*}(S) \tag{B.4.24}$$

There are two sub-cases with this structure:

C1. The basis systems with direct and reversed time span *different representation spaces* of $D^{(i)}$. Then the $2n_i$ elements of the two systems are linearly independent, and

$$\widehat{\mathcal{T}}\,|\,\tau_1 i\mu\rangle = f_{\mu}^{(i)}\,|\,\overline{\tau}_1 i\overline{\mu}\rangle \tag{B.4.25}$$

where

$$\begin{aligned} |\,\tau_1 i\,\mu\rangle &= |\,\rho\, a\, l\, j\, m\, \nu\, i\, \mu\rangle \\ |\,\overline{\tau_1} i,\,\mu\rangle &= (-1)^{l-j+m} f_{\nu}^{(i)}\,|\,\rho\, a\, l\, j\, -m\, \overline{\nu}\, i\, \mu\rangle. \end{aligned}$$

The matrix elements satisfy

$$\begin{aligned} \langle \overline{\tau}_1 i\,\|\widehat{H}\|\sigma_1 i\rangle &= -\langle \tau_1 i\,\|\widehat{H}\|\overline{\sigma}_1 i\rangle^*, \\ \langle \overline{\tau}_1 i\,\|\widehat{H}\|\overline{\sigma}_1 i\rangle &= +\langle \tau_1 i\,\|\widehat{H}\|\sigma_1 i\rangle^*. \end{aligned} \tag{B.4.26}$$

C2. The basis systems with direct and reversed time span *the same representation space* of $D^{(i)}$. In this case only n_i linearly independent basis vectors can be constructed and

$$\widehat{\mathcal{T}}\,|\,\tau_2 i\mu\rangle = f_{\mu}^{(i)} \xi_{\tau_2}^{(i)}\,|\,\tau_2 i\overline{\mu}\rangle \tag{B.4.27}$$

where

$$|\,\tau_2 i\mu\rangle = |\,\rho\, a\, l\, j\, m\, \nu\, i\, \mu\rangle.$$

Writing

$$|\,\overline{\tau}_2 i\mu\rangle = f_{\overline{\mu}}^{(i)} \widehat{\mathcal{T}}\,|\,\tau_2 i\overline{\mu}\rangle \tag{B.4.28}$$

results in the complex phase factor $\xi_{\tau_2}^{(i)}$ defined by

$$\xi_{\tau_2}^{(i)} = \langle \tau_2 i\,\|\overline{\tau}_2 i\rangle / \langle \tau_2 i\,\|\tau_2 i\rangle, \tag{B.4.29}$$

and

$$\langle \tau_2 i\,\|\widehat{H}\|\sigma_2 i\rangle = \xi_{\tau_2}^{(i)} \xi_{\sigma_2}^{(i)*} \langle \tau_2 i\,\|\widehat{H}\|\sigma_2 i\rangle^* \tag{B.4.30}$$

For both cases C1 and C2, the matrix elements coupling symmetry orbitals satisfy

$$\begin{aligned}\langle \overline{\tau}_1 i \,\|\widehat{H}\|\sigma_2 i\rangle &= \xi^{(i)*}_{\sigma_2}\langle \tau_1 i \,\|\widehat{H}\|\sigma_2 i\rangle^* \\ \langle \tau_2 i \,\|\widehat{H}\|\overline{\sigma}_1 i\rangle &= \xi^{(i)*}_{\tau_2}\langle \tau_2 i \,\|\widehat{H}\|\sigma_1 i\rangle^*\end{aligned} \tag{B.4.31}$$

It is possible to determine whether a given irrep $D^{(i)}$ is of type A, B or C by using the Frobenius-Schur test

$$\frac{1}{h}\sum_S \chi^{(i)}(S^2) = \begin{cases} 1 \text{ in Case A,} \\ 0 \text{ in Case B,} \\ -1 \text{ in Case C,}\end{cases} \tag{B.4.32}$$

and cases C1 and C2 can be distinguished by testing for equality of $|\langle \tau i \,\|\overline{\tau} i\rangle|$ and $\langle \tau i \,\|\tau i\rangle$ where

$$\begin{aligned}|\,\tau i\,\mu\rangle &= |\,\rho\, a\, l\, j\, m\, \nu\, i\, \mu\rangle \\ |\,\overline{\tau} i\,\mu\rangle &= (-1)^{l-j+m} f^{(i)}_{\nu}|\,\rho\, a\, l\, j\, -m\,\overline{\nu}\, i\, \mu\rangle\end{aligned}$$

This is satisfied in case C2 but not in case C1.

B.4.5 The TSYM software package

The TSYM package [19] computes the coefficients of a complete set of linearly independent relativistic symmetry orbitals for 45 different point groups. Data are provided for the groups $C_n, C_{nh}, S_{2n}, C_{nv}, D_n, D_{nh}, D_{nd}$ for $n = 1, \ldots, 6$; cubic groups; the tetrahedral groups T, T_h, T_d; the octahedral groups O, O_h; and the icosahedral groups I, I_h. As supplied, the package is limited to systems with no more than 20 atoms and no more than 10 sets of equivalent atoms. The angular quantum numbers are limited by $0 \leq l \leq 5$ and $1/2 \leq j \leq 9/2$. The user has to supply the positions of the atoms and the symmetry group of interest, and the program generates the coefficients $C^{lj\nu i\mu}_{am,a'm'}$ defined by (B.4.10). Full details of program usage are given in the writeup [19].

B.5 Basis sets in atomic and molecular physics

General references Chapter 5, [2], [3], [4] [21], [22]

Conventional texts on quantum mechanics highlight the use of complete orthonormal sets (CNOS) of functions. CNOS have many merits, but they are not always as well suited to calculation on modern digital computers as are certain types of non-orthogonal set. Unfortunately the theory of approximation by sets of non-orthogonal functions is rather more complicated than that of CNOS. We therefore survey some of the issues that must be faced in choosing basis sets for nonrelativistic atomic and molecular physics which will later be generalized for use with Dirac spinors.

We work in a separable Hilbert space $\mathcal{H}$ having a norm $\|\cdot\|$ and inner product $(\cdot,\cdot)$. In nonrelativistic quantum mechanics, the relevant spaces are $L^2(R^s)$ and related spaces, or else Sobolev spaces defined by

$$W_2^{(r)}(R^s) := \{f \in L^2(R^s) \,|\, (c + \mathbf{p}^2)^{r/2}\widehat{f}(\mathbf{p}) \in L^2(R^s)\} \tag{B.5.1}$$

where $\widehat{f}(\mathbf{p})$ denotes the Fourier transform of f, c is a positive real number, and r a non-negative integer. It is clear that

$$W_2^{(r+1)}(R^s) \subset W_2^{(r)}(R^s) \subset L^2(R^s), \qquad r = 1, 2, 3, \ldots \tag{B.5.2}$$

The norms of the spaces $W_2^{(r)}(R^s)$ are stronger than those of $L^2(R^s)$.

Now suppose that we construct a sequence of approximations

$$\Psi_N = \sum_{n=1}^{N} c_n^N \phi_n \tag{B.5.3}$$

where $\{\phi_n\}_{n=1}^{\infty}$ is an infinite set of basis functions. In general the coefficients c_n^N depend on both the number of basis functions N used and the method of determination. However, if the basis set $\{\phi_n\}_{n=1}^{\infty}$ is *complete*, then we can determine these coefficients so that

$$\|\Psi_N - \Psi\| \longrightarrow 0, \qquad N \longrightarrow \infty, \tag{B.5.4}$$

by, for example, a Rayleigh-Ritz process. Thus the capacity to approximate Ψ is settled once we have demonstrated that $\{\phi_n\}_{n=1}^{\infty}$ is complete in the appropriate Hilbert space.

The behaviour of the coefficients $\{c_n^N\}_{n=1}^{N}$ as $N \longrightarrow \infty$ is an entirely different question that is related to the linear independence properties of the basis set $\{\phi_n\}_{n=1}^{\infty}$. Klahn [22] discusses these properties by examining the answers to four questions:

1. If the sequence $\{\Psi_N\}$ satisfies (B.5.4) does it follow that the sequence $c_n^{(N)}, N = n+1, n+2, \ldots$ has a *unique* limit limit c_n, *independent of the method of construction* for each value $n = 1, 2, \ldots$? If so, we can say that Ψ has the *formal expansion*

$$\Psi \sim \sum_{n=1}^{\infty} c_n \phi_n. \tag{B.5.5}$$

2. Does the formal expansion (B.5.5) *represent* Ψ in the sense that

$$\left\| \sum_{n=1}^{N} c_n \phi_n - \Psi \right\| \longrightarrow 0 \text{ as } N \longrightarrow \infty \text{ ?} \tag{B.5.6}$$

   In that case we may write

$$\Psi = \sum_{n=1}^{\infty} c_n \phi_n \tag{B.5.7}$$

in place of (B.5.5).

3. Does Ψ always determine the expansion coefficients c_n uniquely?
4. Do the expansion coefficients c_n determine the function Ψ uniquely?

There are a number of inequivalent criteria of linear independence for infinite sets of functions in Hilbert spaces [22]. We start by considering finite-dimensional unitary spaces $\mathcal{H}_M$ of dimension M with inner product $(\cdot, \cdot)$. Let $\{\phi_n\}_{n=1}^N$ be a set of functions in $\mathcal{H}_M$, and let $[\phi_n]_{n=1}^N$ denote its linear span. Thus any $\Psi \in [\phi_n]_{n=1}^N$ has an expansion

$$\Psi = \sum_{n=1}^{N} c_n \phi_n$$

for suitable complex numbers $c_n, n = 1, \ldots, N$. We say that $\{\phi_n\}_{n=1}^N$ is *complete* if

$$\mathcal{H}_M = [\phi_n]_{n=1}^N$$

so that *any* $\Psi \in \mathcal{H}_M$ has such an expansion. Of course, the expansion coefficients are unique only if the system $\{\phi_n\}_{n=1}^N$ is *linearly independent*; a complete set is linearly independent if and only if $M = N$.

Linear independence of $\{\phi_n\}_{n=1}^N$ in $\mathcal{H}_M$ can be characterized by the following equivalent criteria:

- $\sum_{n=1}^N c_n \phi_n = 0$ if and only if $c_n = 0, n = 1, \ldots, N$.
- There exists a *biorthonormal set* (BOS) $\{\phi_n^*\}_{n=1}^N$ in $\mathcal{H}_M$ such that, for all $m, n = 1, 2, \ldots, N$, $(\phi_m^*, \phi_n) = \delta_{mn}$.
- The lowest eigenvalue μ_1^N of the *Gram* matrix $\mathbf{S}^{(N)}(\phi_1, \ldots, \phi_n)$, with elements given by $S_{mn}^{(N)} = (\phi_m, \phi_n)$, is a positive number[3].
- The *Gram determinant* $\gamma_N(\phi_1, \ldots, \phi_n = \det \mathbf{S}^{(N)}(\phi_1, \ldots, \phi_n)$ is a positive number.

Suppose that $\Psi \in [\phi_n]_{n=1}^N$; then if $\{\phi_n^*\}_{n=1}^N$ is linearly independent and therefore has a BOS, we can write

$$\Psi = \sum_{j=1}^{N} (\Psi, \phi_j^*) \phi_j$$

so that, in particular,

$$\phi_i^* = \sum_{j=1}^{N} S_{ij}^{(N)*} \phi_j, \quad S_{ij}^{(N)*} = (\phi_i^*, \phi_j^*)$$

[3] $\mathbf{S}^{(N)}(\phi_1, \ldots, \phi_n)$ is usually called the *overlap* matrix in the quantum chemistry literature.

Similarly

$$\phi_i = \sum_{i=1}^{N} S_{ij}^{(N)} \phi_j^*, \quad S_{ij}^{(N)} = (\phi_i, \phi_j)$$

The first of these equations shows that the elements ϕ_i^* are always in $[\phi_n]_{n=1}^N$ in the finite dimensional case. It also follows that

$$\mathbf{S}^{(N)*}\mathbf{S}^{(N)} = \mathbf{S}^{(N)}\mathbf{S}^{(N)*} = \mathbf{I}^N,$$

where $\mathbf{I}^N$ is the identity.

Things are a good deal more complicated in infinite-dimensional Hilbert spaces. Let $\{\phi_n\}_{n=1}^{\infty}$ be an infinite set of functions in the Hilbert space $\mathcal{H}$ and let $[\phi_n]$ be its linear span and $\overline{[\phi_n]}$ its closure in $\mathcal{H}$. Then any $\Psi \in \overline{[\phi_n]}$ can be approximated by $\{\phi_n^*\}_{n=1}^{\infty}$ in the sense of (B.5.4). This is obvious if $\{\phi_n^*\}_{n=1}^{\infty}$ is complete in $\mathcal{H}$. However, invoking completeness does not answer any of the questions posed about the expansion coefficients of a given Ψ. There are several linear independence criteria of increasing strength that help to do so:

- **Finite linear independence** The minimum requirement is that $\{\phi_n\}_{n=1}^{\infty}$ has a linearly independent finite subset $\{\phi_n\}_{n=1}^{N} \subset \mathcal{H}$, where $N = 1, 2, 3, \ldots$. Although this permits unique expansions of each $\Psi \in [\phi_n]_{n=1}^N$ it tells us nothing about the convergence of its Fourier coefficients as $N \longrightarrow \infty$.
- **ω-linear independence** A set $\{\phi_n\}_{n=1}^{\infty}$ is said to be *ω-linearly independent* if the relation

$$\left\| \sum_{n=1}^{N} c_n \phi_n \right\| \longrightarrow 0 \text{ as } N \longrightarrow \infty,$$

  where c_n are complex numbers, implies $c_n = 0, N = 1, 2, \ldots$. Clearly ω-linear independence implies finite linear independence, but the converse is false. For example, suppose that $\{\phi_n\}_{n=2}^{\infty}$ is a complete orthonormal set and $\phi_1 \in \mathcal{H}$ but $\phi_1 \notin \{\phi_n\}_{n=2}^{\infty}$. Then $\{\phi_n\}_{n=1}^{\infty}$ is finitely linearly independent but ω-linearly dependent; for if $c_1 = -1$ and $c_n = (\phi_1, \phi_n), n \geq 2$ then $\left\|\sum_{n=1}^{N} c_n\phi_n\right\| \longrightarrow 0$ as $N \longrightarrow \infty$. Similarly, a basis set that is the union of two complete orthonormal subsystems on different atomic centres is also finitely linearly independent but ω-linearly dependent.
- **Minimality** A set $\{\phi_n\}_{n=1}^{\infty}$ is said to be *minimal* (or *strongly linearly independent*) in $\mathcal{H}$ if the relation $\phi_k \notin \overline{[\phi_1, \ldots, \phi_{k-1}, \phi_{k+1}, \ldots]}$ holds for all $k = 1, 2, \ldots$ The following characterizations can be shown to be equivalent:
  A. $\{\phi_n\}_{n=1}^{\infty}$ is minimal in $\mathcal{H}$.
  B. The distance

$$\delta_k = \lim_{N\to\infty} \min_{c_n \in C} \left\| \phi_k - \sum_{i=1, i\neq k}^{N} c_i \phi_i \right\| \tag{B.5.8}$$

  is a positive number $\forall\ k = 1, 2, \ldots$.

C. There exists a BOS $\{\phi_n^*\}_{n=1}^{\infty}$ in $\mathcal{H}$ such that $(\phi_m^*, \phi_n) = \delta_{mn}$.

D. The relation

$$\left\| \sum_{n=1}^{N} c_n^N \phi_n \right\| \longrightarrow 0 \text{ as } N \longrightarrow \infty, n = 1, 2, \ldots$$

where c_n^N are complex numbers implies $c_n^N \longrightarrow 0$.

E. For all positive integers N there exist constants C_N with $1 \leq C_N < \infty$ such that

$$\left\| \sum_{n=1}^{N} c_n \phi_n \right\| \leq C_N \left\| \sum_{n=1}^{N+M} c_n \phi_n \right\|$$

for all integers $M > 0$ and all complex c_n.

The definition of minimality ensures that no one element ϕ_k of a minimal set can be approximated with arbitrary precision by the other elements of the set. Thus deletion of any one element from $\{\phi_n\}_{n=1}^{\infty}$, say the element ϕ_k, makes the set *in*-complete. On the other hand, if a complete set is nonminimal, there exists at least one element ϕ_k such that $\phi_k \in [\phi_1, \ldots, \phi_{k-1}, \phi_{k+1}, \ldots]$; then $\{\phi_n\}_{n=1}^{\infty}$ is said to be *over*-complete, as the remaining subset is still complete. As a corollary of C, we see that a complete set is *exactly complete* if and only if it has a BOS, and is *overcomplete* if it does not.

The characterization C also shows that minimal basis sets are always ω-linearly independent. For if $\Psi = \sum_{n=1}^{\infty} c_n \phi_n = 0$, then the existence of a BOS allows us to conclude that $c_n = (\phi_n^*, \Psi) = 0$. However, it is possible for ω-linearly independent sets to be nonminimal; it is only necessary to demonstrate the existence of an overcomplete ω-linearly independent set. Examples of such sets are given in Klahn's very comprehensive discussion.

Because a minimal set has a BOS, the expansion coefficients can be determined from the equations

$$c_n^N = (\phi_n^*, \Psi_N) = \left(\phi_n^*, \sum_{i=1}^{N} c_i^N \phi_i \right) \;\forall\, n = 1, 2, \ldots$$

independently of the method of calculating the c_n^N. It can also be shown [22, p. 170] that the rate of convergence is given by

$$|c_n^N - (\phi_n^*, \Psi)| \leq \delta_n^{-1} \|\Psi_N - \Psi\| \;\forall\, n = 1, 2, \ldots$$

where δ_n was defined in (B.5.8). Thus if $\{\phi_n\}_{n=1}^{\infty}$ is an exactly complete basis set, the terms of the formal biorthonormal expansion

$$\Psi \sim \sum_{i=1}^{\infty} (\phi_n^*, \Psi) \phi_n$$

are well-defined for all $\Psi \in \mathcal{H}$. When $\{\phi_n\}_{n=1}^{\infty}$ is minimal and incomplete, then the formal expansion remains well-defined for all $\Psi \in [\phi_n]$.

We have already seen in §5.5 that the Rayleigh-Ritz method for nonrelativistic Hamiltonians converges provided the underlying basis set is complete in the appropriate Hilbert space. We now see that this need not guarantee that the expansion coefficients c_n^N converge as $N \longrightarrow \infty$. However, if the basis set is exactly complete in the Hilbert space, it is both complete and minimal, and then the Rayleigh-Ritz method also constructs the formal expansion of Ψ as $N \rightarrow \infty$.

B.5.1 The Coulomb Sturmian functions

The nonrelativistic Sturmian functions were defined by [23, 24] as the normalised solutions of the differential equation

$$\left[-\frac{d^2}{dr^2} + \frac{l(l+1)}{r^2} - 2E_0 + 2\alpha_{nl}V(r)\right] S_{nl}(r) = 0, \quad 0 < r < \infty, \tag{B.5.9}$$

vanishing at the endpoints $r = 0$ and $r = \infty$. The integers $n = 1, 2, 3, \ldots$ and $l = 0, 1, \ldots n-1$ correspond to the usual nonrelativistic quantum numbers and E_0 is a fixed, negative number. The parameter α_{nl} must be adjusted to ensure that the boundary conditions are satisfied. The functions are orthonormal with respect to the weight function $V(r)$ (which must be strictly of one sign, usually negative) so that

$$\int_0^\infty S_{nl}(r) S_{n'l}(r) V(r) dr = -\delta_{nn'}. \tag{B.5.10}$$

The most important case is that in which $V(r)$ is a Coulomb potential

$$V(r) = -\frac{Z}{r}, \quad 0 < r < \infty.$$

We set $E_0 = -\lambda^2/2$, and rewrite (B.5.9) in terms of the independent variable $x = 2\lambda r$, so that

$$\left[-\frac{d^2}{dx^2} + \frac{l(l+1)}{x^2} + \frac{1}{4} - \frac{\alpha_{nl} Z}{\lambda x}\right] S_{nl}(x) = 0, \tag{B.5.11}$$

having the solutions

$$S_{nl}(x) := \mathcal{N}_{nl}\, s_{nl}(x), \quad s_{nl}(x) = e^{-x/2} x^{l+1} L_{n-l-1}^{2l+1}(x), \quad n = l+1, l+2, \ldots, \tag{B.5.12}$$

which vanish at $x = 0$ and $x = \infty$ provided

$$n = \alpha_{nl} Z/\lambda.$$

The $L_k^\alpha(x)$ are Laguerre polynomials [25], and $\mathcal{N}_{nl}$ is a normalization constant. We recover the standard solutions for the radial hydrogenic eigenfunctions when $\alpha_{nl} = 1$, giving $E_0 = -Z^2/2n^2$. Notice that $\lambda = \sqrt{-2E_0}$ (and therefore

E_0) is fixed for the Coulomb Sturmians whereas λ depends upon n for the Schrödinger eigenfunctions.

The properties of the orthogonal polynomials $L_k^\alpha(x)$ are listed in many compilations such as [25]. When $\alpha \geq 1$ they are orthogonal on $(0, \infty)$ with weight function $w(x) = e^{-x}x^\alpha$, such that

$$\int_0^\infty e^{-x}x^\alpha L_k^\alpha(x)L_{k'}^\alpha(x)dx = \frac{\Gamma(\alpha+k+1)}{k!}\,\delta_{k,k'}. \tag{B.5.13}$$

The generating function

$$\Phi^{(\alpha)}(t,s) := \sum_{k=0}^\infty L_k^\alpha(s)t^k = (1-t)^{-\alpha-1}\exp\left(\frac{ts}{t-1}\right), \quad |t| < 1, \tag{B.5.14}$$

can be used to write down explicit representations of the polynomials. It also provides an economical means of evaluating integrals of the form

$$\langle kl|x^p|k'l\rangle = \int_0^\infty s_{kl}(x)x^p s_{k'l}(x)dx$$

for integer values of p for which this integral exists by identifying the coefficient of $t^k u^{k'}$ in the integral

$$\begin{aligned} I_l^{(p)}(t,u) &= \int_0^\infty e^{-x}x^{2l+2+p}\,\Phi^{(2l+1)}(t,x)\Phi^{(2l+1)}(u,x)dx \\ &= \frac{(2l+2+p)!}{(1-tu)^{2l+3+p}}[(1-t)(1-u)]^{1+p}. \end{aligned} \tag{B.5.15}$$

Two cases have immediate application:

A. $p = -1$: then

$$I_l^{(-1)}(t,u) = \frac{(2l+1)!}{(1-tu)^{2l+2}} = \sum_{k=0}^\infty \frac{(2l+k+1)!}{k!}(tu)^k$$

from which we obtain

$$\langle kl|x^{-1}|k'l\rangle = \frac{(2l+k+1)!}{k!}\,\delta_{k,k'}, \tag{B.5.16}$$

which agrees with (B.5.13) if we put $\alpha = 2l+1$ and $k = n-l-1$.

B. $p = 0$: This gives the Gram (overlap) matrix, G, of Coulomb Sturmians. In this case

$$I_l^{(0)}(t,u) = \frac{(2l+2)!}{(1-tu)^{2l+3}}[(1-t)(1-u)]$$

so that there are nonvanishing matrix elements for $k' = k \pm 1$ as well as for $k' = k$. We shall normalize the Sturmians so that

$$\langle kl|k'l\rangle = \delta_{k,k'}, \tag{B.5.17}$$

so that, remembering $k = n - l - 1$, (B.5.16) gives

$$\mathcal{N}_{nl} = \left[\frac{(n-l-1)!}{2n(n+l)!}\right]^{1/2}. \tag{B.5.18}$$

The nonvanishing elements of the Gram matrix, are thus

$$g^l_{nn} = 1, \quad g^l_{n,n+1} = g^l_{n+1,n} = -\frac{1}{2}\sqrt{1 - \frac{l(l+1)}{n(n+1)}}, \quad n = l+1, l+2, \ldots \tag{B.5.19}$$

B.5.2 Completeness and linear independence of Coulomb Sturmians

The classical proof that the orthonormal system $(2/x)^{1/2} S_{nl}(x)_{n=l+1}^{\infty}$ is complete in $L^2(\boldsymbol{R}_+)$ is given by [26, p. 170],[27, p. 95]. It follows that if we define the functions on $\boldsymbol{R}^3 \to \boldsymbol{R}$ by

$$\phi_{nlm}(\boldsymbol{r}) = \frac{S_{nl}(r)}{r} Y_{lm}(\theta, \varphi), \quad n \geq l+1, \quad |m| \leq l,$$

where (r, θ, φ) are spherical polar coordinates of the position $\mathbf{r}$ and $Y_{lm}(\theta, \varphi)$ are spherical harmonics, then the set $\{(2/r)^{1/2}\phi_{nlm}(\boldsymbol{r})\}$ is a complete orthonormal system in $L^2(\boldsymbol{R}^3)$ [21, Lemma 6, p. 31]. Although the weighted Sturmians are therefore a complete orthonormal set in $L^2(\boldsymbol{R}^3)$, it is often more convenient to use the unweighted Sturmians, so that the Gram matrix of the set $\{\phi_{nlm}(\boldsymbol{r})\}$ is tridiagonal in each infinite lm-subset. The completeness and linear independence of the Coulomb Sturmians has to be reconsidered in this context.

The existence of BOS with elements $S^*_{nl}(x) = (2/x) S_{nl}(x)$ and $\phi^*_{nlm}(x) = (2S_{nl}(x)/x^2) Y_{lm}(\theta, \varphi)$ respectively is easily established. Thus the set $\{\phi_{nlm}(x)\}$ is minimal and complete on the Sobolev spaces $W_2^{(p)}(\boldsymbol{R}^3)$ for $p = 1, 2$, showing that we can use L-spinor approximation in Rayleigh-Ritz calculations in atomic and molecular problems.

Gerschgorin's circle theorem [28, 29] allows us to verify that any N-dimensional subset of the Coulomb Sturmians has a positive definite tridiagonal Gram matrix, $G^{(N)}$, when $l > 0$. Each eigenvalue, μ, is located in the union of intervals $1 - \rho_n \leq \mu \leq 1 + \rho_n$, where

$$\rho_n = |g_{n,n-1}| + |g_{n,n+1}|, \quad n = l+1, l+2, \ldots, l+N,$$

so that

$$\rho_{l+1} = |g_{l+1,l+2}| = \frac{1}{\sqrt{2l+4}}, \quad \rho_n \sim 1 - \frac{l(l+1)}{n^2} + O(n^{-3}), \quad n \gg l+1.$$

Thus all eigenvalues lie in the interior of the interval (0,2) so that $G^{(N)}$ is strictly positive. Its condition number is $k_N = (1+\rho_N)/(1-\rho_N) \sim 2N^2/l(l+1)$ when N is large, so that the system is very well conditioned and there is little danger of linear dependence problems in practice. This estimate fails when $l = 0$, as $g_{n,n+1} = -1/2$ independent of the value of n. However the Gram matrix is still positive definite, since $G^{(N)}$ is a tridiagonal matrix with diagonal elements 1 and sub- and super-diagonal elements -1/2.

B.5.3 Basis sets of exponential-type functions

The use of functions of the general form

$$\text{const. } r^l \exp(-\lambda_k r^p) Y_l^m(\theta, \varphi)$$

as primitives for constructing radial wavefunctions is almost universal in *ab initio* electronic structure calculations for atoms and molecules in nonrelativistic quantum chemistry. The case $p = 1$, usually known as Slater-type functions (STF), is most useful for atomic problems, as most schemes for constructing the electron-electron interaction integrals involving more than one nuclear centre are somewhat tedious. The case $p = 2$, the Gaussian-type functions (GTF), have dominated relativistic molecular structure calculations because of the relative ease of evaluating electron-electron interaction integrals.

The quality of this sort of approximation depends upon the choice of the exponents $\{\lambda_\mu\}$. Historically, hardware limitations have dictated the use of basis sets of minimal dimension, and much effort has been devoted to the tedious business of optimizing the exponent set for individual atomic ground states. The results for GTF have been recorded in extensive tables such as those of [30, 31, 32]. These tables do not cover elements heavier than the lanthanides. Completeness issues are essentially irrelevant as far as minimal basis sets are concerned. Linear independence is, on the other hand, essential whatever the size of the basis set. However, if we are to construct a practical computational scheme for relativistic many-body theory and quantum electrodynamics using basis sets of analytic functions, we should aim to use families of functions with suitable completeness properties in an appropriate Hilbert space. This ensures that even if we are compelled in reality to perform calculations with a finite subset of functions, we can in theory expect the calculations to converge to whatever precision we demand by increasing the size of the basis set. The careful review by Klahn [22] of linear independence properties of infinite sets of functions used in quantum chemistry is an invaluable guide to the technical complexities of the subject.

The mathematical properties of expansions in exponential type functions are directly applicable to S- and G-spinors. The tabulated exponent sets [30, 31, 32] for STF and GTF basis sets have usually been carefully and laboriously optimized against a particular test criterion, normally the ground state energy of the target atomic or molecular system. In S-spinor atomic calculations with point nuclei, nonrelativistic STF parameters can be taken over more or less unchanged, as the functional forms (5.9.1) incorporate most of the relativistic changes to the radial amplitudes in the critical region near the nuclei. The linear independence and completeness properties of STF sets then determine the rate of convergence of the approximation as the basis set is enlarged in much the same way as they do in nonrelativistic calculations. For G-spinor basis sets, the position is very similar. Nonrelativistic GTF exponent sets for the same target atom or molecule can often be adopted with little change. However, the nucleus must be modelled as a finite size charge distribution, and it is often desirable to add another G-spinor exponent to describe the rapid variation of the wave function in the nuclear region.

One can often avoid much of the effort of optimizing basis sets by using geometric sequences of exponents of the form

$$\lambda_n = \alpha\beta^{n-1}, \; n = 1, 2, \ldots, N. \tag{B.5.20}$$

These sets are often said to be 'even-tempered', a musical analogy that apparently derives from the observation that exponent sets satisfying the condition

$$\log \lambda_n = A + Bn, \; n = 1, 2, \ldots, N$$

for some constants A and B are found to be optimal for atomic self-consistent field calculations [33]. Such choices have a lot to recommend them as the required integrals can be expressed in terms of functions of the single parameter β with α appearing only in external scale factors. However, the theory of such basis sets is a little complicated by the fact that the set

$$\lambda_n = \alpha\beta^{n-1}, \; n = 1, 2, \ldots, \infty \tag{B.5.21}$$

is *overcomplete* in $\mathrm{L}^2(\mathcal{R}_+)$. This diagnosis rests on a form of the Müntz theorem [22, p. 159]; this states that an STF set with *positive* parameters $\{\lambda_n\}_{n=1}^{\infty}$ is complete in $\mathrm{L}^2(\mathcal{R}_+)$ if and only if

$$\sum_{n=1}^{\infty} \lambda_n/(1 + \lambda_n^2) = \infty. \tag{B.5.22}$$

The divergence of the infinite series is not affected by the deletion of a finite, or even an infinite, number of terms so that if the set is indeed complete it is over-complete in a quite spectacular way. Of course, we should like a *minimal set*: one in which the deletion of even a single member results in an incomplete set. On the face of it, this renders complete STF sets unsuitable for use in variational calculations. However, there is a converse of the Müntz theorem [22, p. 175], which states that an STO set is *minimal* in $\mathrm{L}^2(\mathcal{R}_+)$ if and only if

$$\sum_{n=1}^{\infty} \lambda_n/(1 + \lambda_n^2) < \infty \tag{B.5.23}$$

so that it is then incomplete! This is indeed the case with even-tempered sets, for which (B.5.23) holds for any fixed positive values of α or β. These sets are also Riesz systems [22, p. 191], defined as sets which are both *Hilbertian* (that is the eigenvalues of the Gram matrix are all finite) and *Besselian* (the eigenvalues of the Gram matrix are all positive). In effect, this gives most of what we need for practical calculations, although there remain elements of $L^2(\mathcal{R}_+)$ that cannot be expanded in an even-tempered set.

A convenient procedure for extending even-tempered basis sets has been proposed by Schmidt and Ruedenberg [35] and by Feller and Ruedenberg [36]. The idea is to construct a family of geometric parameters (B.5.21) in which the parameters α and β are chosen so that (B.5.22) is satisfied in the limit $N \to \infty$. They choose two monotone decreasing sequences of parameters $\alpha_N \to 0$ and $\beta_N \to 1$. In the form given by Wilson, [37], these are defined by

$$\ln \beta_N = \left(\frac{N-1}{N} \right)^{1/2} \ln \beta_{N-1}, \quad \alpha_N = \left(\frac{\beta_N - 1}{\beta_{N-1} - 1} \right)^{1/2}.$$

In practice, extrapolation of such systematic sequences of basis sets with N gives results that agree very well with those of well-converged finite difference numerical integration algorithms. However, if the parameter β approaches too close to unity, the condition number of the Gram matrix becomes very large as N increases, and this sets a practical upper bound to the dimension of the basis set as numerical instabilities set in.

Why, then, do the extrapolated results for bound states from such even-tempered sets appear to converge to the same limit as those from finite difference calculations? The largest radial distance that can be represented by the basis set is of order λ_1^{-1}, and the smallest is of order λ_N^{-1}, so that functions with support at greater or lesser distances are not representable within the set. This is of little consequence for most bound state wavefunctions provided the STFs span a wide enough range, and our intuitive feeling that such functions can be well represented is fully justified. Functions in the continuum are non-normalizable, and therefore one expects no convergence of the eigenvalue or of the wavefunction. The main effect of the basis set is to confine the particle within a distance which is some multiple of λ_1^{-1}, and we find that the function closely approximates continuum solutions of corresponding energy at smaller distances. In the finite difference case, the precision of the result depends upon the choice of grid, the size if the region on which it is placed, and the way in which the boundaries are taken into account by the algorithm. Bound state calculations are mainly affected by the grid size provided the region is chosen to be sufficiently large, but highly excited bound states will be less well represented, and continuum solutions are sensitive to the choice of boundary condition at large distances. Continuum solutions with high energy cannot be represented unless their wavelengths are large compared with the grid spacing. We clearly must be alert to the limitations of all methods of approximation.

B.6 Finite difference methods for Dirac equations

General references [38], [39], [40], [41], [42]

B.6.1 An existence theorem

The Dirac equation is a first-order system of differential equations in two dependent variables. It is therefore helpful to begin our discussion with a standard existence theorem on which all methods of numerical integration of ordinary differential equations depend.

Let $w(t)$ be a two-dimensional vector-valued function of t such that

$$\frac{dw}{dt} = F(t; w) \tag{B.6.1}$$

on some interval. In the case of the Dirac equation, we write

$$w(t) := \begin{pmatrix} p(t) \\ q(t) \end{pmatrix}, \quad F(t; w) = \begin{pmatrix} F_{+1}(t; w) \\ F_{-1}(t; w) \end{pmatrix}$$

where if t is a new independent variable such that $r = f(t)$, with $f'(t)$ strictly positive and let $k(t) = f'(t)/f(t)$, $p(t) = P(f(t))$ and $q(t) = Q(f(t))$. Underlying all attempts to construct a solution of this equation is the theorem [39, p. 2]

Theorem B.12. *Let $F : \mathbb{R} \times \mathbb{R}^2 \to \mathbb{R}^2$ be a vector-valued function, $F(t; w)$, with domain $\mathcal{D} := [r_0, R] \times \mathcal{B}$, $\mathcal{B} := \{u \mid \|w\| < \infty\}$, continuous on $\mathcal{D}$ and satisfying there a Lipschitz condition*

$$\|F(t; w) - F(t; x)\| < K\|x - v\|$$

for some positive constant K for all (t, w) and (t, x) in $\mathcal{D}$. Then

(a) the initial value problem

$$\frac{dw}{dt} = F(t; w), \qquad w(t_0) = w_0,$$

has a unique solution $w(t; w_0)$ on $[t_0, R]$;

(b) the solution is Lipschitz continuous in w_0, uniformly in t, that is

$$\|w(t; w_0) - w(t; w_1)\| < e^{K(t-t_0)}\|w_1 - w_0\|$$

for all $(t, w_0) \in \mathcal{D}$ and $w(t; w_1) \in \mathcal{D}$.

Here $\|w\|$ denotes any convenient vector norm in $\mathbb{R}^2$, for example the maximum norm

$$\|w(t)\|_\infty = \max_t(|p(t)|, |q(t)|)$$

or the Euclidean norm

$$\|w(t)\|_2 = (|p(t)|^2 + |q(t)|^2)^{1/2}.$$

The restriction to vectors in $\mathbb{R}^2$ is, of course, inessential, but is all that we require to understand the application of finite difference methods to Dirac's equation.

B.6.2 Initial value methods

The solution of ordinary differential equations by finite difference methods relies on difference equations relating an approximate solution v_j defined at grid point t_j to values at nearby grid points. These difference equations can either be solved by marching along the grid or else by relaxation methods which treat them as a set of simultaneous algebraic equations. Surveys of numerical methods for initial and boundary value problems for ordinary differential equations offer many possibilities [38, 40, 41], many of which have been applied to Dirac's equation. Early work on the DHF method, for example [43], used Runge-Kutta procedures, which have the advantage of being *one-step* methods needing no special attention to start off the marching process but are not particularly efficient, and more recent work has used more efficient *linear multistep* methods.

The grid points of the independent variable t will be taken to be uniformly spaced, $t_j = t_0 + jh$, $j = 0, 1, \ldots, N$; we denote the grid-point values of the approximate solution by $v_j \approx w(t_j)$, $j = 0, 1, \ldots, N$. Consider a *linear k-step formula* with the general form

$$\sum_{j=0}^{k} \alpha_j v_{n+j} = h \sum_{j=0}^{k} \beta_j F_{n+j} \tag{B.6.2}$$

where $F_j = F(t_j; v_j)$, $v_0, v_1, \ldots, v_{k-1}$ are given and α_j and β_j are constants with $\alpha_k \neq 0$ and $|\alpha_0| + |\beta_0| \neq 0$. We associate (B.6.2) with the linear difference operator

$$\mathcal{L}_h[w](t) := \sum_{j=0}^{k} \{\alpha_j w(t + jh) - h\beta_j w'(t + jh)\} \tag{B.6.3}$$

By expanding $w(t + jh)$ and $w'(t + jh)$ in a Taylor series about t, we have that

$$\mathcal{L}_h[w](t) := C_0 w(t) + C_1 h w'(t) + \ldots + C_q h^q w^q(t) + \ldots \tag{B.6.4}$$

We aim to choose the coefficients α_j and β_j in (B.6.2) such that the leading terms on the right of (B.6.4) vanish: $C_0 = C_1 = \ldots = C_p = 0$. The method is then said to be *of order p* so that the first nonvanishing coefficient is C_{p+1}.

The relation between the solutions of the two systems (B.6.1) and (B.6.2) is best described using the notions of *local truncation error*, *consistency* and *stability*.

Local truncation error: The approximation $\{v_j,\ j = 0, 1, \ldots, N\}$ satisfies (B.6.2) exactly. However, if we replace each v_k in (B.6.2) by the exact solution $w(t_k)$ evaluated at the grid point t_k, the result will, in general, be non-zero. The local truncation error associated with the grid point j, denoted $\tau_j[w]$, is defined by

$$\tau_j[w] := \mathcal{L}_h[w]_j,\ j = 0, 1, \ldots, N.$$

Consistency: The scheme (B.6.2) is said to be *consistent of order* $p > 1$ with (B.6.1) provided there exist constants $K_0 > 0$, possibly depending on ϵ, and $h_0 > 0$ such that

$$\|\tau_j[w]\| \le K_0 h^p,\ j = 0, 1, \ldots, N$$

for all t-grids with $h \le h_0$ and for all solutions $w(t)$ of (B.6.1), where $\| \ldots \|$ denotes a suitable vector norm.

Stability: The scheme (B.6.2) is said to be *stable* provided we can find numbers $K_1, K_2 > 0$ and $h_0 > 0$ such that, for any function w defined on the grid and for all $h \le h_0$

$$\|w_j\| \le K_1(\|w_0\| + \|w_N\|) + K_2 \max_{1 \le k \le N-1} \|\mathcal{L}_h[w]_k\|.$$

Theorem B.13. *If the scheme (B.6.2) is stable and consistent of order* p *then, for all grids for which* $h \le h_0$*,*

$$\|w(t_j) - v_j\| \le K_0 K_2 h^p. \tag{B.6.5}$$

The scheme (B.6.2) is then said to be convergent of order p *for the differential equation (B.6.1).*

Proof: Write $x_j = w(t_j) - v_j$; then

$$\mathcal{L}_h[x]_j = \mathcal{L}_h[w]_j - \mathcal{L}_h[v]_j = \tau_j[w] \text{ for } j = 1, 2, \ldots, N-1$$

because the grid function v is a solution of (B.6.2). The result then follows from the stability property.

■

The consistency property ensures that the difference equation (B.6.2) has the differential equation (B.6.1), as its limit as $h \to 0$. However, it is not enough to ensure that the local truncation error(B.6.4) can be made as small as we please[4] as $h \to 0$. We need stability to ensure that random errors do not come to dominate the solution as we step along the grid. The theorem shows that the error depends upon the step length, h and that the error in the solution at each grid point is bounded by

$$\|w(t_j) - v_j\| \le M\, h^p \tag{B.6.6}$$

for some constant M, which will in general depend upon parameters such as the eigenvalue ϵ.

---

[4] Ignoring rounding errors.

B.6.3 Linear multistep methods

A linear multistep method is said to be *explicit* if the coefficient β_k of (B.6.2) vanishes; otherwise the method is said to be *implicit*. The distinction is important for nonlinear differential equations; explicit methods give us the solution v_{n+k} immediately in terms of $v_{n+k-1}, \ldots, v_n$ without any further work whereas v_{n+k} appears in F_{n+k} so that it is necesary to solve a nonlinear equation to obtain v_{n+k}.

The stability of linear differential equations with constant coefficients gives us a good feel for the *local stability* of a linear multistep method for more general systems. Consider the initial value problem on an interval (a, b)

$$w' = A\,w, \quad w(a) \text{ given.} \tag{B.6.7}$$

Let the fixed matrix A have a complete set of linearly independent eigenvectors $\boldsymbol{e}_\lambda$ and eigenvalues λ so that its solution is a linear combination

$$w(t) = \sum_\lambda C_\lambda\, e^{\lambda t}\, \boldsymbol{e}_\lambda$$

for some set of coefficients C_λ. An approximate solution v_j on the grid can be expanded similarly as a linear combination

$$v_j = \sum_\lambda v_j^\lambda\, \boldsymbol{e}_\lambda \tag{B.6.8}$$

where the numbers v_j^λ satisfy the linear algebraic equations

$$\sum_{j=0}^{k} \{\alpha_j - \overline{h}\,\beta_j\}\, v_{n+j}^\lambda = 0, \tag{B.6.9}$$

$\overline{h} = \lambda h$ and $n = 0, 1, \ldots, N-k$. Linear difference equations with constant coefficients such as (B.6.9) have solutions of the form

$$v_j^\lambda = \sum_{i=0}^{k} c_i^\lambda\, {x_i}^j \tag{B.6.10}$$

where the numbers x_i are the roots of the *characteristic polynomial*

$$\pi_{\overline{h}}(x) = \rho(x) - \overline{h}\,\sigma(x) \tag{B.6.11}$$

in which

$$\rho(x) = \sum_{j=0}^{k} \alpha_j x^j, \quad \sigma(x) = \sum_{j=0}^{k} \beta_j x^j$$

are known respectively as the *first* and *second* characteristic polynomials.

We have seen that the solutions of a finite difference method will approach those of the original differential equation as $h \to 0$ if the method is both consistent and stable. The notion of stability has to do with sensitivity of the solution to perturbations: we do not want small initial errors to grow to dominate the solution as we step along the grid.

Whether a linear multistep method is stable for a particular stepsize h depends on $\overline{h} = \lambda h$, and in particular on what happens when $\overline{h} \to 0$. A linear multistep method is said to be *zero-stable* if all roots of $\rho(x) = 0$ lie within and on the unit circle, and those on the unit circle are simple. [5] Dahlquist [44, 45] has proved the following

Theorem B.14. *The solution of a linear multistep method converges to the solution of the associated initial value problem for the differential equation as* $\overline{h} \to 0$ *if, and only if, the method is both consistent and zero-stable.*

A linear multistep method is said to be *absolutely stable* if all the roots of $\pi_{\overline{h}}(x)$ lie inside the unit circle in the complex x-plane. Its *region of absolute stability* consists of all (complex) $\overline{h}$ for which it is absolutely stable. Although a zero-stable method must have all its roots within and on the unit circle, the roots on the unit circle move off it when $\overline{h}$ becomes non-zero, and may therefore move outside the unit circle. In particular, if the dominant root x_0 of $\pi_{\overline{h}}(x)$ approximates an exponential solution, $\mathrm{e}^{\lambda x}$, of the differential equation, then we expect to find $x_0 \approx 1+\overline{h}+O(\overline{h}^2)$ so that if $h = x/n$, then $(1+\overline{h})^n = (1 + \lambda x/n)^n \to \mathrm{e}^{\lambda x}$ as $n \to \infty$ for fixed x. Clearly x_0 will be outside the unit circle for some values of the exponent λ when h is non-zero.

The *strongly stable* methods are more useful. A method is said to be *strongly stable* if it is both consistent and if $\rho(x)$ has one root $x_0 = 1$ with all its remaining roots inside the unit circle. This is certainly the case for the popular class of Adams methods, characterized by $\rho(x) = x^k - x^{k-1}$ for which $k-1$ roots are at the origin. *Explicit* linear multistep methods of Adams type are known as *Adams-Bashforth* methods, whilst *implicit* ones are known as *Adams-Moulton* methods. Predictor-corrector methods use an explicit predictor (P) step to get a first approximation to v_{n+k} followed by an evaluation (E) of F_{n+k} and then using this estimate in an implicit corrector (C) step to obtain a better estimate of v_{n+k}. In principle one could go on to iterate the EC steps, but this is rarely worth the effort. The PEC combination of Adams methods is known as an *Adams-Bashforth-Moulton* method.

The boundary of the region of stability, $\mathcal{R}$, of a zero-stable method is determined by points $\overline{h} = \mathrm{e}^{i\theta}$ on the unit circle. From (B.6.11) we see that the boundary points can be characterized as

$$\partial\mathcal{R} = \left\{\overline{h} \mid \overline{h} = \rho(\mathrm{e}^{i\theta})/\sigma(\mathrm{e}^{i\theta}),\ 0 \le \theta < 2\pi\right\}$$

[5] If the root x_p occurs more than once, then contributions proportional to jx_p^j and so on will appear in (B.6.10). Such terms remain bounded if $|x_p| < 1$, but will lead to instability if $|x_p| = 1$; we therefore want any roots on the unit circle to be simple.

so that locating the region of stability for any particular linear multistep method is straightforward. Books such as those by Lambert [46] or Hall and Watt [40, Chapter 2] survey many popular methods, noting that

- Explicit and PC methods have finite intervals of stability.
- Implicit methods usually have larger intervals of stability than explicit methods.
- Generally the higher the order of a method, the more it restricts h.
- PECE methods restrict h less than PEC or PECEC methods.

Linear multistep methods have been used in a variety of relativistic self-consistent field calculations. Coulthard [47] used Milne's method, Smith and Johnson [48] and Desclaux [49, 50] used a 5-step Adams-Bashforth-Moulton method. The predictor (P) is

$$\begin{aligned} v^{[0]}_{n+5} &= v_{n+4} \\ &+ \frac{h}{720}\left\{1901F_{n+4} - 2774F_{n+3} + 2616F_{n+2} - 1274F_{n+1} + 251F_n\right\} \end{aligned} \tag{B.6.12}$$

and the corrector (C) is

$$\begin{aligned} v^{[1]}_{n+5} &= v_{n+4} \\ &+ \frac{h}{720}\left\{251F^{[0]}_{n+5} + 646F_{n+4} - 264F_{n+3} + 106F_{n+2} - 19F_{n+1}\right\} \end{aligned} \tag{B.6.13}$$

where

$$F^{[0]}_{n+5} = F(t_{n+5}; v^{[0]}_{n+5}).$$

is an E step. Both P and C formulae are consistent of order 5, with principal local truncation errors

$$\tau_P[w](t) = \frac{475}{1440}h^6 w^{(6)}(t), \quad \tau_C[w](t) = \frac{-27}{1440}h^6 w^{(6)}(t).$$

"Milne's device" [46, p. 91] estimates the accuracy of the method by noting that the error in the C formula is $\tau[w]_C(t_n) = w(t_n) - v^{[1]}_n + O(h^7)$, and using the corresponding result for the P formula to eliminate the unknown $w(t_n)$ gives the error estimate

$$\tau_C[w](t) = \frac{27}{502}\left(v^{[1]}_{n+5} - v^{[0]}_{n+5}\right). + O(h^7)$$

This estimate can either be used to monitor the accuracy of the calculation or, following Desclaux [49, 50], used to correct the PEC result.

B.6.4 The nodal structure of Dirac radial wavefunctions

For bound states, $\int_{\mathcal{I}} |P(r)|^2 + |Q(r)|^2\, dr$ must be finite for all $\mathcal{I} \subseteq \mathbb{R}_+$. This condition determines the boundary behaviour in a neighbourhood of $r = 0$,

§5.3, and also the asymptotic behaviour as $r \to \infty$. As in nonrelativistic physics, solutions for which the relativistic momentum p is real represent scattering states whereas any for which p is imaginary must, if they exist, represent bound states. The continuum solutions are normalizable on all finite intervals $[a, b]$ but not on the whole real line.

A phase amplitude representation of the radial functions is useful to understand how the nodal structure of Dirac radial amplitudes depends on the eigenvalue parameter and influences the choice of algorithm for numerical integration. We assume that the function $Z(r) = -rV(r)$ is positive and monotone decreasing over the whole range of r, save possibly near the left endpoint r_0.[6] When $\epsilon < 0$ so that the state is bound, the expression $Z(r) + r\epsilon$ changes sign at some radius r_t. When $0 < r_0 \le r_t \le R < \infty$, this "classical turning point"[7] divides the interval $[r_0, R]$ into an 'inner' and an 'outer' part. We assume, as is reasonable for most bound state problems, that $|(Z(r) + r\epsilon)/2rc^2| \ll 1$ in the outer region. There is no turning point in the upper continuum ($\epsilon > 0$) or in the lower continuum ($\epsilon < -2c^2$).

For a qualitative discussion, it is convenient to introduce a phase-amplitude representation such that

$$P_\epsilon(r) = A_\epsilon(r)\, r^{-\kappa} \sin(\varphi_\epsilon(r)), \quad cQ_\epsilon(r) = A_\epsilon(r)\, r^{\kappa} \cos(\varphi_\epsilon(r)). \tag{B.6.14}$$

where

$$\frac{d\varphi_\epsilon(r)}{dr} = X(r) \cos^2 \varphi_\epsilon(r) + Y(r) \sin^2 \varphi_\epsilon(r) \tag{B.6.15}$$

and

$$A_\epsilon(r) = A_\epsilon(r_0) \exp \int_{r_0}^{r} \left\{ -\frac{1}{2} \sin 2\varphi_\epsilon(s)[X(s) - Y(s)] \right\} ds, \tag{B.6.16}$$

in which

$$X(r) = r^{2\kappa} \left[2 + \frac{Z(r) + r\epsilon}{rc^2} \right]$$

is positive on $[r_0, R]$ under the stated conditions, and

$$Y(r) = r^{-2\kappa} \frac{Z(r) + r\epsilon}{r}$$

is positive for $r < r_t$ and negative for $r > r_t$. Although the amplitude $A_\epsilon(r)$ varies, it is always of one sign, so that the distribution of zeros of the components is determined by the phase angle $\varphi_\epsilon(r)$. The right-hand side of (B.6.15) is strictly positive for $r < r_t$ so that $\varphi_\epsilon(r)$ is a monotone increasing function

---

[6] Whilst this may appear unduly restrictive, it is in practice a good working hypothesis.

[7] The radius at which the classical orbit of a particle moving in the potential $V(r)$ would be an extremum.

of r on this interval. The zeros of $P_\epsilon(r)$ occur where $\varphi_\epsilon(r)$ is an integer multiple of π, and those of $Q_\epsilon(r)$ are located at the *odd* integer multiples of $\pi/2$. Equation (B.6.15) can be rewritten as

$$\frac{d}{dr}\cot\varphi_\epsilon(r) = -Y(r) - X(r)\cot^2\varphi_\epsilon(r)$$

so that for $r_t < r < R$

$$\frac{d}{dr}\cot\varphi_\epsilon(r) \leq -Y(r) \Rightarrow \cot\varphi_\epsilon(r) \leq \cot\varphi_\epsilon(r_t) - \int_{r_t}^{r} Y(s)ds,$$

The right-hand side is never unbounded so that $\tan\varphi_\epsilon(r)$ never has a zero in $[r_t, R]$; all its zeros, the zeros of the large component $P_\epsilon(r)$, must be located to the left of the classical turning point.

By differentiating the relation (5.3.9) with respect to the variable b and rewriting the result in phase-amplitude notation, we find

$$\frac{\partial}{\partial r}\left(A_\epsilon^2(r)\frac{\partial\varphi_\epsilon(r)}{\partial\epsilon}\right) > 0. \tag{B.6.17}$$

so that $\partial\varphi_\epsilon(r)/\partial\epsilon$ is an increasing function of r. Thus $\varphi_\epsilon(r)$ is a continuous increasing function of ϵ at a fixed value of r, so that the successive zeros of both large and small components move inwards (outwards) as ϵ increases (decreases).

It is continuity of the solutions with respect to ϵ that lies at the heart of shooting methods for eigenvalue problems. At the origin, the initial conditions are independent of the value of ϵ. We can choose the boundary condition at $r = R$ so that, say, $\varphi_\epsilon(R) = n\pi$ for some integer n. The continuity of the solutions with respect to ϵ means that if there is an eigensolution for $n = n_0$ with $\epsilon = \epsilon_{n_0}$, then there will also be solutions for $n = n_0 + 1, n_0 + 2, \ldots$ with increasing energies $\epsilon_{n_0} < \epsilon_{n_0+1} < \epsilon_{n_0+2}, \ldots$ For bound states, the location of the classical turning point depends upon ϵ; r_t increases as ϵ increases. The eigensolutions decrease in amplitude on the interval (r_t, R). This interval disappears when ϵ is sufficiently large. When $R \to \infty$ we shall hope to recover the usual bound state spectrum in the region $\epsilon \leq 0$, and bound state eigensolutions must exhibit a square integrable tail when $r > r_t$.

In the continuum, there is no finite classical turning point and the solution is everywhere oscillatory. There are two linearly independent solutions, only one of which is finite at the origin.

By multiplying the inhomogeneous Dirac equation from the left with J^{-1}, where

$$J = \begin{pmatrix} 0 & -1 \\ 1 & 0 \end{pmatrix},$$

we may put it in the form (B.6.1) with

$$F(t; w) = A(t)\,w(t) + B(t) \tag{B.6.18}$$

where, if $r = f(t)$ and $k(t) = f'(t)/f(t)$,

$$A(t) = k(t)\, J^{-1} \begin{pmatrix} -(Y(r)+\epsilon r)/c & \kappa \\ \kappa & -(Y(r)+\epsilon r+2c^2 r)/c \end{pmatrix},$$

and $B(t) = k(t)\, J^{-1}\, X(r)$ where $X(r)$ consists of (two-component) exchange terms in MCDHF calculations together with Lagrange multiplier terms arising from orthogonality constraints. The matrix $A(t)$ has eigenvalues satisfying

$$\lambda^2 = [k(t)]^2 \left(\kappa^2 - (Y(r)+\epsilon r)(2rc^2 + Y(r)+\epsilon r)/c^2\right). \tag{B.6.19}$$

For simplicity, consider the usual grid for which $r = f(t) = r_0 \mathrm{e}^t$, so that $k(t) = 1$. Assume that we can ignore the r-dependence on the right hand side so that (B.6.19) gives local estimates of λ. Then if $Y(r)$ satisfies the conditions of §B.6.4, $\lambda^2 < 0$ to the left of the classical turning point, $r < r_t$, and is positive for $r > r_t$. Thus $\overline{h} = \pm i|\lambda|h$ is pure imaginary in $r < r_t$. The 5-step Adams-Moulton scheme (or PEC iterated to convergence) is then absolutely stable for $|\overline{h}| < 1.3$ locally implying that a practical criterion for the maximum step length to the left of the classical turning point is

$$h < \min_{r_0 < r < r_t} \frac{1.3}{|\lambda|} \tag{B.6.20}$$

To the right of the classical turning point, the roots are both real, one being positive and one negative. There is now one dominant root of the characteristic equation for which, when $\lambda > 0$, $x_1 = \exp \overline{h} + O(\overline{h}^6)$, the remaining roots going to zero for large r. For $\lambda < 0$, the roots of the Adams-Moulton characteristic equation are all inside $\mathcal{R}$ provided

$$h < \min_{r_t < r < R} \frac{1.8}{|\lambda|} \tag{B.6.21}$$

aso that the scheme is relatively stable. It follows that if we step inwards, $h < 0$, the root of (B.6.19) with $\lambda < 0$ dominates in this region. In most applications the choice of h is less restricted by the size of stability domains than by the need for accuracy and the minimization of computational effort.

B.6.5 Discretization of two-point boundary value problems

The inhomogeneous Dirac problem

$$J\frac{dw}{dt} = F(t; w), \quad 0 \le t \le T, \quad w(0) = w_0$$

is equivalent to the integral equation

$$J(w(t) - w_0) = \int_0^T F(t, w(t))\, dt.$$

We can construct a discrete approximation to $w(t)$ on a grid $t_j = jh$, $j = 0, 1, \ldots, N$ spanning $(0, T)$, $T = t_N)$ by replacing the integral on the right hand side by a numerical quadrature. For simplicity consider the *midpoint rule*, which gives us the equations

$$J(v_{j+1} - v_j) = hF_{j+1/2} + h\,\tau_{j+1/2}, \quad v_0 = w_0, \quad j = 0, 1, \ldots, N-1 \tag{B.6.22}$$

connecting consecutive $v_j \approx w(t_j)$, where $F_j = F(t_j, v_j)$. Theorem B.13 shows that if the *local truncation error* satisfies

$$\|\tau_{j+1/2}\| \le K_0 h^2,$$

then the midpoint rule gives the error estimate

$$\| e_j \| = \| w(t_j) - v_j \| \le K_0 K_2 h^2. \tag{B.6.23}$$

Eigenvalue searches require some *characteristic function*, $\psi(\epsilon)$, whose roots are the required eigenvalues. The double shooting method constructs the solution by matching two pieces. Given a trial value of ϵ and initial values P_0, Q_0 from by the power series expansion about the origin, we solve the initial value problem by stepping outwards on the grid $t_0, t_1, \ldots, t_J$, terminating with $P_J^{(-)}, Q_J^{(-)}$. Similarly, we construct a trial solution on $t_J, t_{J+1}, \ldots, t_N$ with one condition $P_J^{(-)} = P_J^{(+)}$ at the join point t_J, choosing t_N is sufficiently large that P_N, Q_N are sufficiently small. These solutions are continuous functions of ϵ, and we can make the jump, $\Delta Q_J = Q_J^{(-)} - Q_J^{(+)}$, vanish by choosing ϵ to be an eigenvalue of the discretized problem. A convenient choice of characteristic function using grid-point values is therefore $\psi_h(\epsilon) = \Delta Q_J$; denote the corrsponding jump using solutions of the differential equation by $\psi(\epsilon)$. Thus for the midpoint rule (B.6.22) we shall have

$$\| e_j \| \le M(\epsilon) h^2, \quad \| \psi(\epsilon) - \psi_h(\epsilon) \| \le \lambda M(\epsilon) h^2 \tag{B.6.24}$$

where $M(\epsilon)$ is a positive number which may depend on ϵ and λ is a positive constant of order unity. For a convergent scheme of order p the eigenvalue and the truncation error both converge like h^p as $h \to 0$ provided the initial guess is sufficiently close:

Theorem B.15. *Let ρ, m, K have the meanings of Theorem B.32 and take $\lambda, M(\epsilon)$ as above. Suppose that the initial eigenvalue estimate $\epsilon^{(0)}$ satisfies*

$$| \epsilon^{(0)} - \epsilon | \le \rho_0 = \rho + M h^p$$

where

$$M = \lambda\, |m| \max_{| \epsilon' - \epsilon | \le \rho} M(\epsilon').$$

Let the grid solution obtained with the trial eigenvalue $\epsilon^{(\nu-1)}$, $\nu = 1, 2, \ldots$ be $v_j^{(\nu)}$ and let $\psi_h(\epsilon)$ be the corresponding characteristic function on the grid.

Then if $w(t)$ is the eigensolution of the differential equation with eigenvalue ϵ, the error at the ν-th iteration is bounded by

$$|\,\epsilon - \epsilon^{(\nu)}\,| \le O(h^p) + O(K^\nu), \quad \|\, w(t_j) - v_j^{(\nu)}\,\| \le O(h^p) + O(K^\nu).$$

Proof: The proof uses the notion of contracting maps: see Appendix B.8. The eigenvalue is obtained as the limit of the sequence

$$\epsilon^{(\nu)} = g(\epsilon^{(\nu-1)})$$

where, (B.8.2), $g(s) = s - m\psi(s)$ for the solution of the differential equation. The corresponding grid sequence differs so that

$$g_h(\epsilon^{(\nu-1)}) = g(\epsilon^{(\nu-1)}) + \delta^{(\nu-1)}, \quad |\delta^{(\nu-1)}| \le \delta = Mh^p.$$

So starting with $\nu = 0$, we have

$$\begin{aligned} |\,\epsilon - \epsilon^{(\nu)}\,| &= |\,g(\epsilon) - g_h(\epsilon^{(\nu)})\,| \\ &= |\,g(\epsilon) - g(\epsilon^{(\nu-1)}) - \delta^{(\nu-1)}\,| \\ &\le K\,|\,\epsilon - \epsilon^{(\nu-1)}\,| + Mh^p. \end{aligned}$$

at each stage. By induction we get

$$|\,\epsilon - \epsilon^{(\nu)}\,| \le K^\nu |\,\epsilon - \epsilon^{(0)}\,| + \frac{1-K^\nu}{1-K}\, M\,h^p,$$

because $0 < K < 1$, establishing part of the theorem. To establish the remainder, we note that

$$\|\, w(t_j) - v_j^{(\nu)}\,\| \le \|\, w(t_j)^{(\nu)} - v_j^{(\nu)}\,\| + \|\, w(t_j)^{(\nu)} - w(t_j)\,\|$$

where $w(t_j)^{(\nu)}$ is the solution of the differential equation using the eigenvalue $\epsilon^{(\nu)}$. Now we have shown previously that

$$\|\, w(t_j)^{(\nu)} - v_j^{(\nu)}\,\| < M(\epsilon^{(\nu)} h^p$$

and Lipschitz continuity of $w(t_j)$ with respect to ϵ is enough to establish that there exists a constant $\mu > 0$ such that

$$\|\, w(t_j)^{(\nu)} - w(t_j)\,\| < \mu |\,\epsilon^{(\nu)} - \epsilon\,|,$$

from which the required bound follows immediately.

■

Rounding errors limit the precision of the calculation, so that there is a practical limit on ν after which there is no improvement in the solution. Note that this process is quite different from the more familiar variational calculation of an eigenvalue using the Rayleigh quotient where a first order error in the wavefunction only contributes to a second order error in the eigenvalue.

B.6.6 Two-point boundary value problems: the deferred correction method

The idea of using simple low order difference equations with relatively poor accuracy but adding difference corrections to the right-hand side using previous trial solutions was first proposed by Fox and Goodwin [51]. This fits well with algorithms for solving eigenvalue problems, especially in the context of self-consistent fields, where practical calculations require iterative determination of the eigenvalue and the difference corrections improve the accuracy of the converged solution. A general theoretical basis for what has become known as the *method of deferred correction* was provided by Pereyra [52, 53, 54].

We need to modify the notation in order to analyse this procedure. We shall see below that each additional term in the local truncation error for the midpoint rule in terms of central differences improves the leading error by a factor $O(h^2)$. Let

$$\mathcal{L}[\,w\,](t) := \mathcal{L}^{(1)}[\,w\,|\,\epsilon\,](t) - \mathcal{L}^{(2)}[\,w\,](t) = 0, \tag{B.6.25}$$

where if $r = f(t)$ and $k(t) = f\,'(t)/f(t)$,

$$\mathcal{L}^{(1)}[\,w\,|\,\epsilon\,](t) = J\frac{dw}{dt} + W(t)\,w(t), \quad w(t) = \begin{pmatrix} P_A(r) \\ Q_A(r) \end{pmatrix}, \tag{B.6.26}$$

with

$$W(t) = k(t)\begin{pmatrix} -[Y(r) + \epsilon r]/c & \kappa \\ \kappa & -(Y(r) + \epsilon r + 2rc^2)/c \end{pmatrix},$$

$$\mathcal{L}^{(2)}[w](t) = \frac{k(t)}{c}\left\{ X[\,w\,] + r\sum_B \epsilon_B w_B(t) \right\} \tag{B.6.27}$$

in its most general form. In the MCDHF equations, $Y(r)/r$ is the direct potential for the orbital A (whose label we have suppressed), and $X[\,w\,]$, the exchange part of (7.4.7), is a linear integral expression in the unknown $w(t)$, $w_B(t)$ are two-component radial functions with the same orbital symmetry whilst κ and ϵ_B are the Lagrange multipliers expressing orthogonality to other orbitals with symmetry κ. After ν iterations we have trial solutions $w^{(\nu)}$ for all orbitals needed to construct the matrix $W^{(\nu)}(t)$ and the right-hand sides, together with estimates of the eigenvalue $\epsilon^{(\nu)}$ and the Lagrange multipliers. We can express the problem of obtaining a new estimate $w^{(\nu+1)}$ as that of solving the inhomogeneous *ordinary differential equation*

$$\mathcal{L}^{(1(\nu))}[\,w\,|\,\epsilon^{(\nu)}\,](t) = \mathcal{L}^{(2(\nu))}[\,w^{(\nu)}\,](t) \tag{B.6.28}$$

with the usual bound state boundary conditions, where $\mathcal{L}^{(1(\nu))}[\,w\,|\,\epsilon^{(\nu)}\,]$ is given by (B.6.26) with

$$W^{(\nu)}(t) = k(t)\begin{pmatrix} -[Y^{(\nu)}(r) + \epsilon^{(\nu)}r]/c & \kappa \\ \kappa & -(Y^{(\nu)}(r) + \epsilon^{(\nu)}r + 2rc^2)/c \end{pmatrix},$$

We now discretize (B.6.28) as in (B.6.22), retaining error terms involving central differences of order up to $2p+1$ on the right hand side

$$\mathcal{L}_h^{(1(\nu))}[\, v \,|\, \epsilon^{(\nu)} \,]_{j+1/2} = \mathcal{L}_h^{(2(\nu))}[\, w^{(\nu)} \,]_{j+1/2} + \tau^{(p)}[v]_{j+1/2} \tag{B.6.29}$$

where

$$h\mathcal{L}_h^{(1(\nu))}[\, v \,|\, \epsilon^{(\nu)} \,]_{j+1/2} = J(v_{j+1} - v_j) + hW^{[\nu]}_{j+1/2}\, v_{j+1/2}, \tag{B.6.30}$$

$$h\mathcal{L}_h^{(2(\nu))}[\, v^{(\nu)} \,]_{j+1/2} = \frac{k_{j+1/2}}{c}\Big\{ X^{(\nu)}[\, v^{(\nu)} \,]_{j+1/2} + r_{j+1/2} \sum_B^{(\nu)} \epsilon_B v^{(\nu)}_{B\, j+1/2} \Big\}. \tag{B.6.31}$$

We obtain the midpoint quadrature rule if we omit the term $\tau^{(p)}[v]_{j+1/2}$ from (B.6.29). The double shooting method needs values of v_0 and v_N to start the integrations on the inner and outer intervals together with matching conditions at their join. The converged eigenvalues and eigensolutions are only accurate to $O(h^2)$.

We have so far defined v_j only at grid points. Equations (B.6.30) and (B.6.31 use values labelled $j+1/2$, which we interpret as evaluating the arguments of each expression at $t_{j+1/2} = (t_j + t_{j+1})/2$. There is no problem about evaluating explicit functions of r at grid points, but what are we to do with $X^{(\nu)}[\, v^{(\nu)} \,]_{j+1/2}$, $W^{(\nu)}_{j+1/2}$ or the unknown $v_{j+1/2}$ which are not necessarily linear in the variable t? For the present, we shall assume that all quantities indexed at $j+1/2$ are defined as the arithmetic means, such as

$$v_{j+1/2} = (v_j + v_{j+1})/2,$$

but we can use actual grid values where they are easy to calculate. We summarize useful finite difference expressions that allow us to correct for this below.

The deferred correction method retains the simplicity of a low order finite difference method, but the error converges faster, like h^{2p}, as $h \to 0$. The reason is indicated by

Theorem B.16. *For each fixed value of $p = 1, 2, \ldots$, the successive iterates $v_n^{[\nu]}$ computed using the deferred correction scheme (B.6.29) approach the required eigensolution asymptotically with error estimates*

$$\|v_j^{[\nu]} - w(r_j)\| \le O\left(h^{2\nu+2}\right) + O\left(h^{2p+2}\right) \tag{B.6.32}$$

and

$$|\epsilon^{[\nu+1]} - \epsilon| \le O\left(h^{2\nu+2}\right) + O\left(h^{2p+2}\right) + O\left(K_n^{\nu}\right), \tag{B.6.33}$$

where K_n is a constant.

Proof: The difference equation (B.6.29) is consistent of order h^{2p+2} because of the presence of the truncation error terms $h\tau^{(p)}[\,v\,]$. Thus we can find a positive constant K_p, independent of ϵ, such that on each interval

$$\mathcal{L}_h^{(\nu)}[\,v\,|\,\epsilon^{(\nu)}\,]_{j+1/2} + \tau^{(p)}[v]_{j+1/2} \le K_p h^{2p+2} \tag{B.6.34}$$

where $\mathcal{L}_h^{(\nu)}[\,v\,|\,\epsilon^{(\nu)}\,] = \mathcal{L}_h^{(1(\nu))}[\,v\,|\,\epsilon^{(\nu)}\,] - \mathcal{L}_h^{(2(\nu))}[\,v^{(\nu)}\,]_{j+1/2}$. Provided the difference scheme generated by $\mathcal{L}_h^{(\nu)}[\,v\,|\,\epsilon^{(\nu)}\,]$, is stable for $h \le h_0$ we can find a positive number $L^{[\nu]}$ such that the error $e_j := v_j - w(t_j)$ satisfies

$$\|e_j\| \le L^{[\nu+1]} \max_j \|\mathcal{L}_h^{(\nu)}[\,v\,|\,\epsilon^{(\nu)}\,]_{j+1/2}\|$$

for $j = 1, 2, \ldots, N$. Now

$$\begin{aligned}
&\mathcal{L}_h^{(\nu)}[\,e\,|\,\epsilon^{(\nu)}\,]_{j+1/2} = \\
&\left\{\mathcal{L}_h^{(\nu)}[\,w\,|\,\epsilon^{(\nu)}\,] - \tau^{(p)}[w]\right\}_{j+1/2} - \left\{\mathcal{L}_h^{(\nu)}[\,v\,|\,\epsilon^{(\nu)}\,] - \tau^{(p)}[v]\right\}_{j+1/2} \\
&\qquad + \tau^{(p)}[\,w\,]_{j+1/2}
\end{aligned}$$

The middle bracket vanishes by (B.6.31) and the last term cancels the local truncation error in the first, which is bounded by (B.6.34). The local truncation error of the midpoint rule is $O(h^2)$, so that if the errors e_j are bounded by some constant m, then we can find a positive constant K_p' such that

$$\|\tau^{(p)}[\,w\,]_{j+1/2}\| \le K_p' m^{[\nu]} h^2$$

where

$$m_n^{[\nu]} := \max_j \|w_{n,j}^{[\nu]}\|.$$

Combining these estimates gives

$$m_n^{[\nu+1]} \le L^{[\nu+1]} \left\{K_p h^{2p+2} + K_p' h^2 m_n^{[\nu]}\right\}$$

Supposing that for all ν we can find a finite positive constant L such that $L^{[\nu]} \le L$, we can choose h so that $K_n = LK_p'h^2 \le 1$ for h sufficiently small. The estimates of the theorem follow by induction as in Theorem B.15.

B.6.7 Construction of difference corrections

The shifting operator, E, relates quantities defined on a uniform grid with stepsize h by

$$Ew(t) := w(t+h), \quad E^p w(t) := w(t+ph),$$

where E^p steps p intervals[8] along the grid. The central difference operator, δ, is defined by

$$\delta := E^{1/2} - E^{-1/2},$$

so that

$$\delta v_{j+1/2} = v_{j+1} - v_j \tag{B.6.35}$$

and, for example,

$$\delta^3 v_{j+1/2} = (E^{1/2} - E^{-1/2})^3 v_{j+1/2} = v_{j+2} - 3v_{j+1} + 3v_j - v_{j-1}$$

Similarly, we can define the averaging operator

$$\mu v_{j+1/2} = (v_{j+1} + v_j)/2$$

and the Taylor series operator by

$$f(x+h) = Ef(x) = e^{hD} f(x)$$

where $D = d/dx$, and the exponential is defined formally by the usual power series expansion

$$\mathrm{e}^{hD} = 1 + hD + (hD)^2/2! + \ldots,$$

which has a meaning for functions $f(x)$ defined as polynomials in x or more generally functions $f(x)$ with continuous derivatives of sufficiently high order. Relations between the various difference operators have been explored in detail in such texts as [55, Chapter 5], [56, Chapter 7]. The derivative operator can be expressed as

$$hD = \log E = 2\log\left[(1 + \frac{1}{4}\delta^2)^{1/2} + \frac{1}{2}\delta\right] = 2\ \mathrm{arcsinh}\ \frac{\delta}{2} \tag{B.6.36}$$

so that, formally, we have the expansion [55, Equation (5.3.120]

$$hf'_{j+1/2} = \left(\delta - \frac{1}{24}\delta^3 + \frac{3}{640}\delta^5 - \frac{5}{7168}\delta^7 + \ldots\right) f_{j+1/2} \tag{B.6.37}$$

Another useful formula expresses a grid function at the mid-point of an interval, $t_{j+1/2}$ in terms of an expression with only grid values [55, Equation (4.6.8)]:

$$f_{j+1/2} = \left(1 - \frac{1}{8}\delta^2 + \frac{3}{128}\delta^4 - \frac{5}{1024}\delta^6 + \ldots\right)(\mu f_{j+1/2}), \tag{B.6.38}$$

where

$$\mu f_{j+1/2} = (f_{j+1} + f_j)/2.$$

A formula that enables us to generate difference corrections in (B.6.29) can be obtained by inverting (B.6.37) after noting that (B.6.35)

[8] p is not necessarily an integer.

$$\delta f_{j+1/2} = f_{j+1} - f_j$$

so that

$$\begin{aligned} f_{j+1} - f_j &= \left(1 - \frac{1}{24}\delta^2 + \frac{3}{640}\delta^4 - \frac{5}{7168}\delta^6 + \dots\right)^{-1} hf'_{j+1/2} \\ &= \left(1 + \frac{1}{24}\delta^2 - \frac{17}{5760}\delta^4 + \frac{367}{967680}\delta^6 + \dots\right) hf'_{j+1/2} \end{aligned} \tag{B.6.39}$$

Finally, by combining (B.6.39) with (B.6.38) we get

$$\begin{aligned} f_{j+1} - f_j \\ = \left(1 - \frac{1}{12}\delta^2 + \frac{61}{2830}\delta^4 - \frac{563}{120960}\delta^6 + \dots\right) \frac{h}{2}\left(f'_{j+1} + f'_j\right) \end{aligned} \tag{B.6.40}$$

which expresses the integral over an interval in terms of central differences involving grid points only. Central difference formulae are convenient for approximating quantities at internal points on the grid. These must be supplemented by equivalent forward or backward difference expressions, [55, Chapter 5], at the ends of the range. Although these are needed at only a small number of grid points, it is important to use expressions that give errors of the same order in h as those used at internal points in order to obtain the full benefit of the deferred correction procedure.

Equation (B.6.40) allows us to express the local truncation error in (B.6.22) in terms of grid-point values of the function F:

$$h\tau_{j+1/2} = \left(-\frac{1}{12}\delta^2 + \frac{61}{2830}\delta^4 - \frac{563}{120960}\delta^6 + \dots\right) \frac{h}{2}\left(F_{j+1} + F_j\right) \tag{B.6.41}$$

Thus the uncorrected algorithm is consistent to $O(h^2)$ and retaining only the leading term of (B.6.41) gives a scheme that is consistent to $O(h^4)$. The additional terms in this case are

$$h\tau^{(1)}_{j+1/2} = -\frac{1}{12}\left(F_{j+2} + F_{j-1} - (F_{j+1} + F_j)\right) \tag{B.6.42}$$

which can be used at all internal points. The overall local truncation error becomes $O(h^4)$ when this is included. Note that if we index the equations using $j + 1/2$, this expression is symmetric about the reference index.

B.6.8 Single stepping algorithms

We have seen that the numerical solution is stable with respect to stepping outwards from the origin as far as the classical turning point, but requires stepping in the opposite direction from the outer boundary to the turning point. The two cases are best treated separately. In particular,we can rearrange the equations so that the outer boundary, which is initially unknown, can be determined automatically by the method of solution.

We start with (B.6.22) for the $\nu+1$-st iteration,

$$J(v_{j+1}-v_j) = hF^{[\nu]}_{j+1/2} + h\,\tau^{[\nu]}_{j+1/2}, \quad v_0 = w_0, \quad j = 0,1,\ldots,N-1$$

where, from (B.6.29) – (B.6.31),

$$hF^{[\nu]}_{j+1/2} = -hW^{[\nu]}_{j+1/2}v_{j+1/2} + hR'_{j+1/2} \tag{B.6.43}$$

where

$$hR'_{j+1/2} = \frac{k_{j+1/2}}{c}\left\{X^{(\nu)}[v^{(\nu)}]_{j+1/2} + r_{j+1/2}\sum_B \epsilon^{[\nu]}_B v^{(\nu)}_{B\,j+1/2}\right\}.$$

and

$$W^{[\nu]}_{j+1/2} = k_{j+1/2} \tag{B.6.44}$$
$$\times\begin{pmatrix} -\left[Y^{(\nu)}+\epsilon^{(\nu)}r\right]_{j+1/2}/c & \kappa \\ \kappa & -\left[Y^{(\nu)}+\epsilon^{(\nu)}r+2rc^2\right]_{j+1/2}/c \end{pmatrix},$$

At this point we have to choose how best to include deferred correction terms into the right-hand side. It is convenient to evaluate $W^{[\nu]}_{j+1/2}$ separately using (B.6.38) as this preserves the symmetry of the difference equation. We then have

$$J(v_{j+1}-v_j) + \frac{h}{2}W^{[\nu]}_{j+1/2}(v_{j+1}+v_j) = hR_{j+1/2} \tag{B.6.45}$$

where, if $T(\delta) = 1-\delta^2/12+\ldots$ is the difference operator appearing in (B.6.41),

$$hR_{j+1/2} = \frac{h}{2}\left(R'^{[\nu]}_{j+1} + R'^{[\nu]}_{j}\right)$$
$$-T(\delta)\frac{h}{2}\left\{W^{[\nu]}_{j+1/2}\left(v^{[\nu]}_{j+1}+v^{[\nu]}_{j}\right) - \left(R'^{[\nu]}_{j+1}+R'^{[\nu]}_{j}\right)\right\}.$$

Ignoring all terms in $T(\delta)$ after the first is equivalent to using the trapezoidal rule for the original quadrature.

B.6.9 Stepping outwards from the origin

Write (B.6.45) in the form

$$\left(J+\frac{h}{2}W^{[\nu]}_{j+1/2}\right)v_{j+1} - \left(J-\frac{h}{2}W^{[\nu]}_{j+1/2}\right)v_j = hR_{j+1/2}$$

which we recast as

$$v_{j+1} = S_{j+1/2}\, v_j + T_{j+1/2} \tag{B.6.46}$$

where

$$S_{j+1/2} = \left(J + \frac{h}{2}W^{[\nu]}_{j+1/2}\right)^{-1}\left(J - \frac{h}{2}W^{[\nu]}_{j+1/2}\right)$$

and

$$T_{j+1/2} = \left(J + \frac{h}{2}W^{[\nu]}_{j+1/2}\right)^{-1} hR_{j+1/2}$$

Because

$$\left(J \pm \frac{h}{2}W^{[\nu]}_{j+1/2}\right) = \begin{pmatrix} \mp\frac{1}{2}hk_{j+1/2}y^{[\nu]}_{j+1/2} & -1 \pm \frac{1}{2}h\kappa k_{j+1/2} \\ 1 \pm \frac{1}{2}h\kappa k_{j+1/2} & \mp\frac{1}{2}hk_{j+1/2}\left(y^{[\nu]}_{j+1/2} + 2r_{j+1/2}c\right) \end{pmatrix}$$

where $y_{j+1/2} = (Y^{[\nu]}_{j+1/2} + \epsilon^{[\nu]}\, r_{j+1/2})/c$, it follows that

$$\left(J + \frac{h}{2}W^{[\nu]}_{j+1/2}\right)^{-1} =$$

$$\Delta^{-1}_{j+1/2}\begin{pmatrix} -\frac{1}{2}hk_{j+1/2}\left(y^{[\nu]}_{j+1/2} + 2r_{j+1/2}c\right) & 1 - \frac{1}{2}h\kappa k_{j+1/2} \\ -1 - \frac{1}{2}h\kappa k_{j+1/2} & -\frac{1}{2}hk_{j+1/2}y^{[\nu]}_{j+1/2} \end{pmatrix}$$

where

$$\begin{aligned} \Delta_{j+1/2} &= \det\left(J + \frac{h}{2}W^{[\nu]}_{j+1/2}\right) \\ &= 1 - \frac{1}{4}h^2k^2_{j+1/2}\left[\kappa^2 - y^{[\nu]}_{j+1/2}\left(y^{[\nu]}_{j+1/2} + 2r_{j+1/2}c\right)\right]. \end{aligned}$$

The behaviour of the outward stepping formula (B.6.46) is determined by the eigenvalues $s_{j+1/2}$ of $S_{j+1/2}$. To simplify the analysis, we temporarily set $k_j = 1$, which is exact for an exponential grid. Then, dropping the subscripts $j + 1/2$, we have

$$s = \frac{1 - h\alpha/2}{1 + h\alpha/2}, \tag{B.6.47}$$

where

$$\alpha^2 = \kappa^2 - (2rc + y^{[\nu]})y^{[\nu]}$$

We can simplify the argument still more by going to the nonrelativistic limit (neglect $y^{[\nu]}$ in comparison with $2rc$) and approximate $y^{[\nu]}$ by writing $Y^{[\nu]}_j = Z - \sigma^{[\nu]}_j$, so that $\sigma^{[\nu]}_j$ is a local screening number. Then

$$\alpha^2 \approx 2r^2\left(\frac{\kappa^2}{2r^2} - \frac{Z - \sigma^{[\nu]}}{r} - \epsilon^{[\nu]}\right)$$

This has a familiar look: the three terms in brackets can be identified as a centrifugal potential, a screened electrostatic potential and an energy parameter. Whilst the details will depend on the variation of $\sigma^{[\nu]}(r)$ with r, we can distinguish at least three separate intervals. The centrifugal term dominates for $r < r_c$, say, so that $\alpha^2 > 0$ and the roots are real numbers $\pm\alpha$. Similarly $\alpha^2 > 0$ when $\epsilon < 0$ for some $r > r_t$. However, $\alpha^2 < 0$ when $r_c < r < r_t$ and the roots are then of the form $\pm i\,|\,\alpha\,|$. It is well-known that further intervals of r may occur in which $\alpha^2 > 0$ for f-shell electrons in rare earths – the so-called double well potentials – but it is sufficient to consider only the simpler case for present purposes.

When α^2 is negative, the roots s of (B.6.47) occur as complex conjugate pairs on the unit circle, so that we expect the solutions of (B.6.46) to be trigonometric in $r_c < r < r_t$. In $r < r_c$ one of the two real roots of (B.6.47) will have $|s| > 1$ and this will give a solution of increasing amplitude as we step outwards. We encounter problems when stepping outwards for $r > r_t$ because we want a normalizable decreasing solution. The unavoidable presence of errors introduces a component related to the root $|s_1| > 1$ that will overwhelm the wanted solution as r increases. We must therefore terminate the inner integration at some grid point r_J near r_t; the precise location is not critical.

B.6.10 Algorithm for the outer region

The process (B.6.46) gives estimates of v_j for all grid points $0 \le j \le J$. If we know the endpoint $t_N = T$, then we can step inwards using

$$v_j = S_{j+1/2}\, v_{j+1} - T_{j+1/2},\ j = N-1, N-2, \ldots J. \tag{B.6.48}$$

where $S_{j+1/2}$ and $T_{j+1/2}$ are the same as in (B.6.46). This equation results by setting $h \to -h$ in (B.6.46) and rearranging. The dominant root of (B.6.47) now the one we want and this process is therefore stable. Unfortunately we may have little idea of the orbital size and an adaptive scheme which automatically determines the boundary $t_N = T$ can avoid waste of memory and computing effort.

Define new 2-component vectors

$$w_{J-1/2} = \begin{bmatrix} 0 \\ P_J \end{bmatrix}, \quad w_{j+1/2} = \begin{bmatrix} -Q_j \\ P_{j+1} \end{bmatrix},\ j = J, \ldots, N-1 \tag{B.6.49}$$

where P_J has been obtained from the outward integration (B.6.39). Write

$$\begin{aligned}
a^{\pm}_{j+1/2} &= 1 \pm \frac{1}{2} h\kappa\, k_{j+1/2},\\
b^{[\nu]}_{j+1/2} &= \frac{1}{2} h k_{j+1/2} y^{[\nu]}_{j+1/2},\\
d^{[\nu]}_{j+1/2} &= \frac{1}{2} h k_{j+1/2} (y^{[\nu]}_{j+1/2} + 2cr^{[\nu]}_{j+1/2})
\end{aligned}$$

along with the matrices

$$
\begin{aligned}
B_{j+1/2} &= \begin{bmatrix} 0 & b^{[\nu]}_{j+1/2} \\ 0 & a^-_{j+1/2} \end{bmatrix}, \\
C_{j+1/2} &= \begin{bmatrix} -a^+_{j+1/2} & -b^{[\nu]}_{j+1/2} \\ d^{[\nu]}_{j+1/2} & a^+_{j+1/2} \end{bmatrix} \\
D_{j+1/2} &= \begin{bmatrix} -a^-_{j+1/2} & 0 \\ -d^{[\nu]}_{j+1/2} & 0 \end{bmatrix}
\end{aligned}
\tag{B.6.50}
$$

Then except at the endpoints, we have

$$
-B_{j+1/2}w_{j-1/2} + C_{j+1/2}w_{j+1/2} - D_{j+1/2}w_{j+3/2} = hR_{j+1/2} \tag{B.6.51}
$$

for $j = J+1, J+2, \ldots, N-2$ whilst

$$
C_{J+1/2}w_{J+1/2} - D_{J+1/2}w_{J+3/2} = h\widetilde{R}_{J+1/2} \tag{B.6.52}
$$

with

$$
h\widetilde{R}_{J+1/2} = hR_{J+1/2} + B_{J+1/2}w_{J-1/2} = \begin{pmatrix} R^+_{J+1/2} + b^{[\nu]}_{J+1/2}P_J \\ R^-_{J+1/2} + a^-_{J+1/2}P_J \end{pmatrix}
$$

and

$$
-B_{N-1/2}w_{N-3/2} + C_{N-1/2}w_{N-1/2} = \widetilde{R}_{N-1/2} \tag{B.6.53}
$$

where

$$
\widetilde{R}_{N-1/2} = R_{N-1/2} + D^{[\nu]}_{N-1/2}w_{N+1/2} = \begin{pmatrix} R^+_{N-1/2} + a^-_{N-1/2}Q_N \\ R^-_{N-1/2} + d^{[\nu]}_{N-1/2}Q_N \end{pmatrix}.
$$

We define $\widetilde{R}_{j+1/2} = R_{j+1/2}$ for $j = J+1, \ldots, N-2$. This has converted the difference equations (B.6.46) into a tri-diagonal form that can be solved by Gauss elimination. The algorithm (Algorithm 5.1 in the text) is simplified by the block 2×2 matrix structure:

(a) With the initial values

$$
E_J := 0, \; x_{J-1/2} := w_{J-1/2},
$$

form the matrices E_j and the vectors $x_{j+1/2}$ for $j = J, \ldots, N-1$ from the equations

$$
E_{j+1} := (C_{j+1/2} - B_{j+1/2}E_j)^{-1}D_{j+1/2} \tag{B.6.54}
$$

and

$$
x_{j+1/2} := (C_{j+1/2} - B_{j+1/2}E_j)^{-1}(\widetilde{R}_{j+1/2} + B_{j+1/2}x_{j-1/2}). \tag{B.6.55}
$$

(b) Starting from the boundary value $w_{N+1/2}$ form the solution for $j = N, N-1, \ldots, J+1$ from

$$w_{j-1/2} := E_j w_{j+1/2} + x_{j+1/2}. \tag{B.6.56}$$

The algorithm is particularly simple to implement as E_j, like $D_{j+1/2}$, has a null second column:

$$E_j = \begin{bmatrix} e_j^+ & 0 \\ e_j^- & 0 \end{bmatrix}.$$

and, if $\left(C_{j+1/2} - B_{j+1/2}E_j\right)$ is nonsingular, then

$$\begin{aligned} &\left(C_{j+1/2} - B_{j+1/2}E_j\right)^{-1} \\ &= \left(\Delta'_{j+1/2}\right)^{-1} \begin{pmatrix} a^+_{j+1/2} & b^{[\nu]}_{j+1/2} \\ a^-_{j+1/2}e_j^- - d^{[\nu]}_{j+1/2} & -a^+_{j+1/2} - b^{[\nu]}_{j+1/2}e_j^- \end{pmatrix} \end{aligned} \tag{B.6.57}$$

where

$$\begin{aligned} \Delta'_{j+1/2} = -e_j^- &\left(a^+_{j+1/2}b^{[\nu]}_{j+1/2} + a^-_{j+1/2}b^{[\nu]}_{j+1/2}\right) \\ &- \left(a^+_{j+1/2}a^+_{j+1/2} - b^{[\nu]}_{j+1/2}d^{[\nu]}_{j+1/2}\right) \end{aligned}$$

Then (B.6.54) gives the non-zero components of E_{j+1} as

$$e^+_{j+1} = -\left(a^+_{j+1/2}a^-_{j+1/2} + b^{[\nu]}_{j+1/2}d^{[\nu]}_{j+1/2}\right)/\Delta'_{j+1/2} \tag{B.6.58}$$

$$\begin{aligned} e^-_{j+1} = \Big\{&\left(a^+_{j+1/2}d^{[\nu]}_{j+1/2} + a^-_{j+1/2}d^{[\nu]}_{j+1/2}\right) \\ &-e_j^-\left(a^-_{j+1/2}a^-_{j+1/2} - b^{[\nu]}_{j+1/2}d^{[\nu]}_{j+1/2}\right)\Big\}/\Delta'_{j+1/2} \end{aligned} \tag{B.6.59}$$

The rest of the algebra is straightforward.

B.6.11 The boundary condition at $T = t_N$

The boundary condition at $T = t_N$ can be determined automatically. For the bound states of isolated atoms, $w_{N+1/2}$ tends to zero when N is sufficiently large, so we can choose

$$w_{N+1/2} = \begin{bmatrix} -Q_N \\ 0 \end{bmatrix} = 0 \tag{B.6.60}$$

and test $\|x_{N-1/2}\|$ to ensure that it is smaller than some specified tolerance for several consecutive steps. Once this condition has been satisfied, we can apply the backwards recursion (B.6.56) to step back to the join point.

The Dirac R-matrix method [57] for electron-atom scattering and photoionization partitions configuration space into two regions: the inner region is atom-centred, a sphere of radius R large enough to contain most of the electrons of the target atom, and it is assumed that exchange between an electron in the outer region and the core can be ignored. On the R-matrix boundary we assume that the target orbitals satisfy the Pauli relation

$$\frac{Q_N}{P_N} = \frac{b+\kappa}{2cR} \tag{B.6.61}$$

where b is an arbitrary constant. The upper line of equation (B.6.56),

$$w_{N-1/2} := E_N\, w_{N+1/2} + x_{N+1/2}$$

is

$$P_N = -e^-_N\, Q_N + x^+_{N+1/2}$$

which with (B.6.61) yields

$$P_N = \frac{2Rc\; x^+_{N+1/2}}{2Rc + (b+\kappa)e^-_N}$$

so that we can start the inward sweep with a finite initial value

$$Q_N = \frac{(b+\kappa)x^+_{N+1/2}}{2Rc + (b+\kappa)e^-_N}. \tag{B.6.62}$$

in equation (B.6.56).

B.6.12 The boundary condition at the origin

The inner boundary at $t = 0$ is close to the origin ($r = 0$). We need only one starting value, v_0, from the power series expansion of §5.4.1. For a point nucleus,

$$P(r) \approx r^\gamma[p_0 + p_1\, r + \ldots], \quad Q(r) \approx r^\gamma[q_0 + q_1\, r + \ldots],$$

where $\gamma = +\sqrt{\kappa^2 - Z^2/c^2}$ and

$$\frac{q_0}{p_0} = \frac{Z}{c(\kappa-\gamma)} = \frac{c(\kappa+\gamma)}{Z}$$

For a finite nuclear model with (direct) potential

$$-Y(r)/r \approx -(Y_1 + Y_2\, r + \ldots)$$

we have

$$P(r) \approx p_0\, r^{l+1} + O(r^{l+3}), \quad Q(r) \approx q_1\, r^{l+2} + O(r^{l+4})$$

for $\kappa < 0$ with

$$\frac{q_1}{p_0} = \frac{\epsilon + Y_1}{c(2l+3)}, \quad p_1 = q_0 = 0$$

or

$$P(r) \approx p_1\, r^{l+1} + O(r^{l+3}), \quad Q(r) \approx q_0\, r^l + O(r^{l+2})$$

for $\kappa > 0$ with

$$\frac{p_1}{q_0} = -\frac{\epsilon + Y_1}{c(2l+1)}, \quad q_1 = p_0 = 0$$

We use the symbol A to denote the leading coefficient, p_0 or q_0; all other coefficients in the series expansion are simple multiples of A, which is fixed by normalization. The power series expansions of the X and Y potentials have to be taken into account in higher order terms to ensure that all expressions are truncated at the appropriate order of h.

B.6.13 Improving a trial solution

We have so far generated an approximate *particular solution* for the target orbital wavefunction $u(t)$ in two parts:

- An inner solution $\{v_j^{(P)}$: at grid points $j = 0, 1, \ldots J$ and
- An outer solution $\{v_j^{(P)}$: at grid points $j = J, J+1, \ldots N$.

At the join point J we have a common value of $P_J^{(P)}$ but the inner and outer solutions have different values of $Q_J^{(P)}$; we denote the difference by $\Delta Q_J^{(P)}$. The corresponding grid solution of the *homogeneous* equation satisfying *the same boundary conditions* will have a jump $\Delta Q_J^{(H)}$ unless ϵ happens to be an eigenvalue of the homogeneous equation. Thus we can construct a solution v_j with $\Delta Q_J = 0$ as

$$v_j = v_j^{(P)} + \alpha v_j^{(H)}, \quad \alpha = -\Delta Q_J^{(P)} / \Delta Q_J^{(H)}. \tag{B.6.1}$$

Because $v_j^{(P)}$ and $v_j^{(H)}$ have the same leading coefficient, A, the result of (B.6.1) is to replace A by $A(1+\alpha)$.

This procedure generates a grid solution v_j that satisfies the finite difference equations of the shooting method at all grid points as long as ϵ is not an eigenvalue of the corresponding homogeneous equation. We now show that these equations can be reformulated in a manner similar to that used in Fischer's nonrelativistic MCHF algorithms [58, §3.11.4] as

$$\left(\boldsymbol{T}^{[\nu]} - \epsilon^{[\nu]}\boldsymbol{S}\right)\boldsymbol{v} = \boldsymbol{R} \tag{B.6.2}$$

where $\boldsymbol{T}^{[\nu]}, \boldsymbol{S}$ are real symmetric $2N+2 \times 2N+2$ matrices, $\boldsymbol{v}$ is a $2N+2$ column vector whose N components are the 2-vectors v_j, and $\boldsymbol{R}$ is formed in a similar way. The matrix $\boldsymbol{T}^{[\nu]}$ is a representation of the Dirac operator on the grid, and $\boldsymbol{S}$ allows for the scale change in the mapping $t \to r = f(t)$. Everything else, including the deferred difference corrections, goes into $\boldsymbol{R}$. This representation underpins the algorithms 5.2 and 5.3. We first rearrange equation (B.6.45)

$$J(v_{j+1} - v_j) + \frac{h}{2} W^{[\nu]}_{j+1/2}(v_j + v_{j+1}) = hR_{j+1/2},$$

and separate $W^{[\nu]}_{j+1/2}$ into two parts,

$$W^{[\nu]}_{j+1/2} = V^{[\nu]}_{j+1/2} - \epsilon^{[\nu]} S_{j+1/2}, \tag{B.6.3}$$

where

$$V^{[\nu]}_{j+1/2} = k_{j+1/2} \begin{pmatrix} -\frac{1}{c} Y^{(\nu)}_{j+1/2} & \kappa \\ \kappa & -\frac{1}{c} Y^{(\nu)}_{j+1/2} - 2cr_{j+1/2} \end{pmatrix},$$

and

$$S_{j+1/2} = \frac{1}{c} k_{j+1/2} r_{j+1/2} \begin{pmatrix} 1 & 0 \\ 0 & 1 \end{pmatrix}.$$

are symmetric matrices. Multiply from the left by the adjoint $(v_j + v_{j+1})^\dagger$ and sum over j. The resulting quadratic expression can be written in the form

$$\boldsymbol{v}^\dagger \left\{ \left(\boldsymbol{T}^{[\nu]} - \epsilon^{[\nu]} \boldsymbol{S} \right) \boldsymbol{v} - \boldsymbol{R} \right\} = 0 \tag{B.6.4}$$

consistent with (B.6.2). The matrix elements of $\boldsymbol{T}^{[\nu]}$ are

$$T_{jj} = \overline{V}^{[\nu]}_j, \quad T_{j,j+1} = +\frac{J}{h} + \frac{1}{2} V^{[\nu]}_{j+1/2}, \quad T_{j+1,j} = -\frac{J}{h} + \frac{1}{2} V^{[\nu]}_{j+1/2}.$$

where the matrices $V^{[\nu]}_{j+1/2}$ are defined at the midpoints of the intervals $(t_j.t_{j+1})$, and the diagonal elements are mean values

$$\overline{V}^{[\nu]}_0 = \frac{1}{2} V^{[\nu]}_{1/2}, \; \overline{V}^{[\nu]}_N = \frac{1}{2} V^{[\nu]}_{N-1/2}, \; \overline{V}^{[\nu]}_j = \frac{1}{2} \left(V^{[\nu]}_{j+1/2} + V^{[\nu]}_{j-1/2} \right), \; 0 < j < N.$$

Similarly, $\boldsymbol{S}$ has elements

$$S_{jj} = \overline{S}_j, \quad S_{j,j+1} = S_{j+1,j} = \frac{1}{2} S_{j+1/2},$$

where $\overline{S}_j$, and also the component vectors $\overline{R}_j$, are defined in the same way as $\overline{V}_j$. The symmetry of the matrix $\boldsymbol{T}^{[\nu]}$ can be demonstrated by noting that $J^\dagger = -J$. All other pieces are clearly symmetric.

B.7 Eigenfunction expansions for the radially reduced Dirac equation

General references [59], [60], [61]

The theory of eigenfunction expansions for the Dirac radial operator on a finite interval is essential for certain applications, especially the R-matrix method of §9.3. Mathematical textbook treatments (for example, Thaller [62]) have generally ignored the subject. Stakgold [61] discusses the problem for second-order differential equations but not linear systems such as Dirac's equation. The present treatment follows closely the analysis of general first-order differential systems given by Atkinson [60, Chapter 9].

B.7.1 The fundamental lemma

In the notation of (5.3.1)–(5.3.3) the radially reduced Dirac equation can be written

$$T_\kappa\, u_{\varepsilon\kappa}(r) = \left\{ cJ\frac{d}{dr} + W_\kappa(r) - \varepsilon \right\} u_{\varepsilon\kappa}(r) = \varepsilon\, u_{\varepsilon\kappa}(r) \tag{B.7.1}$$

where

$$u_{\varepsilon\kappa}(r) = \begin{pmatrix} P_{\varepsilon\kappa}(r) \\ Q_{\varepsilon\kappa}(r) \end{pmatrix}, \quad W_\kappa(r) = \begin{pmatrix} V(r) & c\kappa/r \\ c\kappa/r & -2c^2 + V(r) \end{pmatrix} \tag{B.7.2}$$

and

$$J = \begin{pmatrix} 0 & -1 \\ 1 & 0 \end{pmatrix} \tag{B.7.3}$$

is a skew-symmetric matrix.

First consider the initial-value problem for (B.7.1) on a finite closed interval $[R_1, R_2] \in \mathbb{R}_+$. Assume $V_\kappa(r)$ is real symmetric, integrable and continuous on $[R_1, R_2]$ and that $u(R_1)$ is given. There exists an unique continuously differentiable solution (see, for example, [59, Chapter 3]). We restate Lemma 5.3 and its corollary, which play an important part in this investigation:

Lemma B.17. *Let $u_1(r)$ and $u_2(r)$ satisfy (B.7.1) in some interval $(\alpha, \beta) \subseteq [R_1, R_2]$ with complex parameters ε_1, $\varepsilon_2 \in \mathbb{C}$ respectively. Define*

$$s_{12}(r) = c{u_2}^\dagger(r) J u_1(r) \equiv c\,[Q_2^*(r)P_1(r) - P_2^*(r)Q_1(r)]. \tag{B.7.4}$$

so that $s_{12}(r) = -{s_{21}}^*(r)$ Then

$$s_{12}(\beta) - s_{12}(\alpha) = (\varepsilon_1 - \varepsilon_2^*) \int_\alpha^\beta u_2^\dagger(r)\, u_1(r) dr. \tag{B.7.5}$$

Corollary B.18. *When $\varepsilon_1 = \varepsilon_2 = \varepsilon$ and $u_1(r) = u_2(r) = u(r)$, (B.7.5) reduces to*

$$s_\varepsilon(r) - s_\varepsilon(R_1) = 2i\,\Im\varepsilon \int_{R_1}^{r} u^\dagger(s)u(s)ds, \tag{B.7.6}$$

so that $s_\varepsilon(r)$ is constant on $[R_1, R_2]$ when ε is real.

The scalar $s_{12}(r)$ is the relativistic analogue of the Wronskian of two solutions of the Schrödinger equation because

$$\lim_{c\to\infty} s_{12}(r) \to \frac{1}{2}\left[\frac{dP_2^*(r)}{dr}P_1(r) - P_2^*(r)\frac{dP_1(r)}{dr}\right] \tag{B.7.7}$$

in the nonrelativistic limit. In Corollagry B.18, $s_\varepsilon(r)$ can be interpreted as the radial component of the probability current and, when ε is real, its independence of r expresses particle conservation.

We define the *fundamental matrix solution* of the initial value problem as the 2×2 matrix $U_\varepsilon(r)$ such that

$$U_\varepsilon(R_1) = I, \tag{B.7.8}$$

where I is the identity matrix. The columns of $U_\varepsilon(r)$ are thus two linearly independent solutions of the initial value problem, say $u_\varepsilon^1(r), u_\varepsilon^2(r)$, with the same parameter ε and such that

$$u_\varepsilon^1(R_1) = \begin{bmatrix}1\\0\end{bmatrix}, \quad u_\varepsilon^2(R_1) = \begin{bmatrix}0\\1\end{bmatrix}.$$

Defining

$$S_\varepsilon(r) := c\,U_\varepsilon{}^\dagger(r)\,J\,U_\varepsilon(r),$$

we have

$$S_\varepsilon(r) - S_\varepsilon(R_1) = 2i\,\Im\varepsilon \int_{R_1}^{r} U_\varepsilon{}^\dagger(s)U_\varepsilon(s)ds \tag{B.7.9}$$

so that when ε is real,

$$S_\varepsilon(r) = cJ, \quad R_1 \le r \le R_2. \tag{B.7.10}$$

B.7.2 Boundary conditions: the two-point boundary value problem

When boundary conditions are imposed at both endpoints R_1, R_2, solutions only exist when ε is an eigenvalue of (B.7.1). In order to treat this problem for the Dirac linear system, we consider a class of boundary conditions defined by a nontrivial vector v and square matrices M and N such that

$$u(R_1) = Mv, \quad u(R_2) = Nv, \tag{B.7.11}$$

satisfying two conditions

A $M^\dagger J M = N^\dagger J N$;
B $Mv = 0$ and $Nv = 0$ implies $v = 0$.

Condition A is equivalent to the assertion that $s_\varepsilon(R_1) = s_\varepsilon(R_2)$, consistent with the Corollary B.18 when ε is real. We can satisfy these conditions by choosing *real* rank-1 matrices

$$M = \begin{bmatrix} 0 & m \\ 0 & 1 \end{bmatrix}, \quad N = \begin{bmatrix} 1 & 0 \\ n & 0 \end{bmatrix},$$

This ensures

$$M^\dagger J M = N^\dagger J N = 0,$$

so that

$$P(R_1) = v_2, \quad Q(R_1) = m v_2, \quad P(R_2) = v_1, \quad Q(R_2) = n v_1, \qquad \text{(B.7.12)}$$

where v_1, v_2, the components of v, are *arbitrary complex* numbers.

B.7.3 Boundary conditions at the nucleus

The behaviour of solutions of Dirac radial equations near the nucleus is discussed in §5.3 using power series expansions. The leading terms give values of m and v_2 at a point R_1 close to the origin; equation (5.4.7) applies to the case of a point charge nucleus, and (5.4.8), (5.4.9) to the case of a finite nucleus.

B.7.4 Pauli approximation at R_2

For low-energy electron processes it is often sufficient to assume that the ratio of Dirac radial amplitudes at the outer boundary satisfies the Pauli approximation. This determines the ratio

$$n = (b + \kappa)/2R_2c \qquad \text{(B.7.13)}$$

where b is an arbitrary real constant (often set equal to zero). The value of v_1 need not be specified and depends on the normalization of the solution.

Equation (B.7.13) imposes a value of n that is particularly convenient if the energy parameter is so small that the dynamics can be treated as nonrelativistic for $r > R_2$. It is useful to remember that this boundary condition is still compatible with relativistic electron dynamics in $r > R_2$.

B.7.5 The MIT bag model at R_2

The MIT Bag Model of QCD [63] was devised to confine relativistic quarks within a finite volume, the "bag", whilst avoiding problems with Klein's paradox that occur if one attempts to use a confining potential wall. In atomic physics, it has been used in connection with B-spline representations of electronic states [64].This model is equivalent to setting $n = 1$ with v_1 arbitrary as before.

B.7.6 The eigenvalue spectrum

In this section, we discuss the eigenvalues of the two-point boundary value problem for the radial Dirac equation when the boundary conditions are of the general class of §B.7.2.

Theorem B.19. *The eigenvalues of the 2-point boundary value problem on the interval* $[R_1, R_2]$,

$$T_\kappa u_\varepsilon(r) = \varepsilon u_\varepsilon(r), \quad u_\varepsilon(R_1) = Mv, \quad u_\varepsilon(R_2) = Nv \tag{B.7.14}$$

are all real and have no finite limit point. Let $\mathcal{S} = \{\varepsilon_k, k \in \mathbb{Z}\}$ *be ordered by increasing absolute value and assume, without loss of generality, that no eigenvalue is zero. Then*

$$\sum_{k=0}^{\infty} |\varepsilon_k|^{-1-\delta}$$

converges for any value $\delta > 0$.

Proof: Suppose that (B.7.14) possesses a complex eigenvalue ε associated with some non-zero vector v. By hypothesis, Mv and Nv are not both zero, so that $u_\varepsilon(R_1)$ and $u_\varepsilon(R_2)$ are not both zero and $u_\varepsilon(r)$ must be a non-trivial solution. Condition A implies that $s_\varepsilon(R_1) = s_\varepsilon(R_2)$ and therefore, by Corollary 5.1,

$$\Im\,\varepsilon \int_{R_1}^{R_2} u_\varepsilon^\dagger(r) u_\varepsilon(r) dr = 0.$$

because $u_\varepsilon(r)$ is nontrivial, the integral is strictly positive and $\Im\,\varepsilon = 0$, so that all eigenvalues are real.

The unique solution of the initial value problem with $u_\varepsilon(R_1)$ given can be written in terms of the fundamental matrix solution as

$$u_\varepsilon(r) = U_\varepsilon(r) u_\varepsilon(R_1), \quad r \in [R_1, R_2], \tag{B.7.15}$$

and therefore

$$u_\varepsilon(R_2) = Nv = U_\varepsilon(R_2)\, Mv. \tag{B.7.16}$$

If this equation is to have a nontrivial solution v, the eigenvalue must satisfy the condition

$$\det(N - U_\varepsilon(R_2)\, M) = 0. \tag{B.7.17}$$

Conversely, if ε is a eigenvalue, then the equation $(N - U_\varepsilon(R_2)\, M)v = 0$, has a nontrivial solution v_ε and hence there exists a solution of (B.7.1) with $u_\varepsilon(R_1) = Mv_\varepsilon$ and $u_\varepsilon(R_2) = Nv_\varepsilon$, where v_ε is determined up to a normalization constant. Now each solution of the corresponding initial value problem with the given initial condition is an entire function of ε.[9] Hence $U_\varepsilon(r)$ is also an entire function of ε for each fixed $r \in [R_1, R_2]$ and so is the left-hand side of (B.7.17). As $\det(N - U_\varepsilon(R_2)\,M)$ has no complex zeros, it cannot vanish identically, and the zeros can have no finite limit point. We can therefore order the eigenvalues so that

$$0 < |\varepsilon_0| \le |\varepsilon_1| \le |\varepsilon_2| \le \dots$$

repeating multiple eigenvalues the appropriate number of times. For sufficiently large k, it is easy to show that

$$|\varepsilon_k| \sim \frac{2\pi c}{R_2 - R_1} k, \quad k \gg 1,$$

from which the last part follows. ■

This result implies that the eigenvalue spectrum for this problem has both positive and negative discrete energy values, and is unbounded in both directions. The solutions represent bound states and pseudo-states with energies in the range $(-2c^2, 0)$ and pseudo-states in the intervals $(-\infty, -2c^2)$ and $(0, \infty)$.

B.7.7 The inhomogeneous boundary value problem

Consider the inhomogeneous boundary value problem

$$(T_\kappa - \varepsilon)\,u(r) = -f(r), \quad R_1 < r < R_2, \tag{B.7.18}$$

with boundary conditions $u(R_1) = Mv$ and $u(R_2) = Nv$. This problem has a solution

$$u(r) = \int_{R_1}^{R_2} G(r, s, ; \varepsilon)\, f(s)\, ds. \tag{B.7.19}$$

whenever ε is not an eigenvalue of the eigenvalue problem for the homogeneous equation, where the function $G(r, s, ; \varepsilon)$ is called the *resolvent kernel* or *Green's function* matrix for the problem. To construct the Green's function, we look for a vector $z(r)$ such that

$$u(r) = U_\varepsilon(r)\, z(r), \quad r \in [R_1, R_2]$$

where $U_\varepsilon(r)$ is the fundamental matrix solution defined in §B.7.1. It follows that

$$cJU_\varepsilon(r)\frac{dz}{dr} = -f(r),$$

and multiplying on the left by ${U_\varepsilon}^\dagger(r)$ and using (B.7.10) we get

[9] A complex function $f(z)$ is said to be an entire function if its only singularity is an isolated singularity at $z = \infty$. If this is a pole of order m, then $f(z)$ is a polynomial of degree m.

$$\frac{dz}{dr} = \frac{1}{c} J\, U_\varepsilon^\dagger(r) f(r)$$

so that

$$z(r) = z(R_1) + \frac{1}{c} \int_{R_1}^{r} J\, U_\varepsilon^\dagger(s) f(s)\, ds$$

and

$$u(r) = U_\varepsilon(r)\, z(R_1) + \frac{1}{c} \int_{R_1}^{r} U_\varepsilon(r) J\, U_\varepsilon^\dagger(s) f(s)\, ds \tag{B.7.20}$$

Because $U_\varepsilon(R_1) = I$ this satisfies the initial condition at $r = R_1$, and it also satisfies the second boundary condition $u(R_2) = Nv$

$$Nv = U_\varepsilon(R_2)\, Mv + \frac{1}{c} \int_{R_1}^{R_2} U_\varepsilon(R_2) J\, U_\varepsilon^\dagger(s) f(s)\, ds$$

Thus

$$v = (N - U_\varepsilon(R_2)\, M)^{-1} \frac{1}{c} \int_{R_1}^{R_2} U_\varepsilon(R_2) J\, U_\varepsilon^\dagger(s) f(s)\, ds$$

whenever ε is not an eigenvalue of the homogeneous eigenvalue problem. Substituting back in (B.7.20) gives

Theorem B.20. *Whenever ε is not an eigenvalue of the homogeneous eigenvalue problem (B.7.14) and $f(r) \in [\mathcal{L}^2(R_1, R_2)]^2$, the inhomogeneous boundary value problem (B.7.18) has a solution of the form (B.7.19) with*

$$G(r,s;\varepsilon) = \begin{cases} \frac{1}{c} U_\varepsilon(r) \left[I + M(N - U_\varepsilon(R_2)\, M)^{-1} U_\varepsilon(R_2) \right] J\, U_\varepsilon^\dagger(s), \\ \qquad\qquad\qquad\qquad R_1 \le s < r \le R_2, \\ \frac{1}{c} U_\varepsilon(r) \left[M(N - U_\varepsilon(R_2)\, M)^{-1} U_\varepsilon(R_2) \right] J\, U_\varepsilon^\dagger(s), \\ \qquad\qquad\qquad\qquad R_1 \le r < s \le R_2. \end{cases}$$

Corollary B.21. *$G(r,s;\varepsilon)$ is continuous on the square $[R_1, R_2] \times [R_1, R_2]$ except on the line $s = r$ where it has a jump discontinuity given by*

$$[G(r,s;\varepsilon)]_{s=r-0}^{s=r+0} = -\frac{1}{c} U_\varepsilon(r) J\, U_\varepsilon^\dagger(r) = -\frac{J}{c}. \tag{B.7.21}$$

Theorem B.22. *When $s \neq r$ and neither ε_1 nor ε_2 are eigenvalues of (B.7.14), we have*

$$G(r,s;\varepsilon_1) - G^\dagger(s,r;\varepsilon_2) = (\varepsilon_2^* - \varepsilon_1) \int_{R_1}^{R_2} G^\dagger(t,r;\varepsilon_2)\, G(t,s;\varepsilon_1)\, dt \tag{B.7.22}$$

on $[R_1, R_2] \times [R_1, R_2]$. In particular if $\varepsilon_2^* = \varepsilon_1$, then

$$G(r,s;\varepsilon_1) = G^\dagger(s,r;\varepsilon_1^*).$$

Proof: For the inhomogeneous problem (B.7.18)

$$\begin{aligned} & s_{12}(R_2) - s_{12}(R_1) \\ &= (\varepsilon_1 - \varepsilon_2^*) \int_{R_1}^{R_2} u_2^\dagger(r)\, u_1(r)\, dr - \int_{R_1}^{R_2} \Big(u_2^\dagger(r)\, f_1(r) - f_2^\dagger(r)\, u_1(r)\Big)\, dr \end{aligned} \tag{B.7.23}$$

The left-hand side vanishes by virtue of the boundary conditions and on substituting for $u_1(r)$ and $u_2(r)$ from(B.7.19) we get

$$\begin{aligned} \int_{R_1}^{R_2}\int_{R_1}^{R_2} f_2^\dagger(r)\Bigg\{ & G(r,s;\varepsilon_1) - G^\dagger(s,r;\varepsilon_2) \\ & -(\varepsilon_2^* - \varepsilon_1)\int_{R_1}^{R_2} G^\dagger(t,r;\varepsilon_2)\, G(t,s;\varepsilon_1)\, dt\Bigg\} f_1(s) dr\, ds = 0, \end{aligned}$$

from which the result follows.

■

Define

$$G(r,r;\varepsilon) := \frac{1}{2}\left[G(r,r+0;\varepsilon) + G(r,r-0;\varepsilon)\right] \tag{B.7.24}$$

and the matrix-valued *characteristic function*

$$\begin{aligned} F(\varepsilon) &:= G(R_1,R_1;\varepsilon) \\ &= \frac{1}{2c}\left(M + U_\varepsilon^{-1}(R_2)N\right)\left(M - U_\varepsilon^{-1}(R_2)N\right)^{-1} J^{-1}. \end{aligned} \tag{B.7.25}$$

This enables us to write the Green's function more compactly as

$$G(r,s;\varepsilon) = \begin{cases} U_\varepsilon(r)\left[F(\varepsilon) + (2cJ)^{-1}\right] U_\varepsilon^\dagger(s), & r < s, \\ U_\varepsilon(r)\left[F(\varepsilon) - (2cJ)^{-1}\right] U_\varepsilon^\dagger(s), & s < r \end{cases} \tag{B.7.26}$$

from which

$$G(r,r;\varepsilon) = U_\varepsilon(r) F(\varepsilon) U_\varepsilon^\dagger(r). \tag{B.7.27}$$

From (B.7.22) we have

$$\begin{aligned} G(r,r;\varepsilon_1) - G^\dagger(r,r;\varepsilon_2) &= (\varepsilon_2^* - \varepsilon_1)\int_{R_1}^{R_2} G^\dagger(t,r;\varepsilon_2)\, G(t,r;\varepsilon_1)\, dt \\ &= (\varepsilon_2^* - \varepsilon_1)\int_{R_1}^{R_2} G(r,t;\varepsilon_2^*)\, G(t,r;\varepsilon_1)\, dt. \end{aligned}$$

Setting $\varepsilon = \varepsilon_1 = \varepsilon_2$, and using the definition of the charecteristic function (B.7.25) gives

$$\Im F(\varepsilon) = -\Im\varepsilon \int_{R_1}^{R_2} G(R_1,t;\varepsilon^*)\, G(t,R_1;\varepsilon)\, dt, \tag{B.7.28}$$

where, by symmetry, the integral is positive definite. The fact that $U_\varepsilon(r)$ is an entire function of ε for each value of r means that $F(\varepsilon)$ carries full information about the singularities of the Green's function. This gives the important result

Theorem B.23. *The matrix characteristic function* $F(\varepsilon)$ *is Hermitian for real values of* ε *and is finite except for poles. When* ε *is complex, then* $\Im F(\varepsilon)$ *and* $\Im\varepsilon$ *have opposite sign.*

Corollary B.24. $F(\varepsilon)$ *has only simple poles coinciding with the zeros of* $\det(N - U_\varepsilon(R_2)\,M)$. *Thus there exists a harmonic series in a neighbourhood of each zero* $\varepsilon = \varepsilon_k$ *such that*

$$F(\varepsilon) = \frac{P_k}{\varepsilon - \varepsilon_k} + F_k(\varepsilon) \tag{B.7.29}$$

where $F_k(\varepsilon)$ *is holomorphic in a neighbourhood of* ε_k.

Sketch of proof For full details, see [60].

A scalar function $f(\lambda)$ of a complex variable λ is said to be of *negative imaginary type* if it is holomorphic on the upper half-plane and satisfies $\Im f(\lambda) \leq 0$ for $\Im\lambda > 0$. Suppose that $f(\lambda)$ is meromorphic, real when λ is real (apart from its (real) poles); thus $\Im f(\lambda) > 0$ when $\Im\lambda < 0$. A necessary and sufficient condition that $f(\lambda)$ be a rational function with these properties is that $f(\lambda) = p(\lambda)/q(\lambda)$, where $p(\lambda)$ and $q(\lambda)$ are polynomials with real coefficients and real simple zeros that have no zeros in common. The zeros of $p(\lambda)$ must separate those of $q(\lambda)$. This result can be extended, to all functions $f(\lambda)$ that are holomorphic, save possibly for poles on the real axis that have no finite limit point, for which $\Im f(\lambda)$ and $\Im\lambda$ have opposite signs, Then the zeros of $f(\lambda)$ on the real axis separate, and are separated by, the poles. The final step to extend this to matrix valued functions completes the proof.

■

Theorem B.25. $F(\varepsilon)$ *and (a fortiori)* $G(r, s; \varepsilon)$ *possess discrete simple poles on the real axis. The residues at* $\varepsilon = \varepsilon_k$ *are* $P_k = w_k\, w_k^\dagger$ *and* $u_k(r)\, u_k^\dagger(s)$ *respectively.*

Proof: Equation (B.7.25) can be recast as

$$2c\, F(\varepsilon)\, J\, U_\varepsilon^{-1}(R_2)\, (U_\varepsilon(R_2)\, M - N) = U_\varepsilon^{-1}(R_2)\, (U_\varepsilon(R_2)\, M + N)$$

The right-hand side is regular for all finite ε, so that

$$\frac{P_k}{\varepsilon - \varepsilon_k}\, J\, U_\varepsilon^{-1}(R_2)\, (U_\varepsilon(R_2)\, M - N)$$

is bounded in s neighbourhood of $\varepsilon = \varepsilon_k$, and in particular

$$P_k\, J\, U_\varepsilon^{-1}(R_2)\, (U_\varepsilon(R_2\,)M - N) = 0.$$

Because ε_k is a simple pole, $(U_\varepsilon(R_2)\,M - N)$ has rank 1 and as J and $U_\varepsilon^{-1}(R_2)$ are both nonsingular, P_k (a 2×2 matrix) also has rank 1. Applying this to equation (B.7.26), we see that

$$G(r,s;\varepsilon) = U_{\varepsilon_k}(r)\,\frac{P_k}{\varepsilon - \varepsilon_k}\,U^\dagger_{\varepsilon_k}(s) + G_k(r,s;\varepsilon) \tag{B.7.30}$$

where $G_k(r,s;\varepsilon)$ is holomorphic in a neighbourhood of ε_k.

The eigenvalue equation $(T_\kappa - \varepsilon_k)u_k = 0$ is equivalent to

$$(T_\kappa - \varepsilon)u_k = -(\varepsilon - \varepsilon_k)u_k,$$

with solution

$$u_k(r) = (\varepsilon - \varepsilon_k)\int_{R_1}^{R_2} G(r,s;\varepsilon)\,u_k(s)\,ds. \tag{B.7.31}$$

As $\varepsilon \to \varepsilon_k$, we obtain from (B.7.30)

$$u_k(r) = \int_{R_1}^{R_2} U_{\varepsilon_k}(r)\,P_k\,U^\dagger_{\varepsilon_k}(s)\,u_k(s)\,ds.$$

beacuse $u_k(r) = U_{\varepsilon_k}(r)\,w_k$ this can be written

$$U_{\varepsilon_k}(r)\left\{w_k - P_k\int_{R_1}^{R_2} U^\dagger_{\varepsilon_k}(s)\,U_{\varepsilon_k}(s)\,ds\,w_k\right\} = 0.$$

The bracketed expression must therefore vanish, and we see that

$$w_k = P_k W_k w_k, \tag{B.7.32}$$

where $W_k = \int_{R_1}^{R_2} U^\dagger_{\varepsilon_k}(s)\,U_{\varepsilon_k}(s)\,ds$ is a symmetric positive definite matrix which has a positive definite square root, in terms of which (B.7.32) becomes

$$\widetilde{w}_k = W_k^{1/2}\,P_k\,W_k^{1/2}\,\widetilde{w}_k.$$

where $\widetilde{w}_k = W_k^{1/2}\,w_k$. If the eigenvectors are normalized on $[R_1, R_2]$ so that

$$\int_{R_1}^{R_2} u_k^\dagger(r)\,u_k(r)\,dr = 1,$$

then

$$1 = \int_{R_1}^{R_2} w_k^\dagger\,U^\dagger_{\varepsilon_k}(r)\,U_{\varepsilon_k}(r)\,w_k\,dr = w_k^\dagger\,W_k\,w_k = \widetilde{w}_k^\dagger\,\widetilde{w}_k$$

so that $W_k^{1/2}\,P_k\,W_k^{1/2} = \widetilde{w}_k\,\widetilde{w}_k^\dagger$ from which

$$P_k = W_k^{-1/2}\widetilde{w}_k^\dagger\,\widetilde{w}_k W_k^{-1/2} = w_k\,w_k^\dagger. \tag{B.7.33}$$

Subsituting in (B.7.31) gives the Green's function residue

$$\lim_{\varepsilon \to \varepsilon_k}(\varepsilon - \varepsilon_k)\,G(r,s;\varepsilon) = u_k(r)u_k^\dagger(s) \tag{B.7.34}$$

■

B.7.8 Eigenfunction expansions

Let ε_i be an eigenvalue of (B.7.1), with corresponding eigenfunction $u_i(r)$ and boundary value vector v_i, so that

$$u_i(R_1) = Mv_i, \quad u_i(R_2) = Nv_i. \tag{B.7.35}$$

The fundamental Lemma 5.3 and the boundary conditions (B.7.35) together imply the usual orthonormality condition

$$\int_{R_1}^{R_2} u_i{}^{\dagger}(r)\, u_j(r)\, dr = \delta_{ij}. \tag{B.7.36}$$

In what follows, we assume that there are no multiple eigenvalues, the case of immediate interest; see [60] for a treatment in full generality.

Theorem B.25 suggests that the Green's function can be expanded in the form

$$G(r, s; \varepsilon) \sim \sum_k \frac{u_k(r)\, u_k^{\dagger}(s)}{\varepsilon - \varepsilon_k} \tag{B.7.37}$$

where the sign $\sim$ implies that this is only formal equality at this stage. Substituting in (B.7.19) gives the formal expansion

$$u(r) \sim \sum_k u_k(r) c_k, \quad c_k = \int_{R_1}^{R_2} \frac{u_k^{\dagger}(r)}{\varepsilon - \varepsilon_k} f(r)\, dr \tag{B.7.38}$$

and inserting this into (B.7.18) gives

$$f(r) \sim \sum_k u_k(r) d_k, \quad d_k = (\varepsilon - \varepsilon_k)\, c_k = \int_{R_1}^{R_2} u_k^{\dagger}(r) f(r)\, dr. \tag{B.7.39}$$

We have still to establish in what sense the eigenfunction expansions (B.7.37)–(B.7.39) converge. Using (B.7.15), we can express (B.7.39) compactly in terms of the quantity

$$\psi(\varepsilon) = \int_{R_1}^{R_2} U_{\varepsilon}^{\dagger}(r)\, f(r)\, dr \tag{B.7.40}$$

as $d_k = w_k^{\dagger} \psi(\varepsilon_k)$ with $w_k = Mv_k$. The formal expansion (B.7.39) can therefore be written as a Stieltjes integral

$$f(r) \sim \int_{-\infty}^{\infty} U_{\varepsilon}(r)\, d\rho(\varepsilon)\, \psi(\varepsilon), \tag{B.7.41}$$

where $\rho(\varepsilon)$, the *spectral function*, is a matrix-valued step function, jumping by $(w_k\, w_k^{\dagger})$ at each eigenvalue. An eigenvalue which has multiplicity r is associated with an r-fold jump.

Define the residual error of a partial sum of the eigenfunction expansion of $u(r)$ (B.7.38) by

$$E_K[u](r) := u(r) - \sum_{|\varepsilon_k| \leq K} u_k(r) c_k; \tag{B.7.42}$$

$E_K[f](r)$ is defined in a similar way. The following theorem is based on [60, Theorem 9.6.2]:

Theorem B.26. *Let $f(r)$, $R_1 \leq r \leq R_2$ be measurable and let $\int_{R_1}^{R_2} f^\dagger(r) f(r) dr$, be finite. Let $u(r)$ be absolutely continuous and satisfy almost everywhere the inhomogeneous boundary value problem (B.7.18). Then, for any $K > 0$,*

$$\int_{R_1}^{R_2} E_K[u]^\dagger(r) E_K[u](r) dr \leq \frac{1}{K^2} \int_{R_1}^{R_2} f^\dagger(r) f(r) dr \tag{B.7.43}$$

Corollary B.27. *The right hand side of (B.7.43) vanishes as $K \to \infty$ because $\int_{R_1}^{R_2} f^\dagger(r) f(r) dr$ is finite, showing that the eigenfunction expansion (B.7.38) converges in mean square.*

We shall omit the lengthy proof, which appears in [60, §9.6] in a more general form. It requires the following:

Lemma B.28. *(cf. [60, Lemmas 9.6.1, 9.6.2])*

Let T_κ have no zero eigenvalues and let $v(r)$ be the solution a.e. of

$$T_\kappa v(r) = -f(r), \quad r \in [R_1, R_2], v(R_1) = Mv, \; v(R_2) = Nv,$$

where $f(r)$ is a measurable function in $[\mathcal{L}^2)R_1, R_2)]^2$, and let $f(r)$ be orthogonal to all $u_k(r)$ with $|\varepsilon_k| \leq K$ for any real $K > 0$; then

$$\int_{R_1}^{R_2} E_K[u]^\dagger(r) E_K[u](r) dr \leq \frac{1}{K^2} \int_{R_1}^{R_2} f^\dagger(r) f(r) dr.$$

The result is still valid even if T_κ has a zero eigenvalue.

It is possible to prove that the eigenfunction expansion converges uniformly and absolutely on the whole interval $R_1 \leq r \leq R_2$. This requires some auxiliary results.

Theorem B.29. *For $r \in [R_1, R_2]$.*

(a) If ε is not an eigenvalue of (B.7.14), then

$$\sum_k \frac{u_k(r)\, u_k^\dagger(r)}{|\varepsilon - \varepsilon_k|^2} \leq \int_{R_1}^{R_2} G(r, s; \varepsilon)\, G^\dagger(r, s; \varepsilon)\, ds \tag{B.7.44}$$

(b) If ε is not an eigenvalue of (B.7.14), then

$$\sum_k \frac{u_k^\dagger(r)\, u_k(r)}{|\varepsilon - \varepsilon_k|^2} \le \mathrm{trace} \int_{R_1}^{R_2} G(r,s;\varepsilon)\, G^\dagger(r,s;\varepsilon)\, ds \tag{B.7.45}$$

(c) $\sum_k \dfrac{u_k^\dagger(r)\, u_k(r)}{1+\varepsilon_k^2}$ *is absolutely convergent on* $[R_1, R_2]$.

Proof: When ε is not an eigenvalue of (B.7.14), the 2×2 matrix

$$\int_{R_1}^{R_2} \left\{ G(r,s;\varepsilon) - \sum_{|\varepsilon_k| \le K} \frac{u_k(r)\, u_k^\dagger(s)}{\varepsilon - \varepsilon_k} \right\} \times \left\{ G^\dagger(r,s;\varepsilon) - \sum_{|\varepsilon_k| \le K} \frac{u_k(s)\, u_k^\dagger(r)}{\varepsilon - \varepsilon_k} \right\} ds$$

is non-negative definite. Part (a) follows on using (B.7.31) (valid when ε is not an eigenvalue, and taking the trace gives part (b). Part (c) follows from part (b) by using the fact that i is not an eigenvalue and noting that $|i - \varepsilon_k|^2 = 1 + \varepsilon_k^2$. ■

The next theorem establishes that the eigenfunction expansion converges absolutely and uniformly on $[R_1, R_2]$:

Theorem B.30. *With the assumptions of Theorem B.26, the eigenfunction expansion*

$$\sum_k u_k(r)\, c_k,$$

converges uniformly and absolutely to $u(r)$ on $[R_1, R_2]$.

Proof: It is sufficient to consider only the scalar series

$$\sum_k \left[u_k^\dagger(r)\, u_k(r) \right]^{1/2} |c_k|$$

because each component of the vector $u_k(r)$ is bounded in absolute value by $\|u_k\| = \left[u_k^\dagger(r)\, u_k(r) \right]^{1/2}$. Choose positive integers p, K. By the Cauchy-Schwarz inequality,

$$\left\{ \sum_{k=K}^{K+p} \left[u_k^\dagger(r)\, u_k(r) \right]^{1/2} |c_k| \right\}^2 \le \sum_{k=K}^{K+p} \frac{u_k^\dagger(r)\, u_k(r)}{1+\varepsilon_k^2} \sum_{k=K}^{K+p} (1 + \varepsilon_k^2)\, |c_k|^2. \tag{B.7.46}$$

Because $\int_{R_1}^{R_2} E_K^\dagger[u](r) E_K[u](r)\, dr \ge 0$ and $\int_{R_1}^{R_2} E_K^\dagger[f](r) E_K[f](r)\, dr \ge 0$, we have the Bessel inequalities

$$\sum_0^\infty |c_k|^2 \le \int_{R_1}^{R_2} u^\dagger(r)u(r)\,dr, \quad \sum_0^\infty |c_k\,\varepsilon_k|^2 \le \int_{R_1}^{R_2} f^\dagger(r)f(r)\,dr$$

from which we infer that, for each fixed K sufficiently large,

$$\lim_{p\to\infty} \sum_{k=K}^{K+p} (1+\varepsilon_k^2)\,|c_k|^2$$

is bounded. By Theorem B.29, part (c), the first sum in (B.7.46) is bounded, from which absolute convergence of $\sum_k \left[u_k^\dagger(r)\,u_k(r)\right]^{1/2} |c_k|$ and hence of the eigenfunction expansion of $u(r)$ on $[R_1, R_2]$ follows. The Cauchy convergence principle now assures us that because

$$\lim_{K\to\infty} \sum_{k=K}^{K+p} (1+\varepsilon_k^2)\,|c_k|^2 = 0.$$

for each fixed $p > 0$, then the eigenfunction expansion also converges uniformly as $K \to \infty$ on $[R_1, R_2]$. ■

B.8 Iterative processes in nonlinear systems of equations

General references [39], [42]

Algorithms for solving nonlinear systems of equations by iteration usually depend on the notion of a *contraction mapping* [42, Chapter 3] [39, §1.4]. Suppose that we wish to solve a simple nonlinear equation in one variable,

$$\psi(s) = 0, \tag{B.8.1}$$

iteratively. We consider an associated equation

$$s = g(s), \quad g(s) = s - m(s)\,\psi(s), \tag{B.8.2}$$

where $m(s)$ is bounded and non-vanishing on some interval (s_*, s^*) , but need not be defined more closely at this stage. The corresponding iterative process in $\mathbb{R}^k$ is defined by

Algorithm B.1 *Let* $\boldsymbol{s}$ *denote a point of* $\mathbb{R}^k$.

1. *Set initial estimate* $\boldsymbol{s}^{[0]}$.
2. *For* $\nu = 0, 1, \ldots$ *let* $\boldsymbol{s}^{[\nu+1]} = g(\boldsymbol{s}^{[\nu]})$.
3. *Stop when* $\|\boldsymbol{s}^{[\nu+1]} - \boldsymbol{s}^{[\nu]}\|$ *is sufficiently small.*

The contraction mapping theorem gives conditions for the Algorithm B.1 to converge.:

Theorem B.31. *Let $\boldsymbol{g}(\boldsymbol{s}) : \mathbb{R}^k \to \mathbb{R}^k$ satisfy the Lipschitz condition*

$$\|\boldsymbol{g}(\boldsymbol{s}) - \boldsymbol{g}(\boldsymbol{t})\| \le \lambda \|\boldsymbol{s} - \boldsymbol{t}\|, \; 0 \le \lambda < 1,$$

for all $\boldsymbol{s}$ and $\boldsymbol{t}$ in some neighbourhood $N_\rho(\boldsymbol{s}^{[0]})$, where

$$N_\rho(\boldsymbol{s}^{[0]}) = \left\{ \boldsymbol{s} \, : \, \|\boldsymbol{s} - \boldsymbol{s}^{[0]}\| \le \rho \right\}.$$

Let $\boldsymbol{s}^{[0]}$ be such that

$$\|\boldsymbol{s}^{[0]} - \boldsymbol{g}(\boldsymbol{s}^{[0]})\| \le (1 - \lambda)\rho.$$

Then the points $\boldsymbol{s}^{[\nu]}$ generated by Algorithm B.1 satisfy

(a) $\boldsymbol{s}^{[\nu]} \in N_\rho(\boldsymbol{s})^{[0]}), \; \nu = 0, 1, \ldots;$
(b) $\boldsymbol{s}^{[\nu]} \to \boldsymbol{S}$ as $\nu \to \infty$.
(c) $\boldsymbol{S}$ is the unique root of $\boldsymbol{S} = \boldsymbol{g}(\boldsymbol{S})$ in $N_\rho(\boldsymbol{s}^{[0]})$.
(d) $\|\boldsymbol{s}^{[\nu]} - \boldsymbol{S}\| \le \lambda^\nu (1 - \lambda)^{-1} \|\boldsymbol{s}^{[1]} - \boldsymbol{s}^{[0]}\| \le \lambda^\nu \rho$.

The short proof brings out important features of the iterative process. By hypothesis, $\boldsymbol{s}^{[1]} = \boldsymbol{g}(s^{[0]})$ is in $N_\rho(\boldsymbol{s})^{[0]}$. Suppose that $\boldsymbol{s}^{[2]}, \ldots \boldsymbol{s}^{[\nu]}$ also are in $N_\rho(\boldsymbol{s})^{[0]}$; then

$$\|\boldsymbol{s}^{[\nu+1]} - \boldsymbol{s}^{[\nu]}\| = \|\boldsymbol{g}(\boldsymbol{s}^{[\nu]}) - \boldsymbol{g}(\boldsymbol{s}^{[\nu-1]}\| \le \lambda \|\boldsymbol{s}^{[\nu]} - \boldsymbol{s}^{[\nu-1]}\|.$$

Applying this recursively gives

$$\|\boldsymbol{s}^{[\nu]} - \boldsymbol{s}^{[\nu-1]}\| \le \frac{\lambda^\nu}{1 - \lambda} \rho,$$

and so as $0 \le \lambda < 1$,

$$\begin{aligned} \|\boldsymbol{s}^{[\nu+1]} - \boldsymbol{s}^{[0]}\| &\le \|\boldsymbol{s}^{[\nu+1]} - \boldsymbol{s}^{[\nu]}\| + \|\boldsymbol{s}^{[\nu]} - \boldsymbol{s}^{[\nu-1]}\| + \ldots + \|\boldsymbol{s}^{[1]} - \boldsymbol{s}^{[0]}\| \\ &= (\lambda^\nu + \lambda^{\nu-1} + \ldots + 1)(1 - \lambda)\rho \\ &\le \rho, \end{aligned} \tag{B.8.3}$$

so that $\boldsymbol{s}^{[\nu+1]}$ is also in $N_\rho(\boldsymbol{s})^{[0]}$, proving (a). To prove (b), we need to show that $\{ \, \boldsymbol{s}^{[\nu]} \, \}$ is a Cauchy sequence, and hence has a limit $\boldsymbol{S}$. This follows because

$$\|\boldsymbol{s}^{[\nu+\mu]} - \boldsymbol{s}^{[\nu]}\| \le \lambda^\nu \rho \tag{B.8.4}$$

for all $\mu > 0$, by an argument similar to that used in (B.8.3). Part (c) follows by considering the the limit of $\boldsymbol{s}^{[\nu+1]} = \boldsymbol{g}(\boldsymbol{s}^{[\nu]})$ as $\nu \to \infty$. To prove uniqueness we observe that if there were two roots $\boldsymbol{S}$ and $\boldsymbol{S}'$ in $N_\rho(\boldsymbol{s})^{[0]})$, then

$$\|\boldsymbol{S} - \boldsymbol{S}'\| = \|\boldsymbol{g}(\boldsymbol{S}) - \boldsymbol{g}(\boldsymbol{S}')\| \le \lambda \|\boldsymbol{S} - \boldsymbol{S}'\|.$$

because $\lambda < 1$, this can only be true if $\boldsymbol{S} = \boldsymbol{S}'$. Part (d) follows by a calculation on the lines of (B.8.3) and is consistent with taking the limit of (B.8.4) as $\mu \to \infty$.

The contraction mapping theorem shows that if we can find an expression $\boldsymbol{g}(\boldsymbol{s})$ which is a contraction mapping, and an initial estimate sufficiently close to the root we want, then the algorithm converges. The error decreases by a factor λ in each iteration. The main difficulty for many applications is that it is not always easy to predict a domain containing a fixed point of the mapping so that there is no guarantee that the initial guess $\boldsymbol{s}^{[0]}$ will lead to a successful outcome. Moreover a poorly selected $\boldsymbol{g}(\boldsymbol{s})$ will only converge slowly, so that there are many obstacles to overcome in practice. In the case of the one-dimensional process (B.8.2), we can be a bit more specific.

Theorem B.32. *We seek a root s_n in an interval $\mathcal{D}_n = \{s : |s - s_n| \le \rho_n\}$. Let γ_n and Γ_n have the same sign, $\gamma_n \Gamma_n > 0$, where*

$$|\gamma_n| = \min \left|\frac{d\psi}{ds}\right|, \quad |\Gamma_n| = \max \left|\frac{d\psi}{ds}\right|, \quad s \in \mathcal{D}_n.$$

Choose m_n such that $|m_n| < 2/\Gamma_n$,

$$K_n = \min\{|1 - m_n\gamma_n|, |1 - m_n\Gamma_n|\} < 1,$$

and an initial estimate $s_n^{[0]}$ so that

$$|s_n^{[0]} - s_n| \le (1 - K_n)\rho_n.$$

Then the sequence of iterates defined by

$$s_n^{[\nu+1]} = g(s_n^{[\nu]}) = s_n^{[\nu]} - m_n\psi(s_n^{[\nu]})$$

converges to s_n. The choice $m_n = 2/(\gamma_n + \Gamma_n)$ gives the error estimate

$$|s_n^{[\nu]} - s_n| < K_n^{\nu}|\psi(s_n^{[0]})/\gamma_n|, \ \nu = 1, 2, \ldots$$

Proof: The hypotheses ensure that $d\psi/ds$ does not vanish in $\mathcal{D}_n$, and it therefore has an unique sign on that interval. Now for any two points $s, s' \in \mathcal{D}_n$,

$$g(s') - g(s) = s' - s - m_n[\psi(s') - \psi(s0] = (s' - s)\left[1 - m_n\overline{\frac{d\psi}{ds}}\right]$$

where, by the mean value theorem, $\overline{d\psi/ds}$ is the derivative evaluated at some point $\overline{s} = s + \theta(s' - s)$ where $0 < \theta < 1$. It follows that

$$|g(s') - g(s)| < K_n|s' - s|,$$

so that $g(s)$ gives a contraction map and the iteration converges by Theorem B.31. The choice $m_n = 2/(\gamma_n + \Gamma_n)$ minimizes K_n and leads to the quoted error estimate. ■

This argument shows that the choice of m_n is not critical, although it clearly affects the rate of convergence. It is obviously desirable to choose m_n

to minimize the number K_n by which the error is reduced in each iteration. In fact, convergence can be accelerated by changing m_n at each step. In Newton's method we write

$$\psi(s_n^{[\nu+1]}) = \psi(s_n^{[\nu]}) + (s_n^{[\nu+1]} - s_n^{[\nu]})\frac{d\psi(s_n^{[\nu]})}{ds} \approx 0.$$

Rearranging gives

$$s_n^{[\nu+1]} = s_n^{[\nu]} - \psi(s_n^{[\nu]})\left[\frac{d\psi(s_n^{[\nu]})}{ds}\right]^{-1},$$

or equivalently

$$m_n^{[\nu]} = \left[\frac{d\psi(s_n^{[\nu]})}{ds}\right]^{-1}.$$

Provided $d\psi/ds$ is also Lipschitz continuous in a neighbourhood of s_n we can find a constant M_n such that

$$|s_n^{[\nu+1]} - s_n| < M_n|s_n^{[\nu]} - s_n|^2,$$

giving much faster *quadratic convergence.*

B.9 Lagrangian and Hamiltonian methods

General references [65], [66]

B.9.1 Lagrange's equations

The Lagrangian and Hamiltonian formalism of classical and quantum field theories is modelled on the classical mechanics presented in texts such as [65] or [66]. If q denotes a set of configuration variables $\{q_1, \ldots, q_N\}$ and $\dot{q}$ denotes their velocities at time t, then the *action* associated with the system is

$$S := \int_{t_1}^{t_2} L[q_1(t)\ldots q_n(t), \dot{q}_1(t)\ldots, \dot{q}_N(t)]dt \tag{B.9.1}$$

where $L[q_1(t)\ldots q_n(t), \dot{q}_1(t)\ldots, \dot{q}_N(t)]$ depends on the positions and velocities and may, for open systems, depend on the time variable t as well. The *principle of least action* states that the physical trajectory of the system, expressed by the functional dependence $q(t)$, is that for which S is stationary with respect to small changes $q_n(t) \to q_n(t) + \delta q_n(t)$ in the trajectory subject to suitable boundary conditions, say $q_n(t_1) = q_n^{(1)}$, $q_n(t_2) = q_n^{(2)}$. Considering S as a functional of the coordinates, we have

$$\delta S[q] := S[q + \delta q] - S[q]$$
$$= \int_{t_1}^{t_2} \sum_{n=1}^{N} \left[\frac{\partial L}{\partial q_n(t)} \, \delta q_n(t) + \frac{\partial L}{\partial \dot{q}_n(t)} \, \delta \dot{q}_n(t) \right] dt = 0, \qquad \text{(B.9.2)}$$

where $\delta \dot{q}_n(t) = d\delta q_n(t)/dt$ and $\delta q(t_1) = \delta q(t_2) = 0$. Integrating the last term by parts and using the boundary conditions gives

$$\delta S[q] = \int_{t_1}^{t_2} \sum_{n=1}^{N} \frac{\delta S}{\delta q_n(t)} \delta q_n(t) \, dt \qquad \text{(B.9.3)}$$

where the *variational derivative*, defined by

$$\frac{\delta S}{\delta q_n(t)} := \frac{\partial L}{\partial q_n(t)} - \frac{d}{dt} \left(\frac{\partial L}{\partial \dot{q}_n(t)} \right) = 0, \quad n = 1, \ldots, N, \qquad \text{(B.9.4)}$$

vanishes by (B.9.2) because the variations $\delta q_n(t)$ are independent. The Lagrange equations (B.9.4), one for each degree of freedom, yield the Newton equations of motion for conservative mechanical systems when $L[q(t), \dot{q}(t)]$ is the difference of a kinetic energy expression, $T(q, \dot{q})$ and a potential energy expression, $V(q)$. In the presence of electromagnetic forces, as in (2.6.36), the potential energy term acquires a velocity-dependent character.

The equations are unchanged if we add to $L[q(t), \dot{q}(t)]$ a total derivative with respect to time. In this case, boundary condition terms are added to the action.

B.9.2 Hamilton's equations

The *generalized momenta* of the system are defined by

$$p_n(t) := \frac{\partial L}{\partial \dot{q}_n(t)}. \qquad \text{(B.9.5)}$$

Hamilton's formulation replaces the velocity variables $\dot{q}_n$ of Lagrange's equations by the generalized momenta p_n. This is achieved by a Legendre transformation. The total differential of the Lagrangian is

$$dL = \sum_{n=1}^{N} \frac{\partial L}{\partial q_n} \, \delta q_n + \frac{\partial L}{\partial \dot{q}_n} \, \delta \dot{q}_n$$
$$= \sum_{n=1}^{N} \dot{p}_n \, \delta q_n + p_n \, \delta \dot{q}_n$$

where we have used both (B.9.4) and (B.9.5) to replace the partial derivatives. Defining a new function, the Hamiltonian, by

$$H(q_1 \dots q_N, p_1 \dots p_N, t) := \sum_{n=1}^{N} p_n \, \dot{q}_n - L \tag{B.9.6}$$

we find that H is, as advertised, a function of p and q only:

$$dH = \sum_{n=1}^{N} \{-\dot{p}_n \, \delta q_n + \dot{q}_n \, \delta p_n\} \, . \tag{B.9.7}$$

The equations of motion take the form

$$\dot{q}_n = \frac{\partial H}{\partial p_n}, \quad \dot{p}_n = -\frac{\partial H}{\partial q_n}, \quad n = 1 \dots N, \tag{B.9.8}$$

so that Lagrange's N second order equations (B.9.4) are replaced by the $2N$ first order Hamilton's equations (B.9.8).

The total time derivative of the Hamiltonian is

$$\frac{dH}{dt} = \frac{\partial H}{\partial t} + \sum_{n=1}^{N} \left[\frac{\partial H}{\partial q_n} \dot{q}_n + \frac{\partial H}{\partial p_n} \dot{p}_n \right]$$

and by substituting from (B.9.8) we see that

$$\frac{dH}{dt} = \frac{\partial H}{\partial t}$$

so that H is constant in time (and equal to the total energy of the system) if it has no explicit time dependence, $\partial H / \partial t = 0$.

The total time derivative of a function $f(q, p, t)$ is given by

$$\begin{aligned} \frac{df}{dt} &:= \frac{\partial f}{\partial t} + \sum_{n=1}^{N} \left[\frac{\partial f}{\partial q_n} \dot{q}_n + \frac{\partial f}{\partial p_n} \dot{p}_n \right] \\ &= \frac{\partial f}{\partial t} + \sum_{n=1}^{N} \left[\frac{\partial f}{\partial q_n} \frac{\partial H}{\partial p_n} - \frac{\partial f}{\partial p_n} \frac{\partial H}{\partial q_n} \right] \\ &= \frac{\partial f}{\partial t} + [H, f] \end{aligned} \tag{B.9.9}$$

where the second line uses Hamilton's equations (B.9.8) and the third line introduces the *Poisson bracket* of two expressions f, g

$$[f, g] := \sum_{n=1}^{N} \left[\frac{\partial f}{\partial p_n} \frac{\partial g}{\partial q_n} - \frac{\partial f}{\partial q_n} \frac{\partial g}{\partial p_n} \right] . \tag{B.9.10}$$

In particular, we see that if $f(q, p, t)$ is constant in time then

$$\frac{\partial f}{\partial t} + [H, f] = 0$$

and if also it does not depend explicitly on time, so that $\partial f/\partial t = 0$, then also its Poisson bracket with H, $[H, f]$ also vanishes.

In terms of the Hamiltonian, we see that the action can also be regarded as a functional of the $2N$ variables p and q, so that

$$S[q,p] = \int_{t_1}^{t_2} \sum_{n=1}^{N} p_n \, dq_n - H(q,p)\, dt \tag{B.9.11}$$

If the endpoints $(q^{(1)}, t_1)$ and $(q^{(2)}, t_2)$ are fixed, then

$$\delta S = \int_{t_1}^{t_2} \sum_{n=1}^{N} \left\{ \delta p_n \left(\dot{q}_n - \frac{\partial H}{\partial p_n} \right) + \left(p_n \frac{d\{\delta q_n\}}{dt} - \frac{\partial H}{\partial q_n} \delta q_n \right) \right\} dt$$

Integrating the term in $p_n \, d\{\delta q_n\}/dt$ by parts gives the variational derivatives

$$\frac{\delta S}{\delta p_n} = \dot{q}_n - \frac{\partial H}{\partial p_n}, \quad \frac{\delta S}{\delta q_n} = -\dot{p}_n - \frac{\partial H}{\partial q_n}$$

so that we recover Hamilton's equations when $\delta S = 0$.

B.9.3 Symmetries and conservation laws

The principle of least action selects the path along which the system evolves as the member of a collection of paths that minimizes the action. We can also fix one endpoint and consider the action along the true path as the other endpoint is varied. Then if the Lagrangian L is a function of t as well as of the generalized coordinates and velocities, we see from (B.9.3) that

$$\delta S = \left[\sum_{n=1}^{N} \frac{\partial L}{\partial q_n} \delta q_n \right]_{t_1}^{t_2} + \int_{t_1}^{t_2} \sum_{n=1}^{N} \frac{\delta S}{\delta q_n(t)} \delta q_n(t)\, dt$$

The second term vanishes along the actual path, and we have agreed to set $\delta q_{t_1} = 0$. Using $p_n = \partial L/\partial q_n$ in the first term, and writing $\delta q_{t_2} = \delta q$, we see that

$$\delta S = \sum_{n=1}^{N} p_n \delta q_n,$$

from which we conclude that $\partial S/\partial q_n = p_n$. In the same spirit, if we think of S as a function of the coordinates and time at one endpoint, then the total time derivative of the action satisfies

$$\frac{dS}{dt} = L.$$

However, if we regard S as a function of the coordinates and time at the endpoint, then

$$\frac{dS}{dt} = \frac{\partial S}{\partial t} + \sum_{n=1}^{N} \frac{\partial S}{\partial q_n} \dot{q}_n = \frac{\partial S}{\partial t} + \sum_{n=1}^{N} p_n \dot{q}_n,$$

so that

$$\frac{\partial S}{\partial t} = L - \sum_{n=1}^{N} p_n \dot{q}_n = -H.$$

or, in differential form,

$$dS = \sum_{n=1}^{N} p_n \, dq_n - H dt.$$

If we wish to regard S as a function of both coordinates and time at both ends of the path, then we have

$$dS = \left(\sum_{n=1}^{N} p_n \, dq_n - H dt \right)_2 - \left(\sum_{n=1}^{N} p_n \, dq_n - H dt \right)_1 \tag{B.9.12}$$

General properties of the motion represented by symmetries are often very helpful in simplifying the problem of solving the equations of motion of a system. If the equations of motion are invariant under some symmetry transformation, then we may be able to use this fact to generate a whole family of solutions if we know just one of them. Here we focus on a different aspect: the conservation of quantities such as momentum and charge which emerge from symmetry arguments. The method is applicable to systems with infinite degrees of freedom as well as to the finite systems of this appendix.

We have already encountered one example of this. Suppose we have a nonrelativistic particle moving in force field derived from a time-independent potential. In this case, $H = T + V$, where T is the knetic energy and V is the potential, does not depend explicitly on time, and so $\partial H/\partial t = 0$. Hence $dH/dt = [H, H] = 0$, so that the energy is conserved. Invariance of H under *time translation* is at work here. Alternatively, we can compute the action along the system path connecting $(q^{(1)}, t_1)$ to $(q^{(2)}, t_2)$. If the action is invariant under time translation, we have

$$S(q^{(2)}, t_2 + \eta; q^{(1)}, t_1 + \eta) = S(q^{(2)}, t_2; q^{(1)}, t_1)$$

so that by (B.9.12)

$$\frac{\partial S}{\partial t_2} - \frac{\partial S}{\partial t_1} = -H_2 + H_1 = 0.$$

Again, energy is conserved.

The same kind of argument works for other variations at the end-points. If the action is invariant under a *space translation* in $\mathbb{R}^3$, $\boldsymbol{q} \to \boldsymbol{q} + \boldsymbol{a}$, (B.9.12) gives immediately

$$\left(\sum_{n=1}^{N} \boldsymbol{p}_n\right)_1 = \left(\sum_{n=1}^{N} \boldsymbol{p}_n\right)_2$$

so that *linear momentum* of the system is conserved. Under an infinitesimal rotation in $\mathbb{R}^3$ through an angle $\delta\theta$ about an axis $\boldsymbol{n}$,

$$\boldsymbol{q} \to R\boldsymbol{q} := \boldsymbol{q} + \delta\theta \boldsymbol{n} \times \boldsymbol{q}$$

(B.9.12) gives

$$\left(\sum_{n} \boldsymbol{p}_n \cdot \boldsymbol{n} \times \boldsymbol{q}_n\right)_1 = \left(\sum_{n} \boldsymbol{p}_n \cdot \boldsymbol{n} \times \boldsymbol{q}_n\right)_2,$$

so that if $\boldsymbol{n}$ is an arbitrary unit vector. we see that *angular momentum* is conserved:

$$\left(\sum_{n} \boldsymbol{p}_n \times \boldsymbol{q}_n\right)_1 = \left(\sum_{n} \boldsymbol{p}_n \times \boldsymbol{q}_n\right)_2,$$

B.10 Construction of E coefficients

General references [67], [68], [69]

The relativistic generalization of the McMurchie-Davidson algorithm of §10.8.1 expresses charge density overlap components of §10.7 as a linear combination of products of coefficients $E_q^{\beta\beta'}(\mu\nu;\boldsymbol{k})$ with standard HGTF $H(p_{\mu\nu}, \boldsymbol{r}_{P_{\mu\nu}};\boldsymbol{k})$, enabling straightforward calculation of relativistic interaction integrals. The effectiveness of this construction depends on evaluating the expressions of §10.8.1 in terms of the nonrelativistic E-coefficients defined in [67]. The E-coefficient algorithms of Saunders [68] involve many conditional branches that render them inefficient, and the reformulation described here [69] aims to remove the computational bottleneck to cope with the huge numbers of coefficients needed in practical relativistic calculations.

B.10.1 E-coefficients through Cartesian intermediates

We first consider briefly the generation of E-coefficients using Cartesian Gaussian intermediates (CGTF). This method starts by writing an SGTF as

$$\begin{aligned} S(a,\boldsymbol{r};\boldsymbol{\nu}) &= r^{2n}\,\mathcal{Y}_{lm}(\boldsymbol{r})\exp(-a\boldsymbol{r}^2) \\ &= \sum_{\boldsymbol{k}} s_{\boldsymbol{k}}^{\boldsymbol{\nu}}\, C(a,\boldsymbol{r};\boldsymbol{k}) \end{aligned} \tag{B.10.1}$$

where $\boldsymbol{k} = (i,j,k)$ consists of non-negative integer triplets with $i+j+k = n+2l$, $\boldsymbol{\nu} = (n,l,m)$ and the CGTF $C(a,\boldsymbol{r};\boldsymbol{k})$ is a product of three one-dimensional CGTF,

$$C(a, \boldsymbol{r}; \boldsymbol{k}) = C(a, x; i)\, C(a, y; j)\, C(a, z; k) \tag{B.10.2}$$

with

$$C(a, t; l) = t^l \exp(\mathrm{d} - at^2),$$

As in §10.7, a one-dimensional CGTF product can be expanded in a finite series of one-dimensional HGTF, for example,

$$\begin{aligned} &C(a, x - A; i)\, C(b, x - B; i') \\ &= K_{AB} \sum_{s=0}^{i+i'} e[A, B; p, P; i, i'; s]\, H(p, x - P; s) \end{aligned} \tag{B.10.3}$$

where $p = a+b$, $P = (aA+bB)/p$, $K_{AB} = \exp(-ab(A-B)^2/p)$, and $H(p, x - P; s)$ is defined as a one-dimensional version of (10.8.3):

$$H(p, x - P; s) = \left(\frac{d}{dP}\right)^s \exp[-p(x - P)^2].$$

The expansion (B.10.3) can be obtained, for example, by writing $x - A = (x - P) + (P - A)$ and using Taylor's theorem or elementary algebra. The e-coefficients are non-zero on $0 \leq s \leq i + i'$ where they satisfy a simple recursion

$$\begin{aligned} e[A, B; p, P; i + 1, i'; s] &= \frac{1}{2p}\, e[A, B; p, P; i, i'; s - 1] \\ +(P - A)\, e[A, B; p, P; i, i'; s] &+ (s + 1)\, e[A, B; p, P; i, i'; s + 1] \end{aligned} \tag{B.10.4}$$

with the initial condition $e[A, B; p, P; 0, 0; 0] = 1$. They are set zero when $s < 0$ and $s > i+i'$. Similar equations can be written down for each coordinate direction, and the definition (B.10.2) means that we can generalize (B.10.3) to give

$$\begin{aligned} &C(a, \boldsymbol{x} - \boldsymbol{A}; \boldsymbol{k})\, C(b, \boldsymbol{x} - \boldsymbol{B}; \boldsymbol{k}') \\ &= K_{AB} \sum_{\boldsymbol{s}} E_C[A, B; p, \boldsymbol{P}; \boldsymbol{k}, \boldsymbol{k}'; \boldsymbol{s}]\, H(p, \boldsymbol{x} - \boldsymbol{P}; \boldsymbol{s}) \end{aligned} \tag{B.10.5}$$

where $\boldsymbol{P}$ and K_{AB} are now defined three-dimensionally as in (10.8.2) and $\boldsymbol{s}$ runs over all points with integer-valued arguments in the range $(0, 0, 0)$ to $(i+i', j+j', k+k')$. The final assembly of the E-coefficients of (10.8.2) requires the summation

$$E[\boldsymbol{\nu}; \boldsymbol{\nu}'; \boldsymbol{s}] = \sum_{\boldsymbol{k}} \sum_{\boldsymbol{k}'} s^{\boldsymbol{\nu}}_{\boldsymbol{k}}\, s^{\boldsymbol{\nu}'}_{\boldsymbol{k}'}\, E_C[A, B; p, \boldsymbol{P}; \boldsymbol{k}, \boldsymbol{k}'; \boldsymbol{s}]. \tag{B.10.6}$$

This method retains the computational simplicity of Cartesian Gaussians which are so popular in quantum chemistry, as well as incorporating the vital transformation properties of spherical Gaussians. The small number of coefficients $s_{\boldsymbol{k}}^{\boldsymbol{\nu}}$ needed can be pre-computed and stored, and the other factors in (B.10.6) are easy to generate. However, there are extensive cancellations in the final sum, which limits the numerical accuracy with higher angular momenta. This can be controlled by doing parts of the calculation using integer arithmetic. All of this can be avoided by the method described in §B.10.2 below.

B.10.2 Recurrence relations for E-coefficients

The recurrence relations for E-coefficients presented by Saunders [68, §3.2] are reproduced below, with corrections.[10] The coefficients $E[\boldsymbol{\nu};\boldsymbol{\nu}';\boldsymbol{s}]$ are functions of nine parameters, $\boldsymbol{\nu} = (n,l,m)$, $\boldsymbol{\nu}' = (n',l',m')$, and $\boldsymbol{s} = (s_1,s_2,s_3)$. Only one parameter changes in each recurrence relation and their statement can be made more concise by defining only the active parameters in the equations below. The initial condition is

$$E[\mathbf{0};\mathbf{0};\mathbf{0}] = K_{AB} \tag{B.10.7}$$

There are three relations:
Rule 1: $(l, m=\pm l) \to (l+1, m=\pm(l+1))$

$$\begin{aligned}
&E[l+1,l+1;\boldsymbol{s}] \qquad\qquad\qquad\qquad\qquad\qquad\qquad\qquad\qquad\qquad\qquad\qquad\text{(B.10.8)}\\
&\quad= (2l+1)\left\{\frac{1}{2p}E[l,l;\boldsymbol{s}-\boldsymbol{e}_1] + (P_1-A_1)E[l,l;\boldsymbol{s}]\right.\\
&\qquad\qquad\qquad\qquad\qquad\left. +(s_1+1)E[l,l;\boldsymbol{s}+\boldsymbol{e}_1]\right\}\\
&\quad+i\,(2l+1)\left\{\frac{1}{2p}E[l,l;\boldsymbol{s}-\boldsymbol{e}_2] + (P_2-A_2)E[l,l;\boldsymbol{s}]\right.\\
&\qquad\qquad\qquad\qquad\qquad\left. +(s_2+1)E[l,l;\boldsymbol{s}+\boldsymbol{e}_2]\right\}
\end{aligned}$$

$$\begin{aligned}
&E[l+1,-l-1;\boldsymbol{s}] \qquad\qquad\qquad\qquad\qquad\qquad\qquad\qquad\qquad\qquad\qquad\text{(B.10.9)}\\
&\quad= (2l+1)\left\{\frac{1}{2p}E[l,-l;\boldsymbol{s}-\boldsymbol{e}_1] + (P_1-A_1)E[l,-l;\boldsymbol{s}]\right.\\
&\qquad\qquad\qquad\qquad\qquad\left. +\,(s_1+1)E[l,-l;\boldsymbol{s}+\boldsymbol{e}_1]\right\}\\
&\quad-i\,(2l+1)\left\{\frac{1}{2p}E[l,-l;\boldsymbol{s}-\boldsymbol{e}_2] + (P_2-A_2)E[l,-l;\boldsymbol{s}]\right.\\
&\qquad\qquad\qquad\qquad\qquad\left. +\,(s_2+1)E[l,-l;\boldsymbol{s}+\boldsymbol{e}_2]\right\}
\end{aligned}$$

Here $\boldsymbol{e}_1 = (1,0,0)$, $\boldsymbol{e}_2 = (0,1,0)$, $\boldsymbol{e}_3 = (0,0,1)$, and $\boldsymbol{s} = (s_1,s_2,s_3) \equiv (\rho.\sigma,\tau)$. Similarly $\boldsymbol{P} = (P_1,P_2,P_3)$, $\boldsymbol{A} = (A_1,A_2,A_3)$. Replace $\boldsymbol{A}$ by $\boldsymbol{B}$ to get the relations for $(l',m'=\pm l') \to (l'+1,m'=\pm(l'+1))$.

[10] The notation is slightly different from that of [68]. The right-hand sides of equations (63), (64a,) and (64b) in [68] have each been multiplied by $(2l+1)$; this factor originates from Equation (18a) of that paper.

Rule 2: $(l, m) \to (l+1, m)$. $\Lambda = s_1 + s_2 + s_3$

$$
\begin{aligned}
E[\,l+1;,\boldsymbol{s}] \\
&= \frac{2l+1}{l-|m|+1}\left(\frac{1}{2p}E[\,l;\boldsymbol{s}-\boldsymbol{e}_3] + (P_3 - A_3)E[\,l;\boldsymbol{s}]\right. \\
&\qquad \left. + (s_3+1)E[\,l;\boldsymbol{s}-\boldsymbol{e}_3]\right) \\
&- \frac{l+|m|}{l-|m|+1}\left\{\sum_{i=1}^{3}\left(\frac{1}{(2p)^2}E[\,l-1;\boldsymbol{s}-2\boldsymbol{e}_i]\right.\right. \\
&\qquad \left. + \frac{(P_i - A_i)}{p}E[\,l-1;\boldsymbol{s}-\boldsymbol{e}_i]\right) \\
&\qquad + \left((\boldsymbol{P}-\boldsymbol{A})^2 + \frac{2\Lambda+3}{2p}\right)E[\,l-1;\boldsymbol{s}-\boldsymbol{e}_3] \\
&+ 2\sum_{i=1}^{3}\Big((P_i - A_i)(s_i+1)E[\,l-1;\boldsymbol{s}+\boldsymbol{e}_i] \\
&\qquad \left. + (s_i+2)(s_i+1)E[\,l-1;\boldsymbol{s}+2\boldsymbol{e}_i]\Big)\right\}
\end{aligned}
\tag{B.10.10}
$$

Replace components of $\boldsymbol{A}$ by components of $\boldsymbol{B}$ for $(l', m') \to (l'+1, m')$.
Rule 3: $n \to n+1$

$$
\begin{aligned}
E[\,n+1;\boldsymbol{s}] \\
&= \sum_{i=1}^{3}\left\{\frac{1}{(2p)^2}E[\,n;\boldsymbol{s}-2\boldsymbol{e}_i] + \frac{1}{p}(P_i - A_i)E[\,n;\boldsymbol{s}-\boldsymbol{e}_i]\right. \\
&\qquad + 2(P_i - A_i)(s_1+1)E[\,n;\boldsymbol{s}+\boldsymbol{e}_i] \\
&\qquad\qquad \left. + (s_1+2)(s_1+1)E[\,n;\boldsymbol{s}+2\boldsymbol{e}_i]\right\} \\
&\qquad + \left((\boldsymbol{P}-\boldsymbol{A})^2 + \frac{2\Lambda+3}{2p}\right)E[\,n;\boldsymbol{s}]
\end{aligned}
\tag{B.10.11}
$$

Replace components of $\boldsymbol{A}$ by components of $\boldsymbol{B}$ for $n' \to n'+1$.

B.10.3 Implementation issues

The recurrence relations are quite complicated, and it is necessary to take some care with the allocation of index variables to avoid multiply nested loops and expensive conditional branches. Each polynomial index vector $\boldsymbol{s}$ locates elements of a *block*, labelled by the other parameters, and is associated with an unique *pointer* `tuv`. Each entry $E[b;\boldsymbol{s}]$ in block $b = (\boldsymbol{\nu}, \boldsymbol{\nu}')$, acts a *seed* variable for accumulating contributions to a number of *targets* $E[\widetilde{b};\widetilde{\boldsymbol{s}}]$ according to the pseudo-code

Table B.1. Table of weights $W(b;\boldsymbol{s})$ and targets $\widetilde{\boldsymbol{s}}$

Blocks labelled $b = (\boldsymbol{\nu};\boldsymbol{\nu}')$ with $\boldsymbol{\nu}'$ fixed	
Rule 1: $\boldsymbol{\nu} = (0, l, \pm l)$, $\widetilde{\boldsymbol{\nu}} = (0, l+1, \pm(l+1))$	
$\widetilde{\boldsymbol{s}}$	$W(l, \pm l; \boldsymbol{s})$
$\boldsymbol{s}+\boldsymbol{e}_1$	$(2l+1)/2p$
$\boldsymbol{s}$	$(2l+1)(P_1 - A_1)$
$\boldsymbol{s}-\boldsymbol{e}_1$	$(2l+1)s_1$
$\boldsymbol{s}+\boldsymbol{e}_2$	$\pm i(2l+1)/2p$
$\boldsymbol{s}$	$\pm i(2l+1)(P_2 - A_2)$
$\boldsymbol{s}-\boldsymbol{e}_)$	$\pm i(2l+1)s_2$
Rule 2: $\boldsymbol{\nu} = (0, l, m)$, $\widetilde{\boldsymbol{\nu}} = (0, l+1, m)$	
$\widetilde{\boldsymbol{s}}$	$W(l, m; \boldsymbol{s})$, $\Lambda = l + l'$
$\boldsymbol{s}+\boldsymbol{e}_3$	$(2l+1)/[2p(l-\lvert m\rvert+1)]$
$\boldsymbol{s}$	$(2l+1)(P_1 - A_1)/(l-\lvert m\rvert+1)$
$\boldsymbol{s}-\boldsymbol{e}_3$	$(2l+1)s_3/(l-\lvert m\rvert+1)$
Rule 2: $\boldsymbol{\nu} = (0, l-1, m)$, $\widetilde{\boldsymbol{\nu}} = (0, l+1, m)$	
$\widetilde{\boldsymbol{s}}$	$W(l, m; \boldsymbol{s})$, $\Lambda = l + l' - 1$, $i = 1, 2, 3$
$\boldsymbol{s}+2\boldsymbol{e}_i$	$-(l+\lvert m\rvert)/[(l-\lvert m\rvert+1)(2p)^2]$
$\boldsymbol{s}+\boldsymbol{e}_i$	$-(l+\lvert m\rvert)(P_i - A_i)/[(l-\lvert m\rvert+1)2p]$
$\boldsymbol{s}$	$-(l+\lvert m\rvert)[(\boldsymbol{P}-\boldsymbol{A})^2 + (2\Lambda+3)/2p]/(l-\lvert m\rvert+1)$
$\boldsymbol{s}-\boldsymbol{e}_i$	$-2(l+\lvert m\rvert)(P_i - A_i)s_i/(l-\lvert m\rvert+1)$
$\boldsymbol{s}-2\boldsymbol{e}_i$	$-(l+\lvert m\rvert)s_i(s_i-1)/(l-\lvert m\rvert+1)$
Rule 3: $\boldsymbol{\nu} = (n, l, m)$, $\widetilde{\boldsymbol{\nu}} = (n+1, l.m)$	
$\widetilde{\boldsymbol{s}}$	$W(n; \boldsymbol{s})$, $\Lambda = 2(n+n') + l + l'$, $i = 1, 2, 3$
$\boldsymbol{s}+2\boldsymbol{e}_i$	$1/(2p)^2$
$\boldsymbol{s}+\boldsymbol{e}_i$	$(P_i - A_i)/p$
$\boldsymbol{s}$	$(\boldsymbol{P}-\boldsymbol{A})^2 + (2\Lambda+3)/2p$
$\boldsymbol{s}-\boldsymbol{e}_i$	$2(P_i - A_i)s_i$
$\boldsymbol{s}-2\boldsymbol{e}_i$	$s_i(s_i-1)$

`loop over seed indices` $\boldsymbol{s}$
  `compute target indices` $\boldsymbol{s}'$
  `begin loop over block index` b
    `accumulate` $E[\widetilde{b};\widetilde{\boldsymbol{s}}] \leftarrow E[b;\boldsymbol{s}'] + W(b;\boldsymbol{s})\,E[b;\boldsymbol{s}]$
  `end loop over` b
`end loop over` $\boldsymbol{s}$

$W(b;\boldsymbol{s})$ is referred to as a *weight*. Each sum runs over a set

$$\mathcal{T}_\Lambda = \{\boldsymbol{s} = (s_1, s_2, s_3)\,|0 \le \Lambda = s_1 + s_2 + s_3 \le \Lambda_{max}\}, \tag{B.10.12}$$

with $\Lambda_{max} = \max(2n + 2n' + l + l')$. Let $\Gamma = s_2 + s_3$. Then there are $j_\Gamma = (\Gamma+1)(\Gamma+2)/2$ and $i_\Lambda = (\Lambda+1)(\Lambda+2)(\Lambda+3)/6$ points $\boldsymbol{s}$ for each permissible value of Γ and Λ from which we get the unique pointer

$$\texttt{tuv} = i_\Lambda - j_\Gamma + s_3 + 1. \tag{B.10.13}$$

This scheme eliminates the need to trap target indices that are out of range, a cause of major delays in other computational schemes. Rule 3, which significantly increases the dimension of the calculation, is not needed in nonrelativistic quantum chemistry codes, which only make use of the $n = 0$ relations, but $n = 1$ and $n' = 1$ are required for the calculation of E_q-coefficients when κ is positive. The overheads involved are relatively small as most of the Rule 3 weights can be pre-calculated and the extra coefficients can be calculated on the fly with little extra cost.

The computational algorithm is

- **Initialization**: Compute the root seed $E[\mathbf{0};\mathbf{0};\mathbf{0}] = K_{AB}$, indices $s_i(\texttt{tuv})$, $i = 1, 2, 3$, and components of the vectors $\boldsymbol{A}, \boldsymbol{B}$ and $\boldsymbol{P}$.
- **Stage 1**: Generate $E[(0,l,l);\mathbf{0};\boldsymbol{s}]$ for $0 \le l \le l_A$ on centre $\boldsymbol{A}$ with Rule 1.
- **Stage 2**: Generate $E[(0,l_A,m);\mathbf{0};\boldsymbol{s}]$ for $0 \le m \le l_A - 1$ on centre $\boldsymbol{A}$ with Rule 2.
- **Stage 3**: Generate $E[(0,l_A,m);(0,l',\pm l');\boldsymbol{s}]$ for $0 \le l' \le l_B$ on centre $\boldsymbol{B}$ with Rule 1.
- **Stage 4**: Generate $E[(0,l_A,m);(0,l_B,m');\boldsymbol{s}]$ for $-(l_B - 1) \le m' \le l_B - 1$ on centre $\boldsymbol{B}$ with Rule 2.

  At the end of this stage, we have all $E[(0,l_A,m);(0,l_B,m');\boldsymbol{s}]$ for $0 \le m \le l_A$ and $-l_B \le m' \le l_B$. The complex conjugation rule

  $$E[(n,l_A,-m);(n',l_B,-m');\boldsymbol{s}] = E^*[(n,l_A,m);(n',l_B,m');\boldsymbol{s}]$$

  can then be used to obtain the coefficients with $m < 0$.
- **Stage 5**: The table is completed by inserting the normalization factors s_{lm} from (10.6.4).

The **Stage 1** outputs are stored sequentially in blocks of length i_l labelled by l, keeping the real and imaginary parts separate using floating point arithmetic. The final block gives seed values for the generation of functions centred on $\boldsymbol{B}$. The coefficients $E[(0,l,l);\mathbf{0};\boldsymbol{s}]$ for $0 \le l \le l_A - 1$ act as seed values for **Stage 2**. The first cycle with Rule 2 omits all contributions for which any indices s_1, s_2 or s_3 vanish. The target outputs are transferred to a vector containing $E[(0,l_A,m);\mathbf{0};\boldsymbol{s}]$ with $0 \le m \le l_A - 1$, after which all other data for this value of m can be discarded. The coefficients $E[(0,l_A,l_A);\mathbf{0};\boldsymbol{s}]$ were calculated in **Stage 1**. These outputs are seeds for **Stage 3** in which we generate $E[(0,l_A,m);(0,l',\pm l');\boldsymbol{s}]$ for fixed l_A, m. Each block identified by

l_A and l' is of length $i_{l_A+l'}$ independent of m, m'. These $2l_B + 1$ blocks, one for each value of m' are the first useful results that can be transferred to final storage locations. Finally **Stage 4** generates all the remaining values with $n = n' = 0$. This can be done as a two-dimensional recurrence with the index b varying most rapidly and the second index is segmented in blocks labelled by l. The relativistic coefficients $E_q^{\beta\beta'}[\mu\nu; \boldsymbol{k}]$ can now be generated from (10.8.8)–(10.8.13) as required.

References

1. Jordan T F 1969 *Linear Operators for Quantum Mechanics* (New York: John Wiley).
2. Reed M and Simon B 1972-1979 *Methods of Modern Mathematical Physics* (4 vols.) (New York: Academic Press).
3. Kato T 1976 *Perturbation Theory for Linear Operators* (Berlin: Springer-Verlag).
4. Richtmyer R D 1978 *Methods of Advanced Mathematical Physics* (2 vols) (New York: Springer-Verlag).
5. Elliott J P and Dawber P G 1979 *Symmetry in Physics* (Basingstoke: Macmillan Press).
6. Hamermesh M 1962 *Group theory and its application to physical problems* (Reading, Mass.: Addison-Wesley).
7. Wybourne B G 1974 *Classical Groups for Physicists* (New York: Wiley).
8. Drake G W F ed 1996 *Atomic, Molecular and Optical Physics Handbook* (Woodbury NY: American Institute of Physics).
9. Racah G 1965 *Ergeb. Exakt. Naturwiss.* **37** 28.
10. Stone A P 1961 *Proc. Camb. Phil. Soc.* **57** 460.
11. El Baz E and Castel B 1972 *Graphical Methods of Spin Algebras* (New York: Marcel Dekker, Inc.).
12. Yutsys AP Levinson I B and Vanagas V V 1962 *Mathematical Apparatus of the Theory of Anglar Momentum* (Jerusalem: Israel Program for Scientific Translations).
13. Wigner E P 1959 *Group Theory and its application to the quantum mechanics of atomic spectra* (New York: Academic Press).
14. Judd B R 1963 *Operator Techniques in Atomic Spectroscopy* (New York: McGraw-Hill).
15. Judd B R 1967 *Second Quantization and Atomic Spectroscopy* (Baltimore: The Johns Hopkins Press).
16. Brink D M and Satchler G R 1968 *Angular Momentum* (Oxford: Clarendon Press).
17. Lindgren I and Morrison J 1982 *Atomic Many-Body Theory* (Berlin: Springer-Verlag).
18. Innes F R and Ufford C W 1958 *Phys. Rev.* **111**, 194.
19. Meyer J, Sepp W-D, Fricke B and Rosen A 1996 *Comput. Phys. Commun.* **96**, 263.
20. Atkins P W 1970 *Molecular Quantum Mechanics.* (Oxford: Clarendon Press).
21. Klahn B and Bingel W A 1977 *Theoret. Chim. Acta (Berl.)* **44**, 9, 27.

22. Klahn B 1981 *Adv. Quant. Chem.* **13**, 155.
23. Rotenberg M 1962 *Ann. Phys. (N. Y.)* **19**, 262.
24. Rotenberg M 1970 *Adv. Atom. Molec. Phys.* **6**, 233.
25. Abramowitz M and Stegun I A 1970 *Handbook of Mathematical Functions* (New York: Dover).
26. Szeg'ö G 1959 *Orthogonal Polynomials* (American Mathematical Society Colloquium Publication No. 23).
27. Courant R and Hilbert D 1953 *Methods of Mathematical Physics, Vol. 1* (New York: Interscience Publishers).
28. Gerschgorin S 1931 *Izvestiya Akad. Nauk. USSR, Ser. Math.* **7**, 749.
29. Wilkinson J H 1965 *The Algebraic Eigenvalue Problem* (Oxford: Clarendon Press).
30. Čársky P and Urban M 1980 *Ab Initio Calculations. Methods and Applications in Chemistry.* Springer-Verlag, Berlin.
31. Huzinaga S, Andzelm J, Klobukowski M, Radzio-Andzelm E, Sakai Y. and Tatewaki H (eds.) 1984 *Gaussian Basis Sets for Molecular Calculations* (Amsterdam: Elsevier).
32. Poirier R, Kari R, Csizmadia I G 1985 *Handbook of Gaussian Basis Sets* (Amsterdam: Elsevier).
33. Wilson S *Basis Sets* in [34].
34. Lawley K P (ed.) 1987 *Adv. Chem. Phys.* **69** (New York: John Wiley).
35. Schmidt M W and Ruedenberg K 1979 *J. Chem. Phys.* **71**, 3951.
36. Feller D F and Ruedenberg K 1979 *Theor. Chim. Acta* **52**, 231.
37. Wilson S 1982 *Theor. Chim. Acta* **61**, 343.
38. Press W H, Teukolsky S A, Vetterling W T and Flannery B P 1992 *Numerical Recipes* (2nd edn) (Cambridge: Cambridge University Press).
39. Keller H B 1968 *Numerical methods for two-point boundary value problems* (Waltham, Mass: Blaisdell Publishing Co.).
40. Hall G and Watt J M 1976 *Modern numerical methods for ordinary differential equations* (Oxford: Clarendon Press).
41. Hull T E 1975 in *Numerical solution of boundary value problems for ordinary differential equations* (ed. A K Aziz) pp. 3–26 (New York: Academic Press).
42. Isaacson E and Keller H B 1966 *Analysis of Numerical Methods* (New York: John Wiley).
43. Mayers D F 1957 *Proc. Roy. Soc. A* **241** 93.
44. Dahlquist G 1956 *Math. Scand.* **4**, 33–53.
45. Dahlquist G 1959 *Kungl. Tekniska Hogskolans Handlingen* No. 130 (1959).
46. Lambert J D 1973 *Computational Methods in Ordinary Differential Equations* (New York: John Wiley).
47. Coulthard M A 1967 *Proc. Phys. Soc.* **91** 44.
48. Smith F C and Johnson W R 1967 *Phys. Rev.* **160** 136.
49. Desclaux J P, Mayers D F and O'Brien F 1971 *J. Phys. B: At. Mol. Phys.* **4** 631.
50. Desclaux J-P 1975 *Comput. Phys. Commun.* **9**, 31.
51. Fox L and Goodwin E T 1949 *Proc. Camb. Phil. Soc.* **45** 373 – 388.
52. Pereyra V 1966 *Num. Math.* **8** 376.
53. Pereyra V 1967 *Num. Math.* **10** 316.
54. Pereyra V 1968 *Num. Math.* **11** 111.
55. Hildebrand F B 1956 *Introduction to Numerical Analysis* (New York: McGraw-Hill).

56. *Modern Computing Methods* 1961 (London: H M Stationery Office).
57. Norrington P H and Grant I P 1981 *J. Phys. B: Atom. Molec. Phys.* **14** L261.
58. Fischer C F, Brage T and Jönsson P 1997 *Computational Atomic Structure. An MCHF approach* (Bristol and Philadelphia: Institute of Physics).
59. Coddington E A and Levinson N 1955 *Theory of Ordinary Differential Equations.* (New York: McGraw-Hill).
60. Atkinson F V 1964 *Discrete and Continuous Boundary Value Problems* (New York: Academic Press).
61. Stakgold I 1979 *Green's Functions and Boundary Value Problems* (New York: John Wiley).
62. Thaller B 1992 *The Dirac Equation* (New York: Springer-Verlag).
63. Chodos A, Jaffe R L, Johnson K, Thorn C B and Weisskopf V F 1974 *Phys. Rev. D* **9**, 3471.
64. Johnson W R, Blundell S A and Sapirstein J 1988 *Phys. Rev. A* **37**, 307.
65. Goldstein H 1980 *Classical Mechanics* (2nd edition) (Reading, Mass.: Addison Wesley).
66. Landau L D and Lifshitz E M 1976 *Mechanics*(3rd edition) (Oxford: Pergamon Press).
67. McMurchie L E and Davidson E R 1978 *J. Comput. Phys.* **26**, 218.
68. Saunders V R 1983 in *Methods of Computational Molecular Physics*, Vol. 1, p. 1. ed. G H F Diercksen and S Wilson (Dordrecht: Reidel).
69. Quiney H M and Grant I P 2006 *in preparation*

Index

Springer Series on

ATOMIC, OPTICAL, AND PLASMA PHYSICS

Springer Series on
ATOMIC, OPTICAL, AND PLASMA PHYSICS

1 **Polarized Electrons**
2nd Edition
By J. Kessler

2 **Multiphoton Processes**
Editors: P. Lambropoulos and S.J. Smith

3 **Atomic Many-Body Theory**
2nd Edition
By I. Lindgren and J. Morrison

4 **Elementary Processes in Hydrogen-Helium Plasmas**
Cross Sections and Reaction Rate Coefficients
By R.K. Janev, W.D. Langer, K. Evans Jr., and D.E. Post Jr.

5 **Pulsed Electrical Discharge in Vacuum**
By G.A. Mesyats and D.I. Proskurovsky

6 **Atomic and Molecular Spectroscopy**
Basic Aspects and Practical Applications
3rd Edition
By S. Svanberg

7 **Interference of Atomic States**
By E.B. Alexandrov, M.P. Chaika, and G.I. Khvostenko

8 **Plasma Physics**
Basic Theory with Fusion Applications
3rd Edition
By K. Nishikawa and M. Wakatani

9 **Plasma Spectroscopy**
The Influence of Microwave and Laser Fields
By E. Oks

10 **Film Deposition by Plasma Techniques**
By M. Konuma

11 **Resonance Phenomena in Electron-Atom Collisions**
By V.I. Lengyel, V.T. Navrotsky, and E.P. Sabad

12 **Atomic Spectra and Radiative Transitions**
2nd Edition
By I.I. Sobel'man

13 **Multiphoton Processes in Atoms**
2nd Edition
By N.B. Delone and V.P. Krainov

14 **Atoms in Plasmas**
By V.S. Lisitsa

15 **Excitation of Atoms and Broadening of Spectral Lines**
2nd Edition, By I.I. Sobel'man, L. Vainshtein, and E. Yukov

16 **Reference Data on Multicharged Ions**
By V.G. Pal'chikov and V.P. Shevelko

17 **Lectures on Non-linear Plasma Kinetics**
By V.N. Tsytovich

18 **Atoms and Their Spectroscopic Properties**
By V.P. Shevelko

19 **X-Ray Radiation of Highly Charged Ions**
By H.F. Beyer, H.-J. Kluge, and V.P. Shevelko

20 **Electron Emission in Heavy Ion-Atom Collision**
By N. Stolterfoht, R.D. DuBois, and R.D. Rivarola

21 **Molecules and Their Spectroscopic Properties**
By S.V. Khristenko, A.I. Maslov, and V.P. Shevelko

22 **Physics of Highly Excited Atoms and Ions**
By V.S. Lebedev and I.L. Beigman

23 **Atomic Multielectron Processes**
By V.P. Shevelko and H. Tawara

24 **Guided-Wave-Produced Plasmas**
By Yu.M. Aliev, H. Schlüter, and A. Shivarova

25 **Quantum Statistics of Nonideal Plasmas**
By D. Kremp, M. Schlanges, and W.-D. Kraeft

26 **Atomic Physics with Heavy Ions**
By H.F. Beyer and V.P. Shevelko

Printed in the United States of America